# ENCYCLOPEDIA OF GLOBAL CHANGE

# ENCYCLOPEDIA OF

# Global Change

## Environmental Change and Human Society

Andrew S. Goudie
EDITOR IN CHIEF

David J. Cuff
ASSOCIATE EDITOR

Volume 1

OXFORD
UNIVERSITY PRESS
2002

# OXFORD

UNIVERSITY PRESS

Oxford   New York

Athens   Auckland   Bangkok   Bogotá   Buenos Aires   Cape Town
Chennai   Dar es Salaam   Delhi   Florence   Hong Kong   Istanbul   Karachi
Kolkata   Kuala Lumpur   Madrid   Melbourne   Mexico City   Mumbai
Nairobi   Paris   São Paulo   Shanghai   Singapore   Taipei   Tokyo   Toronto   Warsaw

and associated companies in

Berlin   Ibadan

Copyright © 2002 by Oxford University Press, Inc.

Published by Oxford University Press, Inc.
198 Madison Avenue, New York, New York 10016
www.oup.com

Oxford is a registered trademark of Oxford University Press

Library of Congress Cataloging-in-Publication Data

Encyclopedia of global change : environmental change and human society /
Andrew S. Goudie, editor in chief.
p.   cm.
Includes bibliographical references and index.

ISBN-13 978-0-19-510825-5

1. Global environmental change—Encyclopedias.
2. Nature—Effect of human beings on—Encyclopedias.
3. Global environmental change—Social aspects—Encyclopedias.
I. Goudie, Andrew.
GE149.E47   2000   363.7—dc21   00-058918

The photographs and line drawings used herein were supplied by contributors
to the work, members of the Editorial Board, major museums, and by commercial
photographic archives. The publisher has made every effort to ascertain that necessary
permissions to reprint materials have been secured. Sources of photographs and
line drawings are given in captions to illustrations where appropriate.

EDITORIAL AND PRODUCTION STAFF
*Commissioning Editor:* Sean Pidgeon
*Project Editor:* Norina Frabotta
*Assistant Editor:* Elizabeth Aldrich
*Copyeditors:* Sean Pidgeon, Larry Hamberlin, and Jane McGary
*Proofreaders:* Nancy Bernhaut, Sue Gilad, Gareth Ridout, and Rhoda Seidenberg
*Indexer:* Bonny McLaughlin
*Book Designer:* Joan Greenfield
*Managing Editor:* Matthew Giarratano
*Publisher:* Karen Casey

3  5  7  9  8  6  4
Printed in the United States of America
on acid-free paper

# Contents

Editorial Board

vol. 1, p. vii

Preface

ix

Introduction

xi

Abbreviations, Acronyms, and Symbols

xv

## ENCYCLOPEDIA OF GLOBAL CHANGE
### Environmental Change and Human Society

Directory of Contributors

vol. 2, p. 549

Synoptic Outline of Contents

561

Index

563

# Editorial Board

vii

# Preface

As some of the biographical entries in the *Encyclopedia of Global Change* demonstrate, studies of the diverse impacts that humans have on their environment are by no means new. Equally, there is a long history of studies of the natural changes in the environment that have an impact on humans. However, what has been remarkable in the last two to three decades has been the way in which research on anthropogenic and natural changes has intensified. Even more remarkable has been the way in which concerns with environmental change have proliferated, so that it has become a central focus for academics in a whole range of disciplines, from anthropology to zoology, and involving both social and natural sciences. It has also become a topic of major interest to those outside academics, intriguing and motivating the public at large, causing those in business to rethink some of their approaches, and leading to increased activity by those in government and politics.

The study of natural environmental changes such as the multiple and profound glacial and interglacial fluctuations of the Pleistocene, or the oscillation of climate over shorter time scales such as the El Niño–Southern Oscillation, has demonstrated the frequency, magnitude, and abruptness of change. Our ability to date such changes has been revolutionized by isotopic techniques, while cores through the ocean floors, lakes, bogs, and icecaps have permitted increasingly high-resolution environmental reconstruction. Explanations of past changes have been facilitated by the development of sophisticated models, a process in which the adoption of increasingly powerful computers have played a major role.

The study of anthropogenic changes and their actual and potential impacts has also burgeoned. This is partly because of the growth of human population, of increasing affluence, of new technologies, and of exploitation of new environments. However, new monitoring and analytical techniques have enabled the identification of some potential threats such as the ozone hole, while powerful modeling capability permits the construction of scenarios of impacts and of future change, such as those that might be produced by an enhanced greenhouse effect.

Frequently, it is frustratingly difficult to separate out the effects of human and natural factors in environmental change. Natural and human factors often combine to create a particular effect. A classic example of this is desertification, where natural droughts, exacerbated conceivably by human-induced changes in ground surface reflectivity or atmospheric dust loadings, work with overgrazing, deforestation, and unwise irrigation practices to cause land degradation.

Another feature of recent studies of environmental change is that they have become increasingly global in scope. This is true both of some of the phenomena and of the processes themselves, such as ozone depletion or global warming. It is also true in terms of the ramifications of some processes—human and natural systems are inextricably linked at all scales—and in terms of international business and politics. Environmental processes and changes do not respect national boundaries.

Given the importance of the subject matter, the rapidity of developments in intellectual thought, and the need to appreciate the sheer spread of disciplinary interest, we hope that this interdisciplinary encyclopedia will provide an up-to-date, accessible, and wide-ranging treatment of natural and anthropogenic changes and their interactions at a whole range of temporal and spatial scales.

The conception of this encyclopedia owes a great deal to Sean Pidgeon, who persuaded me to take on the role of editor in chief. Its execution owes a huge debt to the distinguished international advisory board, to the specialist area editors, to the commitment of the associate editor, David Cuff, and to the methodical and meticulous skills of Norina Frabotta. I am grateful to them all. We all hope that the utility of the product of our labors will be enhanced by the large number of illustrations, index, extensive guides to further reading, cross-references, vignettes, list of initials and acronyms, and, a product of the times, the guide to useful Web sites available through the Oxford University Press Web site at http:www.oup-usa.org/reference/globalchange.

Global change is one of the great issues of the new millennium. It is a huge and diverse field. We hope that these two volumes will inform, entertain, and guide all those who seek to understand our ever-changing environment.

—ANDREW GOUDIE
*Oxford*
*September 2000*

# Introduction

The terms "global change" and "global environmental change" are much used but seldom rigorously defined. They are sometimes considered to be synonymous, but the meanings of both have changed (and sometimes diverged) through time. Wide use of the term "global change" seems to have emerged in the 1970s, but in that decade its use was principally, though not invariably, in reference to changes in international social, economic, and political systems (Price, 1989). The term included such issues as proliferation of nuclear weapons, population growth, inflation, and matters relating to international insecurity and decreases in the quality of life. The first use of the term was essentially anthropocentric (Pernetta, 1995, p. 1).

In the early 1980s, the concept of "global change" took on another meaning—one far more geocentric in focus. According to Price (1989), "the concept [was] effectively . . . captured by the physical and biological sciences," and such were the wide range of scales involved, whether spatial or temporal, that "it can be perceived to include almost any topic in the biological, Earth or atmospheric sciences." He also noted that the term has been used to describe worldwide changes in almost any variable. Such variables include Earth's geological evolution; coastal erosion and volcanism; the human-induced changes in atmospheric composition and terrestrial ecosystems caused by urbanization, population growth, and chemical pollution; as well as the resurgence of malaria and the rapid spread of AIDS (acquired immune deficiency syndrome). The time encompassed by the changes in these variables varied billions of years to a single decade.

The geocentric meaning of global change was used to develop the International Geosphere–Biosphere Programme: A Study of Global Change. The program (known as IGBP) was established in 1986 by the International Council of Scientific Unions: "To describe and understand the interactive physical, chemical and biological processes that regulate the total Earth system, the unique environment that it provides for life, the changes that are occurring in this system, and the manner in which they are influenced by human actions."

The term "global environmental change" has, in many senses, come to be used synonymously with the more geocentric use of "global change." Its validity and wide currency were recognized when the journal *Global Environmental Change* was established in 1990 as an international journal to address the human ecological and public policy dimensions of the environmental processes threatening the sustainability of life on Earth. The journal publishes on, but is not limited to, deforestation, desertification, soil degradation, species extinction, sea level rise, acid precipitation, destruction of the ozone layer, atmospheric warming/cooling, nuclear winter, the emergence of new technological hazards, and the worsening effects of natural disasters.

In the United Kingdom, the Research Councils established a U.K. Global Environmental Research Office, and it recognized two scales of issues in its inaugural publication of *The Globe* (Issue 1, p. 1, March 1991): "(a) those that are truly global (e.g., ozone depletion, climate modification) and (b) local and regional problems which occur on a worldwide scale (e.g., acid rain)." The Research Councils also recognized that global environmental change has become synonymous with understanding the complex interacting processes that influence the Earth's environments and the natural and human-induced factors that affect them.

Beran (1995, p. 43), however, has suggested that there is a distinction between the two phrases, that traditionally the term "global change" has referred to a branch of Earth science that deals with Earth's geologic past. To most researchers, "global environmental change" comprises a set of issues that includes climate change, ozone depletion, biodiversity reduction, desertification, eutrophication, and so on. Those who study "global change" address major questions relating to the beginnings of Earth's atmosphere, magnetic field, or salty ocean, and why there are ice ages, soils, and mountains.

In the *Encyclopedia of Global Change*, the term "global change" is synonymous with "global environmental change"—a definition that has been adopted in the United States (Munn, 1996):

> Global change is a term intended to encompass the full range of global issues and interactions concerning natural and human-induced changes in the Earth's environment. The [U.S.] Global Change Research Act of 1990 defines global change as "changes in the global environment (including alterations in climate, land productivity, oceans or other water resources, atmospheric chemistry, and ecological systems) that may alter the capacity of the Earth to sustain life." (See *Our Changing Planet: The FY 1995 U.S. Global Change Research Program*, p. 2, a report of the U.S. National Science and Technology Council, Washington, D.C., 1994.)

That definition is very similar to the one used by the European Commission's European Network for Research in Global Change:

> Global Change research seeks to understand the integrated Earth system in order to: (i) identify, explain and predict natural and anthropogenic changes in the global environment; (ii) assess the potential regional and local impacts of those changes on natural and human systems; and (iii) provide a scientific basis for the development of appropriate technical, economic and societal mitigation/adaptation strategies.

In that context, "global change" encompasses climate change; changes in atmospheric physics and chemistry (e.g., ozone depletion, tropospheric pollution); changes in land use and cover; variability in marine environments and ice cover; changes in biogeochemical cycles; the impact of such changes on natural balances (ecosystems), on natural resources (fresh water, forest products, crops, marine resources), and on socioeconomic balances (sustainable development, industrial development, human health, and the quality of life). Therefore, the importance of both natural and anthropogenic factors has been recognized in promoting environmental changes on a wide range of temporal scales; anthropogenic changes can only be understood in the context of the full range of past, present, and future natural changes.

It is also important to understand what is meant here by "global," and in this encyclopedia there are two components: systemic global change and the cumulative global changes in climate brought about by atmospheric pollution. In the systemic meaning, "global" refers to the full spatial scale; it comprises such issues as the global changes in climate brought about by atmospheric pollution. In the cumulative meaning, "global" refers to the areal or substantive accumulation of localized change. A change is considered "global" if it occurs on a worldwide scale or if it represents a significant fraction of the total environmental phenomenon or global resource. Both types of change are closely related; for example, the burning of vegetation can lead to (1) systemic change, through such mechanisms as carbon dioxide release and albedo change; and (2) cumulative change, through its impact on soil and biotic diversity.

This issue was pursued by Talbot (1989, pp. 25–26), who indicated that environmental issues might be considered as global or might become global in various ways. For example, some activities occur in a few places but may affect the biosphere as a whole: large-scale combustion of fossil fuels may occur in only a few industrial areas but may affect the climate as a whole, and the same is true of the release of other greenhouse gases. Local activities of smaller scale, when repeated worldwide, may also have global effects. Estuaries near coastal cities are heavily polluted and, although each is a case of local pollution, the total impact of so many cases makes it global. Local changes in land use also cumulatively affect the planet's albedo, as does the local clearing of forests, which have resulted in major cumulative effects on species loss and carbon dioxide balance.

Species extinction is an important dimension of the problem, although the loss of species was formerly considered to be a local problem. Today, maintenance of biological diversity has been recognized as being of major global importance.

Environmental problems result in socioeconomic problems for humans. Some global issues, such as climate change, have clear impacts on humans worldwide. Yet the increasing interdependence among nations and peoples has demonstrated that localized environmental problems generate increasingly pervasive economic and social impacts in other world areas. Such socioeconomic impacts become particularly evident in the case of development assistance to emerging nations. With the flow of resources to them now more than 100 billion U.S. dollars per year, much of it in the form of commercial loans, the size of their debt is enormous. In many countries, environmental degradation is reducing their capacity to support themselves—much less to repay the debt—which, in turn, affects other countries and even regions, developing as well as developed.

The *Encyclopedia of Global Change* presents our current knowledge of natural and anthropogenic changes in the Earth's physical, chemical, and biological systems and resources, and it examines the effects of those changes on human society. Its focus is primarily on the changes that affect the human rather than the geologic time scale, although the Earth's natural rhythms and processes are discussed in sufficient detail that present-day or projected trends may be considered against the appropriate background of natural change.

## BIBLIOGRAPHY

Beran, M. "Education and Training of Global Environmental Change Scientists." In *Global Environmental Change in Science: Education and Training*, edited by D. Waddington, pp. 41–47. Berlin: Springer, 1995.

Munn, R. E. "Global Change: Both a Scientific and a Political Issue." In *Policy Making in an Era of Global Environmental Change*, edited by R. E. Munn, J. W. M. La Rivière, and N. van Lookeren Campagne, pp. 1–15. Dordrecht, Netherlands: Kluwer, 1996.

Pernetta, J. "What Is Global Change?" *Global Change Newsletter* 21 (1995), 1–3.

Price, M. F. "Global Change: Defining the Ill Defined." *Environment* 31.8 (1989), 18–20, 42–44.

Talbot, L. M. "Man's Role in Managing the Global Environment." In *Changing the Global Environment: Perspectives on Human Involvement*, edited by D. B. Botkin, M. F. Caswell, J. E. Estes, and A. Orio, pp. 17–33. Boston: Academic Press, 1989.

Turner, B. L., R. E. Kasperson, W. B. Meyer, K. M. Dow, D. Golding, J. X. Kasperson, R. C. Mitchell, and S. J. Ratick. "Two Types of Global Environmental Change: Definitional and Spatial-Scale Issues in Their Human Dimensions." *Global Environmental Change* 1 (1990), 14–22. The most useful discussion of the definition of "global environmental change."

—ANDREW S. GOUDIE

# Abbreviations, Acronyms, and Symbols

| | |
|---|---|
| AABW | Antarctic Bottom Water |
| ABRACOS | Anglo-Brazilian Climate Observation Study |
| AC | alternating current |
| AGCM | atmospheric general circulation model |
| AIDS | acquired immune deficiency syndrome |
| AIJ | activities implemented jointly |
| AMIGO | America's Interhemisphere Geo-Biosphere Organization |
| AMIP | Atmospheric Model Intercomparison Project |
| AMS | accelerator mass spectrometry |
| ANC | acid-neutralizing capacity |
| AOU | American Ornithologists' Union |
| APEC | Asia-Pacific Economic Cooperation |
| AR | autoregressive |
| ASTER | Advanced Spaceborne Thermal Emission and Reflection Radiometer |
| ATP | adenosine triphosphate |
| ATSR | Along Track Scanning Radiometer |
| AU | astronomical unit (1 AU = $1.496 \times 10^8$ kilometers) |
| AVHRR | (NOAA) Advanced Very High Resolution Radiometer |
| BAP | Bureau of Animal Population |
| BASF | Badische Anilin- und Soda-Fabrik |
| BBSRC | (U.K.) Biotechnology and Biological Sciences Research Council |
| BCA | benefit–cost analysis |
| BCE | before the Common Era |
| BCG | bacille Calmette-Guerin |
| BLM | (U.S.) Bureau of Land Management |
| BNC | base-neutralizing capacity |
| BOREAS | Boreal Ecosystem-Atmosphere Study |
| BP | before the present |
| Btu | British thermal unit |
| CAD | computer-assisted design |
| CBD | Convention on Biological Diversity |
| CCAMLR | Convention on the Conservation of Antarctic Marine Living Resources |
| CCGT | combined-cycle gas turbine |
| CCN | cloud condensation nuclei |

| | |
|---|---|
| CDIAC | Carbon Dioxide Information Analysis Center |
| CDM | clean development mechanism |
| CE | Common Era |
| CEOS | Committee on Earth Observation Satellites |
| CEQ | (U.S.) Council on Environmental Quality |
| CERES | Clouds and the Earth's Radiant Energy System |
| CERLA | Comprehensive Environmental Response, Compensation, and Liability Act |
| CET | central England temperature |
| CFC | chlorofluorocarbon |
| CGIAR | Consultative Group on International Agricultural Research |
| CGIS | Canada Geographic Information System |
| CIA | cumulative impact assessment |
| CIESIN | Center for International Earth Science Information Network |
| CIM | computer integrated manufacturing |
| CIMMYT | Centro Internacional de Mejoramiento de Maiz y Trigo (International Maize and Wheat Improvement Center) |
| CIRAN | Centre for International Research and Advisory Networks |
| CITES | Convention on International Trade in Endangered Species of Wild Fauna and Flora |
| CLIVAR | Climate Variability and Prediction Program |
| CMS | Convention on the Conservation of Migratory Species of Wild Animals (or the Bonn Convention) |
| CN | condensation nuclei |
| CNRS | Centre National de la Recherche Scientific (France) |
| CoCP | Council of Contracting Parties |
| COM | cost of mitigation |
| COMECON | Communist Economic Community (or Council for Mutual Economic Assistance) |
| CoP | Conference of the Parties |
| CoP1 | First Conference of the Parties, Berlin, Germany, March–April 1995 |
| CoP3 | Third Conference of the Parties, Kyoto, Japan, December 1997 |

| | |
|---|---|
| CoP4 | Fourth Conference of the Parties, Buenos Aires, Argentina, 2–13 November 1998 |
| CPR | common-pool resource; contraceptive prevalence rate |
| CRED | Centre for Research on the Epidemiology of Disasters |
| CSD | (United Nations) Commission on Sustainable Development |
| CSERGE | Centre for Social and Economic Research on the Global Environment |
| DAAC | distributed active archive center |
| DARPA | (U.S.) Defense Advanced Research Projects Agency |
| DBMS | database management system |
| DDT | dichlorodiphenyltrichloroethane |
| DEM | digital elevation model |
| DGD | Decision Guidance Document |
| DHF | dengue hemorrhagic fever |
| DICE | dynamic integrated climate economy |
| DIF | Directory Interchange Format |
| DMS | dimethyl sulfide |
| DNA | deoxyribonucleic acid |
| DOC | dissolved organic carbon |
| DOD | (U.S.) Department of Defense |
| DSM | distributed shared memory |
| DSS | dengue shock syndrome |
| DU | Dobson unit |
| EA | environmental assessment |
| EBM | energy balance model |
| EC | European Community |
| ECC | electrochemical cell |
| ECE | (United Nations) Economic Commission for Europe |
| ECOSOC | (United Nations) Economic and Social Council |
| ED | electrodialysis |
| EDF | Environmental Defense Fund |
| EDR | electrodialysis reversal |
| EEA | environmental effects assessment |
| EEC | European Economic Community |
| EEZ | Exclusive Economic Zone |
| EIA | environmental impact assessment |
| EIS | environmental impact statement |
| EKC | environmental Kuznets curve |
| EM-DAT | Emergency Events Database: the OFDA/CRED International Disaster Database |
| ENMOD | Convention on the Prohibition of Military or Any Other Hostile Use of Environmental Modification Techniques |
| ENSO | El Niño–Southern Oscillation |
| EOS | (NASA) Earth Observing System |
| EOSDIS | EOS Data and Information System |
| EPA | (U.S.) Environmental Protection Agency |
| ERBE | (NASA) Earth Radiation Budget Experiment |
| ERS | (European) Earth Resources Satellite |
| ESA | Endangered Species Act |
| ESRC | (U.K.) Economic and Social Research Council |
| FAO | (United Nations) Food and Agriculture Organization |
| FCCC | (United Nations) Framework Convention on Climate Change |
| FFA | South Pacific Forum Fisheries Agency |
| FFT | fast Fourier transform |
| FGDC | U.S. Federal Geographic Data Committee |
| FIFE | First ISLSCP Field Experiment |
| FOE | Friends of the Earth |
| FoEI | Friends of the Earth International |
| FRG | Federal Republic of Germany |
| FRONTIERS | Forecasting Rain Optimised using New Techniques of Interactively Enhanced Radar and Satellite data |
| FS | (U.S.) Forest Service |
| FSU | former Soviet Union |
| FWS | (U.S.) Fish and Wildlife Service |
| G-7 | Group of Seven (leading industrial nations) |
| GAC | Global Area Coverage |
| GARP | Global Atmospheric Research Program |
| GATS | General Agreement on Trade in Services |
| GATT | General Agreement on Tariffs and Trade |
| gC | grams of carbon |
| GCDIS | Global Change Data and Information System |
| GCM | general circulation model; global change model; global climate model |
| GCMD | Global Change Master Directory |
| GCOS | Global Climate Observing System |
| GDP | gross domestic product |
| GDR | German Democratic Republic (East Germany; now part of Germany) |
| GEC | global environmental change |
| GECHS | Global Environmental Change and Human Security |
| GEF | Global Environment Facility |
| GEWEX | Global Energy and Water Cycle Experiment |
| GFDL | Geophysical Fluid Dynamics Laboratory (NOAA) |
| GHG | greenhouse gas |
| GIS | Geographic Information Systems; Geographical Information Systems |
| GISS | Goddard Institute for Space Studies |
| GNP | gross national product |

| | |
|---|---|
| GONGO | governmental and nongovernmental organization |
| GPCP | Global Precipitation Climatology Project |
| GPS | Global Positioning System |
| GWP | global warming potential |
| HAP | hazardous air pollutants |
| HAPEX | Hydrological and Atmospheric Pilot Experiment |
| HCB | hexachlorobenzene |
| HCFC | hydrochlorofluorocarbon |
| HFC | hydrofluorocarbon |
| HIV | human immunodeficiency virus |
| HLW | high-level waste |
| HPLC | high-pressure liquid chromatography |
| IAA | International Association of Academies |
| IAEA | International Atomic Energy Agency |
| IAI | Inter-American Institute (for Global Change Research) |
| IARC | international agricultural research center |
| IARIW | International Association for Research in Income and Wealth |
| IAS | Institute for Advanced Study |
| IBAMA | Instituto Brasileiro do Meio Ambiente e dos Recursos Naturais Renováveis (Brazilian Institute of Environment and Renewable Natural Resources) |
| IBM | International Business Machines (Corporation) |
| ICAIR | International Centre for Antarctic Information and Research |
| ICAS | Interstate Council on the Aral Sea |
| ICBP | International Committee for Bird Protection (now BirdLife International) |
| ICCROM | International Centre for the Study of the Preservation and Restoration of Cultural Property |
| ICES | International Council for the Exploration of the Sea |
| ICESat | Ice, Cloud, and land Elevation Satellite |
| ICJ | International Court of Justice |
| ICOLD | International Commission on Large Dams |
| ICOMOS | International Council on Monuments and Sites |
| ICRW | International Convention for the Regulation of Whaling |
| ICSID | International Centre for Settlement of Investment Disputes |
| ICSU | International Council for Science (formerly the International Council of Scientific Unions) |
| ICZM | Integrated Coastal Zone Management |
| IDA | International Development Association |

| | |
|---|---|
| IDEAL | International Decade of the East African Lakes |
| IDGC | Institutional Dimensions of Global Change |
| IDN | (CEOS) International Directory Network |
| IDNDR | International Decade for Natural Disaster Reduction |
| IE | industrial ecology |
| IEA | International Energy Agency |
| IFAD | International Fund for Agricultural Development |
| IFAS | Interstate Fund for the Aral Sea |
| IFC | International Finance Corporation |
| IFOV | instantaneous field of view |
| IFPRI | International Food Policy Research Institute |
| IGBP | International Geosphere-Biosphere Programme |
| IGY | International Geophysical Year |
| IHDP | International Human Dimensions Programme on Global Environmental Change |
| IIASA | International Institute for Applied Systems Analysis |
| IISD | International Institute for Sustainable Development |
| ILW | intermediate-level waste |
| IM | industrial metabolism |
| IMF | International Monetary Fund |
| IMO | (United Nations) International Maritime Organization |
| IMS | Information Management System |
| INCD | Intergovernmental Negotiating Committee for the elaboration of an international convention to combat Desertification |
| INC5 | Fifth Intergovernmental Negotiating Session for a Global POPs Treaty, Johannesburg, South Africa, 4–9 December 2000 |
| INPA | Instituto Nacional de Pesquisas da Amazônia (National Institute for Research in the Amazon) |
| INPE | Instituto Nacional de Pesquisas Espaciais (Brazilian National Institute for Space Research) |
| INSEAD | European Institute of Business Administration |
| IPAT | I [environmental impact] = P [population] $\times$ A [affluence] $\times$ T [technology]; or FI = FP $\times$ FA $\times$ FT |
| IPCC | Intergovernmental Panel on Climate Change |
| IPF | Intergovernmental Panel on Forests |
| IPM | integrated pest management |
| IPP | independent power-producing |
| IPR | intellectual property rights |
| IR | infrared |
| IRC | International Research Council |

| | |
|---|---|
| ISCCP | International Satellite Cloud Climatology Project |
| ISLSCP | International Satellite Land Surface Climatology Project |
| ISO | International Organization for Standardization |
| ISSC | International Social Science Council |
| IT | Industrial Transformation (IHDP); information technology |
| ITCZ | intertropical convergence zone |
| ITO | International Trade Organisation |
| IUCN | International Union for the Conservation of Nature and Natural Resources (the World Conservation Union) |
| IWC | International Whaling Commission |
| IWRB | International Waterfowl Research Bureau |
| IWRM | integrated water resources management |
| JERS-1 | Japanese Earth Resource Satellite-1 |
| JI | joint implementation |
| K/T | Cretaceous-Tertiary |
| LAC | Local Area Coverage |
| LACIE | Large Area Crop Inventory Experiment |
| Landsat | system of U.S. land-surface observation satellites |
| LCA | life cycle analysis |
| LDC | less developed countries; London Dumping Convention |
| LEPA | low-energy precision application |
| LFG | landfill gas |
| LGM | last glacial maximum |
| LIS | Lightning Imaging Sensor |
| LLN | Louvain-la-Neuve |
| LRTAP | (Convention on) Long-Range Transboundary Air Pollution |
| LUCC | Land Use and Land Cover Change |
| LULC | Land Use and Land Cover |
| LWT | Lamb Weather Type |
| MAB | (United Nations) Man and the Biosphere Programme |
| MARC | Monitoring and Assessment Research Centre |
| MARPOL | negotiations of the International Convention for the Prevention of Pollution from Ships |
| MBIS | Mackenzie Basin Impact Study |
| MCA | multicriteria analysis |
| MDC | more developed countries |
| MDR | multidrug resistance |
| ME | multieffect evaporation |
| MIGA | Multilateral Investment Guarantee Agency |
| MIS | marine isotope stages |
| MM | Modified Mercalli intensity scale |
| MNC | multinational corporation |
| MOBILHY | (HAPEX) Mode'lisation du Bilan Hydrique (France) |
| MODIS | Moderate Resolution Imaging Spectrometer |
| MoP | Meeting of Montreal Protocol Parties |
| MOPITT | Measurement of Pollution in the Troposphere |
| MORECS | U.K. Meteorological Office Rainfall and Evaporation Calculation System |
| MPP | massively parallel processor |
| MSF | multistage flash |
| MSS | (Landsat) Multispectral Scanner |
| MSY | maximum sustainable yield |
| MTPE | Mission to Planet Earth (NASA) |
| NADW | North Atlantic Deep Water |
| NAFO | North Atlantic Fisheries Organization |
| NAFTA | North American Free Trade Agreement |
| NASA | (U.S.) National Aeronautics and Space Administration |
| NASA ER-2 | NASA high-altitude research aircraft |
| NASDA | National Space Development Agency of Japan |
| NAT | nitric acid trihydrate |
| NATO | North Atlantic Treaty Organisation |
| NBII | (U.S.) National Biological Information Infrastructure |
| NCB | National Coal Board |
| NCDC | (U.S.) National Climate Data Center |
| NCP | (Montreal Protocol) Non-Compliance Procedure |
| NDVI | Normalized Difference Vegetation Index |
| NEP | net ecosystem production |
| NEPA | National Environment Policy Act |
| NERC | (U.K.) Natural Environment Research Council |
| NGO | nongovernmental organization |
| NHGRI | National Human Genomic Research Institute |
| NIEO | new international economic order |
| NMO | national member organization |
| NOAA | (U.S.) National Oceanic and Atmospheric Administration |
| NPP | net primary production; net primary productivity |
| NPS | (U.S.) National Park Service |
| NSF | (U.S.) National Science Foundation |
| NWT | North West Territories (Canada) |
| OCMIP | Ocean Carbon-Cycle Model Intercomparison Project |
| ODP | ozone depletion potential |
| OECD | Organisation for Economic Co-operation and Development |
| OEM | original equipment manufacturers |

| | | | | |
|---|---|---|---|---|
| OFDA | Office of U.S. Foreign Disaster Assistance | | RMP | (IWC) Revised Management Procedure |
| OGCM | ocean general circulation model | | RMS | (IWC) Revised Management Scheme |
| OGI | old growth index | | RNA | ribonucleic acid |
| OH | free radical hydroxyl | | RO | reverse osmosis |
| OIES | (UCAR) Office for Interdisciplinary Earth Studies | | ROEC | Reed Odorless Earth Closet |
| OPEC | Organization of the Petroleum Exporting Countries | | R-strategists | Short-lived organisms characterized by short generation times and the ability to produce and disperse offspring efficiently and abundantly (e.g., dandelions, cockroaches) |
| OSL | optically stimulated luminescence | | SAGE | Semi-Automatic Ground Environment (U.S. Air Force) |
| OTA | Office of Technology Assessment | | | |
| PAGES | (IGBP) Past Global Changes | | SAR | Second Assessment Report; (JERS-1) Synthetic Aperture Radar |
| PAH | polyaromatic hydrocarbon | | | |
| PAM | plant available moisture | | SBI | Subsidiary Body for Implementation |
| PAN | plant available nutrients | | SBSTA | Subsidiary Body for Scientific and Technological Advice |
| PAT | population, affluence, and technology | | | |
| PC | personal computer | | SBUV | Solar Backscatter Ultraviolet Instrument |
| PCB | polychlorinated biphenyl | | SCAQMD | South Coast Air Quality Management District (California) |
| PCE | perchloroethylene | | | |
| PCH | polychlorinated hydrocarbon | | SCICEX | Scientific Ice Expeditions |
| PDMS | postdepositional modification stratigraphy | | SCOPAC | Standing Conference on Problems Associated with the Coastline |
| PET | polyethylene terephthalate; potential evapotranspiration | | SCOPE | Scientific Committee on Problems of the Environment |
| pH | hydrogen ion concentration | | SD | statistical dynamical (model) |
| PIC | prior informed consent | | SEA | strategic environmental assessment |
| PNA | Pacific North American | | SEEA | system of integrated environmental and economic accounts |
| POC | particulate organic carbon | | | |
| POP | persistent organic pollutant | | SFS | sea floor spreading |
| ppb | parts per billion | | SIA | social impact assessment |
| ppbv | parts per billion by volume | | SMART | Save Money and Reduce Toxics |
| ppm | parts per million | | SNA | system of national accounting |
| ppmv | parts per million by volume | | SOHO | solar and heliosphoric observatory |
| ppt | parts per trillion | | SPM | Summaries for Policymakers; suspended particulate matter |
| pptv | parts per trillion by volume | | | |
| PR | Precipitation Radar | | SPOT HRV-XS | Systeme Pour l'Observation de la Terre (SPOT) High-Resolution Visible (HRV) Multispectral (XS) |
| PRA | probabilistic risk assessment | | | |
| PSC | polar stratospheric cloud | | | |
| PURPA | Public Utility Regulatory Policies Act of 1978 | | SQHW | small quantities of hazardous waste |
| PVC | poly(vinyl chloride) | | SSP | Second Sulfur Protocol (to the Convention on Long-Range Transboundary Air Pollution) |
| PVP | parallel vector processor | | | |
| QA/QC | quality assurance and quality control | | SSS | sea surface salinity |
| QRA | quantitative risk assessment | | SST | sea surface temperature |
| quad | a unit of energy; one quad equals one quadrillion Btu | | SSURGO | Soil Survey Geographic Database |
| | | | STAP | Scientific and Technical Advisory Panel (of the Global Environment Facility) |
| RAID | redundant array of independent devices | | | |
| RC | one-dimensional radiative-convective model | | START | The global change SysTem for Analysis, Research and Training |
| RD | relative dating | | | |
| RECLAIM | Regional Emissions Clean Air Incentives Market (Southern California) | | STATSGO | State Soil Geographic Database |
| | | | STP | sewage treatment plant |
| REE | rare-earth element | | SVD | singular-value decomposition |
| RIS | reservoir-induced seismicity | | SWDA | Solid Waste Disposal Act |

| | |
|---|---|
| SWE | snow water equivalent |
| TAR | (IPCC) Third Assessment Report |
| TCE | trichloroethylene |
| TCP | Technical Cooperation Programme |
| TEK | traditional ecological knowledge |
| TEL | tetraethyl lead |
| TFR | total fertility rate |
| TL | thermoluminescence |
| TM | (Landsat) Thematic Mapper |
| TMI | TRMM Microwave Imager |
| TOA | top of the atmosphere |
| TOC | total organic carbon |
| TOGA | Tropical Ocean and Global Atmosphere program |
| TOMS | Total Ozone Mapping Spectrometer |
| TRMM | Tropical Rainfall Measuring Mission |
| TSCA | (EPA) Toxic Substances Control Act |
| TSS | total suspended solids |
| TVA | Tennessee Valley Authority |
| UARS | Upper Atmosphere Research Satellite |
| UCAR | University Corporation for Atmospheric Research |
| UKMO | U.K. Meteorological Office |
| UN | United Nations |
| UNCED | United Nations Conference on Environment and Development |
| UNCLOS | United Nations Convention on the Law of the Sea |
| UNCOD | United Nations Conference on Desertification |
| UNDP | United Nations Development Programme |
| UNEP | United Nations Environment Programme |
| UNESCO | United Nations Educational, Scientific, and Cultural Organization |
| UNFPA | United Nations Population Fund (formerly United Nations Fund for Population Activities) |
| UNICEF | United Nations Children's Fund |
| UNIDO | United Nations Industrial Development Organization |
| UPS | United Parcel Service |
| URL | Uniform Resource Locator |
| USAID | U.S. Agency for International Development |
| USGCRP | U.S. Global Change Research Program |
| USGS | U.S. Geological Survey |
| UTH | upper tropospheric humidity |
| UTM | Universal Transverse Mercator |
| UV | ultraviolet (radiation) |
| UV-B | ultraviolet-B (radiation) |

| | |
|---|---|
| VC | vapor compression; Vienna Convention for the Protection of the Ozone Layer of 1985 |
| VIP | ventilated improved pit |
| VIRS | Visible and Infrared Scanner |
| VOC | volatile organic compounds |
| WCED | World Commission on Environment and Development |
| WCP | World Climate Programme |
| WCRP | World Climate Research Programme |
| WHO | World Health Organization |
| WIPO | World Intellectual Property Organization |
| WMO | World Meteorological Organization |
| WOCE | World Ocean Circulation Experiment |
| WRAP | Waste Reduction Always Pays |
| WTO | World Trade Organization |
| WWF | World Wide Fund for Nature (formerly World Wildlife Fund) |
| WWW | World Weather Watch; World Wide Web |
| YD | Younger Dryas |

## Measures and Conversions

| | |
|---|---|
| centimeter | 0.39 inches |
| foot | 12 inches (0.3048 meters) |
| gram | 0.0353 ounces |
| inch | 2.54 centimeters |
| kilogram | 1,000 grams (2.2046 pounds) |
| kilometer | 1,000 meters (3,280.8 feet or 0.62 miles) |

## Prefixes for International System of Units

| | |
|---|---|
| yotta (Y) | $10^{24}$ |
| zetta (Z) | $10^{21}$ |
| exa (E) | $10^{18}$ |
| peta (P) | $10^{15}$ |
| tera (T) | $10^{12}$ |
| giga (G) | $10^{9}$ |
| mega (M) | $10^{6}$ |
| kilo (k) | $10^{3}$ |
| hecto (h) | $10^{2}$ |
| deca (da) | $10^{1}$ |
| deci (d) | $10^{-1}$ |
| centi (c) | $10^{-2}$ |
| milli (m) | $10^{-3}$ |
| micro ($\mu$) | $10^{-6}$ |
| nano (n) | $10^{-9}$ |
| pico (p) | $10^{-12}$ |
| femto (f) | $10^{-15}$ |
| atto (a) | $10^{-18}$ |
| zepto (z) | $10^{-21}$ |
| yocto (y) | $10^{-24}$ |

# ENCYCLOPEDIA OF GLOBAL CHANGE

# A

## ACID RAIN AND ACID DEPOSITION

*Acid rain* is a colloquial term that refers to acids and associated compounds in rainfall, cloud water, snow, and airborne particles. *Acid deposition* is a more accurate and inclusive term. Acidity is the concentration of hydrogen ions in a substance; the negative logarithm of the hydrogen ion concentration (pH) is the common measure used for reporting the acidity of natural waters. Lower values correspond to greater acidity. Acid deposition has increased during this century as a result of human activity. Acid deposition has a variety of effects on natural and human systems, some of which have been well documented.

All rain is naturally somewhat acid because the reaction between water and atmospheric carbon dioxide forms some carbonic acid: [*See* Atmospheric Chemistry.]

$$H_2O + CO_2 \leftrightarrow H_2CO_3. \tag{1}$$

The carbonic acid then partially dissociates, contributing hydrogen ions and thus lowering the pH of precipitation to 5.65 (rather than a neutral 7.0):

$$H_2CO_3 \leftrightarrow H^+ + HCO_3^-. \tag{2}$$

Acid rain refers to rainfall that has a pH lower than 5.65. Naturally generated acid compounds (from trees, fires and volcanoes, for example) can in some cases contribute to pH values as low as 5.0. Acid rain derived from human activity can result in rainfall with pH values as low as 2.6.

The precursors of acid deposition generated by human activity are the primary pollutants nitrogen oxide (NO) and sulfur dioxide ($SO_2$). Nitrogen oxides are released during combustion when atmospheric nitrogen ($N_2$) combines with atmospheric oxygen. [*See* Nitrous Oxide.] Most fuels contain nitrogen, which also combines with oxygen to form nitrogen oxide. When a compound containing sulfur is burned, the sulfur combines with atmospheric oxygen to form sulfur oxides. Mobile pollution sources such as automobiles generate one-third to one-half of nitrogen oxides. The combustion of coal and oil and other stationary-source activities (such as the smelting of metals) emit the majority of sulfur oxides. [*See* Fossil Fuels.] Sulfur dioxide emissions vary according to the sulfur content of the fuel and the presence or absence of pollution-control devices in the combustion plant. Nitrogen oxide emissions vary according to the combustion temperature and the presence or absence of pollution-control devices.

*Acid deposition* is a more accurate term than acid rain because acids that are deposited from the atmosphere occur in a variety of forms: rain, snow, clouds, and as dry particles or vapors. In some cases, the amount of acidity contributed by dry deposition (fine particles, coarse particles, and vapors) can be greater than that contributed by precipitation. In a forest in Tennessee, dry deposition accounted for 55 percent of the hydrogen (acidity) and sulfate ions and 64 percent of the nitrate ions. There are two additional measures of the acid status of waters: acid-neutralizing capacity (ANC), which is roughly equivalent to alkalinity, and base-neutralizing capacity (BNC), which is roughly equivalent to acidity. Other substances contained in fuels are released during combustion and contribute to the overall or net acidity. For example, fly ash (noncombustible byproducts of combustion) usually has a high ANC and is released during combustion of oil and coal. In addition, coal and oil contain small quantities of lead, mercury, and other metals that are emitted when they are burned. Although technically not part of acid deposition, fly ash and metals are often associated with the term *acid rain*.

In the presence of atmospheric oxygen and water, oxides of nitrogen and sulfur are transformed (primarily through oxidation) into nitric and sulfuric acid. Simplified equations for these reactions are:

$$2\,SO_2 + O_2 \rightarrow 2\,SO_3 \tag{3}$$

$$SO_3 + H_2O \rightarrow H_2SO_4 \rightarrow H^+ + HSO_4^- \leftrightarrow 2\,H^+ + SO_4^{2-} \tag{4}$$

$$2\,NO + O_2 \rightarrow 2\,NO_2 \tag{5}$$

$$NO_2 + OH \rightarrow HNO_3 \rightarrow H^+ + NO_3^-. \tag{6}$$

The transformation from the primary pollutants sulfur dioxide and nitrogen oxide to the secondary pollutants sulfuric acid and nitric acid occurs over a number of days. Once transformation occurs, the secondary pollutants are more readily washed out of the atmosphere and can contribute to acid deposition. If the reactions take place in a moving air mass, long-distance transport of the pollutants occurs. The reactions involving nitrogen occur more rapidly than those involving sulfur. Thus nitrogen compounds may be transported hundreds of kilometers while sulfur compounds may be transported more than a thousand kilometers before deposition occurs.

In the past three decades, many scientists have focused their research on the acid and sulfate compounds in acid deposition; more recently, nitrate has also been studied.

**History.** Acid deposition occurs whenever the precursors $SO_2$ and $NO$ are present in the atmosphere. Thus acid rain has existed for as long as there have been lightning, forest fires, and volcanoes. But the term *acid rain* generally refers to excess acidity in wet and dry deposition caused by human activity. The earliest documented acid rain events occurred in London and Manchester, England, and Pittsburgh, Pennsylvania. Robert Angus Smith is usually credited with first describing acid rain near Manchester in 1852. Similar events were probably occurring in many other cities around the world at this time. Referring to the contemporary scientific literature, Eriksson (1952) describes the sulfur concentration of rainfall for a variety of locations in Britain, France, Germany, Russia, Japan, and New Zealand during the late 1800s and early 1900s. Since all of these locations except some parts of Russia had sulfur levels above those of New Zealand, Eriksson described them as being impacted by "atmospheric pollution." Herman and Gorham (1957) first reported elevated sulfur and hydrogen ion concentrations in rainfall. Cogbill and Likens (1974) published the first thorough description of acid precipitation in the northeastern United States. They showed that prevailing winds carry acids from midwestern states toward the northeast. They also generated a series of pH isopleths (lines of equal pH) superimposed on a map, which has been widely used since that time (Figure 1).

Later studies showed that transboundary pollution has occurred from the United States to Canada. Similar transboundary transport has occurred in Europe, where the prevailing winds carry pollutants from England, Germany, and the Netherlands to Scandinavia (Figure 2). [*See* Convention on Long-Range Transboundary Air Pollution.]

During the 1970s and 1980s, acid deposition received a great deal of attention from the press, the public and scientists. It was blamed for causing much environmental damage in many parts of the industrialized world. Strong evidence for acid rain damage has been found in many fewer systems than were originally thought, and today acid rain receives much less attention.

**Extent.** The pH isopleth figures shown here (Figure 1) document the extent of acid deposition in Europe and North America. There have been reports of regional acid deposition in West Africa, Japan, South America, and many areas in the former Soviet Union (Figure 2). It is reasonable to expect that there will be acid deposition downwind of any industrial activity or large concentration of motorized vehicles.

**Impacts of Acid Deposition.** There are many hypothetical or confirmed pathways and mechanisms by which acid deposition may affect aquatic ecosystems, forest ecosystems, agricultural systems, human health,

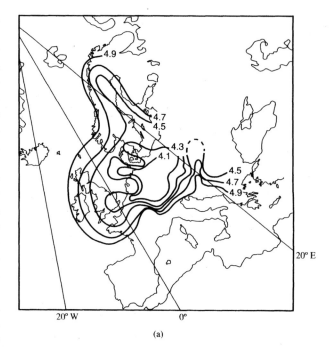

(a)

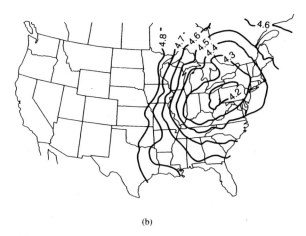

(b)

**Acid Rain and Acid Deposition. FIGURE 1.** pH Isopleths for North America and Europe.

Contour lines of long-term, volume-weighted averages of pH in precipitation in (a) Europe, 1978–1982, and (b) Eastern North America, 1980–1984. (After Spiro and Stigliani, 1996, p. 202.)

buildings and structures, and visibility. For some of these, it is important to distinguish between direct and indirect effects. For example, in aquatic systems, the lowering of the pH of lake water is a primary effect; the mobilization of toxic aluminum that results from the decrease in the pH is a secondary, or indirect, effect.

***Aquatic ecosystems.*** Acid deposition has led to documented acidification (the lowering of pH over time) of lakes and streams in certain portions of northeastern North America, Scandinavia, and the United Kingdom. Decreases in the species diversity of fish and crus-

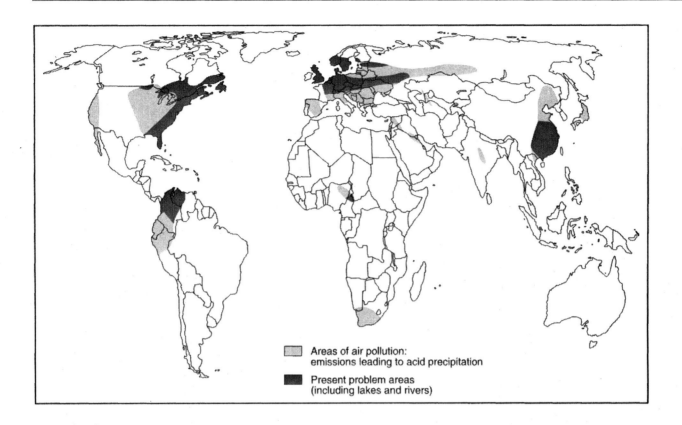

Areas of air pollution:
emissions leading to acid precipitation

Present problem areas
(including lakes and rivers)

**Acid Rain and Acid Deposition.** FIGURE 2. Worldwide Extent of Acid Deposition and Acid Precursors. (Adapted from Kaufman and Franz, 1993, p. 263.)

taceans have been observed in acidified freshwater lakes and streams. Species diversity changes most between pH 3 and pH 6. Lower pH interferes with the osmotic balance of certain aquatic organisms and can also release metals into the water through a process called *mobilization*. These metals are normally bound in complex organic or inorganic compounds in soils and sediments. Since the mobilization of many metals (such as aluminum and mercury) in the water column also has an impact on the physiological function of aquatic organisms, it is not clear whether species changes are a direct or indirect effect of acid deposition.

In many species, differing responses in adults and juveniles have been observed. Furthermore, a decrease in pH can affect the food sources of a fish before it affects the fish itself, further complicating the distinction between direct and indirect effects of acid deposition.

*Forest ecosystems.* Sulfur dioxide, a precursor of acid deposition, can directly affect plants by absorption through stomata. Sulfur dioxide is usually found in close proximity to a combustion source or smelter.

There are few observable effects of acid deposition on individual trees and seedlings at typical levels of acid deposition in North America. However, forest ecosys-

tems are much more complex than individual trees, and we know little about the long-term effects of acid deposition on forests. [*See* Forests.] In North America, there is only one widely accepted effect of acid rain on forests: a component of acid deposition—probably sulfate—has been shown to reduce the cold tolerance of high-elevation red spruce in the Northeast. This has led to a greater incidence of freezing injury to red spruce, seen especially during the period from the 1960s to the 1980s. Acid deposition is believed to have acidified some soils, indirectly affecting forests in central and eastern Europe. A lower pH in rainfall could lead to mobilization of aluminum and accelerated chemical weathering of rock underlying soil. Excess sulfate deposition is believed to lead to excess leaching of both sulfate (a mobile anion) and of positively charged ions such as calcium and magnesium. Some have considered this indirect effect—the mobilization of cation nutrients—to be more serious than any direct effect of acid deposition in forested ecosystems.

*Agricultural systems.* Acid deposition is not one of the major pollutants affecting crop growth or yield. Most crop species are much more sensitive to ozone than acid deposition. Agricultural crops are highly managed and, in most cases, a farmer can compensate for any effects of acid deposition with additional fertilizer. Although there may be a financial cost, ambient levels of acidity are not known to have adverse effects on crop growth or yield, except for crops immediately adjacent to major emission sources. In some cases, nitrogen and

There has been progress in controlling sulfur dioxide ($SO_2$) emissions in many developed countries; but the problem is growing rapidly in Asia, where energy use has surged, and sulfur-rich coal and oil are prevalent. In 1990, $SO_2$ emissions in Asia were 1.4 those in North America. Acid deposition levels were particularly high in southeast China, northeast India, Thailand, and the Republic of Korea. Sulfur dioxide emissions in Asia could soar from 40 million metric tons per year in 1990 to 110 million metric tons per year by 2020 as growth is restored to Asian economies.

In Europe and North America, it is nitrogen oxides that are intractable: their levels, being closely linked to vehicles and electric power generation, have not fallen appreciably since 1980.

**BIBLIOGRAPHY**

Downing, R., R. Ramankutty, and J. Shah. "An Assessment Model for Acid Deposition in Asia." Washington, D.C.: The World Bank, p. 11, 1977.
Organisation for Economic Co-operation and Development. *Environmental Data Compendium.* Paris: OECD, 1997.
Swedish Secretariat. "The Swedish Secretariat on Acid Rain." *Acid News* 5 (1996), 14.
World Resources Institute. *World Resources 1998–99.* New York, pp. 181–183, 1998. The WRI article cites Downing, 1977, and OECD, 1997.

—DAVID J. CUFF

sulfur components of acid deposition may even act as a fertilizer.

***Human health.*** Direct effects of acid deposition occur when asthma sufferers are exposed to sulfur dioxide or sulfuric acid vapor. Only areas in the immediate vicinity of large sources of sulfur pollution are of concern. In such areas, sulfuric acid and acid aerosol exposure can impair the respiratory function of children, the elderly, and those with preexisting health problems. Nitrogen oxide contributes to the formation of tropospheric ozone, which is also a respiratory irritant.

***Buildings and structures.*** Acid deposition has an adverse impact on metal-based and calcium carbonate (limestone and marble) building materials. [*See* Building Decay.] The overall reaction between calcium carbonate rock and acid precipitation is:

$$CaCO_3 + H_2O + CO_2 \rightarrow Ca^{2+} + 2HCO_3^-. \qquad (7)$$

The water and carbon dioxide react as shown in equations (1) and (2) to create acidity ($H^+$). The calcium carbonate consumes $H^+$ and generates $Ca^{2+}$. Acids impact most exposed painted surfaces, possibly including automobile finishes. Some stone structures, such as the Acropolis in Athens, have deteriorated as a result of reactions with gaseous sulfur dioxide ($SO_2$).

***Visibility.*** Sulfuric acid and sulfate aerosols are important contributors to the impairment of visibility in many parts of the industrial world. This impaired visibility, sometimes called *sulfate haze*, occurs in dry climates and can be exacerbated by humidity. Annual haziness in the eastern United States increased by 25 percent between 1950 and 1985. Increased haze trends for the eastern United States are strongly associated with increased sulfur dioxide emission trends. Figure 3 illustrates the extent to which visibility has been reduced in recent times in the eastern United States. Ammonium nitrate, which can form in part from nitrogen oxide emissions, is also a contributor to visibility impairment.

**Acid Deposition in a Changing Global Environment.** Sulfate aerosol is the component of atmospheric acidity that most directly influences global climate. There is strong evidence that sulfate aerosols scatter solar radiation and change the reflective properties of clouds, both of which exert a cooling influence on global climate. Although this cooling effect cannot be compared directly to the warming effect of global greenhouse gases, they may be approximately equal in magnitude (Charlson et al., 1992). However, the continuing reduction in sulfate aerosols associated with the introduction of additional pollution-control measures should cause a steady decrease in the influence of sulfate aerosol on global climate.

While acid deposition has been present in most of the industrialized areas of the world for much of the twentieth century, there is good reason to believe that it has varied in its severity. During the early and middle years of the century, factories and power plants emitted more acid precursors, but they also emitted fly ash containing basic compounds that partially neutralized acidity. As factories increased their pollution-control measures, they emitted considerably less fly ash. The ANC of the

**Acid Rain and Acid Deposition. FIGURE 3.** Clear visibility (a) and impaired visibility (b) over Grand Canyon National Park, Arizona. (Photograph courtesy of Holle Photography.)

emissions was therefore much less, and most acidity was neutralized by its reaction with soil minerals. In addition, as urban areas became more polluted, taller smokestacks were constructed. These taller stacks reduced urban air pollution but increased the amount of nitrogen and sulfur oxides that was transported to distant areas.

In the United States and many other industrialized countries, legislators have focused on sulfur rather than nitrogen emissions, in part because sulfuric acid, with its two hydrogen ions (compared to one in nitric acid) has been deemed to have greater potential for contributing to acidification. As a result, sulfur emissions in the United States (and other developed countries) have dropped since 1970, whereas nitrogen emissions have remained constant or increased during the same time period (Figure 4).

Since the 1980s, investigators in Europe and North America have begun to consider more closely the nitrogen component of acid deposition. Many terrestrial systems are limited by nitrogen, and so it is often thought that the nitrogen added by air pollution may be beneficial. However, there is evidence from a number of ecosystems that too much nitrogen has begun to adversely affect ecosystem function. This nitrogen saturation hypothesis has gained much attention in recent years and is the newest aspect of acid deposition to be studied.

As the global environment changes, acid deposition and its effects may change as well. For example, in a warmer climate, the conversion of sulfur oxides to sulfuric acid in the atmosphere might occur more rapidly, in which case acid rain may be offset and acid deposition distributed more widely. Greater rainfall volumes

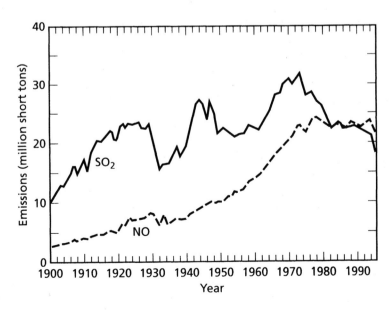

**Acid Rain and Acid Deposition. FIGURE 4.** NO and SO$_2$ Emissions in the United States from 1900–1995. (Modified from U.S. Environmental Protection Agency, 1996, p. 3.)

tend to dilute the acidity in rainwater, so if a region were subject to greater rainfall under a global change scenario, it might receive rainfall with a higher pH. These projections are highly tentative and might be considered third-order effects of global change. To date, they have received little scientific attention.

[*See also* Air Quality; Nitrogen Cycle; *and* Pollution.]

### BIBLIOGRAPHY

Barker, J. R., and D. T. Tingey, eds. *Air Pollution Effects on Biodiversity*. New York: Van Nostrand Reinhold, 1992.

Boyle, R. H., and R. A. Boyle. *Acid Rain*. New York: Shocken Books, 1983.

Charlson, R. J., S. E. Schwartz, J. M. Hales, R. D. Cess, J. A. Coakley, Jr., J. E. Hansen, and D. J. Hoffmann. "Climate Forcing by Anthropogenic Aerosols." *Science* 255 (1992), 423–430.

Cogbill, C. V., and G. E. Likens. "Acid Precipitation in the Northeastern United States." *Water Resources Research* 10 (1974), 1133–1137.

Drabløs, D., and A. Tollan, eds. *Ecological Impact of Acid Precipitation*. Oslo: SNSF Project, 1980.

Erikson, E. "Composition of Atmospheric Precipitation II. Sulfur, Chloride, Iodine Compounds." *Tellus* 4 (1952), 280–303.

Godbold, D. L., and A. Huttermann, eds. *Effects of Acid Rain on Forest Processes*. New York: Wiley, 1994.

Heij, G. J., and J. W. Erisman, eds. *Acid Rain Research: Do We Have Enough Answers?* Amsterdam: Elsevier, 1995.

Herman, F. A., and E. Gorham. "Total Mineral Material, Acidity, Sulfur and Nitrogen in Rain and Snow at Kentville, Nova Scotia." *Tellus* 9 (1957), 180–183.

Irving, P. M., ed. *Acidic Deposition: State of Science and Technology*. Washington, D.C.: U.S. Government Printing Office, 1991.

Kaufman, D. G., and C. M. Franz. *Biosphere 2000: Protecting Our Global Environment*. Dubuque, Iowa: Kendall/Hunt Publishing, 2000.

Lindberg, S. E., G. M. Lovett, D. D. Richter, and D. W. Johnson. "Atmospheric Deposition and Canopy Interactions of Major Ions in a Forest." *Science* 231 (1986), 141–145.

Lucier, A. A., and S. G. Haines, eds. *Mechanisms of Forest Response to Acidic Deposition*. New York: Springer-Verlag, 1990.

National Research Council. *Acid Deposition: Atmospheric Processes in Eastern North America*. Washington, D.C.: National Academy Press, 1983.

Rodhe, H., and R. Herrera, eds. *Acidification in Tropical Countries*. New York: Wiley, 1988.

Schindler, D. W. Effects of Acid Rain on Freshwater Ecosystems. *Science* 239 (1988), 149–157.

Spiro, T. G., and W. M. Stigliani. *Chemistry of the Environment*. Upper Saddle River, N.J.: Prentice-Hall, 1996.

U.S. Environmental Protection Agency. National Air Pollutant Emission Trends, 1900–1995. Research Triangle Park, N.C.: EPA, 1996.

—ANDREW J. FRIEDLAND

# AEROSOLS

Aerosols are technically defined as a suspension of solid or liquid particles in a gas. By this standard, cloud droplets are technically aerosols, but it is common within atmospheric sciences to use the term to refer to the fine particles within the atmosphere. These may be either solid or liquid, and they are considered distinct from drops primarily because of their size. Drops grow to roughly 10–20 microns in radius and exist in regions of the atmosphere that have a relative humidity near and above 100 percent, whereas aerosols are generally less than 10 microns in radius. Larger aerosols can be lifted into the air—soil grains lifted in dust storms, for example, or the bursting of whitecap bubbles or direct injection of drops creating particles of sea salt—but these larger particles are heavy enough to fall back to the surface within a few hours, so that they have no large-scale effects on the atmosphere. [*See* Dust Storms.]

**Role of Aerosols in Cloud Formation.** Aerosols are intimately involved in the formation of clouds. [*See* Clouds.] It is generally quite difficult to form a cloud directly from water vapor—a process termed *homogeneous nucleation*—because water vapor molecules must first form clusters of sufficient size so that further growth is reasonably fast. The formation of these clusters is generally very slow. In the atmosphere, water vapor condenses onto aerosol particles. This process can occur under subsaturated conditions (relative humidity less than 100 percent) if the particle contains compounds that attract water and therefore decrease the equilibrium vapor pressure of water over the particle. However, as the relative humidity of an air parcel is raised above 100 percent, there comes a point at which the growth of the particle by condensation of vapor becomes unstable, allowing the particle to continue to take up water until the relative humidity is reduced back to about 100 percent. This is the mechanism by which warm cloud droplets form in the atmosphere. The particles that facilitate this process are called *cloud condensation nuclei*.

Ice clouds, on the other hand, may form from deposition of vapor directly onto particles (termed *deposition ice nuclei*), from freezing of solution drops (termed *homogeneous ice nuclei*), or from freezing induced by the presence of a solid-phase particle (termed *heterogeneous ice nuclei*). The particles that are able to form ice nuclei are either very rare or require high supersaturations to reach the critical point at which ice nucleation occurs. Large regions of the atmosphere could therefore form ice clouds (that is, they have relative humidity above the saturated vapor pressure over ice), but do not do so because they lack sufficient nuclei of the appropriate type. Because the relative humidity of vapor over ice is less than the relative humidity of vapor over liquid water, once ice begins to form, ice clouds may grow rapidly from supercooled water clouds.

**Sources and Sinks of Aerosols.** Particles may be composed of sulfate, nitrate, organics, compounds associated with soil dust, compounds associated with the sea salt aerosol, ammonia, soot or black carbon, and trace metals. The total trace metal mass is small and is therefore ignored in the following discussion. The sources of these components in the aerosol vary significantly with region. Thus, over the Amazon and in desert areas of the western United States, organic carbon is the most abundant fine aerosol particle component, whereas in rural continental areas of the eastern United States, sulfate accounts for more than 50 percent of the fine-particle mass.

Table 1 presents global average estimates for the sources of the atmospheric aerosol. The fine-particle fraction of the atmospheric aerosol (defined as that portion with diameter less than 2 microns) contains most of the aerosol number and is, therefore, the most important to quantify in order to determine the effects of aerosol on the radiation balance and on clouds. Aerosol sources may be divided into gas phase sources and primary particulate sources. Gas phase sources are aerosol precursor emissions that are converted through photochemical reactions to products that condense to the particulate phase. Primary particulate sources are emitted as particles at or near the source. The gas phase sources include sulfate, nitrate, ammonia, and organic gases. Each of these has sources that originate from either natural or anthropogenic gases. [*See* Air Quality; *and* Atmospheric Chemistry.] Estimates of the magnitude of the anthropogenic and natural sources, based on current estimates in the literature, are included in the table. Estimates of the possible range in source strengths, given current understanding, are shown in

parentheses. Some of the aerosol components are volatile—that is, they exist in both the aerosol phase and the gas phase and adjust their concentrations in the aerosol phase, depending on the local aerosol concentrations of nonvolatile species as well as the local vapor pressure of the vapor compound. These components include nitrate, ammonium, the chloride in sea salt, and some organic compounds. As a result, the table assumes that approximately half of the airborne portion of nitric acid ($HNO_3$) is found as aerosol $NO_3^-$, and half of the sources of ammonia ($NH_3$) become aerosol ammonium ($NH_4^+$). Volatile chloride and volatile organics were not considered in Table 1.

Primary particles are those that are associated with smoke from fires, dust deflation by wind, sea salt from direct injection and bursting of bubbles, and biogenic particles (fungi, spores, epicuticular plant waxes, etc.). [*See* Fire.] Fires mainly originate from anthropogenic activity (burning for agriculture, to clear land, burning of fossil fuels). Dust outbreaks have also been associated with anthropogenic activity, since agricultural processing of the land can lead to more frequent dust outbreaks and because some desertification is associated with land use changes. Tegen and Fung (1995) estimated that about 50 percent of their total dust source strength was associated with land use changes. For dust and sea salt, the table shows an estimate of the total source strength as well as that portion in the fine-particle fraction.

Nearly 50 percent of the total emissions of fine aerosol particles appear to derive from anthropogenic sources. This estimate is a factor of two to five times

**Aerosols. TABLE 1.** Sources of Tropospheric Aerosols (radius less than 2 microns)

| TYPE | Source strength (teragrams per year) | |
| --- | --- | --- |
| | ANTHROPOGENIC | NATURAL |
| Gas-to-particle conversion | | |
| Sulfates | 120 (90–160) | 55 (25–110) |
| Ammonium | 21 (10–30) | 5 (2–10) |
| Nitrates | 30 (15–50) | 22 (5–120) |
| Organic aerosols | — | 30 (10–80) |
| Primary emissions | | |
| Biomass burning for land clearing and agriculture | 50 (10–110) | |
| Fossil fuel burning, industry | 35 (15–75) | |
| Domestic fuels and bagasse burning | 11 (5–25) | |
| Dust | 160 (60–300) (~750 for $r < 10$ microns) | 160 (60–300) (~750 for $r < 10$ microns) |
| Sea salt | | 140 (100–180) (~12,000 for $r < 10$ microns) |
| Biogenic | | 50 (0–100) |
| Total (fine-particle fraction) | 430 (48 percent) (200–750) | 460 (52 percent) (200–1,000) |

larger than previous estimates of the role of anthropogenic aerosols in the aerosol particle budget (see, for example, Prospero et al., 1983), in part because in the current analysis the fine-particle fraction of the aerosol mass has been emphasized. The estimates for all sources in Table 1 clearly demonstrate that atmospheric aerosol sources have a large anthropogenic component and may have changed over time.

When either liquid or ice clouds form precipitation, they also remove the aerosol from the atmosphere. This is the primary sink of atmospheric aerosols, although deposition, wherein aerosols are brought into contact with the surface and remain there, is also a (relatively minor) removal pathway.

**Role of Aerosols in Climate.** Aerosols are important to climate because they scatter and absorb radiation in the atmosphere and because they change the microphysical structure and possibly the lifetime of clouds within the atmosphere. Radiation in the atmosphere is divided into solar and thermal radiation. The former has wavelengths ranging from 0 to about 4 microns, while the latter has wavelengths ranging from 4 to 20 microns. Because most aerosols are smaller than 10 microns and because aerosol particles interact most strongly with wavelengths near their size, most of the scattering and absorption of radiation by aerosols takes place in the solar spectrum. An exception to this rule applies to the larger dust particles that interact strongly with both solar and infrared radiation. The scattering of solar radiation acts to cool the planet, while absorption of solar radiation by aerosols warms the air directly instead of allowing that sunlight to be absorbed by the surface. When sunlight is absorbed by the surface, the heat is transferred back to the atmosphere either by thermal (heat) radiation, by conduction of heat, or by evaporation of water which, when it condenses, releases the heat back to the atmosphere.

As a result of their abundance and the size of droplets, clouds both scatter solar radiation and absorb thermal radiation. Optical depth is a measure of the strength of interaction of clouds with radiation. Small changes in the size of cloud droplets, especially for clouds of intermediate optical depth, greatly increase their scattering of solar radiation while having little effect on their interaction with thermal radiation. The warm liquid clouds at low altitude mainly scatter solar radiation, while the ice clouds at high altitude mainly absorb thermal radiation. Thus, the former act mainly to cool the planet, while the latter act to warm it. Therefore, changes in aerosol concentrations over time have the potential to influence greatly the radiation balance, and therefore the climate.

**Ice-Core Record of Aerosols.** Ice cores provide a proxy record for past climate changes, gas concentrations, and aerosol concentrations. [See Climate Change.]

Aerosol concentrations are recorded directly within the falling snow since these aerosols were either condensation nuclei or ice nuclei. The snow falls and is eventually transformed to ice after the pressure of overlying snow compacts it sufficiently. As this compaction takes place, the porous snow and firn (partly consolidated snow) start to encapsulate air volumes, which then also contain a record of past air composition. Oxygen-18 is the heavy isotope of oxygen, with a molecular weight of 18 instead of the normal 16. D is the heavy isotope of hydrogen, with a molecular weight of 2 instead of the normal 1. Because the heavy-isotope content of precipitation decreases as the condensation temperature decreases, the record of $HD^{16}O$ and $H_2^{18}O$ relative to $H_2^{16}O$ provides a proxy for the temperature at which the precipitation formed, which itself is a proxy for the atmospheric temperature, if the cycle of water evaporation, transport, and removal remained relatively constant over the period of interest.

Ice-core records have been obtained from sites in both Antarctica and Greenland. Each differs in the type of information they provide: Greenland is surrounded by continents, and the proximity of human activity has allowed aerosols associated with the growth of industrial and anthropogenic activities to be recorded. Antarctica, on the other hand, is surrounded by ocean and allows a study of cycles in natural aerosol concentrations. The longest records available are those from Vostok, East Antarctica, covering the past 220,000 years—this ice core therefore covers the last glacial cycle (20,000–15,000 BP) and interglacial cycle (125,000 BP) as well as the penultimate glacial cycle (140,000 BP). Recent drilling projects in Greenland have extended the record back to this same period. Unfortunately, it has been shown that at sites with low snow accumulation rates (such as the continent of Antarctica), the deposition of aerosols such as dust, sea salt, and sulfur is not closely tied to their concentration in falling snow because a variable amount of dry deposition leads to an impure signal in the snow. Thus, a 10°C decrease in local temperature associated with the last ice age may have led to a drier atmosphere and to a reduction in snow accumulation rate. This would record roughly doubled concentrations of aerosols within the ice core without any associated increase in atmospheric concentrations. Nevertheless, concentrations of insoluble species such as dust were more than ten times higher during the Last Glacial Maximum than they are now. In addition, sea salt concentrations were much higher during the last two glacial periods than now. These results have been interpreted as due either to higher wind speeds during glacial maxima or to a more efficient meridional transport in the Northern Hemisphere. In Antarctica, similar increases during the last glacial maximum have been interpreted as being associated with an expansion of arid areas and exposure of the continental shelves. Another aerosol species that was enhanced in Antarctica was sulfate. Because a similar enhancement of methane sulfonic acid (MSA)—which is associated with marine biological sources of sulfate—was also recorded, it is thought that the antarctic increase was due to enhanced biological activity during the glacial maximum. No evidence of an increase in biological activity is evident in the ice cores from Greenland, however.

The increases in sea salt and sulfate aerosols would have led to additional cooling during ice ages, as pointed out by Overpeck et al. (1996). The increased dust, because of its ability to absorb solar and thermal radiation, could actually lead to regional warming events. This absorption might help to explain the abrupt warming events that took place as the glacial era was ending.

Changes in ionic composition found in ice cores have been used as indicators of sudden events of large magnitude. Thus volcanic events, which are a natural source of sulfate aerosol in both the stratosphere and the troposphere, are recorded by large enhancements of sulfate, while forest fires are recorded by enhancements to ammonium, formate, and nitrate. If these events are screened out of the record for the last two hundred years, it can be seen that the Greenland ice core has substantially increased concentrations of sulfate (as well as nitrate), whereas the antarctic ice core does not. The record for sulfate aerosol in Greenland is in broad agreement with the increasing emissions of anthropogenic sulfur dioxide. Decreased concentrations recorded since 1980 have been interpreted as being due to the reductions in emissions that have taken place in Europe and the United States.

**Role of Aerosols in Recent and Projected Climate Change.** Because the sources of anthropogenic aerosols represent a relatively large fraction of the total source strength, it is important to try to understand the possible influence of these recently increasing aerosols on past and future climates. One method for estimating their possible climatic influence is to estimate the radiative forcing associated with the anthropogenic fraction of aerosol. Radiative forcing is the change in outgoing radiation at the top of the troposphere caused by a specific change in composition. It gives a first-order measure of the long-term steady-state response of the climate, because the global average change in surface temperature is related approximately linearly to the radiative forcing. Thus,

$$\Delta T_s \approx \lambda \Delta F$$

where $\lambda$ is defined as the climate sensitivity and $\Delta F$ is the radiative forcing. The climate sensitivity depends on the change in water vapor (the most abundant greenhouse gas in the atmosphere), clouds, sea ice cover and snow, and may vary considerably between different climate models with different representations of physical processes.

It is important to note that radiative forcing is only an approximate measure of climate effects. It relies on the concept that the surface temperature change is effectively communicated throughout the troposphere by convective and radiative processes. However, because emissions into specific regions can alter the vertical profile of temperature and/or the response of clouds in different regions, the concept is only approximately valid. Nevertheless, it provides a first-order estimate of the relative impacts of different emissions. Radiative forcing by aerosols may be compared to the forcing from greenhouse gases and from changes in solar radiation over the past one hundred years. These latter forcings were estimated as about 2.5 watts per square meter (W m$^{-2}$) in a recent scientific assessment.

Recent estimates for the direct forcing by anthropogenic sulfate aerosols and carbonaceous aerosols are between $-0.3$ and $-0.9$ W m$^{-2}$ for sulfate, between $-0.15$ and $-0.25$ W m$^{-2}$ for biomass aerosols (domestic fuels and land clearing), and between $+0.15$ and $+0.2$ W m$^{-2}$ for fossil fuel and industrial sources of organic and soot aerosols. No specific estimates of climate forcing from anthropogenic nitrates or anthropogenic ammonia are yet available, but some authors consider that forcing by these aerosol compounds is small relative to the forcing aerosols. Anthropogenic sources of dust aerosols are estimated to have a global forcing of $+0.09$ W m$^{-2}$ according to Tegen et al. (1996). The climate forcing associated with these estimates could offset a substantial fraction of the warming associated with greenhouse gases, and several climate models have simulated the effect of sulfate aerosols to show that the record and pattern of temperature change are consistent with the fact that the cooling provided by sulfate aerosols is mainly in the Northern Hemisphere. If the ratio of aerosol emissions to greenhouse gas emissions remains constant, however, these impacts should decrease in the future, because the short lifetime of aerosols ensures that the future buildup of concentrations will not keep pace with the buildup of long-lived greenhouse gases.

While estimates of climate forcing by direct reflection and absorption of radiation are reasonably straightforward, estimating the indirect effects of aerosols has proven much more difficult. This is mainly because the indirect effects depend on natural as well as anthropogenic sources, the state of mixing of the aerosol components (i.e., whether the aerosol formed is a new cloud condensation nucleus or whether the aerosol adheres to a preexisting aerosol), and the capability of large-scale models to treat changes in cloudiness in response to increases in anthropogenic aerosols. The state of mixing has been shown to lead to uncertainties ranging from $-0.4$ to $-1.5$ W m$^{-2}$ in forcing by sulfate aerosols associated with changes in drop size distributions (Chuang et al., 1997) and to uncertainties ranging from $-1.1$ to $-4.8$ W m$^{-2}$ in forcing by sulfate aerosols associated with changes in cloudiness (Lohmann and Feichter, 1997). Changes in droplet concentrations due to carbonaceous aerosols from biomass burning, fossil fuels, and industry might also lead to large estimates of negative forcing (Penner et al., 1996). These estimates may be far too large, but if they are proven to be correct, then our explanations of recent climate changes and the consistency of the historical record of temperature change with the record of radiation forcing will need to be altered.

Finally, aerosols from aircraft have been associated with possible increases in cirrus clouds (Jensen and Toon, 1997). If cirrus cloudiness were to increase as a result of aircraft emissions, the likely impact would be a positive forcing leading to a warming, since the net influence of cirrus is to warm the planet. The radiative forcing of cirrus increase caused by aircraft emissions was estimated in the range of 0 to 0.02 W m$^{-2}$.

[*See also* Albedo; Atmosphere Dynamics; Gaia Hypothesis; Global Warming; Nuclear Winter; Pollution; Sulfur Cycle; *and* Volcanoes.]

## BIBLIOGRAPHY

### GENERAL REFERENCES

Chuang, C. C., J. E. Penner, K. E. Taylor, A. S. Grossman, and J. J. Walton. "An Assessment of the Radiative Effects of Anthropogenic Sulfate." *Journal of Geophysical Research* 102 (1997), 3761–3778.

Intergovernmental Panel on Climate Change. *Climate Change 1995: The Science of Climate Change: Contribution of Working Group I to the Second Assessment Report of the Intergovernmental Panel on Climate Change*, edited by J. T. Houghton, L. G. Meira Filho, B. A. Callander, N. Harris, A. Kattenberg, and K. Maskell. Cambridge: Cambridge University Press, 1996. A thorough summary with estimates of radiative forcing and an explanation of the concept with papers current as of 1995.

Jensen, E., and O. B. Toon. "The Potential Impact of Soot Particles from Aircraft Exhaust on Cirrus Clouds." *Geophysical Research Letters* (1997), 249–252.

Legrand, M., and P. Mayewski. "Glaciochemistry of Polar Ice Cores: A Review." *Reviews of Geophysics* 35 (1997), 219–243. A nice summary of what has been learned about aerosols and gases from ice cores.

Lohmann, U., and J. Feichter. "Impact of Sulfate Aerosols on Albedo and Lifetime of Clouds: A Sensitivity Study with the ECHAM-4 GCM." *Journal of Geophysical Research* 102 (1997), 13685–13700.

Penner, J. E. "Atmospheric Chemistry and Air Quality." In *Changes in Land Use and Land Cover: A Global Perspective*, edited by W. B. Meyer and B. L. Turner II, pp. 175–209. Cambridge and New York: Cambridge University Press, 1994. A detailed but concise discussion of sources and sinks of greenhouse gases and aerosols affecting atmospheric chemistry and climate.

Penner, J. E., C. C. Chuang, and C. Liousse. "The Contribution of Carbonaceous Aerosols to Climate Change." In *Nucleation and Atmospheric Aerosols 1996*, edited by M. Kulmala and P. E. Wagner, pp. 759–769. Oxford: Pergamon, 1996.

Prospero, J. M., R. J. Charlson, V. Mohnen, R. Jaenicke, A. C. Delany, J. Moyens, W. Zoller, and K. Rahn. "The Atmospheric Aerosol System: An Overview." *Reviews of Geophysics and Space Physics* 21 (1983), 1607–1629.

Tegen, I., and I. Fung. "Contribution to the Atmospheric Mineral Aerosol Load from Land Surface Modification." *Journal of Geophysical Research* 100 (1995), 18707–18726.

Tegen, I., A. Lacis, and I. Fung. "The Influence on Climate Forcing of Mineral Aerosols from Disturbed Soils." *Nature* 380 (1996), 419–422.

### AEROSOLS AND CLIMATE CHANGE

The following papers discuss climate change associated with aerosols and compare the historical record of temperature change with that from climate models that include radiative forcing by sulfate aerosols.

Hegerl, G. C., et al. "Multi-Fingerprint Detection and Attribution Analysis of Greenhouse Gas, Greenhouse Gas-Plus-Aerosol and Solar Forced Climate Change." *Climate Dynamics* 13 (1997), 613–634.

Mitchell, J. F. B., and T. C. Johns. "On Modification of Global Warming by Sulfate Aerosols." *Journal of Climate* 10 (1997), 245–267.

Overpeck, J., D. Rind, A. Lacis, and R. Healy. "Possible Role of Dust-Induced Regional Warming in Abrupt Climate Change during the Last Glacial Period." *Nature* 384 (1996), 447–450.

Santer, B. D., et al. "Towards the Detection and Attribution of an Anthropogenic Effect on Climate." *Climate Dynamics* 12 (1995), 77–100.

———. "A Search for Human Influence on the Thermal Structure of the Atmosphere." *Nature* 382 (1996), 39–46.

Taylor, K. E., and J. E. Penner. "Response of the Climate System to Atmospheric Aerosols and Greenhouse Gases." *Nature* 369 (1994), 734–737.

—Joyce E. Penner

# AGASSIZ, LOUIS

Jean Louis Rodolphe Agassiz (1807–1873), Swiss–American natural scientist, was born in Motier-en-Vuly, Switzerland, and grew up in the Swiss countryside, where his interest in nature blossomed. On his route to becoming a world-renowned scientist, he had to overcome his family's desire that he enter the field of commerce or medicine. After he received a doctorate of philosophy from the Universities of Munich and Erlangen in 1829, he complemented this in the following year with a doctorate in medicine from the University of Munich. By the time of his medical doctorate, he had already embarked on what was to become a celebrated and varied scientific career, touching many aspects of the natural and biological sciences. Agassiz's early work included a study and classification of Amazon fishes with Karl von Martius and publication over the years

1833–1844 of a five-volume study on fossil fishes, *Poissons fossiles*. The latter contains careful descriptions of more than seventeen hundred ancient species and related zoology to both geology and paleontology. Agassiz is most famous, however, for his formulation and advocacy of a theory that at sometime in the past, during an ice age, glaciers engulfed huge portions of Europe and northern North America. [*See* Glaciation.]

Agassiz was not the first to recognize that individual glaciers at some point had been more extensive. Among his forerunners in this respect were the mountaineer Jean-Pierre Perraudin, the naturalist Johann de Charpentier, and the civil engineer Ignatz Venetz. Perraudin had reached this conclusion as early as 1815, after he decided that many features in the Swiss landscape found well downstream of current glaciers—including granite boulders, piles of rock debris termed *moraines*, polished rock surfaces, and systematic scratches (or striations) on exposed rocks—could most readily be explained as having been deposited or formed by glaciers that formerly extended outward to the locations of these features. (Others felt instead that these features had been deposited or formed by water during Noah's flood.) Perraudin helped convert both Venetz and de Charpentier, who later, in 1836, contributed to convincing the previously skeptical Agassiz. Once convinced, Agassiz became the most prominent advocate of this theory.

By 1837 Agassiz had extrapolated far beyond local evidence and had formulated a full-scale theory of the occurrence of an age in which ice extended over vast portions of the Earth's land cover. On 24 July 1837, he gave his major initial statement of the ice-age theory as part of his presidential address to the Swiss Society of Natural Sciences, meeting in the town of Neuchâtel. Later known as the *Discourse at Neuchâtel*, the address included the sweeping theory that huge ice masses, brought about by markedly lower temperatures than at present, had covered Europe and Asia from their northernmost reaches south to the Mediterranean and Caspian Seas. The theory was received with considerable hostility, but Agassiz was persistent, and his talks, writings, and journeys into the field with influential opponents to examine the evidence firsthand all contributed over the subsequent years to the growing acceptance of an ice-age concept. As early as 1839, new supporters in both North America and Europe had begun amassing evidence of past glaciations on both continents. Agassiz himself gathered evidence from England and Scotland in the early 1840s. Agassiz's best-known work on glaciers, *Études sur les glaciers* (*Studies on glaciers*), was published in 1840 and elaborated this ice-age theory, as well as described the structure, temperature, color, formation, and movement of glaciers, and various glacial processes.

In 1846 Agassiz crossed the Atlantic and began both an eager search for local evidence of former North American glaciation and a popular lecture tour that started in Boston and proceeded to other New England cities. The following year he accepted a professorship of zoology and biology at Harvard University, a career change that came shortly after the death of his wife of fifteen years, Cécile Braun. Agassiz remained a professor at Harvard until his own death in 1873. During the period of his Harvard professorship, he became well known as a teacher and popular lecturer while also gaining further recognition for his research, which now included extensive fieldwork in the Americas. Continuing to seek additional details on the glacial, geologic, and biological history of the Earth, and to collect specimens for museums, he undertook expeditions throughout the eastern and central United States and portions of South America, complementing his earlier fieldwork in central Europe and Great Britain. During much of his Harvard career, he was assisted in his research by his second wife, Elizabeth Cabot Cary.

Agassiz's strong advocacy of the ice-age theory and the evidence he assimilated were central in its growing scientific acceptance through the mid-nineteenth century. By the end of the century, findings from dozens of researchers had led to significant enhancements of the original concept, including the idea of multiple ice ages rather than a single occurrence, quantified estimates of the lowering of sea level due to the existence of the land ice, extensive explicit delineation of the geographic extent of the ice, and theories attempting to explain ice-age occurrences. The basic concept, however, that the Earth has experienced climatic regimes that are vastly different and much icier than today's, remained intact.

Although Agassiz was in the forefront with respect to the ice-age theory, he was known also at the time for being a leading opponent of another noted contemporary scientific theory, namely, the theory of evolution put forward by Charles Darwin in 1859. Agassiz viewed ancient species that are not identical to modern ones as fully separate, with no evolutionary connection. Darwin, on the other hand, accepted major aspects of Agassiz's ice-age theory and used them to explain some of the current geographical distributions of particular plants and animals.

Agassiz's many honors and awards included the Copley Medal of the Royal Society of London and the Wollaston Medal of the Geological Society of London. Complementing his research and teaching, he founded the Museum of Comparative Zoology at Harvard in 1859, founded the Anderson School of Natural History on Penikese Island, Massachusetts, in 1873, and was influential in the establishment of the U.S. National Academy of Sciences in 1863. Agassiz died of a cerebral hemorrhage in Cambridge, Massachusetts, on 14 December 1873, shortly after returning from a research expedition to the southernmost portions of South America.

### BIBLIOGRAPHY

Agassiz, L. *Études sur les glaciers*. Neuchâtel, 1840. Agassiz's major text on glaciers.

———. *Studies on Glaciers Preceded by the Discourse of Neuchâtel*. Translated and edited by Albert V. Carozzi. New York: Hafner, 1967. Translations of Agassiz's major text on glaciers (Agassiz, 1840) and his 1837 presidential address to the Swiss Society of Natural Sciences, laying out his first major statement of the ice-age theory.

Chorlton, W. *Planet Earth: Ice Ages*. Alexandria, Va.: Time-Life Books, 1983. A nontechnical description of ice ages, the impacts of ice ages on humans and other animals, vegetation, sea level, and the landscape, possible causes of ice ages, and the history of the scientific study of ice ages.

Crowley, T. J., and G. R. North. *Paleoclimatology*. New York: Oxford University Press, 1991. An advanced text presenting a modern understanding of the Earth's climate history and the placement of ice ages within it, based on both observations and numerical modeling. Although not referring to Agassiz, the book details how far the concept and elaboration of major climate change has progressed since Agassiz's initial formulation of an ice-age theory less than two centuries ago.

Hallam, A. *Great Geological Controversies*. New York and Oxford: Oxford University Press, 1983. Examinations of five major historical controversies in geology, including, in Chapter 3, the ice-age controversy and Louis Agassiz's role in it. Although written for professional geologists and geology students, the book is also appropriate for others interested in the history of science.

Imbrie, J., and K. P. Imbrie. *Ice Ages: Solving the Mystery*. Hillside, N.J.: Enslow, 1979. An enjoyably readable history of both the concept of ice ages and the attempts to explain them. Agassiz is a central figure in Chapters 1 and 2.

Lurie, E. *Louis Agassiz: A Life in Science*. Chicago: University of Chicago Press, 1960.

———. "Agassiz, Jean Louis Rodolphe." In *Dictionary of Scientific Biography*, edited by Charles C. Gillispie, vol. 1, pp. 72–74. New York: Scribner's, 1970.

—Claire L. Parkinson

# AGENDA 21

Agenda 21 is the lengthy plan of action (approximately five hundred pages) adopted at the United Nations Conference on Environment and Development (Earth Summit) in Rio de Janeiro in 1992. [*See* United Nations Conference on Environment and Development.] It is to serve as a blueprint for national and international efforts extending into the twenty-first century to promote the dual objectives of preserving the natural environment and facilitating the economic development of the poorer nations of the world. As with comparable documents adopted at other United Nations–sponsored conferences, Agenda 21 is not a formal treaty and thus not a legally binding document. Nevertheless, it has been broadly embraced by the international community and

has triggered a myriad of activities around the world aimed at promoting sustainable development.

Agenda 21 drew inspiration from the Brundtland Commission's report *Our Common Future* (1987), which stressed that environmental concerns could not be divorced from economic development, especially in the developing countries. [*See* Brundtland Commission.] The plan was drafted at the PrepCom meetings that preceded the Earth Summit, in which governments as well as many nongovernmental organizations (NGOs) were given opportunities to participate and make proposals, before being finalized in Rio. The document proposes upward of twenty-five hundred specific actions to be undertaken nationally or internationally, which are presented in forty chapters grouped into four major sections. The first section addresses social and economic problems, such as poverty, demographic trends, human health, and human settlements. The second section focuses on issues pertaining to the environment and the management of resources—in particular, those related to atmosphere, deforestation, desertification, biological diversity, biotechnology, the oceans, water, toxics and hazardous wastes, and radioactive substances. The third section draws attention to the roles that may be played by certain groups, such as women, children and youth, NGOs, local authorities, workers, business and industry, the scientific community, and farmers. The fourth section offers proposals for implementing the plan, including how to finance it, capacity building in developing countries, institutional and legal arrangements, transfer of technology, and education and public awareness. It was estimated that implementation of the multifaceted plan would require an annual contribution of U.S.$125 billion by the developed countries, or about 0.7 percent of their collective gross national product.

Later, in 1992, the United Nations General Assembly created the Commission on Sustainable Development (CSD) to oversee efforts to implement Agenda 21. While lacking both the authority to make binding decisions and significant financial resources, the CSD has become the principal international forum for the continuing discussion of issues raised in Agenda 21. The new body has also assumed responsibility for reviewing the reports of states on their efforts to carry out the Agenda 21 recommendations and to further the cause of sustainable development generally. By 1999 upward of 117 countries had established national sustainable development commissions, and a coalition of more than 120 corporations had formed the World Business Council for Sustainable Development. In 1997 the General Assembly convened a special session, called "Rio Plus Five," to review progress in implementing the Agenda 21 plans. The assessment was discouraging and underscored, in particular, the failure of advanced industrialized countries to provide the substantial additional economic and technological assistance needed to carry out the vision.

## BIBLIOGRAPHY

Bryner, G. C. "Agenda 21: Myth or Reality?" In *The Global Environment: Institutions, Law, and Policy*, edited by N. J. Vig and R. S. Axelrod, pp. 157–189. Washington, D.C.: CQ Press, 1999. An overview of Agenda 21 and an assessment of progress toward its implementation.

Sitarz, D., ed. *Agenda 21: The Earth Summit Strategy to Save Our Planet.* Boulder: EarthPress, 1993. A readable summary and analysis of Agenda 21.

United Nations. *Agenda 21 Earth Summit: United Nations Program of Action from Rio.* New York: United Nations, 1992. The full, unedited text of Agenda 21.

—MARVIN S. SOROOS

## AGRICULTURAL REVOLUTION. *See* Human Populations.

## AGRICULTURE AND AGRICULTURAL LAND

The use of land for the production of food, fiber, and fuel is the principal driver of land use and land cover change and, therefore, a key forcing function of global environmental change. The overwhelming majority of agriculture involves the control and enhancement of nature to achieve production, leading to the modification and transformation of land covers (e.g., forests, grasslands, wetlands) and the manipulation of soil and water (e.g., fertilization, irrigation). As a result, changes in agricultural land and the strategies of production contribute significantly to so-called greenhouse gases and to impacts on the hydrologic cycle.

When the meaning of "global change" is expanded beyond climate warming, the reach of agriculture extends even further. Most of the humanly modified and transformed surface of the Earth is connected in some way to agriculture (Table 1). These changes alter the structure and function of ecosystems and reduce biotic diversity, while the production process pollutes the troposphere, primarily through the nitrogen cycle. These types of relationships make agricultural land and production susceptible not only to potential climate change but also to subglobal environmental feedbacks, and they heighten agriculture's role in a full array of environmental issues: to name a few, the migration of biota across disrupted and fragmented landscapes in the face of latitudinal and altitudinal climate shifts, the degradation of marine and lacustrine ecosystems from agricultural runoff, and soil and groundwater depletion.

Beyond these environmental impacts, agriculture draws attention because of its role in a more crowded and consumption-driven world. A doubling of the world's population in the early part of the twenty-first century and the expectation of increasing "average" consumption, including increased emphasis on livestock, will significantly intensify the pressures on agricultural land and production. Adding to these pressures is the competition for urban, industrial, recreation, and preservation land uses, most of which outbid agriculture for land. It is expected, however, that technological and managerial improvements over the long run will increase production on that land remaining in agriculture.

**History of Agricultural Attributes.** The role of agriculture and agricultural land in global environmental change is understood through the history of the trends in key attributes of land use and production: substitution and intensification, expansion, mechanization, crop diversity, and production–consumption linkages.

With few exceptions, agriculture is predicated on manipulating and controlling nature in order to achieve increased production and security over harvest losses. The technologies and strategies used to accomplish

**Agriculture and Agricultural Land.** TABLE 1. Global Areas of Different Land Covers Converted to Cropping, 1860–1978 (thousands of square kilometers)

|  | FORESTS | WOODLANDS | SAVANNAS | GRASSLANDS | SWAMPS | DESERTS | TOTAL |
|---|---|---|---|---|---|---|---|
| Developed regions |  |  |  |  |  |  |  |
| Canada/U.S.A. | 511 | 130 | 107 | 849 | — | ? | 1,597 |
| Europe | 78 | 3 | 18 | 84 | 34 | — | 217 |
| Former USSR | 439 | 136 | 17 | 884 | — | 32 | 1,509 |
| Oceania | 239 | 123 | 133 | 51 | — | 5 | 551 |
| Total | 1,267 | 392 | 275 | 1,868 | 34 | 37 | 3,874 |
| Developing regions |  |  |  |  |  |  |  |
| Africa | 180 | 289 | 239 | 325 | — | 31 | 1,064 |
| Asia | 592 | 628 | 500 | 374 | 177 | 73 | 2,344 |
| Latin America | 384 | 253 | 161 | 401 | 1 | 31 | 1,233 |
| Total | 1,156 | 1,170 | 900 | 1,100 | 178 | 135 | 4,641 |
| World Total | 2,423 | 1,562 | 1,175 | 2,968 | 212 | 172 | 8,515 |

SOURCE: Williams (1990). Reprinted with the permission of Cambridge University Press.

## BIOREGIONALISM AND PERMACULTURE

*Bioregionalism* is the desire to commit oneself to local places and ecosystems and to act locally to improve the environment. *Permaculture* is a movement that aims to devise and recover cooperative and sustainable ways of living with the land: forms of agriculture that do not impoverish and destroy the soil; technologies that work with nature rather than against it; forms of social organization to accompany and support these agricultures and technologies.

**BIBLIOGRAPHY**

Jackson, W. *Rooted in the Land: Essays on Community and Place.* New Haven, Conn.: Yale University Press, 1996.
McGinnis, M. V. *Bioregionalism.* London and New York: Routledge, 1999.
Mollison, J. *Permaculture: A Practical Guide for a Sustainable Future.* Washington, D.C.: Island Press, 1990.
Sale, K. *Dwellers in the Land: The Bioregional Vision.* Athens: University of Georgia Press, 2000.

—ANDREW S. GOUDIE

these goals vary considerably. They may be as simple as burning savanna grasses to enhance new shoots at the onset of the rainy season, thus providing a superior feed for livestock, or they may be as complex as hydroponics, in which computer-controlled cultivation takes place in nutrient-loaded water, not soil. The former strategy relies almost entirely on nature to provide the inputs required for the output (feed), while the latter strategy substitutes fully for nature, even to the extent of controlling climate when undertaken in greenhouses.

Most systems of agriculture, of course, fit somewhere between the extremes in the two examples, from slash-and-burn cultivation and pastoral nomadism to irrigated crops with satellite-informed fertilization and cattle feedlots. Generally speaking, systems that rotate cropping plots and grazing pastures (low-frequency use) tend to rely on natural processes to replenish the critical elements lost in production (e.g., soil nutrients, pasture biomass). Increased frequency of use and increased yields invariably require increased substitutes for nature—soil nutrient and moisture enhancement (fertilizers and irrigation), stabilization of soil on slopes (terraces), reduction of inundation (wetland fields and drainage), and control of nature's vagaries—pests and weeds (pesticides and herbicides), frost and freezes (mounding and mulching), or drought (irrigation). Variants of these and other strategies abound throughout the world, giving rise to many forms of agroecosystems. Yet agriculture historically exhibits a path toward ever-increasing substitutes to increase yields and decrease labor inputs. The overall trend is toward increasing intensification through yields and frequency of cropping.

From the onset of domestication, agriculture expanded to include almost all types of arable land on Earth, and in some cases, such as the polders of the Netherlands, the creation of arable land where none existed. The global scale, therefore, reveals an ever-increasing portion of Earth placed under cropping and grazing production. This global trajectory, however, has not been sequentially uniform by region. Subsequent to about 3000 BP, the most expansive reach of agriculture was found in tropical to midlatitude Eurasia and the Nile drainage. From 3000 BP until 500 BP, significant expanses of agriculture emerged everywhere, including the Western Hemisphere tropics. Subsequently, European colonialism profoundly altered agriculture globally by the huge scale of transcontinental species exchanges that followed, especially the movement of livestock to the Western Hemisphere. The twentieth century witnessed major expansion in the cultivation of grasslands in the Americas, Russia, and Australia, spurred by mechanization and irrigation, and of tropical forest biomes worldwide.

The general global trajectory is replete with counterexamples—regions that have witnessed major declines in cultivation. Salinization led to abandonment of ancient systems of desert irrigation, and population losses and relocation created large-scale abandonment of tropical forest and wetland agriculture in the Americas. More recently, much marginally arable land in western Europe and North America has been taken out of cultivation, and megalopolises have taken over prime agricultural lands.

Mechanization is the dominant trajectory of tool technology in agriculture, serving to reduce labor and increase production. For example, worldwide tractor use per thousand hectares of arable land increased from twelve in the late 1960s to nineteen by the beginning of the 1990s. Mechanization does not necessarily track with intensification per se, however. It is possible to achieve large yields and intensive production in labor-rich systems. Mechanization dominates land-rich and labor-sparse agricultural conditions, such as the Great Plains of the United States. It also appears in land-sparse and labor-rich conditions, such as South Asia, in concert with the use of high-yielding crop varieties. In either of these general cases, research and development aimed at improving mechanized tilling, crop-enhancing applications, and harvesting increases as well, even among relatively impoverished smallholders. For example, the powered "miniplow" has spread from China throughout Asia, diesel pumps lift irrigation water throughout the tropical world, and "backpack" sprayers are increasingly used to apply fertilizers, pesticides, and herbicides. Mechanization, coupled with high-yielding crops, facilitates an increasing size in the number of people fed per farmer. Currently, a Great Plains farmer feeds 75.8 people.

The agricultural revolution some ten thousand years ago led to a proliferation of domesticates, many of them highly localized in their geographic distribution. Subsequently, certain domesticates spread over large areas, becoming dominant staple crops for large regions: rice in lower-latitude Asia; wheat in Europe, North Africa, and Asia Minor; sorghum and millets in sub-Saharan Africa; maize in North and Central America; potato in the Andes; and manioc in Amazonia. Individual localities, however, bred numerous varieties of these and other major crops as well as retaining highly localized species.

Perhaps the first global-level shock to this diversity coincided with the European colonization of the world from the sixteenth through the nineteenth centuries. The impacts of this colonization on population, settlements, land access, and political economies profoundly affected the distribution of biota globally through intentional and unintentional exchange of plants, animals, and pathogens. Species favored by Europeans were privileged in these changes, leading to the spread of wheat, barley, and oats in suitable environments, and sheep, goat, and cattle to almost every agricultural region. And yet many other crops were also spread worldwide during this time: maize, manioc, potato, beans, tomato, and chili, to name a few. These exchanges began the process of lowering the diversity of domesticates everywhere, save in the most isolated locales. The second shock came in the middle of the twentieth century with the emergence of the "green revolution" and the high-yielding varieties emerging from it. The productivity of these crops, and the socioeconomic structures set in place to promote them, continues to reduce the varieties of staple crops everywhere and marginalizes knowledge about the many local species that did not fall into the favored cropping packages. The intensity of focus on an increasingly small number of crops has raised alarm from those concerned about the implications for genetic pools of domesticates.

Finally, the economic structure of agriculture continues to change. The long history of local production and consumption has given way to the spatial separation of the two. Few pure subsistence agriculturalists remain, including pastoralists. Smallholders everywhere engage the market, and the market increasingly is extralocal, serving national and international markets. These same markets, of course, serve the large-holder and corporate and state farms. Produce from anywhere is consumed everywhere, and increasingly the price received by the farmer is affected by international commodity exchanges. Hidden within this complex production–consumption linkage, however, is subsistence production, which persists in part because of the inability of many smallholders and pastoralists to gain sufficient and secure income from commodity production alone.

**Distribution of Major Agricultural Types.** These broader trends in agriculture and their connections to global change and sustainability are more appropriately understood in terms of the specific type or class of systems in question. While no taxonomy of agriculture is accepted worldwide, the attributes on which typologies should be based include the plant or animal focus of production; the intensity of land use; the use of production for subsistence, market, or some mix of the two; and the technomanagerial strategies of production (Table 2). These attributes bleed into one another, making taxonomies difficult and often highly regional in applicability. In general, however, animal and crop production are typically separated, and subdivisions of each are based on the attributes.

*Livestock production.* Large stock (especially cattle, sheep, and goats) are overwhelming important in land use and global change because of the total amount of land devoted to their rearing and of the greenhouse gases emitted by the rise of cattle numbers globally (Table 3). Three major types of large stock systems exist: pastoral nomadism, ranching/dairying, and feedlots. *Pastoral nomadism* is an ancient practice in which the entire populace or a large segment thereof moves seasonally with the livestock in search of natural feed. It is a mistake to underestimate the importance of this practice today, even if production increasingly is sold. Major areas throughout the arid and semiarid world—northern and eastern Africa, Asia Minor, and Central Asia—are dominated by pastoral nomadic activities, as are cold environments in the Hindu-Kush, Himalayas, and northern Eurasia. Grasslands in the Americas, Australia, and southern Africa support *ranching*, characterized by commercial production of meat and stock byproducts and, increasingly, improved pasture. *Dairying*, of course, uses cattle for commercial milk and cheese products and tends to be found in proximity to major urban markets. Its presence is increasing worldwide. The most intensive stock-rearing activities are

**Agriculture and Agricultural Land.** TABLE 2. General Typology and Attributes of Major Agricultural Systems

| TYPE/SUBTYPE | PRINCIPAL BIOTA | INTENSITY | PRIMARY INTENT OF PRODUCTION | TECHNOMANAGERIAL STRATEGY |
|---|---|---|---|---|
| *Livestock* | Large stock | | | |
| • Pastoral nomadism | Large stock | Low | Subsistence | Migration |
| • Ranching/dairying | Large stock | Low–medium | Commercial | Grazing rotation, improved pasture |
| • Feedlots | Large stock | High | Commercial | Supply feed, water |
| *Cultivation: Rain-Fed* | Crops | | | |
| • Shifting | Staple crops | Low | Subsistence | Rotate plots, slash and burn |
| • Permanent | Staple crops | Medium–high | Subsistence/commercial | Tillage, fertilize, weed |
| *Cultivation: Wetland* | Staple crops | Medium–high | Subsistence/commercial | Drainage infrastructure, tillage, fertilize |
| *Cultivation: Irrigated* | Staple crops | | | |
| • Wet rice | Rice | High | Subsistence/commercial | Irrigation and paddy infrastructure, tillage, fertilize |
| • Irrigation | Staple crops | High | Commercial | Irrigation infrastructure, tillage, transplanting, fertilize |
| *Cultivation: Tree and Shrub* | Nonstaple crops | Medium–high | Commercial | Culling, transplanting, pesticides |
| *Horticulture* | Nonstaple crops | High | Subsistence/commercial | Tillage, fertilize, transplant, variable climate control |
| *Agroforestry* | Nonstaple crops, wood fuel, timber | Low–medium | Subsistence/commercial | Culling, planting, fire production |

*feedlots* for cattle. While globally insignificant in area, feedlots are a major source of methane emissions.

**Cultivation.** A significant portion of the Earth's surface has been or is cultivated, the largest portion of which is devoted to rain-fed systems, which are overwhelmingly dominant both spatially and in total crops produced (Table 4). These systems are coordinated with the seasonality of rainfall and focus on staple produce (cereals and root crops). Beyond these general shared qualities, rain-fed cultivation varies significantly. Subdivisions based on the frequency of cropping is significant for global change, however. *Shifting cultivation* refers to that diverse set of systems that rely primarily on the patterned movement of the location of cropped plots. Minimal inputs are employed to enhance or maintain land productivity in the face of soil nutrient depletion and weed and pest invasion. Plots are abandoned, and nature serves to regenerate the land. Shifting cultiva-

**Agriculture and Agricultural Land.** TABLE 3. Changes in Livestock Production, 1961–1991 (million head)

| | Cattle | | Poultry | | Pigs | |
|---|---|---|---|---|---|---|
| REGION | 1961–1963 | 1989–1991 | 1961–1963 | 1989–1991 | 1961–1963 | 1989–1991 |
| More developed | | | | | | |
| North America | 113 | 110.8 | 842.7 | 1447 | 65.2 | 65.1 |
| Europe | 117.4 | 123.1 | 914 | 1289 | 110.6 | 182.9 |
| Oceania | 24.5 | 31.1 | 24.7 | 63.3 | 2.3 | 3 |
| Former USSR | 81.6 | 117.9 | 483 | 1158 | 65.1 | 77.5 |
| Other developed | 16.2 | 18 | 117.7 | 395.3 | 5 | 13.3 |
| Less developed | | | | | | |
| Africa | 102 | 150.8 | 234.7 | 725.3 | 4.2 | 15.1 |
| Near East | 34 | 48.6 | 115 | 642.3 | 0.12 | 0.44 |
| Far East | 288.9 | 359 | 1160 | 4140.4 | 120.8 | 420 |
| Other developing | 0.33 | 0.58 | 2 | 8 | 0.84 | 1.63 |
| World | 956.4 | 1282 | 4257 | 11161.6 | 425.5 | 855.6 |

SOURCE: Heilig (1994).

**Agriculture and Agricultural Land. TABLE 4.** Rain-Fed Agriculture and Projections, 1988/1990–2010

| TYPE OF LAND | MOISTURE REGIME (GROWING PERIOD IN DAYS) | Million Hectares In Use 1988–1990 | In Use 2010 | Balance 1988–1990 | Balance 2010 |
|---|---|---|---|---|---|
| Dry semiarid | 75–119 | 86 | 92 | 68 | 62 |
| Moist semiarid | 120–179 | 148 | 161 | 202 | 189 |
| Subhumid | 180–269 | 222 | 249 | 372 | 344 |
| Humid | 270+ | | | | |
| | | 201 | 232 | 915 | 883 |
| Marginally suitable land in the moist, semiarid, subhumid, humid classes | 120+ | | | | |
| Fluvisols/gleysols | Naturally flooded | | | | |
| | | 64 | 77 | 259 | 246 |
| Marginally suitable fluvisols/gleysols | Naturally flooded | | | | |
| Total with rain-fed potential (of which is irrigated) | | 721 | 812 | 1,816 | 1,725 |
| | | 87 | 108 | | |
| Additional irrigation on nonsuitable land (arid and hyperarid) | | 36 | 38 | | |
| Grand total | | 757 | 850 | 1,816 | 1,725 |

SOURCE: Alexandratos (1995). Copyright 1995. John Wiley and Sons Limited. Reproduced with permission.

tion, therefore, requires access to a considerable land area, most of which is in some stage of regeneration at any given moment. Overwhelmingly associated with subsistence production, shifting cultivation remains significant in the tropical world and is associated with burning forest lands and, in some cases, land degradation. *Permanent rain-fed cultivation*, by contrast, involves systems in which plots are cultivated annually or more, on average, and are rested for only brief periods of time. The depletion of nature's resources are offset by human inputs. In the more intensive systems, land is tilled and weeded, and applications of fertilizers, pesticides, and herbicides are applied to high-yielding varieties of crops. Many of the world's "breadbaskets" rely on mechanized, permanent cultivation, especially in the formerly glaciated soils of the middle latitudes, although irrigation may supplement precipitation.

*Wetland cultivation*, predicated on controlling water within otherwise naturally inundated lands, consumed large areas in the ancient Americas. Today, its significance is marked less by the total area in its employ (e.g., the Netherlands) than by its enormous costs, productivity, and impacts on local hydrologies and methane emissions. Investment in wetlands in the twentieth cen-

tury tended to focus on their drainage, not in the use of wetland systems of production. In contrast, *irrigated cultivation*, also an ancient technology, continues to grow in spatial extent and environmental impacts (Table 5). Predicated on the quality control and delivery of water, it is usually subdivided in types. *Wet rice irrigation* involves significant infrastructures for field flooding and draining in addition to that for water delivery. Subsistence production once dominated wet rice production and remains important, especially in South and Southeast Asia. Increasingly, however, wet rice cultivation has moved to commercial production throughout midlatitude and tropical Asia, as well as in California and Texas, where its cultivation is completely mechanized. The spread of wet rice agriculture contributes significantly to the emission of methane from its flooded fields. Other *irrigation* involves nonflooded cultivation. Often associated with semiarid and arid lands, it involves infrastructures of all sizes, from local streambed runoff to long-distant canals to deep groundwater pumping. Irrigation enhances yields but, if improperly used, can lead to significant land degradation from salinization, as well as to the depletion of aquifers and destruction of downstream ecosystems, illustrated in the death of the Aral

**Agriculture and Agricultural Land.** TABLE 5. World Estimates of Irrigated Land, 1961–2050

| REGION* | Area (Millions of Hectares) | | | | Area Change (Percent) | |
| --- | --- | --- | --- | --- | --- | --- |
| | 1961 | 1989 | 2025 | 2050 | 1989–2025 | 1989–2050 |
| Africa | 7.8 | 11.2 | 12.3 | 12.9 | 1.1 | 1.15 |
| Latin America | 8.2 | 15.8 | 18.2 | 19.8 | 1.15 | 1.25 |
| Middle East | 9.7 | 13.1 | 15.1 | 15.8 | 1.15 | 1.2 |
| China+ | 31.9 | 48.8 | 53.7 | 57.1 | 1.1 | 1.17 |
| South and Southeast Asia | 44.6 | 79.7 | 91.7 | 95.7 | 1.15 | 1.2 |
| North America | 14.4 | 18.9 | 19.9 | 21 | 1.05 | 1.11 |
| Western Europe | 8.3 | 17.8 | 19.6 | 21 | 1.1 | 1.18 |
| Eastern Europe | 1.6 | 5.7 | 6.3 | 6.6 | 1.1 | 1.16 |
| OECD Pacific | 4 | 5 | 5.5 | 5.6 | 1.1 | 1.12 |
| Former USSR | 9.4 | 21.1 | 23.2 | 25.1 | 1.1 | 1.19 |
| LDCs | 102.2 | 168.6 | 190.9 | 201.1 | 1.13 | 1.19 |
| MDCs | 37.7 | 68.6 | 74.5 | 79.4 | 1.09 | 1.16 |
| World | 139.9 | 237.2 | 265.4 | 280.5 | 1.12 | 1.18 |

*LDCs = less developed countries; MDCs = more developed countries; OECD = Organisation for Economic Co-operation and Development.

SOURCE: Leach (1995). Reprinted with the permission of the Stockholm Environment Institute.

Sea. Areas of large-scale irrigation (nonwet rice) include the Central Asian Republics and western China, the Nile and Tigris–Euphrates valleys, south-central Australia, the southern High Plains and western United States, and northern Mexico.

*Tree and shrub cultivation and horticulture* range in kind from house and orchard gardens to plantations and advanced-technology greenhouses. Production emphasizes high-value fruits, oils, condiments, ornamentals, and medicinal species. The house and orchard gardens tend to be subsistence or semisubsistence in orientation, focused on diverse species, and located adjacent to rural settlements and within periurban zones. In contrast, the orchard, vineyard, and plantation (technically a specific labor arrangement) focuses on the commercial monocropping of species requiring special conditions offered by the location. Market gardening, including the greenhouse, were once located adjacent to major markets (e.g., New Jersey and Long Island within the megalopolis of the Atlantic seaboard of the United States). Increasingly, however, the economics of modern transportation relaxes the need for locational proximity, supporting long-distance activities (e.g., the valleys of California, southern Mexico, and Morocco).

Finally, *agroforestry* is a historically important activity that has changed significantly. It once focused on the culling and manipulation of naturally occurring species to ensure a consistent offtake in food, resin, fiber, and wood, much like the present but waning subsistence sector of agroforestry. The "timber plantation"

may be seen as another form of agroforestry, focused on monocropping timber from naturally occurring species or exogenous species. This form of agroforestry is growing significantly in the pinelands of the developed world and, increasingly, in parts of the developing world. Its impacts on the loss of biotic diversity is assumed to be large.

**Major Agricultural Trends.** The significance of agriculture to humankind is attested by the antiquity of domestication and sophisticated means of production. The cultivation of virtually every environment, save those lacking a cropping season or access to water, has long been mastered, as well as biotic means of enhancing production through the use of nitrogen-fixing species, night soils, crop rotations, and mounding and mulching, to name only a few ancient strategies. This knowledge diffused worldwide and was employed and expanded in concert with the longer-term histories of population and settlement change. Prefossil fuel agriculture, however, could not keep pace with the growing demands for produce triggered by the Industrial Revolution, especially with the exponential growth in global population from the middle of the twentieth century onward, accompanied in the latter portion of the century with major increases in global average consumption. These demands refocused the agriculture industry and agricultural science toward enhancing production, giving rise to high-yielding hybrids (genetically manipulated crops) requiring increased applications of industrial-produced fertilizers, pesticides, and herbicides. Yields of major cultivars increased appreciably in the last half of the century, as hybrids of rice, maize, wheat, and potato poured from the Rockefeller-sponsored research and development centers created specifically for this activity. During this time, a staggering 170 percent increase in staple foods was achieved while adding only 1 percent more land (while population doubled). By the end of the century, however, the ability of this "green revolution" to push yields ever higher waned, and increasingly attention is given to the direct manipulation of genetic structure of crops to enhance their resistance to losses resulting from diseases, pests, and droughts.

These advances in the science and technology of production play out within the reality provided by political economic structures. Worldwide, for example, sufficient food is produced to provide for all of humanity at relatively high rates of consumption. The world political and economic systems, however, support highly skewed patterns of distribution and consumption. Not only are global stocks of food skewed in terms of their distribution, but the success of the green revolution in achieving these stocks affected smallholders globally, creating winners and losers in terms of access to technological inputs for production and, ultimately, to land. Thus changes in agriculture and agricultural land use are intimately linked to much broader structural changes, especially in less wealthy economies where large rural labor pools must find new livelihoods, presumably in urban-industrial settings. The success of the green revolution, in part, followed from the realization that a technological solution that increased the total stock of produce globally was easier to achieve than a restructuring of the world political economy to alter distribution and consumption patterns. This reality remains in force, despite international agendas calling it to question and asking for more "sustainable" or environmentally benign policies.

In addition, the first half of the twenty-first century is understood to include a staggering increase in the world's population as well as increases in per capita consumption, urban settlement, and the international flow of people, services, and commodities. Under these conditions, more affluent locales with marginal agricultural lands will continue to follow the twentieth-century trend of de-emphasizing agriculture, in many cases permitting forestation as in New England and western Europe. Less affluent locales, in contrast, will increase agriculture for export, largely through the intensification process, including agriculture within the former Soviet Union. Urban-industrial centers will continue to consume the prime agricultural lands that gave rise to the centers in the first place, as witnessed throughout coastal Asia and exemplified by the changing periurban character of the Pearl River Delta of China, once a breadbasket for rice. The environmental impacts of urban-industrial com-

plexes on regional agriculture will increase, through land competition and tropospheric pollution. Moreover, boom–bust conditions will continue to dominate the remaining "open" agricultural lands of the world, such as Amazonia, leading to significant deforestation and land degradation. These trends, of course, are independent of any impacts that may follow from climate warming or environmental problems.

**Agricultural Implications for Global Environmental Change.** Agriculture and agricultural land use have played an important role in the modification and transformation of the biosphere and promise to do so well into the future. Historically, deforestation for agriculture was the single largest contributor to human-induced carbon dioxide in the atmosphere, and the spread of wet rice cultivation was similarly important as a source of methane. Irrigation in general, especially the drawdown in aquifers, and the drainage of wetlands affected the hydrologic cycle. All these changes played significant roles in regional climate impacts. Future trends of these kinds will likely focus on agricultural expansion at the expense of tropical forests and woodlands, especially in Oceania, Southeast Asia, Africa, and Latin America, and increased irrigation globally. Much of this change will continue to fragment some of the most biotically diverse landscapes in the world, interfering with movements of fauna and generally decreasing biota from edge effects.

Whatever the impacts from expansion, they will likely be matched and exceeded by those of intensification. It is doubtful that major increases in production globally will be accomplished without sustained fossil fuel inputs and further genetic engineering of crops, which will undoubtedly be linked to increased applications of irrigation, fertilizer, herbicides, and pesticides. Nitrogenous fertilizer use worldwide, for example, climbed to a staggering fifty-seven kilograms per hectare of arable land by the 1990s. These inputs will affect trace gas emissions and the hydrologic cycle, just as the intensification of cattle production will increase methane emissions.

Intensification will also significantly affect biotic diversity and local environmental degradation. The enormous diversity in the varieties of domesticated plants continues to be lost in the demand for production of a few major varieties of staples. We now speak of "lost" crops and preserving crop diversity in international seed banks. Poor agricultural practices are the primary cause of surface water pollution in many regions, and agriculture tends to violate one or more pollutant criteria, even where relatively strong enforcement of environmental policies are in place. In the United States alone, it is estimated that farming and ranching are a source of water quality impairment in 60, 50, and 34 percent of impaired river miles, lakes, and estuaries, respectively.

Despite major improvements in soil tillage and grazing practices, agriculture remains a large source of soil erosion and vegetation degradation and is the largest single consumer of fresh water.

**Global Change Implications for Agriculture.** Much uncertainty surrounds the impacts of potential climate change on agriculture and agricultural land use because of the uncertainty in global climate models, the role of enhanced atmospheric carbon dioxide—the fertilization effect—and adaptations that farmers might take. Various modeling efforts suggest little impacts under the low-warming assumption (+1.5°C). Even the high-warming assumption (+4.5°C) may have low-level impacts on global cereal production. Regional variations in the impacts, however, are significant, giving rise to clear winners and losers. The losers appear to be those parts of the tropical world where C4 crops (e.g., maize, sorghum, and millet) dominate. These crops respond less well in experimental conditions to enhanced carbon dioxide conditions. This effect, coupled with increased temperatures and hence increased evapotranspiration, bode poorly for the nonwet rice regions of the tropics—much of Africa and the Americas. In contrast, these same models suggest that C3 crops (e.g., wheat, rice, soybean) will perform well with enhanced carbon dioxide levels, and warmer temperatures will push poleward the use of deeply glaciated soils, such as those on the Canadian plains and in parts of Russia. While the total area of cropped land is likely to increase for reasons that have little to do with climate change, significant changes will take place in lands put into and taken out of cultivation as the relative costs of production change.

Also important for agriculture are the feedbacks from production on tropospheric pollution. Prime crop lands are the source for more than half of the world's emissions of nitrogen oxide to the atmosphere. These lands are also in proximity to major urban-industrial complexes that give rise to ground-level ozone pollution, which reacts with nitrogen oxide to reduce crop yields. This pollution is becoming serious for major breadbasket zones around the world, such as the Ganges basin, and ozone pollution, if unabated, may have significant impacts on various efforts to intensify production.

**Adaptations to Global Change and Degradation.** Perhaps the most important advance in understanding farming behavior and agricultural change in the past quarter century is the recognition that changes in cultivation practices, including the development and use of new technologies, maintain a strong endogenous character. Farmers, from impoverished smallholders to corporate largeholders, are not passive agents but are constantly innovating as conditions change. Agricultural research and development respond similarly. This recognition strongly suggests that, on average and globally, agriculture worldwide will adapt to climate change as

well as to more localized environmental changes, such as land degradation. These adaptations could potentially lower agriculture's impacts on environmental change in general. For example, low- or no-tillage cultivation and organic farming have been shown to reduce air and water pollution, soil nutrient loss, and loss to pests. These practices currently involve increased costs to the farmer, however, and it is not clear how much of these costs can be passed on to the consumer or will be subsidized by society at large. Beyond these methods, conventional costs of inputs remain important to cultivation strategies. For example, the water-inefficient center pivot irrigation used on the southern Great Plains of the United States has given way to more efficient irrigation methods as the depletion of the Ogallala aquifer drives up the cost of pumping water. Herein lies an important issue within environmental and economic debates: Will the market mechanism kick in before irreparable damage to agricultural land or critical production resources takes place?

These kinds of adjustments are taken by the agricultural sector on behalf of itself. Potential climate change, however, suggests that portions of this sector will be reduced or closed down in certain regions or environments and expand in others. In general, where water stress increases and cannot be efficiently alleviated, cultivation will convert to other land uses, perhaps to livestock. Expansion is likely in higher latitudes where temperature currently inhibits the growing season. Regional economic impacts of less favorable agricultural conditions will vary by the political economy in question. In subsistence-oriented and economically poor regions, smallholders will be further marginalized, and movement to already crowded regions and urban areas will be exacerbated. In economically well-developed regions, however, the impacts will be masked by the much larger industrial and service sectors. In these cases, the burden of adjustment to other livelihoods will fall on the individual farmer.

This last scenario assumes, of course, that the agriculturalist confronts the changing physical and economic environment as an entrepreneur alone. There are many reasons to believe, however, that this confrontation will be fought politically, with the agrarian sector seeking support and subsidies to maintain agriculture in place. Confronted with inadequate water and international competition, agriculturalists in the western United States and western Europe have leveraged their political clout for public financed and subsidized irrigation and guaranteed price support, respectively. These kinds of "adjustments" will surely become contested with climate change in the developed world.

[*See also* Animal Husbandry; Biological Diversity; Consultative Group on International Agricultural Research; Deforestation; Desertification; Environmental Economics; Erosion; Fire; Food; Food and Agriculture

In the United States in an average year, over 400,000 hectares (1 million acres) of farmland is lost to urban sprawl. Thus a farmland area roughly the size of the state of Delaware is being turned into housing developments, shopping centers, industrial parks, and roads every year. On an average day, the toll is about 1,100 hectares, or 4.2 square miles. That is a significant portion of the nation's cropland, which totals 166 million hectares. Unhappily, much of the loss occurs on prime farmland: 19 million hectares of Class I soils and 89 million hectares of Class II soils, for a total of 108 million hectares, a large proportion of which is near urban areas because fertile soils and good water encouraged the original settlements that grew to become cities.

The loss of farmland to urban sprawl occurs across the nation because, as an urban area spreads, the value of agricultural land soars. Farmers are faced with rising taxes, while working of the land is made more difficult by encroaching residences. When a developer offers an attractive price for his land, a farmer will often sell and retire. There are some devices, though, that can slow this process.

- Taxing the land at lower rates, as long as it is used for agriculture.
- Zoning laws, if enforced, can stipulate agricultural use of the land.
- Urban growth boundaries can limit the sprawl. The Green Belt around London, England is a well-known application of the idea. In the United States, the state of Oregon leads in implementing growth boundaries.
- Land trusts or conservancies—private organizations that receive gifts of land and also purchase land and development rights to preserve land that has scenic, recreational, or agricultural value.
- In the United States, state, county, or township governments may purchase development rights from the farmer. After appraisals are made, the farmer is paid a per acre easement value, which is the difference between the land's agricultural value and its value to a developer. The farmer remains owner of the property, but he and any subsequent owners must continue farming the land according to a deed of easement that specifies the land use. At county and local levels, funds may be raised by bond issues approved by taxpayers. Such programs have been popular mostly in northeastern states. (In England, private development rights were nationalized in 1947. For any development, other than farm buildings, a landowner must gain permission from the local government.)

### INTERNET RESOURCES

American Farmland Trust. http://www.farmland.org/.
The Land Trust Alliance. http://www.lta.org/.

### BIBLIOGRAPHY

American Farmland Trust. *Farming on the Edge*. Northampton, Mass.: American Farmland Trust, 1997. Identifies the twenty most threatened agricultural regions in the United States and identifies areas in each state where strong development pressure coincides with high-quality farmland.
————. *Saving American Farmland: What Works*. Northampton, Mass.: American Farmland Trust, 1997. A comprehensive guidebook for policy makers, planners, and concerned citizens working to save farmland at the local level.
Daniels, T., and D. Bowers. *Holding Our Ground: Protecting America's Farms and Farmlands*. Washington, D.C.: Island Press, 1997. A general review of the problem and its solutions. Includes extensive notes and bibliography.
*Farmland Preservation Report*. Street, Md.: Bowers, 1990–.

—DAVID J. CUFF

Organization; Forestation; Grasslands; Irrigation; Land Reclamation; Land Use; Nongovernmental Organizations; Pest Management; Salinization; Savannas; Soils; Sustainable Development; Technology; Urban Areas; Water; *and* World Bank, The.]

## BIBLIOGRAPHY

Alexandratos, N. *World Agriculture Towards 2010. An FAO Study.* New York: Wiley, 1995.

Chameides, W. L., P. S. Kasibhatla, J. Yienger, and H. Levy II. "Growth of Continental-Scale Metro-Agro-Plexes, Regional Ozone Pollution, and World Food Production." *Science* 264 (1994), 74–77. The original study to identify the macrospatial scale and global reach of the ozone-urban/industrial agricultural linkage.

Crosby, A. W. *Ecological Imperialism: The Biological Expansion of Europe 900–1900.* Cambridge: Cambridge University Press, 1986. Watershed but not first demonstration of the biological and ecological impacts of the era of European global dominance, focusing on the age of exploration and colonialism.

Darwin, R., M. Tsigas, J. Lewandowski, and A. Raneses. *World Agriculture and Climate Change: Economic Adaptations.* Agricultural Economic Report no. 703. Washington, D.C.: U.S. Department of Agriculture, 1995. Example of modeling impact assessment of climate change of world agriculture.

Economic Research Service, United States Department of Agriculture. Agriculture Information Bulletin no. 710, Washington, D.C., February 1995.

Ervin, D. E., C. F. Runge, E. A. Graffy, W. E. Anthony, S. S. Batie, P. Faeth, T. Penny, and Y. Warman. "Agriculture and the Environment: A New Strategic Vision." *Environment* 40.6 (1998), 9–15, 35–40. An assessment of the ability of the growth in global agriculture to keep up with expected increases in demand under conditions of environmental stress.

Frederick, K. D., and N. J. Rosenberg, eds. *Assessing the Impacts of Climate Change on Natural Resource Systems.* Special issue. *Climatic Change* 28, nos. 1–2 (1994). Reviews of major integrated modeling approaches for understanding climate change impacts on agriculture and other resource uses.

Galaty, J. G., and D. L. Johnson, eds. *The World of Pastoralism: Herding Systems in Comparative Perspective.* New York: Guilford Press, 1990. One of the few recent treatments of pastoralism and pastoralists in the contemporary world.

Grigg, D. B. *The Agricultural Systems of the World: An Evolutionary Approach.* Cambridge: Cambridge University Press, 1974. A geographic view and review of the history and condition of different kinds of agricultural production systems around the world.

Hayami, Y., and V. W. Ruttan. *Agricultural Development: An International Perspective.* Baltimore: Johns Hopkins University Press, 1985. An exemplary review of the neoclassical economics perspective on agricultural change and development embedded within an induced innovation approach.

Heilig, G. "Neglected Dimensions of Global Land-Use Change: Reflections and Data." *Population and Development Review* (1994), 831–859.

Henderson-Sellers, A., ed. *Future Climates of the World: A Modelling Perspective.* Amsterdam: Elsevier, 1995. A state-of-the-art assessment of future-climate modeling illustrating links to such activites as land and agricultural change.

Lamb, H. H. *Climate, History, and the Modern World.* London: Methuen, 1982. Classic assessment of the role of climate and climate change on human history.

Leach, G. *Global Land and Food in the 21st Century:* Trends and Issues for Sustainability. Stockholm: Stockholm Environmental Institute, 1995.

Matson, P., W. J. Parton, A. G. Power, and M. J. Swift. "Agricultural Intensification and Ecosystem Properties." *Science* 277 (1997), 504–509. An example of the linkage between global change or earth system science and socioeconomic conditions.

Meyer, W. B., and B. L. Turner II. "Human Population Growth and Global Land Use/Cover Change." *Annual Reviews in Ecology and Systematics* 23 (1992), 39–61. A review of the current state of understanding of the human-induced changes in terrestrial land covers and uses.

Micklin, P. P. "Dessication of the Aral Sea: A Water Management Disaster in the Soviet Union." *Science* 241 (1988), 1170–1176. One of the initial documentations and demonstrations of the linkage between the demise of the Aral Sea and technologies of production and political economy.

National Research Council. *Lost Crops of Africa.* Washington, D.C.: National Academy Press, 1996. A good example of the array of crops and crop varieties that are decreasingly cultivated in Africa.

Parry, M. L. *Climate Change and World Agriculture.* London: Earthscan, 1990. Assesses various climate change scenarios on global agriculture.

Pingali, P., Y. Bigot, and H. P. Binswanger. *Agricultural Mechanization and the Evolution of Farming Systems in Sub-Sahara Africa.* Baltimore: Johns Hopkins University Press, 1987. Illustrates that induced innovation and intensification can be demonstrated even in regions where agriculture seems to be on the demise.

Richards, J. F. "Land Transformation." In *The Earth as Transformed by Human Action: Global and Regional Changes in the Biosphere over the Past Three Hundred Years*, edited by B. L. Turner II et al., pp. 163–177. Cambridge: Cambridge University Press, 1990. A historical reconstruction of global land use changes by regions of the world.

Rosenzweig, C., and M. Parry. "Potential Impact of Climate Change on World Food Supply." *Nature* 367 (1994), 133–138. A synthesis of state-of-the-art modeling outcomes on global agriculture.

Sen, A. *Poverty and Famines: An Essay on Entitlement and Deprivation.* Oxford: Oxford University Press, 1981. A path-breaking demonstration that malnutrition, hunger, and famine originate in the political economy.

Turner II, B. L., and S. B. Brush, eds. *Comparative Farming Systems.* New York: Guilford Press, 1987. Reviews approaches to agricultural change and intensification largely outside neoclassical economics.

Turner II, B. L., and K. W. Butzer. "The Columbian Encounter and Land-Use Change." *Environment* 43 (1992), 16–20, 37–44. Reviews misconceptions about ancient land and agriculture in the Americas and its meaning for global change.

Waggoner, P. E., J. H. Ausubel, and I. K. Wenick. Lightening the Tread of Population on Land: American Examples." *Population and Development Review* 22.3 (1996), 531–545. An examination of economic and population growth demands on agriculture with a positive spin.

Williams, M. "Forests." In *The Earth as Transformed by Human Action: Global and Regional Changes in the Biosphere over the Past Three Hundred Years*, edited by B. L. Turner II et al., pp. 179–201. Cambridge and New York: Cambridge University Press, 1990.

—B. L. TURNER II AND ERIC KEYS

**AIDS.** *See* Disease.

**AIR POLLUTION.** *See* Atmospheric Chemistry; Convention on Long-Range Transboundary Air Pollution; *and* Pollution.

## AIR QUALITY

Air is a fundamental natural resource, critical to the very existence of life on Earth. The degradation of air quality by human activities therefore constitutes an ominous threat to the environment and to human health and well-being. Indeed, the release of chemical contaminants into the atmosphere has irreversibly altered its chemical composition, causing impacts on all scales, local, regional, and global.

**Historical Overview.** Air pollution is not a recent phenomenon. The remains of early humans demonstrate that they suffered from the detrimental effects of smoke in their dwellings. The classical writers make reference to air pollution in the ancient cities of Rome and Athens, and there is evidence that the major cities of Europe experienced air pollution problems as far back as medieval times.

However, it was with the development during the Industrial Revolution of increasingly large-scale industries based on new sources of energy—the fossil fuels, coal, oil, and gas—that the deleterious effects of air pollutants on human health and well-being really started to become apparent. The problems were most acute in the large cities that sprang up in the 1800s in Europe and North America, especially where industry and workers' dwellings were in close proximity. In spite of public complaint, there was at this time widespread and passive acceptance of air pollution as an inevitable evil. It was the price to be paid for progress and opportunity.

Attitudes began to change in the middle of the twentieth century. Rapid postwar economic expansion in the late 1940s and early 1950s, coupled with tremendous growth in the use of petroleum products, particularly petrol-driven road vehicles, meant that air pollution became an inescapable part of urban life almost everywhere. Severe air pollution episodes occurred with increasingly regularity in cities in both North America and Europe; these episodes coincided with increased rates of respiratory morbidity and higher than usual death rates. Of these episodes, those that occurred in Donora, Pennsylvania, in 1948 and in London in 1952 are without doubt the most notorious. In the case of the 1952 London episode, poor air quality is believed to be the cause of over four thousand excess mortalities over a five-day period. It was these events that lead the public to demand less polluted air in its cities and provided the driving force for the implementation of measures to control pollutant emissions. The U.K. Clean Air Act, introduced in 1956, is especially noteworthy in this regard, as it was the first legal instrument of its kind and a model for similar legislation in many other countries.

In the 1970s it was confirmed that, in addition to impacting human health in the cities, air pollutants were having more subtle effects further afield. As a result, the issue of air pollution control was catapulted into the international arena.

**Sources and Nature of Air Pollution.** Generally speaking, air pollutants can be described as either primary pollutants (when emitted directly into the air by emission sources) or as secondary pollutants (if they were formed in the atmosphere from primary pollutants as a result of physical or chemical processes). Common primary pollutants include carbon monoxide (CO), sulfur dioxide ($SO_2$), the nitrogen oxide (i.e., NO and $NO_2$, which are collectively termed $NO_x$), hydrocarbons (HCs) and other volatile organic compounds (VOCs), particulate matter (dust and smoke) and lead (Pb). Secondary pollutants comprise the photochemical pollutants, the most significant of which are ozone (formed in the lower atmosphere in the presence of sunlight from $NO_x$ and VOCs) and various aerosol particles, which are products of gas-to-particle conversions (e.g., $SO_2$ and $NO_x$ form acidic sulfate and nitrate aerosols, respectively). [*See* Aerosols.]

The combustion of fossil fuels for electrical power generation and domestic space heating, in motor vehicles, and by industry has been, and still is by far, the main source of air pollutant emissions. Fossil fuel combustion in stationary sources leads to the production of sulfur dioxide, nitrogen oxides, and particulate matter. Gasoline-fueled motor vehicles generate nitrogen oxides and carbon monoxide; cars running on leaded gasoline also emit lead; diesel-fueled engines, on the other hand, emit significant quantities of sulfur dioxide and particulates as well as nitrogen oxides.

Fossil fuel combustion also produces carbon dioxide ($CO_2$). Although not strictly an air pollutant in the sense that it has no direct effects on health, its emission is of great significance in view of its greenhouse properties. [*See* Carbon Dioxide.] Emissions of greenhouse gases such as carbon dioxide are responsible for climate change, which along with the problem of stratospheric ozone depletion (caused by industrial emissions of chloride- and bromine-containing chemicals) means that air pollution can be said to have truly global consequences.

*Urban air pollution.* The density of air pollution emission sources is highest in urban areas and other centers of industrial activity; it is here that ambient concentrations of the common air pollutants reach their maximum and thus have the most pronounced impact on human health. In contrast to the situation some twenty-five to thirty-five years ago, when urban emissions were dominated by those of sulfur dioxide, nitrogen oxides, and suspended particulate matter (SPM) from coal combustion (mainly industry and domestic fires), the motor vehicle is now the most important source of air pollution problems in the vast majority of cities, at least in the industrialized world. In contrast, the cities of the less developed countries exhibit a greater variety of air pollution sources. The relative contributions of mobile and stationary sources vary markedly between these cities depending on the level of motorization and the level, density, and type of industry present. Cities in Latin America, for example, tend to have higher vehicle densities than those in other developing regions and are therefore more likely to experience air quality problems associated with the car-related pollutants, that is, nitrogen oxides, carbon monoxide, and ozone.

The contribution from motor vehicles is less important in cities in Africa where levels of motorization are much lower and in cities located in the temperate regions that rely heavily on solid fuels such as coal, wood, or biomass fuels for industry, electricity generation, and space heating and other domestic purposes. Cities that are heavily dependent on solid fuels, such as those in parts of eastern Europe and China, are thus particularly prone to smoke and sulfur dioxide pollution.

The nature and density of emissions sources are, however, not the only factors that govern air quality in urban areas; ambient concentrations of air pollutants at any given location are greatly influenced by local site geography and meteorological dispersion factors—in other words, the weather. In some cases these factors may act to exacerbate air pollution problems by preventing dispersion and dilution, thereby concentrating pollutants within a city. This is certainly the case in cities that have been built in valley locations; the surrounding hills then act as a downwind barrier, effectively trapping pollution close to the city. On a more local scale, a similar effect, known as the "street canyon" effect, is created by buildings and other tall structures. Tall buildings positioned on both sides of a busy road are particularly good at preventing the dispersal of low-level pollutant emissions (i.e., those from cars), giving rise to especially high air pollution levels at curbsides.

In temperate and cold climates, "thermal inversions" can cause buildup of pollution levels over an entire city. Under "normal" conditions, hot pollutant gases rise as they come into contact with colder air masses at higher altitudes. Under certain meteorological conditions, however, the air temperature may increase with altitude; if this happens, an inversion layer forms at anything between a few tens up to a few hundred meters above the ground. This inversion layer keeps pollutants trapped over the city; this in turn acts as a heat cover, prolonging the inversion. Inversions may last several days, or even longer when wind speeds are low. It was inversions of this type that caused the notorious wintertime pollu-

tion episodes of the 1950s and 1960s in London and other cities mentioned earlier.

Whereas higher demands for space heating combined with a tendency toward thermal inversions can often cause problems with wintertime sulfur dioxide and SPM pollution in temperate cities, the summertime is frequently characterized by episodes of photochemical smog and haze. Periods of hot, sunny weather dominated by stagnant, high-pressure conditions are ideal for the formation of ozone from its precursors, the nitrogen oxides and hydrocarbons. Cities such as Los Angeles, where this type of air pollution was first identified back in the 1960s, have almost become synonymous with photochemical smog. Since then, the phenomenon has been observed in most large cities around the world.

In more recent years, the growing realization of the risks to human health and to the environment posed by the presence in the air, albeit at relatively low concentrations, of a large and diverse group of "air toxics" has tended to shift the focus of attention away from the more common or "traditional" air pollutants. These contaminants—many of which are known to be toxic or are carcinogenic and which include several heavy metals (e.g., beryllium, cadmium, and mercury), organics such as benzene, polyaromatic hydrocarbons (PAHs) and polychlorinated biphenyls (PCBs), radionuclides (e.g., radon), and fibers (asbestos)—are increasingly being detected in urban atmospheres. Numerous activities, including the chemical and metal industries, energy production, solvent manufacture and use, waste incineration, and agricultural pesticide use, contribute to emissions of hazardous air pollutants (HAPs). [*See* Chemical Industry.] Specific examples of HAPs and their sources are listed in Table 1.

***Large-scale air pollution.*** During the 1960s and 1970s it became increasingly apparent that the air pollutants generated by human activities in the industrialized, urbanized areas could in fact be transported by prevailing winds for considerable distances. [*See* Convention on Long-Range Transboundary Air Pollution.] The problem of air pollution was thus no longer viewed as a purely local-scale phenomenon, but one that had both regional and even global implications. Problems caused by the long-range transport and subsequent deposition of the acidic gases, sulfur dioxide and the nitrogen oxides, provided the earliest indication of the potential for impacts on the regional scale. Widespread acidification of both aquatic and terrestrial ecosystems was first observed in remote parts of Scandinavia during the late 1960s and early 1970s; since then, similar problems have been observed in other parts of Europe, the eastern portion of North America, and most recently in eastern regions of China. [*See* Acid Rain and Acid Deposition.]

Traditionally, ozone has also been viewed as a predominantly urban pollutant. [*See* Ozone.] In the 1980s,

**Air Quality. TABLE 1.** Selected Hazardous Air Pollutants and Their Sources

| POLLUTANT | MAIN SOURCES OF MAN-MADE EMISSIONS |
| --- | --- |
| Benzene | Motor vehicles (an additive to gasoline) |
| Cadmium | Metal smelting, coal and waste combustion pigments in plastics, fertilizer use, tobacco smoke |
| Formaldehyde | Motor vehicles, plastic and chemical manufacturing, waste incineration solvent use |
| Mercury | Selected mining and smelting operations (including gold mining), coal combustion, selected industrial processes (e.g., the chlor-alkali, electrical, and paint industries), and waste incineration. |
| Polyaromatic hydrocarbons (PAHs) | Coal combustion, motor vehicles |
| Polychlorinated biphenyls (PCBs) | PCBs are released into the environment through their production and use in a wide range of products, such as capacitors, transformers, and other electrical devices, plasticizers, surface coatings, inks and adhesives; further releases arise from the disposal of products containing PCBs (i.e., from waste disposal sites and incinerators). |

however, research has clearly established that ozone and its precursors, nitrogen oxides and volatile organic compounds, can be transported over large areas ranging from several hundred or several thousand square kilometers. Large-scale ozone pollution has been observed throughout the midlatitudes of the Northern Hemisphere, mainly during the summer months and most noticeably downwind of major cities and industrial centers. This is a cause for concern for a number of reasons. First, ecological consequences of high ozone levels in rural areas include damage to all types of vegetation, including agricultural crops and forests. Second, ozone in the troposphere behaves as a greenhouse gas. The observation that levels of ozone in the midtroposphere have roughly doubled in rural parts of Europe since the turn of the century is thus of considerable concern in view of its climatic implications.

Hazardous air pollutants too can exert their effects both locally and, through long-range transport, on regional and global scales. The atmosphere is the predominant pathway for the global dispersion of stable trace pollutants, and wet and dry deposition are the key processes for introducing these pollutants into terrestrial and aquatic ecosystems where they can represent a significant threat to the environment. These threats are of particular concern in the case of the more persistent HAPs, which have a tendency to migrate through different environmental media, that is, to *bioaccumulate*. There is now a growing body of evidence to suggest that all environmental media, but especially wildlife, have become contaminated with industrial pollutants, even in the remotest parts of the world once considered pristine. Elevated levels of heavy metals such as lead, cadmium, and mercury and persistent organochlorines such as di-

chlorodiphenyltrichloroethane (DDT), toxaphene, PCBs, and hexachlorobenzenes (HCBs) have been measured in fish, birds, and marine mammals from both the arctic and antarctic regions.

The arctic region is also of interest because it has become associated with one of the most extensive manifestations of the long-range transport of air pollution—*arctic haze*. During the winter months (January–April), the arctic atmosphere contains a variety of anthropogenic gaseous and particulate species such as sulfates, nitrates, organic trace metals, and carbonaceous particles derived from the midlatitudes of the Northern Hemisphere, that is, Europe, Asia, and, to a lesser extent, North America. The processes governing its formation are now well understood; polluted air builds up over the industrialized regions and is periodically swept poleward by strong, anticyclonic meteorological systems. Upon entering the polar zones, the transport processes slow down and the polluted air masses remain fixed in the arctic region. The stagnant conditions, the polar winter darkness and cold, and the near total lack of precipitation mean that the normal processes by which air is cleansed are not functioning. Consequently, the pollution remains airborne for an unusually long time, breaking up only with the onset of warmer temperatures at springtime.

Arctic haze has built up each year since the Industrial Revolution, but the pace has quickened since World War II. It is of particular concern today because the presence of a sooty haze over the reflecting arctic ice pack may constitute a significant heating mechanism and thus has an influence on climate.

**Human Health Impacts.** The common air pollutants and their main human health effects are summa-

**Air Quality.** TABLE 2. Health Effects of Selected Air Pollutants

| POLLUTANT | MAIN HEALTH EFFECTS |
| --- | --- |
| *Traditional Air Pollutants* | |
| $SO_2$ | Acute health effects include impaired lung function, producing symptoms ranging from coughing and wheezing to bronchitis and asthma. Long-term exposure linked to increased prevalence of respiratory conditions such as chronic bronchitis. |
| Suspended particulates (SPM)/smoke | Similar to $SO_2$; combined exposure to $SO_2$ and SPM associated with increased cardiac-respiratory mortality and morbidity. Particles in the submicron size range (i.e., less than 10 $\mu$m in diameter) are of particular significance with regard to health effects. |
| Nitrogen dioxide ($NO_2$) | Sensitizes lungs to other pollutants and allergies; increases susceptibility to viral infections. Long-term exposure can affect lung function. |
| Ozone ($O_3$) | A lung and eye irritant, causing coughing, choking, impaired lung function, headaches, sore throat, and runny eyes; sensitizes lungs to other pollutants; can aggravate chronic heart disease, asthma, and bronchitis. |
| Carbon monoxide (CO) | Interferes with absorption of oxygen by hemoglobin in red blood cells; impairs perception, slows reflexes, and causes drowsiness, unconsciousness, and even death. |
| Lead (Pb) | A neurotoxin suspected of causing hyperactivity and neurobehavioral effects, particularly in children; affects blood biochemistry and can raise blood pressure. |
| *Hazardous Air Pollutants* | |
| Benzene | Leukemia, neurotoxic symptoms, and bone marrow injury. |
| Cadmium | Acute and chronic respiratory disease, renal dysfunction, and animal carcinogen. |
| Formaldehyde | Chromosome aberrations; irritations of the eyes, nose, and throat; dermatitis; and respiratory tract infections in children. |
| Mercury | Effects on nervous system include deficits in short-term memory and disturbance of sensory and coordination functions; kidney failure. |
| Polyaromatic hydrocarbons (PAHs) | Respiratory tract and lung cancers, and skin cancers. |
| Polychlorinated biphenyls (PCBs) | Spontaneous abortions, congenital birth defects, and bioaccumulation in food chains. |

rized in Table 2. In most cases the impacts are on the respiratory and cardiovascular systems.

In recognition of the ubiquity of air pollution and its potential for human health impacts, most countries in the developed world set about establishing monitoring networks for the routine measurement of air quality. Early efforts focused on sulfur dioxide and suspended particulate matter. By the late 1970s, however, as motor vehicles became an increasingly important source of air pollutants, networks were expanded to cover the traffic-related pollutants: carbon monoxide, nitrogen oxides, and lead. Latterly, more attention has been paid to the need for the monitoring of ozone and volatile organic compounds.

Despite the advances in the scale and scope of urban air quality monitoring, there are still major difficulties in acquiring adequate air quality data for risk assessments. Indeed, quantifying human exposure to air pollution and estimating the risks involved continue to frustrate scientists today. Statistical studies have clearly demonstrated an association between high-pollution days and increased mortality and morbidity (e.g., hospitalization due to a range of illnesses, most notably,

heart attacks and asthma). Causal relationships between increased mortality and morbidity and exposure to specific air pollutants have been much more difficult to establish. This is due, in part, to the practical difficulties in determining the actual exposure of an individual to given air pollutants. Most air quality monitoring, because it takes place at fixed locations outdoors, is unlikely to be representative of the air breathed by an individual, who on average spends up to 80–90 percent of his or her time indoors, where the air quality may be quite different. Moreover, exercise increases the volume of air and hence the pollutant inhaled. A further difficulty arises in attempting to identify the effect of an individual pollutant, which almost invariably occurs in combination with other pollutants and under different weather conditions. It is generally accepted that air pollutants such as sulfur dioxide, suspended particulate matter, nitrogen oxides, and ozone act synergistically; in other words, the effect of exposure to a "cocktail" of pollution is greater than the sum of the individual components.

If it is difficult to assess the exposure to and risks from the traditional air pollutants, then establishing those from the so-called hazardous or toxic air pollutants is even more fraught. This is because for many of these pollutants information on direct health effects is, at best, extrapolated from animal studies or simply not available at all. Assessment is further hampered by the lack of information about ambient concentrations of HAPs; in most countries, the trace air pollutants are not routinely monitored. The health effects of a small selection of toxic air pollutants are summarized in Table 2; effects are varied, ranging from relatively minor symptoms, such as skin and eye irritation, to serious conditions including cancer and birth defects.

In view of the concern expressed by governments and the public about the possibility that hazardous air pollutants can cause cancer, great efforts have been made in recent years to estimate the overall cancer risk from air pollution. Much of this work has recently been reviewed by the Organisation for Economic Co-operation and Development (OECD) as part of its program of work on HAPs. One of the most comprehensive studies reviewed was that conducted in the United States, in which it was estimated that the total nationwide cancer incidence due to outdoor concentrations of air toxics was somewhere between seventeen hundred and twenty-seven hundred excess cancer cases per year (or between seven and eleven annual cases per million population).

Considerable concern has also been expressed in recent years over the relative contribution of indoor sources of air pollution to the overall exposure of individuals. Available evidence shows that in many incidences the air inside homes and inside automobiles may be of considerably lower quality than outdoor air and that people may be exposed to much higher levels of both the traditional and the toxic trace pollutants than was originally believed.

In addition to their health effects, a number of urban air pollutants are responsible for a range of other impacts on the environment. Aside from the problems of a more regional nature that have already been mentioned, sulfur dioxide, the nitrogen oxides, and ozone all have direct adverse effects on plant growth and development, and sulfur dioxide and acidic particulate matter acting together are implicated in the accelerated weathering of buildings and historical monuments. [See Building Decay.] Ozone is also believed to cause damage to materials, especially rubber. The loss of amenity caused by the impairment of visibility is another effect of urban air pollution; it can be observed both in cities and downwind of them. When impacting areas of natural beauty, visibility impairment is considered to be a serious environmental issue.

**National and International Strategies for Combating Air Pollution.** National programs for reducing air pollution in urban areas, launched in the 1960s by most industrialized countries, were based on two fundamental beliefs. First, it was believed that the effects of air pollution occurred only in relatively small geographic areas and that the application of air pollution control technology to local emissions sources would thus cure the problem and ensure that air quality standards were met. Second, it was assumed that the atmosphere had the capacity to assimilate much of the pollution generated by human activity. The solution, based on these assumptions, involved the construction of tall chimneys and stacks for the discharge of pollutants from power plants and industry at higher altitudes to improve dispersal. The result was an increase in the long-range transport of pollution and in adverse impacts on aquatic and terrestrial ecosystems in areas as far as several thousand kilometers away from the original source regions. In effect, the tall-stack policy simply redistributed the problem of air pollution, and it soon became obvious that more stringent controls were needed in order to reduce emission at the source, especially from power stations and motor vehicles.

Over the last twenty to thirty years, a wide range of strategies and techniques have been developed and adopted by many countries, particularly those in the developed world. These control strategies can be grouped according to the following categories:

- Precombustion controls
- Combustion modifications
- Postcombustion control
- New industrial processes
- Energy conservation

Examples of each are given in Table 3. More detailed descriptions of control techniques may be found in the texts listed in the bibliography.

In most countries, the combination of the above control measures, together with structural changes in economics and energy demand, have substantially reduced air pollutant emissions and improved air quality in many urban areas. This applies particularly to the sulfur oxides, particulates, and lead. For example, since the early 1970s, urban sulfur dioxide concentrations have decreased, on average, by between 30 and 75 percent in OECD member countries; similarly, average SPM concentrations, once between fifty and one hundred micrograms per cubic meter, have declined to around twenty to sixty micrograms per cubic meter. In North America, urban lead concentrations have fallen by as much as 85 percent, and in large European cities by about 50 percent. In contrast, volatile organic compounds and nitrogen oxides emissions have generally increased compared to the early 1970s, largely because increases in motor vehicle fleets and distances traveled have outpaced gains made by the implementation of emission controls. Consequently, air quality guidelines for nitrogen oxides and ozone are still frequently exceeded in many countries.

**Air Quality. TABLE 3.** Selected Control Strategies for the Reduction of Air Pollutant Emissions at the Source

| CATEGORY | EXAMPLES |
|---|---|
| Precombustion control | Use of low-sulfur oil and coal; fuel cleaning (reduces the sulfur, ash, and dust content of coal); switching to sulfur-free fuels (natural gas) or nonfossil fuels (nuclear power); reducing lead additives in gasoline. |
| Combustion modification | Changing the methods of fuel burning; e.g., using low nitrogen oxide ($NO_x$) burners and fluidized bed combustion to reduce $NO_x$ and sulfur dioxide ($SO_2$) emissions. |
| Postcombustion control | Removal of pollutants from flue gases; use of catalysts in vehicle exhausts to reduce $NO_x$, carbon monoxide, and hydrocarbon emissions. |
| New industrial processes | Several have been developed in recent years, for example, low-temperature hydro-metallurgical techniques, which are used in the production of nonferrous metals in developed countries to reduce $SO_2$ emissions. |
| Energy conservation | By better use of energy, energy demand and the need to generate as much energy will be reduced. This in turn will reduce the demand for fossil fuels, which will then reduce pollutant emissions. |

Despite the considerable improvement in air quality that has been achieved in many cities—at least with respect to sulfur dioxide, particulates, and lead—it is estimated that approximately half the world's urban residents are still exposed to concentrations of sulfur dioxide and particulate matter in excess of World Health Organization (WHO) guidelines and as a result may be suffering from detrimental effects to their health (United Nations Environment Programme, 1991). Exceedances of WHO guidelines in some cities are far from marginal; cities in northern China, for example, experience levels of sulfur dioxide and particulates three to eight times those of WHO guidelines, and residents of Mexico City breathe ozone levels that are more than 50 percent above WHO guidelines. Nor are the problems confined to large cities in the developing world. Studies in the United States suggest that more than half of Americans live in areas that do not meet national air quality standards (Cortese, 1990). More disconcertingly, many of the fastest-growing cities are situated in tropical regions, where warm, sunny conditions for much of the year, coupled with rapidly growing vehicle numbers, are likely to mean a worsening of air quality conditions in these regions in the future.

It is significant that the impetus for controlling emissions of sulfur dioxide and nitrogen oxides over the past decade or so has stemmed as much from the environmental threats posed by acid deposition as from concerns over human health risks in the urban environment. In 1979, thirty-five European countries signed up to the 1979 United Nations Economic Commission for Europe (UNECE) Convention on Long-Range Transboundary Air Pollution. As part of this convention, three protocols specifying targets for national emission reductions for sulfur dioxide (in 1985), nitrogen oxides (in 1988), and volatile organic compounds (in 1991) have been signed. During the mid-1990s, new protocols to the convention

were negotiated, based on a new approach to selling emission reduction targets—the *critical load approach*. This approach sets emission targets based on environmental benefits that can be achieved; that is, it uses the receptor rather than source as the starting point.

The use of the critical load approach for setting the second generation of abatement strategies within the UNECE convention exemplifies the new thinking that is required for successful air quality management in the future. Increasingly, scientists and policy makers alike are realizing that traditional management strategies, which consider air pollutants individually, are inadequate. Given the complex and interactive nature of air pollution and its potential for major impacts on regional and global scales, it is becoming clear that a more holistic view, in which the combined effects of several pollutants are considered together, is urgently required. Emissions need to be reduced locally, regionally, and globally.

[*See also* Industrialization.]

## BIBLIOGRAPHY

Barrie, L. A. "Arctic Air Pollution: An Overview of Current Knowledge." *Atmospheric Environment* 20 (1986), 643–663.

Cortese, A. D. "Clearing the Air." *Environmental Science and Technology* 24, no. 4 (1990), 442–448.

Faiz, A., K. Sinha, M. Walsh, and A. Varma. *Automotive Air Pollution: Issues and Options for Developing Countries.* Working Paper Series 492. Washington, D.C.: Department of Policy, Research, and External Affairs, The World Bank. Includes much useful information on the relative merits of various technologies for reducing air pollution emissions from cars.

Grennfelt, P., Ø. Hov, and D. Derwent. "Second Generation Abatement Strategies for $NO_x$, $NH_3$, $SO_2$ and VOCs." *Ambio* 23, no. 7 (November 1994), 425–433.

Moseley, C. "Indoor Air Quality Problems." *Journal of Environmental Health* 53 (1990), 19–35.

Murray, F. "Urban Air Pollution and Health Effects." In *The Global Environment: Science Technology and Management*, edited by D. Brune, D. V. Chapman, M. D. Gwynne, and J. M. Pacyna, 1:585–598. Weinheim, Germany: VCH, 1997.

Organisation for Economic Co-operation and Development (OECD). *The State of the Environment.* Paris: OECD, 1991. The chapter on air is a useful, comprehensive summary of all the key air pollution issues facing the OECD member countries.

———. *Control of Hazardous Pollutants in OECD Countries.* Paris: OECD, 1996. Probably one of the most comprehensive documents published to date on the subject of HAPs; includes information on sources, health effects and risk assessment, and control policies, as well as case studies of management strategies currently operating in seven countries.

Pacyna, J. M., and H. Ahmadzai. "Air Pollution Abatement." In *The Global Environment: Science Technology and Management,* edited by D. Brune, D. V. Chapman, M. D. Gwynne, and J. M. Pacyna, 2:724–728. Weinheim, Germany: VCH, 1997. A fairly detailed description—without being overly technical—of currently available technology-oriented methods of air pollution abatement.

United Nations Economic Commission for Europe (UNECE). *Strategies and Policies for Air Pollution Abatement.* Report ECE/EB.AIR/44. New York: United Nations, 1995.

United Nations Environment Programme (UNEP). *Urban Air Pollution.* UNEP/GEMS Environment Library, no. 4. Nairobi: UNEP, 1991. A very readable overview, written in nontechnical language, that provides a useful introduction to the subject; it is based on a UNEP GEMS/WHO assessment of worldwide urban air pollution published in 1988.

Viles, H. A. "Urban Air Pollution and the Deterioration of Buildings and Monuments." In *The Global Environment: Science Technology and Management,* edited by D. Brune, D. V. Chapman, M. D. Gwynne, and J. M. Pacyna, 1:599–609. Weinheim, Germany: VCH, 1997.

World Health Organization (WHO). *Air Quality Guidelines for Europe.* WHO Regional Publications, European Series, no. 23. Copenhagen: Regional Office for Europe, WHO, 1987. The definitive text on health effects of air pollutants, both traditional and trace; summaries of current knowledge about health effects of individual pollutants are published as the WHO's Environmental Health Criteria documents; the air quality guidelines have themselves been revised and updated in 1995.

World Health Organization/United Nations Environment Programme (WHO/UNEP). *Urban Air Pollution in Megacities of the World.* Published on behalf of the World Health Organization, Geneva, and the United Nations Environment Programme, Nairobi. Oxford: Blackwell, 1992.

World Bank. *Air Pollution from Motor Vehicles: Standards and Technologies for Controlling Emissions.* Washington D.C.: The World Bank, 1996. A state-of-the-art review of vehicle emission standards and testing procedures that also attempts to synthesize worldwide experience of vehicle emission control technologies and their application in both developed and developing countries.

—ANN D. WILLCOCKS

# ALBEDO

Albedo (derived from the Latin term *albus,* "white") is defined as the fraction of incident solar radiation reflected by a surface. It can be defined over the entire spectrum of solar radiation (known as *broadband albedo*) or at a specific wavelength (known as *spectral albedo*). Averaged over the globe, the annual broadband planetary albedo at the top of the atmosphere (TOA) is approximately 0.3. Since clouds are highly reflective, with albedos ranging up to about 0.8, and cover approximately 60 percent of the Earth, they have a dominant effect on planetary albedo. [*See* Clouds.] On average, clouds reflect approximately 20 percent of the solar radiation reaching the planet, compared with 6 percent from gas molecules and aerosols and 4 percent from the surface.

Regions where clouds have a large influence on planetary albedo are areas of persistent low stratiform cloud over the eastern subtropical oceans (for example, off California, Angola, and Peru), in the tropics over regions of intense convective activity (Central South America, the Congo Basin, India, and Indonesia), and in the midlatitudes over areas of cyclonic storm formation (eastern coasts of North America and Asia). Near the poles, the relative influence of clouds is not as pronounced since the clear-sky snow or ice background is already highly reflective.

In cloud-free regions, incoming solar radiation is reflected back to space by the atmosphere (gas molecules and aerosols) and surface. The average planetary clear-sky TOA albedo is approximately 0.13. TOA albedos over ocean generally lie between 0.04 and 0.15. Subtropical desert regions (such as the Sahara and Arabian deserts) are much brighter, with TOA albedos between 0.25 and 0.35. As the amount of vegetation over land increases, the albedo decreases. Over forested regions, albedos are commonly between 0.05 and 0.20, while albedos over savannas and grasslands are typically between 0.15 and 0.25. In extreme conditions, such as dust outbreaks over ocean, forest fires, and volcanic eruptions, albedos over even dark surfaces can reach values comparable to those of clouds. By far the largest albedos observed on the planet occur over snow and ice. Just above the surface, albedos generally range between 0.65 and 0.75 for old snow and can reach as high as 0.75–0.90 for fresh fallen snow. At the top of the atmosphere, albedos over snow are reduced slightly (by approximately 10–15 percent) because of absorption by the atmosphere (Warren, 1982).

The wavelength dependence of spectral albedo varies widely with surface type. Most surfaces (except snow) have very low spectral albedos at ultraviolet wavelengths, and higher values in the visible and near-infrared. The chlorophyll in vegetation absorbs strongly in the visible at wavelengths shorter than about 700 nanometers, so that a sharp rise in spectral albedo (often by a factor of three to four) is frequently observed near this wavelength over forest and grassland (Asrar, 1989). Spectral albedos over soils have a more gradual increase with wavelength and are highly dependent upon the mineralogical composition of the soil and whether it is wet or dry. In contrast, spectral albedos over snow are generally high at ultraviolet wavelengths, reach a maximum in the visible (400–500 nanometers), and decrease sharply at near-infrared wavelengths, where ice crystals strongly absorb solar radiation (Wiscombe and Warren, 1980).

Over most surfaces, albedo is also a function of the angle of solar incidence or, equivalently, the angle of the Sun above the horizon. The highest albedos tend to occur near sunrise and sunset, when the Sun is near the horizon. Since the angle of solar incidence decreases with latitude, albedos for a given surface type may also increase with latitude. How albedo varies with the angle of solar incidence depends on the physical properties and composition of the surface. Other factors, such as surface roughness, can also play an important role. For example, albedos over rough snow and wind-driven ocean surfaces are generally smaller than albedos over flat snow and ocean surfaces, and increase less rapidly as the Sun approaches the horizon.

The amount of solar radiation absorbed by the Earth is the principal source of energy that drives the climate system. It is determined from the difference between the solar radiation intercepted by the planet and that reflected back to space. The Earth also emits radiation to space in the form of terrestrial infrared radiation. Over a year, a planetary radiation balance is approached, whereby the absorbed solar radiation is nearly balanced by outgoing terrestrial infrared radiation. A change in planetary albedo could significantly modify this balance and alter climate. For example, an increase in albedo would mean less solar heating and a gradual cooling of the Earth. This cooling would continue until a new balance between emitted terrestrial infrared radiation and absorbed solar radiation was reached at some lower equilibrium temperature. Conversely, a decrease in planetary albedo would result in a warmer climate. Sensitivity studies show that an absolute increase (or decrease) in planetary albedo of 0.01 can potentially lead to a decrease (or increase) in equilibrium surface temperature by as much as 1.75°C (Cess, 1976).

The prediction of global climate change due to natural or man-made perturbations requires an understanding of how the various components of the climate system influence the planetary radiation balance. [See Climate Change.] Clouds have an important modulating role at both solar and terrestrial wavelengths. They reduce the amount of solar radiation absorbed by the climate system by increasing the Earth's albedo, and decrease the loss of terrestrial infrared radiation to space by absorbing and then reemitting part of the radiation back down to the surface. Overall, the net effect

of clouds is to cool the climate system, but on a regional basis their net effect varies considerably, depending on their physical properties—such as cloud fraction, cloud water content, cloud height, and cloud droplet sizes (Arking, 1991). Low stratiform clouds tend to have a strong cooling effect, while high cirrus clouds are believed to have a warming effect.

Overall, the direct radiative effect of aerosols is to cool the climate system, but the magnitude of this cooling is highly uncertain because of the lack of information on the global properties of aerosols (Charlson et al., 1992; Hansen et al., 1995). [See Aerosols.] Perhaps more significant for climate than their direct effect is the indirect effect of aerosols. For water vapor to condense and form cloud particles, aerosols that have an affinity for water vapor are needed. These aerosols, known as cloud condensation nuclei (CCN), are more abundant in polluted areas and over ocean regions of high dimethyl sulfide (DMS) concentrations (Charlson et al., 1987) (DMS is produced by phytoplankton in sea water.). An increase in CCN has been shown theoretically (Twomey, 1977) to alter cloud optical properties and increase the brightness or albedo of clouds of a given liquid water content. Some observational support for this hypothesis is available from satellite measurements over marine stratus clouds in the presence of ship tracks (Coakley et al., 1987; Radke et al., 1989), but, as with the direct effect of aerosols, the magnitude of the indirect effect on cloud albedo and on the radiation budget remains highly uncertain. Recently, the effect of commercial air traffic on cloud cover over industrialized regions has been demonstrated from satellite observations (Minnis et al., 1998). High-level cirrus clouds generated directly from jet aircraft contrails were observed to develop into systems roughly 35,000 square kilometers in extent with lifetimes of several hours. More study is needed to assess the radiative effects of these anthropogenic cloud systems on regional climate.

There is ample evidence of man's influence in modifying major portions of the Earth's land surface. Patterns of land use, such as urbanization, agriculture, overgrazing, and deforestation, have changed the Earth's landscape at an unprecedented rate. Land albedo and water cycling characteristics are among the major components of the surface energy balance, which determines regional climate. Changes in ecosystem structure due to deforestation and desertification generally lead to surfaces that are more highly reflective. Modeling studies show that such increases in land albedo tend to reduce the amount of solar radiation absorbed by the surface, causing a reduction in the amount of energy available for surface evaporation, and a change in regional cloud cover, precipitation, and temperature patterns (Sud and Fennessy, 1982; Chervin, 1979). Depending on how rapidly and by how much the physical characteristics of the surface are altered, these changes may increase the likelihood of drought. To improve model predictions of regional and global climate over land, accurate, global observations of key processes and variables that determine the surface energy budget—such as surface albedo, cloud cover, evaporation, and surface–atmosphere heat exchange—are needed.

Changes in albedo due to variations in regional snow and ice cover are also important climatologically. [*See* Ice Sheets.*] A rise in air temperature over an ice surface can cause melting and replace a bright highly reflecting surface with a dark underlying surface. Provided that cloud cover does not increase substantially, this decrease in surface albedo will allow more solar radiation to be absorbed by the surface and thus enhance warming. This process, known as the "ice-albedo feedback," is one of the main reasons why general circulation models generally predict more warming at high latitudes in response to increases in carbon dioxide concentrations (Mitchell, 1989).

[*See also* Atmosphere Structure and Evolution; Desertification; Global Monitoring; Global Warming; Greenhouse Effect; Pollution; Remote Sensing; Sea Ice; *and* Snow Cover.]

## BIBLIOGRAPHY

Arking, A. "The Radiative Effects of Clouds and Their Impact on Climate." *Bulletin of the American Meteorological Society* 72 (1991), 795–813.

Asrar, G. *Theory and Applications of Optical Remote Sensing.* New York: Wiley, 1989.

Barkstrom, B. R., and G. L. Smith. "The Earth Radiation Budget Experiment: Science and Implementation." *Reviews of Geophysics* 24 (1986), 379–390.

Cess, R. D. "Climate Change: An Appraisal of Atmospheric Feedback Mechanisms Employing Zonal Climatology." *Journal of the Atmospheric Sciences* 33 (1976), 1831–1843.

Charlson, R. J., et al. "Oceanic Phytoplankton, Atmospheric Sulfur, Cloud Albedo and Climate." *Nature* 326 (1987), 655–661.

———. "Climate Forcing by Anthropogenic Aerosols." *Science* 255 (1992), 423–430.

Chervin, R. M. "Response of the NCAR General Circulation Model to Changed Land Surface Albedo." In *Report of the JOC Study Conference on Climate Models: Performance, Intercomparison and Sensitivity Studies*, vol. I, pp. 563–581. Geneva: World Meteorological Organization, 1979.

Coakley, J. A., Jr., et al. "Effect of Ship-Stack Effluents on Cloud Reflectivity." *Science* 237 (1987), 1020–1022.

Hansen, J., et al. "Low-Cost Long-Term Monitoring of Global Climate Forcings and Feedbacks." *Climate Change* 31 (1995), 247–271.

Minnis, P., et al. "Transformation of Contrails into Cirrus during SUCCESS." *Journal of Geophysical Research* 25 (1998), 1157–1160.

Mitchell, J. F. B. "The 'Greenhouse' Effect and Climate Change." *Reviews of Geophysics* 27 (1989), 115–139.

Radke, L. F., et al. "Direct and Remote Sensing Observations of the Effects of Ships on Clouds." *Science* 246 (1989), 1146–1149.

Sud, Y. C., and M. J. Fennessy. "A Study of the Influence of Surface Albedo on July Circulation in Semi-Arid Regions Using the GLAS GCM." *Journal of Climate* 2 (1982), 105–125.

Twomey, S. "The Influence of Pollution on the Shortwave Albedo of Clouds." *Journal of the Atmospheric Sciences* 34 (1977), 1149–1152.

Warren, S. G. "Optical Properties of Snow." *Reviews of Geophysical Space Physics* 20 (1982), 67–89.

Wiscombe, W. J., and S. G. Warren. "A Model for the Spectral Albedo of Snow: I. Pure Snow." *Journal of the Atmospheric Sciences* 37 (1980), 2712–2733.

—Norman G. Loeb

## ALIEN SPECIES. *See* Exotic Species.

## AMAZONIA, DEFORESTATION OF

[*This case study focuses on the results of Amazonian deforestation, which give rise to the destruction of biodiversity, reduced carbon storage, and loss of water cycling, and discusses current and future efforts to contain the problems.*]

One of the most pressing environmental problems today is deforestation in Amazonia. Despite its vast size, the Amazon forest is subject to rapidly increasing pressures, and the environmental consequences of its loss are severe.

**Definitions of Amazonia.** The Amazon River watershed (Figure 1A) totals 7,350,621 square kilometers ($km^2$), of which 4,982,000 $km^2$ (67.8 percent) is in Brazil, 956,751 $km^2$ (13.0 percent) is in Peru, 824,000 $km^2$ (11.2 percent) is in Bolivia, 406,000 $km^2$ (5.5 percent) is in Colombia, 123,000 $km^2$ (1.7 percent) is in Ecuador, 53,000 $km^2$ (0.7 percent) is in Venezuela, and 5,870 $km^2$ (0.1 percent) is in Guyana. Amazonia is between four and seven million square kilometers in area (the precise figure depending on the definition adopted), including, in Brazil, the Tocantins–Araguaia Basin (which drains into the Pará River, interconnected with the mouth of the Amazon) and the small river basins in Amapá that drain directly into the Atlantic. The forested area extends beyond the bounds of the river basin, especially on its northern and southern edges, but a number of enclaves of nonforest vegetation exist within the watershed (Figure 1B). In addition, Greater Amazonia encompasses Suriname (142,800 $km^2$), French Guiana (91,000 $km^2$), and the part of Guyana outside the Amazon River watershed (205,369 $km^2$), bringing the total area of to 7,789,790 $km^2$ (Figure 1C).

In Brazil, the Legal Amazon (Figure 1D) is an administrative region comprising nine states and covering

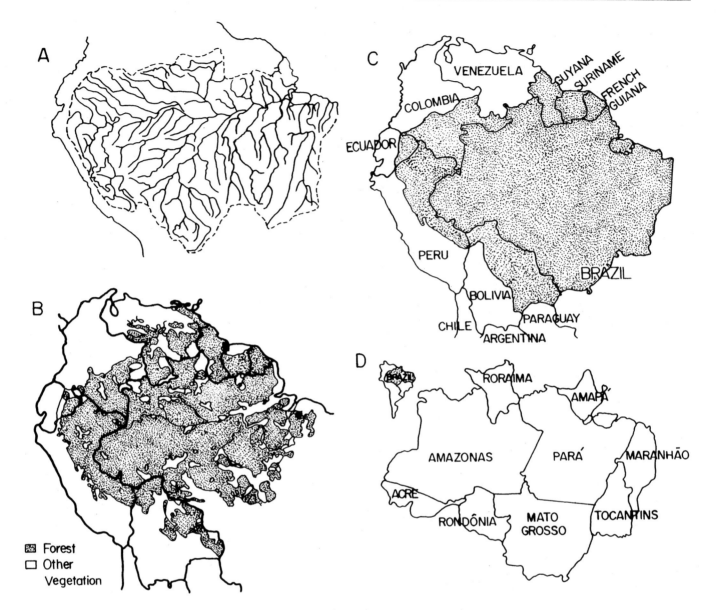

**Amazonia: Deforestation of. FIGURE 1.** The Amazon Region.

(A) Amazon River drainage basin, including Tocantins–Araguaia and Amapá coastal rivers. (B) Amazonian forest vegetation. (After Harcourt et al., 1996, and Daly and Prance, 1989). (C) The greater Amazon region with the addition of the coastal region of Guyana. (After Tratado de Cooperación Amazonica, 1992.) (D) Brazil's legal Amazon region with state boundaries.

five million square kilometers. One million square kilometers of the region was not originally forested, but was covered by various kinds of savanna (especially the *cerrado*, or central Brazilian scrub savanna). The Legal Amazon was created in 1953 and slightly modified in extent in 1977. Because special subsidies and development programs apply within the region, its borders were drawn just far enough south to include the city of Cuiabá

(Mato Grosso), and just far enough east to include the city of São Luís (Maranhão), both of which are outside the portion that is geographically Amazonian.

**Deforestation.** Deforestation refers to the loss of primary (sometimes called *mature, virgin,* or *old-growth*) forest. [*See* Deforestation.] This is distinct from cutting of secondary (successional) forest. In addition to clearing, (such as for agriculture or ranching), deforestation includes forest lost to flooding for hydroelectric dams. It does not include disturbance of forest by selective logging. In Amazonia, virtually all logging is selective because only some of the many tree species in the forest are accepted by today's timber markets.

Wide discrepancies in estimates for deforestation in Amazonia are often the result of inconsistencies in definitions, including the delimitation of the region itself, the inclusion or exclusion of the *cerrado* scrub savanna, classification of secondary forests as "forest" or "deforested," and the inclusion of flooding by hydroelectric dams. Differences in the radiation frequencies recorded and in image resolution for the data gathered by different satellites, and in interpretation of these data, also contribute to discrepancies (see Fearnside, 1990). Operationally, areas are classified as deforested if they are readily recognized as cleared on Landsat imagery.

It is also important not to confuse deforestation with burning: not all land is burned when it is deforested, and many areas are burned that are either not originally forest (especially savanna) or have already been deforested (especially established cattle pastures). Amazonian forest can sometimes burn without being cleared first, as in the case of the Great Roraima Fire of 1998, but these events leave most trees standing and are not considered deforestation.

**Extent and Rate of Deforestation.** Much more complete information for the rate and extent of deforestation exists for Brazil than for the other Amazonian countries because of Brazil's monitoring capabilities at the National Institute of Space Research (INPE). The Food and Agriculture Organization of the United Nations (FAO) compiled estimates for the status of forests in 1990 in all tropical countries. Unfortunately, the FAO definitions of forest types are not entirely consistent with other classifications, particularly with regard to whether the vast Brazilian *cerrado* should be considered a forest. FAO (1993) estimated that, for the period 1981–1990, $36.7 \times 10^3$ square kilometers per year ($\text{km}^2 \, \text{y}^{-1}$) were cleared in Brazil (including *cerrado* and areas outside Amazonia), $2.7 \times 10^3 \, \text{km}^2 \, \text{y}^{-1}$ in Peru, $6.3 \times 10^3 \, \text{km}^2 \, \text{y}^{-1}$ in Bolivia, $3.7 \times 10^3 \, \text{km}^2 \, \text{y}^{-1}$ in Colombia, $2.4 \times 10^3 \, \text{km}^2 \, \text{y}^{-1}$ in Ecuador, $6.0 \times 10^3 \, \text{km}^2 \, \text{y}^{-1}$ in Venezuela, and $0.2 \times 10^3 \, \text{km}^2 \, \text{y}^{-1}$ in Guyana. Deforestation rates in other parts of Greater Amazonia were minimal: Suriname had 130 $\text{km}^2 \, \text{y}^{-1}$ and French Guiana had less than 10 $\text{km}^2 \, \text{y}^{-1}$.

Landsat data interpreted at INPE (Figure 2) indicate that, by 1996, the area of forest cleared in Brazilian Amazonia had reached $517.1 \times 10^3 \, \text{km}^2$ (12.9 percent of the originally forested portion of Brazil's Legal Amazon Region), including approximately $100 \times 10^3 \, \text{km}^2$ of "old" (pre-1970) deforestation in Pará and Maranhão. Over the period 1978–1988, forest was lost at a rate of $20.4 \times 10^3 \, \text{km}^2 \, \text{y}^{-1}$ (including hydroelectric flooding), the rate declined (beginning in 1987) to a low point of $11.1 \times 10^3 \, \text{km}^2 \, \text{y}^{-1}$ in 1990–1991, and climbed to $14.9 \times 10^3 \, \text{km}^2 \, \text{y}^{-1}$ in 1992–1994; the rate then jumped to $29.1 \times 10^3 \, \text{km}^2 \, \text{y}^{-1}$ in 1994–1995, and fell to $13.2 \times 10^3 \, \text{km}^2 \, \text{y}^{-1}$ in 1996–1997 and rose to $17.4 \times 10^3 \, \text{km}^2$ in 1997–1998; a preliminary estimate for 1998–1999 indicates a deforestation rate of $16.9 \times 10^3 \, \text{km}^2 \, \text{y}^{-1}$ (Fearnside, 1997a; Instituto Nacional de Pesquisas Espaciais, 2000). Current values can be obtained from INPE's Web site (see Bibliography). Note,

however, that the official explanations given by INPE as to why deforestation rates rise and fall (decrees affecting incentives and programs for inspection and levying fines) are unlikely to be correct (see below).

**Causes of Deforestation.** Amazonian countries differ greatly in the social factors driving deforestation. In Brazil, most clearing is carried out by owners of large and middle-sized ranches for cattle pasture, whereas the role of small farmers clearing for agriculture is relatively more important in the other countries. Brazil is by far the most important country in tropical forest matters in Amazonia and globally, both in terms of the extent of remaining forest and of the area of forest being cleared each year.

The relative weight of small farmers compared with major landholders in Brazilian Amazonia is continually changing as a result of changing economic and demographic pressures. The behavior of large landholders is most sensitive to economic changes such as the interest rates offered by money markets and other financial investments, government subsidies for agricultural credit, the rate of general inflation, and changes in the price of land. Tax incentives were a strong motive in the 1970s and 1980s. In June 1991, a decree suspended the granting of new incentives. However, the old (that is, already approved) incentives continue to the present day, contrary to the popular impression that has been fostered by numerous statements by government officials to the effect that incentives had ended. Many of the other forms of incentive, such as large amounts of government-subsidized credit at rates far below those of Brazilian inflation, became much scarcer after 1984.

For decades preceding the initiation of Brazil's Plano Real economic reform program in July 1994, hyperinflation was the dominant feature of the Brazilian economy. Land played a role as a store of value, and its value was bid up to levels much higher than what could be justified as an input to agricultural and ranching production. Deforestation played a critical role as a means of holding claim to land (Fearnside, 1987). Deforesting for cattle pasture was the cheapest and most effective means of maintaining possession of investments in land. The extent to which the motive for defending these claims (through expansion of cattle pasture) was speculative profit from increasing land value has been a matter of debate. Hecht et al. (1988) present calculations of the overall profitability of ranching in which the contribution from speculation is critical, while Mattos and Uhl (1994) show that actual production of beef has become increasingly profitable, and that supplementary income from selling timber (allowing investment in recuperation of degraded pastures on the properties) is critical. Obviously, selling of timber can only be depended upon for a few years to subsidize the cattle-raising portion of

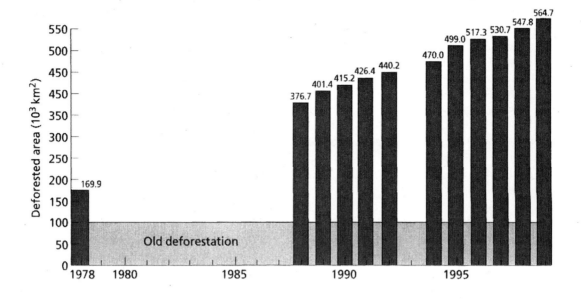

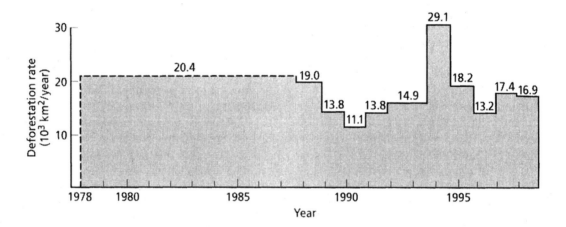

**Amazonia: Deforestation of. FIGURE 2.** Extent and Rate of Deforestation in the Brazilian Legal Amazon.

"Old" deforestation refers to pre-1970 clearing in Pará and Maranhão.

the operations, since the harvest rates are virtually always above sustainable levels. Faminow (1998) has made a more complete analysis of land price trends in Amazonia, and finds that speculative profits cannot explain the attraction of capital to investments in Amazonian ranches.

The decline in deforestation rates from 1987 to 1991 can best be explained by Brazil's deepening economic recession over this period. Ranchers simply did not have money to invest in expanding their clearings as quickly as they had in the past. In addition, the government lacked funds to continue building highways and establishing settlement projects. Probably very little of the decline can be attributed to Brazil's repression of de-

forestation through inspection from helicopters, confiscation of chainsaws and fining of landowners caught burning without the required permission from the Brazilian Institute of Environment and Renewable Natural Resources (IBAMA). Despite bitter complaints, most people continued to clear anyway. Changes in policies on the granting of fiscal incentives also do not explain the decline. The decree suspending the granting of incentives was issued on 25 June 1991—after almost all of the observed decline in deforestation rate had already occurred (see Figure 2). Even for the last year (1991), the effect would have been minimal, since the average date for the Landsat images for the 1991 data set was August of that year.

The peak in 1995 is probably in large part a reflection of economic recovery under the Plano Real, which resulted in larger volumes of money suddenly becoming available for investment, including investment in cattle ranches. The fall in deforestation rates in the years after 1995 is a logical consequence of the Plano Real having sharply cut the rate of inflation. Land values reached a peak in 1995 and fell by about 50 percent by the end of 1997. Falling land values make land speculation unattractive to investors. The association of major swings in deforestation rate with macroeconomic factors such as money availability and inflation rate is one indication that much of the clearing is done by those who invest in medium and large cattle ranches, rather than by small farmers using family labor.

The distribution of 1991 clearing among the region's nine states indicates that most of the clearing is in states that are dominated by ranchers: the state of Mato Grosso alone accounts for 26 percent of the 11,100 square kilometer total. Mato Grosso has the highest percentage of its privately held land in ranches of 1,000 hectares (2,500 acres) or more: 84 percent at the time of the last (1985) agricultural census. A moment's reflection on the human significance of having 84 percent of the land in large ranches (and only 3 percent in small farms) should give anyone pause. By contrast, Rondônia—a state that has become famous for its deforestation by small farmers—had only 10 percent of the 1991 deforestation total, and Acre had 3 percent.

The number of properties in each size class in the census explains 74 percent of the variation in deforestation rate among the nine Amazonian states. Multiple regressions indicate that 30 percent of the clearing in both 1990 and 1991 can be attributed to small farmers (properties less than 100 hectares (250 acres) in area), and the remaining 70 percent to either medium or large ranchers (Fearnside, 1993). The social cost of substantial reductions in deforestation rates would therefore be much less than is implied by frequent pronouncements that blame "poverty" for environmental problems in the region.

The question of who is to blame for tropical deforestation has profound implications for the priorities of programs intended to reduce forest loss. The prominence of cattle ranchers in Brazil (which is not the case in many other parts of the tropics) means that measures aimed at containing deforestation by, for example, promoting agroforestry among small farmers can never achieve this goal, although tools such as agroforestry have important reasons for being supported independently of efforts to combat deforestation (Fearnside, 1995a).

**Impacts of Deforestation.** Deforestation has many significant impacts on local and global ecosystems. Some of the most important are discussed here.

*Loss of biodiversity.* Deforestation results in loss of biodiversity because most tropical forest species cannot survive the abrupt changes that take place when forest is felled and burned, and cannot adapt to the new conditions in the deforested landscape. [*See* Biological Diversity.] The high degree of *endemism*, or presence of species that are only found within a small geographical range, can result in loss of species and loss of genetic variability within species, even when the forest surrounding a cleared area appears to human observers to be identical to the forest that was lost.

The impact of deforestation extends beyond the area directly cleared because of edge effects and the impact of fragmentation of the formerly continuous forest into small islands that are unable to support viable populations of forest species and their biological interactions (Laurance and Bierregaard, 1997). In addition, fire and other disturbance regimes (including logging) are usually associated with the presence of nearby deforestation, thus further extending the impact beyond the edges of the clearings.

The impact of converting forest to another land use depends not only on the patch of land for which conversion is being considered, but also on what has been done with the remainder of the region. As the cumulative area cleared increases, the danger increases that each additional hectare of clearing will lead to unacceptable impacts. For example, the risk of species extinctions increases greatly as the remaining areas of natural forest dwindle.

Biodiversity has many types of value, from financial value associated with selling a wide variety of products, to the use value of the products, to existence values unrelated to any direct use of a species and its products. There is disagreement on what values should be attached to biodiversity, especially those forms of value not directly translatable into traditional financial terms by today's marketplace. While some may think that biodiversity is worthless except for sale, it is not necessary to convince people who hold this view that biodiversity is valuable; rather, it is sufficient for them to know that a constituency of people concerned about biodiversity exists today and is growing, and that this represents a potential source of financial flows intended to maintain biodiversity. Political scientists estimate that such willingness to pay already surpasses U.S.$20 per hectare per year for tropical forest (Cartwright, 1985).

***Reduced carbon storage.*** Storage of carbon, which may in turn reduce global warming caused by the greenhouse effect, represents a major environmental benefit or service of Amazonian forests. [*See* Greenhouse Effect.] The way that this benefit is calculated can have a tremendous effect on the value assigned to maintaining Amazonian forest. As currently foreseen in the Framework Convention on Climate Change (FCCC), maintenance of carbon stocks is not considered a service—only deliberate incremental alterations in the flows of carbon. Even considering only this much more restrictive view of carbon benefits, the value of Amazonian forests is substantial. In 1990 (the year that is the baseline for inventories under the FCCC to assess changes in greenhouse gas emissions), Brazil's 13,800 $km^2 y^{-1}$ rate of deforestation was producing net committed emissions of 263 million metric tons of $CO_2$-equivalent carbon per year (Fearnside, 1997b). The benefit of slowing or stopping this emission is therefore substantial. For comparison, the world's 400 million automobiles emit 550 million metric tons of carbon annually (Flavin, 1989, p. 35). All human activities in the 1980s emitted approximately 7.1 billion metric tons of carbon yearly, 5.5 billion metric tons of which were from fossil fuel combustion (Schimel et al., 1996, p. 79); this means that, while slowing deforestation would be an important measure in combating global warming, it cannot eliminate the need for major reductions in fossil fuel use in industrialized countries.

Although a wide variety of views exist on the value of carbon, already enacted carbon taxes of U.S.$45 per metric ton in Sweden and the Netherlands and U.S.$6.1 per metric ton in Finland indicate that the willingness to pay for this benefit is already substantial. This willingness to pay may increase significantly in the future when the magnitude of potential damage from global warming becomes more apparent to decision makers and the general public. At the level indicated by current carbon taxes, the global warming damage of Amazon deforestation is already worth U.S.$1.6–11.8 billion per year. The value of the global warming damage from clearing a hectare of forested land in Amazonia (U.S.$1,200–8,600) is much higher than the purchase price of land today. (These calculations use U.S.$7.3 per metric ton of carbon as the value of permanently sequestered carbon; this is the "medium" value from Nordhaus, 1991.) [*See* Ecotaxation.]

*Loss of water cycling.* Water cycling is different from biodiversity and carbon loss in that impacts of deforestation in this area fall directly on Brazil rather than being spread over the Earth as a whole. [*See* Hydrologic Cycle.] Several independent lines of evidence indicate that about half of the rainfall in the Brazilian Amazon is water that is recycled through the forest, the rest originating from water vapor blown into the region directly from the Atlantic Ocean (Gash et al., 1996; Shukla et al., 1990). Because recycled water is 50 percent of the total, the volume of water that is recycled is the same amount as one sees flowing in the Amazon River. The Amazon is by far the world's largest river in terms of water flow—over eight times larger than the second largest, Africa's Zaire River, and seventeen times larger than the Mississippi–Missouri system in North America. Some of the water vapor is transported by winds to Brazil's Central–South Region, where most of the country's agriculture is located. Brazil's annual harvest has a gross value of about U.S.$65 billion, and the dependence of even a small fraction of this on rainfall from Amazonian water vapor would translate into a substantial value for Brazil. Although movement of the water vapor

is indicated by general circulation models (Eagleson, 1986; Salati and Vose, 1984), the amounts involved are as yet unquantified.

The role of Amazonian forest in the region's water cycle also implies increasing risk with increasing scale of deforestation: when rainfall reductions caused by losses of forest evapotranspiration are added to the natural variability that characterizes rainfall in the region, the resulting droughts may cross biological thresholds and lead to major impacts (Fearnside, 1995b). These thresholds include the drought tolerance of individual tree species and the increased probability of fire propagation in standing forest. Fire entry into standing forest in Brazilian Amazonia already occurs in areas disturbed by logging (Uhl and Buschbacher, 1985; Uhl and Kauffman, 1990). During the El Niño drought of 1997–1998, over 9,000 square kilometers of undisturbed forest burned in Brazil's far northern state of Roraima. In Amazonia, major El Niño events have caused widespread conflagrations in the forest four times over the past 2,000 years (Meggers, 1994). The effect of large-scale deforestation is to turn relatively rare events like these into something that could recur at much more frequent intervals. [*See* El Niño–Southern Oscillation.]

**Potential Countermeasures.** Strong measures are needed to contain or reverse the deforestation of Amazonia. The continued march of deforestation indicates that current efforts, although well intentioned, are far from sufficient.

*Current efforts.* Current efforts to contain deforestation include the Pilot Program to Conserve the Brazilian Rainforest, financed by the G-7 countries and administered by the World Bank. Components already being implemented as of 1998 include the "PD/A" or "type A demonstration projects" (small projects carried out by nongovernmental organizations), extractive reserves, indigenous lands, and support for scientific research centers and directed research projects. Projects expected to begin soon include natural resources policy (such as zoning), natural resources management (mainly forestry), *várzea* ("flood plain") management, parks and reserves, fire and deforestation control (such as detection of deforestation and burning), and monitoring and analysis of Pilot Program activities in order to learn policy lessons. Activities for which proposals are under preparation (for integration into the PD/A component) include recuperation of degraded lands, environmental education, and indigenous and private-sector demonstration projects.

In addition to the Pilot Program, the Brazilian government has a number of other programs aimed at controlling deforestation. These can be seen on the Web site of the Brazilian Institute for the Environment and Renewable Natural Resources (IBAMA) (see Bibliography).

***Needed policy changes.*** The most basic problem in controlling deforestation is that much of what needs to be done is beyond the purview of agencies such as IBAMA that are charged with environmental problems. Authority to change tax laws, resettlement policies, and road-building priorities, for example, rests with other parts of the government.

The overriding importance of the economic recession means that deforestation rates can be expected to increase again once Brazil's economy recovers, unless the government takes steps now to remove the underlying motives for deforestation. Steps needed include levying of heavy taxes to take the profit out of land speculation, changing of land titling procedures to cease recognizing deforestation for cattle pasture as a *benfeitoria* ("improvement"), removal of remaining subsidies, reinforcement of procedures for Environmental Impact Reports (RIMAs), implementation of agrarian reform both in Amazonia and in the source areas of migrants, and provision of alternative employment in both rural and urban areas (Fearnside, 1989).

Although small farmers account for only 30 percent of the deforestation activity, the intensity of deforestation within the area they occupy is greater than for the medium and large ranchers who hold 89 percent of the Legal Amazon's private land. Deforestation intensity, or the impact per square kilometer of private land, declines with increasing property size. This means that deforestation would increase if forest areas now held by large ranches were redistributed into small holdings. This indicates the importance of using already cleared areas for agrarian reform, rather than following the politically easier path of distributing areas still in forest. Large as the area already cleared is, it has limits that fall far short of the potential demand for land to be settled. Indeed, even the Legal Amazon as a whole falls short of this demand (Fearnside, 1985). Recognition of the existence of carrying capacity limits, followed by maintenance of population levels within these limits, is fundamental to any long-term plan for sustainable use of Amazonia (Fearnside, 1986, 1997c).

***Environmental services as development.*** At present, economic activities in Amazonia almost exclusively involve taking some material commodity and selling it. Typical commodities include timber, minerals, the products of agriculture and ranching, and nontimber forest products such as natural rubber and Brazil nuts. The potential is much greater, both in terms of monetary value and of sustainability, for pursuing a radically different strategy for long-term support: finding ways to tap the environmental services of the forest as a means both of sustaining the human population and of maintaining the forest.

At least three classes of environmental service are provided by Amazonian forests: biodiversity maintenance, carbon storage, and water cycling. Preliminary calculations of indicators of "willingness to pay" for the services lost from 1990 deforestation in the Brazilian Legal Amazon total U.S.$2.5 billion (assuming 5 percent annual discount); maintenance of the stock of forest, if regarded as producing 5 percent per year annuity, would be worth U.S.$37 billion annually (Fearnside, 1997d). The magnitude and value of these services are poorly quantified, and the diplomatic and other steps through which such services might be compensated are also in their infancy. These facts diminish neither the importance of the services nor the need to focus efforts on providing both the information and the political will required to integrate these services into the rest of the human economy in such a way that economic forces act to maintain rather than to destroy the forest (Fearnside, 1997d).

One of the major challenges to finding rational uses for Amazonian forest lies in gathering and interpreting relevant information. The establishment of environmental services of the forest as a basis for sustainable development is, perhaps, the area in which information is most critical. A better understanding of the dynamics of deforestation and of deforestation's impacts on biodiversity, carbon storage, and water cycling is a necessary starting point for the long and difficult task of turning environmental services into a basis for sustainable development in Amazonia.

The term *development* implies a change, usually presumed to be in the direction of improvement. What is developed and whom the improvement should benefit are subject to widely differing opinions. It might be suggested, for example, that in order to be considered "development," the change in question must provide a means to sustain the local population. Infrastructure that does not lead to production is not development, nor is a project that exports commodities from the region while generating minimal employment or other local returns (aluminum processing and export is a good example because it provides almost no employment in Brazil, despite massive monetary, environmental, and social costs of hydroelectric dams built to supply the industry; Fearnside, 1999, 2000).

The production of traditional commodities often fails to benefit the local population. Conversion of forest to cattle pasture, the most widespread land use change in Brazilian Amazonia, brings benefits that are extremely meager (although not quite zero). A high priority must be given to the redirection of development to activities with local returns that are greater and longer lasting. This goal may be achieved by tapping the value of environmental services. The most important challenge in turning these services into development is to reserve the benefits of the services for the inhabitants of the Amazonian interior (Fearnside, 1997d).

[*See also* Albedo; Biomes; Carbon Dioxide; Forests; *and* Human Impacts, *article on* Human Impacts on Biota.]

## BIBLIOGRAPHY

Brazilian Institute for the Environment and Renewable Natural Resources: http://www.ibama.gov.br/. Information on Brazilian programs aimed at controlling deforestation.

Cartwright, J. "The Politics of Preserving Natural Areas in Third World States." *The Environmentalist* 5.3 (1985), 179–186.

Dale, V., R. V. O'Neill, F. Southworth, and M. Pedlowski. "Modeling Effects of Land Management in the Brazilian Settlement of Rondônia." *Conservation Biology* 8.1 (1994), 196–206.

Daly, D. C., and G. T. Prance. "Brazilian Amazon." In *Floristic Inventory of Tropical Countries: The Status of Plant Systematics, Collections, and Vegetation, plus Recommendations for the Future*, edited by D. G. Cambell and H. D. Hammond, pp. 401–426. New York: New York Botanical Garden, 1989.

Eagleson, P. S. "The Emergence of Global-Scale Hydrology." *Water Resources Research* 22.9 (1986), 6s–14s.

Faminow, M. D. *Cattle, Deforestation and Development in the Amazon: An Economic and Environmental Perspective*. New York: CAB International, 1998.

Fearnside, P. M. "Agriculture in Amazonia." In *Key Environments: Amazonia*, edited by G. T. Prance and T. E. Lovejoy, pp. 393–418. Oxford: Pergamon, 1985.

———. *Human Carrying Capacity of the Brazilian Rainforest*. New York: Columbia University Press, 1986.

———. "Causes of Deforestation in the Brazilian Amazon." In *The Geophysiology of Amazonia: Vegetation and Climate Interactions*, edited by R. E. Dickinson, pp. 37–61. New York: Wiley, 1987.

———. "A Prescription for Slowing Deforestation in Amazonia." *Environment* 31.4 (1989), 16–20, 39–40.

———. "The Rate and Extent of Deforestation in Brazilian Amazonia." *Environmental Conservation* 17.3 (1990), 213–226.

———. "Deforestation in Brazilian Amazonia: The Effect of Population and Land Tenure." *Ambio* 22.8 (1993), 537–545.

———. "Agroforestry in Brazil's Amazonian Development Policy: The Role and Limits of a Potential Use for Degraded Lands." In *Brazilian Perspectives on Sustainable Development of the Amazon Region*, edited by M. Clüsener-Godt and I. Sachs, pp. 125–148. Carnforth, U.K.: Parthenon Publishing Group, 1995a.

———. "Potential Impacts of Climatic Change on Natural Forests and Forestry in Brazilian Amazonia." *Forest Ecology and Management* 78 (1995b), 51–70.

———. "Amazonia and Global Warming: Annual Balance of Greenhouse Gas Emissions from Land-Use Change in Brazil's Amazon Region." In *Biomass Burning and Global Change*, edited by J. Levine, vol. 2, *Biomass Burning in South America, Southeast Asia and Temperate and Boreal Ecosystems and the Oil Fires of Kuwait*, pp. 606–617. Cambridge, Mass.: MIT.

———. "Monitoring Needs to Transform Amazonian Forest Maintenance into a Global Warming Mitigation Option." *Mitigation and Adaptation Strategies for Global Change* 2 (1997a), 285–302.

———. "Greenhouse Gases from Deforestation in Brazilian Amazonia: Net Committed Emissions." *Climatic Change* 35.3 (1997b), 321–360.

———. "Human Carrying Capacity Estimation in Brazilian Amazonia as a Basis for Sustainable Development." *Environmental Conservation* 24.3 (1997c), 271–282.

———. "Environmental Services as a Strategy for Sustainable Development in Rural Amazonia." *Ecological Economics* 20.1 (1997d), 53–70.

———. "Social Impacts of Brazil's Tucuruí Dam." *Environmental Management* 24.4 (1999), 485–495.

———. "Environmental Impacts of Brazil's Tucuruí Dam: Unlearned Lessons for Hydroelectric Development in Amazonia." *Environmental Management* 26.2 (2000, in press).

Flavin, C. "Slowing Global Warming: A Worldwide Strategy." Worldwatch Paper 91. Washington, D.C.: Worldwatch Institute, 1989.

Food and Agriculture Organization. *Forest Resources Assessment 1990: Tropical Countries*. FAO Forestry Paper 112. Rome: FAO, 1993.

Gash, J. H. C., C. A. Nobre, J. M. Roberts, and R. L. Victoria, eds. *Amazonian Deforestation and Climate*. Chichester, U.K.: Wiley, 1996.

Harcourt, C., C. Billington, J. Sayer, and M. Jenkins. "Introduction." In *The Conservation Atlas of Tropical Forests: The Americas*, edited by C. Harcourt and J. A. Sayer, pp. 9–16. New York: Simon & Schuster, 1996.

Hecht, S. B., R. B. Norgaard, and C. Possio. "The Economics of Cattle Ranching in Eastern Amazonia." *Interciencia* 13.5 (1988), 233–240.

Instituto Nacional de Pesquisas Espaciais (INPE). *Monitoring the Brazilian Amazonian Forest by Satellite 1998–1999*. Document released via Internet: http://www.inpe.br/Informações_Eventos/Amy 1998–1999/index_amy.htm. São Paulo, Brazil: INPE, 2000.

Laurance, W. F., and R. O. Bierregaard, Jr., eds. *Tropical Forest Remnants: Ecology, Management, and Conservation of Fragmented Communities*. Chicago: University of Chicago Press, 1997.

Mattos, M. M., and C. Uhl. "Economic and Ecological Perspectives on Ranching in the Eastern Amazon." *World Development* 22.2 (1994), 145–158.

Meggers, B. J. 1994. "Archeological Evidence for the Impact of Mega-Niño Events on Amazonia during the Past Two Millennia." *Climatic Change* 28: 321–338.

Nordhaus, W. "A Sketch of the Economics of the Greenhouse Effect." *American Economic Review* 81.2 (1991), 146–150.

Salati, E., and P. B. Vose. "Amazon Basin: A System in Equilibrium." *Science* 225 (1984), 129–138.

Schimel, D., et al. "Radiative Forcing of Climate Change." In *Climate Change 1995: The Science of Climate Change*, edited by J. T. Houghton, L. G. Meira Filho, B. A. Callander, N. Harris, A. Kattenberg, and K. Maskell, pp. 65–131. Cambridge: Cambridge University Press, 1996.

Shukla, J., C. Nobre, and P. Sellers. "Amazon Deforestation and Climate Change." *Science* 247 (1990), 1322–1325.

Tratado de Cooperación Amazonica. *Amazonia without Myths*. Quito, Ecuador: Commission on Development and Environment for Amazonia, 1992.

Uhl, C., and R. Buschbacher. "A Disturbing Synergism between Cattle-Ranch Burning Practices and Selective Tree Harvesting in the Eastern Amazon." *Biotropica* 17.4 (1985), 265–268.

Uhl, C., and J. B. Kauffman. "Deforestation, Fire Susceptibility, and Potential Tree Responses to Fire in the Eastern Amazon." *Ecology* 71.2 (1990), 437–449.

—PHILIP M. FEARNSIDE

# ANIMAL HUSBANDRY

Animal agriculture has evolved with the growth of the human population from 4 million people ten thousand years ago to the 5.85 billion people in the world near the end of the twentieth century. Domestication of a relatively few of the available mammalian species initially occurred in conjunction with climatic, cultural, and societal changes following the end of the last ice age, the Wisconsin deglaciation. Agrarian-based communities developed independently around the world, and generic animal husbandry practices became common. [*See* Agriculture and Agricultural Land.] Today, despite tremendous technological advances in genetics (such as cloning), physiological manipulations (such as artificial insemination), and nutritional provisions (such as high-quality harvested forages), there are still only a few domesticated species used in animal agriculture, and traditional animal husbandry practices are still nearly universally applied. In 1996, the world population of domesticated grazing livestock was approximately 3.3 billion, an increase of 1.1 billion during the last five decades despite a relatively constant total amount of grazeable lands. The bulk of this increase during the last half of the twentieth century has been a near doubling of cattle, sheep, and goats on the continents of Asia and Africa. The unmanaged impacts of grazing animals have been substantial through the course of human history. The modern principles of proper management of grazing lands were developed during this century in response to impacts of unmanaged grazing.

**Domestication.** Domestication is the adaptation by humans of a species, resulting in the modification of traits to the extent that the species is clearly distinguishable from its native origins. There is evidence that humans first domesticated dogs from wolves before 100,000 BP (Vila et al., 1997). However, the history of animal domestication other than canines is generally thought to be much more recent, with beginnings about 10,000 BP. Domestication has been documented to have occurred independently at numerous locations around the world from 10,000 to 5,000 BP (Table 1). In the same period, the world human population has been estimated to have increased from about 4 million (10,000 BP) to 14 million (5,000 BP), with the greatest increases on the European and Asian continents, where domestication (of both plants and animals) was being broadly adopted (McEvedy and Jones, 1978).

As with the beginnings of agriculture during this period, the development of domestication has been attributed to a variety of changing global conditions affecting climate (for example, deglaciation), culture (for example, ritual animal sacrifice), and society (for example, increased sedentary populations). In addition, domestication developed under a wide variety of conditions. For example, domestication occurred in both sedentary and nomadic cultures, in agrarian and hunter–gatherer societies, prior to and subsequent to domestication of plants, and using either monogastric or ruminant species. In addition, domestication did not occur in some situations where both the conditions and the opportunities have been viewed as optimal (Reed, 1977).

Although numerous animal species were tamed during this period (such as elephants in India and hyenas in Egypt), only a few species of mammals were actually domesticated. Ninety percent of domesticated grazing animals in the world in 1996 were either cattle, sheep,

**Animal Husbandry. TABLE 1.** General Timeline (Years before Present) of Domestication of Animals Used as Livestock*

| YEARS BEFORE PRESENT | SPECIES AND LOCATIONS |
| --- | --- |
| 10,000 | Sheep and goats in southwestern Asia |
| 9,000 | Cattle, sheep, and goats in southeastern Europe; pigs in Asia, Europe, and the Far East |
| 7,500 | Alpacas or llamas or both in Peru |
| 6,000 | Wild asses in northern Africa |
| 5,000 | Bactrian camels in Turkmenistan region |
|  | Horses in Asia |
|  | Water buffaloes in southern China |
| 2,500 | Rabbits in Europe |

* Note that numerous species have been tamed and exploited but cannot be regarded as domesticated. Some of these events predated domestication, including reindeer herding fifteen thousand years ago in Siberia.

SOURCE: Adapted from Clutton-Brock (1981).

or goats, and these can be traced to just three species: *Bos primigenius, Ovis orientalis,* and *Capra aegagrus* (Clutton-Brock, 1981). Domestication requires at least thirty generations, a period of one hundred to two hundred years for most of our domesticated animals. The unique demands of domestication have limited its application by humans to only a few species, but the traditions of animal husbandry have been readily conveyed to succeeding human generations. By 5,000 BP, communities were engaged in common animal husbandry practices characteristic of animal agriculture, such as the use of grazing systems, selective breeding practices, and spreading of animal manure onto agricultural fields.

In 1996, grazing livestock numbers were estimated at 3.3 billion (Table 2). Domesticated grazers (primarily cattle, sheep, goats, horses, buffalo, mules, and camels) were 91 percent of the total world population of domesticated animals, including pigs and poultry (Food and Agriculture Organization, 1996). Since World War II, the world populations of grazing livestock—primarily cattle, sheep, and goats—have collectively increased by 1 percent annually. However, this increase of 1.1 billion head during the last five decades has been spatially heterogeneous. Cattle numbers have more than doubled in Africa, South America, and eastern Europe, a rate of increase more than 50 percent higher than the global average. Seventy-seven percent of the global increase in sheep numbers has occurred in Asia. Overgrazing is frequently cited as a principal cause of land degradation on these continents during the twentieth century (Dregne et al., 1991).

**Grazing Animal Impacts.** Grazing refers to the foraging use of herbaceous materials that is such a characteristic feature of livestock use of pastures and rangelands. Grazing animals have well documented direct and indirect impacts upon grazed ecosystems beyond simple plant defoliation. These impacts can be manifested at several scales, from that of the individual plant to landscape areas of thousands of hectares. [*See* Desertification; *and* Erosion.] Direct impacts of grazing include consumption of plant issues, trampling of soil surfaces, disruption of cryptogamic organisms, and removal and excretion of nutrients by the harvester. Indirect effects include spatial and temporal redistribution of nutrients, alteration of numerous plant morphological and physiological attributes, alterations of plant gene frequencies, changes to structural features of plant communities, and alterations of water and energy fluxes within the landscape. Impacts can vary tremendously with variations in environmental conditions. The effects of grazing are frequently compounded by the concurrence of other environmental stresses, particularly drought and fire. The impacts of grazing can often be episodic and reflect the occurrence of mitigating site conditions. Often a general understanding of the effects of grazing will not be useful to anticipating and predicting grazing effects in specific environments (Vavra et al., 1994).

Herbivory is a natural process that can have complex impacts in natural settings (Collins et al., 1998). However, unmanaged grazing by livestock can have extreme and negative effects on ecosystems on local, regional, and continental scales (McNaughton, 1993). On most continents, there is evidence of damage to grazing land resources because of overgrazing by livestock during some prior time periods. Frequently, these damaging effects manifest themselves in a relatively short period, often just within one or two decades. Given the arid and

**Animal Husbandry. TABLE 2.** World Population Numbers (Millions) of the Primary Domesticated Grazing Animals in 1948–1952 and in 1996, by Continent

| | CATTLE | | SHEEP AND GOATS | | HORSES, ASSES, AND MULES | | BUFFALO | | CAMELS | | TOTAL | |
|---|---|---|---|---|---|---|---|---|---|---|---|---|
| | *1948–1952* | *1996* | *1948–1952* | *1996* | *1948–1952* | *1996* | *1948–1952* | *1996* | *1948–1952* | *1996* | *1948–1952* | *1996* |
| World | 759 | 1,320 | 1,059 | 1,722 | 127 | 120 | 81 | 152 | 9 | 19 | 2,035 | 3,303 |
| Western Europe | 100 | 170 | 143 | 141 | 22 | 5 | 1 | <1 | 0 | <1 | 266 | 316 |
| Eastern Europe* | 56 | 118 | 92 | 144 | 14 | n/a | 0† | <1 | 0 | <1 | 162 | 262 |
| North and Central America | 114 | 165 | 51 | 32 | 20 | 22 | 0 | <1 | 0 | 0 | 185 | 219 |
| South America | 135 | 300 | 142 | 110 | 26 | 23 | 0 | 2 | 0 | 0 | 303 | 435 |
| Asia | 242 | 431 | 279 | 795 | 32 | 46 | 79 | 147 | 3 | 5 | 635 | 1424 |
| Africa | 94 | 198 | 205 | 386 | 13 | 19 | 2 | 3 | 6 | 14 | 320 | 620 |
| Oceania | 20 | 37 | 145 | 176 | 1 | <1 | 0 | 0 | 0 | 0 | 166 | 213 |

*Reported as USSR in 1956, and separated as eastern Europe in 1996.

†0 = none reported.

SOURCE: From Food and Agriculture Organization (1956; 1996).

nutrient-poor conditions of many of our global grazing lands, recovery of these lands can be slow or nonexistent (Schlesinger et al., 1990). It is now increasingly recognized that some environments, such as those with highly erodible soils or inherently low primary productivity, may not be suitable for grazing by domestic livestock. However, managed grazing in many environments has been shown to have beneficial or, at most, negligible adverse effects (Heitschmidt and Stuth, 1991).

**Managed Grazing.** Managed grazing is any situation in which the array of grazing behaviors displayed by an animal is under some degree of management control. Most often the level of control is at a pasture scale, which can vary from less than one hectare in humid climates to over 10,000 hectares in arid climates. Key factors for management control are the timing, frequency, intensity, and duration of grazing (Trlica and Rittenhouse, 1993). Each of these controlling factors can have significant effects upon the structures and functions of grazing land ecosystems. For example, given that grazing is a selective process, the effects of grazing intensity can be manifested by both grazed and ungrazed plant species. The most commonly examined controlling factor is intensity of grazing use, but impacts of other factors are well recognized and management principles are well articulated (Holechek et al., 1998).

Grazing methods (systems) are any type of management program in which key factors of grazing are controlled to some degree. Grazing systems fall into four broad categories: true nomadism, semisedentary, transhumant, and sedentary (Williams, 1981). Variations within each category are numerous, but each system is characterized by some level of management over one or more of the main controlling factors. It is important to recognize that grazing systems have evolved over the past ten thousand years. However, on an evolutionary scale, angiosperm–vertebrate interactions have occurred over millions of years. Although grazing systems have been developed either to minimize livestock effects or to mimic historical grazing behaviors of native herbivores, there is no evidence that any one specific combination of control factors (namely, a specific system) has intrinsic advantages over other combinations/systems to either compensate for prior mismanagement or dramatically improve resource conditions. The benefits of particular systems are often in the level of increased human involvement in land management that accompany their deployment. Secondary benefits associated with implementation of a system, such as improved animal distribution associated with stock water development, or restricted growing-season defoliation resulting from construction of new fences to conserve more sensitive areas, are frequently observed.

Today, grazing management is evolving toward achieving desired resource conditions rather than attaining some prior existing states (Westoby et al., 1989). This evolution is occurring both in developed countries that have placed an emphasis on multiple uses of grazing land resources (i.e., functioning to provide forage for livestock, habitat for wildlife, conserved watersheds for high-quality water supplies, and open spaces for recreational adventures) and in less developed nations where livestock production is a key component of subsistence agriculture. A future issue is whether grazing animals can be used as tools for managing landscapes while also providing agricultural products. Management objectives could include a variety of goals such as increased native vegetation diversity, improved watershed functions, decreased use of harvested forages, and biological control of introduced plant species. These objectives require managed control of grazing by livestock to minimize or negate the effects of grazing.

[*See also* Food and Agriculture Organization; Hunting and Poaching; Land Use, *article on* Land Use and Land Cover; *and* Methane.]

## BIBLIOGRAPHY

Clutton-Brock, J. *Domesticated Animals from Early Times.* Austin: University of Texas Press, 1981. A general text summarizing the history of animal domestication from cats to cattle.

Collins, S. L., A. L. Knapp, J. M. Briggs, J. M. Blair, and E. M. Steinauer. "A Modulation of Diversity of Grazing and Mowing in Native Tallgrass Prairie." *Science* 280 (1998), 745–747. A scientific paper summarizing long-term research on the effects of fire and grazing by both cattle and bison on species biodiversity in the tallgrass prairie.

Dregne, H., M. Kassas, and B. Rozanov. "A New Assessment of the World Status of Desertification." *Desertification Control Bulletin* 20 (1991), 6–18. An overview of the amounts of degraded lands by continent by one of the world experts (H. Dregne) on the subject of desertification.

Food and Agriculture Organization. *Production Yearbook*, vol. 2. Rome: Food and Agricultural Organization, 1956. A compilation of agricultural statistics.

———. *Production Yearbook*, vol. 50. Rome: Food and Agriculture Organization, 1996. A compilation of agricultural statistics.

Heitschmidt, R. K., and J. W. Stuth, eds. *Grazing Management: An Ecological Perspective.* Portland, Oreg.: Timber Press, 1991. An excellent overview of information related to rangeland management.

Holechek, J. L., R. D. Pieper, and C. H. Herbel. *Range Management Principles and Practices*, 3d ed. Upper Saddle River, N.J.: Prentice-Hall, 1998. The most current textbook on range management in wide university use in North America.

McEvedy, C., and R. Jones. *Atlas of World Population History.* London: Penguin, 1978.

McNaughton, S. J. "Grasses and Grazers, Science and Management." *Ecological Applications* 3 (1993), 17–20. A brief summary of the literature on differences between grazing by native herbivores and grazing by managed domesticated ruminants.

Reed, C. A., ed. *Origins of Agriculture.* The Hague: Mouton Publishers, 1977. A compilation of both general and specific papers addressing the development of agriculture on different continents.

Schlesinger, W. H., J. F. Reynolds, G. L. Cunningham, L. F. Huenneke, W. M. Jarrell, R. A. Virginia, and W. G. Whitford. "Biological Feedbacks in Global Desertification." *Science* 247 (1990), 1043–1048. This paper contains an excellent conceptual model that describes the processes leading to desertification in deserts.

Trlica, M. J., and L. R. Rittenhouse. "Grazing and Plant Performance." *Ecological Applications* 3 (1993), 21–23. A concise summary of factors influencing effects of grazing on plants.

Vavra, M., W. A. Laycock, and R. D. Pieper. *Ecological Implications of Livestock Herbivory in the West.* Denver: Society for Range Management, 1994. A very readable set of symposium papers reviewing the scientific literature on grazing impacts in the western United States.

Vila, C., P. Savolainen, J. E. Maldonado, I. R. Amorin, J. E. Rice, R. L. Honeycutt, K. A. Crandall, J. Lundeberg, and R. K. Wayne. "Multiple and Ancient Origins of the Domestic Dog." *Science* 276 (1997), 1687–1692.

Westoby, M., B. Walker, and I. Noy-Meir. "Opportunistic Management for Rangelands not at Equilibrium." *Journal of Range Management* 42 (1989), 266–274. A key paper describing a new conceptual model for understanding the dynamics of rangelands.

Williams, O. B. "Evolution of Grazing Systems." In *Grazing Animals*, edited by F. H. W. Morley, pp. 1–12. Amsterdam: Elsevier, 1981. A thorough overview of grazing management terms and the history of different grazing methods.

—KRIS M. HAVSTAD

# ANTARCTICA

[*This entry comprises two articles on natural and anthropogenic environmental changes in Antarctica. The first article focuses on the threats and responses to these changes; the second explores the special rules of environmental protection that have resulted from the implementation of the Antarctic Treaty System.*]

## Threats and Responses

For a long time Antarctica was viewed as a remote continent, the science of which had only local or regional interest and significance. Times change, and with them the general recognition that Antarctica is a fundamental part of earth system science (Hansom and Gordon, 1998). The Antarctic and the Southern Ocean surrounding it now appear to be highly significant elements of the present global climate system, and historical data on past climates suggest that this has been the case for many millions of years. This recognition has ensured an increasing emphasis on the historical record of past climates in rocks, sediments, and ice cores as well as an escalation of previous studies on contemporary processes of great importance (Harris and Stonehouse, 1991). In addition, there is growing concern about pollution and its effects throughout much of the inhabited world. As the only nonindustrialized continent, the Antarctic is now seen as an ideal place in which to measure the global baselines for pollution against which trends elsewhere can be assessed.

Existing models of global climate at present predict that the greatest changes are likely to occur at high latitudes (Kattenberg et al., 1996), with the most recent interpretations suggesting that the Arctic will experience greater short-term changes than the Antarctic. However, because the feedback mechanisms in the south might easily magnify relatively small changes in sea ice, ice-sheet balance, or the link to ocean current systems, the ramifications of any change in the Antarctic are likely to be of global importance. Understanding and documenting these processes at a continental scale is a major focus of current research.

Historical reconstruction of previous climates is one important method for understanding climate change. In interpreting present changes, it is important to relate them to natural variability. Unless the degree of natural variability can be reconstructed accurately, it is difficult to assess if and how present changes are significantly anthropogenically driven. Direct records of climate in Antarctica are limited to this century, are restricted to a few localities, and are only usefully continuous for the last five decades. Yet the records of previous climates that are captured in the ice and sediments make Antarctica a critical area for studies of Southern Hemisphere paleoenvironments. [*See* Aerosols.] Using these proxy indicators, it is possible to reconstruct the frequency and magnitude of previous climate variability over a range of time scales ranging from decades to millennia. [*See* Climate Change.]

**Past Environments.** The Antarctic Ice Sheet is the largest area and volume of ice in the world. [*See* Ice Sheets.] Its maximum depth lies in East Antarctica, where ice over 4.5 kilometers thick is known to exist. The snow falling on the continent accumulates and is gradually turned to ice. There is a slow flow of ice outward from the center of the continent to the coast, but the rate of movement varies greatly from place to place. In areas called *ice domes*, the flow is very slow and ice can accumulate in these regions for up to 500,000 years. Drilling of such areas has provided ice cores from which a range of proxy data on previous climates have been extracted (Wolff and Bales, 1996). There are, of course, numerous caveats on the accuracy and reliability of each data type. The temporal resolution of change can be difficult in cores from such areas since annual precipitation is very low and the annual accumulation layers are difficult to count. It would be dangerous to provide a paleoclimatic model for the whole continent based on a very small number of deep core sites with low accumulation. Hence there are several international initiatives (for example, the European Polar Ice Coring in Antarctica project) to sample ice cores from across the Antarctic in areas with different snow accumulation rates to improve the temporal resolution to annual measurements on a time scale of centuries, and to assess the variability in regional patterns.

The ice cores provide various proxy records of climate change. Variation in mean temperature is measured through analysis of isotope ratios in oxygen and hydrogen, past atmospheric composition is determined by sampling gases trapped in the ice, rates of precipitation are recorded through measurements of the thickness of annual layers of ice, and changes in atmospheric circulation are deduced from examination of dust particles derived from volcanic eruptions and storms. The latest core from Vostok is at present the deepest at 3,623 meters, and has been subjected to the most complete analyses. Two of the most promising other major ice domes for drilling appear to be Dome C, with ice that is at least 3,000 meters deep, and Law Dome, with over 4,000 meters of ice. Such long records span several interglacials and are crucial in trying to understand how glacial/interglacial switching might be triggered.

High-resolution sediment records from the continental shelf, from lakes, and from ice-sheet margins provide a history of changes—again mainly from proxy indicators—in temperature, ice-sheet extent, coastal hydrology, ocean circulation, sea ice variability, and patterns of biological productivity. The lake sediments at present cover only the last fifteen thousand years, but offer opportunities for very precise analyses since the lack of animals and mixing in the lakes means that the annual layers are often very clearly defined. Marine sediments are more heavily disturbed, and so offer poorer resolution but with a time scale extending to millions of years. On some of the antarctic and subantarctic islands there are considerable peat deposits that offer a further insight into the records of terrestrial climate change on time scales of five thousand years or more.

**Contemporary Processes.** While it may be imagined that the significant effects of Antarctica on global change are all related to the stability of the ice sheet, this is not so. The importance of Antarctica in global change resides, at least in part, in the fact that there are several key processes of global importance that are best studied there. The scientific problems are considerable, not least because of the comparative inaccessibility of the area and the high costs of maintaining research on and around the continent. Yet it has become clear that, without the antarctic data, global modeling will be fatally flawed in its attempts to predict future climate.

*The atmosphere.* Undoubtedly, the most well-known antarctic process is the rapid stratospheric depletion of ozone in the antarctic polar vortex during early spring. [*See* Ozone.] This rapid and major change in a key atmospheric constituent, which appears to have

been caused principally by the catalytic activities of anthropogenic chemicals (primarily chlorofluorocarbons), shows little sign of diminishing at present (Jones and Shanklin, 1995) and has had worldwide effects. Its importance lies in the loss of absorption by the upper atmosphere of ultraviolet-B (UV-B) radiation (280–320 nanometers wavelength), which is known to have important mutagenic effects on biological systems. The annual spring deficit in antarctic ozone has been accompanied by a more general decline in mean stratospheric ozone worldwide and the sporadic appearance of similar but more limited depletion in the Arctic (Pyle and Harris, 1991). The political effect of the discovery of the antarctic depletion by British Antarctic Survey scientists in the 1980s was to bring about an international agreement (the Montreal Protocol) to phase out the use of chlorofluorocarbons and replace them with more environmentally benign products. Although this has made great progress, not all countries have implemented the controls, and even the most optimistic forecasts for recovery of ozone concentrations suggest that it will take at least another fifty years.

Perhaps equally important, but less well known, are the measurements that have been made of atmospheric gas composition at the South Pole since 1956. While the extent and accuracy of the data are similar to those from Mauna Loa Observatory in Hawaii (Houghton et al., 1996), the antarctic data are unique in their low signal-to-noise ratio—because of the absence of industry and vegetation in the Antarctic. They provide the most reliable indication yet that key radiatively active natural gases, such as carbon dioxide and methane, have been increasing steadily over the past forty years. More recently, measurements at the South Pole have also demonstrated global trends in other key chemicals linked by many to climate change—nitrogen dioxide, methyl bromide, chlorofluorocarbons, and so forth.

Long-term measurements of the reflective layers in the ionosphere over the Antarctic have recently provided evidence of possible further effects of global change. The height of the F layer, an ionized level in the atmosphere that reflects radio waves, varies both diurnally and seasonally, but has been thought of as characteristically around 300 kilometers. Data published recently (Jarvis et al., 1998) have shown that the height of this layer has been falling for the last thirty-eight years at around 0.4 kilometers per year over the Antarctic Peninsula and that its temperature is decreasing. This is the first evidence linking changes in the lower atmosphere with the middle atmosphere, and is of a similar magnitude to that predicted by earlier models. Its implications remain to be assessed.

*The ice.* Ice and snow have a very high albedo, reflecting a great deal of the heat energy received from the Sun. [*See* Albedo.] The seasonal formation of roughly

16 million square kilometers of sea ice more than doubles the continental area (14 million square kilometers) of high reflectivity and has a major influence on the atmospheric heat balance of the Southern Hemisphere. The summer melting of the sea ice dominates the heat budget of the upper layers of the Southern Ocean. Changes in the spatial extent, thickness, persistence, and annual distribution of sea ice thus have a major influence on climate, a feedback loop that needs accurate characterization for modeling purposes. Satellites are providing increasingly detailed data on sea ice, but with data from only twenty-five years, it is still difficult to extract reliable long-term trends (Hanna, 1996). [*See Sea Ice.*]

The Antarctic Ice Sheet itself (including the floating ice shelves) contains around 75 percent of the world's fresh water. Thus changes in the volume of the ice sheet will result in changes in global sea level, a feature clearly visible in paleoenvironmental reconstructions. Present research aims to provide an accurate assessment of volume to allow two levels of change to be modeled—interannual to interdecadal changes in snow accumulation that produce small changes in sea level, and longer-term changes in mass balance that initiate major changes in the volume of the world's oceans. A 1 percent change in total volume would be a matter for global concern. However, the scientific problems related to making measurements of changes in total volume over an area of 14 million square kilometers to this level of accuracy are considerable. New satellite altimeters are making substantial improvements, but better models of ice dynamics and rates of loss are also essential.

**The Southern Ocean.** The Southern Ocean links the three major Southern Hemisphere ocean basins and accounts for 20 percent of the world's oceans. Of special importance is the role that cold dense water produced by the melting of antarctic ice plays in the fluxes of nutrients and transfers of energy in both the Pacific and Atlantic Oceans. There are complex, and as yet poorly understood, relationships between patterns of large-scale variability in the Southern Ocean and the major oceanographic events farther north, such as the annual monsoon and the periodic appearance of El Niño. [*See El Niño–Southern Oscillation.*]

Measurements of mean sea level have shown that the level has increased steadily for over a century (Titus and Narrayanan, 1995). Most of this is attributed to thermal expansion of the oceans, with as yet no consensus that melting of antarctic ice contributes significantly. One important area of current research is the characterization of air/sea interactions over an annual cycle, where the seasonal formation of sea ice fundamentally changes the responses.

**Current and Future Impacts.** While great reliance has been placed on using direct physical measurements of climate where these exist, the global change scientific community has realized that for many parts of the world this is not possible (Houghton et al., 1996). For those areas it will be necessary to use proxy indicators, and in the Antarctic it has become clear that some of these proxies are significantly better at indicating trends than direct meteorological measurements. This is especially so where the signal-to-noise ratio for the point physical measurements is high. In such cases the integrating ability of a system such as a plant at its distribution limit or a small glacier close to its melting point can provide a more sensitive response.

The longest runs of meteorological data all show warming trends (King and Turner, 1997). Data from Faraday Station on the Antarctic Peninsula cover a fifty-year period and show the largest increase in air temperatures of 2.5°C over that period, while shorter data sets suggest there appears to be a small cooling effect in areas of East Antarctica. This rate of warming is higher than has been measured anywhere else in the world and fits poorly with the estimates of change for the Antarctic Peninsula from existing models. [*See Global Warming.*] The trend in the physical data has also been supported by changes in terrestrial plant communities along the western side of the peninsula, where the frequency of successful reproduction has increased, with consequent changes in plant distribution and biomass. Although it is not yet clear precisely why the change is so large in the peninsula, it does offer useful advance warning of some of the effects of any wider-scale change.

At the other side of the Antarctic, measurements made on ice thickness in lakes in the McMurdo Dry Valleys show a thinning trend, while lake volume is increasing year on year as more meltwater is produced from the local glaciers. Around many lakes there are clear indications from multiple shorelines that this has happened many times in the past.

The partial disintegration of ice shelves along both the western and eastern coasts of the Antarctic Peninsula has been attributed to a localized warming of the climate. The Wordie Ice Shelf in Marguerite Bay has largely disappeared over the last twenty-five years, while more recently there have been major changes in the extent of the Larsen Ice Shelf around James Ross Island (Vaughan and Doake, 1996). One of the suggestions is that this has been due principally to increased air temperature causing increased surface melting; the meltwater then penetrates the ice shelf and weakens its structure. If this is correct, there seems little future for many of the ice sheets around the peninsula that now show extensive summer melt pools. Since they are already floating, the loss of all the Antarctic Peninsula ice shelves would have a negligible effect on sea level. [*See Ice Sheets.*]

There may, however, be more than one type of response to climate change. Some theories suggest that ice-sheet stability may be controlled by a critical threshold, so that, although a roughly linear response may characterize initial trends, different rules come into play at the threshold. This has been seen by some glaciologists as a possible scenario for the collapse of the warmer and inherently less stable West Antarctic Ice Sheet, whose melting would raise mean sea level by around 6 meters. At present the balance of opinion suggests that such a collapse would be extremely unlikely (Bentley, 1998). None of the models at present predicts major changes in the East Antarctic Ice Sheet for centuries to come, although changes in precipitation, which is poorly modeled at present, could affect the balance of the ice sheet in the longer term.

The importance of the linkage between sea ice and climate change has already been indicated. Reduction in sea ice extent or duration would affect the albedo of the Southern Ocean and could have important implications for both oceanography and biology. The primary effects could be on deep convection and ocean circulation, but there would also be immediate impacts on Southern Ocean biological productivity and longer-term potential impacts on biogeochemical processes. The Southern Ocean plays an important role in global cycles of carbon and sulfur. Changes in sea ice could impinge on this in several ways (Hansom and Gordon, 1998). One such mechanism could involve changes in cloud formation. Dimethylsulfoxide and carbonyl sulfide produced by marine organisms are potential precursors of particles that act as cloud condensation nuclei. Sea ice impedes the efflux of these compounds into the atmosphere so that changes in extent and persistence of the ice could impact cloud production with consequent effects on radiative balance.

A second mechanism of potential importance is the removal of excess carbon dioxide from the atmosphere through its sequestration by phytoplankton (which subsequently sediment out). Models suggest this marine carbon sink is of considerable importance in global carbon budgets, but phytoplankton is closely linked to sea ice patterns and nutrient availability so that changes in either affect the rate of carbon uptake.

Organisms are differentially sensitive to increased UV-B, but as yet relatively little is known about what the long-term effects will be on the antarctic and Southern Ocean ecosystems. Of greatest importance is understanding both what protective mechanisms are involved in limiting damage to metabolic processes and nucleic acids, and what species-specific damage will do to community structure and ecosystem functioning in the longer term.

**Pollution.** In making any measurement of change, a reference point is essential. For pollutants, the control needs to be as far away from potential sources as possible. The Antarctic is thus the ideal place to measure the global baseline for the wide range of anthropogenic compounds constantly added to the environment (Wolff, 1992). The monitoring at the South Pole and elsewhere in the Antarctic provides a measure of changes in atmospheric gases. Considerable work is now under way to provide baseline data for persistent organic pollutants (including insecticides and polychlorinated biphenyls [PCBs]), hydrocarbons, and heavy metals from a variety of sources, but care is needed to ensure that any localized sources of pollution (such as the scientific stations) are rigorously excluded. Heavy-metal concentrations in ice cores show global changes in lead associated with increased industrialization. Dichlorodiphenyltrichloroethane (DDT) and other insecticides that have been identified in birds, seals, plants, lake sediments, and snow are a clear indication of transfer into the Antarctic from other continents (Larsson et al., 1992). On the other hand, PCBs found in animals and in sediments may in many instances indicate localized contamination.

## BIBLIOGRAPHY

Bentley, C. R. "Rapid Sea-Level Rise from a West Antarctic Ice-Sheet Collapse: A Short-Term Perspective." *Journal of Glaciology* 44 (1998), 157–163.

Hanna, W. "The Role of Antarctic Sea Ice in Global Climate Change." *Progress in Physical Geography* 20 (1996), 371–401.

Hansom, J. D., and J. E. Gordon. *Antarctic Environments and Resources: A Geographical Perspective.* London: Longman, 1998.

Harris, C. M., and B. Stonehouse, eds. *Antarctica and Global Climatic Change.* London: Belhaven, 1991.

Houghton, J. T., et al., eds. *Climate Change 1995.* Cambridge: Cambridge University Press, 1996.

Jarvis, M. J., et al. "Southern Hemisphere Observations of a Long Term Decrease in $F$ Region Altitude and Thermospheric Wind Providing Evidence for Global Thermospheric Cooling." *Journal of Geophysical Research* 103.A9 (1998), 20774–20788.

Jones, A. E., and J. D. Shanklin. "Continued Decline of Total Ozone over Halley, Antarctica, since 1985." *Nature* 376 (1995), 409–411.

Kattenburg, A., et al. "Climate Models: Projections of Future Climate." In *Climate Change 1995*, edited by J. T. Houghton et al. Cambridge: Cambridge University Press, 1996.

King, J. C., and J. Turner. *Antarctic Meteorology and Climatology.* Cambridge: Cambridge University Press, 1997.

Larsson, P., et al. "PCBs and Chlorinated Pesticides in the Atmosphere and Aquatic Organisms of Ross Island, Antarctica." *Marine Pollution Bulletin* 25 (1992), 281–287.

Pyle, J. A., and N. R. P. Harris, eds. *Polar Stratospheric Ozone.* Commission of the European Communities Air Pollution Research Report 34. Brussels, 1991.

Titus, J. G., and V. Narrayanan. "The Probability of Sea Level Rise." U.S. Environmental Protection Agency, EPA 230-R-95-008. Washington, D.C., 1995.

Vaughan, D. G., and C. S. M. Doake. "Recent Atmospheric Warming and Retreat of Ice Shelves on the Antarctic Peninsula." *Nature* 379 (1996), 328–331.

Wolff, E. W. "The Influence of Global and Local Atmospheric Pollution on the Chemistry of Antarctic Snow and Ice." *Marine Pollution Bulletin* 25 (1992), 274–280.

Wolff, E. W., and R. C. Bales, eds. *Chemical Exchange between the Atmosphere and Polar Snow.* Berlin: Springer, 1996.

—D. W. H. WALTON

## Antarctic Treaty System

The antarctic region is subject to special rules of environmental protection (the Antarctic Treaty System) reflecting the unique physical conditions of these areas and the important role that they play in maintaining regional and global environmental conditions. It comprises four treaties: the 1959 Antarctic Treaty (and its 1991 Protocol on Environmental Protection to the Antarctic Treaty); the 1972 Convention for the Conservation of Antarctic Seals; the 1980 Convention on the Conservation of Antarctic Marine Living Resources (CCAMLR); and the 1988 Convention on the Regulation of Antarctic Mineral Resource Activities (which has been superseded by the 1991 Protocol and is unlikely to enter into force). In addition, the parties to the 1959 Antarctic Treaty and the 1980 CCAMLR have adopted numerous recommendations and conservation measures. In many ways, the antarctic region has played a catalytic and innovative role, contributing to the progressive development of rules and techniques associated with global change, especially in relation to information exchange, scientific advisory processes, environmental impact assessment, observation and inspection, the management of waste streams, liability for environmental damage, enforcement procedures, and institutional arrangements.

The 1959 Antarctic Treaty freezes national claims to sovereignty in the continent and establishes that the area is to be used "for peaceful purposes only," including scientific investigation (Article I). Parties with Consultative status may take additional measures regarding the "preservation and conservation of living resources in Antarctica" (Article IX). In 1964 the Meeting of the Consultative Parties to the 1959 treaty adopted the Brussels Agreed Measures for the Conservation of Antarctic Fauna and Flora, which designate the Antarctic region a "Special Conservation Area."

The 1972 convention applies to the sea area regulated by the 1959 treaty; it limits the number of seals that can be killed or captured, and grants complete protection to certain species. The objective of the 1980 convention is the conservation (including "rational use") of the marine living resources in the Antarctic Treaty area and in the surrounding area that forms part of the antarctic marine ecosystem. Harvesting and associated activities are to be carried out in accordance with principles of conservation adopted under the convention. The 1988 con-

vention was intended to be an integral part of the Antarctic Treaty System to establish the framework for determining whether antarctic mineral resource activities (prospecting, exploration, and development) were acceptable and, if so, under what conditions they could be carried out. It recognized the dangers posed by mineral resource activities for the environment, and elaborated a range of measures designed to ensure environmental protection. These were considered to be insufficient by some of the Consultative parties (France and Australia), which indicated that they would not ratify the Convention, effectively prohibiting its entry into force. Nevertheless, its innovative provisions on institutional arrangements, environmental impact assessment, liability, and noncompliance have influenced developments elsewhere.

In October 1991 the Antarctic Environmental Protocol and its four annexes were signed (a fifth has since been adopted). The protocol establishes a fifty-year moratorium on antarctic mineral resources activities (other than scientific research: Article 7). The protocol and annexes comprise the most comprehensive and stringent regime of environmental protection rules yet established under international law. The objective is the protection of the antarctic environment and ecosystems, based upon the conviction that such a goal is "in the interest of mankind as a whole" (Preamble and Articles 2 and 4). Antarctica is designated a "natural reserve, devoted to peace and science" (Article 2). The protocol includes guiding principles to support environmental protection in the planning and conduct of nonmineral resource activities that will be permitted, principally tourism and especially the scientific research that is essential to understanding the global environment.

### BIBLIOGRAPHY

Bush, W. *Antarctica and International Law.* 3 vols. London and New York: Oceana Publications, 1982–1988.

Heap, J., ed. *Handbook of the Antarctic Treaty System.* 7th ed. Washington, D.C.: United States Department of State, 1990.

Sands, P. *Principles of International Environmental Law,* vol. 1. chap. 14. Manchester, U.K.: Manchester University Press, 1995.

Verhoeven, J., P. Sands, and M. Bruce, eds. *The Antarctic Environment and International Law.* London and Boston: Graham and Trotman, 1992.

—PHILIPPE J. SANDS

# ANTHROPOGEOMORPHOLOGY

Anthropogeomorphology is the study of the human role in creating landforms and modifying the operation of geomorphological processes such as weathering, erosion, transport, and deposition (Goudie, 1993). [*See* Land Surface Processes.] The range of anthropogenic impacts on both forms and process is considerable, and there are very few spheres of human activity that do not, even indirectly, create landforms (see, for example, Brown, 1970; Nir, 1983; Goudie, 1999). It is useful, however, to recognize that some features are produced by direct anthropogenic processes. These tend to be relatively obvious in form and are frequently created deliberately and knowingly (see, for example, Haigh, 1978). They include landforms produced by construction (e.g., spoil tips, bunds, embankments), landforms produced by excavation (e.g., open-cast and strip mines), landforms produced by hydrological interference (e.g., reservoirs and canals), and landforms produced by farming (e.g., terraces). Table 1 lists some of the major anthropogeomorphic processes.

Landforms produced by indirect anthropogenic processes are often less easy to recognize, not least because they tend to involve not the operation of a new process or processes, but the acceleration of natural processes. They are the result of environmental changes brought about inadvertently by human technology. Nonetheless, it is probably this indirect and inadvertent modification of process and form that is the most crucial aspect of anthropogeomorphology. By removing or modifying natural land cover—through the agencies of cutting, bulldozing, burning, and grazing—humans have accelerated rates of erosion and sedimentation. [See Erosion.] Sometimes the results will be spectacular and obvious, for example, when major gully systems rapidly develop; other results may have less immediate effect on landforms (e.g., sheet erosion) but are, nevertheless, of great importance. By other indirect means humans may create subsidence features (Johnson, 1991), trigger off mass movements such as landslides, and even influence the operation of phenomena such as earthquakes (Meade, 1991) through the impoundment of large reservoirs. Rates of rock weathering may be modified because of the acidification of precipitation caused by accelerated sulfate emissions or because of accelerated salinization in areas of irrigation (Goudie and Viles, 1997). [See Building Decay.]

Finally, there are situations where, through a lack of understanding of the operation of processes and the

**Anthropogeomorphology. TABLE 1.** Major Anthropogeomorphic Processes

DIRECT ANTHROPOGENIC PROCESSES

Constructional
  tipping: molding, plowing, terracing, reclamation

Excavational
  digging, cutting, mining, blasting of cohesive or noncohesive materials
  craters
  tramping, churning

Hydrological interference
  flooding, damming, canal construction, dredging, channel modification, draining, coastal
    protection

INDIRECT ANTHROPOGENIC PROCESSES

Acceleration of erosion and sedimentation
  agricultural activity and clearance of vegetation, engineering, especially road
    construction and urbanization
  incidental modifications of hydrological regime

Subsidence: collapse, settling
  mining (e.g., of coal and salt)
  hydraulic (e.g., groundwater and hydrocarbon pumping)
  thermokarst (melting of permafrost)

Slope failure: landslides, flows, accelerated creep
  loading
  undercutting
  shaking
  lubrication

Earthquake generation
  loading (reservoirs)
  lubrication (fault plane)

Weathering
  acidification of precipitation
  accelerated salinization
  lateritization

## HUMANS VERSUS NATURE IN EARTHMOVING

Estimates of human earthmoving show that our efforts are comparable to natural geologic processes. Omitting the indirect effects of actions such as deforestation and cultivation, the following are the magnitudes of deliberate human earthmoving actions in the United States (Hooke, 1994).

|  | BILLION METRIC TONS PER YEAR |
| --- | --- |
| Excavation for housing and other construction | 0.8 |
| Mining | 3.8 |
| Road work | 3.0 |
| Total United States | 7.6 |
| World total (roughly four times the U.S. total) | 30.0 |

For comparison, the following figures represent estimated world totals due to natural earthmoving processes.

|  | BILLION METRIC TONS PER YEAR |
| --- | --- |
| River transport | |
| to oceans and lakes | 14 |
| short-distance transport within river basins | 40 |
| Tectonic forces lifting continents | 14 |
| Volcanic activity elevating sea floor | 30 |
| Glacial transport | 4.3 |
| Wind transport | 1.0 |

**BIBLIOGRAPHY**

Hooke, R. L. "On the Efficacy of Humans as Geomorphic Agents." 4.9 *GSA Today* (1994), 217, 224–225.
Monastersky, R. "Earthmovers." *Science News* (24 December 1994), 432–433.

—DAVID J. CUFF

links between different processes and phenomena, humans may deliberately and directly alter landforms and processes and thereby set in train a series of events that was not anticipated or desired. There are, for example, many records of attempts to reduce coastal erosion by implementing important and expensive engineering solutions which, far from solving erosion problems, only exacerbated them. Examples include dune stabilization schemes in North Carolina (Dolan et al., 1973) and the role of sea walls in causing beach scour (Bird, 1979).

As so often with environmental change, it is not easy to disentangle changes that are humanly induced from those that are natural. There has, for example, been a long-continued debate about the origin of deeply incised valley-bottom gullies, or arroyos, which developed in the southwestern United States over a relatively short period between 1865 and 1915. Some workers have championed human actions as the cause of this erosion

spasm, citing such factors as timber felling, overgrazing, harvesting of grass, compaction along tracks, channeling of runoff from trails and railroads, and disruption of valley-bottom sods by domestic stock. Other workers have championed the importance of natural environmental changes, noting that arroyo incision had occurred repeatedly before the arrival of Europeans in the area. Among the natural changes that could promote the phenomenon are a trend toward aridity (which depletes the cover of protective vegetation) or increased frequencies of high-intensity storms (which generate erosive runoff). Such issues are discussed by Cooke and Reeves (1976) and by Balling and Wells (1990). Similar debates surround the explanation for incision of the Mediterranean Valleys (Vitz-Finzi, 1969; Butzer, 1974) and the deposition of fills of freshwater carbonate (tufa) in the valleys of Europe (Goudie et al., 1993).

Another example of the complexity of causation in

geomorphologic change is posed by a consideration of the potential causes of loss of land to the sea in coastal Louisiana (Walker et al., 1987), something that appears to be proceeding at a rapid rate at the present time. Among the factors that need to be considered are the natural ones of sea level change, subsidence, progressive compaction of sediments, changes in the locations of where deltaic deposition is taking place, hurricane attack, and degradation by marsh fauna. Equally, however, one has to consider a range of human actions, including the role that dams and levees have played in reducing the amount and texture of sediment reaching the coast, the role of canal and highway construction, and subsidence caused by fluid withdrawals.

The possibility that the buildup of greenhouse gases (e.g., carbon dioxide) in the atmosphere may cause enhanced global warming in coming decades has many implications for anthropogeomorphology. [See Global Warming.] While increased temperatures will have a direct impact on some landforms, they will also have an indirect effect because of associated changes in precipitation regimes, rates of moisture loss by transpiration and evaporation, and the distribution and form of vegetation assemblages (Table 2).

**Anthropogeomorphology. TABLE 2.** Some Geomorphologic Consequences of Global Warming

HYDROLOGIC

Increased evapotranspiration loss

Increased percentage of precipitation as rainfall at expense of winter snowfall

Increased precipitation as snowfall in very high latitudes

Possible increased risk of cyclones (greater spread, frequency, and intensity)

Changes in state of peatbogs and wetlands

Less vegetational use of water because of increased $CO_2$ effect on stomatal closure

VEGETATIONAL CONTROLS

Major changes in latitudinal extent of biomes

Reduction in boreal forest, increase in grassland, etc.

Major changes in altitudinal distribution of vegetation types (ca. 500 m for 3°C)

Growth enhancement by $CO_2$ fertilization

CRYOSPHERIC

Permafrost, decay, thermokarst, increased thickness of active layer, instability of slopes, river banks, and shorelines

Changes in glacier and ice-sheet rates of ablation and accumulation

Sea ice melting

COASTAL

Inundation of low-lying areas (including wetlands, deltas, reefs, lagoons, etc.)

Accelerated coast recession (particularly of sandy beaches)

Changes in rate of reef growth

Spread of mangrove swamp

EOLIAN

Increased dust storm activity and dune movement in areas of moisture deficit

SOIL EROSION

Changes in response to changes in land use, fires, natural vegetation cover, rainfall erosivity, etc.

Changes resulting from soil erodibility modification (e.g., sodium and organic contents)

SUBSIDENCE

Desiccation of clays under conditions of summer drought

Increased sea surface temperatures may change the spread, frequency, and intensity of tropical cyclones—highly important geomorphologic agents, particularly in terms of river channels and mass movements. Warmer temperatures will cause sea ice to melt and may lead to the retreat of alpine glaciers and the melting of permafrost (permanently frozen subsoil). The forms of vegetation will change and show latitudinal and altitudinal migration that will also influence the operation of geomorphologic processes. Changes in temperature, precipitation quantities, and the timing and form of precipitation (e.g., whether it is rain or snow) will have a broad range of important hydrological consequences. Some parts of the world may become moister (e.g., high latitudes and some parts of the tropics), while other parts (e.g., the High Plains of the United States) may become drier. The latter would suffer from declines in river flow, lake desiccation, reactivation of sand dunes, and increasing dust storm frequencies.

However, among the most important potential future anthropogeomorphologic changes are those associated with sea level change caused by the melting of land ice. Low-lying areas (e.g., salt marshes, mangrove swamps, deltas, coral atolls) would tend to be particularly susceptible, but rising sea levels could promote more general beach erosion. [*See* Sea Level.]

Indeed, it is because of the existence of potentially highly sensitive landform environments that Goudie (1996) introduced the term *geomorphological hotspots* for areas where the effects of global warming might be especially serious.

Some landscapes will be especially sensitive because they are located in zones where climate is expected to change to an above-average extent. This is the case, for instance, in the high latitudes of Canada or Russia where the degree of warming may be three to four times greater than the global average. It may also be the case in some critical areas where particularly substantial changes in precipitation may result from global warming. For example, various methods of climatic prediction produce scenarios in which the High Plains of the United States will become markedly drier. Other landscapes will be especially sensitive because certain landscape-forming processes are particularly closely controlled by climatic conditions. If such landscapes are close to particular climatic thresholds, then quite modest amounts of climatic change can switch them from one state to another.

[*See also* Erosion; Global Warming; *and* Sea Level.]

### BIBLIOGRAPHY

Balling, R. C., and S. G. Wells. "Historical Rainfall Patterns and Arroyo Activity within the Zuni River Drainage Basin, New Mexico." *Annals of the Association of American Geographers* 80 (1990), 603–617.

Bird, E. C. F. "Coastal Processes." In *Man and Environmental Processes*, edited by K. J. Gregory and D. G. Walling, pp. 82–101. Folkestone, U.K.: Dawson, 1979.

Brown, E. H. "Man Shapes the Earth." *Geographical Journal* 136 (1970), 74–85.

Butzer, K. W. "Accelerated Soil Erosion: A Problem of Man–Land Relationships." In *Perspectives on Environments*, edited by I. R. Manners and M. W. Mikesell. Washington, D.C.: Association of American Geographers, 1974.

Cooke, R. U., and R. W. Reeves. *Arroyos and Environmental Change in the American South-West*. Oxford: Oxford University Press, 1976.

Dolan, R., P. J. Godfrey, and W. E. Odum. "Man's Impact on the Barrier Island of North Carolina." *American Scientist* 61 (1973), 152–162.

Goudie, A. S. "Human Influence in Geomorphology." *Geomorphology* 7 (1993), 37–59.

———. "Geomorphological 'Hotspots' and Global Warming." *Interdisciplinary Science Reviews* 21 (1996), 253–259.

———. *The Human Impact on the Natural Environment*. 5th ed. Oxford: Blackwell, 1999.

Goudie, A. S., and H. A. Viles. *Salt Weathering Hazards*. Chichester, U.K.: Wiley, 1997.

Goudie, A. S., H. A. Viles, and A. Pentecost. "The Late-Holocene Tufa Decline in Europe." *The Holocene* 3 (1993), 181–186.

Haigh, M. J. "Evolution of Slopes on Artificial Landforms—Blaenavon, U.K." Research Paper. Department of Geography, University of Chicago 183 (1978).

Johnson, A. I., ed. "Land Subsidence." *International Association of Hydrological Sciences* 200 (1991).

Meade, R. B. "Reservoirs and Earthquakes." *Engineering Geology* 30 (1991), 245–262.

Nir, D. *Man, A Geomorphological Agent: An Introduction to Anthropic Geomorphology*. Jerusalem: Keter, 1983.

Vita-Finzi, C. *The Mediterranean Valleys*. Cambridge: Cambridge University Press, 1969.

Walker, H. J., J. M. Coleman, H. H. Roberts, and R. S. Tye. "Wetland Loss in Louisiana." *Geografiska Annaler* 69A (1987), 189–200.

—ANDREW S. GOUDIE

## AQUACULTURE. *See* Fish Farming.

## ARAL SEA, DESICCATION OF THE

[*This case study examines the far-reaching environmental consequences of expanded irrigation and the domestic and international efforts to improve the ecosystem of the Aral Sea. The article also outlines lessons to be learned from this ecological disaster zone.*]

One of the planet's most serious examples of large-scale, human-induced alteration of ecosystems is unfolding in the basin of the Aral Sea. Over the past three decades, this huge lake has endured rapid and continuous shrinkage as expanding irrigation has substantially reduced river discharge to it. A spectrum of severe environmental and human problems has accompanied the Aral Sea's drying. National and international efforts are under way to cope with these problems, but even their partial alleviation will be enormously costly and will require many years. The sea serves as an instructive lesson on the results of thoughtless human interference with natural systems and on the need to carefully evaluate contemplated major alterations of the physical and biological environment.

**General Character of the Aral Sea and Its Basin.** The Aral Sea sits amid the deserts of Central Asia (Figure 1). A terminal lake (that is, without surface outflow), the sea's level is mainly determined by the balance between surface inflow from two large rivers (the Amu Dar'ya and Syr Dar'ya) and net evaporation (evaporation from its surface minus precipitation on it). Inflow and net evaporation each averaged near 55 cubic kilometers (13.2 cubic miles) from 1911 to 1960. Hence, the water balance was in long-term equilibrium with a maximum lake level variation over this period of less than one meter. The sea's drainage basin covers 1.8 million square kilometers (0.7 million square miles), with 83 percent located on the territory of Uzbekistan, Kazakhstan, Tajikistan, Kyrgyzstan, and Turkmenistan, newly independent republics that were part of the former Soviet Union until the end of 1991. Most of the remaining part of the basin lies in Afghanistan, with a small portion in Iran. The population of the Aral Sea Basin was near 40 million by 1997. Uzbekistan and Kazakhstan are the nations that border on the sea. However, all of the Aral Sea coastline within Uzbekistan lies within that nation's Karakalpak Autonomous Republic.

The Aral Sea was the world's fourth largest lake (in terms of area) in 1960. A brackish water body (with an average salinity of 10 grams per liter), the lake was populated chiefly by freshwater species. The sea supported a major fishery, directly and indirectly employing 60,000 people, and functioned as a key regional transportation route. Where the Syr Dar'ya and Amu Dar'ya entered, extensive deltas had formed; these sustained a diversity of flora and fauna and had major economic importance for irrigated agriculture, animal husbandry, hunting and trapping, fishing, and harvesting of reeds (which served as a ubiquitous construction material and source of animal fodder).

Since the early 1960s, the sea has steadily shrunk (Figure 2 and Table 1) as expanding irrigation diminished inflow from the Amu Dar'ya and Syr Dar'ya. Lessened precipitation in the mountain zones of flow formation has been responsible for no more than 10 percent of the level decline. Average annual river inflow to the sea during the 1980s was around 7.6 cubic kilometers ($km^3$), or 14 percent of the figure for 1911–1960. For 1990–1996, average annual inflow nearly doubled to 15 $km^3$, a result of heavy natural flow years from 1991 to 1994 and efforts of the basin states to deliver more water to the sea. The Aral Sea divided into two water bodies in 1987: the Small Sea in the north and the Large Sea in the south. The Syr Dar'ya flows into the former and the Amu Dar'ya into the latter.

The irrigated area in the Aral Sea Basin was 4.5 million hectares (11.1 million acres) in 1960. Withdrawals prior to the 1960s had much less effect on discharge to the sea because of compensating factors, particularly downstream diminution of evapotranspiration and deltaic flooding. However, since 1960, the impact of these mechanisms essentially disappeared, and each cubic meter withdrawn has been a cubic meter that did

**Aral Sea. Figure 1.** Location of the Aral Sea.

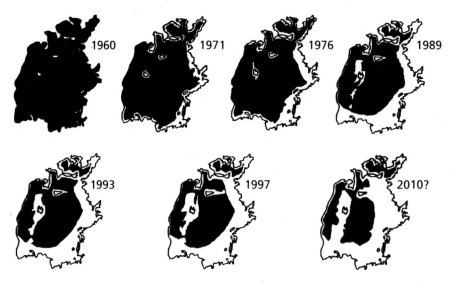

**Aral Sea. Figure 2.** The Shrinking Aral Sea (see Table 1).

not reach the sea. Also, irrigation expansion in the Aral Sea Basin after 1960 required much more water per hectare as long, unlined canals were extended into the desert; flushing requirements rose as more saline soils were brought under irrigation; huge amounts of water went to fill pore spaces in dry soils; giant reservoirs were built, and new irrigation systems discharged their water into the deserts rather than back to the rivers from which it was taken. Thus, as the irrigated area rose from 4.5 to 7.2 million hectares—a 60 percent increase—between 1960 and 1990, consumptive water use rose more than twofold from around 45 km³ to between 90 and 100 km³.

**Environmental, Economic, and Human Consequences.** The consequences of the Aral Sea's drying have been severe. The sea's native fishes disappeared by the early 1980s as salinity rose beyond their ability to adapt, and shallow spawning and feeding areas shrank. This spelled the end of the important commercial fishery and threw tens of thousands out of work. Navigation on the Aral Sea also ceased as efforts to keep the increasingly long channels open to the major ports

**Aral Sea. Table 1.** Hydrographic and Hydrologic Characteristics of the Shrinking Aral Sea (see Figure 2)

| YEAR (JANUARY 1) | AVERAGE LEVEL (M) | AVERAGE AREA (KM²) | % OF 1960 | AVERAGE VOLUME (KM³) | % OF 1960 | AVERAGE SALINITY (G/L) | % OF 1960 |
|---|---|---|---|---|---|---|---|
| 1960 | 53.4 | 66,900 | 100 | 1,090 | 100 | 9.9 | 100 |
| 1971 | 51.1 | 60,200 | 90 | 925 | 85 | 11.2 | 113 |
| 1976 | 48.3 | 55,700 | 83 | 763 | 70 | 14 | 141 |
| 1989 | | 39,866 | 60 | 366 | 34 | | |
| Large Sea | 39.7 | 37,155 | 61 | 345 | 34 | 30 | 303 |
| Small Sea | 39.8 | 2,711 | 45 | 21 | 26 | 30 | 303 |
| 1993 | | 35,089 | 52 | 294 | 27 | | |
| Large Sea | 37.7 | 32,690 | 54 | 276 | 27 | ~35 | 353 |
| Small Sea | 38.5 | 2,399 | 40 | 18 | 22 | ~35? | 353 |
| 1997 | | 31,220 | 47 | 242 | 22 | | |
| Large Sea | 35.9 | 28,459 | 47 | 220 | 22 | >40? | 404 |
| Small Sea | 40.0 | 2,761 | 46 | 22 | 27 | ~30? | 303 |
| 2010* | | 21,609 | 32 | 141 | 13 | | |
| Large Sea | 31.4 | 18,361 | 30 | 113 | 11 | >60? | 606 |
| Small Sea | 42.0 | 3,248 | 54 | 28 | 34 | ~25? | 252 |

*Figures for 1997–2010 are based on an assumed inflow from the Amu Dar'ya to the Large Sea of 10 cubic kilometers per year and from the Syr Dar'ya to the Small Sea of 3 cubic kilometers per year with no flow from the Small to Large Seas.

of Aralsk in the north and Muynak in the south were abandoned.

But damage has stretched well beyond the sea and its immediate shoreline. A 400,000-square-kilometer region around the Aral Sea with a population of nearly four million has also suffered greatly and has been designated an ecological disaster zone. Damage has been particularly severe to the rich ecosystems of the extensive Amu Dar'ya and Syr Dar'ya deltas. This is due chiefly to the reduction of river flow through them and the decline of groundwater levels in them. Desertification is spreading and intensifying in both deltas, leading to the replacement of endemic hydrophyte (water-loving) plant communities with halophytic (salt-tolerant) and xerophytic (drought-tolerant) associations. Ecological simplification is occurring, with examples of floral and faunal extinction. The expanses of unique Tugai forests (communities of phreatophytic [deep-rooted plants that obtain their water from the saturated zone] trees and shrubs) that formerly stretched along the main rivers and distributary channels in the deltas have shrunk significantly, with adverse impacts on the diversity of plants and animals inhabiting them. By the late 1980s, the area of lakes, wetlands, and their associated reed communities had shrunk by 85 percent. Lakes and wetlands provide prime habitat for a variety of permanent and migratory waterfowl, a number of which are endangered, and their loss coupled with increasing pollution of remaining water bodies (primarily from irrigation runoff containing fertilizers, pesticides, and herbicides) has decimated aquatic bird populations. The number of nesting species of birds in the Syr Dar'ya delta is estimated to have fallen from 173 to 38.

Irrigated agriculture, practiced in the deltas of the Amu Dar'ya and Syr Dar'ya for several millennia, has been damaged by the reduced and salinized river flow arriving from upstream. Raised salinity (at times, over 2 grams per liter) has led to lowered crop yields and, in conjunction with inadequate drainage of irrigated fields, to secondary soil salinization. [See Salinization.] Animal husbandry, both in the deltas and in the desert regions adjacent to the Aral Sea, has been damaged by declining productivity of pastures resulting from desertification, dropping groundwater levels, and replacement of natural vegetation suitable for grazing by inedible species.

Salt and dust blown from the increasingly large former sea bottom (over 35,000 square kilometers by 1997) carries as far as 500 kilometers (and perhaps farther) and settles over a considerable area adjacent to the Aral Sea. Total deflated material has been estimated to be as high as 150 million metric tons per year, although salt content is probably not more than 1 percent. Salts, in dry and aerosol form, falling on soils and vegetation retard plant growth and depress crop yields. Natural vegetation and crops in the Amu Dar'ya delta to the south of the sea and pastures in the Ustyurt plateau to the west of the Aral Sea are affected the most. Airborne salt and dust is also suspected as a major factor in increasingly high levels of respiratory illnesses and throat and esophageal cancer.

Climate has been affected significantly in a narrow zone (up to 100 kilometers wide) that formerly bordered on the sea. Maritime conditions have been replaced by more continental and desertic regimes. Summers have warmed and winters cooled, spring frosts are later and fall frosts earlier, humidity is lower, and the growing season shorter. In the northern part of the Amu Dar'ya delta, the growing season has become too short for cotton, forcing a switch from this crop to rice.

The population living near the Aral Sea suffers from acute health problems. Some of these are directly linked to the sea's recession and its consequences (such as respiratory and digestive afflictions and, possibly, cancer from inhalation and ingestion of blowing salt and dust; poorer diets from the loss of Aral Sea fish as a major food source). But others are the result of environmental pollution associated with irrigation and inadequate medical, health, and hygienic conditions and practices. Bacterial contamination of drinking water is pervasive and has led to very high rates of typhoid, paratyphoid, viral hepatitis, and dysentery. Because of this and other factors (high fertility, poor medical care, poor diet, lack of sewage systems), general mortality and morbidity and infant mortality and morbidity are the highest in the former Soviet Union. Karakalpak Autonomous Republic, the republic adjacent to the southern shoreline, reports infant mortality rates in some localities in excess of one hundred per one thousand live births—four times the former Soviet national rate and ten times U.S. levels.

**Rehabilitation Efforts.** In the late 1980s, programs were initiated by the Soviet Union to deal with the key environmental, economic, and human problems arising from or related to the desiccation of the sea. Valuable research was conducted, and plans for coping with the most serious problems were developed. However, only limited implementation was inaugurated before the Soviet Union collapsed in 1991. Responsibility for dealing with the Aral Sea situation then fell on the five newly independent nations in the Aral Sea Basin. In 1992, the five basin states signed a watersharing measure to continue the allocation scheme that existed under the Soviet regime, but added the sea and river deltas as entities with rights to water. Specific allocations among the states and to the Aral Sea and its river deltas are decided by an Interstate Coordinating Water Commission for each hydrologic year (October to October) based on flow forecasts for the Syr Dar'ya and Amu Dar'ya Rivers. In recent years, 12–14 km$^3$ have been allocated to the sea and deltas.

Other agreements to improve the Aral Sea situation have been signed since 1992. The most important established an Interstate Council on the Aral Sea (ICAS) and the Interstate Fund for the Aral Sea (IFAS). The former was given responsibility to develop and implement a program—in coordination with foreign donors, but chiefly the World Bank—for dealing with the Aral Sea crisis, whereas the latter was to raise money from the five member countries for the implementation of this program by levying a fee on each government (originally 1 percent of national income, but now 0.3 percent). In February 1997, ICAS was dissolved and its functions transferred to IFAS.

International help to resolve the problems of the Aral Sea region began in 1990 when the Soviet government invited the United Nations Environment Programme to provide experts to help prepare a diagnostic study of the situation. After the independence of the Central Asian states, major aid from the international community (primarily the World Bank, United States, European Union, and United Nations) started to flow to the region in late 1993. The most ambitious efforts are being coordinated by the World Bank through its Aral Sea Basin Programme. The first of three phases includes 20 projects whose planning and implementation cost is estimated at U.S.$470 million. The program has four major objectives: (1) stabilization of the environment of the Aral Sea Basin, (2) rehabilitation of the disaster zone around the sea, (3) improved management of the international waters of the Aral Sea Basin, and (4) building the capacity of regional institutions to plan and implement the above-mentioned programs. The first phase was scheduled to be completed in three years, but, as of mid-1997, planning activities for many of the projects were still under way and few had reached the implementation stage. Attracting the necessary funding for the projects from donors has proven the most difficult problem.

With a concerted and cooperative effort among the Aral Sea Basin states and help from the international community, there is hope of partially alleviating the most critical problems that beset the Aral Sea region. The key issues of health and medical improvement and the provision of clean drinking water are being pursued as the first priority both by the governments of the region and by the international community. Promotion of measures to slow the region's rapid population growth is also necessary. Improvement of the efficiency of irrigation is a major concern since it is the key to substantial water savings, but will be costly and will take a long time.

Because of its great ecological and economic value, efforts to preserve the delta of the Amu Dar'ya have received considerable attention. Approaches to reducing the frequency and severity of salt and dust storms should also be high on the agenda. A substantial restora-

tion of the Small Aral Sea is feasible with relatively small increases of inflow because its surface area and, hence, its evaporative losses are much smaller than those for the Large Sea. Salinity would drop to levels tolerable by indigenous fishes, allowing restoration of commercial fishing. As noted earlier, partial rehabilitation of the Small Sea is one of the first-phase actions under serious consideration as part of the World Bank's Aral Sea Basin Programme.

It would be extremely difficult to return the entire Aral Sea to its early-1960s size. Average annual inflow from the Amu Dar'ya and Syr Dar'ya would need to reach 50–55 km$^3$, over 4.5 times the 1990–1996 average of 11.5 km$^3$. Such an increase would require huge expenditures on irrigation improvement, a major reform of the agricultural sector (including true land privatization and introduction of meaningful water pricing), and/or a restructuring of the economy away from irrigated agriculture. However, under the most optimistic assumptions, this will take decades and cost billions of dollars. Near-term drastic reductions in irrigation would create economic and social havoc in the region and are neither reasonable nor practical.

Indeed, practical considerations may dictate that the Aral Sea can only partially be revived. It may prove possible to restore late-1970s hydrologic parameters (level, 46–47 meters; area, 52,000–54,000 km$^2$; volume, roughly 700 km$^3$; average salinity, 15–16 grams per liter). This would require an average annual inflow of about 44 km$^3$. Assuming discharge to the sea were reasonably clean, indigenous flora and fauna could be reintroduced to the sea from lakes in the Amu Dar'ya and Syr Dar'ya deltas or from other lakes in Central Asia. This is a compelling argument for preserving remnants of the original Aral Sea flora and fauna, either in deltaic lakes or in parts of the remaining sea in which the habitat conditions of the pre-recession Aral Sea could be maintained artificially. More salt-tolerant fish species could also be introduced, as long as care were taken to select species that would not become major competitors to the indigenous populations.

**Lessons for Global Change and Sustainable Development.** Several lessons may be drawn from the Aral Sea situation. First, the environment is complex and easy to damage, but it is very difficult, costly, and time-consuming to restore, even partially. Hence, it is critically important to carefully evaluate fundamental human interference in large and complicated natural systems and to identify key impacts before proceeding to alter such systems, and to be ready to forego or make major alterations to planned actions if warranted. No such evaluation was performed prior to diversion of most of the flow of the Amu Dar'ya and Syr Dar'ya Rivers, and the consequences for the Aral Sea, its ecosystems, and the people of the surrounding region have been tragic.

Second, special attention needs to be paid to identification of thresholds: sensitivity points in the operation of a system, which, if transgressed by an external disturbing force, cause the system to undergo unexpectedly large and rapid changes. Irrigation development in the Aral Sea Basin crossed such a threshold in the 1960s.

Third, simplistic, and at first glance appealing, proposals for quick correction of complex environmental and human problems (for example, through rapid and massive cuts in irrigation to free water for sea) should be viewed with caution since they are likely to cause severe disruptions and lead to worse problems, including economic collapse. The Aral Sea situation developed over many decades, and a longer period will be required even to partially rectify it.

Fourth, complex problems such as we see in the Aral Sea Basin have multiple causes. The serious problems faced in the Aral Sea region relate not only to the reduced downstream flow of the major rivers and the drying of the Aral, but also to other factors such as poor health and medical conditions, and these too must be dealt with effectively.

## BIBLIOGRAPHY

Micklin, P. P. "Desiccation of the Aral Sea: A Water Management Disaster in the Soviet Union." *Science* 241.4870 (2 September 1988), 1170–1176. A general description of Aral Sea desiccation that first brought this problem to the attention of the Western scientific community.

———. *The Water Management Crisis in Soviet Central Asia.* Pittsburgh: Center for Russian and East European Studies, 1991. A monograph with much detail and figures on the water management situation in Central Asia and the Aral Sea.

———. "Aral Sea." In *McGraw-Hill Yearbook of Science and Technology,* edited by Sybil B. Parker, pp. 22–24. New York: McGraw-Hill, 1993. A brief, general description of the Aral Sea and its problems.

———. "The Shrinking Aral Sea." *Geotimes* (April 1993), 14–19. A brief, popular treatment of key Aral Sea issues.

———. "The Aral Sea Problem." *Civil Engineering,* Proceedings of the Institution of Civil Engineers (August 1994), 114–121. A more technical treatment of Aral Sea desiccation and its consequences.

———. "Aral Sea." *Microsoft Encarta Encyclopedia on CD ROM,* 1998 ed. A short, general article on the Aral Sea and its problems.

Micklin, P. P., ed. "Special Issue on the Aral Sea Crisis." *Post-Soviet Geography* 33.5 (May 1992). Mainly translations from Russian of articles on different aspects of the Aral Sea problem by leading Soviet specialists.

Micklin, P. P., and W. D. Williams, eds. *The Aral Sea Basin,* Proceedings of an Advanced Research Workshop, Tashkent, Uzbekistan, 2–5 May 1994. NATO ASI Series, vol. 12. Heidelberg: Springer-Verlag, 1996. Edited and revised versions of selected papers, covering a wide range of issues, presented at the NATO conference.

World Bank. *The Aral Sea Crisis: Proposed Framework of Activities.* Europe and Central Asia Region, Country Department 3, Country Operations Division I, 29 March 1993. Washington, D.C.: World Bank, 1993. A description of key Aral Sea Basin issues and a proposed plan of action to improve the situation; by the major foreign assistance provider.

———. *Aral Sea Basin Program, Phase 1.* Progress Report No. 3, Aral Sea Basin Unit, Country Department 3, Europe and Central Asian Region, February 1996. Washington, D.C.: World Bank, 1996. Discusses progress and problems of implementing the World Bank program and future plans.

—PHILIP MICKLIN

## ARCTIC HAZE. *See* Aerosols; Air Quality; *and* Atmospheric Chemistry.

## ARID LANDS. *See* Deserts.

## ARRHENIUS, SVANTE

Svante August Arrhenius (1859–1927), Swedish physicist and chemist, was born in Vik, near Uppsala, Sweden. He was awarded his Ph.D. by the University of Uppsala in 1884 and was named docent in physical chemistry the same year. In 1891 he was appointed lecturer at the Stockholm Högskola, later the University of Stockholm, and in 1895 he was promoted to professor in physics. Arrhenius was awarded the Nobel Prize in Chemistry in 1903 for his electrolytic theory of dissociation. In 1905 he was appointed Director of the Nobel Institute for Physical Chemistry, a post he held until a few months before his death. He was a member of the Nobel Committee for Physics, 1901–1927.

Arrhenius first devoted himself to physical chemistry, making major contributions to this field through his theory of electrolytic dissociation (1887) and calculations of the temperature dependence of the rate constants of chemical reactions (the "Arrhenius equation," 1889). Returning to Sweden in the early 1890s after five years spent on the European continent, he turned to cosmic physics. This attempt to develop physical theories linking the phenomena of the seas, the atmosphere, and the solid Earth centered on the Stockholm Physics Society. Debates in the Society concerning the causes of the ice ages led Arrhenius to construct his model of the influence of carbonic acid (carbon dioxide) in the air on the temperature on the ground. [*See* Global Warming.] He presented it in an article published in *Philosophical Magazine* early in 1896. The calculation of the absorption coefficients of water vapor and carbon dioxide that were the key to the construction of the model was made possible through Arrhenius's use of Samuel P. Langley's measurements of heat emission in the lunar spectrum. He constructed the model so that it would show how much the atmospheric carbon dioxide concentration had to decrease to produce the temperature (4°–5°C less than in Arrhenius's time) that could have led to the ice ages, and how much it had to increase to bring about the warmer period that had preceded the ice ages. The general rule that emerged from the model was that, if the quantity of carbon dioxide increases (or decreases) in geometric progression, temperature will increase (or decrease) nearly in arithmetic progression.

In linking the calculations of his abstract model to the actual processes that make for an increase or decrease of carbon dioxide, Arrhenius drew heavily on the pioneering work of Arvid Högbom (professor of geology at the Stockholm Högskola) on the geochemical carbon cycle. The cycle included the burning of fossil fuels as a source of atmospheric carbon dioxide. In a popular lecture early in 1896, Arrhenius predicted that a doubling of carbon dioxide from fossil fuel burning alone would take five hundred years and would lead to temperature increases of 3°–4°C. This is probably what has earned Arrhenius his present reputation as the first to have predicted the effect of this particular carbon dioxide scenario on global warming.

### BIBLIOGRAPHY

Arrhenius, S. "On the Influence of Carbonic Acid in the Air upon the Temperature on the Ground." *Philosophical Magazine* 42 (1896), 237–276.

Crawford, E. *Arrhenius: From Ionic Theory to the Greenhouse Effect*. Canton, Mass: Science History Publications, 1996.

Rodhe, H., and R. Charlson. "Svante Arrhenius and the Greenhouse Effect." *Ambio* 26 (1997). Proceedings of a symposium held at the Royal Swedish Academy of Sciences in 1996 to celebrate the hundredth anniversary of Arrhenius's article in the *Philosophical Magazine*. Includes contributions by R. J. Charlson, E. Crawford, S. Manabe, and V. Ramanathan.

—ELISABETH CRAWFORD

**ASTEROIDS.** *See* Impacts by Extraterrestrial Bodies.

## ATMOSPHERE DYNAMICS

By *atmosphere dynamics* is meant the set of physical processes that sets the atmosphere into motion and causes those motions to evolve. Apart from the gravitational tide, which is not a very significant component of the flow except at the highest levels, motions in the atmosphere are driven by variations of the heat energy absorbed from the sun and reemitted by the atmosphere. The nature of these motions is highly constrained by various factors, especially the rotation of the Earth. Atmospheric motions in the tropics and midlatitudes are very different in character. Low-frequency, long-wavelength *Rossby waves* propagate both horizontally and vertically in the atmosphere and provide the primary mechanism whereby distant forces can modify the local flow. The latent heat released by the condensation of water vapor plays an especially important role in the dynamics of the Earth's atmosphere. The interaction between the dynamics of the atmosphere and the dynamics of the oceans is increasingly appreciated as a major factor governing climate and climate change. [*See* Climate Change.]

**The Thermodynamic Imperative.** The dynamics of the Earth's atmosphere are determined by three basic physical principles: the conservation of matter, the conservation of energy (sometimes called the first law of thermodynamics), and Newton's laws of motion. Applied to the atmosphere, these laws have a complicated mathematical expression, so that the equations governing atmospheric flow are highly nonlinear; that is, they have no solutions in terms of elementary mathematical functions. For this reason, atmospheric dynamics are frequently studied using large computer models. But some general principles can be set out without recourse to such models.

Atmospheric motions are largely driven thermally, by variations of density caused by local heating or cooling of the air. The ultimate source of virtually all the heat input to the atmosphere is radiant energy received from the sun. All hot bodies emit electromagnetic radiation. So-called *black bodies* emit radiant energy at a rate proportional to the fourth power of their temperature, a relationship called Stefan's law. Furthermore, Wien's law states that the mean wavelength of this radiation decreases as the temperature increases. The sun radiates as a black body with a temperature of around 5,700°C, and most of this radiant energy is at the relatively short wavelengths of visible light. The Earth is much cooler; seen from outer space it would appear to radiate longer-wavelength infrared radiation as a black body with a temperature of around −18°C. In the long-term average, the amount of shortwave radiation intercepted by the Earth balances the amount of longwave radiation it emits; that is, there is a state of global radiative equilibrium.

This global radiative equilibrium does not extend to the local radiative balance, however. The solid curves in Figure 1 show that in the tropics the incoming shortwave radiation exceeds the outgoing longwave radiation. At high latitudes the outgoing radiation exceeds the incoming radiation. This local imbalance is explained by the internal heat transport within the atmosphere and oceans. The global circulation of the atmosphere transports heat from the tropics to the high latitudes. The transport of heat by the oceans is not easy to measure directly but is probably comparable to, or a little less than, the atmospheric heat transport.

A similar imbalance holds in the vertical. The dotted curves in Figure 1 show that while some 60 percent of the incoming solar radiation reaches the Earth's surface, only around 10 percent of the outgoing radiation originates at the surface. Again, the circulation of the atmosphere is responsible for transporting heat from the absorbing layers near the surface to the emitting layers in the upper troposphere. The opacity of the Earth's atmosphere to infrared radiation, due to water vapor

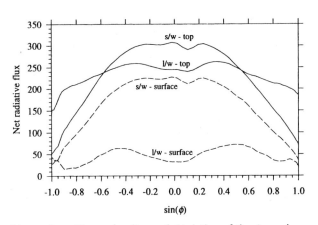

**Atmosphere Dynamics. FIGURE 1.** Variation of the Annual Mean Radiation Budget of the Earth with Latitude.

The solid curves show the net downward shortwave radiation and net upward longwave radiation at the top of the atmosphere. The dotted curves show the corresponding quantities at the Earth's surface. (Data taken from Rossow and Zhang, 1995.)

and other minor constituents, compared with its transparency to visible radiation means that the surface temperature of the Earth is on average 15°C, rather than the −18°C observed from space. This warming is the so-called *greenhouse effect*; without it, life would not be possible on the Earth's surface. The global energy budget illustrated in Figure 1 implies that heat must be transported both poleward and upward by atmospheric motions. [*See* Greenhouse Effect.]

The principles of thermodynamics explain the origin of these motions. Figure 2 is an example of a *thermodynamic diagram* for a parcel of air circulating in the atmosphere. The thermodynamic state of the air parcel—that is, its temperature, pressure, and density, as well as more exotic quantities—can be deduced from the location of the air parcel on this diagram. For example, the curves join points with a particular value of pressure. This particular diagram plots the temperature of the air parcel against a quantity called the *specific entropy*. The change of state of an air parcel is represented by a curve, and the amount of heat added to the air parcel during the change of state is proportional to the area beneath the curve.

The closed curve in Figure 2 represents the changes of state of an air parcel circulating in the atmosphere. Starting from a point near the surface at the cold pole (A), the parcel first moves toward the warmer equator (B). It then rises to the tropopause (C). A considerable amount of heat is added to the air parcel during the journey AB, while rather little heat enters or leaves the parcel during the rapid ascent BC. The air parcel then returns to its starting point but along a different route: the air parcel moves toward the pole at an upper point (D) before sinking back to the surface at point A. Heat must be extracted from the air parcel during its journey along segment CDA, but less heat than was added at segment AB. Thus the parcel has returned to its initial state, but net heat energy, proportional to the area enclosed by the loop ABCD, has been added to it. Because this energy cannot be destroyed, it must have been converted to kinetic energy of the motion of the air parcel. This type of circulation is self-sustaining, with heat energy being converted into kinetic energy of the circulation. The air is heated at low levels, especially close to the equator, and cooled at high levels, especially near the poles, just as the radiation budget shown in Figure 1 requires. The circulation of the Earth's atmosphere is a natural example of a *heat engine*, in which the circulation of air converts a small part of the available heat energy into mechanical energy.

Such circulations, in which net heating takes place at

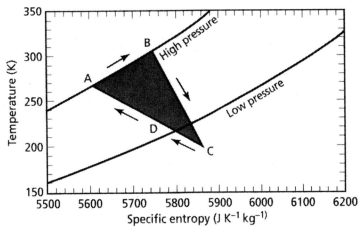

**Atmosphere Dynamics. FIGURE 2.** A "Thermodynamic Diagram" for the Earth's Atmosphere.

In a graph of temperature versus specific entropy, the curves show lines of constant pressure. The closed loop ABCD represents an idealized track of a typical air parcel as it circulates in the atmosphere.

high pressure and net cooling at low pressure, with relatively warm air parcels rising and relatively cool parcels descending, are called *thermally direct circulations*. Thermally direct circulations create kinetic energy of motion from heat. In contrast, *thermally indirect* circulations consume mechanical energy and generate heat. Thermally indirect circulations are sometimes found in the Earth's atmosphere—for example, in the winter stratosphere—but the stronger thermally direct circulations overwhelm their effects in the total circulation.

Eventually, friction prevents the mechanical energy of motion from building up indefinitely. Friction converts mechanical energy back into heat, which is eventually radiated back to space. Although the energy flow into the atmosphere is balanced by energy flow back out of the atmosphere into space, the atmosphere is a net exporter of entropy to space.

**Dynamic Constraints.** A consideration of the thermodynamic imperative alone would suggest that the atmospheric circulation should consist simply of equatorward flow near the surface, with a return poleward flow at upper levels. That this is an oversimplified description of the global atmospheric circulation is due to a number of additional dynamic constraints imposed upon the flow.

The most important of these constraints is imposed by the rotation of the Earth. Imagine a ring of air moving from the equator toward the pole in the upper layers of the troposphere, as shown in Figure 3. Initially, the ring is at rest relative to the rotating solid Earth. But as the ring approaches the pole, its radius contracts, and in order to conserve its angular momentum, it will spin

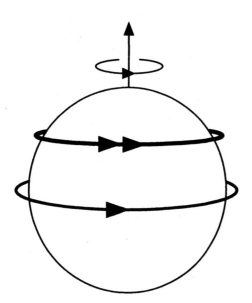

**Atmosphere Dynamics. Figure 3.** A Ring of Air Moves from Low to High Latitudes on a Rotating Planet.

If the ring is initially stationary with respect to the solid planet, it will spin more rapidly than the solid planet at higher latitudes.

more rapidly. From the point of view of an observer on Earth, a westerly wind will develop, rapidly becoming stronger at higher latitudes. Indeed, simple calculations reveal that the wind speed would be more than 100 meters per second (m/s) by the time the ring reached a latitude of 30°. Such wind speeds are already much larger than those observed in the atmosphere and are close to the limit allowed by the thermodynamic considerations of the last section. A simple thermally direct overturning must therefore be confined to a limited range of latitudes.

A ring of air moving equatorward near the Earth's surface would, by the same argument, be expected to develop easterly flow. However, friction between the air and the solid Earth prevents this easterly wind from ever becoming strong. So one effect of rotation is to ensure that the winds in the atmosphere are generally westerly. On average, the Earth's atmosphere rotates some 4 percent faster than the solid Earth, corresponding to westerly winds of about 15 m/s.

A second major effect of rotation is to inhibit vertical motion. The effect is related to the behavior of a rapidly spinning gyroscope, which resists any change in the direction of its rotation axis. Similarly, an air parcel on the rapidly spinning Earth resists the change in direction of its rotation axis that would be implied by any vertical motions. Consequently, vertical motion is inhibited by rapid rotation.

In a fluid of constant density, such as water in a rotating tank, this effect is dramatic. The lack of vertical motion means that the flow is two-dimensional and is exactly the same at every level in the tank. The result is that vertical streaks of dye are pulled out by fluid motions into thin, vertically hanging curtains called *Taylor's ink walls*. The two-dimensionality of the flow is asserted by the Taylor-Proudman theorem.

In the Earth's atmosphere, variations of density mean that the flow is not identical at every level, though the inhibition of vertical motion at middle and high latitudes is still important. Instead, the wind increases or decreases with height in proportion to the horizontal gradients of density. The classical example of this *thermal wind relationship* is the steady increase of the westerly winds with height in the middle latitudes, in balance with the decrease of temperature from pole to equator (and hence increase of density) at every level through the troposphere. The cross section of Figure 4 shows that the westerly wind increases particularly rapidly with height at places where the temperature is decreasing particularly rapidly with latitude.

A second major dynamic constraint on atmospheric circulation is revealed by Figure 4. The temperature decreases with height, at an average rate of 6.5 K/m in the troposphere, considerably less than the 10 K/m that a freely rising parcel of air, expanding and cooling as its

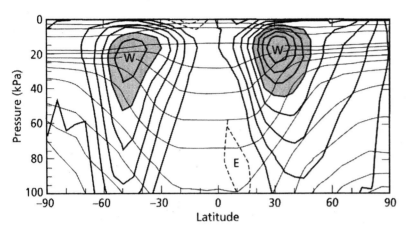

**Atmosphere Dynamics. FIGURE 4.** A Latitude-Height Cross Section of the Temperature and Westerly Wind for the December-January-February Season.

The fine contours indicate the potential temperature, contour interval 10 K, and the thicker contours the westerly wind strength, contour interval 5 m/s, with easterly winds shown by dashed contours and westerlies in excess of 20 m/s shaded.

pressure dropped, would show. As a result, the atmosphere is said to be *stably stratified*. Stable stratification means that if an air parcel is displaced upward, it will find itself colder and therefore more dense than its surroundings. Buoyancy forces will then tend to pull the parcel back toward its original level. As a result of this effect, disturbed air parcels have a tendency to form waves that oscillate in the vertical; these waves are called *buoyancy waves* or *internal gravity waves*, and their period is typically around ten minutes. A particularly graphic example of internal gravity waves is sometimes observed when air flows over a range of hills or mountains, as illustrated in Figure 5. The hill displaces air parcels near the surface, causing them to oscillate downstream of the hill. If the oscillations have a sufficiently large amplitude, they may be marked by lens-shaped *lee wave clouds*, which form at the peaks of the waves.

Such dramatic examples of internal gravity waves are observed only in relatively unusual situations when they are trapped in the lower layers of the atmosphere. But smaller-amplitude internal gravity waves are ubiquitous in the atmosphere. They transmit information relatively quickly from one part of the atmosphere to another and are a primary mechanism by which the midlatitude atmosphere remains close to a state of thermal wind bal-

ance, despite the many heating, friction, and other effects that act to upset this balance. Internal gravity waves also propagate from the surface to higher levels in the atmosphere. It is increasingly recognized that global circulation models need to represent the effects of such vertically propagating gravity waves in order to simulate accurately the structure of the wind and temperature fields near the tropopause.

Because stable stratification also implies a resistance of the fluid to vertical displacements, it acts as a second influence that acts to suppress or reduce vertical motion. The depth over which atmospheric motions extend is governed by the two periods that characterize a rotating stratified atmosphere: the rotation period and the buoyancy period. The ratio of horizontal scales to vertical scales of motion in weather systems is the ratio of the rotation period to the buoyancy period.

**Tropical Circulations.** Motions in the atmosphere have a very different character in the tropics, where the vertical component of the Earth's rotation is small, than in the midlatitudes. In the deep tropics, the components of the motion in a latitude-height (or *meridional*) plane are essentially the simple thermally direct overturning suggested by thermodynamic arguments. Figure 6 shows streamlines of atmospheric flow averaged around latitude circles for the December-January-February period. The flow consists of rising motion in the summer (southern) hemisphere tropics, and descent in the tropics and subtropics of the winter (northern) hemisphere. These motions form an overturning cell, called a *Hadley cell*, after the British scientist who in 1835 first suggested this type of overturning. His theory of the circulation of the Earth's atmosphere accounted for the subtropical trade winds familiar to mariners. The effect of the Earth's rotation is clear when the meridional flow is compared to the westerly winds that also develop, and that are shown by shading in Figure 6. Associated with the poleward flow out of the equatorial regions at upper levels is an increase in the westerly winds. They reach a maximum intensity in the upper troposphere at latitudes of around

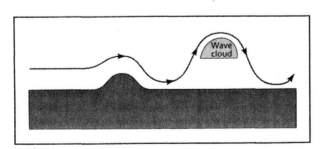

**Atmosphere Dynamics. FIGURE 5.** A Schematic Illustration of Lee Waves, Formed When Air Flows over Hills to Form a Stationary Wake of Internal Gravity Waves Downstream of the Hill.

**Atmosphere Dynamics. FIGURE 6.** Streamlines of the Mean Meridional Stream Function for the December-January-February Period (contour interval $2 \times 10^{10}$ kg/s).

The shading indicates westerly winds stronger than 20 m/s.

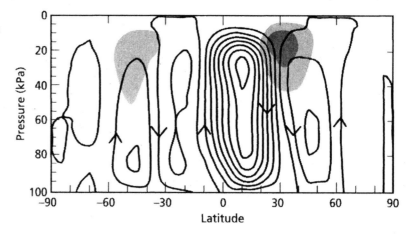

$25°–30°$N in excess of 35 m/s. This maximum is referred to as the *subtropical jet*; its strength is well below the value implied by angular momentum conservation.

The Hadley cell is confined to regions equatorward of about $25°$ of latitude. Poleward of these latitudes, the thermally direct Hadley circulation gives way to weaker thermally indirect circulations. These thermally indirect circulations, called the *Ferrel cells*, are driven by the eddies of the midlatitudes (see Midlatitude Weather Systems, below). They are much less important in the global transport of heat and tracers than the Hadley cells because they are weaker and because the eddies of the midlatitudes overwhelm their transports.

Returning to the tropical latitudes, the distribution of atmospheric heating or cooling is rarely uniform along lines of latitude. Rather, the tropics are characterized by intense centers of localized heating. For example, in most years the most intense heating is located over the Indonesian region, in the western tropical Pacific; the heating is associated with the formation of intense cumulonimbus clouds and heavy rainfall in a region of very warm sea temperatures and high mountains. The latent heat released by the condensation of large amounts of water vapor goes to heat the atmosphere in these regions. How does the tropical atmosphere respond to such localized heating maxima?

The answer to this question can be understood in terms of special types of wave motion that are trapped close to the equator. These large-scale, low-frequency waves are more complicated than the midlatitude Rossby waves discussed in the next section, but their general properties are easily described. In a motionless atmosphere, two kinds of equatorially trapped waves are important. These are *Kelvin waves*, which propagate to the east and are characterized by a single extremum of pressure centered on the equator and dying away within about 2,000 kilometers of the equator. The other group of waves are called *equatorial Rossby waves*; they propagate more slowly to the west, have two extrema of pressure disposed symmetrically about

the equator, and again die away within 2,000 kilometers of the equator. Both types of wave motion are generated by an isolated heating maximum and propagate along the equator away from the heating region. A steady state is achieved when dissipation in the atmosphere balances the excitation of wave activity by the heating. The resulting flow is shown in Figure 7. Because it propagates more slowly, the Rossby-gravity wave extends only a short distance to the west of the heating. The more rapidly propagating Kelvin wave extends a greater distance to the east.

The flow generated by this heating is largely east–west; it takes the form of overturning cells along the equator, with ascent over the heating region and descent at other longitudes. Such cells are called *Walker cells*. The small meridional winds associated with these cells means that they do not play a large role in the poleward transport of heat, which is dominated by the Hadley circulations even when the heating is not zonally uniform.

The same theory can help to account for the flow associated with the Asian monsoon circulations. If the heating maximum is displaced slightly away from the equator, the flow to the east of the heating remains largely unchanged. The flow to the west becomes highly asymmetric about the equator, however, with intense circulating systems developing to the west and poleward of the heating maximum. James (1994, Figure 7.5) illustrates this flow.

**Vorticity and Rossby Waves.** One of the most characteristic features of the motion in the Earth's atmosphere is the dominance of swirling, circulating motions. Even a cursory glance at a weather map reveals that air is generally swirling around various high and low pressure centers. This swirling is measured by a quantity called *vorticity*, which is simply described as a measure of the spin of an air parcel about its vertical axis: vorticity is defined to be twice the angular velocity of an air parcel about its vertical axis. The vorticity of an air parcel can be split into two parts. The first is the spin due to the vertical component of the rotation of the

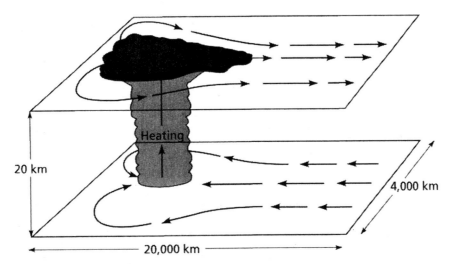

20 km

Heating

4,000 km

20,000 km

**Atmosphere Dynamics. FIGURE 7.** A Schematic Diagram of the Response of a Stationary Tropical Atmosphere to Localized Heating.

The arrows represent the flow toward the heating near the surface and away from the heating at the upper level. The circulation is completed by weak descent away from the heating region.

Earth itself. This component is called the *planetary vorticity*. The planetary vorticity has a maximum value at the North Pole, drops to zero at the equator, and is negative in the Southern Hemisphere. The second component is the spin of the air parcel relative to the solid Earth, called the *relative vorticity*. When the relative vorticity has the same sign as planetary vorticity, it is called *cyclonic vorticity*; when it has the opposite sign, it is called *anticyclonic vorticity*. Cyclonic relative vorticity is associated with centers of low pressure, and anticyclonic vorticity with centers of high pressure.

An extremely important effect concerning the spin of air parcels is illustrated in Figure 8. If an air parcel is displaced to the north or south, if its vertical extent remains fixed, and if frictional forces are unimportant, then the absolute vorticity or absolute spin of the air parcel about a vertical axis remains constant. However, the planetary vorticity decreases as one moves toward the equator. Consequently, the relative vorticity of the displaced air parcel must change, becoming more anticyclonic (or less cyclonic) for a poleward displacement, and more cyclonic (or less anticyclonic) for an equatorward displacement. This concept is readily generalized to cases when the vertical extent of the air parcel varies, through a concept called *potential vorticity*, which will be introduced briefly at the end of this section.

This change of the relative vorticity of an air parcel when it is displaced from its home latitude to another latitude is responsible for *Rossby waves*, a transverse type of wave. Their period is long, often several days, which means that they are readily excited by motions in the large-scale atmosphere. Indeed, they may be thought of as the primary form of wave motion in the Earth's midlatitudes, transmitting information across large distances in the atmosphere; many important atmospheric phenomena can be described qualitatively in terms of Rossby waves. The mean state of the atmosphere can be thought of as providing a wave guide for Rossby waves, and considerable insight into atmospheric dynamics can be gained from understanding the circumstances in which Rossby waves can propagate and their speed and direction of propagation.

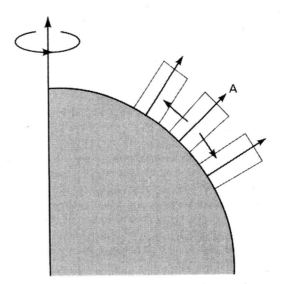

**Atmosphere Dynamics. FIGURE 8.** The Relative Vorticity of Air Parcels that are Displaced in the North–South Direction.

The arrow shows the magnitude of the planetary spin at each latitude. The air initially at latitude A conserves its spin (indicated by the length of the rectangle) when it moves to another latitude. Moved poleward, it spins less rapidly than its surroundings; moved equatorward, it spins more rapidly than its surroundings.

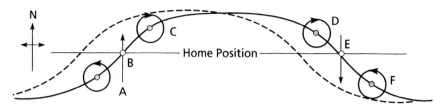

**Atmosphere Dynamics. FIGURE 9.** Schematic Diagram of a Rossby Wave.

Poleward and equatorward, displacements of air parcels generate spins that tend to return parcels to their original latitudes and result in a westward propagation of the entire pattern. The solid line shows the initial position of the displaced parcels and the dashed line their position a short time later.

The essential mechanism for Rossby waves is illustrated in Figure 9, in which the effect of introducing a wavelike displacement of air parcels in an otherwise stationary atmosphere is considered. Those parcels displaced poleward of their initial location acquire a clockwise or anticyclonic spin. Those displaced equatorward acquire cyclonic spin. Consider the flow in the vicinity of the three parcels labeled A, B, and C in Figure 9. The cyclonic flow around A and the anticyclonic flow around C both combine to drive parcel B poleward, thus advancing the wavefront toward the west. In the vicinity of the three points D, E, and F, the flows induced by parcels D and F drive parcel E equatorward. Again, the wavefront moves to the west. Indeed, similar arguments can be applied to any point in the wave, indicating that the entire pattern must move toward the west. A more quantitative discussion reveals that the longer the wavelength, the faster the wave's advance to the west. Unlike many other, more familiar waves, such as sound waves or water waves, Rossby waves can propagate only in one direction, toward the west. Furthermore, the rate of propagation or *phase speed* depends strongly on the wavelength of the wave: Rossby waves are said to be *strongly dispersive*. If the background flow is not still but moving, the Rossby waves still propagate to the west *relative to the flow*. Relative to the Earth's surface, they may propagate either eastward or westward, depending on the strength of the background flow and the wavelength of the Rossby wave.

When the vertical extent of the air parcel varies, it is not the absolute vorticity that is conserved but the absolute vorticity per unit pressure height of the parcel. This quantity is called the *potential vorticity*. Potential vorticity is perhaps the most fundamental quantity governing flow on a rotating planet, and it enables a qualitative understanding of many phenomena to be gained. Potential vorticity is important for two reasons. First, the potential vorticity of an air parcel is conserved when friction and heating rates are small, that is, for time scales of a few days or less. Second, when the distribution of potential vorticity is known, the flow fields and temperature fields can all be deduced: this is called the *invertibility principle.*

In the present context, Rossby waves are supported by strong gradients of potential vorticity. The stronger the gradients, the faster the Rossby waves propagate, and the more rapidly do displaced air parcels oscillate about their home latitude. The gradients of potential vorticity are often considerably stronger than the gradients of planetary vorticity discussed above.

**Midlatitude Weather Systems.** The flow in the Earth's atmosphere is generally unsteady and fluctuates in time. Weather systems form, amplify, and then collapse and are replaced by new weather systems at new locations. The middle-latitude atmosphere is especially unsteady, with large variations of wind, temperature, and other quantities at any given location. The unsteadiness arises because the flow in midlatitudes is unstable, so that any small disturbances amplify rapidly and quickly come to dominate the flow. Subsequently, the unstable disturbances collapse, and a fresh outbreak of instability creates new disturbances. The process is called *baroclinic instability.*

The origin of the instability of the midlatitude flow can be understood in terms of the Rossby waves introduced in the last section. The mean flow in the midlatitudes is westerly, and it increases in intensity with height to values typically close to 40 m/s near the tropopause. This variation of the flow with height can be crudely represented by dividing the atmosphere into two layers: an upper layer with strong westerly winds and a lower layer with weaker westerly winds. Baroclinic instability arises from the interference between Rossby waves in the upper level and Rossby waves in the lower level. Figure 10 shows the two trains of waves. Consider first the upper level. Relative to the average over the whole depth of the atmosphere, the flow here is westerly, and the potential vorticity increases toward the pole. Rossby waves propagate toward the west relative to the flow. The effect of the regions of cyclonic and anticyclonic spin penetrates to the lower level, where it can displace air along a north-south line; this downward penetration is shown by the vertical solid lines in Figure 10. The degree of this penetration depends on the structure of the flow and the temperature field, but generally smaller wavelengths have less effect on the lower layer than longer waves.

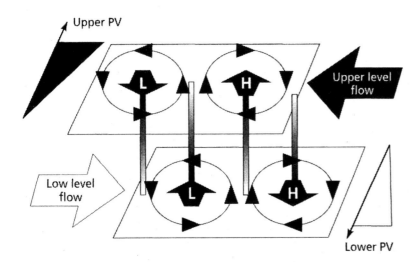

**Atmosphere Dynamics. Figure 10.** Baroclinic Instability, Viewed as Two Interfering Trains of Rossby Waves.

Potential vorticity (PV) increases toward the pole in the upper layer and toward the equator in the lower layer.

Now consider the lower level. Relative to the depth-averaged flow, the wind here is easterly (relative to the Earth's surface, it may be weakly westerly). But the effect of the wind shear and the boundary below means that the effective potential vorticity increases toward the equator. Rossby waves in the lower layer therefore propagate from the west toward the east relative to the local flow. This combination of westerly wind and poleward potential vorticity gradient in the upper level with easterly wind and equatorward potential vorticity in the lower layer means that it is possible to find trains of Rossby waves of the same wavelength in the two layers that do not propagate relative to one another.

With the configuration of the Rossby waves shown in Figure 10, in which the cyclonic centers of the upper wave are to the west of the cyclonic centers of the lower wave, the lower-level wave will be amplified by the circulation penetrating downward from the upper-level wave. At the same time, the upper wave will be amplified by the circulation penetrating upward from the lower wave, an influence indicated by the broken vertical lines. This mutual interference of the two trains of waves ensures that both amplify; in other words, that the vertically sheared flow is unstable.

The essential condition for the instability to operate is that the upper and lower waves should have similar speeds. If the waves are too short, they move only slowly relative to the background flow at their level, and waves will be blown apart. If they are too long, they will propagate too quickly relative to each other to interact. Maximum amplification is obtained for some intermediate wavelength, which for the Earth turns out to be around 4,000 kilometers. This is indeed typical of the spacing between successive troughs in the upper-level flow of the Earth's midlatitude troposphere. The rate at which the waves amplify depends not only on their wavelength but also on the vertical penetration of the influence of the Rossby waves.

This interpretation of the dynamics of baroclinic instability is complemented by a thermodynamic discussion of their effects on the temperature structure of the atmosphere. The wind and the temperature both vary along a growing baroclinic wave. At some longitudes, the wind blows toward the pole, and at the same location, the temperatures tend to be higher. At other longitudes, the wind blows equatorward and the temperature is lower. Thus, the wave systematically moves warm air poleward and colder air equatorward, on average transporting heat toward the pole. These motions are almost, but not quite, horizontal. The actual paths followed by air parcels make a small angle with the horizontal. A typical trajectory slope is around one in one thousand, with warmer air parcels rising gently and colder air parcels sinking. As a result, the baroclinic wave transports heat upward as well as poleward. Here is just the condition required on thermodynamic grounds to convert heat energy into kinetic energy of motion. This is the thermodynamic basis of baroclinic instability, which, because of the gently sloping trajectories followed by fluid elements, is sometimes referred to as *sloping convection.*

In Figure 11 is shown the upward temperature flux due to weather systems, as observed over the Northern Hemisphere in winter. Notice that the vertical temperature fluxes are nearly everywhere upward and that they are largest throughout the middle latitudes. They are particularly concentrated into two elongated regions in the western Atlantic and western Pacific Oceans. These elongated maxima are found to coincide with the tracks followed by surface weather systems, with a predominance of young developing systems at the western end and older, decaying systems at the eastern end. Because of their association with surface pressure depression tracks, these maxima are called *storm tracks.* Their formation is related to the pattern of sea surface temperatures in the western part of the ocean basins, par-

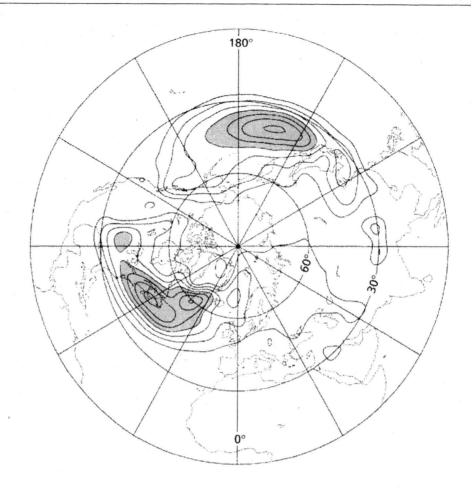

**Atmosphere Dynamics. FIGURE 11.** The Upward Transport of Heat by Weather Systems in the December-January-February Season.

Contours show upward temperature flux, contour interval 0.1 K/Pa/s, with shading indicating values greater than 0.35 K/Pa/s. The concentration of the flux into the two major "storm tracks" is clear.

ticularly the very strong gradients of sea surface temperature that form where warm currents such as the Gulf Stream meet cold water flowing from the north and peel away from the coast. Understanding the location and the length of these storm tracks is an important issue in understanding climate change in the midlatitudes. Changes in the length or vigor of storm tracks could have major impact on regions such as western Europe that border the storm track regions.

**Steady Waves and Teleconnections.** While Rossby waves always propagate from east to west relative to the flow, the winds throughout the midlatitudes blow from the west. It is therefore possible to find a Rossby wave that is propagating to the west relative to the westerly flow, but which is stationary with respect to the Earth's surface. Such stationary Rossby waves can be forced by stationary features such as mountains or patterns of

heating due to variations of sea surface temperature. The effect is comparable to the wake of stationary surface waves that forms over an obstacle in a swiftly flowing river. The wake of Rossby waves formed by localized forcing spreads in the poleward, zonal, and vertical directions, and in favorable circumstances may affect a wide range of latitudes to south and north of the original forcing.

In regions of easterly flow, no Rossby waves are stationary with respect to the ground. The tropical easterlies therefore prevent the patterns of stationary Rossby waves formed in the Northern Hemisphere from spreading into the Southern Hemisphere and ensure that to some extent the waves in each hemisphere are independent of those in the other hemisphere. The line where the zonal wind changes from easterly to westerly is called a *critical line*. Critical lines act as barriers to stationary Rossby wave propagation; they may either absorb or reflect the Rossby waves impinging upon them, depending on the levels of dissipation in the atmosphere and the amplitude of the waves. The evidence of global circulation studies, however, is that the critical lines in the Earth's atmosphere act mainly as absorbers of wave activity; the one-way traffic of Rossby waves out of the midlatitudes into the subtropics is one of the most characteristic dynamic features of the Earth's atmosphere.

The principal sources of steady Rossby waves in the Northern Hemisphere are the major mountain ranges, especially the Rockies and the Tibetan Plateau. As air is forced to rise over these features, it acquires anticyclonic spin, and thus a train of Rossby waves is excited that extends over a large part of the hemisphere. The result is a series of large-scale trough ridge patterns in the winter flow of the Northern Hemisphere that greatly affect the climate at different locations. For example, the winters in the eastern United States are much colder than in the Mediterranean, even though both locations share much the same latitude: the steady wave pattern brings cold air from the north to New York, but warmer air from the south to Rome, during winter. The steady waves also play a substantial role in the poleward transport of heat in the Northern Hemisphere winter.

During the past twenty years or so, it has become increasingly clear that anomalous weather patterns in the middle latitudes can often be traced back to anomalies in the tropics. For example, the severe winter of 1962–1963 in western Europe was associated with abnormally warm sea surface temperatures off tropical West Africa. The mechanism for these long-distance connections or *teleconnections* is the excitation of a poleward propagating train of Rossby waves, excited in the vicinity of a tropical heating anomaly but propagating to high latitudes.

One of the most important of these teleconnection patterns is illustrated in Figure 12. Analysis of forty-five winter months by Wallace and Gutzler (1981) showed that unusually low pressure close to the Date Line in the tropical Pacific is associated with unusually low pressure in the North Pacific, high pressure over northwestern North America, and with low pressure over

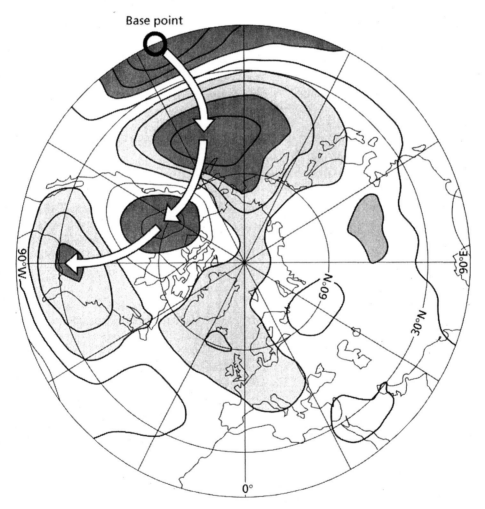

**Atmosphere Dynamics. Figure 12.** The Pacific–North American Pattern, One of the Most Significant and Best-Established Teleconnection Patterns in the Troposphere.

The diagram illustrates the relationship of the midtroposphere pressure with the pressure at a base point, marked with a circle. Dark shading indicates that the pressure tends to be high when the base point pressure is high, and low when the base point pressure is low. Light shading indicates that the pressure tends to be low when the base point pressure is high and vice versa. The pattern can be interpreted as a train of Rossby waves radiating away from the base point, indicated by the arrows. Based on forty-five months of winter data for the Northern Hemisphere.

southeastern North America. On the other hand, unusually high pressure near the Date Line is associated with low pressure over northwestern North America and high pressure over southeastern North America. The high and low centers form a long arc, arching into high latitudes and back toward the tropics, exactly the path that would be followed by a train of Rossby waves excited in tropical central Pacific. This particular teleconnection pattern, one of the most significant observed, is called the Pacific–North American or PNA pattern. Other teleconnection patterns include the North Atlantic Oscillation (NAO), which has an important influence on the Atlantic storm track and hence on low-frequency fluctuations of the climate of western Europe.

**Water Vapor and Dynamics.** Water vapor is the most important variable constituent of the Earth's atmosphere. It can make up as much as 4 percent of the mass of the atmosphere in the moist tropics and as little as a few parts per million in the stratosphere. Motions in the atmosphere transport moisture from source regions to those places where it falls out as rain. The source regions are the warm parts of the oceans over which the trade winds blow, where there is net evaporation of water vapor. Over the continental interiors, rainfall exceeds evaporation and there is a net deposition of water on the surface. This *hydrologic cycle* is closed by the runoff of liquid water in the form of rivers flowing from the continental interiors back to the oceans, and by the flow of glaciers transporting frozen water away from continental interiors at high latitudes. [*See* Hydrologic Cycle.]

The situation is more complex than this transport cycle would suggest, however. Water vapor is not simply advected passively by the atmospheric flow. Rather, there is a strong feedback between the atmospheric dynamics and the evaporation and condensation of water. Atmospheric flow can be strongly modified, or even more or less completely determined, by rainfall and evaporation. The reason for this is the large amount of energy required to evaporate solid or liquid water into its vapor state, energy that is released as heat when the vapor condenses again. The heat required to evaporate water is called the *latent heat* of evaporation. The latent heat released by condensation sufficient to produce just 8 millimeters of rain is equivalent to the average heat received by the Earth from the sun per square meter each day. Indeed, the latent heat of evaporation of water is larger than for any other common substance, and this accounts for the unique role of water in shaping the circulation of the Earth's atmosphere.

Unsaturated air flowing over a warm ocean surface—for example, in the trade winds of the subtropics—is gradually moistened as water from the ocean evaporates into it. Incoming solar energy goes largely into evaporating water in these regions, rather than warming the air, thus moderating the surface temperature of the Earth. When the moistened air rises—for example, when it meets an opposing current of moist air near the equator at the intertropical convergence zone—it cools, and much of its load of water vapor condenses and falls as rain. The latent heat released heats the air, making it more buoyant and intensifying the rising motion. In this way, a feedback exists in which the condensation of water vapor intensifies the atmospheric circulation that led to the condensation. One view of the process is that the heat poured into the climate system over a large area of the tropical oceans is released and concentrated into relatively small regions of strong uplift and rainfall.

The active role of water in atmospheric circulation is one of the principal sources of uncertainty in assessing and predicting climate change. The way in which small-scale clouds are organized to give the large-scale pattern of rainfall and latent heat release is imperfectly understood and consequently is poorly represented in models of the atmospheric circulation. The clouds of water droplets or ice crystals that form as water condenses in the atmosphere have profound consequences for the radiative balance of the atmosphere. They reflect solar radiation, leading to a cooling of the Earth's surface. At the same time, clouds absorb infrared radiation, enhancing the greenhouse warming of the Earth's surface. The balance of these large opposing effects is poorly understood and is another major uncertainty in climate modeling.

**Coupling with the Oceans.** In the previous section, it was shown that the oceans profoundly modify the flow of the atmosphere. The major effect is the strong heating of the atmosphere arising from condensation over regions of large sea surface temperature. But the distribution of sea surface temperature is itself a product of the atmospheric flow. The ocean is both heated and cooled at its surface, and, therefore, unlike the atmosphere, rather little of its motion is generated by a global heat engine effect. Instead, it is set into motion mainly by the stresses due to the wind flow across its surface. There is thus a strong coupling between the atmospheric and oceanic circulations; neither can be understood in isolation from the other. Increasingly, those who endeavor to model climate and atmospheric circulations are being forced to model the coupled ocean–atmosphere system.

The El Niño–Southern Oscillation, or ENSO, phenomenon encapsulates the importance of ocean–atmosphere coupling in the Earth's tropics. Figure 13 illustrates how the ENSO phenomenon is thought to work. Normally, the warmest sea surface temperatures are in the Indonesian region of the Pacific. These warm sea surface temperatures drive intense convection in this region: the high orography of Indonesia further amplifies the convection. The low-level flow that develops in response to such localized heating is illustrated in Figure 7. The westward

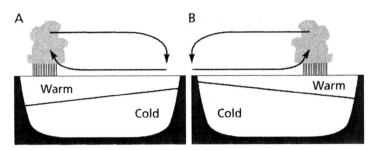

winds to the east of the heating region apply a stress to the ocean surface, which acts to pile up the warm surface waters to the west of the ocean basin in the Indonesian region. The whole picture is therefore consistent. From time to time, however, this normal equilibrium breaks down. The wind stress on the ocean is reduced, allowing the warm water to spread back toward the central Pacific. As the warmest water moves eastward, the atmospheric convection and heating maximum also moves toward the central Pacific. The new configuration is supported by the changing wind stresses. The anomalous circulation regime has profound consequences throughout the tropics, with changed patterns of rainfall leading to floods in some regions and drought in others. At the same time, teleconnection patterns, especially the Pacific–North American pattern described earlier, are excited, leading to anomalous weather patterns throughout much of the subtropics and higher latitudes. [*See* El Niño–Southern Oscillation; *and* Ocean–Atmosphere Coupling.]

What starts the change off is not easily determined. A change in either the atmospheric or the ocean component of the system could trigger the change. But clearly feedbacks between the atmospheric and ocean circulations all act together to drive an irregular low-frequency, nonlinear oscillation. Whether the ENSO cycle is in any sense predictable is an important practical question. Whether the ENSO cycle would have the same character in a warmer world with enhanced greenhouse warming is also an important question.

A small part of the ocean circulation, the so-called *thermohaline circulation*, is generated by a heat engine effect. It relies upon the downward mixing of heat in certain localities in order that the large-scale thermohaline circulation can transport heat upward. The thermohaline circulation is responsible for exchanging surface water and abyssal water in the oceans on time scales of decades to centuries. As well as exchanging water, the thermohaline circulation also exchanges dissolved tracers, such as carbon dioxide, between the upper and deep layers of the ocean. Changes to the thermohaline circulation are thought to play an important role in climate change. [*See the vignette on* Thermohaline Circulation and the Cooling Effects of Warming *in the article on* Climate Change.]

[*See also* Atmosphere Structure and Evolution; Natural Climate Fluctuations; Ocean Dynamics; *and* System Dynamics.]

### BIBLIOGRAPHY

Hartmann, D. L. *Global Physical Climatology*. San Diego: Academic Press, 1994. A comprehensive introduction.

Holton, J. R. *An Introduction to Dynamical Meteorology*. 3d ed. San Diego: Academic Press, 1992. The classic modern text on dynamic meteorology, covering a wide range of topics and taking a quantitative analytical approach.

Hoskins, B. J., M. E. MacIntyre, and A. W. Robertson. "On the Use and Significance of Isentropic Potential Vorticity Maps." *Quarterly Journal of the Royal Meteorological Society* 111 (1985), 877–946. This review article gives a thorough discussion of potential vorticity and its use in understanding Rossby waves, baroclinic instability, and other dynamic effects in the atmosphere and ocean.

James, I. N. *Introduction to Circulating Atmospheres*. Cambridge: Cambridge University Press, 1994. An extensive and quantitative treatment of the material covered by the present article.

Philander, S. G. H. *El Niño, La Niña, and the Southern Oscillation*. San Diego: Academic Press, 1990. Gives a good description of the ENSO phenomenon and discusses the basis of modern models of ENSO.

Rossow, W. B., and Y. C. Zhang. "Calculation of Surface and Top of Atmosphere Radiative Fluxes Based on ISCCP Datasets. II. Validation and First Results." *Journal of Geophysical Research* 100 (1995), 1167–1197.

Wallace, J. M., and D. S. Gutzler. "Teleconnections in the Geopotential Height Field during the Northern Hemisphere Winter." *Monthly Weather Review* 109 (1981), 784–812.

—I. N. JAMES

## ATMOSPHERE STRUCTURE AND EVOLUTION

The term *atmosphere structure* refers to the variation of atmospheric temperature, pressure, density, or chemical composition with altitude. The vertical temperature variation of the Earth's atmosphere is largely controlled by the absorption of solar radiation and by the resulting radiative response of atmospheric constituents. The thermal structure of the atmosphere is so distinctive that it is the basis for dividing the atmosphere into regions. Figure 1 shows the zonal-mean temperature

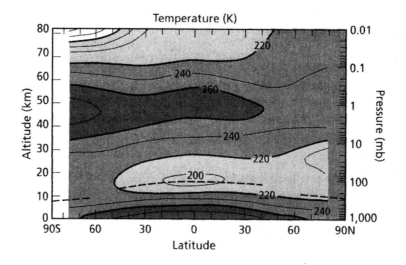

**Atmosphere Structure and Evolution. Figure 1.** Zonal-Mean Temperature (in Kelvin) during Northern Winter as a Function of Latitude and Altitude. (After Salby, 1996. With permission of Academic Press.)

during the northern winter as a function of latitude and altitude. Starting at the surface is a region, known as the *troposphere*, where the temperature decreases with height. Most of what we call "weather" is confined to the troposphere. The top of the troposphere (between 9 and 16 kilometers [km]) is the level at which the temperature reaches a minimum, called the *tropopause* (lower dashed line in Figure 1). The troposphere is characterized by vigorous convection, initiated by solar heating of the Earth's surface. Convection cools the surface and warms the upper troposphere, giving an average lapse rate of 6.5 kelvins (K) per kilometer. Warmer temperatures and enhanced convective activity are responsible for the greater altitude of the tropopause in the tropics (ca. 16 km) as compared with the polar regions (ca. 9 km). Above the tropopause is the *stratosphere,* where the temperature increases to a maximum at the *stratopause* (upper dashed line in Figure 1). The temperature increase, or inversion, in the stratosphere is caused by the absorption of ultraviolet sunlight by ozone and has the effect of inhibiting mass exchange within the stratosphere and between the stratosphere and other parts of the atmosphere. Above the stratopause lies the *mesosphere*. Ozone is less abundant in the mesosphere, and molecules such as carbon dioxide radiate effectively, leading again to a temperature decrease with height that eventually reaches a temperature minimum near the *mesopause* (ca. 80 km). Above the mesopause lies the *thermosphere*, where the temperature rises rapidly to more than 1,000 K, as shown in Figure 2. Radiative processes have a limited effectiveness in the thermosphere owing to low atmospheric densi-

ties and a lack of molecules with substantial dipole moments; therefore, thermospheric temperatures remain high. Instead of being lost by radiative cooling, heat gained by the absorption of shortwave solar radiation and interactions with the magnetosphere in the thermosphere is balanced by conduction to lower atmospheric levels, where the energy can be more effectively radiated away. The temperature in the thermosphere varies from a low value during *solar minimum* (quiet sun) to a high value during *solar maximum* (active sun). Alternatives to this nomenclature are the terms *lower atmosphere* (troposphere), *middle atmosphere* (strato-

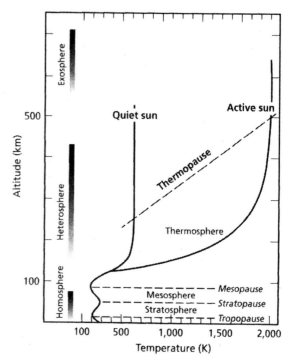

**Atmosphere Structure and Evolution. Figure 2.** Vertical Structure of Temperature in the Atmosphere. (After Banks and Kockarts, 1973. With permission of Academic Press.)

sphere and mesosphere), and *upper atmosphere* (above 80 km), as well as *homosphere* (where molecules are homogeneously mixed) and *heterosphere* (where molecules are diffusively separated).

In addition to the aforementioned regimes, the atmosphere also contains a number of layers characterized by special chemical and physical processes. The lowest 1–2 km of the atmosphere differs from the bulk of the troposphere. This region is known as the *planetary boundary layer* and is characterized by strong interactions (exchange of mass, momentum, and energy) between the surface and the atmosphere. Variabilities (e.g., diurnal changes) are large. Imbedded in the stratosphere is the *ozone layer*. The peak of the ozone layer is around 30 km in the tropics but lowers to about 20 km in the polar region as a result of upwelling motion in the tropical stratosphere and downwelling motion at high latitudes. The well-defined seasonal variation of the ozone layer is driven by a combination of photochemistry and dynamics. Another region, the *ionosphere*, extends from about 80 km to its peak at around 300 km in the thermosphere. The ionosphere is formed by the absorption of extreme ultraviolet solar radiation by the atmosphere, leading to ionization of molecules and atoms. The outermost limit of the thermosphere is the *exosphere*, where molecules may readily escape from the planet if the velocities are larger than the escape velocity and the direction of motion is upward. The thermosphere of the Earth is so hot that atoms in the tail of the Maxwell-Boltzmann distribution can escape.

**Atmospheric Evolution.** A unifying theme that relates our atmosphere to all planetary atmospheres in the solar system is its origin and evolution. All planets share a common origin about 4.6 billion years ago. The subsequent divergence in the solar system may be partly attributed to evolution. The central importance of evolution is illustrated in Figure 3. The box labeled "Origin" represents the source of material to the atmospheres. The sun, through radiation and solar wind, and chemical kinetics determine the rate of change of the composition of the atmospheres. Most chemical reactions in the atmosphere do not result in a net conversion of one species into another, but rather a recycling among different compounds. Some processes do result in a net change of the composition of the atmosphere, as with the box labeled "Evolution" in Figure 3. A change in composition can happen by means of species interconversion, loss of atmospheric species by escape to space, or sequestering in a subsurface reservoir. In the latter case, the material is not permanently lost from the planet and eventually may be recycled through tectonic and volcanic processes. Material can also be gained or lost from space by impact of comets or meteorites. The concept of atmospheric evolution provides a simple way to classify all planetary atmospheres according to their

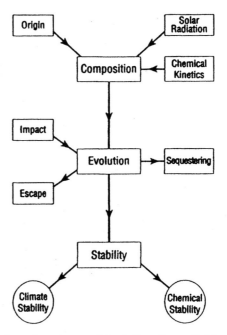

**Atmosphere Structure and Evolution. FIGURE 3.** Schematic Diagram Showing the Relation between Origin and Evolution and Other Parts of Atmospheric Science. (After Yung and DeMore, 1998. With permission of Oxford University Press.)

rates of evolution: (1) little or no evolution (giant planets); (2) moderate evolution (Mars, Venus, Earth); (3) extensive evolution (Titan, Triton, Pluto, Io); and (4) catastrophic evolution (Mercury, Moon, comets).

Volatiles trapped in the Earth's interior during its formation were volcanically outgassed to form the paleoatmosphere. Water was the dominant volatile constituent on all the newly formed terrestrial planets, followed in abundance by carbon dioxide and a much smaller amount of molecular nitrogen. On the Earth, the water condensed to form oceans, and carbonate rock formation within those oceans was responsible for removing the bulk of the atmospheric carbon dioxide. As a result, nitrogen is the Earth's main atmospheric constituent rather than carbon dioxide or water vapor. A mildly reducing, mostly nitrogen atmosphere probably predominated when life originated on the Earth. Life itself inexorably changed the atmospheric composition, as discussed below. *Volcanism* continues to play an important role on our planet by recycling the volatiles that have been sequestered by sedimentary processes.

**Impacts.** The paleoatmosphere would have been subject to frequent bombardments by comets, asteroids, and early planetesimals. To a large extent, the velocity of the impacting objects controls whether volatile material will be added or lost. Slow impacts deliver new volatiles and organic material, while fast impacts cause net erosion of the atmosphere. In the latter process, drag forces on the incoming projectile heat portions of the atmosphere, and the expansion of the impact-generated plumes results in the loss of portions of the atmosphere to space. In addition, it has recently been shown that large impactors (i.e., lunar-size objects) send a strong shock wave throughout the Earth. The resulting ground motion may accelerate the thermosphere to escape velocity. Escape of gases, including the heavier atoms such as oxygen, nitrogen, and carbon, sets a limit to the duration of any dense primitive atmosphere. Geologic and cratering evidence on the Moon and Mars suggest that by 3.5 billion years ago the active period of intense bombardment was over, followed by episodic events after this period. [See Impacts by Extraterrestrial Bodies.]

**Escape Processes.** Recent advances in the study of planetary atmospheric evolution have focused on *nonthermal escape* mechanisms, driven by photochemistry and other energetic processes, as summarized in Table 1. All these processes create energetic particles by photolytic, electron impact, or ionic reactions. Direct momentum transfer may result in knocking atoms off the exosphere, in a process known as *sputtering*. In the case of the Earth, ions may migrate to the polar regions, where the magnetic field lines are open to the tail of the magnetosphere, and they can readily escape along these field lines. For most escape processes, lighter constituents such as atomic hydrogen have the best chance of gaining enough velocity to escape. The current estimate for the rate of hydrogen escape is 300 million atoms per square centimeter per second, with most of the contribution derived from *charge exchange*, followed by smaller contributions from thermal escape and *polar wind*. A general constraint, known as *Hunten's limiting flux theorem*, relates the maximum escape flux of hydrogen to the abundance of total hydrogen in the mesosphere, determined largely by the chemical sources and transport. The escape rates of hydrogen in the past were probably much higher than the present rate. Our planet apparently started in a more reducing state, followed by an inexorable evolution toward a more oxidizing state owing to the loss of hydrogen by escape to space. Thus, a chemical arrow of time permeates every chemical reaction on this planet. Is this the ultimate driving force of biological evolution? Although too small to compete with photosynthesis as a source of oxygen, the dissociation of water vapor followed by escape of hydrogen might have contributed to the initial rise of oxygen (or oxidants) in the prebiotic atmosphere. Oxygen atmospheres arising from such a mechanism have recently been detected on the Galilean satellites of Jupiter, Europa, and Ganymede.

Evidence based on the fractionation pattern of noble gases and their isotopes in the atmosphere (relative to cosmic abundance) suggests that *hydrodynamic escape*, or rapid blowoff, might have occurred in the early atmosphere. The outflow is driven by the massive escape of a light gas (e.g., H or $H_2$). Heavier gases are carried along by the hydrodynamic drag. The energy that drives this escape process is believed to be derived from the early active sun while it went through a *T-Tauri* phase. Alternatively, shock waves in the upper atmosphere excited during impacts by planetesimals might be a source of the energy.

Escape processes can fractionate gases in the atmosphere, preferentially enriching the heavier species and isotopes (e.g., deuterium over hydrogen, heavier noble gases over the lighter species). By studying *isotopic*

**Atmosphere Structure and Evolution. TABLE 1.** Nonthermal Processes Leading to Escape*

| PROCESS | EXAMPLES | PRODUCT[†] | REMARKS |
|---|---|---|---|
| 1. Charge exchange | $H + H^+ * \rightarrow H^+ + H*$ | N | — |
| | $O + H^+ * \rightarrow O^+ + H*$ | N | — |
| 2. Dissociative recombination | $O_2^+ + e \rightarrow O* + O*$ | N | Energy divided equally |
| | $OH^+ + e \rightarrow O + H*$ | N | H takes nearly all the energy |
| 3. Impact dissociation | $N_2 + e* \rightarrow N* + N*$ | N | e* may be a photoelectron or an accelerated electron |
| Photodissociation | $O_2 + h\nu \rightarrow O* + O*$ | N | — |
| 4. Ion-neutral reaction | $O^+H_2 \rightarrow OH^+ + H*$ | N | — |
| 5. Sputtering or | $O + O^+* \rightarrow O* + O^+ *$ | N | Sputtering requires kilovolt or greater energies |
| Knock-on | $O* + H \rightarrow O* + H*$ | N | Knock-on requires much less |
| 6. Solar-wind pickup | $O + h\nu \rightarrow O^+ + e$ | I | Also electron impact |
| | $O^+$ picked up | I | Also magnetospheric wind for satellites |
| 7. Ion escape | $H + *$ escapes | I | Requires open magnetic-field lines (polar wind) |
| 8. Electric field | $X^+ + eV \rightarrow X^+ *$ | I | Generates fast ions and electrons that participate in other processes |

*Asterisks represent excess kinetic energy. [†]N = neutral; I = ion.

SOURCE: Adapted from Chamberlain and Hunten (1987). With permission of Academic Press.

*fractionations*, we can place powerful quantitative constraints on models of the origin and evolution of the atmosphere.

**Imprint of Life.** The earliest geologic evidence for life on Earth is from 3.85 billion years ago. The emergence of life has exerted a profound effect on atmospheric evolution. About 2 billion years ago, the atmosphere made a fundamental transition from an *anaerobic* state to an *aerobic* state, as shown in Figure 4. Today, the global planetary environment is genial and conducive to life. The surface of the planet is protected from harmful ultraviolet radiation by an ozone layer. The existence of the ozone layer, which makes advanced life on Earth possible, is caused by the abundance of atmospheric oxygen, which is in turn a product of the biosphere. The imprint of life on the global terrestrial environment may best be appreciated by comparing the composition of the present atmosphere with a hypothetical atmosphere on an abiotic Earth, as shown in Figure 5. Note that in the absence of life or carbonate formation, carbon dioxide would be the dominant gas in the atmosphere, with total pressure of about 200 millibars (this is a much lower total atmospheric pressure than the present atmosphere). The higher abundance of carbon dioxide (by more than three orders of magnitude compared to the present) is the result of that gas's slower rate of removal from the atmosphere by geochemical processes in the absence of life. Therefore, the composition of the terrestrial atmosphere would, to first order, resemble its sister planets Mars and Venus. Nitrogen and oxygen would be minor constituents, with abundance orders of magnitudes below those at present. The trace gases such as methane, nitrous oxide, ammonia, hydrogen chloride, and hydrogen sulfide would either vanish from the atmosphere or have their abundances greatly reduced. Only the concentrations of carbon monoxide would greatly increase, owing to the greater source from carbon dioxide photolysis. [*See* Atmospheric Chemistry.]

According to accepted theories of the evolution of main-sequence stars, of which the sun is a typical member, the solar luminosity has been steadily increasing by about 40 percent since the sun formed. The mean surface temperature of a planet such as the Earth is determined by energy balance. If the composition of the atmosphere had remained unchanged, the Earth's mean surface temperature would have been below the freezing point of water before about 2 billion years ago. But the sedimentary record shows that liquid water has always been present on Earth. A plausible resolution of the "faint young sun" paradox is that the early atmosphere contained more greenhouse gases (e.g., carbon dioxide). It has been suggested that the existence and evolution of life on this planet may have had a profound impact on the climate by regulating the amounts of carbon dioxide and other greenhouse molecules in the atmosphere. The ability of the biosphere to maintain a global environment that is optimal for life is known as *homeorrhesis*. That the terrestrial atmosphere has been

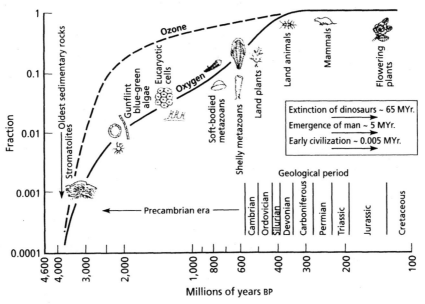

**Atmosphere Structure and Evolution. Figure 4.** Evolution of Oxygen, Ozone, and Life on Earth.

Life could not have become established on land until a substantial ozone layer was formed. The units of oxygen are in fraction of the present atmospheric level (PAL). The abundance of oxygen before photosynthesis is $5 \times 10^{-9}$ PAL. (After Wayne, 1991. With permission of Oxford University Press.)

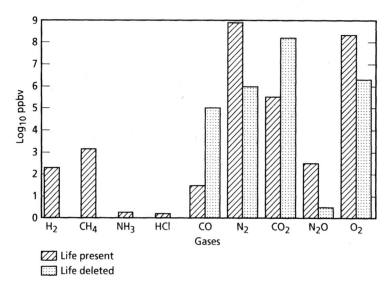

**Atmosphere Structure and Evolution. FIGURE 5.** Partial pressure of reactive gases of the Earth's atmosphere at present (with life) and those calculated to result from reactions were life to be extinguished.

The concentrations are expressed as log base 10 of quantities in parts per billion by volume ($\log_{10}$ ppbv). For example, the concentrations of carbon dioxide in the present atmosphere and in the hypothetical atmosphere with life deleted are 0.0003 and 0.2 bars, respectively. These values correspond to 5.5 and 8.3, respectively, in this figure. (After Lovelock and Margulis, 1989. With permission of Academic Press.)

subject to homeorrhesis by the biosphere throughout its long history is an intriguing hypothesis but difficult to prove.

**Cosmic Perspective.** Recent work has raised the possibility of detecting life on *extrasolar* planets in our galaxy based on the existence of an oxygen atmosphere. The idea is that habitable planets are likely to exist around stars not too different from the Sun if current theories about terrestrial climate evolution are correct. Some of these planets may have evolved advanced life and may have atmospheres rich in molecular oxygen. It is important to keep in mind, however, that the terrestrial atmosphere remained deficient in oxygen until about 2 billion years ago, even though photosynthesis had a much earlier origin (see Figure 4). One should also keep in mind that extraterrestrial life could be vastly different from life on Earth, and molecular oxygen might not be a relevant participant.

It is clear that our planet's environment has evolved from initial conditions that are profoundly different from those today, and it will undoubtedly continue to evolve (with or without human intervention) into physical and chemical conditions that may be drastically different from those at present. Yet life has been sustained and has continued to thrive for at least the last 3.85 billion years. Is there another place in the universe where similar conditions are duplicated, even partially? The ultimate significance of the concept of atmospheric evolution lies not only in connecting us to the origin of the solar system but also in its possible connection to our future on this planet, as well as to other places in the universe where conditions might favor the emergence of life.

[*See also* Atmosphere Dynamics; Carbon Dioxide; Climate Change; Earth History; Earth Motions; Evolution of Life; Greenhouse Effect; Introduction; Methane; Nitrous Oxide; Oxygen Cycle; Ozone; Paleoclimate; Plate Tectonics; Sun; *and* Water Vapor.]

## BIBLIOGRAPHY

Alvarez, L., W. Alvarez, F. Asaro, and H. V. Michel. "Extraterrestrial Cause for the Cretaceous–Tertiary Extinction." *Science* 208 (1980), 1095–1108.

Banks, P. M., and G. Kockarts. *Aeronomy.* New York: Academic Press, 1973.

Chamberlain, J. W., and D. M. Hunten, *Theory of Planetary Atmospheres.* New York: Academic Press, 1987.

Chen, G. Q., and T. J. Ahrens. "Erosion of Terrestrial Planet Atmosphere by Surface Motion after a Large Impact." *Physics of the Earth and Planetary Interiors* 100 (1997), 21–26.

Hall, D. T., D. F. Strobel, P. D. Feldman, M. A. McGrath, and H. A. Weaver. "Detection of an Oxygen Atmosphere on Jupiter's Moon Europa." *Nature* 373 (1995), 677–679.

Holland, H. D. *The Chemical Evolution of the Atmosphere and Oceans.* Princeton, N.J.: Princeton University Press, 1984.

Hsu, K. J. "Is Gaia Endothermic?" *Geological Magazine* 129 (1992), 129–141.

Hunten, D. M. "Atmospheric Evolution of the Terrestrial Planets." *Science* 259 (1993), 915–920.

Hunten, D. M., R. O. Pepin, and J. C. G. Walker. "Mass Fractionation in Hydrodynamic Escape." *Icarus* 69 (1987), 532–549.

Kasting, J. F. "Earth's Early Atmosphere." *Science* 259 (1993), 920–926.

———. "Habitable Zones around Low-Mass Stars and the Search for Extraterrestrial Life." *Origins of Life and Evolution of the Biosphere* 27 (1997), 291–307.

Lovelock, J. E., and L. Margulis. "Atmospheric Homeostasis, by and for the Biosphere: The Gaia Hypothesis." *Tellus* 26 (1974), 1–9.

Melosh, H. J., and A. M. Vickery. "Impact Erosion of the Primordial Atmosphere of Mars." *Nature* 338 (1989), 487–489.

Moses, J. I. "Meteoroid Ablation in Neptune Atmosphere." *Icarus* 99 (1992), 368–383.

Owen, T., and A. Bar Nun. "Comets, Impacts, and Atmospheres." *Icarus* 116 (1995), 215–226.

Salby, M. L. *Fundamentals of Atmospheric Physics.* New York: Academic Press, 1996.

Walker, J. C. G. *Evolution of the Atmosphere.* New York: Macmillan, 1977.

Wayne, R. P. *Chemistry of Atmospheres.* Oxford: Clarendon Press, 1991.

Yung, Y. L., and W. B. DeMore. *Photochemistry of Planetary Atmospheres.* New York: Oxford University Press, 1998.

Yung, Y. L., and M. B. McElroy. "Stability of an Oxygen Atmosphere on Ganymede." *Icarus* 30 (1977), 97–103.

Yung, Y. L., J. S. Wen, J. I. Moses, B. M. Landry, M. Allen, and K. J. Hsu. "Hydrogen and Deuterium Loss from the Terrestrial Atmosphere: A Quantitative Assessment of Nonthermal Escape Fluxes." *Journal of Geophysical Research* 94 (1989), 14971–14989.

—YUK L. YUNG AND JULIANNE I. MOSES

# ATMOSPHERIC CHEMISTRY

There are three essential, interconnected phenomena governing the chemistry of the atmosphere: the presence of sunlight, the presence of life, and the presence of water in three phases (vapor, liquid, and solid). Sunlight is the driver, providing a direct energy source for photosynthesis, providing heat to drive atmospheric circulation and to control the rates of chemical reactions, and providing low-frequency, high-energy ultraviolet radiation capable of breaking chemical bonds, thereby generating highly reactive chemical fragments called *free radicals.* [*See* Atmosphere Structure and Evolution.] Phototropic organisms store energy by converting oxidized carbon (carbon dioxide, $CO_2$) into a reduced form (carbohydrates), whose later chemical oxidation releases the stored energy. The storage and use of this reduced material is necessarily inefficient, and a considerable quantity of reduced material is released into the atmosphere, where it is oxidized by the photochemically produced free radicals. Human activity alters and augments this release—altering it by influencing ecosystems, and augmenting it by releasing carbon from fossil fuels. Water is a key resource for life, but it also profoundly influences atmospheric circulation, controlling the way heat is released and transported in the atmosphere and establishing rates and depths of atmospheric mixing. It serves as a precursor to the dominant atmospheric free radical (hydroxyl,

OH), it serves as a substrate for mixed-phase heterogeneous reactions involving gaseous species and solid particles or liquid drops (collectively known as aerosols), and as rain it eventually cleanses the atmosphere of many of the soluble, oxidized products of the reduced species emitted by living organisms.

The two lowest regions of the atmosphere have distinctly different chemistry. The troposphere extends from the surface to the tropopause at 9–13 kilometers (depending on latitude and season). The stratosphere extends from the tropopause to roughly 30 kilometers. In the troposphere, atmospheric circulation is dominated by solar heating of the surface and subsequent turbulent motions (weather) associated with transportation of heat away from the surface and the tropics. In the troposphere, where temperature decreases rapidly with altitude, mixing is rapid (within one month from the surface to the tropopause, and within two years from pole to pole), and chemistry is dominated by the oxidation of reduced compounds (hydrocarbons, etc.) emitted from the surface. In the stratosphere, where temperature increases with altitude, mixing is much slower, and chemistry is dominated by the absorption of intense ultraviolet radiation at various wavelengths. Light with sufficient energy to break most chemical bonds is absorbed in the stratosphere (see Figure 1), where molecular bonds of oxygen ($O_2$) and ozone ($O_3$) are broken. This simultaneously attenuates the light and warms the stratosphere, causing its stability. Very little high-energy ultraviolet light penetrates into the troposphere or to the surface. The stratosphere thus serves as a shield, protecting living organisms from this potentially damaging radiation.

Atmospheric transport and mixing do not depend on a molecule's weight; light and heavy gases travel through the atmosphere at the same rate. Below 90 kilometers, air is a bulk fluid, carrying gases of any mass as it moves. Heavy gases do not tend to remain near the surface, and light ones do not tend to rise preferentially. Atmospheric mixing occurs when air itself mixes, so the important parameter is the mixing ratio—the ratio of the mass concentration of a particular compound to the mass concentration of air itself (i.e., its density). Mixing ratios are expressed as parts per million (ppm), billion (ppb), or trillion (ppt), as appropriate.

Human activity can influence atmospheric chemistry in three general ways. It can directly influence tropospheric chemistry, both by adding reduced compounds to the atmosphere and by adding compounds that lead to oxidant formation. It can directly influence stratospheric chemistry by altering the photochemical balance associated with ozone. Finally, by changing particulate levels or temperatures, or both, it can indirectly influence the chemical processing rates of atmospheric reactions.

This article will describe the basic chemistry of the

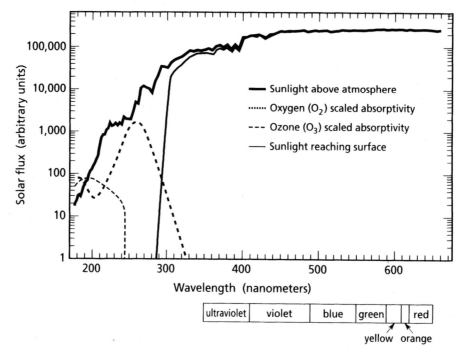

**Atmospheric Chemistry. FIGURE 1.** The Solar Spectrum at the Top of the Atmosphere and at the Surface.

Visible wavelengths are largely unattenuated, while short-wavelength ultraviolet light is absorbed by molecular oxygen and ozone. More energy is required to break the strong double bond in molecular oxygen, so oxygen absorption occurs at shorter wavelengths (higher energy). Most absorption is in the upper atmosphere (stratosphere), but some ozone absorption occurs in the lower atmosphere (troposphere) down to the surface.

stratosphere, the urban atmosphere, and the remote troposphere. It will touch on points where human activity has altered atmospheric chemistry, and it will describe ways in which various influences combine with each other. The stratospheric ozone layer and urban pollution will receive the most attention, both because the effects are dramatic and because they have the longest history of research.

**Oxygen and Hydrogen Chemistry.** Oxygen and water are uniquely important constituents in atmospheric chemistry, in part because they are very common in the atmosphere. [*See* Oxygen Cycle.] Thus, before discussing specific regions of the atmosphere in detail, we shall examine this basic chemistry in general terms.

Atmospheric oxygen exists in four forms: diatomic molecular oxygen ($O_2$), ozone ($O_3$), ground-state atomic oxygen (O), and electronically excited atomic oxygen (O*). Molecular oxygen is the most stable form of oxygen, making up about 20 percent of the atmosphere. Most $O_2$ does not react chemically in the atmosphere; it is a byproduct of life, and it is cycled through living organisms on a time scale of many thousands of years. Ozone has an intermediate reactivity, with a chemical lifetime ranging from a few days to a month or more, and an abundance varying from 0.01 to 10 ppm, depending on location (mostly altitude). [*See* Ozone.] Both O and O* are extremely rare and short lived. Ozone, O,

and O* are often discussed as a chemical family known as odd oxygen ($O_x$).

Photolysis drives atmospheric chemistry (see Figures 1 and 2). It occurs when a molecule absorbs a photon and breaks into two or more fragments. When a chemical process involves absorption of light, we shall use $h\nu$ to denote the photon (Planck's constant, $h$, multiplied by the frequency of the light, $\nu$, gives the energy of a single photon of that frequency). Photolysis of $O_2$ in the upper stratosphere is the ultimate source of stratospheric ozone. Photolysis of nitrogen dioxide ($NO_2$) is the major direct source of tropospheric ozone. Ozone photolysis is the main source of the important free radical hydroxyl (OH) in both the stratosphere and troposphere. Hydroxyl is the main atmospheric oxidant. Hydroxyl and several other radicals containing H atoms comprise a chemical family known as odd hydrogen, or $HO_x$. Hydroxyl radicals are produced when ozone photolysis generates O*, which can in turn react with water vapor to produce two OH radicals. In the stratosphere, O* also reacts with methane ($CH_4$) to produce OH. Other photolysis reactions are also important OH sources.

**Stratospheric Chemistry.** Chemistry in the stratosphere is almost entirely a story of ozone. Ozone is formed photochemically and removed through a series of catalytic reactions. The ozone loss rate has been increased by human activity, mostly by compounds con-

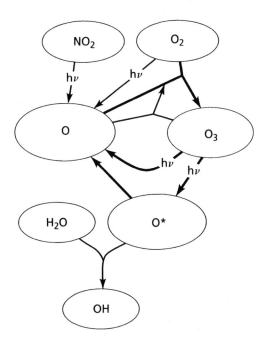

**Atmospheric Chemistry. FIGURE 2.** Basic Oxygen and Hydrogen Chemistry.

Molecular oxygen ($O_2$), nitrogen dioxide ($NO_2$), and ozone ($O_3$) can absorb light and produce O (see Figure 1), although $O_2$ absorption occurs only in the upper stratosphere. Oxygen atoms, O, react rapidly with $O_2$ to form or re-form $O_3$. Some $O_3$ photolysis produces electronically excited O (O*), and some O* reacts with water vapor ($H_2O$) to make hydroxyl radicals (OH). The significance of particular reactions depends greatly on location (stratosphere, remote troposphere, urban atmosphere).

taining chlorine and bromine. Emissions of nitrogen oxides from aircraft also influence ozone chemistry, as may particulate emissions from aircraft and a combination of temperature and humidity changes related to global climate change.

**The Chapman cycle.** Basic ozone chemistry was described by Sidney Chapman (1930). It consists of a series of reactions describing the production, cycling, and loss of odd oxygen. The basic Chapman reactions (also in Figure 2) are as follows:

$$O_2 + h\nu \ (<220 \text{ nm}) \rightarrow O + O \tag{1}$$

$$O + O_2 \rightarrow O_3 \tag{2}$$

$$O_3 + h\nu \ (<305 \text{ nm}) \rightarrow O + O_2 \tag{3}$$

$$O_3 + O \rightarrow O_2 + O_2 \tag{4}$$

$$O_3 + O_3 \rightarrow O_2 + O_2 + O_2. \tag{5}$$

Reaction (1) is the major source of ozone. Reactions (2) and (3) describe the cycling of odd oxygen with no net loss of ozone. Finally, reactions (4) and (5) result in odd-oxygen (and ozone) loss. Ozone is continually produced via reaction (1) and lost via reactions (4) and (5). While the ozone level remains relatively constant in the atmosphere, the flux through the system is enormous; hence it is entirely impractical to consider directly influencing the abundance of ozone by adding it to the atmosphere.

**Ozone catalysts.** While the Chapman cycle describes qualitatively the observed ozone distribution in the stratosphere, it predicts much more ozone than is actually observed. The reason is catalysis of reactions (4) and (5). Catalysts accelerate the rate of a chemical reaction without being consumed. Thus very small concentrations of ozone catalysts can greatly influence ozone levels.

One catalyst is atomic chlorine. Chlorine was first identified as an ozone catalyst when the potential effect of the Space Shuttle's solid rocket boosters was being considered in the early 1970s. Subsequently, Mario Molina and F. Sherwood Rowland (1974) recognized that the common chlorofluorocarbon (CFC) propellants also delivered chlorine to the stratosphere. The chlorine cycle proceeds as follows:

$$Cl + O_3 \rightarrow ClO + O_2 \tag{6}$$

$$ClO + O \rightarrow Cl + O_2 \tag{7}$$

(Net: $O_3 + O \rightarrow O_2 + O_2$; $\Delta ClO_x = 0$).

Oxides of nitrogen are also ozone catalysts; this process was first fully described by Paul Crutzen (1970):

$$NO + O_3 \rightarrow NO_2 + O_2 \tag{8}$$

$$NO_2 + O \rightarrow NO + O_2 \tag{9}$$

(Net: $O_3 + O \rightarrow O_2 + O_2$; $\Delta NO_x = 0$).

These two cycles catalyze reaction (4). A third catalytic cycle of great importance involves odd hydrogen:

$$OH + O_3 \rightarrow HO_2 + O_2 \tag{10}$$

$$HO_2 + O_3 \rightarrow OH + O_2 + O_2 \tag{11}$$

(Net: $O_3 + O \rightarrow O_2 + O_2 + O_2$; $\Delta HO_x = 0$).

This cycle is a catalyst for reaction (5). Because ozone is more common in the lower stratosphere, and atomic oxygen less so, this cycle is very important there. Other reaction sets, including several involving chlorine and bromine compounds, can also catalyze reaction (5). In 1995, Molina, Rowland, and Crutzen shared the Nobel Prize in Chemistry for their work on stratospheric ozone.

Before approximately 1980, natural sources of $NO_x$ and $HO_x$ dominated the budgets of the ozone catalysts in the stratosphere, accelerating the Chapman loss reactions by a factor of five to six. Since then, anthropogenic perturbations have reduced stratospheric ozone by 10–100 percent, depending on time and location.

**Anthropogenic perturbations.** There are two direct human influences on ozone chemistry: the addition of chlorine (and bromine), primarily through halogenated hydrocarbons (CFCs, hydrofluorocarbons [HCFCs], halons, etc.), and the addition of $NO_x$, primarily as an effluent of high-altitude aircraft. In addition, many of the chemical reactions depend strongly on temperature, so that global climate change can influence ozone chemistry profoundly.

To proceed further, we must explore the life cycles of the various ozone catalysts. Each follows a similar pattern. Some long-lived precursor molecule mixes upward from the lower atmosphere into the stratosphere, where it is finally broken apart by the increasingly energetic ultraviolet radiation. At that point, most of the catalyst remains in some relatively inert form, known as a reservoir, but some fraction exists as the reactive form described by the catalytic cycles above. The reservoir compounds are often soluble in water, so that as they mix downward into the troposphere they are eventually removed by rainfall.

The chlorine cycle (Figure 3) provides an excellent example. Ultraviolet light dissociates a chlorine-containing molecule (e.g., CFC-12) in the stratosphere, producing Cl and initiating the catalytic cycle described above. There are two major chlorine reservoirs: Cl can react with methane to produce HCl, or ClO can react with $NO_2$ to produce chlorine nitrate ($ClONO_2$). HCl is converted back into Cl by reacting with OH. Chlorine nitrate is converted back to ClO by photolysis. Typically, of roughly 2 parts per billion of available chlorine, only about 10 parts per trillion are in an active form. Bromine is also a potent ozone catalyst; atom for atom it is more than one hundred times more destructive than chlorine because the reservoir compounds are unstable and a much higher fraction of total bromine remains active. For the opposite reason, fluorine is not an important

catalyst; HF molecules are not converted back to F atoms in the stratosphere.

Note that the various catalytic cycles interact. For example, both the odd-nitrogen and odd-hydrogen cycles influence the chlorine cycle, with nitrogen radicals tending to deactivate chlorine and hydrogen radicals tending to activate it. In a similar manner, both chlorine and hydrogen radicals tend to deactivate odd-nitrogen radicals. The catalytic cycles are in competition and, when one dominates, the other catalysts tend to suppress the dominant catalytic cycle. Thus the net effect on ozone of altering the abundance of a catalyst (in a given part of the stratosphere) depends critically on which cycle is dominant. This is illustrated in Figure 4. An important aspect of this diagram is that it has been directly verified by observations; by exploiting the natural variability of the stratosphere, it is possible to observe the coupled variation of the ozone catalysts and their precursors. A series of missions using the U.S. National Aeronautics and Space Administration (NASA) ER-2 aircraft have provided direct confirmation of the basic mechanisms described here (Fahey et al., 1990; Anderson et al., 1991).

**Heterogeneous chemistry.** We have so far neglected reactions involving aerosol particles in the stratosphere. [*See* Aerosols.] Reactions involving gases and particles, because they involve two or more phases of matter, are known as *heterogeneous* reactions. These heterogeneous reactions tend to enhance the conversion of reservoir species into active catalysts, and thus generally accelerate ozone removal.

**Polar ozone.** In 1985, Joe Farman described an unprecedented loss of ozone over Antarctica (Farman, 1985). For more than a decade, an increasingly pronounced drop in total ozone had occurred during the

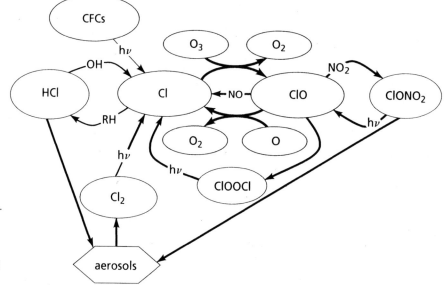

**Atmospheric Chemistry. Figure 3.**
The Stratospheric Chlorine Cycle.

Chlorine radicals (Cl and ClO, or $ClO_x$) catalytically destroy ozone. Most inorganic chlorine is stored in the reservoirs HCl and $ClONO_2$. Heterogeneous reactions involving liquid or solid particles (aerosols in the diagram) can convert $ClONO_2$ and in some cases HCl back into $ClO_x$.

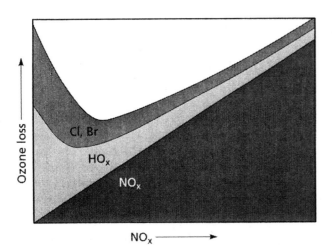

**Atmospheric Chemistry. FIGURE 4.** Total Ozone Removal as a Function of Total Odd Nitrogen for Fixed Odd Hydrogen and Halogens.

The portion directly attributable to $NO_x$ is shaded dark gray. This increases linearly as $NO_x$ increases. The portions due to the halogen (Cl, Br, medium gray) and odd-hydrogen (light gray) cycles are shown by the two curves. These are strongly influenced by total $NO_x$. At low $NO_x$, more $HO_x$ and halogens are activated, so total ozone loss decreases as $NO_x$ decreases. At high $NO_x$, $HO_x$ and halogens are largely sequestered, so total ozone loss increases as $NO_x$ decreases. The effect of changing $NO_x$ thus depends dramatically on actual stratospheric conditions (time, location, etc.). (After Wennberg et al., 1994.)

polar spring (October), just after the Sun emerged from the long polar night. This "ozone hole" has continued to grow, so that by now nearly all of the stratospheric ozone is removed above the entire antarctic continent for one month each year. A similar, although currently less dramatic, phenomenon occurs over the Arctic each spring.

Despite the extensive research conducted on ozone chemistry in the decade preceding this discovery, the polar ozone loss was a complete surprise, and the mechanism was not anticipated in any theoretical model or laboratory experiment. This is a sobering lesson. The cause is a reaction of the two chlorine reservoirs (HCl and $ClONO_2$) on ice or nitric acid ice and liquid particles during the polar winter to produce nitric acid and chlorine gas ($Cl_2$). The $Cl_2$ photodissociates at polar sunrise in the spring. This simultaneously activates all of the 2 parts per billion of available chlorine and sequesters the odd-nitrogen radicals. A similar process combining chlorine and bromine radicals also contributes significantly. Together, these cycles almost completely remove ozone in roughly one week.

This process has been unequivocally proven by observations in the antarctic stratosphere, which show rapid ozone loss almost perfectly correlated with high ClO and BrO concentrations, at a rate consistent with

laboratory measurements of the rate-limiting reactions. Similar conditions in the arctic stratosphere have produced significant ozone losses, but not the near-total removal seen in the Antarctic. The reason is the shorter duration and warmer temperatures of the arctic polar vortex. However, the rates of the heterogeneous reactions responsible for this phenomenon are extremely sensitive both to temperature and to the concentration of water vapor. Thus, future changes in stratospheric temperatures and water vapor concentrations could dramatically change the severity of vernal arctic ozone loss.

*Midlatitude ozone.* Most of the stratosphere (and most human habitation) is not near the poles. Heterogeneous reactions on sulfate aerosols throughout the stratosphere can convert reservoir forms of most of the major catalysts (nitrogen, chlorine, and bromine) into active forms. These reactions are thought to be in large part responsible for the gradual observed ozone loss in the midlatitude stratosphere. Over the past fifteen years, there has been an approximately 15 percent decline in total ozone. The exact cause of this decline is not fully understood.

*Global change issues.* Considerable anthropogenic change has already occurred in the stratosphere: the ozone hole and the midlatitude ozone loss result from halogen-containing compounds of human origin. Emission of these compounds has been sharply curtailed by international agreement, and as a result their levels in the troposphere are now declining. [*See* Montreal Protocol.] Emissions from super- and subsonic aircraft flying in the stratosphere have also been a cause of concern for several decades. Increases in $NO_x$ and increases in sulfate (and thus aerosols) are both under investigation. The $NO_x$ emissions may reduce the efficiency of $HO_x$ and $ClO_x$ ozone depletion cycles, while the particulates may enhance conversion of all catalysts into an active form. An example of changing aerosol loading is provided naturally every decade or so by major volcanic eruptions, which inject aerosols into the stratosphere and produce readily observable decreases in ozone.

**Urban and Regional Pollution.** Pollution in urban areas is also in large measure a story of ozone, particles, and formation of acids. [*See* Pollution.] However, the similarity with the stratosphere ends there. In the urban atmosphere, ozone formation is the issue. Ozone, fine particulates, and acid deposition affect human health directly and also influence agricultural productivity and the health of regional ecosystems. Ozone is a strong oxidant and can damage unprotected tissues in plants and animals. Chemistry driven by sunlight following the emission of hydrocarbons and $NO_x$ produces ozone. During warm summer episodes, urban ozone levels can reach more than ten times the natural background, causing acute respiratory distress. The same chemistry leads to particle formation, causing a haze, or smog, which reduces visibility and may also have direct health effects. Finally, these particles are generally acidic, and their deposition can damage buildings and alter the pH of poorly buffered soils and lakes. Urban and regional pollution is affecting an increasing fraction of the global population and will continue to be an issue of paramount importance as industrialization and urbanization accelerate in the next century. [*See* Air Quality.]

***Urban ozone formation.*** In the troposphere, ozone is formed when hydrocarbons and other compounds are oxidized in the presence of nitrogen oxides. Reduced compounds (mostly hydrocarbons, collectively known as volatile organic compounds, VOCs) are emitted into the atmosphere as a result of incomplete combustion, solvent use, and as a natural effluent from plants. During high-temperature combustion, some of the oxygen and nitrogen molecules in air and in the fuel are broken down and re-formed into $NO_x$, and again some fraction of this $NO_x$ is released into the atmosphere.

The chemistry is driven by photolysis. Photolysis of ozone and aldehydes (partially oxidized hydrocarbons containing a carbon–oxygen double bond, such as formaldehyde, $H_2CO$) produces $HO_x$. The direct reaction of ozone with unsaturated hydrocarbons (those containing carbon–carbon double bonds) can also produce $HO_x$. The hydrocarbon oxidation sequence (Figure 5) then consumes hydrocarbons while producing carbon dioxide, ozone, and more $HO_x$. $NO_x$ catalyzes the process. The crucial step involves the reaction of peroxy radicals ($RO_2$, where R represents an organic fragment) with nitric oxide (NO) to produce RO and $NO_2$. This is followed by $NO_2$ photolysis to produce ozone.

Two significant factors complicate the picture. When $NO_x$ is overly abundant, the reaction of $NO_2$ with OH to produce nitric acid ($HONO_2$) can starve the system of $HO_x$. When, on the other hand, hydrocarbons are overly abundant, the peroxy radicals will tend to react with each other rather than with NO. This bypasses the ozone production step of the main oxidation sequence, oxidizing the hydrocarbons without producing ozone. The first condition is referred to as hydrocarbon limited, because hydrocarbon concentrations are too low for OH radicals to react with them before reacting with $NO_2$. The second condition is known as $NO_x$ limited, because the low NO concentration permits the $RO_2$ self-reactions to occur.

The maximum ozone production occurs when the ratio of hydrocarbons (measured by the total carbon) to the nitrogen oxides is approximately 8:1. The exact value depends on the actual atmospheric composition. When this ratio is smaller than 8:1, reductions in $NO_x$ will increase the ratio, increasing ozone production. When the ratio is larger than 8:1, reductions in hydrocarbons will tend to decrease the ratio, again increasing ozone production. The expected peak ozone as a function of hydrocarbons (VOC) and nitrogen oxides ($NO_x$) is often plotted on an isopleth diagram (Figure 6). Control strate-

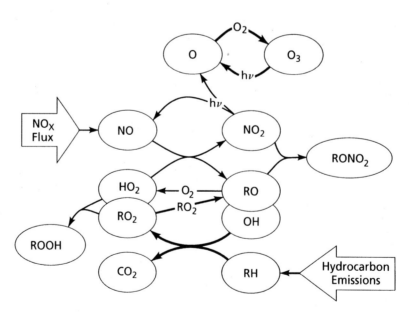

**Atmospheric Chemistry. FIGURE 5.** Ozone Production during Hydrocarbon Oxidation.

The odd-nitrogen radicals, NO and $NO_2$, serve as catalysts for the oxidation of hydrocarbons to carbon dioxide, producing ozone as a byproduct. Organic fragments are indicated by the symbol R (the parent hydrocarbon is RH, the peroxy radicals are $RO_2$). Excessive hydrocarbons or odd nitrogen will drive the cycle out of balance and produce partially oxidized hydrocarbons or nitrates such as nitric acid instead of ozone.

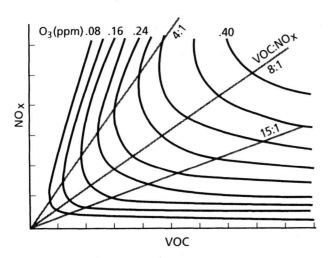

**Atmospheric Chemistry. Figure 6.** An Ozone Isopleth Diagram, Showing Expected Peak Ozone (Isopleths) as a Function of Hydrocarbon (VOC) and Nitrogen Oxide (NO$_x$) Concentrations.

The consequences of reducing VOCs or NO$_x$ depend strongly on the starting point on this surface. (Adapted from National Research Council, 1991.)

gies depend critically on the relative abundance of NO$_x$ and hydrocarbons. This is complicated by the rich array of hydrocarbons typically encountered in urban air and the difficulty of measuring some of them, all of which makes estimating actual hydrocarbon levels difficult. Furthermore, local considerations may differ from considerations downwind; photochemistry in hydrocarbon-limited conditions tends to remove NO$_x$ more rapidly than the hydrocarbons, so that, while hydrocarbon reductions under these conditions may reduce local ozone production, they may actually lead to an increase downwind, once the air mass becomes NO$_x$ limited.

*Particulates.* There are two categories of particulate: primary particulates, such as dust, abrasion fragments, and so on, that are directly injected into the atmosphere, and secondary particulates, such as sulfate and organic aerosols. In areas where sulfur-containing fuels are used with no scrubbing of smokestack gases, sulfate aerosols can be a major issue. These aerosols impinge on visibility, aggravate respiratory problems, and acidify soils and watersheds through dry and wet deposition. Organic aerosols are produced when large hydrocarbons are oxidized. All chemically generated aerosols tend to be much smaller than primary particulates; it is thought that this results in deeper penetration into the lung and thus more severe health consequences, although these findings remain controversial. The U.S. Environmental Protection Agency has recently promulgated regulations requiring control of particles less than 2.5 micrometers in size.

*Acid deposition.* Oxidation processes in the atmosphere often produce acidic products. [*See* Acid Rain and Acid Deposition.] In some cases, especially with sulfur compounds and to an extent nitrogen compounds, sufficient acidity can be produced to alter significantly the acidity of poorly buffered soils and lakes. As with sulfate aerosols, this problem has been effectively controlled in areas where smokestack gases are scrubbed of their acid precursors. Elsewhere, acid deposition remains a serious problem.

*Global change issues.* Increasing urbanization and automobile use throughout the world mean that urban ozone and the associated particulate and acidity issues will be an increasingly conspicuous problem in the next century. Because of the interactions described here, different locations will require different control strategies, but some measure of control will be essential to prevent exposure to dangerously poor air quality for most of the world's population. While emission controls in developed nations, such as the United States, have not greatly reduced urban ozone, ozone pollution has not become markedly worse in spite of continued urbanization and increased fuel use.

**Global Tropospheric Chemistry.** Global tropospheric chemistry is essentially a simplified subset of urban chemistry, occurring under conditions of much lower NO$_x$. The oxidation reactions of methane and CO are for the most part sufficient to describe the system. The evidence we have suggests that the chemical state of the troposphere has changed dramatically over the last millennium (with the bulk of the change occurring in the last one hundred years). Gas trapped in glacial ice at the poles and on mountains reveals a continuous increase in methane, while early assays of atmospheric ozone reveal mean ozone concentrations of less than 30 parts per billion throughout continental Europe. NO$_x$ concentrations are also probably many times higher today than they were a century and a millennium ago.

Several issues relate directly to global change. The increasing methane and NO$_x$ levels suggest that the oxidative capacity of the troposphere (for example, the OH concentration) has changed over the past millennium. These changes continue at an accelerated rate today. However, since the increases in methane and other reduced gases have been accompanied by increases in NO$_x$, it is not yet clear what the overall effect has been. At least one-quarter of the change in radiative forcing (the greenhouse effect) caused by human activities is associated with compounds such as methane that are oxidized in the troposphere. The concentration of these compounds thus depends on the oxidative capacity of the troposphere. Furthermore, their oxidation rate depends on temperature; in this case the feedback associated with this coupling is probably negative (increasing temperatures will increase the oxidation rate of these gases, which will tend to decrease the greenhouse forcing).

Another issue is associated with a naturally occurring sulfur compound, dimethyl sulfide. This gas is emitted from the ocean surface and may be responsible for a large fraction of the aerosol particles in the remote atmosphere. These particles serve as the nuclei of cloud droplets (cloud condensation nuclei, or CCN), and they also directly influence the light-scattering properties of the atmosphere. The number of CCN can influence the properties of clouds, which can influence the light-scattering properties of the clouds. If there are feedbacks between global temperatures and dimethyl sulfide production or oxidation rates, there could be a feedback between the natural sulfur cycle and global climate change. This remains an area of active research.

[*See also* Atmosphere Dynamics; Biogeochemical Cycles; Biomes; Biosphere; Carbon Cycle; Chlorofluorocarbons; Climate Change; Clouds, *article on* Clouds and Atmospheric Chemistry; Deforestation; Greenhouse Effect; Introduction; Land Surface Processes; Methane; Nitrogen Cycle; Nitrous Oxide; Ocean Chemistry; Ocean Dynamics; Phosphorus Cycle; Plate Tectonics; Sulfur Cycle; Volcanoes; Water Vapor; *and the biography of Revelle.*]

### INTERNET RESOURCES

Climate monitoring and diagnostics laboratory of the U.S. National Oceanic and Atmospheric Administration. http://www.cmdl.noaa.gov/. Online graphs of halocarbon trends.

Goddard Space Flight Center. http://www.gsfc.nasa.gov/. Global ozone images and other information.

Harvard University atmospheric chemistry. http://www-as.harvard.edu/chemistry/.

Max-Planck Institute for Chemistry, Germany. http://www.mpch-mainz.mpg.de/.

University of North Carolina. http://airsite.unc.edu/. Many links to other atmospheric chemistry sites.

### BIBLIOGRAPHY

Albritton, D. L., et al., eds. *Scientific Assessment of Ozone Depletion: 1994.* Geneva: World Meteorological Organization, 1994. An authoritative review of stratospheric chemistry. Prior issues (1985, 1989, 1991) are also useful.

Anderson, J. G., et al. "Free Radicals within the Antarctic Vortex: The Role of CFCs in Antarctic Ozone Loss." *Science* 251 (1991), 1.

Barker, J. R., ed. *Progress and Problems in Atmospheric Chemistry.* Singapore: World Scientific, 1995. A collection presenting current research issues.

Chapman, S. "A Theory of Upper-Atmospheric Ozone." *Memoirs of the Royal Meteorological Society* 3 (1930), 103.

Charlson, R. J., and J. Heintzenberg, eds. *Aerosol Forcing of Climate.* Chichester, U.K.: Wiley, 1995. A collection describing many aspects of aerosol chemistry and physics.

Crutzen, P. "The Influence of Nitrogen Oxides on the Atmospheric Ozone Content." *Quarterly Journal of the Royal Meteorological Society* 96 (1970), 320.

Fahey, D. W., et al. "Observations of Denitrification and Dehydration in the Winter Polar Stratospheres." *Nature* 344 (1990), 321.

Farman, J. C. "Large Losses of Total Ozone in Antarctica Reveal Seasonal $ClO_x/NO_x$ Interaction." *Nature* 315 (1985), 207.

Finlayson-Pitts, B. J., and J. N. J. Pitts. *Atmospheric Chemistry: Fundamentals and Experimental Techniques.* New York: Wiley, 1986. A very extensive treatment, especially of urban and regional chemistry.

Goody, R. *Principles of Atmospheric Physics and Chemistry.* New York and Oxford: Oxford University Press, 1995. An overview of global issues in the stratosphere and troposphere.

Molina, M., and F. S. Rowland. "Stratospheric Sink for Chlorofluoromethanes—Chlorine Atom Catalyzed Destruction of Ozone." *Nature* 249 (1974), 810.

National Research Council. *Rethinking the Ozone Problem in Urban and Regional Air Pollution.* Washington, D.C.: National Academy Press, 1991. A comprehensive discussion of the urban ozone issue, including issues of $NO_x$ and VOC control.

Rowland, F. S., and I. S. A. Isaksen, eds. *The Changing Atmosphere.* Chichester, U.K.: Wiley, 1988. Good information on changing concentrations, both recent and longer term.

Seinfeld, J. H., and S. N. Pandis. *Atmospheric Chemistry and Physics: From Air Pollution to Climate Change.* New York: Wiley, 1998. A recent treatment.

Singh, E. B., ed. *Composition, Chemistry, and Climate of the Atmosphere.* New York: Van Nostrand Reinhold, 1995. A collection presenting current research issues.

Walker, J. C. G. *Evolution of the Atmosphere.* New York: Macmillan, 1977. Old, but excellent description of long-term atmospheric evolution.

Wayne, R. P. *Chemistry of Atmospheres.* New York and Oxford: Oxford University Press, 1991. A good overview.

Wennberg, P. O., et al. "Removal of Stratospheric $O_3$ by Radicals: *In Situ* Measurements of OH, $HO_2$, NO, $NO_2$, ClO, and BrO." *Science* 266 (1994), 398.

—Neil M. Donahue

## ATOLLS. *See* Reefs.

## AUSTRALIA

[*This case study discusses the impact of European colonization on the flora and fauna of Australia.*]

The European colonization of Australia is a powerful example of how a human culture can trigger massive environmental change. Since the European discovery and colonization of the continent in the late eighteenth century, there has been a human population explosion. Originally, there were 0.3–1.5 million indigenous hunter-gatherers who inhabited the whole of the continental land mass (760 million hectares); of the current population of over eighteen million people, more than 85 percent live in urban areas comprising less than 1 percent of the total area. European colonization has affected the whole of Australia, with ecological impacts ranging from near complete destruction of ecosystems to support intensive agriculture on less than 10 percent of the continent to more subtle ecological changes such as altered wildfire regimes and introduction of exotic organisms over the remainder. European impacts appear to have been magnified by peculiarities of the Australian environment, such as vast ancient landscapes, extreme rainfall variability, and the geographic isolation of the flora and fauna.

The recent report *Australia: State of the Environment 1996* demonstrates that Australians are beginning to take stock of the environmental impacts of the transition from Aboriginal to European custodianship that commenced in 1788. The great rate of these ongoing changes and imperfect knowledge of the precolonization environment prohibit precise conclusions; however, clear negative trends are apparent.

Europeans have substantially altered the Australia biota by causing both extinctions of native species and the naturalization of numerous exotic animal and plant species (Table 1). Exotic species occur in all Australian ecosystems, although some areas, such as offshore islands, remain relatively free of them. [*See* Exotic Species.] Wild populations of vermin species such as black rats (*Rattus rattus*), brown rats (*Rattus norvegicus*) and house mice (*Mus musculus*) were established by animals escaping from points of introduction such as ships and settlements. Domesticated animals such as cats (*Felis catus*), pigs (*Sus scrofa*), buffalo (*Bubalus bubalis*), camels (*Camelus dromadarius*), horses (*Equus caballus*), donkeys (*Equus asinus*), and goats (*Capra hircus*) have become feral over large parts of Australia. During the nineteenth-century acclimatization, societies actively promoted the establishment of a variety of Northern Hemisphere species to give a European character to Australia's environment and to provide game for recreational hunting. Some of these species have become serious pests, most notably the fox (*Vulpes vulpes*) and rabbit (*Oryctolagus cuniculus*).

Some Australian animals have also been introduced intentionally to areas outside their natural range, such as the laughing kookaburra (*Dacelo novaeguineae*), a medium-sized predatory bird native to eastern Australia, which was released in Western Australia and Tasmania early this century. The only naturalized amphibian, the cane toad (*Bufo marinus*), was introduced in the middle of this century in the mistaken belief that it would control an invertebrate pest of sugar cane. It is still aggressively expanding its range in the Australian tropics. Invertebrates were also introduced, some intentionally, such as honeybees, and others unintentionally, such as ants, wasps, cockroaches, and fruit flies. About 15 percent of the total Australian flora is made up of exotic species, and over two hundred have become serious weeds in agricultural or natural landscapes. Although many weeds were accidentally introduced to Australia, some escaped from gardens—such as the shrub *Mimosa pigra*, which can overwhelm tropical wetlands. Some weeds were spread in the hope of improving pasture production. Only a few proved to be useful, and some of these, like the grass *Cenchrus cilliaris*, have invaded native vegetation. The spread of introduced organisms has been facilitated by transportation systems (namely, roads, tracks, and railway lines) and habitat disturbances such as mineral exploration. For example, the exotic root-rot fungus *Phytophthora cinnamomi* is being effectively spread throughout southern Australia by vehicles, machinery, and boots. A great diversity of Australian species have no tolerance of this pathogen, so it is likely that it will radically transform the species composition of many vegetation types in wetter parts of the continent. [*See also* Exotic Species.]

Of all classes of vertebrates, mammals have suffered the greatest species loss, and this is particularly the case for small marsupials and rodents adapted to the arid zone (Table 1). The extinction of these arid-zone mammals appears to be linked to their inability, unlike that

**Australia. TABLE 1.** Summary Statistics of the Flora and Fauna of the Australian Mainland and Offshore Islands

| GROUP | NUMBER OF NATIVE SPECIES | ENDEMIC SPECIES (PERCENT) | NUMBER OF NATIVE SPECIES EXTINCT SINCE 1788 | NUMBER OF NATIVE SPECIES IN JEOPARDY | NUMBER OF EXOTIC SPECIES NATURALIZED |
|---|---|---|---|---|---|
| Flowering plants | 20,000 | 85 | 76 | 1,009 | 1,500–2,000 |
| Freshwater fishes | 195 | 90 | 0 | 17 | 21 |
| Amphibians | 203 | 93 | 3 | 29 | 1 |
| Reptiles | 770 | 89 | 0 | 51 | 2 |
| Birds | 777 | 45 | 20 | 50 | 32 |
| Mammals | 268 | 84 | 19 | 43 | 25 |

SOURCE: Adapted from Alexander (1996).

of reptiles, to coexist with stock animals and other introduced herbivores and predators, especially during droughts. Cats and foxes have been devastatingly effective predators of native fauna, as evidenced by the survival of otherwise extinct native animals on offshore islands that lack these carnivores. Exotic fish appear to be responsible for the decline and range contractions of over twenty native fish species. The ecological effects of introduced invertebrates are poorly understood, although there is evidence that they have caused the decline of some native invertebrates. Hunting caused the extinction of a number of species, such as the Tasmanian marsupial wolf (*Thylacinus cynocephalus*). Unregulated hunting caused the population collapse of some vertebrate species such as crocodiles (such as the saltwater crocodile *Crocodylus porosus*) in northern Australia and seals (for example, the Australian sea lion *Neophoca cinerea* in southern Australia), although the removal of hunting pressure has allowed populations of most species to recover. The destruction of habitats is now recognized as the most important threat to the survival of a number of plant and animal species. For example, at least thirty-five reptile species, thirty-two bird species, thirteen marsupial species, and three rodent species are in jeopardy because of habitat destruction. Nonetheless, some native species have benefited from agricultural activities. For instance, the galah (*Cacatua roseicapilla*), a native granivorous bird, has been able to expand its range from the arid zone into agricultural landscapes.

It appears that forestry operations threaten the survival of native vertebrates less than land clearance. Logging is thought to threaten the survival of three bird species and two marsupial species. Negative effects of logging arise principally from the destruction of mature trees that serve as animal habitats, the provision of pathways for the spread of exotic organisms, and the alteration of the mix of tree species. Nearly all commercially valuable forest has been logged. About 1 million hectares of native vegetation has been replaced with plantations of exotic conifers (*Pinus* species) or *Eucalyptus* species that have typically been moved outside their natural geographic ranges. [*See* Deforestation.]

Habitat clearance has occurred in an arc from the humid tropics in northeastern Queensland down the eastern seaboard and across to South Australia, with a disjunct area in the southwest of Western Australia. This pattern corresponds to relatively moist climatic zones that form the hinterland of all major Australian cities (Figure 1). National parks, forestry areas, and water catchments provide the largest areas of uncleared vegetation within this agricultural zone. In addition to serving Australia's domestic market, exports from this belt of intensive agriculture provide fifty million people with their annual grain requirements and thirty million peo-

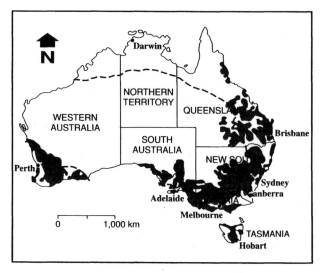

**Australia. Figure 1.** Boundaries of the Australian States and Territories and Their Capitals.

The arid zone that normally receives less than 500 millimeters of rainfall per annum is the area within the dashed line. Shading indicates the areas of land clearance. (Adapted from Alexander, 1996.)

ple with their annual meat requirements, as well as producing large quantities of wool and cotton. European colonization has seen the proportion of the continent covered by forest and woodlands shrink from 30 percent to about 20 percent, with less than one-quarter of the 1788 coverage (8 million hectares) of rainforest remaining. Although habitat clearance peaked after World War II, land is still being cleared, particularly in central Queensland. The rate of clearing across Australia during the last decade was about 500,000 hectares per annum and, given current trends, broad-scale clearing is likely to continue, particularly in the monsoon tropics. At the local scale, habitat clearance has been highly focused, with habitats of little economic potential often being spared from destruction while more productive habitats have been cleared systematically. For instance, less than 0.5 percent of the original coverage (2 million hectares) of native temperate grasslands remain in southeastern Australia, often in tiny fragments such as roadside verges and undeveloped blocks of public lands such as cemeteries and racecourses. It should be noted that gross estimates of habitat destruction are problematic, given disagreements concerning definitions of even broad vegetation types such as forest or woodland, let alone more specific vegetation categories such as rainforest or grassland. Furthermore, there is only a limited appreciation of the extent of the original vegetation in heavily cleared landscapes. Rudimentary data on the distribution of plant and animal species in habitats destroyed by clearing also constrain definitive statements concerning the extinction of species or contraction of their geographic ranges.

Many habitat fragments in cleared landscapes are too small to support viable populations of vertebrates; hence, even in regions where habitat destruction has abated, local extinctions of fauna will not cease immediately. Similarly, there is a widespread and ongoing population crash of scattered eucalyptus trees that escaped destruction during the initial phase of land clearance. The cause of this rural dieback appears to be related to multiple ecological changes associated with intensive agricultural land use, including increased soil nutrients, inhibition of tree regeneration by livestock, and population eruptions of leaf-eating insects. Fragmentation of habitats also affects highly mobile species, such as waterfowl, that require an archipelago of habitats. For example, the drainage of over two-thirds of the wetlands in southern Australia affected the magpie goose (*Anseranas semipalmata*) so badly that it is now restricted to northern Australia, where wetlands are intact. Habitat destruction has also had serious effects on catchments as a result of accelerated soil erosion and dryland salinization that contaminates surface soils, ground water, and streams, hence killing a broad cross section of the native flora and fauna. The latter process is caused by the replacement of deeply rooted woody vegetation with shallow-rooted grasses, which causes the water table to rise by up to 10 meters, thereby mobilizing salt that has accumulated in the soil through geologic time. It has been estimated that about 9 percent of the cleared land in the wheat belt of southwestern Western Australia (14.6 million hectares) is affected by dryland salinization. It is extremely difficult to rehabilitate land damaged by salt. [*See* Salinization.]

Most river systems in the agricultural zone have been modified to control floods and store water for irrigation and hydroelectricity. For example, over 80 percent of the water yield from the Murray–Darling system, Australia's largest catchment (14 percent of the continent), has been diverted for human use. The opposite is true for the catchments in the monsoon tropics that yield over 45 percent of the continent's mean annual runoff. Flood mitigation has adversely affected the ecology of riverine and swamp systems that require periodic flood events for plant species to regenerate and animal species to breed. The aquatic environment of the rivers has been degraded by the combined effects of weed infestation, salinization, sedimentation, pollution with herbicides and fertilizers, modification of stream channels, regulations of flows by dams, and the dominance of introduced fish species such as European carp (*Cyprinus carpio*). In the aggregate, these changes have caused the geographic range contraction of native frog, fish, and invertebrate species, and have facilitated the spread and abundance of introduced fish and weeds. A

symptom of the seriously damaged ecology of river systems such as the Murray–Darling are recurrent blooms of blue-green algae (for example, *Anabaena circinalis*) that cause fish kills and make the water poisonous to livestock, wildlife, and humans. In contrast, only a few river systems in Australia have been seriously polluted by mine tailings.

About three-quarters of the continent is made up of native vegetation suitable for low-density sheep and cattle grazing. Currently, about 60 percent of these rangelands are used for pastoralism. Although pastoralism has not destroyed native vegetation, this land use has a negative effect on biodiversity. Cattle and sheep compete directly for food resources with native herbivores and cause substantial habitat changes such as complete removal of ground cover during drought periods, destruction of riverine vegetation, acceleration of water and wind erosion, fouling of springs and soaks, and trampling of soil. The establishment of bores and water troughs for cattle and sheep has caused increased abundances of feral animals and some large native herbivores such as the kangaroos (*Macropus* species) and the emu (*Dromaius novaehollandiae*), a large flightless bird. The introduction of *Bos indicus*, a cattle species better adapted to the tropics than European cattle (*Bos taurus*), since the mid-1970s has increased the negative effects of grazing in tropical savannas that cover about one-third of the continent. Overgrazing throughout the tropical savanna zone is probably responsible for the range contraction of granivorous birds such as the Gouldian finch (*Erythura gouldiae*). Although precise data are lacking, there is increasing evidence that there has been a marked increase in trees and shrubs on rangelands throughout Australia. This increase in woody species appears to be related to severe overgrazing and also to the associated cessation of Aboriginal landscape burning.

It is clear that Aborigines used fire skillfully to maintain open habitats that maintained high densities of game species. The extirpation of Aborigines from their tribal lands has caused profound changes in fire regimes, with consequent shifts in the distribution of vegetation types. In some environments, the cessation of Aboriginal burning has allowed woody species to increase, while in other environments fires have become more intense and destructive, particularly where fire suppression has caused the buildup of heavy fuel loads. Most of the Australian biota is adapted to recurrent fires, and so the maintenance of biodiversity requires skillful management. However, habitat fragmentation, weeds, and the need to protect economically valuable infrastructure such as buildings and fences frustrates a return to Aboriginal landscape burning in most environments.

European colonists found the Australian continent alien and inhospitable and, rather than adapting to the new environment, they sought to transform it into a facsimile of Europe, primarily by introducing plant and animal species and imposing European agricultural methodologies. This has resulted in the degradation of entire landscapes. Although Australians are learning, it is still standard practice to allow stock numbers to build up during moist periods to densities that are unsustainable when the inevitable drought arrives. Such boom-and-bust cycles cause severe overgrazing and consequent soil loss, breakdown of ecological processes, and loss of biodiversity. Clearly, the conservation of the remaining Australian biodiversity demands amelioration of the environmental changes set in train two centuries ago.

Most Australian environments are biologically unproductive. The peculiarities of the endemic flora and fauna no doubt reflect over thirty million years of geographic isolation, phases of extreme aridity during the ice ages, and adaptations to floods, droughts, and fires (Table 1). Many landscapes are ancient, subdued, and based on sedimentary rocks rich in quartz. Fertile soils are largely restricted to the limited areas along rivers or where there has been volcanic activity in the geologically recent past. Throughout the continent there is considerable variability of annual rainfall. More than two-thirds of the continent receives, on average, less than 500 millimeters per annum (Figure 1). This aridity is a direct consequence of the absence of any great mountain ranges and, because of Australia's midlatitudinal position, the dominance of the subtropical high-pressure system. This climate system causes dry, windy conditions conducive to wildfires in southern Australia during the summer months, and similar "fire weather" during the winter months in northern Australia. El Niño–Southern Oscillation events cause regular cycles of droughts and floods across eastern Australia. These geographic factors rendered Australia particularly vulnerable to the environmental upheavals that followed European colonization.

## BIBLIOGRAPHY

Alexander, N., ed. *Australia: State of the Environment 1996*. Melbourne: CSIRO Publishing, 1996. A useful compendium of information about critical Australian environmental issues.

Conacher, A., and J. Conacher. *Rural Land Degradation in Australia*. Melbourne: Oxford University Press, 1995. An introductory text discussing environmental issues in Australian agricultural landscapes.

Kennedy, M., ed. *Australia's Endangered Species: The Extinction Dilemma*. Sydney: Simon & Schuster, 1990. An illustrated introductory reference manual of Australia's endangered flora and fauna.

Kirkpatrick, J. B. *A Continent Transformed: Human Impact on the Natural Vegetation of Australia*. Melbourne: Oxford University Press, 1994. An introductory text outlining environmental issues surrounding the conservation of Australian native vegetation.

Olsen, P. *Australia's Pest Animals: New Solutions to Old Problems*. Sydney: Kangaroo Press, 1998. A well-referenced overview of feral animal impacts and management options.

Young, A. R. M. *Environmental Change in Australia since 1788*. Melbourne: Oxford University Press, 1996. A well-referenced text outlining European impacts, focusing on the physical rather than the biological environment.

—DAVID BOWMAN

**AUTOMOBILES.** *See* Transportation; *and* Urban Areas.

# B

## BASEL CONVENTION

Adopted in 1989, the Basel Convention on the Control of Transboundary Movements of Hazardous Wastes and Their Disposal constituted an international policy response to scandals involving the dumping of highly toxic wastes from the industrialized world in developing countries during the 1980s. Global in scope, the treaty entered into force in 1992 and had a membership of over 130 countries by 2000. At the time of adoption, its aim was twofold: to promote environmentally sound management of hazardous wastes wherever their place of disposal, and to restrict and monitor the transfer of such wastes beyond the borders of the country of generation through an international control system. In response to growing pressure from developing countries and environmental organizations, the Basel Convention was amended in 1995 to prohibit the export of hazardous wastes from member countries of the Organisation for Economic Co-operation and Development (OECD) to nonmembers of the OECD. As of 2000, this amendment had not entered into force. The discussion as to whether this prohibition should be extended to countries other than OECD members has not led to any result thus far. The Basel Convention also calls for international cooperation between parties in areas related to hazardous waste management, particularly assistance to developing countries, and for exchange of technical information. Protocol on civil liability for damage caused by hazardous wastes was adopted by the Conference of the Parties in December 1999, after nearly ten years of negotiations.

The Basel Convention, as amended, is based on the following principles: (1) The generation of hazardous wastes should be reduced to a minimum. (2) Hazardous wastes should be disposed of as close as possible to the source of generation. (3) Transboundary movements of hazardous wastes are prohibited in the following cases: from OECD to non-OECD countries; to Antarctica; to any state not party to the Basel Convention or to a treaty establishing equivalent standards; and to parties that have unilaterally banned hazardous waste imports. (4) In all other cases, the transaction must take place in accordance with the control system established by the Basel Convention, which is based on the principle of prior informed consent of the prospective countries of import and transit. (5) Hazardous wastes that have been exported illegally, as well as legally exported wastes that cannot be safely disposed of in the importing state, must be returned to the state of origin.

The Basel Convention is the first and thus far the only international legal instrument addressing the problem of hazardous waste management and disposal at a global level. A number of earlier waste management systems dealt with the issue at the regional level. Within the European Union and the OECD, rules on hazardous waste management have been in place since the 1970s. The United States concluded bilateral treaties with both Canada and Mexico in 1986. A number of regional treaties have been adopted since the inception of the Basel Convention, including the 1991 Bamako Convention (Africa) and the 1995 Waigani Convention (Southeast Asia).

The need for an international agreement regulating the transfer of hazardous wastes between different parts of the world arose from a massive increase in the uncontrolled transfer of hazardous wastes from developed to developing countries during the 1980s. This was due to various factors. On the one hand, the development of international trade, communications, and transport systems over the past decades provided interlinkages between countries, making the transport of hazardous wastes over long distances a viable option. On the other hand, the difference in standards between the developed and the developing world favored waste transfer from the developed to the less developed. While a high level of industrialization in developed countries led to the generation of huge amounts of hazardous wastes, environmental awareness as well as the scarcity and costs of disposal options encouraged export to countries where such problems were not likely to be encountered. As a result of a lack of legislation and enforcement mechanisms, infrastructure, and information on the dangers connected with hazardous wastes, poor countries were often willing to accept hazardous wastes in exchange for foreign currency in amounts sometimes exceeding the countries' gross national product. These factors led to a globalization of pollution by hazardous wastes: instead of being limited to the country of generation, this pollution was spread throughout the globe.

Until the 1980s, hazardous waste disposal, for example, in scarcely populated areas in developing countries, was considered by many as a free and unlimited option. A change of attitude occurred after the environmental and health effects of a number of uncontrolled toxic dumps in African countries were widely publicized. It is now almost universally recognized that the path of least resistance is not an acceptable option for waste disposal. The Basel Convention is both a product and a promoter of this change of attitude.

Despite the high profile of the developed–less developed aspect of the problem, the Basel Convention applies to all transboundary movements of hazardous wastes. In fact, the vast majority of such movements take place among the industrialized countries of the OECD. A new angle to the problem was introduced by the geopolitical changes of the late 1980s and early 1990s: whereas the policy debate that led to the adoption of the Basel Convention had mainly opposed the Western industrialized states and the developing countries, central and eastern European countries now emerged as a new group of actors. These countries, as a result of their relatively high levels of industrialization, generate hazardous wastes themselves. At the same time, they were the target of illegal hazardous waste imports from the West during the communist period. In many of these countries, the disposal of stocks of obsolete chemicals remains one of the important environmental problems.

The potential effects of the Basel Convention on international trade in secondary raw materials—such as scrap metal and other wastes for recycling—gave rise to considerable debate, especially after the adoption of the ban amendment. Some countries as well as industry circles feared that the recycling industry would sustain heavy losses, because the Convention could be interpreted to cover recyclable wastes with a market value. To address this problem, two lists containing the wastes covered by the ban and those not covered, respectively, have been included as Annexes VIII and IX to the Convention in 1998. Accordingly, a number of wastes with a value for the recycling industry, which have a low hazard potential, are included in Annex IX, and are thus not covered by the ban.

In accordance with a ministerial declaration adopted in 1999 by the Conference of the Parties, the focus of the Basel Convention during the next decade will be the promotion of environmentally sound waste management and disposal options worldwide.

### BIBLIOGRAPHY

ANALYSIS OF THE BASEL CONVENTION
AND THE NEGOTIATION PROCESS

Kempel, W. "Transboundary Movement of Hazardous Wastes." In *International Environmental Negotiation*, edited by G. Sjöstedt. London: Sage, 1993. An analysis of the negotiation process from a government negotiator's viewpoint.

Kummer, K. *International Management of Hazardous Wastes: The Basel Convention and Related Legal Rules.* Oxford: Clarendon Press, 1995. The 1999 paperback edition includes the most important updates.

Kwiatkowska, B., and A. Soons, eds. *Transboundary Movement of Hazardous Waste in International Law: Basic Documents.* Dordrecht, Boston, and London: Martinus Nijhoff/Graham & Trotman, 1993. A complete collection of international policy and legal documents related to hazardous waste management.

*Review of European Community and International Environmental Law* 7.3 (1998). This issue is devoted to waste management and comprises seven articles on different aspects of the Basel Convention.

REGIONAL REGULATION; THE DEVELOPED–LESS
DEVELOPED ASPECT OF THE PROBLEM

Biggs, G. "Latin America and the Basel Convention on Hazardous Wastes." *Colorado Journal of International Environmental Law and Policy* 5 (1994), 333.

Jones, W. F. "The Evolution of the Bamako Convention: An African Perspective." *Colorado Journal of International Environmental Law and Policy* 4 (1993), 324.

TRADE IMPLICATIONS; THE 1995 BAN
AMENDMENT

De La Fayette, L. "Legal and Practical Implications of the Ban Amendment to the Basel Convention." *Yearbook of International Environmental Law* 6 (1995), 703.

Guevara, M., and M. Hart. *Trade Policy Implications of the Basel Convention Export Ban on Recyclables from Developed to Developing Countries.* Ottawa: Carleton University/ICME, 1996.

Krueger, J. "International Trade and the Basel Convention." The Royal Institute of International Affairs, Trade and Environment Series. London: Earthscan, 1999.

Kummer, K. *Transboundary Movements of Hazardous Wastes at the Interface of Environment and Trade.* UNEP Environment and Trade Monograph No. 7. Nairobi: UNEP, 1994.

—KATHARINA KUMMER

**BEACHES.** *See* Coastal Protection and Management.

# BELIEF SYSTEMS

The study of global change has become so complex in recent years—so dependent on the special skills of chemists, biologists, economists, and others—that the layperson can easily conclude that this is a technical field whose problems can be thoroughly understood and solved only by experts. Yet a large part of the study of global change deals with changes brought about by human beings. Both the scope and magnitude of these changes have increased so enormously in recent centuries that when people speak of global change today, they often mean anthropogenic, or human-induced, global change. This accelerating impact of our species on the earth is largely a consequence of modern science and technology, directed chiefly to satisfying the desires of a consumer society. Dominant as it is today, that consumer society is only one of the thousands of human societies that have inhabited the Earth. It stands out as a uniquely powerful agent of global change and also, not coincidentally, as a society that tends to see the world chiefly as a storehouse of raw materials for the use of human beings. That belief, embedded now for several centuries, drives us to become ever more powerful as agents of change. Thus, an examination of the underlying causes of global change requires an understanding of the belief system responsible for many of those changes.

Whether such an understanding is needed to solve those problems is another matter. Pollution and species extinctions, for example, may be seen as problems created by technology and soluble only through more technology. If that is true, the intellectual history of the society that has generated those problems is irrelevant to their solution. Yet many of the environmental problems created by modern society would be vastly reduced if the underlying values of that society were changed. Thus, to take a simple example, global warming would be a much smaller problem if people abandoned big cars, not to mention cars altogether. A consideration of belief systems is, therefore, in this sense not merely a historical curiosity but part of the solution to the problems of global change.

**The Emergence of the Individual.** The mentality that sustains the emerging global culture can be traced back to the Renaissance. One of the most striking changes that occurred at that time can be seen in the paintings of the era. For centuries, painters had been portraying crowds of angels and Virgins. Suddenly, or so it seems, Jan van Eyck began painting actual living people—and doing so with formidable exactitude. A private devotional portrait such as *Virgin and Child and George van der Paele* (1434) appears, almost without precedent, as a portrait of a real person. Such portraits spread quickly. Van Eyck was from Ghent, but within two generations Giovanni Bellini of Padua was painting portraits with a similar degree of verisimilitude: his portrait reputedly of Giovanni Emo (c.1480) is a good example of the importance newly attached to the individual. Albrecht Dürer, from Nuremberg, would soon paint a portrait of the banker Jacob Fugger (1520) with a detail and accuracy that would please a businessman today. Leonardo da Vinci painted the first realistic portrait of a woman (the Mona Lisa dates from about 1560). Andrea del Verrochio's bust of Lorenzo de' Medici (c.1485) shows that sculptors, too, were achieving an extraordinarily high degree of realism.

This unprecedented accuracy and attention to portraiture was more important than the appearance of photography four centuries later, for it was not only a technical innovation but a conceptual one. It is as though the highly self-conscious Greeks of antiquity had been reborn, with all their pride and competitiveness. It can hardly be immaterial to this rebirth that Erasmus was in these years overseeing the publication of the works of Aristotle and other Greek authors. For the first time, these books were set in the movable type developed a few decades earlier (the Gutenberg Bible dates from 1456). Whatever its ultimate causes, the pride of the new age was evident in Michelangelo, both in his defiant *David* (1504) or in his sternly judicial *Moses* (1515). In literature, Shakespeare puts it best, in *Hamlet* (1600). In a gloomy humor, Hamlet yet considers mankind and says: "How noble in reason! how infinite in faculties! . . . in apprehension how like a God!"

**The Division of Mind and Matter.** Not surprisingly, this new emphasis on the individual was accompanied by a matching tendency to draw a sharp line between humanity and the rest of creation—and to debase the rest of that creation to raw stuff, devoid of soul and awaiting transformation at the hands of mankind. In *The Advancement of Learning* (1605), Francis Bacon praises knowledge when it is acquired "for the benefit and use of men." The aim, he continues, is "to reject vain speculations, and whatsoever is empty and void, and to preserve and augment whatsoever is solid and fruitful." In an anticipation of the almost incredible technological developments of later centuries, Bacon in *The New Atlantis* (1627) writes of a utopia devoted to the "enlarging of the bounds of human empire, to the effecting of all things possible."

The mechanistic aspect of this practical orientation is especially clear in the work of René Descartes. In the *Discourse on Method* (1637), Descartes writes that he would rather put his faith "in the reasonings that each man makes on the matters that specially concern him, and the issue of which would very soon punish him if he made a wrong judgment . . . [and not in the reasonings] made by a man of letters in his study touching speculations which lead to no result." This reasoning faculty, Descartes continues, "this 'me,' that is to say the soul by which I am what I am, is entirely distinct from body." Here is a telling separation: the external world is firmly separated from the spiritual one. Descartes looks at animals and sees only machines. Indeed, he compares them to clocks, which, though without intelligence, tell time "more correctly than we can do with all our wisdom."

**The Scientific Revolution.** This famous Cartesian dualism—the assumption of a gulf between mind and matter, knower and the world, subject and object—stimulated an enormous burst of scientific discovery. Historians often center their narratives on astronomy, especially the work of Copernicus and Galileo, which in turn led to the physics of Isaac Newton. But astronomy was not the only field that grew rapidly in these years, and it was not merely the cosmos whose motions could be understood in mechanical terms. Julien La Mettrie, author of the famous and in its time notorious *L'Homme machine* (1748), drew for his vision of the human body as a machine upon the work of Andreas Vesalius, the pioneering Flemish anatomist, contemporary of Copernicus, and author of *De humani corporis fabrica* (1543).

In Rembrandt's *Anatomy Lesson of Dr. Tulp* (1632), one sees not only Vesalius's objective view of the body but also, in one prominent character who stares at the viewer, a comment from Rembrandt about the revolutionary and disquieting legacy of Descartes, namely, the ability to look upon the world with cool detachment. Another sign of that objectification may be read from the Renaissance and Baroque garden. The medieval garden had been a *hortus conclusus*, an enclosed garden from which one escaped the world. The Renaissance garden, as described by Leone Alberti in *De re aedificatoria* (1452), extended outward across the landscape, controlled it, shaped it with geometric precision. A famous example is the garden built for Louis XIV at Versailles by André le Nôtre.

**An Enlarged World.** Commensurate with the expansion of scientific knowledge came an expanded knowledge of the terrestrial globe. We have only to look at Nathaniel Dance's portrait of Captain James Cook (1776) to see the strength and confidence sustaining this revolution in geographical knowledge. The practical implications of the fusion of science and a planetary stage were immense—and perhaps most immediately dramatic in the case of the botanical world. Linnaeus developed his binomial nomenclature as a practical way of organizing the vast collections of plants that had been collected by the great voyages of discovery. By 1771 Joseph Banks was in charge of Kew Gardens, which he transformed from an ornamental garden into a repository of the plants that would bring vast wealth to Europe. Similar gardens were established in Calcutta in 1786 and in Singapore in 1822. There, five years later, the first Brazilian rubber trees were planted. Soon they would be tapped to demonstrate the feasibility of rubber-tree plantations in Southeast Asia.

**The Enlightenment.** By 1750 Denis Diderot was distilling European knowledge into the famous *Encyclopedia*, testament to the now-mature faith in reason and progress. The very term *enlightenment*, coined about 1660, speaks much about how Europe saw the rest of the world, as well as its own implicitly "dark" past. In 1776 Edward Gibbon began publication of *The Decline and Fall of the Roman Empire*, which dared to view history not through a Christian lens but through a secular and progressive one.

Adam Smith was simultaneously spurring the development of capitalism with the publication of *The Wealth of Nations* (1776). In Germany, Immanuel Kant was liberating mankind from the burden of inherited religious dogma. Arguing that the existence of God could not be demonstrated theoretically or scientifically, he placed the burden of making ethical judgments squarely on the individual.

**Exporting Progress.** The religious faith of Bacon and Descartes was gradually cast aside as superfluous,

and, even before Friedrich Nietzsche in 1882 observed famously that "God is dead," much of Europe's political philosophy had been completely secularized. The British utilitarians, chiefly through Jeremy Bentham, not only espoused a purely rational organization of society but also went a long way toward creating the governing institutions of one. In England such intentions met with competition, but overseas they did not, at least not at official levels. In India, especially, Bentham guided the development of administrative policies for the East India Company, which was then in the hands of his disciple William Bentinck, appointed governor general of India in 1833. No doubt the societies of Asia did not have to be taught by Europeans that wealth was a good thing, but Europeans did introduce to Asia the ideas that enough wealth could be produced to make everyone comfortable and that government was responsible for facilitating this great transformation. These ideas came to India not only through Bentinck but through the indefatigable Thomas Babington Macaulay, who accelerated English-medium instruction and a Eurocentric curriculum in Indian schools. (In a famous and stylistically flamboyant essay, Macaulay also wrote one of the most adulatory of all accounts of Francis Bacon.) In retrospect, it may seem that these Westernizing efforts of the British in India were doomed to failure, but in fact they led directly to the construction of major public works, not only in India but across the imperial domains of the European powers. Of equal or greater importance, they laid the foundations for the development of English as the lingua franca of the modern world, and it is hardly too much to say that they initiated the processes of economic development sweeping across the non-Western world today.

**The Age of Heroic Materialism.** The Victorian era saw not only a rapid acceleration in the globalization of the now-matured European mentality but also its expression in undertakings of industrial gigantism on almost every continent. The achievements of the great nineteenth-century engineers such as Isambard Kingdom Brunel and Ferdinand de Lesseps still seem like the work of titans; so, too, do the achievements of the industrial magnates of the time, such as Andrew Carnegie and Cecil Rhodes. The list can be extended into the twentieth century, with such men as Thomas Alva Edison and Henry Ford, though perhaps the aura of heroism is brightest in the earliest generations, the ones closer to the transformation of a world of wood and muscle into a world of steel and engines.

**The Conservation Movement: An Environmental Policy for Sensible People.** Optimistic as the Victorian age was, a few people, especially in America, began to worry about the unprecedented impact that industrial civilization was having on nature. A recent author, Herman Daly, wrote in 1988 that "there is some-

*Ecofeminism* explores the links between feminist thought and ecological concerns. It has many strands. One of these explores the connections between the oppression of women and on treatment of other animals and of nature as a whole. Another strand considers female connections to nature—a sense for nature's cycles, love, and care—while one contribution of ecofeminism has been to awaken awareness of how much environmental protection around the world is done by women.

**BIBLIOGRAPHY**

Gaerd, G., ed. *Ecofeminism: Women, Animals, Nature.* Philadelphia: Temple University Press, 1993.

Merchant, C. *The Death of Nature: Women, Ecology, and the Scientific Revolution.* San Francisco: Harper San Francisco, 1989.

Southeimer, S. *Women and the Environment.* New York: Monthly Review Press, 1991.

—ANDREW S. GOUDIE

thing fundamentally wrong in treating the Earth as if it was a business in liquidation." A more scholastic tone was adopted by his great predecessor, George Perkins Marsh, author of *Man and Nature* (1864). Abraham Lincoln's minister to Italy, Marsh had studied the historic destruction of the land and water resources of the Mediterranean Basin, and he cautioned that the United States, then in the midst of the almost wanton deforestation of the Lake States, might similarly destroy the physical basis of its own prosperity. It was Marsh to whom the founders of the conservation movement always pointed as the fountainhead—Marsh to whom Gifford Pinchot, Theodore Roosevelt's charismatic forester, pointed in explaining how the United States began taking the steps that led to the creation of the national forests and other conservation institutions created in Roosevelt's time.

The same concern with resource depletion underlay Paul Sears's *Deserts on the March* (1935), a book prompted by the Dust Bowl on the southern Great Plains of the United States. Often it was expressed in messianic terms, like those employed by Walter Lowdermilk, second in command at the U.S. Soil Conservation Service. Speaking in 1939, he offered what he called an eleventh commandment that began with the words "Thou shalt inherit the holy earth as a faithful steward."

Today, the conservation idea is often presented in the new lexicon of "sustainable development," a phrase apparently first used in 1980 by the International Union for the Conservation of Nature. It remains as prudent as ever. Yet it also remains a philosophy conceived in anthropocentric terms, a rational attempt to be sure that we don't destroy nature in our efforts to control it. Thus, the conservation movement, though often seen as the ancestor of contemporary environmentalism, is actually fundamentally different from it, for environmentalism

arises not from a concern with human welfare but from a concern with the welfare of nature itself.

**From History to the Present.** Before turning to the evolution of this remarkable countertrend, it is well to remember that the twentieth century has shaken but not broken the Enlightenment faith in material progress. In an essay whose title is more famous than its message, William James urged that American youth should turn away from militarism and seek instead "The Moral Equivalent of War" (1910). That equivalent, it turned out, was a conscript army working in mines and foundries and constituting an "army enlisted against Nature." During World War II, the Food and Agriculture Organization (FAO) was created, its publicly avowed purpose "to promote the uninterrupted development and most advantageous use of agricultural and other material resources." Today, candidates speak of building bridges to the new millennium, and the world they envision is one of extraordinary health and wealth. Voters, after all, have grown up expecting that they will be more prosperous than their parents and that their own children will be more prosperous still; elected officials have learned, meanwhile, that they will be held responsible if the economy falls into recession.

The world itself has been transformed from manorial fields to monster farms, from walled congestion to sprawling megacities, from muddy tracks to high-speed roads. The burning of ancient forests has become the wholesale extirpation of species, and the erosion that made rivers yellow in antiquity is now complemented by loads of toxic and organic wastes that pollute not only the streams but the atmosphere and the oceans themselves. The ecological dominance of humanity is now so great that Bill McKibben, in *The End of Nature* (1989), argues that "the world outdoors" had come to "mean much the same thing as the world indoors, the hill the same thing as the house." It has been an extraordinary transformation, but all too often it has been seen through culture-bound eyes as the consequence of universal human nature, rather than as an expression of a remarkable and distinctive European culture, one with a history.

**The Countertrend.** For all its power, anthropocentrism does not monopolize European intellectual history. A countertrend, always present, has probably become increasingly important in recent years. Reacting to Cartesian dualism, it seeks to restore the connectedness between human beings and the world. It can appear primitivistic, on the simple grounds that there can be no union with nature except in a state of nature. Yet "connectedness" is a psychological state, not a physical one. Although it almost certainly depends on personal experience of the natural world, it does not require that we abandon the highly artificial world we have made and in which we reside. Balancing the two, indeed, has become a part of everyday life, which daily juxtaposes acts that accelerate our control over nature with acts that embed us back in it.

**Romanticism.** This countertrend arose under the name of Romanticism. Its roots can be traced to the earliest mystics and even through the height of the Renaissance Enlightenment. There is the appalled observer in Rembrandt's *Dr. Tulp*, and there are also the writings of the naturalist John Ray, who wrote in 1691 that many people believe that the world was created for human beings, but that "wise men think otherwise." Jonathan Swift's "A Modest Proposal" (1729) is usually read as an attack on the British, but to anyone familiar with the writings of economists, it is an equally blistering parody of the language of scientific analysis, coolly able to discuss atrocities in morally neutral terms. Jean Jacques Rousseau was still a child when Swift wrote, but his own attacks on the social order and his own praise of primitivism came very nearly to mark the birth of romanticism. For English speakers, the pivotal figures are William Blake and the more immediately influential William Wordsworth. For Germans, the towering presence is Johann Goethe, whose *Faust* (Part 1, 1808) portrays the archetypal modern man, destroyed by his lust for power. An echo of Faust appeared across the Atlantic in Herman Melville's *Moby-Dick* (1851), where Captain Ahab, for reasons that make no sense even to him, is determined to capture the natural world, symbolized by a white whale.

**The Popularity of the Romantic View.** It may seem strange that such radical views would become popular, but the philosophy of Descartes is a cold one, and the warmth of Romanticism has always had greater popular appeal. Once again, painting and garden design are useful indicators. And so one looks to the French artist Claude Lorrain, who in Descartes's time was paint-

ing landscapes inspired by the Virgilian bucolics. Claude stood alone in those years, but by the time of Wordsworth he was inspiring John Constable, whose work drops Claude's classical allusions and reveals instead a love for the preindustrial British countryside. It became the motif for countless magazine pictures, calendars, and the cheapest kind of oil paintings, made and still sold today by the truckload to people who want something pretty on their walls.

Garden design changed radically. At a time when geometric gardens dominated garden aesthetics across Europe, William Temple (1685) praised the Chinese garden as being "without any Order or Disposition." He was wrong, but the judgment was a measure of growing uneasiness with the rigid control of nature. In 1711 Joseph Addison was praising the Chinese garden, and in 1713 Alexander Pope was writing of the "amiable simplicity of unadorned nature." Pope's friend William Kent became, along with Charles Bridgeman, the first garden designer to implement the new "naturalistic" aesthetic; he was followed by Lancelot Brown and Humphry Repton, who influenced the aesthetics of many public parks across England. The same aesthetic came to America in the famous work of Frederick Law Olmsted, and it flourishes in the countless private gardens whose owners lay out curved beds, mound up berms, and seek to create backyards that look "natural."

The treatment of animals constitutes a third domain in which the popularity of the Romantic countertrend can be seen. The Society for the Prevention of Cruelty to Animals was established in Britain in 1824 (the American counterpart was created in 1866). Significantly, the people instrumental in its organization were the same who helped abolish slavery in the British Empire a few years later. The sources of this empathy with living creatures, human or animal, are obscure, but the movement clearly drew strength from literary works. In the case of slavery, the key work was *Uncle Tom's Cabin* (1852), by Harriet Beecher Stowe. In the case of animals, there may not be such a key work, but certainly Anna Sewell's *Black Beauty* (1877) is a candidate. Lesser known, Sarah Orne Jewett's short story "The White Heron" (1886) tells of a girl kind enough to forgo a cash reward, so that a bird may live. The influence of the film version of Felix Salten's *Bambi* (1929) can hardly be exaggerated. Its condemnation of hunters and its sympathy for their victims has now become endemic to popular culture at every age level.

**Reform and Revolution.** At the height of the Victorian era, Charles Dickens's scathing portraits of the human cost of industrialization were not only tolerated but famously popular, even among the classes responsible for the conditions he depicted. Other critics went beyond depiction to prescription. Some were of an artistic or poetic frame of mind, such as John Ruskin and

William Morris, whose love of craftsmanship may be seen as hopelessly archaic until one considers the price of antiques and handmade products today. Others were of a more accommodating or reformist nature, such as the founders of Europe's many Labor Parties. Of all these political expressions of the romantic sentiment, the most important was surely Communism, which arose from the moral outrage of Friedrich Engels and Karl Marx. It is easy to overlook their initially romantic orientation. Engels, for example, wrote in 1844 that buying and selling land was almost as immoral as chattel slavery. In his posthumously published *Dialectics of Nature*, he added that "at every step we are reminded that we by no means rule over nature like a conqueror over a foreign people, like someone standing outside nature."

Just as the Baconian enterprise led to the squalor of Coketown, however, so Marxism became dogmatized and corrupted into tyrannies whose actions, both social and environmental, were catastrophic. Bad enough that idealism should have led to gulags and the Red Guards. It also led to Chernobyl, the plowup of the Virgin lands, and the destruction of the Aral Sea. The Chinese communists undertook to build huge dams on both the Yellow and Yangtze Rivers. So did India, which has had a very difficult time distancing itself from the Soviet-style central planning that so appealed to Jawaharlal Nehru. Not so many years ago, of course, similar projects appealed greatly in the capitalistic West. In fact, it is hardly an exaggeration to say that the model for integrated river-basin development everywhere on Earth was the Tennessee Valley Authority, premised on the idea that rural poverty was best tackled by simultaneously transforming the whole spectrum of land uses. In this sense, the detour of Marxism led in a great circle back to the mentality that Romanticism sought to escape.

**Ecocentrism.** Another avenue proved more fruitful and led from Romanticism to what has been termed *ecocentrism*, which is to say, adopting as a measure of virtue the consequences of human actions upon the biota as a whole, not simply upon the human species. The line from romanticism to ecocentrism runs through the transcendentalism of Henry David Thoreau, who wrote in his journals, "We must go out and re-ally ourselves to Nature every day." It comes fifty years later through the writing of John Muir, the founder of the Sierra Club. It runs through *A Sand County Almanac* (1948), written by a professor of game management who had come around to believing that wolves had as much right to live as people. Aldo Leopold urged his readers to adopt a new ethic, a land ethic based on the principle that ethics rests on communities and that defined *community* as including all living things.

Leopold was ahead of his time, and his book was published only posthumously, after it had been repeatedly rejected by publishers. The real audience for Leopold

developed in the 1960s, after publication of Rachel Carson's *Silent Spring* (1962), which was an indictment of the use made of synthetic poisons in modern agriculture. Oddly, Carson made no mention of Leopold but instead dedicated her book to Albert Schweitzer, the missionary doctor who believed that we would destroy the Earth because we had lost our "reverence for life." Paraphrasing him, Carson wrote that "the control of nature is a phrase conceived in arrogance."

If *Silent Spring* seems destined to live because of its tremendous influence in energizing the ecology movement, we are still too close to most of its successors to judge their staying power. Certainly, there were many popular books written on ecological themes in the years after Carson, among them the phenomenally successful, though critically battered, *The Greening of America* (1969), by Charles A. Reich, and *The Making of a Counter Culture* (1969), by Theodore Roszak. Both were studies of youth as revolutionaries overturning the mechanistic, technocratic world, but both had environmental implications. A few years later, Roszak contributed the foreword to E. F. Schumacher's *Small Is Beautiful: Economics As If People Mattered* (1973), another publishing sensation, which in its advocacy of "intermediate technology" suggested that there were more important things than technological sophistication. "We can," Schumacher wrote, "each of us, work to put our own inner house in order."

Meanwhile, the Norwegian philosopher Arne Naess published an article in *Inquiry* called "The Shallow and the Deep, Long-Range Ecology Movement: A Summary" (1973). It dismissed the efforts of mainstream environmentalism as merely reformist and advocated instead a policy of environmental egalitarianism, in which all species had an equal right to coexist. Here was a full-blown ecocentrism, and it was popularized in the United States by the sociologist Bill Duvall and the philosopher George Sessions, authors of *Deep Ecology* (1985). It was also repudiated as "ecological fascism" by critics who attacked it for its willingness to sacrifice people for the greater good of a supposed ecological community.

A more recent venture into ecocentrism is the "biophilia hypothesis" of Edward O. Wilson; in Wilson's words, biophilia is "the innately emotional affiliation of human beings to other living organisms."

**Superorganicism.** The extreme form of ecocentrism holds that the Earth itself is alive. This is an idea that draws considerable strength from Charles Darwin, who showed that humanity was linked to a community that encompasses all life, past and present. It is an idea that draws strength, too, from the conscious effort to demolish the Cartesian boundary that separates us from the physical world, for, if we concede no boundary, it is difficult to see how we can be alive at the same time as the world around us is not. Such a view is anathema to most scientists, including most practicing ecologists. Thus when the pioneer ecologist Frederick Clements spoke of the "climax communities" of America's prairie grasslands, he ran afoul of British biologists such as the Oxford botanist Arthur George Tansley, who chose to replace the word "communities," redolent of organicism, with the austerely objective term "ecosystem."

Yet superorganicism will not die. The youthful John Muir felt the Earth alive around him. So did Henry David Thoreau, who in fact wrote that "the earth is a body, has a spirit." In 1875 the geologist Eduard Suess coined the term *biosphere*, referring to a layer of the Earth saturated with life, and in 1915 a famous dean of agriculture at Cornell University, Liberty Hyde Bailey, wrote a book called *The Holy Earth* in which he pleaded for what he called a "biocentric" view of the world. Here was a reputable, highly regarded scientist striking a blow against anthropocentrism. A few decades later, the poet Robinson Jeffers wrote "The Answer" (1938), which includes the lines "Integrity is wholeness, the greatest beauty is/Organic wholeness, the wholeness of life and things, the divine beauty of the universe. Love that, not man/ Apart from that."

In 1979 James Lovelock announced what he termed the "Gaia hypothesis," which has given new life to superorganicism by clothing it in science. Lovelock himself has been cautious, stating only that the Earth is "a complex system which can be seen as a single organism." Lynn Margulis, who is often credited as a co-founder of the hypothesis, was even more cautious when she appeared before a meeting of the American Geophysical Union in 1988 and said that the hypothesis held that the Earth's surface conditions are regulated by the activities of life. The public at large, perhaps charmed by the name Gaia, has been far more willing to entertain the notion that the earth is not only an organism but even conscious. Such concepts are only a simple step from the commonplace personification of nature, which led Joseph Wood Krutch in 1929, for example, to write of nature's "blind thirst for life." [*See* Gaia Hypothesis.]

**Implicating Non-Western Belief Systems.** Not surprisingly, the reservoir of non-Western cultures has been explored for helpful insights. This is nothing new: Thoreau ended *Walden* (1854) with a mock-Hindu parable of his own devising, while Schumacher ended *Small Is Beautiful* by saying that the guidance needed to put our houses in order "can still be found in the traditional wisdom of mankind." How different from Macaulay's belief that the corpus of Sanskrit literature was worthless! Contemporary poets such as Gary Snyder have drawn heavily on Zen Buddhism, which in its renunciation appeals to satiated consumers and which in its apprehension of a unified world is perfectly ecocentric. Christianity has been reconsidered, too, perhaps first by

Edward Payson Evans, who noted in 1894 its "anthropocentric character." More recently, Lynn White's "Historical Roots of Our Ecological Crisis" (1967) suggested that we should pay less attention to the biblical injunction to "multiply and subdue the earth" and more to St. Francis of Assisi.

Hinduism and Islam also have been held up as models for ecologically enlightened behavior. So, too, have tribal societies. Until Victorian times their myths were treated as prescientific errors: such was the orientation of James Frazer's *Golden Bough* (first ed., 1890). Now there is a wide audience for studies inclined, as Schumacher was, to see wisdom in traditional ways. Much of this literature has been written by Europeans, often colonials. However sympathetic these authors may be— and certainly many of them have been sympathetic— there is an unfortunate dearth of material by authors writing of their own cultures. There is special value, therefore, in the works of such writers as N. Scott Momaday.

**Regional Planning.** Like anthropocentrism, ecocentrism has transformed the world in tangible ways. One early and easily overlooked case is that of regional planning, which took an early turn toward ecocentrism in the pioneering work of Patrick Geddes, who argued for a species of planning based on careful surveys of physical and cultural character. His intention was to learn so much about the places he planned that the plan itself would be an organic extension of the place. In *Cities and Evolution* (1915) he wrote that "each place has a true personality . . . which it is the task of the planner, as master-artist, to awaken. And only he can do this who is in love and at home with his subject." Geddes's work had considerable impact in Britain, where the Oxford geographer A. J. Herbertson wrote that regions have a spirit "best understood by poets, not scientists." The Geddes outlook came to the United States especially in the work of Lewis Mumford, Clarence Stein, and the other people who in 1923 formed the Regional Planning Association of America. In *The New Exploration* (1928), another member of that association, Benton Mackaye, proposed to temper the growth of what Geddes had called "conurbations" by creating buffers of undeveloped land along ridge crests—"embankments" against the flood, he called them.

Mackaye's ideas overlap a great deal with the "watershed consciousness" that has become prominent in recent years, often blended with the concept of biocentrism. Tony Hiss has written on this subject; so has Peter Berg. In general, they have sought to define geographic provinces by complexes of cultural and physical features, many of which, like vegetation, might be anthropogenically modified. Such regions are not treated only as the units of regional planning but also as natural units of self-sufficiency, around which most economic needs could be met. How this might be accomplished was explored by Ernest Callenbach in his novel *Ecotopia* (1975).

**The Impact of the Environmental Movement.** Ecocentrism has begun expressing itself in other tangible ways, too, although the threads here are complexly interwoven with threads of both superorganicism and anthropomorphic conservation. Indeed, it has been so since the early twentieth century, when the conservation movement was occasionally allied to the preservationist orientation of the Sierra Club.

The combined effect of these belief systems, however, is now very great. Late in the 1960s, photographs of the whole Earth began to be received from the Apollo 8 spacecraft, and it has often been observed that they had a great influence on the public's perception of the Earth. The United States passed a National Environmental Protection Act in 1969, and a raft of litigation soon arose, including the famous "Mineral King" case, in which Justice William O. Douglas held that trees had standing to sue. Among other books on environmentalism, *The Limits to Growth* (1972), by Donella and Dennis Matthews, attracted a great deal of notice. So did the massive *Global 2000 Report to the President* (1982), prepared for Jimmy Carter by his Council on Environmental Quality. The environment continued to grow as a topic of national interest, and in January 1989 *Time* magazine named "the endangered Earth" as its "person of the year." Just as the idea of progress had become part of the air we breathe, so too had the idea that Western civilization is fundamentally destructive. Hence the success of such films as *Dances with Wolves*, in which the whites are portrayed as evil and the Indians as models of civilized humanity.

Ecocentric political parties emerged first in Tasmania and Australia in 1972, as the United Tasmanian Group and Values Party. An Ecology Party was organized in the United Kingdom in 1973, and a Green List for the Protection of the Environment, which came to be known as the Green Party, was organized in Lower Saxony in 1977. In 1983 the Green Party was seated in the Bundestag with 5.6 percent of the vote, but it soon split into radical and pragmatic wings. In 1990 it lost all its representation, apparently because in Germany, as elsewhere, established parties took over the genuinely popular elements of the Green platforms. In the United States, a national Green Party failed to emerge, though Green candidates have sometimes won local elections. In some communities they have been very influential, most strikingly in the politics of college towns and "New Age" communities such as Sedona, Arizona.

International conventions have become an increasingly important arena for ecocentric thought, and their influence on national politics will probably increase. Such pairing of national and international actions can be seen, for example, in the Endangered Species Act, passed by the U.S. Congress in 1973, the same year in which the Convention on International Trade in Endangered Species of Wild Fauna and Flora (CITES) was signed. Similar linkages are likely to affect the speed at which both international and national actions are taken, and the consequences bear not only on the maintenance of biodiversity but also, in the case of climate change, on industrial production.

**The Balance of Power.** The balance of power remains clearly with the forces of development, but it is striking how earnestly voices as varied as Richard Nixon's and Al Gore's have tried hard to ally themselves with the apparently virtuous ground of ecocentric thought. The greed to which they have been opposed is of course as universal as consumer culture. We live, quite simply, in a state of contradiction, now embedded in us as the unlikely offspring of both Francis Bacon and William Wordsworth. We will, it seems, forever vacillate between power and dominion, on one hand, and incorporation and transcendence, on the other. Even so, that vacillation implies a moderating influence on the unalloyed drive to technological control over nature. As such, it will ease the burden our economy places on the Earth. Belief systems, therefore, must be seen as among the most potent ways of dealing with global change.

[*See also* Catastrophist–Cornucopian Debate; Environmental Economics; Environmental Law; Environmental Movements; Ethics; Population Policy; *and* Religion.]

## BIBLIOGRAPHY

Bramwell, A. *The Fading of the Greens*. New Haven, Conn.: Yale University Press, 1994. An account of Green politics, primarily in Europe.

Callicott, J. B. *Earth's Insights: A Survey of Ecological Ethics from the Mediterranean Basin to the Australian Outback*. Berkeley: University of California Press, 1994. A comprehensive review.

Callicott, J. B., and R T. Ames, eds. *Nature in Asian Traditions of Thought: Essays in Environmental Philosophy*. Albany: State University of New York Press, 1989. Essays on Chinese, Japanese, Buddhist, and Indian worldviews.

Chisholm, A. *Philosophers of the Earth: Conversations with Ecologists*. New York: Dutton, 1972. Accounts of meetings with eighteen prominent figures.

Clark, K. *Landscape into Art*. London: J. Murray, 1949. A classic and readable history of European landscape painting. May usefully be read in conjunction with his *Civilization* (New York, 1969).

Clarke, J. J., ed. *Voices of the Earth: An Anthology of Ideas and Arguments*. New York: G. Braziller, 1993. An unusual collection of primary rather than secondary materials; global in scope.

Gottlieb, R S., ed. *This Sacred Earth: Religion, Nature, Environment*. New York: Routledge, 1996. An anthology mostly of secondary writings about the role of nature in various religions.

Hargrove, E. C., ed. *The Animal Rights/Environmental Ethics De-*

bate: *The Environmental Perspective.* Albany: State University of New York Press, 1992.

Joseph, L. E. *Gaia: The Growth of an Idea.* New York: St. Martin's Press, 1990. A history of the Gaia hypothesis, by a cautious and not-quite-convinced believer.

Merchant, C. *The Death of Nature. Women, Ecology, and the Scientific Revolution.* New York: Harper and Row, 1980.

———. *Radical Ecology: The Search for a Livable World.* New York: Routledge, 1992. A textbook.

Mumford, L. *The Myth of the Machine: The Pentagon of Power.* New York: Harcourt, Brace, and World, 1970. An intellectual history that links ideas to their visible consequences.

Nash, R. F. *The Rights of Nature: A History of Environmental Ethics.* Madison: University of Wisconsin Press, 1989. A concise review, restricted to the European culture world.

Parsons, H. L., ed. *Marx and Engels on Ecology.* Westport, Conn.: Greenwood Press, 1977. An anthology, with a lengthy introduction.

Sale, K. *Dwellers in the Land: The Bioregional Vision.* Athens: University of Georgia Press, 1985.

Schneider, S. H., and P. J. Boston, eds. *Scientists on Gaia.* Cambridge, Mass.: MIT Press, 1991. The record of the 1988 American Geophysical Union's Chapman Conference on the Gaia Hypothesis.

Sylvan, R., and D. Bennett. *The Greening of Ethics: From Anthropocentrism to Deep-Green Theory.* Tucson: University of Arizona Press, 1995.

Torrance, J., ed. *The Concept of Nature.* Oxford and New York: Clarendon Press and Oxford University Press, 1992. A collection of essays about nature from Greek to modern times.

Warren, K. J., ed. *Ecofeminism: Women, Culture, Nature.* Bloomington: Indiana University Press, 1997.

—BRET WALLACH

## BENNETT, HUGH HAMMOND

Hugh Hammond Bennett (1881–1960), soil conservationist, often called "Big Hugh," was reared on the badly gully-eroded piedmont of southern North Carolina. Thereafter, he accepted a position as a field soil surveyor in the Bureau of Soils of the United States Department of Agriculture. It was in 1905, while working on a soil survey of Louisa County, Virginia, that he was confronted by the specter of sheet erosion. As Barnes (1960, p. 506) put it, ". . . he mapped eroded and gullied soils over our far-flung corn, cotton and tobacco regions. He decided that this erosion was destroying the resource base for posterity, that it was not something we had to accept and that something ought to be done about it." Thus it was, following in the steps of two American soil erosion pioneers, McGee and Shaler, that Bennett became the world's leading scholar of soil erosion and the father of the soil conservation movement. He made millions understand that topsoil is a valuable but all too finite resource.

In 1909 he was appointed an inspector in the Soil Survey Division of the Bureau of Soils and undertook soil studies not only in the United States but also in Central and South America. This confirmed his concerns and gave him an international and global perspective that he was to develop as the years passed. In 1928 he wrote, for the United States Department of Agriculture, *Soil Erosion: A National Menace.* In that same year he was charged with setting up ten research stations to study soil erosion control. This provided him with a mound of fundamental knowledge, much of which was incorporated in what was his greatest and most magisterial work, *Soil Conservation* (McGraw-Hill, 1939).

Bennett's work struck a chord during the years of the Great Depression, the Dust Bowl, and the Tennessee Valley Authority. In 1933 the Soil Erosion Service was established in the Department of the Interior, and Bennett was made its chief. In 1935 this agency was transferred to the Department of Agriculture and was renamed the Soil Conservation Service, with Bennett at its head. In that same year Bennett addressed a Senate Committee in Washington on the need for a Soil Conservation Act: as he was speaking, the sky darkened with the passage of a dust storm originating in the Great Plains to the west, and so the act was recommended. He encouraged the setting up of Soil Conservation Districts, which subsequently became established over most of the United States.

Although not naturally a good public speaker, "he was a good showman and had a fine sense of humour" (Patrick, 1961, p. 121). He developed this role and "so with speeches and with pictures, he began with his effective brand of earthy eloquence to carry his story to whoever would listen. He realised the dramatic qualities of a gully. Sheet erosion might be the more insidious, but pictures of the road-swallowing, field-slashing gully were what could persuade people that here was a problem of real national concern" (Barnes, 1960, p. 506).

Bennett was awarded the American Geographical Society's Cullum Medal in 1948 for "The scientific conservation of our precious heritage of soil." He was buried in Arlington National Cemetery, "a fighter who served his country no less than those who battled men."

[*See also* Erosion.]

### BIBLIOGRAPHY

Barnes, C. P. "Hugh Hammond Bennett, 1881–1960." *Annals of the Association of American Geographers* 50 (1960), 506.

Bennett, H. H. *Soil Erosion, A National Menace.* Washington, D.C.: United States Department of Agriculture, 1928.

Brink, W. *Big Hugh, the Father of Soil Conservation.* New York: Macmillan, 1951, p.167. Contains a portrait and partial bibliography, but the prose verges on the purple.

———. *Soil Conservation.* Chicago: McGraw-Hill, 1939.

Patrick, A. L. "Hugh Hammond Bennett." *Geographical Review* 51 (1961), 121–124.

—ANDREW S. GOUDIE

## BIOCENTRISM. *See* Belief Systems.

**BIODIVERSITY.** *See* Biological Diversity; Evolution of Life; Fishing; *and* Valuation.

## BIOGEOCHEMICAL CYCLES

Environmental scientists recognize that the chemical elements that compose the Earth are not held in static compartments. The components of the atmosphere and oceans are transported by the fluid movements of these bodies, which determine our daily weather patterns and the major ocean currents such as the Gulf Stream. Even the solid materials of the Earth—its crust and mantle—circulate by the processes that cause mountains to rise from the sea floor and rock materials to break down, delivering dissolved constituents to the sea. For every chemical element of the periodic table, we can outline a global cycle of its movement between the land, the atmosphere, and the oceans. We can compile measurements from studies around the world to estimate the annual transport of materials between these compartments, usually in units of millions of metric tons per year.

There are two broad categories of global cycles: sedimentary cycles and biological cycles. Geologic processes determine the movement of elements in the sedimentary cycles. Elements circulating in sedimentary cycles typically have no gaseous forms; these are the common rock-forming elements, such as calcium, phosphorus, silicon, and most trace metals. Each year, these elements move from land to sea in the process of rock weathering, which stems from the mechanical and chemical breakdown of all crustal materials that are exposed above sea level. About 17,500 million metric tons of material are carried from land to sea each year. Some of the elemental constituents of this material—for example, sodium, may spend a long time as a dissolved constituent of sea water, but eventually they are deposited on the sea floor. The deposition may be physical—for example, the deposition of river-borne sediments—or biological, as when a marine organism, with calcium carbonate ($CaCO_3$) in its skeleton, dies and falls to the sea floor. [*See* Ocean Chemistry.]

If it were not for the internal movements of the Earth's crust and mantle, the movement of elements in sedimentary cycles would be unidirectional, not cyclic. However, tectonic movements of the Earth's crust carry marine sediments deep into the Earth, where they are transformed back into rock. The average age of the exposed sea floor is only about 150 million years—far less than the age of the Earth's oceans at 3.8 billion years. Older sediments have been recycled. Tectonic movements of the crust and mantle result in the uplift of marine sediments to form new mountainous areas on land, renewing rock weathering and rejuvenating the global sedimentary cycle of these elements. Elements moving in sedimentary cycles may spend only a fraction of their life near the Earth's surface, where they may be incorporated in organisms. The average lifetime of an atom of phosphorus in sea water is about twenty-five thousand years. [*See* Phosphorus Cycle.] When this atom of phosphorus is incorporated in a marine sediment, it is removed from the biosphere for about 400 million years.

Elements moving in global biological cycles are typically found in one or more gaseous compounds that are produced by organisms. Carbon, oxygen, nitrogen, and sulfur are good examples. These elements are also found in rocks and also circulate in the sedimentary cycle. For example, a small amount of nitrogen is found in rocks (as $NH_4^+$), and nitrogen derived from rock weathering is carried to the sea by rivers. [*See* Nitrogen Cycle.] But the contrast between the sedimentary and biological cycles is that biotic processes dominate the annual movement of elements with biological cycles. For example, relative to the total amount of nitrogen delivered to the oceans each year, very little is sequestered in marine sediments. The bulk of it, as much as 100 million metric tons per year, is released to the atmosphere as $N_2$ by the biological process of denitrification.

Biological processes dominate the global cycle of carbon. [*See* Carbon Cycle.] Each year the uptake of carbon by plant photosynthesis and the release of carbon dioxide by metabolism result in a movement of carbon to and from the atmosphere that is more than one thousand times greater than its movement in the underlying sedimentary cycle of carbon. The annual additions of oxygen to the Earth's atmosphere by photosynthesis, and its removal by biological metabolism, also dwarf the movements of oxygen in the underlying sedimentary cycle, in which oxygen is consumed by rock weathering and carried to the sea, largely as $SO_4^{2-}$. Thus, elements moving in global biological cycles have a relatively rapid circulation near the Earth's surface as a result of biotic activity.

Geochemists define the *mean residence time* of an element in any compartment of the Earth as the mass of the material in that compartment divided by the annual exchange between that compartment and other sectors of the Earth. For instance, the mean residence time of sodium in the world's oceans is about 75 million years—obtained by dividing the sodium content of all sea water by the annual delivery of sodium to the sea by rivers. In contrast, the mean residence time of bicarbonate ($HCO_3^-$), the major reservoir of carbon in sea water, is about 100,000 years. With its movements primarily determined by biological activity, carbon spends much less time in sea water. As a general rule, the mean residence time of elements moving in biological cycles is much shorter than those moving in sedimentary cycles at the Earth's surface.

One advantage to quantifying the global cycles of the elements, particularly the elements of life, is that we can evaluate the impact of humans on the cycles of individual elements. In the mining of metals from the Earth's crust, humans expose buried rocks much more rapidly than would normally occur by erosion and rock weathering. Thus, humans accelerate the movement of the elements in the global sedimentary cycle. The annual transport of copper in the world's rivers is about three times greater than what one might estimate from the natural rate of rock weathering alone. The difference is largely attributed to the extraction and smelting of copper ores by humans. The mining of coal and the extraction of petroleum accelerate the natural rate at which these carbon-rich sediments would be exposed by rock weathering at the Earth's surface. The current rate of combustion of fossil fuel releases carbon dioxide to Earth's atmosphere about seventy times more rapidly than we would expect in nature. Clearly, humans as a biological species have an enormous capacity to increase the rate of movement of materials in both the sedimentary and biological cycles.

[*See also* Biosphere; Ocean Life; Origin of Life; Oxygen Cycle; *and* Sulfur Cycle.]

### BIBLIOGRAPHY

Bertine, K. K., and E. D. Goldberg. "Fossil Fuel Combustion and the Major Sedimentary Cycle." *Science* 173 (1971), 233–235.
Schlesinger, W. H. *Biogeochemistry: An Analysis of Global Change*. 2d ed. San Diego: Academic Press, 1997.

—WILLIAM H. SCHLESINGER

**BIOLOGICAL CONTROL.** *See* Agriculture and Agricultural Land; Exotic Species; *and* Pest Management.

# BIOLOGICAL DIVERSITY

The term *biological diversity*, or *biodiversity*, refers to the myriad kinds of living things. Most obvious are the "charismatic megavertebrates" that have been the symbolic focus of conservation campaigns—pandas, tigers, or koalas; however, biological diversity includes living things from all taxonomic groups and levels of biological organization. The vast majority are taken for granted: the bacteria, fungi, and invertebrates that decompose organic matter; the phytoplankton that fuel ocean productivity; or the invertebrates and microorganisms that recycle nutrients from sediments. These inconspicuous organisms are rarely the focus of human concern. Even the larger vascular plants—grasses, herbs, and trees—blend into a green backdrop, consuming carbon dioxide and producing oxygen. Many organisms are overlooked or even despised, such as agents of disease, pests and predators of domesticated plants and animals, and animals dangerous to humans. People find it difficult to embrace all living things under the umbrella of biodiversity.

Biological diversity is a huge topic, and the scope of this article is limited by its context. Contemporary global change is brought about mainly by humans, and this article focuses on interactions of humans with other living things: how people perceive and measure biological diversity, why they value it, and how they threaten and protect it.

**Kinds of Biological Diversity.** It is easiest to recognize biological diversity at the level of species, the different kinds of organisms that are regarded as separate biological populations and separate units of evolution and that are given specific scientific names. This focus belies the fact that biological diversity exists on a continuum ranging from genetic variation among individuals in a single population to the various biomes of the Earth, such as tundra, freshwater lakes, and polar oceans.

The term *genetic diversity* usually refers to variation within species populations. Local populations of wide-ranging species may differ genetically; individuals in any single population differ from one another; and a single organism can be endowed with variation from its genetically different parents. Genetic difference among populations of a single species can represent adaptation to local conditions, while genetic variation among individuals in a single population offers the potential for adaptation to changing environmental conditions, which is especially important as the raw material for evolutionary response to global change. Genetic variation within individuals is thought to confer greater fitness in terms of enhanced reproduction and survival over that of individuals who carry two identical copies of the same genetic material. Above the species level, diversity among higher taxa (orders, classes, phyla) represents further differentiation of genetic lineages and correspondingly different ways of making a living. All of these ways of classifying biological diversity rest on genetic connections—presumed evolutionary relationships that range from close (individuals in the same local population) to distant (e.g., animal phyla as different as sponges and arthropods).

Another way of classifying biological diversity at different scales is to define an assemblage of organisms that occur in the same place at the same time but have no close genetic relationship. These are variously termed *communities*, *ecosystems*, or *biomes*, depending roughly on their spatial extent. They are usually named for their most conspicuous organisms (often vascular plants) and sometimes also for the geophysical regions where they occur, such as coral reefs, sea grass beds, or dry tropical forest. Definitions of communities or ecosystems are more arbitrary than those of taxonomic groups, and the boundaries between neighboring communities or ecosystems are more difficult to delineate. The number and distribution of communities within a region contribute to its biological diversity. The fact that a freshwater lake has emergent aquatic vegetation and associated animals along the shoreline, plankton and fish in the open water, and invertebrates and microorganisms in the deepwater sediments makes it more diverse than it would be if the same populations of plants and animals were scattered evenly over the whole lake. The same is true on a global scale, where the major biomes (boreal forest, desert, grasslands) replace one another along altitudinal, latitudinal, precipitation, and soil gradients.

**Measures of Biological Diversity.** Measures of diversity incorporate both *richness*, the number of species or communities in a given area, and *evenness*, the relative abundance of those species or communities. Biodiversity may be expressed in terms of richness alone, or by using indexes that combine richness and evenness. Areas with many species or communities, each about equally abundant, are considered more diverse than areas with fewer species or areas where a few species are very abundant and the rest much less abundant.

**Patterns of Biological Diversity.** Ecologists have detected patterns of biological diversity in space and in relation to biophysical features. One commonly observed pattern is a "species–area" relationship, in which the number of species found in a given area increases as area increases, at first rapidly, then levels off. Species–area relationships themselves show patterns in response to environmental gradients: more species are found in a given area in tropical regions than in higher latitudes; greater animal diversity exists in habitats with greater structural complexity (e.g., forests vs. grasslands); and there are fewer species per area in "extreme" conditions, such as low precipitation. Some of these spatial patterns may reflect a correlation between higher productivity and higher diversity. These patterns can interact with human activities that threaten biodiversity; the conversion of a hectare of tropical forest to pasture may affect far more species than converting an equal area of temperate forest.

Another aspect of spatial pattern in diversity concerns changes in species composition across the landscape. These patterns also interact with human activities. In the United States, national forest managers have sometimes argued that clear cutting enhances diversity at the level of the national forest by adding early successional species to those that inhabit mature forest stands. Opponents of clear cutting, looking at national forest land where most surrounding land is already in early succession, have argued that regional diversity is enhanced by maintaining national forest land in mature stages to provide more habitat for late successional species. *Ecotones*, the zones where two or more communities or ecosystems meet, are often highly diverse, combining species typical of both adjoining systems. Some wildlife managers have argued that creating more such edges—for example, by cutting small patches in mature forest—will increase diversity. Again, taking a broader regional focus suggests that a mix of different types of habitats, some edge and some interior, supports the highest levels of diversity regionwide.

Ecologists have also noted patterns of biological diversity in relation to dynamic and functional features of ecosystems. More diverse systems are thought to exhibit greater stability and greater ability to recover from perturbations, including those caused by humans. Concern over oil spills in the high Arctic is based partly on the expectation that systems with limited diversity in extreme environments will recover slowly, if at all. Biological diversity is thought to be higher in systems with intermediate frequencies of disturbance from sources such as wildfire or hurricanes. Too frequent fires eliminate all but a few very fire-tolerant species, whereas too few fires allow fire-intolerant species to dominate those adapted to survival or recovery after fires. Changing natural disturbance regimes is often a feature of human activities, with persistent consequences for biological diversity; widespread fire suppression in the United States has altered or even eliminated some fire-adapted communities, such as longleaf pine woodland. Some human activities mimic natural disturbance, but often with greater intensity, greater frequency, or both (e.g., clear cutting vs. blowdowns in forests), and this usually pushes ecosystems toward lower diversity. [*See* Fire.]

**Why Biological Diversity Is Important.** Because human activities often affect biodiversity, usually for the worse, questions about its value inevitably arise. What good is it? How much do we need? How should we trade off losses in biodiversity against gains in other areas, such as economic development? Are there ways of valuing biodiversity besides its importance to humans?

*Use by humans.* Utilitarian reasons for valuing biodiversity focus on what it provides to humans, and these have been the mainstay of arguments in favor of conserving it. These arguments view biological diversity as a source of goods, ecological services, and aesthetic and spiritual benefits. An overused example is biomedical prospecting for naturally derived compounds for new medicines. Other examples of the value of raw materials include traditional products like timber or fish and also chemical feedstocks for high-tech materials. In addition, genetic diversity in wild relatives of domesticated plants and animals is a potential source of improved protection against pests and diseases.

Less easily characterized are the ecological functions that biological diversity provides to humans. For example, biological organisms decompose waste products, remove impurities from water, generate oxygen, convert low-quality vegetable matter to high-quality protein, and pollinate crops.

Still less tangible, but not necessarily less important, are the aesthetic, psychological, spiritual, and religious benefits of biological diversity. Humans derive satisfaction from experiencing beauty and order in natural systems. They may draw spiritual strength from wild places where nonhuman forces predominate. Religious and cultural practices in human societies incorporate landscapes, organisms, and biologically derived products from fungi to whales. [*See* Belief Systems.]

*Monetary value.* Questions about how much biodiversity we really need and how biodiversity should be weighed against human activities that threaten it have spurred attempts to express the utilitarian values of biodiversity in monetary terms. For biological products that are already traded in the marketplace, such as timber or ecotourism adventures, market value is a widely accepted way of monetizing worth, although many fear that market prices do not account fully for the side effects of exploiting biological resources. For biological products that might become market goods in the future, such as undiscovered medicines from tropical rainforests, economists have attempted to calculate an option value for preservation of this potential. For ecological services such as water purification, environmental economists have tried to assign monetary value by combining the estimated costs of replacing those natural services with engineered equivalents and the costs of environmental degradation (e.g., the manufacturing costs of using dirtier water for industrial processes). Even the less tangible ways of valuing biological diversity—the aesthetic and spiritual benefits—have been monetized by asking people how much they would be willing to pay to preserve a particular biological feature. These monetary expressions of the value of biological diversity are apt to find their way into cost-benefit analyses, where proposals for human uses of biological systems are weighed against losses of biodiversity, or where opportunities to preserve biodiversity are weighed against expenditures for other social benefits. Many conservation-minded people wonder whether even the most comprehensive efforts to monetize the many uses of biodiversity by humans can capture its full value. [*See* Valuation.]

*Intrinsic value.* Even more skeptical are those who believe that biological diversity has intrinsic value, quite apart from its appreciation and use by humans. In this framework, biological diversity exists in its own right and should continue to do so unless there are compelling reasons for its destruction. This argument follows the same structure as that for the intrinsic value of human life: it is not to be traded for economic gain or convenience, but safeguarded, except in very rare circumstances. The U.S. Endangered Species Act (ESA) works this way in principle (although not in practice); it evaluates actions that jeopardize the survival of endangered species without regard for socioeconomic consequences, unless there are extraordinary human needs that cannot be met without loss of endangered species. Some advocates for the intrinsic value of biodiversity attribute equal value to all biological organisms, regardless of their aesthetic appeal

or level of biological complexity. Under this framework, humans have the same moral obligations toward all living things, and perhaps also to assemblages of organisms like ecosystems, or to ecological phenomena like mass migrations of African wildlife. Other environmental ethicists attribute higher value to complex organisms that behave more like humans—exhibiting emotion, feeling pain. Advocates of the latter framework believe that humans have a greater obligation toward organisms that seem more like us, mainly birds and mammals.

Using any of these methods of valuing biological diversity to decide when to forgo other benefits to protect or restore biodiversity implies an answer to the question, How much diversity do we really need? The answer depends critically on which framework for valuation is used. Those who attribute the same intrinsic value to all living things would answer that we need all the diversity present at some previous time, usually assumed to be prior to the Industrial Revolution and perhaps prior to the Agricultural Revolution as well. Those who focus on the services that biological systems provide to humans would say that we need enough kinds and numbers of organisms to produce oxygen, purify water, or perform other services of concern. Obviously, the latter view could be used to justify considerable losses of biodiversity, particularly over the short term. Thus, the human-dominated landscapes of western Europe or Indonesia, which provide many ecological services and even aesthetic pleasures, have a very depleted flora and fauna compared to prehistoric times.

**Threats to Biological Diversity.** Loss of biological diversity has become a global concern. Although it is true that every species will someday become extinct (just as every individual will die), there is little doubt that industrial humans have accelerated the rate of extinction two to five orders of magnitude above "background" levels. Even early hunter-gatherers extirpated many species of large mammals and birds, particularly on oceanic islands. Today, the decline of one fishery after another to uneconomic levels continues this pattern of uncontrolled exploitation. As devastating as direct killing can be, however, destruction and fragmentation of habitats are undoubtedly the greater threats to biological diversity. Conversion of natural landscapes to agricultural and urban uses causes obvious loss of habitat. Less obvious loss results from human-induced changes in ecological processes, such as changes in productivity through the addition of nutrients from fertilizers or waste water. The spatial pattern of habitat loss can be as important as its extent: the same area of forest habitat broken into small patches sustains a different flora and fauna than do larger patches; barriers to traditional migration routes from road construction can degrade remaining habitat disproportionately to the area actually lost. [*See* Extinction of Species.]

The globalization of modern human society poses special threats to biological diversity. For millennia, humans have been transporting organisms beyond their natural ranges, wittingly and unwittingly, causing extinctions of native species through competition and predation by introduced species. With high-speed transport, pathogenic microorganisms can circle the globe in hours. Even ocean travel homogenizes nearshore ecosystems worldwide by transporting organisms on ship hulls and in bilge water. Once-local pollution has gone worldwide as well: acid emissions from power plants in Europe and the United States circulate in the upper atmosphere and rain down anywhere; heavy metals and pesticide residues appear in the tissues of animals in remote polar regions. Some human actions have global repercussions, changing the chemical composition of the atmosphere and, probably, altering world climates. The interaction of global climate change with habitat destruction and fragmentation is particularly troubling; species that might adapt successfully to gradually changing climates by shifting their ranges will be unable to do so if contiguous habitat has been lost to development. Globalization also has social implications that impinge on biodiversity. Uncontrolled borders facilitate smuggling of endangered species products, such as the bear gall bladders used in Asian folk medicine. The homogenization of human cultures fostered by rapid transportation and electronic communication encourages people in all parts of the world to pursue the same patterns of development and material culture that have already devastated biological diversity in industrial societies. Rapid human population growth exacerbates these threats to biodiversity.

**Protection of Biological Diversity.** Although the impact of humans on biological diversity has been devastatingly negative, there are some encouraging efforts to slow the rate of loss and even to restore biodiversity. Although much can be said about how to conserve biodiversity, this discussion focuses on methods of setting priorities and on examples of institutions and policies for biodiversity protection. [See Conservation.]

*Species conservation.* The job of conserving the world's biological diversity clearly exceeds the resources currently allocated to the task, so hard decisions must be made about which species and ecosystems to protect first. For species, criteria for high priority include uniqueness (species with no close relatives), abundance (species with the lowest numbers and those that have suffered the greatest declines), distribution (species represented in only one or a few locations), threat (species facing the greatest immediate pressures), ecological function (species with functional importance disproportionate to their numbers, such as some key predators), and, sometimes, aesthetic or emotional appeal.

*Habitat and ecosystem conservation.* The futility of evaluating all threatened species in this manner and then implementing protection on a species-by-species basis has led many conservationists to focus instead on habitat and ecosystem protection. A widely used tool for identifying areas with high priority for protection is gap analysis. This method identifies *hot spots* of biological diversity—places where there are high levels of diversity (many species, many habitats) and high levels of endemism (species not found elsewhere). Candidate areas for protection are then selected to form a complementary suite of areas that encompasses the widest variety of resources. Sometimes gap analysis also includes areas where biodiversity is being lost especially rapidly (e.g., dry tropical forest) or where species and habitats are especially vulnerable to extinction (e.g., isolated areas such as oceanic islands). Recently, this kind of analysis has been extended to consider the cost of protecting candidate areas, in order to assemble a suite of areas that protect the most biological diversity for a given cost; this strategy may emphasize protection in developing countries, where land acquisition costs are lower.

*Laws and institutions.* Institutional and legal mechanisms for protecting biological diversity range from local to international and are so numerous that we can only illustrate them with a few examples. The Nature Conservancy (TNC) is a private nonprofit organization whose traditional focus has been species and habitat preservation through the purchase of small areas. Recognizing that biodiversity protection on small parcels is affected by surrounding land uses, and that successful protection must include multiple-use lands as well as those dedicated to conservation, TNC has begun working with both private land owners and public land managers toward conservation on regional scales. [See Environmental Law.]

The ESA constrains actions by public and, to a lesser extent, private entities that threaten the continued existence of species at risk of extinction. Implementation of the act has been controversial, drawing criticism both from those who believe it does too little to list and then protect species in peril, and from those who believe it interferes too often with legitimate activities of public and private land owners. The ESA has been credited with some conservation successes, mainly for species facing single, fairly easily reversed threats, such as dichlorodiphenyltrichloroethane (DDT) contamination of bird eggs.

Internationally, the Convention on International Trade in Endangered Species of Wild Fauna and Flora (CITES) restricts trade in live organisms and products from species listed as endangered. A recent CITES decision continuing restrictions on trading elephant ivory has sparked criticism from a few countries where elephants have been protected successfully. As with most international agreements, compliance is voluntary, and certain countries are notorious for their markets in banned products such as tiger parts and rhino horn. [*See* Convention on International Trade in Endangered Species.]

Another international agreement, the Convention on Biodiversity, highlights the tension between industrialized and developing countries in balancing socioeconomic development and conservation of biological diversity. A major stumbling block to designing and ratifying the convention has been the allocation of financial benefits from commercial development of biologically derived products. [*See* Convention on Biological Diversity.]

***Development and biodiversity.*** The struggle between development and biodiversity is particularly acute in developing countries, where much of the remaining biodiversity is found, but where means to conserve it are most limited and human needs most compelling. One organization concerned with this dilemma is the International Union for the Conservation of Nature and Natural Resources (the World Conservation Union; IUCN), a consortium of governmental and nongovernmental entities formed to protect and sustain the world's resources. [*See* World Conservation Union.] Among its activities are publication of the Red List of threatened species and sponsorship of species survival plans for conservation of selected taxa. Institutional participants in these plans often include conservation organizations, such as World Wide Fund for Nature (formerly World Wildlife Fund; WWF), and zoos from industrialized countries. The role of zoos in protecting biological diversity has been controversial, in part because zoos and countries with wild populations of endangered species may disagree about whether to emphasize captive breeding or protection in the wild, and in part because zoos historically have consumed biological diversity rather than protecting it.

Integrating human needs and biodiversity protection is a new challenge for international aid organizations such as the World Bank, which for many years came down heavily on the side of economic development. [*See* World Bank, The.] Some innovative instruments have been developed for using the wealth of industrialized countries to support conservation in developing countries, including "debt for nature" swaps, in which part of a country's foreign debt is forgiven in return for its dedication of land to the protection of biodiversity.

Another approach to reconciling socioeconomic concerns with conservation is evident in the United Nations Man and the Biosphere (MAB) reserves. These consist of a core area undisturbed by human activities, surrounded by a buffer of low-intensity, biodiversity-compatible human uses (e.g., extraction of nontimber forest products like thatch or nuts). Such systems view human societies—particularly traditional, resource-based economies—as part of nature, to be sustained along with the nonhuman components of the ecosystem. Globalization imperils these traditional societies, which are often linked with intimate knowledge of biological diversity as food, medicine, and religious artifacts. Losing their biodiversity base threatens these cultures, and loss of the cultures extinguishes their rich knowledge of biological diversity. Thus, the preservation of human and nonhuman diversity are closely linked.

[*See also* Agriculture and Agricultural Land; Deforestation; Earth History; Environmental Movements; Exotic Species; Evolution of Life; Global Change, *article on* Human Dimensions of Global Change; International Geosphere–Biosphere Programme; *and* Origin of Life.]

## BIBLIOGRAPHY

Daily, G. C., ed. *Ecosystem Services: Their Nature and Value.* Washington, D.C.: Island Press, 1997.

Leopold, A. *A Sand County Almanac and Sketches Here and There.* New York: Oxford University Press, 1949. A classic source in environmental ethics, proposing a "land ethic" focused on ecosystem integrity rather than on human uses.

Meffe, G. K., and C. R. Carroll. *Principles of Conservation Biology.* 2d ed. Sunderland, Mass.: Sinauer Associates, 1997. A comprehensive textbook, with good attention to human aspects of implementing conservation actions.

Myers, N. *The Sinking Ark.* Oxford: Pergamon Press, 1979. A forceful look at how humans are accelerating extinction.

Noss, R. F., and A. Y. Cooperrider. *Saving Nature's Legacy: Protecting and Restoring Biodiversity.* Washington, D.C.: Defenders of Wildlife and Island Press, 1992. Emphasis on U.S. examples of biodiversity loss and protection.

Western, D., R. M. Wright, and S. C. Strum, eds. *Natural Connections: Perspectives in Community-Based Conservation.* Washington, D.C.: Island Press, 1994. Case studies of community involvement in sustainable development from around the world.

Wilson, E. O. *The Diversity of Life.* Cambridge, Mass.: Belknap Press of Harvard University Press, 1992. A beautifully written overview of biodiversity and its loss.

—Lynn A. Maguire

# BIOLOGICAL FEEDBACK

A *feedback* is the return of output to the input part of a system. The system may be the body of a living organism, a population, or the interacting biotic and abiotic components of an ecosystem. Positive feedbacks amplify a change in the state of the system, thereby destabilizing it. Negative feedbacks act against change and are generally stabilizing provided time lags in response are short. As a result of such feedbacks, nature is frequently nonlinear, with responses to an environmental change being disproportionate to the change itself. The circularity, nonlinearity, and interconnectedness of feedbacks make their consequences difficult to analyze, quantify, and predict.

**Positive Feedbacks.** Figure 1 illustrates a feedback that might lead to rapid global warming. Several such feedbacks may occur in climate change. Warming of soils induced by rising carbon dioxide may increase decomposition in soils, releasing more carbon dioxide. Disruption of climatic zones may kill trees, also releasing carbon dioxide. If the world becomes warmer and wetter, another greenhouse gas, nitrous oxide, may be released in increasing amounts by soils. A cooling stratosphere may deplete ozone faster, harming the plankton that pump carbon dioxide into the ocean depths and so cooling the stratosphere further.

There is evidence that the long-term cooling and drying over the past seven thousand years was triggered by subtle changes in the Earth's orbit, but that abrupt desertification in the Sahara around fifty-five hundred years ago was caused by positive feedback. Reduced vegetation and increased surface albedo result in decreased rainfall, which further reduces the vegetation cover.

Positive feedback may occur in population growth or decline. The more adult organisms there are, the more there are to reproduce, as seen in pest outbreaks and disease epidemics and when invasive species are intro-

duced to new areas (e.g., rabbits and *Opuntia* cactus in Australia). [*See* Exotic Species.] Introduced species may show "ecological release" from the negative feedbacks that naturally check the population. Some species, including our own, can grow at faster than exponential rates through adaptation. Similarly, feral pigs in the Galápagos are spreading their exotic food plants. Some social species (such as several seabirds and whales) breed less readily at lower densities (the "Allée effect"), so generating an "extinction vortex."

"Edge creep" (Soulé, 1986) is a feedback in landscape degradation. Before fragmentation, a habitat may have communities at its edge inimical to those at its core. As habitats are fragmented and penetrated by human agency, so the proportion of edge to core increases, and a positive feedback may ensue. Migrant birds in North America and forest-specialist birds of the tropics may be threatened by such changes.

Other positive feedbacks include overheating in mammals, "arms races" between predators and prey, and the "sexual selection" of ornamentation in mates.

**Negative Feedbacks.** During climate change, negative feedbacks include fertilization of plant growth through increased carbon dioxide and the spread of forests into high latitudes—both increasing uptake of carbon dioxide.

Negative feedbacks in ecosystems are often due to "density-dependent" processes, whereby an increase in the abundance of the population generates pressures reducing further population growth. Such pressures include competition within and between species for food. One of the most common biological feedbacks may be increased disease transmission as host abundance increases.

Negative feedbacks may dampen a population at a stable level ("regulation") or may lead to cycles within stable limits (as with hares and lynx in Canada). However, time lags before the feedback may lead to chaotic population changes.

Models of density-dependent negative feedback are at the core of "sustainable fisheries" management, although the effects on interacting species are often neglected.

**Interactions among Feedbacks.** Positive and negative feedbacks can interact at a variety of spatial and temporal scales. An illustration of a putative linkage of feedbacks is given in Figure 2. Global heat balance is calculated to be sensitive to cloud condensation, such that even small changes in the amount or type of phytoplankton might have large effects on climate.

Feedback loops can be thought of as linked in a complex interconnected web, as any organism or entity within the Earth system is an integral part of many loops. This can lead to "emergent properties"—phenomena that cannot be predicted from knowledge of the constituent components and that are not tractable using linear mathematics.

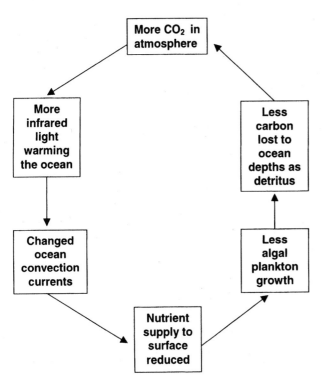

**Biological Feedback. FIGURE 1.** A Positive Feedback: The Plankton Multiplier.

A positive feedback creates rapid global warming after a small increase in warmth at the ocean surface. See Woods and Barkmann (1993) for details.

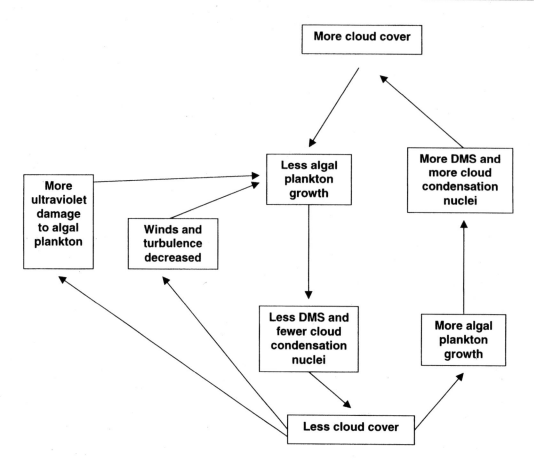

**Biological Feedback. FIGURE 2.** Hypothetical Interlinked Feedback Loops.

Both negative and positive feedbacks may occur after clouds are generated by dimethyl sulfide gas (DMS) released by marine algae. Many features of this system, first proposed by Charlson and colleagues in 1995, have been demonstrated. The overall influence of cloud cover on algae is as yet uncertain.

**Gaia Theory.** One of the most interesting consequences of the consideration of biological feedbacks is the concept of *Gaia*, developed by James Lovelock. Gaia theory portrays the "Earth system" as an intimate coupling between organisms and their environment. Organisms and their environment evolve together as organisms alter their material environment and their environment constrains and naturally selects organisms. These numerous feedbacks stabilize the system and maintain conditions within the range suitable for organic life. The composition of the Earth's atmosphere, for example, is in an extreme state of disequilibrium compared to lifeless planets and yet is fairly stable over periods of time that are much longer than the residence times of the constituent gases. Life on Earth has persisted despite an increase in solar luminosity of 25 percent since its origin 3.8 billion years ago. Gaia theory seeks to explain such phenomena and the mechanisms by which they arise. [*See* Gaia Hypothesis.]

To demonstrate that planetary regulation could emerge from traits developed by natural selection at an individual level, Lovelock and Andrew Watson developed "Daisyworld," a model that has since been elaborated by Stephan Harding. Daisyworld demonstrates the regulation of planetary temperature by coupling temperature and life through the influence of different-colored daisies on the absorption or reflection of heat from the (progressively hotter) sun. These models are useful metaphors to demonstrate principles but are controversial. One criticism is that their stability is dependent on constraints imposed in the assumptions of the model, and they are less stable if evolutionary change is permitted; in living systems, however, there are limits to such adaptation.

An alternative explanation for what appears to be evidence of "self-regulation" is that the conditions of the Earth happened, by chance, to be conducive to the evolution of conscious observers. An intermediate position accepts the existence of stabilizing mechanisms but denies that there is any tendency toward planetary self-regulation: on a planet where destabilizing feedbacks dominate, life is likely to perish before evolution produces conscious observers.

Other approaches including new mathematical models suggest that a number of stable states may emerge spontaneously owing to the complexity of ecological systems. In particular, models in preparation by Peter

Henderson and William Hamilton suggest that the abundance of negative feedbacks such as parasitism may be important to stability.

**Relevance to Human Belief Systems.** Gaia theory is often misinterpreted as implying that humans do not have to worry about their impact on the planet, as the Earth system will adjust to compensate and restore stability. This is a misunderstanding. The Gaian notion of stability is a dynamic stability in which a system can shift to new states of constancy that still result in conditions within the bounds suitable for life. Life persists, but the forms of life change. There is no guarantee that a shift triggered or amplified by human agency will result in conditions conducive to human life, and any sudden shift is likely to impact heavily on human society.

The philosopher Mary Midgely suggests that Gaia and the notion of the interconnected whole reminds us that we are not separate, independent, and autonomous, in contrast to much post-Enlightenment philosophy. The Gaia hypothesis is a scientific theory open to refutation; however, its portrayal of the human species as an integral part of the planet suggests an ethic to fill the recent moral void.

Lovelock has proposed two ethical consequences of the Gaia theory. The first is that stability in ecosystems and on the Earth requires the presence of limits and constraints on human activity and impacts. The second is that there is a need to take care of the environment, as we may harm ourselves by changing the system of which we are part.

[*See also* Belief Systems; Biogeochemical Cycles; Ecosystems; Population Dynamics; Religion; Stability; *and* Sustainable Development.]

## BIBLIOGRAPHY

Charlson, R. J., J. E. Lovelock, M. O. Andreae, and S. G. Warren. "Oceanic Phytoplankton, Atmospheric Sulphur, Cloud Albedo and Climate." *Nature* 326 (1987), 655–661. A classic demonstration of the importance of life in climate change; hypothesizes that locally beneficial adaptations of algae might lead to global phenomena such as cooling and nutrient flow to the oceans.

Claussen, M., C. Kubatzki, V. Brovkin, A. Ganopolski, P. Hoelzmann, and H.-J. Pachur. "Simulation of an Abrupt Change in Saharan Vegetation at the End of the Mid-Holocene." *Geophysical Research Letters* 26 (1999), 2037–2040. An example of the importance of including biological feedbacks in models of climate change.

Hambler, C. "Instability in Daisyworld." *Gaia Circular* 2.1 (1999), 5–6. A general critique of Daisyworld models, suggesting they are not evolutionarily stable against mutation of the preferred conditions of the daisies.

Houghton, J. T., L. G. Meira Filho, B. A. Callander, N. Harris, A. Kattenbury, and K. Maskell. *Climate Change 1995: The Science of Climate Change: Contribution of Working Group I to the Third Assessment Report of the Intergovernmental Panel on Climate Change.* New York and Cambridge: Cambridge University Press, 1996. Chapters in this key review on climate change highlight the importance of biological feedbacks and the need for more research into them.

Kauffman, S. A. *At Home in the Universe.* New York: Oxford University Press, 1995. A highly readable account of the fast-developing mathematical field of complexity theory, which relates the number and strength of interactions between entities in a system with the spontaneous emergence of stable states in the system.

Lenton, T. M. "Gaia and Natural Selection." *Nature* 394 (1998), 439–447. A review of the evidence for global anomaly and stability related to life on Earth. Applies Daisyworld models to rock weathering.

Lovelock, J. E. "Geophysiological Aspects of Daisyworld." In *Biodiversity and Global Change*, edited by O. T. Solbrig, H. M. van Emden, and P. G. W. J. van Oordt, pp. 57–70. Wallingford, U.K.: CAB International, 1994. A good, clear introduction to the Daisyworld models developed by J. E. Lovelock, A. J. Watson, and S. P. Harding. This type of model suggests that abundant life might stabilize a planet's temperature if there are constraints to the range of temperatures that life can endure.

———. *The Ages of Gaia.* 2d ed. Oxford: Oxford University Press, 1995.

———. "A Way of Life for Agnostics." *Gaia Circular* 2.2 (1999), 6–7.

Melillo, J. M., G. D. Prentice, G. D. Farquahr, E.-D. Schluze, and O. E. Salan. "Terrestrial Biotic Responses to Environmental Change." In *Climate Change 1995: The Science of Climate Change: Contribution of Working Group I to the Third Assessment Report of the Intergovernmental Panel on Climate Change*, edited by Houghton et al., pp. 445–481. New York and Cambridge: Cambridge University Press, 1996. An outline of positive and negative feedbacks in climate change.

Soulé, M. E., ed. *Conservation Biology. The Science of Scarcity and Diversity.* Sunderland, Mass.: Sinauer Associates, 1986. See Section 3 for discussion of edge effects.

Woods, J., and W. Barkmann. "The Planktonic Multiplier: Positive Feedback in the Greenhouse." *Journal of Plankton Research* 15 (1993), 1053–1074. A classic example of the vicious circle of biological feedback.

—SUSAN CANNEY AND CLIVE HAMBLER

# BIOLOGICAL PRODUCTIVITY

Biological productivity is the rate of change in biomass per unit area over time (usually a season or a year). Productivity and biomass are not directly related. That is, areas of large biomass (such as boreal forest ecosystems) may not necessarily have high productivity, since the biomass may have been accumulated over long periods of time. [*See* Biomass.] Conversely, areas of high productivity may have low biomass, when biomass is removed seasonally (as is the case in many salt marshes). The majority of plants are green plants and, therefore, are primary producers. [*See* Ecosystems.] Green plants generally produce energy-rich organic matter from solar energy in the visible wavelengths and from carbon dioxide and water. A small subset of organisms can produce new biomass from inorganic chemical energy, as, for example, in chemosynthetic bacteria found around thermal vents in the ocean floor, and from hydrogen sulfide found around sulfurous vents. Beyond green plants and chemosynthetic organisms, virtually all other organisms are "secondary producers" that produce new biomass by consuming, or altering, organic matter produced by green plants and chemosynthetic organisms (Barbour et al., 1987).

Green plants produce simple sugars from carbon dioxide and water. The energy comes from sunlight, captured by chlorophyll and other photosynthetic pigments. The radiant energy is used to excite electrons, and this energy is transferred to reaction centers within chloroplasts, which are chlorophyll-containing organelles found in algal and green plant cells. Transfer pigments allow the transfer of electrons, which produce energy-rich compounds of adenosine triphosphate (ATP) and NADPH (the reduced form of nicotinamide adinine dinucleotide phosphate). Visible radiation, in the presence of the above compounds, the enzyme ribulose bisphosphate, carbon dioxide, and water, produces simple three-carbon sugars. The energy and carbon chains in these simple sugars and inorganic nutrients taken up by the plants are the basic building blocks for the complex organic synthesis of the many organic chemicals present in plants (including sugars, starches, fats, proteins, nucleic acids, and secondary compounds, such as caffeine in coffee).

The organic compounds produced by plants form the basic building blocks needed by herbivores (which feed on plants) and by carnivores. Animals vary in the extent to which they can synthesize new chemical compounds or need them in their diet. All energy and carbon chains used by animals ultimately come from the energy fixed in simple sugars through photosynthesis by green plants or chemosynthetic bacteria. There is a loss of energy at each trophic (food chain) level. Plants are roughly 0.6 percent effective in transferring solar energy to chemical energy in biomass. Each trophic level (e.g., herbivores feeding on primary producers or carnivores feeding on herbivores) is roughly 10 percent efficient. That is, only 10 percent of the energy in biomass consumed at one level is transferred to biological production at the next trophic level. The rest is lost as feeding and assimilation inefficiencies, and through respiration, the process whereby organic matter is broken down to provide energy for the metabolism of the host organism. This limits the possible number of trophic levels. In addition, feeding often takes place at several levels, forming a "food web" that makes calculation of the absolute trophic level difficult.

There are also decomposer or detrital pathways, which break down dead plant and animal material. These pathways may also have a number of steps and be quite complex. Ultimately, the biomass produced by biological productivity is stored above or below ground in living and dead biomass, primarily plant biomass, in lake and ocean sediments, or broken down and released back to the atmosphere as carbon dioxide. Fossil fuel deposits

are the result of plant productivity and biomass altered through time by biological and physical processes.

Key controls on global photosynthesis and productivity include water availability, temperature, solar radiation, carbon dioxide concentration, and nutrient availability. Regional and latitudinal patterns in solar radiation and temperature result in significant gradients of potential primary productivity. Water limitation restricts the primary productivity of over 95 percent of the land surface. However, the limited availability of plant nutrients, including the macronutrient (those nutrients required in large amounts by organisms) nitrogen, also limits primary productivity over large areas. Primary productivity ranges from 10 to 400 grams per square meter per year in arctic and alpine tundra, from 10 to 250 grams per square meter per year in deserts and semidesert scrub, and from 1,000 to 3,500 grams per square meter per year in tropical rainforests. Productivity in temperate forests ranges from 600 to 2,500 grams per square meter per year, while temperate grasslands fall between 200 and 1,500 grams per square meter per year. Extreme deserts, including polar deserts and extreme hot arid deserts, may have productivities of only 0 to 10 grams per square meter per year, whereas swamps and marshes may have productivities of 800 to 6,000 grams per square meter per year (Whittaker and Likens, 1975; Table 1). Productivity is controlled by physiology, vegetation composition, and environment at differing time scales. Changes in any of these environmental variables can affect primary and global productivity.

A number of factors correlate with net or gross productivity. Gross productivity (net photosynthesis before losses due to plant respiration are accounted for; gross productivity minus respiration equals net productivity) and/or net productivity correlates with a number of factors, including leaf area duration (Kira, 1975), mean annual temperature (Lieth, 1973), actual evapotranspiration, mean annual precipitation (Barbour et al., 1987), and spectral absorption (normalized difference vegetation index, or NVDI). These relationships are correlative, however, and do not necessarily indicate causation.

Carbon dioxide is the only major resource limiting primary productivity that does not have a large regional variability, since carbon dioxide in the atmosphere is fairly well mixed and its concentration generally varies by less than 5 percent seasonally and latitudinally. [See Carbon Cycle; and Nitrogen Cycle.] Atmospheric carbon dioxide concentration has varied dramatically over geologic time, and has increased by more than 30 percent since the beginning of the Industrial Revolution, apparently as a result of human activities, including combustion of fossil fuels, forest clearing, and agriculture. This increase in global atmospheric carbon dioxide may have increased global primary productivity by as much as 1–15 percent since the beginning of the In-

**Biological Productivity. TABLE 1.** Annual Net Primary Production, Leaf Area Index, and Chlorophyll Content of Biomes

| TYPE OF VEGETATION | AREA $(10^6 \text{ KM}^2)$ | Net Primary Production RANGE $(\text{KG M}^{-2}\text{A}^{-1})$ | Net Primary Production MEAN $(\text{KG M}^{-2}\text{A}^{-1})$ | AREA TOTAL $(10^{12} \text{ KG})$ | Leaf Area Index RANGE $(\text{M}^2\text{M}^{-2})$ | Leaf Area Index MOST FREQUENT VALUE $(\text{M}^2\text{M}^{-2})$ | CHLOROPHYLL CONCENTRATION MEAN $(\text{G M}^{-2})$ |
|---|---|---|---|---|---|---|---|
| **Continents** | **149.0** | | **0.78** | **117.5** | | **4.3** | **1.5** |
| Tropical rainforests | 17.0 | 1–3.5 | 2.2 | 37.4 | 6–16 | 8 | 3.0 |
| Deciduous woodland (semiarid) | 7.5 | 1.6–2.5 | 1.6 | 12.0 | | 5 | 2.5 |
| Deciduous forests (temperate) | 7.0 | 0.4–2.5 | 1.2 | 8.4 | 3–12 | 5 | 2.0 |
| Evergreen temperate | 5.0 | 1–2.5 | 1.3 | 6.5 | 5–14 | 12 | 3.5 |
| Boreal forests | 12.0 | 0.2–1.5 | 0.8 | 9.6 | 7–15 | 12 | 3.0 |
| Dry scrub and sclerophylls | 8.5 | 0.3–1.5 | 0.7 | 6.0 | 4–12 | 4 | 1.6 |
| Savanna | 15.0 | 0.2–2 | 0.9 | 13.5 | 1–5 | 4 | 1.5 |
| Meadows and steppes | 9.0 | 0.2–1.5 | 0.6 | 5.4 | | 3.6 | 1.3 |
| Tundra | 8.0 | 0.01–0.4 | 0.14 | 1.1 | 0.5–2.5 | 2 | 0.5 |
| Shrub deserts | 18.0 | 0.01–0.3 | 0.9 | 1.6 | | 1 | 0.5 |
| Dry and cold deserts | 24.0 | 0–0.01 | 0.003 | 0.07 | | 0.05 | 0.02 |
| Agricultural crops | 14.0 | 0.1–4 | 0.65 | 9.1 | 4–12 | 4 | 1.5 |
| Swamps, marshes | 2.0 | 1–6 | 0.3 | 6.0 | | 7 | 3.0 |
| Inland waters | 2.0 | 0.1–1.5 | 0.4 | 0.8 | | | 0.2 |
| **Oceans** | **361.0** | | **0.155** | **55.0** | | | **0.05** |
| Open ocean | 332.0 | 0.002–0.4 | 0.125 | 41.5 | | | 0.03 |
| Upwelling zones | 0.4 | 0.41–1 | 0.5 | 0.2 | | | 0.3 |
| Coastal zones | 26.6 | 0.2–0.6 | 0.36 | 9.6 | | | 0.2 |
| Reefs and tidal zones | 0.6 | 0.5–4 | 2.5 | 1.6 | | | 2.0 |
| Brackish water | 1.4 | 0.2–4 | 1.5 | 2.1 | | | 1.0 |
| **Global total (Earth)** | **510** | | **0.336** | **172.5** | | | **0.48** |

SOURCE: Whittaker and Likens (1975); for details see Ajtay et al. (1979); Schulze (1982).

dustrial Revolution. Elevated carbon dioxide may also change the organic matter quality, plant–animal relationships, and plant–microbe relationships. A major area of current uncertainty is the extent to which increasing primary productivity due to global increases in carbon dioxide can be maintained (Intergovernmental Panel on Climate Change, 1996).

Anthropogenic additions to the nitrogen cycle mean that humans now add as much nitrogen to the global biological nitrogen cycle as is produced (or "fixed") naturally (Matson et al., 1997; Vitousek et al., 1997). This nitrogen has the potential to increase the biological productivity over large areas where nitrogen deposition has been augmented. It can also change species composition, biodiversity, and chemical composition of the plant material for animal consumers (Vitousek et al., 1997).

Net terrestrial carbon flux is the difference between carbon uptake in net primary productivity and carbon lost in respiration from decomposer organisms and consumers. Terrestrial net carbon flux is not necessarily in balance. In recent years, terrestrial ecosystems are thought to have been net sinks of carbon dioxide because of regrowth of forests, carbon dioxide fertilization from increasing atmospheric carbon dioxide, increased anthropogenic nitrogen deposition, and warmer global temperatures (Bender, 1996). This increase in net ecosystem carbon dioxide sequestration has apparently slowed the growth of atmospheric carbon dioxide concentration. However, not all ecosystems have responded with increased carbon sequestration. Arctic ecosystems, and some boreal forest ecosystems, have shown a change from net ecosystem sequestration to carbon loss to the atmosphere due to recent warming and drying (Oechel et al., 1993; Goulden et al., 1998).

[*See also* Agriculture and Agricultural Land; Fishing; Forestation; *and* Growth, Limits to.]

### BIBLIOGRAPHY

Ajtay, G. L., P. Ketner, and P. Duvigneaud. "Terrestrial Primary Production and Phytomass." In *The Global Carbon Cycle*, edited by B. Bolin, E. Degens, S. Kempe, and P. Ketner, pp. 129–182. Chichester, U.K.: Wiley, 1979.

Barbour, M. G., et al. *Terrestrial Plant Ecology.* 2d ed. Menlo Park, Calif.: Benjamin Cummings, 1987.

Bender, M. "A Quickening on the Uptake?" *Science* 381 (1996), 195–196.

Goulden, M. L., S. C. Wofsy, J. W. Harden, S. E. Trumbore, P. M. Crill, S. T. Gower, T. Fries, B. C. Daube, S.-M. Fan, D. J. Sutton, A. Bazzaz, and J. W. Munger. "Sensitivity of Boreal Forest Carbon Balance to Soil Thaw." *Science* 279 (1998), 214–219.

Intergovernmental Panel on Climate Change. *Climate Change, 1995: Impacts, Adaptations, and Mitigation on Climate Change: Scientific–Technical Analyses. Contribution of Working Group II to the Second Assessment Report of the Intergovernmental Panel on Climate Change*, edited by R. T. Watson, M. C. Zinyowera, and R. H. Moss. Cambridge and New York: Cambridge University Press, 1996.

Keeling, R. G., et al. "Global and Hemispheric Carbon Dioxide Sinks Deduced from Changes in Atmospheric $O_2$ Concentration." *Nature* 381 (1996), 218–221.

Kira, T. In *Photosynthesis and Productivity in Different Environments*, edited by J. D. Cooper. Cambridge: Cambridge University Press, 1975.

Larcher, W. *Physiological Plant Ecology.* 3d ed. New York: Springer, 1995.

Lieth, H. "Primary Production: Terrestrial Ecosystems." *Human Ecology* 1 (1973), 303–332.

Matson, P. A., et al. "Agricultural Intensification and Ecosystem Properties." *Science* 277 (1997), 504–509.

Myneni, R. B., et al. "Increased Plant Growth in the Northern High Latitudes from 1981 to 1991." *Nature* 386 (1997), 698–702.

Oechel, W. C., et al. "Recent Change of Arctic Tundra Ecosystems from a Net Carbon Dioxide Sink to a Source." *Nature* 361 (1993), 520–523.

Schulze, E. D. "Plant Life Forms and Their Carbon, Water, and Nutrient Relations." In *Encyclopedia of Plant Physiology*, edited by O. L. Lange, P. S. Nobel, C. B. Osmond, and H. Ziegler, pp. 615–676. Berlin and New York: Springer-Verlag, 1982.

Vitousek, P. M., et al. "Human Domination of Earth's Ecosystems." *Science* 277 (1997), 494–499.

Whittaker, R. H., and G. E. Likens. In *Primary Productivity of the Biosphere*, edited by H. Lieth and R. H. Whittaker, pp. 305–328. Berlin: Springer, 1975.

—WALTER C. OECHEL

## BIOLOGICAL REALMS

It has long been recognized that different parts of the world support distinctive assemblages of plants and animals. These general patterns were originally attributed to differences in climate, but this explanation was increasingly questioned as more species were described by natural historians. Thus, in the mid-eighteenth century, George Louis Leclerc Comte de Buffon noted that environmentally similar regions in the Old and New World tropics supported very different fauna. Extensive travels by Alexander von Humboldt and others subsequently showed that regional biotas elsewhere in the world were also quite distinctive. In 1820, Alphonse de Candolle proposed that the world be divided into twenty regions based on characteristic floras. The idea that species can be grouped into distinct regional assemblages was refined in 1858 by P. L. Sclater. He recognized six faunal regions based on the distribution of bird species, and this arrangement is still widely used (Figure 1). The most extensive region is the Palearctic, which includes Europe, the Middle East, and much of Asia. North America and the highlands of Mexico are represented by the Nearctic, which is linked by the isthmus of Central America to the Neotropical region of South America. Most of Africa and Madagascar are included in the Ethiopian region. The Oriental region is bounded on the north and west by the Himalayan mountains and associated ranges; it includes India, Sri Lanka, and Malaysia. The Australasian region, which includes Australia, New Zealand, New Guinea, and several islands in the East Indies, is completely isolated and possesses many endemic species. Subsequently, A. R. Wallace demonstrated that other animal groups have equally distinctive patterns that reflect their adaptation to globally variable environmental conditions.

Although many flowering plant families are widely distributed, cosmopolitan species are comparatively uncommon, and the majority are restricted by general habitat conditions in much the same way as animals. Takhtajan (1986) describes six floristic regions. The temperate and arctic regions of the Northern Hemisphere are termed the *Holarctic*. Central America and much of South America are affiliated with the Neotropical region. The Paleotropics cover most of Africa south of the Sahara, Madagascar, India, and Southeast Asia, with the distinctive flora of southern Africa incorporated into the Cape region. Australia is also recognized for its unusual flora. New Zealand, Patagonia, and other land areas in the Southern Hemisphere are grouped together in the antarctic region, although Antarctica itself is excluded. This scheme recognizes the main divisions proposed by Good (1964) on the basis of endemism. Broad endemic species have similar patterns of distribution and occupy comparatively large ranges within a continent. Conversely, the distributions of narrow endemic species are more restricted; they are usually limited by very specific habitat requirements and are considered to have evolved under conditions of geographic isolation. In his final map, Good (1964) shows thirty-seven floristic regions that range in size from the extensive Euro-Siberian region to smaller, isolated regions such as Hawaii and the islands of Saint Helena and Ascension in the South Atlantic. The recognition of

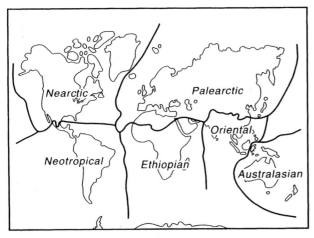

**Biological Realms. FIGURE 1.** Zoogeographical Regions of the World Based on the Distribution of Mammals. (Modified from Pielou, 1979.)

an antarctic temperate floristic region is an important difference between the distributions of plants and animals. The presence of podocarp gymnosperms and evergreen angiosperms (notably, the genus *Nothofagus*) provides the common link between these geographically isolated regions. The Cape flora of southern Africa is also treated as a distinct region by plant geographers. Similarly, the Paleotropical floristic region extends to New Guinea, whereas zoogeographers include many of the islands of Southeast Asia with Australia.

The concept of biological realms is based on the distinctive biota found within each terrestrial region. Movement of organisms between realms is restricted by physical barriers. In many cases, the realms are clearly isolated by the oceans, but mountain ranges and deserts also form insuperable barriers to dispersal. The change in biota can occur quite abruptly; the Palearctic fauna of northern Africa, for example, is very different from that found south of the Sahara. Elsewhere, the boundaries are less distinct. This is the case in Southeast Asia, where the separation of the Oriental and Australian faunal regions has been the subject of much controversy. In 1859, Wallace proposed a line of demarcation running through the Straits of Lombok and coinciding with the deep-ocean region between Borneo and Sulawesi. Wallace's Line separates the placental mammals from the marsupials, although other animal families are more widely dispersed throughout the Malaysian archipelago. This is reflected in the position of the boundaries that have been proposed in this region (Figure 2). Physiographically significant are Huxley's Line, proposed in 1868, which follows the Sunda Shelf, and Lydekker's Line (1896), which marks the limit of the Sahul Shelf. The lowering of sea level during the Pleistocene per-

mitted land animals to move throughout these temporarily emerged marine platforms, although the two faunal realms remained separated by the deep water surrounding Sulawesi.

Zoologists recognize eighty-nine families of terrestrial mammals (excluding bats). Three families are found worldwide because of human dispersal; these include rats and mice (family Muridae), rabbits and hares (Leporidae), and dogs and wolves (Canidae). The remainder vary in terms of their dispersal success. For example, ursids (bears), cervids (deer), and soricids (shrews) are found in all but the Australian region. Conversely, families including the kangaroos and wallabies (Macropodidae), wombats (Phascolomidae), and platypus (Ornithorhynchidae) are restricted to the Australian region. About 58 percent of the families of terrestrial mammals are endemic to a single faunal region. The degree of endemicity ranges from 91 percent in the Australian region to 3 percent in the Palearctic (Cox and Moore, 1993). Although plants are mobile during phases of dispersal, their geographic ranges are ultimately determined by adaptation to the environment. Consequently, many unrelated species exhibit similar patterns of distribution that typically reflect climate, topography, and soil conditions. Botanists recognize more than four hundred families of flowering plants, ranging in size from the Compositae, which contains more than one thousand genera and twenty thousand species, to those that are represented by a single monotypic genus (Good, 1964). Some families—for example, the Compositae and Gramineae—are cosmopolitan in their distribution. Others, such as the Cruciferae and Juncaceae, are predominantly found in temperate regions, while the Rubiaceae and Euphorbiaceae are mainly restricted to the tropics. In contrast, approximately 30 percent of plant families are endemic to specific areas. They include, for example, the Didieraceae of Madagascar, represented by four species, and the Xanthorrhoeaceae of Australia and New Guinea, with twelve species. In contrast, the Cactaceae, a much larger family represented by about two thousand species, has a correspondingly greater geographic range and is endemic to the Americas. Some plant families are notable for their discontinuous distributions in that they are found in geographically separate parts of the world. For example, the Platanaceae are found in North America and western Eurasia, the Strelitziaceae are native to South America and Africa, and the Pittosporaceae are present in Africa and Australia. The origin of such disjunct distributions has been a matter of much speculation.

The distinctive faunas of the Oriental and Australian realms are separated by a distance of only 25 kilometers in the Straits of Lombok and suggested to Wallace that the geography of the Malay Archipelago had changed significantly over time. He accepted the conti-

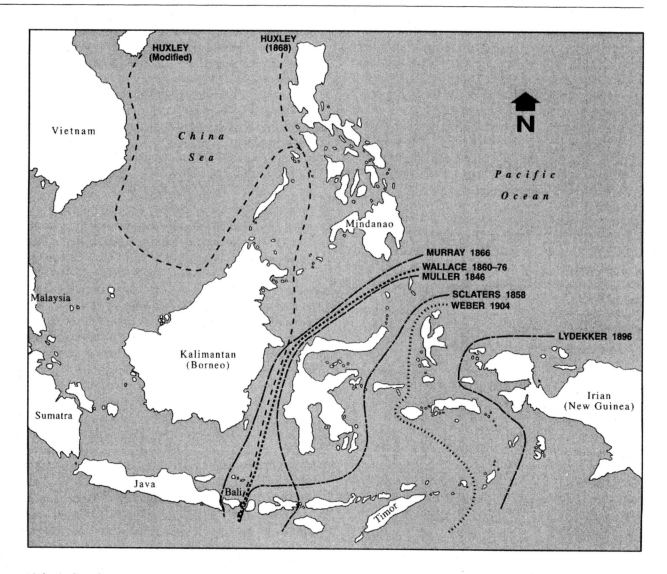

**Biological Realms. Figure 2.** The Various Lines Used to Delimit the Oriental and Australian Zoogeographic Regions. (Adapted from Simpson, 1977.)

nental extensionist tradition, which postulated that islands with faunal similarities to adjacent islands or continents must have become separated in the recent past, whereas a distinct faunal assemblage implies the absence of a previous land connection. The importance of past geologic change was further supported by the fauna of Sulawesi, which has a greater affinity with African species than with those of the adjacent islands. Previously, Wallace had argued that disjunct distributions could only arise from a continuous population and not by accidental transport as suggested by Darwin and others. Thus, he postulated, Sulawesi was a remnant of an ancient landmass covering what is presently the Indian Ocean. Current interpretations of the global patterns of life are now firmly based on the theory of continental drift and plate tectonics.

Fossil evidence suggests that two distinct paleofloral regions had evolved by the early Mesozoic (230 million years ago). North America and Eurasia existed as the single northern landmass Pangea. In the Southern Hemisphere, Gondwanaland united South America, Africa, India, Australasia, and Antarctica. The ancient floras of these regions were distinctly different, with ancestral conifers (Cordaitales) widely distributed across the north, and seed ferns (notably, *Glossopteris*) occurring in the south. The subsequent breakup of Gondwanaland is based on the present distribution of plants, such as the southern beeches (*Nothofagus*), which are disjunct across the southern continents despite poor mechanisms for dispersal. The separation of South America and Africa probably began in the lower Cretaceous (127 million years ago), Madagascar and India broke away in the mid-Cretaceous (100 million years ago), with India becoming isolated in the late Cretaceous (80 million years ago). At about this time, New Zealand began to drift away from Antarctica, with the rift between Australia and Antarctica occurring in the Eocene (49 million years ago). The breakup of Pangea started with the opening of the North Atlantic in the late Triassic (180 million years ago), although links between Europe and North America probably persisted at high latitudes until the Eocene, with connections between the Nearctic and Palearctic subsequently reestablished via Beringia. Africa was the first of the southern continents to join with a northern landmass; a temporary connection with Europe was established in the early Paleocene (63 million years ago), although the present configuration of the Mediterranean Sea did not arise until the Miocene (17 million years ago). Contact between India and Asia occurred in the Eocene, but species movements between these areas would subsequently be limited by associated crustal warping and the formation of the Himalayas. Finally, the connection of North and South America by a land bridge occurred in the Pliocene (6 million years ago).

The isolation and recombination of the continents over time has led to distinctive patterns of species distribution and areas of species richness. The most obvious pattern is that species richness decreases at higher latitudes, so that tropical regions exhibit far greater diversity than temperate or polar regions. Superimposed on this general pattern are the effects of environmental factors such as elevation and precipitation, all of which augment habitat variability and increase the potential for species evolution and survival. Similarly, geographically isolated islands are important centers of species diversity, because invariably they have large numbers of endemic species. It is estimated that about one in six plant species grows on oceanic islands and that one in three of all known threatened plants are island endemics (Groombridge, 1992). Island ecosystems have been altered dramatically since the mid-nineteenth century through the introduction of grazing animals and exotic plants. [See Exotic Species.] The problem of species loss is not confined to islands. Demands for food and natural resources have resulted in extensive loss of habitat. [See Human Impacts, *article on* Human Impacts on Biota.] This, together with accidental or deliberate introductions of exotic species, eradication of competitors, diseases, and other pressures, is rapidly changing the natural patterns of distribution that are the legacy of ancient and complex evolutionary processes.

[*See also* Biological Diversity; Biomes; Ecosystems; *and* Extinction of Species.]

### BIBLIOGRAPHY

Cox, C. B., and P. D. Moore. *Biogeography: An Ecological and Evolutionary Approach.* 5th ed. Oxford: Blackwell, 1993.
Cracraft, J. "Continental Drift and Vertebrate Distribution." *Annual Review of Ecology and Systematics* 5 (1974), 215–261.
Fichman, M. *Alfred Russel Wallace.* Boston: Twayne, 1981.
George, W., and R. Lavocat, eds. *The Africa–South America Connection.* New York and Oxford: Oxford University Press, 1992.
Good, R. *The Geography of Flowering Plants.* 3d ed. London: Longman, 1964.
Groombridge, B., ed. *Global Biodiversity: Status of Earth's Living Resources.* London: Chapman and Hall, 1992.
Pielou, E. C. *Biogeography.* New York: Wiley, 1979.
Simpson, G. G. "Too Many Lines: The Limits of the Oriental and Australian Zoogeographic Regions." *Proceedings of the American Philosophical Society* 121 (1977), 107–120.
Takhtajan, A. *Floristic Regions of the World.* Berkeley: University of California Press, 1986.
Whitmore, T. C., ed. *Biogeographical Evolution of the Malay Archipelago.* New York and Oxford: Oxford University Press, 1987.
Wolfe, J. A. "Some Aspects of Plant Geography of the Northern Hemisphere during the Late Cretaceous and Tertiary." *Annals of the Missouri Botanical Garden* 62 (1975), 264–261.

—O. W. ARCHIBOLD

## BIOMASS

*Biomass* is the quantity of organic matter. It is usually measured as the total for a study area or as a density, either per unit surface area or per unit volume. For studies of life on land, density per unit surface area is the usual measure, and units are typically given in kilograms per square meter, kilograms per hectare, and kilograms per kilometer. In marine and freshwater studies, density per unit volume is the usual measure, but sometimes density per surface area is also used. Density per unit volume is typically given in grams per liter and kilograms per cubic meter. Biomass is typically divided into live and dead organic matter, the latter sometimes referred to as *mortmass*, which is at this time an informal term.

**Importance of Biomass as a Variable.** Biomass is generally considered to be a primary ecological variable. In addition, for specific kinds of tissues of living things, there is a more or less constant ratio of carbon content to total biomass, and energy storage to total biomass. For example, for woody tissue, generally considered to be the primary live-biomass storage on Earth, carbon is 45 percent by dry weight of total biomass. Energy content ranges from somewhat more than 4 kilocalories per gram for woody tissue to approximately 9 kilocalories per gram for fats. These are sufficiently constant so that a measure of total biomass by major tissue type can be used to estimate energy and carbon storage.

The rate of change in biomass over time, also a primary ecological variable, is, by definition, *net biological productivity*—the difference between the amount added and the amount lost. Biomass is added to an ecosystem through photo- or chemosynthesis. It is decomposed and lost by respiration. Photo- and chemosynthetic organisms add organic matter through these processes and lose it through respiration. Other organisms, called *heterotrophs*, add biomass by ingesting, digesting, and incorporating organic matter from other organisms. These also lose organic matter through respiration.

Most live organic matter on Earth is stored as woody tissue, and most dead organic matter is stored in the litter and soils of forests, grasslands, and northern bogs. In terms of global carbon cycling and other biogeochemical cycles, therefore, the biomass of vegetation is most important. The biomass of animals is small by comparison. It has been generally thought that global biomass in bacteria and fungi was small, but this has recently begun to be disputed. Live biomass in the oceans is small in comparison to that in forests and grasslands, but the turnover in biomass (net production) of marine algae and photosynthetic bacteria can be large.

**Methods of Measurement.** There are two basic kinds of measurements of biomass: direct field sampling and remote sensing. The following discussion focuses on estimates of biomass of primary producers (plants, algae, and photosynthetic bacteria) because of the importance of these in comparison to other life forms, as mentioned earlier.

In theory, direct measurements of biomass could be done by direct harvest of large areas, but this is impractical and destructive to ecosystems and can cause significant environmental problems. Therefore, it is rarely done. Exceptions are biomass measurements made as part of other harvest or land-clearing operations, such as forest harvest. More commonly, in land systems, direct measurements typically involve two steps: (1) samples of biomass of individual organisms over a wide range of size so that a statistical relationship can be developed between body size and total carbon storage and (2) statistically valid field sampling of trees by species and size. In some cases, the first step is done by cutting, drying, and weighing entire individual organisms. Often this is not cost-effective, and a method called *dimension analysis* is used. Dimension analysis involves a statistically valid subsampling of individuals. For example, for a tree, a sample of limbs is selected at random. On each of these limbs, a subsample of branches is selected at random. On each selected branch, a subsample of leaves and twigs is selected, dried, and weighed. Statistical regression analysis is then applied to provide a mathematical relationship between simple measures of body size, such as height and diameter, and biomass. Most existing, statistically valid estimates of biomass use this method.

In recent decades there has been considerable work to attempt to estimate biomass by aircraft and satellite remote sensing. There are two premises behind these measures: (1) that there is a direct and statistically significant relationship between light reflected from photosynthetic organs (leaves of land plants, cells of marine algae, and bacteria) and total biomass and (2) that there is a statistically valid relationship between the biomass of the photosynthetic organs and total biomass. These premises hold up better for marine systems than for terrestrial systems. In terrestrial systems, reflection from leaves saturates at a leaf area index (LAI) much below that of an old-growth stand (leaf area index is the number of leaves above a point on the ground). Therefore, remote sensing by reflected sunlight can be used to estimate the biomass of leaves only to a relatively low value. The second premise also is violated for land plants, especially forests. LAI reaches a maximum in most forests while biomass continues to increase. Therefore, after the LAI maximum is reached, LAI no longer is correlated with total biomass.

As a result, the primary utility of remote sensing that depends on reflected sunlight is to determine the lands that are covered by a vegetation type and, where finer distinctions can be made, on major categories within that type, such as recently cleared areas, young stands, and mature stands. These can then be used by estimates derived from direct field measures to develop large area estimates for biomass. Other methodologies are in development, such as the use of radar from aircraft and satellites, but these remain experimental. [*See* Remote Sensing.]

**Estimates of Biomass for Major Areas of the Earth.** Despite its fundamental importance in ecology and to the Earth system, biomass is poorly measured for large areas of the Earth. With a few exceptions, there are no statistically valid estimates of biomass for the major biomes or for the total biomass on the Earth. The two exceptions are the biomass of the boreal forest of North America and the eastern deciduous forest of North America. Despite the lack of statistically valid estimates, biomass estimates are used widely. The problem with existing measures for land biomass is that the estimates are based primarily on studies of old-growth areas, typically near universities, so that there is both a geographic and ecological successional bias in the values, resulting in an overestimate by as much as two to four times, as Table 1 demonstrates. Because these values are used so commonly, a standard set of values for the major biomes of the Earth system is provided here, but the reader is warned that they remain at this time simply rough, first approximations. Even though the estimates are not statistically valid, Table 2 is based on a classic table of standard measures of biomass for all the major biomes. These numbers are still in use today.

[*See also* Biomes.]

### BIBLIOGRAPHY

Ajtay, G. J., P. Ketner, and P. Duvigneud. "Terrestrial Primary Production and Phytomass." In *The Global Carbon Cycle*, edited by B. Bolin et al., pp. 129–182. New York: Wiley, 1979.

Bonnor, G. M. *Inventory of Forest Biomass in Canada.* Petawawa, Ontario: Canadian Forestry Service, Petawawa National Forestry Institute, 1985.

**Biomass. TABLE 1.** Estimates of Above-Ground Biomass in the North American Boreal Forest

| SOURCE | BIOMASS[a] ($KG/M^2$) | CARBON[b] ($KG/M^2$) | TOTAL BIOMASS[c] (BILLION METRIC TONS) | TOTAL CARBON[c] (BILLION METRIC TONS) |
|---|---|---|---|---|
| This study[d] | $4.2 \pm 1.0$ | $1.9 \pm 0.4$ | $22 \pm 5$ | $9.7 \pm 2$ |
| Previous studies[e] | | | | |
| 1 | 17.5 | 7.9 | 90 | 40 |
| 2 | 15.4 | 6.9 | 79 | 35 |
| 3 | 14.8 | 6.7 | 76 | 34 |
| 4 | 12.4 | 5.6 | 64 | 29 |
| 5 | 5.9[f] | 2.7 | 30 | 13.8 |

[a]Values in this column are for total above-ground biomass. Previous studies give total (above- and below-ground) biomass, which is corrected by us assuming that 23 percent of the total biomass is in below-ground roots. (Most references give this percentage; Leith and Whittaker (1975) give 17 percent; we have chosen to use the larger value to give a more conservative comparison.)

[b]Carbon is assumed to be 45 percent of total biomass following Whittaker (1975).

[c]Assuming our estimate of the areal extent of the North American boreal forest—5,126,427 square kilometers.

[d]Above-ground woody plants only.

[e]1: Ajtay et al. (1979); 2: Whittaker and Likens (1973); 3: Olson et al. (1978); 4: Olson et al. (1983); 5: Bonnor (1985).

[f]Value is for all Canadian forests.

SOURCE: Botkin and Simpson (1990). With kind permission of Kluwer Academic Publishers.

**Biomass. TABLE 2.** Abbreviated Version of a Classic Table of Standard Measures of Biomass for All the Major Biomes

| ECOSYSTEM TYPE | AREA (TENS OF THOUSANDS SQUARE KILOMETERS) | WORLD NET PRIMARY PRODUCTIVITY (BILLION TONS PER YEAR) | WORLD BIOMASS (BILLION TONS) |
|---|---|---|---|
| Tropical rainforest | 17.0 | 37.4 | 765 |
| Tropical seasonal forest | 7.5 | 12.0 | 260 |
| Temperate evergreen forest | 8.0 | 6.5 | 175 |
| Temperate deciduous forest | 7.0 | 8.4 | 210 |
| Boreal forest | 12.0 | 9.6 | 240 |
| Woodland and shrubland | 8.5 | 6.0 | 50 |
| Savanna | 18.0 | 13.5 | 60 |
| Temperate grassland | 9.0 | 5.4 | 14 |
| Tundra and alpine | 8.0 | 1.1 | 5 |
| Desert and semidesert scrub | 18.0 | 1.0 | 13 |
| Extreme desert, rock, sand, and ice | 24.0 | 3.0 | 0.5 |
| Cultivated land | 14.0 | 9.1 | 14 |
| Swamp and marsh | 2.0 | 4.0 | 30 |
| Lake and stream | 2.0 | 0.5 | 0.05 |
| Total continental | 149.0 | 115.0 | 1,837 |
| Open ocean | 332.0 | 41.5 | 1.0 |
| Upwelling zones | 0.4 | 0.2 | 0.008 |
| Continental shelf | 26.5 | 9.5 | 0.27 |
| Algal beds and reefs | 0.6 | 1.5 | 1.2 |
| Estuaries | 1.4 | 2.1 | 1.4 |
| Total marine | 361.0 | 55.0 | 3.9 |
| Full total | 510.0 | 170.0 | 1,841.0 |

SOURCE: Whittaker (1975, Table 5.2). Reprinted by permission of Prentice-Hall, Inc.

Botkin, D. B., and L. Simpson. "The First Statistically Valid Estimate of Biomass for a Large Region." *Biogeochemistry* 9 (1990), 161–174.

Olson, J. S., H. A. Pfuderer, and Y. H. Chan. *Changes in the Global Carbon Cycle and the Biosphere*, ORNL/EIS-109. Oak Ridge, Tenn.: Oak Ridge National Laboratory, 1978.

Olson, J. S., I. A. Watts, and L. I. Allison. *Carbon in Live Vegetation of Major World Ecosystems*, ORNL-5862. Oak Ridge, Tenn.: Oak Ridge National Laboratory, 1983.

Whittaker, R. H. *Communities and Ecosystems.* 2d ed. New York: Macmillan, 1975.

Whittaker, R. H., and G. E. Likens. "Carbon in the Biota." In *Carbon and the Biosphere*, edited by G. M. Woodwell and E. V. Pecan, pp. 281–300. Springfield, Va.: National Technology Information Center, 1973.

—DANIEL B. BOTKIN

# BIOMES

There is some diversity in the use of the term *biome.* Begon et al. (1986, p. 608) describe them as "communities characteristic of broad climatic regions." Cox et al. (1976, p. 36) recognize the existence of certain major climatic types, each of which "has a number of characteristic plant and animal communities that have evolved so that they are well adapted to the range of environmental factors in them; such characteristic communities are called biomes." Walter (1985, p. 7), on the other hand, remarks that "the word biome on its own (without a prefix) is used for the fundamental unit of which larger ecological systems are made up," and (p. 2) that it is also a term for "a large and climatically uniform environment within the geo-biosphere." Clapham (1983, p. 244) suggests they are "generalised types of communities comprising several associations with similar community structure."

Some authors have divided biomes into different categories. Notable here is Walter (1985), who has three classes: *zonobiomes, orobiomes,* and *pedobiomes.* Zonobiomes are essentially major climatically defined zones, of which he identified nine:

1. ZBI        Equatorial with diurnal climate, humid
2. ZBII       Tropical with summer rains, humido-arid
3. ZBIII      Subtropical-arid and summer drought, arido-humid
4. ZBIV       Winter rain and summer drought, arido-humid
5. ZBV        Warm temperate (maritime), humid
6. ZBVII      Warm temperate with a short period of frost (nemoral)
7. ZBVII      Arid-temperate with a cold winter (continental)
8. ZBVIII     Cold-temperate (boreal)
9. ZBIX       Arctic (including Antarctic), polar

Walter believed that the large zonobiomes can be further subdivided into *subzonobiomes. Orobiomes* are mountainous environments that can be vertically subdivided into altitudinal belts. *Pedobiomes* are environments associated with a certain type of soil; they are designated according to soil type: lithobiomes (stony soil); psammobiomes (sandy soil); halobiomes (salty soils); helobiomes (moor or swamp soils); hydrobiomes (soils covered with waste water); peinobiomes (soils poor or deficient in nutrients); and amphibiomes (temporarily wet soils).

Biogeographers show no real agreement about the number of biomes in the world: the distinctions that are made are largely arbitrary and biomes grade into one another. Moreover, some biomes have been transformed as a result of land use and land cover changes. It is also important to recognize that there are terrestrial and aquatic biomes, and that aquatic biomes are capable of division into freshwater and marine.

Figure 1 and Table 1 identify the main terrestrial biomes that have been identified by a range of different biogeographers. [*To see the distribution of Earth's biomes, see the maps that accompany the articles* Deserts; Forests; Grasslands; Mediterranean Environments; Savannas; *and* Tundra.]

Cox et al. (1976) identify three principal marine biomes:

1. The oceanic biome of open water, away from the immediate influence of the shore. This can be further subdivided:
   (a) The planktonic sub-biome containing free-floating plankton
   (b) The nektonic sub-biome of active swimmers
   (c) The benthic sub-biome whose fauna is especially adapted for life on the sea floor
2. The rocky shore biome
3. The muddy or sandy shore biome

It is important to recognize that biomes are not the same as biological realms. [*See* Biological Realms.] Biological (or floristic or faunal) realms and kingdoms are defined on the basis of the existence of families or substantial sections of families that are endemic to that particular portion of the globe. The subdivisions are therefore based on flora and not on vegetation types. The distinction between the two is an important one: the flora of an area is the sum total of all the plant species in it; vegetation is the kind of plant cover in that area. Thus, although two floral regions may be similar in their vegetation type (because, for example, they are both examples of tropical rainforest), they do not necessarily have much in common botanically or share more than a few genera.

There is a close connection between biomes and climatic conditions, as suggested in Figure 2. This means

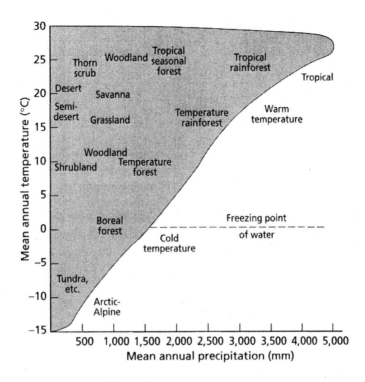

**Biomes. Figure 1.** The Pattern of World Biome Types in Relation to Mean Annual Temperature. (Modified from Gates, 1993, Figure 1, p. 63. With permission of Sinauer Associates.)

that the present distribution of biomes is not the same as the past distribution. For example, the growth and retreat of the Great Laurentide ice sheet over North America in the late Pleistocene and Holocene led to the remarkable changes in major vegetation types illustrated in Figure 3. The tundra–boreal forest biome's boundary has varied by over 20° of latitude. Likewise, at the Late Glacial Maximum (about 18,000 BP), the vegetation of Europe was transformed in comparison to what it was like in an interglacial. Similarly, between about 18,000 and 12,000 BP, the tropical rainforest biome was greatly reduced in extent, while tropical deserts were greatly expanded. Gates (1993) provides a useful

**Biomes. Table 1.** Major Terrestrial Biomes as Identified by Different Biogeographers

| ODUM (1971) | SIMMONS (1979) | COX ET AL. (1976) | CLAPHAM (1983) |
|---|---|---|---|
| Tundra | Tundra | Tundra | Tundra |
| Northern conifer forest | Boreal forests | Northern coniferous forest (taiga) | Boreal conifer forests |
| Temperate deciduous and rainforest | Deciduous forests of temperate climate | Temperate forest | Temperate deciduous forests |
|  |  |  | Temperate rainforests |
| Temperate grassland | Temperate grasslands | Temperate grassland | |
| Chaparral | Temperate sclerophyll woodland and scrub | Chaparral | Temperate evergreen woodland |
| Desert | Deserts | Deserts | Deserts |
| Tropical rainforest | Tropical rainforests | Tropical rainforests | Tropical rainforest |
| Tropical deciduous forest | Tropical seasonal forests |  | Tropical seasonal forests |
| Tropical scrub forest |  |  | Thorn woodland and scrubland |
| Tropical grassland and savanna | Tropical savannas | Tropical grassland or savanna | Tropical savanna |
|  |  |  | Subtropical rainforests |
| Mountains | Mountains | — | — |

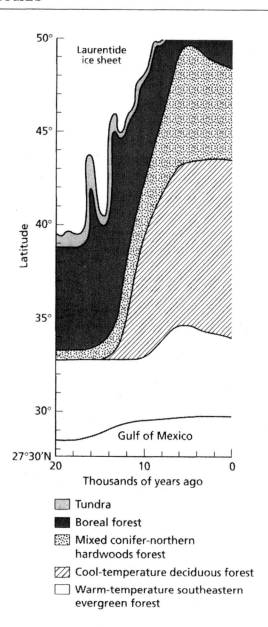

**Biomes. Figure 2.** Schematic Diagram Showing Changes in Major Vegetation Types along a Transect 83° West in Eastern North America over the Last 20,000 Years (i.e., between the Late Glacial Maximum and the Present). (Adapted from Delcourt and Delcourt, 1983.)

introduction to the relationship between climatic change and biome distribution.

It is increasingly true that the nature of biomes has been transformed, particularly in the last few thousands of years, by human activities. [*See* Human Impacts, *article on* Human Impacts on Biota.] As Simmons (1979, p. 92) pointed out, maps are not maps "of contemporary reality. Most of the tundra, for example, is still there and little altered; the deciduous forests of Eurasia have largely gone; the lowland tropical forests of the Congo and Amazon basins are shrinking fast." It is not simply that biomes are being changed in extent because of such processes as deforestation; they are also continually being modified in character because of such processes as the use of fire. Table 2 provides an indication of changes in the area of the major land cover types between pre-agricultural times and the present for the world as a whole. Such changes are reviewed in Meyer and Turner (1994).

Finally, there is increasing interest in the extent to which biomes will be transformed as a consequence of future potential global warming (see Gates, 1993, Chapter 5). [*See* Global Warming.] Some models suggest that wholesale changes in biomes will occur. Using the general circulation model (GCM) developed by Manabe and Stouffer (1980) for a doubling of carbon dioxide levels, and mapping the present distribution of ecosystem types in relation to contemporary temperature conditions,

**Biomes. Figure 3.** A Comparison of Vegetation Conditions of Europe across a North–South Transect in an Interglacial (top) and a Glacial (bottom).

Notice the absence of trees during a glacial cold phase. (*Modified after van der Hammen et al., 1971, Figure 6. Copyright by the American Geophysical Union.*)

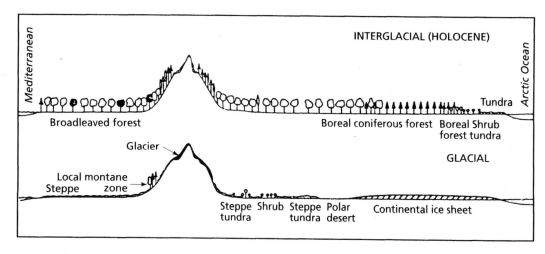

**Biomes. TABLE 2.** Estimated Changes in the Areas of the Major Land Cover Types between Preagricultural Times and the Present*

| LAND COVER TYPE | PREAGRICULTURAL AREA | PRESENT AREA | PERCENT CHANGE |
|---|---|---|---|
| Total forest | 46.8 | 39.3 | −16.0 |
| Tropical forest | 12.8 | 12.3 | −3.9 |
| Other forest | 34.0 | 27.0 | −20.6 |
| Woodland | 9.7 | 7.9 | −18.6 |
| Shrubland | 16.2 | 14.8 | −8.6 |
| Grassland | 34.0 | 27.4 | −19.4 |
| Tundra | 7.4 | 7.4 | 0.0 |
| Desert | 15.9 | 15.6 | −1.9 |
| Cultivation | 0.0 | 17.6 | +1760.0 |

*Figures are given in millions of square kilometers.

SOURCE: From J. T. Matthews (personal communication), in Meyer and Turner (1994). With permission of Cambridge University Press.

Emmanuel et al. (1985) found that the following change would take place: boreal forests would contract from their present position of comprising 23 percent of total world forest cover to less that 15 percent; grasslands would increase from 17.7 percent of all world vegetation types to 28.9 percent; deserts would increase from 20.6 to 23.8 percent; and forests would decline from 58.4 to 47.4 percent. More recently, Smith et al. (1992) have attempted to model the response of biomes to global warming and associated precipitation changes as predicted by a range of different GCMs. These are summarized for some major biomes in Table 3. All the GCMs predict conditions that would lead to a very marked contraction in the tundra and desert biomes, and an increase in the areas of grassland and dry forests. There is, however, some disagreement as to what will happen to moist forests as a whole; though within this broad class the humid tropical rainfall element will show an expansion.

[See also Ecosystems; Heathlands; Mangroves; Mountains; Reefs; Rivers and Streams; and Wetlands.]

### BIBLIOGRAPHY

Begon, M., J. L. Harper, and C. R. Townsend. *Ecology: Individuals, Populations and Communities.* Oxford: Blackwell Scientific, 1986. A particularly wide-ranging and conceptual treatment of all aspects of ecology.

Clapham, W. B. *Natural Ecosystems.* 2d ed. New York: Macmillan, 1983.

Cox, C. B., I. N. Healey, and P. D. Moore. *Biogeography: An Ecological and Evolutionary Approach.* Oxford: Blackwell Scientific, 1976.

Delcourt, P. A., and H. R. Delcourt. "Late Quaternary Vegetational Dynamics and Community Stability Reconsidered." *Quaternary Research* 19 (1983), 265–271.

**Biomes. TABLE 3.** Changes in Areal Coverage of Major Biomes as a Result of the Climatic Conditions Predicted for a Doubling of World Carbon Dioxide Levels by Various General Circulation Models*

| BIOME | CURRENT | OSU[†] | GFDL[†] | GISS[†] | UKMO[†] |
|---|---|---|---|---|---|
| Tundra | 939 | −302 | −5.5 | −314 | −573 |
| Desert | 3699 | −619 | −630 | −962 | −980 |
| Grassland | 1923 | 380 | 969 | 694 | 810 |
| Dry forest | 1816 | 4 | 608 | 487 | 1296 |
| Mesic forest | 5172 | 561 | −402 | 120 | −519 |

[†]OSU = Oregon State University; GFDL = Geophysical Fluid Dynamic Laboratory; GISS = Goddard Institute for Space Studies; UKMO = U.K. Meteorological Office.

*Figures are given in thousands of square kilometers.

SOURCE: From Smith et al. (1992, Table 3). With permission of Academic Press.

Emmanuel, W. R., H. H. Shugart, and M. P. Stevenson. "Climatic Change and the Broad-Scale Distribution of Terrestrial Ecosystem Complexes." *Climatic Change* 7 (1985), 29–43.

Gates, D. M. *Climate Change and Its Biological Consequences.* Sunderland, Mass.: Sinauer Associates, 1993.

Goudie, A. S. *Environmental Change.* 3d ed. New York: Oxford University Press, 1993.

Manabe, S., and R. J. Stouffer. "Sensitivity of a Global Climate Model to an Increase of $CO_2$ Concentration in the Atmosphere." *Journal of Atmospheric Science* 37 (1980), 99–118.

Meyer, W. B., and B. L. Turner, eds. *Changes in Land Use and Land Cover: A Global Perspective.* Cambridge: Cambridge University Press, 1994. A magisterial compendium on the human impact on the environment.

Odum, E. P. *Fundamentals of Ecology.* 3d ed. Philadelphia: Saunders, 1971. A classic consideration of key issues in ecology and biogeography.

Simmons, I. G. *Biogeography: Natural and Cultural.* London: Edward Arnold, 1979. An early analysis of the main controls of biogeographical patterns and processes.

Smith, T. M., H. H. Shugart, G. B. Bonan, and J. B. Smith. "Modelling the Potential Response of Vegetation to Global Climate Change." *Advances in Ecological Research* 22 (1992), 93–116.

van der Hammen, T., T. A. Wijmstra, and W. H. Zagwijn. "The Floral Record of the Late Cenozoic of Europe." In *The Late Cenozoic Glacial Ages,* edited by K. K. Turekian, pp. 391–424. New Haven: Yale University Press, 1971.

Walter, H. *Vegetation of the Earth.* 3d ed. Berlin: Springer-Verlag, 1985.

—ANDREW S. GOUDIE

# BIOSPHERE

*This entry consists of two articles,* Definition *and* Structure and Development. *The purpose of this entry is to define the term* biosphere *and to trace its development. The first article briefly defines the term, distinguishing it from other terms such as* biota *and* biomass, *and traces recent changes in its meaning. The second article discusses the mass and distribution of living things on the Earth through geologic time and in the present.*

## Definition

The term *biosphere* currently has four different meanings: (1) the total quantity of organic matter on the Earth; (2) the biological component of the Earth system and all its associated matter, exchange of energy and information, and internal biological dynamic processes, distinguished from the lithosphere (minerals), hydrosphere (water), and atmosphere; (3) the global habitat for life, or the set of all the places where life exists; and (4) the global system that includes and sustains life and therefore includes all the biota, the lithosphere, the hydrosphere, and the atmosphere. There is no general

scientific agreement about the preferred definition, although the first is most common, followed by the second.

Some scientists prefer the first meaning because it appears to make a coherent set with lithosphere, hydrosphere, and atmosphere. In geology, this is a common meaning of *biosphere.* However, this meaning is redundant with the term *global biomass,* whose definition is clearer and more direct. In usage, it is sometimes unclear whether authors intend the first or second definition.

The third definition, the global habitat, is the region of the Earth that extends from pole to pole and from the depths of the oceans to the tops of mountains. Its upper boundary is actually somewhat ambiguous because some authors assume that it includes the height to which propagules, such as seeds and spores of plants, are transported by the atmosphere. Others limit the global habitat to the region where life can persist and undergo regeneration—birth, growth, and death. Once again, there is no consistency in regard to this distinction.

The fourth meaning, the entire planetary life-support and life-containing system, has gained some popularity in recent decades. *Biosphere 2* was the name given a large, materially closed experimental system in Arizona, which was originally designed to test the ability of people to live within such a system and to sustain their lives and the lives of a diversity of other organisms with no external material exchange. However, it is more common to use the term *Earth system* to mean the same thing, especially in geologic literature and in programs at the U.S. National Aeronautics and Space Administration (NASA) Mission to Earth. This fourth meaning is an important one for our time, and agreement is needed among scientists in geology, climatology, oceanography, and ecology as to whether the preferred term will be *biosphere* or *Earth system.* In this meaning, the biosphere is a system that is materially closed (except for minor transfers of material into the system from cosmic dust, meteors, and asteroids, and transfers out of gases such as hydrogen), but open to the flow of energy and information. Thus defined, the biosphere has sustained life for more than three and one-half billion years. [*See* Modeling of Natural Systems.]

The necessary and sufficient conditions for the persistence of life over a long period are poorly understood, and only in very general terms. For example, it is clear that, for life to be sustained, there must be a flow of energy through the system and a cycling of chemical elements. These are necessary, general conditions, but what the sufficient conditions are remains unknown. The Earth's life-containing and life-supporting system is the only one we know that has sustained life for a long time. At the other extreme are some interesting experiments conducted by Claire Folsom at the University of Hawaii, who sealed small flasks containing ocean water, sediments, bacteria, and algae, along with air. Folsom claims that life persisted in these for thirty years, but it appears that no one has tried to replicate the experiment.

Some argue that the long-term persistence of life is a planetary phenomenon, because only a planet-sized system can have the necessary variety, number, and size of habitats to provide the redundancy required for long-term persistence. The growing evidence that rare collisions between the Earth and asteroids have caused a great decease in biological diversity supports this view. There is no formal theory that describes the necessary and sufficient conditions for the long-term persistence of life, although there are some informal, qualitative theoretical conjectures.

The concept of biosphere or Earth system is essential to understanding what makes life possible, but it is little studied. The earliest known use of the term *biosphere* was by Austrian geologist Eduard Seuss (1831–1914) in *Origin of the Alps*. Although Seuss left the word undefined, he used it to mean "global habitat." A Russian scientist, Vladimir Ivanovich Vernadskii (1863–1945), introduced the concept of the biosphere as it is accepted today in his book *La Biosphère*. Vernadskii used the term to mean "global habitat" and seems to have been the first to use it to mean "Earth system."

[*See also* Atmosphere Structure and Evolution; Biogeochemical Cycles; Biomass; Carbon Cycle; Earth History; Evolution of Life; Modeling of Natural Systems; Ocean Structure and Development; *and* Origin of Life.]

### BIBLIOGRAPHY

Lovelock, J. *The Ages of Gaia: A Biography of Our Living Earth.* New York: Norton, 1988. One of Lovelock's statements about the Gaia hypothesis.

Rambler, M. B., and L. Margulis, eds. *Global Ecology: Towards a Science of the Biosphere.* Boston: Academic Press, 1989. A general introduction to the idea of the biosphere and its scientific study.

Vernadskii, V. I. *The Biosphere.* Oracle, Ariz.: Synergetic Press, 1986. First published in 1926, this classic statement of the idea of the biosphere is available in various English translations.

—DANIEL B. BOTKIN

## Structure and Development

The total amount of living and dead organic matter on Earth—that is, the biosphere—contains about $1.56 \times 10^{22}$ grams of carbon (gC). Nearly all of this carbon is found in sedimentary rocks of the Earth's crust; the other large pool, soil organic matter, contains only $1.5 \times 10^{18}$ gC, or less than 0.01 percent of the total. [*See* Carbon Cycle.]

Nearly all the organic matter on Earth stems from the process of photosynthesis. The receipt of organic materials from outer space delivers only about $3.2 \times 10^8$ gC per year to the Earth, compared with the total photosynthetic production on land and in the oceans of more than $100 \times 10^{15}$ gC per year. The pool of organic matter that has accumulated in soils and marine sediments represents the net long-term storage of photosynthetic materials, after subtracting the efforts of decomposers to metabolize organic matter and return the carbon to the atmosphere as carbon dioxide. On land, more than 99.3 percent of the photosynthetic production is metabolized each year, so only a small fraction—about $0.4 \times 10^{15}$ gC per year—remains to accumulate in soil organic matter or to wash to the sea in runoff. The net accumulation of organic matter in marine sediments is about $0.1 \times 10^{15}$ gC per year globally. Thus, the mean residence time of organic matter in marine sediments is about 150,000,000 years.

Although most of the historical record of the biosphere is contained in sedimentary rocks deep in the Earth's crust, at any particular time the living organisms are confined to a rather narrow zone at the surface of our planet. Here, life extends from the bottom of the deepest oceans (more than 10,000 meters below the surface) to the highest mountain peaks. For example, stable communities of land plants are found in warm volcanic soils at 6,000 meters altitude in the Andes Mountains of Chile. Apart from a few humans orbiting in spacecraft, the record for the greatest vertical extent of the biosphere above the surface of the Earth is held by bar-headed geese, which fly over the Himalayas at altitudes of greater than 9,000 meters during their seasonal migration. On land, the live biosphere may extend several kilometers into the Earth's crust, where communities of microbes metabolize the organic carbon contained in sedimentary rocks and ground waters. Microbial communities have also been isolated at a depth of 500 meters in Pacific ocean sediments, beneath 900 meters of water. Using these extremes as our end points, the live biosphere is confined to a zone about 20 kilometers thick at the surface of the Earth—a thin skin on the Earth's 6,371-kilometer radius.

Within these vertical dimensions, life is found in nearly all habitats on the surface of the Earth. Thermophilic (heat-loving) bacteria tolerate temperatures in excess of 113°C in submarine (hydrothermal) volcanic emissions on the sea floor. Photosynthetic algae tolerate extreme periods of drought and high temperature in deserts, where some species exist in endolithic (inside-rock) environments that offer some modest protection from the arid climate. Three species of higher plants and a variety of soil organisms persist in the McMurdo dry valley of Antarctica, where temperatures during the 30-day growing season may average −3°C. The adaptability of life to such extreme environments is what encourages space scientists to look for primitive life in inhospitable locations on other planets and their moons.

Evidence for photosynthetic organisms on Earth extends to 3.8 billion years ago, when the conditions on our planet were certainly not especially hospitable to life. [*See* Evolution of Life.] Liquid oceans were present on the Earth's surface, but the Earth was still bombarded by large meteorites that may have heated the planet and repeatedly sterilized its surface. Nevertheless, sedimentary rocks of 3.8 billion years ago contain organic matter with a distinctive isotopic composition, similar to that known to be derived from photosynthetic organisms today. [*See* Origin of Life.] The process of photosynthesis discriminates between molecules of carbon dioxide ($CO_2$) that contain different isotopes of carbon—that is, between the most common form of $CO_2$ in which the carbon is carbon-12 and the less common form in which the carbon is carbon-13. This discrimination means that organic matter from photosynthesis typically has about 2 percent less carbon-13 than what is found in the atmosphere. Fossil organic matter with this distinctive, low content of carbon-13 first appears in rocks of about 3.8 billion years ago obtained from Greenland and other areas around the world, and it is taken as strong evidence for the existence of photosynthetic organisms at that time. Small microfossils that resemble modern-day cyanobacteria (blue-green algae) are found in rocks of 3.5 billion years ago.

For most of the Earth's history, life is likely to have been confined to marine environments. Fossil evidence of eukaryotic organisms (cells containing a nucleus) is found in marine sediments of 2 billion years ago, and diverse communities of multicellular animals roamed the oceans of 600 million years ago. Endolithic and other soil algae may have colonized the Earth's land surface by 2 billion years ago, but the first multicellular land plants did not appear until the early Silurian, 430 million years ago. Apparently, the invasion of land was delayed until the Earth had built up a protective shield of ozone ($O_3$) in the stratosphere, shielding the land surface from the Sun's ultraviolet radiation. Thus, for nearly 90 percent of the history of life on Earth, the biosphere was confined to the ocean waters.

The angiosperms—the flowering plants that now dominate the Earth's land surface—arose about 100 million years ago. Today, the highest diversity of angiosperms—sometimes exceeding 300 species of trees per hectare—is found in tropical rainforests. Temperate-zone rainforests dominated by conifers are arguably the mostly highly structured ecosystems on Earth—sometimes exceeding 60 meters in height. Their biomass may hold more than 300 metric tons of carbon per hectare of land surface. At the opposite extreme, deserts and the waters of the open ocean may contain only a few species and less than 1 gram of organic carbon per square meter of surface. [*See* Biomass.]

While we tend to focus on the living portions of the biosphere, it is the much larger pool of organic carbon in marine sediments that provides the best historical record of its past activity—that is, the record of the net production of organic matter by photosynthetic organisms through Earth's history. The rate of accumulation of sedimentary organic matter has not been constant through geologic time. [*See* Ocean Structure and Development.] A large amount of organic matter appears to have accumulated during periods of mountain building, when a large amount of sediment washed from land to sea, burying the products of photosynthesis and isolating them from decomposition. Particularly large amounts of organic matter may have accumulated about 2 billion years ago—coincident with the first appearance of oxygen in the Earth's atmosphere. Large depositions of organic matter and concurrent increases in atmospheric oxygen are also associated with the Permian and Carboniferous periods, 300 million years ago, during which many modern-day coals were deposited (Figure 1).

As recently as 18,000 years ago, at the peak of the last glacial period, the total biomass of plants on land and the rate of land plant photosynthesis were substantially lower, perhaps only 50 percent of today's values. Higher net primary productivity in the world's oceans may have partially compensated for the loss of plant production on land. Studies of glacial climatic fluctuations show that the biosphere is subject to large changes in its activity in relatively short periods of geologic time.

Today, the pool of organic carbon in sedimentary rocks contains about 15 percent of the total amount of carbon on Earth. Most of the organic carbon in sedimentary rocks is found in low concentrations of highly resistant material known as *kerogen*. Only a small amount is found in rich concentrations that are useful as fossil fuels. When we burn coal and oil, we return

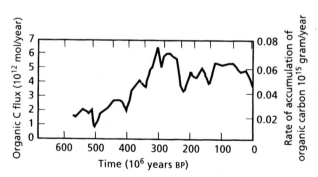

**Biosphere: Structure and Development. FIGURE 1.** Burial of Organic Carbon on Earth during the Last 600 Million Years. (From Berner and Raiswell, 1983. Reproduced with permission of Elsevier Science Ltd.)

carbon dioxide to the atmosphere, removing carbon from the pool of organic carbon that has accumulated in the Earth's crust as a result of photosynthesis during an earlier geologic epoch. The total amount of fossil fuel carbon that is estimated to be recoverable at today's prices is about $4 \times 10^{18}$ grams, or less than 0.4 percent of the organic carbon stored on Earth. Nevertheless, both the total amount of fossil fuel and the rate at which we burn it are very large compared with the carbon content of today's atmosphere ($760 \times 10^{15}$ gC). Thus, carbon dioxide is rising in the atmosphere, with the potential to cause global changes in climate.

[*See also* Atmosphere Structure and Evolution; Biogeochemical Cycles; Earth History; *and* Oxygen Cycle.]

### BIBLIOGRAPHY

Berner, R., and R. Raiswell. "Burial of Organic Carbon and Pyrite Sulfur in Sediments over Phanerozoic Time: A New Theory." *Geochemica et Cosmochimica Acta* 47 (1983), 855–862.
Kenrick, P., and P. R. Crane. "The Origin and Early Evolution of Plants on Land." *Nature* 389 (1997), 33–39.
Schidlowski, M. "A 3,800-Million-Year Isotopic Record of Life from Carbon in Sedimentary Rocks." *Nature* 333 (1988), 313–318.
Schlesinger, W. H. *Biogeochemistry: An Analysis of Global Change.* 2d ed. San Diego: Academic Press, 1997.
Schopf, J. W., and C. Klein, eds. *The Proterozoic Biosphere: A Multidisciplinary Study.* Cambridge: Cambridge University Press, 1992.
Williams, G. R. *The Molecular Biology of Gaia.* New York: Columbia University Press, 1997.

—WILLIAM H. SCHLESINGER

**BIOSPHERE 2.** *See* Modeling of Natural Systems.

# BIOTECHNOLOGY

Multiple definitions now exist for an activity that humans have been involved in for a good portion of recorded history. Fundamental items of the human diet such as bread, cheese, wine, vinegar, soy sauce, bean curd, and pickles are produced through some form of processing, using biological agents. In these instances, the agents are other organisms or groups of organisms that carry out a desirable transformation on the basic substance treated. In most cases, the materials undergoing treatment are themselves edible in a form that has not been biologically or biotechnologically transformed.

With the advent of the molecular biosciences, and the development of tools allowing direct alterations and transformations of the genome, the term *biotechnology* now carries with it a sense of genetic intervention at the molecular level: an implied element of an existing and developing capability to "engineer" organisms according to one's requirements. In large measure, this capa-

bility has been attained. Designer organisms produced to serve a variety of human purposes are now entities that one can generate with some facility through the application of current techniques.

While much of what is considered biotechnology involves the direct manipulation or alteration of the genome, a more broadly based view of this activity would incorporate an array of techniques and processes stemming from the applied molecular biosciences. One might then consider biotechnology as the application of the knowledge, associated techniques, and tools of the molecular biosciences to problems of human import and interest. It is a very broad term for a process with very broad applicability.

While humans have always genetically modified the organisms that they have found of use or interest, current technologies for so doing far exceed in capacity the random approaches of selection for specified genotypes that were utilized in the past. Thus the rate at which one can effect desirable changes in other living systems has increased dramatically. This ability both excites and frightens the general public, given the inherent tendencies of humans to utilize the products of technological advancement for both constructive and destructive purposes (Kappeli and Auberson, 1997; Karp et al., 1997; Carnegie, 1994).

There is much concern that engineered crops and organisms released into the environment will replace natural species and thus reduce the diversity of organisms extant in the biosphere. However, the controlled development and culture of specifically designed organisms as food sources, materials resources, and so forth could reduce the dependence of humans upon organisms that are now harvested from the biosphere at ever-increasing rates: some at such rates that the viability of the species itself is threatened. Organisms engineered with specific and special characteristics offer cost-effective, highly controllable production processes that can in principle be designed to produce minimal environmental impact. High-protein food sources, crops with enhanced nutrient content, and animals and plants with reduced susceptibility to disease or capacities to thrive under marginal conditions are some of the outcomes of biotechnological applications that could markedly reduce current deleterious impacts of human activities on the environment. It is highly probable that, through the application of biotechnology, many of our current problems associated with environmental degradation will be overcome. Environmental insults associated with the cold war are a worldwide legacy whose costs for correction in the United States alone have been estimated at U.S.$147–300 billion (U.S. Department of Energy, 1998). The costs for similar cleanup in the former Soviet Union and eastern Europe are even greater. It is only through development of new analytic, monitoring, and cleanup methodologies that these challenges can be met. The application of biotechnologically derived tools and technologies may provide a means of generating economically practical solutions to these problems.

Applications of biotechnology in pest control are not new. *Bacillus thuringiensis* has been widely applied as a means of controlling forest insect pathogens in several countries. Here, the capacity to conduct direct modification and alteration of the genome provides avenues for directing the organism at specific targets as well as enhancing its overall toxicity to target species. This approach is functional in producing microbiological agents as control systems in general. Insect viruses and viral agents as vectors or carriers of specific messages to target systems provide another example of how current biotechnology can serve multiple purposes (DeVault et al., 1996). [*See* Pest Management.]

While biotechnology's utility in the arenas of medicine and agriculture is self-evident, and numerous examples of the unique benefits obtained through such developments exist, the value of similar approaches in dealing with the deep environmental contamination problems that the world has acquired are as yet less evident. It is this area of application that may in the long run prove of greatest value. It is likely that only through the application of biotechnological approaches will cost-effective means of tracking and treating existing and developing contamination sites be devised (Ramanathan at al., 1997; Selvaraj et al., 1997). [*See* Agriculture and Agricultural Land.]

The overall application potential of biotechnology in a wide array of human activities and the rapidity with which this field is advancing is demonstrated by two documents issued by the Office of Technology Assessment in 1981 (Office of Technology Assessment, 1981) and 1991 (Office of Technology Assessment, 1991). Both documents highlight the areas of beneficial applications visualized across multiple disciplines, as well as areas of societal concern for the negative impacts of such advances. These documents, which were both a result of congressional requests for information on matters of deep interest to a broad political constituency at the time of their initiation, serve as examples of the rapidity with which biotechnological advances have impacted society.

The fundamental areas of biotechnological applications are those that are of significant import to all human societies: the production of foods, the generation of materials, and the maintenance of health and well-being. Biotechnology encompasses processes and procedures that improve environmental conditions and alter entities of import to humans from either a functional or an aesthetic standpoint. Improved health and longevity of companion animals as well as means of re-

covering or retaining species diversity are items that fall within this latter category. Indeed, it is likely that only through the interventions of current biotechnological procedures will certain endangered species be saved. Sewage has long been treated through the applications of microbiological agents, and the process of producing compost for soil enhancement is also an early example of the use of biotechnology to convert waste products arising from human activities into useful products or materials.

Biotechnology has opened new opportunities in the diagnosis, treatment, and prevention of human and animal diseases. The prospects for gene therapy are improving every year (Vile et al., 1998).

Genetic manipulation of metabolic pathways and the production of new microorganisms open avenues for drug exploration and development that were impossible in the absence of current biotechnological techniques. It is now possible to design and develop new materials based upon information generated from genetic and metabolic databases and to model their potential functionality against a host of known receptor types and subtypes. Molecular modeling becomes a part of the standard armamentarium in the war on disease and metabolic disorders. It is now a critical step in the conceptualization and design of therapeutically valuable materials (Kleinberg and Wanke, 1995; Nisbet and Moore, 1997; Frank, 1995).

[*See also* Biological Diversity; Convention on Biological Diversity; *and* Evolution of Life.]

### INTERNET RESOURCE

The Organisation for Economic Co-operation and Development. http://www.oecd.org/ehs/icgb. The OECD follows the applications of biotechnology in the areas discussed and maintains an updated listing of relevant activities and articles on this Web site, as well as a listing of many other relevant sites.

### BIBLIOGRAPHY

Carnegie, P. R. "Quality Control in the Food Industries with DNA Technologies." *Australasian Biotechnology* 4.3 (1994), 146–149.

DeVault, J. D., et al. "Biotechnology and New Integrated Pest Management Approaches." *Biotechnology* 14.1 (1996), 46–49.

Frank, R. "Simultaneous and Combinatorial Chemical Synthesis Techniques for the Generation and Screening of Molecular Diversity." *Journal of Biotechnology* 41 (1995), 259–272.

Kappeli, O., and L. Auberson. "The Science and Intricacy of Environmental Safety Evaluations." *Trends in Biotechnology* 15.9 (1997), 342.

Karp, A., et al. "Molecular Technologies for Biodiversity Evaluation: Opportunities and Challenges." *Nature Biotechnology* 15.7 (1997), 625–628.

Kleinberg, M. L., and L. A. Wanke. "New Approaches and Technologies in Drug Design and Discovery." *American Journal of Health-System Pharmacy* 52.12 (1995), 1323–1343.

Mann, C. C. "Biotech Goes Wild." *Technology Review* (July–August 1999), 36–43.

Nisbet, L. J., and M. Moore. "Will Natural Products Remain an Important Source of Drug Research for the Future?" *Current Opinions in Biotechnology* 8.6 (1997), 708–712.

Ramanathan, S., et al. "Bacterial Biosensors for Monitoring Toxic Metals." *Trends in Biotechnology* 15.12 (1997), 500–506.

Selvaraj, P. T., et al. "Biodesulfurization of Flue Gases and Other Sulfate/Sulfite Waste Streams Using Immobilized Mixed Sulfate-Reducing Bacteria." *Biotechnology Progress* 13.5 (1997), 583–589.

U.S. Department of Energy. *Paths to Closure.* Washington, D.C., 1998. A full-text version is available at http://www/em.doe.gov/closu/.

Vile, R. G., et al. "Strategies for Achieving Multiple Layers of Selectivity in Gene Therapy." *Molecular Medicine Today* 4.2 (1998), 84–92.

—JAMES D. WILLETT

**BIRTH RATES.** *See* Human Populations; *and* Population Policy.

**BOLIDES.** *See* Impacts by Extraterrestrial Bodies.

## BRUNDTLAND COMMISSION

The Brundtland Commission is the unofficial name of the World Commission on Environment and Development, which was chaired by Gro Harlem Brundtland, a former prime minister of Norway. Created as an independent commission in response to a United Nations General Assembly resolution of 1983, the body was given the general mandate of proposing ways in which the international community could achieve sustainable development that would both protect the environment and fulfill the aspirations of the poorer countries for economic development. Following in the tradition of the Brandt Commission on North–South issues and the Palme Commission on security and disarmament issues, the Brundtland Commission was convened at a time of growing concern over global environmental change and the mounting frustrations of the developing countries over their declining prospects for economic development and the failure of the West to enact key provisions of their proposals for a new international economic order (NIEO).

The commission comprised twenty-three prominent individuals from around the world and was supported by an appointed group of expert advisers. The commission held public hearings at locations around the world to hear the views of governmental officials, scientists and experts, industrialists, representatives of nongovernmental organizations, and the general public. The commission's report, *Our Common Future* (1987), covered a wide range of topics, including the international economy, population and human resources, food security, species and ecosystems, energy, industry, the management of commons, and war and peace. The report emphasized that three global crises that arose during the preceding decade—a development crisis, an environmental crisis, and an economic crisis—were integrally related. The poverty and underdevelopment of much of the world's rapidly growing population were seen as factors contributing to a degradation of natural systems, which in turn deepened the economic crises faced by many societies. The report is also notable for promoting *sustainable development*, which it defined as meeting "the needs of the present without compromising the ability of future generations to meet their own needs." It contends that economic growth will be necessary for countries in which the majority are poor, and argues that such nations must also be ensured a fair share of the resources necessary to sustain such growth. The report offers numerous proposals to strengthen the national and international institutions that address economic and environmental problems, to engage the scientific community and nongovernmental organizations, and to foster cooperation with industry. To increase economic resources available to developing countries, the report proposes raising revenue from the use of international commons, such as the oceans and outer space, and taxing certain types of international trade.

*Our Common Future* was widely read and discussed in the years that followed its publication. It informed deliberations in the preparatory meetings for the United Nations Conference on Environment Development that was held in 1992 in Rio de Janeiro and the drafting of Agenda 21, the elaborate plan of action that was adopted at the conference to be a blueprint for furthering sustainable development into the twenty-first century. In a broader sense, it has provided the intellectual foundation for a sustained emphasis in the United Nations during the 1990s on reconciling environmental and developmental objectives. [*See* United Nations Conference on Environment and Development.]

BIBLIOGRAPHY

De la Court, T. *Beyond Brundtland: Green Development in the 1990s.* New York: New Horizons Press, 1990. A critical analysis of the Brundtland report.

World Commission on Environment and Development. *Our Common Future.* New York: Oxford University Press, 1987. The full text of the Brundtland Commission report.

—MARVIN S. SOROOS

## BUDONGO FOREST PROJECT

[*This case study focuses on the Budongo Forest Project, a university-based research and conservation project in western Uganda. It describes the effects of modern hunting techniques on the resident chim-*

*panzee population, and how the project is attempting to work with local people.*]

Because of their close relationship to human beings, first fully understood by Charles Darwin and later documented by fossil evidence and the evidence of genetics, the great apes are among the most important of all living species for conservation. At the same time, threats to their existence have never been greater. There are four species of great ape: the orangutan (the great ape of Asia), the gorilla, the chimpanzee, and the bonobo (the great apes of Africa). Man's closest genetic similarity is with the chimpanzee. Because of the precarious situation of these species, field studies are particularly important in elucidating their behavior in their surviving natural habitat, their interactions with local human populations, and possible approaches to their conservation.

One such study, the Budongo Forest Project, is a field project of the Institute of Biological Anthropology, Oxford University, under the general management and supervision of Professor Vernon Reynolds. The Budongo Forest Reserve is located in western Uganda at 31° east longitude and 1° north latitude, and covers 428 square kilometers of forested area, making it the largest surviving single tract of forest in Uganda (Figure 1).

The Budongo Forest is important because it contains the largest population of chimpanzees in any forest in western Uganda, estimated at around six hundred animals. One community of chimpanzees, the Sonso community, has been studied extensively by the project. In 2000, it numbered forty-nine individuals, including thirteen infants. Besides this research community there are two further communities that have been partially habituated for ecotourism, one at Busingiro in the south of the forest and one at Pabidi in the northeast of the forest. At both of these sites, tourists are encouraged to stop for a while and enter the forest with trained guides who take them along specially constructed trails into the community's range, so that in many cases they are able to observe the chimpanzees at moderately close range. [*See* Tourism.] Forest ecotourism is an up-and-coming revenue earner in Uganda, and the Budongo Forest Ecotourism Project (which is financially separate from the Budongo Forest Project) has become a model of ecotourism in the modern context. It is run by an all-Ugandan committee, and a share of the profits is plowed back into the local communities in the area, so that all can feel the benefits of ecotourism. This in turn has led local people to appreciate the value of the forest and of the chimpanzees in financial terms. One local school has recently been rebuilt with the profits of ecotourism, and another is about to benefit. Work of this kind has brought together the ideas of the conservation movement and the motives of local people, who have attended meetings at which they are introduced to the idea

of great ape conservation, shown how it can benefit them, and later actually given tangible benefits.

The Budongo Forest Project is located in the heart of the forest, off the main roads. It has a staff of twenty-five, and at any time around six graduate students are at work on a variety of research topics. Studies of the chimpanzees have included their diet and feeding habits and the size and constitution of their foraging parties (chimpanzees have a fission–fusion social organization, meaning that they form parties of different size and composition from time to time during each day, and from day to day, so that, even though there are forty-nine animals in the Sonso community, they are never all together at any one time). Other studies have focused on their vocalizations and their intestinal parasites, and, notably for present purposes, the injuries that they have sustained as a result of getting caught in snares or traps set by local farmers to catch duikers (a small antelope species) and pigs. One-third of all the adult members of the Sonso community have such injuries, and this is not an uncommon picture—wherever chimpanzees have been studied.

At Budongo the chimpanzees' injuries consist of a missing foot (one individual), a missing hand (two individuals), or more commonly a deformed hand that is severely flexed (twisted inward) and has lost most of its ability to move (the remaining cases). Despite these injuries, the affected individuals have been shown to be adept at climbing trees and obtaining their food. Indeed, watching the injured chimps make their way up into the trees and move around when aloft is an inspiring sight, for they show very little sign of difficulty, and it is not until a close study is made that we find that they have individualistic and unique ways of feeding, holding branches, moving around, climbing, and nest-making. Some of these features are under study at the present time. It was not known in 1997 whether any chimpanzees died as a result of being caught in snares and traps, although it was suspected that they must have. In May 1997, a dead adult chimpanzee was found in the forest in association with a snare that had tightened around its wrist and led to its death, either from starvation or septicemia and loss of blood. The actual number of deaths from snares remains unknown.

Can anything be done about this waste? Chimpanzees are an endangered species, and the local farmers who set the snares do not eat them (they are traditionally not regarded as edible in East Africa) and do not want to catch them. Once caught in a snare, however, a chimpanzee becomes a dangerous animal and so the farmer will not release it—either it pulls the snare through wrist or ankle and loses a hand or foot, or it manages to pull the snare away from its fixing point on a nearby sapling. It is not known whether chimpanzees, even with their extraordinary intelligence, ever manage to release them-

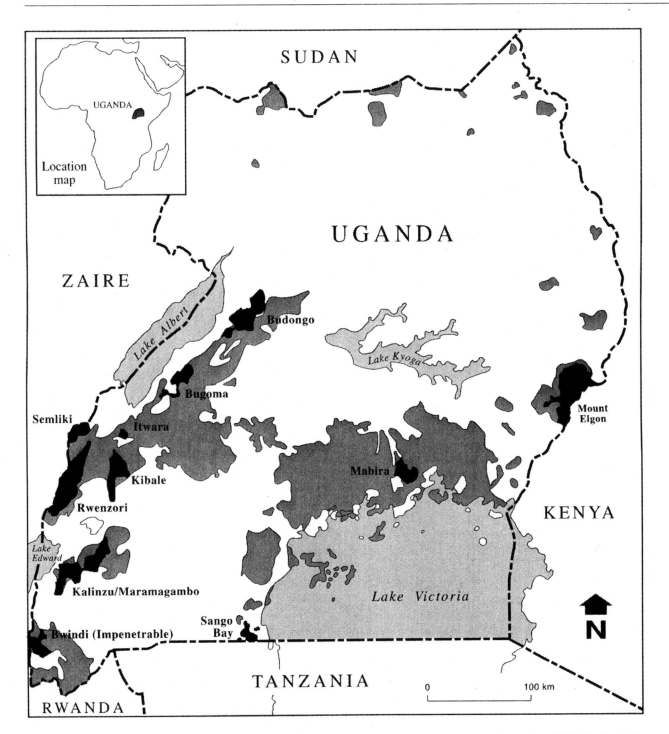

**Budongo Forest Project. Figure 1.** Locations of the principal forest reserves (black) in Uganda and areas of the country where relict forest occurs (dark gray), which were presumably once forested. (After Howard, 1991, p. 3, Figure 1.1.)

selves from a snare as humans do in a similar position. It seems that most often they panic and pull the snare tight, after which release becomes more and more difficult.

The project has been running since 1990. During this time there has been little encroachment on the main forest block, and indeed aerial photographs show that the forest has been slowly expanding, growing outward into surrounding grassland over the last fifty years. This has been possible because the population in this area of Uganda is still below the carrying capacity of the land. That situation cannot last for long. Today, land prices are rising and a big expansion by a local sugar plantation close to Budongo has led to unprecedented pressure on the land, with loss of a number of forest fragments as a result of encroachment. The project has helped the local school with books and a net ball, and will continue to assist local communities in the future. In such ways the project can engender better public relations with local villagers who otherwise see its members as a privileged elite who are on a gravy train that they cannot hope to join. That perception makes them feel different and less willing to listen to pleas for conservation of the forest and its wildlife. If, instead, the project is seen to be willing to part with funds in favor of the local school, church, or other valued treasured communal facilities, friendships and mutual dialogue can develop.

[See also Convention on International Trade in Endangered Species.]

### BIBLIOGRAPHY

Bakuneeta, C., et al. "Human Uses of Tree Species Whose Seeds are Dispersed by Chimpanzees in the Budongo Forest, Uganda." *African Journal of Ecology* 33 (1995), 276–278.
Eggeling, W. J. "Observations on the Ecology of the Budongo Rain-Forest, Uganda." *Journal of Ecology* 34 (1947), 20–87.
Hill, C. M. "Crop-Raiding by Wild Vertebrates: The Farmer's Perspective in an Agricultural Community in Western Uganda." *International Journal of Pest Management* 43 (1997), 77–84.
Howard, P. C. *Nature Conservation in Uganda's Tropical Forest Reserves.* Gland: International Union for Conservation of Nature and Natural Resources, 1991.
Johnson, K. "Hunting in the Budongo Forest, Uganda." *Swara* 19.1 (1996a), 24–26.
———. "Local Attitudes toward the Budongo Forest, Western Uganda." *Indigenous Knowledge and Development Monitor* 4.3 (1996b), 31.
Plumptre, A. J. "The Importance of 'Seed Trees' for the Natural Regeneration of Selectively Logged Tropical Forest." *Commonwealth Forestry Review* 74 (1995), 253–258.
———. "Changes Following 60 Years of Selective Timber Harvesting in the Budongo Forest Reserve, Uganda." *Forest Ecology and Management* 89 (1996), 101–113.
Plumptre, A. J., and V. Reynolds. "The Effect of Selective Logging on the Primate Populations in the Budongo Forest Reserve, Uganda." *Journal of Applied Ecology* 31 (1994), 631–641.
———. "Censusing Chimpanzees in the Budongo Forest, Uganda." *International Journal of Primatology* 17 (1996), 85–99.
———. "Nesting Behavior of Chimpanzees: Implications for Censuses." *International Journal of Primatology* 18.4 (1997), 475–485.
Reynolds, V. "Conservation of Chimpanzees in the Budongo Forest Reserve." *Primate Conservation* 11 (1992), 41–43.
———. "Sustainable Forestry: The Case of the Budongo Forest, Uganda." *Swara* 16.4 (1993), 13–17.

—VERNON REYNOLDS

## BUILDING DECAY

The materials used to construct buildings and engineering structures, such as bridges, suffer from three main types of decay: that caused by atmospheric processes, that caused by contact between different building materials, and that caused by other external agents, such as ground water, soil, and plants. Such decay processes, involving chemical and physical changes to the materials, are entirely natural (and also affect rock outcrops such as cliffs) but may be altered and enhanced by human activity. The decay of building materials leads to color changes, roughening of surfaces, the development of cracks and blisters, and the reduction of the thickness of walls. Such changes in material characteristics are often quite subtle but can be unsightly, expensive to rectify, and even dangerous. Decay processes affect the whole gamut of building materials, from natural stones extracted from quarries, through metals and concrete, to synthetic materials such as plastics. Some building materials are known to be highly durable (for example, resistant to decay processes); whereas others have been found to decay rapidly. Durability is a complex characteristic, influenced by the porosity, structure, and chemical makeup of the material. Water, temperature fluctuations, and the presence of salts all encourage decay. The major concern about the decay of building materials in the modern world comes as a result of the exacerbating effect of a wide range of human impacts that often act to accelerate and alter the natural processes of decay. Air pollution, alterations to groundwater levels and chemistry, and additions of deicing salts to roads in winter, coupled with problems of building maintenance and inappropriate conservation schemes, all accelerate building materials decay. [See Pollution.]

The decay of building materials is a global problem, affecting our cultural heritage, leading to structural failure and costing millions of dollars. In the United Kingdom, for example, many major stone cathedrals are now suffering badly from decay, resulting in crumbling of parts of the stonework and unsightly color changes and crust formation. In India there has been much concern that marble faces of the Taj Mahal may be decaying because of air pollution from nearby heavy industry. Several key archaeological sites, such as Mohenjo-Daro in Pakistan, have suffered accelerated salt weathering damage consequent on changes in ground water. [See Salinization.] The costs of such decay may be very high.

The major natural processes of materials decay can be divided into three types: chemical, physical, and biological. Chemical transformations within materials occur as water or air carrying a range of substances interacts with the material. Rusting of iron is a simple example. Limestone suffers from a range of chemical processes as water acidified with carbon dioxide (as is most rainfall under natural conditions) runs over the surface and reacts with calcium carbonate to form soluble products, which are then removed by the flowing water. Physical processes of materials decay occur as stresses are imposed on and/or within the materials, often as a result of freezing of water or because of salt crystallization in porous materials. External vibrations, such as those caused by heavy traffic or Earth tremors, can have similar effects. Finally, biological decay occurs as a consequence of plant and microorganism growth on or in materials. For example, biofilms composed of a range of microorganisms often grow on reasonably porous materials where water can be retained. These biofilms often cause chemical and physical weathering of the materials surface through their life processes.

Human acceleration and alteration of decay mainly acts through the introduction of greater quantities of reactive chemicals, by affecting the water balance, and by influencing the mechanical stability. Air pollution, for example, increases the loading of sulfur oxides within the atmosphere. [See Acid Rain and Acid Deposition; and Sulfur Cycle.] Sulfur dioxide reacts with calcareous materials (such as limestone and concrete) to produce gypsum (calcium sulfate), which creates a friable weathering crust on the surface of the material. This crust often becomes soiled through the inclusion of sooty particles and other debris common in polluted air, producing an unsightly surface deposit that requires careful management. The stone behind the crust may become softened, leading to catastrophic erosion of the stone once the crust is breached. In areas such as parts of eastern Europe where coal containing high levels of sulfur has been burned for many years, building materials have become badly decayed and soiled as a result.

De-icing salts are commonly applied to road surfaces in winter to prevent the formation of ice, especially in northern Europe and North America, and can often become splashed onto adjacent building surfaces and penetrate into bridges. The salt solutions can easily become trapped within porous building materials. Subsequent crystallization and hydration of the salts can produce mechanical stresses on pore walls of sufficient magnitude to cause physical breakdown. Salts can also be taken up from ground water by porous building materials, and this is an especially serious problem in many arid and semiarid areas where ground water is often naturally salty (because of high evaporation rates); and

human practices such as irrigation often increase groundwater levels and salinities. Groundwater extraction for irrigation purposes can also lead to ground instability and subsidence, which can affect buildings and structures, producing cracking and, sometimes, failure and collapse. [See Ground Water.] Mining may lead to similar subsidence failures.

Bad design, construction, and management practices can also encourage building materials decay. Leakage of water over building surfaces, perhaps because of faulty guttering, can produce accelerated chemical decay as well as encourage the growth of biofilms. The application of inappropriate chemicals for cleaning, preservation, or consolidation of materials can also produce accelerated decay. In porous materials, such chemicals can remain stored for many years, leading to chronic decay problems. However, good management strategies can reduce the threat of accelerated decay and can prolong the service life of many building materials. Allied with other good environmental management practices, such as the improvement of air quality, better management of groundwater resources, and reduction of traffic-related impacts such as de-icing salt applications, such strategies should help to conserve buildings and engineering structures by reducing materials decay.

[See also Air Quality.]

## BIBLIOGRAPHY

Addleson, L., and C. Rice. *Performance of Materials in Buildings.* Oxford: Butterworth Heinemann, 1991. A readable summary of important properties.

Baer, N. S., and R. Snethlage, eds. *Saving Our Architectural Heritage: The Conservation of Historic Stone Structures.* Berlin: Wiley, 1997. A series of state-of-the-art reviews on the processes of decay affecting historic monuments and the potential solutions.

Goudie, A. S., and H. A. Viles. *Salt Weathering Hazards.* Chichester, U.K.: Wiley: 1997. A detailed introduction to how salts cause materials damage, with a focus on building stones. Includes many case studies of problems and solutions.

—HEATHER A. VILES

**BUSH ENCROACHMENT.** *See* Savannas.

# C

## CARBON CYCLE

Carbon circulates among three distinct global reservoirs. These are the atmosphere, the oceans, and the terrestrial system. Of the three reservoirs, the oceans contain by far the greatest amount of carbon (40,000 petagrams; 1 petagram is 1 billion metric tons or $10^{15}$ grams). The terrestrial reservoir contains a much smaller amount of carbon (2,050 petagrams) with a majority in the soil (1,500 petagrams). The atmospheric reservoir contains the smallest amount (currently 775 petagrams), and nearly all carbon in the atmosphere is in the form of the trace gas carbon dioxide. Nevertheless, the atmosphere plays an important role in the global carbon cycle as a conduit between the other two reservoirs. The concentration of carbon dioxide in the atmosphere results from the dynamic exchanges between the oceans, the terrestrial system, and the atmosphere. Over geologic time, the concentration of carbon dioxide in the atmosphere has fluctuated as a result of shifts in the relative surface areas of ocean and land as well as changes in general climate regimes and the influences these have on marine and terrestrial ecosystems. Currently, the burning of fossil fuels is contributing carbon dioxide to the atmosphere at a rate that is increasing the atmospheric carbon dioxide concentration and altering the rates of carbon exchange between the three main reservoirs of the global carbon cycle. [*See Fossil Fuels.*] Figure 1 shows the major reservoirs in the global carbon cycle, their components, and the magnitudes of carbon exchange between them for preindustrial time (before 1860; Figure 1A) and present conditions (Figure 1B). There are other pools and fluxes that do not appear in these diagrams, including soil carbonates, the formation and dissolution of carbonates in the ocean, peat deposition, volcanic emissions, and groundwater and river transport of organic and inorganic carbon. While important on geologic time scales, these pools and associated fluxes do not play a major role during the present, relatively short fossil fuel burning period.

The exchange of carbon between the atmosphere and the ocean is the result of diffusion of carbon dioxide across the ocean surface. The direction and rate of carbon dioxide diffusion vary from place to place and depend on the gas transfer coefficient for carbon dioxide and the difference in partial pressure of carbon dioxide between the atmosphere and the surface ocean waters. The gas transfer coefficient is a function of temperature and is very sensitive to wind speed over the ocean surface (greater wind speed enhances gas transfer by increasing mixed-layer thickness, wave action, and bubble formation). The difference in partial pressure of carbon dioxide between the atmosphere and the surface ocean water determines the magnitude and direction of the exchange. The partial pressure of carbon dioxide in sea water is a function of temperature, salinity, alkalinity, and, most importantly, total carbon dioxide concentration (dissolved bicarbonate and carbonate, as well as dissolved carbon dioxide). The effect of total carbon dioxide on the partial pressure of carbon dioxide in sea water is expressed as the Revelle factor. [*See the biography of Revelle.*] A global mean value is 10. This means that a 1 percent increase in total dissolved carbon dioxide causes a 10 percent increase in seawater carbon dioxide partial pressure. The Revelle factor increases rapidly with increasing total dissolved carbon dioxide. A 10 percent increase in total dissolved carbon dioxide would increase the Revelle factor from 10 to 18. Thus, the more carbon dioxide surface ocean waters take up, the less effective they will become at taking up more carbon dioxide from the atmosphere.

The partial pressure of carbon dioxide in the ocean surface is affected by several ocean processes in addition to carbonate solution dynamics. The large-scale thermohaline circulation that results from the sinking of dense, cold, and salty water at high latitudes and the upwelling of nutrient- and carbon-rich deep waters along continental boundaries makes different ocean regions either sinks or sources of carbon dioxide to the atmosphere. Marine life also affects the uptake of carbon dioxide by the oceans. Photosynthesis carried out by phytoplankton in the well-lit surface layer (euphotic zone) of the ocean converts dissolved inorganic carbon dioxide into organic matter and calcium carbonate. A major portion of this primary production is consumed, decomposed, or remineralized in the surface waters, but a fraction escapes from the well-mixed surface ocean to the stratified deep ocean layers, where most is redissolved and only a minor fraction is deposited in the ocean sediments. This escaping fraction effectively lowers the concentration of dissolved carbon dioxide in the surface ocean, increasing its ability to take up atmospheric carbon dioxide and thereby reducing the atmospheric carbon dioxide concentration. Once carbon is transferred to the deep ocean it remains there, on average, for approximately one thousand years until it is mixed into the surface layer, or wells up as a result of

A

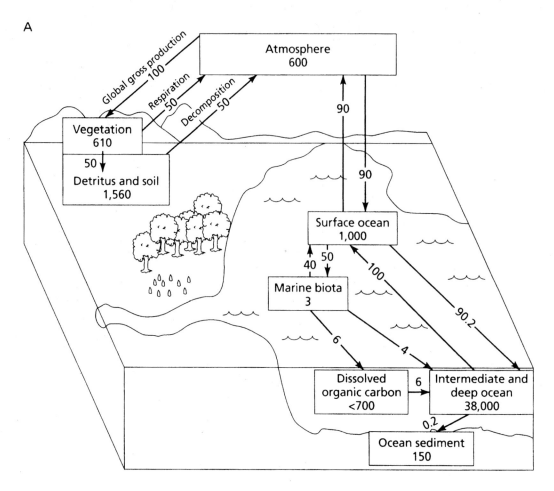

B

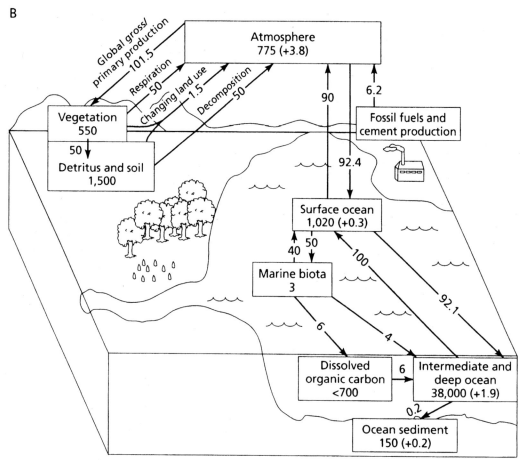

the global thermohaline circulation. [*See* Ocean Chemistry; *and* Ocean Dynamics.]

The terrestrial system, by virtue of the large amount of carbon in soil organic matter, is the second largest of the three global reservoirs. The flux between the atmosphere and the land surface results from the fact that life depends on the incorporation of atmospheric carbon dioxide into organic compounds during plant photosynthesis. Nearly all of this carbon is eventually returned to the atmosphere as carbon dioxide by respiration of living organisms. Some is also returned by fires. [*See* Fire.]

The rate at which carbon dioxide is taken up by plants, also called gross primary production (GPP), depends largely on the availability of light, water, and temperatures suitable for metabolic function. The availability of soil nutrients, particularly nitrogen, also influences the rate of photosynthesis and plant growth. Approximately half of the carbon fixed into organic compounds during photosynthesis is used directly by plants for growth and maintenance and is respired in a short time, returning carbon dioxide to the atmosphere. The remaining fraction, called net primary production (NPP), is the rate at which carbon is incorporated into plant tissues. Some of these tissues persist for only a short time before being shed and decomposed by soil organisms. For example, fine plant roots may last only a few weeks or months, and deciduous leaves less than a year. Other tissues, such as wood, can persist for several decades or even centuries, depending on the forest type and disturbance frequency. Soil organisms return most of the carbon in dead organic matter to the atmosphere as carbon dioxide. A smaller portion of this decomposing material is "humified" or converted by soil organisms into humic compounds that are more difficult to decompose because of chemical resistance or physical protection by soil minerals. Humic compounds can remain in soils for tens to thousands of years before being converted to carbon dioxide. As a result, most of the carbon in the terrestrial system is found in soils, rather than in living plant tissues. The rate of change in carbon storage in terrestrial ecosystems, or what is called net ecosystem production (NEP), is the difference between NPP and the rate of decomposition of shed and dead plant tissues. At equilibrium, NPP is balanced by the rate of decomposition, and NEP is zero. Of course, no single small plot of terrestrial vegetation is ever in equilibrium for long because of natural or human disturbances and subsequent plant regeneration. The notion that NEP may be approximately zero requires an averaging over sufficiently large areas for long enough periods of time to include the effects of disturbances that result in losses of carbon to the atmosphere.

Over the last 150 years, as a result of the burning of fossil fuels, cement making, and the conversion of natural vegetation to agricultural use, there has been a large increase in the amount of carbon dioxide in the atmosphere. [*See* Carbon Dioxide.] Since around 1850, the beginning of the Industrial Revolution, the amount of carbon dioxide added to the atmosphere has increased steadily. The preindustrial atmosphere had 600 petagrams of carbon in the form of carbon dioxide, equivalent to a concentration of 280 microliters per liter. The present atmosphere, with a concentration of 365 microliters per liter, contains 775 petagrams of carbon. The amount of carbon added from fossil fuel burning and cement production between 1992 and 1997 was 6.2 petagrams per year. Land use conversion is estimated to contribute an additional 1.0 petagram per year. (This estimate is a net flux that includes carbon from conversion of native vegetation to cultivation, at present mostly in tropical regions, and carbon uptake by the regrowth of forests on land no longer cultivated, largely in temperate regions.) Of this 7.2 petagrams of carbon per year emitted to the atmosphere, about 2.4 petagrams are taken up by the world's oceans and 3.8 petagrams remain in the atmosphere. Several sources of information (changes in atmospheric oxygen concentration, changes in carbon dioxide concentration in which the carbon is the stable isotope carbon-13, and the latitudinal distribution of atmospheric carbon dioxide) suggest that terrestrial ecosystems take up the remaining 1.0 petagram per year and store it in plant tissues or soil.

Atmospheric carbon dioxide concentration is now 30 percent higher than in preindustrial times. This elevated carbon dioxide stimulates photosynthesis, and higher photosynthesis supports faster plant growth rates. In long-lived perennial species such as trees, increased

---

**Carbon Cycle. FIGURE 1.** Global Carbon Reservoirs.

**(A) Preindustrial Global Carbon Cycle.** The concentration of carbon dioxide in the atmosphere has been constant from the end of the last glacial period, approximately eleven thousand years ago, until the beginning of the Industrial Revolution in the mid-1800s, indicating that the carbon dioxide release from oceans and terrestrial systems was balanced by corresponding uptake. This is shown by the equal exchanges of carbon between the carbon pools in this diagram of the preindustrial global carbon cycle. All pools' units are petagrams of carbon (1 petagram = $1 \times 10^{15}$ grams or 1 billion metric tons) and exchanges are in petagrams of carbon per year. **(B) Global Carbon Cycle (1992–1997).** The burning of fossil fuels and changing land use have resulted in human-induced alteration of the global carbon cycle. The magnitude of the perturbation in carbon fluxes can be determined by comparison with part (A). The rate of change in carbon pool sizes (in petagrams of carbon per year) is indicated by the numbers in parentheses. Land use change has resulted in a decrease of terrestrial vegetation and soil pool carbon content. Currently, losses due to land use change, largely in tropical regions, may be approximately balanced by increased rates of sequestration resulting from carbon dioxide fertilization in relatively undisturbed ecosystems.

growth implies an increase in carbon stored in wood, at least for several decades. In addition, increased plant production, when shed as dead litter, results in a larger amount of decaying organic matter. A portion of this decaying matter becomes humified and enters the soil carbon pool, where it may reside for several decades before undergoing further decomposition. This carbon dioxide fertilization effect may account for most of the current net storage of carbon annually in soil and plant tissues such as wood. For how long, and at what rates, terrestrial and ocean systems will take up carbon dioxide as the concentrations of atmospheric carbon dioxide continue to rise is only partially known and is the subject of considerable scientific investigation.

[See also Biomass; Biosphere; Deforestation; Earth History; Energy; Global Warming; Greenhouse Effect; and Methane.]

Research sponsored by the U.S. Department of Energy, Carbon Dioxide Research Program, Atmospheric and Climate Research Division, Office of Health and Environmental Research, under contract DE-AC05-96OR22464 with Lockheed Martin Energy Research Corp.

### INTERNET RESOURCES

Carbon Dioxide Information and Analysis Center, Oak Ridge National Laboratory, Oak Ridge, Tenn. http://cdiac.esd.ornl.gov/ndps/ndp030.html/. The standard source for annual data on release of carbon dioxide.

### BIBLIOGRAPHY

Barnola, J.-M., D. Raynaud, Y. S. Korotkevitch, and C. Lorius. "Vostok Ice Core Provides 160,000 Year Record of Atmospheric Carbon Dioxide." *Nature* 329 (1987), 408–414. This important paper contains a long history of atmospheric carbon dioxide concentration that provides a basis for current analyses and a glimpse into the dynamic nature of the carbon cycle.
Batjes, N. H. "Total Carbon and Nitrogen in the Soils of the World." *European Journal of Soil Science* 47 (1996), 151–163. An up-to-date analysis of the amount and location of soil carbon.
Broecker, W. S. "Paleocean Circulation during the Last Glaciation— A Bipolar Seesaw." *Paleoceanography* 13 (1998), 119–121. An explanation of the importance of the large-scale thermohaline circulation of the ocean and factors that determine its strength.
Ciais, P., et al. "A Three Dimensional Synthesis Study of $\delta^{18}O$ in Atmospheric Carbon Dioxide, Part I: Surface Fluxes." *Journal of Geophysical Research* 102 (1996), 5857–5872. A technically difficult but clearly written paper demonstrating the significance of terrestrial systems in balancing the current global carbon cycle.
Falkowski, P. G. "The Role of Phytoplankton in Global Biogeochemical Cycles." *Photosynthesis Research* 39 (1994), 235–258. A good general overview of the role of marine biology in the carbon cycle.
Houghton, R. A. "Land-Use Change and the Global Carbon Cycle." *Global Change Biology* 1 (1995), 275–287. A recent and authoritative summary of the impact of land use change on atmospheric carbon dioxide.

Keeling, C. D., et al. "Interannual Extremes in the Rate of Rise of Atmospheric Carbon Dioxide Since 1980." *Nature* 375 (1995), 666–670. A detailed look at variations in atmospheric carbon dioxide concentrations with interesting suggestions on what this variation can reveal about carbon cycle processes.
Keeling, R. F., and S. R. Shertz. "Seasonal and Interannual Variations in Atmospheric Oxygen and Implications for the Global Carbon Cycle." *Nature* 358 (1992), 723–727. A key paper that demonstrates the usefulness of atmospheric oxygen measurements in separating ocean and terrestrial uptake of fossil fuel carbon dioxide.
Marland, G., et al. "Global, Regional and National Carbon Dioxide Emission Estimates from Fossil Fuel Burning, Cement Production and Gas Flaring: 1751–1995." ORNL/CDIAC-25, NDP-30. Carbon Dioxide Information and Analysis Center, Oak Ridge National Laboratory. Oak Ridge, Tenn., 1998. The standard source for annual data on the release of carbon dioxide.
Sarmiento, J. L., and M. Bender. "Carbon Biogeochemistry and Climate Change." *Photosynthesis Research* 39 (1994), 209–234. A good and up-to-date scientific summary of our knowledge of the global carbon cycle with an emphasis on ocean processes.
Schimel, D. S., et al. "Climatic, Edaphic and Biotic Controls over Storage and Turnover of Carbon in Soils." *Global Biogeochemical Cycles* 8 (1994), 279–293. An example of the use of a soil process model to obtain an understanding of soil carbon dynamics at a global scale.
———. "The Global Carbon Cycle." In *Climate Change 1994*, edited by J. T. Houghton et al., pp. 35–71. Cambridge: Cambridge University Press, 1996. A comprehensive explanation of the global carbon cycle that was a key component of the last international assessment of man's impact on climate.

—WILFRED M. POST

# CARBON DIOXIDE

Carbon dioxide ($CO_2$) is classified as a trace constituent of the Earth's atmosphere, accounting for only 0.03 percent of the atmospheric volume. Nevertheless, the presence of $CO_2$ in the atmosphere is of vital importance to the functioning of the planet. Not only does $CO_2$ have a fundamental biological role—$CO_2$ together with water ($H_2O$) and sunlight are the raw materials from which green plants are able to synthesize organic compounds—but it is also a so-called greenhouse gas, and, as such, changes in its atmospheric abundance have the potential to alter the radiative balance of the Earth and therefore the global climate.

**CO$_2$: A Biological Role.** All life on Earth is ultimately dependent on the process by which inorganic carbon (as $CO_2$) from the atmosphere is converted to organic compounds by green plants; this process is called *photosynthesis* and is represented by the equation:

$$6CO_2 + 6H_2O \xrightarrow[\text{chlorophyll}]{\text{light energy}} C_6H_{12}O_6 + 6O_2$$

Through this basic life process, not only are carbon and energy made available to living organisms, but oxygen

($O_2$), vital for all aerobic forms of life, is also produced as waste product. In higher plants the major photosynthetic organ is the leaf; $CO_2$ enters the leaf via open pores (stomata) situated within the leaf surface and diffuses to the leaf's mesophyll cells, which contain chloroplasts and the photosynthetic pigments (chlorophyll). Here $CO_2$ in the presence of sunlight reacts with water that has been transported to the leaf from the roots by the plant's vascular system. The products of photosynthesis, carbohydrates (i.e., sugars), are removed to other parts of the plant or stored, and the waste product, $O_2$, escapes back to the atmosphere via the stomata.

Through the photosynthetic process, plants take up approximately 120 gigatons (Gt) of carbon in the form of $CO_2$ from the atmosphere each year. This uptake is roughly balanced by the reverse process—soil and plant respiration—which releases $CO_2$ back into the atmosphere. This exchange of $CO_2$ between the atmosphere and the terrestrial biosphere is but one component of a larger biogeochemical cycle involving carbon: the carbon cycle. A simplified representation of this cycle,

showing how carbon is circulated between its main reservoirs, the atmosphere, the oceans, the biosphere, and the lithosphere, is shown in Figure 1. [*See* Carbon Cycle.]

**$CO_2$ and Climate: The "Greenhouse Effect."** The Earth's climatic system is a system of great complexity with many interactive components; essentially, however, it can be said to comprise the atmosphere, the oceans, the land and its features, the cryosphere (snow and ice), and hydrology (rivers and lakes). The source of energy that drives this system is radiation from the sun.

Approximately 30 percent of the sun's incident shortwave radiation is scattered or reflected back to space by either microscopic airborne particles (aerosols), clouds and molecules present in the atmosphere, or by the Earth's surface. The remaining radiant energy is absorbed by the atmosphere and by the Earth's surface, thereby warming it. In order to balance this incoming solar energy, the Earth itself must radiate, on average, the same amount of energy that it has absorbed back to

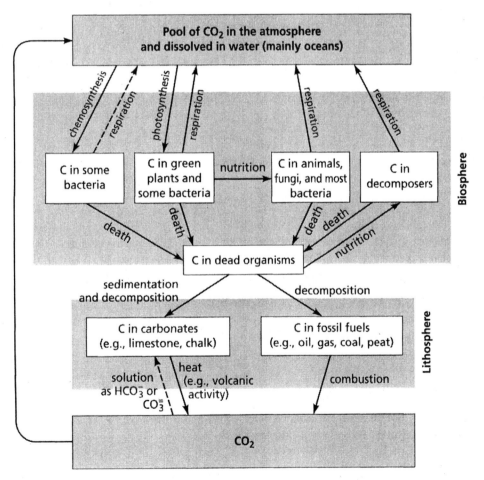

**Carbon Dioxide. Figure 1.** A simplified schematic representation of the carbon cycle showing the main life processes that regulate the flows of carbon between its reservoirs, the atmosphere, the terrestrial biosphere, the lithosphere, and the oceans. (Adapted from Green et al., 1990.)

space. It does this by emitting thermal longwave infrared radiation. Some of this outgoing infrared radiation is transmitted unimpeded through the atmosphere; the bulk, however, is intercepted, absorbed, and then reemitted by certain radiatively active trace gases that are naturally present in the atmosphere. These include water vapor, carbon dioxide, methane, tropospheric ozone, and nitrous oxide. Because radiatively active gases reemit infrared radiation in all directions, the net result is a warming of the lower atmosphere and the Earth's surface. In other words, the presence of these gases in the atmosphere impedes the transfer of thermal energy from the Earth to space. Because the effect is analogous to that of a greenhouse, these radiatively active gases have come to be known as greenhouse gases, and their impact known as the greenhouse effect. [See Greenhouse Effect.]

The operation of the natural greenhouse effect means that the Earth's surface is substantially warmer than it otherwise might be. In the absence of greenhouse gases, the average temperature of the Earth's surface would be $-18°C$; the presence of naturally occurring greenhouse gases in the atmosphere means that the global mean surface temperature is around a habitable 15°C. Water vapor is by far the largest contributor to the natural greenhouse effect, being responsible for around 90 percent of the 33°C overall natural greenhouse warming; $CO_2$ is the next most significant trace gas in this regard, accounting for most of the remaining 10 percent.

Changes in the atmospheric abundance of $CO_2$—or any other greenhouse gas, for that matter—will alter the radiative balance of the Earth and thus have the potential to influence climate. If atmospheric concentrations of $CO_2$ are increased owing to human activity, for example, then an enhanced greenhouse effect is likely. This is because the presence of higher abundances of radiatively active gases means that more of the outgoing terrestrial radiation is absorbed and reemitted by the atmosphere, effectively reducing the efficiency with which the Earth cools to space. This will lead to a warming of the Earth's surface and lower atmosphere. The magnitude of this global warming depends on a number of factors, including the size of the increase in concentration of each greenhouse gas, the radiative properties of the gases involved, and the concentrations of other radiatively active gases and aerosols already present in the atmosphere.

Changes in atmospheric concentrations of greenhouse gases are, however, not the only disturbance that can affect the Earth's radiation balance. Indeed, any factor that alters the amount of radiation that is received from the Sun or lost to space, or that alters the redistribution of energy within the atmosphere and between the atmosphere, land, and oceans, can effect climatic change. Disturbances of this nature are known as *ra-*

*diative forcings*—they "force" the climate system to change in order to restore radiative equilibrium. Radiative forcings are expressed in watts per square meter $(W/m^2)$ and provide a first-order estimate of the potential climatic importance of various forcing mechanisms.

It has been established without doubt that concentrations of the naturally occurring greenhouse gases ($CO_2$, $CH_4$, $N_2O$, and tropospheric ozone) have increased since the late eighteenth century as a result of human activity. In the case of $CO_2$, concentrations have risen from preindustrial levels of around 280 parts per million by volume (ppmv) to the present-day value of 358 ppmv (in 1994), an increase of almost 30 percent. This increase is largely attributed to the combustion of fossil fuels and to land use changes, primarily tropical deforestation. Human activity has also introduced a whole range of new greenhouse gases into the atmosphere, namely, the chlorofluorocarbons, or CFCs. [See Chlorofluorocarbons.] Moreover, the prospect of a substantially altered climate due to an enhanced greenhouse effect is now a real one. Indeed, it has been estimated that atmospheric increases in greenhouse gas concentrations since preindustrial times have already led to a positive forcing of the global climate of the order of 2.5 $W/m^2$. Carbon dioxide is by far the largest single contributor, accounting for just over one-half of this forcing.

**Implications of Increased Atmospheric CO₂.** Over the past two decades, considerable effort has been devoted to the study of the potential effects of increased atmospheric $CO_2$ and other greenhouse gases. A huge body of research has now been amassed on this topic, much of it collated and analyzed by the Intergovernmental Panel on Climate Change (IPCC). [See Intergovernmental Panel on Climate Change.]

Much of the work of the IPCC is concerned with predicting future climate. This is by no means a straightforward task. In the first instance, it involves forecasting the most likely future course of $CO_2$ and other greenhouse gas emissions. To this end, the IPCC has developed a series of emissions scenarios covering the period 1990–2100; these are based on various assumptions about future population and economic growth, land use, technology changes, and energy use. These scenarios, coupled with knowledge about the global carbon cycle and atmospheric chemistry, are then used to project future atmospheric concentrations of the greenhouse gases and thus their radiative forcing. Finally, sophisticated computer models of the climate system are used to develop predictions of future climate in a high-$CO_2$ world.

The latest IPCC projections of future climate, for three emission scenarios, are summarized in Table 1. The most likely course or "best estimate" is an increase in global mean surface temperature of about 2°C and a sea level rise of about 50 centimeters over the period

**Carbon Dioxide. TABLE 1.** Projected Changes in Global Mean Surface Temperature and Global Mean Sea Level According to Various Intergovernmental Panel on Climate Change (IPCC) Emission Scenarios, 1990–2100

| EMISSION SCENARIO*(CM) | $CO_2$ EMISSION IN 2100 (Gt C Yr$^{-1}$) | CLIMATE SENSITIVITY (°C)† | TEMPERATURE INCREASE BY 2100 (°C) | SEA LEVEL RISE (CM) |
|---|---|---|---|---|
| Lowest rate of emissions increase (IS92c) | 4.6 | 1.5 | 0.9 | 13 |
| Midrange rate of emissions increase (IS92a) | 20.3 | 2.5 | 2.0 | 49 |
| Highest rate of emissions increase (IS92e) | 35.8 | 4.5 | 3.5 | 94 |

*IPCC's IS92a emission scenario is generally described as the "best estimate" scenario; it assumes a population growth to just over 11 billion by 2100, average economic growth rates of around 3 percent up until 2025 and declining thereafter, a relatively high continued dependence on fossil fuels, a partial compliance with the Montreal Protocol, and some reductions in emissions of the sulfur and nitrogen oxides, carbon monoxide, and the volatile organic compounds.

†In IPCC reports, "climate sensitivity" refers to the likely equilibrium response of global mean surface temperature to a doubling of atmospheric concentrations of $CO_2$ (or equivalent $CO_2$).

SOURCE: Intergovernmental Panel on Climate Change (1996).

1990–2100 (IPCC, 1996). The IPCC has also tried to calculate what emissions reductions would be required in order to stabilize $CO_2$ concentrations; it has concluded that stabilization of atmospheric $CO_2$ at 450, 650, or 1,000 ppmv would be achieved only if man-made emissions were reduced to 1990 levels by 40, 140, or 240 years from now, respectively, and dropped substantially below 1990 levels thereafter.

Recent improvements in climate models have meant that scientists are becoming increasingly confident about their model predictions. Many unknowns still remain, however, not least of which is the uncertainty surrounding the operation of the various feedback mechanisms that are associated with the behavior of clouds, oceans, sea ice, and vegetation. Feedback mechanisms are processes that either amplify the response of the climate system to radiative forcing (a positive feedback) or act to reduce it (a negative feedback); for example, the amount of water vapor—itself a greenhouse gas—is likely to increase in a warmer, high-$CO_2$ world. This in turn will lead to a further enhancement of the greenhouse effect (i.e., a positive feedback).

**Sources and Sinks of $CO_2$.** In order to project future atmospheric $CO_2$ concentrations, it is essential not only to quantify the sources and sinks of carbon but also to understand the processes that control the global carbon budget. Man-made emissions of $CO_2$ from fossil fuel combustion and cement manufacture are relatively well characterized, being calculated in a systematic fashion for all nations from energy use statistics and the application of appropriate emission factors. According to such computations, $CO_2$ emissions from industrial sources have more than tripled since 1950 and are currently estimated to be around 6 Gt per year (as carbon).

The industrialized nations of the world are by far the greatest producers of $CO_2$. Emissions from North America, Europe, the former Soviet Union, Japan, Australia, and New Zealand collectively account for around 70 percent of the global total; these regions also have the highest per capita emission rates. There are signs, however, that the rate of increase in emissions in certain regions—North America and western Europe— are beginning to tail off; in contrast, emissions are continuing to rise at relatively high rates in some of the rapidly industrializing regions, notably, the Far East and China.

The conversion of large areas of forest to land for agriculture on scales such as those currently practiced in parts of the tropics also lead, through burning and decay, to a net flux of $CO_2$ to the atmosphere. Estimates of emissions of $CO_2$ from this source are less precisely known than those from industrial sources, being derived from patchy information on tropical deforestation rates, biomass inventories, and modeled estimates of regrowth. Although recent satellite data have reduced some of the uncertainties in estimates of deforestation rates in the Amazon, rates for the rest of the tropics are still poorly quantified. Furthermore, for the tropics as a whole, there is incomplete information on initial biomass. Based on the information available, the best estimate for $CO_2$ emissions from terrestrial sources is given as 1.6 ± 1.0 Gt of carbon per annum (C/a); this value represents an average for the ten-year period 1980–1989. As yet, there are no estimates available for the years 1990–1995. [See Agriculture and Agricultural Land.]

Man-made emissions of $CO_2$, totaling some 7 Gt C/a, are summarized in Table 2; also shown are the various reservoirs and sinks for this man-made $CO_2$, together with the most recent estimates of their magnitude. Of these, only one is known with any degree of accuracy, namely, the accumulation of $CO_2$ in the atmosphere, currently put at 3.3 ± 0.2 Gt C/a. The second major sink is

Carbon dioxide is emitted from the combustion of fossil fuels, the flaring of natural gas, and the manufacture of cement, in which limestone (calcium carbonate) is converted to lime (calcium oxide), with the release of roughly half a metric ton of carbon dioxide per ton of cement produced. The relative contributions are as follows (percentage of world annual total):

| | |
|---|---|
| Solid fuels (coal) | 40 |
| Liquid fuels (petroleum) | 37 |
| Gaseous fuels (natural gas) | 18 |
| Cement manufacture | 3 |
| Flaring of natural gas | 1 |

In developing nations, the contribution of cement manufacture is greater than in more industrial nations: in China, for instance, it is 7 percent of the total for the nation, while in Canada and the United States it is roughly 1 percent.

**Carbon Dioxide Emissions by Region and Nation\***

| REGION/NATION | PERCENTAGE OF WORLD TOTAL CARBON DIOXIDE EMISSIONS | PER CAPITA EMISSIONS (TONS) |
|---|---|---|
| Asia | 36.4 | 2.3 |
| China | 14.0 | 2.7 |
| Japan | 5.0 | 9.0 |
| India | 4.0 | 1.0 |
| South Korea | 1.6 | 8.3 |
| North Korea | 1.1 | 0.2 |
| Iran | 1.2 | 3.8 |
| Saudi Arabia | 1.1 | 13.9 |
| Singapore | 0.3 | 19.1 |
| Europe | 27.5 | 8.5 |
| Russia | 8.0 | 12.2 |
| Germany | 3.7 | 10.2 |
| United Kingdom | 2.4 | 9.3 |
| Ukraine | 1.9 | 8.5 |
| Italy | 1.8 | 7.2 |
| France | 1.5 | 5.9 |
| Poland | 1.5 | 8.8 |
| North America | 26.0 | 19.9 |
| United States | 24.0 | 20.5 |
| Canada | 1.9 | 14.8 |
| Mexico | 1.6 | 3.9 |
| South America | 3.3 | 2.4 |
| Brazil | 1.1 | 1.6 |
| Argentina | 0.6 | 3.7 |
| Africa | 3.3 | 1.1 |
| South Africa | 1.3 | 7.4 |
| Algeria | 0.4 | 3.3 |
| Egypt | 0.4 | 1.5 |
| Central America | 2.1 | 3.6 |
| Cuba | 0.1 | 1.5 |
| Oceania | 1.4 | 11.3 |
| Australia | 1.3 | 16.2 |
| New Zealand | 0.1 | 7.7 |

\*Data exclude bunker fuels used in ships and aircraft for international transport.
SOURCE: Calculated from data in World Resources Institute (1998, pp. 344–345).

**BIBLIOGRAPHY**

World Resources Institute. *World Resources, 1998–99.* New York: Oxford University Press, 1998.

—DAVID J. CUFF

**Carbon Dioxide. TABLE 2.** Sources and Sinks of Carbon Dioxide*

| Sources | |
|---|---|
| Fossil fuel combustion and cement manufacture | $5.5 \pm 0.5$[†] |
| Changes in tropical land use (deforestation) | $1.6 \pm 1.0$ |
| Total | $7.1 \pm 1.1$ |
| **Sinks** | |
| Storage in the atmosphere | $3.3 \pm 0.2$ |
| Ocean uptake | $2.0 \pm 0.8$ |
| Other potential sinks[‡] | $1.3 \pm 1.5$ |
|    Northern Hemisphere forest regrowth | 0.5–0.9 |
|    Enhanced growth due to $CO_2$ fertilization | 0.5–2.0 |
|    Enhanced growth due to nitrogen deposition | 0.5–1.0 |
|    Climate anomalies[§] | 0.0–2.0 |

*Data represent average values for the 1980s and are expressed in units of gigatons of carbon per year ($Gt\ C\ yr^{-1}$).

†For comparison, emissions in 1994 were around $6\ Gt\ C\ yr^{-1}$.

‡The estimates of the magnitudes of the "other sinks" are not necessarily additive; for example, the rate of carbon sequestration in midlatitude forests may include contributions from the other processes.

§Various climatic anomalies, for example, such as those caused by the eruption of Mount Pinatubo, may affect carbon storage; this factor is thought to have exerted a positive effect (i.e., taking up carbon) during the 1980s and in 1992 and 1993, but could be either positive or negative during other periods.

SOURCE: Intergovernmental Panel on Climate Change (1996).

uptake by the oceans, estimated to be $2.0 \pm 0.8$ Gt C/a. Together these sinks account for approximately three-quarters of the net emissions of $CO_2$ to atmosphere.

For many years scientists have been uncertain about what happened to the remaining $CO_2$, some 1.8 Gt C/a. More recently, however, considerable progress has been made in elucidating the so-called missing sink in the man-made carbon budget.

Experimental evidence (based on measurements of the relative atmospheric abundances of $^{12}CO_2$ and $^{13}CO_2$) suggests that it is the terrestrial biosphere that is responsible for "mopping up" the missing carbon. In other words, the terrestrial biosphere is currently acting as a net carbon sink. A number of mechanisms have been put forward to account for this observation. First, it is possible that a regrowth of mid- and high-latitude forests, particularly in the Northern Hemisphere, is leading to a net carbon sequestration. Second, fertilization effects—owing to increased atmospheric $CO_2$ and increased nitrogen deposition—may be leading to greater rates of photosynthesis and enhanced net primary production.

It is difficult to estimate the size of these potential carbon sinks with any degree of accuracy, as direct observations of these processes are not available. Analyses of forest inventory data do, however, support the notion that forest area, biomass, and timber volume have been expanding in the temperate latitudes of the Northern Hemisphere. The IPCC estimates that the most likely magnitude of this sink is in the range 0.5–0.9 Gt per annum.

Laboratory studies have clearly demonstrated that the growth of most plant species is stimulated under conditions of high $CO_2$. Furthermore, there is evidence to suggest that $CO_2$ enrichment helps plants use water more efficiently. This has several consequences; it leads to increased productivity for the same moisture conditions and may allow plants to advance into more arid regions. While these direct effects of increased $CO_2$ on plants are well established in controlled environments, their significance in the real world is less certain. Nevertheless, it is considered likely that $CO_2$ fertilization is currently responsible for a net uptake of carbon in the range of 0.5–1.0 Gt C/a (see Table 2). Similarly, increased nitrogen deposition, particularly in North America and Europe, is believed to be enhancing primary production and thus carbon storage by as much as 0.5–0.9 Gt C/a.

Although it is more than likely that the terrestrial biosphere is acting as a net carbon sink, there are no guarantees that it will continue to do so in the future. For example, if atmospheric $CO_2$ is stabilized, the fertilization effect will disappear; alternatively, the $CO_2$ fertilization response may become saturated at higher levels, or plant growth may be limited by other factors, such as the availability of nutrients. Moreover, changes in atmospheric $CO_2$ and climate over the medium term (decades to a century) are likely to produce changes in the structure and pattern of both natural and managed ecosystems. Structural changes will include changes in the local abundance of species and shifts in the global distribution of assemblages of species (biomes). Of particular concern is the possibility that rapid climatic change could shift the competitive balance among species and even lead to dieback of particular species; this could result in a transient release of $CO_2$ into the atmosphere. The magnitude of this feedback is highly uncertain; it could be near zero or as much as 200 Gt of carbon over the next two centuries.

**Atmospheric CO₂: A History of Fluctuations.**
Over the past few decades, growing preoccupation with global warming has led not only to increased efforts to monitor present-day atmospheric $CO_2$ more closely, but also to attempts to chart past changes in atmospheric $CO_2$, particularly those that have occurred over time scales of centuries and more. Routine, direct measurement of atmospheric $CO_2$ actually commenced in the late 1950s at sites at the South Pole and at Mauna Loa, Hawaii. In the 1970s, trace gas monitoring expanded significantly, to the extent that today measurements are made at some thirty or more sites located in remote locations worldwide. The existence of this network of observatory sites means that we now have a reasonably comprehensive picture of the temporal and spatial trends in atmospheric $CO_2$ in the recent past.

In 1994 the global average concentration of $CO_2$ was estimated to be 358 ppmv. During the 1980s globally averaged $CO_2$ concentrations increased on average by $1.53 \pm 0.1$ parts per million per annum (ppm/a), corresponding to an annual average rate of change in atmospheric carbon of $3.3 \pm 0.2$ Gt C/a. In the early 1990s rates of growth in $CO_2$ concentrations dropped to only 0.6 ppm/a in 1991–1992; more recent data indicate that the growth rate has again returned to a value of 1.5 ppm/a. Although the reasons behind these short-term fluctuations in growth rates are not completely understood, it is possible they could be linked to the eruption of Mount Pinatubo in June 1991. [*See* Mount Pinatubo.]

The record of atmospheric $CO_2$ concentrations obtained at Mauna Loa serves well to illustrate the recent rise in $CO_2$ globally (see Figure 2). The pronounced seasonality in the record here is a reflection of the annual withdrawal and production of $CO_2$ by terrestrial biota in the Northern Hemisphere. In addition, there is evidence to suggest that the amplitude of seasonal fluctuations has increased latterly, a feature that is consistent with the notion that Northern Hemisphere forests are currently acting as a net sink for $CO_2$. The recent atmospheric record (i.e., 1958–present) also provides support for the theory that the observed increases are due to man-made emissions of $CO_2$. For example, when seasonal and short-term interannual variations are ignored, the rise in atmospheric $CO_2$ since the late 1950s corresponds to half the size of anthropogenic emissions. Furthermore, the interhemispheric difference in $CO_2$ (concentrations are slightly higher in the Northern Hemisphere owing to a higher density of emission sources) has increased at approximately the same rate as the rate of growth in fossil fuel emissions.

Although direct measurements of atmospheric $CO_2$

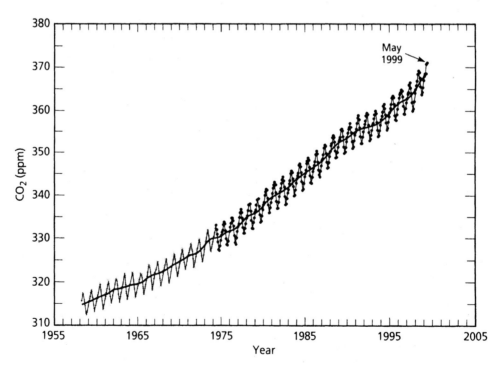

**Carbon Dioxide. Figure 2.** Monthly Mean Concentrations of Atmospheric $CO_2$ as Recorded at Mauna Loa, Hawaii, 1958–1999.

Data prior to 1974 are from the Scripps Institution of Oceanography; data since May 1974 are from the National Oceanic and Atmospheric Administration. A long-term curve is fitted to the monthly mean values. (Courtesy Thomas Conway, National Oceanic and Atmospheric Administration, Climate Monitoring and Diagnostics Laboratory.)

are available only from the late 1950s onward, the analysis of air bubbles trapped in ice cores extracted from both the Arctic and Antarctic regions have extended the record of past atmospheric composition back to well before the Industrial Revolution. Results of studies of this type have firmly established preindustrial atmospheric $CO_2$ as being around 280 ppm; prior to this (i.e., over the last thousand years or so), $CO_2$ concentrations in the atmosphere have fluctuated by no more than 10 ppm around the 280 ppm mark.

A number of deep ice cores, most notably those obtained at Vostok in East Antarctica, have provided a history of atmospheric $CO_2$ concentrations over much longer time scales, as much as 160,000 years BP. This period covers one complete glacial/interglacial cycle. Data obtained from the Vostok core suggest that when going from a glacial maximum (the peak of an ice age) to interglacial conditions, the Earth experienced a change in global average temperature of around $4°$–$5°$C, an increase in atmospheric $CO_2$ of about 80 ppm, and almost a doubling of atmospheric methane ($CH_4$). Other evidence indicates that these changes were accompanied by an increase in precipitation and a sea level rise of the order of 1.2 meters.

There is a close association between temperature fluctuations and atmospheric levels of $CO_2$ over the past 200,000 years; this correlation suggests that changes in greenhouse gas concentrations are likely to have contributed significantly to the glacial/interglacial temperature change. [*See Figure 6 in the article on* Climate Change: An Overview.] It is also significant that changes in greenhouse gas concentrations over the glacial/interglacial cycle are estimated to have caused a radiative forcing of about 2.5 W/m; this is comparable to the present human-induced forcing.

Higher-resolution analyses of paleorecords have revealed the existence of relatively rapid past changes in climate, that is, changes in temperature of several degrees taking place over a decade or so. These discoveries have led some scientists to question whether future climate change can ever be predicted with any degree of reliability. Will, for example, the transition to a warmer world in the twenty-first century be a gradual one, or will there be a series of abrupt or even unexpected changes in climate as the Earth flips between different stable climatic states? Given these uncertainties, it is perhaps not surprising that some scientists have described the discharge of millions of metric tons of $CO_2$ into the atmosphere as the world's greatest geophysical experiment, the results of which still remain to be seen.

[*See also* Atmosphere Structure and Evolution; Atmospheric Chemistry; Biological Feedback; Biomass; Biosphere; Deforestation; Earth History; Electrical Power Generation; Energy; Energy Policy; Fossil Fuels; Geoengineering; Global Warming; Industrialization; International Cooperation; Ocean Chemistry; Ocean Dynamics; *and* Volcanoes.]

## BIBLIOGRAPHY

THE REPORTS OF THE IPCC

The reports of the Intergovernmental Panel on Climate Change (IPCC) are the most comprehensive and authoritative assessments of the climate change issue, representing the consensus view of the worldwide scientific community. They contain comprehensive bibliographies and reference lists. The first scientific assessment report was produced in 1990, the second in 1995. Interim reports were produced in 1992 and 1994. Whereas the 1994 report on the radiative forcing of climate change contains more detailed information about the global carbon cycle, the two reports listed below present the latest information regarding the science of human-induced climate change.

Intergovernmental Panel on Climate Change (IPCC). *Climate Change 1995: The Science of Climate Change: Contribution of Working Group I to the Second Assessment Report of the Intergovernmental Panel on Climate Change.* Edited by J. T. Houghton, L. G. Meira Filho, B. A. Callender, N. Harris, A. Kattenberg, and K. Maskell. Cambridge: IPCC/Cambridge University Press, 1996.

——. *Climate Change 1995: The Science of Climate Change: Summary for Policy Makers and Technical Summary of the Working Group I Report.* Cambridge: IPCC/Cambridge University Press, 1996.

ATMOSPHERIC $CO_2$

Conway, T. J., P. P. Tans, L. S. Waterman, K. W. Thoning, K. A. Masarie, and R. H. Gammon. "Atmospheric $CO_2$ Measurements in the Remote Global Troposphere." *Tellus* 40B (1988), 81.

Lorius, C., and H. Oeschger. "Palaeo-perspectives: Reducing the Uncertainties in Global Change." *Ambio* 23 (February 1994), 30–36.

Raynaud, D., J. Jouzel, J. M. Barnola, J. Chappellaz, R. J. Delmas, and C. Lorius. "The Ice Record of Greenhouse Gases." *Science* 259 (1993), 926–934.

SOURCES AND SINKS OF $CO_2$

Carbon Dioxide Information Analysis Center (CDIAC). Estimates of $CO_2$ Emissions from Fossil Fuel Burning and Cement Manufacturing, Based on United Nations Energy Statistics and the U.S. Bureau of Mines Cement Manufacturing Data. Numeric Data Package 030/R4. Oak Ridge, Tenn.: Carbon Dioxide Information Analysis Center. The CDIAC routinely calculates annual $CO_2$ emissions data for all countries of the world; these data are available upon request in either published form or on diskette.

Houghton, R. A. "Tropical Deforestation and Atmospheric Carbon Dioxide." *Climate Change* 19 (1991), 99–118.

Moore B., III, and B. H. Braswell, Jr. "Planetary Metabolism: Understanding the Carbon Cycle." *Ambio* 23 (February 1994), 4–12.

$CO_2$ AND THE TERRESTRIAL BIOSPHERE

Green, N. P. O., G. W. Stout, and D. J. Taylor. *Biological Science,* vol. 1, *Organisms, Energy and Environment.* 2d ed. Cambridge: Cambridge University Press, 1990. Biology textbook that contains basic details of the biochemistry of photosynthesis.

Leemans, R. "Impacts of Greenhouse Gases and Climatic Change." In *The Global Environment: Science, Technology and Management,* vol. 1, edited by D. Brune, D. V. Chapman, M. D. Gwynne, and J. M. Pacyna, pp. 352–368. Weinheim, Germany: VCH, 1995. A very readable overview of the effects of increased $CO_2$ and climate change on the terrestrial biosphere.

Mooney, H. A., and G. W. Koch. "The Impact of Rising CO$_2$ Concentrations on the Terrestrial Biosphere." *Ambio* 23 (February 1994), 74–76.

Schulze, E. D., and M. M. Caldwell, eds. *Ecophysiology of Photosynthesis.* Berlin: Springer, 1995.

Walker, B. H. "Landscape to Regional-Scale Responses of Terrestrial Ecosystems to Global Change." *Ambio* 23 (February 1994), 67–73.

—ANN D. WILLCOCKS

## CARBON ISOTOPES. *See* Climate Reconstruction.

## CARRYING CAPACITY

That resource limits, including geographical space, set the maximum population size of a species was conventional wisdom at one time in ecology and maintains varying degrees of currency (for example, in conservation biology and ecological economics). Such relationships are referred to as *carrying capacity* and are measured by the number of an organism that can be supported by the requisite renewable resources of the environment. It would be a mistake, however, to assume that the application of the carrying capacity concept to human–environment relationships was borrowed directly from the modern science of ecology. Phrased differently, the antecedents of its applications to human populations are old, traced back in Western thought at least to the seminal population–resource relationships of Thomas Robert Malthus (1798)—that the technology of food production within a given environment (an assumed closed system) sets the limits to which human population numbers can grow. [*See the biography of Malthus.*] Beyond these limits and without a change in technology, the resource base collapses, the environment degrades, and a Malthusian crisis ensues (Coleman and Schofield, 1986).

The application of the carrying capacity concept to humans cycled to the forefront of research attention at least twice in the latter half of the twentieth century. The mid-1960s through the 1970s witnessed the first phase of interest, stimulated by two different research interests—cultural ecology and global population growth—that drew on ideas emanating from ecology, systems theory, and precursors to ecological economics. [*See* Growth, Limits to.] Cultural ecology emerged as a subfield in anthropology and geography and attempted to understand differences in land and resource uses and the level of social and cultural development through the lens of the human–environment relationship. Global population concerns, in contrast, drew attention from diverse communities, including lay ones, focused on the then doubling of the world population to four billion and alarm about the ability of humankind to feed, clothe, and shelter the rapidly growing number of our species. Population growth interests attempted

to assess the carrying capacity of the Earth at large (Meadows et al., 1972) and of regions and countries as well. These assessments projected some areas to exceed their limits and experience Malthusian crises. Cultural ecology, in contrast, used the concept less ambitiously, typically in terms of small areas and in reference to non-Western peoples tied to subsistence livelihoods (Brush, 1975). In either case, several concepts were common to various nomenclatures, definitions, and uses of carrying capacity: the native quality of a bounded unit of land (the environment); the resource response to a given set of technologies and use strategies; and the maximum number of people sustainable in the human–environment relationship. Conceptually, technology and use strategies were thought to determine the quality and quantity of resources from a given environment; if they changed, so did the carrying capacity (Solow, 1970). This nod to human ingenuity was often lost in application, however, and some assessments of critical population densities were reduced to the base environmental qualities of an area, raising the specter of environmental determinism.

Interest in carrying capacity waned during the late 1970s. Cultural ecology exhausted the reach of the concept's usefulness for understanding land use change and cultural development; the fluidity of carrying capacity in regard to changing human–environment relationships was largely recognized; and attention turned to the issue of social structures that shaped the relationships. The global population community's interest momentarily faded for several reasons. Famine (Malthusian crisis) was reinterpreted as having origins in the political economy, not resource technology; the "green revolution" increased food production significantly through hybrid species; and, with the worldwide expansion of family planning, public attention moved toward economic development.

Finally, the concept implied an apparent internal contradiction: carrying capacity expanded or contracted as technology and resource use changed (Cohen, 1995). The principle of the substitutability of resources was difficult to refute, either historically or theoretically (Solow, 1970). If substitutions were exogenous or random in origin, then carrying capacity had a fallback position in terms of momentary capacities. Various works, however, demonstrated the role of endogenous technological change (Boserup, 1965; Turner and Ali, 1995) and the increasing diminution of geographical insularity because of the flows of energy, material, knowledge, and power across regions. Absent this insularity and with endogenous technological change, what was the usefulness of carrying capacity?

The 1990s witnessed a second cycle of interest in carrying capacity, this time as embedded within the complementary concepts of sustainable development and global environmental change. [*See* Sustainable Development.] These linkages purport to change the meaning of carrying capacity by shifting the processes and outcomes of concern and, as a result, the temporal and spatial scales of assessment (e.g., Brown and Kane, 1994; Daily and Ehrlich, 1992). Sustainable development and global environmental change increasingly refocus on the rights of future generations to experience existing nature—for example, biota, landscapes, and air quality—and to use the biosphere without endangering its structures and functions. In this view, worldwide access to high standards of consumption and quality of life must be met without endangering the ability of the biosphere to provide the life-support systems for humankind (Arrow et al., 1995) and, among deep ecologists, for nonhuman species.

In this iteration, carrying capacity focuses on critical flows of the biosphere, such as the carbon and nitrogen cycles, or on potential losses for humankind of biotic and diversity of habitats (Vitousek et al., 1997). Changes in biogeochemical flows disrupt the state of the biosphere, with potential harmful consequences—for example, an atmosphere that permits ultraviolet radiation to penetrate to the Earth's surface unabated threatens biota with excessive levels of radiation, complete with significant health implications for humans. Loss in biotic and ecosystem diversity, it is argued, constitutes losses of genetic stock and the resilience of nature to respond to perturbations of almost any kind.

The problems of substitution and geographical boundedness are challenged in this new version of carrying capacity (Daily and Ehrlich, 1992). Critical states and flows, at least by implication, are thought to be largely unsubstitutable and operate at global rather than local scales. Exogenous flows at the level of the biosphere are reduced by some to incoming solar radiation, funneled through human appropriation of net primary productivity, for example (Vitousek et al., 1997). Places and regions differ in their roles as sources and recipients of global change, and in their abilities to respond to change. They are, however, recognized as connected and their "carrying capacities" calculated through such measures as "ecological footprints" (Rees and Wackernagel, 1994). This change in usage may alleviate the problems inherent in the former meaning of carrying capacity, but it raises new criticisms, largely of a political nature. Reformulated in this way, it is suggested by some that the economically developed world seeks to assert a new kind of authority over the remainder of the world, potentially retarding development elsewhere.

The carrying capacity principle remains in its cycle of interest that began in the 1990s. Other versions will no doubt emerge in future cycles. Its staying power rests in its centrality to the ultimate human–environment questions. Does the environment ultimately set limits on humankind? Or, can human ingenuity restructure nature as needed for human use? These questions and carrying capacity are central to the tensions surrounding the emergence of ecological economics and attempts to calculate the economic value of the biosphere (Costanza et al., 1997; Pearce, 1998)—the ultimate limit?

[*See also* Agriculture and Agricultural Land; Food; Food and Agriculture Organization; *and* Population Dynamics.]

## BIBLIOGRAPHY

Arrow, K., B. Bolin, R. Costanza, P. Dasgupta, C. Folke, C. S. Holling, B. O. Jansson, S. Levin, K. G. Mäler, C. Perrings, and D. Pimental. "Economic Growth, Carrying Capacity, and the Environment." *Science* 268 (1995), 520–521. An interdisciplinary view of carrying capacity applied to resource consumption in general.

Boserup, E. *The Conditions of Agricultural Growth.* Chicago: Aldine, 1965. The watershed argument that population determines the intensity of cultivation and behavior rationale for nonmarket farmers.

Brown, L. R., and H. Kane. *Full House: Reassessing the Earth's Population Carrying Capacity.* New York: Norton, 1994. A reassessment of global carrying capacity including damage to the biosphere.

Brush, S. B. "The Concept of Carrying Capacity for Systems of Shifting Cultivation." *American Anthropologist* 77 (1975), 799–811. A review of the meaning and use of carrying capacity in anthropology and the interdisciplinary field of cultural ecology.

Cohen, J. E. "Population Growth and Earth's Human Carrying Capacity." *Science* 269 (1995), 341–346. A perspective questioning the presumed negative role of population growth on resources.

Coleman, D., and R. Schofield, eds. *The State of Population Theory: Forward from Malthus.* Oxford: Blackwell, 1986. A state-of-the-art review of population–resource relationships in demography and economics.

Costanza, R., R. Dauge, R. De Groot, S. Farber, B. Hannon, K. Limburg, S. Naeem, R. V. Oneill, J. Paruelo, R. G. Raskin, and M. Van Den Belt. "The Value of the World's Ecosystem Services and Natural Capital." *Nature* 387 (1997), 253–260. An incipient attempt to place a monetary value on the biosphere.

Daily, G. C., and P. R. Ehrlich. "Population, Sustainability, and Earth's Carrying Capacity." *Bioscience* 42 (1992), 761–771. An update of the view that population and economic growth stress the Earth's resources.

Malthus, T. R. *An Essay on the Principles of Population, as It Affects the Future Improvement of Society.* London: J. Johnson, 1798.

Meadows, D. H., D. L. Meadows, J. Randers, and W. W. Behrens III. *The Limits to Growth.* New York: Signet Books, 1972. A classic and controversial work attempting to model the global carrying capacity.

Pearce, D. "Auditing the Earth." *Environment* 40 (1998), 23–28. A neoclassical economist responds to attempts to place a value on the biosphere.

Rees, W. E., and M. Wackernagel. "Ecological Footprints and Appropriate Carrying Capacity: Measuring the Natural Capital Requirement of the Human Economy." In *Investing in Natural Capital: The Ecological Approach to Sustainability*, edited by A. Jannsson, M. Hammer, C. Fokle, and R. Costanza, pp. 362–390. Washington, D.C.: Island Press, 1994. A methodology to assess the environmental reach of consumption patterns in cities.

Solow, R. M. *Growth Theory: An Exposition*. New York and Oxford: Oxford University Press, 1970. A neoclassical economic view of the economic implications of growth.

Turner II, B. L., and A. M. S. Ali. "Induced Intensification: Agricultural Change in Bangladesh with Implications for Malthus and Boserup." *Proceedings of the National Academy of Sciences* 93 (1995), 14,984–14,991. An elaboration of the applicability of a Boserupian-type model of agricultural change in a nonintuitive country example.

Vitousek, P. M., H. A. Mooney, J. Lubchenco, and J. M. Melillo. "Human Domination of Earth's Ecosystems." *Science* 277 (1997), 494–499. Ecologists' view of human transformation of the Earth.

—B. L. Turner II and Eric G. Keys

# CARSON, RACHEL

Rachel Louise Carson (1907–1964), U.S. scientist and writer, is widely identified as the "mother of the environmental movement" of the late twentieth century. Thanks to her work, earlier concerns with human encroachment on the natural world, generally referred to as *conservation*, have broadened in the public perception to stress wider ecological implications. [*See* Environmental Movements.] Much of this public understanding came from Carson's books on the ocean, which mustered new knowledge with vivid poetic focus, typified by the title of her first major work, *The Sea Around Us*. Her final book, *Silent Spring*, turned from appreciation of the natural world to the corrosive effect of human activities on the vulnerable Earth.

Carson was of the naturalist tradition, seeing the larger picture while comprehending the details. The desire to bring this broader understanding to everyone, specialists and the general public alike, shaped her writing style. Her education prepared her for this role: undergraduate majors in writing and zoology followed by a graduate thesis in marine biology. Her years as editor in chief of the U.S. Fish and Wildlife Service (1937–1952) sharpened her skill at putting research results into reports significant to scientists and laypeople. By her own writing and editing, she educated a generation of colleagues in scrupulous accuracy and literate presentation.

For her unofficial writing, Carson chose substantial and difficult topics, which were discussed in works such as *Under the Sea Wind* (1941; 50th-anniversary edition, 1991); *The Sea around Us* (1951; revised edition, 1961); and *The Edge of the Sea* (1955). These books brought the latest knowledge of the oceans into syntheses both inclusive and enthralling. Their commercial success enabled Carson to resign from her government job to devote herself to writing, and brought her worldwide recognition that assured prompt response from the experts she sought for later research for *Silent Spring* (1962).

Always alert to developments at the frontiers of science, Carson could put her understanding into universal language. Her remarkable capacity to reach an international audience enabled her to influence the public perception of responsibility to the Earth and its flora and fauna.

The adverse effects of the broad use of new chemical pesticides after World War II were first known from research on fish and wildlife. [*See* Pollution.] Carson saw this evidence early and, as use of active ingredients in the United States increased by 1960 to over 285 million kilograms (627 million pounds) per year, she comprehended the urgency of the threat. (The U.S. Environmental Protection Agency figure for 1997 is about 2 billion kilograms.) The industry promoted the belief that this use of potent poisons was essential to agriculture and daily life; Carson, with her reputation for scientific integrity and her independence from pesticide interests, was able to demonstrate to the public the flaws in these arguments and the grave dangers of uncontrolled pesticide use.

To make her case, Carson assembled scattered and complex evidence, then summarized it in lucid and compelling terms; in so doing, she succeeded both in establishing a new public understanding of the ecology of the Earth and in demonstrating how these new chemical assaults damaged crucial elements of that ecology. *Silent Spring* galvanized people and their governments, and many innovative new laws and agencies followed. These include the Environmental Defense Fund; programs of the Food and Drug Administration; the Conservation in Action series, Fish and Wildlife Service, and related publications; the National Academy of Sciences; the U.S. President's Office of Science and Technology; similar agencies and programs in England and various other countries; and the Rachel Carson Council. The book has remained influential, with sales of 23,200 copies in 1997. It stands today as one of the most powerful works questioning the attitudes and practices of modern industrial society as they relate to ecology and the environment.

Carson gave us a new and critical body of knowledge, and showed by her methods how to recognize and understand new environmental issues. She worked unstintingly to complete her groundbreaking work, which was always accurate and discriminating despite much opposition and difficult personal burdens. She completed *Silent Spring* while battling the cancer that marked her last years.

[*See also* Global Change, *article on* History of Global Change.]

## BIBLIOGRAPHY

Brooks, P. *The House of Life: Rachel Carson at Work*. Boston: Houghton Mifflin, 1972. A literary biography, now out of print but in most libraries, with excerpts from her writing.

Graham, F. *Since Silent Spring*. Boston: Houghton Mifflin, 1970. How the book was written, its immediate tumultuous reception, and longer-term impact. Out of print, but often in libraries.

Lear, L. *Rachel Carson: Witness for Nature*. New York: Henry Holt, 1997. A full, authoritative biography.

—SHIRLEY A. BRIGGS

## CARTAGENA PROTOCOL

Governments adopted the 1992 Convention on Biological Diversity (CBD) as an umbrella agreement that would address all aspects of the decline in biological diversity and exploitation of genetic resources. Written into the treaty was the request to negotiate a special subtreaty—a protocol—on risks posed to biological diversity from trade in living organisms whose genetic code had been engineered, known as living modified organisms (LMOs). Four years later, parties to the CBD began drafting the protocol. In 1999 at Cartagena de Indias, Columbia—at what was slated to be the final negotiating session—delegates failed to reconcile their differences over which LMOs would be included and the protocol's relationship to other international trade agreements. Some countries (led by the United States) wanted few limits on LMOs; others (led by the European Union) wanted the protocol to allow countries to impose tight ("precautionary") limits on LMOs. Delegates from 130 countries resumed talks a year later in Montreal and finally adopted the *Cartagena Protocol*.

Biological diversity, the variety and complexity of living organisms and the ecological systems they occupy, is important because it maintains genetic possibilities for organism adaptation and future human uses. [*See* Biological Diversity.] LMOs—examples of which include corn and soybeans that have been genetically altered to resist pests and diseases as well as salmon that mature twice as fast as natural varieties—could harm natural biological diversity by affecting the evolution of diseases and altering the natural balance of ecosystems, perhaps creating invincible insects and weeds.

The Cartagena Protocol holds the exporter responsible for assessing risks and requesting the importer's permission before transferring seeds, fish, and other LMOs intended for introduction into the environment. It creates an Internet-based Biosafety Clearing House to inform members of risk assessments on all types of LMOs and relay decisions that members have announced about which LMOs they will accept as imports. This Clearing House will help recipient countries, especially in the developing world, to assess risks before agreeing to import. It also better assures that exporters will send shipments according to the wishes of importers. The protocol is more relaxed for LMOs not intended for introduction into the environment, including those for food, feed, or processing and laboratory use. It requires only that shipments containing LMOs be accompanied by labels that indicate the identity of the LMO and any associated health or safety risks. The protocol does not cover pharmaceuticals.

Several provisions in the Cartagena Protocol make its relationship to current World Trade Organization (WTO) law ambiguous. The preamble to the protocol reaffirms that its measures must abide by other existing international agreements, including free trade agreements. Yet the preamble also underscores that the protocol is not subordinate to other international agreements. Using a broad interpretation that is implied in the Cartagena Protocol would suggest that nations have wide freedom to restrict international trade of LMOs. However, a narrow interpretation would suggest that countries must comply with other agreements, in particular the Sanitary and Phytosanitary Agreement (SPS Agreement) and Technical Barriers to Trade Agreement (TBT Agreement) of the WTO. Both those agreements require that countries base their trade restrictions on risk assessment and place other limits on trade measures.

[*See also* Convention on Biological Diversity.]

### INTERNET RESOURCE

Cartagena Protocol on Biosafety. Text of the Cartagena Protocol available at http://www.biodiv.org/Protocol/Protocol.html.

### BIBLIOGRAPHY

Gupta, A. "Governing Trade in Genetically Modified Organisms: The Cartagena Protocol on Biosafety." *Environment* 42.4 (2000), 22.

—VALERIE KARPLUS AND DAVID G. VICTOR

**CATASTROPHES.** *See* Natural Hazards.

## CATASTROPHIST–CORNUCOPIAN DEBATE

*Catastrophist* and *cornucopian* are two relatively neutral and descriptive names for two opposed points of view on the meaning and consequences of global environmental change and on nature–society relations more generally. Other terms often used as synonyms for *catastrophist* are "ecocentric," "pessimist," "doomsday," "neo-Malthusian," and "Cassandra"; for *cornucopian*, "technocentric," "optimist," "Polyanna," and "human exemptionalist."

The catastrophist view maintains that there are definite natural limits to human exploitation and transformation of the Earth that cannot be transgressed without disaster and that the current scale and rate of human activity already exceed the globe's carrying capacity, defined as the maximum demand it can support without lessening the demand that it will be able to support in the future. The cornucopian position tends to dismiss human carrying capacity as a useless and meaningless concept. It holds that human ingenuity and adaptation through technological and social responses, not natural limits, set the bounds to human wealth and well-being. It envisions no resource shortage or pollution problem with which society cannot cope. In interpreting the history of human–environment relations, catastrophists emphasize the rise in human numbers, levels of consumption, and environmental transformation as evidence that pressures on the terrestrial environment are mounting to dangerous levels that threaten humankind's survival. Cornucopians point to rising human numbers, levels of consumption, and life expectancy as evidence of overall and indeed steadily increasing success in the use and management of the Earth. [See Carrying Capacity.]

In academic debate, the catastrophist position is associated most strongly with the natural and especially the biological sciences. Cornucopianism tends to find its home in certain of the social sciences and especially economics, though there are many exceptions to both rules. Orthodox Western economics has long been called "the dismal science" for its insistence on the scarcity of everything and the costlessness of nothing. It is an often-remarked paradox that, in the realm of environment and resources, it is economists who have emerged as the voices of optimism and the questioners of limits to human expansion and well-being. Yet it also was an economic theorist, though a heterodox one from the point of view of the later evolution of the field, who offered the first compelling version of the pessimistic view. One of the few points on which catastrophists and cornucopians can generally agree is in seeing the work of the English clergyman and political economist Thomas Robert Malthus (1766–1834) as a crucial starting point in the history of the controversy. Malthus was not the first to question the adequacy of the Earth's resources to meet human demands, but his classic formulation of the problem has had a profound influence on later debate. [See the biography of Malthus.]

When he published An Essay on the Principle of Population (1st ed., 1798), Malthus was reacting against the confidence expressed by certain thinkers of his time who held the future of the human race to be one of unlimited progress in material well-being. Malthus responded with a model, famous in its own time and ever since, of the relation between population and food supply. It assumed an inherent human drive toward procreation, which ensured that numbers would always tend to increase more rapidly than the means of subsistence could be increased. Whereas population can grow geometrically by multiplication, food supply, Malthus assumed, could grow only arithmetically by the piecemeal cultivation of new lands; hence food supply will always limit the size and material welfare of the hu-

man population. This model formed the basis for a gloomy rather than a cheerful view of the human future. It would consist not of steady progress toward ever higher numbers enjoying ever higher levels of health, wealth, and happiness, but of population always growing to the limits of bare subsistence, its growth ever checked by one or the other of two means Malthus recognized: "misery and vice."

The arguments of the essay differ in several ways from those of modern environmental catastrophists. Malthus saw crises of subsistence not as a looming future consequence of excessive growth but as a factor constantly in operation, something that had always been a part of human life and always would be. He did not directly address environmental change as a constraint on growth in human population and wealth. Land degradation as it might affect food supply was not one of his concerns. In common with other political economists of his time, he treated the power of the soil to produce food as no more subject to degradation than to dramatic improvement. The ultimate problem as he framed the matter was that the globe's land and its productive possibilities existed in a finite supply. His importance for debates over environmental change stems from the broader message of his essay, from the compelling way in which he depicted the Earth as a home of limited resources for its human inhabitants.

Malthusian reasoning was extended to the area of soil fertility and land exhaustion at the end of the nineteenth century by the British physicist Sir William Crookes (1832–1919). In a famous address to the British Association for the Advancement of Science in 1898 and in subsequent writings, he warned that the depletion of soil nutrients, especially nitrates, must eventually curb the world's food supply when the supply of mined nitrate fertilizers gave out. Crookes, however, is an ambiguous figure, as much hopeful cornucopian as worried catastrophist. He looked forward to a solution in the form of an efficient process for synthesizing fertilizers from atmospheric nitrogen, one that was indeed developed within a few years of his warning. By means of this and other advances, including agricultural mechanization and the breeding and dissemination of high-yield grain varieties, Malthus was in a strict sense proven wrong. Population greatly expanded during the twentieth century, but it was fed through dramatic increases in the output of land rather than through an equal expansion of the cultivated area.

New concerns, however, arose over soil erosion, the spread of deserts, and other forms of land decay. A conservation movement that developed around the turn of the century in a number of countries blended concern over land degradation with the fear that a growing scarcity of natural resources such as energy and minerals would emerge as a constraint on economic growth. It argued for regulation to ensure that such resources were not squandered but exploited for the maximum benefit of all future generations. *The Coal Question* (1865) by the English economist William Stanley Jevons (1835–1882) was a classic early statement of this position. Variants of it appeared in the writings of a number of late-nineteenth- and early-twentieth-century theorists who proposed an economics based on energy rather than monetary calculations. Conservationist warnings about growing resource scarcity were countered by an influential set of studies published by the U.S. research institute Resources for the Future in the early 1960s. They showed that the real prices of key nonrenewable resources had not risen over time, as often predicted, but declined. To account for this result, the researchers proposed that early signs of scarcity stimulate a search for new sources, for substitutes, and for more efficient techniques of use, such that over the long run, abundance is likelier to increase than to diminish.

The modern phase of the debate has seen both the catastrophist case and the cornucopian rebuttal extended to cover a wide range of human-induced environmental changes such as air, water, and soil pollution, species loss, and climate change. It has also seen a new emphasis on the possibility of comprehensive global-scale disaster from the transgression of natural limits. *The Limits to Growth* (Meadows et al., 1972) brought the catastrophist perspective to wide public attention. It presented the results of a computer simulation of the world system that showed exponential growth in population and industrial output overwhelming within a century the Earth's capacity to furnish food and non-renewable resources and absorb pollutants. *Models of Doom* (Cole et al., 1973) promptly criticized *The Limits to Growth* for ignoring many processes—such as the adoption of substitutes and remedial and regulatory measures, the identification of new resources, and the more efficient use of resources as they become scarce—as mechanisms by which crises of supply and pollution are averted once they begin to develop.

Disagreements about how well these mechanisms work have been at the center of the debate ever since. The question of how far ingenuity can create substitutes for lost or degraded natural resources has been treated as a fundamental one by the recently arisen field of ecological economics. This field has many affinities with the catastrophist position in seeing many natural resources and services as unique and indispensable. Orthodox neoclassical economic analysis, on the other hand, echoes the cornucopian view of natural resources and services as themselves creations of human ingenuity and replaceable by further ingenuity should they grow scarce in their existing form. Cornucopians tend to regard natural inputs as merely one form of capital, for which nonnatural capital can be substituted. Ecological economists view many forms of natural capital as irreplaceable or at least very costly to replace and condemn as unsus-

tainable growth that does not maintain natural capital undiminished. Differences in methodology reflect these conflicting assumptions. Catastrophists see physical and energy measures of cost, scarcity, and efficiency as the best integrative measuring rods. Cornucopians prefer social measures, particularly market prices, as indicators of abundance or scarcity. Because of the tendency of markets to discount future values heavily in comparison to present consumption, catastrophists far more than cornucopians favor the use of lower-than-market discount rates when calculating the costs and benefits of conservation and preservation, particularly when the possibility exists of irreversible change or loss.

They take similarly opposed positions on many other questions of policy. Both profess concern for the well-being of future human generations while disagreeing as to what course of action in the present will bequeath a better world to them. Catastrophists see unrestrained growth in population, resource demand, and waste emissions as the chief problem. They therefore advocate the limiting of human numbers and of resource use to levels that will be sustainable over the long term. A "steady-state" economy (a concept elaborated upon by the economist Herman Daly) holds out the hope that innovation can indeed make possible further growth in affluence that depends not on greater resource demands but on more efficient and economical use of a fixed flow of material and energy. More rational planning guided by the "precautionary principle" is required in order to keep human activities within the environmental limits that, unchecked, they threaten to violate with catastrophic consequences. Market mechanisms, even when perfectly operating, do not incorporate the possibility of such consequences and cannot alone be trusted to avert them. A key role is thus envisioned for state and society as stewards and custodians of the long-term interests of humankind. Uncontrolled technological development and affluence in the form of higher material consumption are viewed with suspicion.

Cornucopians see the problems, threats, and challenges of resource depletion and environmental degradation as largely self-correcting through the workings of markets and human creativity if those workings are not unduly interfered with. If hubris and unregulated growth are, for catastrophists, the greatest dangers in human–environment relations, from the cornucopian perspective what we have most to fear is fear itself. What the catastrophist sees as the necessary responses to impending crisis, the cornucopian regards as more likely to be the sources of worse problems because they hinder the very processes that can bring solutions at the same time that they raise the general quality of life. Where problems have not corrected themselves, it is because adaptation has been obstructed. The remedy is to remove the obstacles, such as excessive state regulation, that have been placed in the way of its smooth functioning. Planning for sustainability should be viewed with suspicion, an excessive and stifling degree of precaution avoided, individual decisions be left as free as possible, and market transactions and private property rights expanded as far as possible to ensure rational outcomes. Population, termed by the cornucopian economist Julian Simon "the ultimate resource," is part of the solution rather than the problem. Growth in human numbers increases the stock of knowledge and skills; so, too, affluence is viewed positively, as increasing demand for better environmental quality and providing the means to achieve it.

The chief ground of contention between the two perspectives was once the adequacy or the possible exhaustion of the natural resources extracted and used by human society. On this terrain, history to date has arguably borne out the cornucopians over the catastrophists. Critical physical shortages in food production, energy, and minerals have not developed or have not persisted, efficiency has been improved, and substitutes for dwindling resources have been found, despite steadily rising per capita demands by rising populations. Where pollution and more subtle ecological effects are concerned, the record is not so clear. Nor is it clear even where resources are concerned that past experience offers reliable lessons for the future, given the changes of scale and character of human impact today and in the future. Global-scale and rapidly developing crises may tax the adaptive mechanisms that have coped with smaller problems and avoided disaster. As human activity expands in its scale and magnitude, so do its possible disruptive effects. A functioning biosphere itself appears more starkly than ever as a threatened resource for which there are no substitutes, the experience of past successes less reliable a guide to action than before, and the arguments for precaution, restraint, and limits in human interference with the global environment more compelling than ever.

[See also Global Change, article on Human Dimensions of Global Change; Growth, Limits to; Human Populations; IPAT; Sustainable Development; and Technology.]

### BIBLIOGRAPHY

Bailey, R., ed. *The True State of the Planet.* New York: Free Press, 1995. One of two collections (see Simon, below) of essays applying the cornucopian perspective to a wide range of environmental issues.

Cole, H. S. D., et al. *Models of Doom: A Critique of The Limits to Growth.* New York: Universe Books, 1973.

Costanza, R., ed. *Ecological Economics: The Science and Management of Sustainability.* New York: Columbia University Press, 1991. A good collection of readings on the field and its principal themes.

Ehrlich, P. R., and A. H. Ehrlich. *The Betrayal of Science and Reason: How Anti-Environmental Rhetoric Threatens Our Future.* Washington, D.C.: Island Press, 1996. A forceful statement of many of the tenets of the catastrophist position.

Hardin, G. *Living within Limits: Ecology; Economics, and Population Taboos.* New York: Oxford University Press, 1993.

Malthus, T. R. *An Essay on the Principle of Population.* New York: Oxford University Press, 1999. Available in many modern reprints.

Meadows, D. H., et al. *The Limits to Growth.* New York: Universe Books, 1972.

Simon, J. L., ed. *The State of Humanity.* Oxford: Blackwell, 1996. One of two collections (see Bailey, above) of essays applying the cornucopian perspective to a wide range of environmental issues.

—WILLIAM B. MEYER

**CHAOS.** *See* System Dynamics.

## CHEMICAL INDUSTRY

The chemical industry is the modern form of an ancient set of arts and sciences. Neolithic peoples produced copper by reducing minerals such as malachite with charcoal. Bronze, an alloy of copper and tin, perhaps the first synthetic substance invented by humans, appeared around 3,000 BCE, and iron emerged some 1,200 years later. The Egyptians had a rich chemical culture producing plaster, paper, glass, and potash, which they used to make a form of soap. Because they were local arts, based on tacit knowledge, none of these ancient chemical processes created an industry. The first signs of the emergence of what has become one of the world's largest industrial sectors came around the middle of the eighteenth century. As other industries such as textiles, metals, and glassmaking outstripped their traditional sources of raw materials, a new market was created for firms supplying their basic building blocks. This pattern of providing the building blocks for other industries has persisted through much of the chemical industry's history.

The earliest examples of chemical manufacture as an industrial form came in about the middle of the eighteenth century in Europe. Sulfuric acid plants were built in England and France to supply dye-making firms and metal processing. Demand for alkali for glass and soap produced a thriving industry in France at the time of the Revolution. The French early dominance of this embryonic industry waned after the Revolution as the free enterprise model that drove the Industrial Revolution in England spawned a healthy and growing basic chemicals industry. The mainstay was the manufacture of sodium carbonate (sal soda) via the heavily polluting Leblanc process, which produced copious emissions of hydrochloric acid. Regions around the plants were heavily damaged by an early form of acid rain. In perhaps the first instance of formal environmental legislation, the English Alkali Act of 1863 required firms to install absorbing towers to control the acid emissions and established a new bureaucracy to enforce the act.

Very little of the European chemical technology and industry followed the American colonists to the New World. The free trade policies of the new United States made it economical to import their basic chemical stocks from Europe. But for security reasons, one important chemical substance, gunpowder, called for an industry independent of the potentially unfriendly European powers. In 1802, a French émigré, Eleuthère Irénée du Pont de Nemours, built a gunpowder plant near Wilmington, Delaware. DuPont remains today one the world's largest chemical firms and is considered a leader in environmental practices. The deeply acculturated safety consciousness of DuPont, an obvious outcome of the company's traditional product, is considered by many analysts to be the origin of its leadership in the industry. The chemical industry continued to develop during the first half of the eighteenth century as new technologies emerged. Electrochemical processes were discovered but stood in the wings for almost a century until cheap sources of electricity had been developed. Artificial fertilizers, fats, and rubber grew into important parts of the industry, and the earliest synthesis of organic chemicals was made by Wöhler in 1828 when he produced urea from ammonium cyanate and ammonium chloride.

The industry established itself as a major economic sector in the period from about 1850 to the beginning of World War I, following a spate of critically importance scientific discoveries. Bringing back secrets of synthetic dye-making developed in England, German chemists established a powerful organic chemicals industry around firms whose names remain dominant today—Hoechst, Badische Anilin- und Soda-Fabrik (BASF), and Bayer. New organic explosives brought both more power and more danger. Alfred Nobel learned the secret of stabilizing the ultrasensitive but very powerful explosive nitroglycerine, thereby establishing his great fortune. The photographic arts industry arose in Europe and the United States, and the first synthetic fibers from petroleum were made in the United States.

The German industry continued to dominate the world up to the beginning of World War I, but lost this position following Germany's defeat. The industry that grew rapidly after World War II is international in scope, well organized, and highly concentrated. Chemical manufacture continues to be capital intensive with large returns to scale. The industry has not only had to cope with costly and extensive regulations and a poor public image, but, since 1973, with more costly and less reliable feedstocks (such as petroleum). Much of the rapid growth of the global chemical industry following World War II was the direct consequence of abundant, cheap oil supplies that replaced the traditional use of coal. The 1970s saw a series of mergers and investments in new

plant designed to gain market share and dominant positions, but oversupply and higher feedstock prices brought on a long period of poor economic performance. In the late 1980s and later, the global industry recovered and has continued to grow. Many of the giants like DuPont have, however, abandoned their basic commodity chemicals to independent producers and now focus on specialty chemicals and markets.

The industrial hazards of producing chemicals have always been of great concern to laboratory chemists and to plant operators. Nobel's good fortunes were countered by the loss of his younger brother in an explosion in 1864. An early BASF ammonia synthesis plant blew up in 1921 in Oppau, Germany, killing over 600 people. In 1946, a ship carrying ammonium nitrate exploded in Texas City, killing some 500 people. Dioxin, a new industrial hazard, was publicized via an explosion in Seveso, Italy, in 1978, which spread it through the surrounding community. And within recent memory, a 1984 explosion and release of methyl isocyanate at an insecticide plant in Bhopal, India, killed over 2,000 people and injured several hundreds of thousands more. Bhopal set in motion a global outcry that led both to government legislation and to industry self-regulation.

Other forms of hazard emerged as dangerous, unintended side effects were discovered in chemical products and in wastes from their production. [See the biography of Carson.] Thalidomide, marketed by a German pharmaceutical firm, produced anomalies in newborns. It took some three years to track down the cause of these birth defects to the drug that was being marketed as a palliative to the nausea that frequently accompanies pregnancy. An entire Japanese village was afflicted with similar teratogenic problems (i.e., monstrous deformities in newborn children) that were eventually traced to organo-mercury compounds flowing from a company's outfall into Minamata Bay. The local diet depended heavily on fish from the bay, which had become contaminated by the polluted waters. Twenty-three years after fishing was declared off-limits in 1974, local fishermen are again allowed to ply the waters of Minamata Bay. Other chemical disposal practices led to widespread concerns over hazardous waste that had been placed in landfills. In the United States, one community, Love Canal, was relocated following concerns that seepage from an abandoned and covered landfill had led to an unusually large number of childhood cancers and other diseases in the vicinity. Publicity over this event and similar problems at other chemical disposal sites eventually culminated in the passage of the Superfund Law in the United States in 1980.

The chemical industry has been the primary target of much environmental legislation in all of the developed industrial regions of the world. In the United States, the Toxic Substances Control Act (TSCA) was passed in 1976 to control the introduction of new chemicals into

## THE WONDER CHEMICALS

Many of the chemicals now banned or strictly controlled were at one time hailed as wonders that would benefit society.

- DDT had been viewed as the answer to one of humankind's greatest scourges, malaria, until Rachel Carson's *Silent Spring* made its harmful effects on fishing birds a *cause célèbre* in 1962. DDT was eventually banned in all industrial countries, although it still is produced and used in some nations. In the late 1990s, as malaria seems to be making a comeback, some public health officials are advocating the use of DDT as the most effective mosquito control agent ever manufactured.
- PCBs (polychlorinated biphenyls) offered a nonflammable alternative for use as a heat transfer agent and a dielectric filler in electrical devices such as transformers. After decades of use, it was realized that PCBs (in high doses) caused tumors, birth defects, and other abnormalities in laboratory animals. Like DDT, the PCBs are persistent chemicals, remaining unaltered by reaction as they pass through a food chain, becoming more concentrated at each higher level. They were banned in 1976.
- CFCs (chlorofluorocarbons) combine chlorine, fluorine, and carbon. Developed by DuPont in 1930, they have rare properties that make them uniquely suitable for certain applications: they are inert, stable, odorless, nonflammable, nontoxic, and noncorrosive, and they have physical characteristics appropriate for their use as working fluids in refrigerators, air conditioners, and heat pumps. In addition to their use as refrigerants, they are employed as cleaners for electronic components, fumigants for granaries and cargo holds, and propellants in aerosol cans. In addition, the closely related PFCs (perfluorocarbons) are the basis of special emulsions used to transport oxygen to the lungs of premature babies. When the scientific community recognized that CFCs were harmful to ozone in the stratosphere, a compelling irony became apparent. The same stability and inertness that make CFCs safe for use with foods, for instance, also allow their molecules to remain intact for years as they diffuse slowly through the atmosphere and eventually reach ozone-rich layers of the stratosphere, where ultraviolet radiation causes the molecules to dissociate, releasing chlorine atoms that begin the destruction of ozone. If the compounds were less stable, they would react with other chemicals in the lower atmosphere and never reach the stratosphere. The Montreal Protocol of 1987 initiated worldwide cooperation to reduce the use of CFCs and other chemicals harmful to the ozone layer. [See Montreal Protocol.]

—David J. Cuff

commerce. Some 70,000–80,000 chemicals are produced and sold in literally millions of products. TSCA requires chemical manufacturers to provide information on the safety of any new product prior to its market introduction. The original act also banned polychlorinated biphenyls (PCBs). Pesticides are controlled by a related statute. European Community policy followed the U.S. pattern in its broad outlines. The first broad European policy came in 1967 with a Directive on dangerous substances, which set forth a set of requirements for labeling, classification, and handling of hazardous substances. The Directive's 6th Amendment set forth premarket requirements, very similar to those in TSCA.

The history of chemical use and manufacture is full of irony, since many of the chemicals now banned or strictly controlled were originally hailed as wonders that would transform modern life. Dichlorodiphenyltrichloroethane (DDT) had been viewed as the answer to one of humankind's greatest public health scourges—malaria—until Rachel Carson made its harmful effects on birds a *cause célèbre* with the publication of her book *Silent Spring* in 1962. DDT was eventually banned in all industrial countries, but it continues to be produced and used in many developing regions, such as India. In the late 1990s, malaria seems to be making a comeback and some public health officials still advocate DDT's use as the most effective mosquito control agent ever manufactured. [See Pest Management.] Chlorofluorocarbons

As Anastas and Warner (1998, p. v) remark, chemists "possess the knowledge and skills to make decisions in the practice of their trade that can result in immense benefit to society or cause harm to life and living systems and they therefore have responsibility for the character of the decision made." The decision of chemists to minimize adverse impacts on the environment is summarized in the term "green chemistry," the twelve principles of which can be summarized thus:

### THE TWELVE PRINCIPLES OF GREEN CHEMISTRY.
### (AFTER ANASTAS AND WARNER, 1998, P. 30, FIGURE 4.1.)

1. It is better to prevent waste than to treat or clean up waste after it is formed.
2. Synthetic methods should be designed to maximize the incorporation of all materials used in the process into the final product.
3. Wherever practicable, synthetic methodologies should be designed to use and generate substances that possess little or no toxicity to human health and the environment.
4. Chemical products should be designed to preserve efficacy of function while reducing toxicity.
5. The use of auxiliary substances (e.g., solvents, separation agents, etc.) should be made unnecessary wherever possible and innocuous when used.
6. Energy requirements should be recognized for their environmental and economic impacts and should be minimized. Synthetic methods should be conducted at ambient temperature and pressure.
7. A raw material of feedstock should be renewable rather than depleting wherever technically and economically practicable.
8. Unnecessary derivization (blocking group, protection/deprotection, temporary modification of physical/chemical processes) should be avoided whenever possible.
9. Catalytic reagents (as selective as possible) are superior to stoichiometric reagents.
10. Chemical products should be designed so that at the end of their function they do not persist in the environment and break down into innocuous degradation products.
11. Analytical methodologies need to be further developed to allow for real-time, in-process monitoring and control prior to the formation of hazardous substances.
12. Substances and the form of a substance used in a chemical process should be chosen so as to minimize the potential for chemical accidents, including releases, explosions, and fires.

BIBLIOGRAPHY

Anastas, P. T., and J. C. Warner. *Green Chemistry: Theory and Practice.* Oxford and New York: Oxford University Press, 1998.

—ANDREW S. GOUDIE

(CFCs), now banned under the first sweeping global environmental agreement, replaced chemicals that were toxic, flammable, and reactive with a family of compounds that were superior in properties, nonflammable, nontoxic, and virtually nonreactive. The very stability of the CFCs, which are unaffected by normal degradation through tropospheric processes, enabled them to enter the stratosphere, where chlorine was eventually released and reduced natural ozone levels. [*See* Ozone.] The nonflammable family of PCBs replaced flammable compounds for use as heat transfer agents and for di-

electric fillers for electrical devices. But after decades of use, Monsanto, the sole U.S. producer of PCBs, withdrew the product from the market in 1975, anticipating the ban that followed in the next year.

Chlorine-based chemicals have been a particular target since the mid-1970s, when broad limits appeared in directives of the European Community designed to limit emissions into the aquatic environment. By 1990, groups such as Greenpeace, who had published a report entitled *The Product Is the Poison: The Case for a Chlorine Phaseout,* had called for a complete ban on chlorine-based chemicals and hoped to see this issue on the agenda at the 1992 United Nations Environmental Summit Conference in Rio de Janeiro. In North America, the International Joint Commission, an official U.S./Canadian agency, called for the "virtual elimination of the input of persistent toxic substances into the Great Lakes. . . ."

The Superfund Law taxed the chemical and petroleum industries heavily and created a new and very expensive form of liability for actions (such as dumping of hazardous substances) that had been lawful in the past. The law spurred companies to develop new strategies to prevent pollution. Companies such as 3M with its 3P program (Pollution Prevention Pays) had discovered that prevention could avoid investments in control equipment and other regulatory costs. But even this theme could not insulate the industry from continuing regulatory pressures and public scrutiny. A series of newsworthy accidents such as Bhopal, coupled with concerns over the safety of many products, continued to expose chemical firms to legislative scrutiny and to pressures from advocacy groups to limit the marketing of broad classes of chemicals. In 1986, amendments to the Superfund Law required companies to disclose publicly the amounts of a long list of toxic substances being released. Realization of the magnitude of the releases created shock waves both in the public and in high executive levels of the industry.

Faced with continuing regulatory and public pressure, the world chemical industry has created an elaborate self-regulatory system called Responsible Care. In 1988, the U.S. Chemical Manufacturers Association adopted a new self-governing system developed two years earlier by its neighbor to the north—the Canadian Chemical Producers Association. Responsible Care, now mandatory for member firms in the United States and adopted by sister associations around the world, sets forth a set of principles and codes of practice designed to improve environmental performance and to build public confidence in the industry. The codes include requirements for community preparedness, product stewardship, pollution prevention, process safety, distribution, and employee health and safety. Product stewardship sets forth new standards of care reaching forward through the life cycle of the chemicals beyond the traditional plant fence-line.

After almost ten years of experience, it is still too early to evaluate Responsible Care fully. Public opinion has changed only marginally. Investment in environmental management has increased. Public concerns still abound in the late 1990s, having been kept alive by recent data showing anomalous development patterns in aquatic species (e.g., trout) that have been associated with chemicals that disrupt normal sexual traits. The industry's future looks rosy as the global economy continues its rapid growth near the end of the twentieth century, but the potential of future accidents and discoveries of harmful side effects will continue to focus public attention on the chemical industry and maintain the role of environment as an important strategic factor.

[*See also* Hazardous Waste; *and* Prior Informed Consent for Trade in Hazardous Chemicals and Pesticides.]

### BIBLIOGRAPHY

Aftalion, F. *A History of the International Chemical Industry.* Philadelphia: University of Pennsylvania Press, 1991.

Landau, R., et al., eds. *Chemicals and Long-Term Growth: Insights from the Chemical Industry.* New York: Wiley, 1998.

Nash, J., and J. R. Ehrenfeld. "Code Green: Business Adopts Voluntary Environmental Standards." *Environment* 138.1 (1996), 16–20, 36–45.

—JOHN R. EHRENFELD

## CHERNOBYL

[*This case study explains the disaster at the Chernobyl atomic power station and the consequences on human health and the environment.*]

The most catastrophic accident ever to occur at a commercial nuclear power plant took place on 26 April 1986, in northern Ukraine at the Chernobyl (Chornobyl' in Ukrainian) atomic power station. Radioactive fallout of critical intensity covered significant portions of several provinces in Ukraine, Belarus, and the Russian Federation, and lesser amounts fell out with precipitation in numerous other European countries. The resultant health and environmental consequences are ongoing, widespread, and serious.

The Chernobyl atomic power station is one of several such complexes in Ukraine, built to supplement conventional coal-fired electric generating stations. At the time, it was believed that nuclear energy would entail less potential damage to the environment. [*See* Nuclear Industry.] Four other large nuclear power complexes have been constructed near the cities of Zaporizhzhya, Khmel'nyts'kyy, Konstantinovka, and Rivne (formerly Rovno). At these five locations, there are a total of fifteen operating reactors, with four more under construction (and three others shut down). In addition, Ukraine has a major uranium mining center at Zheltyye Viodi, and numerous research facilities.

The Chernobyl reactors utilize a graphite-moderated type of nuclear reactor (Russian acronym: RBMK), with a normal electric output of 1,000 megawatts. These units are water cooled and employ graphite rods to control core temperatures. Each reactor houses 1,661 fuel rods that contain mainly uranium-238 plus much smaller amounts of enriched uranium-235. There are several dangers inherent in the design of RBMK-1000 reactors, including (1) the ability of the operators to disengage safety controls, (2) the lack of a containment dome, and (3) the possibility that, at very low power levels, a rapid and uncontrollable increase in heat can occur in the reactor's core and may result in a catastrophic explosion (Haynes and Bojcun, 1988, pp. 2–4).

This was what happened early in the morning of 26 April 1986. A series of violations of normal safety procedures, committed during a low-power experiment being run on reactor number 4, resulted in a thermal explosion and fire that destroyed the reactor building, exposed the core, and vented vast amounts of radioactive material into the atmosphere. Pieces of the power plant itself were found up to several kilometers from the site of the explosion.

This radiation continued to be released into the atmosphere over a period of nine days, with the prevailing winds carrying the radioactive material initially in a northwesterly direction over northern Europe. The winds later shifted to the southwest, affecting central Europe and the Balkan peninsula. The overall result was significant radioactive fallout (mainly associated with rainfall occurrences) in the countries of Austria, Czechoslovakia, Finland, Germany (mainly Bavaria), United Kingdom, Hungary, Italy, Poland, Romania, Sweden, and Switzerland. Lower levels of radioactive deposition were reported in Denmark, France and the Benelux countries, Greece, Ireland, Norway, Yugoslavia, and several other European nations (Medvedev, 1990, chap. 6). The then-Soviet republics of Estonia, Latvia, and Lithuania were also directly in the path of the initial plume.

Within the Soviet Union, the regions that received the highest levels of radioactive contamination were in the northern Kiev and eastern Zhytomyr provinces in Ukraine, and in the Homyel' and Mogilev provinces of Belarus (then Belorussia). In the Russian Federation, areas situated closest to the Belarus border, such as western Bryansk province, experienced the greatest radiation problems (Nuclear Energy Agency, 1995, p. 32). Lighter fallout, measured in terms of long-lived cesium-137, was recorded in portions of Minsk, Vitsyebsk, Hrodna, Chernigov, Rivne, Cherkassy, Kaluga, Tula, Ryazan', and Orel provinces, as well as in the Baltic republics. Among the capital cities of these provinces, only Homyel' was in a region of high fallout, but other large cities, such as Orel, Mogilev, and Kiev, were right on the border of the danger zone (Bradley, 1997, p. 368).

Large-scale evacuations were conducted in the most heavily affected sections of these provinces. The number of people who had to be resettled totaled around 107,000 in Belarus alone, plus approximately 50,000 more in both Russia and Ukraine (International Atomic Energy Agency, 1996, p. 7). Another source cites a figure of over 84,000 people as having been relocated within the Russian Republic (Savchenko, 1995, p. 76), and other sources speak of over 100,000 evacuees in Ukraine. In many instances the displaced populations had to be moved quickly to places with inadequate housing, social services, and employment. For reasons that have not been fully explained, thousands of people were not relocated from areas of high radiation until many years after the accident.

As would be expected, the effects on human health have been enormous. Around 600,000 people have been "significantly exposed" to radiation from the Chernobyl accident, and thousands of people have developed radiation sickness from exposure to contamination produced by the explosion and subsequent fire (Medvedev, 1990, pp. 129–130). Approximately 270,000 people still live in areas sufficiently contaminated to require ongoing protection measures (Nuclear Energy Agency, 1995, p. 12). Russia, like the former Soviet Union, still reports an official figure of around thirty-one deaths, but the actual number of Chernobyl-related fatalities is often suggested to be in the hundreds, if not the thousands (Marples, 1993, p. 282). In 1996, a senior Russian environmental authority stated that "official [Chernobyl] statistics are incomplete and irreversibly falsified" (Yablokov, 1993).

Medical problems among the general population that are probably attributable to Chernobyl are a serious concern. The first wave of victims were workers associated with the power plant and the thousands of containment and cleanup personnel who needed treatment for radiation sickness. The workers' town of Pripyat' was not immediately evacuated, nor the people even informed of the radiation danger, thereby placing many people, especially schoolchildren, at risk. The 30-kilometer radius "exclusion zone" around the plant was not declared until 3 May, and was inadequate in size. There is general agreement that there will be grave long-term cancer mortality effects from the accident, but also vast disagreement over the magnitude of these carcinogenic consequences. The optimists suggest only a few hundred "extra deaths," all in the former Soviet republics, whereas the pessimists predict as many as 280,000 fatalities worldwide (Medvedev, 1990, p. 166).

The southeastern portion of Belarus received the plurality of fallout from Chernobyl; indeed, between half and two-thirds of all the radioactive fallout from Chernobyl fell on Belarusan territory. The environmental and human toll in this area has been at least as great as that experienced in Ukraine. The city of Homyel', with a population of around half a million, recorded the highest increase in background radiation of any major city in the Soviet Union. Many smaller towns in Belarus, especially those directly across the Pripyat' River from Chernobyl, received more. Approximately 20 percent of the country's agricultural land, possibly totaling in excess of a quarter of a million hectares, as well as 15 percent of the forests of Belarus, are no longer usable (Savchenko, 1995).

Perhaps the most tragic consequence of the accident thus far has been the sharp increase in the incidence of thyroid cancer in children since 1989 in all three republics (Ukraine, Belarus, and Russia). In Minsk (city and province) in 1986 there had been no such cancers, but by 1992, 21 cases had been recorded; in Homyel' province there had been one in 1986 but 97 more from 1987 to 1992. For all of Belarus, there had been 2 cases of thyroid cancer in children in 1986, but a total of 172 cases were recorded in the period 1986–1992 (Marples, 1993, pp. 285–290; World Health Organization, 1995, pp. 20–24). By 1994, the total exceeded 300 (Nuclear Energy Agency, 1995, p. 63). Since the breakup of the Soviet Union, little assistance in dealing with the consequences of Chernobyl has come from Moscow; currently, a significant percentage of the national budget of Belarus must be devoted to dealing with the relocation, environmental, and public health costs of the accident.

The environmental consequences of the accident include large areas of contaminated soil, forests, and water, all of which have major adverse ramifications. [*See* Nuclear Hazards.] Soil and water pollution from the accident have been recorded in twenty-two provinces of the former Soviet Union, as well as in several foreign countries. In places, radionuclides have been measured in the soil at depths up to 25 centimeters, which is the vertical zone in which crop cultivation takes place. Because of radioactive contamination of soil, large areas of fertile farmland have had to be taken out of production, perhaps for decades. For example, in northern Ukraine, over 100,000 hectares of agricultural land, which contain some of the world's richest soils, have had to be abandoned (Savchenko, 1995, p. 53). Not only crops, but also meat and dairy products, have been lost. Deformed calves and pigs have been born on collective farms scores of kilometers away from the power plant. The farmers have had to find work elsewhere, and the losses in agricultural production made up by increased production in other regions. The Nuclear Energy Agency (1995, p. 81) has warned that forest products from the contaminated regions may present long-term radiation exposure problems.

Water supplies were not only contaminated by the immediate fallout, but also by the transport of radioactive sediments. Fishing was prohibited in the portions of the Pripyat' River and the Kiev Reservoir near the accident site, and outside sources of water for Kiev had to be developed quickly. The Pripyat' River flows through the power plant complex (and was the source of its cooling water), and thence flows into the Dnieper River, which runs through Kiev. Several million people living between Kiev and the Black Sea depend upon water from the Dnieper, and thus are potentially exposed to radiation moving through it. In the mid-1990s, though, contamination of drinking water was believed not to be a problem (Nuclear Energy Agency, 1995, p. 81). However, significant amounts of strontium-90 may be imbedded in the banks of the Pripyat' River. Bank collapse and shoreline erosion could eventually release this radiation.

A large concrete containment facility, termed a *sarcophagus*, was completed around the damaged unit number 4 in November 1986, finally halting the release of radiation. However, it was never viewed as a permanent containment structure, and because of the haste of its construction, serious doubts exist about its long-term viability. The sarcophagus covers hundreds of metric tons of nuclear fuel, which continue to produce high temperatures and radiation levels within the ruined reactor building. But the current structural stability of the sarcophagus is very questionable; it contains many cracks, is generally agreed to be deteriorating, and will have to be replaced. The huge cost of this exceeds Ukraine's resources. As a result, at the 1997 economic summit meeting, the major world powers pledged U.S.$300 million to assist in the construction of a second concrete containment facility.

The health and environmental consequences of the accident have been sufficiently great that there has been a widespread call, both from within Ukraine and from other parts of the world, to shut down the Chernobyl complex completely. As of early 1998, only one reactor at the site was still operating. The Ukrainian government has linked the complete closure of the site to the receipt of foreign funding for cleanup assistance that was promised by multinational agreement in 1995.

The consequences of the Chernobyl disaster will remain as significant problems for decades to come. Delayed health effects from radiation exposure will become evident in the coming years. The huge costs of cleaning up the contaminated land and structures, caring for the displaced multitudes, and rebuilding the sarcophagus will have to be funded somehow. Unfortunately, the ability of Belarus and Ukraine to handle the costs and administrative burdens of the clean up and relocations is open to question (Pryde, 1995, chap. 10). In terms of future energy supplies, the accident has had a depressive effect on nuclear power in many parts of the world, most notably the United States, where no new development within the nuclear power industry has taken place since 1986. Russia, on the other hand, plans to continue with its nuclear program and has announced plans to build a number of new units early in the twenty-first century. The ramifications of the Chernobyl accident are extensive and continuing, and it is likely that it will be remembered as one of the defining events of the twentieth century.

[*See also* Nuclear Accident and Notification Convention.]

**BIBLIOGRAPHY**

Bradley, D. J. *Behind the Nuclear Curtain: Radioactive Waste Management in the Former Soviet Union.* Columbus, Ohio: Battelle Press, 1997, pp. 345–370. The most comprehensive work to date on all aspects of commercial and military radioactive wastes in the former Soviet republics.

Haynes, V., and M. Bojcun. *The Chernobyl Disaster.* London: Hogarth Press, 1988. An early review of the disaster and its immediate effects.

International Atomic Energy Agency. *One Decade after Chernobyl: Summing Up the Consequences of the Accident.* Vienna, 1996. Summary report of the Joint EC/IAEA/WHO international conference, "One Decade after Chernobyl," held in Vienna, 8–12 April 1996.

Marples, D. R. "A Correlation between Radiation and Health Problems in Belarus?" *Post-Soviet Geography* 34 (May 1993), 281–292.

Medvedev, Z. *The Legacy of Chernobyl.* Oxford: Basil Blackwell, 1990. A critical evaluation of Chernobyl and its aftermath by a Russian dissident.

Nuclear Energy Agency. *Chernobyl: Ten Years on Radiological and Health Impact.* Paris: NEA, November 1995. A detailed study of the health and environmental consequences of Chernobyl.

Pryde, P. R. *Environmental Resources and Constraints in the Former Soviet Republics.* Boulder, Colo.: Westview Press, 1995. Chapters 9, 10, and 11 are on environmental management problems in Ukraine and Belarus.

Savchenko, V. K. *The Ecology of the Chernobyl Catastrophe.* Casterton Hall, U.K., and New York: UNESCO and Parthenon Publishing, 1995.

World Health Organization. *Health Consequences of the Chernobyl Accident.* Geneva: WHO, 1995. This is the most recent United Nations report on this subject.

Yablokov, A. V. "Facts and Problems Related to Radioactive Waste Disposal in Seas Adjacent to the Territory of the Russian Federation." Report to the President of the Russian Federation by the Government Commission on Matters Related to Radioactive Waste Disposal at Sea, 24 October 1992. Moscow, 1993. Yablokov was senior adviser on environmental matters to President Yeltsin.

—PHILIP R. PRYDE

# CHLOROFLUOROCARBONS

*Chlorofluorocarbons* (CFCs) are volatile compounds of chlorine, fluorine, and carbon that were first synthesized in the 1920s as a substitute for toxic refrigerants such as ammonia and methyl chloride and patented under the trade name Freon. Because of their nontoxic, nonreactive, and noninflammable nature, they became widely used during the following decades not only as refrigerants but also as solvents, propellants, and foaming agents. Those very attributes also meant that they would have very long residence times in the troposphere.

In the mid-1970s, Sherwood Rowland and Mario Molina, at the University of California, showed that

these long-lived CFCs would eventually move up to the stratosphere. There they would be broken down by ultraviolet (UV) radiation, releasing free chlorine and leading to the destruction of ozone in the stratosphere. Since stratospheric ozone absorbs most of the sun's ultraviolet wavelengths below 320 nanometers, this would allow more UV-B radiation to reach the Earth's surface and cause damage to plants and animals. Subsequently, it has been learned that several other substances are capable of destroying stratospheric ozone; these ozone-depletion substances (ODSs) are said to have ozone-depletion potential (ODP).

**Ozone.** Ozone ($O_3$) is a highly reactive form of oxygen having three atoms per molecule instead of the usual two. It occurs in both the troposphere and the stratosphere, but for very different reasons. Over 90 percent of atmospheric ozone is produced in the stratosphere (20–30 kilometers altitude) through the absorption of ultraviolet radiation that would otherwise reach the surface and affect plants and animals. The remainder is produced in the troposphere (0–10 kilometers altitude) through the interaction between sunlight, volatile organic compounds (from natural as well as human sources), and nitrogen oxides (mostly human-made). It is a key pollutant in the photochemical smog that plagues many urban areas and can cause severe injury to plants and animals. [*See* Ozone.]

The decrease in stratospheric ozone was first detected near the South Pole in the 1970s by the British Antarctic Survey, although it was not recognized for what it was until 1985. The phenomenon, called the *ozone hole*, was found to be most pronounced during the polar spring. Subsequently, a decrease was also found in the Northern Hemisphere, as much as 10 percent per decade in Europe and Canada since 1973. Ozone concentration at the South Pole during October dropped from about 325 Dobson units (DU) before 1970 to about 100 DU at the present time. Globally, the decrease since 1980 has averaged 5 percent per decade in the midlatitudes and 10 percent in the polar regions.

**Halocarbons.** CFCs are one of a larger group of substances called *halocarbons*, which are compounds of carbon and one or more of the halogens (fluorine, chlorine, bromine, iodine, and astatine). Although not all halocarbons deplete stratospheric ozone, they are all greenhouse gases having global warming potential (GWP). They include not only CFCs but also low-ODP substances that have been synthesized to replace CFCs and others that have a wide variety of industrial uses. The major halocarbons are listed in Table 1. Almost without exception, these substances are man-made and were not present in the atmosphere prior to their synthesis a few decades ago.

The most potent ODSs are the *bromofluorocarbons* or *halons*, compounds of bromine, chlorines, fluorine,

**Chlorofluorocarbons. TABLE 1.** Halogenated Alkanes and Similar Substances Released to the Atmosphere, with Their Ozone-Depletion Potentials (ODPs) and Global Warming Potentials (GWPs), Atmospheric Half-Lives, and Present Mass Concentrations and Trends

| NAME | CHEMICAL FORMULA | MAIN USE OR SOURCE | OZONE-DEPLETION POTENTIAL | GLOBAL WARMING POTENTIAL* | HALF-LIFE (YEARS) | PRESENT LEVEL (PER $10^{-12}$) | TREND |
|---|---|---|---|---|---|---|---|
| *Chlorofluorocarbons (CFCs)* | | As solvents, refrigerants, propellants | | | | | |
| CFC-11 | $CCl_3F$ | | 1.0 | 4,000 | 60 | 270 | falling |
| CFC-12 | $CCl_2F_2$ | | 1.0 | 8,500 | 100 | 550 | level |
| CFC-13 | $CClF_3$ | | 1.0 | 11,700 | 640 | | |
| CFC-113 | $C_2Cl_3F_3$ | | 0.8 | 5,000 | 90 | 80 | level |
| CFC-114 | $C_2Cl_2F_4$ | | 1.0 | 9,300 | 200 | | |
| CFC-115 | $C_2ClF_5$ | | 0.6 | 9,300 | 400 | | |
| *Hydrochlorofluorocarbons (HCFCs)* | | As substitutes for CFCs | | | | | |
| HCFC-22 | $CHClF_2$ | | 0.05 | 1,600 | 13 | | |
| HCFC-123 | $CF_3CHCl_2$ | | 0.02 | 90 | 2 | | |
| HCFC-124 | $CF_3CHClF$ | | 0.02 | 480 | 6 | | |
| HCFC-141b | $CH_3CFCl_2$ | | 0.10 | 600 | 9 | | |
| HCFC-142b | $CH_3CF_2Cl$ | | 0.06 | 2,000 | 19 | | |
| *Hydrofluorocarbons (HFCs)* | | As substitutes for CFCs | | | | | |
| HFC-23 | $CHF_3$ | | nil | 11,700 | 250 | | |
| HFC-32 | $CH_2F_2$ | | nil | 650 | 6 | | |
| HFC-125 | $C_2HF_5$ | | nil | 2,800 | 30 | | |
| HFC-134a | $CH_2FCF_3$ | | nil | 1,300 | 14 | 1.6 | rising |
| HFC-143a | $CH_3CF_3$ | | nil | 3,800 | 40 | | |
| HFC-152a | $CH_3CHF_2$ | | nil | 140 | 2 | | |
| *Perfluorocarbons (PFCs)* | | From aluminum production and uranium enrichment | | | | | |
| Perfluoromethane | $CF_4$ | | nil | 6,500 | 50,000 | 75 | |
| Perfluoroethane | $C_2F_6$ | | nil | 9,200 | 10,000 | 2.6 | |
| *Halons (Hs)* | | In fire extinguishers | | | | | |
| H-1211 | $CBrClF_2$ | | 4 | | | 3.5 | rising |
| H-1301 | $CBrF_3$ | | 12 | 5,600 | | 2.3 | rising |
| H-2402 | $C_2Br_2F_4$ | | 6 | | | 0.5 | level |
| Carbon tetrachloride | $CCl_4$ | As solvent | 1.1 | 1,400 | 50 | 100 | level |
| Methyl chloroform | $CH_3CCl_3$ | As adhesives and solvents | 0.11 | 110 | 6 | 90 | falling |
| Methyl bromide | $CH_3Br$ | Natural and as soil fumigant | 0.6 | | 1.3 | 10 | level |
| Sulfur hexafluoride | $SF_6$ | For equipment insulation and cable cooling | nil | 25,000 | 3,000 | 33 | rising |

*GWPs are for a 100-year time horizon.

and carbon used as fire-extinguishing agents. Although current concentrations of these substances in the atmosphere are about 1/150 those of CFCs, they are 50 to 100 times (per unit mass) more effective at destroying ozone. The CFCs, along with carbon tetrachloride, methyl chloroform, and methyl bromide, have intermediate ODPs in the range of 0.11 to 1.2. The hydrochlorofluorocarbons (HCFCs), which were synthesized as substitutes for the CFCs, have ODPs between 0.01 and 0.10, while the hydrofluorocarbons (HFCs) and perfluorocarbons (PFCs), which contain no chlorine or bromine, have zero ODP. Nearly all the chlorine and about half the bromine found in the stratosphere come from human sources. Virtually all of the substances in Table 1 have GWPs between several hundred and several thousand times that of carbon dioxide, as well as long atmospheric residence times, making them major threats as greenhouse gases.

**Stratospheric Ozone Destruction.** Ozone is produced in the stratosphere through the photolytic breakdown of oxygen molecules by solar UV radiation, that is:

$$O_2 + UV \rightarrow 2O \text{ and } O_2 + O \rightarrow O_3$$

At the same time, ozone molecules are destroyed naturally by radiation and by collisions with oxygen atoms as:

$$O_3 + UV \rightarrow O_2 + O \text{ and } O_3 + O \rightarrow 2O_2$$

A delicate balance between these production and destruction processes keeps the ozone concentration at a level appropriate for protecting the Earth's surface from harmful UV radiation as well as raising the temperature of the upper stratosphere.

It is believed that a number of man-made substances are capable of interfering with these processes and leading to ozone depletion, including nitrogen oxides from supersonic aircraft flying in the stratosphere and some agricultural emissions. CFCs and their "cousins" appear to be even more dangerous than these, however, because of their large ODPs and very long residence times. While these substances are nonreactive in the troposphere, when they rise to the stratosphere and encounter intense UV radiation, they break down, releasing chlorine or bromine atoms, which react with the ozone in the following manner:

$$Cl + O_3 \rightarrow ClO + O_2 \text{ and } ClO + O \rightarrow Cl + O_2$$

Thus chlorine (or bromine) atoms are recycled through this process, and each one may remove as many as 100,000 ozone molecules before disappearing itself. The chlorine ultimately reacts with nitrogen dioxide or methane to form chlorine nitrate ($ClONO_2$) or hydrochloric acid (HCl), as:

$$ClO + NO_2 \rightarrow ClONO_2 \text{ or } CH_4 + Cl \rightarrow HCl + CH_3$$

The chlorine nitrate and hydrochloric acid in the above equations act as long-lived carriers (or reservoirs) that release their chlorine efficiently on the surfaces of polar stratospheric cloud particles. Nacreous or "mother-of-pearl" clouds are distinctive by being visible after sunset or before sunrise, owing to their high altitude (15–20 kilometers). It was discovered in the 1980s that they are composed of a mixture of sulfuric and nitric acid particles, which form only at temperatures below −80°C, and are involved in ozone depletion in the polar regions. A similar role was found for volcanic emissions in the midlatitudes.

It should not be construed from the above explanation that stratospheric ozone chemistry is well understood, as this is certainly not the case. In reality, it is more complex than indicated above, and there are still some gaps in the understanding, particularly with regard to midlatitude ozone depletion.

**Halocarbon Use and World Reaction.** World production of CFCs rose from 42,000 metric tons per acre in 1950 to almost 1 million metric tons in 1974, when the first theory of ozone depletion was published (Figure 1). Production leveled off for a decade, in response to this warning, but then started to rise again. Increasing evidence of a cause–effect relationship between CFC production and ozone depletion, as well as the deepening ozone hole over Antarctica, finally led to the United Nations Environment Programme's Vienna Convention on the Protection of the Ozone Layer in 1985. This

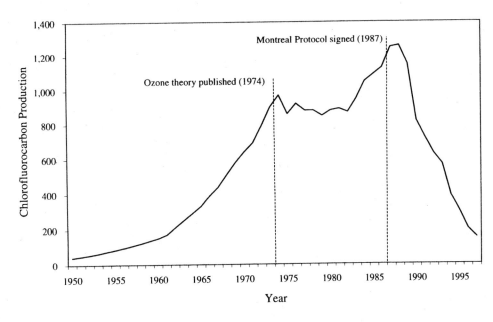

**Chlorofluorocarbons.**
**FIGURE 1.** Global CFC Production since 1950 (kt/a).

prompted forty-seven countries to sign the Montreal Protocol on Substances that Deplete the Ozone Layer in September 1987, encouraging all countries to restrict, and eventually phase out, the production of the CFCs and halons and to replace them with less environmentally damaging substitutes. Its control provisions have been amended and strengthened regularly since then: London (1990), Copenhagen (1992), Vienna (1995), and Montreal (1997). The protocol set deadlines for the phase-out of CFCs, halons, methyl chloroform, carbon tetrachloride, and, more recently, methyl bromide and the HCFCs. The HCFCs and the HFCs (Table 1) were developed as substitutes for CFCs. The addition of hydrogen atoms to the molecules substantially reduced the ODP, GWP, and half-life of the HCFCs, while the subse-

quent removal of the chlorine atoms eliminated the ODP of the HFCs at the expense of increases in GWP. [*See* Montreal Protocol.]

Although production of the major ODSs has dropped dramatically since the Montreal Protocol was signed, their concentrations in the atmosphere, even ten years later, have at best only leveled off (Figure 2). This is because of the long residence times of most of these substances, which indicate that it will be more than one hundred years before there is a significant decline. However, concentrations of ODSs having shorter residence times, such as methyl chloroform, have already begun to decline.

The search for more environmentally friendly substitutes for these products continues. At present, con-

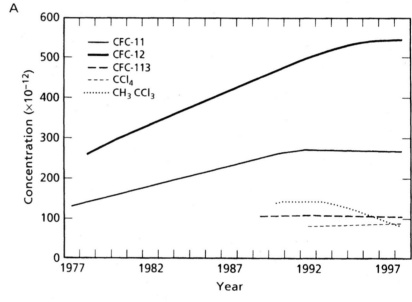

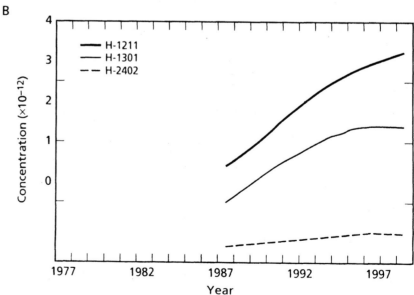

**Chlorofluorocarbons. Figure 2.** Global Mass Concentrations (per $10^{-12}$) of (A) Major CFCs and (B) Other Ozone Depleters in the Atmosphere since 1977.

servative use of zero ODP substances such as the HFCs together with effective recycling appears to offer the best alternative.

[*See also* Atmosphere Dynamics; Atmospheric Chemistry; Chemical Industry; Clouds; Greenhouse Effect; Human Impacts, *article on* Human Impacts on Biota; International Cooperation; Manufacturing; *and* Pollution.]

### BIBLIOGRAPHY

Albritton, D. L., R. T. Watson, and R. J. Aucamp, eds. *Scientific Assessment of Ozone Depletion, 1994*, Geneva: World Meteorological Organization, 1994.

Bolin, B., B. R. Döös, J. Jager, and R. A. Warrick, eds. *The Greenhouse Effect, Climatic Change, and Ecosystems*, SCOPE Report 29, Scientific Committee on Problems of the Environment. Chichester, U.K.: Wiley, 1986.

Cunnold, D. M., P. J. Fraser, R. F. Weiss, R. G. Prinn, P. G. Simmonds, B. R. Miller, F. N. Alyea, and A. J. Crawford. "Global Trends and Annual Releases of CCl$_3$F and CCl$_2$F$_2$ Estimated from ALE/GAGE and Other Measurements from July 1978 to June 1991." *Journal of Geophysical Research* 99D1 (1994), 1107–1126.

Fraser, P., D. Cunnold, F. Alyea, R. Weiss, R. Prinn, P. Simmonds, B. Miller, and R. Langenfelds. "Lifetime and Emission Estimates of 1,1,2-trichlorotrifluorethane (CFC-113) from Daily Global Background Observations, June 1982–June 1994." *Journal of Geophysical Research* 101D7 (1996), 12585–12599.

Houghton, J. T., B. A. Callander, and S. K. Varney, eds. *Climate Change 1992: The Supplementary Report to the IPCC Scientific Assessment.* Cambridge: Cambridge University Press, 1992.

Houghton, J. T., G. J. Jenkins, and J. J. Ephraums, eds. *Climate Change: The IPCC Scientific Assessment.* Cambridge: Cambridge University Press, 1990.

Houghton, J. T., L. G. Meira Filho, B. A. Callander, N. Harris, A. Kattenberg and K. Maskell, eds. *Climate Change 1995: The Science of Climate Change.* Cambridge: Cambridge University Press, 1996.

National Aeronautics and Space Administration. *Report on Concentrations, Lifetimes, and Trends of CFCs, Halons, and Related Species.* Reference Publication 1339. Greenbelt, Md.: NASA, 1994.

—EARLE A. RIPLEY

**CITIES.** *See* Urban Areas.

## CIVILIZATION AND ENVIRONMENT

[*This article examines the environmental background to human development and the impact of human populations on the natural world, especially the effects of early farming and commodity production.*]

Human culture and civilizations have risen and flourished against an ever-changing environmental background. It is no coincidence that the human species emerged during the course of one of the rare periods of continental glaciation and desiccation that have punctuated the history of this planet. [*See* Climate Change.] The earliest stages of this story took place in east Africa, against a background of shrinking forests and fluctuating lakes within the rift valley, and human populations spread from Africa during the massive environmental changes of successive glacial and interglacial cycles of the Quaternary ice age. By the end of the Pleistocene, anatomically and behaviorally modern human populations occupied all of the major continental landmasses except Antarctica. During the current (Holocene) interglacial, radically new developments occurred in at least three separate centers in the middle latitudes (20°–40°N) of the Old and New World landmasses (the Middle East, China, and Central America), in which farming created previously unparalleled densities of population. In each of these areas, urban communities appeared within a few millennia of the beginnings of farming. A complex pattern of spatial diffusion was created, in which farming spread outward from these nuclear areas, to be followed by urban networks extending along the major trade routes around the first cities.

The genesis of this new pattern of human existence can be understood only in terms of the complex transition from glacial to interglacial modes that marked the Pleistocene/Holocene boundary, and the subsequent precession-driven changes in the following five millennia. In conjunction with the behavioral complexity of modern human populations (evident from the late Pleistocene onward, after c.50,000 BP), this massive scale of environmental change caused major changes in human societies. Once initiated, however, these changes (and especially unprecedented rates of population growth) had their own momentum, and environmental changes had a less formative role in human development. They were nevertheless of considerable local or regional importance, since early farming societies were often closely tied to particular environmental zones, where alterations of key parameters could affect a large proportion of the population, as along rivers and on coasts. Farming itself, involving an active restructuring of local ecosystems, was often a significant factor in destabilizing brittle environments and initiating degradation. Semiarid landscapes were particularly vulnerable. Especially where urban societies cultivated cash crops and maintained large herds of animals such as sheep, these impacts had a major effect in initiating soil erosion and in creating present-day vegetational communities such as those of the Mediterranean. In wetter landscapes, deforestation led to the formation of podsols and open moorland, of the kind now characteristic of large areas of northern Europe. Impacts on the tropics, however, were more limited before the major degradation caused by exploitation in the last few hundred years.

Although the course of environmental changes in the Holocene can now be defined with much greater precision than heretofore, these are no longer treated in the simple (and often deterministic) way that was once fashionable. They are now treated as parts of a complex set of interactions between developing societies and their ecological contexts, in which human populations adjusted with great flexibility to alterations in their conditions of life. New forms of evidence, especially ice cores and tree ring sequences, which offer fine resolution in chronology (down to decades or individual years), allow much better correlation between patterns of environmental and cultural change. These have greatly improved the often speculative reconstructions of earlier writers on this topic, and human prehistory and history have emerged as a dynamic interplay between human actors and their changing natural settings, in which human interference has played an increasingly important role.

**Early Holocene Changes and the Beginnings of Farming and Urban Life.** Human populations with the cognitive and behavioral characteristics of modern *Homo sapiens*, including the capacity for representational art, evidence for the exchange of goods, and perhaps language as we know it, emerged only during the last glaciation. In this phase of intense cold and dryness, these propensities were limited by environmental constraints, and it was only as climates became warmer and moister that they found expression in new ways of living. The first signs of climatic amelioration, reflected in the occurrence of sapropelites on the floor of the Mediterranean, began to occur in the late glacial regime around 16,000 BP. The greater productivity of coastal and terrestrial ecosystems allowed an increased density of population where sedentary resources could be harvested, especially along rivers, lakes, and coasts. This increased sedentism is reflected in the occurrence of pottery, used for boiling small, abundant items of food such as shellfish or gathered plant foods in permanent settlements, and first evidenced in the Jōmon culture of Japan around 14,000 BP. Similar adaptations, not yet unambiguously demonstrated by well-dated archaeological finds, were probably widespread in coastal monsoon Asia and have been described in the upper reaches of the Nile (around Khartoum), where they are associated both with riverine resources and the collection of native wild sorghum (and, from 8500 BCE, with the use of pottery).

A critical episode in the development of more interventionist attitudes to nature was the brief reversal to glacial/dry conditions represented by the Younger Dryas interval (11,000–9500 BCE), probably associated with alterations in the pattern of thermohaline circulation in the North Atlantic, which brought renewed desiccation to areas of the Mediterranean where nut-bearing wood-

land had begun to expand in the preceding two millennia. [*See* Younger Dryas.] In the absence of abundant alternative lacustrine, riverine, or littoral resources, foragers in the Levant were forced to turn their attention to the abundant stands of wild cereals that grew extensively in open areas of the Mediterranean woodland belt. Easy to collect, these seeds demanded long hours of grinding to prepare as food. As wetter conditions returned at the onset of the Holocene, around 9500 BCE, certain groups in the Jordan valley deliberately planted these seeds in alluvial fans around springs, probably to support settlements at crucial points on trade routes. Thus began a long-lasting symbiotic relationship between humans and cereals (einkorn and emmer wheat, rye and barley, as well as complementary legumes) whose unforeseen demographic consequences were to result in an explosion of human population numbers, sustained only by bringing larger and larger areas into cultivation.

While these winter rainfall crops came to be grown in western Asia from the beginning of the Holocene, the other centers of indigenous cultivation lay within tropical regions of summer rainfall. Comparable developments occurred only some two to five millennia later, perhaps as a result of declining monsoon rainfall in these areas, at a time when the Sahara was beginning to suffer similar stress. [*See* Holocene in the Sahara.] Farming settlements supported by rice cultivation are known in China from 6000 BCE, and maize-growing villages in Mesoamerica from 3000 BCE. Each of these areas became the center of a major dispersal of crop plants (often carried by expanding human populations), which also gave rise to local supplements or replacements as farming spread to new ecological zones. Moreover, new species were also recruited locally: animals were domesticated and used for meat; tree crops were added to the list of cultigens. Ultimately, animals came to be used also for traction and transport and for the provision of both milk and fibers such as wool, while cereals came to be cultivated by more intensive techniques such as irrigation and the plow. These innovations enabled farming to spread to different parts of the landscape, and together they provided the basis for the local specialization and manufacturing (e.g., of textiles) that provided the economy of the earliest urban centers, which appeared in Iraq from 4000 BCE onward.

Situated on the surface of a prograding deltaic plain and advancing down the graben of the Persian Gulf, the first cities of the lower Euphrates/Tigris river system occupied a landscape radically different in appearance from that of the present day. Changes in sea level and water table, deforestation, deflection of watercourses, and salinization have combined to create the forbidding conditions that exist today: during the fourth millennium BCE it was a more attractive environment for settlement, with better maritime access (because the fan created by the Karun and Karkeh Rivers in Khuzistan had not advanced so far into the Persian Gulf as to create the marshes that exist today below Basra). Even at the time, however, it was notoriously unstable because of catastrophic channel shifts in this area of very low gradients, and the inhabitants of many ancient cities were forced to make periodic moves to new watercourses. The labor of digging canals for irrigation, boat traffic, and water control was a constant demand on local populations. Salinization was undoubtedly exacerbated by irrigation, which used breaches in the levees and distributory canals to spread water over extensive areas, where much of it evaporated. Similar remarks apply to the other early urban centers in Egypt and Mesopotamia, though local irrigation techniques differed (basin irrigation in Egypt, *sailaba* in the Indus), and the silt-laden Nile floods (whose timing reflects the monsoon rainfall regime within its catchment) prevented the kind of salt accumulation that plagued the other alluvial basins. Egypt remained a major grain producer long after it had been eclipsed by the rise of other Mediterranean civilizations, situated in landscapes often more propitious for trade than for cereal production: an annual flotilla of boats carried grain to Rome to feed the inhabitants of the imperial capital at the height of its empire.

**Punctuations: Interruptions and Catastrophes.** Urban centers in alluvial valleys were probably more affected by changes in fluvial geomorphology than by climate. Those supported by dry farming in adjacent areas such as the Syrian steppe, however, were potentially more vulnerable to fluctuations in rainfall, since shifts in the critical isohyet of 200 millimeters, beyond which farming is impossible without irrigation, would have affected large areas of territory. A widespread period of cultural and political regression in the later third millennium BCE, which saw the collapse of the Akkadian Empire in Mesopotamia, the "First Intermediate" period of relative decline in Egypt, and cultural devolution in the Levant, may be indicative of a period of prolonged drought that was sufficiently severe to destabilize the major political units that were economically linked to it. There are some indications from soil micromorphology on sites of this period that this may have been the case, and levels of the Nile at that time were significantly lower. Yet even if this episode does represent a major ecological catastrophe, it was one from which the area rapidly recovered and indeed experienced renewed expansion and prosperity. Human populations have often proved remarkably resilient, at least in the long term.

Another possible example of a casualty of major ecological change is the collapse of the Indus Valley civilization around 1700 BCE, which formed the center of a major regression of urban settlement over an area in-

cluding both the Oxus oases and the Gulf entrepôts in Oman and Bahrain. This was a major retraction in the otherwise expanding pattern of urbanization, and urban life was not resumed in this area for more than a millennium. While the suggestion of catastrophic flooding has been advanced to explain the collapse of the Indus Valley sites (and major changes such as the diversion of the Ghaggar did take place, though at a later date), this seems unlikely to account for the scale of collapse, which affected the entire trading system with which it was connected. Some widespread climatic change might perhaps be postulated, affecting not only hydrology but also (for instance) the strength of monsoonal wind systems, critical for maritime trade; but at the moment this large-scale phenomenon remains a major mystery, in which environmental factors may have played some, though not the only, role. (The other single-factor explanation often proposed—of an Aryan invasion from the north—is equally oversimple.)

Similar doubts about the unique role of climate change in affecting human activities apply to the often-cited instance of the Norse abandonment of Greenland in the Little Ice Age of the twelfth century CE. [*See* Little Ice Age in Europe.] Settled in a major phase of expansion in the tenth century, at a time of major Scandinavian trading and settlement expansion both westward as far as Vinland (Labrador) and eastward to Constantinople and the Volga, the Norse colonies of the North Atlantic dwindled and were ultimately abandoned. Although the increasing severity of climate was no doubt a factor in Viking cost-benefit analysis, the more general withdrawal from long-distance trading activities is equally relevant as an element in the equation, as scattered colonist populations lost touch with their homeland.

The effects of drought on farming populations in marginal areas are undoubtedly well demonstrated by the Pueblo Indian populations of the southwestern United States, practicing floodwater and irrigation farming on a small scale in propitious areas within otherwise desert landscapes. Such groups were naturally vulnerable both to changes in precipitation and to indirectly related phenomena such as arroyo formation, which made farming impossible by incising channels in the small flat areas of silt used preferentially for crop raising. A particular phase of dislocation has been identified in the fourteenth century CE. Similar problems appear to have beset simple farmers in the heartland of western Old World farming development in the Levant during the fifth millennium BCE and may have been a factor in the diversification and intensification of farming practices at this time.

One further hazard to which certain groups were vulnerable was volcanic activity, famous from the eruption of Vesuvius in 79 CE, which buried Pompeii and Herculaneum. The Greek island of Thera (Santorini), occupied by a prosperous Bronze Age population as part of a chain of trading centers across the Aegean in the second millennium BCE, underwent a catastrophic eruption around 1600 BCE that buried the town there. Although tephra have been recovered from contemporary deposits over the adjacent area of the eastern Mediterranean, there seems to be no warrant for the idea that this caused a widespread agricultural catastrophe, or even the temporary end of urban life on nearby Crete (whose Minoan civilization was later to be taken over by Myceneans from the Greek mainland). While eruptions and earthquakes were a constant hazard in this area and often required the reconstruction of buildings, they were no more than temporary setbacks in a continuing story.

**Anthropogenic Degradation.** The spread of human populations to every continent of the world has, like any episode of faunal mixing, brought casualties. Small or isolated populations, like those of Oceania, were especially vulnerable (particularly to human commensals such as rats), and large flightless birds such as the moa were undoubtedly hunted to extinction by humans themselves. [*See* Easter Island.] More doubt attaches to the end of the Rancholabrean fauna in the Americas, which increasingly appears to have been a faunal turnover pulse forced by the rapid changes of the late glacial regime and especially the Younger Dryas: it is less obviously anthropogenic than the effects on the (by then) native human population caused by the later intrusion of white colonists and their diseases.

Whether in North America in the sixteenth century CE or in Europe in the sixth millennium BCE, the initial appearance of farming populations, although accompanied by forest clearance, was usually on too small a scale to cause radical changes. Nevertheless, the continuing process of deforestation, even in temperate areas, initiated irreversible processes of edaphic and geomorphic change. Sandy soils in areas of high rainfall were particularly vulnerable, and large areas of outwash sands in the periglacial regions of both continents were rapidly podsolized. Continued cultivation of these impoverished soils could lead to wind erosion and dune formation. Cultivation of loessic soils on slopes led to slumping and the creation of colluvium in valleys. Some features once attributed to climatic changes are now better seen as long-term anthropogenic impacts, resulting from cumulative disturbance of the landscape by farmers. The "Younger Fill" of the Mediterranean valleys seems to be essentially an artifact of an increasing scale of cultivation and slopewash, not only contributing to valley fills but also accelerating deltaic progradation. Many Mediterranean harbors at valley mouths, even those built as recently as the Hellenistic period (fourth century BCE), are now at considerable distances inland.

**Conclusion.** Although often of critical importance to the groups immediately concerned, environmental changes after the adoption of farming seem to have had remarkably little effect in deflecting the course of human history, by comparison with the role often assigned to them by an earlier generation of geographers. The principal exceptions are those changes caused by human populations themselves and the increasing scale of environmental interference that has accompanied urbanization and interregional trade. Indeed, it could be said that the major anthropogenic impacts have occurred largely since the sixteenth century CE (and most particularly after the discovery of the steam engine), in the context of European expansion and a growing capitalist economy.

The relative stability of the natural environment may, however, be illusory as a long-term prediction, for the Holocene so far appears to have been unusually uneventful by comparison with the kinds of rapid excursions that characterized not just the late glacial regime but the preceding parts of the Pleistocene. There may thus be unexpected magnitudes of change to which now swollen human populations may in the future be exposed: and the degree of environmental intervention by advanced industrial societies may well act as a trigger to such instabilities. That human populations have so far recovered well from environmental perturbations should not encourage complacency. It is an interesting question whether climatic change on the scale of the Younger Dryas event would evoke a comparable response to the invention of agriculture.

### BIBLIOGRAPHY

Butlin, R. A., and R. A. Dodgshon, eds. *An Historical Geography of Europe.* Oxford: Clarendon Press, 1998.
Horden, P., and N. Purcell. *The Corrupting Sea: A Study of Mediterranean History.* Oxford: Blackwell, 2000.
Roberts, N. *The Holocene: An Environmental History.* 2d ed. Oxford: Blackwell, 1998.
Scarre, C., ed. *Past Worlds: The Times Atlas of Archaeology.* London: Times Books, 1998.
Sherratt, A. "Climatic Cycles and Behavioral Revolutions: The Emergence of Modern Humans and the Beginning of Farming." *Antiquity* 71 (1997), 271–287.

—ANDREW SHERRATT

# CLIMATE CHANGE

[*To survey alterations in the climate system, this entry comprises three articles:*

An Overview
Abrupt Climate Change
Climate Change Detection

*The first article presents an overview of climate changes in time scales that range from ice ages to seasonal and daily variations; the second article focuses on an examination of the changes that appear to be rapid and addresses the potential for future abrupt global change; and the third article discusses the observation system and the natural variability associated with the oscillatory behavior of the climate system. See also* Framework Convention on Climate Change; *and* Intergovernmental Panel on Climate Change.]

## An Overview

The Earth undergoes changes in climatic conditions that are observable in a variety of instrumental and other records over time scales ranging from seasons to many millions of years. These changes result from periodic and random adjustments in the intensity and global distribution of solar radiation and the way this energy is redistributed. Atmospheric, oceanic, and cryospheric processes are responsible for this redistribution. Periods in Earth history during which ice is present at high latitudes, and climate changes exhibit pronounced shifts between relatively colder and warmer states, are termed *ice ages.* The present ice age commenced approximately 2.5 million years ago and its occurrence defines the Quaternary period of Earth history. Over long time scales (millions of years), the arrangement of tectonic plates plays a critical role in determining the extent of the cryosphere and whether an ice age state will occur. [*See* Earth History.] When ice ages such as the present one do occur, a series of long-term (tens of thousands of years) shifts between glacial and interglacial conditions results, and these shifts may be further subdivided into short-duration stadial and interstadial events. There is also evidence for climate changes of shorter duration (millennial, centennial, and decadal); these were pronounced in amplitude during the last glacial period and are also present but more subtle in their extent during the present interglacial period. External factors such as quasiperiodic changes in the Earth's orbit around the Sun play a significant role in influencing long-term changes. A combination of internal and external factors is responsible for shorter-duration changes, including variations in solar emissions, volcanic eruptions, and shifts in greenhouse gas concentrations. During the past two centuries, anthropogenic activity has resulted in large increases in the atmospheric greenhouse gas content, which has caused a detectable increase in global temperatures and are predicted to continue to increase for many decades before the climate system reaches a new equilibrium. [*See* Greenhouse Effect; *and* Global Warming.] There is considerable difficulty in separating the intrinsic steady-state equilibrium variability of climate (climate variation) from actual climatic changes, particularly over shorter (namely, human) time scales.

Climate is typically defined to be "average weather" and is often described in terms of mean conditions and their levels of variability over certain preset time periods. It involves processes within the atmosphere, oceans, and cryosphere, caused both by interactions within the climate system and by external factors. These processes act to redistribute energy provided by radiation from the Sun, or insolation. The climate system may be considered as consisting of external and internal components. External components include the Sun and its solar output, the rotation of the Earth, and the geometric arrangement of the Earth in relation to the Sun. Internal components of the system include the atmosphere (its mass and composition), oceans, land and sea ice, and the physical properties of the land surface (including reflectivity or albedo, extent of snow cover, hydrological regimes, vegetation, and biomass). Over short time scales (less than millions of years), the physical components of the Earth system—such as the distribution of land and oceans, continental relief, and the configuration of ocean basins—may be considered as fixed, but over longer time scales these too may be considered as internal variables within the climate system.

**Climatic and Environmental Archives.** Although climate change has become a byword of our time, past instrumental observations of climatic variables, such as temperature and precipitation, have demonstrated that variations in climate are the norm. This pattern of variation is confirmed by a variety of proxy data sources derived from geologic records spanning considerably longer periods of time. Land-based data sources include ecological records from peat bogs, growth patterns of trees with seasonal rings (dendroclimatological data), the orientation of desert dunes, the growth and geochemistry of cave speleothems (calcareous, crystalline deposits, including stalactites and stalagmites), changes in fluvial (river) regimes, the extensive loess deposits that cover vast areas of continental Europe, Asia, and the Americas, and deposits related to high-altitude mountain glaciers. Glacial varves, annual layers that are found in proglacial lakes and result from the summer melting of glaciers, are another significant source of data. The sediments preserved in deep-ocean basins have proved to be an extremely significant source of climate change data. In this case the data relate both to the marine organisms (principally *radiolarians, foraminifera,* and other microorganisms) and the inorganic sedimentary detritus supplied from continental areas as river sediment, eolian (wind-borne) dust, and ice-rafted detritus transported from high latitudes by icebergs. Over the past two decades the records of snow chemistry, quantity, and aerosol components (such as dust, salt, and volcanic ash), and greenhouse gases as recorded in ice cores from the high-latitude and high-altitude cryosphere have provided some of the most detailed records of the past few hundred thousand years of climate change. While such proxy records rarely provide a direct index of climate-related parameters, they are frequently capable of generating qualitative or quantitative indices of climatic characteristics. These records have indicated that the climate changes that have occurred have generally been global in extent and have occurred globally synchronously on time scales ranging from decades to millions of years (Figure 1).

**Earth History and Climate Change.** Geologists have subdivided the 4.5 billion years of Earth's history into a series of eons, eras, and periods, which were first characterized broadly in terms of evolutionary changes in plants and animals (Figure 1). Many of these changes were influenced by changing global environments, and the periods may also be used to describe or classify paleoenvironmental conditions. The most recent era is called the *Cainozoic* or *Cenozoic*. It is subdivided into two periods, the Tertiary and the Quaternary, and it comprises a 60-million-year period of generally cooling global mean temperatures, culminating in the ice age that characterizes the Quaternary. The periods are in turn divided into various series or epochs. The Pleistocene and Holocene series together constitute the Quaternary, which is sometimes simply termed the *late Cenozoic*.

*The quaternary ice age.* Until recently, the date accepted for the onset of the Quaternary was fixed, based on the arrival of cold water marine organisms such as *Arctica islandica* and *Hyalinea balthica* to the sediments of the Mediterranean Sea; their appearance heralded the onset of the cold conditions that characterize the Quaternary. This occurs stratigraphically close to a reversal in the Earth's magnetic field, termed the Olduvian geomagnetic reversal, which has been dated in volcanic rocks to around 1.8 million years ago. Research over the past few decades on deep-ocean sediments indicates that the true date for the onset of cold conditions and extensive Northern Hemisphere glaciation that characterize the Quaternary lies somewhat closer to 2.5 million years ago. The commencement of the relatively stable and warm interglacial Holocene epoch in which we presently live has been fixed arbitrarily at 10,000 years BP.

The Quaternary ice age is by no means the only such cold period of Earth history, and at least five other extensive ice ages have been described. Such major phases of ice-age activity appear to have been separated by periods of around 300–400 million years of nonglacial activity. The ice ages are thought to be controlled by numerous factors, the most important being the orientation, position, and fragmentation of continental plates, the intensity of mountain building, and ocean basin formation. Warm periods are thought to be associated with the formation of large supercontinental plates with

ocean circulation taking place zonally, and confined principally to equatorial latitudes. This results in higher average global temperatures and small equator-to-polar temperature gradients. Ice ages are associated with periods of widespread mountain building and continental fragmentation, which result in the isolation of continental masses at high latitudes (thereby facilitating their refrigeration), meridionally oriented oceanic circulation patterns, increases in weathering rates, and large equator-to-polar temperature gradients.

Given that colder ocean water temperatures facilitate the drawdown of atmospheric gases into the oceans, the ice-age periods are also associated with reduced atmospheric greenhouse gas concentrations. Greenhouse gases increase planetary temperatures by absorbing outgoing longwave radiation, and their reduction during ice-age periods causes a strong positive feedback that accentuates the global cooling. Oceanic drawdown of the greenhouse gas carbon dioxide is further increased by the higher rates of delivery of weathered continental material during mountain-building events, which causes further assimilation of atmospheric gases. Conversely, periods between the ice ages are associated with high levels of atmospheric greenhouse gases. This distinction has led some workers to consider the Earth's history as representing a series of shifts between "icehouse" and "greenhouse" climatic megacycles. For the overwhelming majority of the Earth's history, its climate has been in a greenhouse state.

While the Quaternary ice age commenced 2.5 million years ago, a series of events over the preceding 100 million years caused substantial global climate changes that were critical preconditions for the ice age to develop. Although the overall trend observed is that of a global cooling, the changes frequently occurred as a step function in response to specific events. At around 100 million years ago, Australia separated from Antarctica and allowed the initiation of sea floor spreading, which would eventually lead to the formation of the South Indian Ocean. At much the same time, other plate tectonic reorganizations started the northward migration of India toward the Asian landmass. While these were significant events that were critical to subsequent climate changes, substantial global cooling did not occur until around 50 million years ago, in the early Cenozoic, when the flow of equatorial warm currents was permanently interrupted by the closure of the Tethys Sea as the African plate collided with Europe. [*See* Plate Tectonics.]

During approximately the same period, the Atlantic Ocean was undergoing a series of reorganizations that accentuated ocean circulation and allowed meridional flow from the equator to high northern and southern latitudes. After about 30 million years ago, Antarctica was fully isolated from both Australia and South America, and the resulting development of the cold Circum-

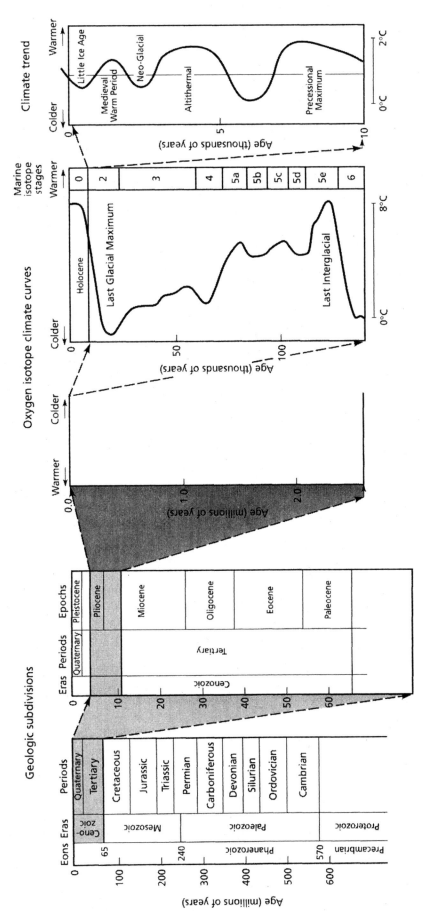

**Climate Change: An Overview. FIGURE 1. Geologic Subdivisions of Earth History and Examples of Scales and Amplitudes of Climatic Variations during the Quaternary Period.** Geologic subdivisions are based on floral and faunal extinction events, many of which are related to global-scale climatic changes. Coarse-scale (glacial–interglacial) Quaternary subdivisions are based on variation in climate as best expressed in the stable isotopic record from deep-sea organisms; finer-scale climatic variations during the Holocene period are based on a variety of evidence, including ice-core records, dendroclimatology, glacial advance/retreat, and instrumental observations.

161

Antarctic Bottom Water current cooled deep or abyssal ocean waters and the surface waters of the southern latitudes. At this time, Antarctica contained only limited land and sea ice, and pollen evidence indicates that abundant forests persisted until around 15 million years ago.

When India collided with Asia around 8 million years ago and formed the uplifted Tibetan Plateau and Himalayan ranges, a further series of important global climate changes occurred. This event caused the deflection of midlatitude Northern Hemisphere wind systems, particularly the trajectories of midlatitude jet streams; allowed the development of the Asian monsoonal circulation system; and accentuated continental-scale chemical weathering and denudation, which further reduced global atmospheric carbon dioxide levels. The formation of the Tibetan Plateau was accompanied by mountain building along the Pacific margin of South America, in western North America, and in the rift system of East Africa, which also contributed to the global cooling trend.

The closure of the Isthmus of Panama around three million years ago caused significant reorganizations of North Atlantic ocean circulation, and the intensification of the Gulf Stream resulted in enhanced levels of atmospheric moisture export to the northern high latitudes. This led, around 2.5 million years ago, to the formation of continental ice sheets in the Northern Hemisphere and the commencement of the ice age proper.

*Interglacial phases.* Over the past few decades it has been recognized that major climate changes were numerous during the Quaternary ice age and involved dramatic shifts between colder glacial conditions and warmer interglacial conditions. The glacial phases are typically long lasting (many tens of thousands of years) in comparison with the interglacial phases, which typically last for roughly ten thousand years. Generally, cold glacial conditions in which polar and alpine glaciers have undergone dramatic expansion and covered large areas in North America (the Laurentide Ice Sheet) and Eurasia (the Fennoscandinavian Ice Sheet) have predominated. During glacial periods, midlatitude temperatures were depressed by around 7°–10°C, and the accumulation of snow and ice at high latitudes resulted in the global lowering of sea levels by up to 120 meters. The intervening interglacials, defined as nonglacial climatic phases, were sufficiently stable and benign to allow the development of deciduous forests similar to those that occurred in northwestern Europe. Another type of interruption to full glacial conditions was the interstadial, a period that was either too cold or too protracted to allow the development of temperate deciduous forest of the full interglacial type. (The term *stadial* refers to an ice advance.)

The concept of multiple glacials, interglacials, stadi-

als, and interstadials during the ice age was proposed in the nineteenth century and further confirmed in the first years of the twentieth century by Penck and Brückner (1909), working in the European Alps. They developed much of the terminology and interpretation of sequences used to this day, and their work is one of the great landmarks in the study of environmental change. They recognized four major glacial advances (Wurm, Riss, Mindel, and Gunz). Similar glacial–interglacial cycles were recognized and correlated elsewhere, as subsequently were stadial and interstadial events of shorter duration and intensity. In North America, the acknowledged glacial periods were also four in number (Nebraskan, Kansan, Illinoian, and Wisconsin), and it was natural to hypothesize contemporaneity and a common cause, namely, variation in the amount and distribution of insolation.

*Stratigraphic evidence.* The main basis for stratigraphic correlation and interpretation of Quaternary environmental records, whether on land or offshore, is via the numbered series of more than twenty glacial–interglacial stages defined principally on the basis of oxygen isotopic evidence from deep-sea cores. There are three stable isotopes of oxygen, of which oxygen-16 is the most abundant; oxygen-17 and oxygen-18 are much rarer. Of the two minority species, oxygen-18 is the more abundant, but even so its ratio with respect to oxygen-16 is only in the range 0.0019–0.0021 in natural materials. Although the isotopes are similar in chemical behavior, there are some processes that discriminate against the heavier isotopes and some that give preference to them, to a degree dependent on temperature. As a net result of such fractionation processes, the ratio of oxygen-18 to oxygen-16 in glaciers is slightly lower than in sea water (that is, the oxygen in the water of glaciers is isotopically lighter). During glacial times, because of the greater amount of water locked up in glaciers, sea water is isotopically heavier. Shells formed in this water are heavier still because there is further fractionation during the formation of shell carbonate, the lower temperature favoring incorporation of oxygen-18.

A record of past oxygen and carbon isotope ratio values is available from calcareous skeletons (commonly called *shells*) of marine organisms (usually foraminifera) in deep-sea sediments. Samples are obtained by means of long coring tubes, of the order of 10 centimeters in diameter and up to 50 meters in length. The continuous sediment cores so obtained carry, in addition, a magnetic polarity record that allows correlation of the climatic variations with the magnetic polarity time scale. [*See* Dating Methods.]

One of the first records of marine oxygen isotopes was obtained by Cesare Emiliani during the 1950s. [*See the biography of Emiliani.*] This showed variations initially interpreted as reflecting primarily the temperature of the water in which the shells had been formed. However, it was later argued that the influence of glacier volume was dominant and that the isotope variations could be considered as a paleoglaciation record. This core showed evidence of thirteen warm and cold phases. These were numbered from the top down, with odd numbers corresponding to warm stages and limited global ice volumes, and even numbers to cold stages with greater global ice volumes (Figure 1). They are now referred to as *marine isotope stages* (MIS).

The stage numbers allocated by Emiliani continue to be used but have been developed so as to include substages. Commonly, these are designated by letters (for example, the warm substages of MIS 5 are named 5a, 5c, and 5e, and the intervening cool troughs are named 5b and 5d). A decimal system has also been developed for the naming of additional layers or horizons within the marine isotope stages in which subtle isotopic variations occur so as to give greater flexibility in dealing with the complexities of the isotope curve. In addition, boundaries between pronounced isotopic maxima (full glacials) and consecutive pronounced minima (peak interglacials) are called *terminations*. These are numbered by roman numerals in order of increasing age. The most recent period of full glacial conditions was at a maximum around 21,000 BP.

Because of the slow sedimentation rate on the ocean floor, and sometimes because of bioturbation (the mixing of sediments on the sea floor by the actions of animals and plants), there is a tendency for any short-term changes to be smoothed out, so that, in most cores, changes persisting for less than a few thousand years are unlikely to be seen and the record is one of long-term changes.

**Astronomical Theory of Climate Change.** The astronomical theory of climate change is by far the most widely accepted theory proposed so far to explain the numerous large-scale glacial–interglacial variations that occurred during the Quaternary period. The theory, also known as the orbital theory or the Croll–Milankovitch theory of climate change, has only in the past two decades gained the widespread support of the scientific community. The theory was first systematically proposed by Milutin Milankovitch, who argued that if the Earth's orbit around the Sun exhibits any degree of cyclic variability, then the subtle variations in incoming solar energy might be responsible for shifts from glacial to interglacial climatic states. [*See the biography of Milankovitch.*] Changes in orbital geometries occur as a result of the gravitational effects of the planetary bodies. [*See* Earth Motions.] There are three major variations in the Earth–Sun orbital configuration, with five primary periodicities that are considered to be significant (Figure 2).

The Earth's orbit around the Sun undergoes changes in its eccentricity from a near-circular to slightly elliptical at periodicities of about 100,000 and 400,000 years. The second major factor is the tilt or obliquity of the Earth's axis relative to the plane in which the bodies of the solar system lie (the ecliptic), which varies between 22° and 25°, with a periodicity of about 41,000 years. The effect of the obliquity variations is to amplify the sea-

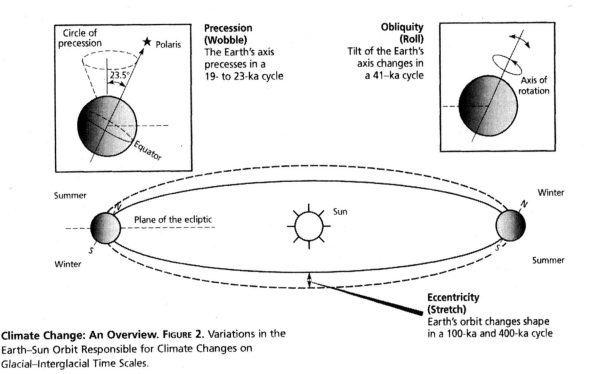

**Climate Change: An Overview. Figure 2.** Variations in the Earth–Sun Orbit Responsible for Climate Changes on Glacial–Interglacial Time Scales.

sonal cycles at high latitudes; the variations have a relatively minor direct effect at lower latitudes. The third key variation, known as the *precession of the equinoxes*, actually consists of two components that change the distance between the Earth and the Sun at any given season. The axial precessional component relates to the wobble of the Earth's axis of rotation and has a periodicity of 26,000 years. The elliptical precessional component has a 22,000-year periodicity and relates to the rotation of the Earth around one of the foci of the orbit. The combination of these two effects results in a shift of the equinoxes through the Earth's elliptical orbit, providing scope for anomalous seasonal patterns that are opposite in direction in the opposing hemisphere. The precessional effects are most pronounced at low latitudes and exhibit periodicities of 19,000 and 23,000 years.

It is only the eccentricity cycle that actually changes the total incoming insolation budget; the other cycles simply serve to redistribute the energy in different spatial and seasonal configurations. Milankovitch hypothesized that the onset of glacial conditions required a minimum in Northern Hemisphere summer insolation, which in turn would require a maximum of eccentricity and a minimum in obliquity, reducing seasonal contrasts and increasing latitudinal energy gradients. Given these conditions, Milankovitch inferred that the summers would remain sufficiently cool to prevent the thaw of winter snow and ice, and the relative mildness of the winter would allow substantial evaporation at intertropical latitudes, which would lead to abundant snowfall at middle and high latitudes. If snow and ice were preserved throughout the year, the land surface albedo would increase, resulting in a positive feedback process ultimately leading to the formation of persistent ice sheets. An additional important positive feedback in this process is the associated cooling of middle- and high-latitude ocean waters, significantly reducing atmospheric greenhouse gas concentrations and therefore reducing the absorption of outgoing longwave (heat) radiation and leading to further cooling.

Spectral analysis of the oxygen isotopic variations in deep-sea sediments demonstrates that the main periodicities of variations in global climate changes closely match those predicted by the astronomical theory (Figure 3). In the absence of internal factors, the astronomical theory would predict simultaneous glacial and interglacial conditions for the Northern and Southern Hemispheres. In fact, climate changes through glacials and interglacials are globally synchronous. Patterns of insolation changes at northern latitudes are most significant because it is here that much of the global surface is covered by land rather than by ocean, and it is the land surface that is capable of accumulating snow and ice and dramatically changing the surface albedo.

The climatic state as influenced by the northern insolation patterns is propagated throughout the globe principally via oceanic thermohaline currents, which transfer energy vertically and laterally in response to gradients of heat and salinity. Although the Milankovitch theory accounts for the periodicities observed in the deep-sea core records, it does not fully explain two important aspects of the record. First, there is a shift at around one million years ago from climatic variability dominated by precessional and obliquity-based (19,000–23,000-year and 41,000-year) cycles to a pattern dominated by variations occurring at periodicities corresponding to the eccentricity cycle. Secondly, the deep-sea record indicates a relatively progressive increase in the amplitude of climatic changes through the Quaternary period.

**Short-Term Climate Variations.** A continuous high-resolution multiproxy climatic record is available in long ice cores, some more than 3 kilometers in length, collected from polar and high-altitude locations. Oxygen isotopic variations, in this case as measured on snow and ice, are again useful for reconstructing past changes. The stable oxygen isotope ratio reflects the ambient temperature at formation, with higher oxygen-18/oxygen-16 values indicating higher temperatures. The long-term pattern of variation is similar to the marine pattern, but there is the possibility of seeing shorter-term (stadial and interstadial) fluctuations since the accretion rate is much greater and there is little opportunity for bioturbation or other disturbances to occur. In the upper part of cores, annual layers are discernible because of seasonal variation of dust and acidity. Further down, estimates have to be made on the basis of models of glacier flow and ice movement or of past accretion rates.

The records collected from both Greenland and the Antarctic indicate rapid fluctuations of climate that must be related to changes in the polar atmospheres. These interstadials are referred to as *Dansgaard–Oeschger events* and are given a numbering system with the prefix *IS*. Their duration was only of the order of a thousand years, and the onset and termination sometimes occurred in a matter of decades. The pattern of cooling and warming during an interstadial–glacial cycle is not symmetrical; instead, cooling occurs gradually and the cycle is completed by rapid warming (Figure 4). A total of twenty-four interstadial intervals have been recognized, lasting from five hundred to two thousand years. The steadiness of climate indicated for the present interglacial—the Holocene, from about 10,000 BP to the present, is in marked contrast to the variability of climate indicated for the last glacial and earlier periods.

The rapid fluctuations during the last glacial period recorded in the arctic ice cores are matched by global variations in mid- and high-latitude sea surface temperature and hence in air temperature. These fluctuations

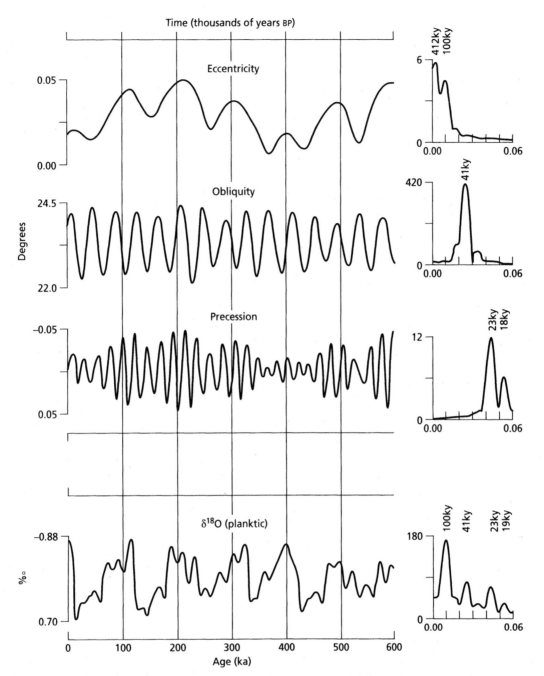

**Climate Change: An Overview. FIGURE 3.** Variations in Orbital Parameters Calculated for the Past 600,000 Years Compared against the Record of Oxygen Isotopic Ratios from Planktonic Deep-Sea Foraminifera.

Spectral analysis of the isotope record indicates significant periodicities of climatic variation occurring at 100,000, 41,000, 23,000, and 19,000 years, which exhibit a good match to those periodicities indicated for the orbital variations. (Modified after Imbrie et al., 1992.)

were initially deduced from the abundance of a temperature-sensitive planktonic (surface-dwelling) foraminifera (*Neoglobigerina pachyderma*) in two cores from the North Atlantic at 50°–55° north latitude and are now recognized within other ocean basins including the Southern Hemisphere via isotopic and other sediment parameters. The fluctuations are grouped into

Bond cycles (Figure 5). In each cycle, there is a gradual decrease in amplitude of the variations as well as a decrease in the temperature of the base level. Near the end of each cycle (that is, at the coldest part), the nature of the ocean-floor detritus indicates the occurrence of a Heinrich event, ascribed to a massive discharge of icebergs into the North Atlantic. Two leading theories

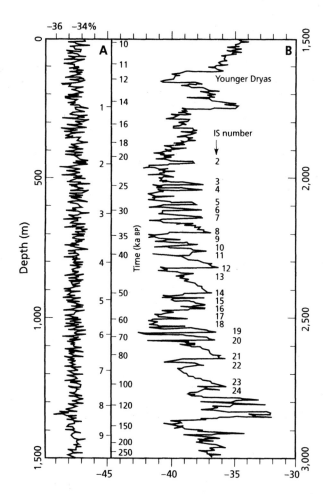

**Climate Change: An Overview. FIGURE 4.** The GISP Greenland Ice-Core Record of Variations in the Isotopic Composition of Snowfall over the Past 250,000 Years.

(A) The Holocene record (0–10,000 years). (B) The late Pleistocene record (10,000–250,000 years). Note that the age scale is not linear. Variations in snow chemistry during the glacial period relating to polar atmospheric changes are pronounced and define numbered, asymmetrical interstadial cycles termed Dansgaard–Oeschger cycles. Each cycle starts as a gradual cooling and terminates in an abrupt warming event. The last significant interstadial cooling event of the last glacial period was the Younger Dryas (labeled on figure). Isotopic records during the Holocene period exhibit relatively little variation. (Modified after Dansgaard et al., 1993.)

have emerged to explain the observed sub-Milankovitch climatic excursions: one proposes catastrophic collapse of large Northern Hemisphere ice sheets caused by internal feedbacks and basal instability (termed *binge–purge cycles*), while the other argues that the global synchrony of the rapid shifts requires an atmospheric mechanism, probably relating to a water vapor feedback process.

As well as recording isotopic variations, polar ice cores also provide a record of greenhouse gas varia-

tions. Analysis of the gas records in combination with estimates of past temperatures derived from the analysis of stable isotopes of hydrogen reveals a close correspondence between global temperature changes and greenhouse gas concentrations (Figure 6). Carbon dioxide exhibits variations that closely match the overall record of insolation variations, while methane abundance exhibits pronounced variations at precessional (19,000–23,000-year) time scales.

Low-latitude environments did not experience cold conditions during glacial phases at higher latitudes. Since the climatic transitions at these latitudes are related more to changes in moisture balance than in temperature, there has been a tendency to consider low-latitude climates to shift from pluvial (wet) to interpluvial (dry) conditions. The main sources of paleoclimatic evidence in these settings are records of past eolian (wind-blown) activity, lake-level data, and analyses of terrestrial dust from marine cores. It is widely recognized that the period of the last glacial was characterized on land by cold, dry, and windy conditions, and that the transition into the Holocene period was accompanied by widespread increases in moisture, at least for the low-latitude regions of the Northern Hemisphere. Glacial-age interpluvial conditions relate to reduced evaporation from oceans, increased continentality caused by lower global sea levels (i.e., the transport distance required to move moist air to inland continental areas), and increased evaporation rates due to enhanced trade-wind circulation patterns which are, in turn, related to increased equator-to-polar temperature gradients. Pluvial conditions such as those that occurred at the start of the Holocene period result in reduced eolian activity, development or expansion of lakes, and widespread reductions in surface albedo. A significant exception to this general pattern is Lake Bonneville in the southwestern United States, which was full during the last glacial period because of the southward deflection of the westerly midlatitude jet stream by the Laurentide Ice Sheet.

One of the most contentious contemporary debates relates to the degree of cooling of tropical sea surface temperatures (SSTs) during glacial periods and its impacts on tropical and extratropical climate. The long-standing view that tropical SSTs have varied little has recently been challenged by a variety of data that collectively indicate that the tropical oceans cooled by about 4°–6°C. Influences on moisture balance in these low-latitude areas include the direct effect of changing monsoon intensity caused by variations in the precessional orbital cycle, and the effects of changing rates of evaporation in adjacent tropical oceans, in part related to sea surface temperatures, which are linked to higher-latitude climates and are more closely related to the eccentricity and obliquity cycles (about 100,000 and 41,000

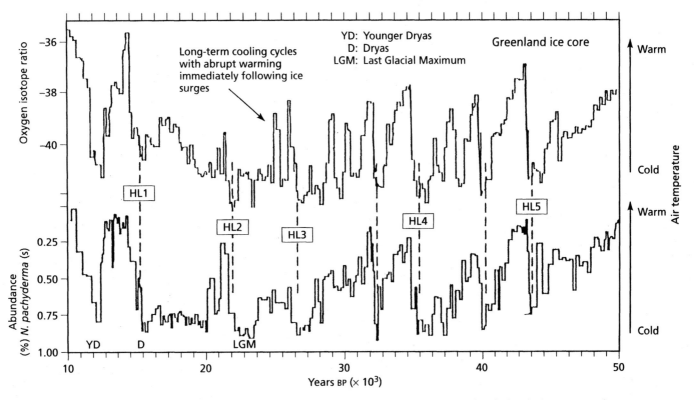

**Climate Change: An Overview. FIGURE 5.** A Comparison between Ice-Core Stable Isotope Shifts (Upper) and Abundances of the Cold-Sensitive Planktonic Foraminifera, *Neoglobigerina Pachyderma* (Lower), during a Portion of the Last Glacial Period. The asymmetrical cooling trends identified in the ice-core records are matched in the patterns of foraminifera abundance. Maximum cooling events correlate with massive releases of icebergs in the North Atlantic, which deposit Heinrich layers (labeled HL1-5 on diagram). (Modified after Bond et al., 1993.)

years, respectively). It has also been noted that dramatic changes in low-latitude moisture may occur over shorter (sub-Milankovitch) time scales.

*The holocene.* Climate changes leading to the present (Holocene) interglacial commenced around 13,000 BP with a series of rapid fluctuations in climate that culminated in the cold Younger Dryas interstadial. [*See* Younger Dryas.] Recovery from the Younger Dryas has been estimated to represent a global warming of the order of 5°–8°C in a period as short as thirty years. Since that time, climate changes have been relatively subtle and related to factors other than the external climatic forcing caused by orbital changes. A range of data sources point to successive variations between warmer and colder periods causing minor retreats and advances of alpine glaciers. The warmest phase of the Holocene period recorded in ice cores was between 10,000 and 8,000 BP, although the traditionally recognized Holocene climatic optimum, termed the *altithermal* or *hypsithermal*, is usually described as occurring between 7,000 and 4,000 BP. Where this optimum of Holocene climate is recognized, it is thought to represent a warming of the order of 1°–2°C above the modern (preindustrial)

levels. Postoptimal global cooling resulted in a neoglacial period of glacier readvance, but in early medieval times there was a return to more favorable conditions known as the *Little Climatic Optimum* or the *Medieval Warm Period*. [*See* Medieval Climatic Optimum.] This phase lasted from around 750 CE to 1300 CE and correlates with the climax of high medieval cultural development and energetic activity. Warmer summer temperatures at this time allowed the development of vineyards in the United Kingdom as far north as York.

The best known of the climate fluctuations of the Holocene is that of the Little Ice Age. Its occurrence is well documented by a range of archival materials (including annals, chronicles, and ship's records) and dendrochronological records. It corresponds to a period of cold and highly variable climate and glacial readvance following the Medieval Warm Period, lasting from around 1300 to 1800 CE. The Little Ice Age had widespread consequences for human populations, including high incidences of crop failure (particularly wheat in the United Kingdom), difficult navigation at higher latitudes, abandonment of settlement in marginal areas (including the final decline of the Anasazi culture in the south-

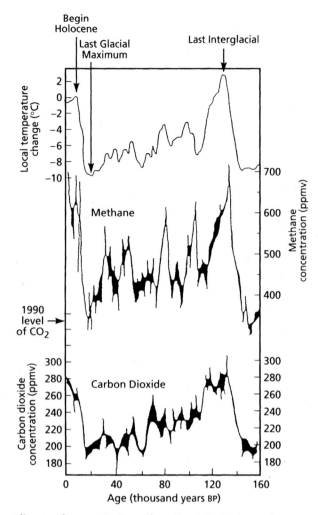

**Climate Change: An Overview. FIGURE 6.** Analyses of Carbon Dioxide ($CO_2$) and Methane Concentrations in Air Trapped in Antarctic Ice Cores.

There is a close correspondence between the greenhouse gas concentrations and past temperature variations. Note the elevated 1990 level of $CO_2$ in comparison to past levels. (Modified after Intergovernmental Panel on Climate Change (IPCC), 1990.)

western United States and abandonment of areas of the Scottish uplands), and extreme winter storms. [*See* Little Ice Age in Europe.]

***Causes of short-term climate change.*** Variations in the climate during Holocene times have been linked with a number of possible causes. The sunspot cycle, which occurs with a 22-year periodicity, has frequently been cited as a potentially significant cause of Holocene and older climate changes. The overall amplitude of the cycles seems to increase slowly and then fall rapidly, with a period of 80–100 years. There also appear to be quasi-cyclic fluctuations on time scales ranging from 180 to 2,200 years. While correlations between solar activity and instrumental and proxy (tree ring and ice core) data have been described, no mechanistic link between

sunspot activity and surface conditions on the Earth has so far been demonstrated. [*See* Sun.] The Little Ice Age has been linked by many workers to the Maunder minimum in sunspot activity.

Collisions of comets with the Earth and very large meteoritic impacts have also been proposed as causes of climatic fluctuations, the best documented being the impact event at the end of the Mesozoic period (about 63 million years ago), which may have resulted in the extinction of the dinosaurs. [*See* Impacts by Extraterrestrial Bodies.] Many of the disturbances that such impacts would cause, such as an increase in stratospheric and tropospheric aerosols, are similar to disturbances internal to the system.

Volcanoes influence climate by projecting large quantities of particulates and gases into the atmosphere. The effect of injected aerosol upon the radiation balance, and whether heating or cooling results, depends largely on the height of injection into the atmosphere. [*See* Aerosols.] Most eruptions inject particulates into the troposphere at heights between 5 and 8 kilometers. These tend to be removed rapidly either by fallout or by rain, and the effect on climate is minimal. More violent eruptions inject debris into the upper troposphere or even into the lower stratosphere (15–25 kilometers). The particulates have a long residence time in the stratosphere (up to years). Nonabsorbing aerosols increase the albedo of the atmosphere and reduce the amount of solar radiation that reaches the surface. Aerosols that absorb in the visible part of the spectrum result in energy transfer directly to the atmosphere. If the aerosol absorbs and emits in the infrared, the greenhouse effect is increased.

Mount Pinatubo injected around 20 million metric tons of sulfur dioxide to heights of 25 kilometers, where it was photochemically transformed into sulfate aerosols. The aerosols generated by the eruption of Mount Pinatubo have been estimated to have resulted in a forcing on the climate system of about −0.4 watts per square meter, with a resultant temporary global cooling of about 0.5°C. Eruptions such as Mount Pinatubo and their effects on the atmosphere are very short-lived compared with the time required to influence the heat storage of the oceans, and they are not likely to initiate significant long-term climatic changes. [*See* Mount Pinatubo.]

The natural variability of the ocean circulation is an important factor for climate. The ocean circulation varies on glacial time scales, when the circulation is known to change markedly, and on interannual time scales, where the El Niño–Southern Oscillation (ENSO) phenomenon is important. [*See* El Niño–Southern Oscillation.]

**Climate Change in the Twentieth Century.** Climate change has also occurred during the twentieth cen-

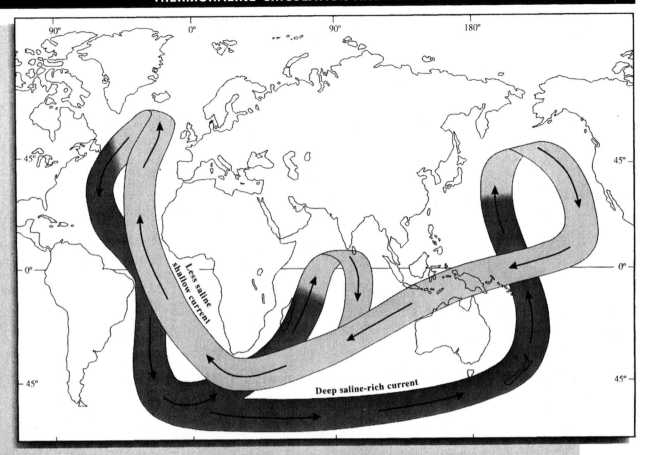

**The Large-Scale Salt Transport System ("Ocean/Global Conveyor") Operating in Today's Oceans.**

This compensates for the transport of water (as vapor) through the atmosphere from the Atlantic to the Pacific Ocean. Salt-laden deep water formed in the northern Atlantic flows down the length of the Atlantic and eventually northward into the deep Pacific. Some of this water upwells in the northern Pacific, bringing with it the salt left behind in the Atlantic due to vapor transport. Flow of the "Atlantic Conveyor" may have been interrupted during cold episodes. (Reprinted from Broecker and Denton, 1990. With permission of Elsevier Science.)

Climate is closely linked to ocean circulation, so that any change in the pattern of currents is likely to have consequences for climate at regional, and even global, scale. Some circulations are driven by water temperature and salinity gradients, notably, the so-called *thermohaline circulation*, which depends upon cooler and more saline waters being more dense and tending to sink.

In modern oceans the most dense waters occur in the North Atlantic, where a combination of low temperature and high salinity leads to large amounts of deep water known as North Atlantic Deep Water. Within the North Atlantic the circulation appears to operate as a *conveyor*, whereby water moves northward in upper levels of the ocean (this includes the Gulf Stream or North Atlantic Drift) ultimately to sink in the vicinity of 60° North. The return limb of the conveyor at depth returns this deep water to the Southern Oceans.

Any warming in high latitudes would increase the melting of glaciers in Greenland and Baffin Island, introducing more fresh water to the North Atlantic, reducing salinity of surface waters, and disrupting the thermohaline circulation. Because Northern Europe is warmed now by westerly winds crossing the Gulf Stream it would have much colder winters in the absence of that circulation.

Some of the abrupt climate changes during the last glacial–interglacial cycle could be explained by the cycling of this circulation.

—ANDREW S. GOUDIE AND DAVID J. CUFF

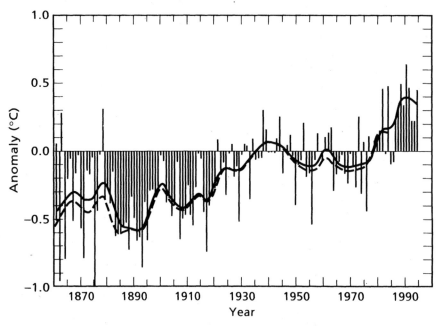

**Climate Change: An Overview.** Figure 7. Annual Global Average Surface Air Temperature Anomalies for Land Areas for the Period 1861–1994.

Bars represent individual anomalies from the 1961–1990 average. The smoothed curves are binomial filters of the individual data. (Modified after Intergovernmental Panel on Climate Change (IPCC), 1990.)

tury. It has been estimated that global (land and sea) surface temperatures have warmed by between 0.3° and 0.6°C, and by between 0.2° and 0.3°C over the last forty years (Figure 7). These temperature levels are close to those of the Northern Hemisphere during the Medieval Warm Period. The observed climate changes are not uniformly spaced, with some areas exhibiting cooling, and the greatest extent of warming occurring over the continents between 40° and 70° North latitude. The warming is thought to be at least partly the result of human activity. Because populations, national economies, and the use of technology are all growing, the global average temperature is expected to continue increasing, by an additional 1.0°–3.5°C by the year 2100. In addition to the warming trend, there have also been changes in precipitation and other climatic factors. Precipitation has increased globally by around 1 percent over the past century. The increased frequency and intensity of ENSO events in the past two decades is also thought to relate to climatic changes relating to tropical ocean–atmosphere interactions.

There are a number of human activities that are believed to be capable of inducing changes in global climate. The most substantial changes have occurred since the Industrial Revolution and relate to the burning of fossil fuels and increased industrialization and pollution.

***The greenhouse effect.*** Increases in atmospheric greenhouse gas concentrations over the past few decades were first detected at the Mauna Loa Observatory in Hawaii. The trend has been confirmed by analy-

ses of gas bubbles preserved in polar ice cores. Levels of carbon dioxide ($CO_2$) have increased in the past century by more than 25 percent since the beginning of the Industrial Revolution. In addition to $CO_2$, the levels of methane ($CH_4$), nitrous oxides ($NO_x$) and halocarbons (chlorofluorocarbons [CFCs] and hydrochlorofluorocarbons [HCFCs]) have also increased. These gases have differing atmospheric concentrations, residence times, and potentials to induce greenhouse warming (Table 1). The relative radiative effects of the various greenhouse gases are measured by their global warming potential (GWP). This is defined as the cumulative radiative forcing caused by a unit mass of gas emitted, relative to a reference gas (usually $CO_2$). Table 1 demonstrates that while $CO_2$ is by far the most abundant of the anthropogenically influenced greenhouse gases, its efficiency as a greenhouse gas is considerably lower than many of the other atmospheric constituents, in some cases by many orders of magnitude.

Increased levels of greenhouse gases will result in a warming of global climate. However, the magnitude of the warming, the time that will be required for the warming trend to reach an equilibrium, and the relative impacts on different regions of the world will depend on the nature of the feedbacks within the climate system.

***Aerosols.*** The influence of tropospheric aerosols associated with industrial pollution and fossil fuel and biomass burning has only recently been identified and, to some extent, quantified. Solid sulfate particles result from the oxidation of sulfur dioxide, emitted when fos-

**Climate Change: An Overview. TABLE 1.** A Summary of Some Key Greenhouse Gases Affected by Human Activities

| SPECIES | CHEMICAL FORMULA | ATMOSPHERIC LIFETIME (YEARS) | PREINDUSTRIAL CONCENTRATION | CONCENTRATION IN 1994 | RATE OF CONCENTRATION CHANGE (PERCENT PER YEAR) | *Global Warming Potential* | | |
|---|---|---|---|---|---|---|---|---|
| | | | | | | 20 YEARS | 100 YEARS | 500 YEARS |
| Carbon dioxide | $CO_2$ | 50–200 | 280 ppmv | 358 ppmv | 0.40 | 1 | 1 | 1 |
| Methane | $CH_4$ | 12 | 700 ppbv | 1,720 ppbv | 0.60 | 56 | 21 | 6.5 |
| Nitrous oxide | $N_2O$ | 120 | 275 ppbv | 312 ppbv | 0.25 | 280 | 310 | 170 |
| CFC-11 | $CCl_3F$ | 50 | — | 268 pptv | 0.00* | | | |
| HCFC-22 | $CHClF_2$ | 12 | — | 110 pptv | 5.00 | 9,100 | 11,700 | 9,800 |
| Perfluoromethane | $CF_4$ | 50,000 | — | 72 pptv | 2.00 | 4,400 | 6,500 | 10,000 |

*Atmospheric emission of chlorofluorocarbons was prohibited under the Montreal Protocol and its amendments.

sil fuels are burned. Other industrial processes and natural and human-initiated biomass burning also contribute particulates, often termed *aerosols*, to the troposphere. These aerosols are localized, being confined mainly to Northern Hemisphere midlatitudes, and have two effects on the climate system. The direct effect of most aerosols is to reflect incoming solar radiation back into space and hence cause cooling. Some particulates, such as soot, are dark in color and have the opposite effect, causing warming. The magnitude of the cooling or warming depends on the nature of the aerosols and their distribution in the atmosphere. There is also an important indirect effect of tropospheric aerosols. They act as additional cloud condensation nuclei and cause more (and smaller) drops to form in clouds, increasing the reflectivity of the clouds and further cooling the planet. The effect of changes in cloud character can have complex repercussions, since the clouds also affect the amount of radiation that escapes from the Earth system.

*Ozone depletion.* The discovery of the Antarctic ozone hole in 1986 and, more recently, a similar but less intense ozone depletion over the Arctic has been a cause of much recent discussion relating to climate changes. The observed ozone destruction now appears to be due to the disturbance of the natural balance of destruction and production that previously existed in the stratosphere. The presence of free chlorine atoms in the stratosphere can now be traced to the photochemical disruption caused by halocarbons when these gases migrate from the troposphere.

Chlorine is the principal cause of the disturbance in ozone chemistry that produces the ozone hole. Although the buildup of CFCs, at least, in the atmosphere is leveling off as a result of the Vienna Convention and the Montreal Protocol, the very long lifetimes of many of these gases mean that they will persist in the atmosphere for perhaps thousands of years. The particular

reactions that act to accelerate the ozone destruction rely on the presence of free chlorine atoms and a solid surface, provided by stratospheric ice clouds. Since CFCs, HCFCs, and the hydrofluorocarbons (HFCs) that are replacing them are radiatively active (they are much more effective greenhouse gases than $CO_2$), they also act to change the atmospheric temperature, and this alters the rate of the chemical reactions. CFCs that remain in the troposphere are effective absorbers of infrared radiation, which would otherwise escape to space. These gases therefore act to enhance the atmospheric greenhouse and to provide a warming influence for the planet. The radiative effect of the reduced stratospheric ozone is to cool the planet. The enhanced levels of tropospheric ozone that have been observed result in a warming. [*See* Ozone.]

*Land surface changes.* Regional changes to the character of the Earth's surface caused by human activities may also cause regional and global changes in climate. These include desertification, changes in levels of forestation, and urbanization. Removal of vegetation and exposure of bare soil during the process of desertification decreases soil water storage because of increased runoff and increased albedo. Less moisture available at the surface means decreased latent heat flux, leading to an increase in surface temperature. On the other hand, the increased albedo produces a net radiative loss.

At present, around 30 percent of the land surface of the Earth is forested and about a third as much again is cultivated. However, the amount of forest land, particularly in the tropics, is rapidly being reduced, while reforesting is prevalent in middle latitudes. As a consequence, the surface characteristics of large areas are being greatly modified. The change in surface character can be especially noticeable when forests are replaced by cropland. The important climatic change after deforestation is in the surface hydrological characteristics, since the evapotranspiration from a forested area can be many times greater than that from adjacent open ground. The largest impacts are the local and regional effects on the climate, which could exacerbate the effects of soil impoverishment and reduced biodiversity accompanying the deforestation.

[*See also* Deforestation; Desertification; Ecosystems; Glaciation; Sea Level; *and the biography of Lamb.*]

### BIBLIOGRAPHY

Aitken M. J., and S. Stokes. "Climatostratigraphy." In *Science-Based Dating in Archaeology*, edited by M. J. Aitken and R. A. Taylor. New York: Plenum Press, 1998.

Bond, G. C., et al. "Correlations between Climate Records from North Atlantic Sediments and Greenland Ice." *Nature* 365 (1993), 143–147.

Bradley, R. S. *Quaternary Palaeoclimatology*. London: Chapman and Hall, 1994.

Broecker, W. S., and G. H. Denton. "The Role of Ocean–Atmosphere Reorganizations in Glacial Cycles." *Quaternary Science Reviews* 9 (1990), 305–341.

Crowley, T. J., and G. R. North. *Palaeoclimatology*. New York: Oxford University Press, 1991.

Dansgaard, W., et al. "Evidence for General Instability of Past Climate from a 250-ky Ice Core Record." *Nature* 364 (1993), 218–220.

Goudie, A. S., and S. Stokes. *Quaternary Environmental Change*. New York and Oxford: Oxford University Press, 1998.

Imbrie, J., A. Berger, and N. J. Shackleton. *Global Changes in the Perspective of the Past*. New York: Wiley, 1992.

Imbrie, J., and K. P. Imbrie. *Ice Ages: Solving the Mystery*. New York and Oxford: Oxford University Press, 1984.

Intergovernmental Panel on Climate Change (IPCC). *Climate Change: The IPCC Scientific Assessment*. Cambridge and New York: Cambridge University Press, 1990.

———. *Climate Change 1995: The Science of Climate Change*, edited by J. T. Houghton, et al. Cambridge and New York: Cambridge University Press, 1996.

Lowe, J. J., and M. J. Walker. *Reconstructing Quaternary Environments*. London: Longman, 1997.

Van Andel, T. J. *New Views on an Old Planet: A History of Global Change*. Cambridge: Cambridge University Press, 1994.

Vrba, E. S., G. H. Denton, T. C. Partridge, and L. H. Burckle, eds. *Palaeoclimate and Evolution with Emphasis on Human Origins*. New Haven: Yale University Press, 1995.

—STEPHEN STOKES

## Abrupt Climate Change

The further back one goes in time, the less easy it becomes to identify abrupt climate change in the geologic record. This is partly because the environmental indicators of change may be less clear, but it is also because the resolution of available dating techniques becomes coarser. When one moves forward into the Pleistocene and Holocene epochs (c. 1.8 million years ago to the present), however, there are a series of high resolutions: precisely and frequently dated cores and sections (from the oceans, icecaps, lakes and loess), which give in-

creasingly clear indications that climate has changed abruptly over scales ranging from decades to centuries and millennial (van Loon, 1999). [*See* Climate Change.]

**Abrupt Changes in Pre-Pleistocene Times.** Although it is difficult to identify abrupt changes of climate in the early geologic record, it is evident that there have been a series of major ice ages in the past [*see* Paleoclimate] and that there have been major spasms of extinction of organisms (Wood, 1999). The Late Ordovician ice age (c. 458–438 million years ago) was associated with a major extinction episode, while the Late Devonian extinction crisis at the Frasnian-Famennian boundary (367 million years ago) may have been associated with a bolide impact or with a decline in oceanic carbon dioxide levels, suggesting that cooling occurred. The end Permian extinctions (245 million years ago) may also have been associated with cooling, especially in tropical latitudes. The great extinctions at the end of the Cretaceous (65 million years ago) have been the subject of great debate and controversy, but it is possible that they were the result of a great bolide impact that could have caused changes in atmospheric gas composition, along with a global drop in temperature, acidification of precipitation, and global wildfires. Major volcanic activity associated with the eruption of the Deccan lavas in India could also have caused an abrupt change in climate at this time, by loading the atmosphere with dust and sulfate aerosols. Many of these events and their causes are discussed by Crowley and North (1991). [*See* Extinction of Species; *and* Impacts by Extraterrestrial Bodies.]

**Abrupt Climate Changes in the Pleistocene.** The availability of well-dated, high-resolution environmental information from cores and sections on land and in the oceans makes the identification of abrupt changes of climate a much easier prospect in the Pleistocene (1.8 million to 10,000 years ago). Indeed, one of the features of paleoclimatic research in the past decade has been the realization of just how abruptly climatic change can occur. An early indication of this came from a deep peat core at Grande Pile in France (Wollard, 1979), where a combination of pollen analysis and precise dating arguably showed that the temperate forest of the last interglacial (Eemian) period was replaced by a pine-spruce-birch taiga within approximately 150 years. This was not a finding that received universal approval (see Frenzel and Blundau, 1987).

Ice cores provide a particularly fine temporal resolution and contain large numbers of valuable environmental indicators. High-frequency swings in isotopic and dust content and other components suggest that dramatic oscillations have taken place in environmental conditions over quite short periods of time. The rapid temperature oscillations that have been identified from ice cores are now known as Dansgaard-Oeschger events. Dansgaard et al. (1993), for example, documented no less than twenty-four interstades (warmer phases) in the last glacial period from the GRIP core. High-frequency shifts are also known from the last interglacial (the Eemian of Europe), but some workers fear that the ice-core records could have been corrupted by deformation within the ice. That said, there is also evidence of a sharp Eemian cold phase in ocean cores (Maslin and Tzedakis, 1996). There is also evidence for early Pleistocene instability (Raymo et al., 1998).

High-frequency abrupt changes have also been identified from ocean cores (Oppo et al., 1998), where the observed sawtooth patterns of climatic variation have been termed *Bond cycles*. Also within the ocean-core sediment record are layers that are rich in dolomite and limestone detritus but poor in foraminifera. Each layer is interpreted as being the result of deposition by a mas-

sive armada of icebergs released from North American (and possibly Fennoscandian) icecaps in the North Atlantic (Bond et al., 1992). The records of these iceberg flotillas are termed *Heinrich events* (Andrews, 1998), and it is evident that they represent cold stadials of short duration (less than one thousand years). The Younger Dryas event, toward the end of the last glacial period, was an example of a very short-lived stadial that came and went with great rapidity (Anderson, 1997). [*See* Younger Dryas.*]

Figure 1 indicates the pattern of changes recorded in ocean and ice cores, including Dansgaard-Oeschger events in the GRIP ice core, Bond cycles, and Heinrich events recorded in ocean cores. The Heinrich events are labeled YD (Younger Dryas) and H1 to H6. This covers the last glacial cycle. Figure 2, which comes from the latest deep Antarctic core from Vostok, extends the

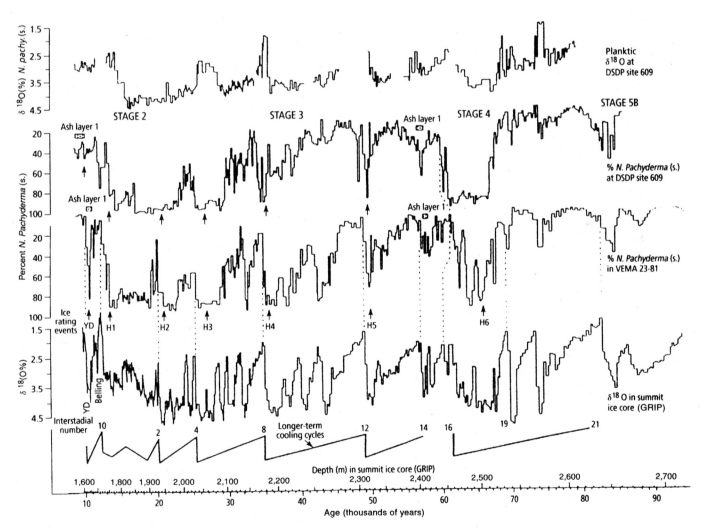

**Climate Change: Abrupt Climate Change. FIGURE 1.** High-frequency climate change events (Dansgaard–Oeschger events, Heinrich events, and Bond cycles) in marine sedimentary records and the GRIP summit ice core (Greenland). The lowermost curve shows the generalized sawtooth structure of the Bond cycles. (After Bond et al., 1993. Reprinted with permission from *Nature*. Copyright 1993. Macmillan Magazines Limited.)

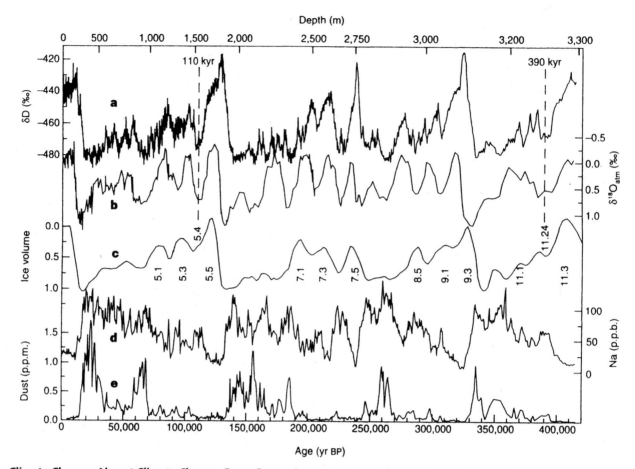

**Climate Change: Abrupt Climate Change. FIGURE 2.** Vostok Time Series and Ice Volume.

Time series (GT4 time scale for ice on the lower axis, with indication of corresponding depths on the top axis and indication of the two fixed points at 110 and 390 kyr) of (a) deuterium profile; (b) $\delta^{18}O_{atm}$ profile obtained combining published data and eighty-one new measurements performed below 2,760 meters; (c) sea water $\delta^{18}O$ (ice volume proxy) and marine isotope stages; (d) sodium profile obtained by combination of published and new measure-

ments with a mean sampling interval of 3–4 meters ($\mu$g per gram or parts per billion); and (e) dust profile (volume of particles measured using a Coulter counter) combining published data and extended below 2,760 meters, every 4 meters on the average (concentrations are expressed in çg per gram or parts per million, assuming that antarctic dust has a density of 2,500 kilograms per cubic meter). (After Petit et al., 1999. Reprinted with permission from *Nature.* Copyright 1999. Macmillan Magazines Limited.)

record of profound variability back even further, to 420,000 years ago (Petit et al., 1999).

**Abrupt Climatic Changes in the Holocene.** Although the Holocene has often been thought of as rather stable, there is growing evidence that abrupt changes of some significance have taken place over the last 10,000 years. One such event has been dated at about 8,200 years ago (Alley et al., 1997), while another occurred at between 3,900 and 3,500 years ago (Anderson et al., 1998). The eolian record of the high plains of North America suggests that dune fields may have flipped repeatedly from a state of vegetated stability to bare mobility during the Holocene in response to severe drought phases (Arbogast, 1996). It is possible that the onset of sharp arid phases was a cause of collapse of civiliza-

tions in Mesopotamia and elsewhere (Weiss et al., 1993). [*See* Civilization and Environment.]

**The Causes of Abrupt Climate Change.** A whole series of possible mechanisms can lead to abrupt climatic changes. These include extreme geographical perturbations, the crossing of thresholds, and the amplification of trends.

The first category could include such phenomena as bolide impacts, which have been implicated in abrupt pre-Pleistocene changes such as those of the late Devonian and end Cretaceous. There is also the possibility that large volcanic eruptions can cause sudden cooling by ejecting large amounts of dust and sulfate aerosols into the atmosphere. It has been suggested, for example, that the great Toba eruption in Indonesia

(c. 73,500 years ago) could have helped to trigger the onset of cooling at the start of the last glacial cycle (Rampino and Self, 1992).

In explaining many of the abrupt changes of the Quaternary, however, most attention has concentrated on the role of triggers and amplifiers that convert a small initial change (brought about, for example, by orbital perturbations) into a much larger shift in temperature and moisture conditions (Adams et al., 1999). The circulation of the North Atlantic probably plays a major role in either triggering or amplifying change, for it seems to be able to flip from one state to another. Triggers that have been postulated are many and include changes in the numbers of melting icebergs entering the ocean, the sudden emptying of large lakes on the margins of ice sheets, and the diversion of meltwater from the Laurentide ice sheet through the Gulf of Saint Lawrence. Such events would alter the temperature and salinity of the ocean and thereby modify the thermohaline circulation.

It is also possible that carbon dioxide and methane concentrations could act as feedbacks in sudden changes. A cooler ocean, for example, would be capable of storing more carbon dioxide, so that more of the gas would be drawn down from the atmosphere, causing further cooling.

Likewise, changes in the albedo (surface reflectivity) of ice, snow, and vegetation could contribute. Cooling could, for instance, produce more extensive snow cover in high latitudes, which, being highly reflective, leads to more heat being reflected back into space, thereby promoting accelerated cooling. Also, ice surges and iceberg armadas could cause the ocean to be more reflective, thereby accelerating cooling. [See Albedo.]

Dust and particulates might also amplify changes. For instance, the drier and colder the world gets, the more desert there is and the higher wind speeds become. More dust gets into the atmosphere, leading to a reduction in convection processes and a further increase in cooling and aridity.

**The Future.** The fact that abrupt climate changes have taken place in the past has implications for the future. As Adams et al. (1999, p. 2) wrote:

> Sudden stepwise instability is . . . a disturbing scenario to be borne in mind when considering the effects that humans might have on the climatic system through adding greenhouse gases. Judging by what we see from the past, conditions might gradually be building up to a "break point" at which a dramatic change in the climate system will occur over just a decade or so, as a result of a seemingly innocuous trigger.

Climate might be an ill-tempered beast that we are poking with sticks.

[*See also* Aerosols; Glaciation; Nuclear Winter; *and* Surprise.]

## BIBLIOGRAPHY

Adams, J., M. Maslin, and E. Thomas. "Sudden Climate Transitions during the Quaternary." *Progress in Physical Geography* 23 (1999), 1–36.

Alley, R. B., P. A. Mayewski, T. Sowers, M. Struiver, K. C. Taylor, and P. V. Clark. "Holocene Climate Instability: A Prominent Widespread Event 8,200 years Ago." *Geology* 25 (1997), 483–486.

Anderson, D. "Younger Dryas Research and Its Implications for Understanding Abrupt Climatic Change." *Progress in Physical Geography* 21 (1997), 230–249.

Anderson, D. E., H. A. Binney, and M. A. Smith. "Evidence for Abrupt Climate Change in Northern Scotland between 3900 and 3500 Calendar Years BP." *Holocene* 7 (1998), 97–103.

Andrews, J. T. "Abrupt Changes (Heinrich Events) in Late Quaternary North Atlantic Marine Environments." *Journal of Quaternary Science* 13 (1998), 3–16.

Arbogast, A. F. "Stratigraphic Evidence for Late-Holocene Aeolian Sand Mobilization and Soil Formation in South-Central Kansas, USA." *Journal of Arid Environments* 34 (1996), 403–414.

Bond, G., et al. "Evidence for Massive Discharges into the North Atlantic Ocean during the Last Glacial Period." *Nature* 360 (1992), 245–249.

Crowley, T. J., and G. R. North. *Paleoclimatology.* New York: Oxford University Press, 1991.

Dansgaard, W., et al. "Evidence for General Instability of Past Climate from a 250 k Yr Ice-Core Record." *Nature* 364 (1993), 218–220.

Frenzel, B., and W. Blundau. "On the Duration of the Interglacial to Glacial Transition at the End of the Eemian Interglacial (Deep Sea Stage 5E): Botanical and Sedimentological Evidence." In *Abrupt Climatic Change*, edited by W. H. Berger and L. D. Labeyrie, pp. 151–162. Dordrecht and Boston: D. Reidel, 1987.

Maslin, M. A., and C. Tzedakis. "Sultry Last Interglacial Gets Sudden Chill." *Eos* 77 (1996), 353–354.

Oppo, D. W., J. F. McManus, and J. L. Cullen. "Abrupt Climate Events 500,000 to 340,000 Years Ago: Evidence from Sub-polar North Atlantic Sediments." *Science* 279 (1998), 1335–1338.

Petit, J. R., et al. "Climate and Atmospheric History of the Past 420,000 Years from the Vostok Ice Core, Antarctica." *Nature* 399 (1999), 429–436.

Rampino, M. R., and S. Self. "Volcanic Winter and Accelerated Glaciation following the Toba Super-eruption." *Nature* 359 (1992), 50–52.

Raymo, M. E., K. Ganley, S. Carter, D. W. Oppo, and J. M. McManus. "Millennial-Scale Climate Instability during the Early Pleistocene Epoch." *Nature* 392 (1998), 699–702.

Van Loon, A. J. "The Meaning of 'Abruptness' in the Geological Past." *Earth-Science Reviews* 45 (1999), 209–214.

Weiss, H., et al. "The Genesis and Collapse of Third Millennium North Mesopotamian Civilization." *Science* 261 (1992), 995–1004.

Woillard, G. "Abrupt End of the Last Interglacial in North-East France." *Nature* 281 (1979), 558–562.

Wood, R. *Reef Evolution.* Oxford: Oxford University Press, 1999.

—ANDREW S. GOUDIE

## Climate Change Detection

Our basic understanding of climate physics, supported by climate model simulations, suggests that the climate is already responding to rising greenhouse gas levels and other anthropogenic pollutants in the atmosphere. Models generally also indicate that this response should, by now, be large enough to be detectable outside the range of natural variability in some, but not all, climate indicators. [*See* Natural Climate Fluctuations]. Detecting evidence of anthropogenic influence in the recent climate record therefore represents a crucial test of our understanding of climate change and our ability to model it correctly.

Climate change detection, much like any other signal-detection exercise, involves the following steps:

1. We identify key indicators in which we have a reasonable chance of observing an unambiguous signal.
2. We simulate how we would expect those indicators to have responded to anthropogenic influence over recent decades, generally (but not always) using a climate model.
3. We compare the simulated signal with the recent observational record and with some estimate of internal climate variability (also usually obtained from a climate model) to establish whether the signal is strong and consistent enough in the observations to be "outside the noise," meaning that there is only a small chance that it could be due to observation error or an internally generated climate fluctuation.
4. Finally, we simulate how these indicators may have responded to other external influences, both natural and anthropogenic, that may have influenced the climate in the recent past, to identify the most plausible explanations of recent climate change that are consistent with the available data.

Various considerations guide the choice of indicator in step 1. Long, consistent records are essential because any signal of anthropogenic climate change is still small relative to year-to-year climate fluctuations. A strong El Niño event, for example, can warm the planet for its duration by more than we expect from thirty years of greenhouse gas emissions. A large volcanic eruption, such as occurred in 1991, can cool it by a similar amount for a year or more. [*See* El Niño–Southern Oscillation.]

Detection studies therefore rely on historical records—mainly air temperatures from meteorological stations and sea surface temperatures from merchant ships—many of which were not originally taken with long-term monitoring in mind. Observing stations move and measurement practices change, creating discontinuities in the record. For example, it is necessary to correct for the use of uninsulated canvas buckets in the late nineteenth and early twentieth centuries, which introduced a spurious cooling of up to half a degree in reported sea surface temperatures (Folland and Parker, 1995). Coverage is very nonuniform, being concentrated in populated regions on land and in the major shipping lanes. Until recently, large parts of the Southern Hemisphere were virtually unobserved.

On the global or continental scale, however, temperature changes over the past century are significantly larger than can be accounted for by observation error or changing coverage, but long-term trends in many regions are much less certain. With the advent of satellite data in the late 1970s, precise global monitoring of surface and tropospheric temperatures is now possible (Spencer and Christy, 1992), but the satellite record is still too short (less than twenty years) for any anthropogenic trend to be distinguishable from natural variability in this record alone (Allen et al., 1994). There is also some debate over the internal consistency of the satellite data sets (Hurrell and Trenberth, 1998; Wentz and Schabel, 1998) but, partly because of the shortness of the record, the significance of the apparent conflict between satellite-based and conventional observations of atmospheric temperature trends has yet to be established.

As well as being the most consistently observed climate variable, the main features of the temperature signal (step 2) are well understood: rising greenhouse gas levels should warm the surface and lower troposphere while cooling the lower stratosphere; land areas are expected to warm faster than oceans because of their lower heat capacity; high-latitude land warming will be enhanced in winter and spring by reduced snow cover while sea surface warming is suppressed over regions of strong heat exchange between the surface and deep ocean (Santer et al., 1996a). Predicting all these aspects of the temperature response is a demanding test of a climate model. Because many key impacts of climate change, such as sea level rise, are directly tied to changing surface temperatures, it is also of practical importance to establish whether the models are simulating these correctly.

The strongest component of the temperature signal is the change in the global mean: near-surface temperatures appear to have risen by 0.3°–0.6°C over the past century, which is a significantly larger rise than we should expect from a simple "red noise" process (Allen et al., 1994). [*See* Global Warming.] It is also larger than the range of century-time-scale trends that climate models predict are likely to occur owing to internal variability—step 3. Unambiguously attributing this change to anthropogenic influence is problematic, however, since the climate system may also be recovering from what may have been a natural cold epoch in the sixteenth to nineteenth centuries, possibly due to reduced solar activity. [*See* Little Ice Age in Europe.] Recent attempts to extend the instrumental record using "proxy" climate observations have confirmed that this twentieth-century warming appears unusual and is unlikely to be solely attributable to changing solar activity. Problems remain, however, with the calibration and interpretation of proxy records.

Comparing the pattern of temperature change over recent decades with the predictions of climate models has several advantages over focusing exclusively on global mean temperature. Models can be used to predict the response to a range of different factors, including, for example, (a) rising greenhouse gases, (b) anthropogenic sulfate and particulate emissions, (c) chlorofluorocarbon (CFC)-induced decline in stratospheric ozone, (d) solar variability, and (e) volcanic activity. The histories of and climate responses to many of these factors are uncertain, but if the patterns of response to factors (b) to (e) are distinct from the pattern of greenhouse-induced climate change, then it may be possible to isolate the component of recent climate change that can be explained only by rising greenhouse gases (step 4). Such attribution is never completely unambiguous because it is impossible in principle to exhaust all possible alternative explanations, but this approach provides a method of evaluating the relative merit of the anthropogenic explanation over a limited number of reasonable alternatives.

Pattern-based detection studies can be categorized into two approaches: pattern correlation techniques (e.g., Barnett and Schlesinger, 1987) and so-called optimal fingerprinting (Hasselmann, 1979; Hasselmann, 1997). Fingerprinting, in principle the most powerful approach, is a variant of linear regression that uses the estimated characteristics of internal climate variability to pick out those aspects of the signal that are most likely to distinguish it from a natural fluctuation. Almost all recent studies report detecting the influence of anthropogenic greenhouse gases on surface and tropospheric temperatures at or above the 95 percent confidence level in the most recent fifty years or so of the record (e.g., Hegerl et al., 1996; Hegerl et al., 1997; Tett et al., 1999). The cooling trend in the lower stratosphere over the forty years for which we have upper-air observations can be attributed to the combined effect of rising greenhouse gases and declining stratospheric ozone and would not be expected if the surface warming were due solely to increased solar activity (Santer et al., 1996b; Tett et al., 1996; Allen and Tett, 1999). The evidence remains ambiguous whether solar variability or volcanic activity has a significant influence on decadal time scales (North and Stevens, 1998), although volcanoes do have a clear short-term impact.

All these studies rely on model simulations of internal climate variability, which are known to be deficient in important respects (Stott and Tett, 1998). Checks for model adequacy can be incorporated into the analysis

(Allen and Tett, 1999), but a natural two-hundred-year climate oscillation that is not represented by the models and whose spatial expression was identical to the pattern of change due to greenhouse gas increase could, inevitably, masquerade as an anthropogenic signal. Such a coincidence may not be as improbable as it sounds. The climate, as a nonlinear system, has a tendency to exhibit *regimes*—that is, preferred modes of behavior that may be excited by various factors, natural or anthropogenic. The pattern of near-surface temperature trends in the Northern Hemisphere, for example, is quite similar to a naturally occurring mode of atmospheric variability (Wallace et al., 1995), although this does not preclude the possibility that human influence may be exciting the interdecadal trend in this mode.

Temperature is only one of a range of variables available for the detection of climate change. Key additional pieces of evidence for an unusual climate change over the twentieth century include the apparent rise in global mean sea level, the widespread retreat of mountain glaciers, and a reduction in Northern Hemisphere snow cover (Santer et al., 1996a). The lack of objective estimates of natural variability in these other indicators, however, complicates their use in quantitative studies.

The Intergovernmental Panel on Climate Change concluded, in 1995, that "the balance of evidence suggests a discernible human influence on global climate." [*See* Intergovernmental Panel on Climate Change.] This cautious wording implies that the "null hypothesis" of no climate change whatsoever can be ruled out at a reasonable ($>$ 95 percent) confidence level. Now that a signal has been detected, the emphasis in detection work has shifted to using the observational record to assess more specific model predictions (Allen et al., 2000). In time, the combined use of multiple variables will be necessary to distinguish between the full range of possible explanations of climate change as well as identifying the influence of key feedback mechanisms. For example, long-term satellite monitoring of the spectrum of infrared radiation emerging at the top of the atmosphere would allow the "spectral signature" of processes such as water vapor feedback to be studied (Goody et al., 1995; Slingo and Webb, 1997), while monitoring of upper-ocean heat content with a combination of altimetry and tomography will allow a more quantitative understanding of feedbacks involving the oceans.

Climate change is a complex process subject to various uncertainties, and different detection results relate to different aspects of the problem. For example, the discrepancy between atmospheric temperature trends and surface temperature trends over the past eighteen years does not necessarily imply that either record is incorrect, but it may be evidence that some of the processes linking surface and atmospheric temperatures are incorrectly simulated in current climate models. It is misleading, therefore, to look for a single smoking gun of anthropogenic climate change: we are trying to understand a process, not solve a crime.

## BIBLIOGRAPHY

Allen, M. R., C. T. Mutlow, G. M. C. Blumberg, J. R. Christy, R. T. McNider, and D. T. Llewellyn-Jones. "Global Change Detection." *Nature* 370 (1994), 24–25.

Allen, M. R., P. A. Stott, J. F. B. Mitchell, R. Schnur, and T. Delworth. "Quantifying Uncertainty in Forecasts of Anthropogenic Climate Change." *Nature* (2000), in press.

Allen, M. R., and S. F. B. Tett. "Checking Internal Consistency in Optimal Fingerprinting." *Climate Dynamics* 15 (1999), 419–434.

Barnett, T. P., and M. E. Schlesinger. "Detecting Changes in Global Climate Induced by Greenhouse Gases." *Journal of Geophysical Research* 92 (1987), 14772–14780.

Folland, C. K., and D. E. Parker. "Correction of Instrumental Biases in Historical Sea Surface Temperature Data." *Quarterly Journal of the Royal Meteorological Society* 121 (1995), 319–367.

Goody, R. M., R. Haskins, W. Abdou, and L. Chen. "Detection of Climate Forcing Using Emission Spectra." *Remote Sensing and Earth Observation* 5 (1995), 22–32.

Hasselmann, K. "On Multifingerprint Detection and Attribution of Anthropogenic Climate Change." *Climate Dynamics* 13 (1997), 601–611.

———. "On the Signal-to-Noise Problem in Atmospheric Response Studies." In *Meteorology of Tropical Oceans*, edited by T. Shawn, pp. 251–259. Royal Meteorological Society, 1979.

Hegerl, G., K. Hasselmann, U. Cubasch, J. F. B. Mitchell, E. Roeckner, R. Voss, and J. Waszkewitz. "On Multifingerprint Detection and Attribution of Greenhouse Gas and Aerosol Forced Climate Change." *Climate Dynamics* 13 (1997), 613–634.

Hegerl, G. C., H. von Storch, K. Hasselmann, B. D. Santer, U. Cubasch, and P. D. Jones. "Detecting Greenhouse Gas-Induced Climate Change with an Optimal Fingerprint Method." *Journal of Climate* 9 (1996), 2281–2306.

Hurrell, J. W., and K. E. Trenberth. "Difficulties in Obtaining Reliable Temperature Trends: Reconciling the Surface and Satellite Microwave Sounding Unit Records." *Journal of Climate* 11 (1998), 945–967.

Mitchell, J. F. B., T. C. Johns, J. M. Gregory, and S. F. B. Tett. "Climate Response to Increasing Levels of Greenhouse Gases and Sulphate Aerosols." *Nature* 376 (1995), 501–504.

North, G. R., and M. J. Stevens. "Detecting Climate Signals in the Surface Temperature Record." *Journal of Climate* 11 (1998), 563–577.

Santer, B. D., K. E. Taylor, T. M. L. Wigley, T. C. Johns, P. D. Jones, D. J. Karoly, J. F. B. Mitchell, A. H. Oort, J. E. Penner, V. Ramaswamy, M. D. Schwarzkopf, R. J. Stouffer, and S. Tett. "A Search for Human Influences on the Thermal Structure of the Atmosphere." *Nature* 382 (1996a), 39–46.

Santer, B. D., T. M. L. Wigley, T. P. Barnett, and E. Anyamba. "Detection of Climate Change and Attribution of Causes." In *Climate Change 1995: The Science of Climate Change*, edited by J. T. Houghton, et al., pp. 411–443. Cambridge: Cambridge University Press, 1996b.

Slingo, A., and Webb, M. J. "The Spectral Signature of Global Warming." *Quarterly Journal of the Royal Meteorological Society* 123 (1997), 293–307.

Spencer, R. W., and J. R. Christy. "Precision and Radiosonde Validation of Satellite Gridpoint Temperature Anomalies. II. A Tropospheric Retrieval and Trends during 1979–90." *Journal of Climate* 5 (1992), 858–866.

Stott, P. A., and S. F. B. Tett. "Scale-Dependent Detection of Climate Change." *Journal of Climate* 11 (1998), 3282–3294.

Tett, S. F. B., J. F. B. Mitchell, D. E. Parker, and M. R. Allen. "Human Influence on the Atmospheric Vertical Temperature Structure: Detection and Observations." *Science* 247 (1996), 1170–1173.

Tett, S. F. B., P. A. Stott, M. R. Allen, W. J. Ingram, and J. F. B. Mitchell. "Causes of Twentieth-Century Temperature Change near the Earth's Surface." *Nature* 399 (1999), 569–572.

Wallace, J. M., Y. Zhang, and L. Bajuk. "Interpretation of Interdecadal Trends in Northern Hemisphere Surface Air Temperature." *Journal of Climate* 9 (1995), 249–259.

Wentz, F. J., and M. Schabel. "Effects of Orbital Decay on Satellite-Derived Lower-Tropospheric Temperature Trend." *Nature* 394 (1998), 661–663.

—MYLES R. ALLEN

# CLIMATE IMPACTS

The accumulation of greenhouse gases in the atmosphere due to fossil fuel combustion, land use change, and other anthropogenic activities may have begun to change the global climate system (Houghton et al., 1996). The changes in atmospheric greenhouse gas (GHG) chemistry and physics are projected to lead to regional and global changes in ambient mean temperature, amount and distribution of precipitation, occurrence of extreme events, and other climate variables. [*See* Global Warming.] The response and feedbacks of the terrestrial biosphere to projected global change are expected to be profound at regional and global scales (Watson et al., 1996). Global change impacts may include shifts in atmospheric, ocean, and land processes such as an increase global mean sea level, greater prospects for severe temperature variability, frequent floods, recurring droughts, changes in soil moisture, and other potentially catastrophic events. Global change will benefit some regions of the world. For example, projected atmospheric warming at high latitudes should lead to improved food and fiber production in Scandinavia, Russia, and Canada. Local, regional, and global costs and benefits associated with climate change have not been fully evaluated, but preliminary analyses suggests global costs may outweigh benefits.

Based on plausible changes in atmospheric GHGs and aerosols, climate system models (e.g., general circulation models, or GCMs) suggest that mean annual global surface temperature will increase by 1°–3.5°C by the year 2100, global mean sea level will rise by 15–95 centimeters, and changes in spatial and temporal patterns of precipitation will occur (Houghton et al., 1996).

The projected average rate of global climate change may be greater than any seen in the past ten thousand years. [*See* Climate Change.] Natural and regional climate variability could substantially differ from the global mean value. For example, these long-term, large-scale changes may interact with natural climate variability of time scales ranging from weeks to decades (e.g., the El Niño–Southern Oscillation [ENSO] phenomenon). Scenarios of future climate change projected by GCMs are uncertain, and analysts continue to search for appropriate data and analytical tools to accurately and reliably conduct biophysical and socioeconomic impact assessments. [*See* Climate Models.]

Human health, ecological systems, and socioeconomic sectors vital to sustainable development—such as freshwater and marine resources, food, fuel, and fiber production, coastal systems, and human settlements—are sensitive to changes in the global climate system (Watson et al., 1996). The terrestrial biosphere is sensitive to the rate and magnitude of not only climate change but also climate variability. Some regions of the world may benefit from climate change while other regions may suffer irreversible declines; for instance, food and fiber production may increase in Russian Siberia but decrease in the agronomic ecosystems of sub-Saharan Africa. Climate change stresses will compound and interact with other anthropogenic impacts, such as acid deposition or stratospheric ozone depletion, influencing the abundance and distribution of terrestrial biosphere resources needed to sustain human economic development and prosperity (Bruce et al., 1996). Global climate change impacts will result in socioeconomic winners and losers, with some regions or countries more vulnerable than others (Strzepek and Smith, 1995). Countries that experience low economic growth rates, rapid increases in human population, and ecological degradation may be particularly vulnerable to global change impacts.

Article 2 of the United Nations (UN) Framework Convention on Climate Change (FCCC) emphasizes the importance of impact assessments in devising appropriate strategies to manage natural ecosystems, resource production, and sustainable economic development in a changing global climate (Bolin, 1998). Impact assessments provide policy makers with estimates of climate change vulnerability for future treaty negotiations (e.g., the Kyoto Protocol) and for national or regional decision making regarding resource management policies and measures to adapt to or mitigate global change. [*See* Kyoto Protocol.] Interpretation of these scientific estimates by policy makers requires careful consideration of the character, magnitude, and rates of global change (Houghton et al., 1996). The 1990, 1992, and 1995 Intergovernmental Panel on Climate Change (IPCC) assessments provided policy makers with summaries of contemporary climate change science (Houghton et al., 1996; Watson et al., 1996). [*See* Framework Convention on Climate Change.]

The objectives of this entry are to (1) introduce science and policy dimensions of contemporary global climate change impact assessments; (2) review assessment tools such as GCMs and analytical methods employed in climate change assessments; (3) present climate change impacts on the Earth's ecosystems; human health; food, fuel, and fiber production; and coastal systems and water resources; and (4) summarize regional impacts on major socioeconomic sectors.

**General Circulation Models.** Researchers have developed a wide range of methods and analytical tools for assessing the impacts of global change on the terrestrial biosphere (Benioff et al., 1996). Plausible scenarios of global change, developed with the assistance of GCMs, define the parameters of biophysical and socioeconomic simulation models that assess climate change impacts at a national and regional scale. These model-driven assessments are illustrative of the potential character and approximate magnitude of impacts that may result from various scenarios of global change (Watson et al., 1996).

GCMs are mathematical simulations of current and future atmosphere, ocean, and land surface processes (Schlesinger and Zhao, 1988; Hansen et al., 1983). These models calculate estimates of present climate at current concentrations of GHGs and aerosols ($1 \times CO_2$ [carbon dioxide]) and future climate resulting from increases in GHGs in the atmosphere ($2 \times CO_2$). The earliest GCM estimates were based on changes in equilibrium climate simulated to result from the equivalent doubling of $CO_2$ concentrations in the atmosphere (Mitchell et al., 1989). Increasingly, GCMs have come to calculate climate changes over time (e.g., decades to centuries), assuming a gradual increase in GHGs (referred to as transient GCM runs). Many impact assessments have used GCMs as the basis for creating scenarios of future climate (Benioff et al., 1996; Watson et al., 1996).

GCMs produce regional estimates of global change that are physically consistent with global atmospheric changes thought likely to occur (Houghton et al., 1996). The GCMs estimate changes in a host of meteorological variables that are consistent with each other within a region and globally. The primary disadvantage of GCMs is that, although they accurately represent global climate processes, their resolution and coarse scale leads to regional climates that are sometimes inaccurate (Boer et al., 1992). In many regions, GCMs significantly under- or overestimate current temperature and precipitation patterns. Moreover, GCMs do not produce output on a geographic and temporal scale of sufficient resolution to adequately drive some biophysical or socioeconomic models (T. M. Smith et al., 1992). Combining GCM output with baseline climate records could mitigate this problem (Benioff et al., 1996).

Assessments of the impacts on various socioeconomic sectors have been completed with some success in countries and regions throughout the world using increasingly comparable and transparent methods and analytical tools (Benioff et al., 1996). Global climate change impacts on agronomic crops, grasslands and livestock, forest systems, freshwater resources, coastal resources, human health, human settlements, and other sectors have been conducted using a wide range of models that link large-scale climatic changes with regional-scale biophysical and socioeconomic responses and feedbacks (Watson et al., 1996; Strzepek and Smith, 1995). A general seven-step method for assessing climate change impacts on socioeconomic sectors was developed by the IPCC. A schematic diagram of this methodology is presented in Figure 1. GCMs that have been historically employed to calculate climate change scenarios for biophysical and socioeconomic simulation models are presented in Table 1. The utility and robust application of simulation models varies by socioeconomic sector, country, and region.

**Climate Change Impacts on Socioeconomic Sectors.** Model simulations predict varied climate change impacts on ecosystems, food and fuel production, water resources, human health, and coastal systems.

*Ecosystems impacts.* Healthy ecosystems and functional biogeochemical cycles not only are the foundation of a sustainable biosphere but also provide many goods and services for human well-being. These goods and services include but are not limited to (1) food, fuel, fiber, fodder, medicines, and energy; (2) storage and processing of key elements such as nitrogen, sulfur, carbon, and other nutrients; (3) freshwater resources such as potable water and fisheries; (4) biological resources and species and genetic diversity; (5) geologic and soil resources; and (6) biogeochemical cycling (e.g., oxygen production, assimilating wastes). Small changes in the Earth's climate system or climate variability may affect the geographic location, structure, function, and reproduction of ecosystem components and the flow of goods and services (T. M. Smith et al., 1992). Climate change impacts will probably be realized through shifts in the rate and magnitude of change in global climate means, as well as the occurrence of extreme events. Terrestrial and aquatic ecosystems may not be able to shift or adapt if climate changes rapidly or if extreme climate change events occur with greater frequency. Increasing atmospheric carbon dioxide will increase the productivity and water use efficiency in some terrestrial plant species but not in others. Indirect climate change impacts may include changes to soil systems or disturbance regimes such as fire, pests, and diseases. Marine and coastal systems may also be affected, but these responses have not been fully quantified. Coastal systems are highly vulnerable to rises in sea level and to increases in extreme events such as hurricanes or typhoons. [*See* Coastlines.]

GCM and biophysical model simulations imply large shifts in the distribution and productivity of vegetation, especially forest and agroecosystems (Dixon et al., 1994). Climate change is projected to occur at a rapid rate relative to the speed at which forest species grow,

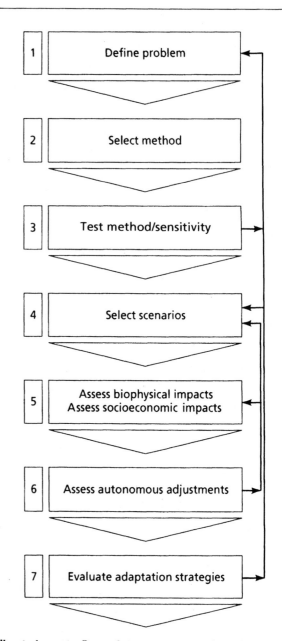

**Climate Impacts. Figure 1.** Intergovernmental Panel on Climate Change's Seven Steps of Conducting Global Change Impact Assessments. (After Carter, 1994).

**Climate Impacts. TABLE 1.** Characteristics of General Circulation Models (GCMs) Employed in Climate Change Impact Assessments

| GCM | WHEN CALCULATED | MODEL RESOLUTION | TEMPERATURE CHANGE (°C) | PRECIPITATION CHANGE (%) |
|-----|-----------------|------------------|-------------------------|--------------------------|
| GISS | 1982 | 7.83° × 10.0° | 4.2 | 11.0 |
| OSU | 1984–1985 | 4° × 5° | 2.8 | 7.8 |
| UKMO | 1986 | 5° × 7.5° | 5.2 | 15.0 |
| GFD3 | 1989 | 2.22° × 3.75° | 3.2 | 10.0 |
| UK89 | 1989 | 2.50° × 3.75° | 2.5 | 9.9 |
| CCCM | 1989 | 3.75° × 3.75° | 3.5 | 3.8 |
| GF01* | 1991 | 4.44° × 7.50° | 3.2[†] | 7.0[†] |

*The model is run for one hundred simulated years assuming a 1 percent annual increase in atmospheric carbon dioxide concentrations.

[†]Transient results are changes between the first decade and the tenth decade of the transient.

SOURCES: Boer et al. (1992); Hansen et al. (1983); Mitchell et al. (1989); Schlesinger and Zhao (1988); Wetherald and Manabe (1988); and Watson et al. (1996).

reproduce, and reestablish themselves (T. M. Smith et al., 1992). For midlatitude forests, an average warming of 1°–3.5°C over the next century is equivalent to a poleward shift of present ecosystems approximately 150–550 kilometers (compared to past migration rates of 4–200 kilometers per century). Thus the composition, distribution, and structure of forest and agroecosystems are projected to change dramatically at national or regional levels (Smith and Shugart, 1993). Globally, the area of low- and midlatitude forests is projected to expand by up to 20 percent, whereas high-latitude forest area could decrease up to 50 percent. Grasslands, savannas, and shrub lands may significantly increase in areal extent, particularly in zones of marginal productivity due to limiting soil factors such as excess salinity or infertility. These shifts in vegetation will be driven by drought-induced declines, changes in extreme ambient temperatures, and disturbance events such as wildfire.

Freshwater aquatic systems, such as lakes, rivers, and wetlands, will be influenced by global change through altered water temperature, flow regimes, extreme events (e.g., flooding and drought), water levels, and thawing of permafrost (Lettenmaier and Sheer, 1991). Major shifts in watershed dynamics may be realized, such as flooding at high latitudes or altitudes, if wide-scale early snowmelt or thawing of permafrost occurs. Lakes and rivers at high latitudes may be greatly influenced by climate change owing to an expansion of the range of warm-water plant and animal species. In contrast, an increase in species loss is predicted in low-latitude aquatic systems owing to loss of habitat. River basins such as the Nile in Africa or the Mississippi in North America will experience greater flow variability, frequency, and duration of extreme events (e.g., floods), which in turn will lead to changes in aquatic habitat, water quality, species diversity, and biological productivity

(Strzepek and Kaczmarek, 1996). The geographic distribution and productivity of wetlands, saltwater marshes, mangrove systems, river deltas, and other sensitive ecosystems will also be at risk to climate change. Changes in all these ecosystems will influence the abundance and distribution of aquatic resources, food supply, tourism, and other goods and services (Markham, 1996).

*Food, fuel, and fiber impacts.* Over the past twenty-five years, irrigation, fertilization, and high-yield crop varieties have allowed global food and fiber production to double. It is not obvious that food, fuel, or fiber production can continue to meet the demands of a growing world population if climate change impacts are substantial in the next century (Bruce et al., 1996). Climate change will interact with other anthropogenic stresses and influence food, fuel, and fiber production on all arable continents, affecting agronomic and forest crop yields and productivity in different ways. Depending on the primary cultivation and management practices, agronomic and forest crop yields may increase or decrease by up to one order of magnitude in response to shifts in regional climate patterns (Rosenzweig and Parry, 1994; Dixon et al., 1994). Direct impacts on agronomic crops will be manifested through shifts in ambient temperature, duration and frequency of precipitation, growing season length, timing of extreme or critical threshold events, and enrichment of the atmosphere with carbon dioxide. Indirect impacts on forest and agroecosystems will result from changes in competing weed species, epidemiology of diseases, occurrence of pests, and soil system processes. Farmers can adapt to climate change by shifting crop-planting dates, changing crop varieties, and altering levels of irrigation and fertilization. Assuming some minimal amount of adaptation by farmers to cope with a changing climate, the net effect of climate change is predicted to be a 5 percent reduction in cereal production worldwide (Strzepeck and Smith, 1995), with larger changes on a regional basis. Cattle, swine, and sheep production may be negatively or positively affected by regional impacts on agroecosystem fodder and feed production (Baker et al., 1992). Extensive or intensive expansion of food, fuel, and fiber production may not be plausible owing to current unsustainable land use practices such as soil erosion, compaction, and salinization.

The effects of climate change on food, fuel, and fiber production and distribution will vary from region to region. In low latitudes, where currently almost 1 billion people are malnourished and lack adequate housing or shelter (Patz and Balbus, 1996), such impacts may affect the socioeconomic conditions of pastoralists and subsistence farmers. In regions where commodity production is well adapted to climate change or where institutions are in place to redistribute in the event of shortfalls, vulnerability to global change is considered to be relatively low. At high and midlatitudes, agroecosystem and forest system yields may increase depending on crop type, growing season, extreme temperature events, and distribution of precipitation (Dixon et al., 1996). In contrast, low-latitude systems, where forest and agronomic crops are at or near maximum temperature tolerance and nonirrigated systems predominate, productivity and quality of yields will generally decline. In China, where a wide variety of ecosystems occur, estimates of climate change impacts on the annual national yield of wheat, rice, and maize range from −20 to +50 percent (Erda et al., 1996).

Marine and freshwater fisheries, a primary source of protein and other goods and services for humans, are sensitive to climate change and are at risk when other stresses such as overfishing, diminishing marshland nursery areas, and aquatic system pollution are considered (Smith and Buddemeier, 1992). Globally, the productivity of marine systems is not predicted to change dramatically owing to shifts in the climate system, although there will be regional winners and losers. Food and fiber production of high-latitude freshwater systems will likely increase with warming ambient temperatures. Impacts on freshwater systems at low latitudes will be manifested at national and regional levels, as institutions for food and fiber production, processing, and distribution shift in response to climate variability (Meisner and Shuter, 1992). In nations and regions strongly affected by the ENSO phenomenon, production and distribution of food, fuel, and fiber in both aquatic and terrestrial systems may be simultaneously influenced by climate change. [See El Niño–Southern Oscillation.]

**Hydrology and water resources impacts.** The impacts of climate change on hydrology and water resources will act concomitantly with other stressors or demands such as a growing world population, technology and infrastructure changes, and socioeconomic patterns. Currently, over 1 billion people to do not have access to potable water, and nearly 3 billion do not have adequate sanitation (Gleick, 1998). The limited technical, financial, and logistical resources of both developed and developing countries will be stressed by adjusting to extremes of water supply and quality. The economic burden of supplying potable water will likely be greatest for developing countries, where human population is rapidly expanding. The burden of extreme events such as flooding will likely be greatest for developed countries at middle and high latitudes.

Periodic and chronic shortages of water, already becoming evident in arid and semiarid regions of the Earth, will likely be exacerbated by climate change (Strzepek and Kaczmarek, 1996). Most GCM scenarios predict increasing frequency and duration of droughts at low- and midlatitude systems (Watson et al., 1996). The water supply of low-latitude developing countries is particularly at risk to climate change because of their dependency on wells and small or isolated reservoirs. Recurrent drought is now a recognized feature of Africa's climate, and nineteen countries on the continent are classified as water scarce by UN agencies (Houghton et al., 1996). At high latitudes, the risk of flooding is predicted to become a more frequent problem, raising concern about the adequacy of the current system of river basin and lake management infrastructure such as dams and levees. [See Dams; and Water Resources Management.]

At high and middle latitudes, one of the more prominent climate impacts will be change in the timing and magnitude of snowmelt (Watson et al., 1996). In basins where winter snow is an important part of the hydrologic cycle, such as the Rocky Mountains of North America or the Himalayas of Central Asia, these changes will dramatically affect watershed dynamics (Strzepek and Smith, 1995). If warming is realized at high and middle latitudes, the ratio of snow to rain will decrease, winter snow accumulation will occur later, and spring snowmelt will occur more rapidly and earlier (Gleick, 1998). Recent impacts associated with early and rapid snowmelt in the Himalayas have been felt in the Ganges River basin of India and Nepal. Similar flooding events have occurred in eastern Europe, South America, and Central Asia.

In recent efforts to evaluate some impacts on water resources, J. B. Smith et al. (1996) analyzed the impacts of climate change on major river basins of the world. River basins such as the Nile, which have low runoff-to-precipitation ratios and high potential evapotranspiration-to-precipitation ratios, are more sensitive to climate change than other basins. In China, annual runoff is predicted to decline more than 10 percent in the Haihe,

Huaihe, and Yellow River basins. The Dnieper River provides about 80 percent of the water supply in Ukraine. Permafrost thawing in Russia, coupled with recurrent droughts in parts of the Dnieper basin, may result in wide swings in future runoff ($-20$ to $+125$ percent). Developing countries (e.g., Egypt and Sudan) and countries with economies in transition (e.g., Kazakhstan, Ukraine) have difficulty in predicting and managing water demand without regarding the added impacts likely to be imposed by future climate change.

**Human health impacts.** Quantifying human health impacts attributable to climate change is just beginning. These impacts depend on many other anthropogenic factors such as human migration, urban water and air pollution, food supply and nutrition, availability of potable water, sanitation, disease vector management, availability of medicines and health care, and myriad other factors. Climate change impacts on human health will be manifested though increases in heat-stress mortality, low-latitude vector-borne diseases, local air pollution (e.g., acid deposition or ozone), and other factors not yet fully evaluated (Patz and Balbus, 1996). At high and middle latitudes, the direct and indirect effects of climate change may not constitute a major hazard to human health. In contrast, impacts on loss of life, community systems, healthcare costs, and loss of human productivity may be greater in low-latitude countries. Quality and distribution of future healthcare delivery systems and basic services such as sanitation will substantially influence the severity of impacts on human health (Watson et al., 1996). [*See* Disease; *and* Human Health.]

Simulation models reveal that the impacts of vector-borne and communicable diseases will shift with a changing global climate. For example, the world geographic distribution of malaria transmission could increase by 15 percent in the next century in response to a $1°$–$3.5°C$ ambient temperature increase (Watson et al., 1996). Malaria infections could increase by 50–80 million cases annually given current plausible scenarios. Increases in nonvector-borne diseases such as salmonellosis, cholera, and giardiasis are also predicted to occur in response to flooding and ambient temperature increases. These impacts have been manifested in some developing countries subject to extreme climate change events such as extended flooding in Bangladesh.

The ENSO phenomenon may provide insights into the impacts of long-term climate change on human health (Nicholls, 1993). Heavy rainfall corresponding to ENSO events has been correlated with outbreaks of mosquito-borne infectious disease. In temperate Australia, rare but severe epidemics of Murray Valley encephalitis occurs after extended rainfall associated with ENSO. Other vector-borne disease outbreaks have been reported in the

United States (equine encephalitis), Australia (Ross River virus), Sri Lanka, Pakistan and Argentina (malaria), and sub-Saharan Africa (West Nile fever). The risk of trypanosomiasis, or African sleeping sickness, increases when tsetse fly habitats expand because of increased precipitation.

In some developing countries, air pollution (e.g., ozone and particulates) and water pollution (e.g., heavy metals) in urban areas already contribute to premature human mortality rates of over 30 percent. The proportion of the developing world's population that resides in urban areas is expected to double by the year 2020. These demographic changes may bring benefits if accompanied by increased access to centralized healthcare systems, potable water, and reliable sanitation. At the same time, climate change effects such as extreme heat events or increases in disease outbreaks could exacerbate problems in rapidly growing urban environments (Patz and Balbus, 1996). Early symptoms of these impacts are being detected in some of the world's large cities such as Cairo, Jakarta, and Mexico City. [*See* Pollution.]

**Coastal system impacts.** Coastal zones are characterized by rich ecosystem diversity and a large proportion of the world's socioeconomic activities (Watson et al., 1996). Beaches, dunes, estuaries, and coastal wetlands are dynamic ecosystems and may adapt naturally to a slowly changing climate system. In contrast, rapid sea level rise, extreme events, or inappropriate coastal zone development will put national and regional coastal systems at risk (Leatherman, 1994). Coastal resources can be protected and managed using dikes, levees, flood walls, and barriers, but increasing population density, short-term infrastructure planning, and inappropriate management of these measures may be risky in a rapidly changing climate (Smith and Lenhart, 1996). The economic impacts of these capital investments by developing countries will be substantial, and some capital cities, such as Banjul, Gambia, may have to be relocated outside high-risk coastal zones (Bruce et al., 1996).

Changes in climate will affect coastal zones through sea level rise, an increase in storm-surge hazards such as hurricanes and typhoon, and possible changes in the frequency or intensity of extreme events (Leatherman, 1994). Coastal systems in many countries face severe sea level rise problems as a consequence of the tectonic or anthropogenically induced subsidence of human settlements. Parts of Bangkok, Thailand, have subsided almost 1 meter over the past one hundred years. Today, approximately 50 million people are annually at risk to flooding from storm surges. Future climate change impacts will exacerbate these problems, affecting ecosystems and human settlements worldwide. Many of the world's largest cities and about one-half of the world's population live in coastal zones and are at risk to climate change impacts (Leatherman, 1991). If a 1-meter increase in sea level is realized, hundreds of millions of people will be affected in low-lying deltas such as the Ganges River of Bangladesh, the Pearl River of China, and the Nile River of Egypt. For some island states, a modest rise in sea level will make storm-surge protection unfeasible, and the retreat or migration of human settlements to other landmasses will be required. [*See* Sea Level.]

Developing countries may be particularly vulnerable to changes in sea level. Inundation of coastal lands (in square kilometers) associated with a 1-meter rise in sea level would affect China (125,000), Bangladesh (25,000), Nigeria (19,000), Malaysia (7,000), Senegal (6,000), and Venezuela (5,000). Over 100 million people in China and Bangladesh may be dislocated by sea level rise. Other impacts on these developing nations may include loss of key wetland systems, coastal zone erosion, and loss of recreational beaches and portions of key cities or ports. These estimates do not include damage associated with storm-surge events.

**Regional Impacts.** As the body of global change science grows, analysts are increasingly able to identify possible local and regional impacts based on GCM scenarios. This summary of regional impacts describes the potential character and approximate magnitude of global change impacts that may result from various scenarios.

*Africa.* Some regions of Africa are particularly vulnerable to global change owing to the fragile nature of ecosystems that are currently subject to multiple stressors, and to an inadequate human and institutional capacity to cope with climate change (Dixon et al., 1996). Terrestrial and aquatic ecosystems, hydrology and water resources, agriculture and food security, coastal systems, human settlements, industry and transportation, and tourism and wildlife are all at risk to climate change. The effects being manifested in sub-Saharan Africa are possible early indicators of long-term impacts in this region. Concomitant widespread poverty, recurrent droughts, inappropriate land use practices, and dependence on rain-fed production of food, fuel, and fiber all exacerbate climate change impacts. The human, institutional, and economic responses required to cope with climate change may exceed the capacity of some countries on this continent. However, local experiences in adapting to a changing climate in Saharan or sub-Saharan Africa may prove valuable in the future.

*Arctic and Antarctic.* GCM scenarios suggest that the biggest changes in temperature will occur at high latitudes, making the Antarctic and Arctic especially vulnerable to climate change impacts. The direct negative influence on humans, however, will be less than on other continents because of low population density. Direct effects will include shifts in marine system ecology, loss of sea and river ice, and permafrost thaw in terrestrial systems (Watson et al., 1996). Indirect effects will include changes in atmospheric, land, and ocean temperatures and precipitation patterns, which could affect climate and sea level processes globally (Houghton et al., 1996). There are major uncertainties regarding the mass balance of Antarctic ice sheets in a changing climate, but small changes could affect the Earth's climate system and the biogeochemistry of the Southern Hemisphere oceans. A large change in the mass balance of polar ice sheets will influence the level of all oceans.

*Asia (high- and midlatitude).* At high and middle latitudes, Asia is a highly populated, rapidly urbanizing region that contains over 20 percent of the world's conifer forest reserves; the productivity and distribution of these resources will shift with a changing climate (Dixon et al., 1994). High-latitude forests will likely be more productive in some regions or countries. The freshwater resources of this region will likely decrease in most major river basins except where permafrost thaws. Estimates of climate change impacts on food crop availability, such as wheat, rice, and maize, vary widely, and regional effects are not easily estimated. In some regions, such as northern Russia, food crop production is predicted to increase. A rise in sea level of 1 meter would severely impact the coastal zones and seaports of China, Russia, and Japan (Watson et al., 1996). Human heat-stress mortality and associated illness may double by the year 2050, particularly in urban areas already affected by other pollutants. The effect of climate change on Asian monsoon and ENSO are major uncertainties, as current simulation models cannot easily estimate impacts based on plausible GCM scenarios (Houghton et al., 1996).

*Asia (low-latitude).* The productivity of high-altitude systems may benefit from climate change, while the impacts on low-altitude systems will be mixed (Erda et al., 1996). The Himalayas play a major role in regional hydrologic cycles such as the monsoon (Houghton et al., 1996). Unfortunately, current simulation models lack adequate resolution to accurately simulate future impacts on precipitation and temperature patterns. Forest and agricultural crops in this region are highly sensitive to changes in temperature and precipitation, and yields may generally decline. Habitat for keystone animal and plant species will also decline. Coastal zones are highly vulnerable to sea level rise, and storm events in this region may increase, particularly in Vietnam, Thailand, the Philippines, Indonesia, and Malaysia (Leatherman, 1994). The incidence and extent of vector-borne diseases affecting humans are expected to increase with a warming climate. Impacts on food security and human health may exacerbate other environmental problems in this diverse region.

*Australasia.* The island states of Australasia will be most dramatically affected by sea level rise and storm events (Erda et al., 1996; Watson et al., 1996). Protection of some small island states from these effects may not be economically or technically feasible in the next century. Australia and New Zealand, with well-developed socioeconomic infrastructures, may be able to adapt to small shifts in the climate system. Fragile ecosystems, hydrology, coastal zones, human settlements, human health, and food security in the Australian exterior are probably vulnerable to a rapidly changing climate. The extensive marginal agricultural systems of Australia's semiarid interior, however, are highly vulnerable to global change impacts.

*Europe.* Climate change impacts in Europe will be significant despite a well-developed infrastructure capable of adapting to climate variability (Dixon, 1997). Ecosystems are fragmented and disturbed, and soils are poor, increasing vulnerability to climate change. Up to 90 percent of alpine glacier mass could disappear by the year 2100. Recurring floods and droughts will affect water-dependent economic sectors, such as agriculture. Permafrost regions and fragile forests of the Nordic countries and Siberia are also at risk, but some models predict greater forest and agroecosystem productivity at high latitudes if warming occurs gradually. Sea level rise may affect the low-altitude countries such as the Netherlands and major river deltas. Negative human health impacts could be realized in urban areas subject to environmental degradation or other stressors.

*Latin America.* Large areas of forest and agroecosystems are projected to be impacted by climate change on this phytogeographically diverse continent, including the Amazon rainforest (Ramos-Mane and Harreau, 1997). Productivity and distribution will be more greatly affected in high-latitude than in low-latitude forests (Figures 2A and 2B). Hydrologic resources are at risk in Central America and in the Andes Piedmont regions, where both recurrent droughts and flooding may occur. The production of grain crops such as wheat, as well as cattle and sheep production, may decline across Central and South America, with the agroecosystems of Brazil, Chile, and Argentina being particularly vulnerable. The geographic distribution of human diseases such as malaria, dengue, and cholera may expand southward and to higher elevations (Patz and Balbus, 1996). Increasing environmental degradation may aggravate current national and international conflicts regarding natural resources and food, fuel, and fiber security.

*North America.* The ecosystems of North America will be subject to both beneficial and harmful impacts of climate change in the next century (Watson et al., 1996). Continental drying will affect ecosystems, hydrology and water resources, and rain-fed agriculture

and livestock (Baker et al., 1992). The distribution and productivity of the long-lived forests of western North America will change with shifts in precipitation and temperature patterns. Coastal zones and associated human settlements such as New York and Miami may be subject to sea level rise and storm events. The technological and economic capability to adapt exists in many parts of North America, but anticipating long-term responses to climate change is complex (Bruce et al., 1996).

**Adapting to a Changing Climate.** The impacts of global climate change on the terrestrial biosphere may be profound, with regional and local impacts being manifested in the next century (Markham, 1996). Local environmental conditions, resource use patterns, economic factors, and human capacity and institutions all influence how future climate changes will affect ecosystems, water resources, food, fuel, and fiber production, human health, coastal systems, and other sectors or systems (Watson et al., 1996). Nonetheless, positive and negative impacts will be felt worldwide, and no country or region will be exempt from the consequences of high-magnitude or rapid climate change. There will likely be socioeconomic winners and losers among nations and regions as these climate change impacts are manifested (Strzepek and Smith, 1995).

Today, many countries and regions do not have coherent or long-term plans or strategies to cope with current climate variability or projected future global change (Watson et al., 1996; Bruce et al., 1996). Fortunately, there are management or technological adaptation options that could be implemented immediately. These options, characterized as "no regrets," provide multiple benefits for sustainable economic development even in the absence of global climate change (Smith and Lenhart, 1996). The challenge for decision makers is to identify sustainable development options and technology deployment opportunities that are sensitive to the climate variability impacts currently being experienced (Bruce et al., 1996). The least expensive adaptation options are generally those that anticipate and plan for a changing climate. Measures and policies that fail to recognize future changes in climate means and frequency of extreme events may prove to be poorly suited to future conditions and unnecessarily expensive. Lessons can be learned from current experiences of coping with climate variability (e.g., El Niño adaptation responses). These experiences can be applied to the design of effective options for adapting to potential changes in the global climate system.

[*See also* Agriculture and Agricultural Land; Ecosystems; Energy; Global Change, *article on* Human Dimensions of Global Change; Integrated Assessment; Natural Climate Fluctuations; Precautionary Principle; Public Policy; Scenarios; Water; *and* Weather Forecasting.]

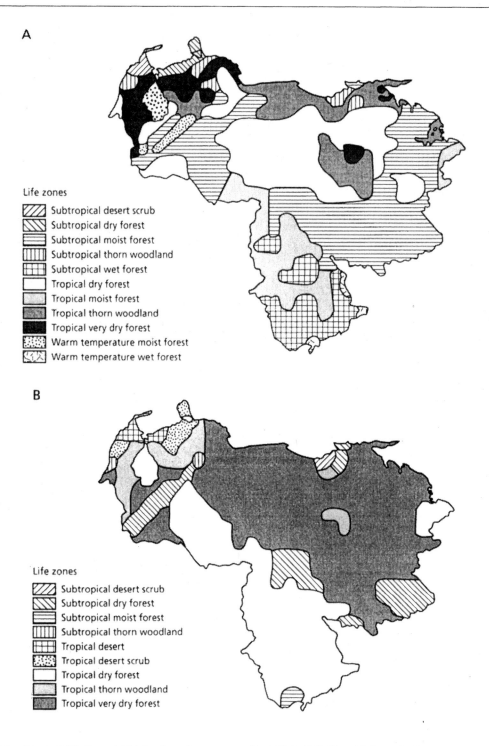

A

Life zones

//// Subtropical desert scrub
\\\\ Subtropical dry forest
≡≡≡ Subtropical moist forest
|||| Subtropical thorn woodland
#### Subtropical wet forest
□□□ Tropical dry forest
▒▒▒ Tropical moist forest
▓▓▓ Tropical thorn woodland
■■■ Tropical very dry forest
∷∷∷ Warm temperature moist forest
⋌⋌⋌ Warm temperature wet forest

B

Life zones

//// Subtropical desert scrub
\\\\ Subtropical dry forest
≡≡≡ Subtropical moist forest
|||| Subtropical thorn woodland
#### Tropical desert
∷∷∷ Tropical desert scrub
□□□ Tropical dry forest
▒▒▒ Tropical thorn woodland
▓▓▓ Tropical very dry forest

**Climate Impacts. FIGURE 2** (A) Current Distribution of Venezuelan Forests. (B) Potential Distribution of Venezuelan Forests on Basis of GCM Scenario.

## BIBLIOGRAPHY

Baker, B. B., R. M. Bourdon, and J. D. Hanson. "FORAGE: A Simulation Model of Grazing Behavior for Beef Cattle." *Ecological Modeling* 60 (1992), 257–279.

Benioff, R., S. Guill, and J. Lee, eds. *Vulnerability and Adaptation Handbook.* Dordrecht: Kluwer, 1996.

Boer, G. J., N. A. McFarlane, and M. Lazare. "Greenhouse Gas-Induced Climate Change Simulated with the CCC Second Generation General Circulation Model." *Journal of Climate* 5 (1992), 1045–1077.

Bolin, B. "The Kyoto Negotiations on Climate Change: A Science Perspective." *Science* 279 (1998), 330–331.

Bruce, J. P., H. Lee, and E. F. Haites. *Climate Change 1995: Economic and Social Dimensions of Climate Change.* New York: Cambridge University Press, 1996.

Carter, T. R., M. L. Parry, H. Harasawa, and S. Nishioka. *IPCC Tech-*

*nical Guidelines for Assessing Climate Change Impacts and Adaptations.* London: University College, Department of Geography, 1994.

Dixon, R. K., ed. "Climate Change Impacts and Response Options in Eastern and Central Europe." *Climatic Change* 36 (1997), 1–232.

Dixon, R. K., S. Brown, R. A. Houghton, M. C. Trexler, A. M. Solomon, and J. Wisniewski. "Carbon Pools and Flux of Global Forest Systems." *Science* 263 (1994), 185–190.

Dixon, R. K., J. A. Perry, E. L. Vanderklein, and H. Hiol. "Vulnerability of Forest Resources to Global Climate Change: Case Study of Cameroon and Ghana." *Climate Research* 6 (1996), 127–133.

Erda, L., et al., eds. "Climate Change Vulnerability and Adaptation in the Asia and the Pacific." *Water, Air and Soil Pollution* 92 (1996), 1–249.

Gleick, P. H. *The World's Water 1998–1999.* Washington, D.C.: Island Press, 1998.

Hansen, J., G. Russell, D. Rind, P. Stone, A. Lacis, S. Lebedeff, R. Ruedy, and L. Travis. "Efficient Three-Dimensional Global Models for Climate Studies: Models I and II." *Monthly Weather Review* 4 (1983), 609–662.

Houghton, J. T., L. G. Meira Fillo, B. A. Callander, N. Harris, A. Kattenberg, and K. Maskell. *Climate Change 1995: The Science of Climate Change.* New York: Cambridge University Press, 1996.

Leatherman, S. P. "Modelling Shore Response to Sea-Level Rise on Sedimentary Coasts." *Progress in Physical Geography* 14 (1991), 447–464.

———. "Rising Sea Levels and Small Island States." *EcoDecision* 11 (1994), 53–54.

Lettenmaier, D. P., and D. P. Sheer. "Climatic Sensitivity of California Water Resources." *Journal of Water Resources Planning and Management* 117 (1991), 108–125.

Markham, A. "Potential Impacts of Climate Change on Ecosystems: A Review of Implications for Policy Makers and Conservation Biologists." *Climate Research* 6 (1996), 179–191.

Meisner, J. D., and B. J. Shuter. "Assessing Potential Effects of Global Climate Change on Tropical Freshwater Fishes." *GeoJournal* 28 (1992), 21–27.

Mitchell, J. F. B., C. A. Senior, and W. J. Ingram. "$CO_2$ and Climate: A Missing Feedback?" *Nature* 341 (1989), 132–134.

Nicholls, N. "El Niño–Southern Oscillation and Vector-Borne Disease." *Lancet* 342 (1993), 1284–1285.

Patz, J. A., and J. M. Balbus. "Methods for Assessing Public Health Vulnerability to Global Climate Change." *Climate Research* 6 (1996), 113–125.

Ramos-Mane, C., and A. Harreau, eds. "Vulnerability and Adaptation to Climate Change in Latin America." *Climate Research* 9 (1997), 1–155.

Rosenzweig, C., and M. L. Parry. "Potential Impact of Climate Change on World Food Supply." *Nature* 367 (1994), 133–138.

Schlesinger, M. E., and Z. C. Zhao. "Seasonal Climate Changes Induced by Doubled $CO_2$ as Simulated by the OSU Atmospheric GCM/Mixed Layer Ocean Model." *Journal of Climate* 2 (1988), 459–495.

Smith, J. B., S. Huq, S. S. Lenhart, L. J. Mata, I. Nemesova, and S. Toure. *Vulnerability and Adaptation to Climate Change: Interim Results from the U.S. Country Studies Program.* Dordrecht: Kluwer, 1996.

Smith, J. B., and S. S. Lenhart. "Climate Change Adaptation Policy Options." *Climate Research* 6 (1996), 193–201.

Smith, S. V., and R. W. Buddemeier. "Global Change and Coral Reef Ecosystems." *Annual Review of Ecological Systems* 23 (1992), 89–118.

Smith, T. M., R. Leemans, G. B. Bonnan, and J. B. Smith. "Modeling the Potential Response of Vegetation to Climate Change." *Advances in Ecological Research* 22 (1992), 13–133.

Smith, T. M., and H. H. Shugart. "The Transient Response of Terrestrial Carbon Storage to a Perturbed Climate." *Nature* 361 (1993), 523–526.

Strzepek, K. M., and Z. Kaczmarek, eds. "Water Resources Vulnerability and Adaptation to Climate Change: The U.S. Country Studies Program." *International Journal of Water Resources Development* 12 (1996), 109–229.

Strzepek, K. M., and J. B. Smith, eds. *As Climate Changes: International Impacts and Implications.* New York: Cambridge University Press, 1995.

Watson, R. T., M. C. Zinyowera, R. H. Moss, and D. J. Dokken. *Climate Change 1995: Impacts, Adaptations and Mitigation of Climate Change.* New York: Cambridge University Press, 1996.

Wetherald, R. T., and S. Manabe. "Cloud Feedback Processes in a GCM." *Journal of Atmospheric Science* 45 (1988), 1397–1415.

—ROBERT K. DIXON

## CLIMATE MODELS

A climate model is an attempt to simulate the many processes that produce climate. It can be considered as comprising a series of equations representing physical, chemical, and biological principles. Any model must be a simplification of the real world. In the case of the climate system, the processes are not fully understood, but they are known to interact with each other to produce feedbacks, so that any solution of the climate model's equations must be an approximation. The approximations made to the laws governing climatic processes can be approached in several ways, leading to a range of different global-scale climate models.

A full general circulation (or global change) model (GCM) is believed to take about twenty-five to thirty person-years to code, and the code requires continual updating as new ideas are implemented and as advances in computer science are accommodated. Most modelers who currently perform experiments with these most complex of models modify only particular components of the models. Computational constraints lead to additional problems. For example, the coarse resolution of global models cannot represent small-scale atmospheric motions (termed *subgrid-scale motions*) such as thundercloud formation. Fine-grid models can be used for weather prediction because the integration time is short. Climate models, however, deal with subgrid-scale processes by parameterization: incorporating a process by representation as a function of other, fully resolved, variables.

The performance of climate models can only be tested against the past or present climate. Usually, when a model is developed an initial objective is to ascertain how well its results compare with the present climate

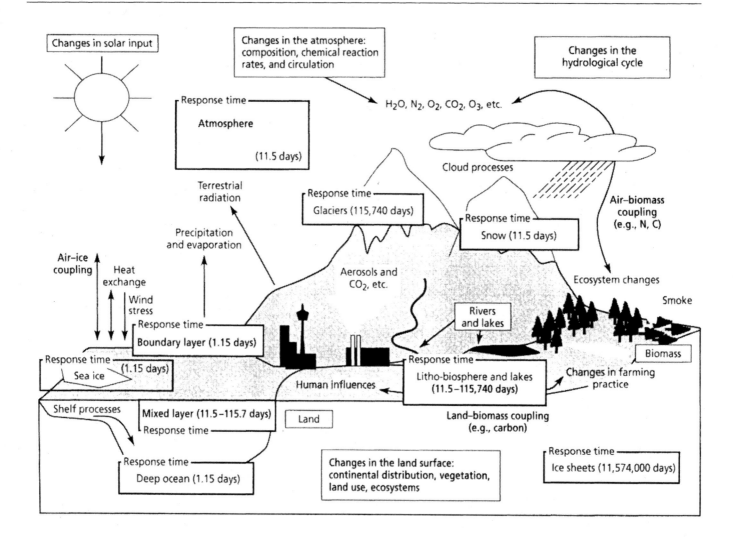

**Climate Models. FIGURE 1.** Components of the climate system (together with their equilibration times—given in seconds) and some of their interactions. (Adapted from McGuffie and Henderson-Sellers, 1996.)

and then to examine the sensitivity of the model to a known forcing such as changes in solar irradiance distribution during a glacial epoch. Although past climates are by no means well known, this comparison provides a very useful step in establishing the validity of the modeling approach. After such tests, the model may be used to gain insight into possible future climates.

The important components to be considered in constructing or understanding a model of the climate system are shown schematically in Figure 1. They include the following.

- *Radiation*—the way in which the input and absorption of solar radiation and the emission of infrared radiation are calculated.
- *Dynamics*—the movement of energy around the globe by winds and ocean currents (specifically from low to high latitudes) and vertical movements (e.g., turbulence, convection, and deep-water formation).
- *Surface processes*—the effects of sea and land ice, snow, and vegetation, and the resultant change in albedo, emissivity, and surface–atmosphere energy and moisture interchanges.
- *Chemistry*—the chemical composition of the atmosphere and ocean and the interactions with other components (e.g., carbon exchanges among ocean, land, and atmosphere).
- *Resolution in both time and space*—the time step of the model and the horizontal and vertical scales resolved.

The relative importance of these processes and the basis for parameterizations employed in their incorporation give rise to a number of different types of climate model. The edges of the "climate modeling pyramid" (Figure 2) represent the basic elements of climate models, and complexity is shown increasing upward. Around the base of the pyramid are the simpler climate models that incorporate only one primary process. Models appearing higher up the pyramid have higher spatial and

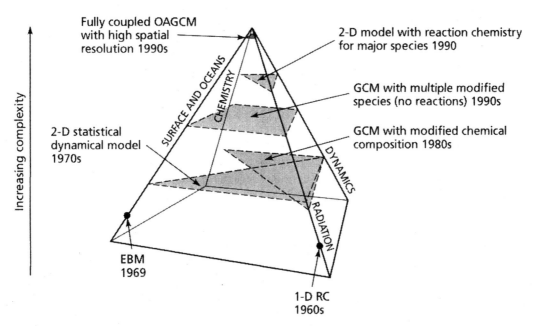

**Climate Models. FIGURE 2. The Climate Modeling Pyramid.**
The position of a model on the pyramid indicates the complexity with which the four primary processes (dynamics, radiation, surface and oceans, and chemistry) interact. Progression up the pyramid leads to greater interaction between each primary process. The vertical axis is not intended to be quantitative. (After McGuffie and Henderson-Sellers, 1996. Copyright John Wiley and Sons. With permission.)

temporal resolutions. There are four basic types of climate models.

- *Energy balance models (EBMs)* are one-dimensional models predicting the variation of the surface (strictly the sea level) temperature with latitude. Simplified relationships are used to calculate the terms contributing to the energy balance in each latitude zone.
- *One-dimensional radiative–convective (RC)* models compute the vertical (usually globally averaged) temperature profile by explicit modeling of radiative processes and a "convective adjustment" that reestablishes a predetermined lapse rate.
- *Two-dimensional statistical dynamical (SD)* models deal explicitly with surface processes and dynamics in a zonally averaged framework and have a vertically resolved atmosphere. These models have been the starting point for the incorporation of reaction chemistry in global models.
- *Global circulation models (GCMs)* incorporate the three-dimensional nature of the atmosphere and the ocean. In these models an attempt is made to represent most climatic processes. They solve fundamental equations including *conservation of energy* (the first law of thermodynamics), *conservation of momentum* (Newton's second law of motion), *conservation of mass* (the continuity equation), and the *Ideal gas law* (an approximation to the equation of state of the atmosphere).

The dynamics of the ocean are governed by the amount of radiation that is available at the surface and by the wind stresses imposed by the atmosphere. [*See* Ocean Dynamics.] The flow of ocean currents is also constrained by the positions and shapes of the continents. Ocean GCMs calculate the temporal evolution of oceanic variables (velocity, temperature, and salinity) on a three-dimensional grid of points spanning the global ocean domain. The formation of oceanic deep water is closely coupled to the formation and growth of sea ice, so that ocean dynamics demand effective inclusion of sea ice dynamics and thermodynamics. As well as acting as a thermal "flywheel" for the climate system, the ocean also plays a central role in the carbon cycle, absorbing approximately half of the carbon that is released into the atmosphere every year. [*See* Ocean–Atmosphere Coupling.]

Energy balance models incorporate the cryosphere, which is the frozen water of the Earth, as if it were a thin, high-albedo covering of the Earth's surface. The solution of the governing equation of an EBM for various values of incoming solar radiation can be used to illustrate the fundamental characteristic of all nonlinear systems. A slow decrease in the solar constant from initial conditions for the present day means a gradual decrease in temperature until a point is reached where a runaway feedback loop causes total glaciation and a rapid drop in temperature. When the solar radiation is then increased, the process is not immediately reversed; the temperature follows a different route until, at a value of the solar radiation greater than that of the present day, temperatures rise again: the modeled climate exhibits hysteresis.

The plant life of the planet has only recently been included in climate models. The first approach has been to delineate geographic boundaries of biomes (vegetation groups characterized by similar species) by using simple predictors available from GCMs such as temperature, precipitation, and possibly sunshine (photosynthetically active radiation). Currently, attempts are being made to evaluate these methods using reconstructions of vegetation distributions during past epochs. These interactive biosphere models are still in their infancy but may provide useful predictions of future responses of the biosphere, including the issue of possible future carbon dioxide fertilization of the biosphere.

As climate models are readily described in terms of a hierarchy (Figure 2), it is often assumed that the simpler models were the first to be developed, with the more complex GCMs being developed most recently. This is not the case. The first atmospheric general circulation climate models were derived directly from numerical models of the atmosphere designed for short-term weather forecasting in the early 1960s, concurrently with the first RC models. On the other hand, energy balance climate models, as they are currently recognized, were not described in the literature until 1969, and the first discussion of two-dimensional SD models took place in 1970.

EBMs have been used extensively in the study of paleoclimates. One common experiment is to introduce the effect of orbital (Milankovitch) variations and altered continental configurations on an EBM. Geochemical data suggest a positive correlation between carbon dioxide and temperature over the last 540 million years. A notable exception to this is the Late Ordovician glaciation (around 440 million years ago), which occurred at a time when the atmospheric carbon dioxide content is believed to have been around fifteen times as high as it is today. Reduced solar luminosity compensated in part for this, but experiments with EBMs have shown that the configuration of the continents was such that the ice sheets could coexist with high carbon dioxide levels. With the benefit of the insight gained from such EBM studies, it has been possible to go on to perform more detailed calculations with a GCM, which have confirmed the hypothesis based on the EBMs. The advantage of EBMs in this kind of problem is the ease with which many different experiments can be performed. Since information on boundary conditions for model simulations is poor, the simple model offers the chance to test a range of situations before embarking on expensive calculations with a GCM.

The climate system is very complex, possessing a very large number of modes of variability. Any attempt to model such a multifaceted system must neglect many aspects, or represent them incompletely. This process, termed *parameterization*, can take many forms. The simplest form is the null parameterization, where a process, or a group of processes, is ignored. The second level, climatological specification, is a form of parameterization that has been used widely in most types of models. In the 1970s, it was not uncommon to specify oceanic temperatures (with a seasonal variation), and in some of these models the clouds were also specified. In climate sensitivity experiments, it is important to recognize all such prescriptions because feedback features of the climate system have been suppressed. The third type of parameterization is only slightly less hazardous. Here, processes are parameterized by relating them to present-day observations: the constants or functions describing the relationship between variables are "tuned" to obtain agreement.

The most advanced parameterizations have a theoretical justification. For instance, in some two-dimensional zonally averaged dynamical models the fluxes of heat and momentum are parameterized via baroclinic theory (in which the eddy fluxes are related to the latitudinal temperature gradient). The parameterization of radiative transfer for clear skies in RCs and GCMs is another example. All that needs to be known is the vertical variation of temperature and humidity. Unfortunately, these parameterizations can lead to problems of uneven weighting because another process of equal importance cannot be adequately treated. In the case of heat and momentum transport by eddies, the contribution to these fluxes from stationary waves forced primarily by the topography and the land/ocean thermal contrast cannot be so easily considered. In radiation schemes, the parameterization of cloudy sky processes is not as advanced as for clear skies.

The interactions between processes in any model of the climate are crucially important. The relative importance of processes and the way that different processes interlink is a strong function of the time scale being modeled. Whether a system is likely to be sensitive to the parameterization used for a particular process often depends upon the response time of that feature compared with other "interactive" features. It is unhelpful to invoke a highly complex, or exceedingly simplistic, parameterization if it has been constructed for a time scale different from that of the other processes and linkages in the model.

The time scale of response is crucially important to all aspects of climate modeling. This time scale is variously referred to as the equilibration time, the response time, or the adjustment time. It is a measure of the time the subsystem takes to reequilibrate following a perturbation. A short equilibration time scale indicates that the subsystem responds very quickly to disturbances and can therefore be viewed as being quasi-instantaneously equilibrated with an adjacent subsystem that possesses a much longer equilibration time. It is common to express equilibration times in terms of the time, called the *e*-folding time, it would take a system or subsystem to reduce an imposed displacement to 1/*e*th of the displaced value. For example, a bowl of soup removed from a microwave oven will reequilibrate with the room environment with an *e*-folding time depending upon the difference in temperature of the bowl contents and the room, and the size and shape of the bowl. A smaller temperature difference, a smaller bowl, or a larger surface-to-volume ratio of the container will result in relatively shorter *e*-folding times. Large *e*-folding times are possessed by subsystems that respond only very slowly. The response time is generally assessed in terms of the thermal response time. Figure 1 shows the equilibration times for many of the subsystems of the climate system. The longest equilibration times are those for the deep ocean, glaciers, and ice sheets ($10^{10}$–$10^{12}$ seconds), while the remaining elements of the climate system have equilibration times nearer $10^5$–$10^7$ seconds.

Climate models have the potential to provide information about future and past climates that have applicability to a wide range of human activities. For example, the search for "safe" disposal sites for nuclear waste materials has involved not only geologic evaluation of possible sites, but also climatological assessment using climate model predictions. The model-predicted threat of a "nuclear winter" following a nuclear war is believed by some commentators to have contributed to the deescalation in weapons development and holdings in the late 1980s.

Mineral exploration companies have examined the results of past climate predictions to try to infer the likely locations of mineral deposits. Evaporites (such as rock salt and gypsum) form near the boundaries of oceans and in shallow basins that are subject to frequent flooding and desiccation. The levels of salinity that are reached in the basins determine the nature of the evaporite deposits. Regions that are amenable to evaporite formation would be indicated in a GCM by regions where the total precipitation minus the total evaporation is negative. In one study, an evaporite basin model consisting of a saline "slab" of water with fixed depth and salinity was run offline (the model was run using output from the GCM as forcing data, rather than being coupled to the GCM). This offers no feedback to the GCM climate, but if the feedback is assumed to be small, this is probably acceptable. The evaporite basin model was used to determine whether evaporite could potentially form rather than to model the process of deposi-

tion. The evaporite model (which computed evaporation based on GCM forcing) was forced with GCM-simulated climate at all model grid points. From such a simulation of the Triassic (roughly 225 million years ago), the locations of major known evaporite deposits in North America, South America (around 120° west latitude), Arabia (10° south, 20° west) and the Western Tethys, Central Atlantic region (10° north, 50°–80° west) are found.

The most widely known current application of climate model predictions is the evaluation of the impacts of greenhouse warming. [*See* Greenhouse Effect; *and* Global Warming.] Evaluation of future climate in this context is prompting the evolution of climate models to take into account the ramifications of climate change for human health; food supply; population policies; national economies; international trade and relations; policy formulation and attendant political processes; national sovereignties; human rights; and international, interethnic, and intergenerational equity. [*See* Climate Change.] In considering the estimated damage due to current emissions of greenhouse gases, the arguments for action are now extending beyond "no regrets" measures: those whose benefits, such as reduced energy costs and reduced emissions of conventional pollutants, equal or exceed their cost. Decisions to be taken in the near future will necessarily have to be taken under great uncertainty. These decisions may be very sensitive to the level at which atmospheric concentrations are ultimately stabilized and to the environmental effects on ecosystems: the net productivity of the oceans, the response of trees and forests to carbon dioxide fertilization and climate change, and methane production by thawing tundra. It is clear that evaluation of this large suite of possible responses to the threat of future climate change must incorporate many issues that are beyond the scope of current climate models but will, perhaps, be encompassed by future models.

[*See also* Atmosphere Dynamics; Modeling of Natural Systems; Natural Climate Fluctuations; *and* Younger Dryas.]

## BIBLIOGRAPHY

Dickinson, R. E. "Climate Sensitivity." In *Issues in Atmospheric and Oceanic Modeling, Part A: Climate Dynamics*, edited by S. Manabe, pp. 99–129. Advances in Geophysics, vol. 28. New York: Academic Press, 1985. This is one of the earliest detailed discussions of climate sensitivity.

Genetic Algorithm for Rule-Set Production (GARP). *The Physical Basis of Climate and Climate Modelling.* GARP Publication Series No. 16, WMO/ICSU. Geneva, 1975. A landmark description of the climate system.

Houghton, J. T., ed. *The Global Climate.* Cambridge: Cambridge University Press, 1984. A full and comprehensive textbook on the global climate.

Houghton, J. T., G. J. Jenkins, and J. J. Ephraums. *Climate Change: The IPCC Scientific Assessment.* Cambridge: Cambridge University Press, 1990. The first IPCC science report describing greenhouse warming.

Houghton, J. T., L. G. Meira Filho, B. A. Callander, N. Harris, A. Kattenberg, and K. Maskell, eds. *Climate Change 1995: The Science of Climate Change Contribution of Working Group I of the Intergovernmental Panel on Climate Change.* Cambridge: Cambridge University Press, 1996. The second IPCC science report identifying the human contribution to greenhouse warming.

Howe, W., and A. Henderson-Sellers, eds. *Assessing Climate Change: Results from the Model Evaluation Consortium for Climate Assessment.* London: Gordon and Breach, 1997. A detailed description of the first industry-funded climate modeling of greenhouse warming and its impact.

McGuffie, K., and A. Henderson-Sellers. *A Climate Modelling Primer.* 2d ed. Chichester, U.K.: Wiley, 1997.

Peixoto, J. P., and A. H. Oort. *Physics of Climate.* New York: American Institute of Physics, 1992. A classic observationally based analysis and description of the climate system.

Schlesinger, M. E., ed. *Physically Based Modelling of Climate and Climatic Change: Parts 1 and 2.* NATO ASI Series C: No. 243. Dordrecht: Kluwer, 1988. A multiauthored graduate-level text on climate modeling.

Trenberth, K. E. *Coupled Climate System Modelling.* Cambridge: Cambridge University Press, 1992. A textbook derived from a graduate-level summer school on the climate system.

—A. HENDERSON-SELLERS

**CLIMATE OSCILLATIONS.** *See* Natural Climate Fluctuations.

**CLIMATE PREDICTION.** *See* Weather Forecasting.

## CLIMATE RECONSTRUCTION

Our knowledge of climate history prior to the availability of instrumental records comes from two main sources: (1) historical observations, where information on past climatic conditions is contained either in direct accounts of the weather or in records that typically relate to agriculture and/or human health; and (2) the paleoclimatic record, where information is preserved by a wide array of physical, chemical, and biological proxies. [*See* Paleoclimate.]

**Historical Records.** The barometer and thermometer were both invented somewhat before the middle of the seventeenth century. During the 1650s, distribution of the first standardized instruments in Europe made possible the collection of accurate instrumental data. The longest continuous time series of weather conditions for a single location is the Central England Temperature Record, which dates back to 1659. Early entries in this record of monthly temperatures for rural sites are largely based on weather diaries, but the record gradually integrated instrumental data as thermometers became more widely available. Several other long in-

strumental records exist, but the data are almost entirely restricted to western Europe. Instrumental records with sufficient coverage to obtain a global-scale picture of climatic conditions only go back to about 1850. For information on climate prior to that time, one must turn to historical records and observations.

The earliest written records of climate come from Egypt, where stone inscriptions detailing Nile flood levels date back to about 5,000 BP. In China, the oldest preserved records date to the Shang dynasty (3,700–3,100 BP) and appear to describe conditions warmer than those of the present. However, for the vast majority of areas on the continents and oceans, historical observations are only available, if at all, for a few hundred years at most.

A great deal of the most useful historical information has come from written accounts of harvest yields and crop prices. An excellent example can be found in records of the harvest dates of grapes in the wine-making regions of northern and central France. A continuous time series of wine harvest dates from this region extends back to at least 1484. While longer-term variations in this record have to be viewed with caution, since tastes in wine have changed over the years and, with them, the date of harvesting, the year-to-year fluctuations in harvest dates have been shown to correlate closely with changes in summer growing temperature, and therefore provide an accurate measure of the weather.

Interpretations of the climatic significance of agricultural records must be made with caution and with an understanding of agricultural principles. Successful cereal growing in northwestern Europe, for example, requires a fine balance between adequate moisture and warmth. Wheat requires a higher summer temperature than barley or oats and grows best when annual rainfall is less than about 900 millimeters. Thus, information on the type of grain as well as the yield is needed when trying to reconstruct summer growing conditions. In parts of the world where summers are typically hotter, such as in North America and India, drought plays a much more important role in the quality of harvests. Hot, dry summers in these regions are bad for most crops, while wetter, cooler summers produce the best harvests.

From feudal Europe there is plenty of evidence to suggest that the eleventh and twelfth centuries were marked by a relatively benign climate, a period often termed the *Medieval Climatic Optimum.* [*See* Medieval Climatic Optimum.] By about 1300, however, it is clear from a variety of historical records that there was a marked deterioration in the weather. Two exceptionally severe winters gripped Europe in 1303 and 1306, followed in the years 1314–1317 by a string of very wet and cool summers. The devastating harvest failures of these years are probably the greatest weather-related disaster ever to hit Europe. In London, wheat prices in the early summer of 1316 were as much as eight times higher than in late 1313, while historical accounts indicate that starvation and pestilence were rampant. Cooler, wet conditions affected grain storage as well, and moldy grain resulted in widespread outbreaks of the dangerous skin condition known as *erysipelas* (Saint Anthony's Fire). In the North Atlantic, the increased incidence of winter storms and the encroachment of sea ice prevented the resupply from Norway of the Viking colonies on Greenland and led to their abandonment by the mid-1300s.

As the discussion here illustrates, shifts in weather and climate have affected agriculture and human health throughout recorded history. Although caution must be used in reading too much into historical records because of the complex interactions between weather, economic activity, and social policies, historical data are a key source of meteorological information in the preinstrumental era.

**Paleoclimatic Records.** A vast wealth of information about past changes in the Earth's climate and environment is recorded in measures as diverse as tree rings, ice, and rocks and sediments that have formed in both marine and terrestrial settings. To use this resource effectively, one must be able to extract information on past climatic conditions from the physical, chemical, and biological components preserved in the geologic record. Such information comes almost entirely from indirect measures, or what are known as *proxies*. Fortunately, a diverse array of proxy types are available to choose from.

In the mid-nineteenth century, observations of the wide distribution of physical deposits and features associated with modern alpine glaciers and ice sheets led to the realization that the Earth had experienced a series of major ice ages in the recent geologic past. [*See* Glaciation.] Some of the best evidence for past glacial activity comes from the debris deposited by moving glaciers. *Drift* is the name commonly used for all types of glacial deposit, while the term *till* is generally applied to deposits laid down directly by ice. The bulk of the material in tills is usually of clay, silt, or sand sizes, but pebbles and large boulders (known as erratics) may be present. A record of changes in the position of glacial fronts is generally derived from ridgelike features known as *moraines* (Figure 1) left at the leading edge of a glacier after its retreat. The problem with the record of moraines and with many other glacial deposits is that they are commonly incomplete, with more recent ice advances obliterating evidence of earlier, less extensive advances. Although careful stratigraphic studies will sometimes reveal evidence of weathering horizons and buried soils (paleosols) that effectively differentiate between successive ice episodes, the problems of accurately dating glacial deposits are acute. Radiocarbon dating of organic material in soils that have developed on or within glacial deposits is most frequently used, but this method can only be used for the last forty thousand

**Climate Reconstruction. Figure 1.** Glacial Moraine at Terminus of Modern Alpine Glacier near Saas-Fee, Switzerland.

Moraines can be used to identify former positions of glacier fronts and usually mark the maximum extent of the ice.

years. Indeed, as recently as the early 1960s it was commonly believed that the most recent series of ice ages in the Pleistocene (the last 1.6 million years) totaled only four in number.

The reconstruction of past climates took a major step forward beginning with work on deep-sea sediments that began in earnest in the 1950s. Typical sediment accumulating on the ocean floor consists of a mixture of the fossil remains of marine microorganisms and terrigenous particles (mainly clays and silts) derived from the erosion and transport of materials from the continents. The biologically produced deposits can be divided into two major categories, depending on the mineralogy of the skeletal material: carbonate "oozes" are dominated by the fossil remains of foraminifera and calcareous nannoplankton, common plankton groups in the ocean that produce shells or skeletal elements made of calcite ($CaCO_3$), while siliceous oozes consist primarily of diatom and radiolarian remains composed of opal ($SiO_2$-$nH_2O$). Species of each of these major plank-

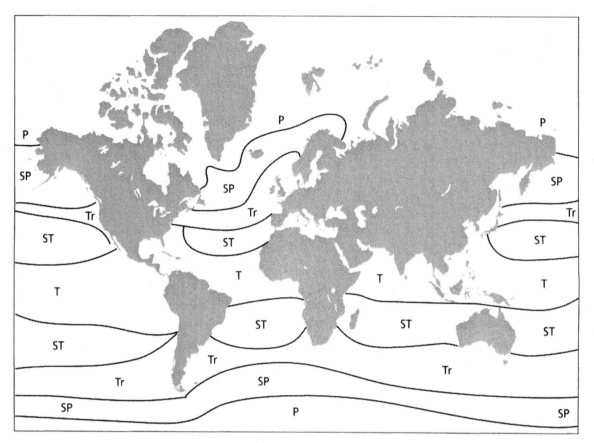

**Climate Reconstruction. Figure 2.** Modern Biogeographic Distributions of Planktic Foraminiferal Assemblages in the Ocean.

The latitudinal distribution pattern of this and other microplankton groups is largely controlled by sea surface temperature (SST). Those groups that leave a record of their distributions in the form of shells in the underlying sediment can be used to infer past changes in SST and other water properties through study of the expansions and contractions of their geographic ranges. (T = Tropical assemblage; ST = Subtropical assemblage; Tr = Transitional Assemblage; SP = Subpolar assemblage; P = Polar assemblage.) (Adapted from Bé and Tolderlund, 1971. With permission of Cambridge University Press.)

ton groups live in the surface waters of the ocean and are distributed according to the physical and chemical characteristics of the upper water column. Many of the taxa show distinct biogeographies, with a distribution by latitude that is clearly related to sea surface temperature (SST) patterns (Figure 2). Past expansions or contractions of the biogeographic ranges of temperature-sensitive species (Figure 3), which are traced and mapped through study of sediment cores, can be used to infer past changes in SST. More sophisticated approaches for estimating SST utilize empirical regression equations (usually known as transfer functions) that relate modern species abundances to historical SST data. Depending on the specific plankton group and ocean, most of the statistical techniques can estimate SST to within 1°–2°C. Problems with this approach can arise from imperfect preservation of the microfossil assemblage and from no-analogue cases in which past assemblages cannot be matched to the modern calibration assemblages. The latter problem becomes exaggerated

as one moves back in time and faunal assemblages change naturally through evolution.

The measurement of stable oxygen and carbon isotope ratios (oxygen-18/oxygen-16 and carbon-13/carbon-12) in the $CaCO_3$ shells of foraminifera from marine sediments has proven to be one of the most powerful tools in the arsenal of the paleoclimatologist. Oxygen isotope records can be used to estimate past water temperatures, the size of ice sheets, and local salinity variations, while carbon isotope records can be used to reconstruct deep-ocean circulation patterns and provide estimates of past oceanic nutrient levels. Isotope ratios for each system are expressed in delta ($\delta$) notation, which relates the isotopic composition of a sample to the known isotopic ratio of a standard. For oxygen isotopes,

$$\delta^{18}O = [(^{18}O/^{16}O_{sample} - {}^{18}O/^{16}O_{standard}) / (^{18}O/^{16}O_{standard})] \times 1{,}000,$$

with units of parts per thousand. Carbon isotope values, $\delta^{13}C$, are reported in the same form.

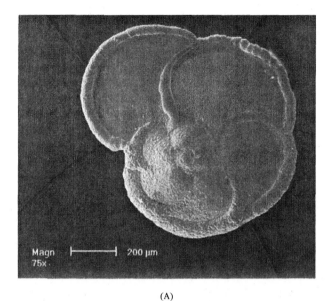

(A)

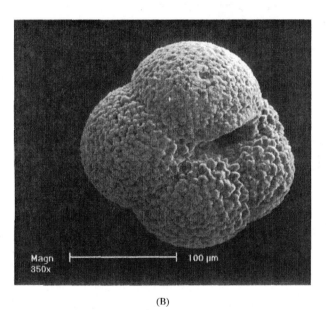

(B)

**Climate Reconstruction. FIGURE 3.** Electron Photomicrographs of Two Common Planktic Foraminifera That Inhabit the Modern Ocean.

(A) *Globorotalia menardii*, a common species found in tropical waters and sediments (magnification = 75×); and (B) *Neogloboquadrina pachyderma*, a dominant component of polar and subpolar foraminiferal assemblages (magnification = 350×).

The use of stable oxygen isotope measurements as a proxy for paleotemperatures began in 1947 with Harold Urey's observation that the oxygen-18/oxygen-16 ratio in calcite should vary as a function of the temperature in which the mineral precipitated. Experimental studies with mollusks confirmed this prediction and led to development of the first paleotemperature equation with the form:

$$T = 16.5 - 4.3 \times (\delta^{18}O_{calcite} - \delta^{18}O_{water})$$
$$+ 0.14 \times (\delta^{18}O_{calcite} - \delta^{18}O_{water})^2,$$

where $T$ and $\delta^{18}O_{water}$ are the temperature (°C) and oxygen isotope value of the water in which the organism lived, and $\delta^{18}O_{calcite}$ is the measured oxygen isotope value of the calcite.

Measurements of $\delta^{18}O$ in foraminifera, corals, and mollusks have now been used routinely for forty years as a paleoclimatic tool and provide most of what is currently known about the evolution of the Earth's climate over the past 100 million years. When used according to the paleotemperature equation, measurements of $\delta^{18}O$ can give reliable paleotemperature estimates. In the absence of changes in $\delta^{18}O_{water}$, a 0.23 parts per thousand increase in $\delta^{18}O_{calcite}$ corresponds to a 1°C decrease in temperature. However, $\delta^{18}O_{water}$ is not uniform and can vary substantially from place to place in the ocean and through time. Water molecules containing the heavier oxygen-18 isotope tend to evaporate less readily than those containing the lighter oxygen-16 isotope. Conversely, fractionation during condensation concentrates the heavier isotope in the precipitation (rain or snow), further enriching the clouds (water vapor) in the lighter $H_2^{16}O$ molecules. In the present hydrologic cycle, water evaporated from the ocean returns there relatively quickly through precipitation, summer melting, and continental runoff. During glacial periods, however, the oceans become enriched in the heavier $H_2^{18}O$ molecules at the expense of $H_2^{16}O$ molecules, which are preferentially stored in the ice sheets. During the recent Pleistocene ice ages, this effect has been shown to be the largest signal recorded at most locations in $\delta^{18}O_{calcite}$ records generated from deep-sea cores. The significance of this observation is that the waxing and waning of ice sheets creates a whole-ocean $\delta^{18}O$ signal that provides a framework for making global correlations based on $\delta^{18}O$ variations preserved in the sediment record. Pleistocene $\delta^{18}O$ stratigraphy is now well established, with chronological control coming from radiocarbon dating, the identification of key paleomagnetic reversals (for example, the Brunhes/Matuyama boundary at 0.78 million years BP), and the presence of marker horizons such as ash layers that can be traced to known volcanic eruptions. [*See* Dating Methods.] Unlike the incomplete continental record, the deep-sea $\delta^{18}O$ record indicates that there have been at least twenty major glaciations throughout the Pleistocene.

As with oxygen isotopes, variations in $\delta^{13}C$ result from fractionation of carbon-13 and carbon-12 during physical or chemical processes. In the ocean, the $\delta^{13}C$ of total carbon dioxide largely reflects the utilization and regeneration of organic matter. Phytoplankton in the surface ocean preferentially utilize isotopically light carbon to build soft tissue during photosynthesis. Remineralization of this soft tissue as the remains settle through the water column after death puts nutrients and light carbon back into the system at depth, a process often referred to as the *biological pump*. In the deep ocean, waters that have been longer removed from the surface typically exhibit higher concentrations of nutrients and lighter $\delta^{13}C$ because of their steady accumulation through time. The fractionation of $\delta^{13}C$ between the surface and deep ocean, and from place to place within the deep ocean, is reliably recorded in the shells of certain planktic and benthic (bottom-dwelling) foraminifera, as well as other calcareous organisms, and has proven to be a useful tool for reconstructing oceanic productivity and deep-water circulation patterns.

The oceanic distributions of certain trace elements reflect processes of climatic interest, and many of these elements substitute readily for calcium in calcareous skeletal materials. Variations in cadmium/calcium (Cd/Ca) ratios in benthic foraminifera have provided important evidence for reorganization of deep-water circulation patterns during Pleistocene glacial–interglacial cycles. In the present ocean, cadmium concentrations are similar to those of the nutrient phosphorus, which increases with water mass age because of the steady rain of organic matter to the deep sea and its subsequent remineralization. Changes in Cd/Ca concentration in foraminiferal shells therefore reflect changes in the age of the water mass, in a manner generally consistent with $\delta^{13}C$. Foraminiferal Cd/Ca and $\delta^{13}C$ measured together in sediment cores from the Atlantic were among the first data to suggest that production rates of North Atlantic Deep Water were greatly reduced during the last glacial maximum.

In corals, the ratio of strontium to calcium (Sr/Ca) is strongly temperature dependent. Measurements of Sr/Ca in corals accurately capture the annual temperature cycle and allow for collection of high-resolution temperature time series in the tropical ocean. A relatively new paleothermometer is the ratio of magnesium to calcium (Mg/Ca). The value of this tool has been demonstrated with benthic marine ostracodes (small crustaceans that secrete a calcified bivalved carapace) and is currently being explored for corals and planktic foraminifera.

Marine sediments also preserve a wide variety of information on other climatic processes. Atmospheric circulation patterns can be assessed from study of the eolian, or wind-transported, material that reaches the deep-sea floor. Wind direction and strength can be assessed through the character and pathways of particles transported offshore by the wind, the size and shape of the grains, and the occurrence of wind-derived higher plant materials and freshwater diatoms from dried lake deposits. In addition to eolian transport, mineral grains can reach the ocean floor by direct riverine input and by an assortment of gravity-driven processes that move materials downslope. The relative abundance of clay minerals and their degree of degradation and alteration provide information on the nature and amount of weathering on the adjacent land surface, while the presence of pollen and terrestrial organic matter can yield clues to continental climates recorded by vegetation types.

While the size of most terrigenous particles entering the ocean is limited by the carrying capacity of wind and water, larger and more poorly sorted mixtures of materials can be delivered to the deep sea by the melting of icebergs that have "calved" and drifted away from land-based ice sheets. In the North Atlantic, sediments deposited during the last glacial are punctuated by distinct layers of ice-rafted materials that can be traced in a broad swath across the ocean. These layers, termed *Heinrich layers* after their discoverer, appear to reflect changes in the dynamics of the Laurentide ice sheet, the large ice mass centered over North America, during the last glacial.

Just as the turn to deep-sea records in the 1950s led to major advances in understanding climate history, the systematic study of ice-core records that began in the 1970s again led to a great advance in knowledge. [*See* Aerosols; *and* Climate Change.] Ice sheets in Greenland and Antarctica have accumulated in place over the last several hundred thousand years and preserve a highly resolved record of climatic conditions; the age of the ice can be resolved in cores to within a few years by methods that include annual layer counting in the uppermost section. The $\delta^{18}O$ record obtained from the ice is very similar to that obtained from deep-sea cores, although $\delta^{18}O$ variations in ice are interpreted in terms of air temperature at the site at which the ice accumulated. One of the most remarkable features of ice-core $\delta^{18}O$ records from Greenland is evidence for large and rapid fluctuations in temperature during the last glacial that are matched by variations in SST proxies in high-deposition-rate North Atlantic sediments. These changes appear to reflect massive reorganizations of the ocean–atmosphere system on time scales of a century or less, and were previously unknown. Another remarkable discovery has come from studies of gas inclusions in ice cores, which provide direct measurements of past atmospheric compositions (Figure 4). Measurements from both Greenland and antarctic ice cores have shown that atmospheric carbon dioxide concentrations during the last glacial were some 75–80 parts per million less than preindustrial levels of about 280 parts per million, and that concentrations during the previous interglacial were as high or higher than preindustrial values. Concentrations of methane, another important greenhouse gas, were also reduced by half during the last glacial. [*See* Methane.]

In addition to their unique record of atmospheric gas

Ice cores provide a wonderful storehouse of stratigraphical and paleoenvironmental information back to the last interglacial period. In the 1990s, two especially deep cores were drilled into the Greenland Ice Sheet. The Greenland Ice-Core Project (GRIP), a European enterprise, reached bedrock at 3,029 meters in 1992, while the North American Greenland Ice Sheet Project 2 (GISP2), located only 30 kilometers away from the GRIP site, reached bedrock at a depth of 3,053 meters in 1993. The GRIP core is at the modern ice divide, whereas the GISP2 core is to its west. The GISP2 core is of large diameter—13.2 centimeters—and a total of forty-two types of measurements comprise the GISP2 research effort. They include the gas content of air bubbles trapped in the ice, concentrations of major ions, cosmogenic isotopes, stable isotopes, dust content, electrical conductivity, and physical properties like crystal characteristics. However, deep cores are not without their problems, of which the disturbances of their lower portions by deformation is perhaps the most serious.

—ANDREW S. GOUDIE

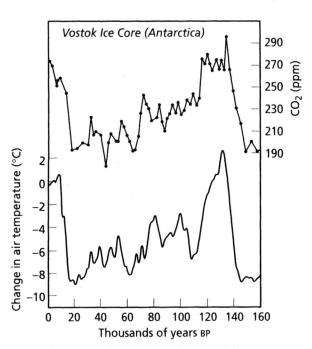

**Climate Reconstruction.** FIGURE 4. Ice-Core Data Spanning the Last Full Glacial–Interglacial Cycle (160,000 Years) from the Famous Vostok Site in Antarctica.

The top panel shows past changes in atmospheric carbon dioxide content determined from measurements of the gas composition of air samples preserved in bubbles trapped in the ice. The bottom panel compares these measurements with a record of air temperature from the same site derived from analysis of the deuterium isotope content of the ice itself. Atmospheric carbon dioxide levels were up to 80 parts per million lower than preindustrial values during the last glacial, but were higher than preindustrial levels during the last interglacial (120,000–140,000 BP). (Data from Barnola et al., 1987; Jouzel et al., 1987.)

levels, the polar ice sheets contain a record of particle fallout from the atmosphere, mainly in the form of dust concentrations and ions such as chloride and calcium. Chloride is derived from sea spray and provides an index of storminess and enhanced wind circulation, while calcium levels are dependent on the source of dust, especially as derived from glacial loess. High dust concentrations in glacial ice from both Greenland and Antarctica indicate generally more arid conditions during glacial maxima.

Although the record of glacial deposits on land is a bit fragmentary, there is an abundance of other well-developed climate proxies for the continents. Pollen grains and spores that accumulate in lakes and bogs provide a record of past vegetation in an area (Figure 5). Abundance data need to be corrected for differences in pollen and spore production between plant taxa, as well as for differing rates of preservation and dispersal, but generally yield reliable estimates of the composition of the surrounding vegetation. Studies of modern pollen rain and modern vegetation patterns indicate that there is a good spatial correspondence between them. Such studies have led to synoptic mapping of inferred vegetation distributions at times in the past, as well as time-series reconstructions from cores at single sites. While such vegetation reconstructions are relatively straightforward, the climatic inferences that can be drawn are often qualitative, with changes interpreted in terms of wetter/drier or warmer/colder conditions.

Variations in tree-ring width from year to year have long been recognized as a major source of chronological and climatic information. The mean width of a ring in any one tree is a function of many variables, including the tree species, its age, the availability of stored nutrients in the tree and surrounding soil, and a host of

**Climate Reconstruction. Figure 5.**
Photomicrograph of a Pine Pollen Grain
Showing Its Distinctive Morphology.

Pollen, such as this specimen of *Pinus ponderosa*, preserved in sediments of lakes and bogs can be used to reconstruct vegetation patterns in the surrounding drainage basin (magnification = 1,500×).

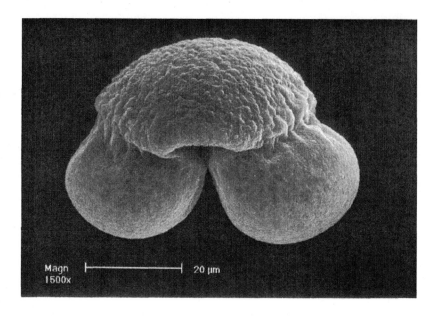

climatic factors, including temperature, precipitation, and availability of sunlight. The extraction of useful climatic data usually follows calibration studies where local relationships are sought between tree-ring data and instrumental or historical data over the period in which they overlap. Tree-ring methodologies have become very sophisticated and involve careful cross-dating, standardization, and analysis techniques. Although climatic information has most often been gleaned from tree-ring width, variations in tree-ring density and in the isotopic composition of the wood itself have proven useful as well.

Animal distributions on the continents, represented both by modern and fossil forms, also provide clues to past climatic states. Insects, for example, occupy virtually every type of terrestrial environment on Earth and produce chitinous exoskeletons that preserve well in sedimentary deposits. Most paleoclimatic work involving insects has focused on fossil beetles. At the larger end of the size spectrum, reptiles are useful because of their ectothermic (cold-blooded) nature. The presence of fossil alligators in sedimentary rocks of early Eocene age (55–50 million years old) on Ellesmere Island, west of Greenland, is a spectacular example of evidence for high-latitude warmth in this earlier geologic epoch.

As one moves back in time, the effects of plate tectonics and the physical rearrangement of the Earth's surface must be considered when one interprets paleoclimate data. The increased difficulty of establishing accurate time correlations inhibits the ability to develop global snapshots of climatic conditions, and biases introduced by physical and chemical postdepositional processes can obliterate or obscure original geochemical signals. Despite such problems, the geologic record contains a rich and diverse archive of information that

offers our only means of deciphering the long history of climatic evolution on Earth.

[*See also* Earth History; Greenhouse Effect; *and* Land Surface Processes.]

### BIBLIOGRAPHY

Barnola, J. M., et al. "Vostok Ice Core Provides 160,000-Year Record of Atmospheric Carbon Dioxide." *Nature* 329 (1987), 408–414.

Bé, A. W. H., and D. S. Tolderlund. "Distribution and Ecology of Living Planktonic Foraminifera in the Surface Waters of the Atlantic and Indian Oceans." In *The Micropalaeontology of Oceans*, edited by B. M. Funnell and W. R. Riedel, pp. 105–149. Cambridge: Cambridge University Press, 1971.

Bradley, R. S. *Quaternary Paleoclimatology: Methods of Paleoclimatic Reconstruction.* Boston: Allen and Unwin, 1985. Contains a thorough and balanced summary of methods used in reconstructing past climates from proxy data series.

Burroughs, W. J. *Does the Weather Really Matter?: The Social Implications of Climate Change.* Cambridge: Cambridge University Press, 1997. An up-to-date analysis of the debate on climate change, with abundant examples of historical climate data.

CLIMAP Project Members. "Seasonal Reconstructions of the Earth's Surface at the Last Glacial Maximum." *Geological Society of America, Map and Chart Series MC-36* (1981), 1–18. Presents detailed maps showing reconstructions of continental and sea surface conditions at the peak of the last ice age, 18,000 BP.

Crowley, T. J., and G. R. North. *Paleoclimatology.* New York and Oxford: Oxford University Press, 1991. An excellent introduction to paleoclimatic data and observations, and the use of models in assessing our understanding of past climate change.

Flint, R. F. *Glacial and Quaternary Geology.* New York: Wiley, 1971. A bit dated but still a classic introduction to the study of glacial deposits on the continents.

Grove, J. M. *The Little Ice Age.* New York: Methuen, 1988. A summary of historical and instrumental data that span the period of cooler climatic conditions between roughly 1450 and 1870 CE.

Imbrie, J., and K. P. Imbrie. *Ice Ages: Solving the Mystery.* Short

Hills, N.J.: Enslow, 1979. An entertaining account of the history of ice-age theory with discussion of evidence that led to an understanding of the role of orbital variations in climate change.

Jouzel, J., et al. "Vostok Ice Core: A Continuous Isotope Temperature Record over the Last Climatic Cycle (160,000 Years)." *Nature* 329 (1987), 402–408.

Lamb, H. H. *Climate, History and the Modern World.* London: Methuen, 1982. Presents an exhaustive compilation and discussion of historical climate records.

Le Roy Ladurie, E. *Times of Feast, Times of Famine: A History of Climate since the Year 1000.* New York: Doubleday, 1971. Translated from the original French, this scholarly work by an economic historian contains detailed analyses of historical records.

Tuchman, B.W. *A Distant Mirror: The Calamitous 14th Century.* New York: Knopf, 1978. Contains an account of the effects of extreme weather events on European social structure in the fourteenth century.

—LARRY PETERSON

# CLOUDS

[*This entry consists of two articles:* Clouds and Atmospheric Chemistry *and* Clouds in the Climate System. *The first article describes the role of clouds in atmospheric chemical processes as well as the effect of chemical processes on the formation of clouds. The second article describes the role of clouds in the climate system, focusing on cloud-radiation feedback.*]

## Clouds and Atmospheric Chemistry

Clouds and atmospheric chemistry have a strong influence on the constituent and energy budget of the atmosphere. The magnitude of this influence on a global scale is difficult to assess, however, in part because of our poor understanding of their complicated interactions. For instance, atmospheric chemical processes may modify the physical and chemical properties of *cloud condensation nuclei*, which strongly control the microphysical and radiative properties of clouds. In turn, clouds influence the atmospheric chemistry by changing the spatial distribution of atmospheric chemicals, by providing sites for liquid phase and heterogeneous reactions, and by altering the actinic flux that drives photochemical activities. Such intricate interactions and feedback mechanisms are important factors that control the global climate and hydrological cycle.

**Effects of Atmospheric Chemistry on Cloud Formation.** Atmospheric chemistry affects cloud formation mainly through its influence on the chemical composition and size distribution of aerosol particles, which in turn have a strong control on the microphysical properties, precipitation efficiency, and lifetime of clouds. [*See* Aerosols; *and* Atmospheric Chemistry.] Because of these effects, anthropogenic emissions and

their chemical reactions may inadvertently affect the formation of clouds and precipitation. Aircraft contrails and ship tracts are the two most visible examples.

*Cloud condensation nuclei.* Cloud condensation nuclei (CCN) are aerosol particles that are capable of initiating the formation of cloud drops and raindrops. The ability of an aerosol particle to activate into a cloud drop is strongly related to the mass and composition of its water-soluble component. Sources of the water-soluble component include the gaseous and aqueous chemical conversions of their precursors (such as the oxidation of $SO_2$ and $NO_x$ into sulfate and nitrate), as well as a direct emission from the Earth's surface (such as ammonia or sea salt). Atmospheric chemistry may therefore influence the microphysical properties of clouds through its effect on CCN.

*Aircraft contrails.* Aircraft contrails are condensation trail clouds formed from jet aircraft engine exhaust. When jet engine exhaust cools down by mixing with the cold ambient air, water vapor and chemically produced condensable gases (mainly sulfuric acid) or chemi-ions contained in it combine to form new particles through the nucleation processes. Further cooling causes cloud drops to form through the condensation of water vapor onto these particles. Some of the cloud drops may also form on soot particles produced in the engine exhaust. When the temperature at cruising altitudes is low enough, these cloud drops freeze instantaneously to form ice particles. Contrails can reflect incoming solar radiation (and thus cool the surface), but at the same time absorb and reradiate terrestrial radiation back toward the surface (and thus warm the surface). Some estimations showed that, if a contrail persists over a twenty-four-hour period, localized surface cooling of a few degrees Celsius might occur. However, this result should vary with contrail thickness and persistency, latitude, time of the year, and other atmospheric conditions.

*Ship tracks.* Ship tracks are cloud trails caused by the entrainment of ship exhaust into marine stratus or stratocumulus clouds. They appear as lines of enhanced brightness in infrared or visible satellite imageries (see Figures 4.1 and 4.2 in Cotton and Pielke, 1995). As a result of processes similar to those occurring with aircraft exhausts, highly concentrated CCN may form in ship exhausts, leading to enhanced droplet concentrations and decreased drop sizes in the ship tracks as compared with the unperturbed clouds. A decrease in drop sizes tends to reduce drizzle formation, such that ship track clouds typically have higher liquid water content than the surrounding clouds. Also, drizzle reduction may impede the *cloud scavenging* effect and thus an increase in CCN lifetimes (see Effects of Clouds on Aerosol, below).

*Polar stratospheric clouds.* Polar stratospheric clouds (PSCs) are ice clouds that form in the polar stratosphere during winter. The formation and consequences of PSCs are strongly related to atmospheric chemistry. Solutes can lower the water activity and thus the freezing ability of haze and cloud drops, an effect particularly strong for stratospheric aerosol particles whose composition by weight is about 75 percent $H_2SO_4$ and 25 percent $H_2O$. So, while the formation of cloud drops in the stratosphere is limited by the low humidity, the formation of ice clouds is hindered by a high solute concentration. During polar nights, however, the stratospheric temperature may fall below about $-80°C$ so that significant condensation of nitric acid (a product of various gas-phase and heterogeneous chemical reactions) may occur to form ternary solutions of $H_2SO_4/HNO_3/H_2O$. A further decrease in temperature then initiates the freezing of these solution drops and thus the PSC formation. The initial growth of PSC particles involves a continuous uptake of $H_2O$ and $HNO_3$ to form crystalline nitric acid trihydrate (NAT) at ice-undersaturated conditions. Growth of PSC particles by condensation of water vapor alone can occur when, with further cooling, the humidity reaches ice saturation.

**Effects of Clouds on Atmospheric Chemistry.** Clouds affect atmospheric chemical processes in many different ways. They can redistribute atmospheric chemicals by processes such as cloud convection and scavenging. Clouds also provide sites for chemical reactions that would not otherwise occur in a clear sky, thus influencing the production and destruction of many trace gases and aerosol particles. Clouds can also reflect or diffuse the solar actinic flux that drives many photochemical reactions. Even high in the stratosphere, clouds reveal their significance by acting as one of the key factors in ozone hole formation.

*Cloud and precipitation scavenging.* Cloud and precipitation scavenging is the removal of gaseous and particulate materials from the atmosphere by cloud and precipitation processes. This may occur both inside and below a cloud, referred to as *in-cloud scavenging* (also called *rainout*) and *below-cloud scavenging* (also called *washout*). The fraction of chemicals being scavenged depends on their solubility, the microphysical processes through which they are incorporated into cloud particles, and whether chemical reactions occur inside or on the surface of cloud particles. Cloud and precipitation scavenging is an important factor in determining the lifetimes of many atmospheric trace chemicals.

*Convective transport.* Convective transport is the transport of trace chemicals in the atmosphere by means of cloud convection. Updrafts and downdrafts associated with clouds play a major role in the vertical redistribution of atmospheric trace chemicals. Pollutants produced near the Earth's surface can reach the upper troposphere rather quickly with the assistance of strong updrafts. Trace chemicals (such as $O_3$) that are richer aloft may also be transported downward by cloud turbulence and precipitation-induced downdrafts. The proportion of chemical being transported by clouds varies from species to species because of chemical fractionation caused by the scavenging processes.

*Acid rain.* Acid rain is rain having a pH lower than 5.6 (the pH of water in equilibrium with typical atmospheric $CO_2$ concentration at standard atmospheric conditions). Acid rain may affect the chemistry of soils, the forests that grow on those soils, and the lakes into which the rainwater drains, and it may cause health risk and environmental impacts. Other forms of acidic atmospheric particles include *acid haze*, *acid fog*, and *acid snow*. The elevated acidity in these particles is caused mainly by the presence of sulfuric and nitric acids, often attributed to anthropogenic sources. One of the main sources of these acids is contained in the cloud condensation nuclei. An additional portion is produced as a result of chemical reactions occurring in cloud or fog water. Many studies have demonstrated that, on a global scale, the production of sulfate through aqueous-phase reactions in clouds exceeds that through clear-sky reactions. Some of the acids are removed from the air in rainfall, known as *acid rain*. [*See* Acid Rain and Acid Deposition.]

*Effects of clouds on gas-phase chemistry.* In addition to transport and removal, clouds have many other effects on gas-phase chemistry in the atmosphere. The selective dissolution of trace gases in cloud water alters the chemical proportions and thus the overall chemical reactions in the gas phase. Also, cloud particles provide a large surface area for heterogeneous chemical reactions. Highly reflective clouds may significantly influence (decrease in the shadow side and increase in reflection sides) the actinic flux that drives the gas-phase chemical reactions. Clouds may also affect atmospheric chemistry through thunderstorm activities because several per cent of the total (natural and anthropogenic) $NO_x$ in the troposphere is produced by lightning.

*Effects of clouds on aerosol.* Because aqueous-phase oxidation of sulfur and nitrogen species adds to the nonvolatile solute mass in cloud drops, larger and more efficient CCNs can be produced if they are not removed through rainout. However, the increase in particle size also leads to a greater deposition velocity. Clouds are thus effective in shortening the lifetimes of atmospheric chemicals not only directly by in-cloud chemical conversion and wet deposition but also indirectly by enhancing dry deposition. Furthermore, the effect of cloud reflection on the actinic flux can indirectly influence aerosol particle nucleation, a process strongly associated with photochemical reactions.

*Ozone hole.* A rapid and accelerating annual decrease in atmospheric ozone during early spring over the Antarctic region has been found to allow an increase of ultraviolet radiation to reach the Earth's surface, which in turn may cause damages to ecosystems and human health. Scientists from the British Antarctic Survey first recognized the ozone hole phenomenon as recently as 1985. [*See* Ozone.] It was then realized that the injection of anthropogenic chlorofluorocarbons into the stratosphere has produced chlorine radicals that act as catalysts in the cyclic process of ozone destruction. One of the key factors in antarctic ozone depletion is the formation during polar nights of stratospheric clouds that allow heterogeneous reactions to occur on the frozen cloud particles. The formation and freezing out of nitric acid in polar stratospheric clouds removes nitrogen-containing molecules that might otherwise be capable of interfering with the ozone loss cycle. These clouds also provide surface sites for the conversion of chlorine reservoir species (e.g., $HCl$ and $ClONO_2$) into chlorine compounds (e.g., $Cl_2$) that are easily broken down into chlorine atoms by the weak ultraviolet light of the polar spring. Chlorine atoms then initiate the catalytic ozone destruction cycles. Owing in part to the exceptional stability of the Antarctic vortex, which limits its mixing with the surrounding air, the ozone depletion trend is most prominent in the Antarctic region, although recent studies have shown a similar but weaker trend in the Arctic.

**Indirect Effects of Atmospheric Chemistry on Climate through Clouds.** Atmospheric chemistry can influence the global energy balance directly through the effects of trace gases and aerosols on solar and terrestrial radiation. The influence may also occur indirectly (and possibly even more strongly) through its effects on clouds. Therefore, any effect that atmospheric chemistry exerts on clouds (or vice versa) eventually influences climate. The *Twomey effect* and *dimethyl sulfide-cloud-climate interactions* are the two most popular hypotheses pertaining to such indirect effects.

*Twomey effect.* Sean Twomey in 1974 first pointed out that increasing anthropogenic pollution would result in greater CCN and cloud drop number concentrations, which in turn enhance the reflectance (albedo) of clouds. [*See* Albedo.] Such clouds tend to cool the atmosphere by reflecting more sunlight. Twomey and others demonstrated theoretically that this cooling effect, called the Twomey effect, has a magnitude comparable, but opposite in sign, to that of greenhouse warming owing to anthropogenic trace gases.

*Dimethyl sulfide-cloud-climate hypothesis.* It has been proposed that biogenic production of dimethyl sulfide (DMS) in oceans and their subsequent transformation into CCNs act as a negative feedback mechanism to counteract global warming. One of the main sources of natural condensation nuclei (CN) over the ocean is believed to be DMS excreted by phytoplankton and lib-

erated into the atmosphere, where it is photochemically oxidized to form sulfate particles. Through mechanisms still not clearly identified to date, these CNs subsequently transform into CCNs to affect cloud formation. It is speculated that the productivity of DMS is temperature dependent such that any increase in ocean surface temperature (an expected outcome of greenhouse warming) would enhance CCN production, which in turn would increase cloud reflectance (albedo) and cause a cooling effect similar to the Twomey effect. Thus, DMS is sometimes referred to as the antigreenhouse gas.

[*See also* Atmosphere Dynamics; Atmosphere Structure and Evolution; Climate Change; Global Warming; Greenhouse Effect; Hydrologic Cycle; *and* Water Vapor.]

### BIBLIOGRAPHY

Charlson, R. J., J. Lovelock, M. O. Andreae, and S. Warren. "Oceanic Phytoplankton, Atmospheric Sulfur, Cloud Albedo and Climate." *Nature* 326 (1987), 655–661.

Coakley, J. A., R. L. Bernstein, and P. A. Durkee. "Effect of Ship-Stack Effluents on Cloud Reflectivity." *Science* 237 (1987), 1020–1022.

Cotton, W. R., and R. A. Pielke. *Human Impacts on Weather and Climate.* New York: Cambridge University Press, 1995. A useful and readable introduction to the anthropogenic factors.

Hobbs, P. V., ed. *Aerosol–Cloud–Climate Interactions.* San Diego: Academic Press, 1993. A useful general survey of current opinion on the subject.

Porch, W. M., C-Y. J. Kao, and R. G. Kelley, Jr. "Ship Trails and Ship-Induced Cloud Dynamics." *Atmospheric Environment* 24A, 1051–1059.

Pruppacher, H. R., and J. D. Klett. *Microphysics of Clouds and Precipitation.* 2d ed. Boston: Kluwer, 1997. An in-depth discussion on the processes occurring in clouds.

Seinfeld, J. H., and S. N. Pandis. *Atmospheric Chemistry and Physics: From Air Pollution to Climate Change.* New York: Wiley, 1998. An in-depth discussion on atmospheric chemistry and climate change.

Twomey, S. "Pollution and the Planetary Albedo." *Atmospheric Environment* 8 (1974), 1251–1256.

—JEN-PING CHEN

## Clouds in the Climate System

Clouds appear in all sorts of shapes and forms. Some clouds are small in horizontal extent, whereas others fill the entire sky. Some clouds are shallow, while others are deep and produce threatening weather. Although individual clouds can be small, they are most often organized into widespread cloud systems. This organization, apparent when the Earth is viewed from space, traces out the large-scale patterns of the major wind streams and giant eddies of the atmosphere (Rossow, 1994).

The excess radiation absorbed by the Earth at low latitudes and emitted back to space at higher latitudes establishes an equator-to-pole heating gradient. This gradient, combined with the dynamic characteristics of fluid motion on a rotating Earth, establishes the currents in the oceans and the eddies in the atmosphere that transport the excess heat at low latitudes poleward to higher latitudes. Clouds are not merely passive tracers of the air motions associated with this transport; they exert a profound influence on our weather and climate and play an important role in regulating climate. A vital component of the hydrologic cycle of the planet, clouds produce precipitation, which is the critical source of fresh water used by humankind. [*See* Hydrologic Cycle.] Clouds also exert a dominant influence on the energy budget of the planet principally through their influence on radiative processes. Specifically, clouds can cool the Earth by reflecting incident sunlight back to space and simultaneously warm the Earth by absorbing upwelling thermal radiation and then radiating lower amounts of this infrared radiation to space. While we are currently able to measure the total incoming and outgoing radiation at the top of the atmosphere (TOA), we do not measure how much heating occurs within the atmosphere versus how much occurs at the surface. This creates some uncertainty in establishing the relative roles of the atmosphere and oceans in transporting heat poleward.

**Cloud-Radiation Feedbacks.** Clouds govern the way radiative heating is partitioned between the atmosphere and the surface, and it is this partitioning of the heating that is vitally important for understanding the Earth's climate (Stephens, 1999). For example, the heating of the atmosphere affects atmospheric circulation, which establishes where clouds form. Clouds in turn affect the heating in the atmosphere and at the surface, thus establishing a cloud feedback loop. Critical to this feedback are processes that govern where and when clouds form and processes that define the cloud's physical properties, which in turn affect how much radiation is absorbed in the atmosphere and at the surface.

Because clouds have such a large effect on the Earth's radiation budget, even small changes in their abundance or distribution, which might occur through the feedbacks mentioned above, could alter the climate more than the anticipated changes in greenhouse gases, anthropogenic aerosols, or other factors associated with global change. These considerations suggest that a detailed understanding of the role of clouds in the climate system is essential for reliable climate simulations. Comparisons of global climate models indicate that "the main uncertainties in climate model simulation arise from the difficulties in adequately representing clouds and their radiative properties" (International Panel on Climate Change, 1995). [*See* Climate Models.]

**Cloud Formation.** Clouds are formed by the ascent and consequent cooling of air to the point of saturation. When air is further cooled, the excess water vapor condenses onto small particles, referred to as *condensation*

*nuclei*, producing cloud water droplets and ice crystals (e.g., Rodgers and Yau, 1996). The size, number, concentration, and even shape of these particles affects the way precipitation is produced in clouds as well as the way radiation is scattered and absorbed. The main mechanisms for lifting air and forming clouds include the following (see, e.g., Ludlam, 1980):

1. Air is heated from below, producing less dense, buoyant air, which rises and cools. The depth of the cloud formed depends on the background atmospheric stability. This mechanism is important, for example, in producing summertime thunderstorms over land in the late afternoon and evenings.
2. Air is lifted as it flows over mountain ranges and slopes.
3. Relatively warm air is lifted over relatively cool air. Midlatitude frontal clouds are examples of clouds formed in this way.
4. Air converges into an area of low pressure, producing ascent that typically occurs over large areas. The type of cloud formed depends to some extent on the nature of the low-pressure system, the strength of the convergence, the background water vapor, and atmospheric stability, among other factors.
5. Air is mixed by turbulence, induced by mechanical or thermal effects. Under certain circumstances, this mixing can cool the air, producing a saturated layer of air and cloud in the upper portion of the layer being mixed. This is the principal mechanism associated with the formation of low-level stratus and stratocumulus clouds.

The mechanisms that lift air and are responsible for cloud formation generally act on a large scale. These mechanisms, referred to as *large-scale forcing*, typically occur in combination. For example, low-level subtropical air is both heated and moistened as it is blown equatorward by the trade winds. There the air converges, producing deep convection as part of the intertropical convergence zone. Air lifted in these deep convective clouds is subsequently mixed with the environment and detrained outward at higher levels, producing extensive anvil clouds.

Large-scale forcing responsible for cloud formation varies considerably in strength. For example, the lifting associated with deep convection is vigorous in comparison with the slow ascent that produces extensive layered clouds. The challenge in understanding and depicting cloud feedbacks lies in correctly determining the magnitude and extent of the large-scale forcing and, in turn, correctly predicting the amount and extent of clouds that form as a result of this forcing. The widespread clouds that form under weak forcing are a special challenge because it is difficult to predict the weak forcing accurately, and the large areas of clouds produced in this way have a particularly important effect on the radiation budget of the Earth.

**Cloud Microphysical Properties.** Clouds form by condensing water vapor onto small condensation nuclei. [*See* Water Vapor.] In perfectly clean air without these nuclei, condensation will not take place unless there is an extreme drop in temperature. In practice, there is a sufficient number of these nuclei present in the air to produce clouds.

The growth of water droplets by condensation is relatively slow, and droplets rarely exceed diameters of 20–30 microns under this mechanism of growth. Ice crystals grow via this condensation mechanism and can reach hundreds of microns in size, partly because saturation of air relative to ice can be substantially greater than saturation with respect to water. The size and shape of ice crystals, the latter being referred to as the *crystal habit*, are determined by the ambient water vapor and temperature. Ice crystals are typically complex in shape because they often experience a variety of conditions as they fall through air.

Precipitation particles are two or more orders of magnitude larger than cloud drops. These larger precipitation particles are formed by collision and coalescence processes that represent the way bigger drops pick up smaller drops after collision. The collision-coalescence process works efficiently when there is a range of particle sizes and when the turbulent motions in clouds promote collisions between drops. Precipitation also forms by a variety of complex ice crystal processes, including collisions with supercooled water drops, collision and collection with other ice crystals, and growth of ice at the expense of evaporating water droplets.

Latent heat that is released in the process of condensational growth is conversely extracted from air when droplets and crystals evaporate. The excess of latent heat released to the atmosphere in condensation over the heat taken from the atmosphere on evaporation is proportional to the precipitation that reaches the surface of the Earth. The excess latent heating in precipitating systems provides an enormous amount of energy that affects the large-scale circulation of the atmosphere.

**The Radiative Properties of Clouds.** The amount of radiation absorbed in the atmosphere and at the surface of the Earth is determined by the process of *radiative transfer*. Radiative transfer is governed by both the amount and distribution of absorbing gases (principally water vapor and carbon dioxide) and by the optical properties of clouds, which in turn are established by the microphysical properties of clouds described above (Stephens, 1999). Three optical properties are important to radiative fluxes:

1. The *optical depth of clouds* is proportional to the vertical path of water and ice in the clouds and inversely proportional to the mean particle size. The tiny amount of mass of water and ice in individual droplets and crystals, when summed up and integrated throughout the depth of the cloud, determines the water and ice paths of clouds. Optical depths of clouds range enormously from values less than 0.1 to values exceeding 100 (Rossow, 1989).
2. The *scattering albedo* is a measure of the fraction of the total of radiation scattered by a small volume of cloud particles versus the fraction absorbed by the same volume.
3. The *particle scattering asymmetry parameter* is a measure of the fraction of the radiation that is scattered forward by a small volume of cloud versus that fraction scattered backward. This parameter has typical values that range between 0.7 and 0.9, which means 70–90 percent of the total radiation scattered by cloud particles is typically scattered forward, and, conversely, approximately 10–30 percent is scattered backward. The backscattering fraction for irregularly shaped ice crystals is typically larger than for spherical water drops.

These three properties combine in a complicated way with the amount of incident radiation to determine how much sunlight is reflected by clouds and how much radiation is absorbed and emitted by clouds. For example, the fraction of sunlight reflected by clouds, referred to as the *cloud albedo*, is proportional to the optical depth. The albedo is also proportional to the backscatter fraction of radiation so that an ice cloud of a given optical depth generally reflects more radiation than a water cloud of the same optical depth. The amount of radiation absorbed and emitted by clouds is also governed by the *cloud emissivity* and *cloud temperature*. The emissivity of the cloud is determined by the combination of optical depth and the scattering albedo. High clouds emit less radiation than warm clouds of the same emissivity because they are colder. [*See* Albedo.]

The combination of cloud emissivity, the height and thus temperature of the cloud, and the cloud albedo determine the overall effect of clouds on the radiative fluxes that leave Earth at the TOA (this is referred to as the Earth's radiation budget). Clouds reduce the net solar input into the planet by reflecting more solar radiation to space; the reflection is defined by the cloud albedo and hence referred to as the albedo effect. Conversely, clouds, as compared to clear skies, reduce the longwave emission to space by effectively raising the altitude of emission to heights characterized by colder temperatures. This is sometimes referred to as the greenhouse effect of clouds.

We are able to determine the net effect of clouds on the radiation budget at the TOA using satellite measurements (see, e.g., Harrison et al., 1994). A convenient way to analyze the overall effect of clouds on fluxes of radiation leaving the Earth is to contrast the fluxes reflected and emitted from cloudy portions of the Earth to fluxes from equivalent clear-sky portions. Satellite flux data can be grouped and analyzed in this way and presented as the difference between clear- and cloudy-sky TOA fluxes. Because clouds reflect more radiation than the clear sky, the difference between clear- and cloudy-sky-reflected solar fluxes is negative everywhere. The magnitude of this difference is largest over the regions of deep convection in the tropics and also in regions of extensive layered clouds that typically occur in the subtropical and midlatitude regions. Conversely, the longwave flux differences are primarily positive, since the radiation emitted from cloudy regions is generally less than the radiation emitted by surrounding clear-sky regions. The highest, coldest clouds associated with the deep convection in the tropics produces the largest longwave differences.

The net effect of clouds on radiation leaving space is determined as the sum of the individual components. The albedo and greenhouse effects cancel each other to a large extent, leaving much smaller net fluxes than are associated with either the longwave or solar fluxes. The largest net flux differences occur in regions of layered clouds in midlatitudes and then only in the Southern Hemisphere. In low latitudes, the albedo and greenhouse effects almost balance, leaving very small net fluxes as a residual. When averaged globally and over the annual cycle, the albedo effect is the more dominant, producing a net flux difference of approximately $-20$ watts per square meters.

The satellite observations cannot tell us about how clouds affect the radiative heating of the atmosphere and surface. It is important to estimate these effects because the nature of cloud feedback depends on how the net effect is distributed between the atmosphere and surface. We are unable to provide definitive estimates of atmospheric and surface heating with the available observations and must resort to use of global models for clues. These models suggest that the longwave effect of clouds observed at the TOA is largely indicative of the effects of clouds on the atmospheric radiative heating. For example, the regions of large positive flux differences at low latitudes represent a heating of the cloudy atmosphere relative to the clear skies in those regions. Conversely, the negative solar flux differences represent a relative depletion of energy reaching the surface under cloudy skies and relatively little effect on atmospheric absorption. How much the solar radiation effect is concentrated at the surface versus how much occurs within the atmosphere continues to be a topic of debate.

**Effects of Aerosol on Clouds.** As mentioned, clouds form on small condensation nuclei, which are some portion of the aerosol particles that remain suspended in air. [*See* Aerosols.] Regional differences in the concentrations and composition of aerosol particles produce regional differences in cloud microphysical and radiative properties. For example, concentrations of nuclei are more abundant over land than over oceans. Clouds formed over land tend to be composed of smaller but more numerous droplets than clouds formed over the remote oceans.

For clouds of the same water content, a cloud formed in air containing abundant nuceli will have smaller characteristic droplets and thus larger optical depth than a cloud formed in air with lower concentrations of nuclei. This observation led to the concept known as the indirect forcing of climate by aerosols. The basic idea is that increasing aerosol concentration leads to increasing cloud optical depths due to decreasing particle size and thus to increasing cloud albedo. This effect is clearly seen at certain times in satellite images of reflected sunlight from clouds above ships on the ocean (see Hobbs, 1993). The effluents emitted from the chimney stacks of ships act as a localized source of increased cloud nuclei.

It has been calculated that the change in cloud albedo introduced by changing the concentration of aerosol can lead to changes in reflected fluxes that are larger than changes expected by direct scattering of sunlight by aerosol when averaged over the globe (Jones et al., 1994). The magnitude of this effect is uncertain, however, and the actual effect of aerosol on clouds may be much more complicated than that described or included in these calculations. For example, both cloud model and observational studies indicate that aerosols may produce even larger influences on clouds through their effects on altering precipitation. Aerosols can alter precipitation, for example, by decreasing particle size and thus reducing the efficiency of the collision-coalescence process in forming precipitation-size particles. Changing precipitation then leads to changes in the water content of clouds and their overall life cycle, thus affecting their optical properties as well as the water vapor content of the atmosphere.

[*See also* Atmosphere Dynamics; Atmosphere Structure and Evolution; Atmospheric Chemistry; Climate Change; Global Warming; *and* Greenhouse Effect.]

## BIBLIOGRAPHY

Barkstrom, B. R., and G. S. Smith. "The Earth Radiation Budget Experiment: Science and Implementation." *Reviews of Geophysics* 24 (1986), 379–390.

Harrison, E. F., P. Minnis, B. R. Barkstrom, and G. Gibson. "Radiation Budget at the Top of the Atmosphere." In *Atlas of Satellite Observations Related to Global Change*, edited by R. J.

Gurney et al., pp. 19–40. Cambridge and New York: Cambridge University Press, 1994.

Hobbs, P. V., ed. *Aerosol-Cloud-Climate Interactions*. San Diego: Academic Press, 1993.

International Panel on Climate Change (IPCC). *Climate Change 1995: The Science of Climate Change*, edited by J. T. Houghton, L. G. Meira Filho, B. A. Callander, N. Harris, A. Kattenberg, and K. Maskell. Cambridge: Cambridge University Press, 1996.

Jones, A., D. L. Roberts, and A. Slingo. "A Climate Model Study of Indirect Radiative Forcing by Anthropogenic Sulphate Aerosol." *Nature* 370 (1994), 450–453.

Ludlam, F. H. *Clouds and Storms: The Effect of Water in the Atmosphere*. University Park: Pennsylvania State University Press, 1980.

Rodgers, R. R., and M. K. Yau. *A Short Course in Cloud Physics*. Oxford: Butterworth-Heinemann, 1996.

Rossow, W. B. "Measuring Clouds from Space: A Review." *Journal of Climate* 2 (1989), 201–203.

———. "Clouds." In *Atlas of Satellite Observations Related to Global Change*, edited by R. J. Gurney et al., pp. 141–163. Cambridge and New York: Cambridge University Press, 1994.

Stephens, G. L. "Radiative Effects of Clouds and Water Vapor." In *Global Energy and Water Cycles*, edited by K. A. Browning and R. J. Gurney. Cambridge: Cambridge University Press, 1999.

Twomey, S. "Influence of Air Pollution on the Short-Wave Albedo of Clouds." *Journal of the Atmospheric Sciences* 34 (1978), 1149–1152.

—GRAEME L. STEPHENS

**CLOUD SEEDING.** *See* Weather Modification.

**COAL.** *See* Carbon Dioxide; Electrical Power Generation; Energy Policy; Fossil Fuels; *and* Mining.

# COASTAL PROTECTION AND MANAGEMENT

It has been claimed that 50 percent of the population in the industrialized world lives within 1 kilometer of the coast, and that roughly 75 percent of the world's population lives in a 60-kilometer-wide strip along the coastline. [*See* Coastlines.] Coasts are by their very nature dynamic areas, with certain hot spots prone to particularly fast changes caused by an interrelated web of natural and physical causes. Such hot spots include sandy coasts (beaches and dunes), coastal wetlands, cliffed coasts in weak sediments, and coral reefs. Coastal change costs money. A recent estimate suggests that some 25 percent of the coastline of the United States is significantly affected by coastal erosion, costing around U.S.$300 million per year in loss of property and building of protective structures. Coastal change constitutes a hazard to human populations. In Bangladesh, for example, cyclone-induced erosion and flooding is a major cause of loss of life and disruption to agricultural pro-

duction. As increasing concern is expressed over the future of the environment, and especially the possibility of sea level rise induced by global warming, it is likely that coastal protection and management will become an increasingly important challenge for many societies.

Coastal change occurs because of natural and/or human factors. The most serious types of change for human use and enjoyment of the coastal zone are erosion and deposition of sediment, deterioration in coastal ecosystems (for example, loss of species), and pollution of coastal land and waters. Increasingly, it is becoming apparent that successful management of these coastal problems must be integrated—that is, all aspects of often multiple problems must be addressed, and the problems themselves must be seen within a wider context of other linked environments.

In the past, coastal protection and management schemes have often been piecemeal, localized attempts to deal with coastal change, often based on a limited scientific understanding of the rates and processes of change. In Britain, for example, coastal protection schemes historically were planned and implemented at local government level, with adjacent parishes potentially instigating conflicting solutions to the same problem. Over the past few decades, local, national, and international coastal management plans have been proposed and there has been a growth in Integrated Coastal Zone Management (ICZM) ideas, with the United Nations Conference on Environment and Development (UNCED) meeting in Rio de Janeiro in 1992 outlining an important role for ICZM in aiding national sustainable development plans. ICZM, as generally practiced, involves three major components: environmental conservation, sea defense, and development planning. Thus coastal protection should be seen as a subset of coastal management. The first comprehensive attempt at coastal zone management was probably the U.S. Coastal Zone Management Act of 1972, which provided for the development of coastal planning and management schemes by state authorities. More recently, in 1995, Australia instigated its Commonwealth coastal policy, designed to facilitate ICZM across the whole country.

Coastal zone management needs to reconcile different uses of the coastal zone through legislation and zoning, while also ensuring that coastal geomorphologic and ecological changes are minimized. A prerequisite of any successful management scheme is thus good scientific understanding of the coastal system involved, how and why it has changed in the past, and how it will react in the future. This is in itself a hugely difficult task. However, once the system is broadly understood, the human ownership and use of the coastal zone also needs determining, before a series of management strategies can be proposed and debated among those involved. A good example is provided by management of the Dorset,

Hampshire, and Sussex coast in southern Britain. Here scientists have been involved for many years in studying the dynamics of the coast, while engineers and managers have been trying to design better coastal protection schemes in an area of great natural beauty. Responsibility for coastal protection and sea defense in England and Wales is shared by district councils and the Environment Agency, which, together with other interested bodies, formed SCOPAC (Standing Conference on Problems Associated with the Coastline) to coordinate expert advice on future coastal management in the area. Scientists provided evidence used to produce a series of scenarios, involving options of retreat, accommodation, and protection against future sea level rise. [See Sea Level.] Such scenarios have been proposed for many other coastal areas facing a prospect of rising sea levels and thus increased coastal erosion and flooding. The retreat option involves developing legislation to prevent further development along vulnerable coasts within a certain distance (the "set-back" line) of the coast and encourage relocation of currently occupied property. This option can be costly and difficult to implement in highly built-up areas. The accommodation option involves adapting to the changes by altering agricultural practices, modifying building construction, and taking out better insurance to live with the coastal changes. Finally, the protection option involves increasing both hard methods (as in structural methods of coastal protection such as sea walls, groins, dykes, and tidal barriers) and soft methods (such as beach nourishment, wetland creation) of protecting the coast from higher sea levels.

All three of these strategies have advantages and disadvantages in any situation, and all should be seen as ongoing rather than one-off solutions. Ownership of coastal land can be very fragmented, making coherent planning difficult. The location of key, highly valuable industrial plants (such as power stations or oil terminals) can bias decisions in favor of protection. Hard engineering schemes are costly and can often fail, or produce knock-on effects down the coast. Soft engineering schemes also often fail and require costly long-term management. In most coastal areas, sea level rise is not the only problem that needs to be dealt with, so such management plans also need to include consideration of issues such as pollution, species loss, and human-induced subsidence (through removal of ground water and other fluids).

[See also Global Warming; and Land Reclamation.]

### BIBLIOGRAPHY

Beatley, T., et al. *An Introduction to Coastal Zone Management.* Washington, D.C.: Island Press, 1994. A simple and clear introduction to the U.S. coastal management situation.

Bray, M. J., et al. "Planning for Sea-Level Rise on the South Coast of England: Advising the Decision-Makers." *Transactions of the Institute of British Geographers* 22 (1997), 13–30. Review of issues and possible solutions along one area of coast.

French, P. W. *Coastal and Estuarine Management.* London: Routledge, 1997. A readable and wide-ranging introduction to coastal management issues and strategies, with lots of British and other case studies.

—HEATHER A. VILES

**COAST EROSION.** *See* Coastlines.

## COASTLINES

Coastlines occupy the zone between land and water (in oceans or lakes) and are often described as a dynamic interface where waves, currents, and tides influence landforms, which in turn affect the movement of water. It is difficult to measure the length of coastline present on the Earth today and, indeed, it is perhaps better to talk about the coastal zone as an area rather than as a mere line running around the edge of continents and islands. The influence of marine or lacustrine waters often extends several kilometers inland, especially where rivers produce estuaries or deltas.

**Types of Coastline.** The world's coastline can be divided into a range of different types, based on varying criteria. The fundamental factor influencing the disposition of coasts at the large scale is plate tectonics. Collision, trailing-edge, and marginal-sea coastal types are the result of plate tectonics. Collision or active-margin coasts are typified by a mountainous hinterland, producing coarse debris in streams, with a narrow continental shelf. Collision coasts are formed along the western side of North America. In contrast, trailing-edge, or passive-margin, coasts are fed by huge river systems that contribute huge volumes of fine-grained sediment to wide, low-angle continental shelves. Examples are found along the eastern side of South America. The major deltas of the world have developed on such passive-margin coasts, where barrier island development is also common. In marginal-sea coasts, such as in the Gulf of Mexico and the South China Sea, although near a collision plate boundary, the coast is protected from its effects by a marginal sea, allowing it to behave like a trailing-edge coast.

At a smaller scale, coasts vary according to the oceanographic setting, sediment size and availability, climate, and sea level history. The latter is a particularly important control on the nature of today's coasts, especially in areas where the vast changes wrought by the last ice age on both land and sea are still continuing to exert an influence. For example, where an ice sheet once depressed a landmass, the land has gradually experienced rebound (a process called isostasy) as the ice has melted and indeed may still be rising (producing an emerging coast characterized by falling sea level). In

## REBUILDING IN THE OUTER BANKS OF NORTH CAROLINA

When hurricanes Fran and Bertha swept through North Carolina in 1996, a state senator observed, "You just can't live in certain parts of the coast . . . these places are just not made for homes." At the same time, the governor said the state should discourage building in vulnerable areas.

But, in that year, the Federal Emergency Management Agency (FEMA) handed out U.S.$115 million to rebuild properties on Topsail Island and two other barrier islands. Despite the Coastal Barrier Resource Act of 1982, which sought to restrict federal spending in undeveloped areas that seemed certain to flood, U.S.$4 million found its way into the north end of Topsail Island: half a million was spent to replace and enlarge a sewer system (which will encourage development there) and another half million was spent to repair roads that wash out as often as once a year.

In August 1998, the area was hit again, but gently, by hurricane Bonnie. This storm washed away more than half the protective dunes that had been built in 1997 at a cost of U.S.$5 million. The dunes were probably gone during the first five minutes of the storm, according to Orrin Pilkey, a geologist at Duke University in Durham, North Carolina.

**BIBLIOGRAPHY**

Jaffe, G., and R. Notoko. "Building in Waves." *The Wall Street Journal* (31 August 1998).

—David J. Cuff

many areas, the major sources of coastal sediment were associated with glacial debris that were combed up from the sea floor as sea levels rose at the end of the last glaciation. Now this source of sediment has been largely used up, leading to sediment-limited coastlines. However, we should not neglect the importance of the present-day setting to the characteristics of today's coasts. Wave environments, tidal range, sediments, and climate are all important in determining whether coasts are characterized by large swathes of beaches backed by dunes, rocky coasts with pocket beaches, muddy coasts such as salt marshes and mangrove swamps, or biogenic coasts such as coral reefs. In many situations, these different coastal types are found close together. Barrier islands, for example, which front much of the coastline of the eastern United States and parts of the Nigerian coast, consist of beach, dune, marsh, and back-barrier lagoon environments. Similarly, along many tropical coasts, coral reefs are found in close proximity to beaches and mangrove swamps. Classification of the spatial diversity of the global coastline is thus difficult, except in the broadest terms; similarly, the coastal zone is characterized by dynamism over a range of temporal scales, producing near-constant changes in topography, vegetation, and water flow.

**Coastal Change.** Coastlines appear to act as systems in which sediments, water, and organisms are linked together by a series of flows and cycles. This means that change in any one factor (for example, removal of vegetation) can also influence other parts of the system. Considerable effort has been expended by coastal scientists in trying to delineate functioning coastal systems. One approach has been to identify coastal sediment cells. Such cells are a convenient way of delimiting areas of coast that function together, linked by flows of sediment along the shore as well as between on- and offshore zones.

Dynamism in coastlines is produced by a range of factors, including, at the large scale, tectonic activity and sea level change (on decadal to century scales, or longer), with El Niño–Southern Oscillation (ENSO) events, storms, and other climatic perturbations producing change on a medium time scale (annual to decadal), and short-term changes being caused by variations in wave energy, tidal height, and currents on a subannual time scale. The dynamic nature of coastlines is essentially determined by sea level, as it is here that the main focus of wave, tidal, and current energy is concentrated and thus where most morphological change occurs. [*See* Sea Level.] Any long- or short-term change in the position of sea level, or of the energy conditions at that level, will produce some change in the coastal morphology. As we have already seen, plate tectonics sets the coastal scene, but it can also produce vertical and horizontal movements along coastlines, influencing the relative sea level. Such changes operate over a range of time scales, from millions of years down to abrupt changes (such as coseismic uplift associated with earthquakes). Sea level also changes as a result of differences in the volume of water held in the ocean basins, which itself is controlled by a range of tectonic and climatic factors. Over the past thousand or so years, sea level

## DELTA POPULATIONS AND RISING SEA LEVEL

If sea levels rise as projected (between 0.3 and 1.0 meters by the year 2100), a large proportion of the world's population will be affected, because they live virtually at sea level on the deltas of major rivers.

One of the most serious situations is in Bangladesh, where a 1-meter rise in sea level could inundate 17 percent of the nation's land area, and displace eleven million people from the densely populated Ganges–Brahmaputra delta. In the Nile delta, a 1-meter rise would displace six million; and in the Niger a similar rise would force half a million to relocate. These effects would be reduced if deposition by the rivers were to elevate the land substantially.

Rising sea level would not only lead gradually to inundation, but would increase risks of flooding and coastal erosion during periodic storms.

Heavily Populated Delta Regions Vulnerable to Sea Level Rise, Listed by River

| | |
|---|---|
| Ganges–Brahmaputra, Bangladesh | Yellow, China |
| Krishna and Godavari, India | Niger, Nigeria |
| Narmada and Tapti, India | Nile, Egypt |
| Irrawaddy, Myanmar | Po, Italy |
| Indus, Pakistan | Rhine, Netherlands |
| Mekong, Vietnam | Rhone, France |
| Hong (Red), Vietnam | Paraná, Argentina |
| Zhu (Pearl), China | Mississippi, United States |
| Yangtze, China | |

**BIBLIOGRAPHY**

The World Resources Institute. *World Resources 1998–99*, pp. 68–69. New York: Oxford University Press, 1998.

—DAVID J. CUFF

has been rising globally at between 0.1 and 0.2 millimeters per year because of continuing (if declining) isostatic crustal adjustments following deglaciation. Over the last one hundred or so years, most studies estimate global sea level to have risen by about 1.0–2.0 millimeters per year, probably as a result of warming of ocean water (producing expansion and thus a volume increase) and the melting of ice stored on land. Recent predictions by the Intergovernmental Panel on Climate Change (IPCC) suggest that, over the next century or so, sea level will rise at an accelerated rate of 4.0–5.0 millimeters per year, as a response to increased global air temperatures. However, much uncertainty surrounds such predictions, and it must be remembered that, locally, sea level may change very differently, depending upon the natural coastal setting and the influence of human activities (e.g., through pumping of ground water, encouraging subsidence).

On the annual to decadal time scale, coasts respond dynamically to changes in oceanographic and climatic regimes. Thus, for example, extreme storm events (such as Hurricane Andrew, which affected the Caribbean in 1992) produce often-catastrophic changes in coastal sediment regimes. The intensely studied ENSO events have also been seen to produce extensive coastal change, especially along Pacific island coral reef coasts where elevated sea water temperatures, coupled with lower sea levels, have wrought considerable damage to the coral ecosystem. [*See* El Niño–Southern Oscillation.] Such extreme events may have a range of impacts on the coastline; and in several cases it is the intensity of local human impacts that determines whether change will be acute or not. On the diurnal (daily) to seasonal scale, coastlines are prone to change as wave energy, climate, vegetation, and sediment volume all vary. Thus, for example, many British beaches have clear summer and winter states—the former characterized by a buildup of sediment supply, often in association with vigorous growth of backshore and dune vegetation. In winter, frequent storm events produce a reduction in sediment volume. It is clear from this brief summary of different scales of coastal change that the causes of change can be difficult to disentangle (even in circumstances in which there is no human impact), assessment of the rates of change is also a complex task, and rates themselves are scale dependent.

**Rates of Coastal Change.** How can we make some assessment of natural rates of coastal change? A range of techniques are available, from stratigraphic methods, which provide information about long-term (thousands of years) variations in the positions of land and sea, through a range of historical sources of information (such as old maps, photographs, and documents of known dates, which can be compared to build up a picture of coastal change over the centuries), to short-term monitoring over a number of months or years of the volume of sediment on, for example, beach and dune coasts. All such methods have their limitations, but used either singly or in combination can provide some picture of the rate and nature of coastal change in a particular area. For example, much work has been carried out by Robert Morton and his co-workers (e.g. Morton, 1979; Morton et al., 1993) on the dynamism of sandy coastlines of the Texas gulf using comparisons of maps and air photographs from 1850 onward coupled with increasingly precise measurements of month-by-month changes. For many areas of the world, however, our knowledge of rates of coastal change is patchy at best.

Rates of coastal change vary hugely depending on both environmental conditions and rock, sediment, and vegetation characteristics. On exposed coasts constructed of weakly cohesive glacial sediments, where wave energies are high and materials are prone to erosion, rates of erosion of up to 10 meters per year (that is, horizontal retreat of the coastline) can be found. Conversely, on plunging clifflines composed of hard rock in low-energy coastal environments, no discernible erosion at all can be produced in a century. Spectacular long- and short-term changes can be produced along delta coasts. The Yellow River delta in China, covering a total area of 15,000 square kilometers, illustrates this well. Between 1128 and 1855 CE it formed a large delta, then the course of the river changed to a more northerly position, and, consequently, erosion has begun to dominate the delta since no new sediment is being delivered. Shoreline retreat here has reached 110 meters per year in places, and around 44 billion cubic meters of sediment has eroded into the ocean since the mid-nineteenth century.

Coastal change can also be positive, of course, with progradation leading to the building out of coastlines. In tropical mangrove swamp environments, for example, increased supply of muddy sediment coupled with colonization of suitable plant species can produce often spectacular accretion. In peninsular Malaysia, for example, a study of maps, aerial photographs, and soil auguring surveys showed that, between 1914 and 1969, the area of mangroves had increased by 26.7 square kilo-

meters, with accretion rates of up to 18–54 meters per year in places. The cause of this accretion seems to have been the development of onshore mining and agricultural activities that resulted in increased river sediment loads. In global terms, however, erosion seems to be a dominant feature. Bird (1985) has coordinated a worldwide survey of coastal change and concludes that of the roughly 20 percent of global coastline that is sandy in type, more than 70 percent has shown net erosion over the past few decades.

Coastal changes can thus be measured or estimated in terms of the relative position of land and sea in both vertical and horizontal dimensions, changes in volume of sediment within the coastal zone, and/or changes in ecosystem characteristics. Many such investigations, especially of short-term coastal change, are carried out within the framework of coastal sediment cells. It is certainly possible to identify coastal hot spots or areas prone to high rates of coastal change, including deltas and other areas prone to subsidence.

**Human Influences on Coastal Change.** Human activities have become an increasingly important component of coastline change within the last hundred or so years, working with natural processes to produce often spectacular change. Human activities can influence physical, chemical, and/or biological characteristics of coastal systems in a variety of direct and indirect, deliberate and accidental ways. [*See* Coastal Protection and Management.] Table 1 gives some examples.

From the table we can see that important deliberate and direct changes can be made through land claim, whereby new land is won from the sea. Such activities have great antiquity in many parts of the world, for example, in the Netherlands where dykes have been constructed from 10 CE onward, and in China where coastal modifications may date back to over 4000 BCE. In Singapore, reclamation has occurred since 1820 CE, with the overall land area of the island increasing from 581.5 square kilometers in 1962 to 641.0 square kilometers in 1992. More widely, coastal engineering schemes also have direct impacts on the coastline, as they aim through a series of "hard" and "soft" alterations to coastal sediment movements to reduce coastal changes. In many areas, however, activities directly within the coastal zone can have largely unplanned impacts on coastal change. Thus, quarrying of rocks and sediments can upset sediment budgets in the local area, a famous example being the erosion of the coast near the village of Hallsands in Devon, England, as a result of mining of offshore sediment for coastal protection elsewhere.

Removal or introduction of vegetation within the coastal zone can also produce coastal change. In many parts of the tropics, mangrove trees are an important component of the local economy and thus are often felled. Furthermore, the mangrove areas are increas-

**Coastlines. TABLE 1.** Human Impacts on Coastline Change

|  | PHYSICAL CHANGE | BIOLOGICAL CHANGE | CHEMICAL CHANGE |
| --- | --- | --- | --- |
| Direct and deliberate action | Land claim<br>Coastal engineering | | |
| Direct/largely unplanned action | Quarrying of material from within the coastal zone | Accidental introduction of species (e.g., *Spartina* grass)<br>Vegetation clearance<br>Aquaculture | Introduction of pollutants onto beaches, or nearshore coastal waters |
| Indirect but deliberate action | Coastal protection and modification schemes influencing nearby coastline | | |
| Indirect/largely unplanned action | Sediment delivery changes consequent on damming rivers<br>Subsidence of coast following removal of ground water or oil | Sediment influx from eroding agricultural land damaging coral reef ecosystems | Pollution from offshore, upstream, and other sources entering a coastal system and producing knock-on biological effects |

ingly being converted to other uses (e.g., agriculture and aquaculture), as in Indonesia where brackish-water fishponds dug into coastal wetland soils now extend over around 269,000 hectares, or 6.5 percent of the former mangrove area. Such conversion can have geomorphologic and ecological consequences, with fishpond areas being prone to erosion and pollution. On the other hand, introduction of species can also have important effects on coastal change. Along much of the British coast, for example, salt marshes have been affected by the spread of the marsh grass *Spartina anglica*, which now covers some 10,000 hectares. This species encourages the accretion of mud, thus increasing the area of coastal wetland, but it reduces biological diversity and is prone to "die-back." Along many coastlines, pollution within the coastal zone can contribute to coastal change through its effects on vegetation and animal life. On fringing coral reefs, for example, sewage is often discharged directly onto the reef, where it often kills off reef-building coral species and encourages the growth of less desirable species, thereby threatening the solidity of the entire reef construction.

Indirect human activities are probably responsible for a vast amount of coastal change, whereby activities upstream, offshore, and alongshore have knock-on effects on coastal systems. Human impacts on river systems, whereby sediment and pollutants are introduced into rivers and debouch into the coastal zone, are particularly important. Of huge impact on many coastal areas has been the reduction of sediment load associated with the damming of major rivers. In the Ebro Delta in northeastern Spain, for example, dams have reduced sediment discharge from roughly 4 million metric tons per year before 1965 to less than 400,000 metric tons per year now, producing shifts in the depositional environ-

ments of the delta with knock-on effects on wetland ecology. Finally, another type of indirect human activity is also a highly potent agent of coastal change, namely, the removal of ground water and oil from coastal aquifers, potentially leading to subsidence. [*See* Anthropogeomorphology; *and* Ground Water.] Venice was particularly prone to such subsidence from the 1950s until groundwater pumping was banned, and it has been estimated that removal of Venetian ground water produced 12–14 centimeters of subsidence.

[*See also* Estuaries; Global Warming; *and* Ocean Dynamics.]

### BIBLIOGRAPHY

Bird, E. *Coastal Changes: A Global Review.* Chichester, U.K.: Wiley, 1985.

Carter, R. W. G. *Coastal Environments: An Introduction to the Physical, Ecological and Cultural Systems of Coastlines.* London: Academic Press, 1988. An extensive and informative survey of coastal processes and management.

Davis, R. A. *The Evolving Coast.* New York: Scientific American Library, 1994. A readable briefing on coastlines and coastal change.

Morton, R. A. "Temporal and Spatial Variations in Shoreline Changes and Their Implications: Examples from the Texas Gulf Coast." *Journal of Sedimentary Petrology* 49 (1979), 1101–1112.

Morton, R. A., M. P. Leach, J. G. Perine, and M. A. Cardoza. "Monitoring Beach Changes Using G.P.S. Surveying Techniques." *Journal of Coastal Research* 9 (1993), 702–720.

Trenhaile, A. S. *Coastal Dynamics and Landforms.* Oxford: Clarendon Press, 1997. An advanced, comprehensive account of coastal geomorphology.

Viles, H. A., and T. Spencer. *Coastal Problems: Geomorphology, Ecology and Society at the Coast.* London: Edward Arnold, 1995. A wide-ranging introduction to coastal environments and human impacts upon them.

—HEATHER A. VILES

# COLORADO RIVER, TRANSFORMATION OF THE

*[This case study discusses the anthropogenic transformation of the Colorado River in the southwestern United States and the geomorphologic and ecological impacts of that transformation.]*

The last 150 years have seen dramatic changes along the Colorado River (Figure 1). Once a major, free-flowing river in an arid region, it has become one of the most controlled rivers in North America, a monument to western expansion and human desires to exploit natural resources.

In 1869, Major John Wesley Powell launched his boats into the Green River near Green River, Utah, intent on exploring and describing a length of the Colorado River only witnessed by pre-Columbian humans. Today, almost twenty-five thousand people raft the Grand Canyon each year, compared with a few hundred prior to river regulation by dams. During the same period of the nineteenth century, shallow-bottom steamboats made their way up the Colorado River from the Sea of Cortez, delivering goods to riverside communities well upstream of Yuma Crossing, a military depot—now Yuma, Arizona. Earlier expeditions up the river reached the Black Canyon of the Colorado, site of Hoover Dam. These steamers had a more than ample supply of fuel for these trips, utilizing extensive cottonwood and mesquite riparian forests along the way. This was the "wild" unregulated Colorado River, a mystery in the upper reaches and a tool for development in the lower.

As one of the most regulated rivers in North America, the Colorado is large for the arid Southwest but does not compare to the Columbia, Missouri, or Mississippi in volume. In addition to the many takeouts for irrigation and urban use (for example, the Central Arizona Project), flows in the Colorado and its major tributaries are controlled by twelve dams, many creating large impoundments; Lake Powell and Lake Mead are the largest. The Colorado River originates in the mountains of southern Colorado, but major inflows come from large tributaries such as the Gunnison River in Colorado, the Green and San Juan Rivers in Utah, the Little Colorado and Gila Rivers in Arizona, and the Virgin River in Nevada. Many of these tributaries are regulated by dams and irrigation takeout structures. The Colorado River, over 2,250 kilometers (1,400 miles) long, drains a watershed of more than 622,000 square kilometers (240,000 square miles) from seven states and northern Mexico terminating at the Sea of Cortez (Gulf of California), where it formed a large delta over the millions of years during which it was free-flowing.

**Controlled Flows.** The hydrology of the Colorado River is now a product of dam releases, which are driven by power and water demands throughout the Southwest and California. [*See* Dams.] Early modifications of river flows were for purposes of irrigation: the Alamo Canal completed in 1901 to irrigate the Imperial Valley is a primary example. Prior to any development of the river, discharge near Yuma, just north of the Mexican border, often peaked at over 3,540 cubic meters per second (125,000 cubic feet per second), with extremes above 5,660 cubic meters per second. Spring peaks in Grand Canyon often exceeded 2,800 cubic meters per second, with low summer and winter flows below 85 cubic meters per second. Prior to dam construction, more than 19 billion cubic meters of water flowed through the Canyon each year. In 1922 the Colorado River Compact divided the flow of the river among the seven basin states. This division, based on flow measurements from wet years of the early twentieth century, allocated more to the upper and lower basin states than the Colorado could deliver later in the century. The Reclamation Act of 1929 initiated major alterations of the Colorado River. This act authorized construction of Hoover Dam, which when completed in 1935 controlled the flow of the river from Nevada to the delta. With few exceptions, peak flows at Yuma dropped below 560 cubic meters per second, and the timing of the peak was changed from May/June to winter months. Construction of Davis and Parker Dams downstream of Hoover Dam further regulated the lower Colorado River flows. These, along with irrigation takeouts, essentially eliminated flows to the delta and Gulf of California that had carried nutrients enhancing agriculture and fisheries. In 1963, Glen Canyon Dam was completed and river discharge through the Grand Canyon was also regulated, driven by hydropower production patterns. Except for wet years, only 10 billion cubic meters of water moved downstream through the Grand Canyon, the remainder being held in Lake Powell behind Glen Canyon Dam. Not only was the amount of water highly controlled, but the pattern of water release from the dam was based on hydroelectric demands, both seasonal and daily. Seasonal peaks now occurred in winter and summer rather than in late spring, and daily fluctuations peaked during the day (seldom exceeding 850 cubic meters per second) and dropped at night. These daily fluctuations, often more than 560 cubic meters per second, created a wave through the canyon, altering the daily river level by as much as 3–5 meters. Water-level fluctuations were a normal occurrence before dams controlled flows, but fluxes occurred seasonally, or in response to occasional storms, and the rate of change, especially the declining limb of a flood, was gradual.

Controlled flows through the lower Colorado River also gave rise to many changes. Some were nature's response to these new conditions, and others were human developments across a highly altered flood plain. Controlled flows reduced the effects of flooding and offered opportunities for channelization of the river. As the

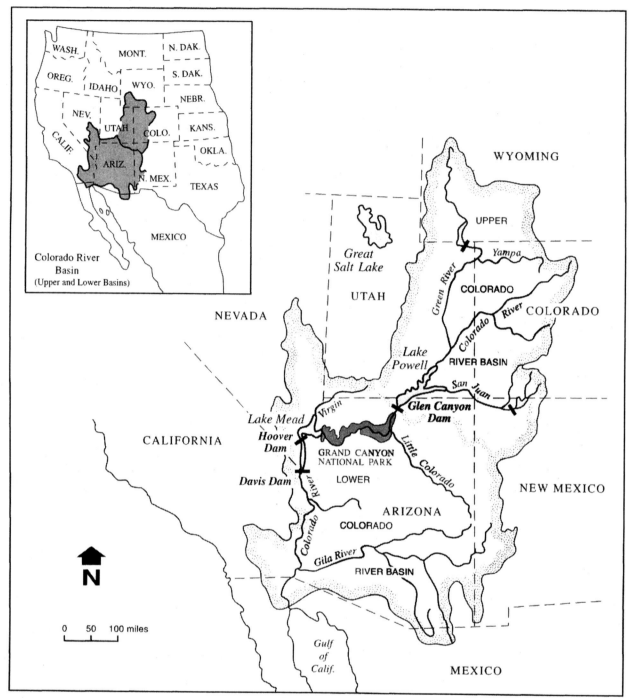

**Colorado River. FIGURE 1. The Colorado River Basin.**

The Colorado River basin drains portions of seven states, including Colorado, Utah, Wyoming, New Mexico, Arizona, Nevada, and California. A small portion of Mexico is also drained by the river before it reaches the Gulf of California. (After Carothers and Brown, 1991. Copyright The Arizona Board of Regents. Reprinted by permission of the University of Arizona Press.)

channel was controlled and natural meanders were eliminated, the remaining flood plain, soon devoid of most riparian vegetation, was used for other purposes, including agriculture and recreational residences.

**Sediment.** Sediment carried by the Colorado River has greatly influenced the riverine ecosystem. The river's name is based on its color and opacity, a consequence of large amounts of transported sediment. For example, prior to completion of Glen Canyon Dam, over 60 million metric tons of sediment were transported past the Grand Canyon gauge near Phantom Ranch in the middle of the Grand Canyon. More than 200 million metric tons were often delivered out of the canyon, of which only 140 million metric tons reached Yuma, the remainder being deposited along the channel and on the flood plain. After dam closure, the sediment load was reduced to about 10 percent of that transported by unregulated flows. Sediment-starved water discharged from the dam suspends much of the accumulated sediment within the canyon and moves it downstream rather than depositing it in eddies and along channel margins. The only sources of sediment below Glen Canyon Dam within the canyon are now the Paria River, the Little Colorado River, and smaller tributaries. The remainder of the sediment is building deltas in Lake Powell, enough to reduce the capacity of the lake by over 850,000 cubic meters per year. Although sediment trapped in Lake Powell extends the life of Lake Mead, it also reduces scour and deposition within the canyon. Elevated sediment deposits form beaches and bars—potential biological habitat—at high flow levels, but formed barren beaches at low flow levels when the river was uncontrolled. This has changed. Scouring is limited because high spring floods no longer occur and sediment deposits now develop only at lower river elevations, if sediment is available.

As sediment was trapped in impoundments behind the ever-increasing number of dams along the Colorado River, river water quality changed. Penstock intakes for generators are sufficiently low that they collect cold, hypolimnetic water (from below the thermocline or steep vertical temperature gradient in lakes). Thus, dam releases are cold and clear, ranging, for example, from 8°–10°C (46°–50°F) below Glen Canyon Dam. This is a major change from a sediment-laden river that had temperature fluctuations from about 27°C (80°F) in summer to near freezing in winter. Because sediment inputs still occur downstream from the dam, and sunshine in the canyon warms the river by a few degrees, the river may function more like an unregulated river as it flows downstream a distance of about 386 kilometers (240 miles) before it enters Lake Mead. Fluctuations in daily river levels have also become truncated downstream. Releases from Hoover Dam and other downstream dams do not flow far before entering another impoundment, so the Colorado River as it extends into the lower basin has little opportunity to recover from dam controls as found within the Grand Canyon.

**Barriers and Aquatic Migration.** Dams not only alter the physical and chemical characteristic of the river; they also create barriers to most aquatic migration along the 2,250 kilometer extent of the Colorado River. The fish community in the Grand Canyon, as well as the lower Colorado River basin, has changed drastically as a result of the presence of Glen Canyon Dam, through other activities of humans attempting to improve commercial or sport fisheries in the area, or

through accidental introduction of nonnative species. [*See* Exotic Species.] Early records in the late 1800s indicate the presence of eight native species in two families in the Grand Canyon: Cyprinidae (chubs and minnows) and Catostomidae (suckers). In recent postdam collections, only two of the Cyprinidae remain (humpback chub, *Gila cypha*, and speckled dace, *Rhinichthys osculus*), and the three Catostomidae remain (razorback sucker, *Xyrauchen texanus*, bluehead sucker, *Catostomus discobolus*, and flannelmouth sucker, *Catostomus latipinnis*). None of these species now uses the whole main stem of the Colorado River within the Canyon as habitat. Some use tributaries (e.g., chub in the Little Colorado River for spawning), while bluehead and flannelmouth are still ubiquitous. Additions to the canyon fish community are nine nonnative fishes. These include rainbow trout (*Salmo gairdneri*), cutthroat trout (*Salmo clarki*), brown trout (*Salmo trutta*), and brook trout (*Salvelinus fontinalis*)—rainbow and brown being most common—as well as common carp (*Cyprinus carpio*), striped bass (*Morone saxatilis*), fathead minnow (*Pimephales promelas*), channel catfish (*Ictalurus punctatus*), and plains killfish (*Fundulus zerbrinus*). Several of these fishes—for example, carp and catfish—were introduced to the lower Colorado River in the 1800s for commercial purposes, and these may be the original source of these fishes throughout the system.

This mixture of native and nonnative fishes is not evenly distributed throughout the Grand Canyon reaches. In all reaches, nonnatives comprise the higher percentage of the community, in some reaches ten to fifty times the native population. The use of tributaries by native fishes as refugia from the changed canyon river environment influences these relationships, but the overall changes that have occurred within the riverine ecosystem have also played an important role in supporting and/or enhancing populations of nonnative species.

Changes in the fish community have been explained primarily by two causes. First, changes in water quality and flooding dynamics from a sediment-laden river with widely fluctuating temperatures and high spring floods to a clear, cool stream with little seasonal fluctuation tend to discourage spawning by natives except in tributaries. Impoundments also improve the survival of nonnative fishes such as striped bass, which prefer lentic (still, nonmoving water, as in lakes) conditions. Second, the high percentage of nonnative fishes that are predatory on natives tends to keep native fish numbers low. Intensive management of the trout fisheries in the 26 kilometers below the dam keeps that reach as a sportfishing mecca, and prevention of the upstream or downstream migration needed by some native species has also been cited as a cause for change in the fish community.

Concomitant with changes in the canyon's fish community has been a change in the aquatic macrophyte and macroinvertebrate communities, the foundation of the aquatic food base. Some aquatic macroinvertebrate species, on emerging from larval stages as winged insects, become a potential food source for the avian community. The energetics of the river below the dam have changed from heterotrophic (pertaining to systems in which most of the organic matter used as an energy source is derived externally, for example, from the river bank vegetation) to autotrophic (pertaining to systems in which all or most of the organic matter is derived internally, for example, through photosynthesis) because clear water below the dam allows the establishment of an extensive stand of photosynthetic green algae (*Cladophora*). These macrophytes are a substrate for periphyton (diatoms and *Gammarus*), which are a major food source for trout and other fishes. Increasing sediment loads downstream reduce the amount of photosynthetic algae and gradually return the river's aquatic food base to a semblance of pre-dam conditions.

**Riparian Vegetation.** Riparian vegetation along the Colorado River has also been greatly altered by river regulation. Prior to regulation, the delta, now waterless and barren, contained the vast majority of riparian vegetation along the Colorado River. Along the lower Colorado River, most original stands of Fremont cottonwood (*Populus fremontii*) and Goodding willow (*Salix gooddingii*) near the river, and honey mesquite (*Prosopis glandulosa*) bosques (forests or wooded thickets) on the floodplain terraces, have been lost, a consequence of river channelization and floodplain development. In a few locations where backwaters along reservoirs and rivers remain, there are dense stands of tamarisk (*Tamarix chinensis*), a small, nonnative tree species that naturalizes rapidly. These now create a wildlife habitat, replacing the role of lost native cottonwood and willow stands. Recent high flows resulting from uncontrolled spills from Colorado River dams in 1983 and 1993 have stimulated cottonwood and willow recruitment in areas near the confluence of the Colorado and Gila Rivers. The response of native riparian vegetation to these uncontrolled spills, which were a result of abnormally high winter snows and spring rains in the watershed, shows the importance of normal hydrology in maintaining a natural riverine ecosystem.

Within the canyons where the zone once scoured by spring floods was usually barren, extensive stands of tamarisk have developed. Only in a few places do native coyote willow (*Salix exigua*) successfully compete with tamarisk. Other nonnative species such as camelthorn (*Alhagi camelorum*) have also taken advantage of the new, less disturbed environments. The original canyon riparian zone, watered by high spring floods, supported species such as honey mesquite, catclaw acacia (*Acacia greggii*), and netleaf hackberry (*Celtis reticulata*). These species are now gradually dying out because no regulated water level reaches their root zone.

The lack of scouring along the river margin and in eddies has allowed the development of marsh ecosystems within backwaters and eddy return channels, a vegetation type essentially nonexistent under pre-dam conditions. Over time, marsh areas gradually fill with sediment and form low riparian zones, but the occasional flood (e.g., that of 1983) scours these areas and reestablishes appropriate conditions for future marsh development. Most marshes form in wide areas of the canyon where backwaters may be extensive.

Regulated flows and the establishment of dense riparian vegetation have enhanced the diversity of the avian community within the Grand Canyon. The lack of wide seasonal swings in river discharge allows greater utilization of the riverine ecosystem by water birds. Waterfowl occur mostly in winter along several reaches through the canyon. They are most common where the canyon is wide, daily river fluctuations are limited, and stream velocity is reduced. Extensive riparian vegetation, albeit mostly nonnative, has increased avian use of this zone. Several breeding pairs of the endangered southwestern willow flycatcher (*Empidonax trailli extimus*) have been found in tamarisk trees along the river, but their existence is threatened by the brown-headed cowbird (*Molothrus ater*). Along the lower Colorado River where the structural diversity of riparian vegetation has been greatly diminished, avian diversity has also been reduced, although tamarisk stands may support large populations of a few species.

Most riverine changes below dams result from regulated flows and reduced sediment transport. Studies have shown the importance of appropriately timed flooding events, since arid-region rivers are disturbance-oriented ecosystems. Removal of the many dams along the Colorado River is highly unlikely; hence the best approach to some semblance of restoration is to attempt to mimic natural hydrological events. In 1996, the Bureau of Reclamation and the U.S. Geological Survey created a controlled flood through the Grand Canyon by releasing high flows from Glen Canyon Dam. The flood did little to scour the nonnative vegetation that has established along the river, or to remove nonnative fishes. It did, however, reestablish many elevated sediment deposits, which will be used as substrate for riparian habitat as well as campsites for whitewater rafters. Management of dams to simulate natural hydrology (pertaining to the properties, distribution, and circulation of water) may be one approach to correcting decades of ecosystem alteration along Western U.S. rivers, the Colorado being only one example.

## BIBLIOGRAPHY

GRAND CANYON REGION

Carothers, S. W., and B. T. Brown. *The Colorado River through Grand Canyon: Natural History and Human Change.* Tucson: University of Arizona Press, 1991. A description of how the Colorado River through the Grand Canyon has changed since construction of Glen Canyon Dam, and comments on past and future directions for its management.

National Research Council. *Colorado River Ecology and Dam Management.* Proceedings of a Symposium, Santa Fe, New Mexico, 24–25 May 1990. Washington, D.C.: National Academy Press, 1991. A series of papers describing changes in the Grand Canyon with emphasis on the 1982–1986 Glen Canyon environmental studies research program, which documented changes caused by operations of Glen Canyon Dam.

Powell, J. W. *The Exploration of the Colorado River and its Canyons.* New York: Penguin, 1987. Powell's description of his 1869 boat trip down the Green and Colorado Rivers through the Grand Canyon from Green River, Utah, to the confluence of the Virgin and Colorado Rivers.

Schmidt, J. C., and J. B. Graf. "Aggradation and Degradation of Alluvial Sand Deposits, 1965–1986, Colorado River, Grand Canyon National Park, Arizona." U.S. Geological Survey Professional Paper 1493. Washington, D.C., 1990. A study of the fluvial and geomorphologic processes that have altered the sand deposits within the Grand Canyon as a consequence of operations of Glen Canyon Dam, which closed in 1963.

Shannon, J. P., et al. "Trophic Interactions and Benthic Animal Community Structure in the Colorado River, Arizona, USA." *Freshwater Biology* 31 (1994), 213–220. Report of a study of the changes taking place in the aquatic food base community in the Colorado River downstream from Glen Canyon Dam, with emphasis on a downstream gradient and effects of dam operations.

Stevens, L. E., et al. "Dam and Geomorphological Influences on Colorado River Waterbird Distribution, Grand Canyon, Arizona, USA." *Regulated Rivers: Research and Management* 13 (1994), 151–169. Report of a study of the distribution and habitat preferences of waterbirds as affected by canyon structure and operations of Glen Canyon Dam.

Webb, R. H. *Grand Canyon, a Century of Change: Rephotography of the 1889–1890 Stanton Expedition.* Tucson: University of Arizona Press, 1996. An analysis of ecological and geologic changes that have taken place in the Grand Canyon based on a comparison of recent photos with those of Stanton's party in 1889–1890.

Webb, R. H., J. C. Schmidt, G. R. Marzolf, and R. A. Valdez, eds. *The Controlled Flood in Grand Canyon.* American Geophysical Monograph 110. Washington, D.C.: American Geophysical Union, 1999. A compendium of many studies undertaken in the Grand Canyon during the controlled high flow release from Glen Canyon Dam in 1996.

LOWER COLORADO RIVER REGION

Fradkin, P. L. *A River No More: The Colorado River and the West.* Tucson: University of Arizona Press, 1984. A history of the Colorado River basin and the ultimate end of a totally controlled river.

Grinnell, J. "An Account of the Mammals and Birds of the Lower Colorado Valley, with Especial Reference to the Distributional Problems Presented." *Zoology* (University of California) 12 (1914), 51–294. An early description of the ecology of the lower Colorado River. Useful for comparison with that of today.

Lingenfelter, R. E. *Steamboats on the Colorado River, 1852–1916.* Tucson: University of Arizona Press, 1978. A history of steamboat activity on the lower Colorado River during a period of early development of the river, when it was still unregulated.

McDougall, D. T. "Delta and Delta Vegetation." *Botanical Gazette* 38 (1904), 44–63. Early ecological descriptions of the Colorado River delta. Useful for comparisons with the barren delta of today.

Ohmart, R. D., et al. "The Ecology of the Lower Colorado River from Davis Dam to the Mexico–United States International Boundary: A Community Profile." U.S. Fish and Wildlife Service Biological Report 85 (7.19). Washington, D.C.: U.S. Department of the Interior, Fish, and Wildlife Service, Research and Development, 1988. A report describing the many ecosystems along the Colorado from Davis Dam to the Mexican border. Data were based on comparisons of early and recent aerial photographs and field reconnaissance.

Reisner, M. *Cadillac Desert: The American West and Its Disappearing Water.* New York: Penguin, 1993. A history of water development in the West, conflicts between bureaucracy and the public, and the resulting altered rivers and ground water of Western arid lands.

Rosenberg, K. V., et al. *Birds of the Lower Colorado River Valley.* Tucson: University of Arizona Press, 1991. Description of the birds of the lower Colorado River with explanations of the ecological habitat preferences of each species.

Sykes, G. "The Colorado River Delta." Publication No. 460, Carnegie Institute. Washington, D.C., 1937. An early description of physiographic and ecological conditions of the Colorado River delta. Published about the same time as the closure of Hoover Dam.

—DUNCAN T. PATTEN

**COMETS.** *See* Impacts by Extraterrestrial Bodies.


# COMMONS

The "tragedy of the commons" has been a famous metaphor for problems related to common-pool resources ever since Garrett Hardin (1968) wrote his evocative article in the journal *Science*. Common-pool resources (CPRs) are natural or man-made resources that share two attributes: (1) substantial difficulty (but not impossibility) of devising ways to exclude individuals from benefiting from these resources, and (2) the subtractability of benefits consumed by one individual from those available to others (Ostrom et al., 1994, p. 6). CPRs share the first attribute with public goods and the second attribute with private goods.

Natural CPRs can range in size from global phenomena, such as the ocean's fisheries and migratory wildlife, to very small-scale local commons such as irrigation systems, small lakes, grazing lands, or inshore fisheries.

## TRAGEDY OF THE COMMONS

Garrett Hardin is a biologist trained in genetics. His 1968 article, bearing the above title, is a frank appraisal of what he calls the population problem. He insists that technological solutions, such as improved agricultural methods, are inappropriate, and he questions the implicit assumption that people have the right to reproduce without restraint in a world of finite resources.

Hardin recognizes that—despite Adam Smith's famous pronouncement in *The Wealth of Nations*—individuals acting to further their own ends do not always advance the public good, but instead deplete common resources. Hardin's best-known illustration, taken from an 1833 pamphlet by William Frank Lloyd, describes a common pasture, open to all and grazed by a number of herdsmen. Each herdsman realizes that any effects of overgrazing will be shared by all, while the benefits of a larger herd will be his alone; so each increases his own flock until the result is a tragic destruction of the grazing land.

Hardin applies the principle to cattlemen grazing in the southwestern United States, to saltwater fishing fleets, to tourists in national parks, and, of course, to individuals who choose to reproduce. He concludes that personal liberty cannot be respected if it threatens the common good. In today's world, regulations and incentives to modify behavior are appropriate and necessary.

**BIBLIOGRAPHY**

Hardin, G. "The Tragedy of the Commons." *Science* 162 (13 December 1968), 1243–1248.
Lloyd, W. F. *Two Lectures on the Checks to Population*. Oxford: Oxford University Press, 1833. Reprinted in part in Hardin, G., *Population, Evolution, and Birth Control*. San Francisco: Freeman, 1964.
Smith, A. *The Wealth of Nations*. London: Ward, Lock, and Tyler, n.d. Originally published in 1776 as *An Inquiry into the Nature and Causes of the Wealth of Nations*.

—David J. Cuff

Man-made CPRs may also range in size from the global (for example, the Internet) to the local (for example, a mainframe computer used by multiple researchers). A key distinction that needs to be made in understanding problems of governing and managing commons is between the resource system itself and the resource units such a system generates. The resource units of a CPR are finite and can thus be overused, leading to many externalities including congestion, higher costs of production, and a change in the attributes of the resource units themselves. For a groundwater basin, the resource system is the physical aquifer that has the potential of storing up to a maximum amount of water. The resource units from a groundwater basin are the quantities of water that are withdrawn for household, agricultural, or industrial uses. Resource systems have their own characteristics, as do the resource units. Resource systems vary in size, regularity, location, ease of access, and ease of monitoring. Resource units vary in quality, mobility, and predictability—and, of course, value. [*See* Resources.]

Because CPRs face problems related to exclusion, potential beneficiaries from their construction, management, improvement, or conservation will be strongly tempted to free-ride on the efforts of others if they can. When a group of fishers agrees to abide by a plan to share a conservatively estimated safe yield from a fishery, all of them benefit. They benefit from the preservation of their fishery over time, from the lowered cost of fishing, and from the increased quality of their catch. They all pay a short-term cost as well. The short-term cost is the immediate income foregone by following the agreed-upon conservative strategy in their fishing efforts. A fishing boat that did not agree to the limit, however, can benefit even more by harvesting at the same level as in the past or even by increasing their harvest rate while others decreased their rates. Unless the fishers who agreed to the management plan find a way of excluding those who did not agree to the plan (as well as monitoring the efforts of those who agreed), the individuals who pay a short-term cost in order to receive long-term benefits may not be persuaded into paying short-term costs and not receiving the long-term benefits. Thus, while the problems of exclusion are difficult to solve, solutions are essential if any highly valued commons are to be governed and managed effectively over the long term.

Because CPRs face problems related to subtractability, users always face potential problems of congestion, higher production costs, and even destruction. These are the obvious problems that fishers who negotiate a management plan try to avoid. It is a much discussed problem for contemporary users of the Internet. Users at key times of the day find that they cannot get information rapidly because so many other users are also accessing the Internet and it has become very congested. Various plans to increase the capacity of the Internet, as well as regulating the amount of use through pricing or rationing, are under discussion.

To govern and manage a CPR effectively, some form of property-rights system must be developed. Property-rights systems can be thought of as a bundle of rights related to who can access the CPR, whether they can withdraw resource units and how much, who can invest in and manage the CPR, who has the right to exclude users, and what type of rights can be transferred to others under what circumstances (Schlager and Ostrom, 1992). When all of these rights are given to private individuals or private corporations, one has a private-property system. When all of these rights are held by a local, regional, or national government, one has a public property-rights system. A common- or communal-property system is involved when a private association of individuals or firms holds at least access, management, and exclusion rights even though they may not be able to alienate these rights to others. In many common-property systems, however, participants may sell their rights if other members of the association approve of the sale. Many CPRs exist without well-defined property-rights systems having been designed. Some of these open-access resources are located within a single country. If demands on these resources grow rapidly, serious degradation can occur before legal actions are initiated. Other open-access resources are larger than a single nation, and the lack of an effective treaty leaves the resource, such as ocean fisheries, unprotected to be exploited ruthlessly. [See Fishing.] The most severe problems of the commons exist for open-access regimes as a result of free-riding and overuse.

Much contemporary policy analysis has presumed that state and private ownership are the only two alternative modes of effectively managing CPRs. Since a major study by a National Academy of Sciences panel was conducted in the mid-1980s, however, common-property regimes of diverse kinds have been found to be as or more effective than some state or private-property systems (Bromley et al., 1992; Feeny et al., 1990). Careful studies of a large number of farmer-managed irrigation systems in Nepal, for example, have found that farmers are able to gain higher yields, use their irrigation systems for three seasons of the year, and have more equitable water distribution systems than government-owned systems (Lam, 1998). For global commons that

cross national boundaries, state ownership is not even an option. Thus, various kinds of international regimes are essential ingredients of any effort to govern and manage a global commons (Keohane and Ostrom, 1995). [See International Cooperation.] Many regimes of all kinds, however, do fail to control access and regulate use so that severe problems of overuse and destruction have occurred.

In an effort to determine the attributes of successful CPR regimes, Ostrom (1990) identified eight design principles. Most successful, robust, long-lasting CPR regimes can be characterized by six or more of these principles. Failed and fragile CPR regimes do not use more than a few of them (Schweik et al., 1997; Morrow and Hull, 1996; Blomqvist, 1996). The particular rules used in effectively managed systems vary substantially. Consequently, it is not possible to develop a set of blueprints that can be used to guarantee successful management. The design principles are as follows.

Design Principle 1: Rules that clearly define who has rights to use a resource and the boundaries of that resource ensure that users can clearly identify anyone who does not have rights and take action against them as well as identify the geographic span of the resource itself.

Design Principle 2 involves two parts. The first is a congruence between the rules that assign benefits and the rules that assign costs so that participants consider the rules to be fair. The second is that both types of rules be well matched to local conditions such as soils, slope, number of diversions, crops being grown, and so on.

Design Principle 3 is concerned with the collective-choice arrangements used to modify the operational rules used to manage a resource. If most users are involved in modifying these rules over time, the information about the benefits and costs as perceived by different participants is taken more fully into account in these efforts to adapt to new conditions and information over time. And, if technological change brings new conditions, participants are able to adjust their rules so as to keep up with new opportunities or new threats.

No matter how high the level of agreement to an initial agreement is, there are always conditions that tempt some individuals to cheat (even when they perceive the overall benefits of the system to be higher than the costs). Without effective monitoring of rule conformance—Design Principle 4—few systems are able to function well at all.

Design Principle 5 focuses on the use of graduated sanctions. The important thing about a sanction for a user who has succumbed to temptation is that their action is noticed and that a punishment is meted out. This tells everyone that cheating on rules is noticed and punished without making all rule infractions into major criminal events. If someone breaks the rules repeatedly and is noticed doing so, the rule breaker eventually faces a severe penalty or expulsion from the set of authorized users.

While rules are always assumed to be clear and unambiguous in theoretical work, this is rarely the case in field settings. Design Principle 6 is the provision of fair, low-cost conflict resolution arenas so that the natural conflicts that occur can be resolved rapidly and in a manner that is considered legitimate.

Design Principle 7 relates to the rights of a group to devise their own institutions and have these rights recognized by international, national, regional, and local governments. When this is the case, the legitimacy of the rules crafted by users will be challenged less frequently in the courts and in administrative and legislative settings.

Design Principle 8 stresses the need, in larger resources with many participants, for nested enterprises that range in size from small to large, to enable participants to solve diverse problems involving economies on different scales. By utilizing base institutions that are quite small, face-to-face communication can be used for solving many of the day-to-day problems in smaller groups. By nesting each level of organization in a larger level, externalities from one group to others can be addressed in larger organizational settings that have a legitimate role to play in relationship to the smaller entities. The last principle is particularly important for governing global commons.

[*See also* Growth, Limits to; Market Mechanisms; Public Policy; *and* Technology.]

## BIBLIOGRAPHY

Blomqvist, A. *Food and Fashion: Water Management and Collective Action among Irrigation Farmers and Textile Industrialists in South India.* Linköping, Sweden: Institute of Tema Research, Department of Water and Environmental Studies, 1996.

Bromley, D. W., D. Feeny, M. McKean, P. Peters, J. Gilles, R. Oakerson, C. F. Runge, and J. Thomson, eds. *Making the Commons Work: Theory, Practice, and Policy.* San Francisco: Institute for Contemporary Studies Press, 1992.

Feeny, D., F. Berkes, B. J. McCay, and J. M. Acheson. "The Tragedy of the Commons: Twenty-two Years Later." *Human Ecology* 18 (1990), 1–19.

Hardin, G. "The Tragedy of the Commons." *Science* 162 (1968), 1243–1248.

Keohane, R. O., and E. Ostrom, eds. *Local Commons and Global Interdependence: Heterogeneity and Cooperation in Two Domains.* London: Sage, 1995.

Lam, W. F. *Governing Irrigation Systems in Nepal: Institutions, Infrastructure, and Collective Action.* Oakland, Calif.: Institute for Contemporary Press Studies, 1998.

Morrow, C. E., and R. W. Hull. "Donor-Initiated Common Pool Resource Institutions: The Case of the Yanesha Forestry Cooperative." *World Development* 24 (1996), 1641–1657.

Ostrom, E. *Governing the Commons: The Evolution of Institutions for Collective Action.* New York: Cambridge University Press, 1990.

Ostrom, E., R. Gardner, and J. Walker. *Rules, Games, and Common-Pool Resources.* Ann Arbor: University of Michigan Press, 1994.

Schlager, E., and E. Ostrom. "Property-Rights Regimes and Natural Resources: A Conceptual Analysis." *Land Economics* 68 (August 1992), 249–262.

Schweik, C., K. Adhikari, and K. N. Pandit. "Land-Cover Change and Forest Institutions: A Comparison of Two Sub-Basins in the Southern Siwalik Hills of Nepal." *Mountain Research and Development* 17 (1997), 99–116.

—ELINOR OSTROM

**COMPLIANCE.** *See* Environmental Law; International Cooperation; *and* Public Policy.

**COMPUTING.** *See* Information Technology.

## CONSERVATION

Conservation is a social movement aimed at altering the attitudes and practices of the human race toward its use of the life-support processes of the planet. Conservation is the antidote to the tendency, found universally in human cultures, to alter their surroundings in order to compete and survive. This is a strong declaratory statement that requires explanation. The term *conservation* is used here as a generic for any act of environmental protection, or behavior, that promotes the sustained well-being of ecosystems and social cohesiveness. In that sense, the interpretation is close to that of Aldo Leopold, the famous U.S. commentator on environmental care, namely, that conservation was an act and a thought process that led to the continuity of the integrity of ecological functioning. This entry points out that ecosystem integrity also requires a democratically functioning society that links social care to ecological care. If Leopold had not been writing as a naturalist, he would surely have linked the two as well. [*See the biography of Leopold.*]

In this very sustaining sense, it can therefore be said that without conservation, the tendency to self-destruct would go unchecked. All societies, to a greater or lesser extent, have exhibited both philosophies and practices of conservation. The actual character of these philosophies alter with the nature of the economy, the character of communication, the state of science, and the educational processes of ethical awareness and concern.

Does the institutionalization of this conservation ethos change anything? The answer depends on the culture of acceptance of the conservation-environmental ideal. If the act of conserving is seen either as altruistic or self-sacrificial, then regulation will always have to act against the social grain. Penalty rather than reward will be the regulatory driver. But if conservation environmentalism becomes accepted as a private and a social good combined, then the enactment of conservation becomes ecologically constructive and socially bonding. We are not there yet, but we could still move in that direction.

Historians and anthropologists have long suggested that no society is immune from a drive to act in a manner that overuses resources and exploits nature. Despite a widespread belief that indigenous people are more or less sustainable, even those societies have to put in place sophisticated cultural taboos to control this objective. In that sense, the taboos, societal regulations, and moral codes are all part of what is properly known as conservation. In certain cases, some societies appear to have succumbed to the ravages of resource scarcity or disease because the conservation readjustment was never adequately put in place. But such conclusions may be too simplistic. The causes of cultural collapse are often as much in the pattern of social order than in the mismanagement of the planet. [See Belief Systems.]

Conservation takes on four main manifestations:

1. The *preservation movement* is anxious to retain the status quo in the protection of particular ecosystems, cultural heritage, or modes of living. This movement is composed of a number of different motives. Some of the justification is self-serving: the desire to protect amenities in one's neighborhood, to maintain property values, or to exclude undesirable intruders. This is the most criticized effect of preservation, for good reason, since it relies on prejudice and political muscle for its operation. But another motive is to protect habitats, ecosystems, or natural processes that are critical for the healthy functioning of a life-supporting planet. In this instance, the advocates are acting more from a global interest than from personal gain. Examples exist in the biodiversity protection strategies aimed at safeguarding species and their habitats that are regarded as essential for the life-support processes of the globe.

2. The *ecomanagement movement* seeks to apply principles of interventionist management in "unruly" ecosystems to harness their functions for the good of humanity. This is largely a utilitarian-based approach seeking to maximize the value of natural resources yet leave them sustainable for future use. The techniques used here include maximum, sustained yield in forestry practices and maximum allowable catch in fisheries practices. In general, these approaches are failing mostly because the state of scientific knowledge is not sufficiently developed to sense the points of tolerance. But more to the point, the managers rarely have the power to control property rights, price signals, and political interference by vested interests. For example, fisheries are being overwhelmed the world over because no governments have the authority or the will to control the fishing industry to the point where it takes only the allowable catch. [See Fishing.]

3. The *precautionary approach* in conservation applies different rules for its persuasion. First, it seeks to create a society that is more ecocentric than technocentric, in that it respects nature and seeks pleasure in giving the natural world room to breathe. Second, it tries to countermand the science and economics of tolerance and limits by applying the *precautionary principle*. This recognizes that science is fallible, that

**Conservation. Table 1.** The Spectrum of Environmentally Related Political Activity*

| *Passive* | | | | | *Active* |
|---|---|---|---|---|---|
| ORDINARY POLITICAL PARTICIPATION | PASSIVE LOBBYING | ACTIVE SUPPORT | ACTIVISM | DIRECT ACTION | REVOLUTIONARY ACTION |
| • Following events in the media<br>• Voting in local and national elections<br>• Responding to surveys and questionnaires | • Letter writing<br>• Signing petitions<br>• Joining groups<br>• Making donations | • Attending meetings and demonstrations<br>• Leafleting and collecting money<br>• Boycotting goods, companies, or institutions | • Organizing events, boycotts, and lobbying efforts<br>• Doing research and writing<br>• Organizing campaigns and fund-raising<br>• Lecturing and public speaking | • Picketing and committing acts of obstruction<br>• Engaging in ethical shoplifting | • Engaging in complete civil disobedience, sabotage, and terrorism |

*These categories are intended to reflect the different degrees of commitment to environmental change. The particular category in which an activity is placed is necessarily somewhat arbitrary. For example, although boycotting could be considered a form of direct action, it is placed under "Active support" because it is essentially nonconfrontational. Also, although some individuals and groups may fall within a single category, many will not.

SOURCE: O'Riordan (1995, p. 28).

outcomes cannot be predicted because the processes being interfered with are not linear or forecastable, and that the interests of future generations are served better if they are given rights in today's management decisions. This reproach is partly ethically biased but is founded more on the basis of a concern for a different approach to justice and the intrinsic rights of natural objects. [*See* Precautionary Principle.]

4. *Direct action.* Modern societies are characterized by a degree of violence, fragmentation, and disobedience to laws, distrust of authorities, especially governments, and a slim sense of do-it-yourself civic activism. Some direct action is illegal, dangerous, and violent. This is still a minority pursuit and is associated with zealots, high-minded and sincere protesters, and genuine troublemakers with criminal intent. It is dangerous to "type" direct action: it comes in many forms for many reasons (see Table 1). Peaceful direct action is more common and is a feature of modern conservation that is growing in significance. The reason for this lies in a more civically active society generally, along with protests led by articulate and well-connected nongovernmental organizations. The rise of the World Wide Web has considerably increased their cohesion and effectiveness. [*See* Environmental Movements.]

Conservation is a profound movement, rooted in the very nature of humanity and triggered by economic change, technological transformation, political fragmentation, and ethical concern. Its very adaptability is its enduring strength.

[*See also* Biological Diversity; Deforestation; Erosion; Industrial Ecology; Land Reclamation; Land Use, *article on* Land Use and Land Cover; Parks and Natural Preserves; Resources; Salinization; Tourism; Water; Wilderness; World Heritage Sites; *and the biographies of Marsh and Thoreau.*]

### BIBLIOGRAPHY

Easterbrook, G. *A Moment of the Earth: The Coming of Age of Environmental Optimism.* London: Penguin Books, 1996.
O'Neill, J. *Ecology, Policy and Politics.* London: Routledge, 1993.
O'Riordan, T. "Framework for Choice: Core Beliefs and the Environment." *Environment* 37.8 (1995), 4–9, 24–29.
Pepper, D. *Modern Environmentalism: An Introduction.* London: Routledge, 1995.

—TIMOTHY O'RIORDAN

# CONSULTATIVE GROUP ON INTERNATIONAL AGRICULTURAL RESEARCH

The Consultative Group on International Agricultural Research (CGIAR), an informal association of governments, private foundations, and international organizations, was established in 1971 to support a global system of international agricultural research centers (IARCs). The CGIAR came about because wise leaders in two private foundations concluded that the best scientists should be mobilized to work on pressing international problems on behalf of the world's poor to alleviate hunger in poor countries.

The CGIAR comprises a network of sixteen autonomous IARCs that seek to promote sustainable agriculture for food security in developing countries. These centers are strategically located in different parts of the globe in response to their respective mandates.

The CGIAR has led the way in making agricultural research a truly global enterprise, involving more than a hundred nations and thousands of scientists in both poor and wealthy countries. The IARCs work on a wide array of formal and informal partnerships with developing-country research organizations and advanced laboratories in developed countries. The CGIAR has been deemed one of the most successful international initiatives of the post–World War II era.

The first two international agricultural research centers, the International Rice Research Institute in the Philippines and the Centro Internacional de Mejoramiento de Maiz y Trigo (CIMMYT, the International Maize and Wheat Improvement Center), gained early recognition for their role in producing new high-yielding rice and wheat varieties that sparked the "green revolution" in Asia beginning in the late 1960s. Since that time, wheat and rice production in Asia has more than doubled, feeding hundreds of millions of people who otherwise would have been threatened with severe food shortages and starvation.

The centers conduct research on crops, livestock, fisheries, and forests, according to research agendas agreed on by their respective Boards of Directors, supported by a Technical Advisory Committee in collaboration with national governmental and nongovernmental organizations, universities, and private industry. They also develop policy initiatives and strengthen national research capacities, mainly through training.

Some of the major achievements of the CGIAR system include genetic varietal improvements in such crops as rice, maize, wheat, cassava, and potatoes, as well as cattle. The result has been increased productivity brought about by yield improvement, genetic resistance to pests and diseases, and improved farming practices that better conserve the natural resource base in dry, semiarid, tropical, and high-altitude regions where most poor farmers routinely struggle to produce their staple food crops and livestock products. The CGIAR also preserves one of the world's largest collections of genetic resources of crops, which are used to develop improved crop strains. Other important contributions include studying food and agriculture policy needs, assisting developing countries to improve national research capabilities, and training of thousands of young developing-country scientists.

Undoubtedly, some of the greatest challenges for developing countries in the twenty-first century relate to food security, poverty alleviation, and natural resources management. Global change may exacerbate the challenge. CGIAR efforts and achievements are a direct response to global change and its consequences.

[*See also* Agriculture and Agricultural Land.]

## BIBLIOGRAPHY

Baum, W. C. *Partners against Hunger.* The Consultative Group on International Agricultural Research. Washington, D.C.: World Bank, 1986.

Consultative Group on International Agricultural Research. *Annual Report, 1997.* Washington, D.C., 1997.

Kabbaj, O. *The Challenge of Development and Poverty Reduction in Africa: 1997 Sir John Crawford Memorial Lecture.* Washington, D.C.: CGIAR, 1997.

—Donald L. Plucknett and Robert B. Kagbo

**CONTINENTAL DRIFT.** *See* Earth Structure and Development; *and* Plate Tectonics.

# CONVENTION ON BIOLOGICAL DIVERSITY

Preservation of nature was one of the first international environmental issues. By the late 1980s, more than two dozen international agreements to protect migratory birds, limit destruction of wetland habitats, and regulate trade in endangered species were on the books. Those agreements were typically fragmented—they protected specific named resources, often denoted on long lists in an appendix to the treaty. But research showed that preservation of biodiversity—the variability among living organisms—would require management of whole ecosystems. [*See* Biological Diversity.] Taken to its logical extreme, this holistic approach would demand an integrated, worldwide effort. That is the goal of the Convention on Biological Diversity (CBD), adopted by governments in 1992 at the United Nations Conference on Environment and Development held in Rio de Janeiro. [*See* Rio Declaration; *and* United Nations Conference on Environment and Development.]

The agreement that became the CBD was initiated by the United Nations Environment Programme (UNEP) and the International Union for the Conservation of Nature and Natural Resources (the World Conservation Union; IUCN) in an effort to promote holistic conservation and management of ecosystems. As negotiations proceeded, many other interests were entrained. The treaty finally adopted in 1992 has three central objectives: (1) conservation of biological diversity, (2) promotion of its sustainable use, and (3) equitable sharing of the benefits of genetic resources.

The treaty establishes biological diversity as the common concern of humankind, while also reaffirming that states control the biological resources within their territories. Implementation of the CBD has required that these sometimes conflicting central objectives and principles be reconciled, which has not been easy.

The first two objectives—conservation and sustainable use of biological diversity—reflect the CBD's origins in the environmental groups and governments, mainly in the industrialized world. The treaty specifically obliges countries to develop national biodiversity plans and to monitor their biological resources. It requires *in situ* efforts to protect biological diversity, such as the establishment of protected areas and rehabilitation of degraded ecosystems. It also obliges countries to complement those efforts with *ex situ* conservation, such as occurs in botanical gardens, zoos, and seed banks.

The obligations of the CBD are loosely worded. Moreover, in most countries, at least some of these efforts to conserve and sustainably use biological diversity were already under way or planned when the CBD was negotiated. Thus many of these commitments codify what governments were already doing (and urge them to do more). At its five-year anniversary it was difficult to identify many areas where the CBD had, itself, resulted in more conservation and sustainable use of biodiversity.

The CBD's third objective—equitable sharing of the benefits of genetic resources—has been the most contentious and ambiguous. The modern capitalist economy is based on private ownership of property, including intellectual property such as inventions. Many inventions, such as novel drugs, are based in part on natural biological diversity. For example, extracts of thousands of wild plants may be screened to discover one that is active against human ailments and thus promising as a commercial drug. Because human effort is needed to identify and refine those parts of biodiversity that are useful, a growing number of legal systems recognize the resulting product as intellectual property that can be owned and protected (for example, with patents). Although many drugs and other products are now entirely synthetic, often their useful properties were first identified in the wild. Examples include cancer-fighting Taxol extracted from the bark of wild (and endangered) yew trees.

Countries rich in biodiversity hoped that, by adding the third objective to the CBD and reaffirming state control over biological resources, they could extract larger royalty payments for inventions based on their raw genetic material. Those payments could make protection of nature more cost-effective. Moreover, many argued that often such material was not "raw"; rather, in many cases its useful properties had been identified and applied over the decades by farmers and other human stewards of biodiversity. They argued that capitalist firms merely took that information, made marginal improvements, and then reaped the full reward by patenting the final product. For example, the Indian Neem tree is the source of more than fifty U.S. patents for products ranging from contraceptives to pesticides. Efforts have been made by India to revoke some patents on the basis that they reward multinational companies for "innovations" that, actually, are derived from age-old Indian techniques.

The allocation of intellectual property rights and other benefits from biodiversity is further complicated by two other considerations. First, the patenting of life forms is a highly controversial issue. Second, many other forms of intellectual property protection have been in existence for a long time, and thus the CBD may conflict with existing practice. For example, plant breeders have long had free access to international seed banks; in turn, they have produced and sold improvements to seed varieties without the obligation to share the commercial benefits. Typically, a new variety incorporates traits from dozens of different strains drawn from private gene banks as well as public collections, notably, the international gene bank network maintained by the Consultative Group on International Agricultural Research (CGIAR). If free access were repealed, it might be a nightmare to work out a scheme for distributing the benefits fairly among the sources of the original material. Moreover, for many staple crops, large public benefits—such as improved food supply and security of the "green revolution"—have flowed from open access to germplasm collections. Open access was enshrined in a nonbinding Undertaking on Plant Genetic Resources adopted (first in 1983 and revised periodically) by the Food and Agriculture Organization (FAO) as well as in several binding international agreements to regulate plant breeding. Efforts are under way in the FAO to reconcile these many agreements and practices with the CBD.

The CBD also created a financial mechanism to transfer resources to developing countries to help them implement the convention's obligations. The task of managing that mechanism was initially given to the Global Environment Facility (GEF) of the United Nations Development Programme, UNEP, and the World Bank. Because the commitments in the CBD are vague, through 1998 the only activities funded through GEF that were unambiguously related to implementation of the CBD were projects for preparing national biodiversity inventories and management plans. In addition, GEF has a closely related funding program on biodiversity that has funded many projects for conservation and use of biological resources. In addition to the financial mechanism, the CBD creates a clearinghouse mechanism intended to aid the transfer of technology.

The CBD is designed to evolve and explicitly envisions the adoption of protocols—legally binding agreements that extend the CBD's commitments. The first protocol—the Cartagena Protocol—adopted in early 2000, addresses the risks to natural ecosystems of trade in genetically modified organisms, also known as biosafety. [See Cartagena Protocol.] However, scientific evidence suggests that other factors, such as the introduction of exotic species into natural environments and the continued destruction of habitat such as forests and wetlands, are greater threats to biodiversity.

[See also Food and Agriculture Organization; Global Environment Facility; United Nations Environment Programme; and World Conservation Union, The.]

### INTERNET RESOURCE

Secretariat of the Convention on Biological Diversity. http://www.biodiv.org/.

### BIBLIOGRAPHY

Raustiala, K., and D. G. Victor. "Biodiversity since Rio: The Future of the Convention on Biological Diversity." *Environment* 38.4 (1996), 16–20, 37–45.

—DAVID G. VICTOR

## CONVENTION ON INTERNATIONAL TRADE IN ENDANGERED SPECIES

The Convention on International Trade in Endangered Species of Wild Fauna and Flora (CITES) is an international agreement that attempts to protect endangered species through regulation of international trade in wildlife. Such an approach was first suggested by the International Union for the Conservation of Nature and Natural Resources (the World Conservation Union; IUCN) in 1963 and supported by a resolution at the United Nations Conference on the Human Environment in 1972. CITES itself was concluded in 1973 in Washington, D.C. and entered into force in 1975. More than 130 states are party to the agreement.

CITES operates to protect species by placing them in one of three appendixes. Species listed in Appendix I are those that are the most endangered. Commercial trade in these species is prohibited, and other trade is permitted only by special permit issued by both exporting and importing countries. These permits verify that the specimen was not taken in contravention of any existing regulations and that export will not be detrimental to the species' overall survival. Species listed in Appendix II are those that are likely to become endangered if trade is not controlled. Trade in these species, or their parts, requires an export permit. Species listed in Appendix III are those that individual countries have indicated are protected within their territory; trade in species originating in the country that listed them requires a permit. Each party to the treaty designates a Management Authority and a Scientific Authority, responsible for issuing permits and determining whether the conditions for trade have been met. Trade in listed species is prohibited with nonparties unless those states present documentation comparable to that required under the treaty.

The biennial Conference of the Parties to CITES votes on species to be listed in Appendix I or II; such a listing requires an affirmative vote by two-thirds of the parties voting. Individual countries may list species within their borders in Appendix III at any time. Once a decision has been made to list a species, any party may decide to lodge a reservation within 90 days, indicating that it will not be bound by the regulations pertaining to that species.

The best-known CITES activities relate to African elephants. The listing of this species in Appendix II failed to prevent a drastic decline in population numbers, due in part to inability to enforce regulated trade, so the species was moved to Appendix I in 1989. While the overall population of African elephants is still endangered, some southern African countries with successful conservation policies argued in favor of managed trade in ivory to provide much-needed income. At the 1997 CITES meeting, an agreement was reached to allow Botswana, Namibia, and Zimbabwe to resume a limited trade in ivory at the end of 1998.

CITES has had mixed success. The generation and inspection of permits for wildlife trade is difficult for countries without well-developed governmental bureaucracies or good control over their borders. A black market thrives in some restricted species. But trade is better regulated than in the past, and the information gathered by the CITES process raises awareness and lowers demand in a way that has helped some species to survive.

[See also Elephants; Extinction of Species; Hunting and Poaching; and World Conservation Union, The.]

### BIBLIOGRAPHY

Burns, W. C. "CITES and the Regulation of International Trade in Endangered Species of Flora: A Critical Appraisal." *Dickinson Journal of International Law* 8.2 (1990), 203–233. Burns addresses the often-overlooked element of protection of flora under CITES and argues that the treaty has not had nearly as much success in protecting these flora as it has fauna.

Favre, D. "Debate within the CITES Community: What Direction for the Future?" *Natural Resources Journal* 22.4 (1993), 875–918.

Hemley, G., ed. *International Wildlife Trade: A CITES Sourcebook.* Covelo, Calif., and Washington, D.C.: Island Press, 1994. In addition to general information about trade in endangered species, this book includes the entire text of the CITES agreement and appendixes and information about parties and particular species.

Lyster, S. *International Wildlife Law.* Cambridge: Grotius Publications, 1985. Chapter 12 addresses CITES specifically. Although this book is somewhat old, it is a classic in addressing issues of wildlife law and contains a particularly lucid and detailed explanation of law under CITES.

—ELIZABETH R. DESOMBRE

## CONVENTION ON LONG-RANGE TRANSBOUNDARY AIR POLLUTION

In the 1960s scientists demonstrated that there was a link between emissions of sulfur in continental Europe (particularly because of fossil fuel burning) and the enhanced acidification of lakes in Scandinavia. Subsequently, it became abundantly clear that air pollutants and acid precipitation could damage sites that were hundreds of kilometers from the points of emission. [See Acid Rain and Acid Deposition.] This appreciation led in 1979 to the signing of the Convention on Long-Range Transboundary Air Pollution under the auspices of the United Nations Economic Commission for Europe (UN/ECE). It came into force in 1983 and now includes forty-three parties among the fifty-five UN/ECE member states. Its geographic spread is large and encompasses a great swath of the old industrialized world, from the United States and Canada in the west to the Russian Federation and the republics of Central Asia in the east.

The objectives of the convention are (1) to protect human beings and the environment against air pollution and (2) to limit and, as far as possible, reduce and prevent air pollution, including that of a transboundary nature. Convention activities initially focused on reducing the effects of acid precipitation through control of sulfur emissions. Subsequently, however, its scope has been widened to address various other types of pollutants: ground-level ozone, nitrogen oxides, persistent organic pollutants (POPs), and heavy metals. Indeed, since the convention came into force in 1983, it has been extended by eight protocols:

1. The 1984 *Protocol on Long-Term Financing of the Cooperative Programme for Monitoring and Evaluation of the Long-Range Transmission of Air Pollutants in Europe (EMEP).*

2. The 1985 *Protocol on the Reduction of Sulphur Emissions or Their Transboundary Fluxes by at Least 30 Percent.*

3. The 1988 *Protocol concerning the Control of Nitrogen Oxides or Their Transboundary Fluxes.*

4. The 1991 *Protocol concerning the Control of Emissions of Volatile Organic Compounds or Their Transboundary Fluxes.*

5. The 1994 *Protocol on Further Reduction of Sulphur Emissions* ("the second sulfur protocol").

6. The 1998 *Protocol on Heavy Metals.*

7. The 1998 *Protocol on Persistent Organic Pollutants (POPs)*.

8. The 1999 *Protocol to Abate Acidification, Eutrophication, and Ground-Level Ozone*.

Three cooperative activities have been identified as being fundamental: collection of data on long-range transport and deposition, determination of possible effects, and increasing our understanding of the cost-effectiveness of abatement strategies. The first of these activities centers on EMEP (the European Monitoring and Evaluation Programme). This plays a crucial role in coordinating an international network for the collection and evaluation of data on the emission, transport, and deposition of air pollutants. The second activity is carried forward by ICPs (International Cooperative Programmes) that look at the effects of air pollution on forests, surface waters, crops, materials, and ecosystems. It also involves the mapping of "critical loads"—a necessary precondition for the second sulfur protocol of 1994, which set emission targets according to the sensitivity of downwind ecosystems. With regard to the third activity, integrated assessment models have been developed that account for the movement of pollutants between countries, costs of reducing emissions from different sources, and environmental impacts based on critical loads and air quality. In addition, a variety of economic mechanisms—joint implementation, burden sharing, economic instruments, and economic assessment of damage—are being explored to gauge their potential to contribute to environmental protection.

[*See also* International Institute for Applied Systems Analysis.]

### INTERNET RESOURCE

Information on the convention and its protocols is provided at http://www.unece.org/env.lrtap.

—ANDREW S. GOUDIE

## CONVENTION ON WETLANDS OF INTERNATIONAL IMPORTANCE ESPECIALLY AS WATERFOWL HABITAT. *See* Ramsar Convention.

## CREATIONISM. *See* Origin of Life.

## CRETACEOUS–TERTIARY BOUNDARY. *See* Impacts by Extraterrestrial Bodies.

## CRYOSPHERE

The cryosphere is that part of the Earth's surface or subsurface environment that is composed of water in the solid state. It contains nearly 80 percent of all the Earth's fresh water. Its six main elements, in diminishing order of area occupied at the surface, are: (1) seasonal snow cover, (2) sea ice, (3) polar ice sheets, (4) mountain glaciers, (5) river and lake ice, and (6) permafrost. Ice in clouds is also an important element in climate but is not considered part of the cryosphere.

**Climatic Roles.** In considering the roles of cryospheric components, it is useful to think in terms of aspect ratio (ratio of lateral extent to thickness) and metabolic rate (annual mass balance/total volume = 1/residence time). Typical orders of magnitude are shown in Table 1. Variations in these quantities, as well as differing response times to atmospheric forcing, determine the differing climatic roles.

Seasonal snow responds rapidly to atmospheric dynamics on time scales of days to weeks. It has a small seasonal heat storage, and its most important climatic effect is its high albedo, affecting surface heat balance. Fresh snow reflects more than 90 percent of the shortwave radiation falling on it. [*See* Albedo.]

Sea ice operates in the climate system at time scales up to seasons or years. It, too, has a high albedo and so its seasonal cycle can also affect surface heat balance in the same way as snow but to a lesser extent, at least in the Northern Hemisphere. Its two most important climatic roles are that it acts as a barrier to the exchange of heat, moisture, and momentum between the ocean and the atmosphere, and that it plays an oceanic role in deep-water formation. River and lake ice behave climatically much like sea ice. [*See* Sea Ice.]

Polar ice sheets, of which the largest cover Antarctica and Greenland, are quasi-permanent features, responding to the atmosphere at long time scales (thousands to hundreds of thousands of years). They have a high elevation and high albedo, and so act as elevated cooling surfaces for atmospheric heat balance. Changes in their volume, although slow, could cause large changes in sea level: complete melt would add more than 80 meters of water. [*See* Sea Level.]

Mountain glaciers are small components of the cryosphere in terms of area and volume. However, their high rates of accumulation and ablation allow rapid changes in response to climatic change (time scales of years to centuries). They are believed to be a major contributor

**Cryosphere. TABLE 1.** Aspect Ratio and Metabolic Rate of Major Components of the Cryosphere

|                  | ASPECT RATIO      | METABOLIC RATE           |
| ---------------- | ----------------- | ------------------------ |
| Mountain glaciers | $10^2$            | $10^{-2}$                |
| Ice sheets       | $10^3$            | $10^{-4}$ to $10^{-5}$   |
| Sea and lake ice | $10^6$            | $10^0$                   |
| Snow             | $10^6$ to $10^7$  | $10^0$                   |

SOURCE: After Untersteiner (1984).

to the current rate of sea level rise, accounting for about half of it, the rest being mainly thermal expansion of the ocean.

Permafrost is permanently frozen subsoil, underlying about 25 percent of the Earth's land surface. [*See* Permafrost.*] It may be continuous (that is, underlying the whole of a region), or discontinuous, occurring in patches. It is a product of present and past climate, changing on time scales of centuries. It affects surface ecosystems and river discharge, and changes in its extent could trigger increased rates of methane emission.

In general, the cryosphere is a very changeable feature of the Earth's surface on all time scales because the freezing point of water lies in the center of the range of terrestrially achieved temperatures. Hence, although it is recognized that the fundamental driving mechanisms for climatic fluctuations occur in tropical and subtropical regions, the most active scene of climatic unrest lies in the polar regions.

**Distribution of Ice in the Cryosphere.** The total volume of water on Earth in all forms is 1.348 billion cubic kilometers. Of this, 97.4 percent is sea water, 0.0009 percent is atmospheric water vapor, 0.5 percent is ground water at depth, 0.1 percent is in rivers and lakes, and 2.0 percent is in the solid state. These categories imply that 77 percent of the fresh water on Earth is in the form of ice and snow. Perennial ice covers 11 percent of the Earth's land surface and 7 percent of the world's oceans.

The area and volume of cryospheric elements, and their seasonal variability, have been estimated by Fitzharris et al. (1996) in the latest IPCC assessment, from which Table 2 is adapted. The table demonstrates some very interesting results. The greatest seasonal variation in area of any cryospheric element (and, indeed, of any geophysical element on the Earth's surface) is that of the seasonal snow cover on land in the Northern Hemisphere. Seasonal snow variability is a feature that would be very obvious to an observer from outer space. Given the high albedo of snow (0.9 when fresh over a flat surface, less when forming a cover over forest) compared with snow-free terrain (typically 0.1), this implies that any retreat of the winter or spring snow line would allow the absorption of more solar radiation by the Earth's surface, causing an anomalously large warming of the lower atmosphere. Thus any warming trend is enhanced at high northern latitudes by a positive ice-albedo feedback mechanism, and this is the main reason why general circulation models (GCMs) of the atmosphere predict a greatly enhanced rate of climatic warming in high northern latitudes under anthropogenic greenhouse forcing.

The seasonal variation of sea ice extent in the Arctic is also significant and contributes to the arctic ice-albedo feedback effect. However, it is much less than the seasonal sea ice variation in the Antarctic. This is because the Arctic Ocean is a truly polar ocean surrounded by land, with only a limited range of longitudes

**Cryosphere. TABLE 2.** Area and Volume Occupied by Major Elements of the Cryosphere

| CRYOSPHERE ELEMENT | AREA ($\times 10^6$ KM$^2$) | VOLUME OF ICE ($\times 10^6$ KM$^3$) |
|---|---|---|
| Seasonal snow | | |
| N. Hemisphere winter | 46.3 | <0.01 |
| N. Hemisphere summer | 3.7 | |
| S. Hemisphere winter | 0.9 | |
| S. Hemisphere summer | <0.1 | |
| Sea ice | | |
| N. Hemisphere winter | 16.0 | 0.05 |
| N. Hemisphere summer | 9.0 | 0.03 |
| S. Hemisphere winter | 19.0 | 0.03 |
| S. Hemisphere summer | 3.5 | <0.01 |
| Polar ice sheets | | |
| E. Antarctica | 9.9 | 25.92 |
| W. Antarctica | 2.4 | 3.40 |
| Antarctic ice shelves | 1.6 | 0.79 |
| Greenland | 1.7 | 2.95 |
| Other icecaps and mountain glaciers | 0.6 | 0.09 |
| River and lake ice | <1.0 | |
| Permafrost | 25.4 | 0.16 |

SOURCE: Adapted from Fitzharris et al. (1996).

within which it can expand seasonally into lower latitudes. There is also a very strong pycnocline (density gradient) separating a low-salinity polar surface-water layer 50–200 meters thick (fed by a vast outflow of fresh water from Siberian rivers) from warmer, more saline, water below (flowing in from the Atlantic and the Bering Sea). The pycnocline hampers heat flow from the warmer layer to the surface, and the ocean heat flux in the Arctic in winter is less than 2 watts per square meter. For these reasons, the Arctic features a permanent sea ice cover of multiyear ice, augmented in winter by a seasonal ice cover of first-year ice stretching into the subpolar seas of the Atlantic, Baffin Bay, Hudson Bay, the Labrador Sea, the Bering Sea, and the Sea of Okhotsk.

Antarctic sea ice has a much larger seasonal variation; in fact, almost all the winter ice cover melts in summer, leaving multiyear ice only in embayments, such as the Weddell and Ross seas that possess semiclosed gyres, and high-latitude seas, such as the Bellingshausen. [See Antarctica.] This is because Antarctica occupies the area around the pole, and the Antarctic Ocean itself does not begin until about 65°–70° south latitude (off East Antarctica), giving less extreme conditions. Furthermore, the Antarctic pycnocline is weak, since the polar surface water is freshened only from icebergs, ice shelf melt, and precipitation, and hence large ocean heat fluxes of 40 watts per square meter or more can occur as a result of heat flow from the underlying circumpolar deep water. In the Southern Hemisphere, the seasonal sea ice cover variation is chiefly responsible for the ice-albedo feedback effect, since seasonal snow (mainly in Patagonia) is relatively unimportant. Since this variation is less than the snow cover variation in the Northern Hemisphere, the predicted enhancement of global warming at southern high latitudes is more modest than in the north.

When we look at polar ice sheets in Table 2 we see the enormous importance of the East Antarctic ice sheet; it is dominant in area and volume and contains in itself more than 60 percent of the Earth's fresh water. In places it is more than 4,000 meters thick. The recently discovered subglacial Lake Vostok, possibly 500 meters deep, underlies part of the East Antarctic ice sheet. The West Antarctic ice sheet is continuous with that of East Antarctica but is generally thinner and lower in elevation, overlying bedrock that is in many places several hundred meters below sea level. It has been hypothesized that this ice sheet may become unstable under global warming, especially if its fringe of ice shelves breaks up or floats away, but these fears are now generally discounted. Nevertheless, local warming at present is indeed causing breakup in the Wordie and Larsen ice shelves that fringe the Antarctic Peninsula.

Ice shelves surround about one-third of the coastline of Antarctica and are the means by which the ice sheet, always slumping outward under its own weight by a creep process, goes afloat and calves off icebergs. Ice shelves are typically 1,000 meters thick in their inner parts, thinning to 500 meters or less at the fronts where they calve, this being the thickness of the tabular icebergs produced. Not all of the flux of fresh water emerging from Antarctica is in the form of icebergs, since there is also a sub-ice circulation often associated with the presence of coastal polynyas at the edge of the ice shelf. These are regions that retain open water through all or part of the winter because katabatic winds (winds that accelerate downslope) from the ice sheet blow seaward and move newly forming coastal sea ice out to sea. The polynyas are sea ice factories responsible for up to 40 percent of antarctic sea ice production, and are also sources of an important water mass called *Antarctic Bottom Water*, which occupies the deepest parts of southern oceans. The formation of sea ice in the coastal polynyas followed by its wind-induced removal causes salt to be deposited in the upper layers (since even new sea ice retains only about a third of the original salt content of the sea water), which helps generate dense water (assisted by a mixing process between water masses). The dense water flows down and outward off the continental shelf as bottom water, but there is also a circulation under the ice shelf, where the sinking and inflowing water helps to melt the ice bottom surface, flowing out again in a diluted state.

The Greenland ice sheet, more than 3,000 meters thick at its center, is the only arctic ice sheet of significance, although there are other smaller icecaps in Ellesmere Island, Devon Island, Svalbard, and some Russian Arctic islands. Together with mountain glaciers, these lesser icecaps take up a very small part of the total ice volume and area.

**The Changing Cryosphere.** Given that the polar regions should exhibit enhanced climatic sensitivity because of ice-albedo feedback, we might already expect to be detecting cryospheric changes due to warming. This is the case in some, but not all, aspects of the cryosphere. The main driving force, air temperature, has certainly shown a strong positive trend since 1966 over northern high-latitude Eurasia and North America (but a slight negative trend over eastern Canada and southern Greenland), making the Arctic warmer than at any time for the past four hundred years (Overpeck et al., 1997). A lesser warming has occurred in the Antarctic.

Seasonal snow extent has been mapped by satellites since 1971, and its thickness on land is now measurable by interpretation of passive microwave sensor data (Parkinson, 1997). Data for the Northern Hemisphere show a large decline in area in the 1980s relative to the 1970s, by about 13 percent in autumn and 9 percent in spring. However, a partial recovery occurred during

1994–1996, and the overall picture appears to be a modest decline superimposed on large fluctuations with periodicities of about five years. [See Snow Cover.]

Sea ice extent can also be mapped by passive microwave from satellites, because of the large difference between the microwave emissivities of ice and open water. By combining the data from two satellites to obtain a seventeen-year record, Bjørgo et al. (1997) found that arctic sea ice extent has shown a diminishing trend from 1978 to 1995, with a decline of 2.6 percent per decade in extent (area enclosed by the ice edge) and 3.4 percent in actual ice area over that period. Antarctic sea ice showed a much smaller and statistically insignificant decline. These data were reanalyzed by Cavalieri et al. (1997) with the addition of data from 1996. They confirmed the arctic trend, but found that the additional recent data gave a small positive trend to antarctic sea ice extent (1.3 percent per decade). Our conclusion so far must be that satellite data sets are still not long enough to identify definite trends in the presence of large fluctuations that themselves have periodicities of three to five years.

Sea ice thickness is much more difficult to measure and is best done from beneath by upward-looking sonar mounted in submarines. The resulting problem is a lack of systematic, frequent coverage. Wadhams (1990) found that, between 1976 and 1987, there was a 15 percent decline in mean ice thickness over the Eurasian sector of the Arctic Ocean. This could represent part of a fluctuation, but more recent data collected in 1996 show a further significant decline in mean thickness. In the Antarctic, where military submarines cannot go, thickness data have come mainly from moored upward sonars, and so far the data sets are not long enough to test for trends.

The Antarctic ice sheet is thought to be in a state of approximately neutral mass balance (i.e., the gain from snowfall each year is matched by the loss from ice shelf melt and iceberg calving). However, disputed evidence from satellite altimetry suggests that the center of the Greenland ice sheet is increasing in elevation while its flanks are shrinking—an effect that might be expected from global warming where warmer, moister winds deposit greater snowfall at high altitude while increased melt occurs at lower altitude. Mountain glaciers have certainly been retreating worldwide throughout the past century—alpine glaciers have lost about half of their mass in 150 years. The general retreat is partly compensated by some advances due to local factors. It has been estimated by IPCC that, over the last one hundred years, the retreat of mountain glaciers has caused global sea levels to rise by 0.2–0.4 meters, approximately equal to the direct contribution from ocean warming due to thermal expansion. The rate of retreat is predicted to increase, causing many small glaciers and icecaps to disappear completely. River and lake ice are diminishing in the sense that breakup dates for river ice have tended to advance (in Siberian rivers) by 7–10 days over the last century.

Finally, permafrost also shows evidence of retreat and melt in Alaska (see, for example, Osterkamp and Romanovsky, 1999) and Siberia, but not in eastern Canada. Much deep permafrost is already relict, that is, it was formed during the past glacial period and has been melting only very slowly since then. One effect of shallow permafrost melt is to increase the emission of methane from the wetter active surface layer, which itself acts as a positive feedback for climatic warming. Another effect of the melt of permafrost where it extends offshore in arctic areas might be the release of trapped gas hydrates from the continental shelf, again increasing greenhouse forcing. [See the vignette on Methane Hydrates in the article on Fossil Fuels.]

The effect of glacier melt on sea level is an instance of the way in which cryospheric changes interact with, or cause, changes in other aspects of the global system. Another example is the effect of sea ice retreat in the Greenland Sea on the global thermohaline circulation (the worldwide "conveyor belt" of currents driven by sinking of surface water in a few parts of the world and upwelling elsewhere). The Odden ice tongue is an area of sea ice that grows out locally from the main East Greenland ice edge in the vicinity of 72°–75° north latitude during winter. Salt released from the growing ice causes convection and makes this one of the few sinking regions in the world's oceans. In recent years, reduced ice growth has caused convection to reduce in volume, and to reach only shallow depths (1,000 meters) instead of the ocean bed (5,000 meters), with probable consequences for the thermohaline circulation of the Atlantic and, ultimately, the world.

[See also Glaciation; Global Warming; Hydrologic Cycle; and Ice Sheets.]

## BIBLIOGRAPHY

Bjørgo, E., et al. "Analysis of Merged SMMR-SSMI Time Series of Arctic and Antarctic Sea Ice Parameters 1978–1995." *Geophysical Research Letters* 24 (1997), 413–416.

Cavalieri, D. J., et al. "Observed Hemispheric Asymmetry in Global Sea Ice Changes." *Science* 278 (1997), 1104–1106.

Fitzharris, B. B., et al. "The Cryosphere: Changes and Their Impacts." In *Climate Change 1995: Impacts, Adaptations and Mitigation of Climate Change; Scientific-Technical Analyses*, pp. 241–265. Contribution of Working Group II to the Second Assessment Report of the Intergovernmental Panel on Climate Change. Cambridge: Cambridge University Press, 1996. The latest consensus view of IPCC on current changes in the cryosphere and their impacts on the global system.

Global Atmospheric Research Programme. "The Physical Basis of Climate and Climate Modelling." GARP Publications Series No. 16. Geneva, 1975.

Hambrey, M., and J. Alean. *Glaciers*. Cambridge: Cambridge University Press, 1992. A popular, well-illustrated account of glaciers and ice sheets and their role in the climate system.

Nesje, A., and S. O. Dahl. *Glaciers and Environmental Change.* London: Edward Arnold, 2000.

Osterkamp, T. E., and V. E. Romanovsky. "Evidence for Warming and Thawing of Discontinuous Permafrost in Alaska." *Permafrost and Periglacial Processes* 10.1 (1999), 17–37.

Overpeck, J. K., et al. "Arctic Environmental Change of the Last Four Centuries." *Science* 278 (1997), 1251–1256.

Parkinson, C. *Earth from Above: Using Color-Coded Satellite Images to Examine the Global Environment.* Sausalito, Calif.: University Science Books, 1997. A popular account of how satellites are used to map properties of the Earth's surface, including the cryosphere.

Paterson, W. S. B. *The Physics of Glaciers.* 3d ed. New York: Pergamon, 1994. The standard work on the dynamics and thermodynamics of glaciers.

Serreze, M. C., et al. "Observational Evidence of Recent Change in the Northern High-Latitude Environment." A new and valuable review based partly on data sets stored at the World Data Center for Glaciology in Boulder, Colo., 1998.

Untersteiner, N. "The Cryosphere." In *The Global Climate,* edited by J. T. Houghton, pp. 121–140. Cambridge: Cambridge University Press, 1984. A useful review of the whole topic with a good bibliography, although now a little outdated.

Wadhams, P. "Evidence for Thinning of the Arctic Ice Cover North of Greenland." *Nature* 345 (1990), 795–797.

Wadhams, P., et al., eds. *The Arctic and Environmental Change.* Amsterdam: Gordon and Breach, 1996. A multiauthor review of current changes in many aspects of the Arctic cryosphere, based on a Royal Society discussion meeting.

—Peter Wadhams

**CYCLONES.** *See* Atmosphere Dynamics.

# D

**DAISYWORLD.** *See* Biological Feedback; *and* Gaia Hypothesis.

## DAMS

Dams and reservoirs have been an integral part of water management from the earliest days of civilization. They store water in wet periods for use in dry periods for agriculture and cities; they produce electricity by tapping the energy of falling water; they reduce the risks of disastrous flooding; they create reservoirs used for recreation and play; and they assist navigation by leveling and increasing flows in low-flow periods. [*See* Electrical Power Generation.] Recently, however, dams have become the focus of intense international debate because of their negative, and often ignored, impacts on both people and natural ecosystems. [*See* Irrigation; Water; *and* Water Resources Management.]

Humans have been building dams since the earliest days of human civilization. There are examples over five thousand years old from Mesopotamia and Egypt. Legend has it that the earliest known dam across a river, the Sadd el-Kafara, was built in the Middle East five thousand years ago (Gleick, 1994). Examples of ancient dams can be found in China, South Asia, Europe from the Middle Ages, and North America before the arrival of Europeans.

Despite this long history, the widespread construction of large dams did not begin until the middle of the twentieth century, when improvements in engineering and construction skills, hydrologic analysis, and technology made it possible to build them safely. Much of the initial large-dam construction occurred in the United States. The government built about fifty large dams between 1900 and 1930. Between 1930 and 1980 it built a thousand more, together with tens of thousands of smaller ones. [*See the vignette for definitions of large dams.*] The first of the truly massive dams was the Hoover Dam, built by the U.S. Bureau of Reclamation on the Colorado River in the 1930s. [*See* Colorado River, Transformation of the.] By 1945, the five largest dams on Earth had been completed, all in the western United States, initiating a surge in dam construction throughout the country. Today, the reservoirs behind U.S. dams store 60 percent of the entire average annual river flow of the United States (Hirsch et al., 1990).

Other countries also saw large dams as vital for national security, economic prosperity, and agricultural survival. In the Soviet Union, Stalin dreamed of transforming that country's massive rivers into controlled projects to provide electricity for Soviet industries and "transform nature" into a machine for the communist state. By the 1970s, the total area flooded by dams in the USSR greatly exceeded that of the United States, and the world's tallest dam—the 300-meter-tall Nurek—was completed in 1980 in Tajikistan. In India, dam construction initiated by the British colonial government was quickly adopted by the independent government after 1947, which built thousands of dams and associated facilities. Following the Chinese revolution in 1949, more than 600 dams were built every year for three decades, with both wondrous and disastrous results.

Today, nearly 500,000 square kilometers of land worldwide are inundated by reservoirs capable of storing 6,000 cubic kilometers of water (Shiklomanov, 1993, 1996; Collier et al., 1996). This redistribution of fresh water is so large that scientists recently reported that it is responsible for a small but measurable change in the orbital characteristics of the Earth (Chao, 1995). Today's dams have nearly 640,000 megawatts of installed hydroelectric capacity and produce nearly 20 percent of the world's total supply of electricity. Table 1 shows the total installed capacity and annual hydroelectric generation for major regions of the world. Figure 1 shows the breakdown of hydroelectric production by region for 1997. Only a small fraction is generated in Africa and Oceania, and much of the hydropower generation is concentrated in a few countries, including Canada, the United States, Brazil, and China. In 63 countries, hydropower supplies more than 50 percent of total electricity supply; it supplies between 90 and 100 percent in 23 countries.

The construction of large dams continues today. In Asia, China is pursuing an ambitious hydropower development program, including the massive and controversial Three Gorges project. India has 10,000 megawatts approved or under construction and another 28,000 megawatts planned. Laos, Nepal, Malaysia, Russia, and the Philippines all have major construction projects under way. In Latin America, Brazil has the most ambitious plans, with more than 10,000 megawatts under construction and plans for another 20,000 megawatts. Honduras, Mexico, and Ecuador all have major projects under consideration. In Europe, projects for new or rehabilitated hydroplants are being pursued in Albania, Bosnia, Croatia, Greece, Iceland, Macedonia, Portugal, Slovenia, and Spain. In North America, Canada plans to build more than 3,000 megawatts of additional

## WHAT IS A LARGE DAM?

There is no single definition of a large dam. According to criteria set by the *International Journal on Hydropower and Dams*, there are more than 300 major dam projects that meet at least one of the following criteria:

- dam height exceeding 150 meters
- dam volume exceeding 15 million cubic meters
- reservoir volume exceeding 25 billion cubic meters, or
- installed electrical capacity exceeding 1,000 megawatts

The International Commission on Large Dams (ICOLD) offers a less restrictive set of criteria for defining a large dam. According to ICOLD criteria, there are roughly 40,000 large dams and over 800,000 small ones (McCully, 1996). A large dam is one whose height is 15 meters or greater; or whose height is between 10 and 15 meters if it meets at least one of the following conditions:

- A crest length of not less than 500 meters
- A spillway discharge potential of at least 2,000 cubic meters per second
- A reservoir volume of not less than one million cubic meters.

—PETER H. GLEICK

capacity over the next 15 years (*International Journal on Hydropower and Dams*, 1997).

Many of these dams may never be built. Serious opposition to large dams is growing throughout the world because of their environmental, social, and cultural consequences. Among the many impacts are land inundation, loss of riparian (river-related) habitat, adverse effects on aquatic species, and reservoir-induced seismicity (RIS), as well as social impacts on local populations and people who must be uprooted and resettled from reservoir areas (Goldsmith and Hildyard, 1986; White, 1988; Covich, 1993; McCully, 1996).

Only a small fraction of land area worldwide has been lost to reservoirs, but this land is often of special value as fertile farmland, riparian woodland, or wildlife habitat. River flood plains are among the world's most diverse ecological systems, balancing aquatic and terrestrial habitat, species, and dynamics. In addition to the area lost to the reservoir behind a dam, there are often secondary impacts on surrounding lands, as humans are required to move to build new homes and farms. Wildlife populations displaced by the dam must seek new habitat. Access roads built in remote areas where many dams are built also bring in loggers and others searching for quick economic gains.

Dams can also destroy dramatic scenery when they are located in mountainous terrain. The Itaipú Reservoir inundated the spectacular waterfall of Sete Quedas on the Paraná River. The dam that provides water for the city of San Francisco flooded the Hetch Hetchy Valley in California, which was considered comparable in beauty to the famous Yosemite Valley. In an ironic twist, the dam that created Lake Powell along the Colorado River was named for a magnificent remote desert

**Dams. TABLE 1.** Hydroelectric Production by Region, 1997

| REGION | INSTALLED HYDROELECTRIC CAPACITY (MEGAWATT) | HYDROELECTRIC PRODUCTION (GIGAWATT-HOURS/YEAR) |
|---|---|---|
| Africa | 20,112 | 66,981 |
| North and Central America and the Caribbean | 153,193 | 668,106 |
| South America | 97,053 | 450,409 |
| Asia | 192,457 | 693,706 |
| Europe | 158,181 | 524,672 |
| Oceania | 12,751 | 41,520 |

SOURCE: Modified from Gleick (1998).

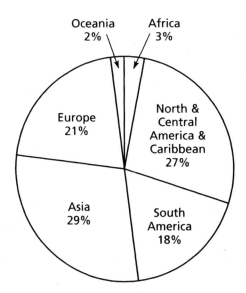

**Dams. Figure 1.** Hydroelectric Production per Region, 1997. (Modified from Gleick, 1998.)

canyon that the rising waters destroyed—Glen Canyon. Indeed, massive public opposition to additional dams along the Colorado resulted in large part from the desire to prevent the destruction of the Grand Canyon, one of the most important symbols of natural beauty in the United States.

Dams radically alter natural free-flowing river systems and put pressure on aquatic ecosystems and species. There is often an immediate decline in the diversity of native fishes after dam construction because the new environment is so different from the original river ecosystem or because nonnative species that outcompete native species are introduced. Dams interrupt the seasonal upstream migration of many species and alter the flow, sediment regime, dissolved oxygen content, and temperature of the habitat, all of which affect reproduction and survival. They alter the hydrology of a basin and the dynamics of riverine ecosystems through changes in nutrient retention, water levels, and water chemistry. Large fluctuations in reservoir levels during droughts or seasonal drawdowns alter the habitat along the margins where fish feed and spawn. Temperature differences in reservoirs often result in completely different aquatic flora and fauna.

The ecological impacts of dams on rivers extend all the way to the sea. Many of the world's major fisheries depend on the volume and timing of fresh water and nutrient flows from large rivers into estuaries or the ocean. Almost all of the fish caught in the Gulf of Mexico, the Grand Banks of Newfoundland, the Caspian Sea, the eastern Mediterranean, and off the western coast of Africa depend on river discharges. The vast salmon fisheries of the eastern Pacific along the coasts of Canada and the United States rely on the ability of the fish to spawn in freshwater streams. Dams in all of these areas, combined with overfishing and fisheries mismanagement, have led to great declines in many anadromous (freshwater-spawning) fish populations.

Partly as a consequence of large dams, more than 20 percent of all freshwater fish species are now considered threatened or endangered, and many of the most severe impacts have been felt by amphibians, insects, waterfowl, and plants. The western coast of the States once boasted 400 separate salmon and steelhead stocks, but a century of habitat loss, dam construction, changes in flow regimes, and overfishing have eliminated nearly 50 percent of them and many of the remainder are at high risk of extinction. Other examples of aquatic species threatened by dams include several species of river dolphins in Latin America and Asia, sturgeon in North America, China, and Russia, and commercial and noncommercial fisheries at the mouth of almost every major river, including the Nile, Colorado, Volga, and Indus, and in the Black, Azov, and Caspian Seas.

Another unusual consequence of some dam construction is RIS, in which earthquakes are caused by the filling of a large reservoir. For most cases of RIS, the intensity of seismic activity increases with the size of a reservoir and the speed the reservoir is filled. The strongest shocks normally occur as the reservoir approaches maximum volume, and earthquakes may continue for a few years after initial filling. RIS has occurred near at least two hundred reservoirs around the world, but the mechanism is not well understood. Of the eleven reported cases of earthquakes greater than 5.0 on the Richter scale associated with filling of reservoirs, ten of them occurred in dams higher than 100 meters.

One of the most serious concerns about large dams has been the involuntary displacement and resettlement of people living in areas to be flooded by reservoirs. Often entire villages are destroyed, with generations of history and culture lost, churches and burial grounds flooded, and homes and farmlands inundated. Limited and often contradictory data are available on populations displaced by dam construction and the creation of reservoirs. According to a World Bank study (1994), approximately 40 million people were displaced by dams between 1986 and 1993, an average of about five million people a year. Others estimate that this many people may have been displaced by dams since the turn of the century (McCully, 1996). The vast majority of people displaced by dam construction are in India and China, two countries with extremely high population densities and aggressive dam-development programs. In China, over 10 million people were officially displaced by water projects between 1960 and 1990, and the massive Three Gorges Dam project now under construction will displace between one and two million more. The resettle-

## ENVIRONMENTAL IMPACTS OF LARGE VERSUS SMALL DAMS

While much has been written about the environmental impacts of large dams, there is still considerable uncertainty about both the impacts of single small dams and the cumulative impacts of many small dams. Moreover, there are many different types of dams, built and operated for many different purposes in a variety of environments, making any generalizations unreliable and misleading.

Single large dams have attracted the most attention from the environmental science and environmental advocacy communities because they often create enormous reservoirs, flood large land areas, displace large populations, and form significant barriers to aquatic species and navigation. Smaller dams may be built on smaller river systems, be designed to either create a reservoir or operate in "run-of-river" mode without storing water, generate small amounts of electricity, or divert small amounts of irrigation water. They thus tend to avoid the more obvious environmental and social disruption of large projects, but they produce fewer benefits as well. Proper comparison of impacts is therefore extremely difficult.

Appropriate measures of the environmental consequences of dams must include not only their size but how they are designed and operated. In a comparison of large and small dam systems in the western United States, multiple small dams were shown to have some impacts that exceeded those of a single large dam (Gleick, 1992). In particular, while many large hydroelectric dams with large reservoirs flood a substantial amount of land and lose water to evaporation, certain kinds of small dams may actually flood more land or lose more water to evaporation per unit of energy produced. Another factor relevant to determining net impacts is how dams are operated. Dams operated in "run-of-river" mode—independent of size—have less impact on natural riverine ecosystems than dams that store water in large reservoirs and alter the timing, quality, and character of streamflow.

This area requires further analysis. No comprehensive study has been done, for example, on the numbers of people displaced by dams as a function of energy produced, land flooded, or water supplied. Few analyses of the net impact to natural ecosystems have been done as a function of either the size of a dam or the number of dams built on a single river. And information is especially limited on impacts in developing countries, where baseline environmental data, reliable social indicators, and environmental assessments are less common or less readily available.

—PETER H. GLEICK

ment issue is complicated to address; it involves restoring and improving living standards as well as trying to resolve a host of problems usually ignored by traditional economic solutions.

**New Developments in the Dam Debate.** The negative environmental and social impacts resulting from past dam projects have given rise to opposition to new ones. Few major dams are now moving forward free of public scrutiny and analysis, and several major projects have been canceled or postponed. Many more have been redesigned to address public concerns and interests.

While many new construction plans are well along and may fill a vital need, serious obstacles will limit the actual level of development and lead to the cancellation of some current projects. Growing public opposition on environmental and social grounds, more comprehensive regulatory procedures, and new complexities in the financing of major dam projects are all working to slow development. The World Bank, long a major international funder in this area, is rethinking guidelines and environmental requirements for major dam projects.

Public debate and opposition to major dam projects have appeared in Nepal, India, South Africa, several countries in Latin America, and even China, where active dissent against government policies is a dangerous and courageous act. Laws in the United States, Norway, and Sweden limit development on pristine rivers. Opposition comes from local grassroots efforts and, internationally, from nongovernmental organizations. Local communities are no longer willing to accept immediate social and environmental disruption for a promise of hypothetical future benefits. Moreover, the frequent disregard for the civil rights of people living in rural areas affected by some dam projects has led to a widespread distrust of all dam projects.

Major projects are likely to be successful only if local

Perhaps the most important anthropologically induced seismicity results from the creation of large reservoirs. Reservoirs impose stresses of significant magnitude on crustal rocks at depths rarely equaled by any other human construction. With the ever-increasing number and size of reservoirs the threat rises. There are at least six cases (Konya, Kremasta, Hsinfengkiang, Kariba, Hoover, and Marathon) where earthquakes of a magnitude greater than 5.0 on the Richter scale, accompanied by a long series of foreshocks and aftershocks, have been related to reservoir impounding. There are many more locations where the filling of reservoirs behind dams has led to appreciable, though less dramatic, levels of seismic activity. Detailed monitoring has shown that earthquake clusters occur in the vicinity of some dams after their reservoirs have been filled, whereas before construction activity was less clustered and less frequent. Similarly, there is evidence that there is a linear correlation between the storage level in the reservoir and the logarithm of the frequency of shocks. Among the reasons why dams induce earthquakes are the hydro-isostatic pressure exerted by the mass of the water impounded in the reservoir, together with the changing water pressures across the contact surfaces of faults. Given that the deepest reservoirs provide surface loads of only 20 bars or so, direct activation by the mass of the impounded water seems an unlikely cause and the role of changing pore pressure assumes greater importance.

However, the ability to prove an absolutely concrete cause-and-effect relationship between reservoirs and earthquakes is limited by our ability to measure stress below depths of several kilometers, and some examples of induced seismicity may have been built on the false assumption that because an earthquake occurs in proximity to a reservoir it has to be induced by that reservoir (Meade, 1991).

**BIBLIOGRAPHY**

Meade, R. B. "Reservoirs and Earthquakes." *Engineering Geology* 30 (1991), 245–262.

—ANDREW S. GOUDIE

populations are fully consulted, involved, and represented in decisions about a project and its alternatives. Yet planning for almost all major dam projects is still usually conducted behind closed doors by a small number of engineers and water experts. In most countries, dam developers solicit little or no public feedback, and decisions are made by a small group of water managers or governmental organizations, together with international funding agencies (Oud and Muir, 1997). Such lack of consideration for the public has also contributed to growing opposition to dams around the world.

While new dams that will contribute to improving the quality of life for many people will be built in the future, there are signs that the philosophy of dam construction is changing. More attention is finally being paid to understanding and mitigating the environmental and social impacts of dams, and to soliciting the participation of local people in making decisions about new facilities. And in some places, some bad decisions made over the past century may finally be reconsidered. In late 1997 the U.S. government for the first time refused to relicense a hydroelectric dam and ordered the structure—the 16-meter-high Edwards Dam on the Kennebec River in Maine—destroyed on the grounds that its costs as a barrier to migratory fish significantly exceed its hydroelectric benefits. Other dams that have environmental or social costs exceeding their benefits are also facing destruction. While none are huge, they symbolize many of the problems facing all dam projects. The precedence of their destruction may herald a new generation of hydrologic engineers and ecological scientists whose field of expertise is the removal of large dams.

[*See also* Ecosystems; *and* Energy.]

**BIBLIOGRAPHY**

Chao, B. F. "Anthropological Impact on Global Geodynamics Due to Water Impoundment in Major Reservoirs." *Geophysical Research Letters*, 22 (1995), 3533–3536.

Collier, M., R. H. Webb, and J. C. Schmidt. *Dams and Rivers: Primer on the Downstream Effects of Dams.* Tucson, Ariz., and Denver, Colo.: United States Geological Survey, 1996.

Covich, A. P. "Water and Ecosystems." In *Water in Crisis: A Guide to the World's Fresh Water Resources*, edited by P. H. Gleick, pp. 40–55. New York: Oxford University Press, 1993.

Gleick, P. H. "Environmental Consequences of Hydroelectric Development: The Role of Facility Size and Type." *Energy* 17.8 (1992), 735–747.

———. "Water, War, and Peace in the Middle East." *Environment* 36.3 (1994), 6.

———. *The World's Water: 1998–1999.* Washington, D.C.: Island Press, 1998.

Gleick, P. H., P. Loh, S. Gomez, and J. Morrison. *California Water 2020: A Sustainable Vision.* Oakland, Calif.: Pacific Institute for Studies in Development, Environment, and Security, 1995.

Goldberg, C. "Fish Are Victorious over Dam as U.S. Agency Orders Shutdown." *The New York Times* (26 November 1997), A12.

Goldsmith, E., and N. Hildyard, eds. *The Social and Environmental Impacts of Large Dams.* Cornwall, U.K.: Wadebridge Ecological Centre, 1986.

Goodland, R. "Ethical Priorities in Environmentally Sustainable Energy Systems: The Case of Tropical Hydropower." World Bank Environment Working Paper 67, 1994.

———. "Environmental Sustainability in the Hydro Industry: Disaggregating the Debates." In *Large Dams: Learning from the Past, Looking at the Future,* proceedings of IUCN and the World Bank Group Workshop, Gland, Switzerland, 11–12 April 1997, edited by T. Dorcey, A. Steiner, M. Acreman, and B. Orlando, pp. 69–102. Washington, D.C.: World Bank, 1997.

Hirsch, R. M., J. F. Walker, J. C. Day, and R. Kallio. "The Influence of Man on Hydrologic Systems." In *The Geology of North America,* vol. O-1, Surface Water Hydrology, pp. 329–359. Boulder, Colo.: Geological Society of America, 1990.

McCully, P. *Silenced Rivers: The Ecology and Politics of Large Dams.* London: Zed Books, 1996.

McPhee, J. *Encounters with the Archdruid.* New York: Farrar, Straus and Giroux, 1971.

Murty, K. S. *Soil Erosion in India, River Sedimentation,* vol. 1. Beijing: International Research and Training Center on Erosion, 1989.

Ortiz, A. *Handbook of North American Indians,* vol. 9: *Southwest.* Washington, D.C.: Smithsonian Institution, 1979.

Oud, E., and T. C. Muir. "Engineering and Economic Aspects of Planning, Design, Construction, and Operation of Large Dam Projects." In *Large Dams: Learning from the Past, Looking at the Future,* proceedings of IUCN and the World Bank Group Workshop, Gland, Switzerland, 11–12 April 1997, edited by T. Dorcey, A. Steiner, M. Acreman, and B. Orlando, pp. 17–39. Washington, D.C.: World Bank, 1997.

Reisner, M. *Cadillac Desert: The American West and Its Disappearing Water.* New York: Penguin, 1993.

Rouse, H., and S. Ince. *History of Hydraulics.* Ames: Iowa Institute of Hydraulic Research, State University of Iowa, 1957.

Scudder, T. "Recent Experiences with River Basin Development in the Tropics and Subtropics." *Natural Resources Forum* 18.2 (1994), 101–113.

———. "Social Impacts of Large Dam Projects." In *Large Dams: Learning from the Past, Looking at the Future,* proceedings of IUCN and the World Bank Group Workshop, Gland, Switzerland, 11–12 April 1997, edited by T. Dorcey, A. Steiner, M. Acreman, and B. Orlando, pp. 41–68. Washington, D.C.: World Bank, 1997.

Sewell W. R. D. "Inter-Basin Water Diversions: Canadian Experiences and Perspectives." In *Large Scale Water Transfers: Emerging Environmental and Social Experiences,* edited by G. N. Golubev and A. K. Biswas, pp. 7–35. United Nations Environment Program, Water Resources Series Volume 7. Oxford: Tycooly, 1985.

Shiklomanov, I. "World Fresh Water Resources." In *Water in Crisis: A Guide to the World's Fresh Water Resources,* edited by P. H. Gleick, pp. 13–24. New York: Oxford University Press, 1993.

Shiklomanov, I. A. "Assessment of the Water Resources and Water Availability in the World." Draft Report to the Comprehensive Assessment of the Freshwater Resources of the World. St. Petersburg, Russia: State Hydrological Institute, 1996.

Stanford, J. A., and J. V. Ward. "Stream Regulation in North America." In *The Ecology of Regulated Streams,* edited by J. V. Ward and J. A. Stanford, pp. 215–236. New York: Plenum, 1979.

United States Committee on Large Dams (USCOLD). *Register of Dams,* 1996. http://www.uscold.org/~uscold/uscold_s.html/.

White, G. "The Environmental Effects of the High Dam at Aswan." *Environment* 30.7 (1988), 4.

World Bank. "Resettlement and Development: The Bankwide Review of Projects Involving Involuntary Resettlement 1986–1993." Washington, D.C.: World Bank, 1994.

———. *Resettlement Remedial Action Plan for Africa.* Washington, D.C.: World Bank, 1995.

—PETER H. GLEICK

# DATING METHODS

The reconstruction of past climates depends heavily on highly refined stratigraphy and dating methods. Paleoclimatology, the study of ancient climates, is a field that is coming into its prime because of the importance of determining the variation and magnitude of climatic change over different parts of the Earth at different times in its history. Stratigraphy includes the composition, sequence, age, and correlation of rock strata or deposits; as a field of study it also includes fossil content, geochemical, chemical, mineralogical, and magnetic properties, mode of origin, and all physical properties and attributes that mark rock or other deposits as strata. Strata are organized or distributed according to superposition (Steno's law) such that younger strata overlie older strata, or are sequentially distributed on the land surface in accordance with episodic geologic activity (for example, relative aged moraines from glaciations, or dune sequence from eolian activity occurring at various times).

All geologic history and paleoclimatic/paleoenvironmental reconstruction is based on the physical, mineralogical, chemical, and biological variations that occur within and between strata in a sequence. Thus lateral changes within strata may be as important as vertical changes within a stratigraphic column in terms of assessing regional variations in paleoclimate. [*See* Climate Change.]

The geologic time scale (Figure 1) is a chronological sequence of geologic events that serves as a measure of the relative or absolute age of a part of geologic time; time is not a material unit or body of strata, and it is not, strictly speaking, a stratigraphic unit. [*See* Earth History.] Geologic time units are subdivided into eons, eras, periods, and epochs; in recent time (less than sixty-five million years ago), where erosion has not removed much chronological evidence, epochs are used to subdivide periods of time (for example, the Pleistocene

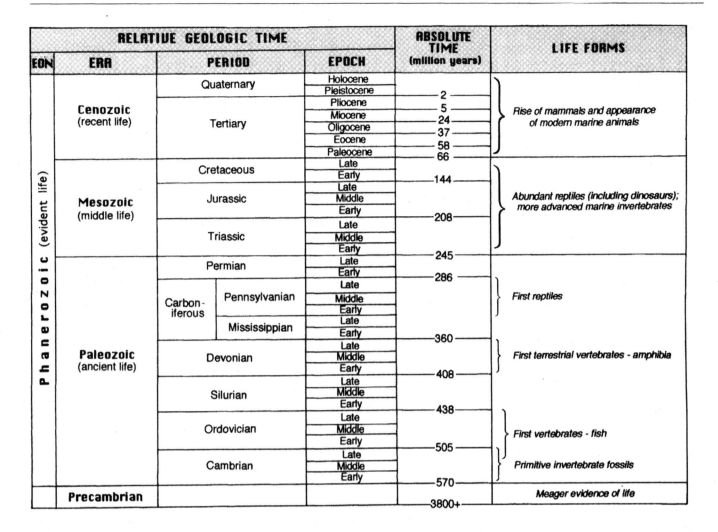

| RELATIVE GEOLOGIC TIME | | | | ABSOLUTE TIME (million years) | LIFE FORMS |
|---|---|---|---|---|---|
| **EON** | **ERA** | **PERIOD** | **EPOCH** | | |
| Phanerozoic (evident life) | Cenozoic (recent life) | Quaternary | Holocene | | Rise of mammals and appearance of modern marine animals |
| | | | Pleistocene | 2 | |
| | | Tertiary | Pliocene | 5 | |
| | | | Miocene | 24 | |
| | | | Oligocene | 37 | |
| | | | Eocene | 58 | |
| | | | Paleocene | 66 | |
| | Mesozoic (middle life) | Cretaceous | Late | | Abundant reptiles (including dinosaurs); more advanced marine invertebrates |
| | | | Early | 144 | |
| | | Jurassic | Late | | |
| | | | Middle | | |
| | | | Early | 208 | |
| | | Triassic | Late | | |
| | | | Middle | | |
| | | | Early | 245 | |
| | Paleozoic (ancient life) | Permian | Late | | |
| | | | Early | 286 | |
| | | Carboniferous — Pennsylvanian | Late | | First reptiles |
| | | | Middle | | |
| | | | Early | | |
| | | Carboniferous — Mississippian | Late | | |
| | | | Early | 360 | |
| | | Devonian | Late | | First terrestrial vertebrates - amphibia |
| | | | Middle | | |
| | | | Early | 408 | |
| | | Silurian | Late | | |
| | | | Middle | | |
| | | | Early | 438 | |
| | | Ordovician | Late | | First vertebrates - fish |
| | | | Middle | | |
| | | | Early | 505 | |
| | | Cambrian | Late | | Primitive invertebrate fossils |
| | | | Middle | | |
| | | | Early | 570 | |
| | Precambrian | | | 3800+ | Meager evidence of life |

Dating Methods. FIGURE 1. Geologic Time Scale. (After Birkeland and Larson, 1989. With permission of Oxford University Press.)

epoch or Holocene epoch during the Quaternary Period). Older units in pre-Cenozoic time are subdivided into periods and occupy much wider time slots based either on relative or absolute dating methods.

**Stratigraphy.** Stratigraphic or time line indicators include any material that provides relative or absolute age control. Tills (glacial sediment), for example, provide evidence for drastic changes in climate but are considered time transgressive (the process, in this case glaciation, does not begin and end at the same time in every place). Tephra (volcanic ash or dust), on the other hand, provides material that can be dated by potassium/argon (K/Ar) or fission track methods and is time parallel (that is, the process begins and ends at the same time everywhere and may be globally synchronous).

Lithostratigraphy is based on a particular lithologic homogeneity that is used to recognize a rock or deposit unit. It works within the framework of both unconsolidated, unweathered, and undifferentiated material (such as deposits) and consolidated, cemented, and indurated material (such as rock). Both are recognized by physical characteristics and/or the presence of certain lithic materials that dominate a unit. A lithostratigraphic unit is named from a type area and should be described from a type section representative of the unit. As with older rocks, it should be identified with a descriptive lithologic name (for example, "Trenton limestone"). The formation is the fundamental unit; other units include group, member, lentil, tongue, and bed.

Lithostratigraphic units are also called rock stratigraphic units. In many glaciated areas, litho- or rock stratigraphic units all have the same lithology, and weathering characteristics and differences in soil expression are therefore used to help differentiate the units (Birkeland et al., 1979; Birkeland, 1984; Mahaney, 1990). Because the beginning and end of deposition may have occurred earlier or later in different areas, the morphological character of the unit may be used in recognition (morphostratigraphy) but should be secondary to its lithic character.

It is common in Quaternary stratigraphy to use the morphology of the deposit as a major criterion in placing it within a sequence. Frye and Willman (1962) defined morphostratigraphic units in the American Midwest as a body of rock identified on surface form. Over thirty morainic units have been named in the Midwest, and they are identified solely on surface form. These moraines may also be grouped together as geologic–climatic units, but then each individual moraine is undifferentiated.

Geologic–climatic units are defined from the rock or soil record, and boundaries within a local area are isochronous. Over a wide region they may not be isochronous. The major unit is a glaciation or climatic episode in which glaciers expanded, developed their maximum volume, and ultimately receded; interglaciations are episodes during which climate was not conducive to expanded ice limits. Subdivisions include stades, which are climatic episodes within a glaciation in which secondary advances occurred; interstades are climatic episodes within a glaciation in which secondary recessions occurred. A geologic–climatic unit may be named after a rock stratigraphic and/or a soil stratigraphic unit. In the type locality of a geologic–climatic unit, the record of its major climatic characteristics should be manifest, for instance, till or record of climatic deteoriation in paleosols below and above the lower and upper limits of the unit (e.g., Pinedale Glaciation with overlying interglacial [Holocene] or interstadial [later Pleistocene] paleosol). Because fossil life forms are found in some but not all rocks, those with fossils may provide a means of differentiation on a biostratigraphic basis. While generally not important for Quaternary rocks or deposits, separation of sedimentary rock is often achieved on the basis of fundamental units called biozones. If fossil content is high and exerts a lithologic importance, biostratigraphic units become, in essence, lithostratigraphic units.

A chemostratigraphic unit is recognized on the basis of its chemical homogeneity; physical and lithic properties often dominate in the field, and chemical properties are discovered after laboratory analysis. Chemostratigraphy often provides considerable information on the source of sediment in lithostratigraphic units, on the chemical and mineral homogeneity of sediments, and on their diagenetic properties. Perhaps most importantly of all, chemostratigraphy provides geochemical profiles of the light and heavy rare-earth elements (REEs) that are important in differentiation of strata (Hancock et al., 1988).

Soil stratigraphic units are formed from weathering of lithostratigraphic, biostratigraphic, and/or chemostratigraphic units over a time interval or hiatus when there is no deposition. A soil stratigraphic unit has physical features and stratigraphic relationships that allow consistent recognition of mappable units. It differs from litho- and rock stratigraphic units in that it forms in them as a result of pedologic processes acting from the surface downward. These units differ from soil orders, great soil groups, and soil series in that they are defined by their stratigraphic relationships. The soil is the single rank of the soil stratigraphic unit and may change laterally in the landscape from a well-leached pedon to a carbonate-rich one. A soil stratigraphic unit should be named after a lithostratigraphic unit using the prefix "post" to avoid a terminological proliferation (Mahaney, 1990). In the past, soil stratigraphic units have been given formal names in accordance with the rules for naming litho- or rock stratigraphic units (Morrison and Wright, 1968).

Postdepositional modification stratigraphy (PDMS) as advocated by Birkeland et al. (1979) is a "hybrid of time stratigraphic and rock stratigraphic units and contains elements of both age and physical properties." Subdivision and correlation of deposits in a geologic sequence is achieved by using multiple relative dating methods, including rock weathering parameters, soils, landform morphometry, and lichens wherever and whenever possible. Despite the number of relative dating (RD) studies in different areas, we still do not know the minimum number of RD parameters that should be used in age assignment (Mahaney, 1990; Birkeland et al., 1979).

Time stratigraphy (chronostratigraphy) consists of material units that comprise all rock formed in an interval of time. Time stratigraphic boundaries are based on geologic time; that is, they are isochronous surfaces based on objective criteria that are nearly time parallel. Such criteria for Quaternary successions include some, but not all, faunal and floral zone boundaries, eustatically controlled shorelines and tephra; evidence for climatic change is excluded because it is time transgressive. Time stratigraphic units for the midwestern United States are based on paleosols representing interglacial stages separating glacial stages, as with the Lake Michigan glacial lobe (Morrison and Frye, 1965).

**Radiometric Dating.** Many dating tools are available to provide absolute or relative age controls. Radiometric methods include radiocarbon, K/Ar, U-series, and rubidium strontium (Rb/Sr); each method can be used to date specific material.

The radiocarbon method, developed by W. F. Libby of the University of Chicago (Libby et al., 1949), is based on the rate of decay of the radioactive carbon isotope carbon-14. There are three isotopes of carbon: carbon-12 and carbon-13, which are stable, and unstable or radioactive carbon-14. The latter is formed in the upper atmosphere by the effect of cosmic-ray neutrons upon nitrogen-14 ($^{14}$N):

$$^{14}N + n \Rightarrow {}^{14}C + p,$$

where n is a neutron and p is a proton (Figure 2). Radiocarbon oxidizes to carbon dioxide ($CO_2$) and enters through photosynthesis and the food chain into plant

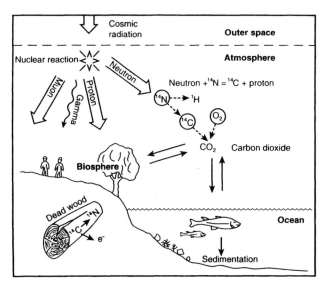

**Dating Methods. FIGURE 2.** Production and Decay of Carbon-14 in Nature. (Adapted from Nydal, 1995. With permission of PACT Belgium.)

and animal pathways, ocean waters, and biogenic carbonate minerals (Nydal, 1995; Figure 2). Since plants and animals utilize carbon in biological food chains, there exists an equilibrium between the carbon-14 concentration in the atmosphere and in organic tissues. When plants or animals die, the carbon-14 uptake ends and only decay continues. The radioactive carbon-14 decays at a constant rate, which is expressed in terms of the half-life value, 5,730±40 years. After every 5,730 years, half of the remaining carbon-14 will have decayed. During the decay, carbon-14 emits a beta particle (b) or electron, and the decay can be written as follows:

$$^{14}C \Rightarrow {}^{14}N + b.$$

The number of b emissions per unit time allows the determination of the remaining amount of carbon-14 in the carbon-containing material. Assuming that the carbon-14 concentration of the atmosphere (and in the food chain) has been constant through the history of the sample, the elapsed time since the start of decay can be established by measuring the amount of carbon-14 left in the sample. All the carbon-14 dates are corrected to 1950 CE. Materials commonly dated through carbon-14 are charcoal, wood, seeds, bone, peat, lake mud, humus-rich soil, and mollusk shells.

Since the carbon-14 concentration is not absolutely stable but has temporal fluctuations, the calibration of radiocarbon dates with historically aged materials becomes important (Stuiver and Pearson, 1986). Radiocarbon dates of sequential dendrochronologically dated trees produce a calendar year/radiocarbon calibration curve that extends back over eleven thousand years, en-

abling radiocarbon dates to be calibrated to calendar dates.

K/Ar can be used to date tephra, many igneous rocks, and some metamorphic rocks. U-series is used to date uranium in peat, in tufas (spring deposits), and in coral; to be effective in peat, the samples must not be subjected to alternating oxidation and reduction cycles. Rb/Sr is used to date volcanic rock. Both U-series and K/Ar dates span the entire age of the Earth, while radiocarbon is limited to 40,000 BP, although accelerator mass spectroscopy (AMS) radiocarbon has the potential to reach to about 90,000 BP.

**Paleomagnetic Dating.** Paleomagnetism is used in Quaternary stratigraphic studies as a tool for correlation and relative dating of equivalent strata, or for the absolute dating of deposits. Magnetic stratigraphy organizes rock strata into identifiable units based on stratigraphic intervals with similar magnetic characteristics. The method is based on the detection of changes in the Earth's magnetic field, and especially changes of polarity that are recorded by ferromagnetic sediments at the time of deposition. The polarity is referred to as *normal* where the north-seeking magnetization gives a Northern Hemisphere pole, as it does today, or *reversed* where the north-seeking magnetization gives a Southern Hemisphere pole.

The dating of Quaternary sediments and rocks by geomagnetic polarity history and by paleomagnetic parameters such as field declination and inclination, secular variation, and susceptibility is now commonplace (Barendregt, 1995). The large-scale features of the Earth's magnetic field have been well worked out for the past five million years or so, based largely on lava flows and deep-sea sediments. These have been used to construct the Global Polarity Timescale (Figure 3). The detailed (albeit short-lived) features for this period are still being discovered and defined through analysis of marine and terrestrial sediments where high sedimentation rates have continued for long periods of time. It is likely that these short-lived events, which provide a record of the excursions and perturbations of the Earth's magnetic field, will ultimately become useful correlative tools.

Fine-grained sediments, lava flows, and baked pottery are the media most frequently used for magnetic dating. Because reversals have occurred repeatedly in the past, their identification within incomplete sedimentary records is only possible through comparison with other stratigraphic or radiometric data collected for similar or related sedimentary sequences. Continuously deposited marine or terrestrial sediments that show a high sedimentation rate provide isochrons that can be used for worldwide correlation. The recent flourishing of research into the secular variation of the Earth's nondipole field also promises to refine the geomagnetic timetable for the Quaternary.

## CONTRIBUTIONS OF DENDROCHRONOLOGY

Bristlecone Pines in White Mountains of Southern California. Some of these trees are over four thousand years old. (From Henry N. Michael, University of Pennsylvania Museum.)

Because the widths of annual tree rings reveal changes in the temperature or moisture in the region where the tree grows, tree-ring data play an important part in the study of recent climate change in one region versus another.

An essential part of such work is assigning absolute calendar dates to specific tree rings that then serve as historical markers. This dendrochronology makes use of periods of distinctive oscillations or patterns in the record of rings. When these appear in the records from both a younger and an older tree (or from a living tree and dead wood), the two records can be overlapped to make a composite history. Using a series of such overlaps, a sequence of dates can be extended backward for thousands of years on the basis of a few trees whose dates of cutting or coring provide the calendar dates.

Dendrochronology has another vital function—correcting the dates obtained from radiocarbon, or carbon-14, dating. While that method can provide a date for charcoal, bone, or other organic materials (or scraps of wood without suitable tree rings), there is some uncertainty about the result because the concentration of the carbon-14 isotope in the Earth's atmosphere has not been constant through time. (It varies with the strength, or intensity, of the Earth's magnetic field.) [See the vignette on Paleomagnetic Evidence of Crustal Movement in the article on Plate Tectonics.] This discrepancy was revealed by carbon-14 dates that did not coincide with known calendar dates at Egyptian sites. Because archaeologists must know how to correct their carbon-14 dates (i.e., radiocarbon dates) for samples of various ages, the solution has been to obtain a large number of wood samples ranging in age from present to thousands of years BCE, to date each one by radiocarbon and dendrochronology methods, and then to assemble the resulting dates in a plot that reveals the discrep-

The only practical way of demonstrating the validity of interpreted magnetostratigraphy is to show that results are reproducible in widely separated sections with different lithologies and sedimentation rates. A long-standing problem of magnetostratigraphy has been the correlation of the terrestrial and marine records (Cooke, 1983). Since geomagnetic polarity reversals are recorded globally, magnetic stratigraphy provides ample opportunity to correlate between the two contrasted sedimentary environments. Recent work (Barendregt and Irving, 1998; Barendregt et al., 1995, 1998) has shown the usefulness of magnetostratigraphy in assigning time lines to glacial sequences where few absolute dates are available. Other studies (Heller and Liu, 1982) have established magnetobiostratigraphies for lake and loess sequences, motivated largely by a desire to understand continental

responses to climatic change, and to late Cenozoic glaciations in particular.

**Varve Chronology.** The clay varve chronology, the first absolute dating method to be developed, was introduced in Sweden by Gerhard De Geer in 1884 and was used to estimate the timing of the disappearance of the Scandinavian ice sheet (Hang, 1997). Laminations in lake sediments are usually referred to as rhythmites or, where the lamination develops because of annual variations in sedimentation, as varves. Annual laminations (O'Sullivan, 1983) in lacustrine sediments are formed as a consequence of seasonal, rhythmic changes in biogenic production, water chemistry, and the inflow of mineral matter (Saarnisto, 1986). Glaciolacustrine varves—alternating layers of coarse and fine sediment—develop in proglacial lakes as a result of summer

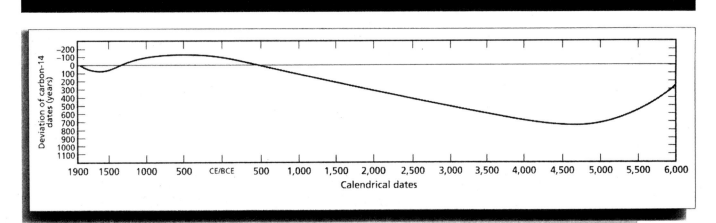

Deviation of Carbon-14 Dates from Absolute Calendrical Dates Established by Dendrochronology. (From Henry N. Michael, University of Pennsylvania Museum.)

ancy or deviation. Researchers from the University of Arizona and the University of Pennsylvania have constructed the sequence of dual-dated samples, collecting wood from Sequoia trees in California, and, more recently, from Bristlecone Pines in the White Mountains of California. These Bristlecones are the world's oldest trees: one living tree is 4,700 years old, and pieces of dead wood from the forest floor have extended the record back beyond 6,000 BCE. Finding the wood samples to bridge all the gaps in the sequence has required many seasons of fieldwork.

The plot shows that for the period 6,000 BCE to 500 BCE, carbon-14 dates are too old; for 500 BCE to 1300 CE they are too young; and for 1300 CE to 1900 CE they are too old. While the graph shows only the pattern, more precise corrections are available in tabular form.

BIBLIOGRAPHY

Cohen, M. P. *A Garden of Bristlecones: Tales of Change in the Great Basin*. Las Vegas: University of Nevada Press, 1998.
Ralph, E. K., and Henry N. Michael. "Twenty-five Years of Radiocarbon Dating." *American Scientist* (Sept.–Oct. 1974), pp. 553–560.
Stokes, M. A., and T. L. Smiley. *An Introduction to Tree-Ring Dating*. Tucson: University of Arizona Press, 1996.

—DAVID J. CUFF

and winter contrasts in sediment input. Counting of varves or annual laminae allows a precise dating of the start, changes, and end of deposition in lakes. By using distinctive (thickness, texture, composition) annual layers or groups of layers (marker layers), it is possible to correlate sediment sections from different lakes and to establish long sequences of varve chronologies. For example, the Swedish Time Scale covers more than 13,300 varve years (Wohlfarth et al., 1995) and is correlated to dendrochronological and ice-core records and high-resolution AMS-dated lacustrine deposits (Björck et al., 1996). However, because they are similar to annually laminated deposits, varves may form as a result of low-viscosity, dense interflows and underflows, as well as from slump-generated surge currents, and the number of rhythms does not necessarily bear any relationship to the number of years during which the deposits accumulated (Miall, 1990).

**Luminescence Dating.** Thermoluminescence (TL), which came into its own as a dating method around 1980, has been superseded by optically stimulated luminescence (OSL). The TL method relies on the release of electrons from traps in quartz minerals by irradiation with a heat source, such as infrared lamps or fire used in kilns to zero samples at some time in the past. As a result, TL has important implications in dating artifacts and is of immense importance in archaeology. The OSL method uses lasers to irradiate individual grains of quartz or feldspars and a photomultiplier tube to collect and count the emitted electrons. The effective use of the methods depends on rigorous relative age controls on the material to be dated, substantial knowledge of the

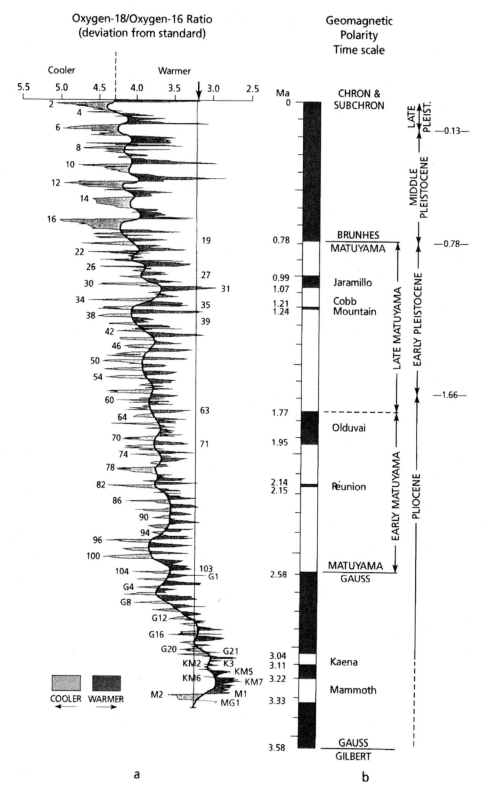

**Dating Methods. Figure 3.** Geomagnetic Polarity Time Scale and Oxygen-18/Oxygen-16 Paleotemperature Record.

Black and white areas (b) are normal and reversed polarity, respectively. Numbers (a) are isotope stages; even numbers are cold, odd numbers are given where they coincide with Chron and Subchron boundaries. Letters and numbers used before isotope stage 104 (Matuyama/Gauss boundary) follow Shackleton et al., 1995. Arrow at top (a) is mean Holocene oxygen-18 value. (Paleotemperature record (a) from Ruddiman et al., 1986; Raymo, 1992. Geomagnetic polarity time scale (b) after Cande and Kent, 1995.)

mineralogy of the material to be dated, and information about the zeroing or bleaching of minerals at present. The range of this method is commonly cited at 200,000 years, although dates as old as 800,000 BP have been reported (Berger et al., 1992). Most eolian (windblown), and lacustrine (lake) sediments, and some fluvial (riverine) sediments, may be dated by both TL and OSL; glacial sediments are excluded because they cannot be zeroed while being transported and encased in ice.

[*See also* Climate Reconstruction; *and* Younger Dryas.]

## BIBLIOGRAPHY

Barendregt, R. W. "Paleomagnetic Dating Methods." In *Dating Methods for Quaternary Deposits*, edited by N. W. Rutter and N. Catto, pp. 24–49. *Geoscience Canada* (Reprint Series Geotest) 2, 1995.

Barendregt, R. W., and E. Irving. "Changes in the Extent of North American Ice-Sheets During the Late Cenozoic." *Canadian Journal of Earth Sciences* 35 (1998), 147–161.

Barendregt, R. W., E. Irving, E. A. Christiansen, E. K. Sauer, and B. T. Schreiner. "Stratigraphy and Paleomagnetism of Late Pliocene and Pleistocene Sediments in Wellsch Valley and Swift Current Creek Areas, Southwestern Saskatchewan, Canada." *Canadian Journal of Earth Sciences* 35 (1998), 1347–1361.

Barendregt, R. W., J. S. Vincent, E. Irving, and J. Baker. "Magnetostratigraphy of Quaternary and Late Tertiary Sediments on Banks Island, Canadian Arctic Archipelago." *Canadian Journal of Earth Sciences* 35 (1998), 580–595.

Berger, G. W., B. J. Pillans, and A. S. Palmer. "Dating Loess Up to 800 Ka by Thermoluminescence." *Geology* 20 (1992), 403–406.

Birkeland, P. W. *Soils and Geomorphology.* New York and Oxford: Oxford University Press, 1984.

Birkeland, P. W., S. M. Colman, R. M. Burke, R. R. Shroba, and T. C. Meierding. "Nomenclature of Alpine Glacial Deposits, or 'What's in a Name'" *Geology* 7 (1979), 532–536.

Birkeland, P. W., and E. Larson. *Putnam's Geology.* 5th ed. New York and Oxford: Oxford University Press, 1989.

Björck, S., B. Kromer, S. Johnsen, O. Bennike, D. Hammarlund, G. Lemdahl, G. Possnert, T. L. Rasmussen, B. Wohlfarth, C. U. Hammer, and M. Spurk. "Synchronized Terrestrial Atmospheric Deglacial Record around the North Atlantic." *Science* 274 (1996), 1155–1160.

Cande, S. C., and D. V. Kent. "Revised Calibration of the Geomagnetic Timescale for the Late Cretaceous and Cenozoic." *Journal of Geophysical Research* 100.B4 (1995), 6093–6095.

Cooke, H. B. S. "Recognizing Different Quaternary Chronologies: A Multidisciplinary Problem." In *Correlation of Quaternary Chronologies*, edited by W. C. Mahaney, pp. 1–14. Norwich, U.K.: Geobooks, 1983.

Frye, J. C., and H. B. Willman. "Morphostratigraphic Units in Pleistocene Stratigraphy." *American Association of Petroleum Geologists Bulletin* 46 (1962), 112–113.

Hancock, R. G. V., W. C. Mahaney, and A. MacS. Stalker. "Neutron Activation Analysis of Tills in the North Cliff Section, Wellsch Valley, Saskatchewan." *Sedimentary Geology* 55 (1988), 185–196.

Hang, T. "Clay Varve Chronology in the Eastern Baltic Area." *Geologiska Föreningens I Stockholm Forhandlingar* 119 (1997), 295–300.

Heller, F., and T. S. Liu. "Magnetostratigraphical Dating of Loess Deposits in China." *Nature* 300 (1982), 161–163.

Libby, W. F., E. C. Anderson, and J. R. Arnold. "Age Determination by Radiocarbon Content: World-Wide Assay of Natural Radiocarbon." *Science* 109 (1949), 227–228.

Mahaney, W. C. *Ice on the Equator.* Ellison Bay, Wisc.: William Caxton, 1990.

Miall, A. D. *Principles of Sedimentary Basin Analysis.* Dordrecht: Springer, 1990.

Morrison, R. B., and H. E. Wright, Jr. *Means of Correlation of Quaternary Successions.* Salt Lake City: University of Utah Press, 1968.

Nydal, R. "The Early Days of $^{14}$C in Scandinavia, and Later Development." *PACT Journal of the European Network of Scientific and Technical Cooperation for the Cultural Heritage* 49 (1995), 9–27.

O'Sullivan, P. E. "Annually Laminated Lake Sediments and the Study of Quaternary Environmental Changes." *Quaternary Science Reviews* 1 (1983), 245–312.

Raymo, M. E. "Global Climate Change: A Three Million Year Perspective." In *Start of a Glacial*, edited by G. J. Kukla and E. Went. Berlin: Springer, 1992.

Saarnisto, M. "Annually Laminated Lake Sediments." In *Handbook of Holocene Palaeoecology and Palaeohydrology*, edited by B. E. Berglund, pp. 343–370. Chichester, U.K.: Wiley, 1986.

Shackleton, N. J., M. A. Hall, and D. Pate. "Pliocene Stable Isotope Stratigraphy of Site 846." *Proceedings of the Ocean Drilling Program, Scientific Results* 138 (1995), 337–353.

Stuiver, M., and G. W. Pearson. "High Precision Calibration of the Radiocarbon Time Scale, AD 1950–500 BC." *Radiocarbon* 28.2B (1986), 839–862.

Wohlfarth, B., S. Björck, and G. Possnert, G. "The Swedish Time Scale: A Potential Calibration Tool for the Radiocarbon Time Scale during the Late Weichselian." *Radiocarbon* 37.2 (1995), 347–359.

—WILLIAM C. MAHANEY,
RENE W. BARENDREGT,
AND VOLLI KALM

**DEATH RATES.** *See* Human Populations.

## DECARBONIZATION

Decarbonization refers to long-term decreases in specific emissions of carbon dioxide per unit of energy. Phrased slightly differently, it refers to the historical decrease in the carbon intensity of primary energy consumption. Figure 1 illustrates historical global decarbonization since 1860 in terms of the average carbon emissions per unit of primary energy. This average has decreased as fuels with high carbon content, such as coal, have been replaced by fuels with less carbon, such as natural gas, or no carbon, such as nuclear energy and most renewable energy sources.

Decarbonization is important because of its impact on potential global warming. When fuels containing carbon—a group that includes all fossil fuels as well as

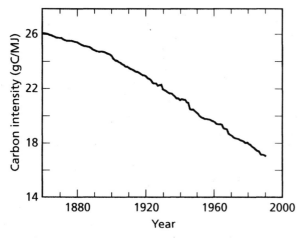

**Decarbonization. Figure 1.** Global Decarbonization of Energy since 1860 (Including Gross Carbon Emissions from Fuelwood) in Grams of Carbon per Megajoule of Primary Energy. (After Nakićenović et al., 1996c. Reprinted by permission of *Daedalus*, Journal of the American Academy of Arts and Sciences.)

biomass—are burned, the combustion process releases, among other things, carbon dioxide. Carbon dioxide is one of the greenhouse gases in the atmosphere. [*See* Carbon Cycle.] Greenhouse gases occur naturally in the atmosphere and make Earth inhabitable to life, but increased atmospheric concentrations of these gases, primarily due to energy use, may lead to enhanced global warming and climate change (See Intergovernmental Panel on Climate Change, 1996). [*See* Global Warming; *and* Greenhouse Effect.]

Of all the greenhouse gases from anthropogenic sources, carbon dioxide is by far the most important, representing more than half of the calculated increase in radiative forcing from anthropogenic emissions. Carbon dioxide emissions from fossil fuels in 1990 were estimated at about 6 billion metric tons of elemental carbon. This represents 70–90 percent of all anthropogenic carbon dioxide emissions (Intergovernmental Panel on Climate Change, 1992). Other sources of anthropogenic carbon dioxide emissions, primarily land use changes such as deforestation, are estimated at about 1.6 billion metric tons of carbon.

Over the past two centuries, energy use has shifted from traditional sources (such as wood and coal) to crude oil and natural gas. [*See* Energy Policy; *and* Fossil Fuels.] The shift has involved the development of elaborate conversion systems for producing higher-quality energy forms, such as electricity, and delivering them to the final consumer. Such structural changes in the energy system, together with improvements in the performance of individual energy technologies, have resulted in significant energy-efficiency improvements and in decarbonization. Compared with specific carbon dioxide emissions of 27–30 kilograms of elemental carbon per

gigajoule of energy (kg C GJ$^{-1}$) for traditional uses of biomass such as fuel wood and peat, emissions are about 26 kg C GJ$^{-1}$ from coal, 20 kg C GJ$^{-1}$ from crude oil, and 15 kg C GJ$^{-1}$ from natural gas (Organisation for Economic Co-operation and Development, 1991, 1995).

The pattern of decarbonization has been different in different countries, and carbon intensities are even increasing in some developing countries. Globally, the rate of decarbonization is about 0.3 percent per year. Primary energy consumption has risen at about 2 percent per year, so that carbon dioxide emissions have increased at approximately 1.7 percent per year.

Decarbonization can be expected to continue over the next few decades as the share of total primary energy supplied by natural gas and nonfossil energy sources increases. Some scenarios anticipate a reversal of decarbonization in the long run as more easily accessible sources of conventional oil and gas become exhausted and are replaced by more carbon-intensive alternatives. Others foresee continuing decarbonization resulting from further shifts to low-carbon energy sources such as renewables and nuclear energy. Virtually all scenarios foresee some increase in the global demand for energy services as the world develops in the future. Even if decarbonization continues at its historical rate, the growth in primary energy consumption implies that, in the absence of policy intervention, carbon emissions will continue to grow.

[*See also* Carbon Dioxide; *and* Scenarios.]

## BIBLIOGRAPHY

Grübler, A., and N. Nakićenović. "Decarbonizing the Global Energy System." *Technological Forecasting and Social Change* 3.1 (1996), 97–110. Reprinted as RR-97–6, International Institute for Applied Systems Analysis. Laxenburg, Austria.

Intergovernmental Panel on Climate Change. *1992 IPCC Supplement.* Geneva, 1992.

———. *Climate Change 1995: Impacts, Adaptations and Mitigation of Climate Change: Scientific-Technical Analyses.* Contribution of Working Group II to the Second Assessment Report of the Intergovernmental Panel on Climate Change. Cambridge and New York: Cambridge University Press, 1996, p. 880.

Marchetti, C. "When Will Hydrogen Come?" WP-82–123, International Institute for Applied Systems Analysis. Laxenburg, Austria, 1982.

Nakićenović, N. "Technological Change and Learning." In *Climate Change: Integrating Science, Economics and Policy*, edited by N. Nakicénović et al. CP-96–1, International Institute for Applied Systems Analyis. Laxenburg, Austria, 1996a.

———. "Decarbonization: Doing More with Less." *Technological Forecasting and Social Change* 51.1 (1996b), 1–17.

———. "Freeing Energy from Carbon." *Daedalus* 125.3 (1996c), 95–112. Reprinted as RR-97–4, International Institute for Applied Systems Analysis. Laxenburg, Austria.

Organisation for Economic Co-operation and Development. *Estimation of Greenhouse Gas Emissions and Sinks.* Final report from OECD Experts Meeting, 18–21 February 1991. Paris, 1991. Prepared for Intergovernmental Panel on Climate Change.

———. *IPCC Guidelines for National Greenhouse Gas Inventories*, vol. 3, *Greenhouse Gas Inventory Reference Manual*. Bracknell: U.K. Meteorological Office, 1995.

Yamaji, K., et al. "An Integrated System for Carbon Dioxide/ Energy/GNP Analysis: Case Studies on Economic Measures for Carbon Dioxide Reduction in Japan." Paper presented at the Workshop on Carbon Dioxide Reduction and Removal: Measures for the Next Century, International Institute for Applied Systems Analysis, Laxenburg, Austria, 19–21 March 1991.

—Nebojša Nakićenović

**DEEP ECOLOGY.** *See* Conservation.

# DEFORESTATION

To deforest is "to clean an area, other than purely temporarily, of forest" (Ford-Robertson, 1971). Removing, or killing, all trees either by natural or human disturbances is not considered deforestation if the area grows back to trees. This temporary removal of trees simply places the forest in a different structural stage—the "open" (stand initiation) structure (Oliver and Larson, 1996). Removal (or killing) of all trees and conversion of the land to desert, agricultural or grazing land, buildings, eroded barrens, or other nontreed conditions is deforestation. For the purposes of this article, the practice of harvesting the trees, cultivating the area for agriculture for several years, and then allowing regrowth to forests in a "shifting agriculture" pattern will not be considered deforestation, since the net effect is a return of the land to a forested condition. Similarly, harvesting native forests and regrowing the area with plantations of exotic tree species will not be considered deforestation, although there may be environmental consequences of this change. Rates of deforestation are difficult to determine because of variations in how data are collected and monitored in different countries; information in this article is based on United Nations Food and Agriculture Organization data, which have some possible shortcomings, as described later.

Deforestation can occur naturally when the climate changes at the forested extremes of cold, dry, or water-saturated conditions so that forests can no longer survive or regenerate. Continental and mountain glacier advances during the Pleistocene (1.5 million to 10,000 years BP) removed forests from vast areas of Europe, North America, and Asia; and rising sea levels, as these glaciers melted, removed forests from the inundated areas. More recent cooling trends within the past few thousands and hundreds of years have caused forests to disappear in high altitudes and polar latitudes. At other times, warming trends have reduced forests as deserts expanded (Lamb, 1988; Gates, 1993). Likewise, sinking land surfaces have caused forests to disappear, such as the forests in southern Louisiana, which are becoming inundated with salt water (DeLaune et al., 1983).

Timber harvesting, land clearing, and/or grazing as well as natural disturbances such as fires, windstorms, or flooding can exacerbate the rate of deforestation that is associated with these geologic changes. In some cold areas, natural or human disturbances can prevent or reverse the change of forests to peat bogs by draining the bogs and/or mixing the mineral and organic soils.

**Rates and Locations of Deforestation.** Recent, rapid deforestation seems to be caused directly by people and has been expressed as a major concern of the international forestry community (Eleventh World Forestry Congress, 1997). The world's forests are estimated to have covered 8 billion hectares in 2000 BCE (Food and Agriculture Organization, 1993) and to have been reduced to 3.45 billion hectares by 1995 (Table 1). Deforestation began in parts of the Mediterranean and Chile about ten thousand years ago (Aschmann, 1973). Humans have accelerated deforestation in the past few centuries in parts of Europe, temperate and tropical Asia, North America, and tropical Central and South America. Following changes in cultures in tropical Asia (e.g., Angkor Wat, Borobodor) and tropical Central and South America (e.g., Mayan, Inca; Turner and Butzer, 1992; von Kooten et al., 1999), forests regrew. Recently, forests have been regrowing in more developed regions such as North America and Europe (Allan and Lanly, 1991; Food and Agriculture Organization, 1997; Table 1).

Globally, deforestation is now occurring at a net rate of 10.8 million hectares per year (0.3 percent per year; 1990–1995 average). The loss of forests is currently higher than this number in less economically developed regions of the world, but is offset by net increases in forest area in more developed regions. The increase in some forest areas compensates statistically for the decrease in others, but this does not necessarily mean that the species diversity, habitats, and socioeconomic benefits are similarly offset.

Currently, the highest rates of deforestation appear to be in tropical forests of Asia, Africa, and Central and South America (Table 1). It is argued that net deforestation is actually lower in these areas because abandoned agricultural land, which regrows to forests, may not be adequately incorporated into the data. It is estimated that 90 percent of the recent tropical deforestation has occurred since 1970 (Skole et al., 1994). In 1990, the tropical forests of the world were estimated to cover about 1.7 billion hectares, compared with 1.9 billion hectares in 1981—an annual tropical deforestation rate of about 0.8 percent (Food and Agriculture Organization, 1993). The rate of deforestation varied substantially by tropical forest type (Table 2). The tropical rainforest experienced the slowest rate of deforestation (0.6 percent annually), but, because of its large area, experi-

**Deforestation. TABLE 1.** Forest Area and Net Forest Change from 1990 to 1995, and Wood Production and Relative Consumption

| REGION | FOREST AREA (MILLION HECTARES) | FOREST AREA CHANGE (MILLION HECTARES) | PERCENT CHANGE | WOOD CONSUMPTION/ PRODUCTION (MILLION CUBIC METERS) | WOOD PRODUCTION (MILLION CUBIC METERS) |
|---|---|---|---|---|---|
| Tropical Africa | 505 | −18 | −3.6% | 537 | 1.0 |
| Nontropical Africa | 15 | −0.3 | −2.0% | 41 | 1.0 |
| Tropical Asia | 280 | −15 | −5.4% | 824 | 1.0 |
| Temperate Asia | 194 | −1.4 | −0.7% | 461 | 1.2 |
| Tropical Oceania | 42 | −0.8 | −1.9% | 9 | 0.6 |
| Temperate Oceania | 49 | 0.3 | 0.6% | 48 | 0.7 |
| North Europe | 53 | 0.04 | 0.1% | 149 | 1.0 |
| West Europe | 59 | 1.8 | 3.1% | 232 | 1.2 |
| East Europe | 34 | 1.9 | 5.6% | 80 | 0.9 |
| Former USSR | 816 | 2.8 | 0.3% | 181 | 0.8 |
| North America | 457 | 3.8 | 0.8% | 903 | 1.0 |
| Central America | 79 | −5.2 | −6.6% | 78 | 1.0 |
| Tropical South America | 828 | −23 | −2.8% | 357 | 1.0 |
| Temperate South America | 43 | −0.6 | −1.4% | 54 | 0.8 |
| Total | 3,454 | −54 | −1.6% | 3,956 | 13 |

SOURCE: Data from Food and Agriculture Organization (1997); data are subject to variation with differing conversion factors and reporting methods.

enced the largest total area of deforestation (4.6 million hectares per year). The highest rates of deforestation were in the tropical uplands, where both moist and dry upland forests experienced a 1.1 percent annual deforestation rate.

Land deforested by poorly managed agriculture or grazing activities in arid, erosive, or otherwise sensitive conditions can result in depletion of the soil structure and nutrient pool, erosion, and/or desertification, so that forests will not grow again. As trees become more scarce for firewood, animal dung is often used instead. This dung and its nutrients are not allowed to replenish soil nutrients, and the soil productivity for agriculture is further depleted. Consequently, other areas are cleared for agriculture or grazing and the deforestation is continued.

**Deforestation. TABLE 2.** Estimates of Forest Cover Area and Rate of Deforestation in Tropical Forests of Different Regions

| | LAND AREA (MILLION HECTARES) | FOREST AREA 1990 (MILLION HECTARES) | PERCENT OF LAND | ANNUALLY DEFORESTED AREA (1981–1990) (MILLION HECTARES) | PERCENT |
|---|---|---|---|---|---|
| Forest | 4,186.40 | 1,748.20 | 42 | 15.3 | 0.8 |
| Lowland | 3,485.60 | 1,543.90 | 44 | 12.8 | 0.8 |
| Tropical | 947.2 | 718.3 | 76 | 4.6 | 0.6 |
| Moist deciduous | 1,289.2 | 587.3 | 46 | 6.1 | 0.9 |
| Dry deciduous | 706.2 | 178.6 | 25 | 1.8 | 0.9 |
| Very dry | 543 | 59.7 | 11 | 0.3 | 0.5 |
| Upland | 700.9 | 204.3 | 29 | 2.5 | 1.1 |
| Moist | 528 | 178.1 | 34 | 2.2 | 1.1 |
| Dry | 172.8 | 26.2 | 15 | 0.3 | 1.1 |
| Nonforest (hot and cold) | 591.9 | 8.1 | 1 | 0.1 | 0.9 |
| Total Tropics* | 4,778.30 | 1,756.30 | 37 | 15.4 | 0.8 |

*Totals may not tally due to rounding.

SOURCE: Data from Food and Agriculture Organization (1993) and von Kooten et al. (1999); data subject to variation with differing conversion factors and reporting methods.

**Deforestation. TABLE 3.** Areas of Degraded Land within the World's Dry Zones

| DRY ZONE REGION | Degree of Degradation (million hectares) | | | | TOTAL DEGRADED | TOTAL NONDEGRADED | TOTAL AREA | PERCENT DEGRADED |
| --- | --- | --- | --- | --- | --- | --- | --- | --- |
| | LIGHT | MODERATE | STRONG | EXTREME | | | | |
| Africa | 118 | 127.2 | 70.7 | 3.5 | 319.4 | 966.6 | 1,286.0 | 25 |
| Asia | 156.7 | 170.1 | 43 | 0.5 | 370.3 | 1,301.5 | 1,671.8 | 22 |
| Australia | 83.6 | 2.4 | 1.1 | 0.4 | 87.5 | 575.8 | 663.3 | 13 |
| Europe | 13.8 | 80.7 | 1.8 | 3.1 | 99.4 | 200.3 | 299.7 | 33 |
| North America | 13.4 | 58.8 | 7.3 | 0 | 79.5 | 652.9 | 732.4 | 11 |
| South America | 41.8 | 31.1 | 6.2 | 0 | 79.1 | 436.9 | 516.0 | 15 |
| Total | 427.3 | 470.3 | 130.1 | 7.5 | 1,035.2 | 4,134.0 | 5,169.2 | 20 |

Note: It is unclear how much of the presently degraded area was once forested.
SOURCE: Odera (1996); United Nations Environment Programme (1992).

In arid areas, some previously forested regions have been changed to "anthropogenic steppes" through deforestation (Boydak and Dogru, 1997). Further degradation can lead to desertification (Aubreville, 1949)—a relatively nonreversible process by which the area loses its ability to sustain forest growth, biological diversity, and most human values. It is uncertain how much of the presently degraded arid land was once forested; Africa and Asia have the most degraded land, while Africa, Europe, and Asia all have high proportions of their drylands in degraded conditions (Table 3).

During desertification, the topsoil and associated soil nutrients, structure, organic matter, and water-holding capacity are lost when the forest is harvested and it and other vegetation are prevented from growing, through a combination of inappropriate cultivation, grazing, and burning. With the loss of topsoil, wind and water erosion often remove the lower soil layers and deposit them in streams, residential areas, and other unwanted locations. The inability of the area to support forests (or other vegetation) means that the area's capacity to provide forest biodiversity and forest products is gone, as is its ability to sequester carbon through forest growth and forest product use. If large areas are converted to deserts, the result may be local climate changes and movement of people to other areas where, if not managed appropriately, they can cause further desertification as well. In areas of sufficient rainfall, vegetation and ultimately forests usually will regrow after the grazing, burning, cultivation, and/or other pressures are removed; however, in arid areas the land may remain as a relatively permanent desert once the vegetation has been removed. [See Desertification.]

Most developed regions have a net increase in forests (Table 1), although within a region there may be deforestation in some places compensated by reforestation in others. Historically, the increase of "civilization" has been associated with expanding deforestation, as en-

larging populations and industrialization consumed more fuelwood and cleared more land for agriculture and grazing (Perlin, 1989). The lack of forests has been suggested as a cause of the decline of some civilizations. Events of the past two centuries, however, are suggesting that a new relationship between economic development and deforestation is emerging. Technological advances in transportation and farming mean that more food can be produced on the very productive areas, fewer people are employed to do the farming, and the products are shipped longer distances than before. As these events occur, the less productive farming and grazing lands are being abandoned and allowed to regrow to trees, and the unneeded farm labor moves to cities. Also, as societies develop, they are now typically substituting fossil fuel for fuelwood and thus decreasing pressure on the forest. The result is a reversal of the deforestation trend, although forests will probably never cover as much area as they potentially could.

As regions develop, deforestation is caused by building construction in expanding urban areas; by clearing of forests on the most productive soils for agricultural use; by construction of roads, reservoirs, and similar structures; and sometimes by poisoning of the air and soil by pollutants (Gordon and Gorham, 1963; Rykowski, 1995). The area deforested is generally less than the reforested area—but the local deforestation can threaten or cause extinction of species that occupy the specific areas and ecosystems being deforested. Deforestation can sometimes have the beneficial effect of providing productive farm and residential areas, thus providing food and shelter and allowing more forests to grow on less productive lands. However, even these cases create negative trade-offs such as loss of species habitats in the productive, deforested areas.

**Causes of Deforestation.** The causes of deforestation are complex and often misunderstood. Commercial timber harvesting in the tropics is rarely a significant

## ACTIVISTS AND THE WOOD INDUSTRY

In the past decade, logging practices and the marketing of wood products have been altered by consumer groups and nongovernmental organizations that monitor forest cutting and the granting of permits to build roads and cut trees. These groups have accomplished at least two kinds of change.

The first change is moving timber companies to actually cease cutting in certain areas. For instance, months of negotiations ended in March 2000 with an eighteen-month moratorium on cutting in selected mountain valleys in the Great Bear cool rainforest of coastal British Columbia. During the moratorium, environmental groups and timber companies will discuss various options for areas of remaining old-growth forests. Industry participants include Weyerhaeuser, International Forest Products Ltd., Western Forest Products, and West Fraser Timber. Environmental groups include Coastal Rainforests Coalition, Greenpeace, and Sierra Club (visit http://www.planetark.org/dailynewsstory.cfm?newsid=6020&newsdate=17-Mar-2000).

The second change is more subtle—convincing retailers to phase out wood products from sensitive regions (a strategy used in the above conflict) and to favor products from properties where harvesting operations are acceptable from an environmental standpoint. Properties and operations that qualify are certified as such, as are the wood and other products derived from those operations.

Today's certified wood and products bearing an FSC label reflect the influence of the Forest Stewardship Council (http://www.fscus.org/aboutfsc), which has an international organization and one in the United States. To paraphrase the council's own statements, it was founded in 1993 as an independent, nonprofit, nongovernmental organization by representatives of environmental and conservation groups, the timber industry, the forestry profession, indigenous peoples, community forestry groups, and forest product certification organizations from twenty-five countries. The FSC supports environmentally appropriate, socially beneficial, and economically viable management of the world's forests. It encourages development of forest management standards and provides public education and information about certification as a tool for ensuring the world's forests are protected for future generations. The international organization has working groups or national contact persons in sixteen nations: Belgium, Bolivia, Brazil, Canada, Colombia, Denmark, Finland, Germany, Ireland, Mexico, the Netherlands, Papua New Guinea, Sweden, Switzerland, the United Kingdom, and the United States. As of September 1999, roughly 37 million acres (15 million hectares) of forest were certified throughout the world, while in the United States, roughly 2 million hectares (equivalent to the area of Massachusetts) were

cause of deforestation, although it can change the forest to the "open" structure until it regrows. Commercial logging, however, can increase road access and remove large trees and so can make forest areas more accessible for conversion to agricultural cultivation and grazing—and hence deforestation.

Shipment of wood (in the form of timber and timber products), from less developed to more developed regions of the world, does not appear to be occurring in such vast quantities that it is a significant cause of deforestation (Table 1). However, it may affect the forest composition, the wood quality, and the socioeconomic conditions of the exporting country, and thereby indirectly affect deforestation. Overseas shipment causes forests and wood products to become valued by the exporting country as a promising economic sector, and the

emphasis on protection, reforestation, and management of forests increases, although this reforestation may be plantations of exotic tree species.

Another common perspective is to attribute deforestation to population growth. As populations rise, the pressures on the forests might be expected to increase; however, it has been difficult to link population and economic growth directly to all deforestation (Skole et al., 1994; Table 4). A population rise, in conjunction with certain lifestyles (e.g., subsistence agriculture and grazing), may account for deforestation in some regions.

Government policies that encourage people to move to rural lands to avoid urban congestion, to strengthen territorial security or claims, or to provide other benefits often promote deforestation. Much of the tropical deforestation is driven by the replacement of forests certified using standards developed for eleven different regions. Standards for certification of forest lands and operations depend upon ten principles (http://www.fscus.org/html). The forester/operator has some flexibility in meeting standards, as long as the forest is regenerated, the biological diversity of the forest is conserved, and the natural cycles, such as nutrient recycling, are maintained in order to protect the long-term health and productivity of the forest. Certification work is carried out by a number of groups, some of which predated the founding of the Forest Stewardship Council. One of these is the Rainforest Alliance, which administers the Smartwood Program (http://www.smartwood.org).

A listing of certified suppliers in the United States (http://www.certifiedwood.org/suppliers) includes approximately three hundred businesses, ranging from sawmills to shops producing cabinetry and furniture. The world's largest distributor of wood products, The Home Depot (874 stores in the United States, Canada, Chile, and Puerto Rico), after being criticized by activists, now gives preference to certified suppliers of wood and seeks to increase the proportion of such wood in its inventory. While this approach appears to be gaining momentum, as of 1999, only about 1 percent of the world's wood supply is certified.

Various industry associations that establish their own forestry policies include the Canadian Institute of Forestry (http://www.canadian-forests.com) and the American Forest and Paper Association Sustainable Forests Initiative (http://www.afanda.org/sfi/menu). The forestry profession in the United States is represented by the Society of American Foresters, founded in 1900 by Gifford Pinchot (http://www.safnet.org). A number of groups promote the sustainable use of forests. Among them are the following:

- World Rainforest Movement (http://www.wrm.org.uy/english)
- Rainforest Action Network (http://www.ran.org/ran)
- Ancient Forests International (http://ancientforests.org)
- Wilderness Committee (http://www.wildernesscommittee.org/ancientforests)
- Native Forest Network Campaigns (http://www.nativeforest.org/campaigns)
- Native Forest Directory (to related groups) (http://www.nativeforest.org/directory)
- Alaska Rainforest Campaign (http://www.akrain.org)
- Tongass Clearinghouse (http://www.tongass.com)

—DAVID J. CUFF

through agricultural activities similar to the deforestation that occurred in temperate North America between 1700 and about 1930 CE. In addition, substantial native tropical forests have been replaced by introducing tree crops such as rubber, palm oil, and coconut. Fertile bottomland forests have gradually been converted to agriculture over many years, and recent water development projects remove forests where water is impounded and allow agriculture to replace forests in areas previously too arid for effective cultivation.

Industrial chemical pollutants released into the air and water can kill trees and soil organism, thus causing deforestation (Gordon and Gorham, 1963; Rykowski, 1995). In most developed countries, pollution-control measures and laws now minimize or prevent this chemical pollution; however, these pollutants can still cause deforestation where these laws are not strict or are ignored, or when accidents happen.

Much of the present deforestation is closely related to subsistence agriculture and pastoral systems and associated large uses of trees for fuelwood (Table 4). Subsistence agriculture and grazing limit areas in which new trees can grow. The net effect is to provide little wood, and so wood is cut over larger areas, which are then cultivated and grazed—and deforestation spreads. Forty-eight percent of the world's harvested wood is used for fuelwood, and 72 percent of the wood harvested in regions with net deforestation is used for fuelwood. Forest growth for fuelwood does not necessarily promote deforestation. For example, a well-managed coppice forest can be harvested for firewood, and new sprouts will grow to a new forest. If, however, the harvesting is fol-

**Deforestation. TABLE 4.** Relation of Fuelwood Use and Population Density to Deforestation by Region

| REGION | FOREST AREA (MILLION HECTARES) | PERCENT CHANGE | FUEL WOOD USE (MILLIONS OF M³) | FUEL WOOD (AS % OF REGIONAL WOOD CONSUMPTION) | REGIONAL POPULATION (MILLIONS) | FUEL WOOD PER CAPITA CONSUMPTION (M³/PERSON) | POPULATION FOREST (PERSONS/ HECTARES) |
|---|---|---|---|---|---|---|---|
| Tropical Africa | 505 | −3.6 | 484 | 91 | 551 | 0.9 | 1.1 |
| Nontropical Africa | 15 | −2.0 | 18 | 44 | 177 | 0.1 | 11.8 |
| Tropical Asia | 280 | −5.4 | 647 | 81 | 1,722 | 0.4 | 6.2 |
| Temperate Asia | 194 | −0.7 | 231 | 42 | 1,663 | 0.1 | 8.6 |
| Tropical Oceania | 42 | −1.9 | 6 | 105 | 7 | 0.9 | 0.2 |
| Temperate Oceania | 49 | 0.6 | 3 | 9 | 21 | 0.1 | 0.4 |
| North Europe | 53 | 0.1 | 9 | 6 | 19 | 0.5 | 0.4 |
| West Europe | 59 | 3.1 | 32 | 11 | 366 | 0.1 | 6.2 |
| East Europe | 34 | 5.6 | 11 | 16 | 122 | 0.1 | 3.6 |
| Former USSR | 816 | 0.3 | 31 | 20 | 292 | 0.1 | 0.4 |
| North America | 457 | 0.8 | 99 | 11 | 293 | 0.3 | 0.6 |
| Central America | 79 | −6.6 | 62 | 78 | 126 | 0.5 | 1.6 |
| Tropical South America | 828 | −2.8 | 239 | 68 | 268 | 0.9 | 0.3 |
| Temperate South America | 43 | −1.4 | 19 | 41 | 52 | 0.4 | 1.2 |
| Total | 3,454 | −1.6 | 1,890 | 48 | 5,678 | 0.3 | 1.6 |

SOURCE: Data from Food and Agriculture Organization (1997); data subject to variation with differing conversion factors and reporting methods.

lowed by grazing and/or fires that prevent new sprouts, local people often dig out the tree roots and cut new forests for fuelwood—and the newly cut areas are then burned and/or grazed. In addition, when trees are cut for fuelwood before their rapid growth stage, large areas are needed to provide a modest wood volume and the trees are kept so small that they cannot be used for other products such as building materials.

**Effects of Deforestation.** Often, and especially in less developed regions of the world, deforestation results in a loss of soil productivity and thus a reduced ability to grow any useful products. This deforestation can eliminate species directly through harvest and indirectly through destruction of habitats. It can eliminate a forested species' habitat or part of the habitat of species that use both forest and nonforest areas, and it can interrupt the migration routes of other species that travel through forests.

Elimination of old-growth forests—forests containing very old or large trees (also known as "late successional forests," "ancient forests," "ancient woodlands," "precolonial forests," and "climax forests")—was a particular concern when it was assumed that old growth was the condition in which all forests existed and that all species survived before being altered by humans. Deforestation does eliminate such forests; however, the old-growth condition was only one of many natural conditions of the forests as they changed through disturbances and regrowth. And there are species that depend

on open forests, savannas, and dense forests, as well as others that depend on old growth (Oliver and Larson, 1996). Deforestation eliminates all of these conditions and threatens the species that depend on them. Elimination of the old forests through deforestation or other harvest is also of concern, since these forests sequester large amounts of carbon. By not harvesting these forests, the carbon stored is much less than the carbon that is added to the atmosphere by using substitute products such as steel, aluminum, brick, and concrete if wood is not available (Kershaw et al., 1993). Although these old-growth forests provide many values, these values must be balanced against the increased carbon dioxide added to the atmosphere if substitute products are used because these forests are not harvested. It must also be remembered that harvesting of old-growth forests is not synonymous with "deforestation" if forests regrow after the harvest. However, certain habitats and other values are lost for a long time in both cases.

Deforestation alters the water flow over and within the soil as well. Elimination of the trees reduces the evapotranspiration and so allows more water to flow into groundwater reservoirs and aquifers as long as the soil structure is maintained. Because there are no trees to regenerate, the soil organic matter and structure degenerate shortly after deforestation. In fine-textured soils such as clays, the soils become compact and easily eroded. Rainfall flows as overland flow rather than through the soil, causing severe erosion and stream deposition, as well as flash floods. In coarse-textured soils such as sands, rainfall can penetrate the soil rather than flow overland, even when the soil profile is ruined. However, the soils often become windblown, and the nutrients are leached into the groundwater reservoir.

There has been an increase in global carbon dioxide during the past century that is believed to affect global climates (Philander, 1998). The increase during the early Industrial Revolution is attributed to burning of wood for energy—and associated deforestation. Since about 1930, the increase is largely attributed to fossil fuel use (Stuiver, 1978). Deforestation affects carbon dioxide in the atmosphere both during the process and afterward. During deforestation, the carbon in the trees is released to the atmosphere as carbon dioxide through burning or rotting. After deforestation, there is no forest to regrow and to absorb carbon dioxide from the atmosphere through photosynthesis (Sampson and Hair, 1992, 1996). In addition, deforestation prevents forest products from being put to use; and so steel, concrete, brick, and other substitute products are utilized; these consume far more fossil fuels in their manufacture and so add much more carbon dioxide to the atmosphere than if wood were used. Energy produced from wood also emits less carbon dioxide than energy produced from fossil fuels; however, burning of wood for energy does not save as much fossil fuel (and store as much carbon dioxide) as using wood for substitute products (Kershaw et al., 1993; Koch, 1991). For most-efficient carbon dioxide sequestration, wood energy should be made from the wood residuals of other wood products of higher carbon substitute value.

Deforestation can also have more direct effects on climate patterns at local levels (Geiger, 1965). Locally, trees reduce wind speeds and direct sunlight to the ground surface and so reduce the surface dryness, extreme heat, and windchill near the ground. Tree shade also reduces extreme daytime heat and nighttime cold, which can occur at the ground surface in deforested places.

Large-scale deforestation may also change global climate patterns, although the ability to discern these effects is confounded by other strong influences on global climate patterns (Gash et al., 1996). Potentially, large-scale deforestation can change global climate patterns by reducing the humidity and causing hotter days and colder nights, thus changing rainfall, precipitation, and wind patterns. The lack of trees means less evapotranspiration and less humid air blowing from a deforested area. The hotter surfaces can also be focal points for convection thunderstorms.

Deforestation may cause drier climates and more extreme temperature fluctuations downwind. There may be less humidity for rainfall, and temperatures may be colder at high altitudes and hotter in low areas in mountainous terrain without moisture to modify the extremes. Forests also release turpenes and other particulates into the atmosphere that form "seeds" for water condensation for clouds and precipitation. These turpenes are less common in deforested areas.

[See also Agriculture and Agricultural Land; Amazonia, Deforestation of; Biological Diversity; Biomes; Carbon Dioxide; Easter Island; Erosion; Extinction of Species; Forestation; Global Change, article on Human Dimensions of Global Change; Global Warming; Greenhouse Effect; Intergovernmental Panel on Forests; International Geosphere–Biosphere Programme; Migrations; and Urban Areas.]

## BIBLIOGRAPHY

Allan, T., and J. P. Lanly. "Overview of Status and Trends of World's Forests." In *Technical Workshop to Explore Options for Global Forestry Management*, edited by D. Howlett and C. Sargent, pp. 17–39. Proceedings of a Workshop in Bangkok, Thailand, 24–30 April 1991. London: International Institute for Environment and Development, 1991.

Aschmann, H. "Man's Impact on the Several Regions with Mediterranean Climates." In *Mediterranean Type Ecosystems, Origin and Structures*, edited by F. Di Castri and H. A. Mooney, pp. 363–371. New York: Springer, 1973.

Aubreville, A. *Climats, Forêts, et Désertification de l'Afrique Tropicale*. Paris: Société des Éditions Géographiques, Maritimes et Coloniales, 1949.

Boydak, M., and M. Dogru. "The Exchange of Experience and State of the Art in Sustainable Forest Management (SFM) by Ecoregion: Mediterranean Forests." Proceedings of the Eleventh World Forestry Congress, 13–22 October 1997, Antalya, Turkey, vol. 6, pp. 179–204, 1997. Published by the Ministry of Forestry of Turkey, Ankara.

DeLaune, R. D., et al. "Relationships among Vertical Accretion, Coastal Submergence, and Erosion in a Louisiana Gulf Coast Marsh." *Journal of Sedimentary Petrology* 53 (1983), 147–157.

Eleventh World Forestry Congress. "Antalya Declaration of the XI World Forestry Congress: Forestry for Sustainable Development: Towards the XXI Century." Antalya, Turkey, 1997. Resolution published, approved, and distributed at the Congress; available from the Ministry of Forestry of Turkey, Ankara.

Food and Agriculture Organization. *Role of Forestry in Combating Desertification*, proceedings of the FAO Expert Consultation held in Saltillo, Mexico, 24–28 June 1985. Rome, 1985.

———. "Forest Resources Assessment 1990: Tropical Countries." FAO Forestry Paper 112. Rome, 1993.

———. *State of the World's Forests*. Rome, 1997. Available via the Internet at http://www.fao.org/waicent/faoinfo/forestry/publclst.htm/.

Ford-Robertson, F. C., ed. *Terminology of Forest Science, Technology, Practice, and Products*. Washington, D.C.: Society of American Foresters, 1971.

Gash, J. H. C., et al. *Amazonian Deforestation and Climate*. New York: Wiley, 1996.

Gates, D. M. *Climate Change and Its Biological Consequences*. Sunderland, Mass.: Sinauer, 1993.

Geiger, R. *The Climate near the Ground*. Cambridge: Harvard University Press, 1965.

Gordon, A. G., and E. Gorham. "Ecological Aspects of Air Pollution from an Iron-Sintering Plant at Wawa, Ontario." *Canadian Journal of Botany* 41 (1963), 1063–1078.

Kershaw, J. A., Jr., et al. "Effect of Harvest of Old Growth Douglas-Fir Stands and Subsequent Management on Carbon Dioxide Levels in the Atmosphere." *Journal of Sustainable Forestry* 1 (1993), 61–77.

Koch, P. "Wood vs. Nonwood Materials in U.S. Residential Construction: Some Energy-Related International Implications." CINTRAFOR Working Paper 36, Center for International Trade of Forest Products, College of Forest Resources, University of Washington. Seattle, 1991.

Lamb, H. H. *Weather, Climate, and Human Affairs.* New York: Routledge, 1988.

Odera, J., ed. "The Present State of Degradation of Fragile Ecosystems in Dry Lands and the Role of Forestry in Their Restoration." International Expert Meeting on Rehabilitation of Degraded Forest Ecosystems Secretariat Note No. 1., Lisbon, Portugal, 1996.

Oliver, C. D., and B. C. Larson. *Forest Stand Dynamics.* New York: Wiley, 1996.

Perlin, J. *A Forest Journey: The Role of Wood in the Development of Civilization.* Cambridge, Mass.: Harvard University Press, 1989.

Philander, S. G. *The Uncertain Science of Global Warming.* Princeton, N.J.: Princeton University Press, 1998.

Rykowski, K. *Sustainable Development of Forests in Poland, State and Perspectives.* Agencja Reklamova-Wydawnicza Arkadiusz Grzegorczyk, Warsaw, Poland, 1995.

Sampson, R. N., and D. Hair. *Forest and Global Change,* vol. 1, *Opportunities for Increasing Forest Cover.* Washington, D.C.: American Forests, 1992.

———. *Forest and Global Change,* vol. 2, *Forest Management Opportunities for Mitigating Carbon Emissions.* Washington, D.C.: American Forests, 1996.

Skole, D. L., W. H. Chomentowski, W. A. Salas, and A. D. Nobre. "Physical and Human Dimensions of Deforestation in the Amazon." *Bioscience* 44.5 (1994), 314–322.

Stuiver, M. "Atmospheric Carbon Dioxide and Carbon Reservoir Changes." *Science* 199 (1978), 253–258.

Thomas, D. S. G., and N. Middleton. *World Atlas of Desertification.* London: Edward Arnold, 1992.

Turner B. L., II, and K. W. Butzer. "The Columbian Encounter and Land-Use Change." *Environment* 34.8 (1992), 16–20, 37–44.

von Kooten, G. C., R. Sedjo, and E. Bulte. "Tropical Deforestation: Issues and Policies." In *International Yearbook of Environmental and Resource Economies: 1999–2000. A Survey of Current Issues,* edited by P. Teitenburg and H. Folmer, vol. 3, pp. 198–249. Cheltenham, U.K.: Edward Elgar, 1999.

World Resource Institute. *World Resources 1996–1997,* p. 204, citing United Nations Food and Agriculture Organization. "Forest Resources Assessment 1990: Global Synthesis." FAO Forestry Paper 112. Rome, 1993.

—CHADWICK D. OLIVER, MELIH BOYDAK, AND ROGER A. SEDJO

## DEMOGRAPHIC TRANSITION.
*See* Catastrophist–Cornucopian Debate; Economic Levels; Growth, Limits to; Human Populations; IPAT; *and* Population Policy.

# DESALINATION

Humans have practiced the desalting of sea water to obtain potable water for centuries. Greek sailors on-board ships used to boil sea water and condense its vapor—a process known as distillation—in the fourth century BCE. Romans, in the first century CE, were reported to have filtered sea water through a clay soil to obtain drinking water. By the fourth century, however, distillation had become the preferred method of desalination (Popkin, 1968). Today, a wide variety of desalination technologies are available on the commercial market.

The first known patent for a desalination process through steam distillation was granted in England in 1869. In the same year, British colonists in Aden built a distillation plant to supply fresh water to merchant and naval vessels stopping at the port. The first large desalting plant may have been the one installed in Aruba, Netherlands Antilles, in 1930, producing 1,000 cubic meters of fresh water per day ($m^3$/day) (Hornburg, 1987). It was not until the mid-1950s, however, that the use of large land-based desalination plants, particularly those using multistage flash distillation, became economically feasible for nonindustrial purposes. This spurred intensive research and development into a variety of desalination processes. By the mid-1960s much of the work in desalination remained experimental, improving on earlier designs of plants that failed to meet expectations.

The arid oil-producing countries of the Middle East stimulated the development of desalination technology and encouraged the growth of the desalination industry. Oil revenues enabled these countries to use desalination as their most reliable and secure solution to the problem of providing fresh water to their populations (WRI, 1993; Ayoub and Alward, 1996).

**Worldwide Desalination Capacity.** Countries such as Saudi Arabia, Kuwait, and the United Arab Emirates are the major users of desalination technology, along with the middle- and high-income regions in the United States, such as the states of Florida and California, whose residents are able to pay for the high costs of desalted water (see Table 1).

The slow adoption of desalination technology in the early 1960s was reflected in the gradual increase in desalination capacity. A sharp rise began taking place toward the end of the 1960s, however, with an average annual increase in cumulative capacity in the order of 125,000 $m^3$/day. This rate jumped threefold in the early to mid-1970s and sixfold by 1990 (Wangnick, 1992). Likewise, since 1960 the worldwide capacity of installed desalination plants has increased enormously. The number of desalination units producing fresh water in excess of 100 $m^3$/day increased from 3,527 in early 1987 to 8,886 units in early 1992, to operating in some 120 countries

**Desalination. TABLE 1.** Total Capacity of Desalting Plants Producing 100 Cubic Meters or More of Fresh Water Daily for Selected Major User Countries, 1900–1997

| COUNTRY | CAPACITY (M³/DAY) | PERCENT* |
|---|---|---|
| Saudi Arabia | 5,373,594 | 23.62 |
| United States | 3,546,872 | 15.59 |
| United Arab Emirates | 2,218143 | 9.75 |
| Kuwait | 1,539626 | 6.77 |
| Japan | 899,140 | 3.95 |
| Spain | 847,435 | 3.73 |
| Libya | 749,414 | 3.29 |
| Qatar | 579,260 | 2.55 |
| Italy | 565,122 | 2.48 |
| Iran | 470,537 | 2.07 |
| Bahrain | 443,329 | 1.95 |
| Korea | 359,353 | 1.58 |
| India | 355,047 | 1.56 |
| Iraq | 332,613 | 1.46 |
| Germany | 267,512 | 1.18 |
| Netherlands Antilles | 232,081 | 1.02 |
| Algeria | 211,721 | 0.93 |
| Great Britain | 204,813 | 0.90 |
| Oman | 199,837 | 0.88 |
| China | 188,357 | 0.83 |
| Egypt | 185,870 | 0.82 |
| Hong Kong | 183,694 | 0.81 |
| Mexico | 169,722 | 0.75 |
| Kazakhstan | 167,619 | 0.74 |
| Netherlands | 159,069 | 0.70 |
| Malta | 148,572 | 0.65 |
| Indonesia | 146,518 | 0.64 |
| Taiwan | 136,255 | 0.60 |
| Virgin Islands | 132,172 | 0.58 |
| Russia | 120,702 | 0.53 |
| Total | 21,133,999 | |

*Percentages indicate share in relation to all other countries.

SOURCE: Wangnick (1998).

with a total or contracted capacity of about 21 million m³/day (Wangnick, 1992; Wangnick, 1998).

**Survey of Desalting Processes.** Desalination processes currently in use are characterized by the quality of energy inputs involved. They fall within two processes: thermal (where a phase change occurs in the feedwater, e.g., distillation) and membrane (where no phase change occurs). In some thermal processes, such as multistage flash (MSF), multieffect evaporation (ME), and vapor compression (VC) distillation, a relatively high thermal energy input is required to bring about a phase change in the sea water. These processes are generally cost-effective in large-scale plants supplying municipal drinking water because the unit cost of product water is lower and thus has high commercial viability (Ayoub and Alward, 1996). Other thermal processes such as freezing, membrane, and solar distillations usually have a lower quality energy requirement (e.g., low-grade thermal energy from such sources as industrial waste heat and solar collectors) and are cost-effective in small- to medium-scale systems used to supply community or family drinking water needs.

In membrane processes such as reverse osmosis (RO), electrodialysis (ED), and electrodialysis reversal (EDR), high-grade electrical or mechanical energy inputs are required to produce potable water without a phase change in the sea or brackish water (Nilsson, 1995). These processes can be scaled to meet various applications, including supplying industrial process water, treating municipal waste water, and supplying potable water for communities (Table 2).

Very large plants, such as the 1 million m³/day plant in Saudi Arabia, typically use the distillation process of seawater desalination. While the major distillation process, multistage flash distillation, has declined in market share from 79 percent in total capacity in 1969 to slightly above 30 percent in 1991, it still occupies a significant share of the very large plants and dual-purpose plants coupled with power generation. Smaller plants typically use the membrane separation process of reverse osmosis. RO plants accounted for about 45 percent of global capacity by 1991, compared with 10 percent in 1969. Approximately two-thirds of all plants worldwide are converting sea water, and the remaining one-third is treating brackish water. Desalination plants are increasingly being used for applications other than the production of potable water, such as for the treatment of effluent water and polluted groundwater sources, and in the production of ultrapure water for the electronics industry (WRI, 1992).

**Desalinated Water Costs.** The costs of desalinated water vary depending on the specific desalination processes chosen, the scale of the operation, and site conditions. In large-scale systems, optimization of energy costs and plant performance are major determinants in the choice of desalination technology, whereas technical, physical, social, and economic factors determine the design of small-scale systems. Although the MSF process has been commercially operating for a much longer time than RO, significant improvements in membrane technology over the last two decades have made the latter process much more economically attractive for large-scale seawater desalination (Malek et al., 1992). Cost studies conducted by various researchers (Ayoub and Alward, 1996; Pappas, 1993; Malek et al., 1992) in-

**Desalination. Table 2.** Commercially Available Desalination Processes

| PROCESS | FEEDWATER | CAPACITY ($M^3/DAY$) | POWER SOURCE | MARKET SHARE (%)* | TOTAL INSTALLED OR CONTRACTED DESALTING CAPACITY (%)* |
|---|---|---|---|---|---|
| *Major Processes* | | | | | |
| Thermal | | | | | |
| MSF | Sea | 4,000 | Natural gas | 32 | 52 |
| ME | Sea | 30,000 | | † | † |
| VC | Sea | 2,000–10,000 | Natural gas | † | † |
| Membrane | | | | | |
| RO | Sea/brackish | 20–2,000 | Electric | 50 | 33 |
| ED | Sea/brackish | Various | Electric | | † |
| EDR | Sea/brackish | Various | Electric | | † |
| *Minor Processes* | | | | | |
| Vacuum freezing | Sea | 38–750 | Electric | Negligible | |
| Solar humidification | Sea/brackish | 1 $m^2$ of collector area produces 4 l/day of potable water | Solar | Negligible | |

*1991 figures.

†ME, VC, ED, and EDR share the remaining 18 percent of market share and 15 percent of total installed or contracted desalting capacity. RO, ED, and EDR are amenable to a wide variety of desalting capacities from 0.10 $m^3$/day to virtually any capacity. However, large-size membrane plants do not have the economies of scale that are achieved by distillation plants.

SOURCE: Ayoub and Alward (1996).

dicate that the unit cost of desalinated water using MSF is approximately in the range of U.S.$1.40–1.90 per cubic meter compared with U.S.$0.80–1.65 per cubic meter for RO systems of similar capacity. The difference in costs is probably due to the greater sensitivity of MSF plants to fuel cost fluctuations. Studies in remote villages in India (Natarjan et al., 1991; Narayanan et al., 1991) of smaller brackish-water RO and seawater ED plants with production capacities of 5–10 $m^3$/day indicate the unit water cost in the range of U.S.$4.00–5.00 per cubic meter. Conversely, the costs of product water from solar stills have been reported to reach as high as U.S.$12.50 for a plant installed in 1986 to supply potable water for the Kalahari dwellers of Botswana (Woto, 1987).

**Conclusion.** There is a growing need to find solutions to the problems of freshwater supply. The technology of seawater and brackish-water desalination for the production of potable water is well established. In terms of large-scale desalination systems, multistage flash distillation and reverse osmosis are presently the dominant processes technically and economically. Newer and more efficient membranes provide real possibilities of reducing the cost of water obtained from large-scale seawater desalination in the future.

[*See also* Deserts; Energy; *and* Water.]

## BIBLIOGRAPHY

Ayoub, J., and R. Alward "Water Requirements and Remote Arid Areas: The Need for Small-Scale Desalination." *Desalination* 107 (1996), 131–147.

Hornburg, C. D. "Desalination for Remote Areas." In *Developing World Water*, pp. 230–232. Hong Kong: Grosvenor Press, 1987.

Malek, A., M. N. A. Hawlader, and I. C. Ho "Large-Scale Seawater Desalination: A Technical and Economic Review." *ASEAN Journal of Science and Technology Development* 9.2 (1992), 41–61.

Narayanan, P. K. et al. "Performance of the First Seawater Electrodialysis Plant in India." In *Desalination and Water Reuse: Proceedings of the Twelfth International Symposium, Malta, 15–18 April, 1991*, edited by M. Balaban, 4: 210–211.

Natarjan, R., W. V. B. Ramalingam, and W. P. Harkare. "Experience in Installation and Operation of Brackish Water Desalination Plants in Rural Areas of India." *IDA Conference on Desalination and Water Reuse: "Water: The Challenge of the 90s," Washington, D.C., 25–29 August, 1991*, 2: 13.

Nilsson, S. *A Review of Desalination Processes and Future Water Needs.* New York: Energy and Atmosphere Programme, BPPS/SEED, United Nations Development Programme, 1995.

Pappas, C. A. "Why Desalination Is Not in Common Use Worldwide." *Desalination and Water Reuse* 3.4 (1993), 34–39.

Popkin, R. *Desalination: Water for the World's Future.* New York: Praeger, 1968.

Wangnick, K. *1992 International Worldwide Desalting Plant Inventory.* Report no. 12. Gnarrenburg, Germany: Wangnick Consulting, 1992.

———. *1998 IDA Worldwide Desalting Plant Inventory.* Report no. 15. Gnarrenburg, Germany: Wangnick Consulting, 1998.

World Resources Institute (WRI). *World Resources 1992–93: Towards Sustainable Development.* New York: Oxford University Press, 1993.

Woto, T. *The Experience with Small-Scale Desalinators for Remote Areas Dwellers of the Kalahari Botswana.* Kanye, Botswana: Rural Industries Promotion/Rural Industries Innovation Centre, 1987.

—JOSEF AYOUB

## DESERTIFICATION

*[To survey desertification, this entry comprises two articles. The first article presents an overview of the causes and results of drought and land degration in arid, semiarid, and dry subhumid areas. The second article describes the objectives and long-term strategies in the management of desertification.]*

### An Overview

*Desertification* refers to land degradation in arid, semiarid, and dry subhumid areas resulting from various factors, including climatic variations and human activities. The term, first introduced by A. Aubréville in his 1949 book, *Climats, Forêts et Désertification de l'Afrique Tropicale,* has been used to refer to a wide variety of ecological problems in dryland habitats throughout the world. Although desertification is usually associated with drought, it is quite common for land degradation to occur without changes in rainfall; for instance, when humans remove vegetation, and bare soil is eroded by wind and water. Desertification is often associated with the loss of desirable plant species and their replacement by species of lesser economic importance. It is associated with a permanent loss in the productive capacity of the land.

The term *desertification* should not be used to describe cyclic phenomena, as when decadal variations in precipitation lead to periods of drought and to losses of vegetation that are fully restored when the rains return. For example, desertification has been used to describe land degradation along the southern border of the Sahara, where a 1975 survey by the United Nations found the desert to be expanding southward at 5.5 kilometers per year in the Sudan. In fact, the southward expansion of the Sahara, reaching its peak in 1984, was effectively reversed with a return to a period of greater rainfall in more recent years (Figure 1), and in most areas there was no long-term loss in the productive capacity of the land. [*See* Drought, *article on* Sahel Drought in West Africa.]

About 30 percent of the world's land surface, 6,150

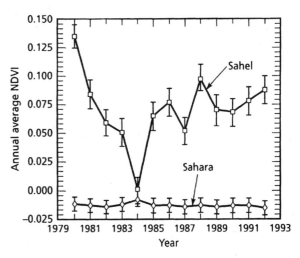

**Desertification: An Overview. FIGURE 1.** Normalized Difference Vegetation Index (NDVI) for Lands Bordering the Sahel Zone and for the Sahara Itself, during a Ten-Year Period. (From Tucker et al., 1994. With permission from Taylor and Francis Ltd.)

million hectares (15,200 million acres), is classified as arid or semiarid. About 16 percent of that area is considered hyperarid (namely, the Sahara) and not subject to further desertification. The remaining 5,200 million hectares, mostly classified as semiarid, is potentially vulnerable to degradation. These areas are mostly grasslands and savannas at the boundaries of arid regions. [*See* Biomes.] The 1991 United Nations assessment of desertification estimated that 3,600 million hectares, or nearly 70 percent of the vulnerable land, was already in some stage of degradation. This was an increase of 117 million hectares over an earlier (1984) UN assessment, or 3.4 percent in seven years. The largest changes were seen in Africa and Asia.

Overgrazing is perhaps the leading cause of desertification worldwide. [*See* Animal Husbandry.] Overgrazing occurs when improper timing, intensity, or frequency of herbivory causes damage to grazed plants. Historically, the intensity of herbivory by livestock has been too high for many arid lands. A survey of grazed lands in South America showed that they often contained about ten times more herbivore biomass than found in adjacent ungrazed lands. The consequences of improperly managed grazing can be devastating to soil resources. In southern New Mexico, overgrazing of semiarid grasslands dominated by black grama (*Bouteloua eriopoda*) has led to the invasion of mesquite (*Prosopis glandulosa*) and creosotebush (*Larrea tridentata*) in less than one hundred years. When shrubs replace semiarid grasses, most of the fertility of the remaining soils is isolated in patches that persist under the shrubs, while the barren soils between shrubs may be nearly devoid of biotic activity. It is often

**Desertification: An Overview. FIGURE 2.**
Crescent Dunes Developing toward the
Front of the Shelter Belt in Gonghe County,
Quinghai, China. (From Ministry of
Geology & Mineral Resources, P.R.C., State
Science & Technology Commission, P.R.C,
and State Planning Commission, P.R.C.,
1991.)

impossible to remediate degradation of these lands by
replanting the native vegetation.

Other human activities also cause desertification. In
many areas, desertification is linked to excessive harvesting of fuel wood from semiarid habitats. Dryland
soils that are used for irrigated agriculture are subject
to degradation as a result of a buildup of salinity in the
soil profile. Lands in both irrigated and rain-fed agriculture are subject to wind erosion during seasonal periods when the soils are left bare. Desertification in the
many areas of China stems from the encroachment of
sand dunes into agricultural areas (Figure 2).

Areas of desertification are vulnerable to wind erosion. As the cover of vegetation is reduced by cattle grazing or agriculture, barren soils are subject to wind
erosion, because the threshold velocity to initiate the
movement of fine soil materials is lower on a barren,
unprotected surface. Wind erosion also increases when
human activities and cattle disrupt the crust of algae and
lichens that often forms at the surface of desert soils.

The infiltration of soil moisture is lower on barren
soils, so that desertified landscapes are subject to high
rates of runoff and soil erosion during infrequent periods of heavy rain. In many construction projects, humans
channelize and reroute the natural drainageways of arid
landscapes, also lowering the effective infiltration of
moisture into the soil and leading to a reduction in the
cover of vegetation. Rain use efficiency—the proportion
of incident rain that is used by plants for growth—is often lower in areas of desertification, and changes in rain
use efficiency are indicative of land degradation.

Areas of reduced vegetation cover often show a
greater reflection of solar radiation. [*See* Albedo.] De-

spite absorbing less incident radiation, barren landscapes have higher soil and air temperatures. Normally
the conversion of water from liquid to vapor consumes
540 calories per gram in the latent heat of vaporization,
which can result in significant evaporative cooling of
the landscape from the uptake and evapotranspiration
of soil moisture by plants. In areas of reduced vegetation, there is less evaporative cooling. In northern Mexico, soil and air temperatures average about 3°C higher
than those in southern Arizona, where grazing has been
less severe. Thus, an increase in the Bowen ratio—the
ratio of the amount of incoming radiation dissipated by
sensible heat to that dissipated by latent heat—is a good
index of desertification. A small proportion, perhaps 1
percent, of the Earth's recent increase in temperature is
thought to derive from increasing desertification.

Changes in the area and borders of arid lands are easy
to monitor by remote sensing. [*See* Remote Sensing.]
Since 1980, NASA scientists have used a high-resolution
radiometer to examine changes in the greenness of the
Sahel zone in Africa. Vegetation greenness is quantified
by the Normalized Difference Vegetation Index (NDVI),
which is based on the differential absorption of red and
infrared radiation by plants. Plants absorb red radiation,
but they reflect infrared wavelengths, whereas bare soils
reflect both forms equally. Thus, an index of greenness
is calculated as:

$$\text{NDVI} = \frac{(\text{Infrared Reflectance}) - (\text{Red Reflectance})}{(\text{Infrared Reflectance}) + (\text{Red Reflectance})}.$$

The index is most useful when the cover of vegetation
is greater than 50 percent. On areas with lesser amounts
of plant cover, the reflectance of bare soils dominates

the radiation received by the satellite, and it is more difficult to see changes in vegetation. The record of vegetation greenness in the Sahel shows a cyclic pattern that is associated with recent changes in regional rainfall, whereas NDVI for the Sahara over the same period shows consistent low values that are unrelated to annual rainfall (Figure 1).

Land degradation often does not follow a smooth progression during desertification. An existing cover of semiarid grassland may persist for many years despite high levels of livestock grazing. Then, within just a few years, perhaps coincident with reduced rainfall, the grasses may disappear entirely from the landscape. In such rapid transitions, ecologists suggest that the landscape has passed a threshold—a level of stress that pushes the ecosystem to a new stable state. Desertification is analogous to a teeter-totter, in which human impact tips the balance of the landscape to a new, degraded state (Figure 3).

The frequency of drought is expected to increase during the next century as global climate change warms the Earth's land surface more rapidly than the ocean surface, where most precipitation is generated. Thus an increasing proportion of the world's land surface is likely to experience greater potential evapotranspiration, with little or no increase in precipitation. Soil dusts generated from desertified areas reflect incoming solar radiation, so a greater area of desertified land may influence global climate. Dust from barren desertified soils in Africa is transported across the Atlantic Ocean and deposited in the southeastern United States (Figure 4). [*See* Dust Storms.] Airborne dusts tend to cool the atmosphere over ocean waters, which otherwise absorb a large proportion of incoming solar radiation. Ironically, an increasing amount of soil dust over the oceans, a likely consequence of an increasing global area of desertified lands, may reduce the rate of greenhouse warming of Earth. Ice-core records show that desert dust was more widespread during the last glacial epoch, when the Earth's temperature was 6°C lower than at present.

Drought is certainly linked to the downfall of great historic civilizations, including the early Mesopotamian civilization in 2200 BCE and the Mayan culture in Mexico around 900 CE. [*See* Civilization and Environment.] Today, about 20 percent of the world's people live in environments at or near the border of desert regions. Concern about the increasing degradation of semiarid lands and an expansion of deserts has led the United Nations to host a number of international workshops to assess and combat desertification. The first, the UN Conference on Desertification (UNCOD), was held in Nairobi, Kenya, in 1977, when the world's attention was focused on drought and famine in central Africa. The most re-

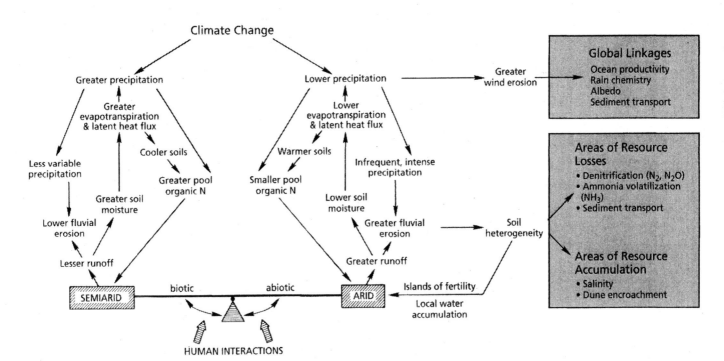

**Desertification: An Overview. FIGURE 3. A Conceptual Model of Desertification.**

The figure shows the properties of semiarid grasslands (left) and desertified shrublands (right), and the balance that is maintained between these lands. (From Schlesinger et al., 1990. Reprinted with permission from the American Association for the Advancement of Science.)

**Desertification: An Overview. FIGURE 4.** Transport of Desert Dust from the Sahara across the Atlantic Ocean. (From Perry et al., 1997. Copyright by the American Geophysical Union.)

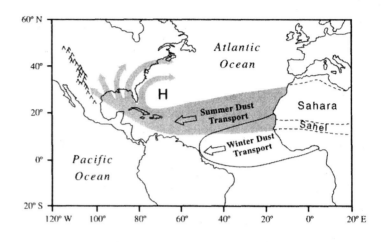

cent convention followed the Earth Summit in Rio de Janeiro in June 1992, producing an international agreement of cooperation that was signed in Paris in 1994. Parties to this convention pledge to combat desertification and reclaim desertified land, with a special focus on Africa, where 66 percent of the continent is dryland.

[*See also* Aerosols; Deforestation; Deserts; *and* Salinization.]

### BIBLIOGRAPHY

Dregne, H. E. "A New Assessment of the World Status of Desertification." *Desertification Control Bulletin* 20 (1991), 6–18.

Laycock, W. A. "Stable States and Thresholds of Range Condition in North American Rangelands: A Viewpoint." *Journal of Range Management* 44 (1991), 427–433.

Le Houérou, H. N. "Rain Use Efficiency: A Unifying Concept in Arid-Land Ecology." *Journal of Arid Environments* 7 (1984), 213–247.

Ministry of Geology & Mineral Resources, P.R.C., State Science & Technology Commission, P.R.C, and State Planning Commission, P.R.C. *Geological Hazards of China and Their Prevention and Control.* Beijing: Geological Publishing House, 1991.

Perry, K. D., T. A. Cahill, R. A. Eldred, D. D. Dutcher, and T. E. Gill. "Long-Range Transport of North African Dust to the Eastern United States." *Journal of Geophysical Research* 102 (1997), 11225–11238.

Rind, D., R. Goldberg, J. Hansen, C. Rosenzweig, and R. Ruedy. "Potential Evapotranspiration and the Likelihood of Future Drought." *Journal of Geophysical Research* 95 (1990), 9983–10004.

Schlesinger, W. H., J. F. Reynolds, G. L. Cunningham, L. F. Huenneke, W. M. Jarrell, R. A. Virginia, and W. G. Whitford. "Biological Feedbacks in Global Desertification." *Science* 247 (1990), 1043–1048.

Tegen, I., and I. Fung. "Contribution of the Atmospheric Mineral Aerosol Load from Land Surface Modification." *Journal of Geophysical Research* 100 (1995), 18707–18726.

Tucker, C. J., W. W. Newcomb, and H. E. Dregne. "AVHRR Data Sets for Determination of Desert Spatial Extent." *International Journal of Remote Sensing* 15 (1994), 3547–3565.

Verstraete, M. M., and S. A. Schwartz. "Desertification and Global Change." *Vegetatio* 91 (1991), 3–13.

West, N. E. "Structure and Function of Microphytic Soil Crusts in Wildland Ecosystems of Arid to Semi-Arid Regions." *Advances in Ecological Research* 20 (1990), 179–223.

—WILLIAM H. SCHLESINGER

## Desertification Convention

On the initiative of the developing countries, the states participating in the United Nations Conference on Environment and Development (UNCED) in Rio de Janeiro in June 1992 agreed to initiate negotiations on an international desertification convention. On 22 December 1992, the General Assembly of the United Nations decided to establish, under its auspices, an International Negotiating Committee for the elaboration of an international convention. After five sessions of the Intergovernmental Negotiating Committee for the Elaboration of an International Convention to Combat Desertification (INCD), the United Nations Convention to Combat Desertification in Countries Experiencing Serious Drought and/or Desertification, Particularly in Africa was adopted on 17 June 1994. The convention includes four annexes concerning regional implementation of the provisions in Africa, Asia, Latin America, the Caribbean, and the northern Mediterranean. However, the annex on Africa includes the most detailed provisions of all annexes, and the convention implies that priority should be given to measures in Africa.

The convention entered into force on 26 December 1996, ninety days after the fiftieth member country had ratified or acceded to the convention, and by the end of March 1998 it had been ratified by 124 states including the European Union, which could—in addition to the EU members states—also join the convention because of its status as a regional economic integration organization.

The major objective of the convention is "to combat desertification and mitigate the effects of drought . . . through effective action at all levels." It includes a "bottom-up approach," since it requires the participation of population and local communities when states decide on the design and implementation of programs to combat desertification or to mitigate the effects of drought. The first Conference of the Parties (CoP), held from 29 September to 10 October 1997 in Rome, accredited 452 nongovernmental organizations (NGOs) to the first and subsequent sessions of the CoP. Most of these NGOs come from developing countries. Such participation of NGOs will be especially necessary when countries affected by desertification or the effects of drought implement the National Action Programs that they are required by the convention to develop.

These National Action Programs should include long-term strategies to manage the problem of desertification and the effects of drought, give attention to the implementation of preventive measures for lands that are not yet or only slightly degraded, promote international cooperation, and provide for effective participation at local, national, and regional levels. One of the most important functions of the convention is to improve scientific and technical cooperation between the member states. In particular, the members of the convention agree to improve the collection, analysis, and exchange of data and information to ensure systematic observation of land degradation and to gather further knowledge about the process and effects of drought and desertification. The convention is also intended to promote scientific and technical cooperation in the relevant areas, the transfer, acquisition, adaptation, and development of technology relevant to combating desertification, and capacity building.

Developed countries are committed by the convention to mobilizing substantial financial resources, including grants and concessionary loans to developing countries and new and additional funding from the Global Environment Facility to support the implementation of the National Action Programs. The convention established a Global Mechanism to promote actions leading to the mobilization and channeling of such resources. The first CoP decided that the International Fund for Agricultural Development (IFAD) should house the Global Mechanism. IFAD, the United Nations Development Programme, and the World Bank established a "Facilitation Committee" for the Global Mechanism to coordinate support from the three institutions.

A Committee on Science and Technology was established by the convention with the task of providing the CoP with information and advice on scientific and technological matters. The first CoP chose the city of Bonn as the location of the permanent secretariat of the convention.

[See also United Nations Conference on Environment and Development.]

## BIBLIOGRAPHY

Corell, E. "The Failure of Scientific Expertise to Influence the Desertification Negotiations." IIASA Working Paper WP-96-165. Laxenburg, Austria, 1996.

Kassas, M. "Negotiations for the International Convention to Combat Desertification, 1993–1994." *International Environmental Affairs* 7.2 (1995), 176–186.

Thomas, D. S. G., and N. J Middleton. *Desertification: Exploding the Myth.* Chichester, U.K.: Wiley, 1994.

Toulmin, C. "Combating Desertification: Encouraging Local Action within a Global Framework." In *Green Globe Yearbook*, edited by H. O. Bergesen and G. Parmann, pp. 79–88. New York and Oxford: Oxford University Press, 1994.

United Nations. *Report of the Conference of the Parties on Its First Session. Part One.* Convention to Combat Desertification held in Rome from 29 September to 10 October 1997. UN Publication CCD/COP(1)/11, 1997.

———. *Report of the Conference of the Parties on Its First Session. Part Two. Actions Taken by the Conference of the Parties at its First Session.* Convention to Combat Desertification held in Rome from 29 September to 10 October 1997. UN Publication CCD/COP(1)/11/Add.1, 1997.

United Nations General Assembly. *Elaboration of an International Convention to Combat Desertification in Countries Experiencing Serious Drought and/or Desertification, Particularly in Africa: Final Text of the Convention.* UN Publication A/AC.241/27, 1994.

—HELMUT BREITMEIER

# DESERTS

Deserts, or drylands, are known to experience significant environmental changes at time scales ranging from the geologic (millions of years) to decadal and annual. Short-term changes (ten years or less) are part of the natural changes associated with climatic variability or, increasingly, with human activities. [*See* Desertification.] Information on significant environmental changes that have occurred in deserts and drylands during the Quaternary (last two million years) geologic period is important not only to identify the response of these areas to past global climate changes, but to use in advancing predictions of changes that might be expected to occur as a result of anthropogenically induced global warming. To understand the nature, timing, and magnitude of changes in deserts, it is necessary to define what these areas are, assess their current extent, examine data gained from proxy sources that allows past changes to be assessed, and then evaluate the status of deserts and drylands into the future.

**Defining Deserts and Drylands.** Most definitions of desert relate directly or indirectly to moisture deficiency. Deserts, however, can and do experience significant, intense rainfall events, albeit on an irregular or unreliable basis, while some have perennial rivers with sources in wetter regions flowing through them—for example, the Nile, which waters areas of Egypt and the Sudan with extremely low rainfall. Few, if any, deserts are totally devoid of plant and animal life, but species that are present tend to have adaptive strategies that permit the accumulation and retention of available moisture, and/or physiologies that allow biological functions to be slowed or shut down at times of acute moisture stress.

The term *desert* is therefore closely allied to the notion of aridity, which has four main causal factors: (1) tropical and subtropical atmospheric stability, (2) continentality (or distance from oceans), (3) topographically induced rain shadows, and (4) in some coastal situations such as Chile and Namibia, cold ocean currents reducing evaporation from the sea surface. Scientific definitions of *desert* have been based on criteria that include the nature and development of drainage systems, the types of rock weathering process that operate, ecological communities, and the potential for crop growth (see Thomas, 1997, chap. 1). Attempts to quantify aridity have focused, in a number of different approaches, on the balance between atmospheric moisture inputs (through precipitation) and losses (through evapotranspiration, determined in part by temperature). Two further factors have been embodied in definitions. First, attempts to define *deserts* and *aridity* since 1950 have tended to be driven by practical concerns related to human use of the environment. Most definitions since that of Perivale Meigs (1953) for United Nations Educational, Scientific, and Cultural Organization (UNESCO) have excluded areas that are too cold for the production of crops. Polar deserts and extreme high-altitude deserts are therefore not included when this approach is taken. Second, as aridity has been investigated, different degrees of moisture deficit have been identified. To this end, hyperarid, arid, semiarid, and now even dry-subhumid areas are all regarded in scientific circles as drylands, with hyperarid areas being viewed as true deserts. However, the term *desert* is applied to regions that do not experience such extreme aridity; for example, the Kalahari Desert in southern Africa embodies areas that range from arid to dry-subhumid. It is therefore difficult to consider deserts without assessing the full range of dryland environments.

***The extent of deserts and drylands.*** The most recent attempts to determine the extent of deserts and drylands on the basis of moisture availability have used an aridity or moisture index in the form *P/PET*, where

*P* is the annual precipitation and *PET* is the potential evapotranspiration. Meigs's moisture index used annually aggregated monthly moisture surplus and deficit data instead of *P*, with *PET* calculated by the Thornthwaite method. The Thornthwaite method only requires mean monthly temperature and daylight hours data, and is therefore suitable for use in drylands where climate data may be sparse. The most recent widely available assessment (by Mike Hulme for United Nations Environment Programme [UNEP]; in Middleton and Thomas, 1997) has used the same method of calculating *PET*. To further rationalize values, Hulme also used meteorological data for a defined time period (1951–1980) to calculate *P/PET*, rather than simply taking mean values from each station supplying data. This overcomes the problem that, in some developing parts of the world, mean values calculated from data runs of a few decades would be treated as equal to those derived from several centuries of data from parts of the developed world. It also gives recognition to an important climatological characteristic of drylands, namely, high interannual and interdecadal climatic variability. This also makes it possible to construct dryland climate surfaces for different decades and to identify spatial changes in their extent (Hulme, 1992).

Table 1 shows the global extent of drylands by different aridity zones, as represented in different sources. Differences between classification schemes represent different data sources and calculation methods, rather than necessarily being actual changes in extent.

***Identifying deserts and drylands.*** Figure 1 shows the global distribution of desert biomes. The identification of deserts and dryland types on the ground is not necessarily simple or something achievable with precision. While plant cover densities in true desert areas may be extremely low and, in extreme cases, limited to microorganisms, the mesophytes (which can tolerate moderate moisture deficiencies) and xerophytes (which can cope with extreme moisture deficiencies) that do occur in different dryland areas have distributions influenced by a range of environmental factors other than

**Deserts. TABLE 1.** The Extent of Global Drylands, Expressed as a Percentage of the Global Land Area, According to Different Classification Schemes

| CLASSIFICATION | DRY-SUBHUMID | SEMIARID | ARID | HYPERARID | TOTAL |
|---|---|---|---|---|---|
| Köppen (1931) | — | 14.3 | 12.0 | — | 26.3 |
| Thornthwaite (1948) | — | 15.3 | 15.3 | — | 30.6 |
| Meigs (1953) | — | 15.8 | 16.2 | 4.3 | 36.3 |
| Shantz (1956) | — | 5.2 | 24.8 | 4.7 | 34.7 |
| United Nations (1977) | — | 13.3 | 13.7 | 5.8 | 32.8 |
| Middleton and Thomas (1997) | 9.9 | 17.7 | 12.1 | 7.5 | 47.2 |

SOURCE: From Thomas (1997, chap. 1). With permission of John Wiley and Sons.

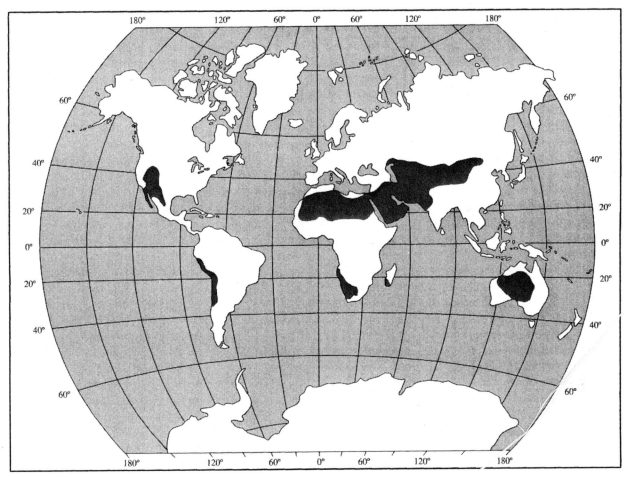

**Deserts. FIGURE 1.** Global Distribution of Desert Biomes.

gross climate. Semiarid and dry-subhumid areas embrace the savanna regions of the Americas, Africa, and Australia, but, in all dryland climatic zones, high interannual and decadal variability in moisture regimes has significant effects on the plant communities present at any particular time. At the intradecadal scale, El Niño–Southern Oscillation (ENSO) events in the Pacific are now known to be linked through the effects of teleconnections (distant changes in the atmosphere that are simultaneous) to significant drought occurrences and aridity intensification in dryland areas as far afield as northeastern Brazil, southeastern Africa, and Australia. Gross seasonal and annual changes in biomass can be loosely identified using satellite-derived Normalized Difference Vegetation Index (NDVI) data, and although there are marked doubts about the precise biological interpretation of NDVI, these data have been used to chart fluctuations in the desert front in response to moisture regime variability (see Tucker et al., 1991), effectively demonstrating the short-term dynamism of desert biomes. [*See* El Niño–Southern Oscillation.]

Deserts and drylands also embrace a wide range of landscape systems, demonstrating that such areas are far more complex than the popular images of seas of sand dunes. Even in the Sahara of North Africa, which in itself embraces many other named desert areas, over 40 percent of the terrain is mountainous and under 30 percent comprises sand seas, according to a survey carried out by Clements et al. (1957).

It is clear that deserts and drylands are complex biomes that are only now receiving the depth of scientific investigation that more temperate and polar regions have received in the past. When past and future global changes affecting deserts and drylands are considered, it is important to recognize three factors that make interpretation complex: (1) the relatively limited understanding of environmental processes that currently affect them, (2) the inherent short-term natural variability that occurs in key climatic forcing factors, and (3) the growing role of humans, with drylands today supporting 17 percent of the global, and nearly 50 percent of the African, population.

**Past Changes in Deserts and Drylands.** The major climatic changes that have affected the Earth in the

Quaternary period are known on theoretical and empirical grounds to have impacted the distribution and extent of drylands. The rock record has also been used to identify the sporadic existence of desert sand seas and dune deposits as far back as the Proterozoic (up to 2,500 million years ago; Glennie, 1987). The distribution and occurrence of such deposits reflect not simply climatic changes that may have affected the Earth in earlier geologic times, but the changing location of landmasses relative to aridity-inducing climate systems, as a function of plate tectonic movements. [*See* Climate Change; *and* Paleoclimate.]

At the scale of Quaternary glacial–interglacial cycles, changes in the extent of deserts and drylands can be expected to have occurred as a result of changes within the partitioning of moisture in the global hydrologic cycle and alterations to the positioning of major climatic systems. Evidence for dryland changes comes from deep-sea core sediments, sections of which may contain dust and eolian (windblown) sands derived from the increased operation of eolian processes under conditions of expanded aridity, or changes in fluvially derived sediments indicative of changed catchment weathering environments, and from terrestrial sediments and landforms. Terrestrial evidence not only points toward larger deserts at times within the Quaternary but, through the existence of paleolake shorelines, significant fluvial deposits, and cave speleothems within present-day desert areas, to times of reduced desert extent as well.

***Reconstructing Quaternary deserts.*** In the last three decades, major advances have been made in understanding past dryland and desert changes, particularly from terrestrial sources of proxy data; however, Stephen Stokes (see Thomas, 1997, chap. 27) notes that reconstruction of Quaternary desert and dryland conditions is hindered by a number of factors, including the following:

- Episodic rather than continuous sediment accumulation as a characteristic of drylands
- The low biomass productivity of the driest areas, limiting the accumulation of pollen and the application of radiocarbon dating
- Groundwater movements contaminating weathering products that otherwise might yield datable precipitates
- The slow accumulation and poor development of soils
- The long exposure of some arid-landscape features at the ground surface, precluding ready determination of the time of development/accumulation

Added to this is the fact that some drylands are deficient in closed sites that preserve longer sediment accumulations that might contain records of fluctuating desert conditions. Thomas and Shaw (1991) note that this is a particular problem in central southern Africa.

Together, these limitations contribute in many parts of the world to difficulties in identifying both more subtle changes that have occurred in the distribution of different dryland zones and the climatic parameters that contribute to the determination of desert conditions. Notable exceptions do exist, and in North America, South Africa, and Australia the pollen preserved within urine-cemented rat middens provides valuable data on biological changes within the arid realm.

One of the most widely used lines of evidence of more extensive Quaternary deserts is fossil, relict, or degraded sand dune systems located in areas that today are not conducive to the operation of eolian processes. Goudie (1993) has discussed some of the principal locations around the world in which these features have been identified: in India, Australia, North and South America and many parts of Africa. One study by Sarnthein (1978) suggested that 50 percent of the land area between 30° north and 30° south latitude comprised active dune fields at the last glacial maximum, compared with only 10 percent today. The resolution and paleoenvironmental utility of Quaternary dune systems is now being enhanced by the application of techniques within the luminescence family of dating methods to date phases of sand accumulation directly, and also by studies that are attempting to gain a greater understanding of the interaction of environmental parameters that control eolian processes and the accumulation of different dune types in dryland environments (see Thomas, 1997, chap. 26). As a result, the coincidence of the greatest late-Quaternary desert expansions worldwide at the last glacial maximum is being questioned, and more complex models of climatic changes affecting drylands are being proposed. New studies are showing that the sediments of dune systems that are inactive today may possess evidence of multiple phases of accumulation in the late Quaternary, and that the focus of enhanced desert conditions was not necessarily at the last glacial maximum. Stronger, more consistent winds, rather than simply drier climatic conditions, may have been important for late-Quaternary dunefield development.

**The Impact of Humans.** The ability of humans to alter desert and dryland environments is not new. Human actions in deserts and drylands are often viewed primarily from a detrimental perspective, with salinization in Mesopotamia around 2,500 BCE attributed by Jacobson and Adams (1958) to agriculture. In central Mexico, O'Hara et al. (1993) have recently shown how significant soil erosion was caused in dryland areas by pre-Columbian societies. These and other studies show that even early civilizations had the propensity to degrade dry environments. During the twentieth century, the growth and changing distributions of dryland populations, and developments in technologies and land uses, have increased this ability, so that while desertification is by no means a new phenomenon, it is one that today

Most of Saudi Arabia is desert, so that climatic conditions are not favorable for rapid large-scale recharge of aquifers. Also, much of the ground water that lies beneath the desert is a fossil resource, created during more humid conditions—pluvials—that existed in the Late Pleistocene, between 15,000 and 30,000 BP. In spite of these inherently unfavorable circumstances, Saudi Arabia's demand for water is growing inexorably as its economy develops. In 1980, the annual demand was 2.4 billion cubic meters. By 1990, it had reached 12 billion cubic meters (a fivefold increase in just a decade), and it is expected to reach 20 billion cubic meters by 2010. Only a very small part of the demand can be met from desalinization plants or surface runoff; over three-quarters of the supply is obtained from predominantly nonrenewable groundwater resources. The drawdown on aquifers is thus enormous. It has been calculated that, by 2010, the deep aquifers will contain 42 percent less water than in 1985. Much of the water is used ineffectively and inefficiently in the agricultural sector (Al-Ibrahim, 1991) to irrigate crops that could easily be grown in more humid regions and then imported.

**BIBLIOGRAPHY**

Al-Ibrahim, A. A. "Excessive Use of Ground-Water Resources in Saudi Arabia: Impacts and Policy Options." *Ambio* 20 (1991), 34–37.

—ANDREW S. GOUDIE

has attained a previously unrealized significance (Thomas and Middleton, 1994). Many concerns about human-induced changes in drylands relate to enhanced rates of soil degradation and total vegetation loss, but studies are now beginning to elucidate the importance at the global scale of desert and dryland biodiversity and the potential use to humankind of specific plant species.

Several environmental factors contribute to the susceptibility of deserts and drylands to degradation by humans. These do not necessarily relate to the sometimes supposed fragility of their ecosystems—which, as Behnke et al. (1993) show, is disputable for many dryland areas—but to the nature of environmental dynamics and human activities. First, the inherent natural variability of dryland climates and ecosystems tends to be overridden today when human activities employ technologies and methods imported from more consistent temperate environments. Second, many geomorphologic processes in deserts and drylands tend to be characterized by significant periods of quiescence, punctuated by abrupt episodes of activity. Thus areas cleared of natural vegetation by grazing pressures, or bare in the period immediately following harvesting, are particularly susceptible to rapid erosion from high-intensity rainstorms or windy conditions. Third, the remarkable growth of urban areas in deserts and drylands since World War II has placed significant pressures on limited water resources and options for waste disposal. In dryland areas of the developing world, rural depopulation resulting from migration to urban centers can lead to the failure of traditional soil and vegetation conserva-

tion techniques. Tiffen et al. (1994) have shown how more people on the land can in certain circumstances halt and reverse environmental degradation.

**Future Changes.** Deserts and drylands will inevitably continue to change and evolve in response to the twin pressures of natural climatic change and direct human impacts. Human impacts on global climate, through anthropogenic global warming, are also likely to have marked impacts. While drylands are themselves net contributors of an estimated 5–10 percent of enhanced greenhouse gases (Williams and Balling, 1995), Ojima et al. (1993) estimate that this proportion will grow dramatically if enhanced management of human activities in drylands is not achieved.

According to the (UCAR) Office for Interdisciplinary Earth Studies (1991), deserts and drylands are expected to be among the biomes that respond most rapidly to global warming. Williams and Balling (1995) have provided a valuable assessment of current modeled predictions of climate changes affecting deserts and drylands under doubled levels of atmospheric carbon dioxide, which are summarized in Table 2. As the sequestration of atmospheric carbon is assessed as a potential route to mitigate the impacts of global warming, deserts and drylands can, according to Squires and Glenn (see Middleton and Thomas, 1997, pp. 140–143), play a vital role. Opportunities exist for both enhanced dryland soil and biomass carbon storage, not least because drylands are a major reserve (about 75 percent) of global soil carbonate carbon, which participates less in global carbon flux than organic carbon. Squires and

Deserts.TABLE 2. Predicted Changes in the Climate of Selected Dryland and Desert Areas, Based on Three Different Climate Models Assuming a Doubling of the Atmospheric Carbon Dioxide Content

| | EUROPE | NORTH AFRICA | SOUTHERN AFRICA | ASIA | MIDDLE EAST AND ARABIA | NORTH AMERICA | SOUTH AMERICA |
|---|---|---|---|---|---|---|---|
| Areas included | Drylands bordering the Mediterranean | Sahara and surrounding areas | Kalahari, Karoo, and Namib | From Iran to China, including Thar | Arabia, and eastern Mediterranean | North Mexico, western United States, and Great Plains | Atacama, Peru, Chaco, Patagonia |
| Summer temperature change (°C) | +4–6 | +4–6 | +2–4 | +6 | +4–6 | +4–6 in United States; +2–4 in Mexico | +2–4 +2–4, +2–6 in Atacama, Patagonia |
| Winter temperature change (°C) | +2–6 | +2–4, +46 in Sahel | +2–4, +46 in Karoo | +4–8 in east; +4–6 in west | +2–4 | +4–6 in United States; +2–4 in Mexico | +2–4 |
| Summer precipitation change | Decrease | Decrease | Decrease | Highly varied; increase in Thar | Decrease | Highly varied | Slight decrease; possible increase in Peru; variable, mainly increased |
| Winter precipitation change | Varied decrease | Decrease | Slight decrease | Slight decrease | Decrease | Highly varied | Slight decrease; increase in east; decrease in west |
| Soil moisture change | Significant decrease | Decrease | Significant decrease | Slight increase (winter); general decrease (summer); increase in Thar | Decrease | Highly varied | General decrease |

SOURCE: From Thomas (1997, chap. 30), using information from Williams and Balling (1995) and other sources. With permission of John Wiley and Sons.

Glenn note that if the antidesertification land-restoration measures proposed by the United Nations were to be implemented, the sequestered carbon would be equivalent to 15 percent of current annual carbon dioxide emissions. This clearly demonstrates the significant linkages between different aspects of dryland environmental changes and their global significance.

[*See also* Drought; *and* Salinization.]

### BIBLIOGRAPHY

Behnke, R. H., et al., eds. *Range Ecology at Disequilibrium: New Models of Natural Variability and Pastoral Adaptation in African Savannas.* London: Overseas Development Institute, 1993. A collection of papers that show developments in understanding the ecological dynamics of drylands.

Clements, T., et al. *A Study of Desert Surface Conditions.* Natick, Mass.: United States Army Environmental Protection Research Division, 1957.

Glennie, K. W. "Desert Sedimentary Environments, Present and Past: A Summary." *Sedimentary Geology* 50 (1987), 135–166. A useful account that includes coverage of evidence of desert environments during different geologic periods.

Goudie, A. S. *Environmental Change.* New York and Oxford: Oxford University Press, 1993.

Hulme, M. "Recent Change in the World's Drylands." *Geophysical Research Letters* 23 (1992), 216–223. A well-explained analysis of climate surface data, exploring decadal-scale variations in the extent of drylands.

Jacobson, T., and R. M. Adams. "Salt and Silt in Ancient Mesopotamian Agriculture." *Science* 128 (1958), 1251–1258.

Meigs, P. "World Distribution of Arid and Semiarid Homoclimates." In *Arid Zone Hydrology,* UNESCO Arid Zone Research Series 1, 1953.

Middleton, N. J., and D. S. G. Thomas, eds. *World Atlas of Desertification.* 2d ed. London: UNEP/Edward Arnold, 1997. An analysis of the United Nations' most recent survey of desertification, supplemented by valuable articles on key social and environmental aspects of change in drylands.

O'Hara, S., et al. "Accelerated Soil Erosion around a Mexican Highland Lake Caused by Prehispanic Agriculture." *Nature* 362 (1993), 48–51.

Office for Interdisciplinary Earth Studies. *Arid and Semiarid Regions: Response to Climate Change.* Boulder, Colo.: OIES, 1991.

Ojima, D. S., B. O. M. Dirks, E. P. Glenn, C. E. Owensby, and T. M. O. Scarlock. "Assessment of Carbon Budget for Grasslands and Drylands of the World." *Water, Air and Soil Pollution* 70 (1993), 643–657.

Sarnthein, M. "Sand Deserts during Glacial Maximum and Climatic Optimum." *Nature* 272 (1978), 43–46.

Stokes, S., et al. "Multiple Episodes of Aridity in Southern Africa since the Last Interglacial Period." *Nature* 388 (1997), 154–158. Uses luminescence dating to show how multiple phases of aridity can be determined from the Quaternary record, and climate modeling to assess the causal factors.

Thomas, D. S. G., ed. *Arid Zone Geomorphology: Process, Form and Change in Drylands.* 2d ed. Chichester, U.K.: Wiley, 1997. A textbook containing thirty chapters by international authorities on the characteristics and environmental processes operating in deserts and drylands, including reconstruction of past environmental changes and accounts of deserts and drylands in each continent.

Thomas, D. S. G., and N. J. Middleton. *Desertification: Exploding the Myth.* Chichester, U.K.: Wiley, 1994. A critical assessment of desertification that includes coverage of the natural variability of deserts and drylands.

Thomas, D. S. G., and P. Shaw. *The Kalahari Environment.* Cambridge: Cambridge University Press, 1991.

Tiffen, M., et al. *More People, Less Erosion: Environmental Recovery in Kenya.* Chichester, U.K.: Wiley, 1994. Challenges the conventional wisdom that more people on the land inevitably leads to degradation in drylands.

Tucker, C. J., et al. "Expansion and Contraction of the Sahara Desert from 1980 to 1990." *Science* 227 (1991), 369–375. Uses satellite-derived data to show how the desert area varies naturally in response to seasonal and annual rainfall variability.

Williams, M. A. J., and R. C. Balling, Jr. *Interactions of Desertification and Climate.* London: WMO/UNEP/Edward Arnold, 1995. Includes coverage of the linkages between climate change and desertification, including assessment of predictions of future climate changes due to global warming.

Williams, M. A. J., et al. *Quaternary Environments.* London: Edward Arnold, 1993. One of several good texts covering Quaternary environmental reconstruction; this one has a chapter specifically devoted to deserts.

—DAVID S. G. THOMAS

## DEVELOPMENT POLICY AND THEORY. *See* Economic Levels; *and* Policy Analysis.

## DINOSAUR EXTINCTION. *See* Impacts by Extraterrestrial Bodies.

## DIOXIN. *See* Pollution.

## DISASTERS. *See* Natural Hazards.

## DISEASE

Infectious diseases have shaped human history. They are dynamic and will continue to influence where and how humans live—and human activities will alter the paths and expressions of infectious diseases. A broad range of societal, biological, and physicochemical factors influence the distribution, incidence, and burden from infectious diseases. In recent years, patterns of infectious disease have changed. These changes include the description of diseases caused by pathogens long present but not previously identified, recognition of seemingly new microbes, changes in old pathogens (for example, changes in distribution, incidence, virulence, resistance to drugs), new disease–disease interactions, and the spread to humans of organisms never previously known to be human pathogens.

Taken together, infectious diseases remain the single

## INFECTIOUS DISEASES

A broad range of societal, biological, and physicochemical factors influence the distribution, incidence, and burden from infectious diseases.

- Infectious diseases are dynamic. They will continue to change.
- Many diseases that appear to be new diseases are not truly new but are only newly recognized and characterized.
- Changes in infections are global in distribution and involve all classes of organisms.
- Infection is a universal phenomenon, affecting plants and animals as well as humans.
- Human activities are a potent force in changing patterns in infectious diseases.
- Infections can cause chronic diseases and are involved as one of the causes in at least 15 percent of human cancers.
- The physicochemical environment affects the distribution, abundance, and dispersal of many human pathogens. Its role is especially prominent in vector-borne and animal-associated infections.

—MARY ELIZABETH WILSON

most common cause of death in the world, causing about one-third of the 51.9 million deaths in 1995. Yet the burden is spread unevenly by population and geographic region. Infectious diseases have a disproportionate impact on the poor and undereducated, and in developing regions of the world. Globally, 98 percent of all deaths in children younger than fifteen years of age are in developing countries—and many of these are from infectious diseases (Murray and Lopez, 1996). The probability of death between birth and age fifteen years is 22 percent in sub-Saharan Africa versus 1.1 percent in established market economies. [See Human Health.]

Although much media attention has focused on tropical, remote locations and on exotic viral infections such as Ebola, the changing pattern of infections involves all geographic regions and all classes of pathogen (such as viruses, bacteria, fungi, helminths [parasitic worms], and protozoa). Most of the major global causes of death from infectious disease are common, widely distributed infections, such as tuberculosis, measles, human immunodeficiency virus (HIV), influenza, pneumococcus, and rotavirus.

Infections kill and disable through many mechanisms. Many pathogens cause acute disease, but infectious diseases also impose a burden through chronic diseases, including cancer. Chronic diseases may be a consequence of persistent infection (for example, hepatitis B virus) or may result from tissue damage caused by past infection. Many cancers have been linked to infections. The World Health Organization (WHO) estimates that more than 1.5 million cancers each year, or about 15 percent of the total, are related to microbes. The three that lead the list are gastric (*Helicobacter pylori*), cervical (papillomaviruses), and liver (hepatitis B and C viruses) cancers.

Many factors are contributing to the changing patterns of infectious disease. Those commonly identified are microbial adaptation and change, human demographics and behavior, environmental changes, technology and economic development, breakdown in public health measures and surveillance, and international travel and commerce (Lederberg et al., 1992). How these influence the appearance, reappearance, and spread of infections will become apparent in the discussion of specific disease examples. Typically, multiple factors interact, leading to changes in a disease. The emergence of a disease may be an unintended consequence of what is viewed as progress: the building of a dam, clearing of lands, mass processing and wide distribution of foods and water, medical interventions (namely, transfusions of blood and blood products, tissue and organ transplantation, cancer chemotherapy), and use of antimicrobial agents.

The burden of disease in humans can increase through increased contact between a pathogen (disease-causing agent) and host, increase in virulence or resistance of the pathogen, increase in the vulnerability of the host, or limited access to effective prevention or therapy. A human or a population can be completely or relatively invulnerable to some infections because of immunity (past infection or immunization), genetic factors, or a whole range of barriers (such as shoes, screens, good housing) or interventions (for example, provision of clean water and adequate waste disposal, control of organisms responsible for transferring pathogens between hosts) that prevent contact between human and pathogen. Good nutrition, including adequate intake of micronutrients, can lead to an improved outcome in at least some infections.

**Microbes.** It is useful to take a broad view of microbial life before focusing on microbes that harm humans. Only a tiny fraction of microbes that exist on Earth have been identified and characterized. Of those identified, most do not infect humans; some are essential for shaping and sustaining life as we know it. Microbes are old, diverse, abundant, and resilient. They live in communities, send signals to communicate with each other, and change in response to changes in the environment. Microbial communities living deep in the Earth metabolize organic materials bound to rocks and sediments and shape the physical environment. Some microbes have short generation times (for example, twenty to thirty minutes for an organism like staphylococcus, compared with twenty to thirty years for a human), and hence can undergo rapid change. Such organisms change via mutation, but also by a variety of molecular maneuvers that involve acquisition and exchange of genetic information (for example, transfer, reassortment, recombination, and conjugation). These can alter microbial traits relevant to human health—virulence, resistance to drugs, and even transmissibility.

The source of microbes causing infections in humans is typically another human, an animal or arthropod, or soil or water. Globally, about 65 percent of infections that lead to death are spread from person to person (such as respiratory tract infections, tuberculosis, HIV, and measles); almost a quarter are carried in food, water, or soil (for example, infections such as cholera and hookworm); and about 13 percent are vector-borne (namely, malaria and leishmaniasis) as per the World Health Organization, 1997.

To cause disease, a microbe must find a way to enter the human host and reach appropriate cells or tissues where it can attach and replicate. Microbes typically enter the body via ingestion, inhalation, or through the skin or mucous membranes. Many important pathogens, including the malaria parasite, are carried to the human host by a mosquito and inoculated through the skin. Some pathogens have complicated cycles that involve one or more intermediate hosts, typically animals that support one stage of development of the pathogen. Other microbes, such as hantaviruses, reside in an animal, known as a *reservoir host*. Humans can become infected if they enter the habitat of the animal or if they come into contact with them or their secretions/excretions through other activities. The human may be irrelevant to the maintenance of the microbe in nature, although the human, if infected, may die as a result of the infection.

**Vectors of Infection.** Many viruses, protozoa (for example, malaria, trypanosomes, leishmania), and a few bacteria (especially rickettsia) and helminths (many filarial parasites) require the assistance of an arthropod to deliver the pathogen to the human host. These arthropods that shuttle pathogens from one host to another are called *vectors*. Vectors are important in carrying

pathogens to animals and plants, as well as to humans. Many animal pathogens can also infect humans—as an essential part of the cycle or as an occasional unhappy event if the human happens to be around when the arthropod takes a blood meal. Sometimes an insect will simply serve as a mechanical carrier of microbes that stick to its feet, hairs, or body, transporting microbes to food, for example. Organisms can also be carried passively on biting mouthparts or in the gut of the insect, but these means of transmission are probably infrequent and do not work for many infections.

In most instances, the arthropod (such as mosquito, tick, sandfly, reduviid bug, louse), in the process of taking a blood meal needed for its own reproduction, will ingest a pathogen along with the blood. The pathogen replicates (sometimes to high levels) or undergoes development to a more mature stage inside the arthropod during a period called *extrinsic incubation*. When the arthropod feeds again, the pathogen can be inoculated into the new host. In a few instances (as with Chagas' disease or American trypanosomiasis), the pathogen is found in the feces of the insect (for example, reduviid bug), which defecates after feeding, allowing the pathogens to enter through the bite site or via other surfaces (such as conjunctivae).

Pathogen–vector interactions may be very specific. Only certain types of mosquito are biologically competent to support the development of a given pathogen. Vectors vary greatly in their capacity to carry pathogens to humans, not only by virtue of their abundance, but also their biting preferences, with some mosquitoes, for example, preferring nonhuman hosts and biting humans only if no animals are available. The likelihood of being bitten is also strongly influenced by human activities, clothing, habits, hobbies, and housing.

**Environment and Climate.** Environmental and climatic factors strongly influence the distribution and incidence of many infectious diseases through direct and indirect mechanisms. In general, infections that are carried by the human host and can be transmitted from one person to another (such as influenza, HIV, measles, chickenpox, and tuberculosis) can and do appear anywhere in the world, unless a population is geographically isolated or immune (because of past infection or immunization). Other infections require certain environmental or climatic conditions or the presence of a specific arthropod vector, animal, or other intermediate host in order to persist in an area. These are the infections that are geographically focal, although the area of distribution can expand or contract in response to a variety of factors, including the weather and environmental changes.

Many infections that are widespread still have striking fluctuations that correspond to the seasons. These may reflect seasonal human activities and also the ambient temperature and humidity as they affect the viability, duration of survival, replication, and transmissibility of the microbe. Influenza, for example, is seen primarily in the winter months in temperate regions, although it occurs throughout the year in tropical environments. Rapid and wide transmission has occurred in special habitats (for example, airplanes, cruise ships, air-conditioned barges, air-conditioned nursing homes) outside of the usual transmission season. In many temperate areas, salmonella, campylobacter, and some other food-borne infections tend to increase in summer months, perhaps reflecting food handling (grilled foods, picnics, higher ambient temperatures, and inadequate refrigeration) and other human activities during those months.

Environmental change, considered broadly for the microbe, includes the application of chemicals and other agents such as antimicrobials. Microbes that have or can acquire resistance mechanisms are more likely to survive. Today, a variety of antiseptic, antibacterial, antiviral, antifungal, and other kinds of antimicrobial agents made by humans permeate the environment—in soaps, hand creams, cutting boards, toys, animal feeds, aquafarming, and antibiotics given to humans to treat and prevent infections. Of the 23 million kilograms of antibiotics used annually in the United States, more than 40 percent is used in animals (Levy, 1998). Increasingly, microbes, including many important human pathogens, are becoming resistant to antimicrobials in common use. Insect vectors also adapt to changes in the chemical environment and can become resistant to insecticides.

The physicochemical environment influences the pathogen, the vector, intermediate and reservoir hosts, vegetation, and human behavior and activities. For each biological species, the physicochemical environment influences the distribution, abundance, and duration of survival. Temperature (not just mean, but also minimum, maximum, and amplitude), humidity, rainfall (pattern as well as amount), types of vegetation, animal life, and distribution of water (such as ponds and rapidly flowing streams) may determine whether a region can support a vector. For an insect vector, the environment also determines its capacity to support the maturation of a pathogen, its biting activity, flight patterns, and shedding or transmission pattern. Hence the epidemiology of many vector-borne infections is characterized by seasonality through influences of weather patterns on host, vector, and microbe.

Seasonal events, such as human and animal migration, vacation, social, and school activities, and occupational work (such as harvesting of crops), can all influence patterns of infectious disease. The epidemiology of infections may also show periodicity or cyclic patterns extending over many years. This may be tied into cyclic weather patterns (for example, periodic climate perturbations such as El Niño–Southern Oscillation, which causes fluctuations in temperature and rainfall) or the accumulation of susceptible people through birth or migration into an area (such as measles in the prevaccination era). [See Migrations.]

**Malaria.** An ancient disease, malaria remains a major killer in many tropical regions. A protozoan infection, it is transmitted by mosquitoes, which require moisture for breeding. Malaria, along with other vector-borne infections, is strongly affected by rainfall and temperature, although the relationship is not a simple one. Increased rainfall can be associated with expansion of malaria breeding sites and an increase in malaria in arid areas. In humid areas, drought can improve breeding conditions for mosquitoes, leading to outbreaks.

Natural predators of vectors influence their abundance. If the reproductive cycle of a predator is longer than that of the mosquito, a mosquito population could surge before the predator population could expand to control it. Temperature, rainfall, and vegetation also affect the abundance of food and the behavior of predators.

Increased rainfall can provide more breeding sites, but excess rain can also destroy breeding sites. Increased temperature, to a point, can speed the maturation of the mosquito through its various stages and may decrease the extrinsic incubation period (the time inside the vector required for the pathogen to develop to an infective stage) for the parasite in the arthropod. Yet excessive temperature may kill or inhibit growth. Very warm temperatures may shorten the survival of adult mosquitoes and immature forms.

Changes in weather patterns or in land use may have varying effects on different vectors. In an area of northern Sarawak (Malaysia) undergoing oil palm development, mosquitoes were surveyed during major changes in the habitat. In association with forest clearing, *Anopheles* mosquitoes decreased and the risk of malaria transmission dropped by 90 percent over a four-year period. During the same period, however, vectors of dengue virus increased (Chang et al., 1997).

Interventions can have unexpected consequences. The use of bed nets has been advocated in highly malarial regions as a way to reduce the number of bites by infective mosquitoes and hence the morbidity and mortality from malaria. In Gambia, West Africa, a national program of impregnated bed nets was introduced in 1992. A follow-up study to assess the impact on malaria found that malaria prevalence was inversely related to vector density. Bed net usage was strongly influenced by the density of biting mosquitoes. Prevalence rates of malaria were reduced in areas where nuisance biting of mosquitoes was sufficient to lead people to use bed nets. The highest rates of malaria were found in villages with low bed net usage.

**Cholera.** A bacterial infection acquired by ingesting food or water contaminated with *Vibrio cholerae*,

## THE WORLD'S TOP TEN INFECTIOUS KILLERS IN 1997

**1. Pneumonia and other lung infections: 3.7 million deaths**

Airborne transmission. Major killer of children in developing countries. Increasing incidence because of poverty-related risks such as malnutrition. Many are treatable with antibiotics.

**2. Tuberculosis: 2.9 million deaths**

Bacteria are airborne in respiratory droplets. Ninety-five percent of cases are in developing countries. Treatable with antibiotics, but multidrug-resistant strains have emerged. Current vaccine widely used but not very effective.

**3. Diarrheal diseases, including cholera: 2.5 million deaths**

Contaminated water and foods are primary causes. Most victims are under five years of age and live in developing countries. Deaths can be reduced with oral rehydration, proper nutrition, and antibiotics.

**4. AIDS: 2.3 million deaths; more than 30 million now infected**

Sub-Saharan Africa thought to have two-thirds of world's cases. Transmissible through body fluids such as blood and semen. No cure or vaccine, but a combination of drugs can prolong life.

**5. Malaria: 1.5 to 2.7 million deaths**

Endemic in 100 countries; highest incidence in sub-Saharan Africa. Transmitted by infected female *Anopheles* mosquitoes. Curable with early diagnosis and prompt treatment, but protozoa are developing increasing resistance to drugs.

**6. Measles: 960,000 deaths**

Infects primarily children; highest incidence in Africa. Deaths usually follow complications such as pneumonia and encephalitis. Virus is airborne in droplets from nose and mouth.

**7. Hepatitis B: 605,000 deaths**

Endemic in Africa, South America, eastern Europe, eastern Mediterranean nations, Southeast Asia, China, and smaller Pacific islands. Transmitted by blood and other body fluids. Majority recover from acute form; chronic carriers risk active hepatitis, cirrhosis, and primary liver cancer. Highly effective vaccine available.

**8. Whooping cough: 410,000 deaths**

Endemic worldwide. Bacteria airborne in respiratory droplets. About one-half of cases are in children under age two. Vaccine available for infants but not adults. Treatable with erythromycin.

**9. Tetanus: 275,000 deaths**

Seventy-five percent of deaths are in Bangladesh, China, India, Indonesia, Nigeria, and Pakistan. Main victims are newborn with infected umbilici. Spores are ubiquitous, especially in animal waste. Preventable by sterile practices and immunization.

**10. Dengue fever and dengue hemorrhagic fever: 140,000 deaths**

Hemorrhagic form is responsible for fatalities, mostly in children under ten years of age. Virus is transmitted by *Aedes* mosquitoes. With increasing urbanization, epidemics are becoming more common, especially in Southeast Asia and Latin America.

SOURCE: World Health Organization, cited in *Natural History*, February 1999, pp. 46–47.

—DAVID J. CUFF

cholera has long caused dramatic epidemics and invokes fear because of its capacity to kill quickly through massive diarrhea. It has a striking seasonal pattern in Bangladesh and in some endemic areas where outbreaks tend to recur.

*Vibrio cholerae* lives in close association with marine life—algae, phytoplankton, copepods, and aquatic plants. It can bind to chitin in crustacean shells. A single copepod can contain up to ten thousand cells of the bacterium. Special techniques can document its continued presence in marine life even when the organism cannot be grown with ordinary culture techniques in the laboratory.

## GLOBAL WARMING AND MALARIA

The links between vector-borne diseases and climatic factors are often close. For example, the *Anopheles* mosquito, which transmits malaria, does not survive easily when the mean winter temperature drops below approximately 15°C. It survives best where the mean temperature is 20°–30°C and humidity exceeds 60 percent. Higher temperatures speed up the malaria parasite's developmental cycle, and the parasite cannot survive below a critical temperature of around 14°–16°C for *Plasmodium vivax* and 18°–20°C for *Plasmodium falciparum*. Thus, by improving the conditions for both the vector and the parasite, unusually hot and humid weather in endemic areas could cause a marked rise in malaria incidence.

Models that seek to relate annual average temperatures and rainfall to malaria incidence worldwide found that for five different climatic change scenarios, the land area affected by malaria would increase by 7 to 28 percent. Cities at a sufficiently high altitude to be situated above the "mosquito line" (e.g., Nairobi and Harare) might be especially susceptible, and the vulnerability of newly affected populations could initially lead to high case-fatality rates because of their lack of natural immunity. On a worldwide basis, some models predict a potential climate-induced increase of around 25 percent in malaria cases.

Climatic change is not the only factor to consider in changing malaria incidence. In sub-Saharan Africa (where 90 percent of all malaria deaths in the world occur today), its incidence has worsened over the past three decades. This can be attributed to a number of factors, including drug and insecticide resistance, declining health service and control infrastructure, increased population movements (migrants and refugees), and land use changes.

### BIBLIOGRAPHY

Martens, W. J. M., L. Niesson, J. Rotmans, T. H. Jetten, and A. J. McMichael. "Potential Impact of Global Climate Change on Malaria Risk." *Environmental Health Perspectives* 103 (1995), 458–464.

Martin, P. H., and M. G. Lefebvre. "Malaria and Climate: Sensitivity of Malaria Potential Transmission to Climate." *Ambio* 24.4 (1995), 200–207.

McMichael, A. J., A. Haines, R. Sloof, and S. Kovats, eds. *Climatic Change and Human Health* Geneva: World Health Organization, 1996.

—ANDREW S. GOUDIE

Under unfavorable conditions the organism can enter a dormant state, in which it requires decreased oxygen and nutrients. In response to changing environmental factors, such as warming of water temperature, increase in pH and nutrients, and decrease in salinity, its population may expand (and become culturable), increasing the risk of human infection, especially in areas where sanitation is poor and drinking water is untreated. Environmental factors such as increase in seawater temperature and influx of organic nutrients can also trigger plankton blooms, which provide a favorable milieu for the proliferation of *V. cholerae*.

History is full of examples of introductions of cholera via ships carrying people infected with *V. cholerae* and excreting it with their feces. Carried in the ballast and bilge of ships, it can also contaminate waters in new ports. Infected people can be a source of infection in a community, but an aquatic reservoir of the bacterium can be a long-term source for periodic introductions into human populations. Shellfish become colonized with *V. cholerae* in their water habitats. They can infect humans if inadequately cooked before being eaten.

The bacterium can survive for up to fourteen days in food and for weeks in shellfish, and can persist indefinitely in an aquatic environment. Although environmental factors can trigger an expansion of bacterial populations, human living conditions and activities determine whether and how often human disease occurs. Discarding of untreated human waste into surface water used by humans for preparation of food, bathing, and so on allows continuing contact between *V. cholerae* and humans. Chlorination of water can kill the bacterium. Filtering of untreated water, which removes copepods, can also reduce the number of viable bacteria.

War and political conflict leading to population displacement and disruption of basic sanitation systems can lead to cholera epidemics. For example, in 1994 when refugees fleeing genocide in Rwanda settled in camps with poor sanitation, a massive outbreak of El Tor cholera in Goma, Zaire, caused seventy thousand cases and twelve thousand deaths. Extreme weather events, such as flooding, can also precipitate cholera epidemics by destroying land and displacing populations. Flooding can also disperse microbes and allow sewage contamination of water sources used by humans.

A new strain of cholera, *V. cholerae* O139 Bengal, appeared in Asia in 1992 and caused explosive epidemics in Bangladesh, India, and other nearby countries, temporarily displacing the resident *V. cholerae* strains. Its virulence was similar to that of typical cholera strains, but it could infect people who were relatively immune to *V. cholerae* O1. This meant that the organism had the capacity to cause high rates of infection, and illustrates how a change in the pathogen can abruptly alter the epidemiology of an old, familiar infection. Such shifts have important implications for vaccine development. Existing cholera vaccines did not protect against this new strain.

**Human Immunodeficiency Virus/Acquired Immune Deficiency Syndrome (AIDS).** Perhaps more than any infection in recent history, AIDS shows the vulnerability of humans to new infection. The appearance of an infection that would claim 2.6 million lives in 1998 was unimaginable a few decades ago. Human behaviors and technology have fueled the spread of HIV throughout the world. Rapid, widespread travel; injections and intravenous drug use; multiple sexual partners (hetero- and homosexual); other sexually transmitted diseases; blood transfusions and use of factor concentrates to treat hemophilia; and, to a much lesser extent, artificial insemination and tissue and organ transplantation have aided the spread of the virus.

The consequences of HIV infection include progressive, profound depletion of the immune system. Death is often a result of opportunistic infection or malignancy. The appearance of HIV and its subsequent study have led to major advances in immunology and also the recognition of new clinical syndromes caused by well-known pathogens. In addition, the presence of HIV has contributed to the characterization of organisms not previously well defined (e.g., cryptosporidium, *Bartonella* species, and microsporidia). HIV has also been an important factor in the resurgence of tuberculosis in many regions of the world.

As HIV has moved into rural and tropical regions, its overlap with the distribution of geographically focal infections, such as Chagas' disease, *Penicillium marneffei*, and leishmaniasis, has made diagnosis and treatment of these infections even more difficult. HIV has altered every aspect of medicine—clinical expression of disease, pathologic findings, interpretation of diagnostic tests, and treatment approaches.

**Tuberculosis.** *Mycobacterium tuberculosis*, the bacterium that causes tuberculosis, infects one-third of the human population. As a single agent, it kills about three million people each year—more than any other single microbe. As with so many infections, tuberculosis is unevenly distributed by geographic region and population. More than 95 percent of the cases and 99 percent of the deaths from tuberculosis are in developing countries.

Tuberculosis spreads from person to person via droplet nuclei, tiny particles less than 10 microns in diameter that are expelled by people with active infection. These droplet nuclei are dispersed widely in the air after coughing, singing, or talking, and remain airborne for hours. Their small size allows them to pass easily into the bronchioles of the lung.

Most people who are infected with *M. tuberculosis* have "silent" infection that may remain unrecognized for a lifetime. In most populations, only 5–15 percent of people infected with *M. tuberculosis* develop active disease. In contrast, HIV-infected people are more likely to become infected and to progress to active disease. Their annual risk of progressing to active tuberculosis is 7–10 percent. The interaction between HIV and tuberculosis is bidirectional. Tuberculosis in HIV-infected people increases viral replication and leads to more rapid progression of HIV disease. There is extensive overlap of the populations infected with HIV and *M. tuberculosis*. HIV is spreading rapidly in many areas of the world where latent infection with tuberculosis is common.

Although drugs that are effective for treating tuberculosis have been available since the middle of this century, prospects for control of global tuberculosis in the near term are poor. About 50 percent of people with untreated active tuberculosis will die from the infection. The death rate is higher in people with AIDS and other conditions associated with poor immune response. Infection with multidrug-resistant (MDR) strains of tuberculosis becomes untreatable if effective drugs are unavailable or inaccessible because of cost.

MDR strains emerge as a consequence of inappropriate treatment of tuberculosis. All bacterial populations of *M. tuberculosis* contain rare bacilli that are resistant to drugs commonly used for treatment. Therapy with too few drugs or for too short a period of time can allow drug-resistant bacilli to emerge and become the predominant infecting strain. Not only can infection persist and progress in the individual, but drug-resistant strains can spread to others who live and work nearby. People with untreated active tuberculosis (whether sensitive or resistant strains) transmit infection, on average, to ten to fifteen people per year.

Many habitats in today's world facilitate the transmission of tuberculosis. These include prisons, homeless shelters, hospitals, refugee camps, and nursing homes. Vulnerable people are brought together and often share the same air circulation system. One habitat that is especially worrisome is the hospital in regions where tuberculosis is common and facilities for respiratory isolation are limited or nonexistent. Hospitals have been the site of many outbreaks of MDR tuberculosis. Hospital workers as well as patients and visitors become infected—although HIV-infected people are especially vulnerable. In Argentina, more than one hundred HIV-infected people acquired MDR tuberculosis in a hospital. The attack rate of MDR tuberculosis among hospitalized HIV-infected people was an extraordinary 37 percent.

A survey published by the World Health Organization in 1997 found that drug resistance to one or more drugs was present in all countries surveyed and exceeded 40 percent in some areas. The median prevalence of MDR tuberculosis was 1.4 percent, but reached 14.4 percent in Latvia. Drugs that are active against many MDR strains do exist, but treatment is difficult (requiring multiple drugs, many with unpleasant and toxic side effects that require careful monitoring) and extremely expensive. For example, treatment of a case of MDR tuberculosis in the United States can cost over U.S.$200,000. In many parts of the world, MDR tuberculosis in practice is untreated tuberculosis.

Crowded living conditions enhance the spread of tuberculosis. Increasing urbanization and poverty, along with the other potent forces of HIV infection and expansion of MDR strains, make for a gloomy forecast. Many researchers are working to develop safe and highly effective vaccines. The currently available vaccine, bacille Calmette-Guerin (BCG), continues to be widely used. Most newborns in the world receive the vaccine. For example, 85 percent of children born worldwide in 1990 received BCG as infants, according to WHO statistics. WHO continues to recommend BCG, preferably at birth, for all infants born in high-risk countries, including those at high risk for HIV infection. A meta-analysis of major studies found an overall protective effect against tuberculosis of about 50 percent and protection against death and disseminated disease in the range of 60–75 percent. While BCG may have reduced deaths from tuberculosis, it is not the answer to global control.

Because of its ease of transmission, tuberculosis anywhere in the world is a concern for everyone. Molecular fingerprinting has allowed the tracing of particular strains of tuberculosis that have been carried across continents and oceans. In western Europe, Canada, and the United States, a significant and rising proportion of the tuberculosis cases are in immigrant populations. In the United States in 1998, 42 percent of all tuberculosis cases were in the foreign born. In some countries, that figure now exceeds 50 percent. Tuberculosis control efforts must be global. In 1996, only an estimated 32 percent of the world population had access to what would be defined by the WHO as high-quality tuberculosis care.

Changes in the microbe can also alter the epidemiology and expression of infection. In 1995, a strain of tuberculosis with high transmissibility was identified in an outbreak in the southern United States. Extensive transmission followed contact, often brief, with people infected with this strain. In two instances, active disease developed after very limited exposure. This strain, when studied in the laboratory, grew much faster than other clinical isolates of *M. tuberculosis*. It is postulated that the more rapid growth could allow this strain to establish infection before an effective immune response could develop (Valway et al., 1998).

**Dengue.** Dengue fever, also known as breakbone fever, is a viral infection transmitted to humans by the bite of an infective mosquito, usually *Aedes aegypti*. In recent years the mosquito vector and the virus have spread into new geographic regions (Figure 1). Dengue virus has caused major outbreaks in the Americas, and it continues to be a major pathogen in Asia. An estimated two billion people live in tropical and subtropical regions at risk for dengue fever, and about 150 million cases occur each year. Although some primates can be infected with the virus, humans are the primary host and the usual source for human infections. An initial infection with the dengue virus typically causes fever, chills, headache, and muscle aches, and is followed by complete recovery. There is no vaccine yet available and no specific therapy, although supportive care in the severe forms of infection can be lifesaving.

Dengue fever is caused by four antigenically distinct ribonucleic acid (RNA) viruses, designated dengue types 1, 2, 3, and 4. Infection with one dengue virus is followed by immunity to that serotype but not to others. Infection with a second dengue serotype can lead to severe disease—dengue shock syndrome (DSS) or dengue hemorrhagic fever (DHF). Mortality in untreated DSS/DHF can reach 20 percent, but falls to less than 1 percent with aggressive supportive treatment. Although an initial infection with dengue can be severe, the relative risk of severe disease is one hundred times higher after a second infection than after primary dengue infection.

In contrast to many vector-borne infections, dengue fever is an urban disease. The primary vector, *Aedes aegypti*, is an urban mosquito, well adapted to human habitats. It rests indoors, sleeps in bedrooms, and prefers human blood. Humans have provided it with many suitable breeding sites—water storage containers, flowerpots, discarded plastic containers, used tires, and tin

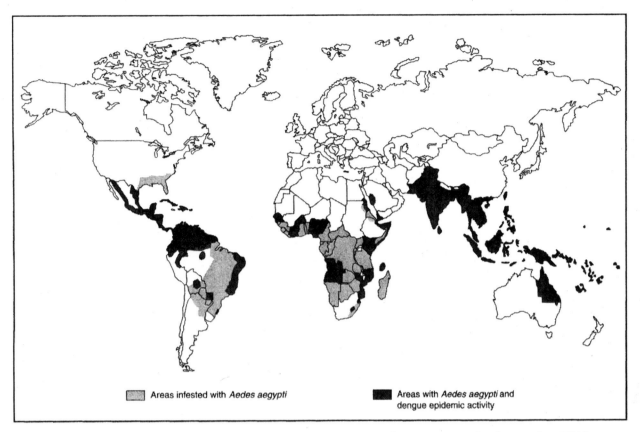

Disease. FIGURE 1. World Distribution of Dengue, 1998. (From Centers for Disease Control and Prevention, 1999.)

cans, among others. It should be noted that *Aedes ae-gypti* is also the principal urban vector for yellow fever virus; hence urban areas with abundant *A. aegypti* are also at risk for the introduction of the yellow fever virus.

Many factors have contributed to the striking increase in dengue. Rapid and frequent travel means that humans who are viremic (have the virus in their bloodstream) regularly carry the virus from one geographic region to another. Expansion of the mosquito populations because of inadequate vector control programs and increasing resistance of the mosquito to insecticides means that many areas will be receptive to the introduction of the virus by a traveling human. Infection can spread rapidly when vectors are abundant and have frequent contact with humans. Poor housing, lack of screens or air conditioning, and dense human populations also favor spread of infection. Epidemics can be huge, sometimes involving a million or more people.

More people live in urban areas today than ever before in history. [*See* Human Populations; *and* Urban Areas.] Seventy-five percent of the world population is in developing countries, and more than 90 percent of the population growth is occurring in these regions. Increasingly, growth of large urbanized centers is taking place in tropical regions. Of the twenty-five megacities (defined as urban areas with more than 8 million people) in the year 2000, nineteen are in developing regions. Huge periurban slums surround many of these cities in tropical and subtropical areas and provide the ideal environment for the appearance and rapid spread of a number of infectious diseases, including dengue fever. It has been estimated that a population of somewhere between 150,000 and 1 million is needed to sustain the cocirculation of more than one dengue serotype, thus setting the stage for severe and complicated dengue infections. More urban centers in tropical regions are reaching this size—and are also increasingly linked to the world via international air travel. With expansion of the populations in which multiple dengue serotypes circulate simultaneously or sequentially, the risk for DSS/DHS increases greatly. Larger human populations can also potentially influence viral evolution. Over the last two centuries, the number of viral lineages has been increasing roughly in parallel with the growing human population. Will an expanding human population lead to increasing rates of viral evolution? Increases in viral diversity could lead to more transmissible or pathogenic lineages (Zanotto et al., 1996).

Mosquitoes typically take some meals from plants (plant sugar can provide energy) but require blood meals for the nutrients necessary for reproduction. *Aedes aegypti* mosquitoes that feed only on human blood have a reproductive advantage and hence can spread viruses more effectively. Dense human populations provide more blood meals for mosquitoes.

Weather patterns also affect the distribution and incidence of dengue through their impact on the mosquito vector. Studies in Thailand in the 1980s showed the strong influence of temperature on vector efficiency for dengue virus. When mosquitoes were incubated at 30°C, the extrinsic incubation period for dengue-2 virus was twelve days but dropped to seven days at 32° and 35°C (Watts et al., 1989).

Interventions must be carried out with clear understanding of the potential consequences. In general, larger mosquitoes are more easily infected with the dengue virus and hence more likely to transmit it. Reducing larval density in breeding sites leads to larger mosquitoes. A control measure that reduced but did not eliminate larvae in breeding sites could have the perverse outcome of increasing dengue transmission to humans.

**Coccidioidomycosis.** The soil-associated fungus *Coccidioides immitis* is found in arid and semiarid areas with alkaline soils, hot summers, and short moist winters. Humans become infected when they inhale airborne arthroconidia of the fungus. Activities and events that disturb the soil (for example, excavation, construction, earthquakes) can lift arthroconidia into the air. Infections are typically acquired only in areas where the appropriate environmental conditions support the growth of the fungus in the soil. Weather patterns can influence risk and can partially explain periodic increases. Heavy rains followed by drought can allow the fungus to thrive—and then to be dispersed during dry, dusty days. In 1977, a major windstorm in the southern extreme of the San Joaquin Valley in California carried aloft soil from this coccidioidomycosis-endemic region and dispersed it over an area encompassing approximately 87,000 square kilometers, an area larger than the country of Austria. This resulted in an epidemic outside the usual endemic area.

**Hantaviruses.** In 1993, in the southwestern part of the United States, a cluster of deaths in previously healthy young adults led to the identification of a hantavirus as the cause of the disease—later named hantavirus pulmonary syndrome. Hantaviruses, a large group of rodent-associated viruses, had long been known to be important pathogens in other parts of the world, but had not been known to cause pulmonary disease in the United States. Rodents infected with hantaviruses shed the virus in urine, feces, and saliva but do not become ill.

Subsequent studies yielded important insights. The virus was not new to the area; infections in humans that had occurred decades earlier could be confirmed retrospectively by studying saved serum and tissues from people who had died of undiagnosed illnesses. Infections were not localized to the area in which the first cases were identified. Once physicians became aware of the infection and had techniques to diagnose it, they identified infections from the same or related hantaviruses from many other parts of the Americas, from Canada to southern South America. As of early 1998, infections had been documented in at least twenty-nine states in the United States. This poses the question, why had an infection that kills about half of the humans who become infected gone unrecognized for so long—and why was it finally recognized and characterized in 1993?

Because of favorable weather conditions that supported abundant food for rodents, the rodent population increased more than tenfold between 1992 and 1993 in the area where the outbreak occurred. An expanded rodent population allowed increased contact between rodents and humans. A cluster of deaths in highly visible, young, healthy adults attracted the attention of astute clinicians. The rapid identification of the hantavirus drew on past research and new molecular techniques. Although the hantavirus pulmonary syndrome has characteristic findings, it shares many features (fever, aches, shortness of breath, pulmonary infiltrates) with many other diseases that are more common. Hence, occasional, sporadic cases in the past did not raise alarms.

Hantavirus researchers expect that these viruses will be found to cause at least occasional human cases in areas where human disease has not yet been reported. Many cases have now been identified in South America. As humans clear lands and change habitats in ways that increase human–rodent contact, the number of human cases will probably increase. A survey of rodents in national parks in the United States in 1994 and 1995 found that antibodies to Sin Nombre virus (the hantavirus found in the southwestern United States) were also widespread in eastern and central regions of the country (Mills et al., 1998). Many factors influence rodent–human contact, including agricultural activities, season of the year, housing, abundance of rodents, and changes in land use, which can change rodent habitats.

[*See also* Climate Change; Global Warming; *and* Risk.]

## BIBLIOGRAPHY

Centers for Disease Control and Prevention. *Health Information for International Travel 1999–2000*. Atlanta, 1999.

Chang, M. S., et al. "Changes in Abundance and Behaviour of Vector Mosquitoes Induced by Land Use during the Development of an Oil Palm Plantation in Sarawak." *Transactions of the Royal Society of Tropical Medicine and Hygiene* 91 (1997), 382–386.

Colditz, G. A., et al. "Efficacy of BCG Vaccine in the Prevention of Tuberculosis: Meta-Analysis of the Published Literature." *Journal of the American Medical Association* 271Z (1994), 698–702.

Colwell, R. "Global Climate and Infectious Disease: The Cholera Paradigm." *Science* 274 (1996), 2025–2031.

Lederberg, J. et al., eds. *Emerging Infections: Microbial Threats to Health in the United States*. Washington, D.C.: National Academy Press, 1992.

Levy, S. B. "Multidrug Resistance: A Sign of the Times." *New England Journal of Medicine* 338 (1998), 1376–1378.

McNeill, W. H. *Plagues and Peoples*. New York: Doubleday, 1976.

Mills, J. N., et al. "A Survey of Hantavirus Antibody in Small-Mammal Populations in Selected United States National Parks." *Amerian Journal of Tropical Medicine and Hygiene* 58 (1998), 525–532.

Morell, V. "How the Malaria Parasite Manipulates Its Hosts." *Science* 278 (1997), 223.

Murray, K. J. L., and A. D. Lopez. *The Global Burden of Disease*. Cambridge, Mass: Harvard University Press, 1996.

Raviglione, M. C., et al. "Global Epidemiology of Tuberculosis: Morbidity and Mortality of a Worldwide Epidemic." *Journal of the American Medical Association* 273 (1995), 220–226.

Reeves, W. C., et al. "Potential Effect of Global Warming on Mosquito-Borne Arboviruses." *Journal of Medical Entomology* 31 (1994), 323–332.

Scott, F. W., et al. "A Fitness Advantage for *Aedes Aegypti* and the Viruses It Transmits when Females Feed Only on Human Blood." *American Journal of Tropical Medicine and Hygiene* 57 (1997), 235–239.

Travis, J. "Africa's Latest Scourge." *Science News* (17 July 1999), 40–42.

Valway, S. E., et al. "An Outbreak Involving Extensive Transmission of a Virulent Strain of Mycobacterium Tuberculosis." *New England Journal of Medicine* 338 (1998), 633–639.

Watts, D. M., et al. "Effect of Temperature on the Vector Efficiency of *Aedes Aegypti* for Dengue 2 Virus." *American Journal of Tropical Medicine and Hygiene* 36 (1989), 143–152.

Whalen, C., et al. "Accelerated Course of Human Immunodeficiency Virus Infection after Tuberculosis." *American Journal of Respiratory and Critical Care Medicine* 151 (1995), 129–135.

Wilson, M. E. *A World Guide to Infections: Diseases, Distribution, Diagnosis*. New York and Oxford: Oxford University Press, 1991.

———. "Travel and the Emergence of Infectious Diseases." *Emerging Infectious Diseases* 1 (1995a), 39–45.

———. "Infectious Diseases: An Ecological Perspective." *British Medical Journal* 311 (1995b), 1681–1684.

Wilson, M. E., et al., eds. *Disease in Evolution: Global Changes and Emergence of Infectious Diseases*. New York: New York Academy of Sciences, 1994.

World Health Organization. *Anti-Tuberculosis Drug Resistance in the World*. Geneva, 1997. The WHO/IUATLD global project on anti-tuberculosis drug resistance surveillance.

Zanotto, P. M. de A., et al. "Population Dynamics of Flaviviruses Revealed by Molecular Phylogenies." *Proceedings of the National Academy of Sciences* 93 (1996), 548–553.

—MARY ELIZABETH WILSON

**DIVERSITY.** *See* Biological Diversity.

# DRIFTNET CONVENTION

Driftnets are large-scale *pelagic* ("ocean-fishing") nets or combinations of nets, intended to be held in a more or less vertical position by floats and weights, whose purpose is to enmesh fish by drifting on the surface or in the water. These nets tend to be indiscriminate, harvesting many more species than those targeted (known as *bycatch*), and are particularly destructive when they are lost and continue to "ghost fish." They can range up to 30 miles (48 kilometers) in length and represent a capital-intensive form of fishing. Concerns over the environmental effects of driftnets rose toward the end of the 1980s and led to the adoption of three UN Resolutions in December 1989 (Resolution 44/225), December 1990 (Resolution 45/197), and December 1991 (Resolution 46/215). Resolution 46/215 called for a global moratorium on all large-scale driftnet fishing, to be implemented on the high seas by 31 December 1992, and was the most strongly worded. (The resolutions specifically targeted large-scale driftnets and did not "address the question of small-scale driftnet fishing traditionally conducted in coastal waters, especially in developing countries, which provides an important contribution to their subsistence and economic development" [Resolution 44/225].) Its provisions indicate a global commitment on the part of the international community to end this practice, but the resolution is by nature legally nonbinding and its implementation has been inconsistent.

In parallel to this global legally nonbinding approach, the states of the South Pacific Forum Fisheries Agency (FFA) negotiated and adopted, in November 1989, the Convention for the Prohibition of Fishing with Long Driftnets in the South Pacific Ocean (the Driftnet Convention or Wellington Convention, which entered into force on 17 May 1991). The Wellington Convention codifies in a legally binding treaty some of the obligations contained in the UN Resolutions, and goes further by restricting access in the area to all vessels engaging in driftnet fishing. The convention specifies that the parties will prohibit the use of driftnets in the Convention Area as well as prohibiting the transshipment of driftnet catches within areas under their jurisdiction. They may also prohibit the landing of driftnet catches, prohibit their processing, prohibit their importation, restrict port access for driftnet fishing vessels, and even prohibit the possession of driftnets on board any fishing vessels under their areas of fisheries jurisdiction. The convention is open for signature to any member of the FFA, any state representing a Territory in the Convention Area, and any Territory in the Area authorized to do so by the state responsible for it. Two protocols to the convention, which refer to the conservation of South Pacific albacore tuna within the Convention Area and contain the final clauses of the convention, were adopted in 1990.

The Wellington Convention is indicative of the strong regional approach that has united the members of the FFA. This cohesive approach was also the key to the adoption of prior treaties on the protection of the marine environment (most notably the 1985 South Pacific Nuclear Free Zone Treaty) and access to marine resources in the Exclusive Economic Zones (EEZs) of the member states (for example, the 1987 Treaty between the Governments of Certain Pacific Island States and the Government of the United States). This regional success also indicates that stronger treaties can often be adopted in a regional context, where the members are united by common goals and concerns. These regional initiatives can then in turn serve as an indication of what could be achieved at the global level.

[See also Fishing.]

### BIBLIOGRAPHY

Burke, W. T. "Regulation of Driftnet Fishing on the High Seas and the New International Law of the Sea." *Georgetown International Law Review* 3 (1990), 265–310.

———. *The New International Law of Fisheries, UNCLOS 1982 and Beyond.* Oxford: Clarendon Press, 1994.

Burke, W. T., M. Freeberg, and E. L. Miles. "The United Nations Resolutions on Driftnet Fishing: An Unsustainable Precedent for High Seas and Coastal Fisheries Management." *Ocean Development and International Law* 25 (1994), 127–186.

de Fontaubert, A. C. "Managing Marine Resources under International Law: Challenges and Opportunities." Presentation to the Seaviews Conference on Marine Ecosystem Management Obligations and Opportunities. Wellington, New Zealand, February 1998.

Food and Agriculture Organization. "Report of the Expert Consultation on Large-Scale Driftnet Fishing." FAO Fisheries Report No. 434.

———. "The Regulation of Driftnet Fishing in the High Seas: Legal Issues." Annex I, FAO Legislative Study, 1991.

Hewison, G. J. "High Seas Driftnet Fishing in the South Pacific and the Law of the Sea." *Georgetown International Law Review* 5 (1993), 313–374.

———. "The Legally Binding Nature of the Moratorium on Large-Scale High Seas Driftnet Fishing." *Journal of Maritime Law and Commerce* 25 (1994), 557–579.

Johnston, D. M. "The Driftnetting Problem in the Pacific Ocean: Legal Considerations and Diplomatic Options." *Ocean Development and International Law* 21 (1990), 5–38.

Northridge, S. "Driftnet Fisheries and Their Impacts on Non-Target Species: A Worldwide Review." FAO Fisheries Technical Paper No. 320, 1991.

Orrego Vicuña, F. *The Changing International Law of High Seas Fisheries.* Cambridge and New York: Cambridge University Press, 1999.

United Nations. "Large-Scale Driftnet Fishing and Its Impact on Living Marine Resources of the World's Oceans and Seas." Report of the Secretary General. UN Doc. No. A/45/663, 26 October 1990.

Wright, A., and D. J. Doulman. "Driftnet Fishing in the South Pacific." *Marine Policy* 15 (1991), 303–337.

—A. CHARLOTTE DE FONTAUBERT

# DROUGHT

[*This entry consist of two articles:* An Overview, *which focuses on the various definitions of drought and the regions affected, and* Sahel Drought in West Africa, *which explains the impact drought has had in the Sahel region of West Africa.*]

## An Overview

Drought is a shortage of water, but little more can be said without qualification. The first challenge is to decide upon an appropriate definition. This is not a simple matter, unfortunately, because there is no universal definition of drought. A shortage of water can be brought about by a reduction in supply (e.g., a decline in rainfall), or by an increase in demand (caused by a change in land use, for example), or by a combination of the two. Drought can be defined, for example, as a period of below-average precipitation (*meteorological* drought); or a shortage of soil moisture in the root zone of crops that affects their productivity (*agricultural* drought); or a low river discharge that adversely impacts upon the ecosystem (*hydrological* drought). There are also *climatological* definitions of drought, in addition to those dealing with ecological and economic aspects.

Drought is described as a creeping phenomenon, with an uncertain start. It is characterized by its intensity, location, impact, and duration. Drought should not be confused with aridity, which is a measure of the lack of water in a region based on observations averaging data over 30 years or more. Warren and Khogali (1992), in their assessment of desertification and drought in the Sudano–Sahelian region, distinguish between:

- *drought*: moisture supplied below average for short periods of one to two years,
- *desiccation*: a process of aridization lasting decades, and
- *land degradation*: a persistent decrease in the productivity of vegetation and soils.

Each of these environmental problems prompts different environmental strategies. Drought requires food storage and short-term relief; desiccation demands more radical measures such as resettlement and changing land use patterns; while land degradation is more amenable to corrective and preventive measures. It is essential to distinguish between a short-term reduction in rainfall (i.e., drought), and the process of climate change occurring on longer time scales. [*See* Climate Change.]

**Drought Impacts and Frequency.** Drought is a worldwide phenomenon. Drylands in India experience drought once every four years, and a similar frequency is observed for Israel, southern Africa, and parts of China, although in the North China Plain there have been thirty-five reported droughts in the last forty years. Australia and the United States have both experienced a number of major droughts this century: California suffered consecutive years of low rainfall between 1987 and 1992, and Australia was completely free of drought for only one year in the same five-year period. In northeastern Brazil, droughts occur eight out of every ten years, often accompanied by floods; such is the variability of the precipitation. Drought has been blamed for death and economic disruption in all continents; for example, 28 million people are believed to have been affected by drought in India in the period 1985–1988, although Africa appears to have been the worst affected. The droughts of the mid-1980s created millions of environmental refugees in Africa and enormous suffering for those left behind. Three major droughts have occurred in the Sahel region this century—during the 1910s, 1940s, and 1970s. The last is of most concern because it appears to have heralded the onset of generally drier conditions that have affected 80 percent of the population of the Sahel. [*See* Fire, *article on* Indonesian Fires.]

**Causes of Drought.** Drought is often considered to be strictly a climatological phenomenon, but to ascribe the problem solely to a lack of rainfall ignores the impacts of human activity—if not in causing drought, then at least in exacerbating the impacts. Publications with titles such as "nature pleads not guilty" and "drought the scapegoat" illustrate the importance attached to nonclimatological causes, a perspective taken up by many writing on drought in Africa. It is therefore necessary when discussing the causes of drought to discuss both the role of climate and the impacts of human activities.

**Drought as a Climatological Phenomenon.** The primary characteristic of meteorological drought is a reduction in precipitation (Figure 1). There are many definitions of the phenomenon, including threshold values (in the United Kingdom, drought is fifteen consecutive days without rain) and statistical values (in South Africa, 70 percent of normal rainfall; in India, twice the standard deviation below the average rainfall). These approaches have the advantage of precision, but they often lack any assessment of impact.

Detailed analyses of the Sahel region with respect to

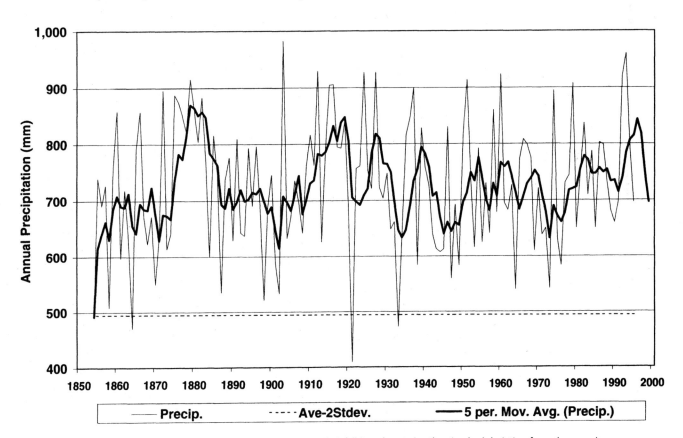

**Drought: An Overview. Figure 1.** Annual Precipitation at Rothamsted Experimental Station, Harpenden, United Kingdom.

This time series (mean 726 millimeters) displays a large range and no clear trend. Note that there are few extreme drought events

(rainfall less than twice the standard deviation from the mean), although recent reported droughts include 1976, 1988–1992, and 1995–1996.

meteorological drought have provided evidence of a persistent decline in annual rainfall since the 1970s. This has been linked to local land use changes, but regional and global phenomena such as El Niño–Southern Oscillation (ENSO) and global warming may offer an alternative mechanism. [*See* El Niño–Southern Oscillation; *and* Global Warming.]

It has been suggested that local causes of rainfall reduction are related to the impacts of land degradation on the energy balance of the atmosphere. This idea has been prevalent since the 1970s, when it was proposed as a *biogeophysical feedback* initiated by overgrazing leading to increased surface albedo, reduced heating of the atmosphere, and hence a reduction in rainfall. [*See* Albedo.] This model has been criticized for ignoring the lack of evidence of extensive overgrazing, the observed increases in surface temperatures in drylands, and latent heat flux changes. More recently, the importance of surface moisture content (and hence vegetation cover) has been demonstrated for areas such as the Sahel, where rainfall is produced by convection, but the latest climate models suggest that there is no clear link between land degradation and rainfall.

On a worldwide scale, the interlinking, or teleconnection, of climate and oceanic circulations has been examined as a possible explanation of drought. Changes in rainfall in Africa have been coupled to sea surface temperature anomalies in the Atlantic, while sea surface temperature changes in the southern Atlantic and southwestern Indian oceans appear to be significant for rainfall in southern Africa. The most widely discussed coupling is that of ENSO with monsoonal rainfalls. During the 1983 El Niño, Australia experienced one of the worst droughts this century, followed by unusually heavy rainfall in 1988 when the Walker circulation was reestablished. Drought in India and northeastern Africa has similarly been related to El Niño. The association with Sahelian rainfall is less clear, however, and for northeastern Brazil there is only a weak correlation between rainfall and El Niño. Caution must therefore be exercised before accepting that an El Niño may herald the worldwide onset of drought conditions, a point made by Wuethrich (1995), who reports that the El Niño phenomenon is becoming increasingly unpredictable and appears to follow a more complex pattern than previously believed.

There are also concerns that drought, especially agricultural drought, will become more frequent as a result of global warming. Higher temperatures may cause an intensification of the hydrological system, but more detailed analyses of seasonal changes suggest an increase in drier periods (England, for example, is expected to experience milder winters but drier summers). General circulation models have indicated that drought will become more frequent in the United States and much of Europe. In the tropics, agriculture is particularly vulnerable in situations where crops are grown near their tolerances of heat and water supply; hence yields may decline under global warming unless the efficiency of water use is increased. Even where an increase in rainfall is predicted, it may not result in higher production. Monsoonal precipitation is expected to increase in northern Australia, for example, encouraging better rangeland production and improvements in cattle ranching. But since plant growth is also limited by low levels of soil phosphorus, a reduction in drought will not necessarily lead to greater productivity unless fertilizers are applied.

**Drought and Environmental Mismanagement.** The changes that have taken place in rural communities during the twentieth century and assumed links between agricultural production, rainfall, and famine have generated much debate over the relationship between environmental mismanagement and drought (Figure 2). Vulnerability is the key issue with respect to mismanagement. Rainfall is highly variable in space and time, and it is inevitable that any region will have periods during which rainfall is significantly below average. Whether this has an impact on human activities or the ecosystem and is regarded as drought will depend upon the resilience or vulnerability of that system. It is argued that rural communities in developing countries have become less resilient with the opening up of their economies to world market forces during the latter half of the twentieth century. The intensification of agriculture and greater water demands for irrigation and urban expansion creates an environment that is more sensitive to rainfall variations (Glantz, 1994). A reduction in rainfall is seen as a trigger mechanism for drought, but the broad phenomenon is explained in terms of social and economic change. Such explanations are not merely the province of the developing countries. The United Kingdom has recently encountered a number of droughts, and blame has been apportioned variously to rainfall, consumer demand, and even the recent privatization of the water supply companies.

**Conclusions.** The important features of drought may be summarized briefly as follows.

1. The appropriate definition for drought in any particular instance must be stated at the outset.
2. Meteorological drought is a normal feature of any climate and should be expected from time to time.
3. Economic or agricultural drought arises through an imbalance between supply and demand and can be the result of a reduction in supply, an increase in demand, or a combination of the two.
4. Because drought may be due to a reduction in rainfall or an increase in demand, solutions to drought may include supply enhancement (desalination, in-

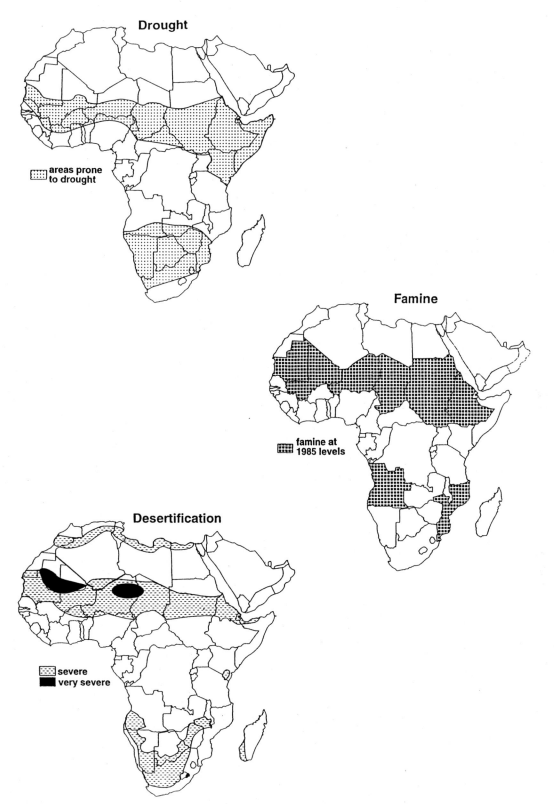

**Drought: An Overview. FIGURE 2.** Concurrent Occurrence of Drought, Famine, and Desertification in Africa.

This suggests a close association between the occurrences of drought, desertification, and famine, yet the maps ignore the complex relationships between these phenomena, and their spatial variability is much greater than the figures indicate. (Adapted from Thomas, 1993. With permission of Royal Geographical Society.)

terbasin transfers, cloud seeding) or demand management (short-term water restrictions or longer-term economic measures and education).

5. There is a need to move away from responding to drought (tactical strategies) to a more proactive approach (strategic measures) for drought management and mitigation.

[See also Desertification; and Famine.]

## BIBLIOGRAPHY

### DROUGHT DEFINITIONS

Agnew, C. T. "Sahel Drought, Meteorological or Agricultural?" *International Journal of Climatology* 9 (1989), 371–382.

Agnew, C. T., and E. Anderson. *Water Resources in the Arid Realm.* London: Routledge, 1992.

Le Houerou, H. N. "Climate Change, Drought and Desertification." *Journal of Arid Environments* 34 (1996), 133–185.

Wilhite, D. A., ed. *Drought Assessment, Management and Planning.* Boston: Kluwer, 1993.

### CLIMATE AND DROUGHT

Glantz, M. H. *Currents of Change: El Niño's Impact on Climate and Society.* Cambridge: Cambridge University Press, 1996.

Glantz, M. H., R. W. Katz, and N. Nicholls, eds. *Teleconnections: Linking Worldwide Climate Anomalies.* Cambridge: Cambridge University Press, 1991.

Gordon, A. H. "The Random Nature of Drought: Mathematical and Physical Causes." *International Journal of Climatology* 13 (1993), 497–507.

Marsh, T. J., R. A. Monkhouse, N. W. Arnell, M. L. Lees, and N. S. Reynard. *The 1988–92 Drought.* Wallingford, U.K.: Institute of Hydrology, 1994.

Nicholson, S. E., and I. M. Palao "A Re-evaluation of Rainfall Variability in the Sahel." *International Journal of Climatology* 13 (1993), 371–389.

Wuethrich, B. "El Niño Goes Critical." *New Scientist* 145.1963 (4 February 1995), 32–35.

### DROUGHT IMPACTS

Binns, T., ed. *People and Environment in Africa.* Chichester, U.K.: Wiley, 1995.

Cannel, M., and C. Pitairn. *Impacts of Mild Winters and Hot Summers 1988–1990.* London: HMSO, 1993.

Downing, T. E. "Climate Change and Vulnerable Places." *Environmental Change Unit Report 1.* Oxford: University of Oxford, 1992.

Glantz, M. *Drought Follows the Plow.* Cambridge: Cambridge University Press, 1994.

———, ed. *Drought and Hunger in Africa.* Cambridge: Cambridge University Press, 1987.

Mortimore, M. *Adapting to Drought.* Cambridge: Cambridge University Press, 1989.

Rozenzweig, C., and M. C. Parry. "Potential Impacts of Climate Change on World Food Supply." *Nature* 367 (1994), 133–138.

Thomas, D. G. "Sandstorm in a Teacup? Understanding Desertification." *Geographical Journal* 159 (1993), 318–331.

Wilhite, D. A., ed. *Drought.* 2 vols. London: Routledge, 2000.

—C. T. Agnew

## Sahel Drought in West Africa

The Sahel region of West Africa has become well known worldwide because of the devastating droughts that it has experienced. Conditions were catastrophic in the early 1970s, with herds decimated, millions of people displaced to refugee camps and cities, and perhaps 100,000 deaths from famine. The drought affected a much larger region, including most of the wetter savanna to the south.

A case study of drought in this region is important for many reasons. The intense impact on the peoples of the Sahel alone demonstrates its importance, but the drought of the 1970s and the subsequent continuation of drought conditions has also triggered an academic interest in many innovative facets of drought research. The drought led to a recognition of the problem of large-scale land degradation, termed *desertification*, and underscored the point that the impacts of drought are closely linked to both environmental and societal factors. The Sahel drought was the impetus for numerous scientific papers suggesting that droughts may have an anthropogenic origin, and it has served as an important case study in the issue of land–atmosphere interaction and land surface feedback on climate. The 1970s drought also prompted an interest in the long-term climate of the region and in long-range forecasting of weather conditions in the region.

**Regional Geography.** The Sahel is a narrow, semiarid expanse of grassland, shrubs, and small, thorny trees, stretching some 5,000 kilometers across northern Africa from the Atlantic Ocean eastward almost to the Red Sea. It includes much of the countries of Mauritania, Senegal, Mali, Niger, Chad, and the Sudan, as well as northern fringes of Nigeria and Burkina Faso. Extending from approximately 14° to 18° north latitude (Figure 1), the Sahel is a zone of climate and vegetation that represents a transition between the desert and the more humid savanna to the south. The Sahel is historically important because centuries ago it was the site of prosperous cities and empires taking part in trans-Saharan trade. Today, most of the population practices some form of agriculture, with pastoralism common in the drier north and sedentary farming common in the south. Millet is the main crop in most of the Sahel.

A precise location for the Sahel is difficult to specify because the environmental characteristics that distinguish it, climate and vegetation, exhibit large and continual north-to-south gradients and also change markedly in time. Because of this, and the extension of the recent droughts to the wetter savanna farther south, the term *Sahel* is commonly (but mistakenly) applied to the entire arid/semiarid region from the southern Sahara to the forest margin at about 10° north latitude.

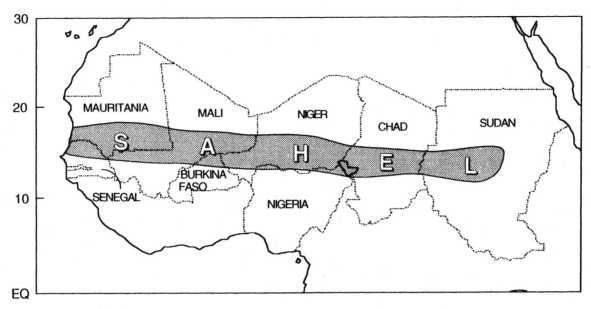

**Drought: Sahel Drought in West Africa. FIGURE 1.** Location of the West African Sahel.

***Climate.*** The Sahelian climate is characterized by high temperatures throughout the year and by a brief rainy season occurring during the high-sun or summer season. The alternation between the few months of rainfall and the dryness during the rest of the year is linked to the shift of the large-scale general atmospheric circulation features that control the climate and weather of West Africa. This shift is illustrated by the conditions in January and August (Figures 2A and 2B).

During January and other months of the low-sun season, the subtropical high-pressure cell of the North Atlantic is displaced far enough southward that it influences much of northern Africa. As a result, the region is generally rainless from October to May or June, and the dry northeasterly Harmattan winds prevail.

During August, the general circulation features are displaced toward the Northern Hemisphere. The subtropical high moves north, and circulation over northern Africa is dominated by a heat low over the west central Sahara. The South Atlantic subtropical high moves northward, and the humid southwesterly flow on its equatorial flank, often termed the "Southwest Monsoon," penetrates far inland into the northern Sahel.

The Harmattan and Southwest Monsoon meet in a region called the "intertropical convergence zone," or ITCZ. Where these air streams converge, air is forced to rise; since the southwesterly flow is warm, humid, and unstable, this convergence leads to cloud formation and convective activity. Thus, rainfall is linked to the ITCZ, and its seasonal migration dictates the seasonal migration of the tropical rains. During the ITCZ's northward

excursion in the Northern Hemisphere summer, it brings rain to the Sahel.

In the Sahel and southward to the Guinea coast of the Atlantic, mean annual rainfall and the length of the rainy season are determined by the length of time the ITCZ is the dominant circulation feature. Thus, both the amount of rainfall and the length of the season increase from north to south. Mean annual rainfall is on the order of 100–200 millimeters in the northern Sahel and 500–600 millimeters in the southern Sahel. The length of the rainy season ranges from about two months in the north to about three months in the south. Throughout the region, the month of maximum rainfall is August.

Although the seasonal cycle of rainfall is linked to the ITCZ, most rainfall is not confined to local thunderstorms embedded in it. Instead, there are organized disturbances, called cloud clusters and easterly waves, that traverse the Sahel from east to west. These systems, linked to a jet stream in the midtroposphere, bring most of the rainfall.

***Water resources, soils, and vegetation.*** Although the Sahel is semiarid and prone to drought, the region's rivers provide some environmental security. The two major rivers, the Senegal and the Niger, originate in more humid regions to the south and provide perennial flow and expansive floods after the rainy season. The flood waters provide areas of cultivation and pasture. The area's largest lake, Lake Chad, is also fed by water from farther south, via the Logone and Chari Rivers.

The vegetation of West Africa exhibits the same longitudinal zonation as the rainfall (Figure 3), changing

A

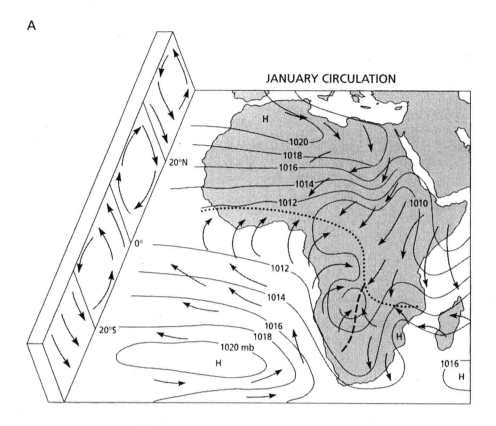

B

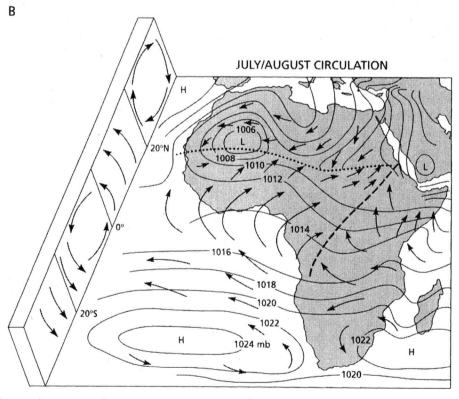

**Drought: Sahel Drought in West Africa. FIGURE 2.** Wind and Pressure Patterns over Africa during (A) January and (B) July/August.

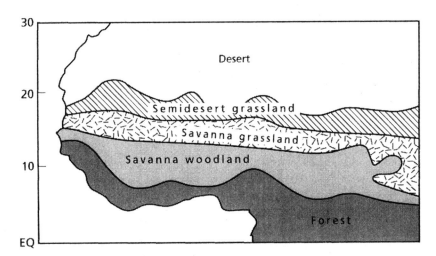

**Drought: Sahel Drought in West Africa. FIGURE 3.** Map of Vegetation Zones in West Africa.

progressively from north to south. In the north along the desert's edge is open grass vegetation, mixed in some areas with occasional woody species of small trees or shrubs. Grasses are perennials, generally not taller than 80 centimeters, and annuals; woody species are often thorny, such as the typical acacia. This is the Sahel proper. Here, the surface is to a large extent bare soil, and the vegetation tends to be clustered in sites of favorable conditions of soil or runoff, creating a mosaic pattern.

The farther south one goes, the taller the vegetation becomes, the greater the proportion of woody species (trees, shrubs, bushes), and the higher the amount of ground cover. The low grass cover of the northern Sahel gives way to savanna grasslands with a taller and more continuous cover. The grasses here tend to be perennials. Much of the savanna is also covered with scattered bushes and trees within the grassland. Farther south is the savanna woodland, a region in which trees become the dominant vegetation and that forms the transition to the forests nearer the coast.

**The Drought of the 1970s.** Drought in the Sahel is generally considered to have commenced in 1968 and extended more or less continuously through 1973 or 1974. Figure 4A shows rainfall deficits during the worst years, 1972 and 1973. Throughout the region, rainfall during these years was generally 30–70 percent below the mean for the period 1931–1960. As the drought began, it encompassed mainly the Sahel and parts of the wetter Soudan zone of savanna woodland to the south; by 1970 or 1971 nearly all of West Africa, down to the humid Guinea coast forest zone, was drought-stricken.

The region suffered not only from the low rainfall, but also from a stark reduction in the amount of water carried by the Sahel's principal rivers. During the years 1972 and 1973, the discharge of the Senegal River and the Chari averaged 60 percent below normal, while that of the Bani was 75 percent below normal. The flow of the Niger was between 25 and 39 percent below normal in 1972 and 1973, respectively.

The loss of water resources resulted in tremendous agricultural and economic losses. In Mali and Chad, fish production dropped dramatically in 1972–1973, falling to about 20 percent of normal in Mali. Rice growing along the river was curtailed in some countries, and irrigated crops such as beans and sorghum were grown on only a fraction of their normal area. In Mali, 40–50 percent of the harvests of staples such as sorghum, millet, and other grains was lost in 1973; in Mauritania, some 70 percent was lost. In Senegal, where peanuts account for about 80 percent of the country's total export earnings, crop production in 1973 was about half that of 1974.

When rainfall increased marginally in the mid-1970s, it was believed that the drought problem had ended. However, the trends of rainfall in the region show that this was only a brief respite in which rainfall did not even return to the long-term mean (Figure 5). By the early 1980s, drought conditions were again severe. During the worst years, 1983 and 1984, rainfall was about 50 percent below the long-term mean (Figure 4B). Thus the meteorological conditions of drought were even worse than those of the early 1970s, and hydrologic resources continued to diminish. The average river discharge for the entire period 1968–1985 was 44 percent below the mean for the Senegal, 39 percent for the Chari, and 22 percent for the Niger. Lake Chad, which still held a considerable amount of water in the 1970s, was almost completely desiccated.

These statistics paint a bleak picture, but a comparison with the more favorable conditions of the 1950s underscores still more strongly the tremendous decline of

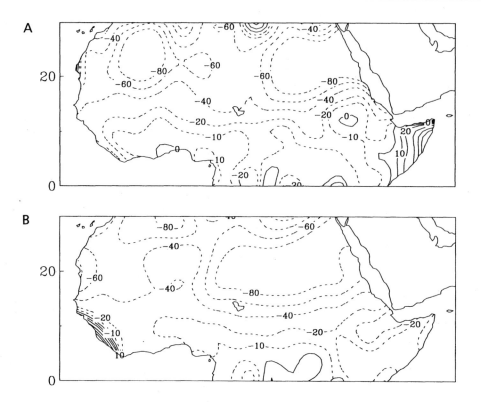

**Drought: Sahel Drought in West Africa. FIGURE 4.** Average Rainfall Deficits, Expressed as a Percentage Departure from the Long-Term Mean for (A) 1972 and 1973 and (B) 1983 and 1984.

rainfall in the region. At many of the stations throughout the Sahel, rainfall during the period 1970–1984 was half that of the 1950s (Table 1).

The two periods of severe drought in the early 1970s and the mid-1980s were part of a multidecadal period of subnormal rainfall. In the Sahel and Soudan zones, rainfall exceeded the long-term mean only marginally in 1968 and 1994; in every one of the twenty-five intervening years it was below the mean. The situation was quite similar in zones to the north and south. Such a long period of continually dry conditions is quite unusual and may be unique to West Africa. It is most pronounced in the central Sahel.

**Causes of the Drought.** As the nature and severity of the Sahel drought became known in the mid-1970s, numerous hypotheses concerning its cause were suggested. Several meteorological explanations suggested natural causal mechanisms, but one very eminent meteorologist, the late Jule Charney of Massachusetts Institute of Technology, proposed that the drought was a result of anthropogenic influences. Even today we do not have an adequate explanation for the drought, particularly its intensity and persistence, but the anthropogenic theory of its origin has almost no support.

However, its ravaging impact was probable exacerbated by human factors.

Early meteorological explanations for drought in the Sahel were quite simplistic: shifts toward the equator of the subtropical highs (especially that over the Atlantic; Figure 2) and of the ITCZ. Subsequent analyses of characteristics of the drought showed the shortcomings of these hypotheses. For one, the ITCZ was as far north in some drought years as during wetter years of the 1950s. Moreover, the drought was part of a continentwide pattern of abnormally dry conditions. Such a pattern, which prevailed throughout the 1980s (Figure 6), is inconsistent with simple displacements of the ITCZ and other circulation features.

More recent investigations have identified links between drought and sea surface temperatures and also between drought and characteristics of the atmospheric general circulation, such as the jet streams prevailing over West Africa and the intensity of other circulation features such as the Hadley cell. It has likewise been shown that drought tends to be associated with reduced intensity of rain-bearing disturbances rather than with fewer rainfall events. This also speaks for factors other than displacements of the ITCZ or subtropical highs.

Charney's theory of anthropogenic origin involved the concept of changes in the land surface resulting from increasing population pressures and concomitant increases in grazing pressure. At the time of the early-1970s drought, numerous satellite and aerial photographs showed changes in the surface albedo, or reflectivity, in heavily grazed areas. Charney hypothesized that the in-

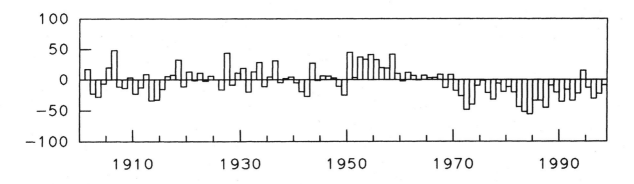

**Drought: Sahel Drought in West Africa. Figure 5.** Annual Rainfall in the Sahelian Zone from 1901 to 1994, Expressed as a Percentage Departure from the Long-Term Mean.

creased reflectivity, which represents a loss of energy to space, would require compensatory changes in the atmosphere. In particular, heat needed to be added to offset the loss, something that could be accomplished by increased atmospheric subsidence, or sinking motion. Since this sinking motion is a major cause of aridity and desert climates, Charney proposed that the baring of high-albedo soils by overgrazing may have triggered the drought via this mechanism. Subsequently, Charney and others suggested additional anthropogenic factors, such as reduced soil moisture and surface roughness and increased dust. [*See* Albedo.]

While all of these mechanisms might influence climate, it can easily be argued that they did not actually cause the drought in the Sahel. This is readily demonstrated by the continental scale of the dry conditions (Figure 6) and by the links to meteorological factors such as sea surface temperatures. Moreover, droughts are an inherent part of Sahel climate. The history of the

**Drought: Sahel Drought in West Africa. Table 1.** Mean Annual Rainfall for Two Multiyear Periods at Selected Stations in and Near the Sahel

| | Mean Rainfall (millimeters) | |
|---|---|---|
| | 1950–1959 | 1970–1984 |
| Bilma | 20 | 9 |
| Atbara | 92 | 54 |
| Nouakchott | 172 | 51 |
| Khartoum | 178 | 116 |
| Agadez | 210 | 97 |
| Timbuktoo | 241 | 147 |
| Nema | 381 | 210 |
| Dakar | 609 | 308 |
| Banjul | 1,409 | 791 |

Sahel is replete with references to long and severe droughts. One quite comparable in magnitude and duration to that prevailing since 1968 occurred in the early decades of the nineteenth century. Like the current arid period, it was a time of continental-scale aridity, long and severe droughts, and regression and desiccation of lakes throughout Africa. Another decimated the Sahelian population during the 1730s–1750s.

**Ramifications of the Study of Sahel Drought.** In the scientific community, the study of Sahel drought has had numerous ramifications relating both to meteorology and to aspects of environment and culture. Meteorologically, the drought prompted an interest in the longer-term climatic picture of this region and elsewhere in Africa. Many researchers worldwide are collecting and evaluating data related to climate time scales of a century and longer, and many of these efforts are being incorporated into international efforts such as PAGES (Past Global Changes), CLIVAR (Climate Variability and Prediction Program), and IDEAL (International Decade of the East African Lakes). A number of researchers and institutions, such as the British Meteorological Office, are producing long-range (seasonal and interannual) weather forecasts for the Sahel based on the data that have been generated through these efforts.

*Interplay of climatic and societal factors in drought.* The recent droughts also made the academic community more aware of the interplay of climatic and societal factors in determining the impact of drought. It has been shown, for example, that traditional organizational structures, drought responses, lifestyles, economic systems, and agricultural practices, common in prior centuries, probably provided some degree of protection from the ravaging effects of severely reduced rainfall. The imposition of new technologies and non-African economic and cultural traditions, largely a consequence of colonial rule, has probably increased the severity of the impact of drought.

All of West Africa was colonized by European countries during the nineteenth century. In "developing" the region, the foreign governments discouraged nomadic ways and promoted urbanization. They subsequently drew West Africa into the international market econ-

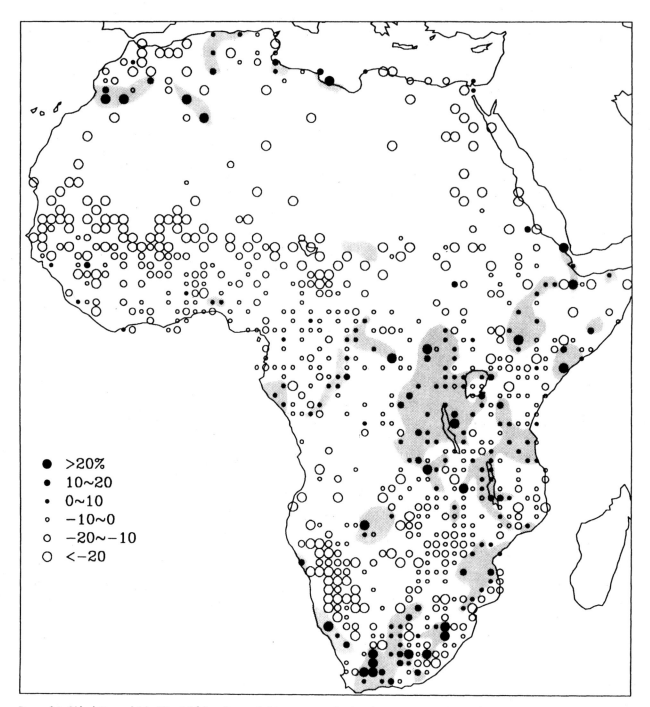

**Drought: Sahel Drought in West Africa. FIGURE 6.** Mean Rainfall for the Decade 1980–1989, Expressed as a Percentage Departure from the Long-Term Mean. Station data are averaged over 1° squares; positive values are shaded.

omy, at the expense of disrupting the traditional subsistence agriculture by favoring cash crops such as peanuts. Much of this regional development occurred during a period of more favorable conditions, particularly during the wet 1950s decade. Moreover, relatively favorable climatic conditions had prevailed for most of the previous hundred years as well, throughout the pe-

riod of colonial rule. The Sahelian nations all became independent around 1960. Following independence came the rapid demise of the favorable climatic conditions, and soon thereafter the onset of drought. The systems established by the colonialists were attuned to the more favorable environment. These systems were ill equipped to deal with the disaster, and traditional coping strategies were no longer viable; this probably served to magnify the effect of the early-1970s drought and the subsequent continuation of drier conditions.

**Desertification.** The drought that ravaged the Sahel in the early 1970s triggered a surge of interest in desertification. Numerous satellite and aerial photos showed evidence of large-scale human impact on the land: international borders separating grazed and ungrazed land were seen vividly from space, as were fenced "exclosures," that is, huge fenced ranches in which the extent of grazing was rigidly controlled. Charney's theory of the origin of the Sahel drought suggested major climatic consequences of desertification.

A U.N. meeting on desertification was convened in 1976, and it was announced that desertification had affected some 35 million square kilometers of land globally and that, overall, 40 percent of the Earth's land surface was at risk of undergoing similar changes. This process was defined as "diminution or destruction of the biological potential of the land [which] can lead ultimately to desert-like conditions." Factors promoting desertification include increasing population of people and animals, sedentarization of the indigenous nomadic peoples, breakdown of traditional market and livelihood systems, innovation of new and inappropriate technology in the affected regions, and, in general, bad strategies of land management such as overcultivation, overgrazing, intensive irrigation, and large-scale deforestation.

In a desertification scenario, the land becomes increasingly impacted by wind and water erosion. Topsoil is eroded away, and vast amounts of land are sometimes washed away to produce huge gullies. The soil's texture, organic matter, and nutrient contents are changed in ways that reduce its fertility. Soil becomes salinized or waterlogged. The land cover becomes more barren or nutrient-rich, and diverse species are replaced by vegetation of poorer quality. The carrying capacity of the land is dramatically reduced.

The interest in desertification had both negative and positive consequences. On the negative side, few rigorous studies were made, claims of the extent and magnitude of desertification were inflated, few data existed to support these claims, and the role of climate was inadequately acknowledged. An exaggerated and simplistic image was evoked of an "advancing desert," a living environment becoming sterile and barren, a desert encroaching on the savanna.

On the positive side, studies of desertification have established many useful facts about the arid and semiarid environments. One is that people, climate, and the environment are intricately linked. In past times, a relatively stable equilibrium was established, with feedbacks between these three components maintaining a healthy and functioning ecosystem. This stability was disrupted only by major climatic fluctuations, such as

the end of the Ice Age, when the savanna dried up to produce the present Sahara. In recent times, however, human pressure on the land has increased, the impact of climatic fluctuations has become more severe, and the stability of the environment is threatened by more moderate climatic fluctuations. Studies of desertification have also demonstrated that the Earth's dryland environments, the arid, semiarid and subhumid lands that support some 20 percent of its population, are truly at risk of severe degradation. At present, prudent land management strategies are needed to prevent or reverse the process. [*See* Desertification.]

**Land–atmosphere feedback.** Although the hypothesis that the 1970s Sahel drought was caused by anthropogenic factors has not received much support, it has led to a related and widely accepted concept concerning the interaction between the land surface and the atmosphere. Central to this concept are the fluxes of energy and mass from the Earth's surface. Because these pose a set of controls on the atmosphere, changes in surface characteristics can potentially influence even large-scale weather patterns. This idea precedes the Sahel drought, but, as a result of that drought, the concept of land–atmosphere feedback has been tested extensively, and much evidence has been produced of its validity.

For example, numerous international field programs have been set up to study the effect of the land surface on the atmosphere. These include HAPEX-MOBILHY (Hydrological and Atmospheric Pilot Experiment–Mode'lisation du Bilan Hydrique) in central France in 1986–1988; FIFE (First ISLSCP—International Satellite Land Surface Climatology Project—Field Experiment) in the Great Plains of the United States in 1987 and 1989; HAPEX-Sahel in Niger in 1992; ABRACOS (Anglo-Brazilian Climate Observation Study) in 1990–1994; and BOREAS (Boreal Ecosystem-Atmosphere Study) in Canada in 1993–1994. These represent a variety of ecosystems and climates. Similar but smaller-scale experiments have been conducted in semiarid regions of the United States, the grasslands of Asia, and elsewhere.

In the case of droughts, there is extensive evidence that a positive feedback can exist such that the changes on the land surface resulting from the drought (for example, reduced surface vegetation cover and soil moisture) act to reinforce the atmospheric anomalies that originally produced the drought conditions. This feedback is probably quite effective in the Sahel and could explain the unusual intensity and duration of droughts in that region. A corollary to this hypothesis is that surface changes induced independently of the drought can also have an impact, so the potential exists for land degradation to influence weather and climate if it is sufficiently extreme and widespread.

The majority of studies concluding that surface conditions affect large-scale weather and climate patterns are simulations with general circulation models (GCMs). Dozens of simulations of various mechanisms and geographic locations have been performed, and nearly all have concluded that large-scale surface changes do influence atmospheric processes. This is not unequivocal proof, because GCMs cannot produce a perfectly accurate picture of global climate and because in most cases the simulations have incorporated unrealistically large surface perturbations. Nevertheless, the vast number of simulations and the relative agreement on major points provide strong support for this feedback hypothesis.

More recently, process models and observations have produced further support. Surface hydrologic feedbacks have been shown to reinforce drought conditions and to have less effect in the case of anomalously wet conditions. Studies have also shown that the atmospheric water vapor over large continental interiors such as West Africa is largely recycled directly from the underlying land surface, so the potential for a feedback effect is strong. In the Sahel, differences in surface energy fluxes between wet and dry years are large enough to invoke changes in atmospheric circulation of the magnitude actually observed in these years. Perhaps most importantly, atmospheric dust in the region has increased steadily as a result of the dry conditions prevailing since the late 1960s. Its continual presence can provide the type of interannual memory required for feedback processes to influence rainfall. Finally, studies using data from the HAPEX-Sahel experiment have demonstrated that areas that randomly receive high rainfall early in the season are more likely than other areas to receive high amounts of rainfall from subsequent storms. This is robust observational evidence of the feedback process in the West African Sahel. More work is needed to understand the mechanisms of the link to the atmospheric processes producing rainfall.

### BIBLIOGRAPHY

Charney, J. G. "Dynamics of Deserts and Drought in Sahel." *Quarterly Journal of the Royal Meteorological Society* 101 (1975), 193–202. A classic paper putting forth the hypothesis that meteorological drought can be a consequence of human activities.

Entekhabi, D. "Recent Advances in Land–Atmosphere Interaction Research." *Reviews of Geophysics* (Supplement, 1995), United States National Report to International Union of Geodesy and Geophysics 1991–1994, pp. 995–1003. A good review of research on land–atmosphere interaction.

Franke, R. W., and B. H. Chasin. *Seeds of Famine: Ecological Destruction and the Development Dilemma in the West African Sahel.* Lanham, Md.: Rowman & Littlefield, 1992. An overview of the societal and economic consequences of drought and famine in the Sahel.

Goudie, A. S., W. M. Adams, and A. Orme, eds. *The Physical Geography of Africa.* New York and Oxford: Oxford University Press, 1995. A detailed, up-to-date overview of various aspects of the physical geography of Africa.

Nicholson, S. E. "African Drought: Characteristics, Causal Theories and Global Teleconnections." In *Understanding Climate Change,* edited by A. Berger, R. E. Dickinson, and J. W. Kidson, pp. 79–100. Washington, D.C.: American Geophysical Union, 1989. A comprehensive review of diverse aspects of Sahel drought, including causes and historical occurrences.

———. "Sahel." In *Encyclopedia of Environmental Biology,* edited by W. A. Nierenberg, pp. 261–275. San Diego: Academic Press, 1995. An introduction to the environment and cultures of the Sahel.

Nicholson, S. E., C. J. Tucker, and M. B. Ba. "Desertification, Drought and Surface Vegetation: An Example from the West African Sahel." *Bulletin of the American Meteorological Society* 79.5 (1998), 815–829. A review for the nonspecialist of the history of the desertification controversy, with emphasis on its meteorological aspects; long-term variations in rainfall are also described.

N'Tchayi Mbourou, G., J. J. Bertrand, and S. E. Nicholson. "The Diurnal and Seasonal Cycles of Desert Dust over Africa North of the Equator." *Journal of Applied Meteorology* 36 (1997), 868–882. Depicts the changes of dust over Sahelian West Africa between the 1950s and the 1980s.

Sircoulon, J. "Les Données Hydropluviométriques de la Sécheresse Récente en Afrique Intertropicale: Comparaison avec les Sécheresses '1913' et '1940.'" *Cahiers O.R.S.T.O.M. Sér. Hydrologie* 13.2 (1976). Summarizes the hydroclimatic aspects of the 1970s Sahel drought and compares it to earlier ones.

Sircoulon, J. H. A. "Variation des Débits des Cours d'Eau et des Niveaux des Lacs en Afrique de l'Ouest depuis le Début du 20ème Siècle." In *The Influence of Climate Change and Variability on the Hydrologic Regime and Water Resources.* International Association of Hydrological Sciences Publication No. 168 (1987), 13–25. A good overview of the century-long hydrologic record of West Africa.

Thomas, D. S. G., and N. J. Middleton. *Desertification: Exploding the Myth.* Chichester, U.K.: Wiley, 1994. A detailed look at the desertification issue, underscoring the weaknesses of numerous studies of this phenomenon.

—SHARON E. NICHOLSON

## DUST STORMS

Dust storms, events in which visibility is reduced to less than 1 kilometer as a result of particle matter being entrained by wind, are important environmental phenomena and have shown great variability in their incidence through time in response to environmental changes (Goudie, 1983). The consequences of dust-storm activity are legion and include the following:

- a reduction in visibility that can create problems for transport infrastructure
- the transport of pathogens and allergens
- disruption of radio communications
- interference with electronic systems
- downwind deposition of deleterious salts (as, for example, in the Aral Sea region)
- modification of regional and local climates
- downwind modification of soils.

Particular interest has arisen recently in the possible role of atmospheric dust loadings in modifying Pleistocene climates (Overpeck et al., 1996) and in contributing to future climatic changes (Andreae, 1996).

**Dust Movements in the Quaternary.** At certain times during the Pleistocene, the world was a very dusty place. This is indicated by extensive deposits of loess, the presence of large amounts of eolian (windblown) dust in ocean-core sediments, and the existence of high quantities of dust in ice cores from the polar regions and elsewhere. This dustiness, especially during cold glacial periods, may relate to a larger sediment source (e.g., areas of glacial outwash), changes in wind characteristics both in proximity to icecaps and in the trade-wind zone (Ruddiman, 1997), and the expansion of low-latitude deserts.

It is possible to obtain a long-term measure of dust additions to the oceans by undertaking studies of the sedimentology of deep-sea cores (Rea, 1994). On the basis of cores from the Arabian Sea, Sirocko et al. (1991) suggested that dust additions were around 60 percent higher during glacials than in postglacial times. Likewise, also working in the Arabian Sea, Clemens and Prell (1990) found a positive correlation between global ice volume and the accumulation rate and sediment size of dust material. Kolla and Biscaye (1977) confirmed this picture for a larger area of the Indian Ocean and indicated that large dust inputs came off Arabia and Australia during the last glacial. In the Atlantic offshore from the Sahara at around 18,000 BP, the amount of dust transported into the ocean was augmented by a factor of 2.5 (Tetzlaff et al., 1989, p. 198)

On a longer time scale there is some evidence that dust activity has increased as climate deteriorated during the last few millions of years. In the Atlantic off West Africa, Pokras (1989) found clear evidence for increased dust input at 2.3–2.5 million years ago, while Schramm (1989) found that the largest increases in mass accumulation rates in the North Pacific occurred between 2 and 3 million years ago. This coincides broadly with the initiation of Northern Hemisphere glaciation. However, no such link has been identified in the southern Pacific Ocean (Rea, 1989). The most comprehensive analysis of dust deposition in the oceans has been undertaken by Leinen and Heath (1981) on the sediments of the central part of the North Pacific. They have demonstrated that there were low rates of dust deposition from 50 to 25 million years ago. This they believe reflects the temperate, humid environment that was seemingly characteristic of the early Tertiary, and the lack of vigorous atmospheric circulation at that time. From 25 to 7 million years ago, the rate of eolian accumulation on the ocean floor increased, but it became greatly accelerated from 7 to 3 million years ago. However, although there is thus an indication that eolian processes were becoming increasingly important as the Tertiary progressed, it was around 2.5 million years ago that there occurred the most dramatic increase in eolian sedimentation. This accompanied the onset of Northern Hemisphere glaciation.

**Dust Deposition as Recorded in Ice Cores.** Another major source of long-term information on rates of dust accretion is the record preserved in long ice cores retrieved either from the polar icecaps or from high-altitude ice domes at lower altitudes. [*See* Aerosols.]

Because they are generally far removed from source areas, the actual rates of accumulation are generally low, but studies of variations in microparticle concentrations with depth do provide insights into the relative dust loadings of the atmosphere in the last glacial and during the course of the Holocene. Thompson and Mosley-Thompson (1981) drew together much of the material that was published at the time they wrote, and pointed to the great differences in microparticle concentrations between the late glacial and the postglacial. The ratio at Dome C Ice Core (East Antarctica) was 6:1, for the Byrd Station (West Antarctica) it was 3:1, and for Camp Century (Greenland) it was 12:1. Briat et al. (1982) maintained that at Dome C there was an increase in microparticle concentrations by a factor of ten to twenty during the last glacial stage, and they explain this by a large input of continental dust. The Dunde Ice Core from High Asia (Thompson et al., 1990) also shows very high dust loadings in the Late Glacial and a very sudden falloff at the transition to the Holocene.

**Loess Accumulation Rates.** By measuring and dating sections in a silky, windblown material called *loess*, it has been possible to estimate the rate at which dust accumulated on land during the Quaternary (see Table 1). The data may somewhat underestimate total

**Dust Storms. TABLE 1.** Loess Accumulation Rates for the Late Pleistocene

| LOCATION | RATE (MILLIMETERS PER THOUSAND YEARS) |
|---|---|
| Negev (Israel) | 70–150 |
| Mississippi Valley (United States) | 100–4,000 |
| Uzbekistan | 50–450 |
| Tajikistan | 60–290 |
| Lanzhou (China) | 250–260 |
| Luochaun (China) | 50–70 |
| Czech Republic | 90 |
| Austria | 22 |
| Poland | 750 |
| New Zealand | 2,000 |

From various sources in Pye (1987), and Gerson and Amit (1987).

dust fluxes into an area because even at times of rapid loess accumulation there would have been concurrent losses of material as a result of fluvial and mass movement processes. Solution and compaction may also have occurred.

The data in Table 1 show a range of values between 22 and 4,000 millimeters per thousand years, but Pye (1987, p. 265) believes that, at the maximum of the last glaciation (at about 18,000 BP), loess was probably accumulating at a rate of between 500 and 3,000 millimeters per thousand years, a rate he suggests was possibly unparalleled in previous Earth history. By contrast, he suggests that during the Holocene, dust deposition has been too low for significant thicknesses of loess to accumulate even though eolian additions to soils and ocean sediments have been appreciable. Pye also hypothesizes that rates of loess accumulation showed a tendency to increase during the course of the Quaternary. Average loess accumulation rates in China, Central Asia, and Europe were of the order of 20–60 millimeters per thousand years during Matuyama time (2.60–0.78 million years ago), and of the order of 90–260 millimeters per thousand years during the Brunhes epoch (since 0.78 million years ago). He also points out that these long-term average rates disguise the fact that rates of loess deposition were one to two orders of magnitude higher during Pleistocene cold phases, and one or two orders of magnitude lower during the warmer interglacial phases when pedogenesis (soil formation) predominated.

**Dust Storms in the Twentieth Century.** The changes in temperature and precipitation conditions in the twentieth century have had an influence on the development of dust storms. Probably the greatest incidence of dust storms occurs when climatic conditions and human pressures combine to make surfaces susceptible to wind attack.

Possibly the most famous case of soil erosion by deflation (wind removal) was the Dust Bowl of the 1930s in the United States (see Figures 1 and 2). In part this was caused by a series of hot, dry years that depleted the vegetation cover and made the soils dry enough to be susceptible to wind erosion. The effects of this drought were gravely exacerbated by years of overgrazing and unsatisfactory farming techniques. However, perhaps the prime cause of the event was the rapid expansion of wheat cultivation in the Great Plains.

Dust storms are still a serious problem in various parts of the United States: the Dust Bowl was not solely a feature of the 1930s. Thus, for example, in the San Joaquin Valley area of California in 1977, a dust storm caused extensive damage and erosion over an area of about 2,000 square kilometers. More than 25 million metric tons of soil were stripped from grazing land within a twenty-four-hour period. While the combination of

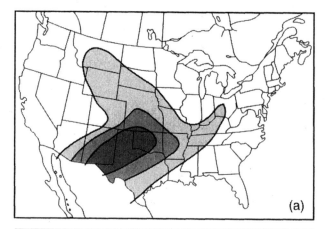

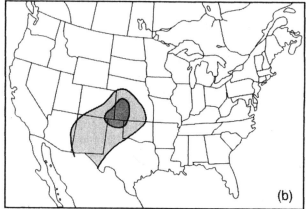

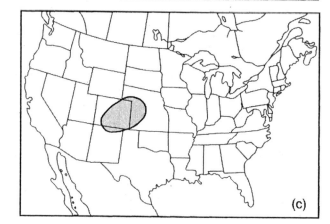

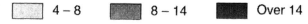

Dust-storm days per month

| 4 – 8 | 8 – 14 | Over 14 |

**Dust Storms. FIGURE 1.** The concentration of dust storms (number of days per month ) in the United States in 1936, illustrating the extreme localization over the high plains of Texas, Colorado, Oklahoma, and Kansas: (a) March, (b) April, and (c) May. (After Goudie, 1983. With permission of Arnold Publishers.)

drought and a very high wind (as strong as 300 kilometers per hour) provided the predisposing natural conditions for the stripping to occur, overgrazing and the general lack of windbreaks in the agricultural land

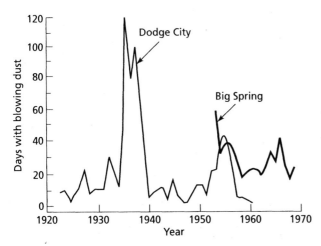

**Dust Storms. FIGURE 2.** Frequency of Dust-Storm Days at Dodge City, Kansas (1922–1961) and for Big Spring, Texas (1953–1970). (After Gillette and Hanson, 1989.)

played a more significant role. In addition, broad areas of land had recently been stripped of vegetation, leveled, or plowed up prior to planting. Elsewhere in California, dust yield has been increased considerably by mining operations in dry lake beds (Wilshire, 1980).

A comparable acceleration of dust-storm activity has occurred in the former Soviet Union. After the Virgin Lands program of agricultural expansion in the 1950s, dust-storm frequencies in the southern Omsk region increased on average by a factor of 2.5 and locally by factors of 5 to 6. Data on trends elsewhere are evaluated by Goudie and Middleton (1992).

Another example comes from the Sahel belt in West

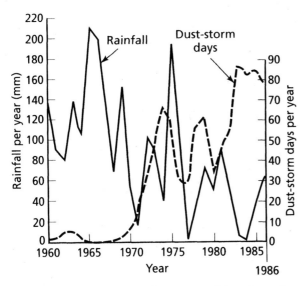

**Dust Storms. FIGURE 3.** Annual Frequency of Dust-Storm Days and Annual Rainfall for Nouakchott, Mauritania, 1960–1986. (After Middleton, 1986. With permission.)

Africa since the late 1960s. Because of the great Sahel drought and the ever-increasing population levels in the area, dust-storm incidence showed a massive increase in the 1970s and 1980s in a great belt from Mauritania in the west to the Horn of Africa in the east. The variation in frequency of annual dust-storm days and annual rainfall totals for Nouakchott, Mauritania, is shown in Figure 3. The increase in dust-storm days after 1968 is dramatic. By the 1980s, rather than having just a few dust storms per year, Nouakchott was receiving over eighty per year.

**The Eolian Environment in a Warmer World.** Given the impact that past climatic variability has had on soil erosion by wind, it is likely that global warming will have a major impact as well. Changes in precipitation and evapotranspiration rates will probably have a marked impact on the eolian environment. Rates of deflation and sand and dust entrainment are closely related to soil moisture conditions and the extent of vegetation cover. Areas that are marginal in terms of their stability with respect to eolian processes will be particularly susceptible; this has been made evident, for example, through recent studies of the semiarid portions of the United States (e.g., the High Plains). Repeatedly through the Holocene, these areas have flipped from states of vegetated stability to states of drought-induced surface instability.

It is apparent from the outputs of general circulation models that, with global climate change, there are likely to be substantial changes in eolian activity (Muhs and Matt, 1993; Stetler and Gaylord, 1996), with dust-storm incidence in the High Plains and the Canadian Prairies (Wheaton, 1990) being comparable to that of the devastating Dust Bowl years of the 1930s.

[*See also* Desertification; Glaciation; Global Warming; *and* Soils.]

### BIBLIOGRAPHY

Andreae, M. O. "Raising Dust in the Greenhouse." *Nature* 380 (1996), 389–390.

Briat, M., et al. "Late Glacial Input of Eolian Continental Dust in the Dome C Ice Core: Additional Evidence from Individual Microparticle Evidence." *Annals of Glaciology* 3 (1982), 27–31.

Clemens, S. C., and W. L. Prell. "Late Pleistocene Variability of Arabian Sea Summer Monsoon Winds and Continental Aridity: Eolian Records from the Lithogenic Component of Deep Sea Sediments." *Palaeoceanography* 5.2 (1990), 109–145.

Gerson, R., and R. Amit. "Rates and Modes of Dust Accretion and Deposition in an Arid Region—The Negev, Israel." In *Desert Sediments Ancient and Modern,* edited by L. Frostick and I. Reid, pp. 157–169. Oxford: Blackwell Scientific, 1987.

Gillette, D. A. and K. J. Hanson. "Spatial and Temporal Variability of Dust Production Caused by Wind Erosion in the United States." *Journal of Geophysical Research* 94D (1989), 2197–2206.

Goudie, A. S. "Dust Storms in Space and Time." *Progress in Physical Geography* 7 (1983), 502–530.

Goudie, A. S., and N. J. Middleton. "The Changing Frequency of Dust Storms through Time." *Climatic Change* 20 (1992), 197–225. A study of meteorological records of dust in the past few decades.

Kolla, V. and P. E. Biscaye. "Distribution and Origin of Quartz in the Sediments of the Indian Ocean." *Journal of Sedimentary Petrology* 47 (1977), 642–649.

Leinen, M., and G. R. Heath. "Sedimentary Indicators of Atmospheric Activity in the Northern Hemisphere during the Cenozoic." *Palaeogeography, Palaeoclimatology, Palaeoecology* 36 (1981), 1–21.

Middleton, N. J. "The Geography of Dust Storms." Ph.D. diss., University of Oxford, 1986.

Muhs, D. R., and P. B. Matt. "The Potential Response of Eolian Sands to Greenhouse Warming and Precipitation Reduction on the Great Plains of the United States." *Journal of Arid Environments* 25 (1993), 351–361.

Overpeck, J., et al. "Possible Role of Dust-Induced Regional Warming in Abrupt Climate Change during the Last Glacial Period." *Nature* 389 (1996), 447–449.

Pokras, E. M. "Pliocene History of South Saharan/Sahelian Aridity: Record of Freshwater Diatoms (Genus *Melosira*) and Opal Phytoliths, ODP Sites 662 and 664." In *Palaeoclimatology and Palaeometeorology: Modern and Past Patterns of Global Atmospheric Transport*, edited by M. Leinen and M. Sarnthein. Dordrecht: Kluwer, 1989, pp. 795–804.

Pye, K. *Aeolian Dust and Dust Deposits*. London: Academic Press, 1987. A comprehensive general review.

Rea, D. K. "Geologic Record of Atmospheric Circulation on Tectonic Time Scales." In *Palaeoclimatology and Palaeometeorology: Modern and Past Patterns of Global Atmospheric Transport*, edited by M. Leinen and M. Sarnthein. Dordrecht: Kluwer, 1989, pp. 842–857.

———. "The Palaeoclimatic Record Provided by Eolian Deposition in the Deep Sea: The Geologic History of Wind." *Reviews of Geophysics* 32 (1994), 159–195.

Ruddiman, W. F. "Tropical Atlantic Terrigenous Fluxes since 25,000 Yrs B.P." *Marine Geology* 136 (1997), 189–207.

Schramm, C. T. "Cenozoic Climatic Variation Recorded by Quartz and Clay Minerals in North Pacific Sediments." In *Palaeoclimatology and Palaeometeorology: Modern and Past Patterns of Global Atmospheric Transport*, edited by M. Leinen and M. Sarnthein. Dordrecht: Kluwer, 1989, pp. 805–839.

Sirocko, F., et al. "Atmospheric Summer Circulation and Coastal Upwelling in the Arabian Sea during the Holocene and the Last Glaciation." *Quaternary Research* 36 (1991), 72–93.

Stetler, L. D., and D. R. Gaylord. "Evaluating Eolian–Climatic Interactions Using a Regional Climate Model from Hanford, Washington, USA." *Geomorphology* 17 (1996), 99–113.

Tetzlaff, G., et al. "Aeolian Dust Transport in West Africa." In *Palaeoclimatology and Palaeometeorology: Modern and Past Patterns of Global Atmospheric Transport*, edited by M. Leinen and M. Sarnthein. Dordrecht: Kluwer, 1989, pp. 1985–2203.

Thompson, L. G., and E. Mosley-Thompson. "Microparticle Concentration Variations Linked with Climatic Change: Evidence from Polar Ice Cores." *Science* 212 (1981), 812–815.

Thompson, L. G., et al. "Glacial Stage Ice-Core Records from the Subtropical Dunde Ice Cap, China." *Annals of Glaciology* 14 (1990), 288–297.

Wheaton, E. E. "Frequency and Severity of Drought and Dust Storms." *Canadian Journal of Agricultural Economy* 38 (1990), 695–700.

Wilshire, H. G. "Human Causes of Accelerated Wind Erosion in California's Deserts." In *Geomorphic Thresholds*, edited by D. R. Coates and J. D. Vitek. Stroudsburg, Pa.: Dowden, Hutchinson and Ross, 1980, pp. 415–433.

Wilshire, H. G., et al. "Field Observations of the December 1977 Wind Storm, San Joaquin Valley, California." In *Desert Dust: Origin, Characteristics and Effects on Man*, edited by T. L. Péwé, pp. 233–251. Washington, D.C.: Geological Society of America, 1981.

—ANDREW S. GOUDIE

# E

## EARTH HISTORY

The Earth and other nearby planetary bodies formed approximately 4.65 billion years ago as a result of the clustering of condensed particles early in the history of the solar system. As the hot, gaseous material of the early solar nebula began to cool, solid mineral particles condensed like snowflakes. These particles collided with one another as they orbited, and they began to clump together as a result of electrostatic forces. Gradually the clumps grew, and they accumulated more material by gravitational attraction. In the innermost part of the solar system, these bodies eventually became the terrestrial planets: Mercury, Venus, Earth, and Mars.

Early in its history the Earth was largely molten, and the surface was completely covered by a magma ocean. The heat that caused melting came from several sources, including the heat released by compression during planetary formation, kinetic energy from constant meteorite bombardment of the surface, and the decay of radioactive materials within the Earth. During or shortly after accretion, while the Earth was still partially molten, dense material sank to its center under the influence of gravity. This process, called *differentiation*, led to the formation of a dense core of iron-nickel metal. The magma ocean eventually cooled and crystallized to form the mantle, composed of silicate rocks of intermediate density. As it solidified, convection within the mantle slowed considerably, and a thin, lower-density rocky crust began to form at the surface. The cooling and thickening of this rigid outer layer made possible a plate tectonic style of convection, that is, thermal convection of the solid mantle driving the movement of overlying cooler, rigid plates. [*See* Plate Tectonics.]

No rocks older than four billion years have yet been discovered on the Earth, although individual mineral grains as old as 4.2 billion years have been found. The processes of meteorite bombardment, melting, metamorphism, and weathering have apparently obliterated any record of rocks older than this. It may be that, prior to that time, the surface was still unstable and any rocks that solidified were constantly recycled into the magma ocean. Lacking the material evidence of rocks older than four billion years, geologists still place the age of the Earth at approximately 4.65 billion years. This is primarily based on the study of meteorites formed from similar materials, in the same part of the solar system, through the same processes of condensation and accretion that were involved in the formation of the Earth.

The ages of such meteorites cluster closely around 4.65 billion years.

When it was first formed, the Earth had an envelope of gases surrounding it. [*See* Earth Structure and Development.] This primordial atmosphere was swept away early in Earth history by a strong solar wind, the flux of ionized material away from the Sun. However, the Earth still contained abundant volatile material, mostly water, in its interior. Through volcanism, volatile elements were degassed, forming a new gaseous envelope that eventually became our atmosphere. Volcanism still occurs today, from which we may conclude that the Earth is still degassing some volatiles from its interior. In Hadean times (prior to 3.9 billion years ago; see Table 1), however, volcanic activity was very active, and degassing was much more vigorous than it is today.

During early Hadean times there was no liquid water on the surface of the Earth—it was still too hot for water to persist in liquid form. The atmosphere was very dense, atmospheric pressure was much higher than it is today, and there was no free oxygen (that is, diatomic oxygen, $O_2$). The surface of the early Hadean Earth did not present a very hospitable environment for life. Eventually, the Earth cooled and the first rain fell. Surface water may have begun to collect as early as 4.4 billion years ago, eventually forming oceans. [*See* Ocean Structure and Development.] The water reacted with chemicals in the atmosphere to form acids that, in turn, reacted with the rocks of the early crust, dissolving some of their chemical constituents. From that point on, the atmosphere, the hydrosphere, and the crust started a history of mutual interaction and constant chemical readjustment, and the different parts of the Earth system evolved chemically toward their present-day compositions. Before long the biosphere, too, joined this interaction.

At some point during the Hadean, life was initiated on the Earth; exactly when, where, and how is still not known. Life may have formed more than once, but failed to persist because of the hostile environment. The origin of life required the presence of simple organic materials. These may have been present on the Earth, or they may have been delivered by meteorites that bombarded the surface during Hadean times; chemical analysis of certain types of meteorites has revealed that they contain organic compounds. For life to form, it was necessary for these organic molecules to be transformed into amino acids and RNA (ribonucleic acid, crucial for cell replication). The exact mechanism through which

**Earth History. TABLE 1.** The Geologic Time Scale

| EON | ERA | PERIOD | MILLIONS OF YEARS AGO |
|-----|-----|--------|----------------------|
| Phanerozoic (from the Greek for "visible life") | Cenozoic (from the Greek for "recent life") | Quaternary | |
| | | Tertiary | 1.6 —— |
| | | | 65 —— |
| | Mesozoic (from the Greek for "middle life") | Cretaceous | 144 —— |
| | | Jurassic | 213 —— |
| | | Triassic | 248 —— |
| | Paleozoic (from the Greek for "ancient life") | Permian | 286 —— |
| | | Pennsylvanian | 320 —— |
| | | Mississippian | 360 —— |
| | | Devonian | 408 —— |
| | | Silurian | 438 —— |
| | | Ordovician | 505 —— |
| | | Cambrian | 545 —— |
| Proterozoic | | | 2,500 —— |
| Archean | | | 3,900 —— |
| Hadean | | | 4,560 —— |

this transformation occurred is not understood, but it must have required an appropriate chemical environment and a source of energy. One form of energy that was common during Hadean times was lightning; another was thermal energy from hot springs or sea floor vents. In any event, isotopic evidence in the rock record indicates biological activity of primitive bacteria (prokaryotes) by the end of the Hadean eon, 3.9 billion years ago (Figure 1). [*See* Evolution of Life; *and* Origin of Life.]

**The Archean and Proterozoic Eons: Early Life.** Single-celled blue-green algae (cyanobacteria) are fossilized in 3.5-billion-year-old rocks from Australia. These organisms probably consumed carbon dioxide and emitted oxygen, forming their own food energy through photosynthesis. Photosynthesis eventually transformed the chemistry of the atmosphere and hydrosphere; most of the free oxygen presently in the atmosphere has come from photosynthesis by plants. However, in early Archean times there was still no diatomic oxygen ($O_2$) in the atmosphere. The Archean atmosphere most likely contained abundant water vapor ($H_2O$) and carbon dioxide ($CO_2$). These are greenhouse gases, which absorb infrared radiation (heat) and cause it to build up near the surface of the Earth. The greenhouse effect resulted in much warmer surface temperatures in the Archean than we experience today, in spite of the fact that the Sun's luminosity was considerably less at that time. Warm, greenhouse-dominated periods of Earth history such as this are referred to as *hot houses*.

Beginning about 2.5 billion years ago, the Earth experienced a prolonged cool period, one of several major cool periods in Earth history that are referred to as *ice houses*. This ice house, which marks the Archean–Proterozoic boundary, was most likely caused by a decrease in the concentration of carbon dioxide in the atmosphere. Lower atmospheric carbon dioxide, in turn, may have resulted partly from a decrease in the vigor of plate tectonic activity as the crust of the Earth thickened and began to stabilize into continental masses separated by ocean basins. With less vigorous tectonic activity, there would have been less active volcanism, and therefore a waning of volcanic emissions of carbon dioxide.

Carbon dioxide is also actively removed from the atmosphere by photosynthetic organisms. [*See* Carbon Cycle; *and* Carbon Dioxide.] Could photosynthesis have been responsible for the removal of significant quantities of carbon dioxide from the Proterozoic atmosphere? Probably not; other, more effective processes were required. Carbon dioxide is removed from the atmosphere quite efficiently through the chemical weathering of rocks. The products of chemical weathering are transported by rivers in the form of dissolved matter, eventually accumulating as ocean-floor sediments. The burial of carbon-bearing sediments sequesters the carbon from further (short-term) interaction with the atmosphere. After the development of life, organic matter accumulated in the same manner on the ocean floor, forming limestones and organic-rich sediments.

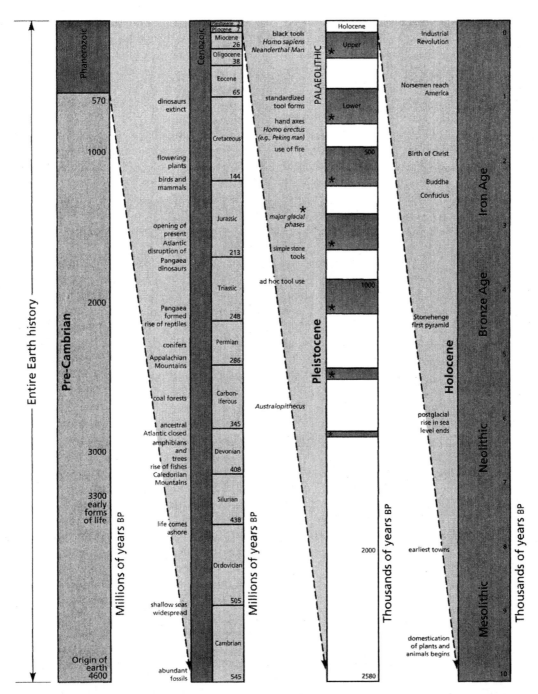

**Earth History. Figure 1.** Entire Earth history, with expanded views of (1) Phanerozoic, all history since beginning of evident life in Cambrian time; (2) Pleistocene, since 2.5 million years ago; and (3) Holocene, the past ten thousand years, essentially postglacial time. (Modified after Barraclough and Stone, 1991, p. 2. With permission.)

Limestones and organic sediments presently function as a reservoir for an extremely large mass of carbon, virtually all of which originated as atmospheric carbon. Its removal and sequestration inevitably led to a decrease in the effectiveness of greenhouse warming, and a lowering of surface temperatures in early Proterozoic times. The mechanisms that remove carbon dioxide from the atmosphere (weathering and the formation of limestones, burial of organic sediments, and photosynthesis) and the mechanisms that add carbon dioxide to the atmosphere (volcanic emissions, respiration, and decay) are part of the long-term carbon cycle. The climatic history of the Earth and the transitions from hot house to ice house and back again have been influenced significantly by shifts in the rates and magnitudes of these processes relative to one another. Thus, in the middle Proterozoic, weathering rates decreased as a result of cool ambient temperatures, shifting the balance back toward hot house conditions once again.

Throughout Archean and early Proterozoic times, much of the oxygen produced by photosynthetic organisms reacted with reduced compounds in the atmosphere or combined with reduced iron dissolved in ocean water; in other words, the oxygen was scavenged before it had a chance to build up in the atmosphere. By mid-Proterozoic times (about two billion years ago), however, the composition of the atmosphere began to resemble more closely its modern-day composition. The ocean, which is coupled to the atmosphere and functions with it in a state of dynamic chemical equilibrium, must also have become oxygenated during this period. Along with the buildup of diatomic oxygen in the atmosphere came an increase in ozone ($O_3$). When ozone was abundant enough to function as a filter, screening out harmful ultraviolet radiation, organisms were able to survive and flourish in shallow waters and, eventually, on land. This critical stage in the evolution of the atmosphere was reached around 600 million years ago. The fossil record shows an explosion of life forms at that time, marking the transition from the Proterozoic eon to the Phanerozoic eon. The end of the Proterozoic was marked by a second major ice house.

**The Phanerozoic Eon: Life Takes Hold.** Much of our detailed understanding of Earth history, and most of what we understand about the evolution of life, comes from rocks and fossils of Phanerozoic age (from the Paleozoic, Mesozoic, and Cenozoic eras) preserved in the geologic record. Note, however, that almost 90 percent of Earth history happened in pre-Phanerozoic times.

The early Paleozoic era was a hot house, caused by an increase in the vigor of plate tectonic activity and carbon dioxide emissions from volcanoes. By the late Paleozoic, all of the continents had drifted together into one great supercontinent known as *Pangaea*. Life evolved rapidly during this time. A great diversity of marine invertebrates dominated Paleozoic seas. In the Silurian period, plants first emerged on land, followed by invertebrate animals (including insects), and finally by vertebrates in the late Devonian period. Fishes, reptiles, and amphibians also diversified greatly during the Devonian. During the Paleozoic hot house, tropical swamps covered vast areas of the land; these were eventually transformed into most of the world's great coal deposits.

The Mesozoic era is characterized by the breakup of Pangaea into a southern supercontinent (called *Gondwana*) and a northern supercontinent (called *Laurasia*), and the subsequent drifting apart of the continents toward their modern-day positions. This allowed for the separate evolution of life-forms on continents that had previously been joined together. Hot-house conditions pertained throughout much of the Mesozoic era, another great coal-forming period. Flowering plants emerged during this time (in the Cretaceous period), and dinosaurs roamed the Earth during the Jurassic and Cretaceous periods.

The beginning of the Cenozoic era signaled yet another ice house, culminating in the great glaciations of the Pleistocene epoch. Modern mammals such as the horse and camel evolved during Cenozoic times, the age of mammals. The oldest known hominid fossils are around 4.4 million years old. The first members of our own subspecies, *Homo sapiens*, appeared around 120,000 years ago.

Embedded in the fossil record of the Phanerozoic eon is a history of rapid adaptation and recovery following catastrophic episodes (*mass extinctions*) in which many species became extinct within a geologically short time. One of the most famous mass extinctions occurred 65 million years ago, at the boundary between the Cretaceous and Tertiary periods. In that extinction, as many as one quarter of all known animal families, including the dinosaurs, became extinct. The most devastating extinction occurred 245 million years ago at the end of the Permian period, when as many as 96 percent of all species died out. On the basis of observational evidence, many scientists argue that some mass extinctions (notably the Cretaceous–Tertiary extinction) may have been caused by giant meteorite impacts. [*See* Impacts by Extraterrestrial Bodies.] Other scientists believe that mass extinctions were more likely to have been caused by extremely voluminous volcanic eruptions, or by rapid climatic or other environmental changes. The study of mass extinctions seems particularly relevant today, given the present rate of extinction of species as a result of human activities, and questions that have arisen concerning the impacts of global climatic change on species survival. [*See* Extinction of Species.]

[*See also* Atmosphere Structure and Evolution; Biological Diversity; Biosphere; *and* Oxygen Cycle.]

### INTERNET RESOURCES

Paleomap Project, Christopher R. Scotese, http://www.scotese.com.
University of California at Berkeley, Geological Time Machine (Museum of Paleontology), http://www.ucmp.berkeley.edu/help/timeform.html.

### BIBLIOGRAPHY

Barraclough, G., and N. Stone, eds. *The Times Atlas of World History.* Times Books, 1991.
Broecker, W. S. *How to Build a Habitable Planet.* Palisades, N.Y.: Eldigio Press, 1985.
Mackenzie, F. *Our Changing Planet.* Upper Saddle River, N.J.: Prentice-Hall, 1998.

—BARBARA MURCK

# EARTH MOTIONS

The motions of Earth, particularly with respect to the Sun, are a fundamental control of many of the environmental changes that have taken place through geologic time. The Earth rotates about its axis and moves around the Sun in an elliptical orbit. Although the gravitational constant is small ($G = 6.670 \times 10^{-11} \, \mathrm{N \, m^2 \, kg^{-2}}$), the law of gravity, mathematically described by Isaac Newton (1643–1727), controls these orbital and rotational motions because large masses and enormous distances are involved.

Broadly speaking, the orientation of the orbital plane (called the ecliptic) is more or less fixed in space. The axis of the Earth's instantaneous rotation is inclined to the ecliptic by about 66°. The elliptical motion of the Earth about the Sun is principally due to the interaction between these two bodies, while the motion of the Moon about the Earth is mainly a consequence of their mutual attraction (as demonstrated by the laws by Johannes Kepler, 1571–1630).

However, the real problem is far more complex. All the parameters that describe these Earth motions vary on many different time scales. A complete discussion of them requires a treatment of the mutual gravitational interactions between the Earth, the Moon, the Sun, and the planets. [See Planets.] In addition, the shapes of the Earth and the Moon must be taken into account because these bodies are not perfectly spherical.

Using theorems on the time variation of the total momentum and of the total angular momentum, it is possible to show that the Earth rotates about its center of mass independently of the displacement of this point. The motion of the center of mass is conditioned by the external forces acting upon the Earth, while the rotation of the Earth about its center of mass is conditioned by the total moment of these external forces about this point. Hence these two motions can be examined separately. The hypothesis that the planets attract each other as if the mass of every one of them were concentrated in its center of mass is valid, except for the Moon's motion around the Earth. In contrast, the flattening of the planets does have a perceptible effect on their rotation.

**Rotation of the Earth.** The rotation of the Earth about its axis, taken for the moment to be fixed in space, mirrors the daily passage of the stars around the sky, which by definition takes one sidereal day. In recent years, this period of rotation of the Earth measured by physical time standards (such as atomic clocks) has been observed to be not constant, showing partly irregular, partly seasonal fluctuations. Typically, the length of the day fluctuates by about one thousandth of a second. On the decadal time scale, the most likely explanation of these fluctuations is related to interactions at the core–mantle boundary between the magnetic field

in the core fluid and the electrically conducting lower mantle (Lambeck, 1981). Of the higher-frequency fluctuations, the annual and semiannual behavior is mainly caused by an exchange of angular momentum in the general circulation of the atmosphere with the angular momentum of the solid Earth (Hide et al., 1997). But there is a wide range of phenomena that may perturb the Earth's rotation, including the secular tidal torques. The Moon's gravitational attraction on the Earth causes two small bulges directed toward the Moon and away from it in the ocean layer and on the solid Earth. Dissipative processes cause a lag in the tidal response and a torque is exerted that does not vanish when averaged over an orbital period of the Moon. The consequence of this torque is a change in the Earth's angular momentum or, equivalently, an increase in the length of the day, currently by about one thousandth of a second in a hundred years. At the same time, the bulge slows the Moon down in its orbital motion and leads to an increase in the Earth–Moon distance of the order of a few centimeters per year (four hundred million years ago, the length of the day was about twenty-two hours, the year consisted of about four hundred days, and the Moon must have been 4 percent closer to the Earth).

Viewed by an observer on the Earth near the North Pole, the stars appear to trace out concentric circles whose center defines the celestial North Pole, the extension of the Earth's rotational axis. The celestial North Pole currently lies close to the star Polaris. As already noted by Hipparchus in about 120 BCE, the rotational axis is observed slowly to trace out a cone, clockwise, with a half-angle of 23.5° about the pole of the ecliptic, which takes about 25,700 years to go full circle around the heavens. This steady motion of the rotational axis in space is termed the precession of the Earth (or general precession in longitude). This is due to the inclination of the major axis of the oblate Earth to the ecliptic. Consequently, the net gravitational force on the Earth due to the Sun does not pass through the center of mass of the Earth. This results in a torque being exerted about the center. The torque attempts to draw the equator into the plane of the ecliptic, but the spinning of the Earth resists this; instead the torque achieves a motion of the spin axis about the pole of the ecliptic. The observed precession is the sum of the solar and lunar torques (because of the large mass of the Sun and the proximity of the Moon) plus a rather minor contribution arising from the other planets.

In addition, the complex interplay of the solar and lunar orbits induces small oscillations in the secular precessional motion of the rotational axis; these oscillations are known as forced nutations. The principal nutation term arises from a nineteen-year periodicity in the inclination of the Moon's orbit.

But observations also show that the Earth as a whole

appears to wobble about its axis. This motion, predicted by Leonhard Euler (1707–1783), is referred to as the polar motion. This consists primarily of two periodic oscillations, one with a fourteen-month period (a free oscillation referred to as the Chandler wobble) and the other with a twelve-month period. All of these precessional and nutational motions provide very important information on the internal structure of the Earth.

**Eccentricity, Precession, and Obliquity.** The incoming solar radiation received over the Earth has a diurnal variation due to the rotation of the Earth around its axis and an annual periodic variation due to the Earth's elliptical motion around the Sun. In addition, the seasonal and latitudinal distributions of this solar radiation have a long-period variation due to the long-term variations in three orbital elements (Berger et al., 1993): the eccentricity $e$, a measure of the shape of the Earth's orbit around the Sun; the obliquity $\epsilon$, the tilt of the equator with respect to the plane of the Earth's orbit; and the climatic precession $e \sin \tilde{\omega}$, a measure of the Earth–Sun distance at the summer solstice ($\tilde{\omega}$ is the longitude of the perihelion measured from the moving equinox, Figure 1). As determined from celestial mechanics, the secular variations of these elements of the Earth's orbit and rotation are due to the gravitational perturbations that the Sun, the Moon, and the other planets exert on the Earth's orbit and on its axis of rotation.

A first approximate solution for the orbit of the planets was given in the seventeenth century by the German astronomer Johannes Kepler. But it was only in the second half of the twentieth century that quite accurate solutions of the complete sets of equations of Joseph Louis Earl de Lagrange (1736–1813) and Siméon Denis Poisson (1781–1840) could be found. At decadal to century time scales, secular planetary theories are the classical

methods built for the ephemerides (tables providing for each day the numerical value of some astronomical parameters, like coordinates of the planets and of the Sun) and used in astronomical observatories (Bretagnon, 1982).

For long-term variations over the last few million years, the equations for the eccentricity, the precession, and the obliquity have been integrated numerically (Laskar et al., 1993a; Quinn et al., 1991) or expressed in trigonometrical form as quasiperiodic functions of time (Bretagnon, 1974; Berger, 1978). In this case one has:

$$e \sin \tilde{\omega} = \Sigma P_i \sin(\alpha_i t + \eta_i) \tag{1}$$

$$\varepsilon = \varepsilon^* + \Sigma A_i \cos(\gamma_i t + \zeta_i) \tag{2}$$

$$e = e^* + \Sigma E_i \cos(\lambda_i t + \phi_i). \tag{3}$$

The amplitudes $P_i$, $A_i$, and $E_i$, the frequencies $\alpha_i$, $\gamma_i$, and $\lambda_i$, and the phases $\eta_i$, $\zeta_i$, and $\phi_i$, are given in Berger (1978) and Berger and Loutre (1991) for three million years.

Let us first stress that the expansions of $e$ and $\varepsilon$ are made in the neighborhood of the constants $e_0$ and $\varepsilon^* = 23.32°$. These average values therefore explain most of the amplitude of the eccentricity and obliquity in absolute terms. This is why the spectrum of $e \sin \tilde{\omega}$ appears to be closely related to the spectrum of $\sin \tilde{\omega}$ and the spectrum of the latitude of the Milankovitch caloric equator,

$$\phi_c = \tan^{-1} \left( \frac{4e \sin \tilde{\omega}}{\pi \sin \varepsilon} \right),$$

is dominated by $e \sin \tilde{\omega}$.

Figure 2 shows the long-term variations of these three astronomical parameters over the past 200,000 years and into the future for the next 100,000 years. Over

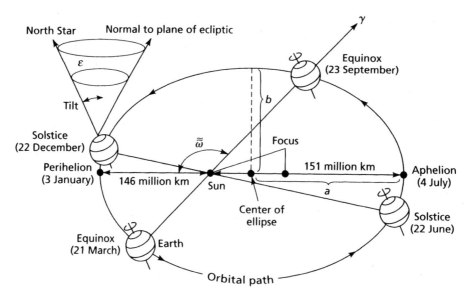

**Earth Motions. FIGURE 1.** Elements of the Earth's Orbit.

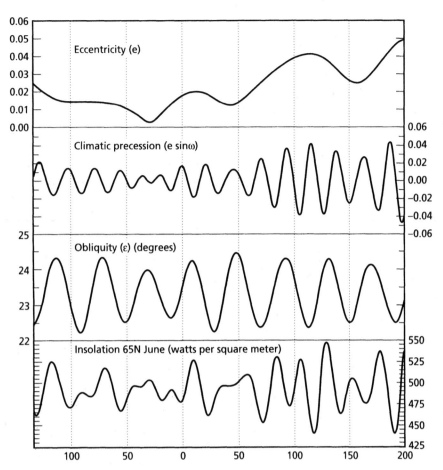

**Earth Motions. FIGURE 2.** Long-term variations of precession, obliquity, eccentricity, and insolation (at 65 degrees north latitude) at the summer solstice from 200,000 years ago to 100,000 years in the future. The data are derived from the astronomical solution given in Berger (1978).

the past three million years, the eccentricity of the orbit varies between near circularity ($e = 0$) and slight ellipticity ($e = 0.07$) at a period whose mean is about 100,000 years. The most important terms in the series expansion occur, however, at 404,000, 95,000, 124,000, 99,000 and 131,000 years and at 2.38 million years in decreasing order of amplitude (Berger and Loutre, 1991). The tilt of the Earth's axis varies between about 22° and 25° at a period of nearly 41,000 years. Although this period corresponds to by far the largest amplitude, there are two other important terms with periods of 54,000 and 29,000 years.

As far as precession is concerned, two components must be considered. The first is the axial precession (the general precession in longitude) with a period of roughly 25,700 years corresponding to the period of the vernal equinox against a fixed reference point. Today, the point around which the stars seem to rotate when viewed from the Northern Hemisphere falls near the star Polaris, which forms the end of the handle of the Little Dipper, but in 2,000 BCE the North Pole pointed to a spot

midway between the Little Dipper and the Big Dipper, and within the next 12,000 years it will point in the direction of Lyra. The second is related to the fact that the elliptical figure of the Earth's orbit is itself rotating counterclockwise in the same plane, leading to an absolute motion of the perihelion whose period, measured relative to the fixed stars, is about 100,000 years (the same as for the eccentricity). The two effects together result in what is known as the climatic precession of the equinoxes, a motion mathematically described by $\tilde{\omega}$ in which the equinoxes and solstices shift slowly around the Earth's orbit relative to the perihelion with a mean period of 21,000 years. This period actually results from the existence of two periods that are close to each other: 23,000 and 19,000 years.

In the insolation formulas used to study past and future astronomical forcing of climate, the amplitude of $\sin \tilde{\omega}$ is modulated by eccentricity in the term $e \sin \tilde{\omega}$. The envelope of $e \sin \tilde{\omega}$ is given exactly by $e$ because the frequencies of $e$ originate strictly from combinations of the frequencies of $\tilde{\omega}$.

Indeed, besides their simplicity and practicality for easy computation, equations (1)–(3) also allow us to obtain directly the main periods associated with the astronomical theory of climate (Berger and Loutre, 1990). The periods calculated in Berger (1978) come from a set of sixteen fundamental periods of the planetary system (divided into two subsets, one related to the frequencies called $g$ and the other to the frequencies called $s$; see, e.g., Berger and Loutre, 1991) and the period of the general precession (which corresponds to a frequency $k = 50.439273$ arc seconds per year). The method used for analytical solution of the differential equations governing the motion of the planets makes it possible to trace each period involved in equations (1)–(3) back to its source. The main astronomic periodicities found in the spectra of geologic records being 100,000, 41,000, 23,000, 19,000, and 54,000 years, their computational origin is given in Table 1.

As we can see, the periods of eccentricity, precession, and obliquity are not independent of each other. For example, the frequencies corresponding to the five most important terms of the eccentricity can be obtained as a linear combination of the frequencies of the most important terms of climatic precession:

$$\lambda_1 = \alpha_2 - \alpha_1$$
$$\lambda_2 = \alpha_3 - \alpha_1$$
$$\lambda_3 = \alpha_3 - \alpha_2$$
$$\lambda_4 = \alpha_4 - \alpha_1$$
$$\lambda_5 = \alpha_4 - \alpha_2$$
$$\lambda_6 = \alpha_3 - \alpha_4$$

As a consequence, we also have

$$\lambda_3 = \lambda_2 - \lambda_1$$
$$\lambda_5 = \lambda_4 - \lambda_1$$

and many other possible combinations of $\lambda$, $\gamma$, $\alpha$, $g$, $s$, and $k$ (Berger and Loutre, 1990).

Although there is not a one-to-one relationship between these periods and the individual planets, it can be shown that the origin of each period, although due to the mutual interactions between all planets, can to some extent be associated with one or a few planets. One might say that Jupiter, Venus, Mars, and Earth are responsible for explaining the eccentricity periods close to 400,000 and 100,000 years. The period of roughly two million years is related to Mars and the Earth. For the obliquity, Mars, Earth, and the Moon explain the 41,000-year period, and Saturn is related to the 54,000-year period. For the climatic precession, the 23,000-year periods given by $\alpha_1$ and $\alpha_2$ are associated with Jupiter, Venus, and the Moon, and the 19,000-year periods (given by $\alpha_3$ and $\alpha_4$) are related to Mars, Earth, and the Moon.

For more remote times, these solutions must be improved. For a few tens of millions of years, time series of these astronomical parameters can be obtained (Laskar et al., 1993a) in which internal geophysical processes such as mantle convection are better taken into account. On the geologic time scale (hundreds of millions of years), the problem is even more complicated, because secular resonances among the inner planets lead to a chaotic behavior of the solar system (Laskar, 1989) and all the terrestrial planets could have experienced large chaotic variations in obliquity. For example,

**Earth Motions. TABLE 1.** Origin of the Most Important Periods in the Astronomical Theory of Paleoclimates

| | | |
|---|---|---|
| Eccentricity | $\dfrac{1}{412885} = \dfrac{1}{176420} - \dfrac{1}{308043}$ | $\lambda_1 = g_2 - g_5$ |
| | $\dfrac{1}{94945} = \dfrac{1}{72576} - \dfrac{1}{308043}$ | $\lambda_2 = g_4 - g_5$ |
| | $\dfrac{1}{123297} = \dfrac{1}{72576} - \dfrac{1}{176420}$ | $\lambda_3 = g_4 - g_2$ |
| Obliquity | $\dfrac{1}{41000} = \dfrac{1}{68829} + \dfrac{1}{25694}$ | $\gamma_1 = -s_3 + k$ |
| | $\dfrac{1}{53615} = -\dfrac{-1}{49339} + \dfrac{1}{25694}$ | $\gamma_3 = -s_6 + k$ |
| Climatic precession | $\dfrac{1}{23716} = \dfrac{1}{308043} + \dfrac{1}{25694}$ | $\alpha_1 = g_5 + k$ |
| | $\dfrac{1}{18976} = \dfrac{1}{72576} + \dfrac{1}{25694}$ | $\alpha_3 = g_4 + k$ |

SOURCE: From Berger and Loutre (1990).

without the Moon, for a spin period ranging from about twelve to forty-eight hours, the obliquity of the Earth would suffer very much larger chaotic variations that might reach 80°–90° (Laskar et al., 1993b). But for the first terms of the eccentricity, the changes in $g$ are sufficiently small that we may assume the 400,000- and 100,000-year periods are not significantly affected—at least when compared with changes in the precessional frequency $k$, which can be much larger over the Earth's history (Berger et al., 1992). $k$ depends indeed mainly upon the rotational angular velocity of the Earth, the Earth–Moon distance, and the Earth's moments of inertia around the equatorial and polar principal axes of inertia. On the assumption that the impact of the chaotic behavior of the solar system on the astroclimatic periods will remain small before 200 million years ago and that the lunar recession rate can be kept constant, the periods of obliquity and precession are reduced back in time, reaching, at 500 million years ago, for example, 16,000 and 19,000 years for climatic precession and 30,000 years for obliquity, with an astronomical precession of 20,800 years.

**Climates of the Past.** During most of the Earth's history the climate has been warmer than it is today. This warm climate has been interrupted by cold periods, the ice ages, which have, on the geologic time scale, been relatively short, covering only perhaps 5–10 percent of the Earth's whole history. Two million years ago, the Earth entered the current Quaternary ice age. [See Climate Change; and Paleoclimate.]

This ice age is characterized by multiple switches of the global climate between glacials (with extensive ice sheets) and interglacials (with a climate similar or warmer than today by a few degrees Celsius). Throughout the past million years, proxy data demonstrate that successive glaciation–deglaciation cycles have occurred with a dominant quasiperiodicity of 100,000 years. [See Glaciation.] The last cycle goes from the Eemian interglacial, centered roughly 125,000 years ago, to the present-day Holocene interglacial, which peaked 6,000 years ago and includes the last glacial maximum (LGM), which occurred 20,000 years ago. In the Northern Hemisphere, the LGM world differed strikingly from the present because of the huge land-based ice sheets, reaching approximately 2 kilometers in thickness and amounting to about 40–45 million cubic kilometers of ice. Sea level was lower by at least 115 meters and the global average surface air temperatures was 5°C below present. Carbon dioxide levels, at roughly 200 parts per million by volume, were much less than two-thirds of their present value (360 parts per million), and aerosol loading was higher than at present (Berger, 1995).

**Long-Term Variations in Solar Radiation.** The combined influence of changes in $e$, $\varepsilon$, and $e \sin \tilde{\omega}$ produces a complex pattern of insolation variations. A detailed analysis of the changes in solar radiation (Berger et al., 1993) shows that it is principally affected by variations in precession, although the obliquity plays a relatively more important role for high latitudes, mainly in the winter hemisphere. [See Sun.]

Changes in incoming solar radiation due to changes in tilt are the same in both hemispheres during the same local season: an increase in $\varepsilon$ leads to an increase in insolation in the summer hemisphere and a decrease in the winter hemisphere. As the strength of the effect is small in the tropics and maximal at the poles, obliquity increases tend to amplify the seasonal cycle in the high latitudes of both hemispheres simultaneously. The precession effect can cause warm winters and cool summers in one hemisphere while causing the opposite effects in the other hemisphere. For example, during current winters, portions of the Northern Hemisphere receive as much as 10 percent more insolation than 11,000 years ago, when the perihelion occurred in the Northern Hemisphere summer. Moreover, because the length of the seasons varies in time according to Kepler's Second Law, the solstices and equinoxes occurred at different calendar dates during the geologic past and will alter in the future. Currently, in the Northern Hemisphere, the longest seasons are spring (ninety-two days nineteen hours) and summer (ninety-three days fifteen hours), while autumn (eighty-nine days twenty hours) and winter (eighty-nine days) are notably shorter. In about 1250 CE, spring and summer had the same length (as did autumn and winter) because the winter solstice was occurring at the perihelion. About 4,500 years into the future, the Northern Hemisphere spring and winter will have the same shorter length and consequently summer and autumn will be equally long (Berger and Loutre, 1994).

Finally it must be stressed that the pattern of solar irradiation at the top of the atmosphere differs significantly from the pattern of radiation absorbed by the surface of the Earth, particularly in high latitudes where the surface albedo is large.

**Milankovitch Theory.** The orbital hypothesis of climatic change was quantitatively formulated by the astronomer Milutin Milankovitch in the 1920s and 1930s. His early calculations provided information on the variations in incident solar radiation, as a function of latitude, for the last million years in winter and summer (Milankovitch, 1941). [See the biography of Milankovitch.] He argued that insolation changes in the high northern latitudes during the summer season were critical to the formation of continental ice sheets. During periods when insolation in the summer was reduced, the snow of the previous winter would tend to be preserved—a tendency that would be enhanced by the high albedo of the snow and ice areas. Eventually, the effect of this positive feedback would lead to the formation of persistent ice sheets.

According to the mathematics of insolation, a minimum in the Northern Hemisphere caloric summer insolation at high latitudes requires a Northern Hemisphere summer occurring at the aphelion, a maximum eccentricity that leads to a large distance between the Earth and the Sun at the aphelion, and a minimum obliquity implying a weak seasonal contrast and an increased latitudinal energy gradient between the equator and the poles. Given these conditions, it is suggested that not only would the northern high latitudes remain cool enough in summer to prevent snow and ice from melting, but also that mild winters would allow a substantial evaporation in the intertropical zone and, thus, abundant snowfalls in temperate and polar latitudes, the humidity being supplied there by an intensified general circulation due to a maximal latitudinal energy gradient.

A simple linear version of the Milankovitch model would therefore predict that the total ice volume and climate over the Earth would vary with the same regular pattern as the insolation; this means that the proxy record of climate variations would contain the frequencies of the astronomical parameters that are responsible for changing the seasonal and latitudinal distributions of the incoming solar radiation. Investigations during the last twenty-five years have indeed demonstrated that the 19,000-, 23,000-, and 41,000-year periodicities actually occur in long records of the Quaternary climate (Hays et al., 1976), that the climatic variations observed in these frequency bands are linearly related to the orbital forcing functions, and that there is a fairly consistent phase relationship between insolation, sea-surface temperature, and ice volume (Imbrie et al., 1992). The geologic observation of the bipartition of the precessional peak, confirmed in astronomical computations by Berger (1978), was one of the first delicate and impressive tests of the Milankovitch theory.

Despite these discoveries in support of the Milankovitch hypothesis, the same investigations over the past million years have identified the largest climatic cycle as being 100,000 years. This eccentricity cycle is very weak in the insolation, and cannot be related to orbital forcing by any simple linear mechanism. The variance components centered near this 100,000-year cycle seem to be in phase with the eccentricity cycle, but its exceptional strength in the climatic record demands non-linear amplification. Moreover, the 100,000-year cycle disappears before one million years ago, at a time when the ice sheets were much less extensive across the Earth, and the shape of the spectra and the phase lags in the climate response to orbital forcing depend upon the location of the deep-sea core and the nature of the climatic parameter determined (Imbrie et al., 1992, 1993).

Since the publication of the papers by Hays et al. (1976), a number of modeling studies have attempted to explain the relationship between astronomical forcing and climatic change. The results are encouraging. For example, the Louvain-la-Neuve 2.5-dimension model, when forced by both insolation and carbon dioxide, reproduces the entrance into glaciation at about 2.75 mil-

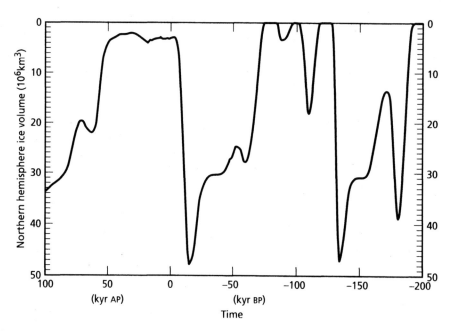

**Earth Motions. Figure 3.** Long-term variations of the Northern Hemisphere ice volume from 200,000 years ago to 100,000 years in the future.

The data are from a simulation by the Louvain-la-Neuve fully coupled climate model. For the future a scenario based upon Vostok $CO_2$ is used (Berger and Loutre, 1996).

lion years ago, the initiation of the 100,000-year cycle at about 900,000 years ago, and the gross features of the last glacial–interglacial cycles. This model predicts that our Holocene interglacial will be exceptionally long, with a high probability of lasting tens of thousands of years into the future (Figure 3; Berger and Loutre, 1996).

## BIBLIOGRAPHY

Berger, A. "Long-Term Variations of Daily Insolation and Quaternary Climatic Changes." *Journal of Atmospheric Science* 35 (1978), 2362–2367.

———. "Modelling the Response of the Climate System to Astronomical Forcing." In *Future Climates of the World: A Modelling Perspective*, edited by A. Henderson-Sellers, pp. 21–69. Amsterdam: Elsevier, 1995.

Berger, A., and M. F. Loutre. "Origine des Fréquences des Éléments Astronomiques Intervenant Dans Le Calcul de L'insolation." *Bulletin de la Classe des Sciences, Académie Royale de Belgique 6e Série, Tome I* 1–3 (1990), 45–106.

———. "Insolation Values for the Climate of the Last 10 Million Years." *Quaternary Science Reviews* 10 (1991), 297–317.

———. "Long-term Variations of the Astronomical Seasons." In *Topics in Atmospheric and Interstellar Physics and Chemistry*, edited by C. Boutron, pp. 33–61. Les Ulis, France: Les Editions de Physique, 1994.

———. "Modeling the Climate Response to the Astronomical and $CO_2$ Forcings." *Comptes Rendus de l'Académie des Sciences de Paris* 323, Série Iia (1996), 1–16.

Berger, A., M. F. Loutre, and J. Laskar. "Stability of the Astronomical Frequencies Over the Earth's History for Paleoclimate Studies." *Science* 255 (1992), 560–566.

Berger, A., M. F. Loutre, and Ch. Tricot. "Insolation and Earth's Orbital Periods." *Journal of Geophysical Research* 98 (1993), 10341–10362.

Bretagnon, P. "Termes Ö Longues Périodes dans le Systäme Solaire." *Astronomy and Astrophysics* 30 (1974), 141–154.

———. "Théorie du Mouvement de L'Ensemble des Planätes, Solution VSOP82." *Astronomy and Astrophysics* 114 (1982), 278–288.

Hays, J. D., J. Imbrie, and N. J. Shackleton. "Variations in the Earth's Orbit: Pacemaker of the Ice Ages." *Science* 194 (1976), 1121–1132.

Hide, R., J. O. Dickey, S. L. Marcus, R. D. Rosen, and D. A. Salstein. "Atmospheric Angular Momentum Fluctuations during 1979–1988 Simulated by Global Circulation Models." *Journal of Geophysical Research* 102.D14 (1997), 16423–16438.

Imbrie, J., E. A. Boyle, S. C. Clemens, A. Duffy, W. R. Howard, G. Kukla, J. Kutzbach, D. G. Martinson, A. Mcintyre, A. C. Mix, B. Molfino, J. J. Morley, L. C. Peterson, N. G. Pisias, W. L. Prell, M. E. Raymo, N. J. Shackleton, and J. R. Toggweiler. "On the Structure and Origin of Major Glaciation Cycles. 1. Linear Responses to Milankovitch Forcing." *Paleoceanography* 7 (1992), 701–738.

———. "On the Structure and Origin of Major Glaciation Cycles. 2. The 100,000-Year Cycle." *Paleoceanography* 8 (1993), 699–735.

Lambeck, K. "The Gravitational Mechanics of the Earth." In *Cambridge Encyclopedia of Earth Sciences*, edited by D. G. Smith, pp. 93–108. Cambridge, U.K.: Cambridge University Press, 1981.

Laskar, J. "A Numerical Experiment on the Chaotic Behaviour of the Solar System." *Nature* 338 (1989), 237–238.

Laskar, J., F. Joutel, and F. Boudin. "Orbital, Precessional, and Insolation Quantities for the Earth from $-20$ Myr to $+10$ Myr." *Astronomy and Astrophysics* 270 (1993a), 522–533.

Laskar, J., F. Joutel, and P. Robutel. "Stabilization of the Earth's Obliquity by the Moon." *Nature* 361 (1993b), 615–617.

Milankovitch, M. *Kanon Der Erdbestrahlung*. Royal Serbian Academy Special Publication 132, Section of Mathematical and Natural Sciences, vol. 33, 1941. Published in English by Israel Program for Scientific Translation for the United States Department of Commerce and the National Science Foundation. Washington, D.C., 1969.

Quinn, Th. R., S. Tremaine, and M. Duncan. "A Three Million Year Integration of the Earth's Orbit." *Astronomical Journal* 101 (1991), 2287–2305.

—A. BERGER

# EARTHQUAKES

An earthquake is defined classically as a shaking of the ground. Such a definition is not particularly useful, however, and can be replaced by the more relevant definition that an earthquake is the sudden rupturing of rock or sudden, rapid movement along a fault that releases stored elastic strain energy. The concept that energy (in the form of stored elastic strain) builds up along a fault zone, to be released catastrophically during an earthquake, is termed the *elastic rebound theory*. This idea that faults have a natural resistance to movement as a result of confining forces and fault friction and can thus accumulate strain, leading to the sudden release of strain and elastic rebound—an earthquake—is attributed to H. F. Reid (1910), who studied the 1906 San Francisco earthquake and saw evidence of this process in the offset of fences, orchards, and other markers along the trace of the San Andreas fault.

The devastation and loss of life from earthquakes can be enormous. Some of the worst natural disasters historically and in modern times have been related to earthquakes. For example, earthquakes in China caused the loss of at least 830,000 lives in 1556, 200,000 in 1927, and perhaps as many as 655,000 in 1976. Even in cases of relatively minor loss of life, the economic impact of earthquakes such as those at Loma Prieta and Northridge (California), and Kobe (Japan) has been enormous. [See Natural Hazards.]

Earthquakes can occur virtually anywhere on Earth, although the majority occur along discrete linear trends (Figure 1). Since earthquakes reflect relative displacements along faults, these regions of high seismicity map the locations of significant relative motion between sections of the Earth's crust and lithosphere. Hence earthquakes are a primary indicator of the location of plate boundaries and the specific way in which the plates are moving with respect to each other across that plate boundary. The relatively aseismic interiors of the regions (plates) indicate that the internal deformation of

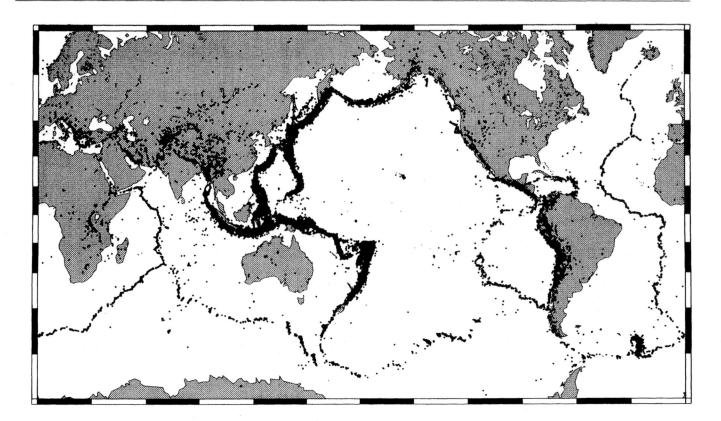

**Earthquakes. FIGURE 1.** Earthquakes of Magnitude Larger than 5.5 Richter during the Period 1973–1998.

Approximately 40,000 earthquakes in this magnitude range occurred during that interval. Plate boundary regions are delineated by the linear trends.

---

plates is small relative to the displacements that occur along the plate boundaries. Not all plates exhibit perfectly rigid behavior. In particular, several regions of the continents (Himalaya–Tibet, the Mediterranean–Alpine zone, and the western United States) exhibit fairly high levels of seismic activity, compatible with other evidence of significant tectonic activity and deformation. [*See* Plate Tectonics.]

The three-dimensional location of an earthquake is called its *focus*. This marks the initiation point of the earthquake. Earthquakes rupture a finite area of a fault zone, while the centroid marks the three-dimensional location of the center of energy release from the earthquake. The hypocenter marks the location of the focus determined by computer earthquake-location programs. The epicenter is the location on the Earth's surface directly above the focus. Most earthquakes occur at fairly shallow levels, with focal depths less than 15–20 kilometers, although earthquakes can occur as deep as nearly 700 kilometers in subduction zones. The inclined plane of seismicity linking shallow earthquakes with deep events in subduction zones is called a *Wadati–*

*Benioff zone* (Figure 2), after the Japanese and American geophysicists who discovered the pattern.

Most shallow earthquakes fit the elastic rebound theory model. They occur along faults where friction resists motion until the strain produces sufficient forces to exceed the fault strength. The maximum depth at which earthquakes occur can be linked to the point at which rocks can more easily deform by ductile (aseismic) rather than frictional mechanisms and thus strain does not accumulate to produce seismic rupture. This transition from brittle (seismic) to ductile (aseismic) deformation is largely controlled by the temperature

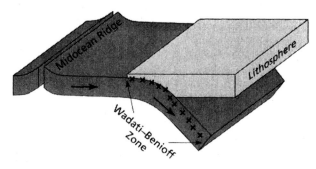

**Earthquakes. FIGURE 2.** Location of Wadati–Benioff Zone in a Subducting Plate.

At shallow levels it marks the boundary between the subducting and overriding plate; at deeper levels seismicity occurs within the subducting slab.

structure; in regions of higher crustal temperatures, earthquakes cease at shallower depths than in regions of colder crust.

Wadati–Benioff-zone earthquakes occurring at depths shallower than 60–70 kilometers are also compatible with the elastic rebound theory model; they occur to greater depths than in most other regions because the subduction process transports cold lithosphere to depth, depressing temperatures and allowing brittle deformation along much of the interface between the two plates. The deeper earthquakes in subduction zones are something of an enigma since they occur within the subducting plate and represent deformation of the plate as it subducts. Many of these earthquakes have been attributed to mineralogical phase changes within the subducting slab as it undergoes a significant thermal and pressure evolution. Recently, deep earthquakes beneath Bolivia and Tonga ruptured fault planes of significant areal extent, seemingly incompatible with the phase-change model for deep earthquakes.

Several approaches are used to measure the size of an earthquake. Intensity measures the local (to the observer) effects of the earthquake. The standard intensity measure is the Modified Mercalli (MM) Intensity Scale, which is a twelve-point scale describing the effects on people and structures. The scale ranges from MM I, which is not felt except by a very few under special conditions, through MM VI, which is felt by all with many frightened, and heavy furniture moves, to MM XII, where damage is total, and objects are thrown into the air. Although the Mercalli intensity measure is easily applied and of particular utility for studying historical, preinstrumental earthquakes, it does not quantify the earthquake itself; a smaller earthquake nearby may have a greater MM intensity than a very large earthquake farther away.

To develop a more objective measure of the size of an earthquake, Charles F. Richter proposed a magnitude scale (now traditionally referred to as Richter magnitude) based on the amplitude of ground shaking recorded on a standardized seismograph (Wood–Anderson) a standardized distance (100 kilometers) from the event. The Richter magnitude was originally applied to southern California, but was later adapted to be applicable to other regions. Since the 1930s, instrumentation has changed and other characteristics of the seismogram have been incorporated into various magnitude scales. All share one characteristic with the Richter scale, namely that the magnitude varies logarithmically with amplitude on the seismogram; however, since they measure different aspects of the seismogram, they do not necessarily produce the same value for magnitude. A unit increase in magnitude represents a factor-of-ten increase in amplitude. The energy in an earthquake increases in a different fashion from ground amplitude. For each unit increase in mag-

nitude, the seismic energy increases by a factor of approximately thirty-two. Finally, the magnitude scale developed in this way tends to become saturated for very large earthquakes, underestimating the true size, and so other systems have been developed to quantify large earthquakes to circumvent this limitation.

The measure of earthquake size generally used by seismologists is the seismic moment. The seismic moment $M_0$ is defined as the product of the fault displacement ($u$), the fault area ruptured ($A$), and the shear modulus ($\mu$):

$$M_0 = \mu\, u\, A.$$

Traditionally this value is given in units of dyne centimeters. The value for $M_0$ can be determined by the above formula, but is more typically determined by analyses of the seismic waveforms recorded by seismographs. In this way the fundamental properties of the earthquake of displacement and area can be obtained directly from the seismogram. The two largest recorded earthquakes (of the instrumental era) are the 1960 Chile earthquake, with $M_0 = 2.5 \times 10^{30}$ dyne centimeters, and the 1964 Alaska earthquake, with $M_0 = 7.5 \times 10^{29}$ dyne centimeters. A magnitude scale derived from seismic moment (moment magnitude $M_w$) has been developed; it is defined as

$$M_w = \frac{2}{3} \log_{10}(M_0) - 10.7.$$

$M_w$ does not suffer from the problems of other magnitude scales for large earthquakes. The two earthquakes described above have moment magnitudes ($M_w$) significantly greater than their Richter ($M_s$) magnitudes (Chile: $M_w = 9.6$, $M_s = 8.5$; Alaska: $M_w = 9.2$, $M_s = 8.3$). Since the size (seismic moment) of earthquakes depends on the area of fault surface ruptured, most great earthquakes occur along subduction zones that can have rupture areas extending 50 kilometers or more down the subducting slab along hundreds of kilometers of plate boundary length. Earthquakes along continental strike-slip faults such as the San Andreas typically rupture only 15–20 kilometers in depth, limiting the maximum size.

The number of earthquakes each year varies with magnitude. Since 1900 there have been, on average, one great earthquake ($M_w$ greater than 8.0), eighteen major earthquakes ($M_w = 7.0$–7.9), and thousands of minor earthquakes ($M_w$ less than 5.0) per year. These rates have been fairly constant this century, and so any apparent increase in the number of earthquakes today compared with the past is a consequence of improved instrumentation rather than any real increase in the rate of seismicity.

Since earthquakes are related to displacements on faults, they are classified according to the type of fault they represent. The three main categories are thrust,

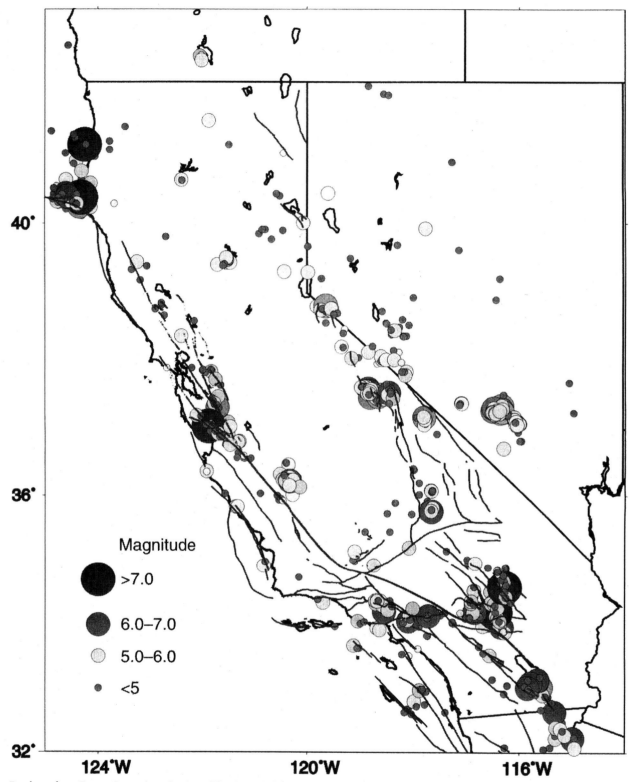

**Earthquakes. FIGURE 3.** Earthquakes in California and the Surrounding Region of Magnitude Greater than 4.5 Richter during the Period 1973–1998.

The trend of the San Andreas is mapped by earthquakes in sections of California, but the pattern is substantially more diffuse in southern California—reflecting the more complex faulting patterns in that region.

normal, and strike-slip earthquakes, with intermediate types that show characteristics of two of the categories. Thrust earthquakes are characteristic of the shallow (depth less than 100 kilometers) parts of subduction zones. Normal earthquakes occur in regions of extension such as midocean ridges and continental rifts. Strike-slip earthquakes occur along the transform faults linking segments of midocean ridges and some of the major faults that cut through the continents, such as the San Andreas in California, the Alpine fault in New Zealand, and the Anatolian fault in Turkey.

Although eight out of the ten largest earthquakes that have affected North America occurred in the Alaska–Aleutian subduction zone, the San Andreas fault is perhaps the most renowned and feared earthquake region in the United States. The 1906 San Francisco earthquake was not large by global standards ($M_w = 7.7$), but its impact on the development of California and the scientific field of seismology cannot be overstated. The combination of a large population and a well developed, seismically active plate boundary fault system make the San Andreas the focus of significant seismic research. Although the majority of earthquakes along the San Andreas (Figure 3) are dominantly strike-slip, the 1989 Loma Prieta earthquake near San Francisco had significant thrust displacement, and the 1994 Northridge earthquake in the Los Angeles region was dominantly thrust (on a previously unmapped fault). Of the fifteen largest earthquakes in the forty-eight contiguous states ($M_w = 7.3$–7.9), eight were associated with the San Andreas, three occurred in the New Madrid region of Missouri in 1811–1812, while the rest occurred within the interior of the western United States.

Most damage that results from earthquakes is a consequence of ground shaking. This shaking and the resulting damage can be amplified as a result of local soil types, geometry of sedimentary basins, and the susceptibility of soils to liquefaction (quicksand-type behavior) during ground shaking. Patterns of damage during the 1989 Loma Prieta earthquake can be largely traced to patterns of soil and susceptibility to liquefaction. The combination of ground shaking and liquefaction leads to substantial landsliding in regions of unconsolidated soils and steep hillsides.

An additional hazard associated with earthquakes in a less predictable way is the tsunami phenomenon. Tsunami (from the Japanese for harbor wave) are generated by undersea earthquakes that rupture the sea floor. These waves can travel long distances across the open ocean virtually undetected (a 1–2 meter amplitude with a 100–200 kilometer wavelength) at great speed (700–1,000 kilometers per hour). As they approach land, the effects of the shallowing seafloor cause a rapid increase in amplitude, leading to devastation of the coast. Tsunami are preferentially generated by large

subduction-zone earthquakes such as the 1960 Chile and 1964 Alaska earthquakes, although some smaller earthquakes (when they rupture slowly) can also produce devastating tsunami.

An ultimate goal of much current earthquake research is to develop ways to predict the occurrence of earthquakes. At present we understand the basic causes of earthquakes and in general we know where earthquakes are most likely to occur; however, we still are unable to accurately predict the time, location, and magnitude of earthquakes. Most attempts at earthquake prediction rely on looking for precursory phenomena, that is, observable events that precede the occurrence of an earthquake. Such potential precursors range from geophysical observations such as changes in the rates of small earthquakes, changes in patterns of ground deformation, changes in water well levels, or changes in the electromagnetic fields in the vicinity of a fault, to observations of anomalous animal behavior. Researchers have instrumented regions around major faults throughout the world in attempts to collect such data in association with earthquakes to search for precursors. The most detailed fault observatory is near the town of Parkfield, California along the San Andreas. That section of the San Andreas experienced a series of $M_s$ 6.0 earthquakes during the twentieth century, and based on that pattern, a similar earthquake was expected to occur about 1990. When that earthquake finally happens, a very detailed set of geophysical, geological, and biological observations before, during, and after the earthquake will be collected. It is hoped that such data will allow earthquake researchers to identify anomalous behavior prior to an earthquake that can be used to improve predictions in other locations.

[See also Land Surface Processes; Risk; and Volcanoes.]

### BIBLIOGRAPHY

Bolt, B. A. *Earthquakes.* New York: Freeman, 1993. A general-purpose, accessible primer on the subject of earthquakes.
Wallace, R.E. *The San Andreas Fault System, California.* Washington, D.C.: United States Geological Survey, 1990. A compilation of articles on the structure and earthquakes of the San Andreas fault system. Contains detailed maps of seismicity and faults.

—KEVIN P. FURLONG

# EARTH STRUCTURE AND DEVELOPMENT

The interior of the Earth is almost completely inaccessible—the deepest mines and boreholes extend only through the upper crust. Yet a surprisingly large amount of information about our planet's deep interior is available. To determine the physical and chemical conditions at depth, Earth scientists employ a variety of indirect investigative approaches, ranging from geophysical studies of the Earth's gravitational and magnetic fields to geochemical studies of lavas and meteorites.

**The Core.** Much of our information about the Earth's deep interior derives from earthquake seismology, the study of seismic waves produced by sudden failure of rock at depth under extreme stress. Seismologists study the travel times, spatial distribution, and other parameters of seismic waves recorded at receiver stations on the Earth's surface to determine the types of material the waves encounter during their transit through the Earth and the depths at which these materials reside. The most useful types of seismic wave for studying the Earth's deep interior are P-waves (primary waves), which can travel through either solids or liquids, and S-waves (secondary waves), which can travel only through solids.

Studies of the distribution of S-waves show that no S-waves are recorded at receiver stations that are at an angular distance greater than about 105° from an earthquake source, creating what is called an *S-wave shadow zone*. Because S-waves cannot travel through liquids, this provides strong evidence that the Earth has a core that is liquid at least at its outer reaches. From the angular distance of 105°, which marks the lower limit of the S-wave shadow zone, the radius of the liquid outer core is calculated to be 3,485 kilometers. P-wave refractions within the core reveal the presence of a discontinuity (abrupt change in seismic properties) within

the core itself. This discontinuity represents the transition from the liquid outer core to the solid inner core, which has a radius of 1,225 kilometers (Figure 1).

The core comprises about 30 percent of the Earth's mass and about 16 percent of its volume, and is believed to be composed predominantly of iron, with lesser amounts of sulfur, nickel, and other elements. A number of lines of evidence support this view. First, from a chemical perspective, it is believed that, with the exception of volatile elements (such as hydrogen, helium, oxygen, and nitrogen), the bulk Earth composition approximates that of the Sun. Compared with the Sun, however, the Earth's crust and mantle are notably deficient in iron, a problem that can be reconciled if much of the Earth's iron inventory is now located in its core. Second, studies of the Earth's gravitational field to determine its mass (about $6 \times 10^{24}$ kilograms), combined with a knowledge of its volume (about $1.1 \times 10^{12}$ cubic kilometers), lead to a calculated mean Earth density of roughly 5.5 grams per cubic centimeter. In comparison, most crustal rocks have densities of about 3 grams per cubic centimeter, and even upon compression deep in the Earth will not yield mean densities as great as that of the mean Earth. This implies that deep within the Earth is a material of much greater density, like that of iron. A third line of evidence comes from studies of meteorites. Meteorites can be divided into two types (with many important subdivisions): meteorites of a rocky composition with silicate minerals called *stones*; and

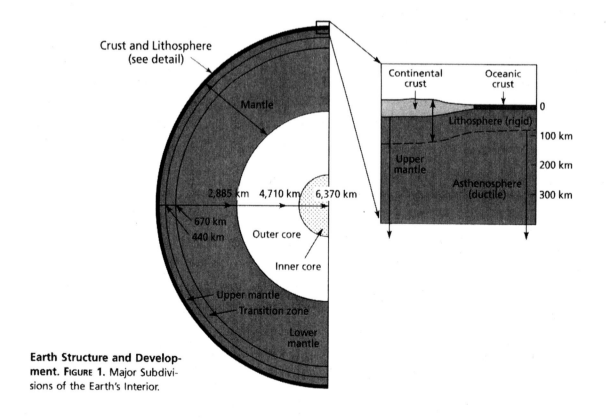

**Earth Structure and Development. Figure 1.** Major Subdivisions of the Earth's Interior.

meteorites predominately of iron–nickel composition called *irons*. It is believed that the irons represent remnants of a now fragmented planetary body, which, like the Earth, developed a metallic iron core. A last and, in some ways, most compelling line of evidence supporting the existence of an iron core is the fact that the Earth has a magnetic field. It is believed that the Earth's magnetic field arises from convection currents in the electrically conducting, molten, outer iron core.

Theories on why the Earth has a molten outer core and a solid inner core rest upon how the physical state of a material (solid, liquid, or gas) varies as a function of temperature and pressure. With increasing pressure (depth) within the Earth, the temperature required to melt a material also increases. At the highest pressures characteristic of the inner core, temperatures are insufficient to melt iron, hence the solid inner core. With increasing distance from the center of the Earth, however, pressure decreases substantially while temperature remains sufficiently high that iron is expected to be in a molten state, hence the liquid outer core. Melting of iron at the depths of the outer core is further favored if the outer core contains other elements, such as sulfur, nickel, and oxygen, which lower the melting temperature of iron.

**The Mantle.** The Earth's mantle extends from a lower boundary with the outer core at about 2,885 kilometers to the base of the crust that forms the thin (5–90 kilometers) outer shell of the Earth. Thus, the mantle comprises the largest component of the Earth, representing about 83 percent of its volume and about 67 percent of its mass. Seismic studies reveal that the Earth's mantle is not homogeneous, but rather exhibits a concentric layering. The boundaries between layers are characterized by abrupt seismic discontinuities that probably represent changes in mantle composition and/or mineral structure. Significant discontinuities exist at depths of about 400 and 670 kilometers, and are used as the basis for subdividing the mantle: the *upper mantle* refers to the region from the base of the crust to 400 kilometers depth; the *transition zone* refers to the region from 400 to 670 kilometers depth; and the *lower mantle* refers to the region from 670 kilometers to the core–mantle boundary.

In contrast to the core, which is entirely inaccessible, the mantle—at least its uppermost 200 kilometers—yields samples for us to study. Mantle xenoliths (foreign rocks) occur embedded in a rare rock type called kimberlite and in some viscous lava flows, which are believed to have been ripped from the upper mantle by the ascending magma or fluid. In addition, at convergent plate boundaries, where tectonic plates collide and can uplift rocks from great depth, portions of the upper mantle are commonly found at the surface. Notable among these are ophiolite complexes, which represent slivers of uplifted upper mantle and oceanic crust. Although highly variable in composition, xenoliths, ophiolites, and other occurrences of uplifted mantle suggest that the upper mantle is composed of peridotite, a rock of ultramafic composition (high in iron and magnesium) consisting largely of the silicate mineral olivine $[(Mg,Fe)_2SiO_4]$, with variable amounts of pyroxenes, spinel, garnet, and other minerals. This view of the composition of the upper mantle is also supported by decades of experimental work showing that partial melting of common peridotites produces melt of basaltic composition, by far the dominant lava type erupted on Earth. With increasing depth in the Earth, minerals that are stable at lower pressure in the upper mantle change in structure and react with other minerals to produce more closely packed mineral structures (e.g., graphite to diamond). These phase changes cause abrupt changes in density and other physical and seismic properties. Mounting evidence suggests that the 400- and 670-kilometer discontinuities, as well as other possible discontinuities, correspond to the predicted depths of such phase changes.

Although the mineralogy of the mantle is certainly related to its elemental composition, important information on the chemical and physical layering of the mantle, as well as on mantle evolution, rests upon studies of trace elements (such as rubidium, strontium, lead) that do not form major constituents of mantle minerals. Much of our understanding of the trace-element composition of the mantle derives from studies of the chemistry of midocean ridge basaltic rocks, from which the composition of their mantle source can be inferred. The vast majority of these rocks are derived from an upper-mantle source that is depleted in incompatible trace elements compared with a calculated primitive (initial) mantle composition. These include a host of elements such as potassium, lead, and the rare-earth elements, and are called *incompatible* because they are not well accommodated in the minerals present and therefore become concentrated in the melt. The depleted nature of midocean ridge basalts contrasts sharply with the enriched trace-element compositions of basalts erupted at hot spots, such as Hawaii or Iceland, which are produced by hot plumes of mantle material rising from the transition zone and/or the lower mantle. Although hot-spot magmas vary in composition, their enriched nature suggests that their deeper source is more primitive in composition and in some cases may have been further enriched by recycling of crustal material to the deep mantle at subduction zones. Thus, the chemical difference between midocean ridge and hot-spot magmas indicates a chemical stratification between the upper mantle and the lower mantle and transition zone.

Since the earliest discussions of continental drift, and more recently with our understanding of plate tectonics, geologists have speculated that the Earth's mantle may convect in the solid state and, in so doing, drive horizontal movements observed at the surface. Convection refers to the buoyancy-driven circulation of a viscous material. A commonly cited example of thermal convection is the heating of a pot of water by a flame below: water heated at the bottom becomes less dense and rises; as it cools along the surface and sides it becomes more dense and descends, establishing a continuous convection cell. Tomography, a branch of seismology that identifies subtle, lateral variations in seismic velocity, suggests that localized hotter and cooler regions that may represent, respectively, hot, less dense, rising convection cell limbs, and cooler, denser, descending limbs are superimposed on the general increase in temperature with depth in the mantle. The dominant heat sources that drive mantle convection are the heat produced by the decay of radioactive elements and continued cooling of the Earth from an initially hotter state.

**The Crust.** The Earth's crust is of two distinct physical and chemical types: oceanic crust and continental crust. The most obvious difference between the two is that the surface of the continental crust is largely above sea level, while oceanic crust is largely below sea level. This seemingly trivial observation masks a fundamental difference in the composition and thickness of the two types of crust, and is governed by the principle of isostacy. Before explaining this principle, it is necessary to define two terms: *lithosphere* and *asthenosphere*. The lithosphere (about 100 kilometers thick, on average) is the rigid outer shell of the Earth and encompasses both the crust (oceanic or continental) and a portion of the upper mantle. In the theory of plate tectonics, it is the effects of the movement of the lithospheric plates that are seen on the surface of the Earth. Beneath the lithosphere lies the asthenosphere, which, although solid, deforms in a ductile way because of its higher temperature and pressure. The concept of isostacy describes the vertical adjustments of the lithosphere, and therefore the elevations of the continents and depths of the oceans, over the yielding asthenosphere in response to variations in the mass of the lithosphere. (This concept is a generalized description of Archimedes' principle, which states that a buoyant body displaces its weight in water.) Because continental crust is both lighter and thicker than oceanic crust, the continental lithosphere has a higher surface elevation than oceanic lithosphere.

*Oceanic crust.* The oceanic crust underlies the deep ocean basins, which comprise some 70 percent of the Earth's surface. Oceanic crust is generally about 6 kilometers thick, but in some areas, such as near fracture zones, it may thin to less than 2 kilometers, and near hot spots it may thicken to as much as 20 kilometers. The average depth below sea level of the oceanic crust is about 5 kilometers, but ranges from above sea level at hot spots to as great as 11 kilometers depth at deep-sea trenches near subduction zones. Because old ocean crust is continuously consumed at subduction zones, the oldest existing ocean crust is only about 200 million years old.

Oceanic crust is formed at midocean ridges, where sea floor spreading causes melting of the mantle. As the plates diverge, mantle material rises to fill the void; mantle that rises adiabatically (that is, no heat enters or leaves the system) retains a temperature similar to its temperature at depth, but, because it is at a reduced pressure, it begins to melt, probably at depths of 50–100 kilometers. Because this type of melting is adiabatic and results from a lowering of pressure as mantle material is drawn upward, it is called *adiabatic decompression melting.* At some point, the melt begins to separate buoyantly from its solid matrix and the overall stress regime acts to focus the melt toward the midocean ridge plate boundary. The melt composition produced by partial melting (5–25 percent) of upper mantle peridotite is basaltic, with about 50 percent silicon dioxide ($SiO_2$), 10–15 percent magnesium oxide (MgO), and 8–12 percent iron oxide (FeO).

Our knowledge of the internal structure of the oceanic crust is based largely on studies of ophiolite complexes. In the ophiolite model, the ocean crust is composed of the following units from top to bottom: a surface layer of marine sediment; an extrusive layer of basalt lava flows 0.3–0.7 kilometers thick; a basalt dike complex 1.0–1.5 kilometers thick that is the feeder system for the overlying flows; a thick (2–5 kilometer) unit of gabbro, a coarse rock of basaltic composition believed to have crystallized in a magma chamber at depth; and finally an ultramafic unit of upper mantle and possibly lower oceanic crust. This neat layer-cake model of ocean crust structure is probably a gross oversimplification, however, since geologic studies of limited exposures of deep portions of the oceanic crust, obtained by submersible, towed camera, or drilling, show a more complex juxtaposition of lithologies, particularly for crust generated at slow-spreading ridges.

*Continental crust.* The continental crust underlies about 30 percent of the Earth's surface, and includes the crust that extends above sea level as well as the continental margin that fringes most shorelines out to a distance of roughly 200 kilometers. The average thickness of the continental crust is approximately 25 kilometers, but ranges from 15 kilometers in regions of crustal stretching such as the East African Rift to as much as 90 kilometers in places of crustal thickening under mountain belts. While the highest elevations (in the Himalayas) are about 9 kilometers above sea level, the average elevation of the continental crust is only some 120 meters. The fact that most inhabited areas are near coastlines and only a few hundred meters above sea level has important implications if global warming melts polar icecaps and raises sea level.

By comparison with the oceanic crust, the continental crust is chemically and physically heterogeneous as a result of the diverse processes that led to its growth and evolution. Physically, the continental crust includes regions of intense folding, faulting, and metamorphism, the most dramatic of which are seen in ancient or modern mountain belts. In other areas, there is little deformation, such as the great accumulations of horizontal marine and near-shore sediments along passive continental margins. In a general sense, each continental land mass has an irregularly shaped, stable, Precambrian (more than 600 million years old) continental "shield" that is surrounded by younger volcanic, sedimentary, and metamorphic rocks. A great deal of effort has gone into estimating the average chemical composition of the continental crust because this information is essential for understanding the chemical evolution of the Earth as a whole. From such estimates, we know that the average composition of the upper continental crust differs from that of the lower continental crust. The average upper crust is generally described as granodioritic in composition with, for example, about 65 percent by weight of $SiO_2$ and less than 10 percent by weight of (FeO and MgO). In contrast, several lines of evidence, including granulitic (high-grade metamorphic) xenoliths recovered from the lower continental crust, suggest that it is more mafic in composition, with about 55 percent by weight of $SiO_2$ and more than 15 percent by weight of (FeO and MgO).

A great deal of research has focused on the question of the growth rate of the continents. In one model, the continents experienced a period of rapid growth early in Earth history (about 3.8 billion years ago) followed by a longer period of zero net growth during which additions to the continents, largely through magmatism at arc volcanoes such as the Andes or Cascades, were balanced by erosion and recycling to the mantle. Other models, in contrast, cite evidence for gradual progressive growth of the continental crust throughout Earth history, possibly punctuated by relatively brief episodes of faster growth and stabilization, particularly between 2.5 and 3.0 billion years ago, because rocks of that age are notably abundant. These and other models have important implications for the geochemical evolution of the Earth as a whole.

**Development of the Earth's Internal Structure.** Radiometric dating of Earth materials and meteorites suggests that the Earth and the other bodies of our solar system formed about 4.6 billion years ago. [See Earth History.] In the nebular theory of solar system formation, our solar system began as a rotating cloud of matter, which collapsed to a rotating disk as a result of gravitational and centrifugal forces. Gravitational attraction of gas and solid "dust" led to a concentration of mass toward the central core (proto-Sun) of the nebula; rising temperatures due to gravitational contraction in turn initiated hydrogen burning, the Sun's nuclear reaction that continues to heat our solar system today.

In the outer regions of the solar system where our planets would form, a temperature gradient existed: hottest nearest the Sun, and progressively cooler with distance. This temperature gradient affected the types of matter that cooled and condensed in each region. Closer to the Sun, in the region that would form the terrestrial planets (Mercury, Venus, Earth, and Mars), temperatures were so high that only refractory materials (materials that remain solid to high temperatures) condensed to solids, whereas volatile elements and compounds (such as hydrogen, methane) remained in a gaseous state and did not accrete significantly. In contrast, in the cooler regions farther from the Sun, such volatiles solidified and accreted to form the giant, or Jovian, planets (Jupiter, Saturn, Uranus, Neptune, and possibly Pluto). In each region, condensed grains began to attract one another gravitationally, colliding and forming larger grains and planetary embryos, and ultimately the planets themselves.

In the case of Earth, it is believed that heating as a result of gravitational collapse, the continued impact of planetesimals, and the radioactive decay of short-lived radioactive isotopes raised temperatures sufficiently to melt the interior of the Earth, producing a "magma ocean." In this largely molten state, iron-rich droplets coalesced and, because of their great density, migrated toward the center of the Earth to form the metallic iron core. As the Earth cooled, the inner core crystallized; as noted above, the outer core remained molten because its temperature is sufficiently high that iron (diluted with other elements) remains in a liquid state. Within a few hundred million years, the region of the Earth that would become the mantle, consisting of oxygen, silica, alumina, and other cations, crystallized to form a silicate (rocky) layer. An important question is how and when the upper mantle became depleted in incompatible trace elements. Some ancient continental rocks were derived from a mantle source that had already suffered depletion by about four billion years ago. One compelling theory is that the upper mantle became depleted through melting (which scavenges incompatible elements) early in Earth's history and this melt formed the initial continental crust.

To this day, the Earth continues to evolve physically and chemically. The upper mantle continues to be depleted further as melting beneath midocean ridges forms the ocean crust. Material continues to be added to the continents through arc volcanism at subduction zones, and removed from the continents through erosion.

Crustal material continues to be recycled to the mantle at subduction zones, possibly foundering to the depths of the core–mantle boundary where, after some period of time, it may rise again and melt to form some hot spots. The Earth is a dynamic, living planet, and the processes that we observe today and its present physical form represent but one stage in its continuing evolution.

[*See also* Planets; Plate Tectonics; *and the biography of Wegener.*]

### BIBLIOGRAPHY

Allegre, C. J., et al. "The Age of the Earth." *Geochemica et Geocosmica Acta* 59 (1995), 1445–1456.

Brown, C. G., and A. E. Mussett. *The Inaccessible Earth.* London: Chapman and Hall, 1993.

Fowler, C. M. R. *The Solid Earth: An Introduction to Global Geophysics.* Cambridge and New York: Cambridge University Press, 1990.

Peltier, W. R., ed. *Mantle Convection.* London: Gordon and Breach, 1989.

Press, F., and R. Siever. *Understanding Earth.* New York: W. H. Freeman, 1994.

—EMILY M. KLEIN

# EASTER ISLAND

[*This case study focuses on the early history of the Rapanui and the role of deforestation in the decline and subsequent fall of their civilization.*]

Easter Island is the most isolated piece of inhabited land on the Earth (Figure 1). It lies in the South Pacific Ocean, 27° south of the equator, 3,747 kilometers from South America. The nearest island is Pitcairn, 2,250 kilometers to the west, home of the descendants of the Bounty mutineers. The island is roughly triangular, with sides of 20, 15, and 15 kilometers (Figure 2). It is entirely volcanic and rises to 510 meters in altitude. There is no recent volcanic activity, but old craters dot the landscape.

Easter Island was so named by the first European to visit it, Jacob Roggeveen, the Dutch mariner, when he first saw it on Easter Day 1722. Roggeveen found that the island was inhabited by pale-skinned people, who appeared to him to be in a rather poor state economically. This was surprising as their settlements of low thatched huts were overlooked by *moai*, giant stone statues (weighing up to 80 metric tons) of human form. Some of these statues were still standing on constructed stone platforms (*ahu*), but others were lying face down. Thus was born the idea of a "collapsed civilization," which became the greatest of several mysteries of Easter Island. Roggeveen also noted, incidentally, that the island bore no forest, but was dominated by grasses and patches of low scrub. This was unusual in a subtropical island with a mild climate. The lack of forest was another mystery of Easter Island.

Later visitors explored the island more thoroughly, and discovered the statue quarries at Rano Raraku, where many unfinished statues were found, including one weighing 200 metric tons. Others, complete apart from eye sockets and carving on the backs, stood nearby partially buried in the ground, apparently waiting to be

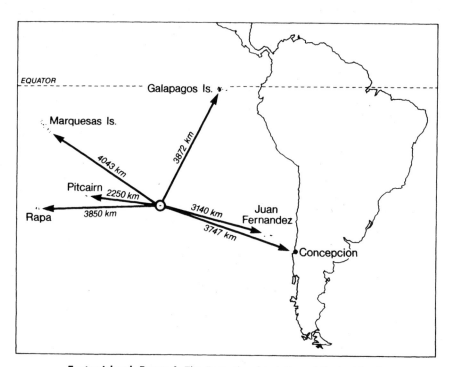

**Easter Island.** FIGURE 1. The Exceptional Isolation of Easter Island.

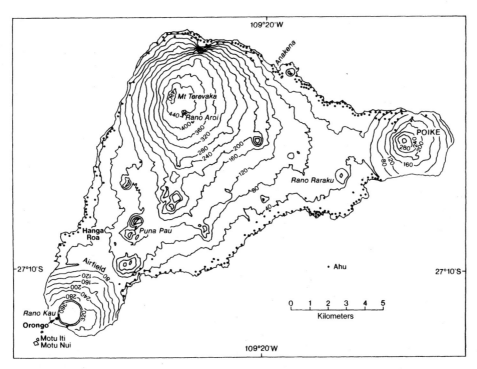

**Easter Island. FIGURE 2.** Topographical Location of Statue Platforms (ahu) on Easter Island. Elevation is given in meters.

transported. But the islanders mostly ignored their statues and were certainly no longer making them. The mystery deepened.

To understand what really happened on the island, we must turn to a combination of archaeology and oral history. According to the legends of the islanders, Rapanui (their name for it) was discovered by their ancestor Hotu Matua, who arrived with a group in a large sailing canoe about 57 generations ago. Where he came from is uncertain. Thor Heyerdahl claimed that it was South America, but most archaeologists and anthropologists agree that the people and their culture are Polynesian, and could well have come from the Marquesas Islands, 4,043 kilometers to the northwest. The date of arrival is uncertain—possibly about 700 CE. The people were soon building crude statues about 2 meters tall (weighing about 10 metric tons) in the local rock on different parts of the island. This was the Early Phase of the island's prehistory.

From about 1100 CE (the Middle Phase) there was a change to building much larger statues (*moai*), almost entirely from the volcanic tuff at Rano Raraku, transporting them distances of up to 10 kilometers or more, and setting them up on *ahu* overlooking each village site. The larger villages had *ahu* with numerous statues, up to 18 at Tongariki. In many cases, the statues were surmounted by huge topknots of reddish stone from another part of the island. We also know now that the statues formerly had eyes, with an iris of white coral and a

pupil of a darker rock. The result must have been most impressive, and it is generally thought that some kind of ancestor worship was being practiced. From the number of *ahu* and village sites, it may be estimated that the population at this time rose to at least 10,000 people. The civilization was at its peak.

And then something went wrong. Major conflicts seem to have broken out. The production of statues at Rano Raraku was terminated—perhaps rather abruptly, judging from the number of unfinished statues in the quarries. In this Late Phase of the island's prehistory, people's time was, instead, turned to the production of obsidian spearheads from the several sources of that rock in different parts of the island. Dating of the obsidian artifacts and waste flakes by measuring the thickness of the hydration rind on them has been carried out by Chris Stevenson. He shows that spear manufacturing increased rapidly from about 1650 CE. Legends about this period speak of famine and warfare. Some of the *moai* were thrown down at this time. They were toppled deliberately, face downward, and frequently a stone was placed so that the head of the *moai* would strike it and thus be snapped off.

The people seem to have changed their religion. The previous ancestor worship was abandoned, and instead the Bird Man cult flourished. This was centered at Orongo, on a cliff top overlooking Motu Nui and Motu Iti, two offshore rocky islets, the breeding site of the sooty tern. Once each year the chiefs of every settlement would assemble at Orongo. Each would dispatch a young champion to Motu Nui to await the arrival of the sooty tern. The chief whose champion found the first egg then became the Bird Man for the ensuing year, and was treated with special deference. This cult has been interpreted by some as being related to the importance of birds and their eggs in the diet, suggesting that there was a shortage of available protein.

This may seem surprising on an island with abundant fish resources offshore. But these were apparently used only in the Early and Middle Phases of the island's pre-history, when large stone fishhooks were used to catch fish up to the size of sharks. In the Late Phase there are only small fishhooks, perhaps mainly suitable for fishing from coastal rocks. Shellfish gathering continued throughout, but the size of the shells diminished through time.

According to legend, the culmination of warfare and famine occurred around 1680 CE, and left the population greatly reduced—perhaps down to around 5,000 people. It was this sorry state of affairs that Roggeveen discovered in 1722. Later eighteenth-century visitors such as Gonzalez, La Perouse, and Cook confirmed this situation. In the nineteenth century things got even worse, the nadir being the slave raid of 1862 in which most able-bodied men were forcibly taken to work in the salt mines of Peru. An international outcry led to the repatriation of some men, who brought smallpox with them and the island population dropped to 111.

Missionary activity eventually flourished on the island, and the Rapanui became Catholics. However, an unfortunate side effect of this was further destruction of *moai*, which the missionaries regarded as evil idols. Wooden artifacts were also destroyed, including many of the famous rongo-rongo tablets, bearing writing of a unique and undeciphered type. Only in 1995 was this writing finally decoded by Stephen Fischer. The tablets prove mostly to bear Rapanui traditional accounts of the creation, but Fischer believes the writing was invented only after European contact.

The population eventually recovered in the twentieth century and is currently about 3,000, though few of these are pure Rapanui. The island is now a part of Chile, and Spanish is the main language, although the Pascuan language (of Polynesian type) survives. Increasingly, the population is dependent on tourism, and there is an airport suitable for large jet planes.

The collapse of the civilization in the period c.1650–1680 CE has been a source of much speculation. William Mulloy attributed it to environmental decline resulting from deforestation by the islanders. Joanne von Tilburg has suggested some kind of psychological decline resulting from the lengthy isolation. Grant McCall has suggested a climatic change—probably a major drought—causing crop failure.

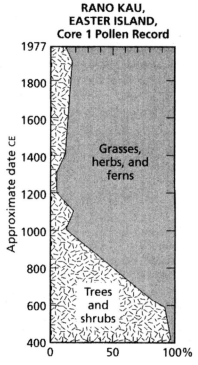

Easter Island. FIGURE 3. Summary of the Pollen Record from Core 1 at Rano Kau.

To test some of these ideas, a program of paleoenvironmental studies was initiated by John Flenley, starting in 1977. This consisted of hand-coring of the sediments in the three main craters (Rano Kau, Rano Raraku, and Rano Aroi), followed by palynology (pollen analysis) and radiocarbon dating of the samples obtained. This program was spectacularly successful. In all three craters, sediments more than 10 meters deep were recovered, and the dates showed that the Holocene period (the last 10,000 years) and part of the Late Pleistocene period were covered. Dates back to before 30,000 BP were obtained from Rano Raraku and Rano Aroi. In Rano Aroi the uppermost sediments had been disturbed, and at Rano Raraku there was also inwashed sediment and a possible hiatus in deposition. But at Rano Kau the sediments appeared to be complete.

The palynology results were very striking. The youngest sediments contained pollen mainly of grasses and sedges—the plants that dominate the vegetation today. But earlier sediments had proportionally less of these, and an abundance of pollen grains of a palm, along with shrubs such as *Sophora*, *Triumfetta*, *Coprosma*, and several species of daisytree. It was impossible from the pollen alone to be sure of the kind of palm involved. Fortunately, palm fruits were recovered from several locations on the island, especially caves, and these were identified by John Dransfield as belonging to the tribe *Cocosoideae*. He named the palm—now extinct—as *Paschalococos disperta*. It appears to have been related to *Jubaea chilensis*, the Chilean wine palm, which is the largest palm in the world. Further evidence that the palm was a sizable tree was provided by the casts of palm trunks found in the lava on the northern cliffs of the island. The palm seems to have been widespread in the lower parts of the island, for root channels of palm type have been found in deep soils at several locations. It was concluded that, throughout most of the last 30,000 years, the island had carried an open palm forest with an understory of shrubs. Even at the peak of the last ice age (around 18,000 BP), there had still been forest, although perhaps of a more open type with more grasses present, and possibly replaced by scrub at the higher altitudes around Rano Aroi.

Clearly, the date of the disappearance of this forest was crucial to any hypothesis relating the collapse of civilization to forest clearance. If the clearance was done by people, it must come after the arrival of people, usually put at around 400–700 CE. If the clearance was a cause of the collapse, it must come before the period 1650–1680 CE. At Rano Kau, it was possible to date the tree and shrub decline in the pollen record rather accurately (Figure 3). It occurred mainly between 600 and 1200 CE, with total disappearance of woody plants by about 1300 CE. In other words, the forest decline fitted the criteria needed to be the explanation of the fall of the civilization.

Of course, this does not necessarily prove that it was a cause, and certainly not that it was the sole cause, of the collapse. Major events rarely have a single cause.

## EASTER ISLAND: ECOLOGICAL DISASTER

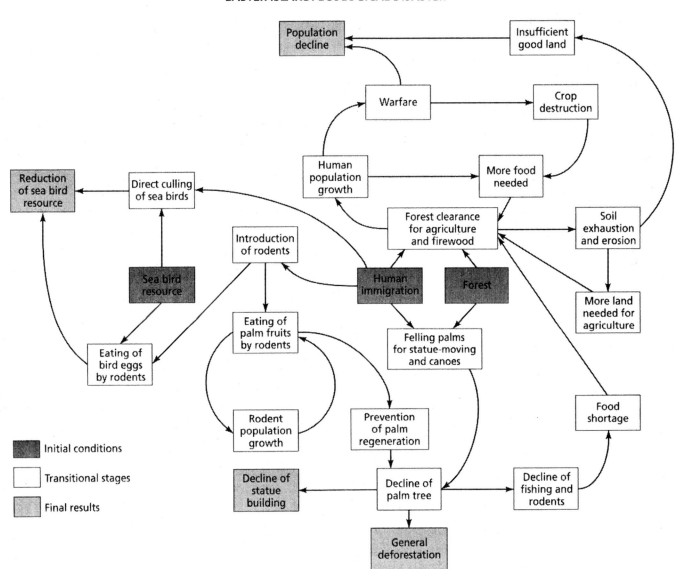

Easter Island. FIGURE 4. Flowchart Showing the Possible Course of Events on Easter Island.

It is entirely possible—even probable—that a major drought triggered the famine that led to the collapse. Droughts happened in many parts of the world in the period 1400–1900 CE, associated with the Little Ice Age of Europe and North America at this time. It seems unlikely there would have been a drought sufficient to destroy the Easter Island forests (which anyway had declined earlier), for even the major Last Ice Age of the Pleistocene had not succeeded in doing this. But there could have been a drought sufficient to cause crop failure, for the island is subject to periodic droughts even today. It seems likely, however, that this would have been only the trigger to set off a collapse of which the underlying cause was deforestation and overpopulation. [*See* Drought; *and* Little Ice Age in Europe.]

We may therefore conclude that what probably happened was an ecological disaster, as illustrated in Figure 4. Reading this outward from the middle, we start with people arriving on a forested island rich in seabirds. They clear land for agriculture and they cut firewood. As agriculture succeeds, population grows, and more land is needed. Soil is lost by erosion and, unlike many more tropical islands, there is no reef to catch the lost soil. The people are eating seabirds and felling palm trees to make canoes and for use as rollers and levers in the moving of *moai*. They have also introduced the Polynesian rat, which eats palm fruits and also seabird eggs. The result is to prevent the palm regenerating, so it becomes a nonrenewable resource. Nesting seabirds—which, according to David Steadman's work on the bird bones in the lava caves, had once made this the most diverse seabird island in the world—declined to only one or two species on offshore islets. The end result of it all was a collapse of the civilization and a human population crash.

One cannot resist drawing the analogy between Easter Island, isolated in the Pacific, and the planet Earth, isolated in space. The Club of Rome's computer predictions for planet Earth—even when revised recently—show in most scenarios a decline of resources and a population crash toward the end of the twenty-first century. The Easter Island model suggests that they could well be correct. Even more disturbingly, Easter Island was a small, simple ecosystem. People must have been able to see what was happening. The person who cut down the last tree could see that it was the last tree, but they still cut it down. The Earth, by contrast, is large and complex. One might do something that appeared harmless, or even beneficial (such as inventing chlorofluorocarbons for refrigeration), and actually damage the Earth ecosystem seriously. Therefore the chances of ecological disaster for the Earth seem even greater than they were for Easter Island.

[*See also* Deforestation.]

## BIBLIOGRAPHY

Bahn, P., and J. R. Flenley. *Easter Island, Earth Island*. London: Thames and Hudson, 1992.

Fischer, S. R. *Rongorongo: The Easter Island Script; History, Tradition, Texts*. New York and Oxford: Oxford University Press, 1998.

———, ed. *Easter Island Studies: Contributions to the History of Rapanuui in Memory of William T. Mulloy*. Oxford: Oxbow, 1993.

Flenley, J. R., A. S. M. King, J. T. Teller, M. E. Prentice, J. Jackson, and C. Chew. "The Late Quaternary Vegetational and Climatic History of Easter Island." *Journal of Quaternary Science* 6 (1991), 85–115.

Heyerdahl, T. *Easter Island: The Mystery Solved*. London: Souvenir Press, 1989.

Irwin, G. *The Prehistoric Exploration and Colonisation of the Pacific*. Cambridge: Cambridge University Press, 1992.

McCall, G. *Rapanui: Tradition and Survival on Easter Island*, 2d ed. Honolulu: University of Hawaii Press, 1994.

Meadows, D., D. Meadows, and J. Randos. *Beyond the Limits*. London: Earthscan, 1992.

Stevenson, C. M. "Archaeological Investigations on Easter Island. Maunga Tari: An Upland Agricultural Complex." *Easter Island Foundation Occasional Paper* 3 (1997).

van Tilburg, J. A. *Easter Island: Archaeology, Ecology and Culture*. Washington, D.C.: Smithsonian Institution, 1994.

—JOHN FLENLEY

# EASTERN EUROPE

[*This case study focuses on the environmental legacy of communism in the region (Figure 1).*]

The communist system of central planning gave rise to major environmental dislocations that had serious consequences in terms of damaged ecosystems, economic inefficiency, and a reduced quality of life. In Poland, pollution costs were reckoned as high as one-tenth of gross national product. This essay attempts to explain the rationale behind the system and indicate some of the ways it impacted negatively on the environment. There is also discussion of the measures being taken to solve these problems now that the system has been replaced.

The communist system of central planning gave priority for production plan targets set in a "war economy," with investment deliberately unbalanced to realize the key strategic objectives. Assets were nationalized, and state-owned enterprises had free use of natural resources. The system rewarded value added rather than sustainable management of the environment and distracted attention away from the negative consequences of such an approach by discouraging criticism of the authorities and insisting that environmental damage could occur only in capitalist countries.

**Environmental Damage under Communism.** High production costs served to increase output "value" (also

**Eastern Europe. Figure 1.** States of Eastern Europe and Locations Mentioned in the Text.

to boost wages and career prospects), but the lack of incentive for conservation made for considerable waste of resources. Oil and gas extraction rates were low and much mining waste could have been subjected to more sophisticated processing. Pipeline leaks meant the loss of 10–15 percent of all oil and gas produced in the former Soviet Union (FSU), and there was considerable risk of explosions, such as the disaster that wrecked trains on the Trans-Siberian Railway near Chelyabinsk in the Urals in 1989. Energy consumption was high in relation to the volume of industrial production, and given the inadequacy of filtering systems, air pollution levels were very high in the vicinity of thermal power stations, especially those burning lignite. Meanwhile, hydroelectric projects involved heavy resettlement costs and losses of agricultural land (21 million hectares in FSU alone), as well as reduced fish catches.

*Air and water pollution.* Various environmental issues can be recognized: excessive noise, inappropriate architecture, and other forms of cultural pollution. But the greatest problems are concerned with air and water pollution in the urban-industrial regions. Pollution associated with the technology of the third and fourth "Kondratieff" economic-development cycles reached unacceptable levels by the 1980s, and the use of environmentally sound technologies for "sustainable development" (first advocated at the Stockholm Conference in 1972) hardly featured on the political agenda. The

problems were exacerbated by the rapid growth of industry without adequate provision for the filtering of emissions and the processing of polluted process water. Coke production was another important source of pollution because of the obsolete methods used, but heavy industry in general was highly damaging: the FSU was responsible for one-fifth of global emissions of carbon dioxide and sulfur dioxide and one-tenth of the chlorofluorocarbons (CFCs) depleting the ozone layer. Yet industry in general was inadequately equipped to filter emissions and process effluent, while the lack of proper sewage treatment increased the scale of river pollution. Transboundary pollution exacerbated local problems involving both air and water: air pollution in Poland was made worse by "imports" from adjacent areas of former Czechoslovakia and East Germany.

Pollution black spots in FSU include the Noril'sk nickel smelter on the Arctic Sea, with its health hazards and landscape damage; also carbon disulfide pollution at Bratsk (also in Siberia), which is injurious to both public health and the local forests. In the case of Lake Baikal (the "Pearl of Siberia"), pollution linked with woodcutting and pulping has constituted a major threat to what is the deepest and most voluminous lake in the world, with six hundred plant and twelve hundred animal species, many of which are unique. Outside the FSU, the fallout of dust and gas (sulfur dioxide) was greatest in the northern industrial areas of Halle/Leipzig, North Bohemia, and Upper Silesia. Farther south there have been notorious black spots such as Copsa Mica in Romania, but the heavily polluted areas are relatively small. Water pollution is also more serious in the north, where untreated sewage is a major consideration. But even modest levels of pollution threaten sensitive ecosystems such as Hungary's Lake Balaton.

*Pollution in rural areas.* Rural areas suffered from intensive farming that involved excessive fertilizer applications, use of heavy machinery, and poorly organized irrigation schemes that made soils salty. Damage also arose from monoculture and indiscriminate use of pesticides, dichlorodiphenyltrichloroethane (DDT), and the defoliant Butifos (leading to a high incidence of cancer and mental illness). Wildlife habitats have been eroded (for the stock of wild animals is falling and there is a threat to rare breeds in Kyrgyzstan and the Far East from hunting), while water supplies have been polluted by excessive fertilizer applications and the accumulation of slurry. Forest damage was severe in the late 1970s and 1980s as the acid rain effect was exaggerated by dry summers, severe winters, and insects. Replanting avoids spruce monoculture by making more use of larch and beech. Large areas of woodland are lost through forest fires in the Russian taiga belt because of inadequate firefighting systems.

*Mining.* Mining made a striking impact on the landscape. In Bohemia, lignite pits were excavated to 150–180 meters, destroying the original environment, including groundwater accumulations. Meanwhile, coal mining in Upper Silesia has an unfortunate side effect not only because of tipping but also through saline waters: 3 million metric tons of salt dumped in rivers linked with the Oder and Vistula basin from coal mines (more than Poland's total salt consumption). This causes huge environmental damage and losses for heating systems and factories that need to use river water. Some mines try to extract salt and create an emulsion with fly ash to fill old passages, but progress is limited because of the high energy input. The tipping of waste material is also a serious problem. For example, the area of East Germany between Hettstedt and Gera is scarred by tipping associated with the mining of copper (Hettstedt/Mansfeld), potash salts (Stassfurt), and uranium (Ronneburg). In the case of uranium mining in Bohemia and East Germany, the environment was endangered by settling ponds in the vicinity of the processing plants: these ponds contain several tens of millions of metric tons of sludge with a high concentration of residual uranium and a number of other detrimental elements.

*Radiation.* Reference should also be made to serious cases of radiation. There was a spectacular accident at Chernobyl nuclear power station in Ukraine in 1986. During the testing of a turbine during a scheduled shutdown, power output rose to one hundred times the normal, thus leading to an explosion and the burning-out of the graphite core of Reactor Four. Radioactive material was released for ten days and contaminated much of Europe. Moreover, radiation hazards associated with the nuclear test site near Semipalatinsk (now Semey) in Kazakhstan, where five hundred tests were carried out between 1949 and 1989, have affected the health of half a million people living in the vicinity. Radioactivity in river water arose through the direct-flow method of cooling at the plutonium factory at Krasnoyarsk, with much more serious problems at the weapons' plutonium/nuclear bomb factory of Mayak ("Chelyabinsk 65") between Chelyabinsk and Yekaterinburg in the Urals. There are further hazards through the abandonment of nuclear submarines and ice breakers at various points around the coast and the dumping of nuclear waste in the seas of the Arctic and the Far East. [*See* Chernobyl.]

**Finding Solutions to Environmental Problems.** In 1987 the Soviet state enterprises had to start paying for resources, and production linked with environmental conservation became a possibility. But laws put in place to limit pollution were not always enforced effectively because of a lack of political will and adequate monitoring equipment. Meanwhile, public opinion became more articulate in dealing with environmental matters. *Glasnost* generated debate by environmental

groups such as Zelenyi Mir (the equivalent of Greenpeace); the Green movement was active in Lithuania as early as 1986. Outside the FSU, too, environmental problems were often major issues of public concern as opposition to communism increased during the 1980s, through the underground press organized by such opposition groups as Charter 77 in former Czechoslovakia. Chernobyl underlined the hazards of transfrontier pollution, and Ekoglasnost in Bulgaria gained considerable influence by campaigning against the pollution crossing the Danube from Romania. Green parties were prominent in the early postcommunist elections, and pressure mounted for closure or modernization of plants responsible for high pollution levels.

*Legislation for environmental standards.* Considerable progress has been made in cleaning up the black spots, but the situation varies between countries. Before 1989 former Czechoslovakia had no central body responsible for environmental problems, but a commission was established in both parts of the country in 1990. And whereas key positions are often held by technocrats with little political standing, substantial clout was eventually mustered in the Czech Republic, with prominent figures in charge at both the Commission and the Ministry. There has been a complete redrafting of legal codes, and environmental impact assessment is now very advanced. It seems that a desire for integration with Europe has made for harmony over environmental policy. Because industrialists wanted integration, European Union (EU) emission standards were accepted even though they were burdensome in terms of the "polluter pays principle" (PPP).

In other countries, too, there is a new realism linked with the EU. And a helpful trend is the greater involvement of local authorities that now have (1) responsibilities for water, sewage, and waste management, and (2) income from taxes to finance environmental projects. Also, the PPP generates fines that go to national and regional ecological funds (though municipalities do not always reserve payments received under national environmental fund disbursements for ecological projects). Environmental issues can be addressed in a more responsible and practical way because self-governments can directly assess the relative importance of different needs expressed by local populations. Hence the beginnings of an effective system of state, regional, and local responsibilities. A national water shortage in Bulgaria makes for pollution control, although the linkage of drainage basins by aqueducts is opposed by some conservationists.

*The case of Poland.* In Poland's top pollution "hot spots," factory closures have been carried out (although only eighteen of the eighty or so polluting industries selected for closure have actually been eliminated), and emissions have been reduced by the installation of abatement equipment. Installations are now in place to reduce coal dust, while desulfurization at major power plants in Poland is bringing sulfur dioxide emissions down. Sewage processing (including the production of compost) and urban recycling are beginning to cope with the growth of domestic waste and the shortage of landfill capacity. Urban bypasses are needed in response to the growth of car traffic, aggravated by the closure of secondary rail routes. Meanwhile, there is a consensus over the protection of sensitive rural areas such as nature reserves or national parks, and more is being done to restore historic towns.

***External assistance.*** Much help has been forthcoming from Finland, Germany, and the Netherlands, also from Japan over the desalinizing of mine water. U.S. World Environment is helping with cleaner technologies for around twenty larger industrial plants, and there is an increased share for gas-based energy (encouraged by World Bank finance for conversion to gas and rural gas distribution). There are EU, Organisation for Economic Co-operation and Development, and United Nations (UN) contacts (with some cancellations of debts to increase scope for environmental spending). In Bulgaria, the European Bank for Reconstruction and Development has financed desulfurization at the huge Maritsa power station, while Japanese loans have reduced sulfur dioxide emissions at the nonferrous metal smelters. The Aral Sea problem is being addressed by UN/World Bank action to improve water supply (though local people generally dislike the chlorinated water and prefer the polluted water to which they are more accustomed). Attention is being directed to irrigation water losses from unlined canals and inadequate distribution of water where fields are uneven (so that water cannot reach all areas without considerable flooding). There is a need for agreement over the division of water among the five republics, previously organized in Moscow. The Aral Sea could eventually be restored, but only through economic restructuring with less emphasis on cotton. Work could be financed by the profits arising from oil and gas if there is a political will based on democratic politics reflecting local opinion.

***Overall assessment.*** The situation remains unsatisfactory despite environmental action and reduced industrial and electricity production. There are still many polluting factories, and additional risks will arise from the selective growth of key industries (such as the Ziar and Hronom aluminum smelter in Slovakia), along with tourist complexes in sensitive areas and infrastructure projects such as nuclear power stations and major highways. War in former Yugoslavia has seen life in rural areas disrupted by land mines, most notably in Bosnia and Herzegovina. And Hungarian and Slovakian approaches to hydropower have generated some high-profile international disputes. Yet the enthusiasm for Green politics seems to have subsided, and there is a general lack of public participation. An environmental ethic is still missing: the idea that a rise in welfare must go hand in hand with environmental improvements. In many cases, nongovernmental organizations concerned with the environment have been unable or unwilling to cooperate with the broader public. The Czech Union of Nature Conservation is quite effective, with around ten thousand members and local organizations in most districts, but in Slovakia the Bratislava Committee (SZOPK) and other environmental groups have been marginalized by the big political questions. Enterprise action has been limited by the tendency for inflation to reduce the value of fines and undervalue the social costs of environmental degradation, while governments have sometimes been slow in setting up adequate inspection services to enforce legislation. But substantial improvements have been made, and pressure from the EU should make for further convergence between eastern and western Europe over the longer term.

## BIBLIOGRAPHY

Bater, J. H. "Natural Resources Management in the Soviet and Post-Soviet Eras." In *Russia and the Post-Soviet Scene: A Geographical Perspective*, pp. 154–175. London: Edward Arnold, 1996. A readable and authoritative review.

Carter, F. W., and D. Turnock, eds. *Environmental Problems in Eastern Europe*. London: Routledge, 1993. Offers comprehensive coverage of all countries outside the former Soviet Union.

DeBardeleben, J., and J. Hannigan, eds. *Environmental Quality and Security after Communism: Eastern Europe and the Soviet Successor States*. Boulder: Westview, 1994. A wide-ranging survey of the whole region.

Fitzmaurice, J. *Damming the Danube: Gabcikovo/Nagymaros and Post-Communist Politics in Europe*. Boulder: Westview, 1994. An excellent survey of a major international dispute.

Frankland, E. G. "Green Revolutions? The Role of Green Parties in Eastern Europe's Transition, 1989–1994." *East European Quarterly* 29 (1995), 315–345. Demonstrates the importance of environmental problems for the development of democracy.

Klarer, J., and B. Moldan, eds. *The Environmental Challenge for Central European Transition Economies*. Chichester, U.K.: Wiley, 1997. A thorough review of national strategies.

Knight, G. C. "The Emerging Water Crisis in Bulgaria." *GeoJournal* 35 (1995), 415–423. A concise study of a major problem.

Lipovsky, I. "The Deterioration of the Ecological Situation in Central Asia: Causes and Possible Consequences." *Europe-Asia Studies* 47 (1995), 1109–1123. Profiles some of the former Soviet Union's most serious ecological problems.

Marples, D. R. *The Social Impact of the Chernobyl Disaster*. London: Macmillan, 1988. A useful and readable introduction to this devastating event.

Massey Stewart, J., ed. *The Soviet Environment: Problems, Policies and Politics*. Cambridge: Cambridge University Press, 1992. An excellent general survey.

Nefedova, T. "Industrial Development and the Environment of Central and Eastern Europe." *European Urban and Regional Studies* 1 (1994), 168–171. Places industry at the center of the pollution problem.

Peterson, D. J. *Troubled Lands: The Legacy of Soviet Environmental Destruction.* Boulder: Westview, 1992. Excellent for communist attitudes toward the environment.

Slocock, B. "The Paradoxes of Environmental Policy in Eastern Europe: The Dynamics of Policy-Making in the Czech Republic." *Environmental Politics* 5 (1996), 501–521. Shows how political issues affect the priority given to environmental problems.

Wiska, A., and J. Hindson. "Protecting a Polish Paradise." *Geographical Magazine* 63.6 (1991), 1–2. A concise study on environmental action in a rural area.

Witkowski, J. "The Quality of the Natural Environment and Demographic Processes in the Large Towns of Poland." *Geographia Polonica* 61 (1993), 367–377. Discusses an important aspect of the environmental situation on which relatively little research has been done.

—DAVID TURNOCK

## ECOLOGICAL ECONOMICS. *See* Environmental Economics; *and* Industrial Ecology.

## ECOLOGICAL INTEGRITY

Recent recommendations have been made to base environmental public policies on concepts of ecological integrity. The recommendations stem from scientists and others concerned about the threats of human activities to ecosystems and species, and from philosophers attempting to derive a more suitable ethic for the relationships between humans and the nonhuman environment. Proponents of "ecological integrity" believe that it offers a better prospect for the management of human and environmental systems compared to concepts of "environmental health" or "sustainable development," which they view as too heavily oriented toward human-dominated systems.

Although ecological integrity has been proposed as a norm for public policies and decision making, the concept is relatively new, and therefore the scientific and philosophical rationales undergirding it have not been developed fully. Hence, proponents of ecological integrity are attempting to relate ecosystem management approaches and the goal to restore integrity to moral principles, that they might serve as a basis for public policy. The following discussion is based on Westra (1994), Lemons and Brown (1995), Westra and Lemons (1995), Lemons (1996), and Lemons et al. (1997).

Despite recent interest in ecological integrity, the concept is not defined succinctly or precisely. Kay (1992) has proposed that ecological integrity encompasses three facets of ecosystems: (1) the ability to maintain optimum operations under normal conditions, (2) the ability to cope with changes in environmental conditions (i.e., stress), and (3) the ability to continue the process of self-organization on an ongoing basis, that

is, the ability to continue to evolve, develop, and proceed with the birth, death, and renewal cycle.

By "optimum operations," Kay means the situation where a balance is maintained between the external environmental fluctuations that tend to disorganize ecosystems—that is, by making them less effective at dissipating solar energy—and the organizing thermodynamic forces that make ecosystems more effective at dissipating solar energy. It is not clear what Kay means by "normal conditions" and hence to what extent ecological integrity can or should refer to human interventions in ecosystems. However, he does believe that integrity includes nonlinear ecosystems whose properties or behavior cannot be explained or predicted by knowledge of lower levels of hierarchical organization within them. These ecosystems have multiple organizational states and processes based on nonequilibrium paradigms that include the following notions: (1) ecosystems are open, (2) processes rather than end points are emphasized, (3) a variety of temporal and spatial scales are emphasized, and (4) episodic disturbances are recognized.

Westra (1994) proposes a definition wherein ecosystems can be said to have ecological integrity when they have the ability to maintain operations under conditions as free as possible from human intervention, the ability to withstand anthropocentric changes in environmental conditions (i.e., stress), and the ability to continue the process of self-organization on an ongoing basis. She argues that concepts inherent in ecological integrity emerge from continuing scientific, legal, and ethical analysis and that while they correspond in her mind to more or less "pristine nature," they cannot be described or predicted precisely because ecosystems are constantly changing and evolving. Westra's definition includes the following: (1) ecosystem health, which may apply to some nonpristine or degraded ecosystems provided they function successfully; (2) ecosystems' abilities to regenerate themselves and withstand stress, especially nonanthropogenic stress; (3) ecosystems' optimum capacity for undiminished developmental options; and (4) ecosystems' abilities to continue their ongoing change and development unconstrained by human interruptions past or present. One of the problems with Westra's definition is that it is impossible either theoretically or practically to determine what constitutes a "natural" ecosystem (Callicott, 1991). Cairns (1977) defines ecological integrity as "the maintenance of the community structure and function characteristic of a particular locale or deemed satisfactory to society." Karr (1992) defines ecological integrity as "the capability of supporting and maintaining a balanced, integrated, adaptive community of organisms having a species composition, diversity, and functional organization comparable to that of natural habitats of the region."

While recognizing that the characterization of ecological integrity in more precise terms is not easy, Noss (1995) nevertheless proposes that the concept can be made operational by selecting measurable and quantifiable indicators that correspond to the ecological qualities we associate with integrity. For Noss, these indicators include: (1) structural measures of patch characteristics; (2) structural measures of patch dispersion; (3) access, flow, and disturbance indicators; (4) structural measures; (5) compositional measures; (6) functional measures; (7) composite indices; (8) measures of genetic integrity; and (9) measures of demographic integrity. There is, however, no *a priori* way to decide which of these attributes should be selected.

According to Kay and Schneider (1995), ecosystems can respond to environmental changes in five qualitatively different ways: (1) after undergoing some initial structural and functional changes, they can operate in the same manner as before the changes; (2) they can operate with an increase or decrease in the same structures they had before the changes; (3) they can operate with the emergence of new structures that replace or augment existing structures; (4) different ecosystems with significantly different structures can emerge; and (5) they can collapse with little or no regeneration. Although ecosystems can respond to environmental changes in one of these five ways, there is no inherent or predetermined state to which they will return. Further, none of the above ways indicates *a priori* whether a loss of integrity has occurred. One problem with this view is that it would not be helpful to decision makers because it accepts all ecosystem responses to change as constituting integrity, with the exception of total collapse, which occurs rarely and is clearly undesirable. Consequently, while in theory science can inform decision makers about the responses of ecosystems to environmental change, it cannot provide a scientific or so-called objective basis for deciding whether one change is more desirable than another. In other words, the selection of criteria to use in such a decision must be based on human judgment regarding the acceptability of a particular change.

Regier (1993) provides an abstract definition stating that ecological integrity exists when an ecosystem is perceived to be in a state of well-being. In part, Regier stresses that a more precise definition of ecological integrity is dependent on people's perspectives of what constitutes complete ecosystems, as many of the aforementioned definitions demonstrate. In addition to reflecting the concerns and values of scientists, definitions of ecological integrity also must reflect various social and ethical values relevant for public policy decisions regarding the protection of ecosystems. One reason for the inclusion of these various values is because there is no a priori scientific definition of ecological integrity,

and therefore the concept encompasses perspectives or ways of viewing the world that inevitably reflect value-laden judgments. The ambiguity of ecological integrity is a recognition that its definition, like many ecological concepts, is determined in part on the basis of value-laden judgments and not solely on so-called value-free or precisely defined scientific criteria.

While apparently being sympathetic to trying to use concepts of ecological integrity as a basis for public policy and decision making, several commentators have questioned the extent to which this can be done at present. Shrader-Frechette and McCoy (1993), Shrader-Frechette (1995), and Lemons (1996) maintain that the methods and tools of ecology are too general and limited in their precision and predictive capabilities. They argue that the value of ecology is primarily heuristic and capable of providing useful information for decision making if studies are based on practical case studies and not on generalizable theories. Unless the problem of scientific uncertainty is overcome, concepts of ecological integrity probably will remain largely philosophical. In any case, before using ecological integrity as a basis for public policies, further work is required to clarify the concept's scientific, philosophical, and practical dimensions.

## BIBLIOGRAPHY

Cairns, J. "Quantification of Biological Integrity." In *The Integrity of Water*, edited by R. K. Ballentine and L. J. Guarraia, pp. 171–187. Washington, D.C.: U.S. Environmental Protection Agency, Office of Water and Hazardous Materials, 1977.
Callicott, J. B. "The Wilderness Idea Revisited: The Sustainable Development Alternative." *The Environmental Professional* 13 (1991), 235–248.
Karr, J. R. "Ecological Integrity: Protecting Earth's Life Support Systems." In *Ecosystem Health*, edited by R. Costanza, B. G. Norton, and B. D. Haskell, pp. 223–238. Washington, D.C.: Island Press, 1992.
Kay, J. "A Non-Equilibrium Thermodynamics Framework for Discussing Ecosystem Integrity." *Environmental Management* 15 (1992), 483–495.
Kay, J., and E. Schneider. "Embracing Complexity: The Challenge of the Ecosystem Approach." In *Perspectives on Ecological Integrity*, edited by L. Westra and J. Lemons, pp. 49–59. Dordrecht: Kluwer, 1995.
Lemons, J., ed. *Scientific Uncertainty and Environmental Problem-Solving*. Cambridge: Blackwell, 1996.
Lemons, J., L. Westra, and R. Goodland, eds. *Ecological Integrity and Sustainability: Concepts and Approaches*. Dordrecht: Kluwer, 1997.
Noss, R. F. "Ecological Integrity and Sustainability: Buzzwords in Conflict?" In *Perspectives on Ecological Integrity*, edited by L. Westra and J. Lemons, pp. 60–76. Dordrecht: Kluwer, 1995.
Regier, H. A. "The Notion of Natural and Cultural Integrity." In *Ecological Integrity and the Management of Ecosystems*, edited by S. Woodley, J. Francis, and J. Kay, pp. 3–18. Delray Beach: St. Lucie Press, 1993.
Shrader-Frechette, K. S. "Hard Ecology, Soft Ecology, and Ecosystem Integrity." In *Perspectives on Ecological Integrity*, edited

by L. Westra and J. Lemons, pp. 125–145. Dordrecht: Kluwer, 1995.

Shrader-Frechette, K. S., and E. D. McCoy. *Method in Ecology.* Cambridge: Cambridge University Press, 1993.

Westra, L. *An Environmental Proposal for Ethics: The Principle of Integrity.* Lanham, Md.: Rowman and Littlefield, 1994.

—JOHN LEMONS

# ECOLOGICAL STABILITY

*Ecological stability* is a set of special cases of system stability. [*See* Stability; *and* System Dynamics.] The classic assumption has been that all ecological systems, including the Earth system, are characterized by classic static stability. James Lovelock proposed a model for a way that vegetation might provide this kind of stability through negative feedback in a simple model he referred to as the "daisy world." Suppose, he said, that the vegetation of the world consisted of only two kinds of plants, white and black daisies, and that both kinds had the same temperature-growth response curve, so that all individuals of both species reached a maximum rate of growth and reproduction at the same optimum temperature. Under any intensity of sunlight, black daisies are warmer than white. If the surface temperature of the Earth falls below the optimum for daisy growth, black daisies will stay warmer, closer to the optimum, and grow and reproduce better than white. The world will then tend to become dominated by black daisies. The better light absorption of a black surface would in turn warm the Earth above the optimum. With the temperature above the optimum, white daisies, which reflect more light, would remain cooler and would then do better, and the Earth would then tend to become populated by white daisies, which would cool the surface. This two-species light-reflecting system would create a negative feedback mechanism and give the Earth system classic static stability. Lovelock proposed that vegetation could have stabilized the Earth's surface temperature in this way. [*See* Gaia Hypothesis.]

The belief that all ecological systems are characterized by classic static stability is obsolete. It is the underlying principle behind the old belief in the balance of nature. It has also been a dominating concept in ecological science. It was believed to be true for populations and ecosystems, as well as for all the biota and the Earth system. For populations, this belief was expressed in the assumption that populations grew according to the Logistic Growth Curve to a maximum known as the *carrying capacity*, a stable equilibrium population level. According to this equation, if the population falls below the carrying capacity, births exceed deaths and the population returns to its carrying capacity. If the population grows above the carrying capacity, deaths exceeds births and the population also returns to its carrying capacity. [*See* Carrying Capacity.]

For ecosystems, the classic idea of succession to a climax state is another statement of classic static stability. For example, according to this classic idea of ecological succession in a forest, a clearing develops through a series of stages into an old-growth forest, which then persists in that old-growth state indefinitely. Under the classic view, the old-growth state had constant and maximum biomass and biological diversity. Before the emergence of Earth system science, it was common to believe that the Earth and all life on it were characterized by this kind of stability, except in cases of human-induced disturbances.

Holling proposed that ecological systems could be categorized by the degree to which they were resistant to change and resilient to change. A system that has high resistance is deflected only a small amount from its equilibrium by a disturbing force of a given magnitude, whereas a low-resistance system undergoes a larger change as a result of the same quantity of disturbance. A high-resilience system is one that returns rapidly to its equilibrium point; a low-resilience system is one that returns slowly to that equilibrium. These two terms, *resistance* and *resilience*, have become common in ecology. It should be recognized that, as defined originally, they referred to systems with classic static stability, and they are therefore another expression of the idea that ecological systems are characterized by this kind of stability.

The degree to which the Earth system is characterized by classic static stability remains debated, but there is a major movement among ecological scientists to reject this idea. It is now clear that this kind of stability is not characteristic of the long-term climate of the Earth. No wild population of organisms has been observed to have this kind of stability, nor are forests characterized by it, because tree species respond to climate change and other persistent environmental changes. Thus, these real ecological systems—populations and ecosystems—lack a fixed equilibrium point, one of the two defining requirements of a system with classic static stability. Instead, environmental and biological variables are characterized by change over time. A system without a fixed equilibrium cannot be characterized by classic static stability. A second problem is that real ecological systems, including the Earth system, appear to be stochastic; that is, they are affected by and/or generate internally random processes, so that the system continually varies. A third problem is that many ecological systems themselves generate changes.

It is necessary to find substitute concepts to describe systems that lack an equilibrium and systems characterized by stochastic processes. Two useful concepts are persistence with defined bounds and the recurrence of "desirable" states.

For example, instead of attempting to maintain a population of an endangered species at a fixed carrying capacity, one could determine the historic range of variation in its numbers in order to characterize a "natural" range of variation, and then seek to conserve the population within that range. Or one might make use of a theory that allows a calculation of a minimum viable population to determine the lower and upper bounds of acceptable variation—acceptable in the sense of maintaining the probability of extinction below some specific value over some specified planning time horizon. In the case of species of interest for commercial harvest, rather than attempting to maintain such a population at a fixed maximum sustainable yield level, one would seek a range that yielded acceptable levels of production.

The second concept, the system as a set of recurrent states, is exemplified by a vegetation community in which fire is a persistent factor, such as chaparral or the many forests dominated by pines growing on sandy soils. In such ecosystems, both plants and animals have evolved and adapted to wildfire and require it to persist. If these systems are forced to a fixed condition—for example, through fire suppression—these species become locally extinct. For these species to persist, fire must not occur too often or too rarely. The desired states—those that allow a species or set of species to persist—occur within a range of wildfire frequencies. Under such wildfire regimes, the desired states become persistent.

With the global Earth system, the issue of the kind of stability or lack of stability is of central importance. It is now clear that the Earth's surface temperature and the carbon dioxide concentration of the atmosphere has varied considerably over the past 150,000 years, yet life has persisted throughout this period. One might conclude that this range of variation leads to sustainability, or persistence of the biota. It is also clear that over geologic time periods, new species evolve and existing species become extinct, and that there are rare catastrophic events that greatly alter the biota. In addition, there have been major biologically induced alterations of the global environment that have, in turn, led to the extinction of some species and the evolution of others. As an example, the alteration of the Earth's atmosphere from anoxic to oxidizing, as a result of the emission of free oxygen by photosynthetic organisms, led to the extinction of some species and made possible the evolution of high-oxygen-requiring species. Thus, over the long term, the biota and the entire Earth system have been characterized by change. This leaves two unresolved questions: Have the biota tended to stabilize the Earth system more than would have occurred otherwise (for example, does vegetation tend to reduce the varia-

tion in surface temperature even if it does not lead to a fixed equilibrium)? And over short time periods, do some attributes of the Earth system in fact exhibit classic static stability?

[*See also* Biogeochemical Cycles; Biological Feedback; Ecosystems; Fire; Population Dynamics; *and* Sustainable Development.]

### BIBLIOGRAPHY

Botkin, D. B., and M. J. Sobel. "Stability in Time-Varying Ecosystems." *American Naturalist* 109 (1975), 625–646. Discusses the ways in which ecological systems differ in their stability characteristics from engineered systems and steady-state systems.
Holling, C. S. "Resilience and Stability of Ecological Systems." In *Annual Review of Ecology and Systematics*, vol. 4, edited by R. F. Johson, P. W. Frank, and C. W. Michener. Palo Alto, Calif.: Annual Reviews, 1973. Applies some of the standard ideas of stability to ecological systems.

—DANIEL B. BOTKIN

# ECONOMIC LEVELS

For millennia, there have been contrasts between the haves and the have-nots, the advanced and the less advanced, or the developed and the less developed. Around 3000 BCE, when the practice of agriculture had not yet spread widely, the advanced nations were those on the eastern fringe of the Mediterranean Sea, while most of western and northern Europe was relatively backward. Much later, the Industrial Revolution triggered the rapid spread of another advance that has affected the world unevenly, bringing wealth to the more industrial nations in contrast to those that have remained more agricultural. At the beginning of the twenty-first century, the world's wealth still is unevenly distributed. The dichotomy between more and less developed nations is a crucial factor in any attempt to deal with world-scale changes in the biosphere and atmosphere, and in any effort to understand the demographic transition toward lower and more stable rates of population growth.

**Defining and Tracking Economic Levels.** To portray differences between developed and less developed nations, the gross domestic product (GDP) is often used as a surrogate for more specific indicators of quality of life. Energy use per capita correlates strongly with that measure, confirming the industrial character of the wealthier nations. The United Nations Development Programme (UNDP) has devised the Human Poverty Index, an aggregate measure that recognizes low life expectancy, illiteracy, and reduced access to health services, safe water, and adequate nutrition. In addition, the UNDP rates nations on the Human Development Index, which combines measures of literacy, life expectancy, and income. To define an income poverty line, the UNDP tabulates two measures: the proportion of the population living on less than U.S.$1 per day, or the proportion living in poverty according to a nation's own definition. World and national statistics present a mixed picture. Since 1965, world average child death rates have fallen by half; and life expectancy has risen by ten years. In developing countries, school enrollment has increased, and adult literacy rates have risen on average from 48 percent in 1970 to 72 percent in 1997 (UNDP, 1999, p. 25). Yet a number of countries have been growing steadily poorer since 1980, contributing to the growing gap between richest and poorest nations. On the basis of GDP, the ratio between the five richest nations and the five poorest was 2.7 in 1820, 7.4 in 1900, 36.6 in 1992, and 152.5 in 1997 (derived from data in UNDP, 1999, p. 38, and Table 7, pp. 155–158).

During the 1990s, there were two trends with regard to income poverty. In percentage terms, the trend has been positive: between 1987 and 1998, the proportion of population in developing nations who lived on less than $1 per day fell from 28 to 24 percent; excluding China, the reduction was from 29 to 26 percent. The absolute number of poor, however, remained roughly constant around 1.2 billion. Excluding China, the number rose from just under 880 million to over 980 million. The absolute increase in the number of poor has been mostly in sub-Saharan Africa, South Asia, Latin America and the Caribbean, and the transition economies of eastern Europe and Central Asia (Wolfensohn, p. 2).

**World Patterns.** In many discussions, less developed countries (LDCs) are contrasted with more developed countries (MDCs). The term *Third World* often is used as a synonym for *poorest nations*; and *the South versus the North* is another expression of the poor versus the wealthy. All these refer to mental maps based on one or more of the measures mentioned above.

For simplicity, we present one such map, based on the Human Development Index for 1994 (Figure 1). That map, using a three-level scale, justifies the term *Third World* for the least-developed nations of intertropical Africa and the cluster of nations in South Asia. Intermediate are parts of South America and much of North Africa and Southwest Asia, the former Soviet Union, China, and Indonesia. The highest category includes North America, western and northern Europe, Australia and New Zealand, Japan, and Malaysia. The term *South versus North* apparently is not validated by the actual map pattern: a better generalization would be *lower versus higher latitude*, for less versus more wealthy.

Any map that uses only a few categories hides significant differences among the members of each group. For instance, oil-rich nations such as Saudi Arabia, Kuwait, and smaller states on the Persian Gulf classify as high or intermediate, but they are quite unlike the Eu-

ropean members of that class, which have more diversified economies (on a map of GNP per capita, those oil-rich nations would stand out as extremely rich). Furthermore, a single map cannot reveal year-to-year changes in the underlying measures. For instance, in the period 1980–1995, some of the world's highest rates of GNP growth took place in some of the poorest nations, including those of East and Southeast Asia. Although an economic collapse in 1998 ended that spurt, the relative wealth of nations continues to change as some participate more successfully than others in world trade. [*See* Global Economy.] The uneven participation by rich and poor nations is revealed by four measures (Table 1) showing that the poorest 20 percent of world nations garner only 1 percent of the activity.

**Poverty and the Environment.** There is still a contrast between the more and the less developed nations. The latter are typically distinguished by problems of water supply and sanitation, and by erosion, desertification, and other threats to agricultural land that result partly from overuse and population pressure. The more industrial nations typically must deal with chemical and other industrial wastes, and with regional and urban air pollution owing largely to vehicle emissions. The distinction between the more and less developed nations is blurring, though; many cities in poor but economically growing nations now have industrial pollution and large fleets of autos, whose unregulated emissions pollute the urban air to levels beyond those in more developed nations. There is little doubt that the poor are disproportionately affected by declines in environmental quality. The rural poor are subjected to water supplies contaminated by agricultural runoff, industrial pollutants, and sewage, and they also live with indoor air pollution from primitive cookstoves.

Populations living on the land and growing their own food are most vulnerable to climate changes, natural or human induced, that alter temperatures or precipitation. In urban areas, the poor live where sanitation is inadequate and collection of solid and toxic wastes is erratic. This theme—that environmental degradation hits hardest those living in poverty—is developed thoroughly in a UNDP Development Report (UNDP, 1998, pp. 66–85). The various environmental threats to health in underdeveloped nations are summarized by one analysis that considers the risks from air (indoor pollution and outdoor, including lead from gasoline), water (lack of sanitation, and exposure to diseases such as malaria), and inadequate nutrition (Figure 2). On these measures, it is clear that populations in Africa and Southeast Asia are most at risk, while those in South Asia and China face risks at the next level.

The converse proposition, that poverty itself contributes to environmental degradation, is less certain. Rural poverty has been seen as both a major cause and

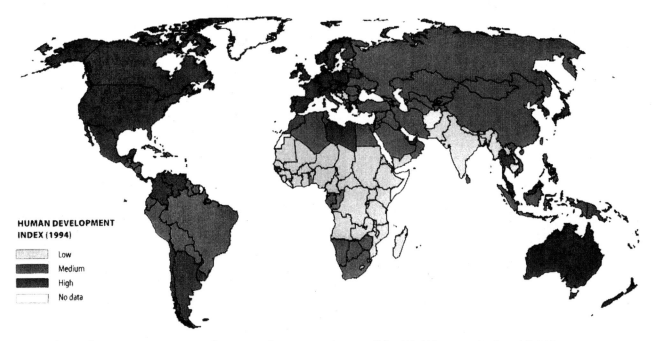

**Economic Levels. FIGURE 1.** Human Development Index (1994).

An index that attempts to capture multiple dimensions of human well-being by combining measures of literacy, life expectancy, and income. (After World Resources Institute, World Resources 1998–99, p. 222, citing United Nations Development Programme, *Human Development Report, 1997.* With permission of World Resources Institute.)

a result of degraded soils, vegetation, water, and natural habitats. The link between rural poverty and environment is viewed as a downward spiral in which past resource degradation worsens today's poverty, which in turn makes it difficult to care for or restore the agricultural resource base. Some farmers are forced to migrate to ecologically fragile lands like steep slopes or semiarid areas vulnerable to desertification. These pressures on the environment intensify as the population grows. But it appears that low income levels are not the only factor driving the degradation: communal ownership of land contributes to the problem, and institutions for managing communal land are missing (UNDP, 1998, pp. 66–67).

Some researchers have addressed the matter directly and have found that the simple downward spiral model may lead to erroneous policies. Scherr (1999) surveys

environmental issues related to agriculture and the poor and recommends a number of policy strategies that could address both poverty and environmental objectives. Prakash (1997), studying areas in the western Himalayan regions of India, concludes that poverty is not so much a cause of environmental degradation as a mechanism by which the true underlying causes (institutional and policy defects) are brought to bear on the environment. Microlevel studies of deforestation have shown that poverty may force people to clear forest cover for cultivation, or to use resources in an unsustainable manner, but they also demonstrate that poor people can and do invest considerable time and resources in forest management. In the absence of a simple causal link between poverty and deforestation, there is need for analysis at the local level (Arnold and Bird, 1999, p. 4).

**Economic Levels. TABLE 1.** Rich versus Poor in Global Opportunity, as of 1997

|  | RICHEST 20% | MIDDLE 60% | POOREST 20% |
| --- | --- | --- | --- |
| Share of world's gross national product | 86% | 13% | 1% |
| Share of exports of goods and services | 82% | 17% | 1% |
| Share of direct foreign investment | 68% | 31% | 1% |
| Share of Internet users | 93.3% | 6.5% | 0.2% |

SOURCE: *United Nations Development Report* (1999, p. 2).

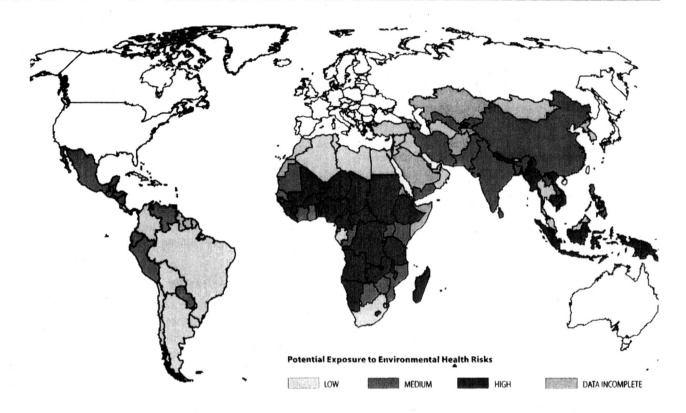

Potential Exposure to Environmental Health Risks

LOW    MEDIUM    HIGH    DATA INCOMPLETE

**Economic Levels. Figure 2.** Environmental Threats to Health in Developing Nations. (After World Resources Institute, 1997, pp. 4–5. With permission of World Resources Institute.)

**Global Warming.** The matter of fuel burning now divides the less from the more developed nations as they try to establish timetables to deal with emissions of greenhouse gases. [*See* Global Warming; Greenhouse Effect; *and* Kyoto Protocol.] Very populous developing nations, like India and China, understandably wish to fuel their growth with fossil fuels, as did the present industrial nations; however, the economic development of their large populations will require extremely large quantities of energy, much of which may be supplied by coal. [*See* Energy; *and* Fossil Fuels.] If that occurs, developing nations will soon rival the industrial nations in their emissions of greenhouse gases. Understandably, many in the developed nations would like to perfect and promote new technologies to allow less developed nations to use fossil fuels efficiently and to move gradually to more benign energy sources.

**Population Growth.** The history of world population growth and the evolving pattern of vital rates (birth rate, death rate, and natural increase) testify to the principle that as a population attains higher economic levels, death rates fall and, sooner or later, birth rates follow. Now, all nations in the most wealthy category have slow growing populations typical of nations that have undergone the European-style transition from high birth and high death rates to low birth and low death rates (with small natural increase) that accompanied the shift to more industrial and more urban societies. When modern medicines were introduced to some less developed regions after World War II, the result was an immediate drop in death rates; birth rates resisted lowering, however, and the demographic transition was incomplete. The resulting growth in less wealthy areas has been a major feature of world population patterns in the past fifty years. It is responsible for a progressive imbalance in numbers, so that roughly 80 percent of the population now lives in the lower two development categories (Figure 1), whose numbers are dominated by a few nations, such as China, India, and Indonesia. [*See* Human Populations.]

Today, not all poor nations have rapid population growth: some are beset by high death rates, especially those African nations afflicted by acquired immune deficiency syndrome (AIDS). Remarkable declines in fertility have taken place recently in rapidly developing countries such as Thailand, the Republic of Korea, Singapore, and Colombia, confirming the association between fertility decline and economic growth and contributing to a slight drop in the world's total annual increase. However, it is not increased income itself that lowers fertility, but social change. Lowered fertility is the result of improved health care, access to family planning, and better education—especially for women.

[*See also* Agriculture and Agricultural Land; Brundtland Commission; Carrying Capacity; Environmental Economics; Food; Industrialization; IPAT; Sustainable Development; Technology; *and* World Bank, The.]

## BIBLIOGRAPHY

Arnold, J. E. M., and P. Bird. "Forests and the Poverty–Environment Nexus." Paper presented at the United Nations Development Programme and Economic Commission Expert Workshop on Poverty and the Environment, Brussels, 20–21 January 1999. New York: UNDP, 1999.

Prakash, S. "Poverty and Environment Linkages in Mountains and Uplands: Reflections on the Poverty Trap Thesis." Campaign for Real Equitable Economic Development (CREED) Working Paper No. 12, February 1997. Available at www.mtnforum.org/mtnforum/archives/reportspubs/librarypraks97.

Scherr, S. J. "Poverty–Environment Interactions in Agriculture: Key Factors and Policy Implication." Paper presented at United Nations Development Programme and the European Commission Expert Workshop on Poverty and the Environment, Brussels, 20–21 January 1999. New York: UNDP, 1999.

United Nations Development Programme (UNDP). *Human Development Report, 1997.* New York: UNDP, 1997.

———. *Human Development Report, 1998.* New York: UNDP, 1998.

———. *Human Development Report, 1999.* New York: UNDP, 1999.

Wolfensohn, J. D. "Rethinking Development: Challenges and Opportunities." Address to Tenth Ministerial Meeting of United Nations Conference on Trade and Development (UNCTAD), 16 February 2000. Available at www.rrojasdatabank.org/wolfen.

World Bank. *World Development Indicators 1997.* Washington, D.C.: World Bank, 1997.

World Resources Institute. *World Resources 1997.* Washington, D.C.: World Resources Institute, 1997.

—DAVID J. CUFF

## ECOPOLITICS. *See* Environmental Movements.

## ECOSYSTEMS

An *ecosystem* is an ecological community (the living organisms inhabiting an area) and its associated physical (nonliving) environment. The term was originated in 1935 by A. G. Tansley (1935). Prior to this time, scientists had begun to approach the idea through such concepts as the "biocoenose" (Mobius, 1887); limnologists S. A. Forbes (1887) and A. Thienemann (1931) had discussed food chains, trophic structure, and organic nutrient cycling in lakes; A. J. Lotka (1925) had applied thermodynamic principles to food webs and chemical cycles; and E. Transeau (1926) had studied primary production and energy budgets in land plants. Since 1935, ecosystem ecology has developed into an important subdiscipline of ecology, with direct linkages to the fields of biogeochemistry and global change biology (Golley, 1993).

Ecosystems perform two basic functions: (1) they capture and process energy, and (2) they cycle and regenerate nutrients. Each function involves interactions between the ecological community (biotic component) and the physical environment (abiotic component). Accomplishing these functions requires an interacting group of organisms including *producers* (autotrophs), who capture energy, and *consumers/decomposers* (heterotrophs), who degrade wastes and regenerate nutrients. Because no single species does both, ecosystems are the smallest ecological units capable of sustaining life (Morowitz, 1979). Recognition of the importance of ecosystems to the sustainability of life on Earth is the basis of *ecosystem management*, a management approach in which biodiversity conservation is achieved by conserving ecosystems. Relationships among producer, consumer, and decomposer organisms constitute an ecosystem's *trophic structure* (Figure 1).

The ecosystem concept is independent of scale. At one extreme, model ecosystems (called Folsom bottles) can be created by sealing algae, bacteria, water, and sediments inside small flasks. Exposed only to sunlight as an energy source, these bottles have been known to sustain life for twenty years (Biosphere 2 is an application of this approach on a larger, more complex scale). At the other extreme, the planet Earth can be viewed as one large ecosystem (Lovelock, 1979). Despite the disparity in size, Folsom bottles and planet Earth have a number of common features as ecosystems. In particular, both are sustained by the same energy source—the sun. Because all ecosystems require an external energy source, no ecosystem is totally self-sustaining. Similarly, interconnections between ecosystems are the rule rather than the exception, and for that reason, defining ecosystem boundaries can be problematic (Reiners, 1986).

Most ecosystems use solar energy to drive ecosystem processes. Through photosynthesis, primary producers (plants) capture light energy (by means of photosynthetic pigments), using it to split water molecules and generate chemical energy in the form of free electrons, and releasing molecular oxygen as a byproduct. The free electrons are attached to carbon dioxide, converting inorganic carbon to organic (reduced, electron-rich) form. Biological organisms access the energy in organic carbon by metabolizing it through respiration (effectively the reverse of photosynthesis), using this energy to drive biological processes, and regenerating inorganic carbon dioxide and water as byproducts. At the ecosystem level, the total amount of inorganic carbon converted to organic form per year is called *annual gross primary production* (GPP).

Primary producers use a portion of this fixed carbon to support their own biological processes. The remainder accumulates as biomass or *net primary production* (NPP; equal to GPP minus primary producer respiration). At the ecosystem level, NPP represents energy available to support consumer populations. Energy

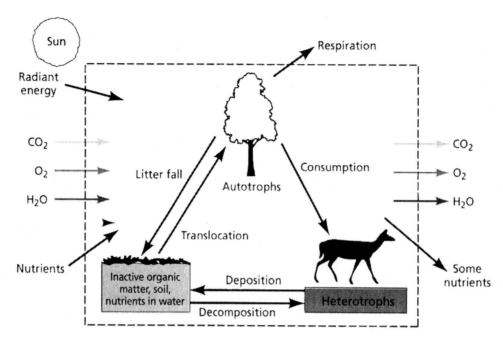

**Ecosystems. FIGURE 1. A Simple Ecosystem and Its Food Web.**

Primary producers (plants) use radiant energy from the sun to drive photosynthesis, fixing inorganic carbon ($CO_2$) in organic (reduced) form. Both producers and consumers use organic carbon to drive biologic processes through metabolism, resulting in the respiration (oxidation) of organic carbon back to the atmosphere as carbon dioxide. As carbon cycles from producer to consumer organisms, energy is dissipated as heat. Thus an external energy source is required for the continued operation of ecosystem processes. Nutrients, in contrast, cycle continually among ecosystem compartments. Additionally, nutrients may be imported or exported from the ecosystem via atmospheric and hydrologic pathways. (Adapted from O'Neill, 1976, as modified in Smith, 1996. Reprinted by permission of Addison Wesley Longman Publishers, Inc.)

transfer is accomplished by the consumption and respiration of organic carbon by consumer/decomposer populations. Organisms are functionally classified by the trophic level on which they feed. Herbivores feed on primary producer (plant) biomass, carnivores feed on animal biomass, omnivores feed on both, and detritovores feed on dead organic matter. Diagrams of the feeding relationships within ecosystems, called *food chains* or *webs*, identify linkages between producer and consumer organisms and can be used to track the flow of matter and energy through the ecosystem. Ultimately, fixed carbon either accumulates within the ecosystem as *net ecosystem production* (NEP; equal to GPP minus community respiration) or is recycled back to the atmosphere through respiration. [*See* Carbon Cycle.]

As carbon cycles through ecosystems, energy flows from one trophic level to the next. The transfer of energy across trophic levels has been estimated to be 5–20 percent efficient at best. Thus productivity and trophic complexity (i.e., the number of links in the food chain) are intimately associated ecosystem properties. Keeping ecosystems operating requires a constant input of energy from an external source (e.g., the sun). In contrast, carbon and other chemical elements cycle constantly within the ecosystem. Both inorganic and organic forms can be assimilated by living organisms;

metabolism of organic matter by consumer/decomposer organisms releases chemical elements back to the environment in inorganic form. This second function of ecosystems (nutrient regeneration) is closely linked to NEP, because rates of energy capture and carbon assimilation can be limited by the availability of other chemical elements, particularly nitrogen and phosphorus. Nitrogen is a key component of the photosynthetic enzymes that control carbon fixation. By means of adenosine triphosphate, phosphorus serves as the energy currency of the cell. Both nitrogen and phosphorus are also important components of nucleic acids (e.g., deoxyribonucleic acid [DNA]). [*See* Phosphorus Cycle.]

There is some controversy over the use of the term *ecosystem*. Some ecologists prefer to think of the community as the highest level of ecological organization (e.g., Begon et al., 1996). Ecosystem processes are viewed simply as interactions between the community and the physical environment. In the *reductionist* view, the whole is explained simply as the sum of its parts. In the alternative *holistic* view, ecosystem properties and processes transcend those of ecosystem components. An intermediate view defines ecosystems in a simple, operational manner, as the list of species in an area and their nonbiological environment. Regardless of which view one holds, the utility of the systems approach to ecological analysis lies in the fact that organisms are grouped according to function rather than taxonomy. Functional analysis can provide insights into competitive relationships among organisms with similar functional roles. Quantifying rates of energy flow and nutrient cycling can identify factors that control energy assimilation by ecological communities. The ecosystem concept may be most appropriate to understanding the dynamics of Earth's environment, where numerous and varied component ecosystems interact with a constantly changing physical environment over long time scales to produce the phenomenon known as global change.

[*See also* Biogeochemical Cycles; Biological Diversity; Biological Realms; Biomes; *and* Population Dynamics.]

### BIBLIOGRAPHY

Begon, M. J. J. L. Harper, and C. R. Townsend. *Ecology: Individuals, Populations and Communities*. Oxford: Blackwell, 1996.
Botkin, D. B., and E. A. Keller. *Environmental Science: Earth As a Living Planet*. New York: Wiley, 1995.
Forbes, S. A. "The Lake as a Microcosm." *Bulletin of the Peoria Scientific Association* 52 (1887), 77–87.
Golley, F. B. *A History of the Ecosystem Concept in Ecology*. New Haven, Conn.: Yale University Press, 1993.
Likens, G. E. *The Ecosystem Approach: Its Use and Abuse*. Vol. 3 of *Excellence in Ecology*. Oldendorf-Luhe, Germany: The Ecology Institute, 1992.
Lotka, A. J. *Elements of Physical Biology*. Baltimore: Williams and Wilkens, 1925.
Lovelock, J. E. *Gaia: A New Look at Life on Earth*. Oxford: Oxford University Press, 1979.
Mobius, K. *Die Auster und die Austernwirthschaft*. Berlin: Wiegandt, Hempel and Parey, 1887.
Morowitz, H. J. *Energy Flow in Biology*. Woodbridge, Conn.: Ox Bow Press, 1979.
O'Neill, R. V. "Ecosystem Persistence and Heterotrophic Regulation." *Ecology* 57 (1976), 1244–1253.
Reiners, W. A. "Complementary Models for Ecosystems." *American Naturalist* 127 (1986), 59–73.
Ricklefs, R. E. *Ecology*. 3d ed. New York: Freeman, 1990.
Smith, R. L. *Ecology and Field Biology*. 5th ed. New York: Harper Collins College, 1996.
Stiling, P. D. *Ecology: Theories and Applications*. 3d ed. Upper Saddle River, N.J.: Prentice-Hall, 1999.
Tansley, A. G. "The Use and Abuse of Vegetational Concepts and Terms." *Ecology* 16 (1935), 284–307.
Thienemann, A. "Der Produktionsbegriff in der Biologie." *Archiv für Hydrobiologie* 22 (1931), 616–622.
Transeau, E. "The Accumulation of Energy by Plants." *Ohio Journal of Science* 26 (1926), 1–10.

—MARK R. WALBRIDGE

# ECOTAXATION

*Ecotaxation*, also known as pollution, pigovian, or green taxation, is an economic instrument using taxes or subsidies to encourage behavior that reduces pollution or otherwise directs private activities toward actions that are deemed environmentally and ecologically favorable.

Taxation of polluting activities is designed to give firms and individuals an economic disincentive to create externalities. An externality exists when an activity of a firm or individual imposes a cost on another party for which the firm or individual is not charged by the price system of a market economy. The sufferer of the externalized cost can be a private citizen, firm, or society at large.

A common example of an externality is pollution experienced by residents on a river downstream from a polluting industrial production facility. The residents downstream of the industry might suffer from decreased water quality for drinking, swimming, or fishing as a result of the pollution emitted from the factory. The industry is externalizing the cost of pollution to another party.

The occurrence of an externality is illustrated in Figure 1. The solid curve represents a firm's cost in creating a product to sell to consumers. The cost of environmental degradation caused by pollution and the costs related to nuisances the residents neighboring the production facility must endure are not included in this cost curve. The dashed curve represents the total cost of the firm's production to society. It is the sum of private cost (the cost of production to the firm) and social cost (the pollution and nuisance costs). The amount of externality is the difference between the two curves.

Without public intervention, the polluting industry has no incentive to include the cost of pollution in its production price. The firm will produce and sell goods for any price above *a*, which is driven to quantity *q* by supply-and-demand forces. The actual price at which this supply-and-demand equilibrium would occur at quantity *q*, if the price of pollution were included, is price *b*. Without being forced to *internalize* the costs of pollution, the polluting firm or individual has no economic incentive to curtail its production of pollution. Without intervention, the air and water may be treated as waste receptacles for pollution, with the cost being

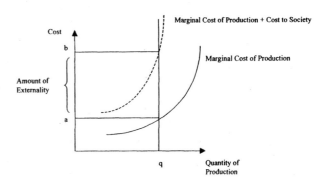

**Ecotaxation. FIGURE 1.** The Occurrence of an Externality.

shifted away from the polluter to a private or public entity.

Although ecotaxation has its foundations in theories proposed by the famous British economist Arthur C. Pigou in his 1920 book *The Economics of Welfare*, the externality tax did not gain practical significance until the 1970s. Pigou observed that polluting enterprises create external costs to third parties and proposed that an externality tax be imposed on the polluter in order to internalize the costs. His work expounded the "polluter-pays" principle, or the belief that the polluter should compensate the sufferer for any economic loss created by the pollution. The complete allocation of rights to the sufferer, however, has been challenged in economics literature, which questions if this paradigm will provide for a system that provides for the maximum social welfare.

Despite the economic arguments against it, the "polluter-pays" principle is generally accepted internationally as a verifiable rule. The Organisation for Economic Co-operation and Development adopted the principle in 1972 as foundation for its pollution policy, calling it a "fundamental principle of allocating costs of pollution prevention and control measures."

Ecotaxation methodology entails setting a price per unit of pollution emitted. This *marginal price* is either charged for each unit of pollution emitted or rewarded as a subsidy for reducing pollution by one unit. The marginal unit tax can be applied directly to pollution emissions or indirectly by taxing the use of products that have polluting consequences. Direct taxes are linked "backward" to the producer of the product, and indirect taxation is linked "forward" to the consumer.

Carbon dioxide, a cause of the greenhouse effect and global warming, is taxed both directly and indirectly. Governments often tax firms according to the number of units of carbon dioxide pollution they emit. The direct tax is based on the amount of the pollutant that is emitted from the production facility. Observing Figure 2A, we see that a direct tax on pollution emissions causes the price of production to increase. The producer reacts to increased production costs by reducing production or installing a pollution abatement technology to avoid the tax. The producer's supply curve will shift back, resulting in increased consumer prices and cost of consumption, and therefore decreased pollution.

Carbon dioxide pollution is indirectly taxed when energy production that creates carbon dioxide byproducts carries a surcharge to the consumer. Each kilowatt-hour of energy that carries a tax increases the price of energy to the consumer. As represented in Figure 2B, the tax increases the price to the consumer. The price increase causes the consumer to demand and use less energy, represented by a movement along the demand curve. The decreased quantity of energy demanded re-

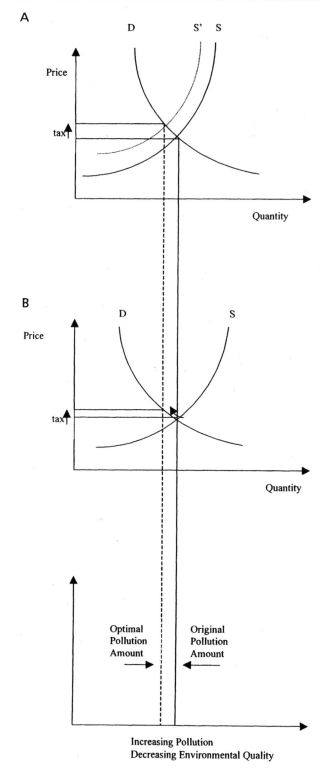

Ecotaxation. FIGURE 2. The Effects of (A) Direct and (B) Indirect Pollution Emissions Taxes on Pollution and Price.

duces the carbon dioxide pollution resulting from energy production and its use.

Specific examples of ecotaxation are product charges, pollution taxation, and energy taxation. Many countries tax products such as soft drinks and other beverages in order to internalize the cost of disposal of the used containers in which they are sold. Other examples of product charges are taxes on water consumption, automobile tires, and batteries. Product charges generally exemplify indirect taxation. Governments place direct taxes on the generation of hazardous wastes, household wastewater, chlorofluorocarbons, halons, carbon dioxide, and sulfur. As illustrated above, taxation can be applied as either a direct or indirect tax and has been a major focus of ecotaxation. Energy taxes include transportation fuel and household energy-use taxes.

Examples of ecotaxation subsidies are recycling tax credits, direct payments for pollution abatement, and reduction in real estate taxes to encourage environmentally friendly land use, such as preserving stream banks with vegetative cover or wetlands protection. Subsidies are often used in the agricultural sector. Governments reward farmers for reducing pesticide and fertilizer runoff, fencing cattle out of streams, and reducing soil erosion. Agricultural subsidies are more prevalent in more developed countries. This is due to the high importance those countries place on maintaining inexpensive, viable food supplies. Less developed countries often do not have sufficient funds in their treasury to subsidize environmentally sound agricultural practices and are reluctant to raise food prices to do so. These countries are hesitant to tax farmers because large percentages of the population work in agriculture, often earning extremely low incomes. Thus, ecotaxation has not been largely implemented in the agricultural sector of less developed countries.

The effects of pollution often cross political boundaries. One region will often suffer from pollution produced in other regions. Therefore, the environmental policy of one country has a large impact on the environmental quality of another. There are economic implications for inconsistent ecotax policies as well. For example, one country may place a large tax on its industries in order to diminish the production of carbon dioxide greenhouse gases. The environmental quality the country's industries are paying for is shared by the entire planet. Another country may not tax greenhouse gases at all, counteracting the benefits of the efforts of other countries' environmentally friendly policies. Furthermore, if industries in both countries produce the same commodity, the industry that isn't taxed will probably be more competitive on a global market.

Governmental intervention in externalities has historically comprised "command and control" regulation. This form of pollution control entails a pollution standard mandated by the government, either by specifying the exact pollution control technology required to comply with the policy using the "best available technology," or alternatively, by allowing individual firms to find technologies that meet the pollution standard at the lowest cost possible. While "least cost compliance" provides incentive for technological innovation of pollution reduction, governmental enforcement costs tend to be higher.

A major criticism of standard-setting policies is the disincentive for polluters to reduce emissions as low as possible, but instead to emit exactly the mandated amount of pollution. Furthermore, command-and-control regulation has not been effective in addressing nonpoint source pollution (pollution that cannot be directly attributed to a single source because many sources are responsible for emissions, such as pesticide runoff from farmland).

Many economists believe that a centralized regulatory policy structure will not provide for economically efficient levels of pollution. An economically efficient policy would reduce pollution emissions to the desired level at the lowest cost to society. Current economic theory suggests that a market can produce an efficient mix of pollution abatement and environmental compensation based on individual polluter situations. Ecotaxation is one such market mechanism. Ecotaxation is an economic instrument that has emerged from natural resource and environmental economics policy. [See Market Mechanisms.]

Like command-and-control regulation, ecotaxation is derived from a governmental pollution standard. A goal is established for total pollution emissions in a specific region. The market mechanisms differ from command-and-control regulation. The market, rather than a central authority, determines the allocation between users of the total pollution quota.

For example, household energy usage varies as a result of tastes, preferences, incomes, available technology, standards, and design practice. A pollution policy based on command-and-control regulation requires large expenditures on information in order to allocate pollution limits efficiently to firms and individuals. Some households find it cheaper to decrease energy usage than others do. The most cost-effective pollution policy would concentrate energy usage decreases in households where the reductions are achieved at the least cost rather than relying on command-and-control regulations.

Using market mechanisms, the households that can most efficiently reduce usage will be "self-selecting" in their response to the market signal provided by the energy tax. The same is true for polluting industries. Some industries will find it cheaper to reduce pollution emissions than pay the tax. Their pollution reduction will make up for the industries that find economic advantage in continuing to pollute. The environmental goal is met in a more cost-effective manner.

Failures of ecotaxation can stem from lack of foresight regarding the outcome of the tax. For instance, if petroleum-based fertilizers were taxed in order to decrease nitrogen pollution, economic forces might cause farmers to substitute the use of manure for the synthetic fertilizers. If the manure is not applied correctly, it may exhibit higher leaching and runoff characteristics than fertilizers, thus compounding the nitrogen pollution. Another example is the European refuse bag tax, in which garbage bags carry a tax. Instead of increasing investment in recycling and reduction of consumption, the tax induces many citizens to dump their refuse illegally, causing much worse pollutive consequences.

As concern over environmental degradation has grown in the last twenty years, countries are trying to ameliorate the damage being caused by various forms of pollution. Countries are generally moving away from centralized regulatory systems and toward market mechanisms. Ecotaxation is proving to be a viable means of pollution reduction and externality compensation. Empirical evidence suggests that ecotaxation could be economically viable in less developed countries as well. However, many of these countries are currently lacking regulatory institutions that allow for the effective implementation of market mechanisms in pollution control.

[See also Environmental Economics.]

### BIBLIOGRAPHY

Andersen, M. S. The Use of Economic Instruments for Environmental Policy. Available at http://www.iisd.ca/linkages/consume/skou.html. Last consulted 5 August 1998. Although advocatory, an insightful essay detailing the international linkages of economic instruments.

Cornwell, A., and J. Creedy. Environmental Taxes and Economic Welfare: Reducing Carbon Dioxide Emissions. Cheltenham, U.K.: Edward Elgar, 1997. Mathematical models illustrating sector responses to public policy implementation of pollution taxes.

Organisation for Economic Co-operation and Development (OECD). Climate Change: Designing a Practical Tax System. Paris: OECD, 1992. International collection of OECD representatives' papers on tax policy design.

———. Taxation and the Environment: Complementary Policies. Paris: OECD, 1993. Concise text on international implications of ecotaxation.

———. Environmental Taxes in OECD Countries. Paris: OECD, 1995. Collection of tables listing specific taxes and their annual revenue.

———. Agriculture and the Environment: Issues and Policies. Paris: OECD, 1998.

Pearce, D. W., and R. K. Turner. Economics of Natural Resources and the Environment. Baltimore: Johns Hopkins University Press, 1990. Encompassing text on the subject.

Pigou, A. C. The Economics of Welfare. London: Macmillan, 1920. Pioneering work on the subject of pollution taxes, yet of predominantly historic significance.

Sullivan, T. F. P. Environmental Law Handbook. 14th ed. Rockville, Md.: Government Institutes, 1997. General environmental law text.

—L. LEON GEYER AND JOSEPH K. SOWERS

**ECOTOURISM.** See Tourism.

## ELECTRICAL POWER GENERATION

In the more than a century since Thomas Edison introduced central station power in New York City in 1882, electric power has grown to become a major industry essential to a modern economy. From electric lights, elevators, and air conditioning to CD players, faxes, and computers, economical and reliable supplies of electricity are essential to support a wide range of services and activities in our society. Connecting almost every home, office, and factory in the developed world, the electric power system has fundamentally transformed the growth, productivity, living standards, and expectations of twentieth-century society. However, at the beginning of a new century, we are witnessing the largest restructuring in the history of the electric power industry, affecting many nations, including both developed and developing countries. This change will have far-reaching implications for the future development of the electricity industry in the new millennium. More fundamentally, as we look beyond the horizon, this change will further power the information revolution and increasing global interdependence of the twenty-first century.

**The Traditional Structure.** The electric power industry has traditionally been considered a natural monopoly industry. This means that the electricity supply to customers in a given service area can be provided at a lower cost by a single large firm than by several smaller, but less efficient, competing firms. An important argument for natural monopoly is that it prevents needless duplication of expensive facilities such as transmission and distribution networks. Typically, therefore, the provision of electricity takes the form of public enterprise or private franchise subject to government regulation on pricing, entry, investment, and reliability standards. Regulation is generally viewed as a substitute, albeit an imperfect one, for competition as an institutional device for assuring good performance. In exchange for the exclusive right to serve a specified geographic area with an assured rate of return, an electric utility accepts the obligation to serve at a uniform price. Under this obligation, all customers must be served without preference or discrimination. Unregulated industries, by contrast, have greater latitude on charging customers differentiated prices or turning away unprofitable customers.

The electric power industry is one of the most capital-intensive in most countries; in the United States, for instance, the investor-owned utilities had almost U.S.$595 billion in assets in 1996, amounting to almost 5 percent of the gross capital stock of all industries. Electric power systems consist of three distinct segments: generation, transmission, and distribution. Traditionally, electric utilities are vertically integrated through common ownership of facilities in all three segments, and horizontally coordinated across different geographic areas through power pools, long-term contracts, or cooperative planning activities. To understand this traditional structure, it is useful to review briefly the characteristics of the three segments.

Generation represents the most important segment of the electric power system in terms of asset value, accounting for over 50 percent of physical assets. Generation converts stored energy fuel, hydropower, or sunlight into electric power. Depending on the energy conversion process, there are four main types of generation: (1) thermal power (coal, oil, and natural gas), (2) hydropower, (3) nuclear power, and (4) nonhydro renewables (geothermal, wind, solar, biomass, etc.). Figure 1 shows the generation mix for all countries in the world from 1981 to 1996. In 1996, the world consumed 13,142 billion kilowatt hours of electricity with 61.3 percent from thermal, 19.3 percent from hydropower, 17.3 percent from nuclear, and the remaining 2.1 percent from nonhydro renewables.

Most electricity today is generated from the combustion of fossil fuels. When fossil fuels are burned, particulates, $NO_x$, sulfur dioxide ($SO_2$), and other emissions are released from the smokestack into the atmosphere, with local or regional effects. To minimize or prevent these emissions, electric utilities can use "cleaner" fuel sources or technologies, install low-$NO_x$ burners or otherwise modify combustion processes, or install filters, precipitators, scrubbers, selected catalytic reduction devices, or other postcombustion controls. Control of air pollution from fossil power plants is not new and will remain a primary environmental concern of every government for many decades. However, issues that increasingly dominate environmental policy debates will be international, involving environmental damage in-

flicted by one country, or a group of countries, on other countries or on the planet. [*See* Air Quality; *and* Fossil Fuels.]

From a global perspective, the most potentially serious problem is the emission of carbon dioxide. A buildup of carbon dioxide in the upper atmosphere might cause greenhouse effects that could change the global climate system, altering temperature and precipitation patterns, storm frequency and severity, oceanic currents and sea levels, natural ecosystems, and economic systems. Research on climate change, its causes, and its impacts is still at a very early stage. In December 1997, an international conference with delegates from an unprecedented number of countries was convened in Kyoto to discuss a global strategy to mitigate greenhouse effects. At the conclusion of the conference, the international delegates announced the Kyoto Protocol, which set greenhouse gas emission targets for major developed countries. If the global climate risk proves real, fossil fuel consumption will have to be cut back significantly, and alternative energy sources, such as nonhydro renewables, will have to be developed very rapidly. [*See* Kyoto Protocol; *and* Renewable Energy Sources.]

The transmission system delivers power from generation to local distribution systems. Most transmission systems use alternating current (AC), because the voltage level can be increased easily so that less energy is lost in the wires during transmission across long distances. The most efficient way to supply power to many customers is to transform power up to very high voltages at the boundary of the power plant and then transform it back down at the other end of the grid, as close as possible to the actual user of the power. The transmission system feeds substations that reduce voltage and spread the power from each transmission line to many successively smaller distribution lines that deliver electricity to the ultimate consumer.

The transmission system plays an important role in maintaining reliability, a hallmark of a modern electric power system. Reliability management in an interconnected AC power network involves some technical complexities. Certain unforeseen events, such as failures of key transmission equipment or generator facilities, may cause cascading outages throughout the network. For instance, when a line failure occurs in an interconnected AC power system, power flows throughout the network redistribute themselves according to physical laws (Kirchhoff's laws). As a consequence, this may cause overload on other lines, triggering additional failures that can eventually bring the entire system to collapse. To ensure the reliability of the power system, contingency analysis is regularly conducted so that power flows are kept within safety levels. For these reasons, centralized operation is necessary to ensure the reliability of a power system.

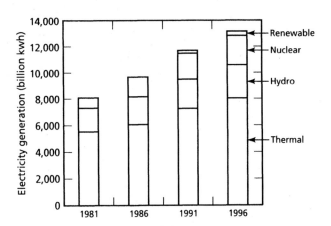

**Electrical Power Generation. FIGURE 1.** World Electrical Generation, Showing Types.

The distribution system, through a small number of connections to high-voltage transmission lines, delivers electric power at relatively low voltage to a large number of geographically dispersed customers. It is generally thought that distribution systems have important natural monopoly characteristics within limited geographic areas and, therefore, should be regulated or operated by a public utility in the future.

In distribution, information technology is emerging as a powerful force of change. High-speed power line networks, automated real-time meters and other "gateway to the home" devices, and the ubiquitous Internet, for example, have enabled new types of entities to enter the electric power industry. Power marketers and providers of electricity and related services can now be different from the provider of electrons. With the introduction of interactive remote meters (meters with two-way communication features), the rigidity in electricity would largely disappear, and the feasibility of continuous spot markets for electricity would be greatly enhanced.

**Structural Underpinnings.** Several features that distinguish the electric power industry from other industries contribute to the traditional institutional structure:

1. Nonstorability
2. Intertemporal and random variability of demand
3. Necessity to balance in an interconnected transmission network
4. Direct connections to customers
5. Capital intensity and economies of scale

The demand for electricity varies continuously and unpredictably from hour to hour, day to day, and season to season. Because of technological limitations, however, electricity cannot be stored. In an interconnected network, the demand and supply of electricity must be balanced instantaneously over time at every point of the network to maintain frequency, voltage, and system stability and to avoid power outages. Unlike other types of network, electricity in an AC electric transmission network flows in directions determined by physical laws rather than by contracts and is therefore more difficult to control. A local variation of demand and supply of electricity affects the power flows throughout the interconnected network. An equipment failure in one part of the network can cause the entire system to collapse. Efficiently meeting a new demand may involve coordinated adjustments of generators located far from the source of the demand. Hence the transmission system does not simply provide a physical connection between electricity generation and consumption facilities; it also involves active coordination of a large number of generating units dispersed throughout the network to support vigorously varying demand.

The centralized organization afforded under the traditional vertically integrated industry structure facilitates the essential task of coordinating efficient system operation and balancing the supply and demand of electricity continuously in response to changing system conditions. However, this structure effectively precludes a competitive market organization for the generation segment, even if competition in generation is otherwise feasible. To develop a competitive market for electricity generation, a necessary early step is to unbundle the vertically integrated structure. A challenge is to design a market organization that will accommodate system coordination in transmission without suffocating market competition in generation.

In a vertically integrated utility, two distinct types of economies of scale have been widely noticed. The first, observed at the plant level, is the reduction in average or incremental cost with increasing size of the generation unit. This type of economy was prevalent in the 1950s and 1960s—from 1949 to 1965, the incremental cost of generation capacity fell 37 percent as plant size increased. This can be important in countries or regions served by a relatively small system undergoing rapid growth. The second economy of scale occurs at the system level. As the size of the system increases, through either growth or combination, the system load shape tends to become relatively flatter and less volatile, a phenomenon known as the portfolio effect. A utility can then substitute for peak load units with more cost-effective baseload units and, at the same time, reduce the amount of reserve capacity (in percentage terms) required. To exploit this economy of scale, electric utilities commonly participate in power pools, interchange agreements, and other cooperative planning activities.

**Forces of Change.** Dramatic changes are now taking place in the institutional structure of electric power sectors in most countries around the world. In countries such as Argentina, Australia, Canada, Chile, New Zealand, Norway, Peru, the United Kingdom, and the United States, these changes generally feature greater competition in the generation segment of the industry. However, in several developed and developing countries such as France, China, and India, it may be a very long time before real competition is introduced. In the United States, reforms are being introduced most quickly in California and the northeastern states. These changes are driven by a combination of institutional, economic, and technological forces described below.

*Institutional forces.* In the United States, the restructuring in the electric utility industry is part of a broad movement of deregulation that has swept across the country. During the 1970s, many economists and others came to believe that the regulatory framework established since the Great Depression was no longer tenable, and that the introduction of competition could reduce economic inefficiencies resulting from traditional regulation. In industry after industry, including airlines, natural gas, petroleum, telephone, trucking, railroads, banking, and securities brokerage, deregulation was introduced, and a new industry structure emerged. The demonstrated success of market mechanisms has further boosted public support for the deregulation movement.

In the United States electric utility industry, the 1970s marked the breakdown of a regulatory consensus among regulators, utilities, consumers, and the public that had existed since the turn of the century. The foundation of this consensus was a long period of low inflation, cheap fuels, declining electricity prices, and high growth in electricity demand. It began to crumble with the rise in inflation rates in the late 1960s; it was exacerbated by the 1973 oil price shock, as imported oil shot from $1.50 to about $12 per barrel in a year, raising energy bills and public concern. Utilities, saddled with higher costs (partly due to uneconomical nuclear power plants), increased rates considerably. In response to public outcry, regulators in many states were forced to hold down rates, which, in turn, led utilities to halt projects adding generation capacity. To attract alternative suppliers to the market, Congress passed the Public Utility Regulatory Policies Act of 1978 (PURPA). By the 1980s, a whole new independent power-producing (IPP) sector had been formed, one of the principal forces pushing for restructuring today. While restructuring is occurring at a different pace in each state depending on local economic and political conditions, one thing seems clear: there is a broad disillusionment with the institution of regulation. The existing regulatory framework in the United States electric utility industry is increasingly untenable in light of market and technological realities.

In the United Kingdom, the election of the Thatcher government in 1979 marked a major watershed in British politics and economic policy. Privatization became an important element in the Thatcher government's overall economic program. For instance, the Electricity Act of 1983 was introduced to encourage the growth of IPPs. It led to the Electricity Act of 1989, which restructured the electric power industry prior to its sale.

In a number of other countries over the past few years, the electric power industry has undergone a substantial degree of privatization. While the reasons vary from country to country, some of the more evident ones include raising revenues for the state, acquiring investment capital, facilitating technology transfer, and improving industry performance. Electricity demand is expected to grow fastest in the developing nations, for which privatization is one means of obtaining badly needed foreign capital.

***Economic forces.*** The economic forces for change are coming from large consumers and IPPs. Even though most states do not yet permit retail wheeling, which entails buying and selling electricity across jurisdictions, consumers have increasingly found ingenious ways to avoid regulatory barriers. Industrial consumers have been most aggressive. Since energy costs are approximately 5 percent of most manufacturers' total operating costs (and as much as 30 percent in such energy-intensive industries as aluminum processing and steel making), manufacturers have significant incentives to reduce energy bills. One of the most important economic stimuli for consumers to support regulatory reform in the United States is the gap between the current regulated electricity price and the projected competitive market price. In California, for instance, electricity prices are currently about 10 cents per kilowatt hour, which is about 3–4 cents above what a consumer could buy directly from the wholesale market plus the local delivery costs. This price gap reflects historical investments in nuclear power and high-priced contracts that utilities were required to sign under PURPA. Competition offers electric utilities an opportunity to escape from the burden of setting rates based on these inefficient "stranded" investments (assuming they can negotiate some sort of return or write-off during the deregulation process). This explains why regulatory reform efforts in the United States have been concentrated in the states where the price gap is largest.

***Technological forces.*** In recent years, electric generation has experienced some significant technological improvements that have contributed to changes in the structure of the electric power industry. New generation technologies such as combined-cycle gas turbines (CCGTs) have been producing power at efficiency levels of 50–60 percent, with advanced technology targeting a level of 70 percent or better by 2005. This compares to a level of 30–40 percent for conventional power plants. CCGTs offer potential entrants a technology that can be introduced at modest scale (300–600 megawatts) and with a short construction time (two to three years). In addition, the price of natural gas has fallen to a level that has made new CCGTs competitive with some older, less efficient, coal-fired generation stations. The new combined-cycle generating technology has effectively ended the economies of scale in generation and significantly increased the feasibility of competitive generation markets.

A major force for future change in generation is the recent technological advance in distributed resources, which are small, geographically dispersed generation technologies such as fuel cells and microturbines. Such

## FUEL CELLS

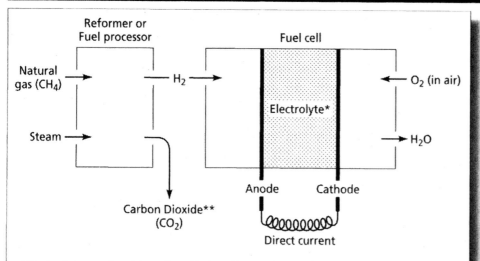

Reformer or
Fuel processor

Fuel cell

Natural gas ($CH_4$) → $H_2$ → → $O_2$ (in air)

Electrolyte*

Steam →

→ $H_2O$

Anode    Cathode

Carbon Dioxide**
($CO_2$)

Direct current

*Electrolyte may be either phosphoric acid or molten carbonate.

**Only 65–75% of the carbon dioxide emitted from fuel combustion.
In large operations the $CO_2$ could be collected, then sequestered
in exhausted natural gas fields.

First demonstrated in 1839, then refined in the 1960s for use in spacecraft, these devices convert the chemical energy of a fuel like hydrogen directly into electricity without combustion—a valuable trait at a time when combustion of fuels is responsible for the majority of greenhouse gases and air pollutants.

One unit, the size of a tractor trailer, is installed at the international airport in Pittsburgh, Pennsylvania, where its generating capacity of two hundred kilowatts meets the needs of a cluster of buildings, while hot water from the process is piped into radiators to heat a nearby hanger.

Similar units can be installed in apartment complexes and in commercial or industrial buildings. A larger unit, with nominal generating capacity of two thousand kilowatts was tested in 1996 in Santa Clara, California, as a demonstration funded by a group of utility companies and the U.S. Department of Energy.

Smaller fuel cells are being used in experimental electric vehicles that readily meet zero-emission requirements because the main byproduct of the process is water. Finally, tiny cells are being developed as alternatives to batteries in applications such as laptop computers and cell phones. Supplanting batteries in vehicles and electronic devices could benefit the environment because disposal of batteries is wasteful and difficult to accomplish safely.

While methanol may be the best feedstock for miniature and specialized cells, hydrogen for a stationary fuel cell usually is derived from natural gas, which essentially is methane ($CH_4$). Using methane as feedstock is attractive because natural gas is abundant in many parts of the world; and furthermore, methane can be produced sustainably from sewage and other organic wastes. At least one major oil company, however, is experimenting with a process that uses gasoline as the raw material, or feedstock. Gasoline on board the vehicle would be fed to a fuel processor, which would supply a fuel cell with hydrogen. Electricity from the cell would drive an electric motor.

In the absence of methane or gasoline, hydrogen can be produced directly from water by hydrolysis, using electrical energy wherever it is abundant. If hydrogen were to become a major fuel, produced at hydroelectric or photovoltaic installations, an essential part of the energy system would be fuel cells to convert the hydrogen to electrical energy.

### BIBLIOGRAPHY

Hayashi, A. M. "Taking on the Energizer Bunny." *Scientific American* (April 1998), 32.
Ingersoll, J. G. "Energy Storage Systems." In *The Energy Sourcebook*, edited by R. Howes and A. Fainberg, pp. 345–349. New York: American Institute of Physics, 1991.

—DAVID J. CUFF

distributed generators can be combined with advanced storage devices to provide flexible, customized service offerings that will meet a variety of customer needs. For instance, microturbines in the range 25–100 kilowatts can provide standby power, remote power, or commercial cogeneration economically, and fuel cells of similar sizes can provide premium power to residential customers. Competitive markets are likely to foster the penetration of distributed generation, which will further diminish the economies of scale in generation.

Institutional, economic, and technological forces are somewhat intertwined. For instance, the deregulation of the natural gas industry has been a clear case of success for market mechanisms. It led to the secular decline of gas prices, which contributed to the price gap between regulated and unregulated wholesale electricity markets, creating economic pressures for change in the electric power industry. Moreover, combined with the technological improvement in CCGTs, low gas prices helped effectively end the economies of scale in generation.

**Future Challenges and Opportunities.** Following the basic model of the previously deregulated network industries such as telephones and natural gas, the competitive segment (the generation of electricity) is being separated structurally or functionally from the natural monopoly segments (the transmission and distribution of electricity). The generation of electricity is being deregulated, and customers are being given the opportunity to choose among competing suppliers. The geographic scope of wholesale power markets is likely to expand in ways that transcend traditional jurisdictional boundaries, as evidenced in Europe by the emergence of the Nordic market, which now includes Norway and Sweden and is being extended to Finland and Denmark. However, a fundamental challenge in the design of new regulatory mechanisms is how to price the use of the transmission system.

Transmission pricing is a highly challenging issue because of the presence of network externalities, a problem that arises from the unintended effects a transaction between two parties may exert on other parties in the transmission network. It is important to get transmission pricing right to enable efficient competition of electricity markets. Two pure approaches are being pursued: (1) the tradable physical rights approach and (2) the nodal pricing approach. The first approach involves defining physical transmission rights and determining the prices of these rights through market trading. The second approach involves using an optimal dispatch algorithm to calculate a set of nodal prices from the submitted energy bids. In general, the tradable physical rights approach maximizes the freedom individual energy traders have to structure transactions and

minimizes the role of the transmission system operator, while the nodal pricing approach entails a more active and central role for the transmission system operator. The choice between these approaches or some combination of them will have major impacts on the institutional structure of the electricity generation market.

A predictable consequence of the changes in the electric power industry is that the complexity and volume of transactions will increase dramatically, demanding a significantly greater communication and information processing capability. Moreover, competition is likely to spur further demand for information technologies, because firms must respond quickly and reliably to customers' diverse needs. The increased demand will stimulate the development of advanced control, computing, and metering technologies. These developments, in turn, will constitute a powerful driving force for further changes in industry structure toward greater decentralization.

Electric restructuring will fundamentally change every aspect of the electric power business. But the shape of the long-term, deregulated future will be largely determined by technological innovation, the most powerful force of progress that we know. Indeed, there are reasons for optimism that the changes caused by restructuring will stimulate innovations that will power the world into the information age of the twenty-first century.

[See also Acid Rain and Acid Deposition; Energy; and Hydrogen.]

**BIBLIOGRAPHY**

Chao, H.-P., and H. Huntington, eds. *Designing Competitive Electricity Markets.* London: Kluwer, 1998.

Einhorn, M., ed. *From Regulation to Competition: New Frontiers in Electricity Markets.* London: Kluwer, 1994.

Gilbert, R. J., and E. P. Kahn, eds. *International Comparisons of Electricity Regulation.* Cambridge: Cambridge University Press, 1996.

Hart, D., and A. Bauen. "Fuel Cells: Clean Power, Clean Transport, Clean Future." London: *Financial Times Energy,* 1998.

Joskow, P. L., and R. Schmalensee. *Markets for Power: An Analysis of Electric Utility Deregulation.* Cambridge, Mass.: MIT Press, 1983.

—HUNG-PO CHAO

# ELEPHANTS

[*This case study focuses on the threats to elephants as a result of heavy hunting, human density, and land cultivation and considers steps for the conservation of African and Asian elephants.*]

There are two extant species of elephant, the African elephant (*Loxodonta africana*) and the Asian elephant (*Elephas maximus*). They are the largest land mammals, and the only living members of a large family of elephants, the *Elephantidae*, which fossil studies reveal as dominating the Pleistocene (1,600,000 to 10,000 years BP) faunas of Africa, Eurasia, and North America. Both genera originated in sub-Saharan Africa but, while *Loxodonta* species never ranged outside the African continent, *Elephas* expanded its range into Asia on two separate occasions, a mid-Pleistocene form culminating in the present-day Asian elephant. There are considered to be two subspecies of African elephant, the savanna or bush elephant (*L. africana africana*) and the smaller forest elephant (*L. africana cyclotes*) of central Africa. The geographic range and numbers of both species have been greatly reduced during the second half of the twentieth century, and the Asian form, in particular, represents one of the most seriously endangered large mammals in the world today.

The African elephant was once distributed throughout virtually the entire continent, including North Africa, but today its range is restricted to some 4.5 million square kilometers in thirty-seven countries, the largest populations occupying the forests and savanna woodlands of central and East Africa (Figure 1). In the nineteenth century there were an estimated 5 to 10 million elephants in Africa, and perhaps 3 million as recently as 1970, but by 1979 the numbers were reduced to 1.2 mil-lion, and by 1989 the rough estimate was a mere 600,000. Greater attention to the quality of surveys on which estimates have been based has recently (1999) produced qualified estimates of "definitely" 301,773 and "speculatively" 487,345. These figures may not reflect real differences with earlier estimates but probably represents a realistic picture of the current total.

Within historical times the Asian elephant's range included Syria, Iraq, and Iran, as well as Afghanistan and Pakistan; it also extended northward into China's Henan Province. Today it is confined to thirteen countries, extending from India and Sri Lanka in the west to Sumatra and Borneo in Southeast Asia (Figure 2). The habitat is greatly fragmented and totals an area of some 500,000 square kilometers, mostly isolated patches of relic forest. The most optimistic estimates put the species population of the Asian elephant at around 55,000, a little more than one-tenth that of its African cousin. Its dense tropical forest habitat makes counting of the Asian species very difficult and is responsible for a lack of reliable data on earlier population sizes. Thus it is impossible to quantify the rate of decline in numbers but, from the destruction of habitat in recent decades, it can be confidently assumed that the overall population size has decreased very substantially during this period.

The continually declining numbers and loss of habitat of both types of elephant give rise to widespread concern regarding the eventual survival of these charismatic large mammals. Because of this, much effort is currently

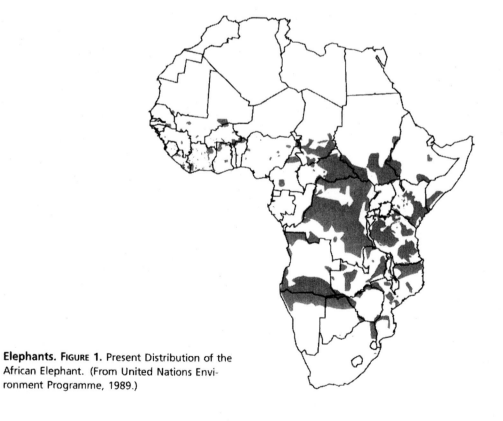

**Elephants. FIGURE 1.** Present Distribution of the African Elephant. (From United Nations Environment Programme, 1989.)

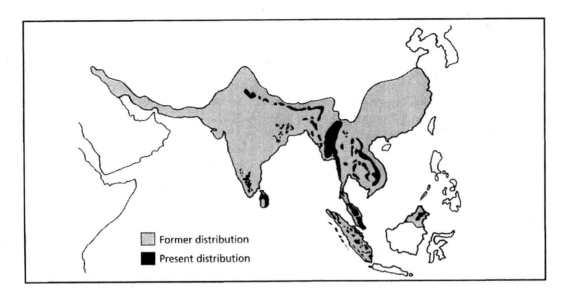

**Elephants. Figure 2.** Past (within Historical Times) and Present Distributions of the Asian Elephant. (From Santiapillai and Jackson, 1990.)

going into the identification of the precise nature of threats to their populations and the design of effective conservation strategies that will guarantee their long-term presence as significant components of the wild faunas of Africa and Asia.

**Threats to the African Elephant.** Numbers of African elephants have been in decline for a very long time. They were reduced at about 2 percent per annum during the nineteenth century, partly as a result of widespread agricultural expansion under colonial rule, which destroyed considerable areas of their savanna habitat. At the same time, there was large-scale slaughter of elephants for ivory and, by 1890, they were virtually exterminated in South Africa. During the first half of the twentieth century, there was a lull in both agricultural development and hunting for ivory, which was restricted by newly introduced game laws. Most elephant populations remained fairly stable, while others actually increased, with the result that elephants became an agricultural nuisance in some countries. However, from 1950 onward, elephant numbers started declining with increasing rapidity, initially because of both habitat loss and illegal hunting, but after 1970 hunting mortality increased dramatically and became the dominant influence on the rate of decline. In the decade 1979–1989 the continent-wide population of elephants, estimated from censuses and informed observations, dropped by 50 percent, from 1.2 million to 600,000. Destruction of the herds by poaching was particularly severe in the countries of East and central Africa, while, by contrast, a number of southern African countries,

with more efficient protection, reported stable or rising populations. Significantly, much of the slaughter took place within protected areas such as national parks and reserves, which in East Africa lost 77 percent of their elephants.

For much of this period, customs records show that the mean annual export of raw ivory from Africa was in excess of 700 metric tons, equating to an annual kill of around 8 percent of the population and confirming that the observed losses were due almost entirely to hunting mortality. Steep rises in the price of raw ivory (from U.S.$5 per kilogram in 1968 to over U.S.$150 per kilogram in the mid-1980s) contributed greatly to elephant decline, persuading whole villages in Africa to abandon farming in favor of elephant poaching. The total value of ivory exported from Africa during 1979–1989 is believed to have exceeded U.S.$500 million. The rising value of ivory was paralleled by an increase in the availability of automatic weapons, allowing entire herds to be killed easily without skilled marksmanship. This was a byproduct of the arms race, civil strife, and war which, during 1971–1980, saw Africa's arms imports increase from U.S.$500 million to U.S.$4.5 billion and general military expenditure grow by 6.6 percent annually. A partial halt to large-scale organized poaching and massive export of ivory from Africa was brought about in 1990 by a worldwide ban on trade in African elephant products imposed by the Convention on International Trade in Endangered Species of Wild Fauna and Flora (CITES), to which the majority of countries trading in ivory are signatories. More recently (1999), as a result of a modification of the CITES ban, several southern African countries with well-managed elephant populations exported a quota of ivory to Japan. However, some countries held this loosening of controls responsible for an alarming upsurge in poaching and ivory smuggling; in response to this, CITES agreed (April 2000) to a two-year hold on ivory sales to allow improved monitoring to be developed. [*See* Convention on International Trade in Endangered Species.]

Pressure from settlement, agricultural expansion, or heavy hunting in unprotected portions of habitat often results in the congregation of large numbers of elephants in the protection of parks and reserves. Unnaturally high densities resulting from this compression effect may lead to overuse of the elephants' food plants, leaving woodland and bushland seriously degraded, which in turn causes the elephants physiological stress or even high mortality. The problem was illustrated in the 20,000 square kilometer Tsavo National Park in Kenya, when, during a severe drought in 1970–1971, a total of six thousand elephants out of a park population of twenty-three thousand died from starvation. In some protected areas, such as Kruger National Park in South Africa, overpopulation is counteracted by periodic culling of elephants to bring total numbers within the area's carrying capacity. Either way, compression resulting from pressures on the species' habitat throughout much of its range is a threat to elephant survival.

The greatest long-term threat to the African elephant is the ever-expanding human population, particularly in those areas in which rainfall is high enough for cultivation, an activity that is incompatible with the presence of elephants. With the human population increasing by as much as 4 percent per annum in some areas, the future of the elephant looks bleak, and in the long term there will probably be few of them left outside protected areas—which presently constitute only 24 percent of the species' total range.

**Threats to the Asian Elephant.** The greatest threat to the Asian elephant lies in the fact that 20 percent of the world's human population lives within or close to its range and is escalating at 3 percent per annum. Unlike its African cousin, however, there has been no dramatic decline in numbers of the Asian elephant as a result of poaching, but rather a steady attrition resulting from the loss and fragmentation of its forest habitat. Throughout its range, large tracts of forest have been cleared to provide timber and make way for settled agriculture and plantation crops such as rubber and oil palm. Prime valley habitats have been submerged by dams and reservoirs serving hydroelectric and irrigation projects. Some forests of Southeast Asia were seriously damaged by prolonged military conflict, particularly by the use of chemical defoliants and napalm (although more forest has been destroyed since the Vietnam war ended than was destroyed during the war). In many areas, settlement and development projects have cut off the elephants' seasonal migration routes, leaving herds pocketed in small patches of forest that cannot supply their year-round needs. Subsequent invasion of newly planted crops brings these displaced animals into conflict with farmers, leading to both human and elephant fatalities. In India alone, some three hundred people are killed annually by elephants.

Many remaining Asian elephant populations are too small to be viable. In Thailand, for example, twenty-nine protected areas hold between thirteen and seventeen hundred elephants, but only thirteen of these reserves hold more than twenty-five individuals. Of the larger populations, some are endangered by being in politically unstable areas, such as the eight to eleven thousand animals inhabiting the states of northeastern India, where local unrest frequently results in weakened enforcement of wildlife laws. War has also taken its direct toll, as in Vietnam, where United States forces bombed elephants because they were being used as transport animals by the Viet Cong guerillas.

Only about one-third of the Asian elephant's habitat is in some category of protected area. However, many of these reserves are of less than optimal size and do not protect the year-round range of the local elephant population. In some cases, annual ranges are more than 50 percent larger than the area under protection, thus exposing elephants to hostile influences for much of the time. Furthermore, a majority of Asian protected areas suffer from human activities within their boundaries that are illegal and inimical to the well-being of elephants and other wildlife. These include damage to trees by the collection of timber, fuelwood, and fodder, settlement, cultivation, grazing of domestic stock, and fires. The construction of dams and other development works has also seriously degraded a number of reserves.

Poaching, mainly for ivory but also for meat, hide, and bones, constitutes a further threat in some areas. A 1989 survey of ten proposed elephant reserves in India indicated some degree of poaching as a problem in eight of them. Because only males bear tusks in the Asian elephant, heavy poaching for ivory leads to a skewed sex ratio: reported as 1:23 in favor of females in one southern Indian reserve. Such a decrease in the proportion of males in a population increases genetic drift, which leads to inbreeding and eventually to high juvenile mortality and poor breeding success. Some populations have a high proportion of naturally tuskless males (known as *maknas*), and here the sex ratio is more balanced, with less adverse genetic effects.

The centuries-old practice of capturing and taming Asian elephants for domestic use poses an unusual threat to the dwindling wild populations. Some countries have banned live capture, but in Myanmar (Burma) considerable numbers are still caught for logging work.

**Elephant Conservation.** Current human demographic trends in Africa and Asia, coupled with the expansion of settlement and cultivation, are the key factors in the survival of both species of elephant.

Even if all forms of illegal offtake can be brought under control, pressure on elephant habitats will continue and is likely to intensify, especially in Asia. There is a need for conservation programs based on a careful consideration of biological, social, political, and economic factors that reconcile the needs of elephants and people competing for the same living space.

There have been several continent-wide action plans for both elephant species, but national governments are not bound by the prescriptions that they contain. The urgent need is for national elephant conservation plans to which individual governments are committed and that can be used as the basis for seeking international funding. Political will and plans exist in a few countries, but adequate public finance is frequently lacking. Recent action plans recognize the impossibility of saving all remaining populations, and incorporate criteria for the identification of those with intrinsic value on which conservation effort should be concentrated.

Fundamental to good conservation planning is research aimed at the acquisition of reliable data on such matters as elephant population dynamics and interaction with vegetation, as well as patterns of human density and land use leading to the identification of areas of potential conflict. Some large populations of the African elephant are still poorly known, especially in the tropical forests of central Africa.

The problem of conflict between elephants and humans must be taken into account in the planning of land use in areas adjacent to elephant ranges, encouraging an approach in which local people manage their resources sustainably and in a manner that minimizes impacts on elephant habitat. In Asia, a concept of Managed Elephant Ranges is emerging that, while giving priority to elephant conservation, permits human activities compatible with it except in core areas that are linked by corridors of protected habitat. Education programs are essential for communities living in association with elephants, to acquaint them with elephant behavior and point out the value of the species to humankind and potential benefits from their preservation. In many African and Asian countries there is also a need for the introduction of land use planning procedures and legislation that strictly limit the impacts of forestry, mining, mineral extraction, and other development activities on elephant habitat. Political support for existing protected-area systems containing elephants needs to be strengthened and requests for downgrading of the legal protection of such areas resisted. The establishment of new reserves in regions of low coverage, such as the central African forests, is to be encouraged.

In addition to these habitat protection considerations, actions to control poaching and illegal trading in ivory must remain major objectives. There is a need for a review of conservation legislation in some countries, so as to ensure adequate powers for enforcement personnel and the imposition of realistic penalties. Studies on cost-effectiveness of field enforcement in Africa indicate a relationship between numbers of elephant killed and expenditure on anti-poaching per unit area of habitat, with money spent on intelligence gathering being especially critical. A figure of U.S.$200 per square kilometer is required to guarantee the integrity of reserves and the safety of elephants, and yet the majority of African conservation budgets are way below this figure and have been cut by more than 50 percent in recent years as a result of economic structural readjustment programs. Good field equipment, sound leadership, staff morale, and the cooperation of local people are further key factors that require strengthening in both Africa and Asia. People living around reserves often feel excluded, but many authorities now understand the wisdom of sharing tourist revenues with them in the form of employment, healthcare, and education, which encourages local participation in the conservation effort.

Although there is now very little market for ivory in the West, there is still a demand in the strong economies of the Far East, notably Japan and China. Significant seizures of illicit ivory still take place in Africa in spite of a dramatic price drop following the 1990 CITES ban, and vigilance against ivory smuggling will always remain a necessity in the conservation of both elephant species. Whether the controlled export of ivory recently commenced by some southern African nations with healthy elephant populations will ultimately eliminate the illegal trade remains an open question. However, the burgeoning human population and the subsequent reduction in wilderness available for elephants is likely to ensure a steady long-term decline in their numbers in Africa and Asia, even if illegal killing for the ivory trade is brought under control.

## BIBLIOGRAPHY

Barnes, R. F. W., G. C. Craig, H. T. Dublin, G. Overton, W. Simons, and C. R. Thouless. African Elephant Database 1998. IUCN/SSC African Elephant Specialist Group. Gland, Switzerland and Cambridge, U.K.: IUCN, 1999. A detailed country-by-country analysis of the latest data on numbers and distribution of the African elephant.

Chadwick, D. H. The Fate of the African Elephant. San Francisco: Sierra Club Books, 1992. A very readable account of elephant life and the efforts of people involved in their conservation throughout Africa and Asia.

Cumming, D. H. M., et al. African Elephants and Rhinos: Status Survey and Conservation Action Plan. Gland, Switzerland: IUCN, 1990. Contains prescriptions for the conservation of the African elephant as of 1990.

Ivory Trade Review Group. *The Ivory Trade and the Future of the African Elephant*, vol. 1, *Summary and Conclusions*, vol. 2, *Technical Reports*. Oxford, 1989. A very detailed technical analysis of the ivory trade and its impact on the African elephant.

Kemf, E., and P. Jackson. *Asian Elephants in the Wild*. WWF Species Status Report. Gland, Switzerland: Worldwide Fund for Nature, 1995. Provides a concise summary of issues surrounding the conservation of the Asian elephant, highlighting the contribution of the WWF.

Santiapillai, C. "The Asian Elephant Conservation: A Global Strategy." *Gajah* 18 (1997), 21–39. A preview of the latest international proposals for the conservation of the Asian elephant.

Santiapillai, C., and P. Jackson. *The Asian Elephant: An Action Plan for its Conservation*. IUCN/SSC Asian Elephant Specialist Group. Gland, Switzerland: IUCN, 1990. Country-by-country status and conservation recommendations on the Asian elephant, as of 1990.

Thouless, C. *Review of African Elephant Conservation Priorities: A Working Document of the IUCN/SSC African Elephant Specialist Group*. Nairobi: IUCN/SSC African Elephant Specialist Group, 1997. Up-to-date recommendations for conservation authorities, but readable by the informed layman.

United Nations Environment Programme. *The African Elephant*. UNEP/GEMS Environmental Library No. 3. Nairobi, 1989. A concise factual account of elephant numbers and distribution in 1989 and their relationship to human demography and economics.

—JOHN B. SALE

# EL NIÑO–SOUTHERN OSCILLATION

Throughout recorded history (and undoubtedly before) there have been droughts in Indonesia, excessive rains in the southeastern parts of North America, floods in the regions of what is now southeastern Brazil, and other climatic anomalies throughout the world. These anomalous episodes have come every few years and until recently seemed to have no apparent pattern, connection, or reason for occurring. [*See* Natural Climate Fluctuations.] Only in the last decade or so have we come to realize not only that these episodes are connected, but also that they are far-flung aspects of a single climate phenomenon known as El Niño–Southern Oscillation (ENSO for short). We have come to understand that this phenomenon of ENSO arises in the tropical Pacific from interactions between the atmosphere and the ocean. These interactions are now well enough understood that they can be modeled by computer and predicted many months in advance. These predictions of ENSO in the tropical Pacific can then be used to provide guidance on the probability of climatic anomalies in many (but not all) regions throughout the world. [*See the biography of Walker.*]

This article describes the phenomenon of ENSO in the tropical Pacific, briefly indicating how we have come to know what we know, what some of its impacts are within and outside of its immediate region, what determines its predictability, and how these predictions have already been put to use.

**The Normal Tropical Pacific.** The tropical Pacific is approximately fifteen thousand kilometers wide from the coast of Ecuador to the islands around New Guinea, Sulawesi, Borneo, and the other islands of the so-called maritime continent. The western tropical Pacific is very warm (with persistent sea surface temperatures of 29°C) and has heavy precipitation (on the order of one centimeter per day throughout the year), and the sea level pressure is low. By contrast, the eastern Pacific is several degrees cooler and has very little rainfall, and the sea level pressure is high.

These properties are linked by an atmospheric circulation connecting the eastern and western parts of the tropical Pacific. At the surface, the trade winds blow from east to west, from the high to the low pressures, bringing moisture to the warm western Pacific. The air rises in deep precipitating cumulonimbus clouds, thereby lowering the surface pressure, and proceeds to flow out aloft. At high levels in the atmosphere, the air flows eastward toward the eastern Pacific, where it then sinks, suppressing rainfall and raising the surface pressure. The westward-moving trades cause an upwelling of cold ocean waters near the equator (the Coriolis force requires the water to flow to the right of the winds in the Northern Hemisphere and to the left in the Southern Hemisphere), thereby cooling the eastern ocean. The circuit is closed by the simple observation that in the tropics, regions of persistent precipitation tend to lie over warmer water, usually at least 28°C. The region of persistent precipitation thus lies far to the west, drawing the westward low-level trades, which cool the east, ensuring that the warmer region consistently lies to the west.

The asymmetry between east and west is enforced by the *thermocline*, the thin region of large temperature gradient near the surface of the ocean (about one hundred meters deep in the tropics) that demarcates the transition between warmer water near the surface and the cold water below. The westward-moving trades keeps the thermocline deep in the west and shallow in the east at the same time that it causes upwelling on the equator. The upwelling acts on a shallow thermocline in the east and is therefore able to bring up cold water very easily, while the same winds, acting on a deeper thermocline, are incapable of cooling the warm water in the west.

The conditions around the normal tropical Pacific Ocean stay roughly the same with the seasons: the west stays warm, has rising motion into deep precipitating clouds, and has low sea level pressure; the low-level winds stay to the west. The east stays cold at the surface and has descending motion and high pressure. The atmosphere above the Pacific thus constitutes a vast circulation system spanning fifteen thousand kilometers and maintaining the conditions at the surface of the ocean favorable for its own existence. The system seems cooperative: the ocean and the atmosphere interact in such a way as to maintain normal conditions in the tropical Pacific. The normal system, as described above, is portrayed by Figure 1A. [*See* Ocean Dynamics.]

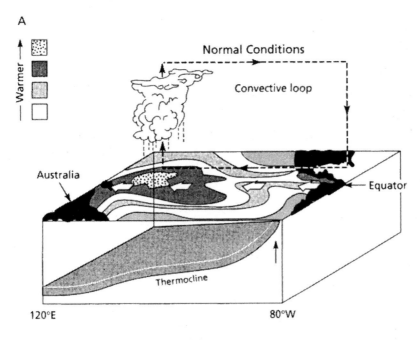

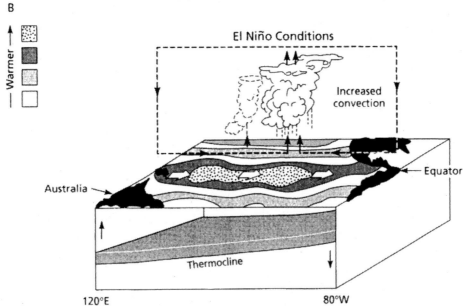

**El Niño–Southern Oscillation. FIGURE 1.** (A) Normal Conditions in and over the Tropical Pacific Ocean.

The sea surface temperature in the west is high, rain is falling over the warm pool of water in the west, and the sea level pressure is low. In the east, the water is cool, no rain is falling, and the pressure is high. The surface trade winds blow from east to west and bring moisture from the east to precipitate in the west. Below the surface, the thermocline, a region of sharp vertical temperature change that separates the relatively warm water near the surface from the cold water below, is deep in the west and shallow in the east. (B) **El Niño Conditions in and over the Tropical Pacific Ocean.** The warm water in the west expands eastward, and the precipitation similarly expands. The pressure in the west rises and in the east falls as the precipitation moves eastward. The trade winds relax and become weaker (or reverse to eastward), and the thermocline flattens.

**The Anomalous Tropical Pacific: ENSO.** Every few years, anywhere from three to four years, the coupled atmosphere–ocean system either strengthens or weakens. When it weakens, the trades relax; in response, the thermocline shoals in the east and deepens in the west. Because the weaker trades no longer cause as much upwelling, they cool the east less, and the eastern sea surface temperature rises. The warm water in the west is no longer so confined to the far west and moves eastward, bringing the precipitation with it. The pressure rises in the west and falls in the east as the conditions for maintaining the trades weaken. These weakened conditions are collectively known as the warm phase of ENSO. Since there are now warm anomalies in the eastern Pacific, it is sometimes called El Niño. Figure 1B shows the conditions characteristic of the warm phase of ENSO.

Similarly, when the system strengthens, the trades in-

crease in intensity, and the thermocline shoals still further in the east. The stronger trades, acting on a still shallower thermocline, cool the east even more strongly, and the sea surface temperature decreases. The precipitation retreats even further to the west with the retreating warm water. The pressure falls still further in the west and rises in the east as the conditions for maintaining a stronger trades strengthen. These conditions collectively are called the cold phase of ENSO.

The pressure signal of this ENSO cycle is easily seen. Long sea level pressure records in the western Pacific (Darwin, Australia) and in the eastern Pacific (Tahiti) may be represented in terms of anomalies—deviations of pressure from its normal value. The two records display an out-of-phase relationship (Figure 2A). The degree of difference (Tahiti minus Darwin) is often taken as an index to represent the cycle; this Southern Oscillation Index (SOI) is negative during warm phases of

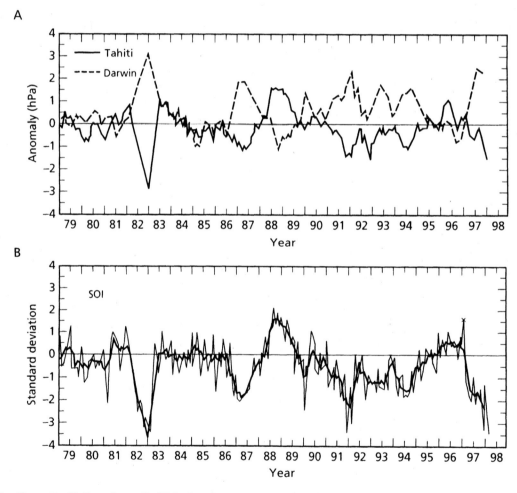

**El Niño–Southern Oscillation. Figure 2.** (A) Surface Pressure Anomalies in the Western Pacific (Darwin, Australia) and in the East-Central Pacific (Tahiti).

The Darwin anomaly is the deviation of the sea level pressure from its normal value at that time of year. The out-of-phase rela-

tionship of sea level pressure ten thousand kilometers apart is apparent. (B) **The Southern Oscillation Index (Tahiti–Darwin Sea Level Pressure) Expressed as Standard Deviations from Normal.** One standard deviation corresponds to an excursion of about 1.5 hectopascal (i.e., 1.5 mbar).

ENSO and positive during cold phases. Figure 2B shows the record of the SOI: note especially the very large warm phases of ENSO during 1982–1983 and 1997–1998. The warm phase of ENSO in 1997–1998 was so strong that the water in the eastern Pacific was not much colder than the water in the western Pacific so that water of temperature 29°C filled the entire tropical Pacific. Precipitation moved away from the west and into the central and eastern tropical Pacific. The pressure difference between east and west was almost zero, and the trades hardly blew. In sum, the system was running as weakly as it could, a highly abnormal situation.

**Effects of Warm and Cold Phases of ENSO.** We may distinguish two types of effects—those local to the tropical Pacific and those remote from this region. The effects local to the tropical Pacific are easier to understand and more robustly connected to warm and cold phases of ENSO.

*Effects local to the tropical Pacific.* The local effects of ENSO are of two main types: those that effect local land regions, mostly through the effects of the movement of precipitation when sea surface temperature changes, and those that directly effect the ecosystems in the ocean.

The grand movement of precipitation into the eastern tropical Pacific as sea surface temperature warms during warm phases of ENSO generally brings precipitation to places in the central and eastern ocean where it otherwise rarely occurs and removes rainfall from the western Pacific, where it normally is heavy. The arid coastal plains of Peru normally get little rainfall, and life is relatively well adapted to this normal lack of rainfall. During strong warm phases of ENSO, torrential rainfalls occur, and agriculture, roads, buildings, and infrastructure in general are all adversely affected. Conversely, in the western Pacific, where life is well adapted to large amounts of rainfall, forests and agriculture suffer from lack of rainfall, and forest fires are much more abundant. The Indonesian forest fires of 1982–1983 and 1997–1998 were devastating to large areas of hardwood forests, with the resulting smoke affecting large regions of the maritime continent. [*See* Fire, *article on* Indonesian Fires.]

The eastern tropical Pacific is an exceptionally fertile part of the ocean. Because the upwelling is so large, and the thermocline is so close to the surface, not only cold water but also plentiful nutrients and carbon, characteristic of the deep ocean, are brought to the surface. The net result is that the normal tropical Pacific maintains an extraordinarily abundant and diverse ecosystem, ranging all the way from plankton to grazers to fish to seabirds. During warm phases of ENSO, this ecosystem is disrupted both by the changes of temperature and by the deepening of the thermocline and the weakening of upwelling. The net effect of the removal of the nutrient source to the upper ocean is a much less productive environment, with fisheries off the coast of Peru, Ecuador, and Chile suffering and mass die-offs of seabirds. The effect on fisheries is somewhat countered by the warmer water—warm-water fish species not normally native to the cold coastal waters appear during warm phases in an attempt to seek waters to which they are normally adapted. Since the coastal fisheries tend to be specific to cold-water fish, they have seldom been able to take advantage of this substitution.

The weakening of upwelling to the surface waters also has an effect on the carbon dioxide content of the atmosphere. The normal tropical Pacific is a source to the atmosphere of about one gigaton (1 trillion kilograms) of carbon a year—a significant part of the global carbon cycle. During warm phases of ENSO, carbon-rich waters are no longer upwelled from the deep to the surface, and the disruption of the aquatic ecosystems means that organic carbon in particulate form is no longer being packaged by surface biota for return to the deep. Consequently, the release of carbon to the atmosphere decreases significantly. During the large warm phase of ENSO of 1982–1983, the source of carbon dioxide to the atmosphere in the tropical Pacific dropped to near zero.

*Effects remote from the tropical Pacific.* The remote effects during a warm phase of ENSO are summarized in Figure 3. These effects are objectively determined by correlating temperature and precipitation around the world with the Southern Oscillation Index. While it is believed that movements of convection with sea surface temperature in the tropical Pacific emit planetary waves around the Earth and thereby affect other regions of the globe, the precise mechanism is not known and is a subject of intensive research. The connections, whatever their mechanisms, can be captured by numerical atmospheric models so that the effects of a known sea surface temperature in the tropical Pacific can be estimated in regions far from the tropical Pacific.

**Predicting ENSO and Using the Predictions.** Numerical models are now being used to predict the sea surface temperature anomalies in the tropical Pacific six months to a year in advance. It is important to understand the nature of this prediction, which requires a suitable model for making the prediction and sufficient data to initialize the model.

Because ENSO intrinsically involves the atmosphere and the ocean, the correct model for prediction is a coupled atmosphere–ocean model. The atmosphere forces the ocean, mostly though wind acting on the sea surface, and thereby changes the ocean's surface temperature. In turn, the sea surface temperature affects the atmosphere.

A major component of the World Climate Research Program is the Tropical Ocean and Global Atmosphere

A

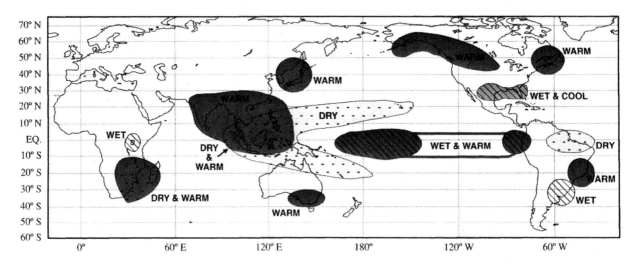

B

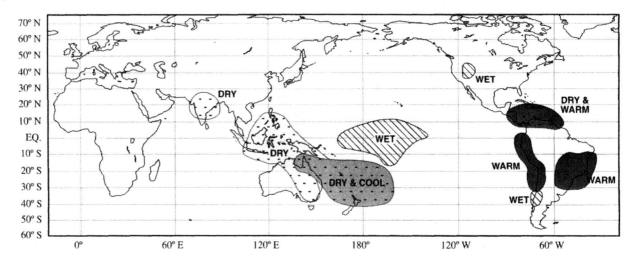

**El Niño–Southern Oscillation. FIGURE 3.** (A) Global Effects of Warm Conditions in the Tropical Pacific during Northern Winter. (B) Global Effects of Cold Conditions in Tropical Pacific during Northern Summer.

Because the global effects occur predominantly in the winter hemisphere and are correlations, they usually, but not always, occur in conjunction with warm or cold conditions in the tropical Pacific.

(TOGA) program. The so-called TOGA Observing System, a network of moored buoys across the entire tropical Pacific basin, measures winds and humidity at the surface of the ocean and temperatures within the top five hundred meters of the ocean. The results are telemetered to satellites and are available almost immediately for research and predictions. The data for the previous day may be freely accessed at http://www.pmel.noaa.gov/toga-tao/realtime.html.

These data are then used to estimate the state of the entire tropical Pacific ocean as input to the prediction. (The state of the atmosphere is known from the analysis performed for weather prediction.) When the model, having been initialized, is then allowed to run for a pe-

riod of time corresponding to a given number of months, the result is a prediction of sea surface temperature the same number of months into the future. Comparing the prediction to the actual sea surface temperature at the predicted time serves to verify the prediction's accuracy. In practice, it would be extraordinarily time consuming to wait many years for a sequence of forecasts to verify. Instead, prediction accuracy is developed by forecasting events that have already occurred and can be verified by comparing to past data; this process of retrospective forecasting is called "hindcasting." The accuracy so developed is a property of a *series* of forecasts and not that of a single forecast. All individual forecasts have an a priori probability of being correct, the probability being higher when the accuracy of a series of forecasts is higher.

In general, the quantity forecast by such models is sea surface temperature in the tropical Pacific, whereas the quantities of most use are air temperature and precipitation. Values for air temperature and precipitation, both local and remote, are obtained from the sea surface temperature either directly, by the same model making the forecast, or indirectly by use of a different model or by statistical inference. Once the tropical sea surface temperature is forecast, the accuracy of the air temperature or precipitation forecast depends on the region affected, the robustness of the signal from the tropical Pacific to the region in question, and the natural variability upon which the ENSO signal is imposed. The forecast is thus likely to be best in the tropics near the Pacific, where the ENSO signal is direct and the background variability is low, and poorest far from the equatorial Pacific, in midlatitudes where the natural variability, especially in winter, is largest.

**What Next?** The study of ENSO progressed considerably in the last decade of the twentieth century. The world was totally unaware of the development of the 1982–1983 warm phase of ENSO owing to an almost complete lack of measurements. The warm phase of 1997–1998, in contrast, was monitored throughout its evolution and predicted at least six months in advance, and preparations based on those predictions were made in various parts of the world. Other parts of the world, however, remained totally unprepared (or perhaps unaware) of the largest ENSO anomaly ever recorded. New knowledge takes time to disseminate and to affect old ways of doing things.

It is only a matter of time before the advanced nations of the world begin to use the forecast information for both private and public gain. The less developed nations of the world need help and guidance to use climate information effectively. Institutions have arisen, especially the International Research Institute for Climate Prediction, to do the physical and social science research needed to apply climate forecast information successfully in less developed parts of the world.

[*See also* Atmosphere Dynamics.]

## INTERNET RESOURCES

A number of Web sites are now available for learning about El Niño and for watching the evolution of El Niño. A general site with links to most major El Niño sites is maintained by the National Oceanic and Atmospheric Administration (NOAA) Office of Global Programs: http://www.ogp.noaa.gov/enso/.
The state at the surface of and below the surface of the tropical Pacific may be found at: http://www.pmel.noaa.gov/toga-tao/realtime.html.
A current guide to predictions and the use of predictions may be found at: http://iri.ldeo.columbia.edu/.

## BIBLIOGRAPHY

Battisti, D. S., and E. S. Sarachik. "Understanding and Predicting ENSO." *U.S. National Report to IUGG (1991–1994): Contributions in Oceanography*. Special issue. *Reviews of Geophysics* supp. (1995), 1367–1376.
Cane, M. A. "El Niño." *Annual Reviews of Earth and Planetary Science* 14 (1986), 43-70.
Fagan, B. *Floods, Famines, and Emperors: El Niño and the Fate of Civilizations*. Boulder: Perseus Books, 1999.
Grove, R. H., and J. Chappell, eds. *El Niño: History and Crisis*. Cambridge: White Horse Press, 2000.
Latif, M., D. Anderson, T. Barnett, M. Cane, R. Kleeman, A. Leetmaa, J. J. O'Brien, A. Rosati, and E. Schneider. "A Review of Predictability and Prediction of ENSO." *Journal of Geophysical Research* 103 (1998), 14375–14393.
National Research Council. *Learning to Predict the Climate Variations Associated with El Niño and the Southern Oscillation: Accomplishments and Legacies of the TOGA Program*. Washington, D.C.: National Academy Press, 1996.
Neelin, J. D., D. S. Battisti, A. C. Hirst, F-F. Jin, Y. Wakata, T. Yamagata, amd S. E. Zebiak. "ENSO Theory." *Journal of Geophysical Research* 103 (1998), 14261–14290.
Trenberth, K. E., G. W. Branstator, D. Karoly, A. Kumar, N.-C. Lau, and C. Ropelewski. "Progress during TOGA in Understanding and Modeling Global Teleconnections Associated with Tropical Sea Surface Temperatures." *Journal of Geophysical Research* 103 (1998), 14291–14324.
Wallace, J. M., E. M. Rasmusson, T. P. Mitchell, V. E. Kousky, E. S. Sarachik, and H. von Storch. "On the Structure and Evolution of ENSO Related Climate Variability in the Tropical Pacific: Lessons from TOGA." *Journal of Geophysical Research* 103 (1998), 14241–14259.

—EDWARD S. SARACHIK

## ELTON, CHARLES

Charles Sutherland Elton (1900–1991), British ecologist, made significant contributions to both animal and community ecology. Elton studied zoology under Julian Huxley at Oxford University from 1918 to 1922. After graduating, he began teaching as a part-time instructor and had a long and distinguished teaching career at Oxford from 1922 to 1967. At Oxford, he continued to work closely with Julian Huxley. In addition to being influenced by Huxley, Elton was affected by animal ecologist Victor Shelford and his seminal book, *Animal Communities in Temperate America* (1913). After a series of arctic expeditions with Huxley and work with the Hudson Bay Company to study animal populations in Canada, Elton wrote his first and most important book, *Animal Ecology*, in 1927. Encouraged by Huxley, he wrote it in eighty-five days. Noted biologist George Evelyn Hutchinson later said that *Animal Ecology* was "one of the greatest biological books of the century." In this work, Elton brought together many concepts in the emerging field of ecology. He introduced four principles that served as the basis for integrating population and community ecology: (1) food chains, (2) food size,

(3) niche, and (4) the pyramid of numbers. Elton's *Animal Ecology* helped establish ecology as the "quantitative study of living organisms in relation to their environments." He insisted that ecology focus on what animals and plants were doing in their biological communities. Elton argued that animal ecologists should study the "general mechanism of animal life in nature, and in particular to obtain some insight into the means by which animal numbers are controlled."

In 1932, Charles Elton helped found the Bureau of Animal Population (BAP) at Oxford. From 1931 to 1967, under Elton's leadership, the BAP was an international center for research on the dynamics of animal populations and ecology. It was merged into the Oxford zoology department in 1967 after Elton's retirement. Many alumni of the BAP later took prominent jobs in ecology throughout the world. Some ecologists consider Charles Elton's most important contribution to animal ecology to have been his leadership and support for the Bureau of Animal Populations.

Elton also helped found the *Journal of Animal Ecology* in 1932. As editor of this journal, he helped shape and influence the emerging field of animal ecology. After his groundbreaking book *Animal Ecology*, he wrote five other books: *Animal Ecology and Evolution* (1930), *The Ecology of Animals* (1933), *Voles, Mice, and Lemmings: Problems in Population Dynamics* (1942), *The Ecology of Invasions by Animals and Plants* (1958), and *The Pattern of Animal Communities* (1966).

Charles Elton had a long and distinguished career as an animal ecologist. Through his work as a teacher at Oxford, his leadership of the BAP, his editing of the *Journal of Animal Ecology*, and his six books, Elton had a major impact on twentieth-century ecology. Few other ecologists have played such a major role in shaping the modern science of ecology as Charles Elton.

[*See also* Exotic Species.]

### BIBLIOGRAPHY

Elton, C. *Animal Ecology and Evolution*. Oxford: Clarendon Press, 1930.
———. *Voles, Mice, and Lemmings: Problems in Population Dynamics*. Oxford: Clarendon Press, 1942.
———. *The Ecology of Invasions by Animals and Plants*. London: Methuen, 1958.
———. *Animal Ecology*. London: Science Paperbacks and Methuen, 1966.
———. *The Pattern of Animal Communities*. London: Methuen; New York: Wiley, 1966.

—CHRIS H. LEWIS

## EMILIANI, CESARE

Cesare Emiliani (1922–1995), Italian paleontologist, was born in Bologna. He obtained a D.Sc. in Geology at the University of Bologna in 1945, and a Ph.D. in Geology at the University of Chicago in 1950. From 1950 to 1956 he was Research Associate with the Enrico Fermi Institute for Nuclear Studies, then moved to the Department of Geological Sciences at the University of Miami as an Associate Professor from 1957 to 1963 and full Professor and Chairman from 1963 to 1993.

Emiliani was the author or coauthor of 175 papers and thirteen books in the fields of micropaleontology, stable isotope stratigraphy and paleotemperature calculation, uranium/thorium dating, and human evolution. His most influential paper, "Pleistocene Temperatures," was published in 1955 (Emiliani, 1955). The paper presented for the first time (1) variations in the composition of oxygen isotope ratios in carbonate tests of marine planktonic foraminifera, (2) temperature changes of ocean surface water during the late Pleistocene, (3) an oscillation of warm and cold stages, that is, the oxygen isotope stratigraphy, (4) ocean temperature fluctuations in relation to glaciations on the continents, and (5) the application of Milutin Milankovitch's orbital theory to calculate the ages of the ice ages. [*See* Earth Motions.]

These five approaches are today the foundations of marine and terrestrial paleoclimatic research. The absolute ages assigned by Emiliani to the warm and cold stages have changed, but "Pleistocene Temperatures" was extremely influential in establishing the oscillating nature of past climate variations. Emiliani was one of the first to apply a new technique (oxygen isotope ratios), based on physical fundamentals, to quantify past ocean properties. He crossed the border between nuclear physics and geology, applied the latest technical innovations to high-quality samples, and provided a grand vision for the interpretation of his results. He spent his entire scientific life on the further validation and improvement of this concept, supporting it with sound scientific papers and initiating large programs such as the Deep-Sea Drilling Project.

Almost forty years after his pioneering study, he published his last book, *Planet Earth* (1992). The book "presents a global picture of our world—how it originated, how it evolved, how it works—and provides the background necessary to assess ways to stabilize it." The book touches on almost all aspects of modern Earth sciences: the planetary system, the Earth's interior, its atmosphere, evolution of life, and the environment. *Planet Earth* is a physically oriented textbook, comprehensive for the young student, and still challenging for the scientist. It appears to be typical for Emiliani that this book again presents a vision: to understand quantitatively our modern industrialized world in the context of its geologic history. Cesare Emiliani was awarded the Swedish Vega Medal and the Agassiz Medal of the United States National Academy of Sciences.

[*See also* Climate Change; *and the biography of Lamb.*]

**BIBLIOGRAPHY**

Emiliani, C. "Pleistocene Temperatures." *Journal of Geology* 63 (1955), 538–578.
———. *Planet Earth.* Cambridge: Cambridge University Press, 1992.

—FRANK SIROCKO

**EMISSIONS.** *See* Air Quality; Market Mechanisms; *and* Pollution.

## ENDANGERED SPECIES CONVENTION.
*See* Extinction of Species.

## ENERGY AND HUMAN ACTIVITY

Modern economies rely on oil, natural gas, and coal (known collectively as fossil fuels) to provide most of the energy needed to operate businesses, supply transportation, and provide comfortable living spaces. Industrialization is almost synonymous with the substitution of these fuels for human and animal power and wood combustion. While supplies of fossil fuels are plainly not limitless, world supplies are adequate to support all forecast global needs for many decades to come. New technologies for finding and extracting energy sources have led to real declines in the price of fossil energy during the past few years, and it is unlikely that real prices will begin to increase for over a decade. In spite of what may be a long period of comparatively stable prices, however, it is clear that current patterns of energy use are not sustainable. The emissions of carbon dioxide and other gases associated with extracting and using fossil fuels may be responsible for about 70 percent of human impact on the world's climate—85 percent in industrialized countries such as the United States. World population is expected approximately to double before stabilizing during the next century, and energy use per person will increase by a factor of two to three in the absence of major technological change. Energy use and energy-related emissions could therefore increase by 400–600 percent. A shift to sustainable sources of energy will clearly be needed before the end of the next century simply because sources of low-cost fossil fuel will be exhausted. Given the enormous investment in an economic infrastructure keyed to fossil energy resources, at least a century will be needed to invent and deploy an energy system capable of supporting ten billion prosperous people. There are no technological reasons why this task cannot be achieved and, given determined action, changes integrated into the natural turnover of energy plant and equipment. Building a political consensus to act before a crisis is evident is, however, another matter.

**Energy Basics.** The precise physical definition of energy is the ability to do work. Energy can take a variety of forms and when we say that energy is used or consumed what we really mean is that it has been changed to a more useful form. The proposition that energy is neither created nor destroyed by these transformations has not been shaken by centuries of investigation. An automobile traveling down a highway has kinetic energy, or energy of motion, which results from converting the chemical potential energy in the car's engine in a chemical reaction that releases heat and drives a mechanical process. Water resting in high mountain lakes has potential energy that can be converted to useful work through water wheels and their modern equivalent. This mechanical energy can be converted to energy in the form of moving electric charges and sent down wires to heat the filaments in light bulbs, which release energy in the form of light and heat. In the end, such energy becomes heat in rooms or in heated tires and highways—heat that is eventually radiated into space.

Photosynthetic processes store vast amounts of energy from the sun in the form of chemical energy in plant materials that may be preserved in fossil fuels. We have enjoyed an economy built on finding and extracting this energy—and in the process releasing much of the carbon captured from ancient atmospheres. Our problems start when we need to pay for creating stored energy in the first place.

A sustainable economy must find ways to convert energy efficiently to do useful work and also find a long-term energy source that places a minimal burden on the environment.

**Trends in Demand.** Energy demands change rapidly with development. Figure 1 indicates the astonishing range in energy use per person worldwide. Each citizen of an affluent, sprawling economy like the United States consumes five times as much as the average person on the planet and ten times more than people in poor nations. This difference is due in part to the absence of traditional fuels such as animal and wood power in these accounts, but most of the difference results from increased consumption of energy-demanding services. Increased affluence leads to larger homes with more appliances, more travel (especially in individual cars and by air), more spacious stores and office environments, and more industries to produce the prodigious range of goods required by such economies. It is clearly possible that some of these amenities may need to be sacrificed if energy costs increase in response to future shortages or to reflect the environmental costs of energy use. A more attractive outcome, however, would be to discover ways for the world to enjoy increasing levels of amenity in the face of these rising prices by finding more clever ways to supply needed services.

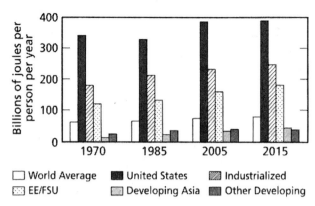

**Energy and Human Activity. FIGURE 1.** Per Capita Energy Use. (From U.S. Department of Energy, 1998b and 1998c.)

***Long-Term trends.*** The rate of growth in per capita energy in industrial nations is expected to slow over the coming decades even if no action is taken to reduce energy demand for environmental or other reasons. Over 40 percent of the investment in new plant and equipment in the United States is in the information industry, for example, not in equipment for manufacturing products with high inputs of energy and materials. The U.S. gross domestic product increased nearly eight percent between 1996 and 1998, but energy use increased less than one percent, largely because of productivity gains possible from new technologies. Standard forecasts suggest, however, that developing nations will invest heavily in the automobiles, mills, offices, and highway infrastructure that generated rapid growth in demand for energy in the industrialized nations a generation earlier. As a result, most standard forecasts suggest that about two-thirds of overall world energy growth is likely to occur in developing nations during the next two decades. Creative development strategies may, however, let many of these countries leap directly to a high-technology, energy-efficient future and avoid the high energy needs of mid-twentieth century industrialization.

***Energy use.*** Most of the energy needed even by the most modern economy is used to achieve very familiar objectives. About two-thirds of U.S. energy is used to provide heating, lighting, and other services in buildings and to move people and freight.

The large differences in patterns of energy use in nations with roughly similar per capita income result in part from the effect of prices (U.S. energy prices are among the lowest in the world) and in part from changes in the structure of national economies and lifestyle choices. It appears, however, that many nations are shifting to U.S. styles of energy use. In the United States, for example, there are 0.6 cars for every person, compared with about 0.4 for France and 0.3 for Japan. Japanese and French car ownership have grown dramatically in the past two decades. Rates of car ownership in Japan

have increased threefold since 1970. Americans also drive an average of 12,500 kilometers each year, a rate that continues to increase. This is four times more distance than an average Japanese citizen travels by car, and over twice the average distance in France. In spite of significant differences in fuel costs, the fuel efficiency of new U.S. automobiles is only about 10 percent better than new cars sold in Japan and 27 percent better than French vehicles (many of which use diesel engines that do not meet U.S. air quality standards).

Larger houses and a harsher climate also contribute to higher U.S. demand. In the United States the average home provides about 60 square meters per person, while the typical Japanese house provides roughly half this space, and typical European homes provide only about 10 percent more space than Japanese houses. The average size of housing in all three groups has increased steadily during the past two decades. Perhaps because of a harsher climate in the United States, the energy efficiency of U.S. homes is better than that of most European homes, but worse than that of Japanese homes, where central heating is often not used. The United States also provides nearly twice as much space per person for office environments, shopping facilities, and other service buildings than other industrial nations.

**The Potential for Efficiency Improvements.** Much more energy is actually used to deliver energy services—such as air conditioning buildings, moving freight, or operating industrial equipment—than would be needed by theoretically perfect machines.

Take the example of automobile travel. Only about an eighth of the energy in automobile fuel, for example, is put to use in propelling the car. The rest is lost in the process of converting heat to mechanical power, to friction, and in idling. The energy needed to propel the car can also be reduced significantly by cutting the weight of the vehicle, reducing the rolling resistance of tires, and designing the car's shape so that air resistance is reduced. In addition, the energy used to accelerate the vehicle can be recaptured when brakes slow the vehicle. Finally, there is the issue of whether the trip was necessary to begin with. Well-designed communities can reduce the need for trips by car and shorten the ones that are taken. Studies in Paris and San Francisco found that people in suburban communities with low population densities traveled twice as far each day as people who live in communities with ten times the population density. Carbon dioxide emissions can, of course, be further reduced by using fuel that releases no net carbon—such as a fuel derived from a renewable resource. Taken together, changes in vehicle design, changes in community design, and changes in fuels can lead to order-of-magnitude reductions in the greenhouse gas emissions associated with personal transport without affecting the quality of the transportation services provided. Similar gains are possible throughout the economy. There is, of

**Other (13.4%)** Largely oil and gas pipelines.

**Aircraft (13.1%)** Demand expected to grow 4%/year for 20 years. Efficiency gains possible in improved engines, air-frames, and air traffic control.

**Heavy Trucks (15.4%)** Heavy diesel engine efficiency can increase by at least a third; improved dispatching can reduce traveling with partial loads.

**Light Trucks (23.2%)** Research goals in the United States and elsewhere would increase efficiencies of these vehicles through improved engines.

**Automobiles (34.9%)** Efficiency gains of at least a factor of two are possible using light-weight materials as substitutes for steel, low air and rolling resistance designs, and use of hybrid drive trains. Proton exchange fuel cells which convert fuels directly to electricity may prove to be the best long-term propulsion system.

**Energy and Human Activity. FIGURE 2.** U.S. Energy Use in Transportation. (Energy use data from U.S. Office of Technology Assessment, 1991.)

course, no guarantee that these potential gains can be captured. At a minimum, a major investment in research will be needed to achieve them.

Many options are available for reducing energy de-

mand in almost every part of the economy. The following discussion reviews some of the major concepts available in each major sector.

***Transportation.*** Transportation consumes between a quarter and a third of the energy used in industrial economies. Compact nations with good rail service rely less heavily on highway transportation than nations with sprawling suburbs and a continental economy, such as the United States, Canada, and Australia. The convenience and privacy of automobile travel clearly has high appeal, however, and automobile transportation is increasing around the world—often in the face of national measures to encourage other forms of transit. A study of seven advanced industrial nations suggests that a doubling in income results in a tripling of car ownership rates. In the United States, energy use in personal transportation has increased largely because nearly half of all new personal vehicles sold now are minivans, sport utility vehicles, and other "light trucks." Air travel is also growing rapidly worldwide. Figure 2 summarizes the potential for reducing greenhouse gases in the U.S. transportation system.

***Buildings.*** Nearly a third of U.S. energy, and a growing fraction of the energy in most industrial economies, is consumed in residential and commercial buildings (Figure 3). Buildings are the major consumers of electric energy worldwide, taking over 75 percent of U.S. electricity, if industrial buildings are included. Rising demand for electricity in developing nations can be traced primarily to demand growth in space-conditioned

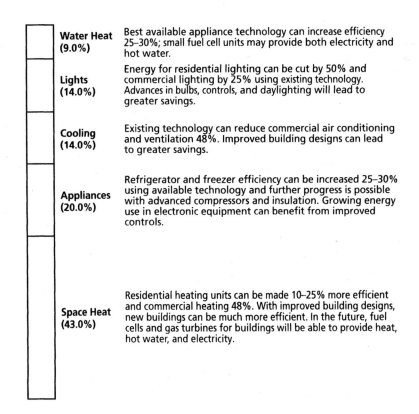

**Water Heat (9.0%)** Best available appliance technology can increase efficiency 25–30%; small fuel cell units may provide both electricity and hot water.

**Lights (14.0%)** Energy for residential lighting can be cut by 50% and commercial lighting by 25% using existing technology. Advances in bulbs, controls, and daylighting will lead to greater savings.

**Cooling (14.0%)** Existing technology can reduce commercial air conditioning and ventilation 48%. Improved building designs can lead to greater savings.

**Appliances (20.0%)** Refrigerator and freezer efficiency can be increased 25–30% using available technology and further progress is possible with advanced compressors and insulation. Growing energy use in electronic equipment can benefit from improved controls.

**Space Heat (43.0%)** Residential heating units can be made 10–25% more efficient and commercial heating 48%. With improved building designs, new buildings can be much more efficient. In the future, fuel cells and gas turbines for buildings will be able to provide heat, hot water, and electricity.

**Energy and Human Activity. FIGURE 3.** U.S. Energy Use in Buildings. (Energy use data from U.S. Office of Technology Assessment, 1991.)

buildings and appliances. Also, in recent years, electronic appliances such as advanced televisions, computers, and fax machines have emerged as major sources of growth in demand for electricity.

***Industry.*** Industrial energy use in a nation depends both on the efficiency of its industrial processes and the mix of industries present. Materials industries such as aluminum, chemicals, forest products, glass, metal casting, petroleum, and steel use much more energy per dollar of value added than industries like computer manufacturing. In the United States these industries are responsible for 80 percent of industrial energy use (Figure 4). Some industries also use fossil fuels as feedstocks. Modern production methods make it possible to increase the efficiency with which energy and materials are used—often by using improved design tools to minimize materials in basic designs and to ensure that production processes operate continuously at peak performance points. Changes in the mix of industries—away from traditional materials industries—have reduced the energy intensity of industry by 10–20 percent over the past two decades. The exception is Denmark, which appears to have shifted to more energy-intensive industries.

**Energy Supplies.** World energy supplies were dominated by renewable sources until the early twentieth century, when coal provided more than half the world's energy. Since World War II, oil and natural gas together have provided an increasing share, while nuclear energy has made an impact only since 1970 (see Figure 5).

***Petroleum.*** Petroleum provides a very useful way to store energy. It is comparatively inexpensive to produce and is inexpensive to transport and store since it is a liquid at room temperature. Filling an automobile gas tank in two minutes with a gasoline hose is equivalent to connecting the car to a 40-megawatt power line for the same length of time—and a lot safer. While petroleum was once a major source of electric generation and a major source of industrial power, most new plants use natural gas or coal if they are available. Petroleum, however, dominates world transportation energy needs, providing over 97 percent of all transportation energy in the United States.

There is enormous unmet demand for transportation worldwide: increasing world income is expected to increase petroleum demand by 3.5 percent annually for the next decade and a half, with total demand reaching ninety-five million barrels per day by 2015. South Korea provides a spectacular example of the link between development and petroleum use. During the Korean economic boom between 1985 and 1995, oil consumption grew an average of 13.1 percent per year. Demand grew rapidly throughout the Southeast Asian region.

New discovery and production techniques are likely

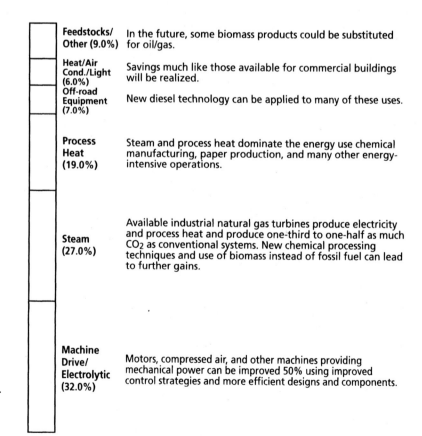

**Energy and Human Activity.** FIGURE 4. U.S. Energy Use in Manufacturing. (Energy use data from U.S. Office of Technology Assessment, 1991.)

Feedstocks/Other (9.0%)    In the future, some biomass products could be substituted for oil/gas.

Heat/Air Cond./Light (6.0%)    Savings much like those available for commercial buildings will be realized.

Off-road Equipment (7.0%)    New diesel technology can be applied to many of these uses.

Process Heat (19.0%)    Steam and process heat dominate the energy use chemical manufacturing, paper production, and many other energy-intensive operations.

Steam (27.0%)    Available industrial natural gas turbines produce electricity and process heat and produce one-third to one-half as much $CO_2$ as conventional systems. New chemical processing techniques and use of biomass instead of fossil fuel can lead to further gains.

Machine Drive/Electrolytic (32.0%)    Motors, compressed air, and other machines providing mechanical power can be improved 50% using improved control strategies and more efficient designs and components.

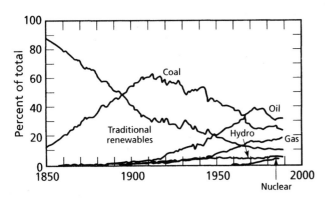

**Energy and Human Activity.** FIGURE 5. World Energy Use. (From Nakićenović, 1998, p. 13. Reprinted with the permission of Cambridge University Press.)

to ensure that petroleum prices do not increase by more than 2 percent per year during the next two decades. Forecast price increases range from no increase to a high of about 2 percent per year. The political crises of the early 1970s and 1980s drove prices to very high levels, but the collapse of production cartels and the pressure of competing energy sources have kept prices low for nearly a decade. Prices in 1998, however, were at near-record lows. Proven oil reserves and a reasonable estimate of undiscovered oil are three to four times greater than the total volume of oil that has already been extracted and used (see Figure 6). The Persian Gulf oil region still has reserves of prodigious size—enough to last eighty years at current production rates—and production costs are between U.S.$1.00 and $1.50 per barrel.

The likely stability of oil prices also results from new finds in many parts of the world and production methods that have sustained production in the United States and other mature fields. This production has permitted world supplies to remain constant in spite of enormous declines in production from the former Soviet Union and Iraq.

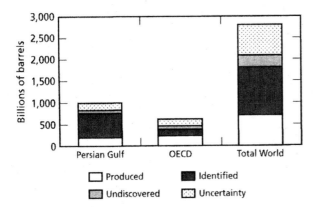

**Energy and Human Activity.** FIGURE 6. World Oil Reserves in 1993. (From U.S. Department of Energy, 1998b.)

New discoveries in the North Sea have made the combined United Kingdom and Norwegian production in 1997 equivalent to Russia's. Italy has discovered more than a billion barrels of oil in the Apennines. Production increases are expected in North Africa and South America, and rapid recovery in production is likely throughout the former Soviet Republics. Very large discoveries near the Caspian Sea could hold reserves larger than those in the United States.

This optimistic view must, of course, be tempered by several cautionary notes. First, stable prices assume a stable political environment. The United States and other large economies face declining domestic oil production and the prospect of sharply increasing imports. It is likely that U.S. imports will double by 2020. History suggests that massive disruptions in world supply are possible—probably likely—given the volatility of many oil production regions. World sensitivity to disruption will increase by the first part of the next century since an increasing fraction of world oil reserves will be concentrated in the Middle East and transported through a small number of shipping lanes. Second, in spite of many false cries of alarm about reaching the end of world oil supplies, the end must inevitably come. If world demand for oil continues to increase as expected, real supply constraints—and high price pressures—are likely by the middle of the next century. In spite of constant improvements in technologies for refining and extracting oil, the International Energy Agency of the Organisation for Economic Co-operation and Development now estimates that world oil production is likely to peak some time between 2010 and 2020 as we begin to approach the real limits of world petroleum resources. World consumption of petroleum will need to decline steadily after this peak is reached—something that could trigger price increases and dramatic price fluctuations. The need to control greenhouse emissions from petroleum may lead to constraints before scarcity begins to increase petroleum prices. [*See* Fossil Fuels; *and* Greenhouse Effect.]

***Natural gas.*** Like petroleum, natural gas is also inexpensive to produce. Since it is a gas, however, greater investments are required to transport and store it. Pipeline transportation can be inexpensive per unit of energy delivered, but pipeline projects are large, require major capital investment, and often require the resolution of difficult political and environmental issues. Pipelines in North America have provided inexpensive, reliable fuels—indeed, over 90 percent of all new U.S. electric generation added between 1996 and 2000 is expected to be powered by natural gas; 85 percent of new U.S. capacity between 2000 and 2020 is expected to be built to use natural gas.

Use of gas is expected to increase rapidly as new gas pipelines are completed. Demand will be driven in large part by the low price of gas electric generation. Coal prices will continue to decline while natural gas prices increase slowly, but coal in the United States will not be cheaper than gas—measured in dollars per unit of energy—until about 2005. This price advantage, however, is offset by the extremely high efficiency of modern gas plants. Combined-cycle devices (which burn gas in a Brayton cycle engine—much like the engines used in modern jet aircraft—and use the hot exhaust to drive a steam turbine) are capable of very high efficiencies in comparatively inexpensive facilities. Commercial plants will soon achieve efficiencies exceeding 54 percent (higher heating value) compared with the 46 percent efficiency typical for new gas turbines sold in 1988. Modern coal–steam plants achieve only about 35 percent efficiency.

Fuel cells offer an entirely new approach for converting chemical energy to electricity. Presented with hydrogen fuel, they do not require combustion but instead use a membrane that in some ways mimics the membranes used in living cells to convert chemical energy to work. In this sense, fuel cells are a first step beyond using fire as the primary method of exploiting chemical energy in fuels. Molten carbonate fuel cells can achieve efficiencies comparable with combined-cycle plants. Proton exchange membrane units can achieve efficiencies nearly as great, but operate at much lower temperatures and can be installed in much smaller sizes—including residential-scale equipment. Fuel cells can operate from natural gas using chemical processors called reformers, which strip hydrogen from methane, sending the hydrogen to the fuel cell and exhausting the residual carbon dioxide. Enormous advances have been made in fuel cell technologies in the past few years, and large-scale commercial introduction of small units is likely in the near future—most likely in systems designed for commercial buildings. The price of proton exchange membrane fuel cells would be cut enormously if these devices were also produced to power automobiles. [See Hydrogen.]

Fuel cell or other natural gas systems can enjoy further advantages if they can be located close to sites where heat can be used for manufacturing processes or to supply building heat. Cogeneration units that produce useful heat and electricity can convert 70 percent or more of the energy in the fuel to a useful product. Fuel cells, which can be built economically in a wide range of sizes, would enjoy an advantage in such markets.

The high efficiency of combined-cycle gas plants, the comparatively low capital costs of gas plants, and the speed with which these plants can be built, is likely to give gas-fired electric plants a price advantage over other fossil plants for at least a generation in areas where gas is available. This advantage will be reflected in markets as deregulation of electric power spreads in the United States, more pipelines are built worldwide, and European nations become more comfortable with foreign gas supplies and remove political restrictions on gas use. Increasing interest in clean energy sources is likely to increase national investments in natural gas. Demand for gas is expected to grow rapidly in Europe and the former Soviet Union.

While about half of the increase in natural gas use in the United States is expected to result from electric generation, low costs, environmental concerns, and ease of use are expected to result in growing demand for gas in industry and commercial and residential buildings. Nations around the world are investing heavily in the pipeline infrastructure needed to capture the benefits of natural gas. Nearly 20,000 kilometers of new pipelines were completed in 1996 alone, and estimates suggest that about 34,000 kilometers were installed between 1998 and 2000. About half of this new construction is in South America. Major construction projects are also under way throughout Southeast Asia and Europe. Natural gas can also be liquefied by cooling and shipped in specially built ships when conventional pipeline shipments are not practical, but the cost of the process is high, making it difficult for the liquefied gas to compete with other fuels in most markets. Over the long term it may be more profitable to convert natural gas chemically into a liquid fuel or another value-added product that can be more easily transported.

Known world reserves are large, and production costs continue to fall as advanced drilling techniques make it possible to tap into many reservoirs from a single platform (Figure 7). Prices are also stabilized by competition among producers; gas reserves are distributed more evenly than oil reserves. The large reserves in Eastern Europe and the former Soviet Union will be able to produce for more than eighty years at current rates, and the reserves in the Middle East may produce for one hundred years.

*Coal.* With 81 percent of world production coming from the United States, China, the former Soviet Union, Australia, Germany, and South Africa, coal is very un-

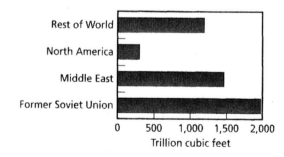

**Energy and Human Activity.** FIGURE 7. World Natural Gas Reserves in 1997. (From U.S. Department of Energy, 1998b.)

evenly distributed throughout the world. It is bulky and expensive to ship and produces enormous amounts of ash and waste products. It also is dirty, creating many potential local and regional air pollution hazards because of $SO_x$ and $NO_x$ production, and presents major global environmental challenges since coal combustion results in 80 percent more carbon dioxide per unit of energy delivered than natural gas. Where it is available, however, coal can be extremely inexpensive. [See Pollution.]

In most of the industrial world, coal demand is expected to grow slowly or, in the case of the former Soviet Union, decline as natural gas replaces older, inefficient industrial equipment and inefficient coal–electric systems. Sharp competition and privatization have led to investments in coal mining, resulting in sharply increased productivity and—in many cases—declining costs. Coal mining productivity has increased 5–10 percent annually in most major coal production nations. Real coal prices at the mine are expected to decline by 1.5 percent annually in the United States. These cost reductions will, however, only serve to slow the shift to noncoal fuel sources in most of the world. More than 82 percent of forecast growth in coal during the next fifteen years is expected to occur in China and India—areas that have limited access to natural gas.

While conversion of coal to liquid fuels does not appear attractive in the near future, coal may be assisted by the investments made to improve the efficiency of natural gas combined-cycle units. Units that convert coal to gas and use the gas in a combined-cycle plant with a net efficiency of 41 percent are being tested in the United States and the Netherlands. Efficiencies of over 50 percent may be possible.

***Nuclear fission and fusion.*** The community of scientists who built the weapons dropped over Japan during World War II emerged with an ardent hope that these same nuclear forces could be harnessed for peaceful purposes—foremost among them the generation of electricity. Successful nuclear reactors are in fact able to take the energy released by fission of uranium and use it to generate heat, which in turn produces steam sent to a turbine generator. Nuclear power is attractive because large resources of uranium fuels are available worldwide. During the next few decades, fuel removed from dismantled nuclear weapons could provide additional fuel supplies. In 1996 the United States signed a contract with Russia for materials from dismantled warheads that could supply a third of U.S. commercial nuclear fuel needs. It is also possible to recover more useful fuel from the byproducts of nuclear combustion than were in the fuel originally.

Large national programs have made nuclear power a major source of electricity, supplying about 20 percent of electricity in the United States, 45 percent in Europe overall, and over 70 percent in France.

Nuclear technology is, however, unlikely to fare well during the next two decades because of concerns about safety resulting from the major accident in Chernobyl, environmental concerns about disposal of nuclear waste and decommissioning of plants when they reach the end of their useful lives, and fear that nuclear material from reactors could be diverted to manufacture nuclear weapons. The greatest immediate problem faced by nuclear power, however, is poor economics. High construction costs more than offset the lower fuel costs of nuclear power. New combined-cycle gas plants produce electricity at costs far lower than those of nuclear facilities. Moreover, gas plants can be built much more rapidly and in smaller sizes than nuclear facilities—an attribute that is very attractive to investors unwilling to take risks on long-term energy forecasts. In the past these risks were minimized by public regulation, but when the United Kingdom and several U.S. states introduced competition into electric generation, special provision had to be made to provide shelter for owners of nuclear facilities.

Without a major change in policy, nuclear generating capacity in the world is likely to decline during the coming decades. Japan, China, South Korea, and India still have vigorous nuclear construction programs under way, but Belgium, Finland, Germany, the Netherlands, Spain, Sweden, Switzerland, and the United Kingdom have frozen new construction of nuclear facilities. No new nuclear facilities have been ordered in the United States since the late 1970s, and the number of operating nuclear plants in the United States in 1998 is the same as the number operating in 1989. Nearly half of the U.S. plants will reach the end of their forty-year licenses by 2020. If problems arise that shorten lifetimes to 30 years, nearly all United States plants would need to be retired before that date. Research may find ways to extend lifetimes safely.

A long-term hope is that it may be possible to harness nuclear fusion—the reactions that produce the sun's heat—for electric power generation. Two hydrogen atoms colliding at high energy fuse to become a helium atom, releasing energy in the process. Teams of scientists in the United States, Russia, Japan, and Europe have worked for decades to find ways to heat hydrogen to the temperatures needed to create fusion reactions, but even the optimists suggest that practical systems are many decades away. Funding for the programs continues in national energy budgets, but ambitious plans for a major international test facility have met major setbacks as nations around the world have reexamined budget priorities. [See Nuclear Industry.]

***Renewables.*** The world relied almost entirely on renewable energy forms for the majority of its energy needs until the beginning of the twentieth century (see Figure 5). Advanced technology raises the possibility that a large fraction of world energy needs can once again be provided from energy resources continuously replenished by the sun. The productivity gains possible through these new technologies mean that the renewable energy technologies of the twenty-first century can provide the electricity and high-quality fuels needed by a sophisticated economy. Studies by the Shell International Petroleum Company and by the United Nations Intergovernmental Panel on Climate Change (IPCC) suggest that, by 2025, renewable resources could be an economic substitute for half to two-thirds of fossil fuel energy in use today (Johansson et al., 1992).

Increased reliance on renewable fuels is clearly attractive in the context of climate change since these technologies produce no net carbon dioxide. Achievement of the IPCC targets, however, will require major investments in research and development, and care must be taken that the renewable technologies do not create other environmental hazards. There must also be sensitivity to the unique characteristics of these technologies, not all of which will make sense for all sites. Desert areas are comparatively attractive for technologies such as photovoltaics, which convert sunlight directly into electricity. Windy areas can extract the least expensive electricity from wind. Areas with large volumes of scrap from lumbering or food processing operations and areas where energy crops can be grown efficiently can be good sites for biomass energy production.

Renewable resources in the United States, primarily hydroelectricity and biomass, supplied 7.9 percent of U.S. energy in 1996.

*Hydroelectric power.*   In use since the time of the Romans, hydroelectric power still provides a higher fraction of world electric needs than nuclear power (19.4 percent compared with 17.2 percent). About 2,000 terawatt hours (1 terawatt = $10^{12}$ watts) of electricity are generated annually from hydroelectric resources, but another 13,000 terawatt hours could be developed economically—particularly in South America, the former Soviet Union, and southern Asia. The development of these resources would, of course, result in the flooding of large areas—an act that itself can create severe environmental disruption. The impact on regions with large amounts of undeveloped hydroelectric resources would be comparable to that of the Tennessee valley or Pacific Northwest regions in the United States during the 1930s.

Technologies are also available for increasing the efficiency, and therefore the output, of existing sites. U.S. research has focused on advanced turbine designs to achieve these improvements as well as methods for mitigating the impact of dams on fish migration paths.

*Biomass.*   About half of the commercial biomass resources used in the world are in North America. Biomass resources supplied about 3 percent of U.S. energy

needs in 1996, primarily in the form of wood waste used to raise steam and provide electric power in lumber mills and paper mills. Worldwide resources of organic waste materials are on the order of 100 exajoules (1 exajoule = $10^{18}$ joules), and there are many opportunities to produce energy crops on land not suitable for conventional agriculture without exploiting areas now in virgin forests or grasslands. Expanded use of biomass, however, depends on progress in three key areas:

1. *Improved yields for biomass crops.* Worldwide research on crops has focused almost exclusively on food crops and fiber production. Crops designed to be used as a fuel source have very different characteristics. It seems likely that high yields are possible even in areas where periodic drought or flooding make them incompatible with conventional food production. Research should be able to increase yields by 50–60 percent from the current 9 dry metric tons per hectare per year.

2. *Improved techniques for producing liquid fuels from biomass.* Heavy domestic subsidies have resulted in robust ethanol industries in the United States and Brazil. The United States produces about 4 billion liters of ethanol annually from corn. Corn is, of course, optimized as a food crop. Its starches can easily be converted to sugars and distilled using techniques known since ancient times. New biotechnologies make it possible, however, to break the cellulose and hemicellulose in woody materials into starches that can be converted to alcohols. These technologies are not economic in the United States and other areas where gasoline prices are low (the value of ethanol in the United States is U.S.$0.18–0.26 per liter as a fuel additive and U.S.$0.13–0.15 per liter as a neat fuel). Ethanol costs today are a quarter of what they were in 1980, but at U.S.$0.32 per liter, are still too expensive to compete without subsidies. U.S. research goals are to reach U.S.$0.18 per liter in 2010 and U.S.$0.15 per liter in 2030.

3. *Improved techniques for converting biomass to electricity.* In the near term, the simplest use of biomass is to mix it with coal in conventional steam plants in areas where the resource is very inexpensive. Mixes of up to 15 percent appear feasible with modest additions to existing plants. Over the long term, biomass conversion can take advantage of new combined-cycle and fuel cell technologies developed to improve the performance of electric generation from natural gas and coal. Biomass can be converted to a gas and introduced into these systems, and actually offers some advantages because of its low sulfur content and easy gasification.

*Wind power.* The central United States, northern China, and many other parts of the United States have enormous wind resources that can be exploited at costs comparable to today's low marginal electric costs. Worldwide there is over 7,200 megawatts of wind power capacity installed. While sales in the United States have slowed, worldwide investment remains strong.

Technology available by 2005 should deliver electricity at U.S.$0.034 per kilowatt hour in moderate-quality winds and U.S.$0.028 per kilowatt hour in good sites. Costs can be cut another 20 percent by 2030, making wind power among the cheapest sources of electricity. High wind towers and improved turbine and control designs can both increase efficiency and load factors.

The exploitation of wind potential requires that the variable nature of winds be overcome and that transmission infrastructure be available to transport power from wind sites—which are often remote—to areas where power is needed. One way to do this is to locate inexpensive compressed-air storage facilities near the generation sites. A recent analysis suggests that wind systems equipped with such systems could deliver electricity with a 90 percent system capacity factor to a destination 2,000 kilometers distant for a total cost of about U.S.$0.05 per kilowatt hour.

*Photovoltaics.* Photovoltaic electric generators are solid state devices that convert solar energy directly into electricity with no moving parts. They can be installed in a wide range of applications since there are few economies of scale in the installations. At present, their high costs limit them to markets where the costs of alternative electric supplies are very high. Applications include remote microwave repeaters, buoys, highway emergency telephones, and remote villages. As prices fall, however, markets are expanding rapidly. In 1996, 98 megawatts of photovoltaic arrays were produced.

Research goals in the United States call for reaching U.S.$1.5 per watt by 2010 and U.S.$1.0 per watt by 2020. Most techniques for achieving these low prices involve increasing the efficiency and lowering the production costs of thin-film systems based on amorphous silicon, cadmium telluride, and copper indium diselenide.

The 2020 price translates into a delivered cost of electricity that is about U.S.$0.07 per kilowatt hour for corporate financing and U.S.$0.05 per kilowatt hour for units financed with a residential mortgage. Since photovoltaic systems are typically installed on rooftops or other sites close to major consumption sites, these rates should not be compared directly with the marginal cost of generating electricity from a large generating system that must supply customers through a large transmission and distribution system.

*Other renewable technologies.* A variety of other renewable technologies are also under active investigation. Geothermal energy can provide economic power in sites where it is available. Worldwide production is 7 gigawatts, of which 2.7 gigawatts is produced in the United States. Solar thermal technologies use mirrors to concentrate sunlight on units that convert the heat to electric power or heat for industrial uses. Costs have been falling, but companies working on these systems were badly hurt in the United States when subsidies were removed in California and other areas.

*Issues of scale.* An obvious feature of most modern energy systems is their immense scale. Huge hydroelectric sites provide the cheapest power, and there are clear economies of scale in coal and nuclear electric generating facilities. These economies of scale led to regulation based on an assumption that there was a social advantage in eliminating competition among providers so that these scale economies could be enjoyed—hence a natural monopoly. It is not at all clear that these scale economies will continue into the next century. Many of the key technologies—fuel cells, photovoltaic cells, and many advanced gas turbines—show little or no scale economies over a wide range of production sizes.

The larger systems suffer several inherent disadvantages that have been exposed during the past few decades. Large systems require huge capital investments, and their profitability depends on accurate demand forecasts, whereas forecasts have been notoriously unreliable. The large systems also require extensive transmission and distribution systems, which typically cost as much as the generating system itself and have higher maintenance costs. In contrast, a system that connects many small systems through a distributed grid is much less vulnerable to the failure of any single element. The distributed systems also allow more flexibility in scheduled maintenance. Energy systems may well continue to evolve in this way, linking many distributed generation sites instead of depending on a large central site. Fusion reactors are likely to be large, but could change the logic of energy scale economies if the energy proves to be extremely inexpensive.

Most nations have abandoned the concept of natural monopoly and have begun to abandon regulations that prevent nonutility generators from competing with publicly regulated units. These regulatory changes will make it easier for the market to discover the real economy of scale in energy production.

**Policy Issues.** Taken together, the combined impact of investment in energy efficiency and renewable energy on greenhouse gas emissions could be considerable. A recent U.S. study concluded, for example, that by investing wisely in these innovations, the United States could reduce its emissions to 1990 levels at a cost of U.S.$50 per metric ton of carbon saved. It is clear, however, that these opportunities will not be captured without some policy intervention.

The challenge of reducing greenhouse gas emissions from energy use is made easier by the fact that most innovations that increase productivity also increase energy productivity and reduce waste that appears as emissions. Quality in modern production systems is measured by improved design, more precise controls, lighter and stronger materials, and tailoring to the needs of specialized markets—attributes that typically add information and not mass to the products. Encouragement of innovations that produce both productivity and environmental benefits lies at the heart of national proposals that must make political and economic sense.

***Regulations and pricing.*** The most fundamental way to ensure that investors have an incentive to develop energy-saving inventions or substitutes for conventional fossil fuels is to ensure that market prices reflect real costs. In many nations, prices for favored groups are kept far below market prices as an act of social policy. The United States and many industrialized nations have moved away from price regulation of fuels and are in the process of deregulating the generation of electricity. The process is neither pretty nor painless since many investments that made sense in a period of heavy regulation are nearly valueless in an open market. Moreover, while there is good reason to believe that, over the long term, elimination of electric power regulation will achieve more competition and more innovation, there is a real danger that perverse incentives may be given for operating older, more polluting plants. The details of deregulation proposals will be debated heatedly worldwide for some time to come.

Perfect markets, of course, still give investors no incentive to reduce greenhouse gas emissions; this will require a different level of intervention. The most straightforward approach would be a simple fuel tax, but such taxes are politically difficult in most nations. A more modest approach is to find a way to charge energy users a modest fee to provide funds for research and programs to encourage energy efficiency and renewable energy. Such funds are already operating in most utility areas, and provision can be made for them to operate even under a completely market-driven electric power system. Provision can also be made to require that some fraction of energy be produced from renewable resources.

Another approach is to provide subsidies to low-carbon energy sources through tax relief or other methods. Several such proposals are being actively considered in the United States. Unless funds are adequate to provide subsidies for a broad range of efficiency and low-carbon energy investments, great care must be taken to target these subsidies in areas where they can accelerate needed technological innovation. Provisions for ending these subsidies when they have served their purpose are also essential.

Systems for establishing tradable permits in greenhouse gases would have an effect similar to taxes or subsidies, but would not require a political process to select the levels either of taxes or of subsidies. The price of an emission permit would, in theory at least, come to equilibrium at the lowest possible price consistent with meeting the reduction goals.

***Research and development.*** National programs to encourage technological innovation are justified on the grounds that many crucial innovations that benefit the nation as a whole do not pay individual investors an adequate return to justify the expense of developing and introducing the innovation. The benefits to individual investors may be distant and very uncertain, and the concept could move quickly into common practice. Public investment in research is justified by the gap between the benefits to the nation and the benefits that can be captured by individual firms. This gap is easy to see in the case of basic mathematics or science, but it clearly also exists for the investments in basic engineering needed to develop products such as improved materials for fuel cells or improved crop yields for biomass production. In the case of climate change, there is another kind of uncaptured public benefit—the benefit of a stable, clean environment. Returns measured in the form of an improved environment benefit the entire community, but can pay no direct returns to investors. In the absence of a high energy tax, investors will underinvest in areas that have large benefits for climate change. This gap provides additional incentives for public intervention and investment.

Two key issues must be confronted in designing an effective research program: how to establish priorities, and how to manage the research process. There is the ever-present danger that high-priority areas may be missed, while those with political support will be funded. There is also a risk that public funds will be used to pay for research that would have been funded by industry in the absence of the program. These are not theoretical fears: history is littered with examples of how not to conduct energy research programs.

Most national research enterprises have learned some painful lessons from these experiences and are relying increasingly on research partnerships that provide competition and require significant matching from private investors. Policy makers can identify areas where progress is needed and establish competitive programs to design the research priorities. One major example of this is the partnership in the United States that aims to develop an automobile with three times the fuel efficiency of today's typical car. The policy goal was accepted by government and industry leaders, and a group of technological specialists in industry, universities, and governments established the research priorities. An annual review of the program is conducted by the National Academy of Engineering.

**Conclusion.** One of the few things that is certain about the world's economic future is that prosperity will be linked to the management of energy. The twentieth century is likely to be the only century that derived the bulk of its energy resources from fossil fuels. The next century must find alternatives. It is also likely that energy will be used with great sophistication through systems applying precisely the right amount of power at the right place and at the right time. This kind of elegance characterizes the use of energy in biological systems. What is not in doubt is that the application of energy will always be critical if humans are to master the environment they inhabit. We can dare to hope that this power will be well used.

[*See also* Carbon Dioxide; Electrical Power Generation; Energy Policy; Global Warming; Growth, Limits to; Industrialization; Renewable Energy Sources; *and* Resources.]

### INTERNET RESOURCES

*Federal Energy Research and Development for the Challenges of the Twenty-First Century*, Report of the Energy Research and Development Panel. http://www.whitehouse.gov/WH/EOP/OSTP/Energy/.

International Energy Agency. http://www.iea.org/energy.htm/.

U.S. Department of Energy, Energy Information Administration. United States Federal Government data can be found in a series of publications on this site. http://www.eia.doe.gov/.

U.S. Department of Energy Interlaboratory Working Group. *Scenarios of U.S. Carbon Reductions: Potential Impacts of Energy Efficient and Low Carbon Technologies by 2010 and Beyond*. A good review of options for using renewable energy and energy efficiency to control greenhouse gases. http://eande.lbl.gov/EE.html.

### BIBLIOGRAPHY

International Energy Agency. *Indicators of Energy Use and Efficiency*. International comparative statistics and an analysis of differences in national patterns of energy use. Paris, 1997.

———. *World Energy Outlook*. London: IEA Publications, 1998. Provides a detailed review of energy forecasting information, including a new estimate that world petroleum production will peak sometime between 2010 and 2020. Also available on the Internet (see Internet Resources above).

Johansson, T. B., et al., eds. *Renewable Energy: Sources for Fuels and Electricity*. Washington, D.C.: Island Press, 1992. A comprehensive review of renewable energy technology.

Johansson, T. B., et al. "Options for Reducing Carbon Dioxide Emissions from the Energy Supply Sector." *Energy Policy* 24 (1996), 985. A useful summary of the careful review of international options for using renewable energy resources prepared by the Intergovernmental Panel on Climate Change.

Nakićenović, N., A. Grübler, and Alan McDonald, eds. *Global Energy: Perspectives*, Cambridge: Cambridge University Press, 1998.

President's Commitee of Advisors on Science and Technology. In *Federal Energy Research and Development for the Challenges of the Twenty-First Century*, Report of the Energy Research and Development Panel. Washington, D.C., 1997. A review and critique of U.S. research priorities for energy. Also available on the Internet (see Internet Resources above).

U.S. Department of Energy. *Nuclear Power Generation and Fuel Cycle Report*. Washington, D.C., 1997. One of several Energy Information Administration publications with useful material on individual technologies. Also available on the Internet (see Internet Resources above).

———. *Annual Energy Outlook 1997*. Washington, D.C., 1998a. Annual publication. Particularly useful for historical series and forecasts. Also available on the Internet (see Internet Resources above).

———. *International Energy Outlook 1998*. Washington, D.C., 1998b. Annual publication. Also available on the Internet (see Internet Resources above).

———. *International Energy Review 1997*. Washington, D.C., 1998c. Annual publication. Also available on the Internet (see Internet Resources above).

———. *Monthly Energy Review, January 1998*. Washington, D.C., 1998d. Monthly publication. Particularly useful for historical series and forecasts. Also available on the Internet (see Internet Resources above).

———. *Renewable Energy Annual*. Washington, D.C., 1998e. Annual publication, also available on the Internet (see Internet Resources above).

U.S. Office of Technology Assessment. *Changing by Degrees: Steps To Reduce Greenhouse Gases*. Washington, D.C., 1991.

—HENRY KELLY

## ENERGY POLICY

Energy is an essential and pervasive component of modern life. It is not surprising, then, that a broad spectrum of government policies impact on energy. These include mineral law, tax policy, environmental policy, industrial policy, foreign policy, and defense policy. Policies affecting energy reach down as far as local building codes and household appliances.

**History.** Texas passed a law to regulate the sale and use of petroleum in 1889. The next year it established the Texas Railroad Commission, which was to be a powerful regulator of oil and gas production for the next century. In 1928, the United States Supreme Court held that the Texas Railroad Commission regulation was a proper exercise of the police power of the state in controlling the development of natural resources. Energy policy existed before 1973, but was generally fragmented. Policies for different fuels or different components of the energy system were addressed independently, if at all.

Much of early energy policy related to war or preparation for war. Shortly before World War I, Winston Churchill said, "If we cannot get oil, we cannot get corn, we cannot get cotton, and we cannot get a thousand and one commodities necessary for the preservation of the economic energies of Great Britain." Churchill's arguments led the British Government to take control of the private Anglo-Persian Oil Company. During that war, the allies established an organization to control oil supplies. In the United States, the Fuel Administration was established.

After World War I, energy policy focused primarily on oil. French energy policy in the 1920s, for example, focused on containing the oil trusts, building domestic refineries, bringing order to the market, and developing foreign oil leases. When the oil market fell into anarchy after the discovery of oil in eastern Texas, the Texas Railroad Commission was given power to control output and remained a force in government involvement in energy policy. In 1933, under President Franklin Roosevelt, the Secretary of the Interior was given extraordinary powers to set monthly oil production quotas for each state. This was chosen instead of government fixed prices. The courts struck down the law the following year, but the value of the system in controlling the market was recognized and continued on a quasi-voluntary basis. Also in the 1930s, the Tennessee Valley Authority was established as an experiment in energy and social policy.

Oil again proved of vital concern during World War II. Japan was highly dependent on oil imported from the United States for shipping and for its navy. When Japan attacked China, the United States and Britain were faced with a difficult policy decision. Cutting off American oil to Japan was likely to precipitate a Japanese takeover of the oil fields in the Dutch East Indies. Finally, only months before Pearl Harbor, a de facto embargo was put in place. Like Japan, Germany had no domestic oil. Production of synthetic oil from coal was a major technological effort within Germany throughout the war. Protection of the Romanian oil fields and capture of the larger Russian fields in Baku was an important reason for Germany's invasion of Russia.

By the mid-1960s, the need for more a comprehensive energy policy was beginning to be recognized throughout much of the world. In 1972, the Conference Board organized a conference on energy policy. Key policy areas identified were the short supply of natural gas, increasing dependence on foreign oil, a patchwork approach to the energy and environment relationship, the need for more emphasis on developing new energy technology and resources, and the absence of public concern about energy issues. The latter ended abruptly in the next year.

For much of the world, the importance of energy policy crystallized during the oil embargo of 1973–1974 by the Organization of the Petroleum Exporting Countries (OPEC). The public became acutely aware of the need for energy policy. In the United States, the federal government consolidated several independent agencies involved with energy. Individual states formed energy offices or commissions to address energy issues more broadly. The United Kingdom created a Department of Energy. Sweden formed the Secretariat for Future Studies, whose first effort was a project on Energy and Society. The International Energy Agency (IEA) was established to coordinate energy policy among the countries belonging to the Organisation for Economic Co-operation and Development (OECD). Numerous energy research institutes were formed around the world. The idea emerged that energy analyses needed to focus on the entire energy system rather than on its parts individually. Government and academic studies were initiated to provide a basis for formulating energy policy. Governments promoted innovation of new energy technologies as part of their energy policy. Within a decade, the new policy emphasis appeared to work. Expectations of higher energy prices and concerns over energy supplies led to increases in energy efficiency. For example, the average consumption of heating oil in single-family houses in West Germany fell from 35–40 liters per square meter per year in 1970 to 18–25 liters per square meter per year in 1982. Sources of oil were diversified and concerns over shortages decreased. Energy prices did not skyrocket as had been predicted; instead, they fell. What had appeared to be a severe shortage of natural gas turned out to be an abundance.

During the 1980s and into the 1990s, policies in industrialized countries moved from a strong government role in planning and regulation to policies of deregulation and privatization of government-owned energy businesses. Concern for energy conservation declined with falling energy prices. Mergers among electric utilities resulted in international electric generation companies.

The transition of the eastern European countries toward market-based economies created expectations for improved energy efficiency and reduced pollution. Initial results were disappointing, showing increased energy intensity and slow implementation of energy conservation. Subsidies, especially for residential energy markets, were difficult to eliminate.

Increasing concern with climate change in the late 1980s created new implications for energy policy. The important role of the energy system in the production of carbon dioxide, the principal component of anthropogenic greenhouse gas emissions, meant that energy policy was closely linked with climate policy. [*See* Carbon Dioxide.]

**Policy Goals.** A government's energy policy goals are shaped by many factors. These include the availability of domestic energy resources; revenue generation; the importance of a secure energy supply and the opportunity of increasing energy supply (or increasing efficiency or conservation measures) to meet expanding needs; stable energy prices; availability of energy technology and the capital needed to build these technologies; consumer preferences and the character of the market; and attitudes toward social welfare and the social and political feasibility of the policy. In addition, energy policy must accommodate interacting goals from other areas such as protection of human health and the environment, economic growth, foreign policy, and national security.

**Strategies for Policy Intervention.** Policies can take many forms. They may be explicit energy policies or policies in other areas that affect the energy system. The latter include tax policy, transportation policy, land use policy, and research policy.

An environmental policy to reduce greenhouse gas emissions has significant implications for energy policies since such a large part of those emissions originate in the energy sector. Other environmental policies, such as reduction of atmospheric concentrations of sulfur, particles, ozone, and mercury, also have energy implications. Indeed, measures to reduce greenhouse gas emissions will result in other air quality improvements. [See Air Quality.]

Because of the importance of energy, the security of energy supplies is an important part of energy policy. Continued access to the oil resources of the Dutch East Indies, Malaya, and Indochina was vitally important to Japan in the years before World War II. Access to oil supplies at reasonable prices continues to be a vital consideration of oil-importing countries today.

*Tax policy.* Tax policy may be aimed at maximizing government revenue from a lucrative source, such as oil production. Alternatively, tax policy may focus on obtaining economic or environmental benefits. Firms involved in extracting resources may be subject to various taxes: income, royalties, and profits or excess profits. There are often complex methods to depreciate investments. Many of these taxes are in place for historical reasons and have little to do with current energy policies.

The oil depletion allowance is an especially controversial tax policy. It is based on a more general minerals policy that allows the owner of mineral deposits to deduct from other taxes the loss due to depletion of the mineral resource as minerals are extracted. Since it is difficult to evaluate the resource in the ground, the deduction is generally based instead on an arbitrary percentage of revenue. Percentage depletion deductions can exceed the cost of investment. It is argued that this deduction reduces the cost of oil and thus distorts the market. Since renewable resources are not depleted, producers of hydro, solar, or wind power cannot claim a depletion allowance. One suggestion is that, instead of a tax deduction, the depletion allowance be converted to a charge for the depletion of nonrenewable natural resources.

Oil and gas companies may also benefit from other tax benefits, such as faster depreciation of expenses. Sometimes the costs of exploration and drilling can be expensed in the year that they occur, rather than spreading the deduction over a longer period as is typical for industrial firms. Various tax incentives have also been used to encourage the growth of renewable energy sources. These have also included faster depreciation rates.

Other tax benefits for energy also exist. In the United States, profits on sales of coal under some agreements can be treated as capital gains rather than regular income. Municipally owned electric and gas distribution facilities can issue tax-free bonds, reducing the cost of their services. Since price is an important factor in demand, this tends to increase the demand for energy services. There are other more subtle tax benefits. For example, allowing corporations to deduct the cost of free parking for employees encourages driving rather than the use of more energy-efficient mass transit.

Taxes are also used for revenue generation. Motor gasoline is commonly taxed, often to provide revenue to maintain roads. Within the OECD, gasoline taxes range from 33 to 81 percent of the total cost of gasoline.

*Regulation.* Sometimes referred to as "command and control," regulation has been used widely in the energy industry: for example, to specify fuel quality for specific applications, to limit environmental emissions, to prohibit use of certain fuels, to ration fuels for certain uses, and to control fuel prices. Most governments are experienced in managing regulatory policy, but this approach runs counter to the current deregulatory trend. Economic incentives may work better, especially in market-driven economies. Regulation can be effective, but is inefficient, allows little flexibility, and often leads to lengthy litigation. Nonetheless, it remains an effective tool in the policy toolbox and in many cases is necessary to back up market-driven policies. For example, regulation may be used to require measurement and reporting of emissions to avoid cheating.

*Market mechanisms.* The new trend in regulatory policy is to make use of the market mechanism. Economists have long argued for such an approach, especially as a means of controlling environmental pollution. Others, however, have seen this as a license to pollute. Only recently have serious experiments been attempted. The willingness to experiment with market mechanisms may stem partly from the failure of a regulatory approach to achieve desired goals and broader knowledge and acceptance of economic tools. There are two basic approaches: taxation and "cap and trade." One development has been for the government to fix the amount of emission reduction to be achieved (the "cap") rather than to attempt a pure benefit-cost analysis.

Two methods of reaching the cap are then available. In cap and trade, tradable permits are issued for each allowable unit of emission based on an allocation scheme. The number of permits may be equal to or less than existing emissions and may decrease in subsequent years. Polluters either take action to decrease emissions, or buy permits from others. Firms that can reduce emissions at a cost less than the value of the permits will sell permits, while those whose cost to reduce emissions exceeds that of the permits will buy. The market determines the value of the permits. This has been applied with apparent success in the United States for sulfur emissions from electric power plants, and new efforts are being developed for nitrogen dioxide emissions. While cap and trade may work well for electric generation plants and industry, it would be more difficult to implement for homes and businesses, where emissions from individual buildings are not measured.

In the alternative approach, an emissions tax is levied. The tax might apply to measured emissions or to fuels, based on projected emissions. The tax raises the cost of the fuel or energy to the industry, business, or consumer. The effect of the price signal would be to reduce emissions, to switch to a cleaner fuel, or to implement energy conservation. In the late 1980s and early 1990s, a carbon tax was much discussed. Some countries instituted such a tax. By the mid-1990s, it became clear that, at least in the United States, an additional tax on fuels was not politically acceptable and the focus shifted to cap and trade, with increasing interest in international trade.

There are several concerns associated with the use of economic mechanisms. One is that poor people will suffer inordinately from higher energy prices. There may be several issues of this sort that may need to be addressed through subsidies that would increase the complexity of a carbon tax. While most studies on the impact of reducing greenhouse gases recognize this, little attention has been paid either to the social implications or to the effect that such subsidies may have on economic projections. Farms are another example. Farmers use relatively small amounts of energy, but the price of that energy is an important factor in the viability of the farm. Agriculture contributes to greenhouse gases in ways other than fuel use. Governments may come under pressure to protect domestic industries that use large amounts of energy, imposing a greater burden on others. Finally, especially for international trading of permits, methods to verify that reductions have actually been made are important.

An important consideration of a carbon tax is the large amount of revenue it can generate. In the United States, for example, typical estimates are on the order of U.S.$100 tax per metric tons of carbon. If a target of 1,300 million metric tons of carbon emissions per year were achieved, this might produce annual revenues of over U.S.$100 million. How this revenue is distributed back into the economy is very important to the overall effect of the policy on the economy. Some studies have suggested that the revenue can be returned in a way that changes the tax structure, resulting in an increase in the rate of economic growth. Such comparisons must not be restricted to different policies for redistribution of the revenue. For strategies that involve changing the tax structure, comparisons must also be made with a situation in which the tax structure is changed in a similar way without a carbon tax. For example, a study might find that using the carbon tax revenue to eliminate corporate income taxes or capital gains taxes would increase growth, while recycling the revenue in a neutral way would slow economic growth. But both are compared with the present situation. The former should also be compared with a base case in which the tax structure is changed to eliminate corporate income taxes in the absence of a carbon tax. The result is that the effect on the economy is lower with the carbon tax.

***Privatization.*** Energy privatization is a worldwide trend placing greater reliance on market forces and less on government, reversing a nationalization trend earlier in the century. This has stemmed from a recognition that government-managed enterprises are inefficient, from the increasingly global economy, from the need to raise investment capital, and from a conviction that free enterprise advances the wealth of nations better than nationalized industries. The collapse of communism in eastern Europe and the former Soviet Union has led to the privatization of many previously state-run industries. Most of Latin America has adopted some form of privatization. Chile was the first to privatize energy facilities. The United Kingdom was the first major industrialized country to do so on a large scale. Further reasons for privatization include raising revenue for the state by selling assets, acquiring investment capital, improving managerial performance, moving toward market-determined prices, reducing costs, and reducing the frequency of power outages.

The methods of privatization vary. Argentina, the United Kingdom, Chile, and New Zealand auctioned off state-run companies to the public. Formerly communist countries (such as the Czech Republic) have distributed ownership to the public by issuing stock certificates directly to individual members of the public, simultaneously creating a stock exchange. In some instances, workers and management at the facilities were allocated a greater share in ownership. In some cases, the share of government ownership is reduced gradually. British Petroleum gradually reduced state ownership from 66 percent to zero over the course of almost two decades. The Russian company Gazprom, the world's largest gas company, was converted to a state-owned joint stock company in 1993 and began to be privatized in 1994. In the United States, the trend was to allow independent power companies to compete with traditional utilities, then to unbundle generation, transmission, and distribution of electricity.

Privatization has led to a substantial increase in the number of multinational energy companies. The handful of such companies a decade ago has increased to scores, and there are increasing numbers of multinational coal companies, pipeline companies, electric utility companies, and power generation equipment and construction companies. Privatization has also led to growing interconnection of companies previously focused on gas, oil, or electricity.

**Deregulation.** Over the course of decades of regulation of energy, the regulatory environment became cumbersome and inefficient. Deregulation aims to return a regulated monopoly or oligopoly to more market-driven processes.

In the United States, deregulation of natural gas began in 1978. Gas was in short supply because regulated prices could not keep up with increasing oil prices. Initially, the price of natural gas at new wells was deregulated. Wellhead price controls were not eliminated until 1989. In 1985, gas supply and pipelines were unbundled for local distributing companies and, in 1992, supply and transportation were completely unbundled. Industry was the first to take advantage of this. Individual states are experimenting with approaches for residential customers.

Several U.S. states are now in the process of deregulating electricity. This involves separation of generation, transmission, and distribution, allowing the consumer to choose among generating entities. The principal argument in favor of deregulation has been that competition among generators will lead to lower prices and better service. There have also been proposals for "green pricing," that is, consumers may be willing to pay a higher price for electricity if they know it comes from a low-pollution source. [See Electrical Power Generation.]

Since the distribution system is a natural monopoly, deregulation affects the generation and transmission stages. The latter is the most complex part of deregulation. Electric utilities routinely "wheel" wholesale electricity from different sources to maintain least-cost operations and to assure service during peak load periods. Deregulation involves wheeling retail electricity from a larger number of generators. Argentina began a process of deregulation of oil production in 1991, leading to a doubling in production of crude.

Remaining questions as to how deregulation will work relate to its effects on reliability, quality (maintenance of voltage and frequency), and service. Also to be determined is who will be responsible for assuring that these attributes are maintained, and the extent to which deregulation will work for residential consumers. Based on recent experience in California, there is some question as to whether generators will compete for the residential market.

**Voluntary action.** Government energy policies can encourage voluntary actions on the part of individuals and corporations. The Energy Star and Green Lights programs in the United States are highly successful in promoting more energy efficiency in buildings, appliances, and lighting. Contests have been devised to promote development of energy-efficient designs of buildings and of high-efficiency appliances.

For a time, electric utilities were encouraged through regulatory incentives to sponsor voluntary actions such as the development of a highly efficient refrigerator. Deregulation may reduce the incentive to participate in such programs.

**Research and development.** Governments sponsor research, development, and demonstration of new energy technologies. This can be done either by funding the work directly or by providing tax credits or other forms of subsidy to industry. A government can use its own facilities to demonstrate new technologies. Subsidies or tax credits can be used to encourage consumers to buy or install energy-efficient or renewable energy technologies. Research and development possibilities include supply-side and end-use technologies, including more efficient electric generation, home appliances, industrial equipment, and motor vehicles, as well as more basic research on materials such as superconductors. Innovative approaches have been tried, such as the "Golden Carrot" refrigerator and government–industry partnerships such as the Partnership for a New Generation Automobile.

**Information.** A decision to provide information on energy supplies and technologies is a policy. Information on energy technologies, supplies, and projections allows industries and consumers to make more intelligent choices. Governments can provide information directly, for example by advertising the benefits of more efficient lighting. Governments can require that utilities provide consumers with information, for example, showing a comparison of energy use with similar-sized homes in the same area, along with suggestions on how to reduce energy use. Manufacturers can be required to provide information on the energy consumption of appliances.

**Energy policy targets.** There are numerous mechanisms by which energy policy can be implemented. These can address energy resources and supply, energy conversion, or energy use. They may encourage voluntary action or they may be mandatory. Experience in the United States with the *Climate Change Action Plan* (U.S. Congress, 1993) demonstrated that voluntary policies, while helpful, were not enough to achieve the levels of change in the energy system that the government sought.

Deregulation of wellhead prices, changes in tax policy, elimination of subsidies to coal mines, and elimination of restrictions on output are examples of policy action on resources. Extraction of energy resources is also strongly affected by environmental policy.

Environmental regulation to prevent oil spills and lawsuits over the damages and cleanup of spills have had an effect on the energy system. Legislation in many countries has provided opportunities for government review, unbundling of gas supply and transportation, provision of information, and lawsuits on pipeline siting and construction methods. Concerns over possible health effects of electromagnetic fields have impacted transmission line placement and construction.

Governments have generally either owned or provided close oversight to public utilities. Deregulation and privatization is changing this. Independent power producers are now common. Ownership of public utility companies and independent power producers has become internationalized. Regulation of environmental emissions from electricity generating plants and oil refineries has increased. Concerns with toxic emissions and rules requiring public notice of toxic emissions have led to efforts to reduce those emissions.

*Transportation.* Transportation is closely tied to land use patterns. These are slow to change and in many cases are changing in directions that increase the need for transportation rather than reduce it. The social feasibility of a policy is an important consideration. For example, Europe imposes gasoline taxes at a level that would not be socially acceptable in the United States or Canada. Policies can encourage increasing fuel efficiency by encouraging more efficient engines or by shifting to more efficient modes of travel. Policy tools, primarily appropriate to industrialized countries, include the following:

- Promotion of ride sharing by providing information on the need for ride sharing, provide mechanisms that facilitate the forming of ride-sharing groups, implement or increase parking fees at workplaces, provide special lanes for vehicles with multiple occupants.
- Expand use of telecommuting by providing information on ways in which telecommuting can help business while decreasing commuting.
- Research fuel-efficient vehicles through development of vehicles that run on alternative fuels, including electricity, natural gas, methanol, ethanol, and hydrogen. Develop more efficient engine types. Develop fuel cells for vehicles.
- Aim to increase overall efficiency by moving people from automobiles to other modes of transportation,

especially for commuting to work. The possibilities include trains, buses, bicycles, and walking. To achieve this on a voluntary basis requires policies that make these modes more attractive, faster, more comfortable, and more convenient. It requires a substantial commitment in upgrading capital stock and, over the longer term, changing land use patterns.
- *Mandatory tools,* such as a ban automobiles from certain sections of cities and limiting parking in certain sections of cities to local residents.

*Efficiency standards.* Frequently buildings are constructed with insufficient attention to energy efficiency. Building codes in many areas are beginning to require minimum levels of insulation and energy-efficient windows, for example, in new construction. Retrofitting of existing buildings is a more difficult issue, but can be addressed through upgrading requirements at the time of sale.

District heating and cooling systems can be an attractive alternative in densely populated areas. These require community involvement. Community energy analyses, as required in Sweden, can allow the planning and implementation of such possibilities. Cogeneration (supplying heat and electricity from the same device) reduces energy demand substantially. Tax credits for significant energy efficiency upgrades to buildings can be offered. The state or energy utilities can provide energy audits of buildings as a service as well as comparisons of energy usage with other similar buildings in the area. In Norway, opportunities for increasing energy efficiency of buildings have been placed on an Internet Energy Kiosk. Building managers could be trained in energy efficiency.

This policy has proved very effective in both voluntary and mandatory programs. Computer manufacturers, even beyond the United States, have voluntarily adopted the U.S. Environmental Protection Agency's Energy Star program. The Energy Star label can be displayed on a device only if specific efficiency standards are met.

Since consumers tend to make purchase decisions based on first cost rather then life cycle cost, manufacturers may ignore operating and maintenance costs. An individual manufacturer who tries to do better may lose sales. Appliance standards require all manufacturers to meet specific energy efficiency standards.

**Difficulties in Implementing Energy Policies.** New energy policies often cause disruption to producers, consumers, or both. Special interest groups may fight for or against a new policy. Policies that involve increased prices or taxes or that have potential adverse economic impacts may be politically difficult to implement. Policies that restrict the use of certain fuels, such as coal or wood, may cause hardships to individuals.

Policies that require new equipment may have difficulties because of a shortage of capital and a perception that the new equipment may be insufficiently tested and thus pose a financial risk.

Countries that are in the transition to a market economy face particular difficulties. New policies to allow the market to set prices on energy services that were once provided free or were highly subsidized must be implemented gradually to avoid hardship and political backlash.

Some energy conservation programs have been highly successful, but others have faced severe difficulties. A residential or commercial tenant has little incentive to make energy-saving capital improvements to the building since he may move, while the landlord has no incentive since the tenant pays the energy bill. In many businesses and industries, energy conservation does not come to the attention of top management, since the energy bill, although perhaps large, is a small fraction of total costs. Individuals and builders are usually reluctant to use new, energy-conserving materials because they are not familiar and may require new approaches to construction.

A change in the pattern of living, from compact cities and villages to more dispersed housing and industries, has led to an increase in the use of individual motor vehicles and, in many areas, a decrease in more efficient public transport. Initiated in the United States, but spreading to much of the world, this new pattern has led to a situation of high energy use that essentially defies improvement by any policy.

**Energy Policy in Developing Countries.** The problems facing developing countries are substantially different from those of industrialized countries, and developing countries therefore have different policy objectives. In many such countries, a substantial fraction of the population, especially in rural areas, does not have access to electricity. Expanding electric service is often a key policy goal.

Oil, natural gas, and electricity are frequently state monopolies. Oil-producing countries frequently subsidize the domestic price of oil products substantially. Over the past decade, however, there has been a trend toward reducing or eliminating subsidies, privatization, and "unbundling" of the components of the energy system, that is, separating exploration, production, refining, transport, and sales of oil and gas and separating generation, transmission, and distribution of electricity. This has come about in part as a means of obtaining needed capital, to increase efficiency of operation, to reduce smuggling, and by pressure from the International Monetary Fund, the World Bank, and other funding organizations.

Argentina is a good example. Production and distribution of natural gas and electricity was privatized, and deregulation and privatization of the oil and gas industry spurred a significant increase in production. China and India focus on better use of local coal resources. One example is the large number of fluidized bed industrial boilers introduced in China. In Ethiopia, policy has focused on switching from fuelwood to kerosene as a means of reducing deforestation.

A policy of cooperation among neighboring countries has improved the energy sector in many countries. Examples include the Mercosur regional trade pact in South America and the Chad–Cameroon oil pipeline.

**Energy-Producing Countries.** OPEC was created in 1960. Its stated objective is to coordinate and unify petroleum policies among member countries to secure fair and stable prices for petroleum producers, an efficient, economic, and regular supply of petroleum to consuming nations, and a fair return on capital to those investing in the industry. The OPEC member states are Algeria, Indonesia, Iran, Kuwait, Libya, Nigeria, Qatar, Saudi Arabia, United Arab Emirates, Venezuela, and Iraq. Although organized as a cartel, OPEC limited itself to negotiation with the Western oil companies throughout the 1960s. The West refused to take OPEC seriously. At the time, there was a surplus of oil available. The 1967 Arab–Israeli war seemed to confirm the validity of this approach, since even the one-hundred-day closure of the Suez Canal did not seriously interrupt the flow of oil. This was in part due to increased oil from then pro-Western Libya. This changed two years later when the Libyan king was overthrown by Muammar al-Qaddafi. From 1970 to 1973, oil prices were ratcheted upward and the oil-producing nations gained control of their oil. In 1973, Saudi Arabia told the United States that increased oil sales depended upon reduced support for Israel. Further price increase were introduced. From the late 1960s to 1973, the price of oil rose from U.S.\$1.80 to U.S.\$11.65 per barrel. With the 1973 Arab–Israeli war, Arab shipments of oil to the United States were embargoed.

Higher oil prices resulted in a phenomenal transfer of money from the West to the oil-producing nations. Western economies were disrupted, and developing countries with no oil production of their own were devastated. The rising prices, however, encouraged new production in other countries, and by the early 1980s the non-OPEC countries were producing more oil than OPEC. Competition forced OPEC to lower prices.

**Technology Transfer.** Substantial technological advances have been made over the past two decades in energy production and use. These include a doubling of the efficiency of fossil power plants using combined-cycle technology, similar increases in efficiency of many end-use devices such as refrigeration, and improvements in energy conservation. Many industrialized countries have programs to transfer new energy technologies to developing countries. One difficulty is that these new technologies are only partially implemented in the industrial countries themselves. Developing countries are often reluctant to be the testing ground for new technologies. Another difficulty is that the new technologies are often unsuitable for use in developing countries. Finally, the new technologies are seldom provided as a gift. The developing countries usually have to pay for the technologies either by expending scarce capital or by borrowing and increasing national debt. By careful selection of technologies appropriate to the country and its available resources, there are opportunities for developing countries to use technology transfer to make rapid improvements in their energy infrastructure. Already, the implementation of mini-steel mills in developing countries is changing the global structure of that industry. In some cases, the infrastructure of the industrialized countries can be bypassed: an example is the use of cell phones.

Several industrialized countries have transferred energy models to developing countries and provided training. The International Atomic Energy Agency also has a long-term commitment to training member countries in energy modeling.

[See also Economic Levels; Energy; Environmental Economics; Fossil Fuels; Greenhouse Effect; and Renewable Energy Sources.]

### BIBLIOGRAPHY

Carnegie Commission on Science, Technology, and Government. Task Force on Environment and Energy. *Organizing for Environment, Energy, and the Economy in the Executive Branch of the U.S. Government.* New York: The Commission, 1993. Discussion of linkages of energy, environment, and economic and international policy, with recommendations.

Dasgupta, P., et al. "The Taxation of Exhaustible Resources." In *Public Policy and the Tax System*, edited by G. A. Hughes and G. M. Heal. London: Allen and Unwin, 1980.

Goldemberg, J. "Leapfrog Energy Technologies." *Energy Policy* 26 (1998), 729–741.

International Atomic Energy Agency. "Incorporation of Environmental and Health Impacts into Policy, Planning and Decision Making for the Electric Sector." Key Issue Paper 4. Vienna, 1991. Discusses decision-making processes, policy options, instruments and constraints, and implementation of policies.

International Energy Agency. *Energy Efficient Communities.* Paris, 1994.

Kræmer, T. P. and L. Stjernström. *Energy Policy Instruments: Description of Selected Countries.* Copenhagen: AKF Forlaget, 1997. Summaries of energy policies implemented in thirteen countries.

Starling, G. *Strategies for Policy Making.* Chicago: Dorsey Press, 1988. A primer on policy, strategic thinking, and policy analysis; includes case examples.

United States Congress. *Climate Change Action Plan.* Washington, D.C.: United States Government Printing Office, 1993.

United States Department of Energy. *A Compendium of Options for Government Policy to Encourage Private Sector Responses to Potential Climate Change.* DOE/EH-0102/3. Washington, D.C., 1989.

——. *Policies and Measures for Reducing Energy Related Greenhouse Gas Emissions, Lessons from Recent Literature.* DOE/PO-0047. Washington, D.C., 1996. A practical compendium covering buildings, manufacturing, transportation, electricity generation, and economy-wide options. A key reference source.

——. *Privatization and the Globalization of Energy Markets.* DOE/EIA-0609. Washington, D.C., 1996.

World Commission on Environment and Development. *Our Common Future.* New York and Oxford: Oxford University Press, 1987. Frequently referred to as the "Brundtland Report" after the chairman of the commission, which was created by the General Assembly of the United Nations to propose long-term environmental strategies.

Yergin, D. *The Prize.* New York: Simon and Schuster, 1991. The history of oil.

—S. C. MORRIS

**ENSO.** *See* El Niño–Southern Oscillation.

## ENVIRONMENTAL ACCOUNTING

Environmental accounting is the systematic effort to assess the ecological significance of economic activity and the economic worth of natural resources. Standard measures of national economic welfare, such as gross domestic product (GDP), calculate the monetary value of market transactions. The environmental impact of these transactions is largely disregarded. Environmental accounting provides indices for a sustainable economics. It is grounded in the premise that measures of national (or regional) welfare should include assessments of: (1) available natural capital (stocks of natural resources), (2) the depletion or degradation of natural resources and the welfare effects of such depletion or degradation, and (3) expenditures for environmental protection and restoration.

Promoting "green businesses" and eliminating government subsidies for polluting industries are two of the more common practical goals of environmental accounting. The use of environmental labeling is one means of promoting green business practices. Numerous national "ecolabeling" efforts have been developed, such as Germany's "Blue Angel" and the United States' "Green Seal" programs. In 1992 the European Union developed a regulatory framework for such programs. Ecolabeling programs identify for consumers products whose production, use and disposal are less environmentally harmful than noncertified products. The success of these programs varies from product area to product area and from country to country. [*See* Greening of Industry.]

Efforts to curtail the subsidization of environmentally harmful business practices is another practical goal of environmental accounting. Beginning in 1993, Friends of the Earth and the National Taxpayers Union spearheaded the "Green Scissors" campaign. The Green Scissors Report annually challenges the U.S. Congress to cut $30 billion to $40 billion from the federal budget by terminating between thirty and sixty environmentally harmful federal programs. In addition to the recommended cuts in spending and subsidies, the report also exposes environmentally destructive tax loopholes, mostly given to extractive industries, that cost taxpayers over $4 billion annually.

A more ambitious goal of environmental accounting is *full-cost pricing*. Full-cost pricing is grounded in the premise that the price consumers pay for goods and services should factor in long-term environmental costs. Businesses often "externalize" the environmental costs of production. Financial benefits from resource exploitation may then be immediately realized while the environmental costs of this exploitation are distributed across time and space. Businesses may not clean up the waste streams produced by their operations, for example, effectively passing along the costs of abating or living with pollution to residents living "downstream." Alternately, natural resources may be depleted without recompense to future generations that would otherwise benefit from their use. Were full-cost pricing implemented, goods and services would be priced to include the environmental costs associated with their production, distribution, use, and disposal, effectively eliminating these externalities.

Some of the more straightforward proposals for full-cost pricing are geared toward the levying of "green taxes." Many environmental groups endorse a tax on gasoline that would pay for the costs of building and maintaining highways and abating the pollution caused by vehicle exhausts. Pollution taxes that would achieve similar results for industrial facilities are also widely endorsed, as are taxes on resource and energy use and waste production. [*See* Ecotaxation.]

**The History of Environmental Accounting.** As environmental awareness grew in the late 1960s, scholars began investigating the deficiencies of the current systems of economic accounting. In the early 1970s E. F. Schumacher observed that the modern industrial system was treating its irreplaceable natural capital as if it were income. Viewing the capacity of the biosphere to absorb pollution as a type of natural capital itself, Schumacher maintained that the saturation of the biosphere with pollution constituted an unsustainable depletion of resources. Schumacher's concerns were shared by many environmental scholars, including Herman Daly and other economists. Only in the late 1980s, however, did environmental accounting gain widespread attention and undergo extensive conceptual clarification, empirical measurement, and practical implementation. Today

many scholars, nongovernmental organizations, national and regional governments, and international agencies engage in and promote environmental accounting.

In 1991 a Special Conference on Environmental Accounting organized by the International Association for Research in Income and Wealth (IARIW) was held for scientists and practitioners in Baden, Austria. The conference produced a draft manual, later issued by the United Nations (UN) Statistics Division as the handbook *Integrated Environmental and Economic Accounting*. That same year, a colloquium held by the National Academy of Sciences investigated the relation of environmental accounting to the emerging field of "industrial ecology," which employs lifecycle analysis of products and "design for environment" strategies. The 1992 UN Conference on Environment and Development held in Rio de Janeiro, also known as the "Earth Summit," produced a consensus view on the merit of public policies oriented toward sustainable development. The conference action plan, Agenda 21, endorsed environmental accounting as a component of sustainable development strategies.

In 1993 the findings of the Baden conference were partially incorporated into the revised edition of the globally adopted system of national accounting (SNA). A system of integrated environmental and economic accounts (SEEA) was produced to accompany the SNA. In 1996 a second IARIW conference on environmental accounting that focused on the implementation of the SEEA was held in Tokyo. This system of environmental accounting was conceived as a "satellite" account that would supplement rather than replace existing national accounting measures, such as the GDP.

Certain components of national environmental accounting—the valuation of natural resource stocks and use, for instance—are becoming mainstream. Other features that cannot easily be measured in market transactions and require difficult estimations, such as the valuation of externalities, remain controversial.

**Greening Business.** Support within the business community for environmental accounting has been reserved. Relatively few corporations have attempted to integrate its principles into their business practices. Full-cost pricing, in particular, is seldom embraced. In part, this is because it is perceived as a threat to profit margins. Were the full-cost pricing implemented, for example, the cost of monocropped fruits and vegetables produced by agribusiness and sold on supermarket shelves, despite industrial efficiency and economies of scale, might rise well above that of organic produce purchased at farmers markets or local grocers. Full-cost pricing is also a complex, time-consuming exercise that entails the discovery and disclosure of the environmental costs of the complete lifecycles of products. Determining the full-cost price of commercially produced fruits or vegetables would entail assessing: (1) the environmental costs of their agricultural production, including soil erosion, the eutrophication of water sources caused by the use of synthetic fertilizer, and the bioaccumulation of contaminants caused by pesticide applications; (2) the environmental costs of their delivery to consumers, including the pollution caused by trucking to market and the energy depletion caused by refrigerated transport and storage; and (3) the environmental costs of their waste streams, including spoilage, discarded packaging, and noncomposted remains. The present total unpriced environmental costs of the U.S. food system are estimated at $150 billion to $200 billion per year. Whereas traditional accounting methods of agricultural production estimate an average profit of $80 per acre, full-cost pricing studies have estimated a loss of $26 per acre (Faeth et al., 1991). Full-cost pricing of imported foodstuffs may yield an even greater deficit.

The full-cost pricing of automobile use would also increase its costs significantly, making public transportation much more economical. Current gasoline taxes, vehicle taxes, and road tolls in the United States cover less than two-thirds of the total capital and operating costs of highways. Operating costs not covered each year by these user fees generally come out of local taxes. In turn, the indirect costs associated with automobile travel, such as highway law enforcement, tending to accident victims, and smog abatement, amounts to billions of dollars each year. Were these expenditures figured into the price of motor vehicle fuel, the cost of gasoline would rise by $3 to $7 a gallon (MacKenzie et al., 1992). If other collateral effects were assessed, the price would rise even higher. The destructive effect of car emissions on the nation's wheat, corn, soybean, and peanut crops yields losses of $1.9 billion to $4.5 billion annually. Motor vehicle pollution, in turn, adds between $40 billion and $50 billion to annual healthcare expenditures, and causes as many as 120,000 unnecessary or premature deaths. With these expenditures included, the full-cost price of gasoline could mount to as much as $11 per gallon. Environmental accounting, advocates conclude, would have the effect of "taking cars off welfare."

Along with public transport, alternate forms of energy production would become more economically feasible were environmental accounting applied to fossil fuel consumption. Some sixty-four thousand deaths from heart and lung disease are caused each year by particulate air pollution in the United States. If the cost of caring for these patients and the cost of lost productivity and premature death were added to the price of burning coal or oil, these fossil fuels would become more expensive than many forms of renewable energy, such as solar power.

**Greening the GDP.** For a sustainable economics grounded in full-cost pricing to become feasible, the common measures by which standards of living are assessed would have to change. Currently, quality of life is typically measured in terms of GDP. Such measures do not take the collateral ecological effects of economic activity into account. Environmental organizations were quick to publicize the fact that the 1989 Exxon *Valdez* disaster, which dumped 11 million gallons of oil into Alaska's coastal waters, actually raised the GDP. That is because the cleanup efforts were formally recorded as contributions to the nation's economic productivity. Less controversial but no less significant examples abound: walking, biking, or taking mass transit to work contributes less to the GDP than the private use of an automobile. Wearing a sweater on winter nights contributes less than raising the thermostat. In short, the environmental imperative to "reduce, reuse, and recycle" is undermined by standard measures of economic productivity. Greening the GDP would necessitate counting the depletion and degradation of natural resources as a debit, not a credit to national accounts. A more comprehensive measure of quality of life would include various indicators of environmental welfare, such as human morbidity and mortality rates; ecological health indicators for forests, rivers, lakes, coastal regions, and agricultural land; levels of air and water purity; presence of wildlands; and biological diversity.

The World Bank, the International Monetary Fund, and a number of UN agencies have explored means of recalculating GDP to account for environmental losses and gains. The feasibility of environmental accounting at the regional level has been explored in a number of countries, including Indonesia and the Philippines. In the United States, green economic indices have not been widely accepted by the business community, nor have they been implemented by the government. In France, Germany, Japan, the Netherlands, New Zealand, and Norway, alternate measures, such as the Net National Welfare or the Net National Product, are more widely employed. These indices are calculated by subtracting various environmental costs from the GDP, namely: (1) the money spent protecting against and repairing environmental damage; (2) the monetary equivalent of the degradation remaining after such protective and reparative expenditures have been made; and (3) an allowance for the depletion of natural resources.

[*See also* Environmental Economics.]

## BIBLIOGRAPHY

Ahmad, Y., S. El Serafy, and E. Lutz. *Environmental Accounting for Sustainable Development.* Washington, D.C.: World Bank, 1989. A brief account with an international focus.

Chertow, M. R., and D. C. Esty, eds. *Thinking Ecologically: The Next Generation of Environmental Policy.* New Haven, Conn.: Yale University Press, 1997. A very informative collection of essays.

Daly, H. E., and J. B. Cobb Jr. *For the Common Good: Redirecting the Economy toward Community, the Environment, and* a *Sustainable Future.* 2d ed. Boston: Beacon Press, 1994. A readable yet comprehensive introduction.

Dryzek, J. S. *Rational Ecology: Environment and Political Economy.* New York: Blackwell, 1987. A critical look at standard economic practices and an endorsement of sustainable economics.

Faeth, P., et al. *Paying the Farm Bill.* Washington, D.C.: World Resources Institute, 1991.

Friends of the Earth and the National Taxpayers Union Foundation. *The Green Scissors Report: Cutting Wasteful and Environmentally Harmful Spending and Subsidies.* Washington, D.C., 1997. An example of environmental accounting in practice.

Hardin, G. *Living within Limits: Ecology, Economics, and Population Taboos.* New York: Oxford University Press, 1993. A hard-nosed assessment of ecological constraints to economic and population growth.

Hawken, P. *The Ecology of Commerce: A Declaration of Sustainability.* New York: HarperCollins, 1993. An overview of environmental business practices and possibilities.

Hecht, J. E. "Environmental Accounting: Where We Are Now, Where We Are Heading." *Resources* (Resources for the Future) (Spring 1999).

MacKenzie, J. J., R. C. Dower, and D. Chen. *The Going Rate: What It Really Costs to Drive.* Washington, D.C.: World Resources Institute, 1992. An informative, empirical assessment.

Roodman, D. M. *Paying the Piper: Subsidies, Politics, and the Environment.* Worldwatch Paper 133. Washington, D.C.: Worldwatch Institute, 1995. An informative, empirical assessment.

Schmidheiny, S. *Changing Course: A Global Business Perspective on Development and the Environment.* Cambridge, Mass.: MIT Press, 1992. A useful overview of corporate environmentalism.

Schumacher, E. F. *Small Is Beautiful: Economics as If People Mattered.* New York: Harper and Row, 1973. A classic introduction to environmentally grounded economic thought.

United Nations. *Integrated Environmental and Economic Accounting.* New York: United Nations, 1993. A useful summary account.

Uno, K., and P. Bartelmus, eds. *Environmental Accounting in Theory and Practice.* Dordrecht: Kluwer Academic, 1998. A well-organized and informative anthology.

Welford, R., and R. Starkey, eds. *Business and the Environment.* Washington, D.C.: Taylor and Francis, 1996. A helpful introduction.

—LESLIE PAUL THIELE

# ENVIRONMENTAL ECONOMICS

In a general sense, the last one hundred years have witnessed the turning of a complete economic cycle. The twentieth century began against a background of globalization with the international economy and economic doctrine buttressed by free movement of capital, free trade, economic market liberalization and stable money. Following World War I, periods of extreme political and economic instability alternated with long runs of more stable and regulated conditions as material well-being in the industrialized countries increased rapidly. Beginning in the early 1970s the modern version of neoliberal market-based economics, with characteristics such as deregulation, privatization, tax cuts, and curbs on public spending, returned to dominate the political scene. At the international level, the 1990s saw the new deregulation, information technology, financial markets integration, and other globalization trends leave the world community increasingly prone to "local" economic crises, with consequent widespread ramifications and with environmental pressures and stresses. These environmental risks threaten both local and regional environmental resources and ambient quality, as well as global environmental systems such as the climate and the stratospheric ozone layer.

Having defeated socialist central planning philosophy and practice, market capitalism and economics in the twenty-first century will need to meet the challenges posed by environmental change on an unprecedented scale. The subdiscipline of environmental economics came to prominence during the 1960s at the time of the first wave of modern popular "green" thinking and policy perceptions within developed countries, known as environmentalism (O'Riordan, 1981). Nevertheless, this branch of economics shares with its parent discipline a common history and an overlapping but not identical set of fundamental ideas. Historically, economics has been strongly influenced by developments in science, as early on, formal economics followed the mechanistic models of Newtonian physics. More recently developments in population biology and community systems ecology have stimulated new thinking in resource and environmental economics and management. At the core of environmental economic thinking is the recognition that our economic system (which provides us with all the material goods and services necessary for a "modern" standard of living) is underpinned by and cannot operate without the support of ecological systems of plants and animals and their interrelationships.

A "materials balance" perspective of the economy is fundamental to environmental economics thinking. The laws of thermodynamics imply that an economic production-consumption system inevitably sucks in "useful" low-entropy matter and energy and pushes out "useless" high-entropy matter and energy such as low-temperature heat, gases, and particulate matter. Much analysis and debate has therefore been devoted to the nature and severity of a range of natural resource supply problems ("source limits"), as well as to pollution/waste assimilation problems ("sink limits"). In 1966 Kenneth Boulding wrote an essay on "Spaceship Earth" that combined economics and some science in order to bring together the view of the economy as a circular resource flow system, and of the environment as a set of limits (Boulding, 1966). Economists have explored the dynamics of the natural systems with which the economy interacts in order to understand optimal rates of resource use and have treated pollution (known technically as externality effects) as a pervasive characteristic of the system.

The notion that material growth in the economic system necessarily increases both the extraction of environmental resources and the volume of waste deposited in the environment was highlighted in the mass-balance general equilibrium models of Ayres and Kneese (1969). These models, following earlier insights by Leontief (1966), Daly (1973), and Georgescu-Roegen (1971), yielded important insights into pollution (waste) externalities, and helped in the development of pollution control instruments and policy. But they did not become a focus for further extensive work in environmental economics. Indeed, such work as was done on both mass-balance and entropy models was out of the mainstream. The so-called regional environmental quality models of the 1970s, for example, never really gained full entry into the core of standard economics (Basta et al., 1978), and Georgescu-Roegen's work on entropy-based models was definitely on the fringe. Environmental economists would counter that "materials balance constraints" thinking has influenced work on marketable permit schemes for control of air pollutants such as sulfur and nitrogen oxides and their practical implementation, as well as underpinning recent integrated modeling research in the context of, for example, climate change. The mass-balance work has generated two ecological-economic axioms. First, since perfect recycling of resources is probably precluded on thermodynamic grounds, the potential growth of physical output is finite. Second, since the waste generated in the process of production is seldom inert, higher rates of physical growth imply higher rates of change in the processes of the environment (Perrings, 1987).

The interrelationships between population growth, increased material output and wealth, and the state of environmental systems and their natural resources have become increasingly complex as globalization and the scale of economic activity have expanded (Daly, 1992). In the future, an approximate doubling of world population and gross world product per capita by around 2030 would impose significant additional stress and shock on both local and global ecosystems. Water resources in particular would become scarce commodities, with critical geopolitical consequences. Local resource allocation problems have proved difficult to mitigate even when global supplies are adequate, because of nonuniform spatial distribution of natural resources (in economic terms, "natural capital") and because the political and purchasing power of the poor is low (Dasgupta and Mäler, 1998). These local environmental deteriorations are often masked by global statistics (macroaverages) that show a positive trend despite the distributional inequities.

Economic growth and endogenous technical innovation result in evolutionary changes in the structure of consumption and in the mix of goods and services produced. Overall, both complexity and organization are increased. Among the factors thought to influence the relationship between income levels and the state of the environment are rates of economic growth; the energy processes and flows involved as dissipative economic activities develop while retaining their reliance on nonrenewable resources; structural change and the composition of production and consumption; international trade; the density of economic activity; power and income inequalities; and external shocks. According to some interpretations of the so-called environmental Kuznets curve (EKC) relationships, further economic growth brings with it environmental improvement. The hypothesis is that as industrializing economies with low levels of per capita income continue to grow, the emissions of certain pollutants also increase. After per capita income reaches a threshold level, however, more economic growth brings environmental improvements with it, if appropriate policy responses are made (Grossman and Krueger, 1992; Selden and Song, 1994).

Tests of the EKC hypothesis have produced mixed results, with EKC relationships holding for some but not all environmental indicators. The best fits have been found for some air pollution indicators and a small number of water pollution indicators, with a bias toward local and short-term pollution, rather than indirect, longer-term global pollution. The income levels that provide turning points for most EKC curves are quite high, so the prospects in the short to medium term for global pollution are gloomy (Barbier, 1997; Cole et al., 1997). Institutional factors (the quality of policies, enabling property rights, and other institutions) can reduce environmental damage at low income levels and speed up environmental improvement at higher income levels. It also seems to be the case that a more equal distribution of power and income contributes to less environmental pressure; that is, more equitable income distribution, wider literacy, and greater political liberties and civil rights improve the general quality of the environment (Torras and Boyce, 1998).

Because environmental economics has accepted the hypothesis of an extensive interdependence between the economy and the environment, some of its analysts have also pointed out that the design of economies offers no guarantee that the life support functions of natural environments will persist. We do not have what we could call an "existence theorem" that relates the scale and components of an economy to the set of environment-economy interrelationships underlying that economy (Pearce and Turner, 1990). Without this theorem there is a risk of degrading and perhaps destroying environmental functions. If we are interested in sustaining our economy over time, it becomes important to establish some principles and then practical rules for sustainable economic development.

There is, then, a very real but ultimate sense in which economic activity is limited or bounded by the capacities of natural environments. Thus the "limits" concept and debate, which started in the 1970s and has simmered on ever since, has it origins in the work of thinkers such as Thomas Malthus, David Ricardo, and Karl Marx. Malthus worried about absolute limits or scarcity. Ricardo took a more sophisticated and slightly more optimistic perspective when he argued that relative limits or scarcity was the real problem for a growing economy. In Ricardian analysis, limits are set by rising costs as the highest-grade resources, which are exploited first, become exhausted and have to be substituted for by successively lower-grade resources. The costs of exploitation, including pollution costs, escalate as the "grade profile" of resources declines.

Marx highlighted, among many other things, the possibility that economic growth might be limited because of social and political unrest. The "social limits to growth" theme was picked up again by some economists during the development of environmental economics in the 1970s. At that time, opinion poll evidence in the rich countries seemed to indicate that despite huge absolute increases in the material standard of living, people on average said they did not feel much happier with their lives, the Easterlin Paradox (Easterlin, 1974). It turned out that the "feel good factor" was a complex phenomenon influenced as much by relative income and social status as by absolute quantities.

The "social limits" theme was further extended and elaborated on during the 1970s with the addition of moral concerns connected with economic growth and development. Ethical issues surfaced on the potentially negative impact of the fast-growing and competitive modern economic system on the prospects for future human generations (intergenerational equity concerns) and nonhuman nature (bioethics debate), as well as on the already declining moral standards in contemporary society.

Between the 1970s and the 1990s, there has been a shift in environmental concern away from absolute limits to growth (source constraints) and toward waste assimilation (sink constraints) and related global environmental change (climate change and biodiversity loss). In the 1970s the "limits to growth" debate asserted a Malthusian viewpoint in its espousal of the physical limits to economic growth based on resource constraints. Economic critiques of this position stressed the role of technological change and the price effects of absolute resource scarcity leading to increased conservation measures and substitution. The Brundtland Commission Report (World Commission on Environment and Development, 1987) could be said to be following

in this philosophical tradition. In the 1980s the interdependence of the world economic system was recognized in several major reports, including that of the Brundtland Commission, but the latter's particular focus was on environmental interdependency. [See Growth, Limits to.]

The Brundtland Report is portrayed as a direct precursor to the specific focus of the 1992 United Nations Conference on Environment and Development conference on global environmental issues (the "Rio Summit"). What are the similarities and differences between the "limits to growth" and "global environmental change" concepts? Both phenomena, as generally perceived, have been based on scientific analysis and are equally widely (although not unanimously) accepted in the contemporary scientific communities and in environmental movements. Both concepts have also led to prescriptions on the consumption/population debate that are neo-Malthusian in philosophical outlook. The essential difference lies in the general acceptance by capitalist economies and governments of the concept of global change (and of environmental awareness in general) where "limits to growth," interpreted as a zero economic growth objective, was shunned. Global environmental change, by contrast, is perceived as a problem that is amenable to mitigation in ways that further the agendas of developed countries—for example, by creating markets for new products and less-polluting technologies. We are therefore still waiting for the creation of an effective international financing organization to deliver new policies and practice that will benefit developing countries. [See Brundtland Commission.]

In 1982, for example, global income (gross domestic product) stood at approximately U.S.\$11,000 billion; by 1997 this had increased to U.S.\$29,000 billion. But is this process, or something like it, sustainable over the long run? This conundrum is made more difficult when it is further realized that between 1960 and 1991 the richest 20 percent of all nations had their share of world income rise from 70 percent to 83 percent, while the poorest 20 percent of countries suffered a fall from 2.4 percent to 1.4 percent (United Nations Development Programme, 1992, 1994).

The trend toward an ever-widening gap between the richest countries and the poorest has been reinforced by a number of factors, which in combination are ensuring that the disparity will persist into the future. Only the rich countries have the capacity to invest in more technological innovations, while the poorest countries have to rely on "trickle down" technology transfers, which often prove to be ill suited to local environmental and labor market conditions. Savings rates are low in poor countries, inhibiting the investment and growth process. Debt burdens mean that too great a proportion of what is generated in poor countries (usually through the export of income-inelastic goods) in poor countries is transferred out in the form of interest payments. Many poor countries are plagued by political instability and strife, which inhibit inward investment and divert scarce domestic resources away from community wealth creation activities. Finally, population growth in some countries is so rapid that infrastructure investment is a permanent diversion from more direct income generation (Sandler, 1998).

**Sustainable Economic Development.** Sustainable economic development may be characterized as a process of change in an economy that ensures that welfare is nondeclining over the long term. It is nevertheless the case that the concept is often difficult to pin down in operational terms. The dominant view among both environmental and ecological economists is based on capital theory and defines sustainable development in terms of the maintenance of the value of a capital stock over time. The definition of capital used encompasses both natural capital (the functions, goods, and services provided by the environment), and manufactured, human, and institutional capital (with the latter taken to include ethical or moral capital and cultural capital). [*See* Sustainable Development.]

While the capital theory approach itself is criticized by some analysts, who identify problems in aggregating natural and produced capital, the main difference between analysts concerns the problem of the substitutability of produced and natural capital. The ecological economic concepts of weak and strong sustainability are, for example, defined in terms of the degree to which various capital stocks may be substituted for each other. Weak sustainability assumes perfect substitutability between natural and other forms of capital. Under weak sustainability, the maintenance of an aggregate capital stock over time is both a necessary and a sufficient condition for sustainable economic development. Economic growth can continue indefinitely according to this perspective, as long as the "Hartwick Rule" is observed, that is, that the economic rents derived from the exploitation of exhaustible natural resources (fossils fuels, etc.) are invested in other forms of capital capable of yielding an equivalent stream of income in the future.

Strong sustainability assumes well-defined limits to substitution. Under strong sustainability a minimum necessary condition for sustainability is that separate stocks of aggregate natural capital and aggregate "other" capital must be maintained. Keeping the natural capital base intact over time has been interpreted to mean conserving all "critical" natural capital (e.g., life-support functions and services and supporting environmental attributes), which by definition is subject to irreversible loss. There are no plausible technological substitutes for climatic stability, stratospheric ozone, topsoil, or species diversity. Technological optimism, in this regard,

is simply misguided. These system functions should be regarded as finite resources, vulnerable to human interference. While such functions are potentially renewable, they can be irreversibly destroyed, if not actually consumed. The maintenance of critical natural capital could be achieved through the imposition of environmental standards or regulations mandating lower bounds on appropriate natural capital stocks, including the environment's waste assimilation capacity. It seems better to err on the side of caution and conserve natural capital in order to maintain the options of future generations.

The role that technology and innovation might play in any future sustainable-development strategy is a substantive question. A growing body of analysis adopts the evolutionary perspective, in which the direction of technical progress and economic activity is seen not as an autonomous phenomenon but as an endogenous process conditioned by the structure of economic incentives, technological opportunities, and prevailing institutions. Future outcomes are relatively unpredictable because of the complex interaction taking place between different actors who are endogenous to the process of change. Technical innovations are the outcome of leaning processes inside firms, but progress is hindered by inertia in the firms' organizational routines and by path dependency, with the attendant risk of becoming locked into an unsustainable technological trajectory. [*See* Technology.]

The older notions of innovation processes driven by the push and pull of technology and demand have therefore been supplanted by the concepts of technological paradigms and trajectories. Technical change within a paradigm is biased in certain directions, relies heavily on learning-by-doing, and is a cumulative process. Improvement in techniques is conditioned by input prices, the institutional structure (monopoly, etc.), and government regulation (including environmental regulations). But "localized technical change" ("lock-in" effects) is also common within paradigms and requires public investment or grants to release new innovative activities. Sufficiently large changes in the external context in which a firm or industry operates, including environmental pressures, can force a paradigm shift in the technology used (e.g., from end-of-pipe to radical "clean" technologies). Firms that can anticipate changes in environmental regulations can benefit from "first-mover" advantage (Porter and van der Linde, 1995) and appropriate economic returns from new technology that complies with future standards, gaining internal cost reductions, new markets, and enhanced public image.

Firms that improve their resource use efficiency (by capital substitution) reduce input costs per unit of output and also minimize waste disposal costs. The trigger can be environmental regulations and changes in wider social values and concerns. The impact of environmental compliance costs on overall competitiveness also appears so far to be insignificant. Even the latest Club of Rome report now claims that resource productivity (reducing waste at all stages in the production-consumption cycle) can, and should, grow fourfold (Von Weizsacker et al. 1997). But critics of the ecoefficiency strategy warn that it is not a panacea, even if it did prove feasible to get resource productivity gains of a factor of four or more. There may be "rebound effects" such as reduced pressure on resources, price falls, and increased aggregate consumption.

**Pollution Control Policy Instruments.** It was Arthur Pigou (1920) who first formalized the impact of pollution on the working of the economy. He distinguished between the private costs of production and consumption (encapsulated in fuel, raw materials, labor costs, etc.) and the full social costs (i.e., on society as a whole) of such activities. Thus pollution gives rise to external costs, which drive a wedge between private and social costs. The socially optimal level of external costs is unlikely to be zero (zero pollution) because of the natural capacity of the environment to absorb some waste and the costs of controlling pollution. Zero pollution is desirable, however, when the predicted damage from the disposal of certain toxic and hazardous substances is thought to be catastrophic in some sense. Until the 1960s the economics literature, focusing on externalities between two parties, still dealt with externality as if it was an occasional problem causing a deviation away from Pareto optimality in competitive markets. Coase (1960) argued that even if markets may not secure the optimum amount of externality, they can be pushed in that direction without the necessity for full-scale regulatory activity involving taxes or standard-setting. The so-called Coase theorem laid down that bargaining (with compensation) between polluters and polluted in the market, given an established system of property rights, will lead to the socially optimal level of pollution. Regardless of which party holds the property rights, there is an automatic tendency to approach the social optimum. If this is correct, we have no need for government regulation of externality, for the market will take care of itself.

But since pollution externality effects are in reality pervasive and can involve large numbers of gainers and losers, some form of government intervention to control the rate and extent of pollution is required. For many environmental resources, markets fail to operate properly or are simply nonexistent. The reasons for this have been identified in terms of a variety of factors such as high transactions costs (caused by temporal and spatial distances); missing, ill-defined, or unprotected property rights; and inadequate or difficult-to-perceive information about the effects of environmental change. Through some combination of these factors, market prices, when they exist, fail to signal real social scarci-

ties and can mislead policy. Corrections can be made by imposing regulations covering resource use and pollution levels and by imposing taxes and other incentive instruments (Cropper and Oates, 1992).

Baumol and Oates (1975) formulated the least-cost theorem for pollution charges, pointing out that real-world pollution problems involve a combination of environmental quality standards and enabling instruments. Given certain assumptions, pollution charges are the least-cost method of achieving an exogenously set standard. More recently, other economic incentive instruments have been championed. The idea of marketable permits was first formulated by Dales in the 1960s and was extensively analyzed subsequently. When some of the Baumol-Oates assumptions are relaxed, however, and criteria such as distributional equity and ethical considerations are introduced, the case in favor of the incentive approach is much less clear cut (Bohm and Russell, 1985).

The political economy of public finances and fiscal regimes is complex and mirrors the various stakeholders and political interests present in contemporary societies. In principle, market-based incentive instruments offer efficiency gains over direct regulation measures, although the magnitude of the efficiency advantage is conditioned by the real-world context and application. Moreover, economic efficiency is only one of at least six not necessarily complementary principles (the others being environmental effectiveness, administrative cost effectiveness, fairness, institutional concordance and revenue raising) that are thought to be relevant in any policy instrument choice situation. So while economic instruments are inherently efficient and effective, the social gains they offer are limited by a policy process typically driven by multiple conflicting objectives.

Following Coase the "property rights" approach has been extended. Key neoclassical assumptions about human behavior in the marketplace (e.g., self-interested utility maximization) have been extended to cover the activities of bureaucrats in the public sector (borrowing from public choice theory), and notions of extended rationality (more than mere self-interest) have been resisted. It is argued that in an economy with well-defined and transferable property rights, individuals and firms have every incentive to use natural resources as efficiently as possible. Increased government intervention should be resisted because public ownership of many natural resources lies at the root of resource conflicts: governments may fail. The misallocation of environmental resources is not, therefore, just a question of market failure. Environmental resource depletion and degradation can also be caused by government (institutional) failures such as inefficient or uncoordinated resource management policies and inappropriate tax exemptions or subsidies of various sorts. The world's

## PERVERSE SUBSIDIES

Subsidies for various types of human activity can be useful. They can overcome deficiencies of the marketplace, they can support disadvantaged segments of society, and they can promote environmentally friendly technologies. But there are many subsidies—termed *perverse subsidies* by Norman Myers and Jennifer Kent—that are adverse in the long run to both the economy and the environment. As they pointed out, subsidies for agriculture can lead to soil erosion, pollution from agricultural chemicals, and release of greenhouse gases. Subsidies for fossil fuels contribute to acid rain, the enhanced greenhouse effect, and urban smog. Road transportation subsidies promote pollution and cause a loss of scenic amenity. Water subsidies encourage waste of an increasingly scarce commodity, while subsidies for the fishing industry lead to the depletion of fish stocks.

**BIBLIOGRAPHY**

Myers, N., and J. Kent. *Perverse Subsidies: Tax $s Undercutting Our Economies and Environments Alike.* Winnipeg: International Institute for Sustainable Development, 1998.
—ANDREW S. GOUDIE

coastal zones, for example, are now under severe pressure from multiple resource demands. In many places this environmental pressure is exacerbated by natural geophysical factors and climate change so that zones have become more vulnerable (less resilient) to further stress and shock. A marked feature of the pollution and resource overexploitation problems in coastal zones is the significance of "out of zone" activities and their effects. Most of the damage occurring in these places is related to activities located in the wider drainage basin areas and beyond (Bower and Turner, 1998).

If environmental resource property rights do not exist or are easily challenged, then too rapid a rate of resource exploitation will almost certainly result. The open-access problem (which applies to a range of resources, fisheries, wetlands, some forests, and the waste assimilation capacity of oceans and atmospheres) has, unfortunately, been confused in the environmental economics literature by frequent references to the "common property" problems and the "tragedy of the common" problem. In fact, common property is property owned by a community and is often subject to usage rates or social norms. The term "tragedy of open access" is therefore a better term and relates both to the problem of the optimal rate of resource exploitation and to the problem of pollution and the rate at which the environment's assimilative capacity could itself be depleted or destroyed.

**Ecological Economics.** Environmental economics is not a static body of knowledge but an ongoing process of change, refinement, and debate. Most recently, over the last ten years or so, a split has occurred that has led some analysts to comment that a potentially separate subdiscipline called "ecological economics" has

begun to emerge. Several writers have sought to show that despite differences in emphasis between resource, environmental and ecological economics (Turner et al., 1997), and in experimental methodologies and philosophies between economics and ecology, the joint research and experimentation agenda offers significant opportunities for the integration of economics and ecology. Such an integration is a key element in any strategy that seeks to improve the understanding and management of coevolving interrelated complex ecological and socioeconomic systems (Norgaard, 1981). Indeed, ecological economics has recently been defined as an overarching mode of inquiry encompassing both resource economics with its foundations in population ecology, and environmental economics, with links to systems ecology (Dasgupta, 1996). It is then possible to integrate problems of resource management with problems of environmental pollution and degradation, within the context of the underpinning environment as the provider of an extensive and diverse set of services and capital stock.

Nevertheless, the hypothesis that fundamental differences do exist between ecological and mainstream economics finds stronger support in the realms of philosophy, ethics, and social policy. The differences between ecological economics and the mainstream school of thought seem to be at their widest in the context of philosophy, ethics, human psychology, and social welfare conditions, as well as policy prescriptions that are supported by the contending camps. Some ecological economists strongly favor the at least partial substitution of a collectivist in place of an individualistic perspective. This has important implications for, among others, the neoclassical principle of consumer sovereignty, for the monetary valuation of environmental benefits, and for moral codes of conduct and "social limits to growth" arguments. The "social limits" position was given prominence in the 1960s and 1970s by orthodox economic writers such as Scitovsky, Hirsch, Mishan, and Thurow. Questioning the social desirability of the economic growth society (the "zero-sum society") centered on market exchanges, they put forward an array of more and less radical policies to restructure the market economy along more communal and egalitarian lines. This line of thought has strongly resonated within ecological economics and has been used to reinforce arguments for the establishment of humanistic "steady-state" and "bioeconomic" economic and social systems (Lutz and Lux, 1988).

Ecological economists lay great stress on the need for a historical perspective on socioeconomic–natural systems interactions, in order to progress the analysis and debate on the long-run dynamics of human-nature interactions (Faber et al., 1996). They tend to favor an evolutionary perspective and believe that the concept of open systems, far from equilibrium, is a useful heuristic for describing technological and socioeconomic change. Such change is seen to be sensitively dependent on small historical events and characterized by path dependency ("lock-in" effects) and the unpredictability of outcomes. An attitude of openness is advocated because of the existence of irreducible ignorance and the related concepts of surprise and novelty. In other words, some systems changes may not, in principle, be predictable, but a proper recognition of that unpredictability will still be important and useful for policy responses. Given contexts in which combinations of irreversibility effects, surprise outcomes, and irreducible ignorance exist, the appropriate policy response should be a flexible one. Policy should be conditioned by the precautionary principle and notions such as safe minimum standards, with due regard for the cost effectiveness of option choices and social opportunity costs.

More fundamentally, many ecological economists believe that the key to the mitigation of environmental problems (and in particular problems of pollution and waste assimilative capacity limits) lies outside the realm of science and technology. They have focused on the realm of ethics and philosophy. Policy analysis must, in their view, fully incorporate the concept of full and actual compensation for pollution sufferers and should concentrate on the consequences of the distribution of costs and benefits among multiple stakeholders. But justice is not to be seen as exclusively a matter of income distribution; it is also a question of procedures that define the resulting distributional outcome. This stance necessarily involves questions about community, social norms and collective preferences. The analytical framework should also be extended to cover the ethical analysis of intertemporal and interspecies choice (Costanza et al., 1997).

Ecological economists are also preoccupied with issues of macroenvironmental scale, threshold effects, and ecosystem "health" and "integrity." Ecological economics accepts the principle of separating scale and allocation decisions in environmental policy (Daly, 1991). Mainstream economics by and large provides no answer to the issue of optimum scale. One of the most frequently cited ecological economic arguments is that current levels of economic activity are such that further growth threatens to overwhelm the carrying capacity of the environment (both as a source of natural resources and in particular as a sink for wastes), and that this could result in environmental collapse. The argument is easily misconstrued as a resurrection of the "limits to growth" arguments of the 1970s. It is more subtle than that.

Economic activities are seen as being embedded within open ecological systems, which themselves are developing and maintaining their own self-organization over time. Therefore, the type and overall extent of economic activity is judged acceptable or not according to the severity of their effect on the "health" (the capacity for self-organizing activity) of the larger ecological system within which the economy operates. Ecological economics predicts that pollution and waste disposal ("sink") limits to growth will prove more recalcitrant than resource availability limits, because of the different types of novelty and ignorance they engender. While resource depletion can be forecasted and adaptive strategies can be encouraged by means of incentive mechanisms, pollution problems may prove to be more intractable. They may endanger the health of the biosphere itself and limit adaptive and innovating response strategies.

Relieving these constraints depends on the extent to which it is possible to decouple growing economic systems from their underpinning ecological foundations. The ecological economics view is that the economy and the environment are jointly determined systems and that the scale of economic activity is now such that this matters (the carrying and assimilative capacity of many ecological systems are binding constraints). Nevertheless, the biophysical carrying and assimilative capacity of ecological systems are not static but vary with the preferences and technology of the user, as well as with changes in the nature of the ecosystem itself. The real problem appears to lie in the phenomenon of dynamic feedback effects. Ecosystems do not always clearly signal when some carrying or assimilative capacity has been breached, and the feedback effects of so doing are often indirect and long delayed. They may also be irreversible. In the absence of private indicators of the scarcity of environmental resources—the externality and public goods problems—there is no reason to believe that private resource users will recognize such effects on the global commons. Hence there is no reason to believe that resource users will respect the constraints imposed by the environment's carrying and assimilative capacity. The level of economic activity is currently regulated by reference to a very short-run set of indicators, mainly the stability of market prices. It is argued that it is necessary to take account of the longer-term environmental consequences of current levels of activity precisely because it has important consequences for the future potential of the system, to sustain its provision of valuable functions and services.

Ecological processes function in such a way as to provide an array of ecological services on which most economic activity relies. Ecological services include maintenance of the composition of the atmosphere, amelioration of climate, operation of the hydrologic cycle (including flood controls and drinking water supply), waste assimilation, recycling of nutrients, generation of soils, pollination of crops, and provision of food, as well as the maintenance of particular species and landscapes. The value of ecological services and the patterns in which they are available given different institutional contexts are important components of the economic and ecological research agenda.

Ecological economists argue that a characteristic feature of the ecological economic system is that its dynamics are discontinuous around critical threshold values for species and their habitats and for ecosystem processes and functions (just as they are discontinuous around economic and institutional thresholds). Such thresholds are the boundaries between locally stable equilibria.

One consequence of discontinuous change is the novelty of many effects, and this is related to the scale issue. Current rates of human population growth and consequential rates of growth in the demand for environmental resources have increased the interconnectedness of ecological and economic systems. This is not just a problem for the decoupling of the economy from its environment. It has also moved societies and ecosystems into such novel and unfamiliar territory that the future evolution of both has become more unpredictable. The lack of information about thresholds and the precise consequences of breaching them has led ecological economics to advocate a precautionary approach. This favors conservation and environmental protection measures, unless the social opportunity costs are "unacceptably" large. More importantly, it favors protection of the joint system's resilience.

These concerns have also led to a questioning of cost-benefit thinking, which underpins the conventional economic approach. Conventionally the idea has been to compare all the relevant benefits from a project, policy, or course of action with the costs of such activity. Both benefits and costs are translated, as far as is feasible, into monetary terms and discounted over a given time horizon. Only projects with benefits greater than costs are economically acceptable.

A large environmental economics literature has grown up, since the late 1960s, encompassing a range of monetary valuation methods and techniques designed to price the spectrum of environmental goods and services provided by the biosphere. Because many environmental goods and services are nonmarketed commodities, the valuation methods have utilized involved market-adjusted, surrogate, and simulated-market approaches.

As far as conventional economic theory is concerned, the value of all environmental assets can be measured by the preferences of individuals for the conservation and use of these commodities. Given their existing preferences, individuals will possess a number of held values, which in turn result in objects being given various assigned values. In principle, economists begin to arrive at an aggregate measure of value (total economic value) by distinguishing user values from nonuser values.

By definition, user values derive from the actual use of the environment. Slightly more complex are option values—values expressed through options to use the environment in future. They are essentially expressions of preference (willingness to pay) for the conservation of environmental systems or components of systems against some probability that the individual will make use of it at a later date. A related form of value is bequest value, a willingness to pay to preserve the environment for the benefit of one's descendants. It is not a user value for the current individual valuer, but a potential future user value for his or her descendants.

Nonuser values are more problematic. They suggest non-instrumental values that are in the real nature of the thing but unassociated with actual use or even the option to use the thing. Instead, such values are taken to be entities that reflect people's preferences but include concern for, sympathy with, and respect for the rights or welfare of nonhuman beings. These values are still anthropocentric but may include a recognition of the value of the very existence of certain species or whole ecosystems. Total economic value is then made up of actual use value plus option value plus existence value.

Some ecological economists are concerned that the total economic value concept does not capture the full value of ecosystems and that a number of difficult theoretical and philosophical questions remain to be answered in the context of nonuser values. Other analysts claim that while philosophers debate, the real environment is deteriorating, particularly in developing countries, and much useful valuation analysis is relatively uncontroversial and should be deployed within a cost–benefit approach to aid decision makers as a matter of priority (Turner and Pearce, 1993).

Clearly, while there are limits to the economic calculus (i.e., not everything is amenable to meaningful monetary valuation), economic valuation methods and techniques can and should play a significant role in the project, program, and policy appraisal process that leads to the setting of relative values (including environmental assets values). Constanza et al. (1997) controversially estimated the current economic value of seventeen ecosystem services on a biosphere-wide basis at between U.S.$16–54 trillion per year, giving an average annual value some 1.8 times the current global gross national product.

Costanza et al.'s study has engaged environmental scientists and policy makers, but the global, biome-scale, economic value calculations risk criticism from both scientists and economists. On the basis of the data and methods cited in the article and supporting inventory, the conclusion that the value of the biosphere services really is, on average, U.S.$33 trillion per year is not supportable. Apart from raising policy maker, scientist, and citizen awareness about the environment's economic value and the possible significance of the loss of that value over time, the global value calculations do not serve to advance meaningful policy debate in efficiency and equity terms, in practical conservation-versus-development contexts. Such calculations, with their "single number" outcomes, shroud a number of fundamental scaling problems having to do with valuation contexts, that is, the temporal, spatial, and cultural specificity of economic value estimates. Such values can meaningfully be assigned only to relatively small ("marginal") changes in ecosystem capabilities (functions and services). The practical problem is that determining precisely what is and what is not a discrete and marginal change in complex ecological systems is not straightforward.

[*See also* Agriculture and Agricultural Land; Catastrophist–Cornucopian Debate; Coastal Protection and Management; Commons; Ecotaxation; Ethics; Fishing; Food; Forests; Human Impacts; Industrial Ecology; Joint Implementation; Land Use, *article on* Land Use Planning; Pollution; Resources; Valuation; *and* Water Resources Management.]

## BIBLIOGRAPHY

Ayres, R. U., and Kneese, A. V. "Production, Consumption and Externalities." *American Economic Review* 59 (1969), 282–297.

Barbier, E. B. "Introduction." *Environmental Kuznets Curve* Special issue. *Environment and Development Economics* 2 (1997), 357–367.

Basta, D. J., J. L. Lounsbury, and B. T. Bower. *Analysis for Residuals: Environmental Quality Management.* Washington, D. C.: Resources for the Future, 1978.

Baumol, W. M., and W. Oates. *The Theory of Environmental Policy.* Englewood Cliffs, N.J.: Prentice-Hall, 1975.

Bohm, P., and C. Russell. "Alternative Policy Instruments." In *Handbook of Natural Resources and Energy Economics,* edited by A. Kneese and J. Sweeney, pp. 395–460. Amsterdam: North-Holland, 1985.

Boulding, K. E. "The Economics of the Coming Spaceship Earth." In *Environmental Quality in a Growing Economy,* edited by H. Jarret, pp. 3–14. Baltimore: Johns Hopkins University Press, 1966.

Bower, B. T., and R. K. Turner. "Characterising and Analysing Benefits from Integrated Coastal Management." *Ocean and Coastal Management* 38 (1998), 41–66.

Coase, R. "The Problem of Social Cost." *Journal of Law and Economics* 3 (1960), 1–44.

Cole, M. A., A. J. Rayner, and J. M. Bates. "The Environmental Kuznets Curve: An Empirical Analysis." *Environment and Development Economics* 2 (1997), 401–416.

Costanza, R., et al. "The Value of the World's Ecosystem Services and Natural Capital." *Nature* 387 (1997), 253–260.

Costanza, R., C. Perrings, and C. J. Cleveland, eds. *The Development of Ecological Economics.* Cheltenham, U.K.: Edward Elgar, 1997.

Cropper, M. L., and W. E. Oates. "Environmental Economics: A Survey." *Journal of Economic Literature* 30 (1992), 675–740.

Dales, J. H. *Pollution, Property, and Prices.* Toronto: University of Toronto Press, 1968.

Daly, H. E. "Ecological Economics and Sustainable Development: From Concept to Policy." Environment Department Divisional Working Paper 1991-24. Washington, D.C.: World Bank Environment Department, 1991.

———. *Steady State Economics.* 2d ed. London: Earthscan, 1992.

———, ed. *Toward a Steady State Economy.* San Francisco: W. H. Freeman, 1973.

Dasgupta, P. "The Economics of the Environment." *Environment and Development Economics* 1 (1996), 387–428.

Dasgupta, P., and K-G. Mäler. "Analysis, Facts and Predictions." *Environment and Development Economics* 3 (1998), 504–511.

Easterlin, R. A. "Does Economic Growth Improve the Human Lot?" In *Nations and Households in Economic Growth,* edited by P. David and R. Weber, pp. 89–125. New York: Academic Press, 1974.

Faber, M., R. Manstellen, and J. Proops. *Ecological Economics: Concepts and Method.* Cheltenham, U.K.: Edward Elgar, 1996.

Georgescu-Roegen, N. *The Entropy Law and the Economic Process.* Cambridge: Harvard University Press, 1971.

Grossman, G. M., and A. B. Krueger. *Environmental Impacts of a North American Free Trade Agreement.* Princeton, N.J.: Princeton University, Woodrow Wilson School of Public and International Affairs, 1992.

Hawkin, P., A. Lovins, and L. H. Lovins. *Natural Capitalism: Creating the Next Industrial Revolution.* New York: Little, Brown, 1999.

Leontief, W. *Input-Output Economics.* New York: Oxford University Press, 1966.

Lovins, A. B., and L. H. Lovins. *Climate: Making Sense and Making Money.* Snowmass, Colo.: Rocky Mountain Institute, 1998.

Lutz, M. A., and K. Lux. *Humanistic Economics: The New Challenge.* New York: The Bootstrap Press, 1988.

Norgaard, R. B. "Sociosystem and Ecosystem Coevolution in the

Amazon." *Journal of Environmental Economics and Management* 8 (1981), 238–254.

Oates, W. E. "Forty Years in an Emerging Field: Economics and Environmental Policy in Retrospect." *Resources* 121 (Fall 1999).

O'Riordan, T. *Environmentalism.* London: Pion Press, 1981.

Pearce, D. W., and R. K. Turner. *The Economics of Natural Resources and the Environment.* Hemel Hempstead, U.K.: Harvester Wheatsheaf, 1990.

Perrings, C. *Economy and Environment: A Theoretical Essay on the Interdependence of Economic and Environmental Systems.* Cambridge: Cambridge University Press, 1987.

Pigou, A. C. *The Economics of Welfare.* Oxford: Oxford University Press, 1920.

Porter, M., and C. van der Linde. "Towards a New Conception of the Environment Competitiveness Relationship." *Journal of Economic Perspectives* 9 (1995), 97–118.

Sandler, T. "Global and Regional Public Goods: A Prognosis for Collective Action." *Fiscal Studies* 19 (1995), 22–1247.

Selden, T. M., and D. Song. "Environmental Quality and Development: Is There a Kuznets Curve for Air Pollution Estimates?" *Journal of Environmental Economics and Management* 27 (1994), 147–162.

Skea, J. "Environmental Technology." In *Principles of Environmental and Resource Economics*, edited by H. Folmer, H. Landi-Gabel, and I. B. Opshcoor, pp. 389–412. Cheltenham, U.K.: Edward Elgar, 1995.

Torras, M., and J. K. Boyce. "Income, Inequality and Pollution: A Reassessment of the Kuznets Curve." *Ecological Economics* 25 (1998), 147–160.

Turner, R. K., and D. W. Pearce. "Sustainable Economic Development: Economic and Ethical Principles." In *Economics and Ecology*, edited by E. B. Barbier, pp. 177–194. London: Chapman and Hall, 1993.

Turner, R. K., C. Perrings, and C. Folke. "Ecological Economics: Paradigm or Perspective." In *Economy and Ecosystems in Change*, edited by J. van den Bergh and J. van der Straaten, pp. 25–49. Cheltenham, U.K.: Edward Elgar, 1997.

United Nations Development Programme. *Human Development Report, 1992.* Oxford: Oxford University Press, 1992.

———. *Human Development Report, 1994.* Oxford: Oxford University Press, 1994.

Vatn, A., and D. Bromley. "Externalities: A Market Model Failure." *Environmental and Resource Economics* 9 (1997), 135–151.

Von Weizsacker, E., A. B. Lovins, and L. H. Lovins. *Factor Four: Doubling Wealth, Halving Resource Use.* London: Earthscan, 1997.

World Commission on Environment and Development. *Our Common Future.* Oxford: Oxford University Press, 1987.

—R. KERRY TURNER

## ENVIRONMENTAL IMPACT. *See* IPAT.

## ENVIRONMENTAL IMPACT ASSESSMENT

Environmental impact assessment (EIA) was first introduced in the United States through the 1969 National Environmental Policy Act (NEPA), which came into effect on 1 January 1970. A previous attempt to introduce EIA failed after U.S. President Dwight D. Eisenhower opposed the Resources and Conservation Bill of 1959.

Put simply, EIA is a process of systematically identifying and assessing expected environmental impacts prior to the implementation of a proposed project, policy, program, or plan. The identification of significant negative impacts may prevent the proposal (which, to date in practice, has often been a project) from going ahead. However, the original proposal is generally modified, or mitigation measures are introduced, to ameliorate the anticipated negative environmental impacts. While often focusing on negative environmental impacts, the EIA process should also consider the possibility that a proposal may generate positive environmental impacts, particularly on sites that are currently degraded.

EIA is now a legal requirement in the European Union, in many countries and provinces, including sixteen U.S. states, and sometimes at the level of individual cities. Since the late 1980s, international institutions such as the United Nations Environment Programme (UNEP), the Asian Development Bank, the World Bank, and international aid agencies have required EIA for particular development proposals. [*See* World Bank, The.] The focus within these institutions is usually on capacity building within individual countries so that the countries can develop structures, skills, and motivation to undertake EIA themselves. To this end, in 1992 the World Bank released a three-volume sourcebook designed to assist people involved in the EIA process, and to promote sustainable development by identifying negative environmental impacts at the earliest possible stage.

Thomas (1998) defined EIA as an environmental impact statement (EIS) plus an Assessment Report. While EIA requirements vary between jurisdictions, Thomas's definition encapsulates the core of the process. However, different countries use different terminology, which often causes confusion beyond this simplified understanding of EIA. For example, in some places EIA is simply known as "impact assessment" because it is broader than a narrow definition of the physical environment. In other locations it is known as environmental assessment (EA) or, in the Australian state of Victoria, as environmental effects assessment (EEA). The word *impact* is sometimes perceived to have negative connotations. However, in the United States, an environmental assessment is a preliminary study undertaken within the EIA process to identify the likelihood of significant impacts, which then require the preparation of a full EIS (Burris and Canter, 1997).

EIA has expanded beyond NEPA's initial coverage of United States federal government department projects. Wood's (1995) comparison of EIA systems in six countries and the State of California shows that EIA may include provincial and/or state development proposals and private development proposals, depending upon the specific legislation in a particular location. The introduction of EIA in many countries has been based upon the NEPA model, although attempts have been made both in Canada and in the European Union to move away from the legal focus in the NEPA approach.

The European Economic Community (EEC) adopted EIA for all member countries in 1985, through Directive 85/337/EEC. This Directive covered the types of project that required assessment, the scope of the assessment, and the public consultation requirements for all EEC member countries. Prior to this time, some European countries were advanced in EIA while other countries lagged in the introduction of this process. The 1985 Directive left member states to implement the requirements of the Directive in legislation that they considered to be appropriate. This has resulted in significant variations in the number of EIA reports prepared annually, ranging from over five thousand in France to below one hundred in Belgium, Denmark, Ireland, Italy, and the Netherlands (Wood, 1995). Recent changes to the 1985 Directive include extending the assessment requirements to cover more projects and expanding the notion of public consultation to include consultation in a neighboring member country, if people there are likely to be impacted by a proposed project, policy, plan, or program.

EIA has been extended to consider the importance of transboundary impacts. At Espoo, Finland, in 1991, the United Nations Economic Commission for Europe (ECE) adopted the ECE Convention on EIA in a Transboundary Context. It was signed by twenty-eight countries, and became effective in September 1997. It is premature to gauge the effectiveness of the Espoo Convention, but it is compatible with the principles of sustainable development and with the European Union's emphasis on developing internal links between countries through EIA, rather than the more legally based approach of NEPA in the United States.

EIA has also been adapted from its initial focus on environmental protection to include the idea of sustainable development, which was introduced in the World Conservation Strategy in 1980 and popularized by the World Commission on Environment and Development (the WCED or Brundtland Commission) in 1987. The WCED's expert group on environmental law reported that states should undertake, or require, EIA (referred to as "prior environmental assessments") on proposed activities that may have significant effects on the environment or the use of natural resources.

In addition to the above changes, forms of impact assessment have developed either as an extension from, or a critique of, EIA. Social impact assessment (SIA) considers the likely social and cultural impacts of a proposed project, plan, policy, or program. Strategic environmental assessment (SEA) recognizes the importance of decisions taken at the level above individual projects: that is, at policy, plan, and program levels. A new European Union Directive on SEA is currently in draft form and is being debated, while the World Bank adopted Operational Directive 4.00 in 1989, requiring SEA of regional and sectoral activities (Therivel and Partidario, 1996). Cumulative impact assessment (CIA) recognizes the existence of a range of cumulative impacts, such as time crowding, space crowding, time lags in experiencing impacts, compounding effects on a single site, and so on, which result in the whole being far greater than the sum of the individual parts. While SIA and CIA are now frequently incorporated into an expanded form of EIA, the treatment of social and cumulative impacts is often unsatisfactory.

Overall, it can be said that there is a range of opinion about the effectiveness of EIA. Supporters may see EIA as an important process that discourages the environmentally worst projects from entering the public arena, while also preventing the most undesirable aspects of some submitted proposals from being implemented. However, most supporters recognize that EIA by itself does not necessarily guarantee high-quality development, environmental protection, or sustainable development.

In contrast, some people perceive the EIA process as a way of legitimizing controversial development proposals and doing little to maintain environmental quality. They argue that few projects are refused because of the findings of the EIA process. However, this is not a good measure of effectiveness because projects may be substantially revised during the EIA process, and the mere existence of a process may prevent some environmentally damaging projects from being advocated. Furthermore, opponents of EIA argue that many key decisions have previously been taken at the policy level, and that individual project-based EIA also fails to consider the cumulative impacts of each development. These concerns are partly being addressed by SEA at the policy and program level, and by CIA.

Between these extremes, a range of opinion suggests that EIA has both strengths and weaknesses, and that the weaknesses could potentially be addressed given political will. Recent studies of the effectiveness of NEPA in the United States over a twenty-five-year period conclude that its major benefits were the framework for collaboration that it provided for federal government agencies, that it encouraged agencies to acknowledge potential environmental impacts to the public, and that it encouraged decision makers to think about potential environmental impacts before committing resources. The major weaknesses of NEPA included inadequate strategic planning, inadequate monitoring and adaptive environmental management, and the need to address cumulative impacts more effectively (Canter and Clark, 1997; Council on Environmental Quality, 1997). Importantly, many of the above issues surrounding NEPA in the United States also apply to EIA in other countries.

EIA has changed significantly since its introduction in NEPA. Such evolution is evidence that this traditional project-based form of impact assessment is capable of being extended to newer forms of assessment (for example, SEA, CIA, SIA) and to embrace ideas such as sustainable development. However, the effectiveness of the EIA process in practice, as opposed to an EIA legislative framework, is dependent upon the values, motivation, and skills of people involved at all stages in this process.

[*See also* Environmental Law; Human Impacts, *article on* Human Impacts on Earth; *and* IPAT.]

### BIBLIOGRAPHY

Beattie, R. "Everything You Already Know about EIA (but Don't Often Admit)." *Environmental Impact Assessment Review* 15.2 (1995), 109–114. A short, frank article that should be read by anyone interested in EIA.

Burris, R., and L. Canter. "Cumulative Impacts are not Properly Addressed in Environmental Assessments." *Environmental Impact Assessment Review* 17.1 (1997), 5–18. A study of thirty EIAs in the United States to assess their handling of cumulative impacts.

Canter, L., and R. Clark. "NEPA Effectiveness—A Survey of Academics." *Environmental Impact Assessment Review* 17.5 (1997), 313–327. An accessible summary of leading academics' opinions on the effectiveness of NEPA over twenty-five years.

Council on Environmental Quality. *NEPA: A Study of Its Effectiveness after Twenty-five Years*. Washington, D.C.: Office of the President, 1997.

Fearnside, P. "The Canadian Feasibility Study of the Three Gorges Dam." *Impact Assessment* 12.1 (1994), 21–57. A solid critique of the EIA process for this massive dam project that identifies issues relevant to this specific EIA process, and issues pertinent to EIA generally.

Gilpin, A. *Environmental Impact Assessment (EIA), Cutting Edge for the Twenty-first Century*. Cambridge: Cambridge University Press, 1995. A wide coverage of EIA in Europe, the Nordic countries, North America, Asia, and the Pacific, and in international institutions.

Thérivel, R., and M. R. Partidario, eds. *The Practice of Strategic Environmental Assessment*. London: Earthscan, 1996. A good overview of SEA, with five sectoral case study chapters and three regional case study chapters drawn from Europe, the United States, and Nepal.

Thomas, I. *Environmental Impact Assessment in Australia: Theory and Practice*. 2d ed. Sydney: Federation Press, 1998. A useful guide to EIA in Australia, but suitable for an international audience because it includes chapters on the history of EIA, forms of impact assessment, public participation, and EIA systems in a range of countries.

Wood, C. *Environmental Impact Assessment: A Comparative Review*. Harlow, U.K.: Addison Wesley Longman, 1995. A comprehensive coverage and comparison of many facets of EIA in the United States, the United Kingdom, the Netherlands,

Canada, Australia (Commonwealth system), New Zealand, and the State of California.

—Phil McManus

# ENVIRONMENTAL LAW

*Law* refers to a body of rules accepted as binding on the members of a given society. Sophisticated legal systems have existed since early times and include the Code of Hammurabi in Babylonia in the eighteenth century BCE and the Laws of Manu in India (200 BCE). In modern societies, law is the primary instrument through which policies are implemented; it can express minimum standards of morality or reflect the aspirational goals of a society. Law reflects a society's values and adapts as they change, while at the same time being an important instrument that allows such changes to occur in an ordered fashion.

A diversity of legal traditions coexists in the world today. The common-law tradition evolved from judge-made law in medieval England and is now used by most of the former British possessions including the United States. Another tradition, based on civil law, is characterized by the enactment of "codes" that integrate and organize systematically all legal rules pertaining to a particular subject. This tradition has its roots in Roman law, was revitalized by the enactment in France of the Napoleonic Code in 1804, and is now used in most countries in western Europe and Latin America, as well as in some Asian countries such as Japan. Other legal traditions are based on religious beliefs and include Jewish, Hindu, Muslim, and Canon law.

Sources of law vary among different legal traditions. They usually, however, include judicial decisions rendered by courts and tribunals. Another important source is the body of statutes or acts that are adopted by the state—in democracies, by an elected legislature. These statutes and acts are in turn implemented by detailed regulations promulgated by the national administration. Additional sources of law include custom, general principles of law, and legal writing by scholars. The science of law is divided into different branches. The basic branches are contract, tort, property, criminal, administrative, and constitutional law. Others include admiralty, corporate, intellectual property, and environmental law.

**Domestic Environmental Law.** *Domestic* (or *national*) *law* refers to the legal system applicable to a defined territory over which a sovereign power has jurisdiction. *International law*, on the other hand, regulates the conduct of states and other international actors. Over the years, domestic and international law systems have evolved in parallel. In certain fields and regions of the world, international law has shaped and significantly contributed to the development of domestic environmental law. Yet international environmental law also reflects domestic experiences considered successful by the community of nations. The result is a complex relationship where both levels of environmental law mutually contribute and reinforce each other. This section deals with domestic environmental law. The next section will discuss international environmental law.

**Early developments.** Environmental law is a relatively new field; other branches of law have historically been used to remedy environmental problems. In the common-law system, tort law—which provides remedies for harm caused by one individual to another—provided the necessary legal foundation in early cases. Nuisance actions were the most popular because they allow a successful claimant to receive not only compensation but also a court order to abate the nuisance, such as a smell or smoke. In the civil law system, claimants invoked tort and property law in generally the same way. Historically, however, tort law, based as it is on the protection of individual's rights and the need to prove specific injury, has not been a significant means of preventing environmental degradation.

This inadequacy of tort and property law convinced governments, including local authorities, to adopt measures to tackle the most pressing environmental problems. There is some debate regarding the true nature of the first local ordinances regulating odors, smoke, and waste water. Some scholars argue that they are early environmental statutes. On the other hand, others consider that they are simply health-based policies having the effect of regulating environmental problems. As a matter of fact, most of these early measures were enacted after sporadic crises that endangered public health.

**Modern environmental law.** Since the 1970s, environmental law has experienced an unprecedented growth in many countries. This growth was essentially made possible through the enactment of new statutes and regulations that provide for higher standards of environmental protection. The level of government that has enacted these instruments varies from one country to another. In federal states such as Canada, jurisdiction over the environment is shared between the provincial and federal governments. In the United States, the federal government has adopted most of the important environmental statutes, but their implementation is delegated to the states through a complex system of incentives and responsibilities. The European Union (EU) has a developed system of environmental law, the legal basis for which is now to be found in the 1992 Maastricht Treaty on European Union. Although implementation is the responsibility of the individual EU member states, European law permits individuals as well as other member states and the European Commission to pursue actions for breach of these rules before the European Court of Justice in Luxembourg. [See European Community.]

Most countries have created institutions that handle environmental matters and enjoy varying degrees of independence, power, and jurisdiction. The primary function of these institutions is to coordinate domestic efforts aimed at the protection of the environment. This normally involves statute and regulation development, environmental law enforcement, integration of environmental concerns in governmental decision making, and general environmental education. The nature of the institutions also differs greatly from one country to another, and there is no ideal arrangement. Many countries have created an independent environment ministry or have established a specialized agency, such as the U.S. Environmental Protection Agency, which was created by an executive order and reports directly to the president. Some countries, such as the United Kingdom, have both. Another approach, adopted widely in Latin America, is the creation of an environmental commission that regroups representatives of many other ministries and departments.

*Organization of environmental statutes.* Environmental statutes have traditionally been drafted and organized around important themes such as nature conservation and protection of the principal natural media—air, water, and soil. This organization allows the elaboration of rules of limited application that are easier to manage and enforce, but may fail to acknowledge the importance of a holistic approach and to deal with important natural relationships such as the effects of air pollution on water quality. Other countries have adopted different approaches. New Zealand has a seminal 1991 Resource Management Act, which integrates all sectors and relevant activities, while Canada has consolidated five of its main environmental statutes into one single act of general application. A similar technique is also used in other countries such as Chile, which has adopted environmental framework laws under which sectoral laws can be promulgated in an integrated way.

*Legislative techniques.* Despite the particular organization of a country's environmental laws, a lawmaking body will resort to a number of legislative techniques to attain its policy objectives. Such techniques include prior authorization, environmental standards, liability, and environmental impact assessments.

A general *prior authorization* requirement prohibits any person from engaging without prior permission in any activity that could harm the environment. This essentially establishes a permit or license system whereby any activity that constitutes a potential source of pollution requires the permission of a central authority. This technique can be adapted to serve different policy goals. The scope of the permitting system can be broad enough to cover almost any component of the biosphere, or limited to regulate only certain types of activities.

*Environmental standards* are mostly "command and

control" measures where a central authority mandates specific requirements to be followed by the regulated community. As such, authors distinguish them from "economic instruments," which rely on market-based approaches.

The objective of standards is to prescribe specific quantitative and qualitative limits to be followed by the regulated community. They may take at least five different forms. First, health standards are normally based on risk assessment analysis that identifies safe tolerance levels. These are used to control pesticides and other similar substances and may be enacted without taking into account the compliance costs for the regulated community. Second, ambient environmental standards are used widely in the control of water and air pollution. These standards prescribe specific limits on the concentration of certain designated pollutants that will be tolerated, for example, in the ambient air or water. They may be used for the control of nonpoint or diffuse pollution sources such as the nitrate content of runoff from agricultural land. Compliance with such standards may require major changes of agricultural or commercial practice. Third, emission and discharge standards are also used to combat air and water pollution. Instead of specifying limits applicable to the ecosystem, the standards place limits regarding the composition of the actual emissions or discharge by a specific source.

Two further forms of standards relate to technology. The most commonly used standard is based on technology. A statute may prescribe the use of the "best available technology." Through a cost-benefit analysis, the environmental agency will then specify for each class of industry the technology that it considers the "best available" and that is therefore mandated. Such standards may be relatively easily upgraded. More progressive are "technology forcing" standards that cannot be met by the regulated community under the current state of technology. The intention is that the obligation to meet this type of standard will stimulate and "force" technological innovation. This technique has been used in the United States to regulate motor vehicle emissions.

*Liability* refers to the condition of being actually or potentially subject to a legal obligation. Under civil liability, a person will be liable if he or she was negligent, that is, if his or her conduct fell below the objective standard of a reasonable person. Criminal liability is more serious and requires proof beyond reasonable doubt of an unlawful act as well as of the defendant's specific intent. Strict liability is an intermediary concept that is commonly used in environmental laws. It relieves the state of the obligation to prove that the unlawful act resulted from negligence (civil liability) or that the defendant's conduct was intentional (criminal liability). In other words, the state needs only to prove that the particular defendant committed an unlawful act, for exam-

ple, the discharge of waste water. Another important liability concept consists of joint and several liability, where violators will be held liable together and individually. In this case, the government can sue both violators or either one of them to recover, for example, cleanup costs. This technique is useful when it can be proved that each defendant contributed to an unlawful activity but the exact contribution of each is difficult to demonstrate, or when the injury is simply indivisible.

Retroactive liability is the hallmark of modern soil statutes and constitutes an exception to general principles of law. Under these principles, one should not be held liable for the acts of another, or for actions that were lawful when they were taken. Many governments have invoked this exception as a solution to the contamination of land by hazardous wastes. In urban areas, land contamination often results from decades of intensive industrialization that has occurred without any meaningful preexisting environmental standards. Under some soil statutes, current and past owners of contaminated land may be held liable for cleanup costs even if they have not personally contributed to the contamination. Under certain circumstances, operators, transporters, and, to a limited extent, lenders can also be held liable. Retroactive liability is still controversial and has raised some problems. It has important economic consequences, as the value of such land may drop to become negative in cases where cleanup costs exceed the property's value. In the long term, retroactive liability can also result in new investments going only to pristine "greenfield" sites to avoid contaminated areas that are often situated in disadvantaged communities. Despite these difficulties, the harshness of the liability provision has in some countries coerced industries into better environmental behavior and has substantially minimized major health risks.

Among modern environmental statutes, *environmental impact assessment* (EIA) laws crystallize a preventive approach to environmental protection because they integrate environmental considerations in decision-making processes. Generally, EIA laws require the preparation of an environmental impact assessment for any proposed development activity to review and assess its environmental impacts. The requirement can be applicable to a broad spread of actions and include issuance of a permit or prior authorization, the funding of a project, and the adoption of a new statute or policy. The first step under EIA laws (known as screening) is the determination of whether the proposed activity is likely to cause environmental impacts beyond a certain threshold. If such a determination is positive, the proposer must proceed with the preparation of a formal assessment. Depending on the apprehended impacts, the general public will be notified and public consultations will be held. The environmental assessment may be required to identify appropriate mitigation measures or alternatives to the proposed action that minimize environmental impacts. The key issue is whether EIA statutes oblige the propose to implement the mitigation measures and alternatives previously identified. Without such a mitigation requirement, EIA laws may render decision making more transparent, but they do not provide any effective safeguards to protect the environment. [*See* Environmental Impact Assessment.]

*Enforcement of environmental law.* Enforcing environmental law is critical to ensure that the regulated community actually complies with the policies embodied in a statute. The goals of a good enforcement program are for a government to achieve general environmental compliance through deterrence, identify environmental violators efficiently, and prosecute them diligently. Compliance can be achieved through general education and outreach to the regulated community, backed by effective prosecution procedures. In addition, government bodies may conduct inspection activities periodically or on the basis of probable cause. In some countries, the regulated industry is obliged to make its monitoring data publicly available. This information may also allow nongovernmental organizations to play an important role in identifying violators.

The government, through its administrative agency, is normally the entity responsible for prosecuting violations of environmental law. In some countries, individuals or nongovernmental organizations can also sue violators and recover a share of the awarded penalty as a reward for their initiative, through procedures known as citizen suits or public-interest actions. In addition, national constitutions or environmental statutes may protect the right of an individual to a clean environment. In India, for example, such provisions have allowed the courts to take a highly proactive role in the protection of the environment.

*New trends in environmental law.* Two new trends are currently shaping environmental lawmaking. The first is integrated pollution control (IPC), which allows for the regulation of an ecosystem as a whole, instead of approaching it on a sector-by-sector basis. This specifically avoids the transfer of pollution from one medium (such as water) to another (such as air) and helps in controlling pollution from nonpoint or diffuse sources. This approach has been pioneered in the United Kingdom and now in the European Union.

The second trend is the use of economic instruments that complement command and control measures. Under this approach, the government sets out targets and allows members of the regulated community to allocate among themselves the burden of compliance. Theoretically, if the price of noncompliance is set at an appropriate level, the desired abatement of pollution will be achieved. The advantage is that sources with lower com-

pliance costs will overcomply and receive economic benefits from those with higher compliance costs. The result is the attainment of pollution abatement at a lower net cost to society than in comparison to strict command and control measures. Other economic instruments include the use of taxes, environmental auditing, ecolabeling (to reassure consumers that a product meets certain environmental standards), and the reduction of subsidies that allow the regulated community to play a role in shaping new practices.

**International Law.** Modern international law has its roots in the sixteenth- and seventeenth-century public law of Europe—law that was created to govern the diplomatic, commercial, military, and other relations of the society of Christian states. With the growing penetration of Europe into Asia in the late eighteenth and early nineteenth centuries, other subjects were included in the community of states, but it was only with the formation of the League of Nations in 1920, to which "any state" could be a member, that the international system began to aspire to be truly global.

*Doctrinal foundations of international law.* International law rests on the doctrine of sovereignty and equality of states. This doctrine enshrines the principle that national states are sovereign and have equal rights and duties as members of the international community, notwithstanding differences of an economic, social, political, or other nature. This fundamental feature of international law has created systemic limitations—the absence of an established central legislative authority comparable to a nation system and the absence of a compulsory, or even widely used, judicial system, coupled often with the absence of effective enforcement machinery for breaches of international law. Little surprise then that, even after a few centuries of the existence of international law, many still ask what has been described as "the standard sherry party question" (Harris, 1991, p. 5)—Is international law really law? Despite its systemic limitations, international law does exist. States make it, they follow it, and on occasions they break it. Certain breaches are spectacular but overshadow the general everyday pattern of compliance.

*Sources of international law.* The international community, in the face of the rudimentary character of international lawmaking institutions, has developed its own system of norm creation and of international lawmaking. The system has two basic parts: treaties and customary international law. Treaties—solemn binding agreements between subjects of the international legal order, principally states—can be binding only on those that consent to them. They originate in a framework of international negotiation over matters of common interest and result in an agreement, in the form of a text, that usually reflects mutual advantage. Once the text is agreed upon (and at that stage often signed), the process of ratification commences. This is the process by which the parties ensure, by their various constitutional means, that when the treaty comes into force, the legal, financial, and administrative mechanisms by which the parties will be able to honor their new obligations are in place. Only after these national measures have been put in place will the state be in a position to notify the depository (the state or institution formally holding the list of parties) that it wishes to be bound by the treaty—this is the act of ratification. (It is possible, if the parties wish, for treaties to come into force immediately upon signature. Similarly, states that have not participated in the negotiation process or that have not signed the treaty may accede.) Once the treaty has received the agreed number of ratifications, it then comes into force. This is not an easy process. Pressures of government time, changing priorities, or simple second thoughts can cause dramatic delays. The larger the enterprise, the more apparently intransigent the problems often are. For example, on 16 November 1994, the Law of the Sea Convention, signed in Montego Bay, Jamaica, in December 1982, finally came into force. It had taken some twelve years, as well as considerable legal ingenuity in the negotiation of an amending agreement, for this major international legislative act to receive the sixty ratifications it required to enter into force. [*See* Law of the Sea.]

International customary law is defined by the Statute of the International Court of Justice as general practice accepted by states as law. In simple terms, it is something that states do because they regard themselves as legally obliged to do it.

Treaties and custom constitute *hard law*, law that nation states are obliged to follow under pain of sanction from the international legal system and community. There exists in contrast a category of law that is termed *soft law*, which comprises nonbinding instruments that lay down guidelines or desiderata for future action, or through which states commit themselves politically to meeting certain objectives. Soft law is largely based on international diplomacy and customs, dependent on moral suasion or fear of diplomatic retribution. The 1972 Stockholm Declaration and the 1992 Rio Declaration, which embody a series of widely revered environmental principles, constitute good examples of soft law, although a number of those principles may be said to have crystallized into "harder" obligations representing customary law (Freestone, 1994). There also exist subsidiary sources of international law such as doctrine, judicial decisions, general assembly resolutions, and opinions of international jurists.

**Development of international environmental law.** International environmental law refers to the corpus of international law relevant to environmental is-

sues (Birnie and Boyle, 1992). While the status of international environmental law as a discipline in and of itself is disputed by a few international scholars who believe that no autonomous "international law" exists apart from the general international law (Brownlie, 1995), it appears well established that environmental perspectives and concerns have stimulated and catalyzed international legal development. The growth of international environmental law is premised on the globalization of environmental problems and concerns, attributable to two crucially interlinked factors, ecological and economic interdependence.

Huge conceptual leaps were made in international environmental law in the last quarter of the twentieth century. Environmental problems progressed from being tackled within a bilateral, coexistence framework to a multilateral, cooperation framework. Further, international environmental law traversed the path from being merely reactive, such as in the negotiation of treaties to address the known threats of marine oil pollution, to being proactive, as in the case of the UN Framework Convention on Climate Change (UNFCCC), which is an anticipatory response to the possibility of future anthropogenic global climate change. [*See* Framework Convention on Climate Change.]

The development of international environmental law can be traced through two main phases: from 1972 to 1992, which was the period of burgeoning international environmental consciousness surrounding and after the UN Conference on the Human Environment in Stockholm in 1972, and from 1992 onward. [*See* United Nations Conference on the Human Environment.] This latter period, initiated by the negotiations leading up to the 1992 UN Conference on Environment and Development in Rio de Janeiro, is distinguished by concerns for sustainable development and includes the current phase of experimentation with economic market-based instruments to achieve environmental compliance. [*See* United Nations Conference on Environment and Development.]

*From Stockholm to Rio (1972–1992).* The Stockholm Conference, held in 1972, acted as a catalyst for several environmental initiatives. It resulted in a declaration containing a series of normative environmental principles, a 109-point environmental action plan, and a resolution recommending institutional and financial implementation by the United Nations. The result of these recommendations was the creation of the United Nations Environment Programme (UNEP), established by UN General Assembly resolution and based in Nairobi. UNEP plays an active role in convening and organizing meetings to negotiate global environmental treaties. The Convention on the Control of Transboundary Movements of Hazardous Wastes and Their Disposal, signed in Basel, Switzerland, 22 March 1989 is a case in point. The Basel Convention is built around two basic principles—the principle of proper waste management and the principle of prior informed consent (Kummer, 1995). UNEP was also directly responsible for the development of the important Regional Seas Program, which has resulted in a network of regional framework conventions protecting the marine environment, each with protocols developed to meet the special requirements of the region.

This era also witnessed the birth of several other international environmental treaties. Of particular significance is the 1985 Vienna Convention for the Protection of the Ozone Layer. The very real and apparently imminent threat of depletion of the ozone layer by commercially produced chemicals, principally chlorofluorocarbons (CFCs), prompted the convening of a conference in 1985 to negotiate the Vienna Convention. The format chosen was a framework convention: general obligations and institutional framework were laid down by the treaty, to be made more specific in the future by the negotiation of detailed protocols (or subtreaties open to the parties to the main convention). The discovery of the ozone hole over Antarctica led to intense intergovernmental negotiations resulting in the Montreal Protocol on Substances That Deplete the Ozone Layer in 1987. The protocol called for a freeze on the production and consumption of CFCs and halons at 1986 levels, followed by a 50 percent reduction in CFC use by industrialized countries over a ten-year period. Developing countries were allowed to increase their CFC consumption for a period of ten years. The protocol was deliberately designed as a flexible and dynamic instrument—countries were allowed to select the most economic mix of reductions with incentives to reduce the most harmful chemicals. [*See* Montreal Protocol.]

A follow-up to the Stockholm Conference, held in 1982 in Nairobi, spurred the United Nations to set up the World Commission on Environment and Development, chaired by Gro Harlem Brundtland, then prime minister of Norway. Its 1987 report *Our Common Future* placed the concept of *sustainable development* into the realm of international environmental law. At the suggestion of the commission, preparations began for the Rio Summit, officially the Conference on Environment and Development, thus marking the end of the era of emphasis on the "human environment" and the beginning of the era of emphasis on "environment and development." [*See* Brundtland Commission; *and* Rio Declaration.]

*Rio and beyond.* The United Nations Conference on Environment and Development, held in Rio de Janeiro twenty years after the Stockholm Conference, was popularly perceived as an attempt at environmental planning on a grand scale. In addition to a tremendous surge in environmental consciousness, the Rio Summit resulted in

- Agenda 21, an action plan for the next ten years and into the twenty-first century
- The Rio Declaration on the Environment and Development
- The 1992 United Nations Framework Convention on Climate Change, which was to provide a framework for the negotiation of detailed protocols on further issues such as controls on the emissions of greenhouse gases, particularly carbon dioxide
- The 1992 Convention on Biological Diversity, which was aimed at arresting the alarming rate at which species were disappearing through pollution and habitat destruction
- A nonlegally binding declaration on forests

Despite the obvious significance of these environmental initiatives, perhaps the enduring legacy of the Rio Summit lies in its contribution to the development of a framework of international environmental law principles. If indeed the maturity of international environmental law is to be assessed by the development of "discrete discipline specific" principles (Freestone, 1994, p. 210), then the Rio Declaration heralded the coming of age of international environmental law. [*See* Agenda 21; Framework Convention on Climate Change; *and* Convention on Biological Diversity.]

***Principles of international environmental law.*** Several principles of international environmental policy, some first enunciated in the Stockholm Declaration, were crystallized through the Rio process. Among them were the principles of *precaution, polluter pays, sustainable development, common but differentiated responsibility,* and *environmental impact assessment.* Some of these concepts, such as the polluter pays and environmental impact assessment, have their roots in domestic environmental law. Environmental impact assessment, for instance, was first established in the domestic law of the United States under the 1972 National Environment Protection Act. Other principles, such as that of common but differentiated responsibility, are products of international thought and action. International lawyers still dispute whether any or all of these concepts remain policy principles or have hardened into binding principles of customary international law (Boyle and Freestone, 1999).

*Precaution.* Enshrined in Principle 15 of the Rio Declaration, the precautionary principle postulates that where serious harm is threatened, positive action to protect the environment should not be delayed until irrefutable scientific proof of harm is available. It represents an important tool for decision making in uncertainty, which a significant body of opinion argues is now a legal principle. In its strongest formulations, this principle can be seen to require a reversal of the normal burden of proof, so that a potential actor would need to prove that a proposed activity will not cause harm before it can be sanctioned. It has been endorsed by virtually all recent environmental treaties, including regional treaties such as the 1992 Maastricht Treaty on European Union, the 1992 Paris Convention on the North East Atlantic, and the Helsinki Convention on the Baltic, as well as global environmental treaties such as the UNFCCC, the Convention on Biological Diversity, and the 1995 United Nations Agreement on Straddling Fish Stocks and Highly Migratory Fish Stocks. [*See* Straddling and Migratory Stocks Agreement; *and* Precautionary Principle.]

*Environmental impact assessment, public participation, and access to information.* Related to the precautionary principle is the concept of *environmental impact assessment.* It is based on the premise that rational planning constitutes an essential tool for reconciling development and environment. EIA provides an important modality for the implementation of the precautionary principle. Though first debated at Stockholm, the concept of environmental impact assessment found place only in the Rio Declaration. Agenda 21 calls on countries to assess the suitability of infrastructure in human settlements, ensures that relevant decisions are preceded by environmental impact assessments, takes into account the costs of any ecological consequences, and integrates environmental considerations in decision making at all levels and all ministries. The EIA requirement is embodied in several international instruments notably the 1991 UN Economic Commission for Europe (ECE) Convention on Environmental Impact Assessment in a Transboundary Context, the 1992 Biodiversity Convention, and the World Bank Operational Policy 4.01 (1999). The value and legitimacy of the EIA process has, in recent times, been strengthened by the evolution of the right of access to information on the environment and the right of public participation. The Rio Declaration recognizes in Principle 10 that environmental issues are best handled with the participation of all concerned citizens. It has recently been validated in the UNECE Convention on Access to Information, Public Participation in Decision-Making and Access to Justice in Environmental Matters, signed on 25 June 1998 by thirty-seven countries. The convention recognizes not only that "every person has the right to live in an environment adequate to his or her health and well-being, and the duty . . . to protect and improve the environment," but also that "citizens must have access to information, be entitled to participate in decision-making and have access to justice in environmental matters." In order for people to fulfill these rights and responsibilities, it obliges signatory states to, among other provisions, (1) make envi-

ronmental information available "as soon as possible," and "without an interest having to be stated" by the solicitor; (2) take specific measures to ensure complete public participation in decisions of specific activities, plans, programs, policies, and other regulations related to the environment; and (3) ensure that any person who feels the state has not met specific environmental commitments has access to a review procedure before a court. The value of such participation is enhanced by the right of access to information, a right that has found its way into various international instruments. The European Community Directive 90/313 on Access to Environmental Information assures the public free access to and dissemination of all environmental information held by public authorities throughout the EU.

*Common but differentiated responsibility.* Articulated as Principle 7 in the Rio Declaration, this principle requires states to cooperate in a spirit of global partnership to protect the environment. Yet, because states have contributed differently to global environmental problems, the principle recognizes that they should have common but differentiated responsibilities. A ready example is afforded by Article 4 of the 1992 UNFCCC, which places on developed countries an obligation to take the lead in meeting the required reductions in greenhouse gas emissions, and on developing countries an obligation only to implement these commitments to the extent that developed countries have met their commitments to provide financial resources and to transfer technology. (A similar provision can be found in Article 20(3) of the Convention on Biological Diversity.) As a general principle, sure to govern further negotiations on the UNFCCC, the principle of common but differentiated responsibility is highly significant. The 1997 UNFCCC Kyoto Protocol in its structure mirrors the philosophy of common but differentiated responsibility. The developed countries are committed to reducing their overall emissions of greenhouse gases by at least 5 percent below 1990 levels between 2008 and 2012. The developing nations have no such commitments. Although every nation state has the responsibility of reducing global greenhouse gas emissions, only the OECD and economies-in-transition countries are required to make specific quantified emission limitations. The limitations, even among these countries, vary to take into account differing domestic circumstances. Developing countries are provided with an opportunity to participate through the Clean Development Mechanism, which allows countries to cooperate on specific projects to reduce greenhouse gas emissions. [*See* Kyoto Protocol.]

*The polluter pays.* The *polluter-pays* principle requires that the costs of pollution be borne by the party responsible. The practical implications of this principle lie in its allocation of economic obligations in environmentally damaging activities. This seemingly intuitive principle has not received the kind of broad support that the precautionary principle has in recent times. Principle 16 of the Rio Declaration, for instance, supports the "internalization of environmental costs," taking into account the polluter-pays principle but only "with due regard to the public interest and without distorting international trade and investment." An example of an international instrument that refers expressly to the polluter-pays principle is the 1972 OECD Council Recommendation on Guiding Principles Concerning the International Economic Aspects of Environmental Policies, which endorses the polluter-pays principle to allocate costs of pollution prevention and control measures so as to encourage rational use of environmental resources.

*Sustainable development.* Defined by the 1987 Brundtland Commission report as "development that meets the needs of the present without compromising the ability of future generations to meet their own needs," the *sustainable development* principle is at the heart of many environmental initiatives. It recognizes the need for intergenerational equity, sustainable and equitable use of resources held in common by the current generation, and the integration of environmental considerations into economic and other development initiatives. This principle also finds itself reflected *inter alia* in the Framework Convention on Climate Change in Article 3(4). Although specifically recognized as a legal principle in the separate opinion of Judge Weeramantry in the 1997 *Gabcikovo-Nagymaros* case before the International Court of Justice (ICJ), the very breadth of the concept means that considerable controversy still surrounds this argument.

**Compliance and enforcement of international environmental law.** The term *enforcement* in the context of international environmental law refers to the measures taken to ensure the fulfillment of international legal obligations or to obtain a ruling by an appropriate international body that obligations are not being fulfilled.

Initially, only the general principles of state responsibility and dispute settlement guided efforts at enforcement of international environmental law. As the principal subjects of international law, states assume the obligation to enforce international environmental law. Enforcement by states arises primarily in situations of transboundary environmental harm and involves a determination by an international body, such as the ICJ, in The Hague. The ICJ, the United Nations' principal judicial organ, rules on questions of international law, including issues of international environmental law. In fact, however, its contribution to the development of international environmental law principles has been very slight.

A range of techniques and a panoply of international actors are today involved in the enforcement of international law (Boisson de Chazournes, 1995). Enforcement includes a wide array of forms, including diffusion of information, monitoring, verification, and inspection. For example, it is increasingly common for international law agreements to mandate their conference of parties, the plenary body of environmental agreements, to conduct *implementation reviews*. This review mechanism monitors national compliance with the obligations undertaken under the environmental agreement. Such a review is based primarily on national self-reporting, though some conventions do provide for independent means of gathering information.

Other conventions may use incentives or disincentives—that is, adopt the "carrot and stick" approach to obtain participation and ensure compliance. For example, under the Montreal Protocol, trade restrictions can be imposed on imports to and exports from nonparties to the protocol, and a fund has been created to assist countries to comply with their obligations under the protocol, thereby encouraging participation. Recently negotiated conventions use creative, dynamic, and flexible means to obtain environmental compliance. The UN-FCCC Kyoto Protocol provides a number of "flexibility mechanisms" (including cooperative implementation, emissions trading, and technology transfer) to assist parties in meeting their commitments.

Among the concerned actors are also international organizations and nongovernmental organizations. *International organizations* have a small but useful role to play in the enforcement of international environmental obligations. States have traditionally been reluctant to endow international organizations with enforcement powers, but some recent instruments do provide certain bodies with limited enforcement authorities. For instance, the 1982 UNCLOS (United Nations Convention on the Law of the Sea) provides the International Sea Bed Authority with the power to supervise implementation of parts of the convention, invite attention of the assembly to cases of noncompliance, and institute proceedings for noncompliance. *Nongovernmental organizations* often play the role of self-appointed "watchdogs" against the national governments and can thus help in the enforcement of international law through political means or public-interest litigation to ensure that governments keep to their international environmental commitments. The individual as an actor on the international arena also deserves mention. With the increasing emphasis on public participation and provision of access to environmental information, in international discourse, the individual's role in ensuring international environmental compliance is becoming increasingly relevant.

## BIBLIOGRAPHY

Ball, S., and S. Bell. *Environmental Law: The Law and Policy Relating to the Protection of the Environment.* 2d ed. London: Blackstone, 1991. The first book to provide a systematic exposition of English environmental law, incorporating also European Union policies and requirements.

Birnie, P. W., and A. E. Boyle. *International Law and the Environment.* Oxford: Oxford University Press, 1992. An outstanding work of scholarship and reference, providing an invaluable map for the emerging field of international environmental law.

Boisson de Chazournes, L. "La mise en oeuvre du droit international dans le domaine de la protection de l'environnement: Enjeux et défis." *Revue générale de droit international public* 99 (1995), 37. Analysis of the role played by traditional means and new procedures for ensuring compliance and enforcement of international environmental law.

Boisson de Chazournes, L., R. Desgagné, and C. Romano. *Protection internationale de l'environnement: Recueil d'instruments juridiques.* Paris: Pedone, 1998. The first comprehensive French-language collection of international legal materials on the protection of the environment. Contains useful thematic summaries and analytical bibliographies.

Boyle, A. E., and M. Anderson, eds. *Human Rights Approaches to Environmental Protection.* Oxford: Clarendon Press, 1996.

Boyle, A. E., and D. Freestone, eds. *International Law and Sustainable Development.* Oxford: Oxford University Press, 1999.

Brownlie, I. "International Law and the Fiftieth Anniversary of the United Nations." *Recueil des cours* 255 (1995), 181.

Chertow, M. R., and D. C. Esty, eds. *Thinking Ecologically: The Next Generation of Environmental Policy.* New Haven, Conn.: Yale University Press, 1997.

Churchill, R., and D. Freestone, eds. *International Law and Climate Change.* London: Graham and Trotman/Martinus Nijhoff, 1991.

David, R., and J. E. C. Brierley. *Major Legal Systems in the World Today.* 3d ed. London: 1985.

Freestone, D. "The Road from Rio: International Environmental Law after the Earth Summit." *Journal of Environmental Law* 6 (1994), 193. An assessment of the impact of the UN Conference on Environment and Development on the development of international environmental law.

Freestone, D., and E. Hey, eds. *International Law and the Precautionary Principle: The Challenge of Implementation.* London: Kluwer Law International, 1996.

Harris, D. J. *Cases and Materials on International Law.* 4th ed. London: Sweet and Maxwell, 1991.

Huglo, C., and J. de Malafosse. *Juris-classeur Environnement.* 3 vols. Paris: Ed. du Juris-classeur, 1998. The most comprehensive source on French domestic environmental law.

Kamto, M. *Droit de l'environnement en Afrique.* Vanves, Frances: Internationales Francophone-Association des Universités Partiellement ou Entièrement de Langue Française (EDICEF/AUPELF), 1996. Very good analysis on domestic and regional environmental law in Africa.

Kiss, A., and D. Shelton. *International Environmental Law.* New York: Transnational, 1991. The first and still classic exposition of the new subject area of international environmental law.

Krämer, L. *Focus on European Environmental Law.* London: Sweet and Maxwell, 1992.

———. *European Environmental Law Casebook.* London: Sweet and Maxwell, 1993.

Kummer, K. *International Management of Hazardous Wastes: The Basel Convention and Related Legal Rules.* Oxford: Clarendon Press, 1995.

Lyster, S. *International Wildlife Law.* Cambridge: Grotius Press, 1985. Although dated, this is still an excellent and highly read-able account of the various international treaty regimes that aim to protect wildlife.

Reitze, A. W., Jr. *Air Pollution Law.* Charlottesville, Va.: Michie Butterworth, 1994.

Rodgers, W. H., Jr. *Environmental Law.* 2d ed. St. Paul, Minn.: West, 1994. Provides a comprehensive overview of domestic environmental law in the United States.

Sand, P. H., ed. *The Effectiveness of International Environmental Treaties.* Cambridge: Grotius Press, 1992. This volume brings together expert reports prepared for UNCED assessing the success of international environmental treaty law in meeting its declared objectives.

Sands, P. *Principles of International Environmental Law.* Manchester: Manchester University Press, 1995. First of a three-volume collection provides a stimulating scholarly and highly readable exposition of international environmental law. The remaining two volumes, edited jointly with Richard Tarasofsky and Mary Weiss, provide an invaluable collection of primary materials on international and European environmental law.

United Nations Environment Programme (UNEP). *Handbook of Environmental Law.* Nairobi: UNEP, 1992.

—DAVID FREESTONE AND
LAURENCE BOISSON DE CHAZOURNES

# ENVIRONMENTAL MOVEMENTS

Environmental movements are networks of groups and individuals that exhibit common ecological values. Movement members seek to shape sensibilities, commitments, and practices within the general public as well as influence business practices and government policy. To achieve these ends, many members of environmental movements organize themselves into political parties or form nonprofit organizations.

There are currently over fifty green political parties worldwide. The earliest ecology-focused political parties were organized in the early 1970s in New Zealand and the United Kingdom. The United States Green Party formed in the mid-1980s, by which time West German Greens had already gained significant representation in government (Figure 1). The vast majority of green groups are not political parties but nongovernmental organizations (NGOs). It is difficult to estimate the total number of environmental NGOs worldwide, but clearly there are many tens of thousands. Environmental NGOs engage in scientific research, public education, political advocacy, and hands-on environmental protection, preservation, or restoration. They exploit strategically the resources of the media, government, the legal system, and public opinion to further their goals. While most environmental groups are organized locally, regionally, or nationally, many have global orientations and international memberships or affiliations. In June 1992, representatives from fourteen hundred NGOs participated in the Earth Summit in Rio de Janeiro in an effort to identify environmental problems, demand governmental action, and help craft political, economic, so-

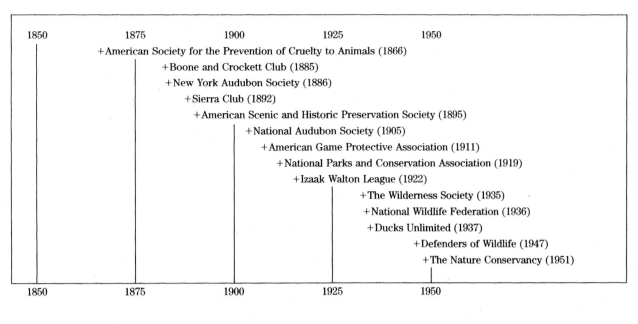

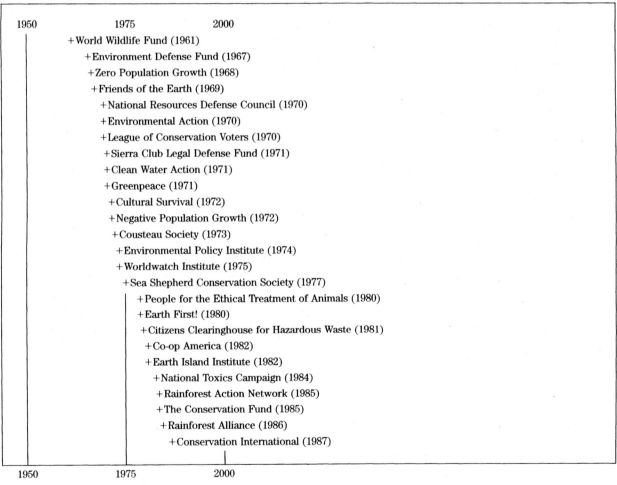

**Environmental Movements. FIGURE 1.** Founding Dates of Selected Environmental Organizations.

cial, and technological solutions. For reasons of brevity, this article focuses primarily on North American environmental movements. [See Nongovernmental Organizations.]

**The History of Public Environmental Protection.** As the environmental dangers of industrialization became evident in the nineteenth century, the popularity of nature writing and natural history grew in America and Europe. In the late 1820s, John James Audubon began publishing his *Birds of America*. Ralph Waldo Emerson wrote his famous essay "Nature" in 1836. In 1854, Henry David Thoreau published *Walden*, America's most famous tribute to the harmony of humanity and nature. George Perkins Marsh, in 1863, wrote of the destructive potential of human beings in *Man and Nature*. By the early 1870s, the first popular magazines advocating nature conservation were being published. At this time, John Burroughs' collections of nature essays were also widely employed in secondary-school curricula.

By the mid- to late 1800s, the protection of nature and the conservation of natural resources in Europe and America were gaining both public and governmental support. A conservation movement formed. Early conservationists are often divided into the two categories: resource conservationists and nature preservationists. Resource conservationists promoted the efficient management of natural resources. Preservationists focused on the protection of nature for its own sake.

Borrowing from European practices, Gifford Pinchot (1865–1946) championed resource conservationism in his role as chief of the U.S. Forest Service. President Theodore Roosevelt, an avid outdoors man, strongly supported Pinchot's efforts. Roosevelt also vastly expanded the national forest system, created national parks, and inaugurated the National Wildlife Refuge System. The first citizen conservationists were members of regional alpine or birdwatching clubs. These citizen groups concerned themselves not only with the efficient management of natural resources for human use, but with the preservation of wildlife for its own sake and for the aesthetic, spiritual, and recreational benefits it afforded humans. They worked both in tandem with and in critical opposition to the early resource conservationists within government. George Bird Grinnell, editor and publisher of *Forest and Stream*, was the founder in 1886 of America's first popular conservation organization, the Audubon Society. Its primary mission was the protection of plumage birds from the millinery industry and the protection of certain game birds from unregulated sport hunting. Many women, including former suffragette Rosalie Edge, played an important role in maintaining the organization's preservationist orientation. In Europe, similar efforts to protect wildlife, forests, and wilderness were well under way, particularly in Britain.

The most famous preservationist of the time was John Muir (1838–1914), a Scottish-born immigrant who spent much of his life hiking and climbing in America's wilds. Muir founded the Sierra Club in 1892 and remained its president until his death. The Club promoted the enjoyment and preservation of the forests, canyons, and mountains of California's Sierra Nevada. Muir and his fellow Sierrans opposed extensive logging and livestock grazing in the Sierra Nevada. After a protracted campaign, Muir lost the battle to halt the damming of the Hetch Hetchy valley. The flooding of the valley for use as a reservoir had been endorsed by Gifford Pinchot as an efficient utilization of natural resources.

Resource conservationists, maintaining anthropocentric or human-centered values, had the upper hand in the first wave of environmentalism. They formed the backbone of the early conservation movement. Yet preservationists, with biocentric or nature-centered perspectives, were not without influence. Moreover, because the natural resources that early conservationists aimed to conserve often included wildlands and wildlife, their policies, if not their principles, frequently dovetailed with those of preservationists.

An important figure of this period was Aldo Leopold (1887–1948), who exemplified the tension between anthropocentric and biocentric perspectives within the early conservation movement. Leopold worked under Pinchot in the Forest Service. After observing the pitfalls of shortsighted forestry and agricultural practices and predator extirpation programs, Leopold came to understand the human economy as a subset of an overarching ecological balance. He cofounded the preservationist Wilderness Society with Bob Marshall in 1936. In *A Sand County Almanac*, published posthumously in 1949, Leopold developed the first formalized environmental ethic. Leopold's "land ethic" extended moral concern to the biotic community as a whole.

By the 1950s, preservationist goals were becoming integrated into the agendas of most conservation groups. With the passage of the Wilderness Act of 1964, largely due to the lobbying efforts of Bob Marshall, the U.S. government officially recognized its duty to protect the integrity of wilderness for present and future generations.

In the 1960s, and more markedly by the early 1970s, a second wave of environmentalism arose. Previously, both resource conservationists and nature preservationists placed their activities under the rubric of nature conservation. The words *environment* and *environmentalism* were not yet in general circulation. By the mid-1970s, talk of environmental protection was widespread in the media, in schools, and in the halls of government.

Not unlike their predecessors, second-wave environmentalists were concerned with managing natural resources efficiently to satisfy human needs. But they were also troubled by the growth of these needs and the ecological costs of satisfying them on a global scale. With mounting unease, the general public learned that humankind was becoming the victim of its own environmental abuses. Consumerism and the mass production of goods had yielded a tremendous increase in litter. Waste disposal and energy resource problems were mounting. Rapid suburban development led to the paving over of green spaces. Urban air quality was noticeably deteriorating; many cities suffered from deadly inversions that trapped heavily polluted smog. Oceans were often used as dumping grounds, and many streams and rivers were clogged with effluent that made their waters undrinkable and frequently unfit for swimming or fishing. In this context, the public began to worry about the planetary effects of accelerating human production and reproduction. This global sensibility was aided by the photographs of the Earth taken by Apollo astronauts. For the first time, the Earth could be seen as a closed system with distinct and quite visible limits.

The rapid growth of second-wave environmentalism was sparked by the publication of Rachel Carson's *Silent Spring* (1962). In scrupulous detail, Carson documented the widespread use of pesticides and their devastating effects upon bird populations. America, Carson predicted, would soon face a spring wholly deprived of its beloved avian singers. *Silent Spring* also foretold a time when chemicals recklessly introduced into nature in the pursuit of profit would take a significant human toll. After reading *Silent Spring*, President Kennedy appointed a special panel of the Science Advisory Committee to study pesticide use. The panel largely corroborated Carson's findings. In turn, a group of private citizens inspired by Carson's work demonstrated that the use of dichlorodiphenyltrichloroethane (DDT) to control mosquitoes was the cause of a sharp decline in the osprey populations on Long Island. They founded the Environmental Defense Fund (EDF). By 1972, the EDF had succeeded in having the use of DDT banned nationwide.

In 1968, Paul Ehrlich published *The Population Bomb*. It quickly became a bestseller. Ehrlich argued that overpopulation was the chief obstacle to resolving many of the world's most pressing economic and ecological problems. He prophesied that humankind would breed itself into oblivion. After publishing his book, Ehrlich founded Zero Population Growth, an organization with a mandate of stemming the tide of human numbers.

Widespread concern about environmental degradation in the United States and the growth in the world's population burst into a true mass movement in 1970. On 22 April of that year, the first Earth Day was celebrated. An estimated twenty million Americans participated. In the following year, biologist Barry Commoner accomplished for the issue of technology what Ehrlich had achieved for population. In *The Closing Circle* (1971), Commoner raised concerns about the social and ecological effects of a centralized, technological way of life. Commoner argued that ecological devastation was directly tied to the way society was organized and the manner in which its productive capacities were designed.

Employing complex computer-aided analyses, the authors of *The Limits to Growth* (1972) brought overconsumption, the third agent of environmental degradation, to the public eye. This detailed study described the accelerating rate of natural resource depletion that modern technology, increasing human numbers, and rapid resource consumption had produced. If growth trends in world population, industrialization, food production, pollution, and resource depletion continued unabated, its authors predicted, the planetary limits to growth would be reached within the next one hundred years, with catastrophic results. While these predictions proved exaggerated, public concern for the environment grew steadily in light of such dire forecasts and in the wake of the widely observed degradation of the air, land, and water.

Between 1901 and 1960, an average of three conservation groups formed each year in the United States. Between 1961 and 1980, an average of eighteen new groups were founded each year. These organizations expanded the environmental agenda and radicalized its operations. In 1971, Greenpeace was formed. Originally organized to oppose nuclear testing carried out by the United States off the coast of Alaska, Greenpeace quickly became involved in a wide array of environmental issues, including well-known campaigns to end seal hunting and whaling. Greenpeace inaugurated an era of environmental "direct action." Although eschewing violence and the destruction of property, Greenpeace members frequently engaged in civil disobedience and put themselves in harm's way in efforts to publicize or prevent environmental misdeeds. Activists climbed smokestacks to release banners decrying pollution, positioned their rubber dinghies between whales and harpoon-laden whaling ships, and sailed into nuclear testing zones. The organization symbolized the willingness of people to risk their own welfare in the interests of environmental protection. More radical groups such as the Sea Shepherd Society and Earth First! engaged in ecological sabotage—sinking whaling ships or destroying tree-harvesting machinery—to protect wildlife and wilderness.

The vast majority of environmental groups adopted legal means to protect the environment. They gained widespread public support for their efforts. Indeed, by the early 1980s certain segments of the environmental movement were actively courted by business interests and the political establishment. Environmentalism was no longer a "fringe" movement. It had become mainstream. Many of the larger environmental groups, with memberships in the millions and annual budgets ranging over U.S.$100 million, were now run by professional administrators who controlled entire departments of scientists, lobbyists, lawyers, public relations personnel, communications and media consultants, fundraisers, membership recruiters, and business liaisons. These large organizations often worked in tandem with local volunteer groups. Yet many grassroots environmentalists objected to the national organizations' professionalism, commercialism, and reliance on corporate donations. Beginning in the late 1980s, the mainstream groups were also criticized for catering exclusively to the middle and upper classes. In the United States and many other countries, an "environmental justice" movement formed to challenge the disproportionate suffering of minorities and the poor from the effects of environmental degradation. In India, for example, a Chipko ("tree-hugging") movement was initiated by peasants and rural women in the late 1980s to protest the destruction of their forests by business interests.

**Public Values and the Antienvironmentalist Backlash.** Support for environmentalism rises and falls periodically as the public's economic, security, or other concerns wax and wane. Support for movement organizations typically swells in the wake of large-scale environmental catastrophes. The disaster of leeching toxic wastes at Love Canal (1978), the evacuation of Times Beach, Missouri as a result of dioxin contamination (1982), and the massive pesticide leak at the Union Carbide plant in Bhopal, India (1984) that killed many thousands of people and injured as many as half a million, vastly heightened environmental concerns and stimulated the development of a "toxics" movement. The near-catastrophe at the nuclear facility at Three Mile Island in Pennsylvania in 1979 and the massive explosion and radiation leak at the Chernobyl nuclear power plant in the Ukraine in 1986, which exposed millions of people to radioactive fallout that drifted as far as Ireland and Sweden, kept the antinuclear movement strong while demonstrating the global interdependence of nations in ecological affairs. The public's concern with the problem of solid waste disposal was likewise heightened in 1987, as a barge loaded with Long Island garbage was sent in search of a dumping ground. After spending 164 days at sea and being rejected by four foreign countries, the barge returned to New York to unload its cargo. Public concern grew again the following year, when medical waste that had been dumped at sea washed up on New York and New Jersey shores, undermining tourism.

Massive environmental degradation caused by oil and fuel spills also worried the public in the 1980s and bolstered the environmental movement. In March of 1989, the *Exxon Valdez* oil tanker ran aground on Bligh Reef in Alaska's Prince William Sound. Eleven million gallons of oil were spilled, creating an 80-kilometer-wide slick that contaminated over a thousand miles of formerly pristine coastline. Movement membership rose dramatically in the wake of the *Valdez* spill. National outrage and extensive lawsuits forced the oil company to spend billions of dollars trying to repair the damage.

As many as four out of five Americans claim to be environmentalists and support increased environmental protection. The environmental movement's success in gaining adherents is significantly offset, however, by the general public's reluctance to translate its proclaimed environmental values and beliefs into environmentally responsible behavior. Organized resistance to environmentalism also poses a threat to the movement. This resistance first arose in the late 1970s, when the Sagebrush Rebellion began in the western United States. Supported chiefly by energy companies and large public-lands ranchers, the Sagebrush Rebels pitted themselves against the federal government in its role as the caretaker of public lands and, to a lesser extent, against environmentalists who encouraged the protection of national forests and other federal lands. Identifying himself as a Sagebrush Rebel, President Ronald Reagan encouraged the selling of public lands to private owners. In the late 1980s, a broader coalition of antienvironmental forces initiated the Wise Use movement and the Property Rights movement. These movements sought to undo or prevent environmental legislation and regulation; open up public lands and wilderness areas to increased mining, drilling, logging, commercial development, and motorized recreational use; and ensure the right of industrialists, land developers, and property owners to receive compensation whenever environmental regulations threatened or harmed their economic interests.

The need to defend the environmental agenda against the Wise Use and Property Rights movements prompted many environmental organizations to close ranks in the 1990s. The antienvironmental backlash also underlined the failure of the environmental movement to branch out into rural areas and secure grassroots support from land-based peoples. In response to these shortcomings, many environmental organizations promoted "community-based" environmental protection and attempted to demonstrate how local economic needs could be met in tandem with the protection of the environment. [*See* Belief Systems.]

**Prospects for the Future.** For the foreseeable future, environmental movements will be defined by many different voices. The movement remains divided, for example, on the benefits and dangers of free trade; animal rights activists continue to stand at odds with environmentally oriented hunters and fishermen; biocentric preservationists maintain sharp differences with mainstream resource conservationists in philosophic principle and practical policy. Such diversity bears the potential of fragmenting the environmental movement. At the same time, environmentalists are learning to value diversity within their movement for the same reason that they value diversity in nature: it lends resilience to a complex web of relations. Diversity within the environmental movement may foster the resilience it needs to protect the environment in a quickly changing world.

[*See also* Conservation; Friends of the Earth; Greenpeace; International Cooperation; Public Policy; *and the biographies of Carson, Marsh, Nicholson, Pinchot, and Thoreau.*]

### BIBLIOGRAPHY

Brown, M., and J. May. *The Greenpeace Story*. New York: Dorling Kindersley, 1991. A readable, illustrated account.

Bryant, B., ed. *Environmental Justice: Issues, Policies, and Solutions*. Washington, D.C.: Island Press, 1995. A concise summary.

Dalton, R. *The Green Rainbow: Environmental Groups in Western Europe*. New Haven, Conn.: Yale University Press, 1994. A detailed, social scientific examination.

Dowie, M. *Losing Ground: American Environmentalism at the Close of the Twentieth Century*. Cambridge, Mass.: MIT Press, 1995. A critical challenge to mainstream environmentalism.

Dunlap, R., and A. Mertig, eds. *American Environmentalism: The U.S. Environmental Movement 1970–1990*. New York: Taylor and Francis, 1992. A comprehensive empirical assessment.

Echeverria, J. D., and R. B. Eby, eds. *Let the People Judge: Wise Use and the Private Property Rights Movement*. Washington, D.C.: Island Press, 1995. A critical yet balanced account of antienvironmentalism.

Fox, S. *John Muir and his Legacy: The American Conservation Movement*. Boston: Little, Brown, 1981. A classic history.

Hays, S. P. *Conservation and the Gospel of Efficiency: The Progressive Conservation Movement*. Cambridge, Mass.: Harvard University Press, 1958. A detailed assessment of the early resource conservationists.

Kamieniecki, S., ed. *Environmental Politics in the International Arena: Movements, Parties, Organizations, and Policy*. Albany: State University of New York Press, 1993. A useful summary of global environmental groups.

Kempton, W., et al. *Environmental Values in American Culture*. Cambridge: MIT Press, 1995. A useful analysis of environmental opinions and commitments.

Leopold, A. *A Sand County Almanac, with Essays on Conservation from Round River*. New York: Ballantine Books, 1966. A classic of ecological thinking and environmental ethics.

Manes, C. *Green Rage: Radical Environmentalism and the Unmaking of Civilization*. Boston: Little, Brown, 1990. An insider's view of Earth First!

Nash, R. *The Rights of Nature: A History of Environmental Ethics*. Madison: University of Wisconsin Press, 1989. A useful summary and historical analysis.

Sale, K. *The Green Revolution: The American Environmental Movement 1962–1992*. New York: Hill and Wang, 1993. A brief, readable account.

Thiele, L. P. *Environmentalism for a New Millennium: The Challenge of Coevolution*. New York: Oxford University Press, 1999. A detailed history and analysis of environmentalism.

Wapner, P. *Environmental Activism and World Civic Politics*. Albany: State University of New York Press, 1996. An informative exploration of transnational environmental groups.

—LESLIE PAUL THIELE

**EPIDEMIOLOGY.** *See* Disease; *and* Human Health.

## EROSION

Erosion is a natural process by which the flow of water, wind, and ice removes weathered material from the Earth's land surface. By acting on rocks of different resistance, erosion shapes the landforms of hills and valleys that make up our landscape. The study of the nature and evolution of landforms is known as the science of *geomorphology*, and erosion is a geomorphologic process. [*See* Land Surface Processes.]

Erosion comprises three phases: the detachment of individual soil particles from the soil mass, the pick-up or entrainment of those particles by the flow, and the transport of the particles over the land. When the energy of the eroding agent is no longer sufficient to move the particles, they are deposited. Erosion thus connects the source areas of sediment, where the weathering takes place, to areas of sediment sinks where the deposition occurs. Studies of sediments in partially enclosed environments such as lakes can often be used to interpret erosion rates on the surrounding land. Erosion rates are normally expressed as a weight of soil removed from a unit area of land over a given period of time. Typically, units of metric tons per hectare per year are used in erosion studies.

The main impact of erosion is on the surface material or soil. Under well-vegetated conditions such as forests and grasslands, a balance exists between the rate of soil erosion and the rate of soil formation. The vegetation cover protects the soil from erosion. It intercepts rainfall, reducing the amount that reaches the ground surface by some 20–30 percent annually. It increases the amount of water that infiltrates the soil because the plant roots open up passages in the soil through which water can move. Overall, the amount of water flowing as runoff over the land surface is reduced. The percentage of the annual rainfall contributing to runoff varies from less than 1 percent in densely vegetated areas to nearly 60 percent in urban areas and on bare soil. By determining the fate of the rainfall, the vegetation cover directly influences the erosion by raindrop impact and surface runoff.

**Importance of Raindrop Impact.** Raindrop impact on bare soil is a powerful agent of particle detachment. Professor Norman Hudson (1934–1996), a leading international consultant in soil conservation, carried out a simple but classic experiment in the 1950s in Zimbabwe to show the importance of covering the soil against falling raindrops. Soil loss was measured from two experimental plots over a ten-year period. Both were kept free of vegetation by hand-weeding. One plot was open to the rainfall, while a double layer of fine-mesh wire gauze was suspended over the other plot. The gauze allowed rain to pass through but broke up the falling raindrops so that they reached the soil surface in small droplets. Over a ten-year period, an average annual soil loss of 127 metric tons per hectare was recorded on the bare plot, but only 0.9 metric tons per hectare was recorded on the plot covered with the gauze. The experiment thus demonstrated the importance of cover in controlling the rate of erosion by reducing the rate of soil particle detachment.

The maximum effect of a vegetation cover in controlling raindrop impact occurs when the cover is at the ground surface. When rainfall is intercepted by a plant canopy, its properties are altered. First, the rainfall at the ground becomes concentrated beneath individual drip points. Second, the raindrops coalesce on the leaves to form larger drops. Typically, natural rainfall has a median drop size (the value at which 50 percent of the drops are larger and 50 percent smaller) by volume of about 2.0 millimeters; in contrast, the median drop size of leaf drips is about 4.8–5.0 millimeters. Although the energy of the rainfall is absorbed by the plant canopy, the drips gain energy again as they fall from the leaves. With vegetation canopies on or close to the ground surface, this gain is very small, but once the canopy rises to 0.5 meters or more above the ground, it is sufficient to detach soil particles. Detachment rates on bare soil under forest canopies can be two to four times higher than those on open ground. Fortunately, in most forested areas the soil is protected by a litter layer of decaying plant matter. On agricultural land, however, no such protection exists, and detachment rates under crops such as maize and cassava can be two or more times that of uncropped bare land. Since detachment is the first phase of the water erosion process, its control is important. If few or no soil particles were detached, there would be no material for the eroding agents to transport and the erosion rates would therefore be low.

**Runoff.** A ground-level vegetation imparts roughness to flowing water, reducing its velocity and therefore its ability to erode the soil. Runoff will detach soil particles from the soil mass and entrain them in the flow when its velocity exceeds a critical value that depends upon the resistance of the soil material. The critical velocity is higher for coarse materials (sands, gravels), which weigh more, and for fine particles (clay), which are held together by cohesion. It is lowest for silts and fine sands, which explains why silty and loamy soils are more prone to erosion than other soil types. Once a soil particle has been picked up by the flow it can be carried at much lower velocities than those required for detaching it. This is particularly true for clay particles which, once detached, can be transported considerable distances (tens of meters to kilometers) before coming to rest at a point in the landscape where velocity is reduced and deposition occurs.

Surface runoff initially takes the form of a shallow (1–2 millimeter) sheet-like flow spreading out over the slope surface. Its velocity is low because of the rough nature of the ground surface and any vegetation present, and it is not able to detach soil particles. It can, however, transport material splashed into it by raindrop impacts. As the water flows downslope it becomes concentrated into flow paths, its velocity increases, and it is able to cut small channels or rills into the land surface. Dense rill networks can be formed on bare farmland during intense rainstorms; they are also a feature of slopes left bare of vegetation for several months at a time, such as road cuttings, embankments, industrial sites, and residential areas left bare while construction work is in progress.

In many situations, rill channels do not cut very deeply because the underlying material is less weathered, more compact, and therefore more resistant to erosion. Sometimes, however, the reverse is the case, and once the rill has cut through the surface soil it can continue to develop, forming gullies 20–30 meters deep and 20–40 meters wide. Such features are common in many tropical and subtropical areas where the bedrock is chemically weathered to form a layer of saprolite (which is much weaker than the overlying soil) 5–20 meters or more thick. The spectacular gullies or *lavakas* of Madagascar originate in this way, and similar features are found in the southeastern United States, southern Brazil, West Africa, Zimbabwe, Swaziland, and South Africa.

**Wind.** Some of the most catastrophic examples of soil erosion are associated with wind acting on bare soil, following clearance of the vegetation cover. Wind is able to detach and transport large quantities of material, carrying the finer particles in suspension high into the atmosphere. Clouds of dust can affect people hundreds or even thousands of kilometers away, blotting out the sun, creating respiratory problems, and polluting the air. The best known example is the Dust Bowl that occurred in the Great Plains of the United States in the 1930s. During the 1920s, rainfall was above average and the prairie grasslands were plowed up for wheat cultivation, but cereal production was not sustainable when drought conditions returned. The first great "blow" occurred on May 11, 1934, and within a few hours the sun was obscured over a large area including Chicago, New York, and Washington, D.C. Over the next five years some 350 million metric tons of soil were eroded, and in some areas over 1 meter of soil was removed. [*See* Dust Storms.]

A similar episode occurred in Kazakhstan in the 1950s, following the extension of agriculture on to marginal lands in Kazakhstan under the Russian-operated Virgin Lands program. Wind erosion is also a major problem in the drier parts of Argentina, the Middle East, the Sahel, India, Pakistan, China, and Australia. Its association with semiarid areas means it is often associated with desertification. In reality, however, it is the lack of vegetation cover rather than climatic conditions that controls its distribution. For example, in Iceland the rainfall is more than sufficient to support vegetation, yet wind erosion is a major environmental issue. Largely as a result of overgrazing, vegetation occupies only some 25 percent of the country today, compared with 65 percent when the first settlers arrived in the eighth century CE.

**Historical Perspectives.** Given the protective effect of vegetation, it is not surprising that erosion rates are sensitive to the ways in which the type and amount of plant cover have changed over time. [*See* Land Use.] Sedimentation studies in lakes show that over much of northern Europe, mean annual erosion rates were relatively high (greater than 1.3 metric tons per hectare) in the immediate postglacial period (up to 10,000 BP). As climate became warmer and forest cover more extensive, mean annual erosion rates fell between 10,000 and 2,500 BP to less than 0.5 metric ton per hectare. Since then, erosion rates have been locally highly variable, depending on the level of human impact. Where deforestation occurred and the land was plowed, rates increased by about 500 CE to around 1 metric ton per hectare, but elsewhere they remained low. The Middle Ages saw a further extension of the land under farming as a result of demographic pressure as well as an increase in the use of spring-sown crops, so that land was bare for the autumn and winter seasons. This, coupled with clearing of the forests for charcoal burning and shipbuilding, led to an increase in erosion which, locally, reached 10 metric tons per hectare. Fluctuations in erosion over time and space were common, however, with rates decreasing wherever war, famine, and disease reduced population numbers. In the late Middle Ages (1350–1400), very high rates were recorded in parts of central Europe as a result of catastrophic rainfall events; for example, deep gullies developed in parts of Lower Saxony in Germany, where rates locally exceeded 2,000 metric tons per hectare. In contrast, the period 1400–1750 was less extreme and the gullying ceased.

The late eighteenth and early nineteenth centuries saw the introduction of new crops on the fallow land, improved manuring, and overall improved agricultural practices. As a result, erosion declined further until the late nineteenth century, when a second phase of gullying took place, also related to a greater frequency of intense rain events. Erosion rates increased locally to 100–200 metric tons per hectare before falling again in the early part of the twentieth century. Since the 1950s, the intensification of agriculture, the concentration on cereals, oil seeds, and root crops with consequent decline in rotational grass, and the removal of hedgerows and field banks to enlarge fields have created unstable bare soils depleted in organic matter, and higher erosion rates have been the result.

This brief history of erosion (Figure 1) demonstrates the close relationship between erosion rates and human impact. In studies of soil erosion, a distinction is often made between erosion as a natural process and accelerated erosion as a result of human impact. Rates of accelerated erosion can be between ten and one thousand times the natural rate, depending on local climate, soil, slope, and land cover. The study of the role of human impact on geomorphologic processes is known as anthropogeomorphology. [*See* Anthropogeomorphology.]

**Consequences of Erosion.** Since soil, along with water and air, is one of the resources needed to support human life, the rate at which it is removed is important. Soil is required for growing crops for food and industrial use and fodder for livestock. The loss of soil by erosion has effects both on site, where the erosion occurs, and off site, at the places downstream or downwind where sedimentation takes place. On site, the main effects are loss of soil and, more importantly, decline in soil fertility. The two effects are separated because the first relates to the mass of soil material, the second to its composition. When erosion occurs, it is a selective process in which the finer particles are removed preferentially and the coarser material left behind. The fine particles (clays) contain the nutrients, particularly nitrogen and phosphorus, required for plant growth. The material that remains is not only less fertile but retains less water. Erosion has caused a decline in wheat yields in Australia of between 6 and 52 percent in the decade 1973–1983, equivalent to A\$13–155 per hectare. In Sierra Leone, a typical 3 hectare holding loses U.S.\$58 per annum as a result of declining yields due to erosion; this is some 20 percent of the annual farm income. It has been estimated (Stocking, 1988) that the annual cost to Zimbabwe of nitrogen and phosphorus loss due to erosion is U.S.\$1.5 billion.

Off-site effects relate to silting of rivers, canals, and reservoirs, damage to roads and property as a result of burial by sediments, and pollution of water bodies through increased concentrations of sediment particles and transfer of chemicals, particularly phosphorus, adsorbed to the clays. The off-site costs of erosion are considerable, amounting to A\$2–4 per hectare in northwestern New South Wales.

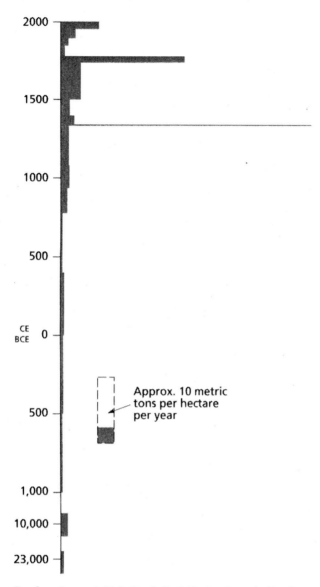

Erosion. Figure 1. Variation in Peak Erosion Rates in Northern Europe in Historical Times. (Adapted from Bork, 1988 and van Vliet-Lanoë et al., 1992.)

Over time, if no remedial measures are taken, accelerated soil erosion can lead to a self-perpetuating system in which the loss of soil fertility results in poorer vegetation cover, hence reduced soil protection, more erosion, and, in turn, less vegetation. Under these extreme conditions, land can become desert-like. In areas of low but intense rainfalls, erosion can be a major contributing factor leading to desertification. [See Desertification.]

**Soil Conservation.** Soil erosion has long been recognized as one of the leading environmental issues threatening our ability to meet world needs for food. Professor W. C. Lowdermilk, a leading American archaeologist of the 1930s, documented evidence that erosion was a major factor contributing to the decline of the ancient civilizations in Mesopotamia and the Middle East. The remains of terracing from these times illustrate a very early understanding of erosion control. The foundation of soil conservation as a worldwide movement dates from the establishment in 1935 of the United States Soil Conservation Service, with H. H. Bennett as its first director. [See the biography of Bennett.] The basis of soil conservation is wise land use and good land management. The Soil Conservation Service pioneered a system of land evaluation to determine the suitability of the land for agriculture. All land was assigned to one of seven classes, depending on the type and severity of limitations for arable farming. With increasing severity of limitation, a greater level of soil conservation is required.

Soil conservation measures can be divided into three broad groups, defined as agronomic, soil management, and engineering. Agronomic measures relate to the way the land cover is managed and include the growing of grasses and legumes with other crops in rotation, the use of cover crops, and agroforestry. Soil management relates to the way the land is tilled. Soil conservation tillage retains the residue from previous crops to help provide a surface mulch to protect the soil. Engineering measures include various forms of terracing and bunding, contour grass barriers, shelterbelts, and purpose-built waterways to convey excess runoff to safe outlets. In many cases, these measures are integrated into a coherent system of soil and water management. For example, systems to control gully erosion usually include planting of grasses and shrubs to help reduce runoff and erosion on the hillsides and construction of check-dams to help stabilize the channel.

Throughout the 1950s and 1960s, soil conservation was promoted in many countries through local equivalents of the United States Soil Conservation Service. This resulted in a top-down approach in which conservation staff proposed and designed erosion-control measures and the farmers were expected to implement them. Although there were examples of successful conservation projects, they often had to be supported by financial incentives to farmers from national and local governments. Implementation of the measures usually increased a farmer's costs without any appreciable short-term gain in income, and it was the community rather than the individual farmer who benefited. The need to subsidize conservation in this way is unsustainable, and over the last decade a new "bottom-up" approach has been adopted. Farmers are encouraged to be involved in decisions on how their land is managed. The emphasis is on improving all aspects of land husbandry, including erosion control, rather than dedicating resources specifically to soil conservation, and on recognizing the role played by indigenous conservation methods.

[*See also* Agriculture and Agricultural Land; Deforestation; *and* Soils.]

## BIBLIOGRAPHY

Bork, H. R. "The History of Soil Erosion in Southern Lower Saxony. "*Landschaftgenese und Landschaftsökologie* 16 (1989), 135–163.

Hudson, N. W. *Land Husbandry*. London: Batsford, 1992. Modern approaches to land use and land degradation issues in developing countries.

———. *Soil Conservation*. 3d ed. London: Batsford, 1995. A practical guide to modern soil conservation.

Johnson, D. L., and L. A. Lewis. *Land Degradation: Creation and Destruction*. Oxford: Blackwell, 1995. Case studies from the developing and developed world emphasizing interactions between physical and human environments.

Morgan, R. P. C. *Soil Erosion and Conservation*. 2d ed. Harlow, U.K.: Longman, 1995. A general introduction to the subject.

Pimental, D., ed. *World Soil Erosion and Conservation*. Cambridge: Cambridge University Press, 1993. Papers on soil erosion in different countries and regions of the world.

Stocking, M. "How Eroding Soils Lose Money." *International Agricultural Development* 8.1 (1988), 11–12.

van Vliet-Lanoë, B., M. Helluin, J. Pellerin, and B. Valadas. "Soil Erosion in Western Europe: From the Last Interglacial to the Present." In *Past and Present Soil Erosion: Archaeological and Geographical Perspectives*, Oxbow Monograph No. 22, edited by M. Bell, and J. Boardman, pp. 101–114. Oxford: Oxbow Books, 1992.

—R. P. C. MORGAN

# ESTUARIES

Estuaries are places where rivers meet the sea. They can be defined as the lower portion of the river or an arm of the sea, where saline sea water is diluted with fresh water. Estuaries are inherently dynamic as a result of short-term fluctuations in water level and water masses due to tides, storm surges, and runoff from tributary streams as well as long-term changes in water level and sediment inputs that change the size of the basins and the rate of migration of the shorelines. The characteristics of many present-day estuaries result from a slowing down of the rate of sea level rise several thousand years ago that allowed sediment to accumulate and create beaches, shallow water subtidal flats, and intertidal land that developed into salt marshes. These diverse environments and the tidal creeks that traverse them provide valuable habitat and feeding areas for fauna. Estuaries also provide fish and shellfish for human consumption and sheltered harbors and navigation corridors for boating and shipping. The favorable location of estuaries for human activity causes them to be foci of human settlement. Many of the world's largest cities are located on estuaries, and many are subject to the most intensive levels of use applied to any marine water area.

**Recent Changes.** The expansion of industrial and commercial activity in the nineteenth century and recreational activity in the twentieth century greatly modified the character of natural estuaries, and human development is still increasing in intensity and spatial coverage. Human modifications affect biological processes by altering photosynthetic production, nutrient cycling, food supply, activity patterns, and mortality rates. The introduction of exotic or cultured species has changed biota. The construction of dams and levees and the alteration of tributaries for flood control or use of water for human consumption or agricultural production alter the migratory patterns of fish, change the distribution of salinity and nutrients, change the range of predators, and change the amount of sediment delivered to the estuaries—which, in turn, alters siltation rates and shoreline locations.

The size and shape of estuaries is altered by land reclamation for agricultural and residential use and for the building of causeways and bridges, leading to loss or fragmentation of habitat. The dredging of channels for navigation and disposal of the unwanted dredge spoil changes turbidity levels, circulation patterns, and zones of sediment erosion and accretion. The construction of generating plants for electricity has caused thermal pollution of the waters. The use of fertilizers and inputs of human waste from ever-increasing populations contributes to organic enrichment in estuaries (known as *eutrophication*), which leads, in turn, to excessive growth of algae, increasing metabolism, causing changes in the structure of biological communities. Toxic materials such as heavy metals, petroleum products, and pesticides cause direct chronic or lethal effects to primary ingestors, and are passed through the food chain to predators. To these negative impacts can be added positive impacts, including the restoration of natural habitat and the creation of recreational beaches using artificial fill.

**Sensitivity and Resilience of Estuaries.** Estuaries are resilient natural features because their physical environments and biota are adapted to continuous inputs from rivers and from the sea that renew the supply of water, food, larvae, and other essential elements to small damaged areas, aiding recovery and protecting long-term net stability. The organisms that are adapted to estuaries have great tolerance to the rapid changes in temperature and salinity that are helpful in resisting external forces, although some adverse human inputs are exceeding the capacity of some estuaries to absorb them.

One of the greatest threats to the future viability of estuaries is the effect of accelerated sea level rise, coupled with human attempts to reduce the threat of flooding or to retain a stable shoreline position by using protective structures. A rise in sea level will tend to cause the land/water boundary of an estuary to shift inland, bringing about more frequent inundation of land that was less affected previously. The extent to which estuaries can adapt by creating new natural shoreline environments will depend on the rate of sea level rise, the amount of sediment available, and the actions of humans to protect developed land. Coastal marsh systems are one of the landscape types most threatened by sea level rise. Valuable salt marshes will be lost in places because sediment inputs will be insufficient to allow marshes to build up as fast as they are inundated. Estuaries may become deeper and more marine as well.

**Estuarine Management.** Programs have been implemented in many countries at national, state and provincial, and local levels of government to improve water quality and enhance resources that have been threatened by pollution, development, and overuse. [*See* Coastal Protection and Management.] Regulations include waterfront development laws, wetland and tideland management acts, water quality planning and maintenance programs, health regulations, dredge-and-fill permit programs, critical area laws requiring site plan reviews, erosion control setback lines, shore protection programs, and parks and recreation programs. Environmentally oriented restrictions on development have prevented the filling of marshes in many countries, but other habitats that are not protected by legislation, including beaches and naturally functioning upland margins, are still being eliminated through human development. Much of this development has occurred at locations where new marsh would otherwise be created as a result of sea level rise. The amount of salt marsh will be reduced where losses due to wave erosion and inundation cannot be replaced by the creation of new marsh on the landward side because the land is developed for human use. This loss of estuarine habitat has been called the *coastal squeeze*. Management plans for overcoming the coastal squeeze include adding sediment to existing marsh surfaces to allow them to keep pace with inundation, or introducing managed retreat, where low-lying upland that is not intensively developed will be allowed to revert to marsh. Creation of marsh by reshaping the land, reintroducing tidal flow, and planting marsh vegetation is also possible, and a number of projects have been attempted with varying degrees of success.

[*See also* Coastlines; Ecosystems; *and* Ocean Life.]

### BIBLIOGRAPHY

Cronin, L. E. "The Role of Man in Estuarine Processes." In *Man's Impact on Environment*, edited by T. R. Detwyler, pp. 266–294. New York: McGraw-Hill, 1971. A landmark article in a landmark volume.
———, ed. *Estuarine Research*, vols. 1 and 2. New York: Academic Press, 1975.
Day, J. W., Jr., et al. *Estuarine Ecology*. New York: Wiley, 1989. A comprehensive synthesis that serves as a text and standard reference.
Kennedy, V. S., ed. *Estuarine Perspectives*. New York: Academic Press, 1980.
Kennish, M. J. *Practical Handbook of Estuarine and Marine Pollution*. Boca Raton, Fla.: CRC Press, 1997. A recent, comprehensive review of problems of pollution and programs for monitoring and controlling them.
Lauff, G. H., ed. *Estuaries*. Washington, D.C.: American Association for the Advancement of Science, 1967. The first of a series of edited volumes representing the state of the art in estuarine research. Selected representative volumes in this series, taken from research results presented at meetings of the Estuarine Research Federation, include Cronin (1975), Kennedy (1980), Wiley (1976), and Wolfe (1986).
Nordstrom, K. F., and C. T. Roman, eds. *Estuarine Shores: Evolution, Environments and Human Alterations*. London: Wiley, 1996. Contributions in this book focus on horizontal changes to estuarine shores, in contrast to the many existing evaluations of vertical changes or processes and biota in deeper waters.
Reed, D. J. "The Response of Coastal Marshes to Sea-Level Rise: Survival or Submergence." *Earth Surface Processes and Landforms* 20 (1995), 39–48. A good review of the literature. Companion articles in the same issue assess effects of sea level rise on other coastal environments.
Wiley, M., ed. *Estuarine Processes*, vols. 1 and 2. New York: Academic Press, 1976.
Wolfe, D. A., ed. *Estuarine Variability*. Orlando, Fla.: Academic Press, 1986.

—KARL F. NORDSTROM

# ETHICS

[*This entry consists of three articles that explore the meaning and significance of several ethical concepts and positions relevant to the human implications of global change and their significance to human welfare.*]

An Overview
Intergenerational Equity
Environmental Bioethics

*The first article discusses technocentric and ecocentric world views as well as a range of supporting ethical positions. The second article reviews, at a more technical level, the concept of intergenerational equity. The final article focuses on the definition and origins of environmental bioethics. For a related discussion, see* Sustainable Development.]

## An Overview

Environmental decisions are ethical choices. Different ethical theories offer conflicting views of the values to which decisions should appeal, of equity in the distribu-

## ECOTHEOLOGY

Ecotheology seeks to explore the links between religious thought and ecological concerns. It has sometimes been argued that Judeo-Christian thought is at the base of the environmental crisis, though others argue that Christian thought includes the need for stewardship rather than domination. Other religions have different attitudes to the environment and may, for example, promote a respect for animal life and for the conservation of trees.

**BIBLIOGRAPHY**

Gottlieb, R. *This Sacred Earth: Religion, Nature, Environment.* New York: Routledge, 1996.
Kinsley, D. *Ecology and Religion: Ecological Spirituality in Cross-Cultural Perspective.* Englewood Cliffs, N.J.: Prentice-Hall, 1995.

—ANDREW S. GOUDIE

tion of benefits and burdens, and of the extent of the ethical reference class—who should count.

The utilitarian holds that we should aim at the decision that maximizes the welfare of affected parties. Hedonist utilitarianism takes welfare to consist in pleasure and the absence of pain, preference utilitarianism in the satisfaction of preferences. The utilitarian approach can allow direct preferences for environmental goods to be a component of welfare. It need not involve a narrow definition of welfare such that nature is just a resource for the production of consumer goods.

Objections to utilitarianism include the following: distributions of benefits and burdens matter as such, not just as a means of increasing total welfare—a decision that placed all burdens, for example health hazards, on a particular minority might be unequitable even if it maximized the total welfare; the rightness of an action is not determined solely by its consequences— some acts, for example torture, are wrong no matter what the consequences.

Rights-based theories assert that beings have certain claims and rights, which are not open to being traded against increases in the total welfare. There are two distinct justifications for this position. The first appeals to respect for the intrinsic value of individuals: individuals possess a special value that is not reducible to their value to others. The second appeals to contract: ethical rules are those that would be agreed between rational persons about the conditions under which they will cooperate, rather than compete in pursuing their interests; rational contractors would agree to their essential interests being guaranteed from interference through the institution of rights.

These different approaches have implications for the constituency of ethics—who is to count. For the utilitarian, if a being has welfare interests then it counts— as such it classically involves an extension to incorpo-

rate all sentient beings, human and nonhuman, present and future. Intrinsic-value rights theorists allow any being that is an end in itself to be included. They differ on whom that incorporates: Kantians extend rights to persons, animal-rights theorists to sentient beings, biocentric rights theory to all living things. Contract theory has more difficulty in extending the ethical constituency beyond those capable of entering into contracts: future generations and nonhumans can be included if rational contractors have an interest in their well being.

For utilitarian and rights theorists, only individuals count. Ecocentric holist positions include collective entities such as biological communities as direct objects of ethical concern. Typical is Aldo Leopold's land ethic, according to which actions are right if they tend "to preserve the integrity, stability, and beauty of the biotic community."

Different approaches have implications for the size of human populations. If each new individual has a positive welfare, for the utilitarian, paradoxically, population increase may be one way of augmenting total welfare. In contrast, some ecocentric positions entail a radical reduction in human population.

[*See also* Belief Systems; Conservation; Environmental Law; Gaia Hypothesis; Human Impacts, *article on* Human Impacts on Biota; Policy Analysis; Valuation; *and the biography of Leopold.*]

### BIBLIOGRAPHY

Attfield, R. *The Ethics of Environmental Concern.* New York: Columbia University Press, 1983. A clear overview of basic positions in environmental ethics.
Leopold, A. *A Sand County Almanac.* New York: Oxford University Press, 1949 [1987]. The most influential source for many recent ecocentric perspectives.
O'Neill, J. *Ecology, Policy and Politics: Human Well-Being and the Natural World.* London and New York: Routledge, 1993. A

defense of a virtues-based approach that is critical of utilitarian, rights-based, and ecocentric positions.

Singer, P. *Practical Ethics*. New York: Cambridge University Press, 1979. An overview of ethical positions and defense of a utilitarian perspective.

Taylor, P. W. *Respect for Nature: A Theory of Environmental Ethics*. Princeton, N.J.: Princeton University Press, 1986. A biocentric rights-based approach to environmental ethics.

—JOHN O'NEILL

## Intergenerational Equity

The central question posed by this ethical concept is what, if any, are the current generation's obligations to future generations. Underpinning the answers to the question are beliefs about whether or not future generations have interests and/or rights. Arguments in favor of such an assignment of interests and rights sometimes seek support via the analogy with contemporary strangers. Thus future generations and contemporary strangers are entitled to broadly similar ethical treatment. As long as it is feasible to assume that some people will definitely exist in the future and will be in no relevant way unlike current people (right-holders), then they are worthy of equal moral consideration.

Arguments against giving equal moral consideration to future generations have been based on some combination of the following points:

- the temporal location of future generations, and the asymmetry of power between generations that do not overlap;
- uncertainty or complete ignorance of distant future peoples' wants and needs; and
- the contingency of future people, that is, future people may not exist at all and their actual number depends in large measure on current actions and decisions.

Some versions of utilitarian philosophy favor, at best, a weak form of intergenerational equity and support the argument that the current generation's moral duties are limited to assignable people and the requirement not deliberately and knowingly to harm future prospects without good cause. The difficulty for these utilitarians is that, on the basis of an individualistic philosophy, the distinction between possible or potential people in the future and actual individuals who will exist at some subsequent time is problematic. Thus it is argued that the assignment of rights to all possible people is an inadequate basis for an environmental ethic focused on sustainable development. If all possible people have a right to life, or to other rights hypothetically, then a current policy based on environmental conservation and growth moderation may not be morally supportable. Some utilitarians will go as far as concluding that the present generation has no obligation to future generations, arguing that a person has not been wronged by another unless he or she has been made worse off by the other's act. Nothing the current generation can do can wrong future actual people (brought about by our actions), unless what is done results in such a "poor" future world that these actual future generations turn out to wish they had never been born at all.

If, however, the individualistic viewpoint is moderated by a more collectivist approach, then a number of points in support of intergenerational equity obligations can be made. While there will always be uncertainty about the precise form and extent of future individual preferences, it seems reasonable to conclude that basic needs (clothing, food, shelter, community etc.) will exist and will not be radically different from contemporary ones for the foreseeable future.

A strategy that satisfies these basic needs will be a prerequisite of the satisfaction of most other wants and interests of future generations, regardless of their precise configuration. Further, it does not necessarily follow that our obligations to the future require us to know the identity of individual members of succeeding generations, only that there will be future generations of humans. If we accept the existence of future generations as given, then the contingency problem alone does not provide sufficient ground for subordinating future generations' interests to those of existing contemporaries. A "total utility" variant of utilitarianism could therefore be deployed to support a normative ethic in favor of positive obligations to future generations. Members of the current generation have a collective obligation to ensure that human existence continues at a fairly high level of general (total) utility, by maximizing basic-needs provision. If population can be stabilized, this ethical position would require equal provision for the basic needs of each generation. The "generalized obligations" serve to maintain an inheritance of at least a constant flow of resources into the future to ensure ongoing human life.

A number of attempts have been made by philosophers to utilize a contractarian approach to suggest criteria for intergenerational equity. In particular, John Rawls's (1972) "Theory of Justice" has, in amended form, been used as the basis for a number of theories of intergenerational equity and resource-use strategies. But the Rawlsian case that can be made out for the formulation of a principle governing the just distribution and rate of resource usage over long periods of time requires the bringing together of different philosophical principles (e.g., in the tradition of John Hume but also Immanuel Kant) that are not necessarily complementary. The dual aspects of the theory—justice as rational cooperation (Hume) and justice as hypothetical universal assent (Kant)—may conflict as soon as the analysis moves away from the self-contained society of contemporaries.

The contractarian approach is based on actual or hypothetical negotiations that are said to be capable of yielding mutually agreeable principles of conduct, which are also binding upon all parties. Rawls has formulated an idealized decision model to produce procedural rules for a just society peopled by contemporaries. Rational and risk-averse individual representatives form contemporary society in an "original position" (the negotiations) and operating from behind what is called the "veil of ignorance" (individuals are assumed not to know to which stratum of society they themselves belong) choose the principles of justice. According to this model, among the agreed rules would be equal opportunity for all individuals as well as the "difference principle" (or "maximin criterion"), which requires an acceptable standard of living for the least well off in society.

A number of analysts have sought to apply the difference principle intertemporally to guarantee future generations an adequate natural resource endowment and a habitable environment. If the representatives were deprived of the knowledge about which generation they were part of, it could be argued that a maximin rule would again be rationally chosen. The rule could translate into a contract that mandated a bequest of a constant capital stock (human, reproducible, and natural capital elements) across generational time. It turns out, however, that "veil of ignorance" conditions may need to be drawn even more tightly, and, strictly speaking, should leave the individual representative unsure about the total number of generations, as well as about his or her status as a possible or actual person and which species he or she represents.

Pushing the Kantian approach to intergenerational fairness further, the "justice as opportunity" argument can be deployed. The preservation of "opportunities" for future generations becomes a minimal notion of intergenerational fairness. The present generation does not have a right to deplete the economic and other opportunities afforded by the biospherical resource base since it does not "own" it. Sustainability criteria should be imposed as prior constraints on the maximization of social preferences concerning the distribution of welfare across generations. Each successive generation may be charged with a duty to ensure that the expected welfare of its offspring is no less than its own perceived welfare. The passing-on over time of the resource base "intact" does not necessarily mean literally intact; in cases where depletion occurs, this must be compensated for by capital investment in alternative options and/or technological innovation to increase resource productivity, the combined impact of which is to offset the impacts of depletion (both source and sink resources capacity).

This moral rule also supports a bias against actions that generate current benefits while simultaneously imposing the risk of irreversible future losses if the maintenance of options would allow for improved discussions as new information becomes available. It suggests that precautionary action, such as the protection of biological diversity and the ozone layer, has much to recommend it. Nevertheless, from a weak sustainability position, some depletion of natural resource stocks can be compensated for via capital substitution and is not necessarily inconsistent with the sustainable development policy objective. But, given the level of socioeconomic and scientific uncertainty that pervades our current and foreseeable future societies, it may equally be better to work toward strong sustainability, conserving the maximum feasible amounts of natural capital to maintain the options for the future.

[*See also* Belief Systems; Conservation; Environmental Law; Gaia Hypothesis; Human Impacts, *article on Human Impacts on Biota*; Policy Analysis; Valuation; *and the biography of Leopold.*]

## BIBLIOGRAPHY

Attfield, R. *The Ethics of Environmental Concern.* Oxford: Blackwell, 1983.

Brown, P. G., and D. MacLean. *Intergenerational Justice in Energy Policy.* Totowa, N.J.: Rowman and Littlefield, 1983.

Howarth, R. B. "Sustainability Under Uncertainty: A Deontological Approach." *Land Economics* 71 (1995), 417–427.

Norton, B. G. "Intergenerational Equity and Environmental Decisions: A Model Using Rawls' 'Veil of Ignorance.'" *Ecological Economics* 1 (1989), 137–159.

Page, T. *Conservation and Efficiency.* Baltimore: Johns Hopkins University Press, 1977.

Parfit, D. "Future Generations: Further Problems." *Philosophy and Public Affairs* 11.2 (1982), 113–172.

Rawls, J. *A Theory of Justice.* New York and Oxford: Oxford University Press, 1972.

—R. Kerry Turner

## Environmental Bioethics

Environmental bioethics, or environmental ethics, is the theory and practice relating to values in and duties to, or concerning, the natural world. This involves an environmental philosophy, or philosophy of nature. A recurrent issue is whether ethics is applied to the environment analogously to its application in business, medicine, engineering, law, and technology; or whether it is more radical, revising traditional ideas about what is of moral concern. The applied ethic is *anthropocentric*; the central concern is whether humans are helped or hurt by the condition of their environment.

A *naturalistic* ethics is more radical, holding that animals, plants, endangered species, ecosystems, and even the Earth as a whole ought to be the direct objects of moral concern, at times at least. Others than humans also count morally—whales who are slaughtered, or ancient forests, or Earth as a biotic community disrupted by human-introduced global changes. If so, environmental bioethics differs from bioethics more generally, which has previously been largely medical, with human health and welfare its concern. Environmental ethics is unique in moving outside the sector of human interests.

Philosophers have formed philosophies of nature for millennia, although in the West the meaning of human life has usually been their main focus. Environmental ethics and philosophy arose in the second half of the twentieth century. The impetus came from a dramatically escalating power of humans to affect nature, associated with, for example, the loss of species or global warming. Industrialization, advanced technologies, global capitalism, consumerism, ecological and evolutionary sciences, and exploding populations have all contributed to a concern that humans are not now in a sustainable relationship with their natural environment. Nor have they distributed the benefits derived from natural resources equitably, nor have they been sensitive enough to the welfare of the many other species.

A parallel concern for animal welfare also arose, sometimes advocating *animal rights*. These ethicists typically cared for domestic animals, such as food animals or pets, but they also took wild animals into consideration. They found common interests with environmentalists, but also many divisions. One may be vegetarian, while the other approves of hunting. An animal welfare ethics seeks to be "humane"; an *ecocentric* ethics has more concern for ecosystems, soil and water conservation, or wilderness preservation. The American forester Aldo Leopold termed this a *land ethic*, where "land" is the regional biotic community of life. [*See the biography of Leopold.*]

*Biocentrism* is sometimes used as a general synonym for any naturalistic or nonanthropocentric ethics. Biocentrism refers more specifically to an ethics of respect for life, focused on organisms or individuals. In contrast to an animal welfare ethics, biocentrism holds that any living thing can count morally, although plants, insects, animals, and humans need not count equally. Animals and plants can and must use each other as resources, which justifies some human uses of the environment. But human uses can be unjustified, failing to consider the worth, or intrinsic value, of living things—cutting old-growth forests, for example, instead of recycling or using tree plantations to meet needs for paper and wood.

*Axiological environmental ethics* identifies various values carried by nature. Humans value nature as their life-support system, economically, recreationally, scientifically, aesthetically, for genetic diversity, as cultural symbol, religiously, and so on. Such values may be assigned to natural things by humans, or they may come into existence in human interactions with nature. Further, many values are found in nature that are present there independently of human valuations—autonomously intrinsic values. Such values are discovered, not placed.

Plants and animals alike defend their own lives; they are members of species lineages perpetuated over millennia. Ecosystems are the sources and systems of life, having generated numerous species over evolutionary time. An adequate ethics will need to optimize all relevant values, humanistic and naturalistic. Moral concern needs to focus on the relevant survival unit, not always the individual, often the species, the ecosystem, or, ultimately, the planet Earth.

Some values carried by nature are said to be "subjective," meaning that they are found only in the experiences of psychological subjects resulting from their interactions with the world, as contrasted with "objective," meaning present in the world independently of such experiences in subjects, particularly those of humans. No one denies that humans enjoy such experiences of value. The issue is whether, apart from humans, animals and plants embody any values on their own, and whether these too can and ought to figure in environmental ethics (animal pain, the flourishing of a sequoia tree, or an endangered insect species). Some, invoking *sentient environmental ethics*, think that the higher animals (vertebrates, mammals) with developed central nervous systems are capable of such valuing experiences (subjects), but that lower animals and plants are not, much less other mere objects, such as mountains or rivers. All the latter have no values of their own.

*Deep ecology* emphasizes the ways in which humans, although individual selves, ought to extend their selves through a web of connections, taking a model from ecology where living things are what they are in their environments. In this view, humans have such entwined destinies with the natural world that their richest quality of life involves a larger identification with the communities, human and natural, by which they are surrounded. Such transformation of the personal self will result in an appropriate care for the environment.

*Human ecology* or *social ecology* taken as an environmental ethics argues that the focus needs to be on the institutions of society, its politics, business, development plans, population policies, legal systems, patterns of resource distribution, and so on. What is needed is not some revised metaphysics of nature so much as criticism and reformation of human patterns of social behavior regarding nature. Although these follow from the worldviews of individuals, an environmental ethics must be corporate, because action must be taken in concert. Accordingly, an environmental policy, a political ecology, a green politics is especially needed. *Bioregionalism* argues that such human–nature relationships can best be worked out at regional levels—as in the United States Southwest, or the Scandinavian Arctic. Bioregionally, people can retain a strong sense of place; government, business, lifestyles, and environmental integrity can be kept in a coherent regional whole.

Since the United Nations Conference on Environment and Development (UNCED) at Rio de Janeiro in 1992, much interest has gathered under the umbrella of *sustainable development ethics*. Developing nations argued that environmental conservation could not and should not be separated from development, which involved environmental justice as much as environmental preservation. About four-fifths of the world's population, including most nations, produce and consume only about one-fifth of the world's goods. About one-fifth of the world's population (the Group of Seven [leading industrial] nations and a few others) produce and consume four-fifths of the goods. That, they say, is neither fair nor sustainable. Consumption by the rich is more of a problem than increasing population in the developing nations. The global warming threat is produced by industrial nations, not developing ones. They call for an *ethics of ecojustice*.

Sustainable development, however, has proved a rather elastic concept, sometimes meaning sustainable growth, or profit, or resource base, or sustainable opportunity, or sustainable communities. Those inclined to a more ecological ethics argue that ultimately humans ought to have a *sustainable biosphere ethics* (which global climate change might seriously threaten).

An *ecofeminist ethics* finds an alliance between the forces that exploit nature and those that exploit women, often also finding a care for nature present among women, contrasting with an attitude of dominion among men. Such patriarchal bias has been present in many societies but has especially characterized the modern West.

Attitudes toward nature include religious convictions, and many argue that a *religious environmental ethics* must be part of the answer; indeed it is essential to an answer, since religion, globally, is more influential than philosophy; and religion, misunderstood, has helped cause the crisis. Monotheistic religions, such as Christianity, Judaism, and Islam, urge an ethics of the *stewardship of creation*; or they may prefer to speak of caring for a sacred creation, or *reverence for life*. A *creation spirituality* has a strong sense of the divine presence in nature. Others argue that Eastern religions have something to offer, such as the *yang* and *yin* of Taoism in harmonious balance, or the *ahimsa*, noninjury and respect-for-life traditions in Hinduism and Buddhism. Native Americans and indigenous peoples in Africa, Australia, and South America have claimed that their traditions respect the natural world better than the modern West.

A *postmodern environmental ethics* doubts whether humans can know nature independently of the various cultural schemes used to interpret nature, plural as these are. A worldview is a social construction more than a realist account of nature as it is in itself. These views can be better or worse (judged by their sustainability, or eq-

uitable distribution of resources, or quality of life), and that is all that is needed for an environmental ethics, which is, after all, about relating persons to places: *an ethics of place*. Such environmental ethics may differ between various peoples: a *pluralist environmental ethics*. A *communitarian ethics* locates humans in both human and biotic communities, with values and duties at multiple levels and scales. These ethics may join, however, as all humans see themselves as Earthlings, with their home planet as a responsibility.

[*See also* Belief Systems; Conservation; Environmental Law; Gaia Hypothesis; Human Impacts, *article on* Human Impacts on Biota; Policy Analysis; *and* Valuation.]

### INTERNET RESOURCE

An extensive Web site bibliography is at http://www.cep.unt.edu/ISEE.html/. This site also includes the *International Society for Environmental Ethics Newsletter*, updating the current bibliographic literature on a quarterly basis.

### BIBLIOGRAPHY

Attfield, R. *The Ethics of Environmental Concern*. 2d ed. Athens: University of Georgia Press, 1993.

Callicott, J. B. *In Defense of the Land Ethic*. Albany: State University of New York (SUNY) Press, 1989.

———. *Earth's Insights: A Survey of Ecological Ethics from the Mediterranean Basin to the Australian Outback*. Berkeley: University of California Press, 1994.

Devall, B., and G. Sessions. *Deep Ecology*. Salt Lake City, Utah: Peregrine Smith Publishers, 1985.

Elliot, R., ed. *Environmental Ethics*. New York: Oxford University Press, 1995.

Gruen, L., and D. Jamieson, eds. *Reflecting on Nature: Readings in Environmental Philosophy*. New York: Oxford University Press, 1994.

Johnson, L. *A Morally Deep World: An Essay on Moral Significance and Environmental Ethics*. Cambridge: Cambridge University Press, 1991.

Norton, B. G. *Toward Unity Among Environmentalists*. New York: Oxford University Press, 1991.

Regan, T., ed. *Earthbound: New Introductory Essays in Environmental Ethics*. New York: Random House, 1990.

Rolston, H. *Environmental Ethics: Values in and Duties to the Natural World*. Philadelphia: Temple University Press, 1987.

———. *Conserving Natural Value*. New York: Columbia University Press, 1994.

Taylor, P. *Respect for Nature*. Princeton, N.J.: Princeton University Press, 1986.

—HOLMES ROLSTON III

# ETHNOBIOLOGY

Ethnobiology is the comparative study of human beliefs about, and perceptions, categorizations, uses, conservation, and management of, the visible and invisible worlds ("bios") that provide a context for "life." The term is roughly synonymous with *ethnoecology*, although some practitioners use ethnobiology as the more general term referring to all "living things," leaving ethnoecology as the study of how humans understand interactions between elements of ecosystems and landscapes (Toledo, 1992). Hardesty (1977, p. 291) defined ethnoecology as "the study of systems of knowledge developed by a given culture to classify the objects, activities and events of its universe." These differences are semantic and reflect variations in the different use of "ecology" and "biology" in different languages. Generally, both terms are used interchangeably and in the broadest sense (namely, the human perception of all living things). Subcategories of ethnoecology/ethnobiology are common and mirror the fragmented subdisciplines of Western science (for example, ethnobotany, ethnozoology, ethnoentomology, ethnoornithology, ethnopedology, ethnoastronomy, ethnomedicine, and so forth) more than they do the holistic cosmovisions of other traditions (Ellen, 1993).

The continuum of research and methodologies that comprises ethnoecology extends from the perception of phenomena (the "cognition"), to the ordering of cognitive categories (the "ethnotaxonomy"), the social expression of the system (the "culture"), the impact of cultural systems on physical surroundings (the "anthropogenic and cultural landscapes"), the management of those landscapes (the "conservation and resource management"), the exploitation of knowledge and resources for wider use and application (the "applied research"), and the moral concerns of how to apply such research equitably while protecting the rights of knowledge holders (the "ethics and intellectual property rights"). Thus, ethnobiology and ethnoecology inevitably draw from disciplines such as cognitive psychology, linguistics, anthropology (including ethnology and biological and social anthropology), geography, ecology, biology, history of science, philosophy, ethics, and additional areas of study such as forestry and plant sciences, zoology, agriculture, and medicine.

Each phase of the continuum evokes problems of epistemology, terminology, and methodology, since there is no general agreement between disciplines on how to view this spectrum of concerns (Atran, 1990). In practical terms, ethnobiology requires crossdisciplinary and interdisciplinary methodologies that are sensitive to profound differences between cultures, especially in relation to "sacred" categories, places, or domains that may restrict or prohibit access to researchers. Thus, ethnobiologists are constantly torn between maintenance of the rigors of their Western scientific "discipline" and respect for and understanding of the metaphysical, aesthetic, and spiritual dimensions that define and encompass the societies of the peoples they study.

Conklin (1957) became a pioneer in ethnobiological research when he described the extensive knowledge of natural history of the Hanunoo people of the Philippines, as did Bateson (1979) in his thesis on the "unity of mind and nature." Levi-Strauss (1962) provided the neoteric framework for understanding the links between the "savage mind" and the environment, and drew inspiration from the early (1900) work of Barrows on the Coahuila Indians' knowledge and use of the edible and medicinal plants of the Southern California desert. Berlin et al. (1966) set the current standard for comparative studies of ethnobiological classification systems based on their studies of the Tzeltal Maya of Mexico (also see Berlin, 1992). Majnep and Bulmer provided another classic standard on the Kalam of New Guinea in *Birds of My Kalam Country* (1977). Debates still rage over whether cognitive, or utilitarian, factors underlie folk classification systems (Hunn, 1993). More recently, ethnobiological studies have tended to focus on more practical aspects of ethnobiological knowledge, especially its application to conservation, natural resource management, and environmental change and management (Berkes, 1989; Martin, 1995; Warren et al., 1989).

One goal of ethnobiology is to stimulate dialogue between scientists and local knowledge specialists so that a common scientific language can eventually emerge to discuss and study what is generally known as *traditional ecological knowledge*, or TEK (Slikkerveer, 1998). Historically, ethnobiology has concentrated, but not been limited to, studies of indigenous and traditional peoples (such as peasants, hunters, gatherers, fisherfolk, small farmers, and so forth). Today, however, there is increasing interest in knowledge systems of urban dwellers, suburban and periurban communities, and peoples in industrial societies.

Since the 1992 Convention on Biological Diversity (CBD), global interest in the "wider use and application" of traditional peoples' "knowledge, innovation and practices" has accelerated. The search for new products and genes for biotechnology (bioprospecting) has highlighted ways in which the knowledge and genetic resources of indigenous and local peoples have been exploited without compensation (known as *biopiracy*), despite requirements of the CBD for equitable benefit sharing with and protection of the "holders" of TEK. Thus, a debate on the ethics of science and industry in their study and exploitation of "traditional resources" has ensued (Posey, 1996). It is no longer possible to undertake ethnoecological research without dealing with full disclosure of research intentions and outcomes, prior informed consent from affected communities, equitable benefit sharing from any products or derivatives of research, and other ethical issues in research and dissemination (such as publication, films, photos, and so forth). The application of intellectual property rights (IPRs) has become a particularly contentious ethical issue, since industrial "discoveries" can be protected by law, but collective knowledge of indigenous, traditional, and local communities cannot. The CBD has initiated a process, in conjunction with the Food and Agriculture Organization (FAO) and the World Intellectual Property Organization (WIPO), to investigate *sui generis* ("unique" or "specially developed") options for the protection of traditional genetic resources and the TEK associated with them. This process builds upon a long-standing attempt within FAO to develop farmers' rights to guarantee just compensation for traditional farmers who have developed and conserved for millennia the folk varieties and land-races from which high-yielding domesticated crop plants were developed—and upon whose future improvement and adaptation these crops still depend (Posey and Dutfield, 1996).

Ethnobiologists and ethnoecologists have become leaders in the struggle to prevent extinction of plant and animal species and the loss of languages and cultures that encode knowledge about biodiversity and management strategies that conserve it. [*See* Biological Diversity.] The International Society for Ethnobiology meets biennially and has developed, together with indigenous and traditional peoples, a model Code of Conduct to guide "equitable partnerships" between experts in TEK and Western scientists. Many countries have their own societies or associations of ethnobiologists (namely, the United States, India, Mexico, Brazil, Thailand, and China). Several journals that deal specifically with ethnobiology and ethnoecology are published, including *Journal of Economic Botany* and *Journal of Ethnobiology* (both published in the United States), *Etnoecologica* (Mexico), and *Indian Journal of Ethnobotany*. There is also a global network of indigenous knowledge resource centers coordinated by the Centre for International Research and Advisory Networks (CIRAN/Nuffic), based in The Hague (Netherlands), which publishes the *Indigenous Knowledge and Development Monitor*. Ethical and benefit-sharing issues are dealt with by the international Working Group on Traditional Resource Rights (see Internet Resource below) and the Indigenous Peoples' Biodiversity Network.

[*See* Convention on Biological Diversity; Ethics; *and* Extinction of Species.]

### INTERNET RESOURCE

Working Group on Traditional Resource Rights. http://oxford.users.ac.uk/~wgtrr/.

### BIBLIOGRAPHY

Atran, S. *Cognitive Foundations of Natural History.* Cambridge: Cambridge University Press, 1990. A fundamental study of the underlying perceptual elements of ethnobiology.
Bateson, G. *Mind and Nature: A Necessary Unity.* New York: Dutton, 1979. Sequel to *Steps to the Ecology of the Mind*, which established the link between cognition, society, and nature.
Berkes, F., ed. *Common Property Resources: Ecology and Community-Based Sustainable Development.* London and New York: Belhaven Press, 1989. A broad collection of studies on the relationship between property systems and environmental management by communities.
Berlin, B. *Ethnobiological Classification: Principles of Categorization of Plants and Animals in Traditional Societies.* Princeton: Princeton University Press, 1992. Provides an outline of the general principles of ethnobiological classification, followed by an exemplary case study of indigenous Mayans of Chiapas, Mexico.
Berlin, B., et al. "Folk Taxonomies and Biological Classification." *Science* 154 (1966), 273–275. One of the foundation works on the universal principles of folk classification.
Conklin, H. "Hanunoo Agriculture: Report of an Integral System of Shifting Cultivation in the Philippines." Forestry Development Paper No. 5. Rome, 1957. The historic study of traditional agriculture that helped fuel the early interest in traditional knowledge and ethnobiology.
Ellen, R. *The Cultural Relations of Classification.* Cambridge: Cambridge University Press, 1993. One of the primary theoretical studies on ethnobiology and classification, building upon extensive field data and case studies.
Hardesty, D. *Ecological Anthropology.* New York: Wiley, 1977. A foundation study on the role of culture and environment in molding human societies.
Hunn, E. "The Ethnobiological Foundation for TEK." In *Traditional Ecological Knowledge: Wisdom for Sustainable Development*, edited by N. Williams and G. Baines. Canberra: Centre for Resource and Environmental Studies, 1993.
Levi-Strauss, C. *La Pensée Sauvage.* Paris, 1962. (English translation: *The Savage Mind*, Chicago: University of Chicago Press, 1996.) A classic work that lays out the importance of biology and nature in shaping human society and the universal perceptions of nature that are influenced by biological function.
Majnep, I., and R. Bulmer. *Birds of My Kalam Country.* Auckland: Auckland University Press, 1977. One of the most influential studies in ethnoornithology and ethnobiology.
Martin, G. *Ethnobotany.* London: Chapman and Hall, 1995.
Posey, D. A. *Traditional Resource Rights.* Gland, Switzerland: IUCN, 1996.
Posey, D. A., and G. Dutfield. *Beyond Intellectual Property Rights.* Ottawa: IDRC Books, 1996.
Slikkerveer, J. "Traditional Ecological Knowledge; An Introduction." In *Cultural and Spiritual Values of Biodiversity*, edited by D. Posey. London: UNEP and IT Press, 1998.
Toledo, V. "What is Ethnobiology? Origins, Scope, and Implications of a Rising Discipline." *Etnologica* 1.1 (1992), 5–21.
Warren, D. M., et al. *The Cultural Dimension of Development.* London: Intermediate Technology Publications, 1989.

—DARRELL ADDISON POSEY

## EUROPEAN COMMUNITY

The European Community (EC) is a supranational organization composed of fifteen countries; it has some features of an international organization and some features of a federation. It was originally established in 1957 by six countries as the European Economic Community, a common market based on a customs union.

Since then, its competencies have gradually been increased and new members have joined. Community policies include such fields as competition, trade, agriculture, fisheries, transport, environment, and research. In 1992, the Treaty on European Union was signed, establishing an Economic and Monetary Union and a Common Foreign and Security Policy in addition to the existing Community structure. In 1999 the Treaty of Amsterdam entered into force, transferring additional powers to the European Union (EU). The EU is now composed of the following three branches, each with different institutional arrangements: the European Community, the Common Foreign and Security Policy, and the Police and Judicial Cooperation in Criminal Matters. The main features of the EC have remained the same. Policies in the area of global change fall within the scope of the EC. The second branch is based on political cooperation between the member states. The third pillar proposes common action for the member states.

The most important political institutions of the Community are the Council, the Commission, and the Parliament. The Council is an intergovernmental body composed of delegates of the member states represented by government ministers. The Commission and the Parliament represent the interests of the Community as a whole. The powers of the Parliament have increased in recent years. Most measures are now adopted jointly by the Council and the Parliament in the cooperation and co-decision procedures. The Commission formulates proposals for new measures and supervises their implementation. It represents the Community at international negotiations with a mandate of the Council. It is the task of the European Court of Justice to ensure the uniform interpretation and the enforcement of Community law.

The binding legislative acts of the Community are Regulations, Decisions, and Directives. Regulations and Decisions are directly applicable; Directives must be transposed into national law by a national act, although the European Court has recognized a direct effect of Directive provisions that are sufficiently clear and unconditional. A Regulation applies generally, whereas a Decision concerns an individual case. Most environmental measures are adopted in the form of Directives, usually implemented with great delay by the member states. International environmental treaties are normally ratified with a Decision.

As far as global change is concerned, the EC has taken action in the areas of climate change, ozone depletion, wildlife and habitat protection, biodiversity and endangered species, acid rain, and genetic engineering. In most cases, measures are adopted as a result of an international commitment. In recent years, the EC has ratified most of the important international conventions on the above-mentioned topics. The Commission, as the EC's representative, often plays an important role during the negotiations. Usually both the EC and the member states participate in treaties to establish a mixed agreement. Among these are the Vienna Convention for the protection of the ozone layer and the Montreal Protocol on substances that deplete the ozone layer, the United Nations (UN) Convention on Biodiversity, the UN Framework Convention on Climate Change, the Convention on International Trade in Endangered Species, and the Convention on the Conservation of European Wildlife and Natural Habitats.

In its proposals on global change issues, the Commission generally pushes for a high level of protection, but proposals are very often watered down during discussions with the member states in the Council. A pertinent example is the Commission proposal on the adoption of a combined carbon and energy tax, which the Council—reluctant to accept EC competence in the sphere of taxation—has never adopted. The most far-reaching binding measure in the area of climate change currently in force is a community-wide monitoring mechanism of greenhouse gas emissions. The Environment Council agreed in March 1997 on a reduction target of 15 percent by 2010 for carbon dioxide, methane, and nitrous oxide as the community position in the international talks under the Framework Convention on Climate Change and in particular at the meeting in Kyoto in December 1997. The parties who signed the Kyoto Protocol committed themselves to reduce greenhouse gas emissions by at least 5 percent below 1990 levels during the period 2008–2012. The EU member states must reduce their emissions by 8 percent between 2008 and 2012. The action plan adopted in 1998 in Buenos Aires during the fourth session of the Conference of the Parties to the FCCC provides for a work program to continue the implementation of the convention. The Commission has published a large number of policy documents which, although nonbinding, tend to have a stimulating effect on national and international policy development in the area of climate change (see bibliography).

## BIBLIOGRAPHY

Commission Proposal on the adoption of a combined carbon and energy tax (COM (92) 226; Official Journal 1992 Nr C 196/1). The proposal was never adopted by the Council, which was reluctant to accept EC competence in the sphere of taxation.

Commission Proposal for a Council Decision Concerning the Signature by the European Community of a Protocol to the United Nations Framework Convention on Climate Change (COM (98) 96 final). The Proposal has still not been adopted by the Council.

Communication of 3 June 1998 from the Commission to the Council and the European Parliament: Climate Change—Towards an EU post-Kyoto strategy (COM (98) 353 final).

Communication of 19 May 1999 from the Commission to the Council and the European Parliament: Preparing for implementation of the Kyoto Protocol (COM (99) 230 final).

Communication of 8 March 2000 from the Commission to the Council and the European Parliament on EU Policies and Measures to Reduce Greenhouse Gas Emissions: Towards a European Climate Change Programme (COM (2000) 88 final).

Council Decision 93/389 for a Monitoring Mechanism of Community $CO_2$2 and other Greenhouse Gas Emissions (OJ No L 167/31); amended by Decision 99/296, OJ No L 119/35. The Decision was adopted to ensure consistency with the monitoring mechanism to be established under FCCC, and concerns all greenhouse gas emissions not controlled by the Montreal Protocol.

Council Decision 94/69 Concerning the Conclusion of the UN Framework Convention on Climate Change (OJ No L 033/11). States the Community target that was agreed prior to the negotiations at UNCED: stabilization of $CO_2$ emissions by 2000 at 1990 levels in the Community as a whole. Annex B to the Decision lists all the instruments adopted at the Community level, including the above-mentioned monitoring mechanism.

Council Decision 647/2000/EC of the European Parliament and of the Council of 28 February 2000 adopting a multiannual program for the promotion of energy efficiency (SAVE) (1998 to 2002) (OJ No L 079/6).

Hartley, T. C. *The Foundations of European Community Law.* 4th ed. Oxford: Clarendon Press, 1998. A comprehensive overview of EC law, including a discussion of the institutions and their functioning, the decision-making procedures, agreements with third countries, the doctrines of the European Court, and references to relevant case law.

—IDA J. KOPPEN

**EUROPEAN UNION.** *See* European Community.

**EUSTASY.** *See* Sea Level.

**EUTROPHICATION.** *See* Marginal Seas; *and* Pollution.

# EVOLUTION OF LIFE

The world of living organisms is remarkable for its extreme variation, from bacteria one-thousandth of a millimeter long, through extinct dinosaurs, to the blue whale, weighing in at one hundred metric tons and the one-hundred-meter-high giant redwood tree; from wormy denizens of the deep-sea volcanic trenches to birds soaring to thirty thousand feet above the Earth's surface. The only concept that makes scientific sense of this bewildering diversity of perhaps as many as thirty million different species of organisms is evolution. [*See* Biological Diversity.] For beneath the differences lies an equally astonishing basic similarity. All organisms are made of the same essential molecular building blocks of proteins, carbohydrates, lipids, and so on. And all make use of the same macromolecular polymer, de-

oxyribonucleic acid (DNA), to code their genetic information so they can pass on to future generations the instructions needed for developing into the right kind of organism. Evolution, in biological terms, is the process of irreversible descent with modification. It explains the similarity by the idea that all organisms have descended from a common ancestor. It explains the differences by the idea that different lineages diverged in different ways from those ancestors. All living organisms fit onto their respective branches of the single tree of life.

Even before Charles Darwin, a number of scientists had proposed rudimentary theories of evolution to explain this pattern of similarities and differences between different species. But it was Darwin who first made a detailed, extensive, and systematic compilation of all those aspects of the diversity of the living world for which evolution is a satisfactory explanation. First of all, comparative anatomy pointed to the existence of characters of different degrees of generality. What this means is that some characters are found in a large number of different species, for example, the vertebral column in all those animals called vertebrates. Other characters are only found in subsets of these larger groups, such as four legs, which occur only in those vertebrates known as tetrapods. Yet further characters subdivide the subdivisions, so that only certain tetrapods possess, say, moist, naked skin and are known as the amphibians. A very large number of characters of organisms therefore point to a hierarchical arrangement of species as groups within groups within groups. If descent with different modification in different lineages did indeed occur, then predictably this hierarchical arrangement of groups supported by a hierarchical distribution of characters would indeed exist.

Other kinds of evidence adduced by Darwin included the slightly different species found on islands compared to those on the nearest mainland, as witnessed famously by the Darwin's finches of the Galápagos Islands. He also noted the existence of vestigial organs, of no apparent purpose, but explicable by evolution as reductions of parts of the organisms no longer required for a modified way of life. Finally, the existence of fossilized remains of animals and plants no longer living, and often of quite unfamiliar appearance, can be accounted for by the belief that in the course of evolution, earlier kinds of organisms become extinct and thereby give way to later kinds.

It was not difficult to persuade the mid-nineteenth-century scientific community of the truth of evolution. But Darwin's even greater genius lay in his discovery of a mechanism that would explain why evolution actually occurred and why it followed the paths that it took. This is the theory of natural selection.

**Natural Selection.** The single most remarkable feature of living organisms is their capacity for adaptation. Every individual is able to live in its own particular complex environment and carry out its basic functions, such as gaining energy, avoiding predation, and ultimately reproducing, because it is well designed to perform those functions. It might have teeth suitable for processing its particular foodstuff, coloration matching its background to avoid being too visible, ways of attracting a mate, and so on, depending on exactly what kind of organism it is and how it makes its living. *Adaptation* is the term applied to this matching of structure to function, and it was once the strongest argument presented in favor of a Creator: How could the fine design of the human eye or the horse's limb have come about without some foresight of what it was going to be used for? A little ironically, however, adaptation is now the overwhelmingly strongest argument in favor of evolution, because it is the key to understanding how and why evolutionary change occurs.

Darwin argued from three commonplace observations about living organisms. First, all organisms tend to produce many more offspring than the single one in its lifetime that would be required to maintain a constant size of population. Second, the individuals in a population all differ slightly from one another: no two members of a species have exactly the same size, shape, color pattern, and so on. Third, offspring tend to resemble their parents in these variable characteristics. Therefore, since there will not be enough resources or space for all the offspring to survive, what he termed a "struggle for existence" will arise among them. Any one of them that has some variation in character that by chance happens to increase its ability to survive will be more likely to reach reproductive age and produce offspring of its own. And those offspring will inherit the favorable variation. As time passes, a higher and higher proportion of the favorable variant will occur and a lower and lower proportion of the less favorable ones. Eventually, the population will consist entirely of organisms with the new characteristic. The species will have evolved to a new kind that is better able to cope because it is better adapted, or fitter, than the old.

The essential point about this process is that the initial variation itself comes about entirely by chance, and not by any reference to the organism's needs. Most of this random variation may well cause reduced fitness, and only a very small amount confers a benefit. Nevertheless, the latter will tend to increase and the former to be lost, thus increasing the level of adaptation to the particular environment. Darwin likened the process he had discovered to the process that stockbreeders had been using for centuries to improve the quality of their stock. Here the breeder allows to breed only those animals or plants that show the characteristics regarded as desirable, for example, fatness in pigs. After a time, the herd as a whole consists of pigs with increased average fatness. It is by analogy with this well-known practice of artificial selection that the expression *natural selection* was coined.

Darwin announced his theory in 1858, along with Alfred Russel Wallace, who had independently hit upon the same idea. However, it was the publication of Darwin's book *On the Origin of Species by Means of Natural Selection* the following year that saw the beginning of the establishment of evolution by natural selection as the most important single idea in biology. At first, not all went well. While the general concept of evolution was widely accepted, a number of flaws in the notion of natural selection as the driving mechanism were perceived. The main problem was that natural selection can work only if an appropriate mechanism for the inheritance of characters from parents to offspring exists, and this was simply not known. For instance, it was generally believed at that time that inheritance was blending, which is to say that offspring were the average between paternal and maternal characters; if true, then it is difficult to see how novel, favorable variation in one individual could persist in future generations. It was also accepted that characters acquired by an organism during its life could somehow be passed on to the next generation, in which case evolution ought to be readily reversible. There was also incredulity that the relatively slight degree of variation found in a typical population of a wild species could possibly affect how likely particular organisms were to survive and reproduce. Could a millimeter or so difference in the length of a bird's beak really matter? And finally there was the persistent failure to discover the long sequences of fossils showing gradual adaptive change with time that the theory predicted should exist.

It was not until the turn of the century that the underlying mechanism of heredity was discovered. In fact, it was the rediscovery of the principles that the Austrian monk Gregor Mendel had found almost half a century earlier. By a series of breeding experiments with pea plants, Mendel had shown that the characters of an organism are controlled and transmitted by means of "particles" that are neither blended with each other nor altered by direct environmental effects on the organism. Spontaneous, random alterations to these particles could therefore be the basis of variation. Once formed, the variation could be passed on to future generations with no blending or loss. Nor was there any difficulty with the idea that natural selection could sort out the different particles, or *genes*, as they became called, with those causing greater fitness accumulating at the expense of those causing lesser.

At this point in the development of evolutionary theory there was still one outstanding bar to complete acceptance of natural selection, namely, the apparent triviality of the level of variation seen in nature. A number of mathematically minded evolutionary biologists, notably Sir Ronald Fisher in Britain and Sewell Wright in America, were able to calculate that even very small differences in fitness between individuals in a population could, given enough time, produce significant evolutionary change. And reference to the fossil record indicates that there has indeed been plenty of time. Meanwhile, many observations since then of natural selection occurring in the wild have confirmed beyond any possible doubt that the level of variation in nature is perfectly adequate to fuel the required rates of evolution.

By the middle years of the twentieth century, all this information on variation, genetics, and mathematical modeling came together in what is termed neo-Darwinism or the synthetic theory of evolution. With its stress on randomness of variation, upon which selection acts through the greater rate of production of offspring by some variants than others, natural selection finally came to offer a powerful explanation for the diversity of organisms and the origin of their adaptations. Yet there was still a huge area of ignorance: it was completely unknown at the molecular level what genes were and how they worked. The next step toward full understanding of evolution began in 1953, when James Watson and Francis Crick described the structure of the DNA molecule. The ensuing molecular revolution has affected the whole of biology, and no area more so than evolutionary theory. It is now known that a gene is a polymer of nucleotides forming a DNA molecule, and that mutations causing variation are spontaneous mistakes in DNA replication that are completely random with respect to the selection forces acting upon the organism. It has also become clear why acquired characters cannot be inherited by offspring: there is no possible molecular means by which some modification to an organism caused by a direct effect of the environment could become coded for in a DNA sequence. Thus, in the same way that Mendelian genetics had earlier shown how Darwinism can work, so molecular genetics has in turn shown how Mendelism can work: the three together form the contemporary theory referred to as the new evolutionary synthesis.

**Patterns of Evolutionary Change.** Natural selection is the force behind evolutionary change, explaining why it happens and how it brings about adaptation. But it does not of itself predict what kind of patterns evolutionary change will create. To discover this, evidence from currently evolving species and observations from the fossil record are necessary.

*Speciation* is the process by which one species splits into two or more, and it is a fundamental step in the pattern of evolutionary change over time. A species is defined as a population of organisms that can interbreed among themselves but cannot successfully interbreed with members of another species. Thus species are re-

productively isolated from each other, and the moment a new species has formed, an irreversible step in evolution has occurred: there can be no going back to the single ancestral species. Most speciation is believed to occur initially by the physical, geographical separation of different parts of a hitherto continuous population. The barrier might be a river, mountain, or stretch of sea. The consequence is that the now separate populations undergo their own separate evolutionary changes, due to different mutations and gene combinations arising by chance, as well as by different selection pressures acting in the different areas. The result will be increasing genetic divergence from one another until there is sufficient dissimilarity to prevent successful interbreeding, even if the two populations should subsequently come into contact with one another. There are other possible methods by which new species might come about without necessarily having complete geographic separation, but these are more controversial ideas and have their own difficulties.

Whatever the actual mechanism of speciation is, one of the most astonishing things that the fossil record illustrates is that the history of life on earth has consisted of a complex, uninterrupted pattern of origins and extinctions of species. Species, and the higher groups that they constitute, such as families and orders, are forever originating, flourishing, declining, and being replaced by yet new species and groups, in a never-ending kaleidoscope of taxonomic turnover. The life span of a species is extremely variable, but is typically of the order of one million to ten million years. This may seem a long time, but compared to the several hundreds of millions of years of life on Earth, it indicates that the vast majority of species have become extinct, and that those existing today are but a relatively few tips of the branches of the overall phylogenetic tree of life. And they too, in a geological sense, must be seen as ephemeral, due eventually to be replaced by as yet unformed new species.

The fossil record contains another unpredictable surprise, which is the phenomenon of mass extinction. [*See* Extinction of Species.] From time to time through paleontological history there have been brief occasions when a very large number of the existing species have disappeared. On the geologic time scale these events appear to be virtually instantaneous, although given the poor time resolution of stratigraphic rocks, this could actually encompass a period as long as a few hundred thousand years. Many different kinds of organisms are affected, across many geographical areas and environments. Of the twenty or so such biotic crises in life's history, five particularly large mass extinctions saw the loss of an estimated 75 percent or more of species. Of these, the event at the close of the Permian period, 250 million years ago, was associated with the disappearance of as many as 96 percent of the world's species.

The cause of mass extinctions is unclear. A number of triggers to massive environmental deterioration have been suggested, such as impacts of meteorites, volcanoes, changing continental dispositions, falling or rising sea levels, and global cooling and warming cycles. Bits of evidence for each of these categories, in different combinations, are associated with different mass extinctions. Perhaps the answer is that any one of several possible combinations of them can be the ultimate cause of a breakdown in the global ecosystem.

At any event, whatever the cause, mass extinctions have played a major role in the pattern of evolution. The species that survive a mass extinction do not have any obviously favorable characteristics. Rather, they appear to be the few species that were fortunate not to succumb to whatever was happening. But they are nevertheless the species that initiate the recovery in biodiversity. Thus mass extinctions have the effect of breaking up long evolutionary trends and removing many hitherto successful groups. To name but one example, after the end-Permian extinction, the very successful brachiopods (lampshells) barely recovered, while the superficially but previously less diverse bivalve mollusks proceeded to dominate the shelly fauna of the sea floor. This effect has been termed, somewhat dramatically, "resetting the evolutionary clock."

**History of Life on Earth.** While it is not, and perhaps never will be, precisely understood how life originated, there is much evidence to suggest how it might have come about. The Earth began its existence as a condensation from a hot cloud of dust about 4.7 billion years ago. [*See* Earth History.] By around 4 billion years ago an atmosphere of gases such as nitrogen, carbon dioxide, and methane had formed from outgassing volcanoes, while water vapor had condensed to form the early oceans. Experiments show that if a source of energy such as electric sparks or ultraviolet light acts on such a mixture, many of the basic molecules of life, such as amino acids, simple sugars, and precursors of the nucleotides that constitute nucleic acids, form spontaneously. If there were some way of combining these units into larger biomolecules, then perhaps an elementary molecular system capable of simple replication would eventually arise. It has been variously suggested that this might have occurred in such locations as the surfaces of mineral crystals, drying-up shorelines, or hot deep-sea vents. Whatever the details may have been, it can be imagined that even a very simple, primitive, and inaccurate replication system might start to undergo natural selection. If some variant of the system was a little more stable, or a little more likely to replicate, then it would tend to survive and accumulate at the expense of less effective versions. At some stage a lipid membrane enclosing and stabilizing the system would be selected for, after which internal organization could

increase. Thus an essentially cell-like entity, capable of absorbing and utilizing energy from the environment and of replicating accurately, can be imagined. By definition such an entity would be a living cell.

Such simple, isolated cells as these were the sole form of life for the next 2.5 billion years or so. They constituted the world of the prokaryotes, which are still very much part of every corner of the present biosphere in the form of bacteria and the blue-green algae.

About one billion years ago the most important revolution in the whole of life's history since its origin occurred, although at the time it might not have seemed all that remarkable. A number of different kinds of prokaryotic cells formed an association, or symbiotic relationship. Some sort of actively feeding prokaryotic cell appears to have ingested a small kind of aerobic cell, but instead of being digested, the latter remained alive and functional. The host cell now had inside it an inclusion capable of gaining energy from the aerobic respiration of organic molecules. At least some of these hosts went on to ingest a second kind of cell, this time a photosynthetic form able to use light energy to manufacture is own organic molecules from carbon dioxide and water. The result of this extraordinary process of endosymbiosis was the eukaryotic cell. These are larger than prokaryotes and have internal structures called organelles: mitochondria are the sites of aerobic respiration derived from the first of the endosymbionts, and chloroplasts are the sites of photosynthesis derived from the second kind of endosymbiont. In some cases, a third kind of endosymbiont became the whiplike flagella used for locomotion in a number of eukaryotes. Furthermore, eukaryotic cells evolved a special vesicle called the nucleus to house and protect the DNA of the genes, and the DNA molecules themselves became incorporated into highly organized structures called chromosomes, which greatly increase the precision of cell division.

The eukaryotic cell is larger, more complex, and capable of greater levels of activity than the prokaryotic cell, and there are many kinds alive today, such as the familiar amoeba and the planktonic algae of the oceans. More significantly still, the cells of a species of eukaryote had the potential to cooperate with each other in ways that prokaryotes cannot, and this ability was to lead to the next great stage in the history of life. Simple multicellular animals, consisting of large numbers of eukaryotic cells, first appear in rocks about 700 million years old. They were an assortment of peculiar forms, superficially resembling jellyfish, sea pens, and crinkly matlike creatures referred to as the Ediacaran fauna after the locality in Australia where they were first discovered. They were all soft-bodied, achieved relatively little diversity, and had disappeared by about 570 million years ago. This moment in time is the beginning of the Cambrian period, a period marked by the Cambrian "explosion." Within a geologically very short period of time of about 10 million years, practically all the main fossil animal groups appeared in the fossil record. There were many kinds of jointed-limb arthropods, such as crustaceans and the now extinct trilobites. There were corals, mollusks, echinoderms (starfish relatives), brachiopods (lampshells), and a number of less familiar groups. All are characterized by having hard and fossilizable external skeletons, the reason, no doubt, for their abundant appearance in the fossil record. But why they appeared so relatively suddenly is not at all clear. Possibly they reflect an increase in oxygen levels at the time, which allowed the evolution of increased body size. Animals with large bodies are more in need of the support of a skeleton than small organisms. At any event, these fossil groups constituted what is in essence the modern marine ecosystem. New groups and subgroups replaced the old, but the general arrangements would have been perfectly familiar to a visiting modern ecologist.

Abundant and diverse as life had become in the seas, the land remained unconquered for a long time. It may be that oxygen levels had to reach a certain level, or perhaps ultraviolet light levels were intolerable. Moreover, land is a difficult environment to attain. The loss of the support of the water demands the evolution of particularly effective skeletal elements, while some sort of waterproofing, special air-absorbing organs, and an ability to avoid or withstand a large diurnal temperature range are necessary. Despite these problems, primitive land plants with simple root systems and internally supported stems are found in rocks about 400 million years old. At about the same time terrestrial invertebrates, such as wingless insects, mites, and spiderlike animals, occur.

The conquest of land by the vertebrate animals, with all that that event foreshadowed, followed in due course. There was a group of fishes called lobe-fins or rhipidistians, many of which lived in shallow fresh water, breathed air, and had their paired fins supported by bones. Around 370 million years ago, one lineage of lobe-fins evolved limbs in place of fins and other adaptations that allowed them to spend at least periods of their life on the wet marshy land surrounding the watercourses. These early tetrapods had a completely new habitat at their disposal, and the great radiation of tetrapods duly commenced. Some remained relatively dependent on freestanding water; their modern representatives are the amphibians. At least one lineage soon evolved more elaborate water-conserving devices, particularly a waterproof skin and a shelled, protecting egg that can be laid on land without drying out. This lineage, the amniotes, includes the great reptile group, with forms such as dinosaurs on the ground, pterosaurs in the air, and

ichthyosaurs and plesiosaurs reinvading the seas. For the whole of the Mesozoic era, which lasted from 250 to 67 million years ago, reptiles dominated the tetrapod habitats.

Two modest groups of tetrapods had also evolved during this time, but they both consisted of mostly small animals. What they had achieved was a high level of activity and an independence of fluctuations in the air temperature. These were the mammals and the birds respectively, the homeothermic or "warm-blooded" animals. At least in the case of the mammals, they were nocturnal and apparently unable to compete with the dinosaurs. The end of the Mesozoic is marked by the end-Cretaceous mass extinction, possibly caused by a meteorite impact, but certainly responsible for the extinction of the dinosaurs and other dominant reptile groups of the Mesozoic. The mammals and birds survived, and within a matter of a few million years had radiated to become the new dominant terrestrial animal life.

It required only the origin of humans, the most recent significant evolutionary event, to complete the story of the evolution of life. A mere four million years ago, a member of the relatively obscure mammalian order called the primates evolved a new mode of locomotion. Bipedalism had certain advantages for an apelike species living in the African savannah. It also had the initially unforeseen effect that it allowed the subsequent evolution of the forelimbs for carrying objects and using tools, and the evolution of a hugely enlarged brain carried on top of the erect torso. Quite why these interconnected anatomical modifications happened is hotly debated. That they were to have an enormous impact on the world's biota in the years to come is beyond dispute.

[See also Biological Realms; and Exotic Species.]

## BIBLIOGRAPHY

### NATURAL SELECTION AND NEO-DARWINISM

Darwin, C. On the Origin of Species by Means of Natural Selection or the Preservation of Favoured Races in the Struggle for Life. London: Murray, 1859. There were six editions altogether, each containing revisions and additions. Uniquely for a book published so long ago, it is still an important primary reference on the subject.

Dawkins, R. The Blind Watchmaker. Harlow, Essex: Longmans, 1986. A very readable account of Darwinian theory, with many good examples.

Depew, D. J., and B. H. Weber. Darwinism Evolving. Cambridge, Mass.: MIT Press, 1996. Quite advanced, but a good account of the contemporary incorporation of molecular genetics into evolutionary theory.

Dobzhansky, T. Genetics and the Origin of Species. New York: Columbia University Press, 1937. The defining point in the combination of Darwinism and Mendelism to create the neo-Darwinian theory.

Gould, S. J., and S. C. Morris. "Showdown on the Burgess Shale." Natural History (December 1998–January 1999), 48–55.

Marshall, C. R., and J. W. Schopf, eds. *Evolution and the Molecular Revolution*. Sudbury, Mass.: Jones and Bartlett, 1996. A series of rather easier essays on the impact of molecular genetics on a number of aspects of evolutionary study.

Mayr, E. *The Growth of Biological Thought: Diversity, Evolution, and Inheritance*. Cambridge, Mass.: Harvard University Press, 1982. Brilliant, readable analysis of the history of the development of evolutionary thought from Darwin to the present.

PATTERNS OF EVOLUTION

Doolittle, W. F. "Uprooting the Tree of Life." *Scientific American* (February 2000), 90–95.

Kemp, T. S. *Fossils and Evolution*. Oxford: Oxford University Press, 1999. Contains an up-to-date discussion of the processes of speciation, taxonomic turnover, and mass extinction as revealed by the fossil record.

HISTORY OF LIFE

Benton, M. J. *Vertebrate Palaeontology*. 2d ed. London: Unwin Hyman, 1996. A standard textbook account, particularly good on the evolutionary radiation of the tetrapods.

Clarkson, E. N. K. *Invertebrate Palaeontology and Evolution*. 3d ed. London: Chapman and Hall, 1993. The clearest standard account of the fossil metazoan invertebrates.

Gordon, M. S., and E. C. Olson. *Invasions of the Land: The Transitions of Organisms from Aquatic to Terrestrial Life*. New York: Columbia University Press, 1995. Comprehensive coverage of all the groups, plants, invertebrates, and vertebrates.

Loomis, W. F. *Four Billion Years: An Essay on the Evolution of Genes and Organisms*. Sunderland, Mass.: Sinauer, 1988. A fairly technical review of the origin and life and living systems, including quite detailed chemistry.

Margulis, L., and K. V. Schwartz. *Five Kingdoms: An Illustrated Guide to the Phyla of Life on Earth*. 3d ed. New York: Freeman, 1998. An excellent overview of the main groups of living organisms, including prokaryote and simple eukaryote groups.

Ward, P. D., and D. Brownlee. *Rare Earth: Why Complex Life is Uncommon in the Universe*. New York: Springer-Verlag, 2000.

—TOM S. KEMP

## EXOTIC SPECIES

An exotic species is one that is found in a habitat in which it did not evolve. Also called *alien species, nonnative species,* or *invaders,* exotic species can refer to all forms of life, including plants, higher animals, invertebrates, and lower forms that include parasites and pathogens. Throughout the history of life on Earth, the movement of species into new niches or habitats has been intimately connected with the spread of life forms and the creation of new species and new assemblages of species and ecosystems. Exotics may colonize empty habitats or niches (for example, following major disturbances such as storms, volcanic activity, and ice ages), or may move into already occupied areas and create new conditions through such processes as predation, competition, and coevolution. Throughout the history of life on Earth, most exotics, along with other associated life forms, have become extinct, but often not before they have contributed genetically and eco-logically to the biota and ecosystems that followed. In a sense, the Earth's present biodiversity is at least in part the end result of over three billion years of exotic species.

One of the best known examples of this process is provided by the Galápagos Islands finches, particularly famous because Charles Darwin's study of them gave him insights into evolution that later appeared in his historic *Origin of Species* (1859). Following the arrival of one species of finch on an island, the birds spread to other islands. On each island the finches changed dramatically in form and habit to enable them to survive under the different islands' varying supplies of food and other environmental conditions. In the process, they created a series of new finch species and modified species associations and ecosystems. [*See* Evolution of Life.]

While the invasions of exotic species have been an integral part of the development of life on Earth, the process has been altered significantly and the rate of invasions greatly accelerated by human activities. As a result, exotic species in large part have ceased to be a "natural" part of the development of biological diversity and instead represent a growing ecological disaster and a major threat to that biological diversity. [*See* Biological Diversity.]

Today there are still some invasions of exotic species that are not necessarily due to human intervention. In the United Kingdom, for example, around 80 percent of the invading species of birds during the past century apparently arrived without help from humans. In North America, again without apparent help from humans, the nine-banded armadillo is rapidly extending its range northward. In both of these cases, human alteration of the environment may be playing some role by creating suitable habitats, but the invasions are basically a natural process. In the case of the armadillo, at least, it appears that the species is in the process of following a warming that has been proceeding since the last ice age. Other species of animals and plants are also gradually extending their range northward for the same reason.

**Introduction of Exotic Species.** By far the greater part of exotic invasions now are due to human activities. Some of these introductions are intentional, because of the perceived benefits to be gained from the introductions. The examples are legion. Crop plants and commercial lumber trees have been introduced intentionally on a global scale. Some of these have indeed proven beneficial, while others have had dramatically negative impacts. Throughout the world, especially in developing nations, indigenous forests have been cleared to plant fast-growing eucalyptus or pine trees. This process is still actively promoted by international development agencies. These plantations have often proven uneconomic or, at best, short lived, and usually fare worse than indigenous species in terms of erosion control, wildlife habitat, and maintenance of biodiversity and ecological services.

To provide sport fishing, trout have been introduced into mountain streams, rivers, and lakes worldwide. These introductions have often provided the desired fishing. In the rare cases where no native fishes were present, these exotic trout represented a net benefit. In most cases, however, the trout preyed on or outcompeted indigenous fishes, depleting the biodiversity and changing the aquatic ecosystem.

Fish hatcheries are a major source of exotic species. In North America, hatcheries have been developed for decades as a way to restock streams that had been overfished and to augment the fish catch for both commercial and sport fisheries. The hatchery programs have been politically popular and many millions of dollars have been spent on them. In recent years, however, research has increasingly shown that exotic hatchery fish have a host of negative effects and often no demonstrable benefits. Among other problems, they dilute the genetic makeup of the native fish, in many cases leading to depletion or loss; they introduce diseases; and, when subjected to careful scientific analysis, most have not proven to increase the fishery yields.

Deer have been introduced in many areas for sport hunting and have often become a major threat to indigenous plants and the habitats of some forms of native wildlife. Efforts to protect indigenous biodiversity have often pitted conservation authorities against the hunters. The impacts of the exotic deer and the resultant controversies are most acute in island situations such as Hawaii and Mauritius.

Exotic game birds have also been very widely introduced for sport hunting, but in general their impact has been more benign than that of introduced mammals. Some have been introduced to fill a habitat niche that was not being used by native game birds. The ring-neck pheasant (*Phasianus colchicus*), introduced in 1881 from Asia, has become successful as a game bird in much of the United States. Since it usually occupies agricultural lands, it appears generally to coexist rather than compete with native wildlife.

Kudzu (*Pueraria thunbergiana*), a very fast-growing vine, was introduced from China into the United States as an ornamental plant in the 1870s, and it was reintroduced widely in the 1930s for soil erosion control. It can grow a foot a day, and it covers anything in its path with a green blanket, blocking out the sun and killing trees and other vegetation in its way. It is now regarded as a major menace to most forms of land use and to biological diversity conservation. The U.S. Forest Service estimates the kudzu infestation in the United States at about 2.8 million hectares (7 million acres), from southern Florida to Massachusetts, and expanding rapidly. In a similar manner, rhododendrons introduced as orna-

mental plants into Ireland from the Himalayas now rate as a major threat to many native plants, especially in southwestern Ireland where they are taking over forest areas and shading out smaller plants and tree reproduction.

Goats and mongooses are intentional exotic introductions that play havoc with island environments worldwide. In a process that started centuries ago, fishermen and other sailors have put goats on islands to provide a source of meat. This is still being done, for example, in the Galápagos Islands. The voracious goats eat virtually anything, removing native vegetation and preventing its regeneration, causing erosion, and damaging the habitat for indigenous plants and animals. Goats are also regarded as one of the major agents by which humans denuded the vegetation of much of the Middle East. Mongooses were introduced onto islands in the Pacific and the Caribbean to control the rats and, in some cases, snakes. However, they found the indigenous birds, mammals, and reptiles easier prey, and they have depleted the native biota on many islands, without much affecting the rats.

Many exotics have been imported for specific uses, such as pets, but have escaped and become established. The state of Florida has been particularly afflicted by tropical birds, fishes, and plants that have escaped from intended captivity. The Asian walking catfish is an example, recently introduced as a pet but now escaped, evading control efforts, and threatening native Floridian freshwater fishes. The African "killer" bee was introduced into Brazil by a beekeeper for breeding experiments with domestic bees, but it escaped. This particularly aggressive bee has killed or interbred with more placid native or domesticated bees, spreading north through South and Central America, and is now entering some of the southern states of the United States. The gypsy moth was introduced into Cape Cod a century ago as the basis of an American silk industry, but it escaped and has become a major pest, threatening forests throughout the eastern United States.

**Unintentional Introductions.** Probably the greatest number of exotic species problems stem from unintentional introductions. Ocean-going ships have carried dirt as ballast. On arrival at port they drop the ballast over the side, and the dirt, with all its associated organisms, is introduced into the local environment. Many ships carry water as ballast, which is also released in or near port, introducing whatever organisms are carried in it. In this manner, for example, the zebra mussel (*Orisena polymorpha*) was introduced to the Great Lakes in about 1988. It has spread throughout most of the Great Lakes and many rivers, including most of the Hudson River basin, and in the decade since its introduction it is estimated to have cost industry over U.S.$15 billion.

## THE AFRICANIZED HONEYBEE

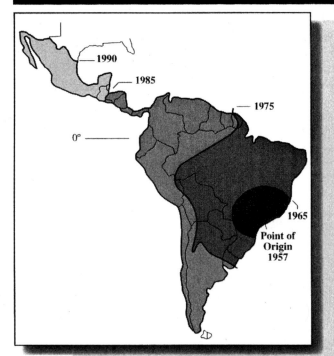

The spread of the Africanized honeybee in the Americas between 1957, when it was introduced in Brazil, and 1990. (Modified after Texas Agricultural Station, in *Christian Science Monitor*, September 1991.)

The growth of world trade and communications has caused a great increase in the number of accidental introductions and subsequent explosions of fauna and flora. Indeed, as C. S. Elton (1958, p. 31) put it, "We are seeing one of the great historical convulsions in the world's fauna and flora."

A recent example of the spread of an introduced insect in the Americas is provided by the Africanized honeybee (*Apis mellifera scutellata*). A number of these were brought to Brazil from South Africa in 1957 as an experiment, and some escaped. Since then, they have moved northward to Central America and the United States, spreading at a rate of 300–500 kilometers per year, and competing with established populations of European honeybees. They reached Texas in 1990, Arizona and New Mexico in 1993, California in 1994, and Nevada in 1998.

The Africanized honeybee, popularly known as the "killer bee," defends its nest (stings) far more than does the European honeybee, swarms more frequently, and is far less selective about where it will nest.

—ANDREW S. GOUDIE

Wherever they have gone people have carried pests, parasites, and diseases. Europeans coming to the Americas brought smallpox, chicken pox, and other European diseases that caused fatal epidemics among the indigenous inhabitants. Rats (*Rattus* species) and the house mouse (*Mus musculus*), carried as stowaways on boats, have been introduced throughout the world.

Very serious damage has been caused by parasites and diseases carried by exotic species that were intentionally introduced. Rinderpest, a viral disease, was introduced to Africa in the late 1800s in cattle brought from Asia, where the disease was endemic. In less than four years, it swept through the continent, devastating the herds of domestic cattle and many wild ungulates. A few years later, some nursery stock brought from Asia to the United States carried a fungus parasite that attacked the American chestnut trees. Although the chestnut had been a dominant tree in eastern U.S. forests, by the 1930s it had virtually disappeared.

**Impacts of Exotics.** Exotics that are predators can wreak particular havoc on native prey species, particularly when no such predator existed prior to the exotic's introduction, so that the native species have not developed defenses against the predation. Cats, dogs, rats, and mongooses, particularly when introduced into is-

land habitats, have decimated native prey species, especially small mammals and ground-nesting birds. In Mauritius, for example, dogs were a major factor in the extinction of the dodo, whose eggs were laid on the ground. The brown tree snake (*Boiga irregularis*) was accidentally introduced onto Guam Island after the World War II, presumably in shipments of derelict vehicles and equipment being salvaged from the Pacific war zone. This snake is a formidable predator, and by 1985 most of Guam's bird species were either isolated in small pockets or were exterminated, the native lizards and mammals were decimated, poultry and other small livestock had been reduced, and over 160 cases of serious snake bites had been reported and, presumably in an effort to eat them, the victims were mostly babies who were bitten while they slept. [*See* Extinction of Species.]

**Competition.** The principle of competitive exclusion states that complete competitors (that is, species that compete directly for the same resource or niche) cannot coexist: one will always exclude the other. Where the introduced species is a superior competitor, this principle operates to the detriment of the native species, with the exotic expanding its numbers and range rapidly and, in the process, threatening the native competitors with extinction. The zebra mussel, noted above, is one such example. Another is the sea lamprey (*Petromyzon marinus*), which invaded the Great Lakes through the new seaway canal system in the 1960s. The Mediterranean annual grasses that were introduced into California in packing materials at the time of the gold rush so completely replaced the native perennials that they created a whole new ecosystem. The exotic kudzu, also noted above, is rapidly outcompeting native vegetation in much of the eastern and southeastern United States, as are the rhododendrons introduced to Ireland.

***New herbivore niches.*** Another form of competition comes from herbivorous species that fill a new niche. As noted, deer and goats have been introduced into islands where no such hoofed herbivore previously existed. The native plants have no defenses against such an animal, and these animals' feeding has decimated native species. The same thing occurred when, in 1839, the Asian water buffalo was introduced into northern Australia as a beast of burden. There were no native grazers of this magnitude, nor any predators of such a large animal, and the buffalo population increased very rapidly. It soon exhausted the native vegetation and there were periodic die-offs from starvation. This process of population eruption followed by a crash is characteristic of introductions of exotic ungulates. [*See* Australia.]

***Diseases and parasites.*** The disastrous spread of exotic diseases such as smallpox in humans and rinderpest in ungulates has been discussed above. In the case of diseases such as smallpox, the population in which the disease was endemic, in this case the Europeans, had developed significant resistance to it. But, when it was introduced to human populations that had no such resistance, the results were catastrophic. The same thing happened with rinderpest. The Asian cattle had developed resistance or even immunity to the disease, but, when it was introduced to African livestock and wildlife that had no such resistance, the results were devastating.

In the same way, parasites such as the chestnut fungus, gypsy moth, and hemlock adelgid were of little concern in their native lands. The chestnut fungus, for example, had little impact on its native European chestnut trees because they had developed resistance to it. But when it was introduced to the United States, where the American chestnut had no such resistance, it virtually exterminated that species.

**What Determines the Impact of Exotics?** Some exotics are successful invaders, usually causing serious impacts, while others apparently never get a foothold or appear to be benign. Several factors combine to create conditions in which an exotic is a successful invader. The species must have characteristics that fit the conditions into which it is introduced, and the recipient ecosystem must be in some way vulnerable to such introductions. The vulnerability of ecosystems varies over time, and stress of some sort can predispose an ecosystem to allowing invaders in. A closed forest may be resistant to exotics that require sunlight until the necessary open light conditions are created by a forest fire, a storm blow-down, or even a falling tree. The sea lamprey invaded the Great Lakes almost as soon as the seaway canal from the ocean provided an entry route, while the zebra mussel did not establish itself for many years, even though the opportunities presumably were there at the same time. Apparently the factors that determine the success or failure of invaders include the interrelations between the biology of species, the properties of the ecosystems at the time, and perhaps pure chance.

In their native habitats, most species have existed for a long period and have evolved together so that the predators, parasites, prey, competitors, and symbionts (species that provide each other mutual advantages) have adjusted to one another. There are limiting factors in such an ecosystem that assure that none of the species involved gets far out of balance. The prey species develop defensive strategies and adaptations to protect themselves from the predators, so that there is a dynamic balance between them. Similar mechanisms are developed between plants, animals, diseases, and parasites. However, when exotic species are introduced into a new habitat where such limiting factors are not present, they have the opportunity to increase without their previous limitations, usually with devastating results on the native biota.

In general, the exotics that have been benign or beneficial have been domestic species of plant or animal that required human care to survive, such as most field crops, or that fitted into an unoccupied niche. But, even in these cases, domestic exotics have escaped and become pests, or their diseases or parasites have done so. People still believe that introductions can bring substantial benefits, and the process of intentional as well as unintentional introductions continues to expand. With the exception of certain domestic species, however, exotic introductions continue to cause far more problems than they solve and, on a global basis, exotic species represent one of the greatest threats to biodiversity and to related human welfare.

[*See also* Biological Diversity; Biological Realms; Biomes; *and the biography of Elton.*]

### BIBLIOGRAPHY

Elton, C. S. *The Ecology of Invasions by Animals and Plants.* London: Chapman and Hall, 1958.

Mooney, H. A., and J. A. Drake, eds. *The Ecology of Biological Invasions of North America and Hawaii.* New York: Springer-Verlag, 1986.

Office of Technology Assessment. "Harmful Non-Indigenous Species in the United States." Publication No. OTA-F-565. Washington, D.C., 1993.

Simberloff, D. "Why Do Introduced Species Appear to Devastate Islands More than Mainland Areas?" *Pacific Science* 49 (1995), 87–97.

U.S. Geological Survey. *Status and Trends of the Nation's Biological Resources.* Reston, Va.: U.S. Department of the Interior, 1999.

Wilcove, D. S., et al. "Quantifying Threats to Imperiled Species in the United States." *Bioscience* 48.8 (1998), 607–615.

—LEE M. TALBOT

# EXTINCTION OF SPECIES

Other than some archaebacteria that may still survive from life's origin on Earth, most organisms that evolved on this planet (more than 99 percent) have become extinct. Extinction is the rule rather than the exception. Massive species extinction is best attributed to the dynamic condition of the Earth. Continents continually move in all directions (up, down, and sideways), global polarity (i.e., the location of the magnetic poles) reverses at roughly 100,000-year intervals, and our planet is constantly (in geologic time) bombarded with projectiles from space. Earth's rotation is gradually slowing (one second every 50,000 years) and the moon is spiraling away from us at a rate of about one and a half inches per year. Subject to such unexpected and irregular assaults and changes, it is no wonder that so many of Earth's organisms do not adapt quickly enough to new conditions to survive.

Organisms that apparently survived for millions of years are those that could tolerate a range of temperatures and changing physical conditions. Thus when extreme climate change occurred after a long, stable environment, some species survived by moving to habitats protected from severe changes. For example, the deep-dwelling coelacanth, a fish discovered in 1938 and living around the Cormoro Islands in the Madagascar Channel, is the sole survivor of at least sixty coelacanth species that thrived during the Mesozoic era, about 245 million years ago. Today's coelacanth has hardly changed from its closest fossil ancestor of 60 million years ago. Evidently coelacanths were initially shallow-water dwellers whose remains were quickly covered with sediments, thereby ensuring good fossilization. At the end of the Cretaceous period (65 million years ago), some coelacanths moved to deeper water where conditions for fossil formation do not exist, and hence there are no coelacanth fossils since that period. Living conditions at a depth of three hundred meters evidently stayed uniform enough for the present line to survive for 60 million years.

If a species cannot find a safe habitat as the coelacanth did, it can survive by being flexible in habitat requirements. Fossil records show that during the Pleistocene epoch in North America (the last one million years) there existed about fifteen species of carnivores, coyote-size or larger; today there are only seven. Gone are the saber-toothed tigers, dire wolves, short-faced bears, huge American lions, and bear-dogs. The flexible coyotes, however, are still expanding across North America in a variety of habitats from California suburbs to New England forests. Some animal species, particularly short-lived and prolific insects, can adapt to change by exploiting through breeding those morphological characteristics that allow them to survive long enough to develop into a new species. The natural extinction of individuals with unsuitable physical characteristics occurs rapidly, in a few centuries or even decades.

Extinctions thus occur when habitat conditions change too rapidly for the occupants to adapt. Species disappear for reasons as varied as the loss of their regular food source and an inability to switch to other diets, or by being unable to adapt to variations in sunlight, temperature, nutrient regimes, parasites, predators, diseases, salinity, and a host of other critical factors. The scale of such extermination ranges from declining single species units succumbing to a species-specific disease or parasite, to whole orders of fauna such as the dinosaurs. Although humans watched the death of the last known passenger pigeon, human-observed extinctions are relatively rare events. For evidence of prehuman extinctions, scientists must depend on fossil dating.

Fossils are the critical evidence of prehistoric life, but conditions necessary to create and preserve them are limited. Plant and animal remains are preserved in ancient tar pits, at the bottom of sinkholes, and in the compacted sediment of sea and lake beds. The latter are brought to the surface by tectonic upheaval. Only the hard parts (bones, teeth, shells, etc.) are preserved; soft parts, except for occasional outlines of soft tissue, almost always disappear. Despite this limitation, however, paleontologists can reconstruct the general appearance of long-extinct animals and plants from their scattered remains.

**Major Extinctions.** The more fossils collected and identified, the clearer becomes the image of the Earth's past life. [*See* Earth History.] The fossil record shows five major extinction events: (1) at the end of the Ordovician, about 440 million years ago; (2) toward the end of the Devonian, about 365 million years ago; (3) late Permian, about 250 million years ago; (4) near the end of the Triassic, about 215 million years ago; and (5) late Cretaceous, about 65 million years ago (Claeys et al., 1992). The greatest extinction occurred at the end of the Permian; more than 90 percent of species in the oceans and about 70 percent of terrestrial vertebrate families disappeared. Insects were also reduced; eight of twenty-seven known orders of Permian insects disappeared. Plant fossils, however, showed relatively little evidence of mass extinctions. At this time, Pangea, the super continent, still existed, but scientists can only speculate on what caused such a catastrophic event and how long it went on: Two million years? One million years? Or less?

This extinction calamity was probably the result of a combination of factors, all with complex interactions. One hypothesis proposes a three-phase sequence: climate changes caused marine basins to dry out, thereby reducing the habitat area of shallow-water invertebrates; concurrently, massive volcanic eruptions may have darkened the daylight; and global warming from a buildup of $CO_2$ in the atmosphere may have depleted $O_2$ in the ocean depths, causing anoxia and further disruption of life (Raup, 1986; Isozaki, 1997).

The next great extinction evidently occurred near the end of the Triassic (roughly 215 million years ago). Recent evidence indicates that contributing to this extinction event could have been a meteorite impact near the border of Botswana and South Africa, which produced a crater about 75 kilometers wide at the start of the Cretaceous, about 145 million years ago (Hecht, 1997a; Renne et al., 1992). Although extinctions were not as great as those of the Permian/Triassic boundary, nevertheless about 20 percent of all groups of species were wiped out. A meteorite, when approaching the Earth at a steep angle, compresses the air before it so densely that at contact the energy released is so great that the meteorite evaporates instantaneously, resulting in a huge explosion that carries the resulting debris high into the stratosphere. There the dust soon circles the Earth, blocking a portion of normal sunlight for several years. This reduction in solar radiation evidently had such a direct effect on photosynthesis and other solar-controlled processes that global climate was severely altered until the atmospheric debris settled.

Another massive global extinction reoccurred almost a million years later at the boundary of the Cretaceous and Tertiary periods (65 million years ago). Strong geologic evidence indicates that this event also coincided with a meteorite striking the Earth near Mexico's Yucatan Peninsula; the impact would inject sixty times the object's mass into the atmosphere as pulverized rock and a fraction of this dust would stay in the stratosphere for several years (Alvarez, 1980). The resulting, mostly submarine crater is about 300 kilometers wide. Other geologic evidence indicates increased volcanic activity during this period, as well as falling sea levels. This extinction event is perhaps best known for the demise of the dinosaurs, which died out because, we surmise, they could not adapt to the changed global climate (Hecht, 1997b). [See Impacts by Extraterrestrial Bodies.]

Other abrupt extinctions occurred among marine organisms. Charles Marshall of University of California, Los Angeles believes that 50 to 75 percent of ammonite species, which resembled the spiral-shelled chambered nautilus of today, disappeared along the west coast of Europe about the time of the Yucatan meteorite impact. Until then they were one of the dominant invertebrate species in the shallow waters of the world's oceans.

Plants, too, were affected by the meteorite impact, particularly in the western interior of North America where there was an almost 80 percent turnover among the large plants and trees across the Cretaceous–Tertiary boundary. Further north, in what is now Canada, the extinction levels were low (Wolfe, 1986; Johnson et al., 1989; Nichols and Fleming, 1990).

The most recent of Earth's major extinctions occurred only ten to twelve thousand years ago. With the rapid improvement in dating fossils, the interval when forty or more large mammal species disappeared in the New World has been cut to as little as five hundred years. Among those lost were giant camels, various species of horses and bison, wooly mammoths, and mastodons.

The cause of these late Pleistocene extinctions is controversial. Some have attributed it to indiscriminate slaughter by human immigrants to North and South America; others to a change in global climate. The role of human hunters will be discussed later, but the timing of their arrival with the disappearance of this megafauna seems to support a direct connection.

The climate changes that most likely ended the dinosaurs' reign 65 million years ago also created new habitats for mammals. These animals soon evolved elaborate strategies for cold-weather survival, including hibernation, caching mast (acorns, beechnuts, etc.) for winter food, and changing winter diet to include the foliage of evergreens and the shoots of hardwoods. Many mammals developed thick coats for insulation against low temperatures. Their rapid evolution was perhaps most remarkable in the New World, where, until the end of the Pleistocene, there was an impressive assemblage of birds and mammals free from human competition on both continents.

**Local Extinctions.** Although global catastrophes happen infrequently, local extinctions of species and replacement by others continues. For example, eastern U.S. hardwood forests were dominated by American chestnut trees (*Castanea dentata*) through the nineteenth century. They grew straight and occupied a large percentage of the canopy. They were valuable as lumber and for their mast crops, an important food for wildlife. Chestnuts became infected with a particularly virulent form of fungal pathogen that arrived in 1904 on a shipment of European chestnut (*C. sativa*) logs. Within thirty years, native chestnuts were infected and ceased to flower, causing certain wildlife species dependent on its nuts to decline or disappear. The canopy gaps were eventually filled with other hardwood species, which in time produced enough mast for the eastern wild turkey to reoccupy its New England range.

The turkey's return illustrates the ebb and flow of plant and animal species following local extirpation.

Human-caused extinction of competitive carnivores often has unexpected results. The killing of all eastern timber wolves in the mid-nineteenth century allowed coyotes from the Midwest to extend their range eastward, so that within roughly one hundred years from the death of the last wolf in New England, coyotes had become an important predator throughout the Northeast. This pattern is now being reversed in Yellowstone National Park, where human-assisted timber wolf reintroduction has reduced the competing coyote population there. It is more difficult to determine the long-term ecological effects of the now absent bison and elk in the hardwood forests east of the Mississippi and how their disappearance influenced the rate at which white-tailed deer occupied their niche. The conversion of hardwood forests to agriculture undoubtedly hastened the local extinction of bison and elk, just as it encouraged the proliferation of deer.

In a similar fashion, but just as difficult to quantify, is the effect of the decimation of the Antarctic blue whale population on the apparent population increase in other krill (*Euphausia* species) feeders such as the minke whale (*Balaenoptera acutorostrata*), crabeater seal (*Lobodon carcinophagus*), Antarctic fur seals (*Arctocephalus gazella*), and krill-feeding penguins. From studying such examples of population fluctuations, scientists confirm that any niche vacated by an organism, for whatever cause, is eventually reoccupied by other adaptable species. It seems that there are always more organisms seeking niches to occupy than vacant niches waiting to be filled.

Many extinctions can occur without humans even being aware of it, especially in the case of bacteria, viruses and other micro-organisms that can disappear unwitnessed. When microorganisms threaten the health of humans or their livestock, such as the brucella bacteria (cause of brucellosis), nations combine resources to develop antibacterial drugs and vaccines to neutralize such assaults or even to eradicate the pathogen. The latter effort is rarely successful but was achieved when the smallpox virus attacked its last human host in Somalia in 1977. This virus exists now only in secure containers at government laboratories.

The examples heretofore cited are evidence that the extinction of one group of species or of an individual species allows other existing sympatric ones to evolve relatively rapidly and expand their ranges, as mammals did when the dinosaurs died out. Similarly, when ammonites disappeared in the ocean at the end of the Permian (245 million years ago), along with about 90 percent of marine species then existing (Raup, 1979), other marine invertebrates replaced them. What is not known is how long these replacement species took to reoccupy the vacant marine niches.

From a geologic perspective, hominids evolved relatively recently, with the genus *Homo* appearing about 1.6 million years ago. Anatomically modern humans appeared quite suddenly about 120,000 years ago, and as social equivalents of contemporary humans only about 40,000 years ago. By then they were skilled hunters who may have contributed to the extinction of scores of large animals in Australia and then in the New World. We do not know why or even if these early humans indulged in such slaughter. In historical times people's motives for overkill were clearer: wolves, for example, were eliminated because they competed with humans for their livestock and game. Bison in North America were virtually extirpated by European immigrants to eliminate a major food source for a competing human population. Smallpox was eliminated to protect human health.

It is difficult even for humans to eliminate a species completely, and they have generally succeeded only locally. Human-assisted total extirpation usually occurs when the species is already a relict, one whose population was already declining. In the New World, the Steller's sea cow and the Labrador duck (now extinct), like the Franklinia tree (no longer found in the wild), were probably all surviving marginally when first confronted by humans.

Hard evidence of human-caused extinction is rare. Paleolithic humans (stone-tool producers from as far back as a million years) were undoubtedly successful hunters, and the remains of animals we assume they successfully killed are found in caves they occupied. Some of their prey animals still survive, but scientists do not know whether those that vanished did so because of direct human intervention or from a combination of causes.

Not until mesolithic times (about fifteen thousand years ago) is there evidence that humans could have had a direct effect on animals' extinction. Subfossil mammoth bones have been found intermixed with stone spearheads, indicating that the beast may have been killed directly by humans. Experimental butchering of an Asian elephant carcass by Smithsonian paleoanthropologists demonstrated that stone blades attached to wood handles were quite adequate to the task.

As the large grazers and browsers were increasingly hunted by humans, the competition between other mammalian predators also intensified. An interesting study reported by Blaire van Valkenburgh showed that the incidence of broken teeth in carnivores such as sabertooths, dire wolves, American lions, and even coyotes was three to five times greater than tooth breakage observed today. In fact, the percentage of broken teeth in sabertooths of forty thousand years ago was relatively less than it was only ten thousand years ago, when the saber-tooths and their prey were dying out. One reasonable conclusion is that there was much fiercer competition for prey animals at the end of the Pleistocene and that the increase in the number of humans at that time competing with other mammal carnivores may have been a primary reason for what is sometimes called the Pleistocene overkill.

There are arguments opposing this viewpoint, however. Some scientists postulate that even gracile human hunters were not so numerous or efficient at killing prey that they alone could have extirpated a whole spectrum of mammal species and that there must have been other factors to account for such a massive die-off. Besides climate change, which occurred when the last continental glaciers retreated north, and hunting, a third possible cause could have been disease introduced to an animal population with no or an impaired immunity (Stevens, 1997). In 1994 about a third of the lion population in the Serengeti died from canine distemper, which normally attacks only dogs. An estimated one thousand lions succumbed to a strain of this disease that crossed from dogs to the cat family (Pain, 1997). Distemper is spread by a morbillivirus in droplets from the mouth of an infected mammal; as it would be unusual for village dogs to get close enough to lions to infect them, there must have been other intermediaries. The spotted hyena, for example, often mixes with lions at a kill site and could have been an intermediate carrier. The Serengeti epidemic died out by the end of 1994 and the lions recovered, but a lethal pathogen introduced by humans or their domestic dogs from the Old World might indeed have been able to eliminate 120 to 130 mammal species more successfully than hunting could have. To accomplish this daunting task, however, the lethal pathogen would have had to (according to Dr. Preston Marx at the Aaron Diamond AIDS Research Center in New York):

1. Kill all age groups of its host rapidly
2. Live on an immune independent host that advanced with or on humans across North America
3. Be lethal to a broad range of species without killing its host, especially if the host was human

Some viruses, such as the one causing distemper, attack a wide range of mammals—lions in Africa or foxes and raccoons in North America—but historically the spread of distemper virus in a wild mammal population generally fades away without eliminating all its hosts. Today's distemper virus, therefore, does not meet the conditions listed.

The search for evidence of paleolithic pathogens is exciting, but surviving evidence is rare if indeed extant. Results of disease or dietary deficiency can be discovered in skeletal material and if the evidence from such bones matches that of current diseases, scientists can use such results to extrapolate the history of those diseases.

As technology has become more sophisticated, human ability to exterminate or reduce animal populations improves. For example, repeating rifles and shotguns dramatically increased the firepower of hunters, just as the advent of powerful winches and large nylon nets increased the catch of commercial fishers. Intensive study of animal life cycles furthermore enables harvesters to uncover weak links in the quarry's reproductive strategy to exploit, control, or even to eliminate species. For example, blowfly larvae cause serious health problems to domestic stock. The female fly lays her eggs on any open wound of the host. When the eggs soon hatch, the larvae grow rapidly by eating the flesh around the wound, thus preventing healing. Careful observation showed that females mated only once, after which they were unreceptive. Discovery of this fact led to raising millions of flies, sterilizing the males by radiation, and then releasing them to mate. Using this procedure, blowflies were virtually eliminated from most of the continental United States.

The indiscriminate use of chemical pesticides to control malaria-carrying mosquitoes was rampant during World War II, when planes sprayed whole islands in the southwest Pacific with dichlorodiphenyltrichloroethane (DDT). Insect control through such spraying, however, is dying out. Only a tiny fraction of the targeted species ever receive a lethal spray droplet, but myriad beneficial insects are also killed in the process. Pesticide-resistant strains of insects soon evolve. Even more damaging than the indiscriminate insect kill was the realization in the 1960s that DDT entered the food chain through ingestion by invertebrates, so that when they were eaten by birds and fish, DDT accumulated in the fatty tissues of the consumers. Ospreys, eagles, and peregrine falcons, at the top of the food chain, became so contaminated with DDT from what they ate that they could not metabolize calcium properly and produced eggshells too thin for successful incubation.

Pesticides improperly or too heavily applied also affect human health, and the developed world restricts their application on crops and around human habitation. It is doubtful whether application of any herbicide or insecticide has totally exterminated a species, unless it was already a relict one. [See the biography of Carson.]

Efforts to control the desert locust (*Schistocerca gregaria*) have been ongoing for centuries in Central Africa's Sahel. Should it ever be extirpated, an unpredictable insect herbivore might replace it and cause even more damage than the original pest. What would then be the fate of those animals dependent on locusts? The planned elimination of any plant or animal, no matter how noxious to humans, is risky, and most scientists are reluctant to endorse such efforts because of the unknown long-term consequences.

The necessity to control a natural population and thereby put it at risk of extinction poses a dilemma. Historically most organism extinctions, caused partially or totally by human activity, were unplanned. John James Audubon and his contemporaries, for example, never imagined that wild passenger pigeons could vanish by the end of their century. A combination of excessive pigeon slaughter and the clearing of these birds' hardwood forest habitat contributed to their extermination; once the pigeon population fell below a critical number, the species was doomed.

Although humans face a dilemma over which of several similar species to save and which to let die, there is little agreement on the criteria for making such a selection. Those chosen for salvation are generally large enough to be easily recognized and ideally possess a cultural identity, for example, the American bison, bald eagle, peregrine falcon, whooping crane, and even such mammalian carnivores as the red wolf and the black-footed ferret.

The extinction of a species is irrevocable. It cannot be recreated, although similar-looking animals can be selectively bred to resemble an extinct one, as the Hecks did in Germany in the 1930s to "recreate" the extinct tarpan, a wild forest horse. Just as the death of an individual represents the total loss of its particular combination of genes, the loss of a whole species, especially when witnessed by humans, is a traumatic event. An opposing attitude holds that no species is eternal and its very existence therefore is transitory. Thus any extinction is regrettable but not earthshaking. These conflicting viewpoints are debated, but most people seem indifferent to a species loss unless it directly affects their well-being. Even when an industry such as cod fishing is threatened by a rapid decline in cod, the fishers tend to blame nonhuman competitors, including the weather, for the decline of their catch. If the catch species becomes "commercially extinct," so be it. The industry then tries to find another fish species still plentiful enough to harvest. [See Fishing.]

**Conservation Efforts.** Fortunately, there are politically powerful conservation groups. With common goals they have pressured governments to ratify international treaties to protect endangered species or threatened sites. The overharvest of ducks and geese, particularly following a series of dry breeding seasons, led to the Migratory Bird Treaty, ratified by Canada and the United States in 1918, Mexico in 1936, Japan in 1972, and the Soviet Union in 1976. So many waterfowl breed in the northern United States and Canada and winter in Mexico and Central America that protecting a species in only one country is ineffective to conserve it. Even in Franklin Roosevelt's administration, conservationists worried about the destruction of tropical forests in Central and South America. As a result, the Convention on Nature Protection and Wildlife Preservation in the Western Hemisphere was signed in 1940 and has now been ratified by about a score of countries that have estab-lished parks and wildlife reserves under its authority. [See Conservation.]

With the end of World War II, international conservation efforts expanded. Although commercial whaling had virtually ceased during the war, the threatened renewal of large-scale commercial whaling by Japan, Norway, Iceland, and the Soviet Union led to the formation of the International Convention for the Regulation of Whaling. In 1946 the fourteen major whaling nations under this convention appointed members to the International Whaling Commission, whose task was to set harvest quotas. Despite an increase of signatory nations to about forty, the enforcement of the convention remains contentious. For example, whales may be harvested above the internationally agreed quota by a signatory if the whales are for "scientific research." Evidence is strong that meat from specimens so designated has ended up in the lucrative Japanese market for this commodity. Great political pressure is exerted by whaling nations to increase quotas on the few remaining commercially harvestable species. At least twelve whale species are already endangered to some degree, including four so rare that it is not even worth looking for them. Nonetheless, despite the rarity of whales, harvesting continues, with yields remarkably close to the quotas set. Quota enforcement has been marginal, and it was not until 1966 that blue whales and humpbacks were given full protection. The North Pacific grey whale has had a harvesting prohibition since 1947 and is thus one of the few whale species whose numbers have increased to a level at which they could be downgraded from endangered to threatened under the Convention on International Trade in Endangered Species of Wild Fauna and Flora (CITES). The process to protect whales by international agreement has been painstakingly slow, but conservation groups, even in Norway and Japan, have begun to be heard by their governments, which eventually may ban all whaling. Even with complete protection for decades, some rare species such as the northern right whale is barely holding its own. With an estimated population of only 350 animals, there might already be too few left to sustain them.

After about a decade of work by the World Conservation Union, which was started in 1948, the CITES was signed in Washington, D.C., on November 3, 1973. It went into effect in July 1975 when the tenth nation ratified it, and now over 130 countries have done so. This convention, if properly enforced, could control the trade in endangered species and their products. Each species so designated is listed in appendices to the treaty and is classified as I, II, or III according to its vulnerability to extinction. CITES signatories meet periodically to assess the condition of listed species and change their classification in the appendices according to their increase or decrease in numbers. The widespread adoption of CITES has made it today the world's most powerful legal tool for protecting endangered wildlife by strict regulation of the market for endangered plants and animals. [*See* Convention on International Trade in Endangered Species.]

With the recent increase in ecotourism, conservation measures have become economically beneficial to developing nations. The Charles Darwin Foundation for the Galápagos Islands (begun in 1959) was one of the first organizations of its kind to concentrate its efforts on a whole endangered ecosystem. It operates in Ecuador as an international organization, and its board includes prominent Ecuadorian scientists and government officials. The foundation works with the nation's park service in studying and protecting the endemic flora and fauna of the Galápagos Islands and seeks to control or eliminate such introduced feral stock as dogs and goats. The governance of the Darwin Foundation was used as a model for the Seychelles Islands Foundation, established in 1979, to protect the endemic flora and fauna of these islands, especially the giant tortoises and the flightless rail on Aldabra. The King Mahendra Trust for Nature Conservation in Nepal followed in 1982, and other private (nongovernmental) conservation associations proliferated throughout Asia, Africa, and Central and South America. Many have been moderately successful in achieving goals in local conservation. During the next millennium, any success in renewing the face of the Earth will ultimately depend on the concern of the local citizenry. No longer will many developing nations be willing to follow dutifully the recommendations of outsiders on how to manage their own resources. Ecotourism, once touted as a significant economic incentive to conserve natural resources, has in some places grown so fast that it has destroyed the very landscape needed to attract foreign visitors. Galápagos tourism, a major Ecuadorian revenue source, for example, is being overwhelmed with fifty thousand tourist visits annually.

In a major effort to protect wildlife, Congress in 1973 passed the Endangered Species Act (ESA), which sought to preserve the habitat on which an endangered species depended. Enforcement of this law has led to considerable controversy between developers and certain federal agencies that are required to confirm that construction projects will not be detrimental to the habitat and survival of an endangered species.

The most notorious controversy engendered by this act was that involving the Telleco Dam and the snail darter. In 1971 the Tennessee Valley Authority (TVA) planned to build a dam in a valley of the Little Tennessee River. TVA was enjoined from starting construction on the grounds that they had not filed an environmental impact statement, as required by law; the U.S. Fish and Wildlife Service believed there to be fish in the river whose habitat would be destroyed by the dam. While construction was delayed, the ESA was passed and about that time a three-inch-long fish that ate snails was discovered in the fast-flowing part of the river above the dam site. Building the dam would flood this portion of the river and destroy the fish's habitat. The battle continued in the courts, where a decision to allow the dam to proceed was overturned by a higher court. In June 1978 the Supreme Court upheld the injunction on the basis of the ESA. However, the Supreme Court ruled that Congress could amend the act to allow exceptions. The dam was ultimately built owing to enormous political pressure. It was a sad day for conservation, but many valuable lessons were learned that were later put to use in such cases as that of the spotted owl, whose habitat was threatened in old-growth Douglas fir forests of the Pacific Northwest.

In the United States, the ESA has become a symbol of federal government interference in the "right" of property owners to do whatever they want with their land. The act has become a rallying point used to defend what its opponents perceive to be an erosion of citizens's rights. As the world becomes more crowded, however, traditional individual rights will have to give way to insure the safety and well-being of the majority. The ESA has had an important effect on development, with most conflicts being resolved through negotiation. The mandatory use of turtle excluder devices on shrimp trawls is one example of how the act was applied to save endangered sea turtles from being caught in shrimp nets. Although large, widely recognized animals receive the most attention when threatened, the ESA makes no size distinction. Small, obscure denizens can be saved for both emotional and practical reasons. Some believe that all organisms hold a right to existence that is ethically not subject to human actions that threaten them. Similarly, all organisms have their own particular kind of beauty or value and should therefore be maintained for everyone's good. Endangered species are worthy of preservation regardless of their direct or even indirect benefit to humans. This concept hinges on the assumption that we are an integral part of the global ecosystem and thereby have an obligation to keep it in balance in order to insure our own survival.

**Extinction Rates among the Five Kingdoms.** In considering the unglamorous snail darter, it is worth comparing the extinction rates of the five kingdoms of life. The monera are the simplest form and have cells without a nucleus. This group includes bacteria, blue-green algae, and viruses. We normally do not think much about monera unless they multiply beyond control. For example, two common bacteria, streptococcus and staphylococcus, are on our bodies almost continually, but our immune system normally keeps them under control. The same applies to the bacteria that live inside us. Viruses are somewhat harder to control because they are so small and can mutate rapidly. The virus that causes Newcastle's disease in poultry, for example, is thought to penetrate the shell of a bird's egg. The ability of the human immunodeficiency virus to alter the character of its protein coat has made it extremely difficult to develop a vaccine to control it. The fossil record is generally not accurate enough to record the disappearance of monera species.

Protista are one-celled organisms with a nucleus that also do not readily fossilize. Because scientists thus cannot determine their distinctive species characteristics, their extinction record is barely known.

Fungi do fossilize, but because they are dependent organisms and are small and structurally simple, they have never been a dominant life form. Paleomycology, still in its infancy, only in recent decades has begun to trace the evolution of fungi. Although fungi have been found on fossil plants as far back as the Devonian (more than 400 million years ago), they seem to have changed little morphologically through time. It is presently extraordinarily difficult to estimate extinction rates among fungi, but as technology improves, scientists will learn more of fungal evolution and consequently their extinctions (Stubblefield and Taylor, 1988).

Plants are the fourth kingdom of life, and because their cells have cellulose walls, their fossil record is relatively complete, although scattered and sketchier than that of most animals, including shallow-water marine invertebrates. Early plants evolved in the Devonian and were slender-stalked rhyniphytes that grew from horizontal rhizomes. They were gradually dominated by more advanced plants, trimerophytes and progymnosperms, which by the end of the Devonian (350 million years ago) had competitively displaced their predecessors by having longer, stronger stalks and roots as well as branches and leaves. They were followed by pteridosporus, which had evolved a better reproductive system. By the late Carboniferous (270 million years ago), conifers appeared and spread because of their efficient adaptation to xeric conditions. The middle Triassic represented a decline in the diversity of vascular plants, but plants in general have suffered considerably fewer extinctions than animals from catastrophic events such as the meteorite impact at the end of the Cretaceous (65 million years ago), which eliminated the dinosaurs but sparked an increase in plant diversity.

According to Knoll (1984), plant extinctions are generally caused by competition from other plants better adapted to changing conditions. Plants are more vulnerable than mobile animals to rapid climate change because they can only migrate slowly to more favorable sites. However, plants and trees do migrate in ecological time as they did during the continental glaciers' ebbs and flows in North America and Europe. Tree species richness today, for example, is much greater in North America, where mountain chains run north–south, than in Europe,

where they run east–west; the tree seed of European hardwoods could not cross the Alps or the Carpathians when squeezed by the advancing glaciers from the north. Plants have evolved elaborate strategies to survive after catastrophic fires, floods, windstorms, and defoliation. They often drop their leaves to conserve moisture, sprout when cut, and alter the soil in which they grow both physically and chemically (Challinor, 1968).

Plant extinctions follow a pattern different from that of animals. Unlike the mass extinctions animals suffered at the end of the Permian (225 million years ago), terrestrial flora changed relatively mildly. There were again only moderate losses at the Cretaceous–Tertiary boundary (65 million years ago), caused primarily by climate change and the competitive effect of the rapid spread of flowering plants. There is no clear fossil record of simultaneous global extinction of land plants, but as Knoll points out, human destruction of tropical forests throughout the world may lead to floral extinctions greater than any natural causes have accomplished so far.

Extinctions in the animal kingdom are better recorded in the fossil record than that of the first three kingdoms, and perhaps even more precisely than that of plants. Although the animal extinction record may be the most detailed, it does not allow good comparison with the other kingdoms, whose fossil records are less complete.

Extinction is the rule: virtually all species eventually die out. The causes are complex and not clearly understood. The role of humans in the process is undoubtedly significant but hard to quantify, and humans are by no means immune from extinction themselves.

[*See also* Biological Diversity; Biological Realms; Biomes; Catastrophist–Cornucopian Debate; Deforestation; *and* Ethnobiology.]

**BIBLIOGRAPHY**

Alvarez, L. W., W. Alvarez, F. Asaro, and H. V. Michel. "Extraterrestrial Cause for the Cretaceous-Tertiary Extinction." *Science* 208 (1980), 1095–1108.

Challinor, D. "The Alteration of Surface Soil Characteristics by Four Tree Species." *Ecology* 49 (1968), 286–290.

Claeys, P., J. G. Casier, and S. V. Margolis. "Microtektites and Mass Extinctions: Evidence for a Late Devonian Asteroid Impact." *Science* 257 (1992), 1102–1104.

Grzimek, B. "18 Equines." In *Animal Life Encyclopedia*, vol. 12, p. 565. New York: Van Nostrand Reinhold, 1975.

Hecht, J. "Big Blast Brought on the Cretaceous." *New Scientist* (March 29, 1997a), 21.

———. "Did Sea-Level Fall Kill the Dinosaurs?" *New Scientist* (April 5, 1997b), 19.

Isozaki, Y. "Permo-Triassic Boundary Superanoxia and Stratified Superocean: Records from Lost Deep Sea." *Science* 276 (1997), 235–238.

Johnson, K. R., D. J. Nichols, M. Atrep, and C. J. Orth. "High-Resolution Leaf-Fossil Record Spanning the K/T Boundary." *Nature* 340 (1989), 708–711.

Knoll, A. H. "Patterns of Extinction in the Fossil Record of Vascular Plants." In *The Extinctions*, edited by M. H. Nitecki. Chicago: University of Chicago Press, 1984.

Nichols, D. J., and R. F. Fleming. "Plant Microfossil Record of the Terminal Cretaceous Event in the Western U.S. and Canada." In *Global Catastrophes in Earth History*, edited by V. L. Sharpton and P. D. Ward. Special Paper 247. Boulder, Colo.: Geological Society of America, 1990.

Pain, S. "The Plague Dogs." *New Scientist* (April 19, 1997), 32–37.

Phillips, M. K., and D. W. Smith. *Wolves of Yellowstone*. Stillwater, Minn.: Voyageur Press, 1996.

Raup, D. M. "Size of Permo-Triassic Bottleneck and Its Evolutionary Implications." *Science* 206 (1979), 217–218.

———. "Biological Extinction in Earth History." *Science* 231 (1986), 1528–1533.

Renne, P. R., et al. "The Age of Paraná Flood Volcanism, Rifting of Gondwanaland and the Jurassic-Cretaceous Boundary." *Science* 258 (1992), 975–978.

Stevens, W. K. "Disease is New Suspect in Ancient Extinctions." *New York Times* (April 29, 1997), pp. C1, C7.

Stubblefield, S. P., and T. N. Taylor. "Recent Advances in Palaeomycology." Tansley Review no. 12. *New Phytologist* 108 (1988), 3–25.

Ubelaker, D. H. *Skeletal Biology of Human Remains*, Smithsonian Contributions to Anthropology no. 41. Washington, D.C.: Smithsonian Institution, 1997.

Van Valkenburgh, B. "Tough Times in the Tar Pits." *Natural History* 103.4 (1994), 84–85.

Wolfe, J. A., and G. R. Upchurch. "Vegetation, Climatic and Floral Changes at the K/T Boundary." *Nature* 324 (1986), 148–152.

—DAVID CHALLINOR

# EXTREME EVENTS

Among the possible consequences of global climate change are changes in the severity and frequency of extreme events such as floods, droughts, tropical cyclones, and severe thunderstorms (Figure 1). [*See* Drought.] These events entail large societal costs, including injury, loss of life, property loss, and impairment of business; and changes in their intensity and frequency are therefore of special concern. In developing nations, loss of life from such events can be extreme. For example, a tropical cyclone in Bangladesh in 1970 killed nearly half a million people, while more than three million have perished in single flooding events in China (Southern, 1979). In developed nations, sophisticated warning technology has greatly reduced loss of life from extreme weather, but at the same time the rapidly expanding technological infrastructure has vastly increased economic vulnerability to extreme events (Landsea and Pielke, 1998). For example, Hurricane Andrew of 1992 was the second costliest national disaster in United States history, with insured losses in excess of U.S.$15 billion (in constant 1990 dollars), while the cost of the Florida hurricane of 1926, normalized to 1990 dollars, was in excess of U.S.$70 billion (Landsea and Pielke, 1998). In this article, the present understanding of the response of the frequency and intensity of extreme events to global climate change is reviewed. [*See* Natural Hazards.]

**Floods.** There are two broad categories of flood: river floods that result from weeks or months of anomalously large rainfall over large portions of drainage basins, and flash floods that typically result from a single convective storm of a few hours' duration. Many considerations enter the estimation of how such events are related to climate.

It should first be recognized that globally integrated precipitation is intimately related to the atmospheric greenhouse effect. When water condenses and falls out of the atmospheric column, latent heat of vaporization is released to the air. Over long periods of time over the globe, this heat release must be compensated by radiative cooling of the atmosphere. If we ignore, for the moment, absorption of sunlight by atmospheric constituents and dry turbulent heat fluxes, then the integrated cooling of the atmosphere is just the difference between the net outgoing longwave radiative flux at the surface and at the top of the atmosphere. This is, in turn, a gross measure of the greenhouse effect. Thus an increasing greenhouse effect implies increasing precipitation and a more active hydrologic cycle. The effect is quantitatively small, however: doubling carbon dioxide, for example, is thought to result in about a 7 percent increase in global precipitation.

Quite apart from changes in global precipitation, changes in the distribution of precipitation in space and time are of concern. Such changes may be brought about by changing atmospheric circulation patterns and changing degrees of atmospheric stability to convection. There are several reasons to suppose that global warming might lead to concentration of precipitation into more localized and intense events. [*See* Global Warming.] First, it has been shown by Bister and Emanuel (1996) that the average degree of convective instability in the atmosphere depends on the thermodynamic efficiency of the atmospheric heat engine. Very simply, the rate of conversion of thermal energy into the energy of convection depends on the difference between the surface temperature and the average temperature at which the atmosphere loses heat to space. Greenhouse warming increases this temperature difference and so increases the kinetic energy density of convective clouds. Precipitation from such clouds should therefore be more intense even though, on average, precipitation episodes would be less frequent. Quantitatively, however, the effect is slight. For example, a radiative–convective equilibrium model that experiences a 2°C rise in surface temperature caused by greenhouse warming shows an increase in the average vertical velocity in clouds of about 5 percent.

There is some indication that global precipitation is becoming increasingly concentrated in more episodic, high-intensity events (Tsonis, 1996). It has not been established, however, that this trend is a consequence of anthropogenic global warming.

**Extreme Events. FIGURE 1.** Hurricane Bret approaching the southern coast of Texas on 22 August 1999, as seen from a U.S. National Oceanic and Atmospheric Administration (NOAA) geostationary satellite. (Courtesy Dave Santek, Space Science and Engineering Center, University of Wisconsin-Madison.)

Theory and several numerical modeling studies suggest that major atmospheric circulation systems in the tropics would increase in intensity in response to global warming. These circulations greatly affect the distribution of precipitation in the tropics and subtropics, enhancing precipitation in equatorial regions and decreasing it in subtropical regions such as the Sahara Desert. Increases in the intensity of such circulations would have the effect of increasing rainfall in regions where it is already plentiful and decreasing it elsewhere. Taken together, these results point to an increase in the occurrence of flash floods, river floods, and drought in association with global warming, but the magnitude of such an increase is difficult to estimate at present, as is the magnitude of global warming itself. [*See* Atmosphere Dynamics.]

**Tropical Cyclones.** We are concerned about three separate aspects of tropical cyclone activity: frequency, intensity, and geographic distribution. Changes in any of these aspects would have major consequences for society.

There is no clear understanding of the many factors that control the intensity of individual tropical cyclones, and consequently there is virtually no skill in forecasting storm intensities even over only twenty-four hours (DeMaria and Kaplan, 1997). On the other hand, there is a definite upper bound on tropical cyclone intensity that can be calculated from knowledge of the temperature of the ocean surface and of the overlying atmosphere. Examination of long records of tropical cyclones worldwide reveals that their intensities obey a nearly universal distribution function such that the rate of occurrence of tropical cyclones of maximum wind exceeding $V$ is given by

$$C(1 - V/V_m),$$

where $V_m$ is the potential maximum wind speed and $C$ is a constant that depends on location and time of year (Emanuel, 1999). The form of this equation appears to be the same in all ocean basins, and so it is reasonable to suppose that it might be invariant with climate change. Then only the parameters $C$ and $V_m$ vary with climate. We can associate $C$ with cyclone frequency and $V_m$ with intensity. Studies of the variations of tropical cyclone frequency and potential intensity have been summarized recently by Henderson-Sellers et al. (1998). There is a general consensus that $V_m$ increases with anthropogenic global warming. The physics behind this increase is straightforward: increasing the greenhouse effect entails a reduction of net upward longwave radiative flux at the surface. To balance the heat budget of the ocean there must be a compensating increase of sensible and latent heat flux from the ocean, implying an increase of either the average surface wind speed or the thermodynamic disequilibrium between ocean and atmosphere. The latter is directly proportional to $V_m$. Estimates of the magnitude of possible increases of $V_m$ vary widely, however. A doubling of atmospheric carbon dioxide is estimated to increase $V_m$ by between 5 percent and as much as 20 percent. This could have serious consequences, assuming that the aforementioned distribution function is universal, even if $C$ remains constant. For example, an increase in the upper bound on intensity from 60 to 70 meters per second would increase the frequency of storms with winds greater than 50 meters per second by almost 70 percent, keeping the total frequency of all storms constant.

Estimates of the overall change in frequency of occurrence of all tropical cyclones have been made principally on the basis of integrating global climate models with increased greenhouse gases. They range from modest decreases to modest increases, depending on the model and on how the model's climate is altered. Because of this result, it is not at present possible to provide useful estimates of how total storm frequency might change. This also reflects a comparative ignorance of the factors determining global tropical cyclone frequency in the present climate.

There is one issue that has been settled rather decisively on the basis of both theory and modeling: the total geographic area afflicted by tropical cyclones would neither expand nor contract appreciably as a result of climate change. This is because tropical cyclones occur almost exclusively in parts of the tropical atmosphere undergoing very slow, large-scale ascent, and are largely absent in regions of large-scale sinking. The dynamics of the tropical atmosphere dictate that roughly half of the tropical atmosphere will experience ascent, with the other half undergoing descent. This will not change appreciably with modest climate change. This result counters the naive view that, since tropical cyclones occur at sea surface temperatures in excess of 26°C and since global warming would expand the area of water with temperatures above this value, so too would the area affected by tropical cyclones expand. (In essence, the 26°C threshold value also increases with global warming; it is not a universal threshold.)

Notwithstanding the absence of change in total surface area affected by tropical cyclones, one might expect some change in the distribution of tropical cyclones within those regions currently affected by them. Unfortunately, global climate models reveal no consistent picture of what such changes might be, since they may depend on comparatively subtle changes in the general circulation of the atmosphere.

Finally, it must be pointed out that individual ocean basins, especially the Atlantic, experience large decadal fluctuations in tropical cyclone activity, quite independent of global climate change (Landsea and Gray, 1992). The magnitude of such decadal changes is so large as altogether to eclipse any changes that might be attributable to global warming, and so policies and procedures related to tropical cyclone preparedness must naturally be more concerned with such decadal fluctuations.

**Severe Winter Storms.** As in the case of tropical cyclones, we are interested in possible changes in the frequency, intensity, and distribution of winter storms, which provide much of the precipitation that falls outside the tropics and which are associated with rainstorms, blizzards, ice storms, and severe coastal and marine windstorms. Winter cyclones are powered by the energy that resides in the general pole-to-equator temperature gradient in the lowest 10 kilometers or so of the atmosphere, although they are also influenced by atmospheric stability, condensation of water vapor, and surface fluxes of heat, momentum, and moisture. Cyclonic storms are often concentrated in a few relatively narrow belts, or storm tracks.

One universal feature of global-warming simulations in climate models is the relatively large warming that takes place at the poles compared with the tropics. Thus the average pole-to-equator temperature gradient is reduced. While this is true at the Earth's surface, the opposite is the case high in the atmosphere, where equatorial warming exceeds polar warming in most simulations. Theory also suggests that this should be the case. Since the growth rate of baroclinic cyclones (winter storms) depends on temperature gradients at the surface and in the upper troposphere, it is not obvious whether they should increase or decrease with greenhouse warming.

A closer look at the energy budget suggests that the velocities associated with baroclinic eddies should scale both with some tropospheric average pole-to-equator temperature difference and with the pole-to-equator moist static energy gradient. The moist static energy is a measure of the total nonkinetic energy content of air. The gradient of this quantity increases with greenhouse warming. This would have the effect of increasing the velocities associated with equator-to-pole heat transport, offsetting to some extent the effect of the possible reduction of the equator-to-pole temperature difference.

The idea that the intensity of winter storms might increase as a result of greenhouse warming is borne out by the few studies that have been performed using general circulation models. Lambert (1995) found that the frequency of intense winter storms increased in a global climate model subject to doubling of carbon dioxide, even though the frequency of all cyclones diminished. Lunkeit et al. (1996) found an increase in extreme wind events in a coupled ocean–atmosphere climate model subjected to rising greenhouse gas concentrations. Neither study found appreciable shifts in storm tracks.

**Severe Thunderstorms, Hail, and Tornadoes.** Hail, tornadoes, and local damaging winds are all associated with atmospheric convective systems whose horizontal dimensions are far too small to be simulated by global climate models. Thus such models are incapable of making direct predictions of the frequency, intensity, or geographical distribution of such events. We must at present rely on indirect inferences about changes in such phenomena, based on explicit simulation of the large-scale conditions known to be conducive to them.

We know both from direct observation and from numerical simulation with specialized, cloud-resolving numerical models that the severity of convective storms depends on certain combinations of atmospheric stability and the vertical shear of the horizontal wind. Large amounts of instability to convection coupled with strong vertical shear, particularly in the lowest 2–3 kilometers of the atmosphere, are conducive to severe thunderstorms. These conditions arise from regional circulations, and so it is unlikely that tabulations of their global averages will prove useful in assessing changes in severe convective storm activity. Rather, one needs to know about changes in regional circulations and thermodynamic conditions in certain regions such as central North America, which is unusually prone to severe thunderstorms.

Unfortunately, global models are very inconsistent with one another in predictions of regional responses to anthropogenic climate change. Thus, even if one comes to accept predictions of changes in global average quantities, models cannot yet be regarded as credible in their predictions of regional change. This greatly hinders projections of changes using models. It is also difficult to detect changes in their frequency from past observations. Thunderstorms are too small to be detected systematically with the standard observational network, and one must rely on meteorological radar for such systematic detection. In the United States, a comprehensive network of radars has been in place since the 1950s, but little attempt has been made to detect trends in convective radar echoes. Over the past decade, these radars have been replaced by modern Doppler radars, which are more accurate and can also detect air motion within thunderstorms.

Severe thunderstorms, hail, and tornadoes in the United States are heavily concentrated in the Midwest and Plains states. Changing demography and levels of education have had such strong effects on reporting of tornadoes and hail that inferences about any long-term trends in such phenomena are not possible. The advent of satellites, Doppler radar, and trained storm spotters over the past two decades, however, now make it certain that virtually every significant tornado or hail event in the United States will be detected, so that trends in such activity will henceforth be observed.

[*See also* Climate Impacts; Scenarios; *and* Sea Level.]

## BIBLIOGRAPHY

Bister, M., and K. A. Emanuel. "Moist Convective Velocity and Buoyancy Scales." *Journal of the Atmospheric Sciences* 52 (1996), 3276–3285.

DeMaria, M., and J. Kaplan. "An Operational Evaluation of a Statistical Hurricane Intensity Prediction Scheme (SHIPS)." Preprints of the 22d American Meteorological Society Conference on Hurricanes and Tropical Meteorology. Boston, 1997, pp. 280–281.

Emanuel, K. A., Influence of Climate Processes on Tropical Cyclone Potential Intensity. Preprints of the 23d American Meteorological Society Conference on Hurricanes and Tropical Meteorology. Boston, 1999, 1073–1076.

Henderson-Sellers, A., et al. "Tropical Cyclones and Global Climate Change: A Post-IPCC Assessment." *Bulletin of the American Meteorological Society* 79 (1998), 19–38.

Lambert, S. J. "The Effect of Enhanced Greenhouse Warming on Winter Cyclone Frequencies and Strengths." *Journal of Climate* 8 (1995), 1447–1452.

Landsea, C. W., and W. M. Gray. "The Strong Association Between Western Sahelian Monsoon Rainfall and Intense Atlantic Hurricanes." *Journal of Climate* 5 (1992), 435–453.

Landsea, C. W., and R. A. Pielke, Jr. "Trends in U.S. Hurricane Losses, 1925–1995." Preprints of the American Meteorological Society Symposium on Global Change Studies. Boston, 1998, pp. 210–212.

Lunkeit, F., et al. "Cyclonic Activity in a Warmer Climate." *Contributions to Atmospheric Physics* 69 (1996), 393–407.

Southern, R. L. "The Global Socio-Economic Impact of Tropical Cyclones." *Australian Meteorological Magazine* 27 (1979), 175–195.

Tsonis, A. A. "Widespread Increases in Low-Frequency Variability of Precipitation over the Last Century." *Nature* 382 (1996), 700–702.

—Kerry A. Emanuel

# F

## FAMINE

Famine is a protracted total shortage of food in a restricted geographical area, political unit, or cultural group, causing widespread disease and mass deaths from starvation. In most instances, a famine cycle (crop failure, lack of food, social disruption, moral degeneracy, migration, epidemics, and death from starvation) is triggered by restriction of food imports to food deficit areas or crop failures in traditional food surplus areas (Figure 1). The primary human factors have been food denial related to war, political decisions, poverty, and inadequate transportation infrastructure. The primary natural factors have been crop failures resulting from frost, floods, drought, violent natural events, insects, and plant diseases. Throughout history, famine has rivaled pestilence and war as a scourge of humankind.

Specific and general locations of high-frequency famine regions have shifted globally as civilizations or nations emerged, flourished, and declined and as food demands in certain places exceeded food production or the ability to import food. Case studies on critical regions (Figure 2) have proven that acute food crises, starvation, and famine are not short-term critical events but must be understood as the results of longer-term processes. Seeds of famine require time to mature. The threat to certain groups of humans increases slowly, and the possibility for successful food crisis management becomes more and more limited. The famine process displays site-specific stages of development that are characterized by the growing vulnerability of social groups to food shortages, up to the point where mass starvation becomes famine.

Famine affects people of all ages, sexes, and nationalities. The first to die are those who live in poverty, in a society lacking equity, in a culture that does not encourage education, and in a place with unsustainable population or population growth. Famines have appeared in the world's best agricultural regions, in all natural zones and climatic regions, and have not been restricted to one cultural area, nation, or racial group. The map of world famine regions in Figure 2 shows the general areas affected by more than eight hundred famines spanning four millennia. Although famines occurred throughout the famine zones in all time periods, the highest percentage of total famines in each time period occurred in very specific regions. The map identifies the location of six major famine regions and the "Future Famine Zone."

**Modern "Four Horsemen."** Concerns over world food production, chronic hunger-undernutrition-malnutrition problems, and unstable world food reserves reflect the best instincts of many individuals in affluent societies. These concerns are compounded by fear of the modern "Four Horsemen" of famine: uncontrolled population growth in food deficit regions, abrupt climatic perturbations (El Niño) and climatic change, revolutionary dietary expectations, and misguided political decisions by self-centered and culturally biased individuals, political parties, or religious leaders. Famines have claimed the lives of more people between 1981 and 2000 than at any other time since World War II. Famines have taken place in the Sahel and Ethiopia, Mozambique, Bangladesh and northeastern India, Kampuchea and Vietnam, Central America, and the highlands of eastern Africa. With high population growth in areas of political and economic instability, illiteracy, disease, pollution, massive rural-to-urban migration, and gross inequalities in food production and food purchasing capability, humankind is facing a never-ending struggle against famine.

**Seeds of Famine.** It is conservatively estimated that at least 800 million people in developing countries are chronically undernourished, and possibly 200 million people in developed countries suffer from mild undernourishment and aspects of malnutrition. Per capita daily caloric intake in developing countries averages about twenty-two hundred (up from two thousand per day in 1980), compared with thirty-four hundred calories in developed nations. The average minimum daily requirements vary according to physical size, climate, type of work performed, and sex of the individual. Per capita annual food supplies for direct human consumption in developing countries are 235 kilograms (kg) of cereals, 16 kg of meat, and 36 kg of milk. Per capita annual food supplies for direct human consumption in developed countries are 146 kg of cereals, 81 kg of meat, and 199 kg of milk. People in developing countries are more dependent on cereal grains than people in developed countries. Cereal production is about 30 percent of total agricultural production in developing countries and less than 22 percent in the developed countries. The International Food Policy Research Institute (IFPRI) estimates that 8 percent of those who suffer from chronic undernutrition live in the Near East and North Africa, 13 percent in Latin America and the Caribbean, 16 percent in East Asia, 24 percent in South Asia, and 37 percent in sub-Saharan Africa.

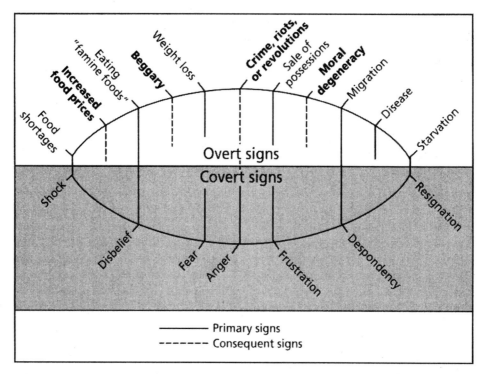

**Famine. FIGURE 1.** The Modern Starvation Cycle. (Modified after Dando, 1980. With permission of Arnold Publishers.)

**World Food Production.** As a result of improved agricultural technology, practices, and policies, the world's agriculturalists produce more than twice as much food in the 1990s as they did in the 1950s. World grain production increased 60 percent between 1969 and 1990, and grain supplies could provide every person on Earth with at least twenty-eight hundred calories per day. World food production per capita also increased but not equally in all countries of the world; production per capita was lower in underdeveloped countries than in developed countries. The amount of arable or potentially arable land that is under cultivation varies markedly from one continent or one country to another, and there are opportunities for increased productivity. World food trade and food imports have tempered food problems in many underdeveloped countries and, as such, are not evidence of internal national agricultural failures. For many developing countries, it is economically efficient to allocate portions of total agricultural investments for the production enhancement of high-value indigenous export commodities that command hard currency and import low-value, high-bulk supplemental food commodities. Moreover, both developed and underdeveloped countries that import and export food commodities are capable of substantial increases in food production and reductions in food losses. What has caused the present and potential famine crises in many countries of the world is not a world shortfall in food. Famine in the future will be related to decisions by political leaders,

poverty, a worldwide unprecedented and accelerating demand for food, civil strife, religious prejudice, and ethnic antagonisms, all combined with a pervasive urban bias in national food policies.

**A Modern Political Famine: China 1959–1961.** Agriculture in China was socialized in the 1950s and organized to serve the interests of the state. Application of the "Stalin Model for Agricultural Development" was achieved in three phases: (1) land, animal, and tool redistribution stage, 1949–1952; (2) cooperative-socialist stage, 1953–1957; and (3) commune stage, commencing in 1958. Social disruption associated with the collectivization of agriculture, the commune system, and abuses associated with the zeal generated by the "Great Leap Forward Campaign" of 1958 upset China's intricate traditional farming system. Peasant apathy and indifference, diversion of nearly 40 percent of the agricultural workforce to nonagricultural activities, removal of one-third of the agricultural land from food production, and a severe drought created an agricultural crisis of catastrophic proportions.

Grain production dropped from more than 200 million metric tons to 150 million, and the vast country experienced the horrors of its first socialist, almost national famine. Food shortages in urban areas, serious in late 1959, reached a critical point during April and May 1961. Twenty million urban residents were ordered into the countryside to ease food shortages in urban areas; food rationing (below human requirement) was ordered

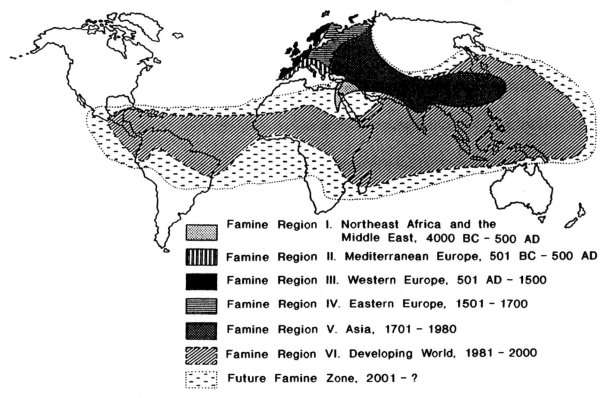

**Famine. FIGURE 2.** World Famine Regions, 4000 BCE–2001 CE. (Modified after Dando, 1980. With permission of Arnold Publishers.)

Famine Region I. Northeast Africa and the Middle East, 4000 BC – 500 AD

Famine Region II. Mediterranean Europe, 501 BC – 500 AD

Famine Region III. Western Europe, 501 AD – 1500

Famine Region IV. Eastern Europe, 1501 – 1700

Famine Region V. Asia, 1701 – 1980

Famine Region VI. Developing World, 1981 – 2000

Future Famine Zone, 2001 – ?

for urban dwellers in noncritical occupations. Threatened with the first nationwide famine in Chinese history, action was taken by the Chinese Communist Party (not Mao) to rectify the errors in national agricultural decision making, to conserve food, and to save as many lives as possible.

Domestic food supplies were able to provide less than eighteen hundred calories per day to each working citizen. Grain imports were increased from 2.6 million metric tons in 1960–1961 to 5.8 million in 1961–1962. Stringent food rationing was enforced, military food rations reduced, and the quality of the food provided lessened. Malnutrition and undernutrition caused widespread night blindness, famine edema, famine diarrhea, liver disorders, tuberculosis, amenorrhea, beriberi, and a general rise in mortality. The natural population increase in China in 1960 dropped to zero, famine-related deaths exceeded 6.8 million, and the calculated total loss of population as a result of the 1959–1961 famine was 32.1 million.

In the millennia of Chinese history, famines have always been regional or local, never nationwide. Of the three major causes of massive loss of life in China (famine, pestilence, and war), famine remains the most subject to social control and elimination. Mao Zedong stressed that the state could organize the nation's re-

sources for the common good. It was only when the agricultural disaster of 1959 was followed by a second monumental setback in 1960 and millions of people had died, that the Communist Party leadership modified Mao's plan, implemented drastic measures to feed those in need, and restored agricultural productivity. The Chinese famine of 1959–1961 was a state-created, unintentional social catastrophe. This famine was the precursor model for future "political" famines in developing nations led by a dictator or driven by an ideology.

**A Modern Media Famine: Ethiopia 1984–1985.** In the mid-1980s millions of Africans living within the Sahel—a narrow semiarid zone stretching three thousand miles from Senegal to Sudan—were confronted with death, and hundreds of thousands of inhabitants of countries outside this zone experienced a similar fate. What began as a series of local food shortages became a horrible multinational human catastrophe. The Great African Famine of 1984–1985 claimed the lives of at least 300,000 people in Ethiopia, and millions there have not been accounted for. At the same time, at least 200,000 people died in Mozambique, 50,000 in Senegal, and thousands in Chad. An estimated 2.5 million suffered from malnutrition and related diseases in Mali, and 1 million Sudanese suffered from acute food deprivation. Precise dimensions of this human tragedy are unknown because governments are unwilling or unable to provide accurate statistics. The triggering physical factor that led to this international famine was two vast belts of drought that spread across the continent; the basic cultural factors were overpopulation in marginal agricultural areas, pervasive poverty, abuse of a delicate physical environment, government mismanagement, and civil strife. Warnings of impending disaster were issued by many scholars, international development agencies, and national planners. These warnings were ignored by government leaders in most countries. [See Drought.]

Thousands of emaciated and destitute refugees from famine-stricken Ethiopia streamed across the border into neighboring Sudan each day, placing an intolerable strain upon the Sudanese government and people. More than 1 million Ethiopian refugees sought food, fuel, and medicines in Sudan. Sudan soon faced a threat of famine among its own people—a horrendous price to pay for an open-door policy and a concern for those in need. The threat of a continent-wide famine was averted by the actions of national and international relief agencies stimulated by television media reports. Yet insensitive national leaders, inexperienced government administrators, political expediency, cultural pride, and religious intolerance have created the setting for more famines in Africa.

**A Modern Silent Famine: North Korea 1995–1999.** A claustrophobic political system, a faltering state-controlled food production system, a crippled industrial economy, and a series of devastating natural disasters have produced a modern "silent" famine in the Democratic People's Republic of Korea. Out of a population of 23 million, it is estimated that 16 million suffered from malnutrition and undernutrition, 2–5 million were considered to be at high risk of dying, and at minimum 800,000 were killed by famine. Many of those who died were elderly, children, individuals or groups considered nonessential to the communist government, and urban inhabitants who had no access to foraging off the bleak land. The famine reached its most devastating stage in late 1996 and early 1997. It led to acute malnutrition and stunted the growth of two out of every three children under the age of five. North Korean authorities were successful in maintaining order and preventing large refugee flows, yet at least 400,000 fled the famine-torn nation for China. The model for the North Korean famine is the great Ukrainian famine of 1946–1947 under Stalin's rule and the Chinese Famine of 1959–1961 (or Mao's famine) during the Great Leap Forward, when political control prevented movement of refugees, external fact finders were denied entry into the country, and the state assumed control of famine amelioration activities. Because the world knew little of these two famines, they are called "silent" famines. This silent famine can also be termed a national "carpet" famine, for 90 percent of North Koreans survived on a state-determined food ration insufficient to maintain human health. Ubiquitous suffering and chronic malnourishment threaten millions.

The Democratic People's Republic of Korea retains a Stalin-model communist system of government under the leadership of Kim Jong Il. Since its war with South Korea in the 1950s and the collapse of communism in the former Soviet Union and eastern Europe, North Korean leaders have chosen a path of national isolation and self-reliance, or chuch'e. It maintains 1.04 million troops, and the military accounts for 25 percent of the nation's annual budget. Causes of the famine were multifaceted and were related to human-made and natural disasters. Food shortfalls were related to years of economic mismanagement, failed collective farming practices, political posturing, and redirection of resources toward the military. Coupled with antiquated political decision making, one natural disaster after another has plagued the country. Hail destroyed much of the food crop in 1994, floods reduced yields in 1995 and 1996, and a severe drought and typhoon destroyed most of the corn crop and much of the rice crop in 1997. After four consecutive years of reduced food crop production, of loss of life, and of human suffering, the leaders of North Korea realized the long-reaching repercussions and the gravity of the food supply situation. The government increased efforts to secure food imports from Thailand, Vietnam, and China, and from relief agencies in the developed world; 3.5 million metric tons of donated food raised national diets to subsistence nutritional levels. Those who have survived, weakened and malnourished, will have difficulty resisting disease and infections. The extent of death and suffering in North Korea may never be known; the famine's victims have died in silence, out of view of the world's media.

**How Many People Can the Earth Feed?** Thomas Robert Malthus, in his *First Essay on Population* (1789), wrote, "the power of population is indefinitely greater than the power in the earth to produce subsistence for man." Malthus concluded that unless family size was regulated, famines would eventually become a global epidemic and over time result in the destruction of civilization. He could not foresee the agrotechnical revolution, transportation innovations, global migrations, and a one-world food economy. Nevertheless, world population estimates and projections to the year 2050 give rise to both guarded optimism and deep concern. Population in the developed countries of the world is increasing slowly and in some cases declining. A number of developing countries have made great strides toward population stability. Unfortunately, in most developing countries of the world, population growth continues at a level that threatens quality of life; in these countries hunger is a way of life for millions, poverty exists to the detriment of family life, women's rights are minimal, and large numbers of citizens are migrating internally and internationally. The world's population is increasing by, at minimum, 90 million people annually. Demographers believe that the world's population in the year 2025 will be about 8.3 billion, approximately 34 percent more than the estimated 6.2 billion people on Earth in 2000. To the dismay of all who are concerned about human physical, mental, and economic development, most of the population growth in the future will occur in developing countries. [See the biography of Malthus.]

Demographers disagree about the maximum number of people the Earth can support. Predictions of the world's carrying capacity depends on assumptions of future food productivity levels. The world's carrying capacity is subject to uncertain physical and biological conditions, technical and logistical deficiencies, ecological constraints and feedbacks, economic problems and limitations, and political and cultural restrictions. The United Nations Food and Agriculture Organization estimated in the 1980s that under optimal conditions the world could support 33 billion people. A more recent estimate, considered more reasonable, is that sufficient food could be produced to support 10 to 15 billion people—if political and socioeconomic conditions are favorable. [See Population Dynamics.]

**Famine Mitigation.** To eliminate famine, the world's leaders must accept the concept that an adequate diet is a basic human right, strive to implement means to continually increase food supplies, insure that all households have adequate incomes to purchase necessary foods, develop regional and international systems to provide early famine warning and emergency amelioration activities, and maintain resilience and flexibility in famine relief programs. No social or economic problem facing the world today is more urgent than famine. While this distressing event is not new, its persistence despite the remarkable technical and productive advances of the twentieth century is difficult to understand.

A successful program to reduce famine incidence includes an international early warning system. Four key components of a successful system are: (1) detecting and predicting widespread food shortages; (2) forecasting or informing decision makers in regions, nations, or areas of the impending threat; (3) communicating to the world a warning message of an impending famine and needed relief; and (4) creating global food reserves and a means to quickly deliver food to needy famine victims. Unfortunately, any early-warning system is involved in symptoms rather than causes. Effective famine mitigation is always problematic in societies that lack re-

sources and those in which politics or religion dictate who will live or who will die.

**Twenty-first-Century Urban Famines.** The United Nations currently estimates that 45 percent of the world's population consists of urban dwellers. Half of the world's population will live in urban areas by 2005, and by 2025, 60 percent of the world's population will be urban. Seventy-five percent of those who live in more developed countries of the world and 37 percent of those who live in less developed countries are urban dwellers. In 1994, 2.6 billion people lived in urban areas, .9 billion of these in more developed countries and 1.7 billion in the less developed countries. The world's urban population is growing at a rate of 2.5 percent per year, three times that of the world's rural population. [See Urban Areas.]

Living in large cities within the developing world are millions of new urbanites who exist on the edge of survival—a phenomenon that can be termed "subsistence urbanization." A small reduction in the consumption of food replaces, for many, acute malnutrition with death by starvation or disease. Traditional local foodstuffs are becoming scarce and expensive. The rural elite generally have moved to cities, and the country's agricultural sector has lost a critical portion of its capability to feed those who live in cities. Urbanization without rural development eventually forces urban leaders to rely upon imported food to maintain the urban population. Any disruption in food deliveries from developed countries could result in food shortages, food riots, revolution, mass starvation, or urban famine.

**Prospects of Future Famines.** Most of those who died of famine in the 1980s and 1990s did not die because the world lacked food or the capacity to produce food. Famines in the 1980s and 1990s were caused by complex political and religious conflicts that resulted in war, national policies that discouraged food producers, or political leaders who would not seek or accept food relief. In Africa, portions of Latin America and the Caribbean, and segments of South Asia, food consumption per capita has declined in the past decade. Growth in world food production is also being hindered by environmental degradation, climatic changes and weather perturbations, and diminished investments in agricultural research and rural infrastructure. If the leaders of the world continue to ignore the population and food issues that confront them, famines are inevitable in the next century. Seeds of famine, particularly dreaded political and urban famines, are being sown in countries of the world once believed to be exceptional places for people to live and once perceived as being without hunger and famine problems.

[See also Agriculture and Agricultural Land; Carrying Capacity; Catastrophist–Cornucopian Debate; Climate

Change; Desertification; Food; Global Warming; Human Populations; *and* Migrations.]

### BIBLIOGRAPHY

Avery, D. T. *Global Food Progress 1991.* Indianapolis: Hudson Institute, 1991.

Bhatia, B. M. *Famines in India: A Case Study of Some Aspects of the Economic History of India.* Bombay: Asia Publishing House, 1967.

Brown, L. *Full House.* New York: Norton, 1994.

Cartledge, B. *Population and the Environment.* Oxford: Oxford University Press, 1995.

Dando, W. A. "Man-Made Famines: Some Geographical Insights from an Exploratory Study of a Millenium of Russian Famines." *Ecology of Food and Nutrition* 4 (1976a), 219–234.

———. "The Soviet Famines of 1946–47." *The Great Plains-Rocky Mountain Journal* 5.3 (1976b), 15–21.

———. *The Geography of Famine.* London: Arnold, 1980.

———. "Famine in China, 1959–1961: Some Geographical Insights." In *China in Readjustment,* edited by C. K. Leung and S. S. K. Chin, pp. 231–249. Hong Kong: Centre of Asian Studies, University of Hong Kong, 1983a.

———. "Biblical Famines, 1850 B.C.–A.D. 46: Insights for Modern Mankind." *Ecology of Food and Nutrition,* pp. 231–249, 1983b.

———. *World Hunger and Famine.* Indiana State University, Department of Geography, Geology, and Anthropology, Professional Paper no. 20. Terre Haute: Indiana State University Press, 1995.

Dando, W. A., and C. Z. Dando. *A Reference Guide to World Hunger.* Hillside, N.J.: Enslow, 1991.

Food and Agricultural Organization. *Urbanization and Hunger in the Cities.* Madrid: FAO, 1986.

Greenough, P. R. *Prosperity and Misery in Modern Bengal: The Famine of 1943–44.* New York: Oxford University Press, 1982.

Lanagan, K. "Hungry Nations in a World of Plenty." *Farmline* 4.8 (1983), 16.

Maass, W. B. *The Netherlands at War: 1940–1945.* New York: Abelard-Schuman, 1970.

Malthus, T. R. *First Essay on Population, 1798.* London: Macmillan, 1926.

Ruowang, W. *Hunger Trilogy.* London: M.E. Sharp, 1992.

World Bank. *Population and Development.* Washington, D.C.: World Bank, 1994.

World Resources Institute. *World Resources: A Guide to the Global Environment.* Oxford: Oxford University Press, 1996.

—WILLIAM A. DANDO

## FIRE

[*This entry comprises two articles on fire. The first article provides an overview of fire characteristics, how humans have used fire, the chemical reactions caused by fire, the ecological effects of fire, and addresses the use of fire as a management tool. The case study explores the environmental consequences of fires caused by drought and forest degradation in Indonesia.*]

## An Overview

Fire has been a conspicuous feature of the Earth's environments for a very long time. Both volcanoes and lightning have been, and continue to be, important causes of ignition of vegetation. Present-day evidence of this can be seen in the volcanic flows on Hawaii igniting *Metrosideros polymorpha* communities, lightning strikes associated with dry thunderstorms in many parts of the world (Figure 1), and charcoal present in sediments both of recent origin and of great age. Interestingly, some of the ignitions of recent large-scale forest fires in Indonesia have been attributed to exposed coal seams that have been smoldering for decades.

Today, fires in natural vegetation occur in parts of every continent except Antarctica, and fires have touched almost every biome (a major regional ecological community) to a greater or lesser extent. Some regions are, however, more prone to fires than others (see Chandler et al., 1983). Northern European temperate de-

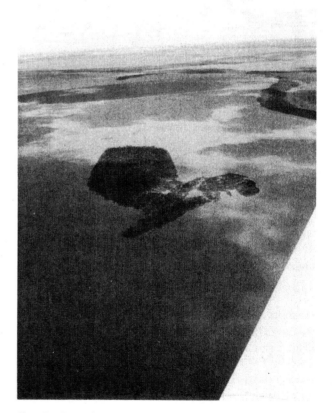

**Fire: An Overview. FIGURE 1.** One of many small fires resulting from lightning strikes during the wet season in Cerrado (savanna) at Emas National Park, Brazil. (Photograph courtesy of Mario Barroso.)

- Temperate forests across Europe typically experience few fires per year (167 per million hectares), but they average less than a hectare each, giving an estimated return period of six thousand years.
- In the Alaskan Taiga, there are few fires per year (an average of 2.7 per million hectares), but they average 1,800 hectares each.
- The wet eucalypt forests of Australia have an intermediate occurrence (sixty-six fires per million hectares) and an intermediate average size (167 hectares), giving an estimated return period of forty-three years.

(From Chandler et al., 1983.)

### Ignition

The climate of a region determines the typical season and frequency of ignitions, because lightning storms are the principal source of fires. Human-caused ignition interferes with this natural pattern. For example, in savannas, extensive human-caused fires early in the dry season leave little vegetation unburned by the time lightning storms occur early in the wet season.

### Fuel

Climate, soils, and plant productivity combine to determine the amount of biomass available as fuel for a fire. These factors also determine the rate of recovery of biomass after a fire. The amount of fuel and the rate of its recovery can, in turn, determine fire intensity, season, and frequency.

The plant community contributes to flammability. The fuel type (e.g., abundant small twigs and leaves) and distribution (e.g., well-aerated litter layer, abundant dry matter suspended in shrubs) can determine the likelihood of ignition and the subsequent rate of spread and intensity. The essential oils present in many plants can increase the flammability even of green leaves and stems.

Climate controls flammability directly by determining the rate of desiccation of biomass and litter, hence determining the season of burning. Climate at the time of fire (e.g., strong winds) affects fire intensity and rate of spread.

### Topography

The rate of spread and intensity of fire can be determined by topography because fires spread more rapidly uphill than downhill. Also, discontinuities such as rivers, lakes, and even small rock outcrops can cause patchiness of fire.

—ANDREW S. GOUDIE

ciduous forests rarely experience fires; when they do occur, they are small in area and of low intensity. Further south in Europe, oak-dominated Garrigue communities in the Mediterranean region burn regularly. Savannas in the seasonal wet–dry tropics across the world are noted for high fire frequencies (Figure 2), with large-area, fast-moving fires typically occurring late in the dry season; the Australian eucalypt forests and the Californian chaparral are well known for high fire intensities that often threaten suburban houses; grasslands and shrublands are noted for rapid rates of spread of fire; and the extensive northern taigas experience fire only infrequently but it can occur over massive areas. [See Biomes.]

Fires differ greatly from one another. They can differ in many respects, which, together, define the typical fire regime of an area (see the vignette, Fire Regimes). Components of the fire regime include: intensity; time between fires (frequency); season; type (crown, ground, peat); and extent (area covered) and patchiness within the burned perimeter. Typical fires in different regions may differ in some or all of these characteristics. Within a region, or even within a small local area, there can be considerable variation in fire regime. In the vast savanna-like areas of South America, closed-canopy gallery forests alongside watercourses burn less frequently than the surrounding savanna-like vegetation

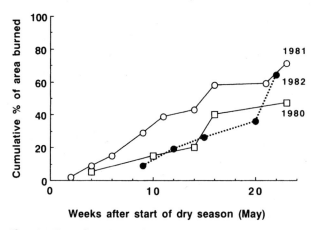

Fire: An Overview. FIGURE 2. Cumulative area of savanna grassland burned through each of three dry seasons (1980–1982) in the northern seasonal tropics of Australia. (Data from Andersen, 1996.)

(e.g., Kellman and Meave, 1997). Similarly, monsoon rainforest alongside rivers and watercourses in northern Australia resists the frequent fires that burn through the surrounding savanna (Andersen et al., 1997). In the United States, savanna-like longleaf pine (*Pinus palustris*) and wiregrass (*Aristida* species) communities burn frequently in low-intensity fires, while adjacent sand pine "scrub" (*Pinus clausa*) requires infrequent but high-intensity fires for its perpetuation (Myers and Ewel, 1990). Within a single area, fire characteristics will vary from one fire to the next.

What causes variations in fire regime? Many factors interact to determine the range of fire regimes that occur in a particular area. A source of ignition is clearly needed before any fire can occur. Once ignition happens, sufficient fuel is required to allow a fire to "catch" (fuel is typically plant biomass that is fine, well aerated, and dry enough to burn). The amount of fuel, the moisture levels, the humidity, the wind speed and variability, and the topography then all combine to determine the nature of the fire as it burns.

**Human Interactions with Fire.** Fire has been elegantly described as "the first great force employed by man" (Stewart, 1956), reflecting both the widespread use of fire by humans for millennia in some parts of the world, and also the impact that this burning has had on the environment. Reasons for burning were (and are) many, including clearing of undergrowth for ease of travel and removal of pests, facilitation of hunting by attracting concentrations of game into recently burned areas, and stimulation of flowering and fruit production by food plants. The range of possible human associations with fire has been comprehensively described by Pyne (1982, 1991, 1995). It is important to note that the time span of human habitation varies enormously

among continents, from more than one million years in Africa (Pyne, 1995) to more than fifty thousand years in Australia (Roberts et al., 1990) to only tens of thousands of years in some areas in the Americas (Coutinho, 1990).

These sorts of aboriginal burning activities undoubtedly led to an increased frequency of ignitions on native shrublands, grasslands, and forests and hence a fire regime that was much altered from the prehistoric situation. This alteration would inevitably have produced changes in the vegetation, and some authors have argued that some changes may have been massive (e.g., Flannery, 1994; Pyne, 1995). While the magnitude of the effects of aboriginal burning on vegetation change will continue to be debated, it is clear that frequent burning activities would not have been distributed uniformly across a region, but would have been concentrated in sites of frequent habitation and movement.

The complex interactions between human activities, fire, and plant communities that must have occurred historically are well illustrated today by large-scale fires in tropical regions. In recent years (1997–1998 in Southeast Asia and in Amazonia), human-set fires have been able to burn into rainforest areas because human activities such as selective logging or slash-and-burn agriculture have altered otherwise fire-resistant forest to make it more flammable. When these activities coincide with particularly dry climatic conditions associated with El Niño events, the resulting fires can affect large areas.

An important lesson that comes from studies of the deliberate introduction of burning to a plant community previously not exposed to fire is that it can rapidly create and maintain a more flammable group of plant species, thereby increasing the likelihood of subsequent burning and substantially altering the previous plant community. For example, clearing for forestry and for shifting agriculture in Indonesia can result in the domination of a site by *Imperata cylindrica* (a grass species), which outcompetes colonizing tree seedlings, dries out rapidly, and burns readily, carrying fires frequently into adjacent fire-resistant forest (Makarim et al., 1998).

Fire was adopted as a potent tool in broad-scale land clearing associated with European settlement in many continents (Africa, North and South America, Australia). Initially, this would certainly have increased the fire frequency in areas being settled, as deliberately lit fires escaped into surrounding forest. However, in all continents, the European attitudes to fire led to policies of effective containment of deliberate fires and widespread suppression of other fires.

One important consequence of successful suppression of natural fires, especially in dry and seasonal environments in which decomposition rates of fallen litter and wood are slow, is the buildup of flammable plant material. Subsequent fires, when they inevitably do oc-

Fires are one of the most important forces in the human environment and can cause loss of life and property. Humans have often attempted to control and suppress fires and this has frequently been a major feature of forest management policies. However, it can be argued that while fire suppression can reduce the frequency of fires, it can also cause especially fierce and hazardous fires by allowing stocks of inflammable material to buildup. The wisdom or otherwise of fire-suppression policies reached public prominence in 1988 because of devastating fires in Yellowstone National Park. Fires began in June and did not die out completely until the onset of winter in November. Somewhere between 290,000 and 570,000 hectares burned in by far the worst fire since Yellowstone was established as the world's first national park in the 1870s.

Was this inferno the result of a policy of fire suppression? Without such a policy the forest would burn at intervals of 10 or 20 years because of lightning strikes. Could it be that the suppression of fires over long periods of, say, one hundred years or more, allegedly to protect and preserve the forest, led to the buildup of abnormal amounts of combustible fuel in the form of trees and shrubs in the understory? Should a program of burning be carried out to reduce the amount of available fuel?

Fire suppression policies at Yellowstone did indeed lead to a critical buildup in flammable material. However, other factors must also be examined in explaining the severity of the fire. One of these was the fact that the last comparable fire had been in the 1700s, so that the Yellowstone forests had had nearly 300 years in which to become increasingly flammable. In other words, because of the way vegetation develops through time (a process called *succession*), very large fires may occur every 200–300 years as part of the natural order of things. Another crucial factor was that weather conditions in the summer of 1988 were abnormally dry, bringing a great danger of fire.

Romme and Despain (1989, p. 28) remarked in the conclusion to their study of the Yellowstone fires that the severity of the conflagrations was caused by favorable weather conditions occurring at a time when the forest was at a very flammable stage in the succession cycle. They suggested that if fires occur naturally at intervals of a few hundred years, then a few decades of effective fire suppression could hardly be sufficient for excessive amounts of combustible fuel to accumulate.

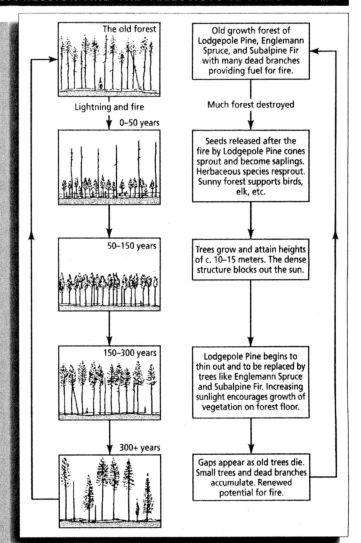

Ecological Succession in Response to Fire in Yellowstone National Park. (After Romme and Despain, 1989, pp. 24–25.)

**BIBLIOGRAPHY**

Romme, W. H., and D. G. Despain. "The Yellowstone Fires." *Scientific American* 261 (1989), 21–29.

—ANDREW S. GOUDIE

cur, can therefore be of very high intensity and a serious threat to human life, properties, and installations. In many environments, these high-intensity fires are quite likely to be far more threatening to humans and human values (e.g., property, farms, timber resources, water quality) than to native plant and animal species and to the ecology of the native communities.

The human response to these postsuppression, catastrophic fire events was generally to devise and apply methods of hazard reduction, mostly by reducing the amount of fuel available to a wildfire. By the 1950s, hazard-reduction burning was becoming accepted in some parts of the world as a most effective tool for large-area reduction of fuel loads. This technique involves deliberate burning of vegetation at a time of year when the fires can be controlled. A mosaic of blocks of vegetation is deliberately burned each year, with a rotation sufficiently short to ensure that fuel loads remain low enough to permit containment of a future wildfire.

In summary, the history of human use of fire in many fire-prone regions has been one of altering the fire regimes. Aboriginal use of fire will generally have increased the fire frequency and perhaps broadened the range of seasons of burning from the prehistoric regime, at least in some areas of the landscape. The area affected by each fire is likely to have declined, on average. Early European settlement will have produced a further increase in fire frequency and broadening in the range of seasons of burning, associated with land clearing, followed soon after by a decrease in fire frequency as fire suppression was practiced. More recently still, broad-scale hazard-reduction burning programs in some regions of the world have shifted the fire regime yet again so that high-frequency, cool-season burning predominates. As discussed below, there is growing concern that the widespread application of hazard-reduction burning may be a greater threat to biodiversity than occasional high-intensity fires, and more sensitive techniques of hazard reduction are being sought.

**Physical Effects of Fires.** Fire is basically a chemical reaction by which energy is released from a stored form in biomass. It is essentially the reverse of photosynthesis, in which energy is captured and stored as carbohydrates in plant tissues as plants grow. Fire consumes biomass, releasing the stored energy as heat and releasing carbon, nitrogen, oxygen, and other elements such as sulfur and phosphorus to the atmosphere. Elements are released in two forms: as gases, or contained within particulate matter in the smoke (Raison et al., 1985). The release of these elements by fire represents a significant source of carbon and trace gases that is of concern in relation to global climate change. However, post-fire plant recovery is rapid in many fire-prone ecosystems. Post-fire photosynthesis may therefore quickly compensate for losses of carbon caused by a fire, but this is yet to be quantified in many ecosystems and would be expected to vary both among different types of fire and among different plant communities.

Fire causes net losses of some nutrients from ecosystems as these are converted to gas and "exported" in smoke. In calculating these losses, it is important to note that the soil (and its roots and other organic material) represents a considerable nutrient store. Hence the losses of some nutrients in a fire might be only a small proportion of the total, even if it is a large proportion of the above-ground biomass (e.g., Kauffman et al., 1994).

Fire can play an important role in the redistribution of some nutrients within an ecosystem. [See Biogeochemical Cycles.] Phosphorus, for example, is stored in plant biomass, but is normally withdrawn from dying leaves before they drop to the ground. By scorching leaves and causing premature leaf drop, burning redistributes the phosphorus to the soil surface. Thus, after fire, phosphorus levels in the soil surface can be substantially increased and become available for plant growth, even if there has been an overall net loss of this element from the ecosystem.

The role of fire in the nitrogen cycle is somewhat different and more complex. [See Nitrogen Cycle.] Much nitrogen may be lost from an ecosystem during burning (Frost and Robertson, 1987, reported a range from 4 to 33 kilograms per hectare for tropical savannas). However, in many fire-prone plant communities, fire also stimulates the germination of legumes, which fix nitrogen through the rhizobium symbiosis. Thus, in some areas, high-intensity fires may both cause an immediate loss of nitrogen from the system and also stimulate the process for recovery of nitrogen. Post-fire nitrogen fixation may compensate for losses in a fire within a year (Medina, 1982).

Another physical consequence of fire is the removal of the protective cover for the soil that is provided by leaf litter (which impedes surface water flow and absorbs the direct impact of raindrops) and a shrub and tree canopy that intercepts raindrops. As a consequence of the removal of this protective cover, massive soil erosion can follow fires, and there are some awe-inspiring examples of this in the western United States (see Chandler et al., 1983, chap. 7). In other sorts of soil, such as sandstones, fire can be an agent of weathering, causing exfoliation, pitting, and fracturing of the rock surface.

The increase in the amount of phosphorus available in the soil surface layers after fire, described above, can also have deleterious effects. If there is heavy rain soon after a large-scale fire, the pulse of nutrients running off into rivers, lakes, and reservoirs can cause algal blooms, with a resultant decline in water quality. [See Phosphorus Cycle.]

Figure 3 summarizes the multiple possible effects of fire on soil properties, hydrology, and geomorphology.

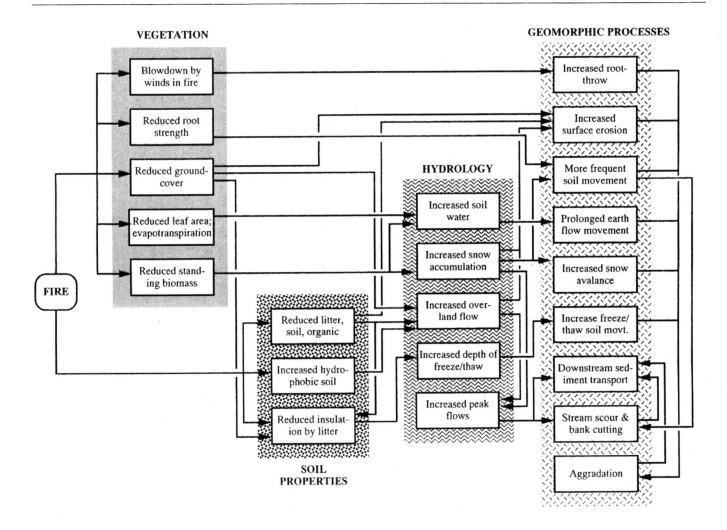

**Fire: An Overview. FIGURE 3.** Effects of fire on soil properties, hydrology, and geomorphology, both directly and indirectly through effects on vegetation. (Modified from Swanson, 1981.)

**Ecological Effects of Fires.** The ecological effects of a particular fire depend upon the characteristics of the fire (e.g., intensity, rate of spread, ground versus canopy, time since the previous fire, season) interacting with the adapted characteristics of the organisms (Whelan, 1995; Bond and van Wilgen, 1996).

*Fire regimes and adaptations.* Evolution is expected to have provided organisms in a particular region with features that allow them to tolerate a type of fire that might cause extinction elsewhere. Thus a particular fire that is typical of the type of fire that occurred throughout the evolutionary history of the biota in the area would not be expected to have devastating effects. In contrast, a high-intensity, canopy fire that burns a community that has never experienced fires (or has experienced only low-intensity, patchy, ground fires) would be expected to cause substantial ecological disruption.

Many characteristics of plants and some features of animals in fire-prone ecosystems clearly facilitate survival through even the most spectacular fires. For plants, these characteristics include (1) thick, insulating bark, (2) specialized underground stems, (3) resprouting from roots, (4) burial of a seed store in the soil, and (5) protection of a seed store in woody fruits in the canopy (see Gill et al., 1981; Wein and MacLean, 1983; Booysen and Tainton, 1984; Whelan, 1995; Bond and van Wilgen, 1996).

Some examples of resprouting from stems that are insulated by thick, corky bark are striking indeed. Most of the woody tree and shrub species in the Brazilian *cerrado* (woody savannas) resprout in some way, and many produce large new leaves soon after fire. Likewise, many trees and shrubs in Australian heathlands and eucalypt forests resprout from trunks and branches within days of even high-intensity canopy fires.

Even if above-ground parts of plants are killed by the heat of a fire, the soil is a good insulator (and most of the heat of a fire is directed upward, not downward). Many plants therefore resprout again from the roots, much as they would after severe grazing or pruning. In

A

B

C

**Fire: An Overview. FIGURE 4.** Survival of Fire by Plants Can Take Many Forms. (A) Resprouting from lignotuber of *Persoonia laevis*, a woody shrub (family Proteaceae) in eastern Australia. (B) The large (5 centimeters long) woody follicle of *Xylomelum pyriforme* (family Proteaceae) in southeastern Australia protects the seeds from the heat of a fire. (C) Seedlings germinating soon after fire, from seeds that were released from follicles opening as a result of being burned.

Mediterranean climate regions, in particular, resprouting from lignotubers or burls is very common (James, 1984; Figure 4A). This structure is below ground but is a specialized stem rather than a root. Many monocotyledons (groups of flowering plants, including, for example, orchids, grasses, and lilies, characterized by seedlings having only a single primary leaf—cotyledon) that have their growing points deeply buried in leaf bases or below ground simply keep growing after fire.

Whether or not established plants are killed in a fire,

potential offspring can be protected either in insulating fruits or by being buried in the soil. Some plants protect their seeds in woody fruits, such as the "closed-cone" pines in the United States and the Mediterranean region and several genera of angiosperms, especially in Australia and South Africa (Figure 4B). Finally, seeds that are incorporated into a soil-stored seed bank are well protected since soil temperatures do not generally rise above the lethal temperature for seeds, even 2 centimeters below the surface. In many ecosystems, it is therefore common to see a flush of germination soon after a fire (Figure 4C).

Adaptations to fire are more difficult to define for animals. Although animals can survive the passage of even a fast-moving, high-intensity fire in refuges such as burrows, crevasses, and rock outcrops, the use of such areas can hardly be interpreted as evolved responses to fire. Some forms of animal behavior have been seen as adaptive in fire-prone environments. Rather than exhibiting panic and undirected fleeing, some animals clearly seek out safe areas within their habitat, sometimes doubling back through the flame front to the relative safety of the already burned vegetation (Christensen, 1980; Whelan, 1995). This is the case not only for mammals. Insects, too, can seek out small areas that provide safety from the heat of a fire. Gandar (1982) found that grasshopper densities in unburned patches of African savanna increased enormously immediately after a fire, as individuals fleeing the fire front found the unburned refuges.

***Reestablishment of populations.*** Mortality of individual plants and animals can be substantial in a fire. Although there have been few attempts to quantify the proportion of animals killed in a particular fire, corpses are found, and a large number of species may be represented among the fatalities.

The finding of many individual animals dead and injured as a direct result of a fire does not necessarily equate to a decline or disappearance of the population. There are various means by which populations of plants and animals can reestablish after fire: (1) continued (or even increased) reproduction by survivors within the burned area, (2) spread from refugia within the burned area, and (3) recolonization from outside the burned area. Which of these reestablishment mechanisms occurs after a particular fire will depend not only on the type of fire but also on the evolved characteristics of the organisms. For example, a high-intensity crown fire can eliminate some tree species from North American conifer forests, but stimulate the release of viable seeds from closed cones of other species, whereas a lower-intensity ground fire may permit both species to survive. Population reestablishment by recolonization from outside a burned area appears to be a rare strategy in many fire-prone ecosystems (Whelan, 1986, 1995).

Many of the species that are able to tolerate fires have also evolved to exploit the favorable post-fire conditions. Fire-stimulated flowering, with seeds released soon afterward, and fire-stimulated germination (Figure 4B) are the most obvious forms of exploitation by plants. These characteristics ensure that seedlings appear at a time when they are most likely to become established plants, when light and nutrient availability are high, competition is reduced, and herbivore densities are perhaps low, during the immediate post-fire period.

An extreme form of this sort of fire response is seen in plants that have been termed *fire weeds*, which appear from seeds or tubers immediately after fire, grow rapidly, flower, reproduce, and then disappear again—to reappear only after the next fire. Such species do not actually disappear from the site during the period between fires, they are simply hidden from view.

The characteristics of some plants have even been interpreted as facilitating fire, using it for their own benefit. Some of the pines in North America, for example, produce large amounts of well-aerated, highly flammable litter—pine needles and branchlets. This type of fuel certainly encourages the ignition and spread of fires, which are easily tolerated by the pine trees, facilitate germination and establishment of pine seedlings, and eliminate understory hardwoods that threaten to outcompete the pines.

Some animals, too, have evolved to depend on fires. A most interesting example is a group of beetles known as the *fire beetles*. The larvae of these animals burrow in dead branches, and fire is an obvious creator of this resource. The adult beetles detect forest fires from a distance and fly in to lay their eggs in still-smoldering twigs and branches (Evans, 1966; Wikars, 1995).

***Survival in the post-fire environment.*** Surviving the fire itself is just the first step for many organisms. The post-fire environment provides many challenges for survival, growth, and reproduction. For plants, the climate during the post-fire period can cause mortality of seedlings. The seedling is perhaps the most vulnerable phase in a plant's life cycle, and even a brief hot and dry period soon after germination can cause mortality. This is most significant for those species in which fire kills the adult plants and releases seeds from dormancy (obligate seeders). Maintenance of the population in the site therefore depends entirely upon the survival of some of the seedlings.

Seedlings are also susceptible to herbivory—a small mouthful can kill a seedling, whereas it would not affect an established plant. A number of studies have shown that herbivory can have potent effects on plant populations and communities in recently burned vegetation. For example, Figure 5 shows that mortality is roughly equivalent in sites that are burned but not grazed, grazed but not burned, and neither burned nor grazed, but it is substantially increased in sites that are both burned and grazed. A marked change in the plant community can be a consequence of this sort of interaction, with palatable plant species declining in abundance and being replaced by unpalatable species. Many decades of small-area burning on Rottnest Island (Western Australia) have converted a community characterized by almost impenetrable *Acacia rostellifera* tall shrubs to a prickly, low *Acanthocarpus preissii* scrub. With each fire, the small wallaby that is abundant on the

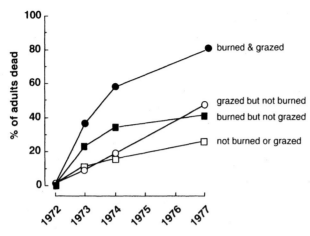

**Fire: An Overview. FIGURE 5.** Herbivory can have a much greater impact on individual plants and plant communities after fire than in unburned vegetation. Leigh and Holgate (1979) showed that mortality of the shrub *Daviesia mimosoides* was greatest when plants had been burned but were exposed to grazing. (Reproduced from Whelan, 1995. With permission of Cambridge University Press.)

island (*Setonyx brachyurus*) grazed selectively on the acacia seedlings stimulated to germinate by the fire.

For those animals that survive the passage of a fire, there are additional challenges to be met. Fire changes the structure and composition of the habitat for animals through its effects on plant individuals, populations, and communities. There is typically increased predation after fire, and appropriate foods may have been eliminated. Thus many studies of small mammals have found that individuals may survive the fire well, and be recorded living in recently burned areas, only to succumb soon afterward to predation and/or starvation.

Survival in the post-fire environment may require a change in diet. For the woylies (small rat-like kangaroos [marsupials]) studied by Christensen in Western Australia, the animals that survived the experimental fires were feeding extensively on fruiting bodies ("truffles") of fungi, which are apparently facilitated by fire, whereas they fed mostly on seeds and leaves prior to the fires. As another example, the common, mouse-sized dasyurid marsupial *Antechinus stuartii*, which occurs throughout eastern Australia, is predominantly an insectivore. Populations are typically not eliminated by fire but decline soon afterwards, probably for a combination of reasons including increased predation and lack of food during the post-fire winter. In areas where fire stimulates mass flowering of plants over extensive areas in winter (see Figure 6), the *Antechinus* has been observed to feed extensively on the nectar (Figure 7).

***Inappropriate fire regimes.*** It is tempting to conclude that communities containing species with obvious fire-tolerance characteristics are adapted to fire, and that fire can therefore be used as a management tool with no risk of detrimental ecological effects. This is a dangerous, simplistic conclusion. Even with fire-tolerance characteristics, inappropriate fire regimes can nevertheless cause local extinctions of plants or animals. Obligate seeder species die in a fire but their seeds are protected in the soil or in canopy-stored fruits—a well-known characteristic that is commonly seen as a fire adaptation. However, if a second fire were to follow before the seedlings had matured and accumulated their own seed bank, the species would disappear.

Even for resprouter species, too-frequent fires can prevent recruitment, while numbers of adult plants are slowly and perhaps imperceptibly depleted. For example, savanna fires in northern Australia cause mortality rates of canopy trees ranging from 1 to 15 percent. When fires are frequent, they prevent newly germinated indi-

**Fire: An Overview. FIGURE 6.** Stimulation of Flowering by Fire is a Common Occurrence. The monocotyledon *Xanthorrhoea resinosa* in sedgeland in eastern Australia flowers en masse in the year following fire but rarely without fire. (Photograph by P. Tap; reproduced from Whelan, 1995. With permission of Cambridge University Press.)

Fire: An Overview. Figure 7. The small, insectivorous marsupial *Antechinus stuartii* exploits the fire-stimulated flowering of *Xanthorrhoea* in the winter after fire, revealed by the footprints left on the flowering stalk.

viduals from recruiting and becoming mature trees (Andersen et al., 1997).

The mortality of established plants may vary according to the season of burning. The physiological state of a tree at the time of burning appears to determine its fire response. Of course, season of burning interacts with fire intensity, because fires in the hot-dry season are likely to be more intense than cool-season fires. The intensity that is apparent to humans may be misleading, however, as a tree species with thick insulating bark may tolerate the hottest of fast-moving fires but succumb to a much lower-intensity, slow-moving fire that burns through the organic layers of the soil and kills roots.

In some environments, fire-prone and fire-intolerant plant communities intergrade. Examples of this include monsoon rainforest along streams and the fire-prone savannas of northern Australia; hardwood hammocks and the surrounding fire-prone longleaf pine sandhill com-munity in Florida; patches of rainforest and surrounding fire-prone eucalypt forest in eastern Australia; and closed-canopy gallery forest and the surrounding fire-prone *cerrado* in subtropical Brazil. Inappropriate fire regimes can alter the balance between the two vegetation types in examples such as these, with too-frequent fires in the more fire-prone community pruning back the boundary of the fire-intolerant community.

**Fire and Land Management.** Humans have been using fires for millennia, and some of these fires have clearly been set with the intention of altering the environment. The deliberate use of fire, and attempts to prevent fire, continue today in many parts of the world. Why? There are two main reasons. First, fires can have beneficial effects in themselves. For example, fire may be used to stimulate flowering of plants used in export wildflower industries, it can be used to maintain habitat required for the conservation of particular animal species, or it can be used to maximize numbers of game animals used in hunting (e.g., quail in the southeastern United States and grouse on Scottish moorlands; see Hobbs and Gimingham, 1987). Second, deliberate fires can be used to reduce the likelihood of occurrence and/or the intensity of a subsequent wildfire that would otherwise damage resources such as timber quality, water quality, livestock, human properties, and lives in rural and urban areas.

*Objectives for applying fire in management.* The first question that should be aimed at any proposal to use fire in management is "What for?" What are the objectives of this form of management? Like any management tool, the deliberate application of fire (or deliberate suppression of fire, or deliberately doing nothing) should be aimed at achieving a clearly stated objective, and the degree to which the objective is attained needs to be monitored and assessed.

There are many possible objectives, because there are many land uses in areas prone to burning. Table 1 summarizes some of these. The prevention of uncontrolled, high-intensity wildfires is important for achieving many of these objectives. For example, large areas of eucalypt forest in Western Australia have been managed for many decades so as to maximize the production of high-quality timber. It was observed in these, and many other, forests that high-intensity wildfires reduced timber quality, even though the trees themselves could survive fire, because of fire damage to trunks. One objective of management is therefore to reduce the incidence of trunk-damaging wildfires. As described above, the strategy of suppressing all fires as soon as possible after detection was found not to be a sustainable long-term solution, because wildfires eventually did occur and they were very intense indeed. It was in this context that a management prescription of applying frequent, low-intensity, cool-season, controlled fires was devised.

**Fire: An Overview.** TABLE 1. Summary of Some of the Possible Objectives of Management for Which Fire Might be Employed as a Tool

| MANAGEMENT OBJECTIVE | USE OF FIRE IN ACHIEVING OBJECTIVE |
| --- | --- |
| Forestry | Rotational hazard-reduction burning to prevent widespread high-intensity fires that would damage the resource (sawlog-quality timber)<br>Removal of tree species in competition with desired timber species<br>Control of soil-dwelling pathogens<br>Stimulation of regeneration of desired, commercial tree species |
| Wildflower harvesting | Maximization of production of inflorescences on woody perennial shrubs |
| Water resources | Maximization of runoff without erosion<br>Rotational hazard-reduction burning to prevent widespread high-intensity fires that would cause erosion and nutrient runoff into reservoirs |
| Primary production | Stimulation of "green pick" for stock in rangelands<br>Control of parasites and pathogens of livestock |
| Urban | Reduction of fuel loads around houses, subdivisions |
| Nature conservation | Perimeter hazard-reduction burning to protect neighbors<br>Maintenance of particular species/assemblages that require a specific fire regime<br>Control of soil-dwelling pathogens<br>Creation of wildflower displays |

SOURCE: From Whelan (1995), with permission of Cambridge University Press.

Termed hazard-reduction burning, this prescription was designed to ensure that the amount of flammable biomass (fuel load) in the forest would never build up to sufficient levels to sustain a damaging wildfire.

This management prescription for the deliberate use of fire is also appropriate for achieving other objectives, where prevention of large-scale, high-intensity wildfires is important, such as the protection of rural and urban properties, protection of installations such as water pumping stations and electricity substations, and protection of water quality in dams and reservoirs.

For other management objectives, this prescription of frequent, cool-season burning may be quite inappropriate. For example, where conservation of biological diversity is the primary management objective, a number of theoretical and experimental studies have shown this applied fire regime to be inappropriate. Such a short fire-return interval can cause local extinction of those plant species that do not resprout after fire and whose seedlings are still not mature by the time the next fire occurs. Furthermore, the patchy fires produced in cool-season fires can maximize herbivore pressure on the plants regenerating in the burned patches, leading to an altered plant community, as described above for the *Acacia rostellifera* community on Rottnest Island in Western Australia. High-frequency fires can also alter the structure of the plant community by removing shrubs and fallen, hollow logs and branches, which are important components of the habitat of many small mammal and bird species.

These sorts of problems with fire regimes that are designed to provide adequate protection from wildfires offer special challenges for management of areas such as native forests that are designated multiple-use areas, with multiple management objectives. Examples are native forests for sustainable timber production, recreation, and conservation of biodiversity, or reservoir catchment areas designated for protection of water quality and also for conservation of biodiversity. How can a set of fire regimes be designed and applied such that species are not threatened by too-frequent burning in an inappropriate season of the year, while fuel loads are modified sufficiently to provide adequate protection from wildfires? Future fire management will find creative solutions to dilemmas such as these.

***Fire regimes for conservation.*** What is the appropriate fire regime to be applied if conservation of biodiversity is an objective of management? How can we decide what fire regime to apply to a large national park or wilderness area, for example? There have been several approaches to this problem. Two of these are based on the expectation that the biota have evolved to tolerate and exploit the prehistoric and historic fire regimes, so that recreating these regimes would ensure adequate conservation. The "let wildfires burn" policy argues that fires starting from lightning in very large reserved natural areas can be allowed to burn unrestrained, thus producing a "natural" fire regime. The debates following the large wildfires in the Greater Yellowstone area in the United States in 1988 illustrate the difficulties with this approach (see Christensen et al., 1989). There are very few areas of wilderness large enough for this strategy to be achievable, and even very large areas have boundaries with other land uses that would be threatened by an intense wildfire. Further, in many areas of the world, human burning has been a significant part of the environment for tens of thousands of years.

A similar argument is that the pre-European burning patterns should simply be mimicked, and that this "natural" fire regime would ensure adequate conservation of biodiversity. There are two main problems with this approach. First, even with techniques such as dendrochronology (dating of fire scars contained in tree rings) and assessment of pollen and charcoal in dated sediments, it is virtually impossible to reconstruct all aspects of a prehistoric fire regime (e.g., season, intensity). Second, even if an accurate reconstruction could be achieved, the landscape is different now. Many conservation areas are now bounded by rural or urban development, dissected by roads, powerlines, and railway lines, and threatened by invasions of weeds and feral animals. Applying a "natural" fire regime today may have "unnatural" consequences.

As a result of these problems with identifying and applying natural fire regimes, many ecologists have argued that fire management for conservation needs to be based on knowledge of the effects of particular fire regimes on the biota to be conserved. This is a challenge for management, because of a vast lack of knowledge about the ecological effects of fire. A great deal is known about the fire responses of only a very limited number of species of plants and animals, a little is known about a few species, and almost nothing is known about most species. Studies that have been done clearly illustrate that the ecological effects of fire vary from one species to another and also, for a given species, from one area to another. Thus, a fire regime prescribed (and shown to be appropriate) for a species in one area is not necessarily portable to another area.

Modern fire management is therefore designed according to the principles of "adaptive management" (Walters, 1986; Whelan and Baker, 1999). This approach recognizes that existing knowledge may never be sufficient to permit the a priori formulation of an appropriate fire regime. Instead, a management prescription is devised, based on the best available information, and treated as a hypothesis. The application of the fire regime is designed to be an experiment, with monitoring and data collection as an integral part of the prescription. The results of the management process are used to modify the prescription in the next cycle of its application.

[See also Forests; Grasslands; and Savannas.]

## BIBLIOGRAPHY

Andersen, A. N. "Fire Ecology and Management." In Landscapes and Vegetation Ecology of the Kakadu Region, Northern Australia, edited by C. M. Finlayson and I. van Oertzen, pp. 179–195. Dordrecht: Kluwer, 1996.

Andersen, A. N., et al. "Fire Research for Conservation Management in Tropical Savannas: Introducing the Kapalga Fire Experiment." Australian Journal of Ecology 23 (1997), 95–110.

Bond, W. J., and B. W. van Wilgen. Fire and Plants. London: Chapman and Hall, 1996.

Booysen, P. de V., and N. M. Tainton, eds. Ecological Effects of Fire in South African Ecosystems. New York: Springer, 1984.

Chandler, C., et al. Fire in Forestry, vols. 1 and 2. New York: Wiley, 1983.

Christensen, N. L., et al. "Interpreting the Yellowstone Fires of 1988: Ecosystem Responses and Management Implications." Bioscience 39 (1989), 678–685.

Christensen, P. E. S. "The Biology of Bettongia penicillata (Gray, 1937), and Macropus eugenii (Demarest, 1817), in Relation to Fire." Forests Department of Western Australia Bulletin 91 (1980), 90.

Coutinho, L. M. "Fire in the Ecology of the Brazilian Cerrado." In Fire in the Tropical Biota, edited by J. G. Goldammer, pp. 83–105. Berlin: Springer, 1990.

Evans, W. G. "Perception of Infrared Radiation from Forest Fires by Melanophila acuminata De Greer (Buprestidae, Coleoptera)." Ecology 47 (1966), 1061–1065.

Flannery, T. The Future Eaters. Sydney: Reed, 1994.

Frost, P. G. H., and F. Robertson. "The Ecological Effects of Fires in Savannas." In Determinants of Tropical Savannas, edited by B. H. Walker, pp. 93–140. Oxford: IRL Press, 1987.

Gandar, M. V. "Description of a Fire and Its Effects in the Nylsvlev Nature Reserve: A Synthesis Report." South African National Science Report Series 63 (1982), 1–39.

Gill, A. M., et al., eds. Fire and the Australian Biota. Canberra: Australian Academy of Science, 1981.

Hobbs, R. J., and C. H. Gimingham. "Vegetation, Fire and Herbivore Interactions in Heathland." Advances in Ecological Research 16 (1987), 87–159.

James, S. "Lignotubers and Burls: Their Structure, Function and Ecological Significance in Mediterranean Ecosystems." Botanical Reviews 50 (1984), 225–266.

Kauffman, J. B., et al. "Relationships of Fire, Biomass and Nutrient Dynamics along a Vegetation Gradient in the Brazilian Cerrado." Journal of Ecology 82 (1994), 519–531.

Kellman, M., and J. Meave. "Fire in the Tropical Gallery Forests of Belize." Journal of Biogeography 24 (1997), 23–34.

Leigh, J. H., and M. D. Holgate. "Responses of Understorey of Forests and Woodlands of the Southern Tablelands to Grazing and Burning." Australian Journal of Ecology 4 (1979), 25–45.

Makarim, N., et al. "Assessment of 1997 Land and Forest Fires in Indonesia: National Coordination." United Nations FAO/ECE/ILO Joint Committee on Forest Technology, Management and Training. International Forest Fire News 18 (1998), 4–12.

Medina, E. "Physiological Ecology of Neotropical Savanna Plants." In Ecology of Tropical Savannas: Ecological Studies, edited by B. J. Huntley and B. H. Walker, pp. 308–335. Berlin: Springer, 1982.

Myers, R., and J. J. Ewel, eds. *Ecosystems of Florida*. Orlando, Fla.: University of Central Florida Press, 1990.

Pyne, S. *Fire in America: A Cultural History of Wildland and Rural Fire*. Princeton: Princeton University Press, 1982.

——. *Burning Bush: A Fire History of Australia*. New York: Henry Holt, 1991.

——. *World Fire: The Culture of Fire on Earth*. New York: Henry Holt, 1995.

Raison, R. J., et al. "Mechanisms of Element Transfer to the Atmosphere during Vegetation Fires." *Canadian Journal of Forest Research* 15 (1985), 132–140.

Roberts, R. G., et al. "Thermoluminescence Dating of a 50,000-Year-Old Human Occupation Site in Northern Australia." *Nature* 345 (1990), 153–156.

Romme, W. H., and D. G. Despain. "The Yellowstone Fires." *Scientific American* 261 (1989), 21–29.

Stewart, O. C. "Fire as the First Great Force Employed by Man." In *Man's Role in Changing the Face of the Earth*, vol. 1, edited by W. Thomas, pp. 115–133. Chicago: University of Chicago Press, 1956.

Swanson, F. J. "Fire and Geomorphic Processes." In *Proceedings of a Conference on Fire Regimes and Ecosystem Properties*, pp. 401–420. USDA Forest Service, General Technical Report WO–26, 1981.

Walters, C. J. *Adaptive Management of Renewable Resources*. New York: Macmillan, 1986.

Wein, R. W., and D. A. Maclean, eds. *The Role of Fire in Northern Circumpolar Ecosystems*. New York: Wiley, 1983.

Whelan, R. J. "Seed Dispersal in Relation to Fire." In *Seed Dispersal*, edited by D. R. Murray, pp. 237–271. Sydney: Academic Press, 1986.

——. *The Ecology of Fire*. Cambridge: Cambridge University Press, 1995.

Whelan, R. J., and J. R. Baker. "Fire in Australia: Coping with Variation in Ecological Effects of Fire." *Proceedings of the Bush Fire Management Conference: "Protecting the Environment and Land, Life and Property."* Nature Conservation Council, Sydney, 1999.

Whelan, R. J., et al. "Responses of Heathland *Antechinus stuartii* to the Royal National Park Wildfire in 1994." *Journal of the Linnean Society of New South Wales* 116 (1996), 97–108.

Wikars, L.-O. "Clear-Cutting before Burning Prevents Establishment of the Fire-Adapted *Agonum quadripunctatum* (Coleoptera, Carabidae)." *Annales Zoologici Fennici* 32 (1995), 375–384.

—Robert J. Whelan

## Indonesian Fires

Indonesia is the largest country in Southeast Asia and has the third largest area of tropical rainforest of any country in the world. Its rainforests are among the most species-rich. A single hectare of Bornean lowland rainforest commonly supports more than 150 species of trees. Indonesia also has the largest area of tropical peat swamp forest in the world, with major forests in coastal Sumatra and Kalimantan and unique highland peat swamps in central Kalimantan. Indonesia is home to many highly endangered large mammals, including the Sumatran rhinoceros (*Dicerorhinus sumatrensis*) and the orangutan (*Pongo pygmaeus*). The country has a tropical climate with in excess of two meters of rainfall per annum and little or no dry season.

In 1997–1998, much of Southeast Asia and Australia experienced a prolonged drought. In July 1997 the world was alerted to a growing environmental disaster as a vast blanket of smoke spread over Southeast Asia from thousands of wildfires burning on the islands of Borneo, Sumatra, Java, Sulawesi, and New Guinea. From August until the beginning of November, Malaysia, Singapore and parts of Indonesia choked under a thick pall of smoke that closed airports, ruined the tourist industry, and affected the health of an estimated 50 million people (Figure 1). The Air Pollution Index in Kuching, Sarawak, reached an all-time high of 839 at the end of September. Readings between 50 and 100 are deemed moderate, and levels up to 200 are unhealthy. There were substantial increases in acute respiratory ailments, including pneumonia and asthma, across the region and it is likely that there will be long-term effects from carcinogens that will not become apparent for decades.

A report by the World Wide Fund for Nature (formerly World Wildlife Fund, WWF) and the Singapore-based Economy and Environment Program for Southeast Asia concluded that economic losses from smoke and fire to Malaysia, Singapore, and Indonesia totaled over U.S.\$1.3 billion. This estimate does not include any appraisal of the loss of Indonesia's natural forest resources such as timber and rattan, and the damage to biodiversity. As well as natural forest, at least 15 thousand hectares of palm oil and rubber plantations were destroyed. There is considerable difficulty in compiling and reconciling data from the many thousands of separate fires. Accurate assessments of the total land area burned by early 1998 are not available, but crude estimates range from 0.5 to 2.5 million hectares (1.2–6.2 million acres).

A fierce debate has raged over the extent to which Indonesia was the victim of a natural disaster or culpable of mismanaging its forests and failing to implement an environmentally sound land use policy. Undoubtedly, drought, forest degradation, and pressures to convert land to agriculture and plantations all played an important role in the disaster.

**Drought.** In 1997–1998, large areas of Indonesia experienced the worst drought for fifty years. This was a consequence of a change in the circulation patterns over the Pacific Ocean that occurs approximately once every four to six years. Normally, cold sea surface temperatures in the eastern Pacific encourage high-pressure air masses to form and winds to flow in a westerly direction toward areas of warm low pressure over Southeast Asia. The ascent of moist air masses in the western Pacific releases heavy convective rainfall. For reasons that are still not fully understood, this pattern breaks down occasionally in what is known as an El Niño–Southern

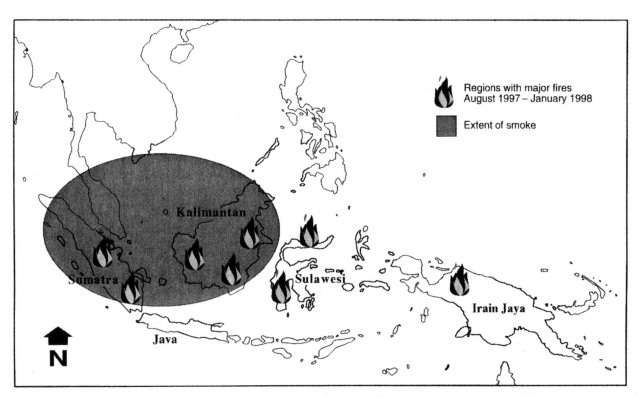

Fire: Indonesian Fires. FIGURE 1. Regions with Major Fires, and Showing Extent of Smoke from August 1997–January 1998.

Oscillation (ENSO) event. Warm sea surface temperatures in the eastern Pacific encourage a reversal of the normal circulation pattern, bringing heavy rainfall to the western seaboard of the Americas and drought conditions to Southeast Asia and Australasia. [See El Niño–Southern Oscillation.]

The 1997 El Niño event was slow to develop, but by August was one of the most severe on record. As the drought worsened, the number of wildfires in South Kalimantan, Sumatra, and Java increased daily. By September, fires had spread to large areas of Sulawesi and Irian Jaya, and the pall of smoke stretched over 2,500 kilometers to Malaysia, Thailand, and the Philippines. The unusual block of high-pressure air over Southeast Asia prevented smoke from forest fires from dispersing or being damped down by rainfall. Monsoon rains finally started to fall in mid-November and gave some respite from the acrid smoke. However, El Niño persisted through the monsoon season, and by early January the drought had reasserted itself in East Kalimantan, one of the driest areas on the island of Borneo. Wildfires started again and by February 1998 the WWF estimated that there were in excess of one thousand fires burning on the island of Kalimantan once more.

The human consequences of the drought were severe. Water supplies in the East Kalimantan capital, Samarinda, were cut off for over a month because salt water was creeping up the local river. Only the wealthy could afford to buy bottled water at twice the usual price, and cholera cases escalated as the poor used dwindling river water for consumption and sanitation. The drought is thought to have cut the Indonesian coffee crop by up to 120,000 metric tons. There was widespread failure of the rice crop, and the important Javanese corn crop was severely reduced. Although the smoke-bound cities of Malaysia, Singapore, and Indonesia were the focus of media attention, rural populations fared far worse. The drought conditions created a crisis for food and water supplies in Central Kalimantan. Many rural communities were cut off from the outside world as rivers dried up and airplanes were grounded by low visibility. Without access to medical care and emergency food supplies, and with bush meat and forest fruits unavailable, many rural communities suffered from malnutrition and consequent high levels of infant mortality.

**Fire in Tropical Rainforests.** Droughts associated with El Niño events are not uncommon in Indonesia, occurring on average twice a decade. In the recent past they have also been strongly associated with forest fires. The largest forest fire ever recorded occurred during the last severe El Niño episode in 1982–1983. An area of approximately four million hectares of forest burnt in Kalimantan and Sabah (Malingreau et al., 1985; Beaman et al., 1985). Big forest fires were also recorded in the less intense El Niño years of 1991 and 1994. Contrary to popular perceptions, fires appear to have occurred repeatedly in the forests of Borneo during the Pleistocene. Wirawan (1991) reported evidence for extensive fires in the coastal area of East Kalimantan in 1915 following an extensive drought, and Goldammer and Seibert (1990) dated extensive charcoal deposits up to 17,500 years old from the same region.

The magnitude and frequency of wildfires are a function of two factors, namely, fire hazard and risk of ignition. Fire hazard depends on both the amount of fuel available and its flammability. The amount of woody debris and leaf litter on the forest floor of an undisturbed lowland rainforest is not significantly less than that in other forest ecosystems where fires are common (Kauffman et al., 1988). The low frequency of forest fires can therefore be attributed primarily to the high moisture content of the fuel biomass, resulting in low flammability (Uhl et al., 1988).

A dense tree canopy buffers the forest microclimate against brief dry periods. Measurements made over a period of two years beneath closed-canopy forest in northern Borneo indicate that relative humidity is at 100 percent for at least nineteen hours per day and very rarely drops below 70 percent (Brown, 1993). Uhl et al. (1988) have suggested that fire only occurs in rainforest in the Venezuelan Amazon when relative humidity drops below 65 percent. During periods of extreme drought, relative humidity in the forest understory may drop to the point where woody debris becomes flammable. Measurements during March 1998 in northeastern Borneo, after a period of two months with no significant rainfall, indicate relative humidity dropping to levels below 60 percent (D. Bebber, personal communication, 1998). Even in undisturbed forest, fuel biomass is sufficient to allow the forest to burn at this point. Without doubt, the El Niño event of 1997–1998 created a significant fire hazard, even in undisturbed rainforest.

**Forest Disturbance and Fire Hazard.** Forest degradation has unequivocally exacerbated the fire hazard in Indonesia. In the fires of 1982–83, six times more logged than unlogged forest burned (Beaman et al., 1985). The explanation for this is twofold: a change in forest microclimate, and an increase in fuel biomass.

The microclimate of a tropical rainforest is strongly influenced by forest structure. Where the forest canopy has been opened by either natural or human disturbance, direct sunlight penetrates to the forest floor and microclimatic conditions become hotter and much drier. Even with in excess of 100 millimeters of rainfall per month, Brown (1993) recorded relative humidity dropping to less than 40 percent in the center of a small clearing in northern Borneo. The degree to which logging opens up the structure of tropical rainforest depends on

the skill and care of the logging company. Figures reported from logging concessions in Indonesia vary between 30 and 65 percent of trees removed, damaged, or killed (Brown, 1998). Relative humidity is therefore much more likely to drop to levels at which the forest becomes flammable in a logged forest.

Logging of tropical rainforests not only makes the forest flammable, but it also increases the amount of woody debris on the forest floor. Tree crowns and bark are never extracted from the forest and add to the large biomass of trees incidentally damaged or killed during forestry operations. Fuel loads in logged forest are thus many times higher than those in undisturbed forest.

Clearance and drainage of peat swamps for agriculture have been other major causes of increased fire hazard. In 1996 the Indonesian government sanctioned the conversion of one million hectares of primary and logged peat swamp forest in Kalimantan to irrigated rice fields. Work started at the beginning of 1997 on the construction of over 20,000 kilometers of drainage ditches. Such developments elsewhere have caused significant areas of the unique shallow ombrogenous peat swamps of Kalimantan and Sumatra to dry out and become a fire hazard. As these domed swamps are above the water table, fires will only be extinguished once the peat is saturated by rainfall. Fires burning beneath the surface have proved almost impossible to extinguish: peat swamps that caught fire in 1982 were still alight a decade later. Peat domes have accumulated slowly over the last ten thousand years, and support large numbers of unique plant and animal species (Rieley and Page, 1997).

**Sources of Ignition.** Although periodic drought may occasionally create a fire hazard in undisturbed rainforest, sources of ignition are rare. Lightning strikes are known to set fire to trees, but they are most common during periods of heavy monsoon rainfall and fires are therefore unlikely to spread. Fire is a traditional tool of shifting cultivators, but until recent decades their population densities have been low. The situation is now changing rapidly. Approximately 1.26 million families practice shifting cultivation in Indonesia (Fraser, 1998), with up to half a million of these in Kalimantan alone, many of them brought in by the government from overcrowded Java. In drought years their slash-and-burn practices create a serious risk of wildfire. Landless migrants have been reported to burn plots in the forest to establish a land claim.

Of equal significance in terms of the area of forest cleared each year are plantations. In a period of economic collapse in most areas of the Indonesian economy, industrial plantations have been booming. Oil palm prices in particular have been very buoyant, encouraging a rush of investment. Indonesia aspires to be the world's largest producer of oil palm and paper pulp, and the total area of plantations is set to double to 6 million hectares by 2000 (Fraser, 1998). In 1997, 300,000 hectares of forest were approved for conversion to plantations.

Once the commercially valuable trees have been extracted from a natural forest, the residual stand is burned and a plantation crop established. Fire is used for land clearance because it is cheap, fast, and efficient. After serious wildfires in 1994, the government banned the use of fire for land clearance in the dry season. However, enforcement of this legislation is almost impossible and the regulations were almost universally ignored. The Indonesian government was remarkably candid in laying most of the blame for wildfires at the feet of plantation and timber companies. In September 1997, the Forestry and Environment ministries published a list of 176 company concessions where fires were burning. Of these, 133 were plantation companies, 28 logging concessions, and 15 transmigration land clearance contractors. Ninety-one of the concessions were in Kalimantan, 85 of them in Sumatra.

**Environmental Consequences of the Fires.** Most fires in undisturbed tropical rainforests are of low intensity and burn only litter and coarse woody debris (Uhl and Kauffman, 1990). Wirawan (1991) reported that after the 1982–1983 fires most understory seedlings and saplings were killed but many large canopy trees survived. The density of pioneer tree regeneration was sufficiently great to inhibit invasion of grasses, and long-term impacts on forest structure and composition are likely to be small (Woods, 1989). In contrast, in logged forest, fires were of such intensity that very few plants survived. In many parts of South Kalimantan, fires are now so frequent that forests have given way to a pyrophytic vegetation dominated by coarse grasses such as *Imperata cylindrica*.

The numbers of animal and plant species lost directly to fire are likely to be small when compared with the ongoing losses due to forest degradation. Much of the forest area that burned had already suffered a progressive spiral of decline. However, the accelerated rate of habitat loss due to fires in 1997–1998 will have had important consequences for many species reaching critical population sizes.

Although, in the long term, regeneration of lowland rainforest after a light fire is possible, in peat swamp forest the peat itself is destroyed, resulting in a fundamental change in the habitat. Fires have also turned peat swamps from sinks for atmospheric carbon into a major source. Six months of fires in Indonesian peat swamps may release more carbon than is released from all of Western Europe in one year (Rieley and Page, 1997).

**Natural or Anthropogenic Disaster?** The 1997–1998 fires in Indonesia are a powerful example of how human mismanagement of the environment can turn a natural hazard into a disaster. Prolonged drought clearly exacerbated the fire hazard, but the risk of wildfire was very much greater in degraded forest. Most fires were deliberately started to clear land. Drought provided an opportunity for uncontrolled acceleration of forest clearance. A great deal of scientific effort was devoted to predicting the drought hazard, but very little to assessing the vulnerability of forests and human communities to drought and fire.

Both Indonesia and Malaysia were very much against a forests convention in the first round of negotiations held prior to the United Nations Conference on Environment and Development (1990–1992), arguing that forest management was entirely a sovereign issue. However, at the negotiations hosted by the Intergovernmental Panel on Forests (1995–1997), both countries moved to support negotiations for a convention. [*See* Intergovernmental Panel on Forests.] Ironically, the 1997–1998 fires have emphasized the very considerable international responsibilities of countries such as Indonesia to manage their forest resources wisely.

## BIBLIOGRAPHY

Beaman, R. S., J. H. Beaman, C. W. Marsh, and P. V. Woods. "Drought and Forest Fires in Sabah in 1983." *Sabah Society Journal* 8.1 (1985), 10–30.

Brown, N. "The Implications of Climate and Gap Microclimate for Seedling Growth in a Bornean Lowland Rainforest." *Journal of Tropical Ecology* 9.2 (1993), 153–168.

———. "Degeneration Versus Regeneration: Logging in Tropical Rainforests." In *Tropical Rainforest: A Wider Perspective*, edited by F. B. Goldsmith. London: Chapman and Hall, 1998.

Fraser, A. I. "Social, Economic and Political Aspects of Forest Clearance and Land-Use Planning in Indonesia." In *Human Activities and the Tropical Rainforest*, edited by B. K. Maloney. Dordrecht: Kluwer, 1998.

Goldammer, J. G., and B. Seibert. "The Impact of Drought and Forest Fires on Tropical Lowland Forest of East Kalimantan." In *Fire in the Tropical Biota: Ecosystem Processes and Global Challenges*, edited by J. G. Goldammer. Berlin: Springer, 1990.

Grove, R. H., and J. Chappell, eds. *El Niño: History and Crisis.* Cambridge: White Horse Press.

Kauffman, J. B., C. Uhl, and D. L. Cummings. "Fire in the Venezuelan Amazon 1: Fuel Biomass and Fire Chemistry in the Evergreen Rainforest of Venezuela." *Oikos* 53 (1988), 167–175.

Malingreau, J. P., G. Stephens, and L. Fellows. "Remote Sensing of Forest Fires: Kalimantan and North Borneo in 1982–83." *Ambio* 14.6 (1985), 314–321.

Rieley, J. O., and S. E. Page, eds. *Biodiversity and Sustainability of Tropical Peatlands.* Cardigan, Wales: Samara Publishing Ltd., 1997.

Uhl, C., and J. B. Kauffman. "Deforestation, Fire Susceptibility and Potential Tree Responses to Fire in the Eastern Amazon." *Ecology* 71.2 (1990), 437–449.

Uhl, C., J. B. Kauffman, and D. L. Cummings. "Fire in the Venezuelan Amazon 2: Environmental Conditions Necessary for Forest Fires in the Evergreen Rainforest of Venezuela." *Oikos* 53 (1988), 176–184.

Wirawan, N. "The Hazard of Fire." Paper presented at the conference "Towards a Sustainable Environmental Future for the S.E. Asia Region," Yogyakarta, Indonesia, May 6–10, 1991.

Woods, P. "Effects of Logging, Drought and Fire on Structure and Composition of Tropical Forest in Sabah, Malaysia." *Biotropica* 21.4 (1989), 290–298.

—NICK BROWN

**FIREWOOD.** *See* Deforestation; Energy; *and* Forests.

**FISHERIES.** *See* Ocean Life.

## FISH FARMING

Fish farming, also commonly referred to as *aquaculture*, is believed to have originated over two thousand years ago in China. Fish farming is a generic term used for the cultivation of aquatic organisms, primarily for food; such organisms include finfish (as in salmonids, tilapias, and carps), molluscs (for example, oysters and mussels), crustaceans (such shrimp, crayfish, and crabs), and seaweeds. Globally, nearly 156 species or species groups are cultured. However, only about 30 percent of these are farmed commercially.

In the latter part of the twentieth century, fish farming has gradually transformed from an art form to a science with a complementary industry. The 1996 aquaculture industry was worth U.S.$46.5 billion, and the volume of production was 34.1 million metric tons (Food and Agriculture Organization, 1998). Indeed, this industry is considered as one of the fastest-growing primary industries globally, and especially in the developing world.

The reasons for the upsurge in this sector are twofold. First, the realization since the late 1950s that the supply of aquatic products that could be harvested from the wild, in particular the oceans, is exhaustible and has an upper limit, even under the best management practices. [*See* Fishing.] Second, and consequently, research and development in this sector aimed at closing the increasing gap between demand and supply for aquatic products for a growing world population has surged since the 1970s. As a result, in the last three decades, commercial fish farming has seen major advances in induced breeding, in genetic improvement of farmed species, in diet formulation and manufacture, in disease control and prevention, in recirculation engineering, and in husbandry. The single most important development is perhaps the technique of artificial propagation, which has enabled commercial fish farming to become independent of wild stocks, with a few exceptions such as the culture of eels.

The output from the aquaculture industry has more than doubled over the period 1985–1994, from 11.2 million metric tons to 25.5 million metric tons. If aquatic plants (only a small proportion of which are used for direct human consumption) are excluded, the increase over the same period is from 7.7 million to 18.5 million metric tons, or nearly 140 percent. The ten major species and/or species groups that lead current global aquaculture production, together with their 1984 and 1995 production figures, are given in Table 1.

It is important to note that most of the species in Table 1 feed low down in the food chain; they are mostly *planktivorous* ("plankton-eating") and/or detritivorous and none of them is a higher-order carnivore (such as

**Fish Farming. TABLE 1.** Production Increases for Leading Species and Species Groups in the Decade 1985–1994

| SPECIES/SPECIES GROUP | 1985 (1,000 METRIC TONS) | 1994 (1,000 METRIC TONS) | FACTOR 1994/1985 |
|---|---|---|---|
| Silver carp* | 1,079.5 | 2,210.2 | x2.2 |
| Grass carp* | 373.1 | 1,819.4 | x4.9 |
| Common carp* | 651.4 | 1,526.6 | x2.3 |
| Indian carps* | 388.9 | 1,155.6 | x3.0 |
| Oysters† | 919.8 | 1,096.8 | x1.2 |
| Bighead carp* | 493.8 | 1,072.5 | x2.2 |
| Clams† | 338.4 | 1,061.1 | x3.1 |
| Scallops† | 122.5 | 1,038.3 | x8.5 |
| Mussels | 758.7 | 991.1 | x1.3 |
| Marine shrimp‡ | 213.0 | 920.6 | x4.3 |

SOURCE: Modified from New (1997).

*Finfish.

†Mollusc.

‡Crustacean.

trout or salmon). However, if the farming of finfish and crustaceans is to develop further and continue to contribute significantly to human food needs in the future, all farming activities need to be intensified particularly in view of the increasing limitations on the potential farming area and water resources. In contrast, in the early phases of growth of the industry it was often believed that fish farming required only marginal land and that it did not overly degrade the environment. The intensification of farming activities will involve use of higher stocking densities, application of better husbandry techniques and water quality management, and provision of artificial feeds when most, if not all, the nutrition of the stock is provided for externally.

An increasing use of external feed input will result in increasing amounts of nitrogen and phosphorus in fish farm effluent, resulting in increased environmental perturbations. It is generally accepted that environmental viability is a prerequisite to the sector's economic sustainability. Future developments in aquaculture must minimize environmental degradation while intensifying production.

Equally, the provision of nutritionally balanced feed to a growing fish-farming industry is problematic, especially in view of the potential limitations of fishmeal, which is a key ingredient in fish feeds. It is believed that with conventional feeds only about 25 to 35 percent of the nitrogen made available in the feed is incorporated into fish mass. The rest is released into water as uneaten and undigested food and metabolic waste. If fish farming is to develop and be sustained in the long term, two problems must be addressed: that of feed resource and that of utilization efficiency of nitrogen and phosphorus. In this context, high-energy diets with a relatively low protein content appear to have much potential from both cost-effectiveness and environmental viewpoints. The best example of this emerges from the Norwegian salmon-farming industry. In 1972 this industry used 1.9 kilograms of animal protein to produce one kilogram of salmon (0.18 kilograms of protein), with a nitrogen retention of only 9 percent. By 1996, diets with 0.4 kilograms of animal protein were used to produce one kilogram of salmon, which is equivalent to a 45 percent increase in nitrogen retention. Under experimental conditions, retentions of up to 66 percent have been achieved (Grisdale-Helland and Helland, 1995). Improvements in nitrogen and phosphorus retention, or efficient conversion of dietary nitrogen and phosphorus into the farmed product, will ensure minimal environmental perturbations from fish farming. This, together with decreasing use of fishmeal in fish diets, will increase the long-term sustainability of fish farming globally, with minimal impact on eutrophication and associated water quality problems.

[*See also* Coastlines; Environmental Economics; Estuaries; Fishing; Food; *and* World Bank, The.]

### BIBLIOGRAPHY

Food and Agriculture Organization. *Aquaculture Production Statistics 1987–1996.* Rome: FAO, 1998.

Grisdale-Helland, B., and S. J. Helland. "Methods to Assess Optimal Nutrient Composition of Diets for Atlantic Salmon." *Journal of Applied Ichthyology* 11 (1995), 359–362.

New, M. B. "Aquaculture and the Capture Fisheries: Balancing the Scales." *World Aquaculture* 28 (1997), 11–30.

—SENA S. DE SILVA

# FISHING

Before the Industrial Revolution, fisheries were practiced from small boats close to the shore, much as happens today in parts of the Third World. Industrialization started with the use of steamboats with purse seines in the mackerel and menhaden fisheries off the eastern seaboard of the United States in the 1870s. In the 1890s the steam trawler, with the otter trawl, was introduced in the North Sea. As a consequence, the catches per unit of effort (where effort is time spent fishing) were reduced so far that the trawlermen were distressed enough to demand help from the government.

In the 1930s steam was replaced by diesel, and after World War II freezers were introduced in factory ships. The older side travelers were replaced by stern trawlers. The most important later steps in industrialization were made by the Russians and Japanese from the 1960s onward. The Russians used large factory ships supported by trawlers. The Japanese used very long fleets of drift nets to catch salmon on the high seas and used pole and line gear and purse seines to catch tuna. These fleets ranged all over the world ocean, working far from their home ports. By the 1970s their activities had infuriated Third World countries to such an extent that the de facto Law of the Sea became effective in 1977, allowing countries to declare extended economic zones out to two hundred miles. They could then exclude the distant-water fleets.

In addition to the general trends of industrialization, particular fisheries appeared. The Japanese invented a method of catching saury with the aid of very strong lights. In the southern North Sea, plaice and sole fishermen returned to the older beam trawls with heavy chains called ticklers to stir the seabed, which demanded the use of large engines of up to five thousand horsepower. Tuna were caught with pole and line, with purse seines, and with drift nets. Herring fisheries, which used to provide all Europe with winter protein, now yield meal for mink in Denmark and eggs for caviar from British Columbia for the Japanese. Tuna fishermen from San Diego work for skipjack far away in the western equatorial Pacific. Drift netters off west Greenland catch salmon from French and British rivers. Russian

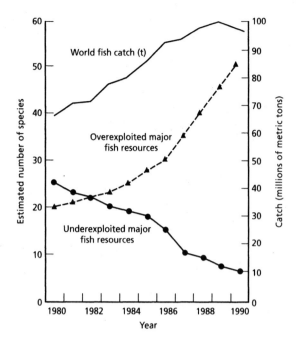

**Fishing. FIGURE 1.** As world catches increased in the 1980s, the proportion of overexploited stocks rose sharply. (After Garcia and Newton, 1995. With permission.)

trawlers work the Antarctic Peninsula for krill and *Neotenus*.

The long development of exploitation from the 1880s to the present day has caused worldwide exploitation. Figure 1 shows how, with increasing catches throughout the world, the proportion of overexploited resources rose, despite the de facto Law of the Sea. By the 1990s it was clear that current management practices were not effective. Further, Garcia and Newton (1995) showed that by 1989 total costs exceeded revenue by 43 percent. This implies that the fisheries are often subsidized, perhaps to sustain employment, but it is still poor management.

**International Control of Fisheries.** The control of fish stocks in the sea has long been international, because the stocks are inevitably shared between the nations that exploit them. [*See* Straddling and Migratory Stocks Agreement.] An early example was the International Fish Commission in Seattle, which regulated the stock of Pacific halibut shared between Canada and the United States from the early 1930s onward. The present structure is complex. In the Northwest Atlantic, the North Atlantic Fisheries Organization (NAFO) coordinates the science and management of stocks between Cape Hatteras, North Carolina, and west Greenland. The fishermen come from Canada, the United States, and other countries, including Japan and Europe. In addition, there are bilateral agreements between Canada and the United States. In the Northeast Atlantic the pre-

dominant organization, the European Commission, is advised scientifically by the International Council for the Exploration of the Sea, based in Copenhagen. Countries such as Norway negotiate with the European Union.

In the Pacific, in addition to the Pacific halibut of the International Fisheries Commission, there is the Inter-American Tropical Tuna Commission based in La Jolla, California. There is also the International North Pacific Commission in Seattle, which manages the Pacific herring stocks, the king crab, and others. Many regional commissions have been established by the Food and Agricultural Organization in Rome. For example, the Southeast Atlantic Fisheries Commission is concerned with the intense fisheries in the Benguela Current off the west coast of Namibia and South Africa.

The International Whaling Commission was established in 1946 to sustain the industry. [*See* Whaling Convention.] In 1963 the Committee of Four recommended that the capture of blue whales be banned, and after two years catches stopped. Since then, the commission has been concerned with a number of issue, principally the moratorium on catches in the Southern Ocean. The Commission on the Conservation and Management of Living Marine Resource, which is responsible for the stocks in the Southern Ocean, has as its purpose the management of those animals, such as leopard seals, that became more abundant with the decline of the whale stocks. It is, however, concerned with more general questions in the Antarctic that need international action.

**Fisheries Science.** The aim of fisheries science is to describe the populations of fish, the annual recruitment of the youngest fish, and their growth and death by fishing and by natural causes. From such information, yield (or catch in weight) per recruit can be estimated as a function of fishing mortality. It has to be expressed in this way because recruitment is so variable. Fish such as cod or plaice can grow by an order of magnitude during their exploited lives, and there is a maximum in their yield per recruit curves. At high fishing rates the fish are caught before they have had time to grow very much. Growth overfishing is that at which the fishing mortality is greater than that at which is found the maximal yield per recruit. Another form of overfishing, recruitment overfishing, occurs when the magnitude of recruitment declines at high fishing mortality.

In the North Atlantic, most estimates are made by virtual population analysis. Because the earstones, or otoliths, of many fishes can be used to age them, catches in numbers can be expressed in catches at age. At the oldest age, an estimate of natural mortality starts the reconstruction of the age distribution of the population back up the year class to the age of recruitment; then, fishing mortality by age can be estimated. Over a num-

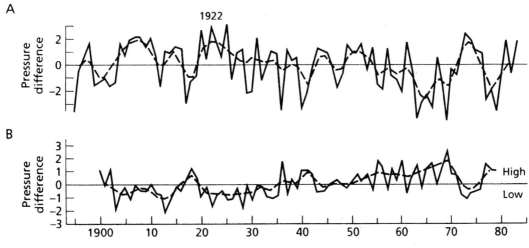

**Fishing. Figure 2.** (A) The trend in time of the North Atlantic Oscillation winter pressure index from 1894 to 1984. (B) The rise of the Greenland High pressure anomaly between the late 1920s and the late 1960s and early 1970s. (After Rogers, 1984. With permission.)

ber of years, matrices of age distributions can be constructed that yield estimates of recruitment of each year class and of fishing mortality and catchability each year; catchability is the coefficient relating fishing mortality to time spent fishing or fishing effort. With measures of growth in weight, yields per recruit can be calculated. The magnitude of recruitment in numbers can be plotted on stock in weight to give the stock recruitment relationship, which can reveal whether the stock is secure from recruitment overfishing and possibly whether it has occurred. A disadvantage of this method is that the measures in the present year are not well estimated, although devices are available to overcome this.

In the North Pacific, a different method is used that does not depend on aging. Catch in weight is plotted on the number of fishing days, and a logistic curve is fitted to the data for which there is a maximum. The fishing effort should be restrained to somewhat below that at which the maximum is obtained. The decline in yield beyond the maximum is attributed to unspecified density-dependent effects, which can include the effects of recruitment overfishing.

In recent years it has become clear that because fishermen increase their efficiency continuously, the catchability coefficient also rises continuously. The coefficient is estimated directly by means of virtual population analysis. Originally, with the logistic curve, an average estimate of catchability was used, but in recent years annual estimates of recruitment have appeared from which catchabilities can be derived.

**Management.** The basis of management is that fishing mortality should not be too high. [*See* Driftnet Convention.] The solution to the problem of growth overfishing in the North Sea was to increase the mesh sizes in the trawl codends to allow the little fish to escape and grow before being caught again. In this way the greatest yield per recruit was obtained without making fishermen leave the sea. This happened once when fishing mortality again reached high levels and no further increase in mesh size was feasible.

There are many indirect ways of achieving the same end: for example, by designating closed areas and closed seasons and, of course, by limiting catches. Total allowable catches have been in use in the North Atlantic for nearly two decades. Fishermen tend to distrust them, and there is possibly a misreporting of catches. In the North Pacific, the Pacific halibut has been caught off Alaska and British Columbia for a day or two each year for some decades.

Stocks of cod have been under pressure throughout the North Atlantic in recent years. That at Iceland has been well sustained. When the northern cod off Newfoundland collapsed in the early 1990s, thousands of fishermen lost their jobs. The North Sea cod was in danger of recruitment overfishing, but it might survive because fishing mortality has been reduced by paying fishermen to leave the sea.

**Climate and Fisheries.** The response of fisheries to climatic change has been long known. For example, for centuries the stocks of the Atlanto-Scandian herring off Norway have alternated in abundance with those of the Swedish herring in the Skagerak. The Norwegian period occurs at the start of each century. The causes remain unknown, but when there is ice north of Iceland, the herring return to Norway.

Figure 2 shows the trend of the North Atlantic Oscillation (1895 to 1984) and the Greenland High (1900 to 1979). The North Atlantic Oscillation is portrayed by the

## CANADA'S COD FISHERY

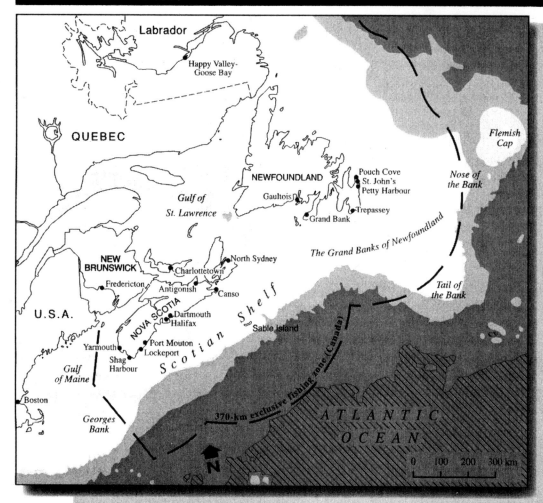

(After Cameron, 1990. With permission of *Canadian Geographic*.)

When European explorers encountered the coast of Newfoundland around 1500 they reported cod so abundant they could be taken ". . . not only with the net, but in baskets let down with a stone." In the five hundred years since, the cod have been treated as an international resource, harvested by fishing fleets from Spain, Portugal, Britain, and Russia in addition to Canadian ships.

Through the nineteenth century, there had been a number of alarming drops in cod landings (in 1817, 1868, 1885, and 1894), but the technical revolution that ultimately threatened the cod population began in 1906 when the French built the first large steam trawlers. By 1909, thirty of them patrolled the Grand Banks, soon to be followed by a similar fleet of Norwegian trawlers. The next technical advance was in 1954 with the advent of the British Fairtry, the world's first factory-freezer trawler. This new breed of ships, and its modern counterparts, can fish around the clock in virtually any weather, using advanced electronic gear to locate and track the fish. Soon other nations had built fleets of similar ships—all trawlers, or draggers that scoop up a variety of species while scouring the ocean floor and disrupting breeding grounds.

These big ships patrol the Grand Banks, the offshore fishery, while hundreds of smaller vessels owned largely by Nova Scotia fishermen, work the inshore fishery on the Scotian Shelf. Recent trouble began when the inshore catch dropped from the usual 150,000 to 35,000 metric tons between 1958 and 1974. It was suspected that a great assault by foreign trawlers in the late 1960s had reduced the cod stocks that normally migrate to the Scotian Shelf in the spring of each year. [See Straddling and Migratory Stocks Agreement.] In defense, Canada established a two-hundred-nautical-mile exclusive fishing zone (in accord with new international law) in 1977. This embraced most of the Grand Banks, but left three outlying shelf areas unprotected.

After establishment of the exclusive zone, federal and provincial subsidies to fishermen encouraged the Canadian industry to expand: between 1974 and 1981 the number of licensed fishermen grew from 36,500 to 53,000, while processing capacity was more than doubled. Expansion of the fishing industry was seen as a way to deal with unemployment in the region, and the federal government set allowable catch limits at levels dictated more by social policy than by sound fishery science.

By 1981, in the face of greatly reduced cod landings, there was wide recognition of trouble in the cod stocks. After a series of commissions and studies, it was determined in 1987, and again in 1989, that the federal fishery surveys had been faulty and had failed to recognize that large landings (or catches) of cod may not reflect abundant populations of fish, but just the effectiveness of modern trawlers that can find, track, and catch large numbers, even in the face of declining fish stocks.

Meanwhile, foreign fleets continued to harvest cod from the edges of the Grand Banks not protected by the exclusive zone established in 1977: the Tail of the Bank, the Nose of the Bank, and the Flemish Cap. At times, there were over two hundred vessels fishing just beyond the zone boundary, and half of those were gigantic factory-freezer ships. Assuming that large catches beyond the boundary would impact cod populations elsewhere in the fishery, Canada began negotiations with members of the European Community and worked to reach some agreements at the Rio de Janeiro Earth Summit in 1992.

In 1992, surveys of the cod stock showed it to be at extremely low levels; and the federal government closed the Canadian cod fishery. Still, foreign trawlers continued to work the outer banks, while Canadian fishermen collected unemployment insurance. On March 9, 1995, after a series of warnings, Canada fired warning shots upon a Spanish fishing ship, eventually boarded her and escorted the captured vessel to port in St. John's, Newfoundland.

The closing of the cod fishery was considered a moratorium on fishing while cod stocks were reassessed. It was hoped the fishery would open again in 1994; but in 1998, commercial fishing for cod still was banned in virtually all parts of the fishery. In May of 1998, the Canadian Minister of Fisheries and Oceans announced, "The outlook for the Atlantic cod fishery remains bleak. The Department's stock status reports and the Fisheries Resource Conservation Council are consistent in warning that very little has changed since the collapse and subsequent closure of this fishery." In June of 1988, the federal government announced its latest measures, worth $730 million, to assist individuals and coastal communities in adjusting to opportunities beyond the fishery, and to plan for a sustainable cod fishery. These measures include buying out fishing licenses and various payments for retraining or retirement of those who worked in the fishing industry.

BIBLIOGRAPHY

Cameron, S. D. "Net Losses." *Canadian Geographic* (April–May 1990), 29–37.
Harris, M. *Lament for an Ocean.* Toronto: McClelland and Stewart, 1998.

—DAVID J. CUFF

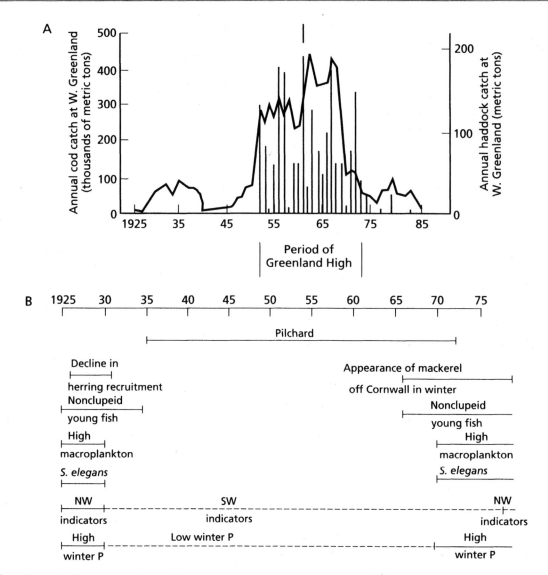

**Fishing. Figure 3.** (A) Catches of Cod off West Greenland,. 1925–1985.

The bars indicate catches of haddock that do not spawn there and so must have drifted there from Iceland. (After Dickson and Brander, 1993.) (B) The structure of the Russell cycle in the western English Channel, showing periods of abundance. (After Cushing, 1995. With permission.)

winter surface air pressure difference between Iceland and the Azores and expresses the tension between the high or low over Greenland and the corresponding low or high over the North Atlantic to the south. [*See* Natural Climate Fluctuations.] From the late 1920s the Greenland High persisted in winter until it collapsed in the late 1960s and the early 1970s. Figure 3a shows the catches of cod off west Greenland from 1925 to 1985; the period of high catches was that of the Greenland High. The Greenland stock is supported by the migration of larvae and juveniles from Iceland, so long as winds from the east prevail over the Denmark Strait (the condition of the Greenland High) and there is relatively

warm water off west Greenland under winds from the south. Figure 3b displays the changes in the western English Channel, known as the Russell cycle, when pilchards replaced herring between 1925 and 1935; profound changes occurred in the macrozooplankton, which reverted in the late 1960s and early 1970s. Again the pilchard period was that of the Greenland High, when southwesterly winds predominated in the western English Channel, characteristic of a low in winter over the North Atlantic.

As the Greenland High collapsed, a large mass of cool and relatively fresh water was generated off east Greenland. It drifted to the region north of Iceland and thence

**Fishing. TABLE 1.** During the passage of the Great Salinity Anomaly of the 1970s, the recruitments to a number of fish stocks were reduced, as tested by a Wilcoxon rank test, comparing the anomalous years with the rest

| STOCK | ANOMALOUS YEARS | NUMBER OF YEAR CLASSES | PROBABILITY* |
|---|---|---|---|
| Iceland summer herring | 1965–1971 | 35 | 0.01 |
| Iceland spring herring | 1962–1971 | 45 | 0.01 |
| East Greenland cod | 1965–1971 | 13 | 0.05 |
| West Greenland cod | 1969–1972 | 15 | 0.01 |
| North Grand Bank cod | 1971–1973 | 20 | ns |
| South Grand Bank cod | 1971–1973 | 22 | 0.05 |
| West Scotland saithe | 1974–1978 | 16 | 0.01 |
| North Sea saithe | 1975–1977 | 18 | 0.05 |
| Faroe saithe | 1975–1977 | 18 | 0.01 |
| Faroe Plateau cod | 1975–1977 | 18 | ns |
| Faroe Plateau haddock | 1975–1977 | 18 | ns |
| Northeast Arctic saithe | 1978–1981 | 19 | ns |
| Northeast Arctic cod | 1978–1981 | 21 | 0.01 |
| Northeast Arctic cod, 0 group | 1978–1981 | 15 | ns |
| Northeast Arctic haddock | 1978–1981 | 22 | 0.01 |
| Northeast Arctic haddock, 0 group | 1978–1991 | 16 | ns |
| Blue whiting | 1978–1981 | 6 | 0.01 |

*ns = not significant.

to the Grand Bank and across the North Atlantic to the Barents Sea for a period of fourteen years. Table 1 shows the reduction in recruitment of a number of "deep water" stocks across the North Atlantic during the passage of the Great Salinity Anomaly of the 1970s. The water was cooler, which may have delayed production, but it also may have been stratified earlier, which may have advanced production. In either case, the time of onset of production would have become mismatched with time of spawning of the fish.

In Figure 4 are shown the times of onset of the spring outburst (with spring species of phytoplankton). Between the 1950s and the 1970s, the wind stress in the North Sea increased and there were more gales in the western North Sea. Because fish in temperate waters tend to spawn at the same time each year, change in the time of onset of the spring bloom may bear on the processes of recruitment to the stocks of fish. In the west central North Sea, production in the 1970s was delayed by a month or six weeks.

In 1962 the year class of North Sea haddock was twenty-five times larger than any of its predecessors since 1925. Stocks of other codlike fishes increased by a factor of three. Figure 5 displays the catches of cod in the North sea from 1903 to 1993. The value of the gadoid outburst amounted to $12 billion. As the Greenland High collapsed, the wind stress in the central North Sea in-

creased and the spring outburst was delayed. It is possible that the magnitude of the cod recruitment was linked to the delay in production of the preferred food of the larvae.

In the upwelling areas of the world ocean, anchovies and sardines are the predominant small pelagic fishes. Since the 1950s they have been heavily exploited, and as the sardine stocks were reduced they were replaced by anchovies. Figure 6 shows an example of such an exchange, or regime shift, off Japan; sardine catches were high in the 1930s, to be succeeded by anchovies between the 1950s and the late 1970s. Then the sardine catches recovered to record levels. Sardine larvae grow quickly in persistent but weak upwelling, whereas anchovies, which grow slowly, can tolerate strong but intermittent upwelling. Such may well be examples of recruitment overfishing, triggered by changes in the nature of the upwelling as the wind stress changes in strength and direction.

It has been long known that the temporal variation in fisheries, appearances and disappearances, were linked to climatic change. Recruitment is possibly determined during the larval stage and depends on the variation in time of onset of the spring outburst, determined by the Sverdrup mechanism. Simply, this is the dependence of the time of onset (in temperate waters) on the progress of stratification, which itself depends on wind

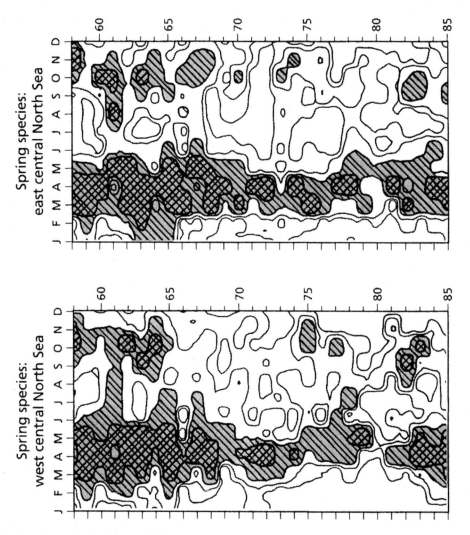

**Fishing. Figure 4.** The changes in the timing of the spring outburst (as numbers of spring phytoplankton species) in the west and east central North Sea, 1958–1985.

In the west central North Sea, the time of onset in the 1970s was delayed by a month or six weeks (but not in the east central North Sea). (After Dickson et al., 1988. With permission.)

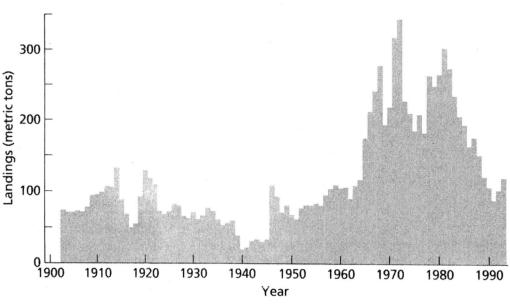

**Fishing. Figure 5.** Catches of Cod in the North Sea, 1903–1993.

The effect of the gadoid outburst is clearly shown; the gain of all codlike fishes was worth $12 billion. (After Daan et al., 1994. With permission.)

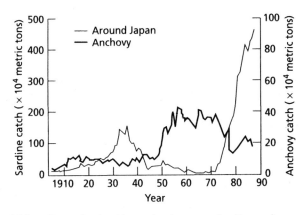

**Fishing. Figure 6.** The Alternation between Sardine and Anchovy Catches off Japan. (After Kondo, 1988. With permission.)

stress and direction and cloudiness. Such is a description of events in the spring outburst, analogues of which appear in the upwelling areas where there are great fisheries. The changes in the fisheries over the centuries have been most dramatic. If the science of recruitment were more fully understood, not only would the condition of the fisheries be much improved but so also would our understanding of climatic change.

[*See also* Marginal Seas.]

### BIBLIOGRAPHY

Cushing, D. H. *The Provident Sea.* Cambridge: Cambridge University Press, 1988.

————. *Populations, Production and Regulation in the Sea.* Cambridge: Cambridge University Press, 1995.

————. *Towards the Science of Recruitment in Fish Populations.* Oldendorf-Luhe, Germany: Ecology Institute, 1996.

Daan, N., H. J. C. Heesen, and J. G. Pope. "Changes in the North Sea Cod Stock during the Twentieth Century." *ICES Marine Science Symposium* 198 (1994), 229–293.

Dickson, R. R., and K. M. Brander. "Effects of a Changing Wind Field on Cod Stocks in the North Atlantic." *Fisheries Oceanography* 2 (1993), 124–153.

Dickson, R. R., P. M. Kelly, J. M. Colebrook, W. S. Wooster, and D. H. Cushing. "North Winds and Production in the Eastern North Atlantic." *Journal of Plankton Research* 10 (1988), 151–169.

Garcia, S. M., and G. Newton. "Current Situation, Trends, and Prospects in World Capture Fisheries." In *Conference on Fisheries Management.* Seattle: Global Trends, 1995.

Kondo, K. "General Trends of Pelagic Fish Population: Study of the Relationship between Long-Term Fluctuations of the Japanese Sardine and Oceanographic Conditions." In *Proceedings of the Twenty-fifth Anniversary Symposium Fisheries and Fisheries Oceanography in the Coming Century* (Tokyo, 10–13 November 1986), 178–184.

Rogers, J. C. "The Association between the North Atlantic Sea Surface Temperature and the Southern Oscillation in the Southern Hemisphere." *Monthly Weather Review* 112 (1984), 1999–2015.

—D. H. CUSHING

**FLOODS.** *See* Extreme Events; Insurance; *and* Natural Hazards.

**FLUOROCARBONS.** *See* Chlorofluorocarbons.

# FOOD

Food is anything that humans eat or drink that provides nourishment and enables them to survive, grow, and remain healthy. Whether by scavenging, gathering, hunting, fishing, or cultivating, humans have sought that which is necessary to maintain vital processes, through the cycle of seasons and the vagaries of years. Human assiduity, over the millennia, has basically consisted of devising means to secure food and developing efficient ways to ensure a stable year-round food supply. Adequate and reliable food sources, secured with a reduction in time expended in the food quest, enable humans to devote more attention to other endeavors. Producing adequate and nutritious food with minimal human labor, however, has not been constant over the centuries nor uniform among continents, countries, or regions. Periodic shortages of food have been characteristic of most places in the world from time immemorial. [*See* Famine.] Because food production is highly dependent upon whether and climate, is affected by episodic surges of plant and animal diseases, and is impaired by ecologically based fluctuations in insect pest population, foodstuffs at any particular place in any particular year vary. Concomitantly, national priorities to the detriment of agriculture likewise affect agricultural production and retard the rate of agricultural advance. Any global change that impacts the basic parameters of food production will affect the diets and survival of millions.

**Food from Plants.** Food consumption patterns vary greatly between and within cultural realms on the surface of the Earth. People associate a specific food or foods with places, regions, nations, and continents. Traditional foods are reflective of what biota thrive in a particular climate, landform, and soil region. Most of the basic staple foods, however, come from a small number of plants and animals. The most important globally are (by tonnage produced):

1. Wheat
2. Corn (maize)
3. Rice
4. Potatoes (white and sweet)
5. Manioc (cassava)
6. Barley
7. Soybeans
8. Sorghums and millets
9. Beans, peas, and chickpeas
10. Peanuts

Wheat and rice are by far the most important staples, with over one-third of all the cultivated land given to them. [See Agriculture and Agricultural Land.] Wheat, the leading bread and noodle grain, is the dominant food staple in North America, Europe, North Africa, northern China, the Commonwealth of Independent Nations (former Soviet Union), and the Near East. Rice is the dominant food staple in most of South and East Asia. Asia has historically produced and consumed nine-tenths of the world's annual rice crop. Although world production of corn (maize) exceeds that of rice, it directly supplies much less of human food calories. Corn is used primarily as animal food in developed nations but is a food staple in some developing countries. It is eaten in South America and Mexico in the form of baked cakes or tortillas, and in southeast Africa as doughy balls, gruel, or paste. Potatoes rival wheat and rice as a source of human food. White potatoes provide more nutrients per hectare than wheat or rice. They are a staple food in the highlands of South America and in eastern and central Europe. Sweet potatoes, manioc, and yams are tropical starch-producing root crops of special importance in portions of lowland South America, along the west coast of Africa, and in the Far East. Barley is used primarily as animal food, but it is consumed by humans in bread form within selected African, Asian, and South American countries. Sorghums and millets are staple subsistence foods in the savanna regions of Africa and India. They are consumed as bread and gruel and fed to food-producing animals. Soybeans, beans, peas, chickpeas, and peanuts are a few of many edible leguminous seeds whose direct consumption is highest in parts of tropical Africa, Southeast Asia, and Central and South America. Those living in areas of the world where plant foods are the principal sources of proteins, vitamins, and minerals suffer most from undernutrition, malnutrition, hunger, and famine.

**Food from Animals.** Animals have been used for food from the time humans learned to scavenge, hunt, and fish. Of the many animals originally hunted or fished and found suitable for food, only a few were domesticated. Those considered staple food sources include:

1. Cattle
2. Pigs
3. Chickens and turkeys
4. Sheep
5. Goats
6. Water buffaloes
7. Camels
8. Rabbits and guinea pigs
9. Yaks
10. Llamas and alpacas

Fish and other aquatic species are excellent food sources, but few have been domesticated.

Foods of domesticated animal origin are more expensive to produce than foods of plant origin, but animals are able to thrive on plant materials unsuitable for human consumption. [See Animal Husbandry.] Cattle, as producers of meat and milk, are the most valued staple animal food source. Dairy farming is very important in the United States and northwestern Europe. Major commercial beef cattle operations exist in the grasslands of the North American Great Plains, northern Argentina, Uruguay, and eastern Australia. Zebu, water buffalo, and other humped cattle are of great importance in the food production tasks of those who dwell in tropical Africa and Asia. Sheep and goats, closely related ruminant animals used as a source of meat and milk, are especially important to the diets of those who live in the Middle East, eastern Europe, Central Asia, Australia, and selected western European nations. Sheep and goats can forage on sparse vegetation and thus make extremely marginal rangeland a producer of high-quality food. Unlike cattle, sheep, or goats, pigs are omnivorous animals that cannot digest cellulose. Their food requirements are similar to those of humans. The main food products secured from pigs are pork, bacon, and lard. Pigs are bred and reared for food on a large scale in the People's Republic of China, the United States, Brazil, the Commonwealth of Independent Nations (former Soviet Union), and Europe. Chickens are the most important domesticated fowl used as a human food source. Other domesticated birds used for human food include turkeys, ducks, geese, and guineas. Chickens have extraordinary biological adaptability and are important food sources for most of the inhabitants of the world. Water buffalo, camels, yaks, llamas, alpacas, guinea pigs, and rabbits are staple food sources in regions of the world where traditionally there is low consumption of meat for religious or cultural reasons. Fish include many water animals that are classified as "seafood," for example, fish, crabs, shrimp, lobsters, turtles, oysters, and frogs. Fish are an inexpensive and wholesome sources of food. [See Fishing.] There are three main classes of fish: oily (such as salmon, mackerel, and herring); lean (flat fish and turbot); and shellfish (lobsters, crabs, oysters, scallops, clams, and shrimp). Seafood is an important component of the diets of those who live in Japan, South and East Asia, Europe, the Commonwealth of Independent Nations (former Soviet Union), Iceland, and the United States.

**Human Food Needs and Diets.** Humans devote much thought, time, and effort to producing, processing, preserving, distributing, preparing, and eating food. The human body must secure approximately fifty different nutrients from food ingested. Each nutrient serves a specific body function, individually or synergistically with other nutrients. Some furnish the body fuel; others provide materials for the building and maintenance of body tissue, and still others act to regulate body processes. [*See* Human Health.] The six general classes or kinds of nutrients are (1) carbohydrates, (2) fats and other lipids, (3) proteins, (4) vitamins, (5) minerals, and (6) water. Carbohydrates are a major source of energy for work or heat. Cereals, high in carbohydrates, are a universal staple food, for they provide the human body with the greatest amount of energy per unit area of land cultivated, are easy to store and transport, and are inexpensive. Fats and lipids are the most concentrated source of energy for body needs. They are the source of essential fatty acids, carry fat-soluble vitamins A, D, E, and K, promote efficient use of carbohydrates and proteins, and make food appetizing and satisfying. Food fats of animal origin include milk, butter, lard, fatty fish and meats, egg yolks, cream, and cheese. Food fats of plant origin include olive oil, coconut oil, peanut oil, and soybean oil. Proteins are used by the human body to (1) build new tissue, (2) maintain tissue and aid in the replacement of regular tissue loss, (3) serve as a regulatory substance and influence water balance, (4) make enzymes essential for digestion and metabolic processes, (5) form milk, (6) supply energy, and (7) make hormones and antibodies. Meat, milk, fish, poultry, eggs, cheese, nuts, peanuts, peas, and beans are protein-rich foods. Vitamins promote growth, maintain health, sustain long life and mental alertness, facilitate use of amino acids and energy source metabolisms, and assist in developing resistance to infections. Vitamins are classified on the basis of their solubility. Vitamin C and the vitamins of the B complex are water soluble; vitamins A, D, E, and K are fat soluble. Minerals are found in the body in numerous combinations. Calcium, phosphorous, sodium, potassium, sulfur, and magnesium are important macro inorganic minerals and are present in the body in relatively large amounts. Iron, manganese, copper, zinc, cobalt, iodine, and others are vitally important but present in minute quantities. People who eat a mixed plant and animal diet seldom experience deficiencies or toxicities from lack of mineral elements. Water is second only to oxygen in importance for human survival. A person can survive for weeks without food but only a few days without water. Acute water deficiency or dehydration produces weakness, lassitude, thirst, loss of weight, and mental confusion. The body's water content must be maintained, for water is the medium for transporting dissolved nutrients and wastes and for regulating body temperatures.

Undernutrition, malnutrition, and hunger describe diets and food intake imbalances. Undernutrition or undernourishment refers to an inadequacy in the quantity of a diet, specifically to insufficient calories. Malnutrition refers to an inadequacy in the quality of a diet, specifically to insufficient essential nutrients. Hunger is acute and persistent undernourishment that causes physical discomfort and pain. An adequate diet must supply the body with all the energy needs, with substances for body building and repair, and with protective regulators to maintain body health and vigor. Many qualitative nutritional deficiencies have no external symptoms, while others result in open and visible deficiency diseases. Providing individuals and cultural groups with the knowledge of basic dietary requirements and the ability to secure the necessary quantity of high-quality foods is the only permanent solution to undernutrition, malnutrition, and hunger. Seven general dietary recommendations can help healthy people, age two and older, achieve and maintain good health: (1) eat a variety of foods; (2) balance food intake with physical activity; (3) choose a diet with plenty of grain products, vegetables, and fruits; (4) consume a diet low in fat, saturated fat, and cholesterol; (5) use fats, oils, and sweets sparingly; (6) select a diet moderate in salt and sodium; and (7) drink alcoholic beverages in moderation, only at mealtime and when there is no associated risk.

**World Food Production and Demands.** As a result of improved agricultural technology, practices, and policies, the world's food production has more than doubled in the past fifty years. In 1986 to 1996 alone, world grain production increased 12 percent; root crops, 18 percent; sugar crops, 23 percent; pulses, 6 percent; tree nuts, 32 percent; oil crops, 46 percent; fruit, 27 percent; meat, 31 percent; milk, 3 percent; eggs, 47 percent; and seafood, 18 percent (Table 1). World per capita production actually increased during this time but not equally in all regions or countries. Production per capita was lower in underdeveloped countries than in developed countries and may have decreased for sub-Saharan Africa. The greatest increase in efficiency of food production has resulted from scientific and technological advances, primarily in the areas of nutritional quality, plant and animal genetics, disease and pest control, and enhancement of plant and animal environments. Because of these advances in agriculture in many countries, there is a substantial world food surplus, while at the same time more people than ever are undernourished and malnourished. This world food surplus is largely due to millions of undernourished and malnourished people not having the financial wherewithal to purchase needed food.

Global demands for cereals are projected to increase from 1,886 million metric tons in 1996 to 2,490 million metric tons in 2020. Meat demands will increase from 208 million metric tons in 1996 to 306 million metric tons in 2020, and roots and tuber demands will increase from 666 million metric tons in 1996 to 855 million metric tons in 2020. Most of the increases in food demands are projected to occur in developing countries, accounting for at least 80 percent of the increase in grain, 90 percent

**Food. TABLE 1.** World Food Production, 1996 and 1986 (per 1,000,000 metric tons)

| PRODUCT | 1996 | 1986 |
|---|---|---|
| Cereals | 1,886 | 1,680 |
| Starchy roots | 666 | 564 |
| Sugar crops | 1,494 | 1,215 |
| Sweeteners | 158 | 127 |
| Pulses | 55 | 52 |
| Tree nuts | 6 | 5 |
| Oil crops | 425 | 292 |
| Vegetable oils | 79 | 53 |
| Vegetables | 526 | 378 |
| Fruit, excluding wine | 448 | 385 |
| Stimulants | 13 | 10 |
| Spices | 5 | 3 |
| Alcoholic beverages | 217 | 185 |
| Meat | 208 | 159 |
| Offals | 14 | 11 |
| Animal fats | 29 | 30 |
| Milk, excluding butter | 540 | 522 |
| Eggs | 50 | 34 |
| Fish, seafood | 109 | 92 |
| Aquatic products, other | 8 | 5 |

SOURCE: United Nations Food and Agricultural Organization (FAOSTAT Database, 1999).

of the increase in meat, and more than 90 percent of the increase in roots and tubers. In response to strong consumer demands for animal products in developing countries, use of grain for feeding livestock will increase considerably by 2020. Grain for animal food will increase by 200 percent, while demands for grain for direct human consumption are projected to increase only 50 percent. In developed countries, grain use for animal food will exceed grain demand for food in both absolute and relative terms. Because of the substantial increase in consumer demand for livestock products, the market for maize is projected to increase faster than for wheat or rice in both developed and developing countries. In the early 1990s, developing countries accounted for between 45 and 50 percent of the world meat consumption; by 2020 developing countries will consume between 60 and 65 percent. Direct human consumption of fish will increase from 109 million metric tons in 1996 to 130 million metric tons in 2020. Much of the increase in fish consumption will be in East Asia, North America, and Australia. Supplies from capture fisheries have stagnated at about 90 million metric tons. Aquaculture, however, has become the fastest growing food production system in the world. Share of global fish production from aquaculture rose from 13 percent in 1990 to 21 percent in 1998. Global food requirements for the year 2050 will depend on the size and nature of the world's population, on the efforts to eliminate undernutrition and malnutrition, and on the extent to which diets change. These variables must be considered when world and national policy makers plan for the future in such areas as food production, trade, and development assistance.

***World dietary transition.*** The well-recognized global economic and demographic transitions currently under way are matched by an equally important dietary transition (Table 2). [*See* Human Populations.] Economic factors have a huge influence on global eating habits. As income rises and populations become more urban, societies enter into different levels of the nutritional continuum. Generally, diets high in complex carbohydrates and fiber give way to more varied diets with a higher proportion of meats and sugars, that is, a Westernization of global eating habits. Whereas high-fat

**Food. TABLE 2.** World Production of Rice, Wheat, Corn, and Meat (percent of reported totals, ranked by country)

| RICE | % | WHEAT | % | CORN | % | MEAT | % |
|---|---|---|---|---|---|---|---|
| China | 34 | China | 19 | United States | 36 | China | 27 |
| India | 22 | India | 12 | China | 22 | United States | 16 |
| Indonesia | 9 | United States | 11 | Brazil | 7 | Brazil | 5 |
| Bangladesh | 5 | France | 6 | Mexico | 3 | France | 3 |
| Thailand | 4 | Russia | 6 | France | 3 | Germany | 3 |
| Burma | 4 | Turkey | 3 | Argentina | 2 | Russia | 3 |
| Japan | 2 | Germany | 3 | India | 2 | India | 2 |
| Brazil | 2 | Australia | 3 | Romania | 2 | Other | 41 |
| Philippines | 2 | Pakistan | 3 | Italy | 2 | | |
| Other | 16 | Other | 34 | Other | 21 | | |

SOURCE: United Nations Food and Agricultural Organization (FAOSTAT Database, 1999).

(meat and milk) and high-sugar diets were once restricted to rich industrial nations, now even low-income societies have access to Western diets. The proportion of energy from protein will probably remain constant, but diets of rich nations will continue to be largely based on animal rather than vegetable proteins. Rapid world urbanization, usually associated with greater per capita incomes and substantial economic growth, has a strong effect on diet structure and food demand. People who live in urban areas consume a variety of different foods and enjoy food combinations distinct from those of their rural counterparts. Urban complexes of 5 million to 30 million dominate many lower-income countries and are growing faster in the less developed than in the more developed regions of the world. The most explosive growth of these megacities is in Asia, where urban conglomerates with more than 10 million residents have grown annually from 2 percent in 1970 to an estimated 20 percent in 2020.

*World dietary diversity.* Major components of the Westernization of global eating habits are a more diverse and varied diet, greater use of processed foods, and increased frequency of restaurant and fast-food dining. Diets of the poorest of the poor, however, are based on a small number of foods and often consist of little more than coarse grains and starchy roots. As personal incomes increase, diet combinations from the five food groups (meat, milk, grain, vegetable, and fruit) are served in daily meals. The case of China exemplifies the early stages of dietary transition. Diets of the Chinese poor were largely based on rice, millet, sorghum, cabbage, salted vegetables, soybean sauce, and some meat. As incomes and overall affluence increased, fresh and pickled vegetables were replaced by more meat, eggs, dairy products, and fresh fruit. As the diets became more diversified, portion size generally dropped for all foods except meat and fresh fruit. In contrast to China, Japan witnesses the late stages of dietary transition. Japan experienced a period of accelerated economic growth from 1950 to 1990, and its gross national product in 1994 exceeded that of the United States. During that period, traditional diets of large servings of rice accompanied by soybean soup, pickled vegetables, fish, and shellfish changed to one of small servings of rice accompanied by red meat, poultry, milk, eggs, fruit, and fresh vegetables. Many components of the Japanese diet of today were inaccessible to the average consumer in 1950, when, for example, 66 percent of daily energy was derived from rice. In 1996 this figure had declined to 35 percent. The Westernization of eating habits in China and Japan is generally viewed as beneficial and is influencing what food items are being produced in major food-exporting countries.

**World Food Consumption Trends.** In the current period of global food surplus and dietary transition,

more than 80 percent of the world's total supply of calories is derived from plants, as is 67 percent of protein and 55 percent of fats. These global figures, however, mask the large variations between the developed and developing nations. Plant foods provide in excess of 90 percent of the calories in developing countries and less than 70 percent in developed countries. Livestock products provide a greater proportion of calories in developed countries than in less developed countries. Animal foods are more expensive per calorie than plant foods. Meat consumption has increased in the less developed nations of the world, while fears of the effects of certain animal foods on human health have led to a decline in their importance in some developed countries. Plant foods such as sugar, vegetables, fruit, and vegetable oils are more important foodstuffs in the diverse diets found in developed areas of the world than in developing areas. Also, absolute consumption of bread and potatoes has declined in most rich industrialized countries, and starchy staples now account for less than 25 percent of all calories consumed. The poorer, less industrialized countries derive a greater proportion of calories from starchy staples, with cereals and roots their cheapest form of food.

Rice is the single most important starchy staple, providing approximately 40 percent of the world's food calories. In many countries of East, South, and Southeast Asia, rice supplies 70 percent of all calories. Wheat is the second most important starchy staple and is the leading plant crop in most of the world. Wheat flour provides eight times as many calories per hectare as beef or lamb. Maize makes up a surprisingly small proportion of the calories derived from starchy staples. Oats, barley, and rye have ceased to be of much significance as world food crops. Potatoes, manioc (cassava), yams, sweet potatoes, and taro provide only 10 percent of the world's supply of calories from starchy staples.

Meat is the most important livestock product, closely followed by milk. Neither fish nor eggs are of world significance, for they account for less than 1 percent of all calories. In the richest regions of the developed nations, calories derived from meat are substantially greater than those derived from milk. Calories per capita derived from meat are very low in tropical Africa, the Near East, and South and Southeast Asia. This may be partly explained by various food taboos related to animal products.

Pork provides nearly 50 percent of the world's meat calories, beef over 25 percent, and poultry at least 10 percent. A pig is the most efficient converter of food into meat calories and is suited to be produced in densely populated areas with a shortage of arable land. Cattle that primarily graze for their fodder are most profitably produced in areas of low population density. Poultry consumption has increased rapidly in the developed countries since the late 1950s, in part because of the intensive feeding practices of poultry factories and a concomitant reduction in the price of poultry meat. Sheep provide a significant amount of meat in Australia, New Zealand, North Africa, and Southwest Asia.

International variations in plant and livestock consumption per capita are large and are influenced by many factors, including (1) religious taboos, gender differences, age, and biological differences between people; (2) the cost of food and in particular the price per calorie; (3) family income for poor households spending a higher proportion of their low incomes on food (for example, in the United States only about 12 percent of a family income is spent on food, but in poorer developing countries more than 70 percent of family income is spent on food); (4) physical environment opportunities and limitations, particularly climate, soils, and landforms; and (5) limitations in choice of food for those who live in countries that are extraordinarily dependent upon external sources and have little control over what foods can be imported and when. There has been a great increase in world food trade, but in 1996 only about 14 percent of all the world's food production entered world trade.

**Constraints to Food Production.** Environmental degradation, aspects of climatic change, insect and disease proliferation, soil nutrient depletion, and the impact of weather perturbations such as those associated with El Niño have a negative effect on locally produced foodstuffs. Overall, critical agroparameters for food production cannot be controlled with technology currently available, but risks can be tempered. Numerous countries of the developing world are at risk of being marginalized from the world food economy because of an increase in unsustainable agricultural practices regardless of the reasons. [*See* Sustainable Development.] Efficient use of cropland necessitates soil conservation and soil building; efficient use of grazing land necessitates restricting the number of livestock to fodder-sustainable capacities. [*See* Carrying Capacity.] Rapid population growth and exploitation are the root cause of natural resource degradation and are linked to failures in national agricultural and population policies. Human errors in the use of natural resources are compounded by natural events such as the climatic phenomenon El Niño. [*See* El Niño–Southern Oscillation.] Every three to seven years, the El Niño cycle has far-reaching effects on world food crop production, livestock and fish production, and forest and natural vegetation growth. In the late 1990s, El Niño severely decreased food production in Asia, Africa, Australia, and North and South America. In 1998, floods reduced food production in forty-one countries, drought reduced food production in twenty-two countries, and forest and grass fires reduced food production in two countries.

Apart from crop and livestock losses attributed to El Niño, ideal conditions have also been created to produce destructive outbreaks of animal and human diseases. Indonesia in 1998 experienced a record food deficit as a result of reduced harvests attributed to one of the worst droughts of this century; this drought was attributed to El Niño. Approximately 7.5 million poor Indonesians in fifteen provinces experienced acute food shortages. This food production loss coincided with the country's reduced ability to buy imported foods in the wake of horrible financial and political crises. The price of rice in Indonesia increased by 50 percent in a ten-month period. The number of countries facing food emergencies increased to thirty-eight in 1999, including twenty countries in Africa, eight in Asia, six in Latin America, four in the Commonwealth of Independent States, and one in central Europe.

*Sustainable food production in Africa.* Nowhere in the world has a combination of environmental degradation and cultural animosities had as adverse an effect on locally produced foodstuffs as in Africa. Rapid rural population growth forced millions of subsistence farmers in the 1960s and 1970s to expand cultivation and herding into areas recognized as drought prone and unsuitable for sustainable food production. Drought, desertification, and locust infestations, along with genocide, terrorism, religious antagonisms, and political uncertainty, reduced food production and food availability in the 1980s and 1990s. The end result was a series of horrendous famines and the loss of millions of lives. Africa, currently the world's continent with the highest proportion of undernourished and malnourished people, is expected to triple its population by 2050; population will increase from 800 million in 1998 to 2 billion in 2050. In 1998, forty-three of the low-income, food-deficit countries of the world were in Africa. Migrations of destitute pastoralists and subsistence farmers into towns and cities where food shortages have already claimed lives have magnified the food–population problem. An estimated 20 percent decrease in grain production in the past two decades has left millions hungry and physically weakened and has resulted in the deaths of millions. Unless corrective measures are taken, Africa will have at minimum 300 million chronically malnourished by 2010.

Nowhere in Africa has the paradox between increased food supplies and continuation of starvation manifested itself more than in Sudan. The Sudanese government reported a 6.2 percent increase in economic growth in 1998 and more than a 50 percent expansion of land devoted to agriculture in the 1990s. Sudan, the largest producer of livestock in Africa, has become a major exporter of meat to North Africa and the Middle East. Yet over 1 million Sudanese are lingering on the brink of starvation. A religious-cleansing civil war and a scorched-earth policy have destroyed crops and displaced 5 million people. An estimated 250,000 died of hunger in 1998, and hundreds of thousands more starved in 1991 and 1993. In 1998, approximately 94 percent of the population lived below the national poverty level. The disparity between those who can afford to buy food and those who cannot is especially pronounced in Khartoum, the capital. Shops and outdoor market stalls have an abundance of food. Those who can afford to buy food complain that the cost of food has increased 50 percent in 1998. Those who cannot afford to buy food are undernourished and malnourished, dull and listless, and many survive on a diet of bread and tea.

Critical factors in resolving the life-taking food problems in Africa include tempering ethnic and religious conflicts, restoring political stability, and implementing plans to ensure sustainable food to all citizens. Food provisioning will be a major determinant of Africa's future population distribution and ethnic composition. Compounding an acute everyday food–population problem, social conflicts and misguided political decisions lead to the loss and spoilage of millions of tons of meat, grain, fruits, roots, and tubers. In the low-income, food-deficit countries, the loss of food to military activities, poor storage, insects, rats and other pests, rudimental transportation, and archaic marketing systems is estimated to exceed 50 percent of production.

**World Food Security.** More than 800 million people throughout the world, and particularly in developing countries, do not have enough food to meet reasonable nutritional needs. By most conservative estimates, 400 million people are suffering from acute lack of the necessary calories, protein, vitamins, and minerals needed to sustain their bodies and minds in a healthy state. Children are the worst affected. It has been estimated that 14 million children under the age of five die every year from the combined effects of undernutrition, malnutrition, and disease. In portions of the underdeveloped world, food scarcity and infection kill 50 percent of the children before they reach their fifth birthday. This is occurring in a world of 5.9 billion people who have available 15 percent more food per person than 4.0 billion people had twenty years ago. Food supplies have increased substantially, but constraints on more equitable access to food, as well as natural and man-made disasters, prevent all citizens of the world from having their basic food needs fulfilled. Given the anticipated increase in world population and stress on natural resources, problems of hunger and famine are likely to persist in the future. Every country in the world has vulnerable and disadvantaged individuals, households, and groups who cannot meet their food needs. In 1998, 70 percent of the world poor were women. Even when overall food supplies are adequate, poverty impedes access to the quantity and velocity of food needed to maintain life. Rapid

population growth and excessive rural migration to urban areas have resulted in negative social, economic, environmental, and nutritional impacts. Large portions of the population in developing countries could face chronic undernourishment by the year 2010. This undernourishment will be exacerbated by period shortages of most basic foods attributed to climatic change, natural hazards, and civil strife. [See Growth, Limits to.]

The world's agriculturalists have the ability to produce more food than the world's population can consume, at least for the next few decades. World food production has increased steadily for years at a rate greater than population growth. Biotechnology, improved agrotechniques, and aquaculture are three promising tools to meet future food demands and to maintain a safe global food supply. A great concern of those who research food and famine issues is the global population in 2050, projected to be 9.8 billion people. The challenge to world food security is to balance food supply and demand while improving equity, reducing poverty, and ensuring adequate quantities of high-quality food for all the world's inhabitants.

[*See also* Brundtland Commission; Catastrophist–Cornucopian Debate; Consultative Group on International Agricultural Research; Food and Agriculture Organization; Land Use, *article on* Land Use and Land Cover; Technology; *and* World Trade Organization.]

## BIBLIOGRAPHY

Bender, W., and M. Smith. *Population, Food, and Nutrition.* Washington, D.C.: Population Reference Bureau, 1997.

Brown, L. R. *Population and Affluence: Growing Pressures on World Food Resources.* Washington, D.C.: Population Reference Bureau, 1973.

Dando, W. A. "Wheat in Romania." *Annals of the Association of American Geographers* 64, no. 2 (1974), 241–257.

Dando, W. A., and C. Z. Dando. *A Reference Guide to World Hunger.* Hillside, N.J., and Hans, U.K.: Enslow, 1991.

Dando, W. A., and J. D. Schlichting. *The New Soviet Food Program, 1982–90: Problems and Prospects.* Washington, D.C.: National Council for Soviet and East European Research, 1988.

Drewnowski, A., and B. M. Popkin. "The Nutrition Transition: New Trends in Global Diet." *Nutrition Reviews* 55, no. 12 (1997), 31–43.

Durham, D. F., and J. C. Fandrem. "The Food Surplus." *Population and Environment* 10, no. 2 (1988), 1–4.

Grigg, D. "International Variations in Food Consumption in the 1980s." *Geography* 78 (1993), 251–266.

Islam, N. *Population and Food in the Early Twenty-First Century: Meeting Future World Demand of an Increasing Population.* Washington, D.C.: IFPRI, 1995.

Pinstrup-Andersen, P., R. Pandya-Lorch, and M. W. Rosegrant. *The World Food Situation: Recent Developments, Emerging Issues, and Long-Term Prospects.* Washington, D.C.: IFPRI, 1997.

Popkin, B. M., A. M. Siegra-Riz, and P. S. Haines. "A Comparison of Dietary Trends among Racial and Socioeconomic Groups in the United States." *New England Journal of Medicine* 335 (1996), 716–720.

Sloan, A. E. "Food Industry Forecast: Consumer Trends to the Year 2020 and Beyond." *Food Technology* 52, no. 2 (1998), 37–43.

Stillings, B. R. "Trends in Food." *Nutrition Today* 29, no. 5 (1994), 6–13.

—WILLIAM A. DANDO

# FOOD AND AGRICULTURE ORGANIZATION

The Food and Agriculture Organization (FAO) is a specialized agency of the United Nations (UN) as defined in Article 57 of its charter. One of the agencies that were created at the same time as the United Nations or have their roots in institutions associated with the former League of Nations, the FAO had a predecessor entity, the International Institute of Agriculture in Rome. The FAO's origin lies in a conference held in Hot Springs, Virginia, in 1943, where forty-four countries decided that a permanent organization for food and agriculture should be established. The actual birth of the FAO occurred at a founding conference in Quebec City, Canada, in 1945.

The FAO is one of the largest specialized agencies, both in membership by states and in its staff. It has multiple roles, providing services to governments through the compilation of data and a variety of information and support services. Drawing upon its networks and the skills of its technical staff, it also is a source of advice to governments on agricultural policy and planning as well as administrative and legal planning. Linked to this is its function as a neutral forum on many issues related to its mandate, such as commodities, fisheries, forestry, agriculture, and food security. The FAO is also the convener of international meetings at all levels. These meetings range from global summits, such as the 1996 World Food Summit, to the regular meetings of its technical committees and a multitude of expert meetings.

Through its field program, the FAO gives assistance to a large number of developing countries. Most of this program is financed by voluntary contributions by member states, while 16 percent comes from the FAO's regular budget through the Technical Cooperation Programme.

At the time of its creation, the primary focus of the FAO was on those aspects of agriculture that have a clear link to improving the production and consumption of food and agricultural products. The main objective of the organization, in the language of its charter's preamble, is "to promote the common welfare . . . for the purposes of raising levels of nutrition and standards of living of the peoples . . . securing improvements in the efficiency of the production and distribution of all food and agricultural products, bettering the condition of rural populations; and thus contributing toward an expanding world economy and ensuring humanity's freedom from hunger." The emphasis on agriculture as such chimed well with the prevailing philosophy of international development until the 1970s. The FAO was fully in line with the concepts and the strategic orientations of the "green revolution."

In the course of the 1980s and 1990s the organization was subject to a crisis similar to that of other components of the UN system. Member states began increasingly to question not only the efficiency, effectiveness, and soundness of the internal structures and management but also the value of the agency's activities for the outside world. The FAO was particularly criticized for being top-heavy, centralist, and out of touch with the mainstream of creative thinking and analysis in agriculture and forestry. The agency was rapidly losing influence and status among UN agencies and was not a determining factor in the considerations of the followup to the major UN conferences of the 1990s, which are the basis of today's International Development Agenda.

Two factors have helped to put the agency back on a better track: the hosting of the successful World Food Summit in 1996 and the tough exercise of streamlining the organization and agreeing with its member states on a "Strategic Framework" for the years 2000–2015. It can be said that the FAO now has created the necessary preconditions for reinventing itself as an effective player in the multilateral system. Internal reforms and the necessary reorientation of activities following the adoption of the Strategic Framework are, however, still to be done.

[*See also* Agriculture and Agricultural Land; Animal Husbandry; Carrying Capacity; Famine; Fishing; Food; Land Use; Pest Management; *and* Prior Informed Consent for Trade in Hazardous Chemicals and Pesticides.]

### BIBLIOGRAPHY

Food and Agriculture Organization. *Reforming FAO: The Challenge of World Food Security.* Rome: FAO, 1997a.

———. *FAO's Emergency Activities.* Rome: FAO, 1997b.

———. *The Basic Texts of FAO.* Rome: FAO, 1998.

———. *The Strategic Framework for FAO.* Rome: FAO, 1999.

———. *Reforming FAO into the New Millenium.* Rome: FAO, 2000.

More information can be found on the Web site of the FAO at http://www.fao.org.

—HUGO-MARIA SCHALLY

**FORECASTING.** *See* Future Studies; Scenarios; *and* Weather Forecasting.

# FORESTATION

Forestation refers to the renewal of the trees on a forest area from which the trees have recently been harvested or removed by natural disturbance (e.g., fire, insects, wind, disease, or landslide). The term implies forest replacement by some means other than natural regeneration (natural seed fall, germination, and seedling establishment). This term is similar to *afforestation*, which refers to forestation on lands that have not previously supported forest plant communities, or not for a long time. A third term, *reforestation*, has the same meaning as *forestation*.

Forestation is a practice that is undertaken when the natural processes of forest renewal are not working, are expected to work unacceptably slowly, or will result in a new forest that does not have the desired species mixture or genetic constitution. These processes may not be working for a variety of reasons. It may be because there is no source of seed close enough for natural seed dispersal processes to deliver seed of desired species to the site in sufficient density and at an acceptable speed. This may be the result of large wildfires, very extensive clear-cut logging that left no seed trees, or simply because there are no seed trees of the desired tree species in the surrounding forest. The natural processes of reforestation may not work because the surface of the soil (the seedbed) is not in a suitable condition for seed germination and seedling establishment. Consumption of seed by insects, rodents, or other seed-eating animals, or the killing of seed by fungal pathogens in the soil may prevent natural generation or make it very unreliable. Unfavorable microclimatic conditions (temperature, moisture, and exposure to frost) that kill newly germinated seedlings will also influence natural processes of forest renewal. Finally, competition from grass, other herbs, and shrubs may prevent new seedlings from becoming established. The purpose of forestation, then, is to assist or replace natural processes that are not achieving forestation objectives.

Forestation is mainly associated with even-age forestry that uses clear-cutting as a silvicultural system, but it may also be used in uneven-age forestry and with other silvicultural systems. Forestation may be used in partially harvested forests to change the tree species composition or to sustain species that have been removed by harvesting or that do not have reliable natural regeneration.

**Stages in the Development of Forestry.** Forestry generally develops over time through a series of rather predictable stages. Very different forestation strategies have been associated with these different stages.

***Unregulated exploitation before the development of forestry.*** The earliest relationship between humans and forests is simply the use of the values of the forest without any explicit activity to ensure their renewal (i.e., without any management of those values). When people are few and lack powerful technology, humans are generally unable to take more from the forest than natural processes can replace. This level of ex-

ploitation has little or no effect on the forest and the future availability of values. But as the numbers of humans and the power of their technology increase, the demands on the forest begin to exceed nature's regenerative capacity; forest alteration and depletion of resource values begin to occur. If this continues unchecked, it may eventually lead to deforestation. [*See* Deforestation.]

***Administrative forestry.*** In an effort to ensure continued supplies of wood for fuel, commerce, war, and other uses, societies that have experienced forest depletion generally institute taboos, religious edicts, laws, policies, or regulations concerning forest use and renewal. Similar mechanisms have been used to protect and conserve the many other values and services forests provide. These early attempts to ensure forest sustainability, or the sustainability of specific values, were generally not based on knowledge of the ecological character of the forest; consequently, they frequently fail. Rigid administrative approaches to the conservation and management of dynamic and spatially variable forests almost never succeeds in sustaining the functions and structures of local forest ecosystems or of forested landscapes. Forest alteration and sometimes degradation continues.

***Ecologically based forestry.*** In contrast to "ecoforestry," which frequently involves more of a spiritual than a scientific approach, ecologically based forestry takes into account both the ecological diversity of the forest landscape (climates, soils, geology, topography) and the associated biological diversity (the interaction between the physical environments and the region's animals, plants, and microbes, as well as the processes of ecosystem disturbance and recovery). Once this overall ecosystem diversity has been recognized, policies and regulations are developed that respect the ecology of the forest values and services that are to be sustained. [*See* Biological Diversity.]

***Social forestry.*** Where ecologically based forestry is designed to sustain a wide range of values, it may provide the foundation for the sustainable forest management that today's society wants. In many situations, this has not been the case. Aesthetic, spiritual, recreational, and various nonwood products and nonconsumptive values have often received less attention and may have been impaired, even when overall ecosystem function and integrity have been conserved and sustained. Consequently, ecosystem-based forestry may evolve into "social forestry," a stage that is based on a strong ecological foundation but that recognizes and sustains a much wider range of forest values.

In the exploitative, preforestry stage, nature is eventually unable to renew forests and their values as fast as they are exploited. In the administrative stage, an initial reliance on natural regeneration is often unsuccessful because the bureaucratic approach to management and harvest does not respect the ecology of the natural regeneration process. The predictable result is an increasing dependence on planting, seeding, or other methods of assisted forestation. This in turn encourages the use of harvesting methods, such as clear-cutting, that are easier and generally more economical. Such methods may be biologically superior for the planting and establishment of a new forest of light-demanding species that are adapted to ecosystem disturbance and are often the desired timber species.

As forestry evolves into the ecologically based stage, clear-cutting and planting may continue to dominate forestation strategies in ecosystems where they are ecologically appropriate. However, there is a closer matching of the tree species planted to the ecological conditions of the site, and greater care is taken that the planted seedlings are in appropriate physiological condition. There is also increasing use of modified clearcutting and of other silvicultural systems that make natural regeneration both more feasible and more successful. In the social forestry stage, there is often strong public pressure to move away from clear-cutting and planting. Most of the general public favor the visually more acceptable minimal disturbance, partial harvesting. Where forestry is economically marginal, there may also be economic pressure to accept "free" natural regeneration, and to use more economic partial harvesting systems. Unfortunately, this often regresses into exploitive "selective" harvesting—"high grading." Thus unless forestation remains based on the ecology of forest renewal, social forestry can return back to the earlier, unsustainable stages of forestry.

**Planted Forests and Naturally Regenerated Forests.** When most people hear the term *plantation forest* (i.e., a forest that has been planted), they think of straight rows of a single species of tree, all of the same height and similar diameter, in dense dark stands with little or no understory vegetation. Forests established primarily for short-rotation timber production on previously agricultural land, or on land deforested long ago, often look like this. The public tends to compare this almost agricultural tree production system with "natural," unmanaged forests, which they think of as consisting of many species of different sizes and ages, arranged in somewhat random patterns of dense patches interspersed with gaps, and with a well-developed understory of herbs, shrubs, possibly mosses, and tree seedlings. [*See* Agriculture and Agricultural Land.]

In reality, where forestation occurs by planting in sustainable, managed, naturally mixed-species forests, the planting is often done merely to ensure that the planted species will be present in the developing "seminatural" forest, which is created as much by natural regeneration of the other species in the surrounding forest as it is by planting. "Plantations" in this type of forest are

quite different from the agriculture-like stands normally associated with the term.

There is thus a continuum of variation in the ecological character of the forest that results: from plantation afforestation, to planted seminatural forest, to natural, unmanaged forests. The last will differ ecologically from either of the former in a number of ways, but this generally has little to do with whether or not the forest was renewed by planting or natural processes.

**Can Planted Forests Substitute for "Old Growth"?** There is a lot of confusion about plantation forests and "old-growth" forests. This is understandable because of the confusion about what an old-growth forest is and what a planted forest is.

With growing public interest in clear-cutting and old growth, an evaluation was made of the differences between recently harvested areas, young forests established by planting, mature forests established by natural regeneration, and old-growth coastal forests in Vancouver Island, western Canada. The comparison was made for a variety of measures of biological diversity, species composition, and forest structure. The result suggested far fewer statistically significant differences than had been expected. This is not because there are no ecological differences between these different forest ages and conditions, but because any one of these stages in forest development can exhibit such a wide range of ecological conditions. There is no simple, easy-to-recognize division of forests into either "old growth" or not "old growth," and no single forest ecosystem condition described by the term *planted forest*.

"Old growth" is a forest condition that has been defined by tree age, tree size, variability in age and size, the presence of large dead trees (snags) and logs on the ground in various stages of decay, patchiness (gaps in the canopy), many different canopy layers, and a variety of other characteristics. Because so many different characteristics are attributed to this condition, few forests conform perfectly to the definition. Instead, we should think of forests as having an index of "old-growthness," or simply an old-growth index (OGI); the degree to which they resemble the ideal "old-growth" condition. A particular forest can be evaluated for each of the many characteristics that define this condition for that kind of forest ecosystem. By summing the individual evaluations, one can assess the degree to which that forest exhibits the "ideal" condition.

Forests vary greatly in their OGI. Agricultural-type tree plantations have a very low OGI. Ancient forests that have not been affected by natural disturbance or human impacts for many centuries often have a very high OGI. Many forests planted following the harvest of a forest with a high OGI have quite high OGI values themselves, owing to a biological legacy carried over from the preharvest forest to the managed forest. The extent of this legacy depends on the OGI of the original forest, how it was harvested, the type of ecosystem, and how the new forest has been managed. The legacy declines over successive rotations of managed forest, unless they are managed to maintain the stand structures that are associated with high-OGI forest. Intensively managed, short-rotation, single-species plantation forests grown to maximize timber values generally have few "old-growth" characteristics. Consequently, they do not provide habitat for those plants, animals, and microbes that depend on these characteristics. In contrast, a planted forest that is managed to have a medium to high OGI may provide habitat for, and support healthy populations of, many species that people normally associate with unmanaged old-growth forest.

**Natural Processes of Succession.** The ecological process of succession is the successive occupancy of a particular area by different plant communities and their associated animal and microbial communities following the removal or modification of the previous communities by natural disturbance or forest harvesting. Herbs may invade open areas created by disturbance, which in the course of time may be replaced by shrubs. These may then lose out to invading light-demanding tree species, which may in turn eventually be replaced by shade-tolerant tree species. The different living communities in this sequence are called *seral stages.*

Several different mechanisms are involved in this process of replacement of one living community by another over time. One of the key processes is the arrival and establishment of the seeds or spores of the plants of the next seral stage. Forestation affects succession by establishing seedlings or providing seeds. Where these are of species that are characteristic of the next seral stage, forestation will result in natural patterns of succession. Where they are of a much later seral stage, it accelerates succession. Where they are of the same or an earlier seral stage, forestation may cause a particular seral stage to be repeated or create an earlier seral plant community.

In some environments, the sequence of seral stages is quite predictable, and succession may actually require particular sequences of stages if it is to proceed. This is especially true of succession in harsh environments. In most forests the sequence of stages is naturally more variable and less predictable. One or more stages may be omitted. Consequently, there is no single "natural" successional pattern to follow. This natural variation has its limits, however, and the long-term health and productivity of the forest may depend on there being a sequence of seral stages over time. Forestation policies should respect this need in those forests where it exists. The biological diversity of a forested region will also depend on the existence of a range of seral stages and forest ages across the landscape.

Where forest managers wish not only to sustain a particular forest condition over several tree crop rotations but also to sustain the biodiversity and other benefits provided by a series of seral stages, they may be able to do this by managing mixed-species forests. By using forestation practices that sustain mixtures of species from different seral stages, some of the ecological attributes of several stages may be combined in a single seral stage. However, this may also require intensive stand management that may not be economically feasible. It will also generally require the retention at time of harvest of permanent or semipermanent patches of old trees and standing dead trees (snags) to provide in the developing new forests some of the structural elements and associated habitat values of older seral stages.

**Forestation, Sustainable Forest Management, and Ecological Rotations.** One of the most fundamental characteristics of forests is that they change over time by the processes of disturbance and succession. Sustainability in most forests is therefore related not to unchanging conditions but to a nondeclining pattern of change. The sustainability of this pattern is defined by three factors (Figure 1): (1) the degree of ecosystem change caused by disturbance, (2) the frequency of this disturbance, and (3) the rate of ecosystem recovery from the disturbance.

A high degree of disturbance may be sustainable because the ecosystem recovers rapidly (is resilient), or because the disturbance is infrequent, or both. A short tree crop rotation (frequent harvests) may be sustainable if the harvest does not change ecosystem conditions greatly, or the ecosystem is resilient, or both. Frequent, severe ecosystem disturbance is unlikely to sustain ecological conditions that develop under lower-intensity or less-frequent disturbances. The combination of intensity and frequency of disturbance and rate of ecosystem recovery that will sustain a nondeclining pattern of change in some ecosystem condition is called the *ecological rotation* for that condition.

Forestation can promote the sustainability of tree growth and timber production by speeding succession and thereby increasing the resilience of the forest ecosystem. By ensuring prompt establishment of a new forest, it may render combinations of rotation lengths and silvicultural systems sustainable that are unsustainable when based on unreliable natural regeneration. Prompt forestation, however, may shorten or eliminate seral stages, that are needed for both long-term site productivity and for nontimber values. The shortened rotations that may be possible with forestation may result in the loss of certain desired ecosystem structures and conditions. Ecological rotations should be evaluated for all the different values that are to be sustained.

Forestation is a key component of sustainable forest management, but it should complement rather than re-

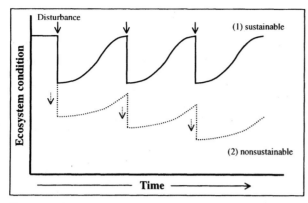

A

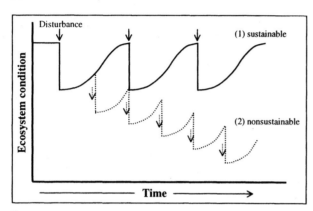

B

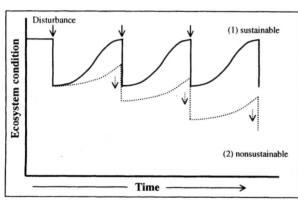

C

**Forestation. FIGURE 1. The Concept of Ecological Rotation.**

Stand level sustainability is a nondeclining pattern of change. Sustainability can be achieved with various combinations of the degree of ecosystem disturbance, the frequency of disturbance, and the rate of ecosystem recovery from the disturbance. (A) Disturbance is too great given the frequency of the disturbance and the rate of recovery. (B) Disturbance is too frequent given the degree of the disturbance and the rate of recovery. (C) Rate of ecosystem recovery is too slow for the degree and frequency of disturbance.

place natural regeneration. It should be used to help achieve management objectives, rather than as a means to shorten rotations, especially where this results in the loss of desired ecosystem values.

[*See also* Ecosystems; Forests; *and* Intergovernmental Panel on Forests.]

### BIBLIOGRAPHY

Nyland, N. D. *Silviculture: Concepts and Applications.* New York: McGraw-Hill, 1996.
Smith, D. M., B. C. Larsen, M. J. Kelty, and P. M. S. Ashton. *The Practice of Silviculture.* 9th ed. New York: Wiley, 1997.

—J. P. (HAMISH) KIMMINS

# FORESTS

The present distribution of world forests has evolved through time, owing some characteristics to continental movements in the distant geologic past and others to the slow reestablishment of forest boundaries during the fifteen thousand postglacial years in Northern Hemisphere areas affected by Pleistocene ice sheets (Pielou, 1991). In future decades, more rapid changes in forest boundaries may occur as forests respond to the changing temperature, precipitation, and soil moisture that accompany global climate change.

Early in human history, forests provided food and shelter for their inhabitants, whose lives were attuned to forest resources and to seasonal cycles. For most of modern humanity, however, forests provide timber and paper, while forest lands themselves in some regions are being converted rapidly to agricultural or grazing lands. [*See* Deforestation.] For many urbanites in more wealthy nations, forests are now revered as refuges, where camping, hiking, and hunting provide recreation in a style that hints of an earlier aboriginal existence. [*See* Wilderness.]

In general, forests are the ultimate result of plant succession in a moist and relatively mild environment. If moisture and temperature allow trees to prosper, ultimately they will prevail over grasses and shrubs because they block their sunlight. Forests are distinguished from woodlands (scattered trees with intervening areas of grasses or sedges), which are found in regions whose climate does not support a continuous cover of trees. Each of the various forest types discussed below is an eloquent expression of its climate; numbers of species and the tendency toward evergreen or deciduous types, for instance, logically reflect the temperature and precipitation regimes in the regions they inhabit.

**Major Forest Types.** The world's forests may be classified both by region (tropical, midlatitude, or boreal) and by species composition (deciduous, evergreen, or mixed).

*Tropical forests.* There are three major tropical forest regions: in South and Central America, in Africa, and in Southeast Asia (Figure 1). The tropical rainforest is the most diverse biome on earth; covering only 6 or 7 percent of all land surface, it nevertheless is home to over 65 percent of all species. In near-equatorial re-

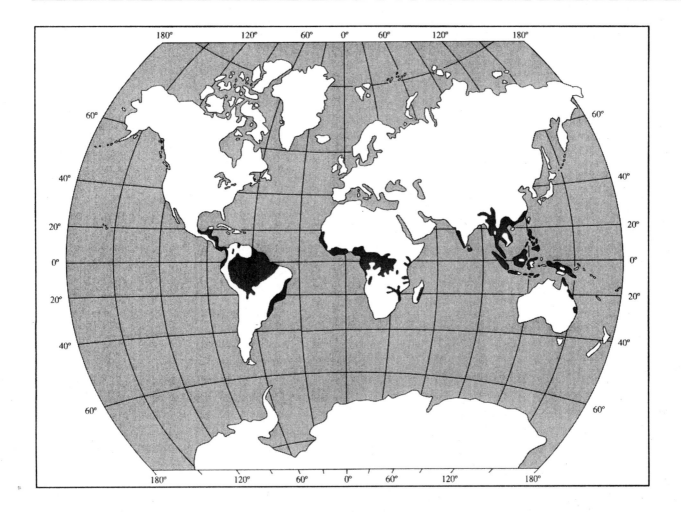

**Forests. FIGURE 1.** Natural Extent of Tropical Forests. (After Archibold, 1995, p. 15, Figure 2.1. With permission of Kluwer Academic Publishers.)

gions where there is no winter and no significant drought grow hundreds of species of trees, all broad-leaved and evergreen. The species usually are scattered, although stands in some variants, such as monsoon forest, are dominated by one species. The trees, and the life that inhabits the forest, are stratified into layers, the uppermost being those trees that require most sunlight, the lowermost being shade-tolerant species.

Among the valuable woods unique to this biome are mahogany, teak, ebony, and the dipterocarps of Southeast Asia. These forests have been extensively harvested for lumber and cleared for grazing lands, so that less than 75 percent of the original forest remains.

This forest occurs either near the equator or on low-latitude windward coasts, the common element being year-round abundant precipitation. Farther from the equator, in areas that are dry in the low-sun season a woodland holding broad-leaved deciduous trees occurs:

the leaves fall during the drought and grow again when rains return in the season of high sun. [*In the entry on* Biomes, *the woodland is mapped with savannas.*]

***Midlatitude broad-leaved deciduous and mixed forest.*** Most extensive in eastern Asia, eastern North America, and Europe, this type occurs in latitudes 25° to 50° in both hemispheres (Figure 2). A distinct winter season is responsible for the deciduous habit of the broad-leaved trees and for the presence of needle-leaved species mixed with broad-leaved. Economically valuable species include the hardwoods—oak, beech, hickory, maple, and walnut—and pine and spruce among the needle-leaved trees.

In North America and Europe, much of this forest was harvested for lumber and fuel and cleared for agriculture; but in many areas it has regrown, as marginal farms have been abandoned and fossil fuels have displaced wood. In some areas of North America and Europe, this forest has been affected significantly by acid rain and acid deposition.

***Coniferous (needle-leaved) forests.*** These are far more prevalent in the Northern Hemisphere than the Southern, partly for climatic reasons, but also because of the Late Triassic breakup of the supercontinent Pan-

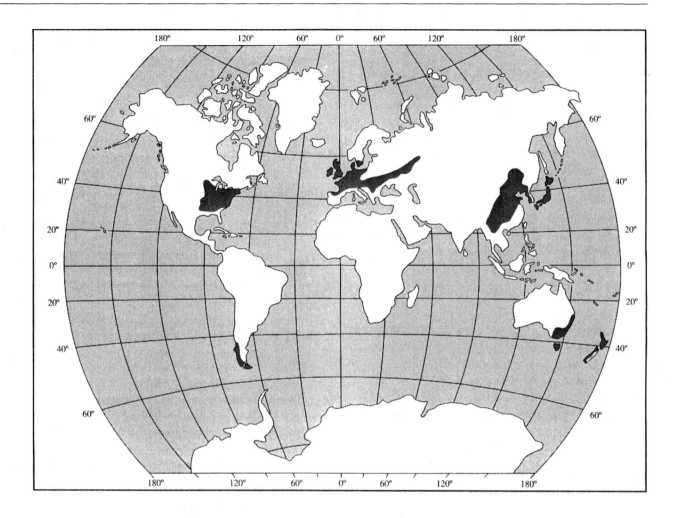

**Forests. Figure 2.** Natural Extent of Temperate Broad-Leaved Deciduous Forests. (After Archibold, 1995, p. 165, Figure 6.1. With permission of Kluwer Academic Publishers.)

gaea into northern and southern landmasses, Laurasia and Gondwanaland. [*See* Plate Tectonics.]

*Temperate rainforest.* Although there are small areas of comparable forest in Japan and New Zealand, this lush forest is fully developed only on the west coast of North America (Figure 3). There the forest is distinguished by a few spectacular long-lived evergreen species—giant sequoia, redwood, western cedar, Sitka spruce, and Douglas fir—which in this environment of heavy rainfall grow to heroic dimensions that make them irresistible to timber companies and to naturalists. Because of ambitious cutting in the past few decades, only a small portion of old-growth forest remains. It is in this coastal forest that the spotted owl and marbled murrelet have been identified as species threatened by commercial logging. These majestic trees, in their cathedral-like groves, inspire vigorous conservation efforts and movements to set aside sizable tracts to be protected from logging.

*Boreal forest.* The boreal forest is a forest of cold-winter continental climates and is extensive only in North America and Eurasia: in the Southern Hemisphere, no comparable land masses exist in middle to high latitudes (Figure 3). Although there are some birch and aspen, the trees are mostly needle-leaved evergreen (spruce, fir, hemlock, pine), with conspicuous needle-leaved deciduous (the larch or tamarack) in mountains and through great expanses in Eurasia.

This forest is limited at its warm edge by lack of moisture: it grades into cool grasslands of the interior. At its poleward edge it is limited by reduced energy during the growing season; there it grades through taiga woodland into grasslike tundra vegetation near sea level, and in its highland expression (montane forest) it grades into alpine meadow. The cold-edge limit is called treeline, whether it is in a mountain range at latitude 40° north or near the Arctic Ocean at latitude 65° north. The sea-level treeline coincides roughly with the 10°C isotherm of the warmest month. In warmer, southern areas the trees are valuable for timber. Farther north, the trees are more slender and stunted and are harvested mainly for paper production.

Recent records suggest that warming of climates is most pronounced in middle- and high-latitude areas.

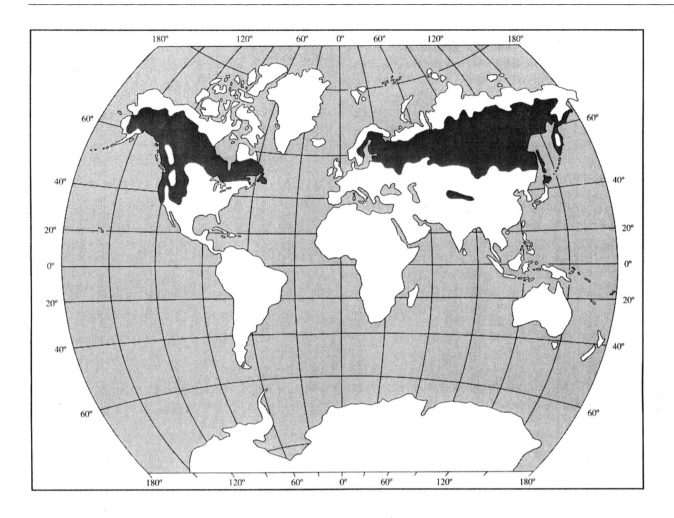

**Forests. FIGURE 3.** Natural Extent of Coniferous (Needle-Leaved) Forests. (After Archibold, 1995, p. 238, Figure 8.1. With permission of Kluwer Academic Publishers.)

[*See the vignette on* Global Hot Spots *in the entry on* Global Warming.] This implies that the boreal forest is likely to be among the forests most affected.

**The Role of Forests in Local, Regional, and Global Change.** The importance of forests in environmental change was appreciated by George Perkins Marsh (*Man and Nature*, 1864), who argued that some of the prime differences between landscapes in the Old World and the New were caused by the long history of forest clearing in the former. He suggested that forest removal could set in train a cascade of changes that included the denudation of slopes, generation of floods, alluviation of bottomlands, and silting-up of estuaries.

*Soils, hydrology, and habitat.* Forest removal can lead to loss of nutrients from soils, especially in tropical forests where much of the nutrient is cycled through trees and the upper humus layer of the soil. Forest removal can also cause a rise in groundwater levels that lead to salinization of low-lying areas, as in the wheatbelt of Western Australia or the prairies of Saskatchewan. In addition, there is some evidence that tree removal can hasten the process of laterite hardpan formation in the tropics, though this is true for only a proportion of tropical soils. Likewise, in cool, temperate areas forest removal can contribute to soil acidification, podzol formation, and peat development.

With regard to soils, however, the most important role of forests is controlling rates of erosion and mass movement. A tree canopy protects the soil from direct rainfall, shortening the fall of raindrops, decreasing their velocity, and thus reducing kinetic energy. The humus in forest soils helps to absorb the impact of raindrops and confers an improved soil structure that increases infiltration and reduces runoff. Furthermore, roots bind slopes and reduce the incidence of mass movement. [*See* Anthropogeomorphology; *and* Land Surface Processes.]

From a hydrological standpoint, forests transpire large amounts of moisture; they intercept rainfall; and because they promote infiltration rather than runoff, they enhance the recharge of groundwater reservoirs and lend streams lower peak flows and less flashy hydrological records. The most important implication is flooding: in a region where highlands have been stripped of forest, heavy rains will be followed by extreme flow volumes, gullying, erosion of stream banks, and loss of farmland; the resulting pools of stagnant water can promote diseases such as malaria. Even stream chemistry is influenced by forests, because removal of trees releases nutrients that increase the concentration of dissolved ions such as nitrates in river water (Anderson and Spencer, 1991).

It is widely understood that forests, especially tropical forests, are habitat to countless species. The removal or fragmentation of forests is a crucial threat to biodiversity. [*See* Extinction of Species.]

***Climate change.*** Forests play a role in climate at a variety of scales. For example, deforestation leads to the release of carbon dioxide into the atmosphere, thereby contributing to the enhanced greenhouse effect. The amount of atmospheric carbon derived from deforestation has increased from about 0.4 gigatons per year in 1850 to 1.6 gigatons per year at present. This compares with roughly 5.5 gigatons per year from burning fossil fuels (Adger and Brown, 1994). More locally, forests serve to lower surface albedo (reflectivity) and absorb more solar radiation; they increase surface roughness which lowers wind speeds; and they increase air humidity by transpiring moisture (Myers, 1991).

While they exert an influence on world, regional, and local climates, forests, their boundaries, and their vertical zonation on mountains are likely to be affected profoundly by any persistent global changes in climate (Gates, 1993; Peters, 1988; Smith et al., 1992; Zabinski and Davis, 1989).

Four different general circulation models all suggest that if atmospheric carbon dioxide is doubled, boreal forest will shift poleward, reducing the extent of subarctic woodland (taiga), but itself will be reduced in area as temperate, broad-leaved deciduous forest and cool grasslands encroach from the south. This shift in middle to high latitudes probably will be mirrored by an expansion of tropical forests at the expense of neighboring savanna lands (Archibold, 1995, pp. 425 and 428).

[*See also* Biomass; Carbon Cycle; Forestation; Global Warming; Intergovernmental Panel on Forests; *and* Trade and Environment.]

### BIBLIOGRAPHY

Adger, W. N., and K. Brown. *Land Uses and the Causes of Global Warming.* Chichester, U.K.: Wiley, 1994.

Anderson, J. M., and T. Spencer. "Carbon, Nutrient, and Water Balances of Tropical Forest Ecosystems Subject to Disturbance." *MAB Digest* 7 (1991).

Archibold, O. W. *Ecology of World Vegetation.* London: Chapman and Hall, 1995.

Brown, P. *Climate, Biodiversity, and Forests: Issues and Opportunities from the Kyoto Protocol.* Washington, D.C.: World Resources Institute, 1998.

Gates, D. M. *Climate Change and Its Biological Consequences,* Sunderland, Mass.: Sinauer, 1993.

Hunter, M. L. *Wildlife, Forests, and Forestry: Managing Forests for Biodiversity.* New York: Prentice-Hall, 1990.

Myers, N. "Tropical Forests and Climate." *Climate Change* 19.1–2 (1991), 1–265.

Perlin, J. *A Forest Journey: The Role of Wood in the Development of Civilization.* New York: W. W. Norton, 1989.

Peters, R. L. "The Effect of Global Climatic Change on Natural Communities." In *Biodiversity,* edited by E. O. Wilson, pp. 450–461. Washington, D.C.: National Academy Press, 1988.

Pielou, E. C. *After the Ice Age: The Return of Life to Glaciated North America.* Chicago: University of Chicago Press, 1991.

Smith, T. M., H. H. Shugart, G. B. Bonan, and J. B. Smith. "Modelling the Potential Response of Vegetation to Global Climate Change." *Advances in Ecological Research* 22 (1992), 93–116.

Zabinski, C., and M. B. Davis. "Hard Times Ahead for Great Lakes Forests: A Climate Threshold Model Predicts Responses to $CO_2$-Induced Climate Change." In *The Potential Effects of Global Climate Change in the United States,* edited by J. B. Smith and D. Tirpak. Washington, D.C.: U.S. Environmental Protection Agency, 1989.

—DAVID J. CUFF AND ANDREW S. GOUDIE

# FOSSIL FUELS

A number of significant changes taking place at the end of the twentieth century involve fossil fuels. They are famously implicated in urban air pollution, regional air quality problems such as acid deposition, and possible global warming due to rising levels of greenhouse gases. [*See* Air Quality; Global Warming; *and* Greenhouse Effect.]

As industrialization and growing affluence spread through developing nations, fossil fuels remain the energy foundation—still inexpensive, compact, convenient, transportable, and generally irresistible, were it not for their serious effects on the environment. Our dependence upon fossil fuels is so ingrained both institutionally and commercially that it is difficult to see an early end to their dominance, especially if new technologies prove able to ameliorate their impacts on the environment. Alternative sources of primary energy will make significant inroads only if the fuels become much more expensive, either through deliberate energy policies or through the decline of fossil fuel resources. It will be shown below that, while that decline is inevitable, even crude oil, the least robust of the fossil fuels, will not be exhausted until well into the next century. [*See* Renewable Energy Sources.]

While advancing industrial technology may be blamed ultimately for our large-scale burning of fuels, there are some technical advances that offer environmentally friendly ways of using fossil fuels. Other changes in technology now improve our access to certain fuels, expanding the resources available and altering our answers to the perpetual questions about how much of the resource remains and how long it will last.

As with any mineral resource, our estimates of fossil fuel resources are constantly being revised and, despite the fixed endowment provided by nature, are subject to changing technologies, prices, and continued exploration. [*See* Industrialization; *and* Resources.]

**Fossil Fuels Defined.** Coal, crude oil, and natural gas, in all their various forms and occurrences, are called fossil fuels because they originated in plant or animal life that grew millions of years ago. Unlike firewood, which can be replaced by tree growth in a few decades, these fuels accumulated and formed in sedimentary rock over periods of millions of years of geologic time and will not be replaced in the foreseeable future. Because they are derived from organic remains, the fossil fuels have saved, indirectly, vast quantities of solar energy from the distant geologic past. Whereas the fuels accumulated slowly through millions of years, humankind promises to dispose of that inheritance in a few hundred years.

The fossil fuel era began with large-scale use of coal in the late eighteenth century. Coal and petroleum fueled the process of industrialization, and now throughout the world support virtually all space heating, industrial processes, private transportation, and over 60 percent of electrical generation.

In 1850 the major fuel in the United States was wood, although coal had made an impact with the advent of steam power. By 1900, coal had displaced wood from most applications and accounted for three-quarters of the nation's raw energy, 9 percent coming from the young oil industry. By 1950, oil and gas had assumed 55 percent of all raw energy needs, while coal's share had shrunk to about 35 percent. At the end of the twentieth century, over 85 percent of raw energy for the United States came from coal, oil, and natural gas. A similar dependence upon fossil fuels is seen in all the world's regions; the dependence is greater in the developing nations of Asia, Africa, and the Middle East, where nuclear power does not contribute a large proportion of total energy.

At the end of the twentieth century, when energy use is growing rapidly in both industrial and developing nations, there are two major issues relating to fossil fuels. The first is how the burning of fossil fuels affects the atmosphere, land, and waters, and how to deal with that problem through energy policies and the application of new technologies. The second is the matter of supply and how it will be impacted by present and future use rates. Both these issues must be viewed in the context of how fossil fuel resources are distributed across the world, and how energy needs are growing in the world as a whole and in industrial versus developing nations.

**Fossil Fuels and the Environment.** In addition to the impact of coal mining on the land surface and on the quality of nearby streams, the combustion of coal, petroleum products, and natural gas yields an array of emissions that directly affect the atmosphere and indirectly affect lakes, other water bodies, and forests.

Because these fuels are various combinations of hydrogen and carbon, their combustion produces carbon dioxide—the foremost greenhouse gas implicated in global warming. While there is some uncertainty about the link between temperature trends and man-made (anthropogenic) emissions, there is no doubt about the rise in carbon dioxide concentrations in the atmosphere. Nor is there any doubt that the increase is due to burning of fossil fuels—coal at the beginning of the industrial period, supplemented massively by petroleum products and natural gas in the twentieth century. A second greenhouse gas, methane, is added to the atmosphere in small amounts by coal-mining operations, and by oil refineries and natural gas production. And a third, nitrous oxide, is added in small amounts by the combustion of all fuels.

Although all fossil fuel combustion contributes to emissions of carbon dioxide, the various fuels are not the same in chemical makeup and are not equivalent in their contribution of carbon dioxide. Coal, because it is so rich in carbon, contributes the most carbon dioxide per unit of energy delivered in a process such as firing a boiler for electrical power generation. Petroleum products such as gasoline and fuel oil are intermediate. Natural gas, predominantly methane, is much richer in hydrogen than carbon and contributes the least carbon dioxide per unit of energy delivered. In some comparisons the emissions are expressed in kilograms of carbon added to the atmosphere: for example, a coal-fired generating plant will add roughly 0.3 kilograms of carbon for every kilowatt-hour of electrical energy generated, while a gas-fired two-stage turbine and boiler combination will—because of the efficiency of the design as well as the different fuel—contribute only 0.1 kilograms of carbon for the same electrical output. [*See* Electrical Power Generation.]

The prevalence of carbon in world fuel burned (amount of carbon per unit of energy) has actually diminished through the industrial period, as wood has been displaced by coal, and coal has been displaced by oil, gas, and alternatives such as nuclear fuels and hydropower. While total carbon emissions have increased along with increases in world energy use, the carbon emission per unit of primary (raw) energy has declined as an average. This trend may or may not continue: future inroads by renewable sources of energy would displace fossil fuels, but rapid growth in the use of coal in developing regions could offset that decline. In any case, decarbonization of the world's composite or average fuel supply on a percentage basis has less impact on the atmosphere than the steady increase in energy use and in total raw amounts of carbon emissions. [See Carbon Cycle; Carbon Dioxide; *and* Decarbonization.]

## TECHNOLOGICAL CHANGE AND FOSSIL FUELS

Predicting the future of fossil fuels has been hazardous because of rapid changes in technology. Advances in recent decades have enabled oil companies to explore more effectively, to drill and produce in areas previously inaccessible, and to wring more petroleum from reservoir rock in known fields. Taken together, these advances have allowed companies to deal with world oil prices that fluctuate and have stayed remarkably low in the last decade of the twentieth century as production capacity has exceeded demand. Changes now on the horizon could make the use of certain fossil fuels more acceptable environmentally, and could make some unconventional resources accessible.

### Drilling and Production

- Enhanced recovery methods, such as fracturing, and the use of steam, detergents, and miscible fluids (including, oddly, carbon dioxide) have progressively improved the proportion of in-place crude that can be extracted from a reservoir rock. These methods are partly responsible for upward revisions in reserve estimates.
- Refinements in mining and separation techniques have in recent decades made both the Canadian tar sands and the Venezuelan heavy-oil deposits amenable to production, thereby making it possible for these resources to play a significant role as conventional crude oil resources decline in the twenty-first century.
- Whereas seismic surveys have been used extensively since the 1950s to map the thickness of reservoir rock and to delineate structures that could cause traps, recent refinements make them more potent tools for exploration and provide detailed three-dimensional portraits of the reservoir in a known field, revealing the shape and extent of oil-saturated beds that could not be traced and mapped from drilling records alone.
- At the same time, new methods have refined directional drilling, so that a drill bit can be guided toward a specific section of reservoir rock, and even driven horizontally through a thin saturated pay zone that would be uneconomic if penetrated by a vertical hole.
- Advanced drilling and production platforms, along with installations on the sea floor, have allowed exploration and production in ever-deeper ocean waters, expanding the exploration of continental shelf areas in the North Sea, the Gulf of Mexico, and Indonesia. Furthermore, directional drilling from a single platform can reach multiple drilling targets economically.
- A combination of deep-water capability, directional drilling techniques, and unconventional methods is likely to be needed if methane hydrates are to be exploited. Successful techniques would open up this resource and drastically expand the natural gas resource picture.

Global warming and carbon dioxide aside, air and water pollution due to combustion of fossil fuels is a significant problem. Sulfur in coals leads to emission of sulfur oxides, which ultimately form sulfuric acid, harmful to the lungs and a key constituent of acid rain and acid deposition. Lead in gasoline finds its way into the air and water. Nitrogen oxides add significantly to acid rain and acid deposition and contribute to urban smog. If our response to the threat of global warming is to slow the burning of fossil fuels, the reduced pollution will be a significant dividend.

**Coal.** Coal is quite varied in character. The nature and depth of a deposit dictate whether it can be mined by surface or underground methods; and the coals themselves differ in their energy content and purity, both of which can be explained by processes that occur in the evolution of a coal bed. Certain ancient bogs where plant materials collected and became peat were, by virtue of their location and surrounding bedrock, apt to gather sulfur or other impurities such as arsenic and selenium. Coals derived from these peat accumulations will be correspondingly rich in those impurities and of lower grade.

### Applications

- Small gas-fired turbines are gaining ground in cogeneration electrical dynamos, whose rejected heat is used for space heating or industrial heat.
- Natural gas can now be used as a feedstock for fuel cells, battery-like devices that convert hydrogen directly into electricity. [*See the vignette* Fuel Cells *in the article on* Electrical Power Generation.] It can also be a source of hydrogen for use as a fuel in engines or power plants.
- While low-temperature, high-pressure liquefaction of gas for shipping by special tanker has been done for years, research now seeks a way to liquefy the gas chemically. The result would be a liquid fuel more versatile than that shipped at high pressure and low temperature in a tanker. It could be transported by pipeline; and, because of its higher energy density, it would be cheaper to transport than the natural gas itself (Fouda, 1998).
- Some natural gas fields produce carbon dioxide as an impurity, along with the desirable methane. Carbon dioxide produced in this way at a field in the North Sea is being separated at the surface and injected back into the reservoir rock to avoid adding to greenhouse gases.
- Coal can be converted to liquid methanol for internal combustion engines; and it can be converted to methane gas, as at the U.S. Department of Energy's Great Plains Gasification Plant in North Dakota. Carbon dioxide, a byproduct of that process, is now piped across the border to be used for enhanced recovery in an oil field in the province of Saskatchewan.
- In a process being developed at Utrecht University, the Netherlands, coal can be converted into hydrogen and carbon dioxide. The hydrogen can be burned in a power plant, while the carbon dioxide is sequestered in exhausted natural gas reservoir rock. This process is more expensive than simply burning the coal as a fuel, but not so expensive as removing carbon dioxide from the flue gases. If hydrogen is to be used in combustion or in fuel cells, much of it may be derived from coal, rather than from electrolysis of water. [*See* Hydrogen.] Such a conversion method, if refined and made economically attractive, could mean that use of coal in future decades need not affect the atmosphere profoundly. If industrial nations were to perfect such conversion methods for their own use, they would be available for developing nations that have abundant coal resources.
- Carbon dioxide released by combustion during the decades of transition from fossil fuels can be captured and sequestered by a number of methods. [*See* Geoengineering.]

—DAVID J. CUFF

Quite separate from grade is rank, which identifies how drastically the original peat material has been transformed or metamorphosed into coal. Through millions of years of heat and the pressure of overlying younger sediments (and ultimately sedimentary rock), peat gradually loses moisture and gains in energy content in a maturing process often called *coalification*, a process that, incidentally, releases methane. Coals that are immature and low in energy content are grouped as *lignite*. More mature are the *subbituminous* coals. Next in rank are *bituminous*. And the most completely altered and generally highest in energy content by weight are *anthracite* coals. For simplification, the lower two ranks may be called *soft coals*, and the higher two, *hard coals*. Worldwide, the most extensive deposits of coal materials took place during the Carboniferous period, 250–300 million years ago, when Northern Hemisphere continents were in warmer climates that favored prolific growth and widespread swamps where peat accumulated. Continents now in the Southern Hemisphere did not benefit as much from this period, but did accumulate coal materials during the Cretaceous period, roughly 100 million years ago. Because the large Carboniferous deposits are so old, their coals are mature and of higher rank;

hence over 60 percent of the world's coal is bituminous or anthracite, while less than 40 percent is subbituminous or lignite. [*See* Earth History.]

Because grade and rank are independent, the all-important trait of low sulfur content does not always occur in any particular rank of coal; however, an anthracite with low sulfur content is extremely desirable. For the production of coke, bituminous coals are ideal. Soft coals are less valuable from the standpoint of energy content; but that characteristic may be offset in some deposits by a low sulfur content. One safe generalization is that the coals of lower rank usually occur in thick near-horizontal beds that are easily mined (often by surface mining methods), whereas anthracites are in beds, contorted and tightly folded by the same compressive forces that carried the coalification process beyond the bituminous stage.

The applications of coal are changing. Bituminous coal continues to be a source of coke, but new methods of steel production require less of it. Coal as a boiler fuel for heating buildings has largely been displaced by oil and natural gas: its major use now is in electrical generation. At the end of the twentieth century, coal supplied roughly 25 percent of all the world's primary, or raw, energy and roughly 35 percent of all the energy used in electrical generation.

While coal use has been displaced by nuclear power and by natural gas in parts of Europe, total use continues to rise in the United States and Japan. Any declines in industrial countries have been more than offset by growth in coal use in China and other Asian countries (see Figure 1). From 1980 to 1995, China's share of world coal use grew from 17 to 28 percent, and in the developing countries of Asia as a group it grew from 23 to 39 percent. It is projected that, in the period 1995–2015,

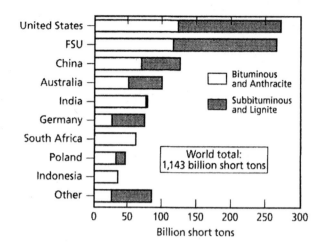

**Fossil Fuels. Figure 2.** World Recoverable Coal Reserves. (From Energy Information Administration, 1997, p. 65.)

world coal use will rise by 45 percent, and China and India together will account for almost half of that rise. In the developing nations of Asia alone, coal consumption is projected to double between the years 1995 and 2015, and China and India will be responsible for 80 percent of that growth. India's use will probably mirror that in most world regions, the growth being for electrical generation. But in China, where alternative energy sources are scarce, more than half the growth may be in nonelectrical uses, such as steam and heat for industrial processes and making coke for steel plants.

If total world coal use is expected to grow by 45 percent from a reference year of 1995 to the year 2015, this portends serious impacts on the atmosphere, at a time when the world community is grappling with how to regulate greenhouse gas emissions. It is conceivable, though, that as oil and gas resources decline, rising prices will encourage novel applications of coal that do not release of large amounts of carbon dioxide. [*See the vignette* Technological Change and Fossil Fuels.]

World reserves of all ranks (i.e., well defined and economically recoverable amounts of all ranks) are enough to last more than 200 years at current rates of production and use. Over half the reserves are in three regions—the United States, the former Soviet Union, and China. Four other countries, Australia, India, Germany, and South Africa, account for an additional 27 percent (Figure 2). In 1995 those seven regions accounted for over 80 percent of world coal production. Bituminous coals suitable for coking do not occur in every region: Australia, the United States, and Canada supplied 85 percent of the world's coking coal in 1995. Between now and the year 2015, the major exporters of coal are expected to be Australia, the United States, and South Africa.

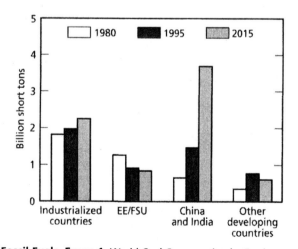

**Fossil Fuels. Figure 1.** World Coal Consumption by Region, 1980, 1995, and 2015. (From Energy Information Administration, 1997, p. 66.)

**Crude Oil and Natural Gas.** Collectively known as petroleum, crude oil and natural gas occur in deposits (or accumulations) of various kinds. A comprehensive view includes both conventional and unconventional deposits; the latter are of very substantial magnitude and could significantly extend the petroleum era.

As with coal, crude oil and natural gas originated in ancient plant and animal life that accumulated along with sediments and was eventually transformed into oil or gas, which in turn became trapped and preserved in sedimentary rock such as sandstone or limestone. Unlike coal, though, much of our petroleum originated in marine rather than continental sediments, and the plants and animal remains were of microscopic scale, while the materials in coal show obvious leaves and stems. Another difference is the timing: coal deposits did not occur until after land plants colonized the continents, the first economic coal beds being in rock of late Devonian age (370 million years old); petroleum accumulations, though scarce in older rocks, have been found in rock of Cambrian age (500 million years old) at the beginning of the Paleozoic era. Despite these few very old accumulations, roughly 70 percent of the world's crude oil occurs in rock of Cenozoic age, that is, less than 100 million years old. With regard to exploration for and definition of fuel resources, the major difference between petroleum and coal is that coal beds are more continuous and more predictable than are accumulations of crude oil or natural gas.

Conventional crude oil and natural gas fields are ac-cumulations of these fluids in the pore spaces and fractures of rock, where the accumulation depends upon some form of trap that has interrupted the upward migration of oil or gas from a source rock. It is generally accepted that the source is a shale rich in organic matter, thick beds of which are usually found in the stack of sedimentary rock that includes the oil and gas accumulations.

Known oil and gas fields demonstrate traps of various kinds, but all of them entail a porous and permeable reservoir rock that is saturated with the fluid, and an overlying dense impermeable rock which, along with a convenient structure, provides the necessary seal (Figure 3). Some traps (and hence some fields) hold just crude oil, some just natural gas, and others a combination of the two. Natural gas alone usually forms at greater depths and higher temperatures than does crude oil, although the separation into oil accumulations in one area and gas in another can occur during the migration of fluids.

For a sizable oil accumulation to form, all the necessary factors must coincide: there must be a source rock buried deeply enough to provide the temperature and pressure needed to convert organic material to petroleum; there must be porous and permeable rock to allow migration and accumulation of oil or gas; a suitable trap must block the upward migration and cause the fluids to accumulate; and, finally, this trap must form before the migration occurs. Oil or gas may escape the trap if it is disturbed by subsequent folding, faulting, or

**Structural Traps**

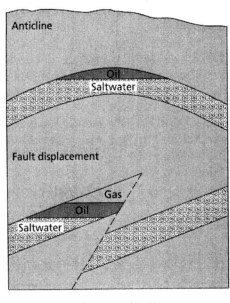

**Stratigraphic Traps**

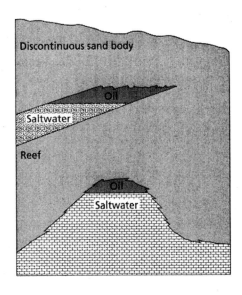

Potential Reservoir Rocks

Sandstone    Limestone    Impervious Rock, such as shale

**Fossil Fuels. FIGURE 3.** Some Traps and Reservoir Rocks for Crude Oil and Natural Gas. (From Cuff and Young, 1985, p. 70. Copyright 1985. Reprinted by permission of the Gale Group.)

tilting. This is one reason that most of the world's oil and gas resources are in relatively young rock.

Exploration for oil and gas begins by identifying regions with thick layers of marine sedimentary rock, preferably with evidence of rocks such as sandstone or ancient coral reefs that are porous and permeable to fluids. Then, drilling sites are selected on the basis of geologic and geophysical evidence that suggests structures that will serve as traps. Whether the structure is there, and whether the potential trap holds petroleum or just ancient salt water, can be determined only by drilling.

When a field is discovered, and then delineated by development wells, the thickness and areal extent of rock saturated by petroleum can be estimated from drilling records. Core samples from the reservoir rock reveal how much pore space exists to hold oil or gas. These measures allow an estimate of the volumes of oil-in-place or gas-in-place. On the basis of the rock's permeability, it is then possible to estimate the proportion that will be recovered: for natural gas the proportion may be as high as 85 percent; for crude oil, however, the proportion is much lower, 35 or 40 percent being a historical average. New technologies can greatly enhance the recovery, and revise upward the estimates of how much oil a given field may produce, given certain assumptions about prices. These recoverable amounts, calculated for each field and summed for a state or a region, constitute reserves. As with any mineral resource, the reserves are in the rock awaiting production; they are not a stockpile at the surface. One exception is the U.S. Strategic Petroleum Reserve, which is oil already produced, then purchased by the U.S. government and stored in underground caverns that serve as natural storage tanks.

Amounts of oil or gas that qualify as reserves are those recoverable from known fields with current technology and at current prices. The most certain are proved reserves. A more comprehensive assessment that assumes fields will expand with time and also assumes older fields will yield more oil through application of enhanced recovery techniques is called *indicated reserves*.

*Conventional* crude oil includes all those occurrences in which the oil is a liquid sufficiently thin (of low viscosity) that it will flow from its reservoir rock into a borehole and can be pumped to the surface. *Unconventional* crude oil sources include the Athabasca oil sands of Alberta, Canada, the Orinoco heavy oil of Venezuela, and smaller or poorly defined deposits of oil sands in Colombia, Jordan, and the United States. The Athabasca oil sands have already produced over one billion barrels of oil (at a recent rate of 250,000 barrels per day), while production in the Orinoco heavy-oil region of Venezuela is now expanding and expected to reach one million barrels per day by the year 2000 (Edwards, 1997). Whereas the Athabasca deposit comprises beds of bituminous sand that must be mined and processed at the surface, the Orinoco deposit is thick tarry oil that can be dissolved in solvent and pumped into a borehole.

Another quite different source is oil shales, or organic shales, which are rich in a substance called kerogen that can be converted to a crude oil if the rock is mined, crushed, and heated. These appear to be potential source rock that was never subjected to extreme depth and heat, and therefore still contains the original raw material. The world's largest deposits are in the United States and Brazil, and in the former Soviet Union where some oil has been produced from the shales. Generally, oil shales require very expensive production methods, so that large-scale exploitation will await crude oil prices of U.S.$40–50 per barrel. These resources are excluded from the world summary of crude oil that follows.

Conventional natural gas is in permeable rock that allows gas (or oil containing the gas) to flow readily into a borehole in response to pressure within the rock. Some accumulations in the United States considered unconventional are in tight sands or organic shales that must be fractured artificially before they will yield gas. The amounts are relatively small compared with resources of conventional gas, and are roughly comparable to the methane gas that could be recovered by drilling coal beds before mining. Larger amounts of unconventional gas reside in sandstones deeply buried in the Gulf Coast region of the United States. These geopressured zones will be tapped only if a use can be found for the hot brines that would flow to the surface along with the gas.

The foregoing examples of unconventional gas are North American accumulations that exist in consolidated rock but are excluded from the usual tabulations of natural gas resources. While published assessments are more readily available for occurrences in the United States, there are comparable possibilities for coal bed methane and for tight gas sands in other regions of the world. Only geologic occurrences have been considered here as unconventional sources of crude oil or natural gas. Excluded are the possibilities of producing synthetic oil, gas, or alcohols from various crops or wastes, usually referred to as biomass.

**Oil and Gas Resources.** For many decades some authors have been predicting an early end to the petroleum era: after all, there is a fixed endowment of oil and gas in the Earth, and our production and use of oil and gas has been growing. But the end has not come rapidly, nor has the price of crude oil climbed steadily, as would be expected in the face of tightening supplies. There have been a number of reasons.

First, some predictions were based on published es-

## METHANE HYDRATES

Recently, an entirely different unconventional source of natural gas has been recognized. It may require drastically different exploration and production methods, and may therefore never be exploited; but the potential resource is so large that it must be included in any survey. In two very different environments—permafrost regions of the Arctic, and sediments of the modern sea floor—methane and other constituents of natural gas occur in a crystalline form bound to ice molecules and named *gas hydrates*, or the more specific *methane hydrates*. (The most general term for the occurrence of gases in this form is *clathrates*.) Whether this gas is geologically old enough to be called a fossil fuel is not clear.

In the marine setting, the origin of the gas may be decomposition of organic material in the sediments; but these occurrences are not the same as conventional gas in offshore fields. That gas has been formed at high temperature from decomposition of organic material and then trapped in gaseous form in deeply buried rock. Gas hydrates, though, are buried less than 600 meters below the ocean floor, in sediments, not sedimentary rock; furthermore, they are ice and occur as a crystalline solid in which gas molecules are surrounded by water molecules in the form of ice. Their thickness appears to be limited by the local temperature gradient, the lower edge being controlled by the depth where above-freezing temperatures begin (temperature rises with greater depth in the earth) while the upper edge is where the water and sediment depth provide enough pressure for the hydrates to form and persist. In the continental Arctic, the maximum depth is, again, roughly 600 meters, near the bottom limit of permafrost.

Studies to date suggest that the larger part of the total resource is on the ocean floor, not in the arctic permafrost. Some hydrates in the ocean sediments occur in discrete nodules and patches that would be difficult to exploit, but they also occur in thicker and more continuous beds that could be tapped by modern drilling and production methods, although water depth in some areas may be prohibitive. Some estimates suggest that the gas in all the deposits combined (not recoverable, but in-place amounts) may have greater energy content than all the world's recoverable energy from identified coal beds and the reserves in conventional crude oil and natural gas fields.

Methane trapped in these hydrates may be significant in another way. As a potent greenhouse gas, methane would promote global warming if it were released by the melting of hydrates, either in the Arctic or in the ocean floor. It may, in fact, be implicated in past episodes of climate warming. [See Methane.]

—DAVID J. CUFF

timates of reserves and neglected the unproven or undiscovered resources and the fact that reserve estimates tend to grow as fields are developed. Second, oil prices did not soar steadily, because, with prices at U.S.$35–40 per barrel in the early 1980s, industries and other users found ways to conserve, and to switch to natural gas and coal for some applications. These same high prices encouraged continued development work in high-cost areas such as the Alaskan North Slope, the North Sea, and western Siberia, and stimulated exploration that led to new or increased production in Mexico and China. Nevertheless, there are signs now that abundant cheap crude oil will not be available after the first decade of the twenty-first century.

Since the 1980s, the rate of crude oil production in the world has exceeded the rate of discovery of new resources. For natural gas, the discovery rate has kept pace with production (Figure 4). Any growth in world crude oil reserves has been due to upward revision of the estimates of reserves in new fields (a phenomenon that depends partly on new technologies and partly on the early estimates being conservative).

Although analysts differ in their estimates of the total conventional crude oil remaining in the world, using an accepted method for predicting the future of oil production, they agree that production will peak very early in the twenty-first century (see below). After production peaks, efforts will be directed toward the second half of

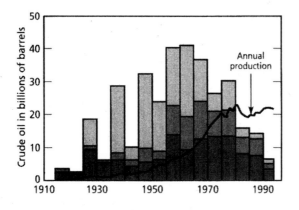

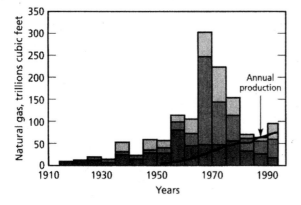

Middle East

Eastern Europe, Former Soviet Union, Asia, Oceania

Western Hemisphere, Africa, Western Europe

**Fossil Fuels. FIGURE 4.** World Annual Discovery and Production Rates for Conventional Crude Oil and Natural Gas.

Discovery rates are averaged over five-year periods. (From Masters et al., 1994, p. 537.)

mer Soviet Union the demand for oil, while growing, is not expected to return to the 1990 level by the year 2015.

According to the same forecast, world use of natural gas is projected to rise at three times the rate of crude oil increase because it is more abundant, cheaper, and less harmful to the atmosphere. It will displace other fuels in industrial applications, in electrical generation, and in residential and commercial uses. The world average growth in demand between 1995 and 2015 is 5 percent per year. In developing Asia it is expected to be 8 percent. In Central and South America, the growth in demand is expected to be over 5 percent, largely in electrical generation and industrial uses. Among industrial nations, the greatest growth in gas demand is projected for western Europe, where it will be used mostly for electrical generation. The slowest growth in gas demand is expected in eastern Europe and the former Soviet Union: although its growth rate will be only 2–3 percent per year, in the period 1995–2015 the demand will be up by 70 percent. One factor that encourages gas consumption is the growth in shipments of liquefied natural gas by tanker. Liquefaction and loading facilities are being built in Oman, Qatar, Nigeria, and Trinidad (adding to those in North Africa) to serve potential customers in Japan, South Korea, Taiwan, and Thailand.

***Oil and gas resources remaining.*** As with all other mineral resources, the remaining resources of oil and gas can be understood completely only if geologic certainty and economic feasibility are both considered. [*See* Resources.] Thus, proved reserves of oil or gas are the amounts well defined geologically and recoverable at today's prices and with current technology, while indicated reserves are larger amounts that assume the growth of existing fields as they mature, and also that

the world's ultimate crude oil resources, which will be more difficult to attain than the first half, despite the fact that a large proportion resides in the prolific fields of the Middle East. The largest and most obvious accumulations have been discovered, leaving the smaller and more remote occurrences for future exploration.

***Demand for oil and gas.*** One forecast of demand for crude (Energy Information Administration, 1997) shows a continuing increase in the next century (Figure 5). Among industrialized nations, the slowest growth of crude oil demand is expected in western Europe, where oil for power generation and home heating is being displaced by natural gas. Among developing countries, the demand in Asia is expected to grow at 4 percent per year, while in China the growth rate will be closer to 5 percent, with oil consumption tripling between 1995 and 2015 as China's transportation sector modernizes. Slightly lower growth in oil use is expected in Central and South America, while in eastern Europe and the for-

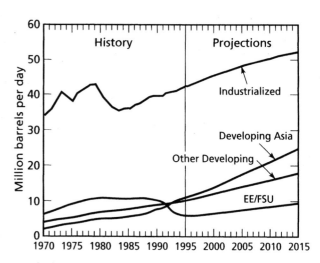

**Fossil Fuels. FIGURE 5.** World Oil Consumption by Region, 1970–2015. (From Energy Information Administration, 1997, p. 31.)

known fields will yield additional oil in response to new recovery methods. Although different nations use somewhat different definitions, it is possible to compile a nation-by-nation list in which indicated reserves may be compared directly. For a realistic assessment of all resources, it is then necessary to estimate the amounts of oil and gas economically recoverable from accumulations not yet discovered. Daunting as this may seem, there are established methods that depend on knowledge of the volumes of suitable sedimentary rock that have not yet been sufficiently explored (see Masters et al., 1994). Those two categories may be added to yield total remaining resources. Then, some assumed annual production (use) rate may be applied to estimate the life of the resource. That approach is handicapped by ignorance of future production rates and by the unrealistic assumption that all the remaining resource will be produced, whereas much of it may prove to be economically unattractive.

Nonetheless, it is useful to summarize all the crude oil remaining (Figure 6A). World reserves of crude oil are more than twice the undiscovered estimate, which underlines the maturity of the world's petroleum enterprise. For completeness, the total estimated resources of natural gas liquids are added to the illustration, al-

though these will become available only during production of natural gas: only the two major oil sands and heavy-oil deposits are represented there; and none of the world's oil shale deposits are included because their recoverable amounts are difficult to assess. Nevertheless, it is clear that the amounts thought to be recoverable from well-defined deposits in the Athabasca and Orinoco regions are equivalent to all the reserves (recoverable amounts) of conventional oil in the Middle East. If all the conventional crude were produced at the current rate of around 25 billion barrels per year, the 1,574 billion barrels remaining would last sixty-two years. The 567 billion barrels of unconventional oil, if all were consumed, would add another twenty-three years. For natural gas, undiscovered amounts are thought to be almost as large as reserves (Figure 6B), the sum of the two being 9,817 trillion cubic feet. This total resource remaining, if produced at the current rate of roughly 90 trillion cubic feet per year, would last 109 years. Such arithmetic does not yield a realistic scenario, but it does make the resource numbers more meaningful.

***Future rates of conventional crude oil production.*** A different and widely accepted approach to the future of oil or gas is to focus not only on the amounts

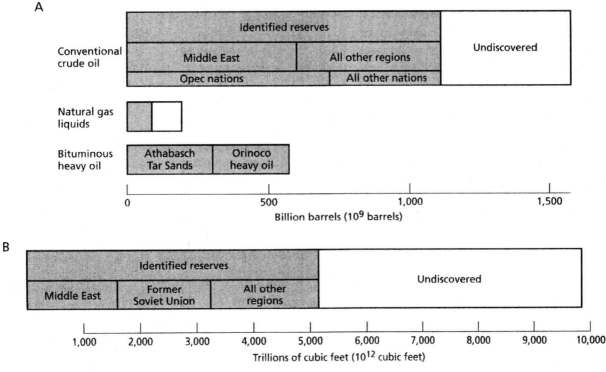

**Fossil Fuels. Figure 6.** (A) World's Remaining Crude Oil Resources as of 1 January, 1993. Bituminous heavy oil amounts are estimated recoverable from identified resources (Meger and Dewitt, 1990). Liquid petroleum gases (LPGs) such as propane are separated from natural gas at the wellhead. Organization of the Petroleum Exporting Countries (OPEC) nations are Venezuela, Algeria, Gabon, Libya, Nigeria, Iran, Iraq, Kuwait, Neutral Zone, Qatar, Saudi Arabia, United Arab Emirates, and Indonesia. (B) World's Remaining Conventional Natural Gas Resources as of 1 January, 1993. (From Masters et al., 1994, pp. 530–534.)

**Fossil Fuels. TABLE 1.** Recent Estimates of World Ultimate Conventional Crude Oil Resources (In Billions of Barrels)

| | CUMULATIVE PRODUCTION | IDENTIFIED RESERVES | UNDISCOVERED RECOVERABLE | ULTIMATE RECOVERY | YEAR OF PEAK PRODUCTION |
|---|---|---|---|---|---|
| Campbell and Laherrere (1998) | 800 | 850 | 150 | 1,800 | 2005 |
| Masters et al. (1994) | 800 | 1,103 | 471 | 2,272 | 2005–2020 |
| Edwards (1997) | 800 | 1,111 | 1,005 | 2,836 | 2020 |

remaining, but on the projected production history of the total resource endowment, part of which has been exhausted, and part of which remains. This is referred to as ultimate resources, or ultimate production.

Three estimates of ultimate resources have been published recently (Table 1). The first component, cumulative production for the world, can be agreed upon easily as 800 billion barrels. Identified reserves, though, include amounts that may be recoverable as known fields mature and the latest technologies are applied. The table shows two estimates near 1,100 billion barrels and one of only 850 billion barrels, which deliberately excludes 300 billion barrels of questionable reserves added in a revision made by six Organization of the Petroleum Exporting Countries (OPEC) members in 1987 (Campbell and Laherrere, 1998). The three authorities cited here understandably show a wide range in their estimates of oil recoverable from undiscovered accumulations, the most pessimistic being 150 billion barrels and the most optimistic being 1,005 billion. The matter of probabilities makes these estimates somewhat difficult to compare. Campbell and Laherrere believe their 150 billion to be the most realistic number. Masters et al. (1994) consider their 471 billion to be the most probable, citing 1,005 billion as their most optimistic number, consistent with a probability of only 5 percent. Edwards (1997), though, chooses that high-end number as his favorite, implying that future technologies and higher prices will lead to the greater recoverable amounts.

Experience with various mineral resources has shown that production of a resource increases gradually until half the ultimate resources are gone; then production declines abruptly. The process can be quantified, and an appropriate bell-shaped curve fitted to the history of production to date. This points to the year of maximum production and the beginning of decline (Ivanhoe, 1996). The numbers required are past production, estimated reserves, and estimated recovery from undiscovered accumulations as given in Table 1.

Applying the production curve methodology to ultimate recovery of only 1,800 billion barrels of oil, Campbell and Laherrere predict that world production will peak "in the first decade of the twenty-first century" (Table 1). Using the much larger ultimate recovery of 2,836 billion barrels (resulting from his more generous assessment of undiscovered recoverable amounts), Edwards predicts the production peak for crude oil in the year 2020. The resource estimate of Masters et al. leads to an intermediate date for the production peak.

***The fossil fuel era.*** On the basis of his generous estimate of crude oil resources, Edwards provides a graphic portrait of the beginning, peaking, and decline of the petroleum era (Figure 7). After crude oil production peaks in 2020, gas, being more abundant, peaks near 2030. As production of conventional crude declines after the year 2020, production of Canadian and Venezuelan crude expands in response to higher prices. Two coal futures are illustrated: in one the use of coal declines slightly from a peak in 2030; but in the alternative projection it expands to a maximum in the year 2100.

As production of total fossil fuel declines after 2030, the growing demand for energy (rising with population growth and increase of per capita use) will have to be met by some mixture of renewable energy sources and nuclear power (Edwards, 1997, p. 1302).

***Distribution of resources.*** Oil and gas fields occur only in areas with thick sedimentary rock. But, beyond that, some sedimentary basins, or areas where sediment accumulated and rock was formed, are richer in petroleum than others. Oil and gas fields of the Middle East, the U.S. Gulf Coast, Mexico, Venezuela, and North Africa are in regions that are now subtropical in climate. According to one interpretation, this is explained by a long view of geologic history: these favored areas were, at a critical period during the migration of continental plates, in equatorial or near-equatorial latitudes, a setting that favored the accumulation of organic material in source rock and also the deposition of carbonate rock and reefs, to serve eventually as reservoir rock (Klemme and Ulmishek, 1991). Furthermore, these conditions led to deposition of evaporite rocks, such as salt and gypsum, which serve as cap rocks for reservoirs and also deform under pressure to cause structural traps such as those in the U.S. Gulf Coast. [*See* Plate Tectonics.]

Whatever the mechanisms, the world's remaining conventional crude oil is now concentrated in the Middle East, which holds roughly half the identified reserves. OPEC nations hold 64 percent of identified reserves (Figure 8A). This region-by-region comparison, like that for natural gas below, relies upon a 1994 U.S.

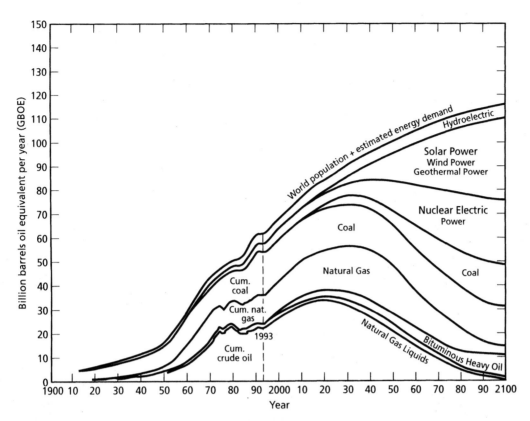

**Fossil Fuels. FIGURE 7.** The Fossil Fuel Era, Showing Production in Billion Barrels of Oil Equivalent. (Adapted from Edwards, 1997, p. 1301. AAPG Copyright 1997. Reprinted by permission of the American Association of Petroleum Geologists, whose permission is required for future use.)

Geological Survey analysis that was unique because its authors adjusted the reserve reports from over fifty nations to arrive at estimates of identified reserves that were roughly comparable across all nations (Masters et al., 1994). Unfortunately, there is no updated version of this study; but annual listings of the more conservative proved reserves reported by each nation are published each year in the *Oil and Gas Journal* and *World Oil*. The graph reveals that not all regions have been equally exploited: North America, for instance, has produced half its original endowment of conventional crude oil, while the Middle East has produced only 20 percent. This reflects the early beginning of the oil industry in the United States, and a period early in the twentieth century when the United States exported to many parts of the world. While undiscovered resources in the Middle East may not be large in relation to reserves, those massive reserves guarantee that this region will dominate crude oil supplies in the next century as resources dwindle in other regions.

A similar review of all conventional natural gas remaining, by region, shows an extreme concentration in the Middle East (especially Iran) and the former Soviet Union (Figure 8B). Once again, North America has produced nearly half its ultimate endowment of gas, while the two dominant regions have barely begun to exploit their resources. Given the attractiveness of natural gas, those regions will play a very important role in the decades ahead. Natural gas, possibly bolstered by unconventional sources, will be an important energy source as a fuel or as a feedstock for various processes.

**Energy Content of Coal, Oil, and Natural Gas Resources.** When the three fossil fuels are tabulated separately and in their own peculiar units—imperial tons for coal, barrels for oil, and cubic feet for natural gas—their relative importance as energy sources is not evident; but they can be compared by translating resource amounts into British thermal units (Btu) or into quads (one quad is one quadrillion Btu). For simplicity, undiscovered amounts of oil and gas are ignored, as are hypothetical resources of coal: only recoverable reserves of coal are compared with indicated reserves of conventional crude oil and natural gas (Table 2).

Crude oil and natural gas reserves appear to be roughly equivalent in energy content; but coal reserves contain five times the energy in crude oil reserves, and more than twice the energy in oil and gas combined.

[*See also* Conservation; Economic Levels; Energy; Energy Policy; Methane; Mining; Technology; *and* Transportation.]

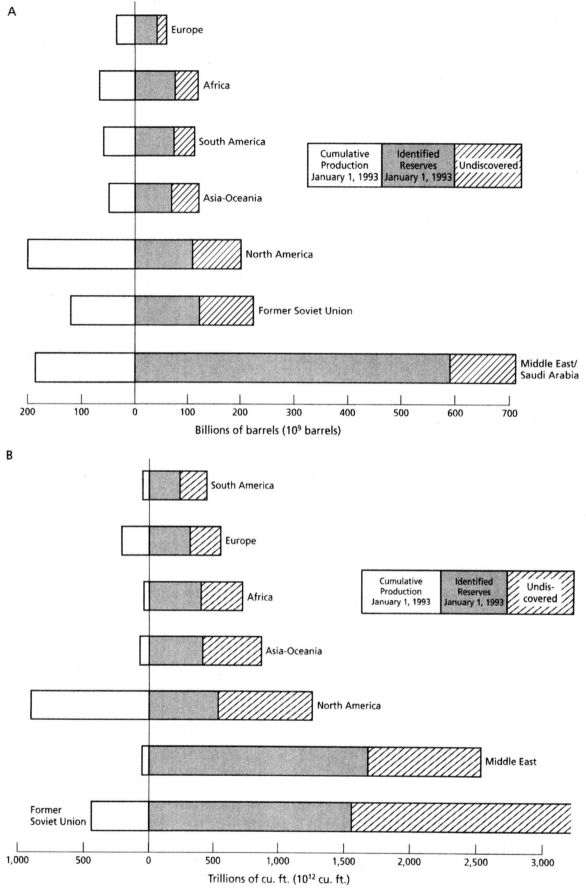

**Fossil Fuels. FIGURE 8.** Regional Distribution of Resources Produced and Remaining (Ultimate Resources) as of 1 January, 1993. (A) Crude Oil and (B) Natural Gas. (From Masters et al., 1994, pp. 530–534.)

**Fossil Fuels. TABLE 2.** Energy Content of Coal Versus Conventional Crude Oil and Natural Gas Reserves

| | | |
|---|---|---|
| Coal | Estimated recoverable reserves: | 1,143 billion short tons |
| | Average energy value per ton: | 22 million Btu |
| | Total energy content: | 25,146 quads |
| Crude Oil | Indicated reserves: | 1,103 billion barrels |
| | Average energy value per barrel: | 5.56 million Btu |
| | Total energy content: | 6, 132 quads |
| Natural Gas | Indicated reserves: | 5,136 trillion cubic feet |
| | Average energy content per cubic foot: | 1.035 thousand Btu |
| | Total energy content: | 5,315 quads |

SOURCE: "Recoverable Reserves of Coal," Energy Information Administration (1997, p. 65), "Indicated Reserves of Conventional Crude Oil and Natural Gas," Masters et al. (1994, pp. 530–534).

### INTERNET RESOURCE

International Energy Agency publications. http://www.caer.uky.edu/iea/ieapubs.htm/.

### BIBLIOGRAPHY

Anderson, R. N. "Oil Production in the 21st Century." *Scientific American* (March 1998), 86–91.

Bartsch, U., and B. Muller. *Fossil Fuels in a Changing Climate: Impacts of the Kyoto Protocol and Developing Country Participation.* Oxford: Oxford University Press, 2000.

British Petroleum Company. *BP Statistical Review of World Energy 1996.* London, 1996. [This review is issued annually.]

Campbell, C. J., and J. H. Laharrere. "The End of Cheap Oil." *Scientific American* (March 1998), 78–83.

Cuff, D. J., and W. J. Young. *The United States Energy Atlas.* New York: MacMillan, 1985.

Edwards, J. D. "Crude Oil and Alternate Energy Production Forecasts for the Twenty-first Century: The End of the Hydrocarbon Era." *American Association of Petroleum Geologists Bulletin* 81.8 (August 1997), 1292–1305.

Energy Information Adminstration. *International Energy Outlook 1997: With Projections to 2015.* DOE/EIA-0484(97). Washington, D.C.: U.S. Department of Energy, 1997.

Greinert, J., and E. Lausch. "Flammable Ice." *Scientific American* (November 1999), 76–83.

International Energy Agency. *Coal Information 1995.* Paris, 1996. Issued annually.

Ivanhoe, L. F. "Updated Hubbert Curves Analyse World Oil Supply." *World Oil* 217.11 (November 1996), 91–94.

Klemme, H. D., and G. F. Ulmishek. "Effective Petroleum Source Rocks of the World: Stratigraphic Distribution and Controlling Depositional Factors." *American Association of Petroleum Geologists Bulletin* 75.12 (December 1991), 1809–1851.

Masters, C. D., E. D. Attanasi, and D. H. Root. "World Petroleum Assessment and Analysis." In *Proceedings of the 14th World Petroleum Congress,* Stavanger, Norway. Chichester, U.K.: Wiley, 1994.

Oak Ridge National Laboratory. "1995 Estimates of $CO_2$ Emissions from Fossil Fuel Burning and Cement Manufacturing, Based on United Nations Energy Statistics and United States Geological Survey Cement Manufacturing Data." ORNL/CDIAC-25, NDP-030. Oak Ridge, Tenn., 1997.

Skinner, B. J. *Earth Resources.* 3d ed. Englewood Cliffs, N.J.: Prentice-Hall, 1986.

United Nations Statistical Division (UNSTAT). *1995 Energy Statistics Yearbook.* New York, 1997.

*World Oil.* See August issue for annual listing of oil and gas reserves and production by nation and region.

"Worldwide Look at Reserves and Production." *Oil and Gas Journal* (December 30, 1996). Annual listings by nation and region always appear in the last weekly issue of the year.

TECHNOLOGICAL CHANGE AND FOSSIL FUELS

Fouda, S. A. "Liquid Fuels from Natural Gas." *Scientific American* (March 1998), 92–95.

Schneider, D. "Burying the Problem." *Scientific American* (January 1988), 21–22.

—DAVID J. CUFF

## FRAMEWORK CONVENTION ON CLIMATE CHANGE

Adopted in 1992 under the auspices of the United Nations, the Framework Convention on Climate Change (FCCC) establishes the basic system of governance for the emerging global climate change regime (Table 1). Given the substantial uncertainties and complicated politics of the greenhouse warming issue, the FCCC requires very little by way of actual mitigation or adaptation measures. Instead, it creates a long-term evolutionary process to encourage further research, promote national planning, increase public awareness, and help create a community of states. As of 11 May 2000, the FCCC had 182 parties, making it one of the most widely accepted international environmental agreements. Pursuant to the FCCC, on 11 December 1997, the Third Conference of the Parties adopted the Kyoto Protocol, which establishes more stringent emission limitation commitments for developed countries. [*See* Kyoto Protocol.]

**History.** In 1985, an expert meeting held in Villach, Austria, sponsored by World Meteorological Organization (WMO) and United Nations Environment Programme (UNEP), concluded that significant anthropogenic climate change is highly probable and first suggested the development of a global convention to address the problem. In 1990, following the First Assessment Report of the Intergovernmental Panel on Climate Change (IPCC) and the Second World Climate Conference, the UN Gen-

**Framework Convention on Climate Change. TABLE 1.** Landmarks of the Emerging Climate Change Regime

| EVENT | DATE | ORGANIZER | OUTCOMES, CONCLUSIONS, RECOMMENDATIONS |
|---|---|---|---|
| Villach Conference | 1985 | WMO and UNEP | Significant climate change highly probable<br>States should initiate consideration of a global climate convention |
| Toronto Conference | 1988 | Canada | States should cut global $CO_2$ emissions by 20 percent by 2005<br>States should develop a comprehensive framework convention on the law of the atmosphere |
| UN General Assembly | 1988 | UN | Climate change a "common concern of mankind" |
| Hague Summit | 1989 | Netherlands, France, Norway | New institutional authority should be developed within the UN involving nonunanimous decision making |
| Noordwijk Conference | 1989 | Netherlands | Industrialized countries should stabilize greenhouse gas emissions as soon as possible<br>"Many" countries support stabilization of emissions by 2000 |
| IPCC First Assessment Report | 1990 | WMO and UNEP | Global mean temperature likely to increase by 0.3°C per decade under business-as-usual scenario |
| Second World Climate Conference | 1990 | WMO and UNEP | Countries need to stabilize greenhouse gas emissions<br>Developed states should establish targets and/or national programs or strategies to limit emissions |
| UN General Assembly | 1990 | UN | Establishment of INC/FCCC |
| INC 5 | 1992 | UN | Adoption of FCCC |
| UNCED | 1992 | UN | FCCC opened for signature; 154 signatories |
| FCCC's entry into force | 1994 | | Entry into force on 21 March, 1994, ninety days after the fiftieth ratification |
| Berlin Conference (COP-1) | 1995 | FCCC | Berlin Mandate for negotiations to strengthen FCCC commitments |
| Kyoto Conference (COP-3) | 1997 | FCCC | Kyoto Protocol adopted |
| COP-4 | 1998 | FCCC | Buenos Aires Plan of Action adopted. |

SOURCE: From Bodansky (1995).

eral Assembly established an intergovernmental negotiating committee (INC/FCCC) with the mandate of negotiating a convention containing "appropriate commitments" in time for signature at the June 1992 Earth Summit in Rio de Janeiro.

The INC/FCCC held five negotiating sessions between February 1991 and May 1992, when it finalized the text of the Framework Convention on Climate Change. The two principal issues were (1) quantified targets and timetables to limit greenhouse gas emissions, which were advocated by, among others, the European Union and the Alliance of Small Island States (AOSIS), but opposed by the United States; and (2) financial assistance. On the latter issue, developing countries sought new and additional funds for their incremental costs of implementation, as well as the creation of a new financial mechanism, while developed countries wished to give more limited assistance, administered by the recently established Global Environment Facility (GEF). The final text of the FCCC reflects compromises on both issues.

The Convention was opened for signature at the 1992 Earth Summit, where it was signed by 154 nations. It entered into force on 21 March 1994, following its ratification by 50 nations. One year later, the first Conference of the Parties (COP-1) concluded that the commitments contained in the FCCC were inadequate and adopted the Berlin Mandate, which called for the negotiation of additional commitments by industrialized countries. The Berlin Mandate negotiations concluded in December 1997 at COP-3, with the adoption of the Kyoto Protocol, which establishes emissions limitation commitments for each industrialized country, as well as market mechanisms that parties can use to meet these commitments, including emissions trading and the Clean Development Mechanism. Detailed rules elaborating the provisions of the Protocol are being negotiated pursuant to the 1998 Buenos Aires Plan of Action, which was scheduled for completion at COP-6 in November 2000. As of June 2000, the Kyoto Protocol had not entered into force.

**Framework versus Regulatory Convention.** As its name suggests, the FCCC establishes a general framework of governance rather than a detailed regulatory regime. Because of the difficulty of gaining agreement on substantive commitments, framework conventions have been widely used in international environmental law. Agreement proved particularly difficult in the climate change context given the uncertainties about the science of climate change, the potentially high economic stakes, and divergent national interests—for example, between big emitters of greenhouse gases such as the United States, rapidly developing countries such as China, oil-producing states, and small island developing states. Other examples of framework conventions include the 1979 Long-Range Transboundary Air Pollution Convention (LRTAP), which addresses the problem of acid rain in Europe, and the 1985 Vienna Convention on the Protection of the Ozone Layer. The rationale of the framework convention approach is to proceed in steps, beginning with the creation of a framework of institutions and mechanisms, which draws nations in, without requiring them to make detailed commitments, and only later elaborating more substantive regulations, generally in separate protocols such as the Montreal Protocol on Substances that Deplete the Ozone Layer or, in the case of the climate change regime, the Kyoto Protocol.

**Key Provisions of the FCCC.** The framework of governance set forth in the FCCC and the Kyoto Protocol include the provisions set out in Tables 2 and 3, respectively.

*Objective and principles.* To guide the evolution of the climate change regime, the FCCC defines its ultimate objective as the stabilization of atmospheric concentrations of greenhouse gases at safe levels (that is, levels that would "prevent dangerous anthropogenic interference with the climate system"). The elaboration of the regime will require the determination of (1) what concentrations levels are safe and (2) what emission reductions are necessary to achieve these concentrations. The FCCC also sets forth several general principles, including the principles of "common but differentiated responsibilities" (for example, different countries may have different obligations to regulate their emissions that cause global warming), inter- and intragenerational equity, precaution, cost-effectiveness, and sustainable development.

*Commitments.* As a framework convention, the FCCC imposes rather limited obligations, although industrialized (and in particular western) countries have somewhat more stringent obligations than developing countries. All parties to the FCCC have general obligations aimed at promoting (1) long-term planning through the development of national programs to mitigate and adapt to climate change and (2) international review of national actions. In addition, industrialized country parties identified in Annex I of the Convention) agreed to a nonbinding working target and timetable to return greenhouse gas emissions to 1990 levels by the year 2000, and western industrialized countries (identified in Annex II) agreed to provide financial assistance to de-

veloping countries, primarily to fund the preparation of national inventories and reports. The 1997 Kyoto Protocol sets forth legally binding emission limitation commitments for each industrialized country, ranging from a reduction of 8 percent to an increase of 10 percent compared to 1990 emission levels. These commitments apply to a basket of six greenhouse gases, for a five-year commitment period from 2008 to 2012.

*Joint implementation.* A controversial issue has been whether states must implement their commitments to limit greenhouse gas emissions at home, or whether they can do so through actions in other countries. The Kyoto Protocol includes mechanisms that allow industrialized countries to meet their commitments through emission reductions in other countries, including: (1) trading of emission allowances; (2) undertaking emission reduction projects in another industrialized country, an approach known as *joint implementation;* and (3) undertaking emission reduction projects in developing countries through the Clean Development Mechanism. The Kyoto Protocol states that reductions through these mechanisms must supplement domestic action, but does not specify any quantitative limits on their use.

*Institutions.* The Conference of the Parties (COP) meets yearly and is the principal decision-making body of the FCCC. It is assisted by the Subsidiary Body on Scientific and Technological Advice (SBSTA) and the Subsidiary Body on Implementation (SBI). The Convention also establishes a secretariat and a financial mechanism, currently operated by the Global Environment Facility (GEF). The IPCC has no formal role under the FCCC, but has continued to serve as the main source of scientific information relating to climate change.

*Reporting and review.* To promote transparency, provide information needed for the elaboration of the regime, and build trust among the parties, nation parties must submit periodic reports on their greenhouse gas inventories and national policies and measures. The reporting requirements are differentiated: developed countries have more stringent requirements than developing countries with respect to the content and timing of reports. The national reports are reviewed by experts, who provide in-depth analyses of individual reports as well as synthesis reports. The Kyoto Protocol sets forth more stringent requirements to monitor and report on emissions, as well as an expert review process and the development of a procedure to determine compliance with its binding emissions limitation commitments.

[*See also* Climate Change; Global Environment Facility; Global Warming; Intergovernmental Panel on Climate Change; Joint Implementation; *and* United Nations Conference on Environment and Development.]

**Framework Convention on Climate Change. Table 2.** Key Provisions of the FCCC

| | |
|---|---|
| Objective | Stabilize atmospheric greenhouse gas concentrations at a level that would prevent dangerous anthropogenic interference with the climate system, within a time-frame sufficient to (i) allow ecosystems to adapt naturally, (ii) protect food production, and (iii) allow sustainable economic development (art. 2) |
| Principles | Intra- and intergenerational equity; differentiated responsibilities and respective capabilities; right to promote sustainable development; precaution; cost-effectiveness; comprehensiveness; and free trade (art. 3) |
| Commitments | *All countries*—general commitments to: develop national greenhouse gas inventories; formulate national mitigation and adaptation programs; promote and cooperate in scientific research, education, training, and public awareness (arts. 4(1), 5, 6)<br>*Developed countries* (listed in Annex 1)—recognize that a return to earlier emission levels of $CO_2$ and other greenhouse gases by the end of decade would contribute to modifying long-term emission trends, and will report with aim to return to 1990 emission levels (art. 4(2))<br>*OECD countries* (listed in Annex 2)—commitments to: fully fund developing country inventories and reports; fund the incremental costs of agreed mitigation measures; provide assistance for adaptation; and facilitate, promote, and finance technology transfer (art. 4(3)-(5)) |
| Institutions | Conference of the Parties (art. 7), Secretariat (art. 8), Subsidiary Body for Scientific and Technological Advice (SBSTA) (art. 9), Subsidiary Body for Implementation (SBI) (art. 10), financial mechanism (art. 11) |
| Reporting ("communication of information") | *All countries*—communication of information on national greenhouse gas inventories and on steps taken to implement the Convention (art. 12(1))<br>*Developed countries* (listed in Annex 1)—detailed description of policies and measures to limit greenhouse gas emissions and enhance sinks, and a specific estimate of their effects on emissions |
| Adjustment mechanism | Reviews of the adequacy of commitments every three years, based on the best available scientific information (art. 4(2)(d)) |

SOURCE: From Bodansky (1995).

**Framework Convention on Climate Change. Table 3.** Key Provisions of the Kyoto Protocol

| | |
|---|---|
| Emission reduction commitments | Specific emission limitation commitments for each industrialized country set forth in Annex B. Commitments apply to basket of six greenhouse gases (carbon dioxide, methane, nitrous oxide, and three trace synthetic gases.<br>First five-year commitment period runs from 2008–2012. Negotiations on second commitment period to begin no later than 2005. |
| Sinks | Emissions and removals due to afforestation, reforestation, and deforestation since 1990 count toward emission targets. Other sink activities can be added by decision of the Parties. |
| Emissions trading | Industrialized countries may trade their emission allowances. |
| Joint implementation | Industrialized countries may receive credit toward their emission reductions, resulting from projects undertaken in another industrialized country. |
| Clean Development Mechanism | Industrialized countries may receive credits toward their targets for emission reductions resulting from projects undertaken in developing countries. CDM governed by an executive board, with specific projects overseen by "operating entities" (for example, multinational accounting firms). |
| Institutions | Generally the same as FCCC institutions. Conference of the Parties serves as meeting of the Protocol Parties (COP/MOP). |
| Reporting and reviewing | Industrialized parties must have "national systems" to monitor and report on their greenhouse gas emissions. Emission inventories must follow IPCC inventory guidance. National inventories reviewed by expert review teams, which can recommend adjustments to inventory numbers that fail to follow IPCC guidelines. |
| Compliance | As of June 2000, compliance institutions and rules under negotiation pursuant to Buenos Aires Plan of Action. |

## BIBLIOGRAPHY

Bodansky, D. "The Emerging Climate Change Regime." *Annual Review of Energy and Environment* 20 (1995), 425–461. General overview through April 1995.

———. "The United Nations Framework Convention on Climate Change: A Commentary." *Yale Journal of International Law* 18 (Summer 1993), 451–558. A detailed history and legal analysis of the FCCC.

Churchill, R., and D. Freestone, eds. *International Law of Global Climate Change*. London: Graham & Trotman, 1991. An older but useful collection of essays on legal issues relating to climate change.

Grubb, M. *The Kyoto Protocol: A Guide and Assessment*. London: Royal Institute of International Affairs, 1999. Overview of the Kyoto Protocol.

Mintzer, I. M., and J. A. Leonard, eds. *Negotiating Climate Change: The Inside Story of the Rio Convention*. Cambridge: Cambridge University Press, 1994. Interesting accounts of the FCCC negotiations by key participants.

Oberthür, S., and H. E. Ott. *The Kyoto Protocol: International Climate Policy for the 21st Century*. New York: Springer, 1999.

Patterson, M. *Global Warming and Global Politics*. London: Routledge, 1996. An excellent survey of political science analyses of the FCCC negotiations.

Susskind, L. E. *Environmental Diplomacy: Negotiating More Effective Global Agreements*. New York: Oxford University Press, 1994. Criticisms of the framework convention/protocol approach.

Victor, D. G., and J. E. Salt. "From Rio to Berlin: Managing Climate Change." *Environment* 36 (1994), 6–15, 25–32. A perceptive analysis of post-Rio developments.

—DANIEL M. BODANSKY

# FRIENDS OF THE EARTH

Established in the United States in 1969, Friends of the Earth (FOE) links the century-old preservationist tradition associated with Sierra Club founder John Muir to a wide-ranging, contemporary global agenda. Its founder, the noted conservationist David Brower, had been forced to resign as Sierra Club Executive Director after his actions led to loss of the Club's tax-exempt status. His new organization would have an international outlook, an activist orientation, and an expanded set of core issues. Yet there would be constant tensions over FOE's agenda and management style, ultimately leading to Brower's departure in 1984.

In the 1970s, FOE turned its attention to new environmental issues, such as nuclear energy and the Supersonic Transport. It was also instrumental in the development of the League of Conservation Voters, an overtly political organization that monitors politicians and supports environmentally friendly candidates. FOE established its first international offices in Paris (1970) and London (1971). Currently, Friends of the Earth International (FOEI) has fifty-eight national members. Although autonomous, they operate under guidelines established by FOEI, whose secretariat is located in Amsterdam. The national organizations are highly diverse and are active, to varying degrees, in global, national, and local affairs. Some groups, such those in the United States and the United Kingdom, are very active in domestic politics; others, such as FOE-Japan, focus mainly on international issues.

Beyond its broad mission of protecting the Earth and repairing past damages, FOEI promotes citizen participation, democratic decision-making, social, economic, and political justice, and environmentally sustainable economic development. FOEI is particularly active in monitoring the activities of, and advocating reforms for, such institutions as the International Monetary Fund, the World Bank, the International Tropical Timber Organization, the Asian Development Bank, and the U.S. Agency for International Development. Among the principal issues that engage FOEI are global climate change, ozone depletion, trade liberalization, toxic chemicals, ocean pollution, tropical and temperate forest protection (the Rainforest Action Network is an FOE affiliate), management of the Antarctic and Arctic, and the environmental impacts of war. FOEI has been an active participant in international treaty negotiations, including the Montreal Protocol on ozone depletion and the United Nations climate change convention.

Much FOE activity is directed toward economic issues. Rather than advocating radical economic change, most FOE campaigns are oriented toward regulation of market economics in the service of environmental and social sustainability. In contrast with their adversarial origins, the U.K. and U.S. organizations are now regarded by some environmental activists as too accommodating of established corporate interests. The U.S. organization, for example, participates in national and local discussions with the steel industry and engages in efforts such as the Green Scissors campaign to reduce wasteful spending that appeal to politically conservative elements of society. At the same time, however, FOE has developed cooperative relationships with local citizens' groups and labor unions to work on issues ranging from opposition to hazardous waste facilities to river restoration.

[*See also* Environmental Movements; *and* Nongovernmental Organizations.]

## INTERNET RESOURCE

http://www.xs4all.nl/~foeint/. Provides up-to-date information on FOE activities and publications, with links to Web sites of national groups.

## BIBLIOGRAPHY

Burke, T. "Friends of the Earth and the Conservation of Resources." In *Pressure Groups in the Global System: The Transnational Relations of Issue-Oriented Non-Governmental Organizations*, edited by P. Willets. London: Frances Pin-ter, 1982. Describes the origins of FOE and its activities in the United Kingdom, Europe, and the rest of the world.

Jordan, G., and W. Maloney. *The Protest Business? Mobilizing Campaign Groups.* Manchester and New York: Manchester University Press, 1997. Examines public support for FOE in Britain and critically analyzes relevant social-movement theories.

Lamb, R. *Promising the Earth.* London: Routledge, 1996. A "celebratory" account of FOE-UK's first twenty-five years.

—ROBERT J. MASON

## FUTURE STUDIES

Future studies refers to the study of forces creating the future—both their possible outcomes and their implications for present decisions.

Three methodologies predominate in future studies. In conventional forecasting, rigorous mathematical models predict future conditions. Other methods focus on gathering and synthesizing expert opinion about the future. Finally, scenario planning methods utilize speculative, but carefully constructed, descriptions of future worlds. All three methodologies arose in the United States in the aftermath of World War II—a time of relatively stable geopolitics and industrial structures. This originating postwar context shaped the early development of the discipline and, later, contributed to its crisis.

At the outset of the cold war in the early 1950s, the U.S. military created a number of units dedicated to forecasting future conditions in the Soviet Union. Faced with the prospect of huge defense expenditures, leaders hoped that such forecasts could guide future defense investments, especially in intercontinental ballistic missiles and long-range bombers. (One of these units was installed by the Air Force at Douglas Aircraft; it later became famous as an independent research institute, the Rand Corporation.) Paralleling military efforts, such major U.S. corporations as General Electric, U.S. Steel, and General Motors were also beginning to experiment with new planning techniques. The war effort had taught large industrial enterprises a host of new methods of management and control, but the application of these complex practices demanded longer-term planning by managers. Business leaders sought methods for predicting future market conditions—demand for products, optimum manufacturing capacity—to steer the explosive growth of their organizations.

By the late 1950s, an emerging discipline of future studies combined military and business planning techniques. It was mathematical forecasting that would dominate this discipline for its first twenty years, as models became ever more ambitious and sophisticated. One of the most prominent was called *industrial dynamics*, a model of the evolution of whole industries. Developed by Jay Forrestor, this model grew into modern systems dynamics. Meanwhile, funding continued to flow from big business and the military, and academic departments of future studies and systems dynamics were established at Houston, Sussex, Hawaii, and Massachusetts Institute of Technology. A wave of substantial books on future studies received wide attention, and numerous journals on the subject appeared, including *Technological Forecasting and Social Change*, *Future Studies*, *Future*, and *Futuribles*. Independent institutions dedicated to future studies flourished: The Rand Corporation, The Futures Group, the Institute for the Future, and the Hudson Institute, among others.

Academic future studies reached its peak in the late 1960s and early 1970s. However, there had been as yet little consideration of the limitations of predictive modeling. Were there social phenomena that could not be forecast using conventional methods? Forecasting methods assumed that social phenomena were systems whose structural dynamics could be identified through the analysis of historical data and then extrapolated into the future. But conventional forecasting was in reality constrained by the specific priorities of its postwar founders (priorities such as economic and business indicators, military capabilities, and technology), and by the necessity that relevant data be tightly defined (easy to measure and quantify). This may help to explain the persistent inattention of future studies to the more heterogeneous issues of social transformation and ecological crisis that the public considered urgent by the late 1960s.

Conventional forecasting was always aimed at prediction. Some commentators, such as Donald Michael (then of the University of Michigan), questioned the viability of this goal in the 1960s. But it was the series of domestic and global crises in the early 1970s, unanticipated by most conventional planners, that most seriously damaged the reputation of predictive future studies. The environmental movement, the oil and energy crisis, inflation and economic volatility, the war in Vietnam, and political upheaval in China, coming in succession, led many to question whether forecasting was capable of anticipating major historical changes.

The crisis of the 1970s spurred work in alternative branches of future studies that had previously remained obscure. One was the Delphi study, a multiple-round survey of experts that sought a convergence of opinion on a central question. However, Delphi studies did not have much predictive success, and soon fell out of use. Another tool was the scenario—a broad, qualitative story about a future world. In contrast to theoretical forecasting, scenario planning was an essentially empirical method with pragmatic aims. Herman Kahn had pioneered scenario planning in the 1950s at the Rand Corporation, but while Kahn had developed one master scenario, planners now started from the premise that alternative scenarios could represent divergent futures. At institutes like the Stanford Research Institute and Battelle, and in the planning department of Royal Dutch/Shell, the multinational oil company, planners developed multiple scenarios as planning tools. Proponents believed that the exercise of constructing several plausible scenarios could alert planners to unexpected changes in political, social, or economic environments.

A divide had now appeared in future studies, between the predictive methods of forecasting and the scenario-based methods that assumed an inherently unpredictable future. In the late 1970s and 1980s, however, diminished expectations for conventional forecasting continued to undermine the academic side of the field. Scenario-based future studies, lacking a rigorous methodology or a mathematical component, was not readily adopted by academics trained in forecasting; scenario planning was developed privately in a few corporations and research institutes. Meanwhile, conventional future studies attracted fewer young people. Many university departments were eliminated; future-studies journals, learned societies, and book publishing all suffered as well. Mathematical forecasting was still used, applied to more limited questions, and indeed continued to grow in sophistication, but the field of future studies as a whole lay fallow in the 1980s.

Recently, scenario planning has won new adherents and is beginning to resuscitate popular interest in future studies. The success of corporations that engaged in scenario work—notably Royal Dutch/Shell, which prospered through two oil crises—attracted considerable interest among business and military planners. Scenario planning is now being taught in many business schools, and is generating a methodological literature. At the same time, the scenario school itself has split. On one side are practitioners who seek to develop rigorous methodologies for constructing scenarios; on the other, practitioners who believe that scenarios are most effective as flexible, pragmatic learning exercises.

[*See also* Modeling of Natural Systems; *and* Scenarios.]

## BIBLIOGRAPHY

HISTORICAL SOURCES ON FUTURE STUDIES
Most of the seminal early works on forecasting and scenario planning are now out of print or hard to find. Examples include the following:

Kahn, H., and A. J. Wiener. *The Year 2000: A Framework for Speculation on the Next Thirty-three Years*. New York: Macmillan, 1967.
Martino, J. P. *An Introduction to Technological Forecasting*. New York: Gordon and Breach, 1987.
Meadows, D., and Meadows, D. *The Limits to Growth*.
Michael, D. *The Next Generation*.

CURRENT SOURCES ON SCENARIO PLANNING
Schwartz, P. *The Art of the Long View*. New York: Doubleday, 1991. An introduction to scenario planning for the general reader.
Van der Heijden, K. *Scenarios: The Art of Strategic Conversation*. Chichester, U.K.: Wiley, 1996. A detailed overview of the scenario-planning process.

—PETER SCHWARTZ AND MILES B. BECKER

**FYNBOS.** *See* Heathlands.

# G

## GAIA HYPOTHESIS

The Gaia hypothesis postulates that the Earth's surface environment is maintained in a habitable state by self-regulating feedback mechanisms involving organisms tightly coupled to their environment. The Earth's atmospheric composition, climate, much of the chemical composition of the ocean, and the cycling of many elements essential to life are hypothesized to be regulated. The idea arose from the involvement of the British independent scientist and inventor James Lovelock in the 1960s space program, and was developed during the early 1970s in collaboration with the American microbiologist Lynn Margulis.

Lovelock was employed by the U.S. National Aeronautics and Space Administration (NASA) as part of the team that aimed to detect whether there was life on Mars. Lovelock's interest in atmospheric chemistry led him to seek a general, physical basis for detecting the presence of life on a planet. He recognized that most organisms shift their physical environment away from equilibrium. In particular, organisms use the atmosphere to supply resources and as a repository for waste products. In contrast, the atmosphere of a planet without life should be closer to thermodynamic equilibrium, in a state attributable to photochemistry (chemical reactions triggered by solar ultraviolet radiation). Thus, the presence of abundant life on a planet may be detectable by atmospheric analysis.

Such analysis can be conducted from Earth using an infrared spectrometer (which detects the characteristic absorption due to specific gases) linked to a telescope. Using this technique, it was discovered that the atmospheres of Mars and Venus are dominated by carbon dioxide and are relatively close to chemical equilibrium, suggesting that they are lifeless. In contrast, the atmosphere of the Earth is in an extreme state of disequilibrium because of the activities of life, in which highly reactive gases such as methane and oxygen coexist many orders of magnitude from the photochemical steady state. Remarkably, despite this disequilibrium, the composition of the Earth's atmosphere was known to be fairly stable over geologic periods of time. Lovelock concluded that life must regulate the composition of the Earth's atmosphere.

Interestingly, the composition of the Earth's atmosphere is particularly suited to the dominant organisms. For example, nitrogen maintains much of the atmospheric pressure and serves to dilute oxygen, which at 21 percent of the atmosphere is just below the level at which fires would disrupt land life. Yet oxygen is sufficiently abundant to support the metabolism of large respiring organisms such as humans. Both oxygen and nitrogen are biological products—oxygen is the product of past photosynthesis, while the gaseous nitrogen reservoir is largely maintained by the actions of denitrifying organisms (which use nitrate as a source of oxygen and release nitrogen gas). Furthermore, the oxygen content of the atmosphere has remained within narrow bounds for over 350 million years. [*See* Atmospheric Chemistry.]

It is also remarkable that life on Earth has persisted despite major changes in the input of matter and energy to the Earth's surface. Most notably, the Sun is thought to have warmed by about 25 percent since the origin of life on Earth, over 3.8 billion years ago. This increase in solar output alone should raise the Earth's surface temperature by about 20°C. Yet the current average temperature is only 15°C. The continuous habitability of the Earth in the face of a warming Sun suggested to Lovelock that life may have been regulating the Earth's climate in concert with its atmospheric composition.

The idea was named "Gaia" after the Greek goddess of the Earth, and was first published in 1972. Lovelock then sought an understanding of the organisms that might be involved. Lynn Margulis was already developing the theory of symbiogenesis—that eukaryotic cells (those with genetic material contained within a distinct nucleus) evolved from the symbiotic merger of previously free-living prokaryotes (organisms, including bacteria, whose genetic material is not enclosed within a cell nucleus). Margulis contributed her intimate knowledge of microorganisms and the diversity of chemical transformations that they mediate to the development of what became the Gaia hypothesis of "atmospheric homeostasis by and for the biosphere."

The Gaia hypothesis was used to make predictions, for example, that marine organisms would make volatile compounds that can transfer essential elements from the ocean back to the land. Lovelock and colleagues tested this ancillary hypothesis on a scientific cruise between England and Antarctica in 1972. They discovered that the biogenic gases dimethyl sulfide and methyl iodide are the major atmospheric carriers of the sulfur and iodine cycles. Later, the Gaia hypothesis was extended to include regulation of much of the chemical composition of the ocean.

**Daisyworld.** The Gaia hypothesis was greeted with hostility from many scientists and leading scientific journals, partly because of its mythological name. The first scientific criticism of the hypothesis was that it implies teleology, some conscious foresight or planning by the biota. Most subsequent criticisms have focused on the need for evolutionary mechanisms by which regulatory feedback loops could have arisen or been maintained. As Richard Dawkins pointed out (Dawkins, 1983) the Earth is not a unit of selection, so Gaian properties cannot be "adaptations" in a strict neo-Darwinian sense since they cannot be refined by natural selection. This poses a challenge to explain how such properties could arise.

The Daisyworld model (Figure 1) was formulated to demonstrate that planetary self-regulation does not necessarily imply teleology. It provides a hypothetical example of climate regulation emerging from competition and natural selection at the individual level. Daisyworld is an imaginary gray world orbiting, at a similar distance to the Earth, a star like our Sun, which gets warmer with time. The world is seeded with two types of life, black and white daisies. These share the same optimum temperature for growth, 22.5°C, and limits to growth of 5°C and 40°C. When the temperature reaches 5°C, the first seeds germinate. The paleness of the white daisies makes them cooler than their surroundings, hindering their growth. The black daisies, in contrast, warm their surroundings, enhancing their growth and reproduction. As they spread, the black daisies warm the planet. This further amplifies their growth and they soon fill the world. At this point, the average temperature has risen close to the optimum for daisy growth. As the sun

warms, the temperature rises to the point where white daisies begin to appear in the daisy community. As it warms further, the white daisies gain a selective advantage over the black daisies and gradually take over. Eventually, only white daisies are left. When the solar forcing gets too high, regulation collapses.

Daisyworld illustrates that both positive and negative feedback are important for self-regulation. While the solar input changes over a range equivalent to 45°C, the surface of the planet is maintained within a few degrees of the optimum temperature for daisy growth. The initial spread of life is characterized by positive feedback—the more life there is, the more life it can beget. This is coupled to an environmental positive feedback—the warming due to the spread of black daisies enhances their growth and reproduction rates. The long period of stable, regulated temperature on Daisyworld represents a predominance of negative feedback. If the temperature of the planet is greatly perturbed by the removal of a large fraction of the daisy population, then positive feedback acts rapidly to restore comfortable conditions and widespread life. The end of regulation on Daisyworld is characterized by a positive feedback decline in white daisies—solar warming triggers a reduction in their population that amplifies the rise in temperature.

The modeling approach pioneered in Daisyworld provided the beginnings of a mathematical basis for understanding self-regulation. It was soon adapted to study the regulation of atmospheric composition and climate on the early Earth. With this work, Lovelock began to refer to Gaia as a theory, in which self-regulation is understood as a property of the whole system of life tightly coupled to its environment. This replaced the original

**A**

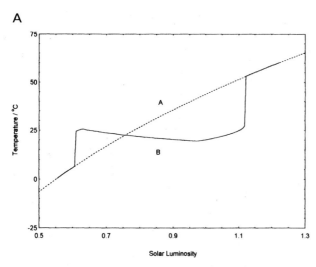

**B**

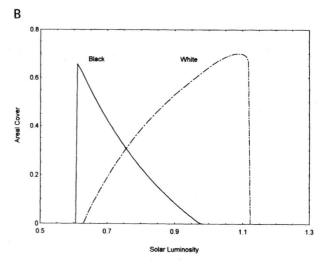

**Gaia Hypothesis. Figure 1.** The Daisyworld Model.

A thought experiment to demonstrate that planetary self-regulation can emerge from natural selection at the individual level between types of life with different environment-altering traits. In this case the traits are "darkness" (albedo = 0.15) and "paleness"

(albedo = 0.65) of black and white daisies on a gray planet (albedo = 0.4). (A) Planetary temperature as solar luminosity increases. The dashed line (A) is without daisies and the bold line (B) is with daisies. (B) Areal cover of black and white daisies.

suggestion that regulation is "by and for the biota" (which is often interpreted as teleological, although never intended as such). The term *homeostasis*, which refers to regulation around a fixed set point, has also been revised. More appropriate is Margulis's suggestion of *homeorrhesis*, which describes regulation around an evolving point.

The Gaian view of Earth history is one of periods of stability—of the environment and life, coupled together—interspersed with periods of rapid change. Such a pattern of punctuated equilibria appears consistent with the geologic record of both environmental proxies and species. Furthermore, evidence that the Earth has remained habitable despite major, periodic disruptions, including the impact of planetesimals and massive volcanic outbursts, supports the notion that the Earth is self-regulating. These events appear to have caused mass extinctions and climate change and yet, in all cases, diverse, widespread life and a tolerable climate returned within a short period of geologic time.

**Climate Regulation.** Climate feedbacks somewhat analogous to those in Daisyworld can be found in the real world. For example, the trees of the boreal forests can be likened to the dark daisies. They possess traits of shedding snow and darkness that give them a low albedo (reflectivity) and make them warmer than their surroundings. The presence of forest warms the high northern latitudes by approximately 4°C in winter. Over much of the surface of the Earth and over longer time scales, the Gaia theory predicts, from the relatively high solar input at present, that the predominant effect of organisms should be to cool the planet.

A Gaian mechanism for long-term climate regulation, involving the biological amplification of rock weathering, was proposed in the early 1980s. Over million-year time scales, the carbon dioxide content of the Earth's atmosphere and the resulting greenhouse effect on the Earth's temperature is determined by the balance of carbon dioxide input and removal fluxes. Removal occurs in the process of weathering of silicate rocks on land and the subsequent formation of carbonate rocks in the ocean. A chemical negative feedback mechanism exists whereby (for example) increases in planetary temperature are counteracted by increases in the rate of rock weathering and the uptake of carbon dioxide. However, the rate of rock weathering is greatly enhanced by the activities of soil microbes, plants, and lichens. This offers the potential for more responsive stabilization of the Earth's temperature. For example, rising temperature may trigger increased plant growth and microbial respiration, which reduces the carbon dioxide content of the atmosphere. The evolution of biological amplification of rock weathering is estimated to have progressively reduced the level of carbon dioxide in the Earth's atmosphere and counteracted the

warming Sun, such that this process now cools the Earth by roughly 20°C.

In the mid-1980s, Gaian thinking led to the hypothesis that production of dimethyl sulfide (DMS) by marine phytoplankton also cools the climate. DMS is oxidized in the atmosphere to form sulfate aerosol particles that can grow, often in combination with ammonium (another biological product), to become cloud condensation nuclei (CCNs). Increases in the number density of CCNs make clouds more reflective, increasing the scattering of solar radiation back to space and thus causing cooling. Temperature affects phytoplankton growth directly and also determines the degree of stratification in the ocean water column and hence the supply of nutrients to the surface layers. Hence there is potential for climate feedback involving the growth of DMS-emitting phytoplankton. The nature of this feedback is the subject of intensive ongoing research. [*See* Sulfur Cycle.]

A regional example of self-regulation is the Amazon rainforest, where the trees, through generating a high level of water cycling, maintain the moist environmental conditions in which they can persist. Nutrients are also effectively retained and recycled, in contrast to the nutrient-poor soil. If too much forest is removed, the water-regulation system can collapse, the topsoil is washed away, and the region reverts to arid semidesert. Such change may be irreversible.

**Current Work.** In recent years, the implications of Gaian feedback for ecology and evolution have been explored. The Daisyworld model has been extended to include different types of "daisy" as well as herbivores and carnivores. Studies of biodiversity within this context (of plant life tightly coupled to climate) suggest that the potential for biodiversity is an essential part of an ecosystem's capacity to respond to perturbation. Different herbivore-feeding strategies have been found to have different effects on the self-regulating capacity of the system. Increases in the number of connections in a Daisyworld food web have been found to increase the stability of both population dynamics and climate, as has the introduction of carnivores to a model with only herbivores and plants. Simulations of habitat fragmentation in Daisyworld have revealed that a critical threshold exists at which the plants become isolated in separate "islands" and regulation breaks down. This emphasizes the importance of spatial interaction for self-regulation.

The challenge of reconciling the theories of Gaia and natural selection is now being readdressed, and evolutionary biologists are showing renewed interest in Gaia. When random mutation of the albedo of the daisies is incorporated in Daisyworld, this extends the range of temperature regulation. However, Daisyworld only represents one special case of a direct connection between the effect of a trait on its bearer and on the global environment. To test further the effect of natural selection on environmental regulation, new models are being developed that incorporate the random generation of environment-altering traits.

Fresh emphasis is also being placed on the importance of life in increasing the cycling of nutrient elements, both on land and in the ocean. For example, it has long been recognized that phosphate and nitrate are available in ocean waters in just the ratio required by phytoplankton. The effects of nitrogen fixation, denitrification, preferential recycling of phosphorus from ocean sediments, and the resultant feedbacks are being modeled to test whether they can account for such regulation. This is generating a new focus on the molecular biology of Gaia—the enzymes responsible for the regulation and the trace elements crucial to their functioning. [See Biogeochemical Cycles.]

Gaian hypotheses concerning the mechanisms responsible for gradual onset and rapid termination of ice ages have been put forward and modeled. These may help us to understand and predict the biosphere's response to global change. Contemporary observations indicate that members of both the marine and terrestrial biota are involved in removing a significant fraction of the excess carbon dioxide released to the atmosphere by human activities. However, this negative feedback is not sufficient to prevent the carbon dioxide content of the atmosphere from rising. Furthermore, Lovelock has predicted that increasing temperature and resultant stratification of the ocean will trigger a decline in phytoplankton and their cooling effect via DMS emissions, providing a positive feedback on global warming. [See Global Warming.]

The Gaia hypothesis has contributed greatly to our understanding of the Earth as a whole system. The previous view that life is merely a passenger on a dead planet has largely been replaced with recognition of the coevolution of organisms and their environment. The degree to which organisms are involved in regulating conditions at the surface of the Earth remains a subject of controversy, but the concept has proved its worth in stimulating valuable research. It may thus offer a new paradigm for environmental science.

[See also System Dynamics.]

## BIBLIOGRAPHY

### GENERAL REFERENCES

Bunyard, P., ed. Gaia in Action: Science of the Living Earth. Edinburgh: Floris Books, 1996. Collected papers from three meetings on the scientific and philosophical implications of the Gaia hypothesis.

Lovelock, J. E. Gaia: A New Look at Life on Earth. New York and Oxford: Oxford University Press, 1979. The classic exposition of the hypothesis, born out of the frustration of censorship from scientific journals. A rare and inspiring blend of science and poetry.

————. The Ages of Gaia: A Biography of our Living Earth. New York and Oxford: Oxford University Press, 1988. At present, the definitive scientific exposition of what the author now describes as the Gaia theory. A thorough response to the criticisms triggered by the original hypothesis, seeking consistency with natural selection, and replete with suggested and (to a lesser degree) tested regulatory mechanisms.

————. Gaia: The Practical Science of Planetary Medicine. London: Gaia Books, 1991. An accessible and well-illustrated introduction to the subject for the general reader.

Margulis, L., and D. Sagan. Microcosmos: Four Billion Years of Microbial Evolution. London: Allen & Unwin, 1987. The microbial history of life and evolution. Stresses the importance of microbes in regulating the Earth's surface environment.

————. Slanted Truths: Essays on Gaia, Symbiosis and Evolution. New York: Springer, 1998. Provides a varied introduction to the work and thought of Lynn Margulis and colleagues.

Schneider, S. H., and P. J. Boston, eds. Scientists on Gaia. Cambridge, Mass. and London: MIT Press, 1991. Edited scientific and philosophical papers from the 1988 Chapman Conference on the Gaia hypothesis (sponsored by the American Geophysical Union). Covers a broad spectrum of views on the subject.

Volk, T. Gaia's Body: Toward a Physiology of the Earth. New York: Springer, 1998. Emphasizes the greatly amplified cycling of essential elements resulting from the existence of life.

Williams, G. R. The Molecular Biology of Gaia. New York: Columbia University Press, 1996. Focuses on the biochemistry of the enzymes that catalyse matter transfers between organisms and their environment, as a route to understanding the regulation of global biogeochemical cycles.

### SCIENTIFIC PAPERS

Charlson, R. J., J. E. Lovelock, M. O. Andreae, and S. G. Warren. "Oceanic Phytoplankton, Atmospheric Sulphur, Cloud Albedo and Climate." Nature 326 (1987), 655–661. Much cited paper, proposing both a mechanism of climatic cooling due to phytoplankton and (more speculatively) a resulting regulatory feedback.

Kump, L. R., and J. E. Lovelock. "The Geophysiology of Climate." In Future Climates of the World: A Modelling Perspective, edited by A. Henderson-Sellers, pp. 537–553. Oxford: Elsevier, 1995. An accessible review of postulated climate feedback mechanisms involving life.

Lovelock, J. E. "Gaia as Seen through the Atmosphere." Atmospheric Environment 6 (1972), 579–580. The first "Gaia" paper.

Margulis, L., and J. E. Lovelock. "Biological Modulation of the Earth's Atmosphere." Icarus 21 (1974), 471–489. One of a collection of jointly authored papers that clarified the Gaia hypothesis and proposed regulatory mechanisms.

Watson, A. J., and J. E. Lovelock. "Biological Homeostasis of the Global Environment: The Parable of Daisyworld." Tellus 35B (1983), 284–289. Gives the equations, mathematical analysis, and an interesting variant of the Daisyworld model.

Whitfield, M. "The World Ocean: Mechanism or Machination?" Interdisciplinary Science Reviews 6 (1981), 12–35. A comprehensive review that extends the Gaia hypothesis to include biological control of aspects of the chemical composition of the ocean.

### CONTRASTING VIEWS

Dawkins, R. The Extended Phenotype. New York and Oxford: Oxford University Press, 1983.

Doolittle, W. F. "Is Nature Really Motherly?" *The CoEvolution Quarterly* (Spring 1981), 58–63. A thoughtful critique of Lovelock's first book (Lovelock, 1979). The author argues that the Gaia hypothesis is inconsistent with natural selection. For a response to this, see Watson and Lovelock, 1983; Lovelock, 1988.

Holland, H. D. *The Chemical Evolution of the Atmosphere and Oceans*. Princeton: Princeton University Press, 1984. A thorough textbook on the Earth's geochemical history. The author also argues that the Gaia hypothesis is not necessary to explain the continuity of life on Earth for 3.8 billion years.

RECENT WORK

Lenton, T. M. "Gaia and Natural Selection." *Nature* 394 (1998), 439–447. A review and synthesis; clarifies the types of environmental feedback and explores their implications at levels from the individual to the global.

—TIM LENTON

## GENERAL AGREEMENT ON TARIFFS AND TRADE. *See* World Trade Organization.

## GENETIC DIVERSITY. *See* Biological Diversity; *and* Evolution of Life.

## GENETIC ENGINEERING. *See* Biotechnology; *and* Technology.

## GEOCHEMICAL CYCLES. *See* Biogeochemical Cycles.

## GEOENGINEERING

Geoengineering is the intentional large-scale manipulation of the global environment. The term has usually been applied to proposals to manipulate the climate with the primary intention of reducing undesired climatic change caused by human influences. These geoengineering schemes seek to mitigate the effect of fossil fuel combustion on the climate without abating fossil fuel use—for example, by placing shields in space to reduce the sunlight incident on the Earth.

Possible responses to the problem of anthropogenic climate change fall into three broad categories: (1) abatement of human impacts by reducing the climate forcings, (2) adaptation to reduce the impact of altered climate on human systems, and (3) deliberate intervention in the climate system to counter the human impact on climate—geoengineering.

It is central to the common meaning of geoengineering that the environmental manipulation be deliberate, and be a primary goal rather than a side effect. This distinction is at the heart of the substantial moral and legal concerns about geoengineering. For example, while it may be argued that modern agriculture constitutes geoengineering, the global-scale transformation of the nitrogen cycle that it causes is a side effect of food production, and is usually viewed differently from the deliberate modification of the global environment.

Explicit consideration of human modification of the global climate dates back at least to the Swedish scientist Svante Arrhenius, who was among the first to analyze the role of carbon dioxide ($CO_2$) in regulating climate. In 1908 he suggested that warming resulting from fossil fuel combustion could increase food supply by allowing agriculture to extend northward (Arrhenius, 1908). [*See the biography of Arrhenius.*]

Sporadic analysis of the potential for global climate modification continued through the first half of the century. The 1950s and 1960s saw increasing interest in the possibility of control of weather and climate for human benefit. Discussion of climate engineering as a means to counteract destructive human influences began in the 1970s at a time of increased concern about the negative effects of technological change.

**Examples of Geoengineering Proposals.** Proposals to engineer the climate may be usefully classified by their mode of action. Most approaches to mitigating climate change propose to alter global energy fluxes through one of two strategies: increasing the amount of outgoing infrared radiation through reduction of atmospheric carbon dioxide, or decreasing the amount of absorbed solar radiation through an increase in albedo. The few proposals that fall outside this categorization typically involve modification of ocean currents (e.g., Johnson, 1997). Geoengineering has also occasionally been proposed for nonclimatic problems such as ozone depletion.

Albedo modification schemes aim to offset the effect of increasing carbon dioxide on the global radiative balance, and thus on average surface temperatures. An albedo change of roughly 1.5 percent is needed to offset the effect of doubled carbon dioxide. Even if perfect compensation of the radiative balance could be achieved, the resulting climate may still be significantly altered. The climate changes could result from changed vertical and latitudinal distributions of atmospheric heating. In addition, the increase in carbon dioxide would have substantial effects on plant growth, independently of its effect on climate. These effects cannot be offset by an increase in albedo. Table 1 summarizes various geoengineering schemes. [*See* Albedo.]

*Stratospheric aerosols.* Aerosols influence radiative fluxes either directly by optical scattering and reradiation, or indirectly by increasing the albedo and lifetime of clouds. It appears that anthropogenic sulfate aerosols may currently influence the global radiation

**Geoengineering. TABLE 1.** Summary Comparison of Geoengineering Options*

| GEOENGINEERING SCHEME | COM[†] | TECHNICAL UNCERTAINTIES | RISK OF SIDE EFFECTS | NONTECHNICAL ISSUES |
|---|---|---|---|---|
| Injection of carbon dioxide into the ocean | 30–80 | Costs are much better known than for other geoengineering schemes. Moderate uncertainty about fate of carbon dioxide in ocean. | Low risk. Possibility of damage to local benthic community. | Like abatement, this scheme is local with costs associated with each source. Potential legal and political concerns over oceanic disposal. |
| Injection of carbon dioxide underground | 30–80 | Costs are known as for carbon dioxide in ocean; less uncertainty about geologic than oceanic storage. | Very low risk. | Is geologic disposal of carbon dioxide geoengineering or a method of emissions abatement? |
| Ocean fertilization with phosphate | 1–3 | Uncertain biology: Can ecosystem change its P:N utilization ratio? | Moderate risk. Possible oxygen depletion may cause methane release. Changed mix of ocean biota. | Legal concerns: Law of the Sea, Antarctic Treaty. Liability concerns arising from effect on fisheries; N.B. fisheries might be improved. |
| Ocean fertilization with iron | 0.3–3 | Uncertain biology: When is iron really limiting? | As above. | As above. |
| Intensive forestry to capture carbon in harvested trees | 3–100 | Uncertainty about rate of carbon accumulation, particularly under changing climatic conditions. | Low risk. Intensive cultivation will impact soils and biodiversity. | Political questions: how to divide costs? Whose land is used? |
| Solar shields to generate an increase in the Earth's albedo | 10–100 | Costs are large and highly uncertain. Uncertainty dominated by launch costs. | Very low risk. However, albedo increase does not exactly counteract the effect of increased carbon dioxide. | Security, equity, and liability if system used for weather control. |
| Stratospheric $SO_2$ to increase albedo by direct optical scattering | <<1 | Uncertain lifetime of stratospheric aerosols. | High risk. Effect on ozone depletion uncertain. Albedo increase is not equivalent to carbon dioxide mitigation. | Liability: ozone destruction. |
| Tropospheric $SO_2$2 to increase albedo by direct and indirect effects | <1 | Substantial uncertainties with respect to transport of aerosols and their effect on cloud optical properties. | Moderate risk: unintentional mitigation of the effect of carbon dioxide already in progress. | Liability and sovereignty because the distribution of tropospheric aerosols strongly affects regional climate. |

*While based on current literature, the estimates of risk and cost are the author's alone.

†Cost of mitigation (COM) is in dollars per metric ton of carbon dioxide emissions mitigated.

budget by about 1 watt per square meter—enough to counter much of the effect of increased carbon dioxide. Budyko (1977) was the first to suggest increasing the albedo by injecting sulfur dioxide into the stratosphere, where it would mimic the action of large volcanoes on the climate. The injection of about 10 teragrams per annum into the stratosphere would roughly counteract the effect of doubled carbon dioxide on the global radiative balance. Several technologically straightforward alternatives exist for injecting the required sulfate into the stratosphere at a trivial cost compared with other methods of climate modification (see National Academy of Sciences, 1992).

The most serious problem with this scheme may be the effect of the aerosols on atmospheric chemistry. The Antarctic ozone hole has clearly demonstrated the complexity of chemical dynamics in the stratosphere and the resulting susceptibility of ozone concentrations to aerosols. Recent elaborations of this scheme have focused on tailoring the scattering properties of the particles, and on choosing particles that might be chemically inert. Depending on the size of particles used, the aerosol layer might cause significant whitening of the daytime sky. Such whitening is one of the classic valuation problems posed by geoengineering: How much is a blue sky worth? [See Aerosols; and Ozone.]

***Space-based shields.*** The possibility of shielding the Earth with orbiting mirrors is the most technologically extravagant geoengineering scheme. While expensive, it has clear advantages over other geoengineering options. Because solar shields effect a "clean" alteration of the solar constant, their side effects would be both less significant and more predictable than those of other albedo modification schemes. Assuming that the shields were steerable, their effect could be eliminated at will. Additionally, steerable shields might be used to direct radiation at specific areas, offering the possibility of weather control. [*See* Weather Modification.]

Most discussion of solar shields has assumed that they would be placed in low Earth orbit; however, such shields act as solar sails and would be rapidly pushed out of orbit by the sunlight they were designed to block. This problem was recognized by Seifritz (1989), who proposed using a single shield of 2,000 kilometer radius near the Lagrange point between the Earth and the Sun. (Objects at the Lagrange point can have quasi-stable orbits that remain on a line between the Earth and the Sun.) Such a shield would be stable with weak active control.

A rough estimate of the cost of such a scheme can be made by assuming that it is dominated by the cost of lifting the required mass to orbit. Detailed estimates of the minimum required mass densities can be found in the solar sail literature. They range from 2 to 10 grams per square meter (including support structures). The mass of a system required to reduce solar flux by 1.5 percent is 1–5 teragrams (note that a recent proposal aims to reduce the required mass radically by use of fine mesh with tailored optical scattering properties). The current cost of launching payloads to orbit is about U.S.$20 per gram. However, given economies of scale—which would certainly apply here—it is argued that launch costs could be substantially lower.

***Sequestration of carbon dioxide from fossil fuel combustion.*** The climatic impact of fossil energy use may be reduced by capturing the resulting carbon and sequestering it away from the atmosphere. Carbon can be captured from fossil fuels by separating carbon dioxide from the products of combustion or by re-forming the fuel to yield a hydrogen-enriched fuel stream for combustion and a carbon-enriched stream for sequestration. The linked technologies of separation and sequestration are often called "carbon management." During the 1990s, a broad-based research program in carbon management emerged, and demonstrated substantial progress in the necessary technologies and improved understanding of the potential for geologic and oceanic carbon dioxide sequestration. Driven by concerns over climatic change, large-scale sequestration of carbon dioxide has already begun; Statoil of Norway is injecting carbon dioxide separated from natural gas into an aquifer beneath the North Sea. Other projects are planned, as are various pilot-scale sequestration experiments.

As carbon management emerges as a plausible near-term option for reducing carbon dioxide emissions, the degree to which it constitutes geoengineering is becoming controversial: proponents contend that it is abatement, while opponents contend that it is geoengineering. In fact, carbon management occupies an ambiguous place in the conventional abate/adapt/geoengineer taxonomy outlined above. The term geoengineering was coined in the early 1970s by Marchetti (1977), who proposed that carbon dioxide from combustion could be disposed of in the ocean. Oceanic sequestration would constitute a deliberate intervention in the carbon cycle. Thus is seems reasonable to label it geoengineering. However, proposed "zero emission" power plants, which would emit nothing to the atmosphere and would sequester their carbon dioxide emissions in stable geologic formations, may be seen as a novel form of mitigation.

Many recent engineering studies have addressed the technical feasibility of capturing carbon dioxide from power plants and compressing it for sequestration in the ocean or underground. The rough consensus of the studies is as follows:

- For new power stations, the amortized additional cost of carbon dioxide capture and sequestration would raise electricity prices by 30–150 percent, a cost that is less than the current costs of nonfossil alternative energy sources such as solar power.
- The costs would be substantially higher for retrofitting existing power plants.
- Most costs arise from the separation of carbon dioxide from other exhaust gases, rather than from its compression and sequestration.

Carbon may also be captured from fossil fuels by re-forming them to produce hydrogen and carbon dioxide. If hydrogen was used as a primary energy carrier—a proposed route to large-scale decarbonization of the energy system—then the existing cost advantage of carbon management over nonfossil alternatives is augmented by the technical advantages of thermochemical over electrochemical hydrogen production.

Carbon from fossil fuel combustion may be sequestered in geologic formations or in the ocean. Three types of reservoir have been seriously considered: depleted oil and gas fields (global capacity roughly 200–500 billion metric tons [gigatons] of carbon), deep coal beds (100–200 gigatons of carbon), and deep saline aquifers ($10^2$–$10^3$ gigatons of carbon). Questions remain about the long-term stability of these reservoirs, particularly oil and gas fields. The remainder of this discussion will focus on oceanic sequestration because it most clearly constitutes geoengineering.

One may view $CO_2$-induced climate change as a problem of mismatched time scales. It is due to the rate at which combustion of fossil fuels is transferring carbon from ancient terrestrial reservoirs into the comparatively small atmospheric reservoir. When carbon dioxide is emitted to the atmosphere, atmosphere–ocean equilibration transfers roughly 80 percent of it to the oceans with an exponential time scale of about three hundred years. The remaining atmospheric carbon dioxide is removed on much longer time scales. Injection of carbon dioxide into the deep ocean accelerates this equilibration, reducing peak atmospheric concentrations. The efficiency of equilibration depends on the location and depth of injection. For example, injection at about 700 meters depth into the Kuroshio current off Japan would result in much of the carbon dioxide being returned to the atmosphere in about one hundred years, whereas injections that formed "lakes" of carbon dioxide in ocean trenches would more efficiently accelerate equilibration of the carbon dioxide with the deep-sea calcium carbonate reservoirs.

The dynamic nature of the marine carbon cycle precludes defining a unique static capacity, as may be done for geologic sequestration. Depending on the increase in mean ocean acidity that is presumed acceptable, the capacity is of order $10^3$–$10^4$ gigatons of carbon, much larger than current anthropogenic emissions of about 6 gigatons of carbon per year.

In considering the implications of oceanic sequestration one must note that—depending on the injection site—about 20 percent of the carbon returns to the atmosphere on the three-hundred-year time scale. Supplying the energy required for separating, compressing, and injecting the carbon dioxide requires that more fossil fuel be used than would be needed if the carbon dioxide were vented to the atmosphere. Thus, while oceanic sequestration can reduce the peak atmospheric concentration of carbon dioxide caused by the use of a given amount of fossil-derived energy, it increases the resulting atmospheric concentrations on time scales greater than about five hundred years.

*Ocean surface fertilization.* Carbon could be removed from the atmosphere by fertilizing the "biological pump" that maintains the disequilibrium in carbon dioxide concentration between the atmosphere and the deep ocean. The net effect of biological activity in the ocean surface is to bind phosphorus, nitrogen, and carbon into organic detritus in a ratio of about 1:15:130 (this includes the carbon removed as calcium carbonate, $CaCO_3$) until all of the limiting nutrient—usually phosphorus—is exhausted. The detritus then falls to the deep ocean, providing the pumping action.

A simple interpretation of this ratio suggests that adding phosphate to the ocean surface should remove carbon dioxide from the atmosphere–ocean surface system in a molar ratio of about 130:1. This first-order model of the biology ignores the phosphate–nitrate balance. Adding phosphate to the system without adding nitrate would only remove carbon in this ratio if the ecosystem shifted to favor nitrogen fixers.

In some areas of the southern oceans the limiting nutrient may be iron (Fe), for which the molar ratio Fe:C in detritus is about 1:10,000, implying that iron may be a very efficient fertilizer of ocean surface biota. This idea has received considerable attention and has stimulated some valuable research. Iron fertilization has been demonstrated *in situ*, but it is not clear that sustained carbon removal is realizable.

Ocean fertilization would have significant side effects. For example, it might decrease dissolved oxygen with consequent increased emissions of methane—a greenhouse gas.

*Afforestation.* Large-scale forest management or afforestation for the purpose of removing atmospheric carbon dioxide is a form of geoengineering. (Note that this definition is unavoidably fuzzy; for example, it may not be appropriate to consider afforestation for a mix of purposes as geoengineering.)

It appears that temperate-zone Northern Hemisphere forests already capture a significant amount (roughly 10–20 percent) of fossil fuel carbon. Uncertainty about the dynamics of carbon in forest ecosystems limits our ability to predict their response to climatic change and increasing carbon dioxide; in particular, it is uncertain whether such changes would accelerate or reverse the sequestration of carbon in forests. Capturing a substantial fraction of fossil fuel carbon would require intensive management of forests on a very large scale. For example, fast-growing forests of young trees can capture about 5 metric tons of carbon per hectare per year under optimal conditions. To capture the full anthropogenic carbon dioxide emissions, about 1 billion hectares would be required—roughly the current global area of managed forest. To capture carbon continuously at this rate, it would be necessary to dispose of the trees so that their carbon could not return to the atmosphere. Fertilization would be required to replace the nutrients removed with the trees. Intensive forest management on this scale would have a substantial impact on forest ecosystems. [*See* Forestation.]

**Evaluating Geoengineering.** Most discussions of geoengineering have focused on assessments of technical feasibility and approximate cost. [*See* Risk.] However, it is probable that issues of risk, politics, and ethics will prove more decisive factors in real choices about implementation. This is true both because of the strong negative reactions often provoked by most geoengineering proposals, and because many geoengineering schemes are inexpensive relative to abatement or adaptation.

***Economics and risk analysis.*** The simplest economic metric for geoengineering is to compute the "cost of mitigation"—the ratio of the cost to the amount of mitigation effected (typically measured in U.S. dollars per metric ton of carbon emission mitigated). This measure permits comparison between geoengineering schemes and between geoengineering and the abatement of emissions. Table 1 includes the cost of mitigation for various schemes. The costs are highly uncertain. For albedo modification schemes, additional uncertainty is introduced by the somewhat arbitrary conversion from albedo change to equivalent reduction in carbon dioxide.

Examination of the cost of mitigation reveals that it varies by more than two orders of magnitude between various schemes, and that for some (e.g., stratospheric aerosols) the costs are very low compared with either abatement or adaptation. However, such direct cost comparisons have little meaning given the very large differences in the nonmonetary aspects of these responses to climate change; for example, risk of side effects, certainty of effect, and social distribution of cost.

*Geoengineering as a fallback strategy.* Focusing on the marginal cost of mitigation permits a more meaningful comparison between geoengineering and abatement. Although the cost of mitigation is uncertain, there is much less doubt about how the cost of mitigation scales with the degree of mitigation required. While econometric and technical methods for estimating the

cost of moderate abatement differ radically, both agree that costs will rise steeply if carbon dioxide emissions are to be abated by more than 50 percent. In sharp contrast, some geoengineering schemes (e.g., albedo modification) have marginal costs that, while highly uncertain, are roughly independent of, and may even decrease with, the amount of mitigation effected. Other schemes (e.g., carbon dioxide sequestration) have marginal costs that are initially higher than abatement, but rise more slowly. These relationships are illustrated in Figure 1.

Geoengineering may serve as a fallback strategy by putting an upper bound on the costs of mitigation should climate change be more severe than we expect. In this context, a fallback strategy must either be more certain of effect, faster to implement, or provide unlimited mitigation at fixed marginal cost. Various geoengineering schemes meet each of these criteria. The notion of geoengineering as a fallback option provides a central—or perhaps the only—justification for taking large-scale geoengineering seriously. A fallback strategy permits more confidence in adopting a moderate response to the climate problem: without fallback options, a moderate response is risky given the possibility of a strong climatic response to moderate levels of fossil fuel combustion.

*Risk assessment.* Questions about the advisability of geoengineering revolve around risk: risk of failure and risk of side effects. Climate prediction is too uncertain to allow quantitative assessment of risk. However, if a

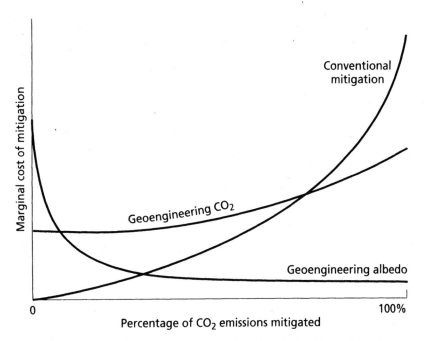

**Geoengineering. Figure 1.** Schematic Comparison between Modes of Mitigation.

Conventional mitigation means any method other than geoengineering; for example, conservation, fuel switching, or use of nonfossil energy sources. Albedo modification schemes (e.g., solar shields) have high initial capital costs, but can provide essentially unlimited mitigation at fixed marginal cost. Geoengineering by disposal of carbon dioxide costs more than conventional mitigation for small amounts of mitigation, but less than conventional mitigation if we require mitigation of all carbon dioxide emissions.

geoengineering scheme works by imitating a natural process, we can make a qualitative risk assessment by comparing the magnitude of the engineered effect with the magnitude and variability of the natural process, and then assume that similar perturbations entail similar results. For example, the amount of sulfate released into the stratosphere as part of a geoengineering scheme and the amount released by a large volcanic eruption are similar. We may estimate the magnitude of stratospheric ozone loss by analogy.

Even crude qualitative estimates of risk can give insight into the relative merits of various geoengineering schemes when considered in conjunction with other variables. Table 2 illustrates this with a comparison of risk and cost.

*Political considerations.* The cardinal political reality of geoengineering is that, unlike other responses to climate change (e.g., abatement or adaptation), geoengineering could be implemented by one or a few countries acting alone. Various political concerns arise from this fact with respect to security, sovereignty, and liability.

Some geoengineering schemes raise direct security concerns; solar shields, for example, might be used as offensive weapons. A more subtle but perhaps more important security concern arises from the growing links between environmental change and security. Whether or not they were actually responsible, the operators of a geoengineering project could be blamed for harmful climatic events that could plausibly be attributed—by an aggrieved party—to the geoengineering scheme. Given the current political disputes arising from issues such as the depletion of fisheries and aquifers, it seems plausible that a unilateral geoengineering project could lead to significant political tension.

In general, international law has little bearing on geoengineering. However, Bodansky (1996) points out that several specific proposals may be covered by existing laws; for example, the fertilization of Antarctic waters would fall under the Antarctic Treaty System, and the use of space-based shields would fall under the Outer Space Treaty of 1967.

As in the current negotiations under the Framework Convention on Climate Change, geoengineering would raise questions of equity. In this case, geoengineering might simplify the politics. As Tom Schelling (1996) pointed out, geoengineering ". . . totally transforms the greenhouse issue from an exceedingly complicated regulatory regime to a simple—not necessarily easy but simple—problem in international cost sharing."

One must note that not all geoengineering schemes are amenable to centralized implementation. For example, carbon management requires diffuse implementation at the manifold sources of fossil fuel combustion.

***Ethics.*** Discussion of geoengineering commonly elicits strong negative reactions. Within the policy analysis community, for example, there has been vigorous debate about whether discussion of geoengineering should be included in public reports that outline possible responses to climate change. Fears have been voiced that its inclusion in such reports could influence policy makers to take it too seriously, and perhaps to defer action on abatement given knowledge of geoengineering as an alternative (see Schneider, 1996, for discussion of the debate over geoengineering in the 1992 National Academy of Sciences panel). While these concerns are undoubtedly serious and substantive, it is difficult to disentangle their various roots and, in particular, to separate pragmatic from ethical concerns. [*See* Ethics.]

Many of the objections to geoengineering that are cited as "ethical" have an essentially pragmatic basis. Three common ones are:

- *The slippery-slope argument.* If we choose geoengineering solutions to counter anthropogenic climate change, we open the door to future systematic efforts to alter the global environment to suit humans. This is a pragmatic argument, because in the future we will be as free as we are now to choose to what extent we wish to geoengineer. An ethical argument

Geoengineering. TABLE 2. Costs versus Risks of Geoengineering Schemes*

| | Cost | | |
|---|---|---|---|
| RISK | LOW | MEDIUM | HIGH |
| Low | — | Intensive forestry for carbon sequestration | Solar shields carbon dioxide disposal |
| Medium | Tropospheric sulfur dioxide Ocean fertilization with iron | Inert stratospheric aerosols Ocean fertilization with phosphate | Balloons in the stratosphere |
| High | Stratospheric sulfur dioxide | — | — |

*This kind of systematic intercomparison is useful in setting geoengineering research priorities. (Cost and risk estimates are qualitative estimates informed by current knowledge.)

must define why such large-scale environmental manipulation is bad, and how it differs from what humanity is already doing.

- *The kluge argument.* Geoengineering is a "technical fix," "kluge," or "end-of-pipe solution." Rather than attacking the problems caused by fossil fuel combustion at their source, geoengineering aims to add new technology to counter their side effects. Such solutions are commonly viewed as inherently undesirable—but not for ethical reasons.

- *The unpredictability argument.* Geoengineering entails "messing with" a complex, poorly understood system: since we cannot reliably predict results, it is unethical to geoengineer. Because we are already perturbing the climate system with consequences that are unpredictable, this argument depends on the notion that intentional manipulation is inherently worse than manipulation that occurs as a side effect.

One may analyze geoengineering using common ethical norms; for example, one could consider the effects of geoengineering on intergenerational equity, or on the rights of minorities (e.g., the inhabitants of low-lying countries). However, these modes of analysis say nothing unique about geoengineering, and could be applied in a similar manner to many other technological choices. Some people would argue that such analysis fails to address a particular ethical abhorrence that they feel about geoengineering, and that we should look for an ethical analysis that addresses geoengineering in particular; for example, an environmental ethic.

The simplest formulations of environmental ethics proceed by extension of common ethical principles that apply between humans. A result is "animal rights" in one of its variants (see, for example, Regan, 1983). Such formulations locate "rights" or "moral value" in individuals. When applied to a large-scale decision such as geoengineering, an ethical analysis based on individuals reduces to a problem of weighing conflicting rights or utility. As with analyses that are based on more traditional ethical norms, such analysis has no specific bearing on geoengineering. Alternative, and more controversial, formulations of environmental ethics locate moral value in systems of individuals, such as a species or a biotic community (see, for example, Callicott, 1989). It is plausible that such a formulation of environmental ethics could more directly address the ethics of geoengineering.

[*See also* Carbon Dioxide; Deserts; Global Warming; Greenhouse Effect; *and* Water Resources Management.]

## BIBLIOGRAPHY

### HISTORICAL WORKS CONCERNING DELIBERATE CLIMATE MODIFICATION

Arrhenius, S. *Worlds in the Making*. New York: Harper, 1908. Published twelve years after Arrhenius first calculated the effect of carbon dioxide on climate, this extraordinary general exposition of planetary science includes discussion of deliberate climate modification.

Budyko, M. I. *Climatic Changes*. Washington, D.C.: American Geophysical Union, 1977. This wide-ranging treatment of the interrelationship of humans and climate discusses geoengineering using sulfates in the stratosphere (originally published in Russian in 1974; based on work from the mid-1960s).

Kellogg, W. W., and S. H. Schneider. "Climate Stabilization: For Better Or for Worse?" *Science* 186 (1974), 1163–1172. An early summary of various geoengineering schemes that discusses their uncertainty and their political ramifications.

Marchetti, C. "On Geoengineering and the Carbon Dioxide Problem." *Climate Change* 1 (1977), 59–68. The term geoengineering was coined in this paper.

### GENERAL WORKS ON GEOENGINEERING

Bodansky, D. "May we Engineer the Climate?" *Climatic Change* 33 (1996), 309–321. A review of the legal implications of geoengineering.

Jamison, D. "Ethics and Intentional Climate Change." *Climatic Change* 33 (1996), 323–336. The only paper on the ethics of geoengineering.

Keith, D. W, and H. Dowlatabadi. "Taking Geoengineering Seriously." *EOS* 73 (1992), 289–293. A general review of geoengineering.

National Academy of Sciences. *Policy Implications of Greenhouse Warming*. Washington, D.C.: National Academy Press, 1992. The chapter on geoengineering contains many detailed cost estimates.

Office of Technology Assessment. *Changing by Degrees: Steps to Reduce Greenhouse Gases*. Report OTA-O-482. Washington, D.C., 1991. This report contains substantial treatment of geoengineering, particularly afforestation.

Schelling, T. C. "The Economic Diplomacy of Geoengineering." *Climatic Change* 33 (1996), 303–307.

Schneider, S. H. "Geoengineeering: Could—or Should—We Do It?" *Climatic Change* 33 (1996), 291–302. Contains a summary of the intellectual history of geoengineering.

Watts, R. G., ed. *Engineering Response to Global Climate Change*. Boca Raton, Fla.: CRC Press, 1997. This book includes an up-to-date chapter on geoengineering.

### WORKS ON SPECIFIC GEOENGINEERING METHODS

Dickinson, R. "Climate Engineering: A Review of Aerosol Approaches to Changing the Global Energy Balance." *Climatic Change* 33 (1996), 279–290.

Herzog, H., B. Eliasson, and O. Kaarstad. "Capturing Greenhouse Gases," *Scientific American* (February 2000), 72–79.

Johnson, R. G. "Climate Control Requires a Dam at the Strait of Gibraltar." *EOS* 78 (1997), 277–281.

Martin, J. H. "Testing the Iron Hypothesis in Ecosystem of the Equatorial Pacific Ocean." *Nature* 371 (1994), 123–124.

Monastersky, R. "Iron Versus the Greenhouse." *Science News* 148 (1995), 220–222. A good review of iron fertilization for the nonspecialist.

Parson, E. A., and D. W. Keith. "Fossil Fuels without Carbon Dioxide Emissions: Progress, Prospects, and Policy Implications." *Science* 282 (1998), 1053–1054. A summary of recent developments in carbon sequestration.

Seifritz, W. "Mirrors to Halt Global Warming?" *Nature* 340 (1989), 603. Describes a stable solar shield scheme that involves a large shield at the Lagrange point.

OTHER CITED WORKS

Callicott, J. B. *In Defense of the Land Ethic.* Albany: State University of New York Press, 1989.

Regan, T. *The Case for Animal Rights.* Berkeley: University of California Press, 1983.

—DAVID W. KEITH

# GEOGRAPHIC INFORMATION SYSTEMS

Geographic information systems (GIS) apply computer technology to the tasks of capturing, storing, manipulating, analyzing, modeling, and displaying information about the surface of the Earth, and the phenomena distributed on it. They have emerged over the past three decades as a distinct form of computer use, with its own software industry and array of products, directed at applications ranging from management of the resources of utility companies to support for global change science. Worldwide sales of GIS software in the late 1990s were in the region of U.S.$600 million annually, with much larger investments in associated digital geographic data.

GIS deal with information that is geographically or spatially explicit, representing the spatial variation of phenomena over the Earth. Although many forms of software are capable of handling such information in limited ways, GIS is the only form designed expressly for this purpose, with a full range of necessary data structures and functions. Global change science is also inherently geographically explicit, dealing with spatial dynamics and differentiation over the surface of the planet. Thus GIS is uniquely suitable as a tool for the computing functions needed to support global change science.

It is often helpful to think of GIS as a computer that contains maps. One of the simplest reasons for manipulating maps with computers is to make them easier to construct and draw, and GIS are often used for this purpose. By computerizing the map-making process it is possible to edit easily, manipulate the map's contents without the labor-intensive task of redrafting, communicate maps electronically, and create output in any convenient form. The advent of GIS has made it possible for anyone to be a cartographer, provided he or she is in possession of the necessary software, a computer to run it on, and a suitable printing device.

This view of a GIS as automated mapping systems is much too simplistic, however. The first GIS were generally agreed to have been the Canada Geographic Information System (CGIS), a project developed in the Canadian Government in the 1960s under the direction of Roger Tomlinson. At the time there were no printers capable of making acceptable maps, even in black and white, and the design of CGIS did not include map output. Instead, the project was justified entirely on the basis of the need to analyze geographic information obtained from maps. Its original design included a thorough cost-benefit analysis that is still a model for the industry, and the design found substantial net benefits to computerization, despite the high costs and crude technology of the time.

The case for computerizing the analysis of geographic information rests on two propositions: first, that the few traditional tools that exist are very labor intensive and crude; and second, that once geographic information is in digital form there are massive scale economies because of the many forms of analysis that are possible. For example, there are two traditional ways of measuring area from a map: use of a mechanical planimeter, and counting of dots on a transparent overlay; and both are incredibly tedious and inaccurate. But measurement of area from a digital representation of a map is trivial, and virtually as accurate as the representation. Once a representation has been created, it is easy to add functions to the software to perform almost any analysis imaginable.

Today's GIS software include functions to support the creation of digital data by manual or automated digitization of paper maps; functions to convert between map projections; functions to integrate data from different sources by converting formats or removing spurious differences; functions to make mapped output more publishable and cosmetically pleasing; and links to specialized software for modeling physical processes or conducting statistical analyses. All of these are relevant to global change science, and GIS have become one of that science's most valuable analytic tools.

**Principles.** The most conspicuous distinction in GIS is between two competing forms of digital representation. The vector approach builds a database from digital representations of points, lines, and areas. The location of each primitive object is recorded, using an appropriate combination of coordinates referenced to the Earth's surface, often in latitude and longitude but also in standard coordinate systems such as UTM (Universal Transverse Mercator, a world standard initially developed for military applications). The characteristics of each object are termed its "attributes," and a vector GIS will accommodate large numbers of these, in the form of names or measurements of various kinds. Objects are grouped into classes, each member of a class having the same dimensionality (all points, for example) and the same group of characteristics. It is convenient to think of the attributes of a class of objects as forming a table, and many vector GIS incorporate relational database management systems to handle the tables.

The raster approach covers the relevant part of the Earth's surface with an array of rectangular cells, and describes variation by allocating a value to each cell. Almost all designs allow only one value per cell, so that representations of multidimensional variation are built by constructing several layers of cells, each describing the variation of one variable. A cell can contain a digital representation of a number, as in digital images of the Earth from space (remote sensing) or representations of the variation of elevation (digital elevation models, or DEMs); or a digital representation of a class, as in layers of land cover, vegetation classification, or land use. The fixed cell size gives the raster representation the appearance of constant spatial resolution, whereas vector representations have resolutions that are unlimited in principle, but limited in practice by the nature of the data.

Vector GIS originated in applications where this representation is most appropriate. For example, it is clearly much more reasonable to represent the links in a connected river or water supply network as lines than as collections of cells. This is particularly apparent in the management of telephone networks, where the connectivity in the system is apparent only at the most detailed spatial resolution. Vector data are also dominant in social, economic, and demographic data, and thus in the human aspects of global change science, because of the practice of collecting such data for irregularly shaped regions. On the other hand, it is clearly much more reasonable to process information gathered by remote sensing in a raster GIS, because the data are collected in that form. Raster GIS are similar in many ways to the image-processing systems developed for handling remotely sensed data, but add functions that allow these data to be integrated with other, possibly vector, data. Vector GIS are similar in many ways to computer-assisted design (CAD) systems, but add functions and capabilities that reflect the special needs of users of Earth-referenced data. Modern GIS attempt to integrate raster and vector approaches, although with only partial success, and many systems still reflect their earlier roots in one camp or the other.

But this distinction between raster and vector is only one instance of a much more general issue: that geographic reality is infinitely complex, and there are always many possible ways of representing the same phenomena. One very important manifestation of this issue concerns scale. To have any chance of creating a representation of real geographic complexity in the limited space of a digital computer, it is always necessary to generalize. Cartographers know this as the scale problem, and define scale as the ratio between distance on a paper map and the equivalent distance on the ground. Traditionally this measure, the representative fraction, has defined the level of detail of a map. Unfortunately there is no easy translation between it and what a scientist recognizes as spatial resolution, because a small object can always be shown on a map of any scale using an appropriate symbol. But useful rules of thumb exist, and it is often assumed that the spatial resolution

of a data set obtained from a map at a given scale can be estimated by computing the ground distance corresponding to 0.5 millimeters on the map. Thus a map at scale 1:24,000 will tend not to show features smaller than 12 meters across, although there are clearly exceptions.

In addition to scale, builders of geographic databases must deal with the distinction between two different conceptualizations of geographic variation. Many scientific laws, including those of fluid motion, are written in terms of variables that change continuously in space, forming what a mathematician would call a *field*, a function of the two spatial dimensions. But maps are more usually conceived as showing collections of discrete entities, embedded in an otherwise empty space. The distinction between fields and discrete features is apparent in weather forecasting: atmospheric models are written in field variables such as pressure, and predict fields of precipitation or temperature; but weather is often understood in terms of the behavior of discrete highs, lows, or fronts. [*See* Weather Forecasting.] Both field and discrete-entity conceptualizations can be represented in either raster or vector form, and yet the meaning and legitimacy of operations carried out on them can be entirely different. An area of homogeneous vegetation may be treated as a discrete patch in a landscape model, but as a continuous field when variation within the patch becomes important. A model of ant behavior may deal at one scale with individual ants as discrete features, then at another scale with the population as a continuous density, related perhaps to the continuous variation of some resource. Unfortunately, then, a single digital representation can have very different meanings; and a single phenomenon can have many different representations. The large number of possible mappings between reality and its digital representation is a source of endless and profound challenges for GIS.

**The Geographic Information Technologies.** GIS are only one of a number of new technologies for handling geographic information in digital form. Another is remote sensing, which provides an increasingly important source of data, with the powerful advantage that it is global in scope. The Global Positioning System (GPS) is a constellation of satellites designed to allow position on the Earth's surface to be determined to a known level of accuracy. Each satellite transmits signals that are precisely timed by on-board atomic clocks; a receiver on Earth is able to resolve position in three dimensions if at least four satellite signals are being received. A simple hand-held receiver, available for a few hundred dollars, is capable of determining position to an accuracy of tens of meters, and there are versions of the technology that are capable of determining relative position to centimeter accuracy, and to meter accuracy in vehicles moving at speed. A technically distinct Russian system can be used in conjunction with GPS to improve positioning. [*See* Remote Sensing.]

One unexpected consequence of the widespread use of GIS for ocean and air navigation and for positioning of scientific observations is that it exposes the inaccuracies in much of the world's published mapping, and the inconsistencies between the mathematical figures of the Earth that support mapping. Unfortunately for global change science, mapping remains an activity of strategic importance in many parts of the Earth, and there are very significant gaps in scientific access to good base mapping.

GIS make increasing use of electronic communication. There are now many archives of geographic information of value to global change science that are accessible to anyone connected to the Internet. The EOS (Earth Observing System) Data and Information System (EOSDIS) is one prominent example. With no more than a standard World Wide Web browser, a user is able to access the archive, browse its contents, and in many cases download large and potentially valuable data sets. Such archives of geographic information are being sponsored by national and state governments, nongovernmental organizations, universities, and research libraries. It is also increasingly possible to manipulate the data in an archive at its source, allowing full use of GIS functions without the necessity to download what is often a very large set of data.

**Recent Trends.** GIS are a new and rapidly developing technology, positioned to take advantage of broader trends in computing. At the same time, its origins in the paper map have established a legacy that is on the one hand an advantage, because it allows the user of a GIS to understand its potential in terms of a familiar metaphor, but on the other hand limiting, because it fails to acknowledge the true potential of GIS. In recent years much work has gone into developing GIS in directions that go beyond the metaphor of a map inside a computer. Today, GIS users can expect functions that process data that have significant temporal elements (maps are inherently static); that deal effectively with uncertainty (maps present the world as simpler than it really is); and that handle the third dimension (maps are inherently two-dimensional). A development of particular significance to global change science is the ability to analyze data distributed on the curved surface of the Earth; early GIS took the distorting projections of maps for granted, and processed information as if it were planar. Such distinctions are moot over small areas, but become very important if GIS are used to estimate the mean of the global temperature field, for example.

Of particular significance for global change science is the role of GIS in facilitating integration of data. Integrated assessments require access to data relating to both social and physical themes, ranging from demo-graphic and economic statistics to climate, soil class, and land cover. Often the only way of integrating such data is within a common geographic framework implemented in a GIS. Inevitably the data of one discipline adopt that discipline's conventions, which may be very different from those of another discipline. Point records of climate must be merged with raster databases from remote sensing and statistics for the irregularly shaped reporting zones commonly used by statistical agencies. Scales vary widely, as do map projections. The impact of GIS on global change science has been greatest in two areas: first, where data must be integrated across such disciplinary differences; and second, where the results and predictions of global change science must be integrated with other concerns in support of policy decisions. Thus GIS are as likely to be used in a policy agency as in a global change science laboratory. [*See* Integrated Assessment.]

The explicit representation of geographic distributions and spatial relationships in a GIS has led to a much broader recognition of the issues and problems involved in working with spatial data. Besides scale, and the difficulties associated with scaling up and scaling down, these issues include data quality and the assessment of the accuracy of maps of such variables as land cover or soil type; the complex ways in which data errors propagate in models of spatially distributed systems, such as global climate models; and the uncertainties inherent in data that must be generalized from spatially limited samples. Recently, the term *geographic information science* has emerged as an umbrella for the fundamental issues raised by the use of GIS; the field has an active international research community and a rapidly growing literature.

[*See also* Global Monitoring; Information Management; *and* Modeling of Natural Systems.]

## BIBLIOGRAPHY

Bugayevskiy, L., and J. Snyder. *Map Projections: A Reference Manual.* London: Taylor & Francis, 1995. An excellent compendium on the geometric basis of mapping.

Burrough, P., and A. Frank, eds. *Geographic Objects with Indeterminate Boundaries.* London: Taylor & Francis, 1996. A collection of recent research papers on dealing with geographic uncertainty.

Burrough, P., and R. McDonnell. *Principles of Geographical Information Systems.* 2d ed. Oxford: Clarendon Press, 1998. An influential text on GIS from a physical resource management perspective.

Chrisman, N. *Exploring Geographic Information Systems.* New York: Wiley, 1997. A very readable and personal view of GIS.

Clarke, K. *Getting Started with Geographic Information Systems.* Upper Saddle River, N.J.: Prentice-Hall, 1997. An excellent recent introduction.

DeMers, M. *Fundamentals of Geographic Information Systems.* New York: Wiley, 1997. Another recent introduction.

Ehleringer, J., and C. Field, eds. *Scaling Physiological Processes:*

*Leaf to Globe.* San Diego: Academic Press, 1993. An ecological perspective on the issue of scale in global change science.

Goodchild, M., B. Parks, and L. Steyaert, eds. *Environmental Modeling with GIS.* New York: Oxford University Press, 1993. A compendium of applications of GIS to the modeling of various elements of environmental systems at local to global scales.

Longley, P., M. Goodchild, D. Maguire, and D. Rhind, eds. *Geographical Information Systems: Principles, Techniques, Management, Applications.* 2d ed. New York: Wiley, 1998. A comprehensive, commissioned state-of-the-art review of GIS.

Martin, D. *Geographic Information Systems: Socioeconomic Applications.* London: Routledge, 1996.

Quattrochi, D., and M. Goodchild, eds. *Scale in Remote Sensing and GIS.* Boca Raton, Fla.: Lewis Publishers, 1997.

Rhind, D., ed. *Framework for the World.* Cambridge, U.K.: GeoInformation International, 1997. A recent collection on the issues impeding comprehensive global mapping.

Star, J., J. Estes, and K. McGwire, eds. *Integration of Geographic Information Systems and Remote Sensing.* New York: Cambridge University Press, 1997. A review of the issues at the interface of these two geographic information technologies, of particular interest to global change science.

Worboys, M. *GIS: A Computing Perspective.* London: Taylor & Francis, 1995. An excellent introduction to GIS from a computer scientist.

—MICHAEL F. GOODCHILD

## GEOMORPHOLOGY. *See* Land Surface Processes.

## GEOPOLITICS. *See* Middle East; Security and Environment; *and* Water.

## GLACIATION

Geological fieldwork has established that the Earth's long and complex history has been punctuated by glaciation on regional and local scales. The fact that glaciers grew only episodically and that glacial ice was not always present is of importance not only for deciphering past and future climates but also for increasing our understanding of the interactions among the entire array of factors contributing to climate change. The main variations in glacial activity in the past have resulted from the complex interactions of mountains, oceans, atmosphere and the biological realm.

**Mass Balance, Ice Movement, Erosion, and Deposition.** Recognition of past episodes of glaciation depends on identifying of the erosive and depositional effects of the moving ice. These differ in the two main types of ice accumulations: mountain glaciers and continental ice sheets. Glacial ice forms as the result of burial and compaction of snow. An imbalance between accumulation and ablation is required before ice will flow, erode, and deposit sediment (Figure 1). Once a glacier is born, through the delivery of precipitation to a

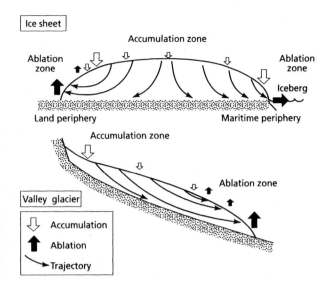

**Glaciation. FIGURE 1.** Mass balance in an ice sheet, indicating areas of accumulation, types and location of major ablation areas, and flowlines of the moving ice. (Modified after Bennett and Glasser, 1996.)

site of year-round low temperatures, any motion is in part governed by the amount of accumulation relative to the amount of ablation. The rate of motion is dependent on the development of shear planes within the ice as governed by the degree to which accumulation and ablation are out of balance. Ablation is caused by melting both at the surface through solar heating and at the base through geothermal heat flow, and by minor wind erosion. In ice sheets, iceberg calving and sublimation are important additional mechanisms of ablation. Accumulation and ablation are concentrated in different parts of a glacier. Accumulation dominates in the elevated zone of ice sheets where temperatures are low, while ablation is more active in the warmer distal reaches. When there is excess accumulation, the potential energy of the ice mass increases and the glacier advances, thus maintaining the mass balance between accumulation and ablation. An excess of ablation correspondingly leads to retreat. When the two are in mass balance the glacier will neither advance nor retreat, though unusual basal melting can lead to surges of the ice. Normally, advances or expansions of ice masses are equated to direct cooling, and this is often explained by a decrease in ablation owing to lower temperatures in the distal parts of the glacier. But given the variety of factors influencing glacier flow, it is apparent that the connections between glacial advance and climate are too complicated to allow firm conclusions. Sufficiently large advances of ice sheets can contribute to widespread and even to global cooling through a regional increase in albedo.

Glacial features can be distinguished in both erosional and depositional landforms of the younger glaciations, and sometimes in those of ancient glaciations. The early stages of ice growth in elevated terrains result in the carving out of semicircular depressions at the heads of valleys (cirques) and the creation of truncated drainages in tributaries (hanging valleys). Seeking the low points, the ice tends to follow valleys and, because its mass operates on valley sides as well as on the valley bottom, alters their cross-sectional profiles from the normal V-shape to the typical U-shape seen in many mountainous regions and their seaward extensions in fjords. Small-scale features sculpted by ice include linear bedrock striae, sometimes of the nailhead type or accompanied by crescentic gouges that yield information on the direction of past flow, and humpbacked masses of bedrock (*rôches moutonées*), with smoothed stoss sides and leeside surfaces showing abundant evidence of ice plucking in the form of irregularly fractured bedrock. On the regional scale, glaciers can create enormous erosional depressions, such as the basins of the Great Lakes in the north-central United States and southern Canada. Typically, glaciated terrains are bordered by regions where periglacial features, such as frost wedges and polygonal ground, are common.

Erosion rate apparently increases with increasing glacial cover. In ice sheets, direct erosion of bedrock is the principal source of incorporated rock fragments, while those in the various types of mountain glaciers usually include a high proportion of material weathered from valley walls by frost action. In ice sheets, clasts are often transported as discrete basal layers reflecting erosive advances of the ice. In montane glaciers, clasts fall or roll onto the ice surface and through continuing ice accumulation are buried in deeper parts of the ice mass. On melting, all material carried by the ice, in size ranging from boulders to rock "flour" (fine particles originating from the grinding action on bedrock) is left as either primary or reworked sediment. There are several types of primary deposits, all showing the poorly

mixed characteristics of the poorly sorted rock debris (till): ground moraines (carpets of debris over large areas); end moraines (resulting from the final melting of the most distal ice); and lateral moraines (derived from the letting down of trains of valley wall detritus). These materials can be modified by two main processes: readvances of the ice and the action of running water.

Erosion is also active when a glacier readvances, the difference being only that soft till or other sediments left from the previous retreat, as well as resistant bedrock, is affected. Till pellets derived from such erosive readvances are commonly incorporated within the younger till. But in most cases meltwater has the greatest impact, in extreme cases removing essentially all primary traces of glaciation. Meltwater may issue from below, within, or on top of the ice mass and may be a factor even while the glacier is advancing. Meltwater streams, being under high hydrostatic pressure, can erode into bedrock and scour subglacial and englacial tunnels as well as superglacial channels, in all of which sediment can accumulate. These structures are referred to as *esker deposits*, some of which have lengths of tens of kilometers and are cut tens of meters deep into rock. Many such channels debouch into standing bodies of water, both marine and freshwater, where they form subaqueous fans; freshwater sites are favorable for the generation of alternating summer (coarse) and winter (fine) bands called *varves*. Much till is reworked by meltwater into glacial fluvial sediments, which include highly variable lithologies and environments. Readvancing glaciers are also thought to be responsible for the formation of some sets of elongate hills (drumlins) by molding of till into parallel mounds up to twenty meters high.

The record of past glaciations is preserved in sedimentary rocks. On a global scale, the rocks reveal strong evidence of distinct glacial episodes ranging from more than 2.3 billion years ago to the present (Figure 2). The

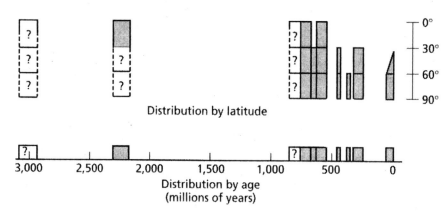

**Glaciation. FIGURE 2.** The Distribution of Glacial Deposits in the Geological Record, by Age and by Latitude.

Paleolatitude determined by paleomagnetic studies and continental reconstructions. The question marks indicate uncertainties in dating and positioning.

best-documented cases are based on till or its consolidated equivalent, tillite, together with evidence of the scribing action of clasts within the moving ice, seen in the form of glacial striae. The poorly mixed sediment of the till or tillite is derived from the mixed materials carried by or moved along beneath the ice and deposited on melting. Glacial striae occur on the underlying bedrock and also on clasts in till and tillite. These features are reliable indicators of glacial activity, as can be the associated sedimentary facies (lacustrine varves with dropstones, glaciofluvial deposits, ice-wedge casts, eskers, etc.). Thus, recognition depends on a variety of criteria.

A distinction is made, however, between typical tillites and better-sorted and stratified fine sediments containing outsize clasts. The latter may result from glaciation but may also be indicative merely of rafting of clasts by seasonal ice (river ice, marine shore ice). These rafted clasts (dropstones), while they are good guides to cold winters, cannot be attributed to glacial icebergs in the absence of till.

**Pre-Quaternary Glaciations.** Though the record of early climates is dominated by evidence of global warmth, there were possibly two intervals of glaciation during Paleoproterozoic time, the first, a problematic one, at sometime between 3.1 billion and 2.9 billion years ago, and the second a tripartite composite with ages clustered at around 2.3 billion years ago. Evidence for the first episode, in South Africa, is debatable in that verifiable tillites are not known and there is no striated floor; however, some rafted clasts are striated. Tillites have not been discovered elsewhere in strata of this age. The second interval is represented by glacial sediments in restricted areas of North America, Africa, Australia, and Europe and is often collectively referred to as the Huronian glaciation. The sequence of three well-documented deposits in the Lake Huron region of Canada is roughly correlated with other accumulations located farther north and with others in the north-central United States. Deposits recently examined paleomagnetically appear to have originated in low latitudes. The possible age range of all these deposits is from about 2.5 billion to 1.6 billion years ago but all can be encompassed within the range of the best dated, between about 2.2 billion and 2.3 billion years ago.

In one of Earth's main glacial periods, parts of all continents experienced glaciation in the Neoproterozoic, within the time range of about 1,000 to 570 million years ago. At least three major episodes are recognized in some regions, but not all deposits can be correlated in time. The deposits record glaciations that spread over large areas and persisted for long periods. Again, low latitudes of formation have been determined. For both the Paleoproterozoic and Neoproterozoic low-latitude episodes, it may be that there was a large increase in the orbital obliquity of the Earth relative to that of the present, such that solar insolation received in the low latitudes was less than that at the high latitudes. An alternative explanation suggests that the globe was everywhere covered in ice, the "Snowball Earth" hypothesis. It has also been suggested that localization of Precambrian glacial strata can be attributed to tectonic elevation of areas marginal to both active and passive continental margins.

The first glaciation of the Paleozoic Era is recorded in unusually well preserved latest Ordovician deposits in North Africa and in glaciomarine strata in formerly adjacent terranes now located in Europe and North America. This glaciation appears to be of considerably smaller scale and shorter duration than earlier and later ones. Importantly, paleomagnetic investigations show that glaciation had shifted to high latitudes, a situation that characterizes all Phanerozoic glaciations. Slightly younger early Silurian glacial deposits of northern Brazil may be explained by relative movement between the Gondwana continents and the South Pole, such that Africa was first affected and later South America. A short and restricted glacial episode in the late Devonian of Brazil may also reflect this polar motion/continental drift.

Possibly the most extensive series of glaciations in Earth history affected the Gondwana continents in the late Paleozoic interval (early Carboniferous-late Permian, 345–230 million years ago), including the Indian subcontinent and drifted terranes now found in Southeast Asia. The best evidence from age dates and paleomagnetic studies suggests that the growth and decay of ice sheets tracked the relative motion of the South Pole across the supercontinent, from the South America/Africa segment to Antarctica and Australia. The final traces of late Paleozoic ice, likely to have been seasonal, not glacial, are in late Permian dropstones of southeastern Australia. Coeval dropstone strata (and possible tillites) occur in northeastern Russia. The initiation of some Gondwana glaciations probably was due in part to topographic uplifts near continental margins, as well as the regional effects of tectonic loading of the lithosphere. But the construction of massive ice sheets in central parts of continents may instead have been related to the assembly and thermally generated uplift of the central part of the amalgamated Laurentia and Gondwana (Pangaea).

Given the low incidence of solar radiation near the poles, it is hard to understand why glaciation has not been perennial in Earth history. As an example of this paradox, glaciation did not again affect the globe until the early Cenozoic Era, some 230 million years after the late Paleozoic episode. Although there are indications of high-latitude ice rafting in the Jurassic record and during the early Cretaceous in strata from Australia,

Siberia, the Canadian Arctic, and Svalbard, definite effects of glaciation are not known from this time. High concentrations of carbon dioxide in the atmosphere and increased transport of heat by the oceans have been invoked in various models in efforts to explain the intermittently warm polar zones.

The Cenozoic cooling 65 million years ago first resulted in early Eocene glacial deposition in the Antarctic Peninsula area, and by the early Oligocene (37 million years ago) the ice sheet had begun to form in Antarctica. In contrast, the earliest known glacial deposits of the Northern Hemisphere (from southeastern Alaska) are of middle Miocene age, some 12 million years ago. Support for enhanced ice growth in the early Oligocene comes from the enriched oxygen isotope composition of benthic foraminifera, which reflects the buildup of isotopically depleted Antarctic ice. Global ice volume greatly increased again beginning in the Middle Miocene, possibly as a result of thermal isolation of Antarctica by the Circumpolar Current, newly formed as continents moved away from the polar zone. While the East Antarctic ice sheet formed early in this history, the marine-based West Antarctic ice sheet probably originated in the late Miocene. The late Pliocene (2 million years ago) saw the birth of the major Northern Hemisphere ice sheets, and Antarctic ice again expanded, as did mountain glaciers in midlatitude highlands.

Tectonic uplifts can cause drawdown of carbon from the atmosphere through enhanced weathering of newly exposed silicate rocks. Through this process, late Miocene uplift of the Tibetan plateau and possibly other mountain belts may have intensified the glaciation. The initiation of Cenozoic glaciation apparently took place much earlier.

Of the several possible contributors to glaciation outlined above, each may have played the key role at particular times. For all of the glacial times, the concentration of greenhouse gases in the atmosphere is inferred to have been lower than in the immediately preceding warm times. Whether these decreases in themselves caused the glaciations or whether they were instead a consequence of global cooling is as yet uncertain. If large enough on the global scale, the effect of biological utilization of carbon can cause significant drawdown of atmospheric carbon dioxide and lead to or intensify glaciation. The variable release of carbon dioxide by volcanic action, particularly in the zones of explosive eruptions, also seems to have altered atmospheric composition over time. Tabulation of the secular variation in volcanism reveals that in some instances glaciation shortly followed or coincided with decreased activity.

While the causes of glaciation are poorly understood and much hypothesized about, their terminations are even more controversial. A marked change in obliquity, the impetus for which is unknown, might explain the end of the Paleoproterozoic glaciation, but the high obliquity/low obliquity cycle would need to be repeated to explain the Neoproterozoic low-latitude and subsequent high-latitude glacial events. Moreover, confounding the strict uniformitarian interpretation that there will be glaciation whenever a continent lies over the pole, the late Paleozoic glaciation ended while western Gondwana still occupied the polar zone. Regarding the tectonic models of late Paleozoic glaciation, not all glacial deposits of this age are known to be positioned adjacent to tectonically active zones involving major uplifts, and far-field subsidence of sedimentary basins, leading to the collapse of nearby ice sheets, does not explain why known smaller ice bodies also terminated. As for the termination of late Paleozoic glaciation, the breakup of Pangaea could not have been responsible, because separation took place only much later, in the second half of the Mesozoic Era. It is possible that changes in oceanic circulation caused this glaciation to end, but just what those changes might have been is unknown. However, terminations of glacial stages during the Quaternary Period, as they are seen in the Northern Hemisphere, can be related to changes in atmospheric and oceanic circulation in the North Atlantic region that diminished the evaporative moisture supply to the ice masses or may have warmed them.

**The Last Million Years: Ice Volume, Oxygen Isotopes, and Sea Level.** Because it is not possible to determine even relative global ice volumes directly from examination of sequential glacial deposits, it has been necessary to devise a method of indirect measurement. One such method is oxygen isotope analysis. Because evaporation favors oxygen-16, the lighter of the element's two principal stable isotopes, both atmospheric water vapor and the glacial ice formed from its condensation are relatively deficient in the heavier oxygen-18. Conversely, sea water at times of high glaciation is relatively rich in oxygen-18. The ratio between oxygen-16 and oxygen-18 in an ocean sediment sample can thus be read as an indication of global ice volume and hence temperature. Such readings are generally expressed as $\delta^{18}O$, that is, the deviation, in parts per thousand, between a sample's ratio and a standard ratio.

The relative ice volume record of the last million years is preserved in the $\delta^{18}O$ of marine carbonate (Figure 3), and isotopes measured in sequence have the added advantage of providing a well established time scale. Spectral analysis of $\delta^{18}O$ data for the last million years and earlier intervals reveals the presence of three cycles related to the Earth's orbit, which in turn affect the latitudinal distribution of incident solar radiation. The shortest of these Milankovitch cycles, at 23,000 years, records variations in the time of year when the Earth-Sun distance is at the minimum (precession), and

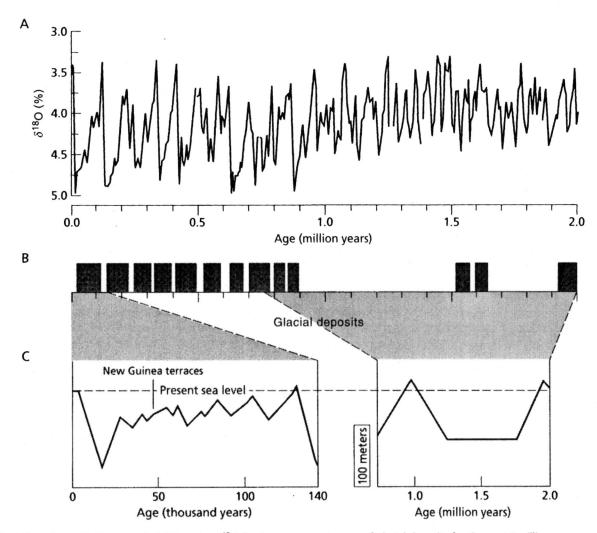

**Glaciation. FIGURE 3.** The record of (A) marine $\delta^{18}O$ for the past 2 million years (larger values of $\delta^{18}O$ signify oceanic enrichment in oxygen-18 relative to oxygen-16); (B) approximate ages of glacial deposits for the past 2 million years; and (C) short- and long-term variations of sea level relative to the present (at 0) over the past 2 million years.

the second shortest (41,000 years) relates to variations in the tilt of the Earth (obliquity). The third cycle, at 100,000 years, is due to the changing eccentricity of the Earth orbit.

Well-dated sedimentary sequences show that the dominant variability over the last 700,000 years coincided with the eccentricity cycle. Prior to that, the obliquity and precession cycles dominated the record. The precession and obliquity components of the isotope record are considered to be largely linear responses; that is, they indicate changing ice volumes that corresponded in time with and were forced by the orbital factors. Variations in the calculated solar radiation for eccentricity, however, are too small to explain the large amplitudes in the $\delta^{18}O$ signal seen at 100,000-year intervals and moreover at times are out of phase by about 15,000 years. Amplification of the input from the eccentricity cycle is therefore required, and the behavior

of massive ice sheets may yet provide an answer. When correlation of glacial deposits with peaks in the filtered 100,000-year sequence is attempted, the relationship is unclear, largely because of poor dating control for tills. But it is likely that through their long-term growth, their rapid decay, and lags resulting from ice-lithosphere (isostatic) feedbacks, the ice sheets of the Northern Hemisphere themselves exert indirect controls on ice growth through a combination of changes in sea level, atmospheric composition, and deep oceanic circulation.

Buildups of polar ice are also reflected in the lowering of sea level (glacio-eustasy). Reconstruction of sea level history has been made from dating and measuring the elevation of individual shorelines in flights of terraces in places such as Papua New Guinea and Barbados, where tectonic uplift is assumed to be constant. The record is known only back to about 140,000 years and can be correlated to the $\delta^{18}O$ record; the relation-

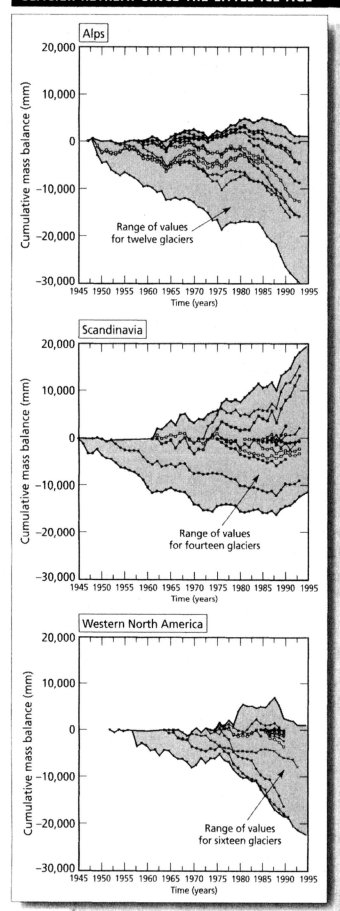

Since the nineteenth century, many of the world's alpine glaciers have retreated up their valleys as a consequence of the climatic changes, especially warming, that have occurred in the last hundred or so years since the ending of the Little Ice Age (Oerlemans, 1994). Studies in the changes of snout positions obtained from cartographic, photogrammetric, and other data therefore permit estimates to be made of the rate at which retreat can occur. The rate has not been constant, nor the process uninterrupted. Indeed, some glaciers have shown a tendency to advance for some of the period. However, if one takes those glaciers that have shown a tendency for a fairly general retreat (see table), it becomes evident that, as with most geomorphologic phenomena, there is a wide range of values, the variability of which is probably related to such variables as topography, slope, size, altitude, accumulation rate, and ablation rate. It is also evident, however, that rates of retreat can often be very high, of the order of 20–70 meters per year over extended periods of some decades in the case of the more active examples. It is therefore not unusual to find that, over the last hundred or so years, alpine glaciers in many areas have managed to retreat by some kilometers.

Current glacier tendencies for selected regions are shown in the accompanying figure. These are expressed in terms of their mass balance, which is defined as the difference between gains and losses (expressed in terms of water equivalent). In the European Alps (top figure) a general trend toward mass loss, with some interruptions in the mid-1960s, late 1970s, and early 1980s, is observed. In Scandinavia (middle), glaciers close to the sea have seen a very strong mass gain since the 1970s, but mass losses have occurred with the more continental glaciers. The mass gain in western Scandinavia could be explained by an increase in precipitation, which more than compensates for an increase in ablation caused by rising temperatures. Western North America (bottom) shows a general mass loss near the coast and in the Cascade Mountains (Hoelzle and Trindler, 1998).

**BIBLIOGRAPHY**

Goudie, A. S. *The Changing Earth*. New York and Oxford: Oxford University Press, 1995.
Hoelzle, M., and M. Trindler. "Data Management and Application." In *Into the Second Century of Worldwide Glacier Monitoring: Prospects and Strategies*, edited by W. Haeberli et al., pp. 53–64. Paris: UNESCO, 1998.
Oerlemans, J. "Quantifying Global Warming From the Retreat of Glaciers." *Science* 264 (1994), 243–245.

—ANDREW S. GOUDIE

**Retreat of Glaciers in Meters per Year in the Twentieth Century**

| LOCATION | PERIOD | RATE |
|---|---|---|
| Breidamerkurjökull, Iceland | 1903–1948 | 30–40 |
|  | 1945–1965 | 53–62 |
|  | 1965–1980 | 48–70 |
| Lemon Creek, Alaska | 1902–1919 | 4.4 |
|  | 1919–1929 | 7.5 |
|  | 1929–1948 | 32.9 |
|  | 1948–1958 | 37.5 |
| Humo Glacier, Argentina | 1914–1982 | 60.4 |
| Franz Josef, New Zealand | 1909–1965 | 40.2 |
| Nigardsbreen, Norway | 1900–1970 | 26.1 |
| Austersdalbreen, Norway | 1900–1970 | 21 |
| Abrekkbreen | 1900–1970 | 17.7 |
| Brikdalbrenn | 1900–1970 | 11.4 |
| Tunsbergdalsbrenn | 1900–1970 | 11.4 |
| Argentière, Mont Blanc | 1900–1970 | 12.1 |
| Bosson, Mont Blanc | 1900–1970 | 6.4 |
| Oztal Group | 1910–1980 | 3.6–12.9 |
| Grosser, Aletsch | 1900–1980 | 52.5 |
| Carstenz, New Guinea | 1936–1974 | 26.2 |
| **Region** | | |
| Rocky Mountains | 1890–1974 | 15.2 |
| Spitzbergen | 1906–1990 | 51.7 |
| Iceland | 1850–1965 | 12.2 |
| Norway | 1850–1990 | 28.7 |
| Alps | 1850–1988 | 15.6 |
| Central Asia | 1874–1980 | 9.9 |
| Irian Jaya | 1936–1990 | 25.9 |
| Kenya | 1893–1987 | 4.8 |
| New Zealand | 1894–1990 | 25.9 |

SOURCE: Goudie (1995, Table 6.3).

ship is 0.11 $\delta^{18}O$ to 10 meters of sea level change. In most isotope measurements, however, the contribution due to the temperature effect is not known, and there are limitations to using this equation far back in time owing to the likely contributions both from local tectonics and from global tectono-eustasy. The late Quaternary oxygen isotope curves show what appear to be gradual rises but abrupt falls, and Quaternary ice sheets are thus thought to have undergone slow accumulation and rapid disintegration. Such features are not readily discernible in the short sea level curve because only the stillstands are recognized.

Glaciation in the Quaternary Period has been a continuation of conditions established earlier in the Cenozoic Era. Ice caps of North America and Europe, together with other glaciers, advanced and retreated periodically, mostly in phase with the orbital motions of the planet. The effects of some of these fluctuations (ice ages, including the Younger Dryas, ca. 13,000–11,500 years ago) are reflected in ocean sediments. The so-called Heinrich Events, consisting of abundant ice-rafted debris in the North Atlantic, record episodes in the glacial history of the Quaternary North American ice sheet. Also, the thermally generated North Atlantic Deep Water appears to have switched on and off during the Neogene (25 million to 1.8 million years ago). The "Little Ice Age," including advances of alpine glaciers, occurred during our present interglacial between about 450 and 150 years ago.

At present, the Earth is in an interglacial stage of the

Quaternary glacial phase (the last 1.8 million years); that is, ice is present only in polar zones and at high elevations. This interglacial began about 10,000 years ago and is merely the latest of several such stages. The last glacial stage, when ice sheets were much larger and reached to 40°N latitude, began at about 120,000 and culminated at about 21,000 years ago. The climate history of this time interval is well known from studies of ice cores from Antarctica and Greenland. Oxygen and deuterium isotopes and carbon dioxide and methane abundances have been measured in air bubbles trapped in the ice. A striking feature of these records is the sudden changes in surface air temperature, spaced at on the order of a millennium and taking only a few decades to complete, as recorded by the Greenland isotopes. In Greenland during the last glacial maximum, the mean annual surface air temperatures were as much as 21°C lower than today. The carbon dioxide concentrations in bubbles, governed by past atmospheric compositions, are low when temperatures are low and high when they are high, indicating the strong relationship between the two variables, with the final large rise occurring be-

tween about 20,000 and 10,000 years ago. Both ice cores and oceanic sediment cores reveal cyclicities related to the orbital Milankovitch cycles. The oxygen isotope stratigraphy developed from ocean sediments can be correlated with the shorter-term ice core records, with the glacial deposits on land, and with the sequence of windblown loess (glacial times) and soil (interglacials) in central China and elsewhere in the high northern latitudes.

The last glacial maximum (ca. 21,000 years ago) saw the development of Northern Hemisphere ice sheets to thicknesses of three kilometers. The decline of this ice has importance as an indication of how, and why, global warming might proceed in the future. At the peak, the Laurentide ice sheet covered the northeastern quadrant of North America as far south as the region of the Great Lakes, at about 42° latitude (Figure 4). To the west, the Cordilleran ice sheet occupied mountainous regions and extended to the Gulf of Alaska., but reached only as far south as the U.S.–Canadian border. In Eurasia, the Scandinavian, Barents, and Kara ice sheets attained similar latitudes as those of North America and also occupied

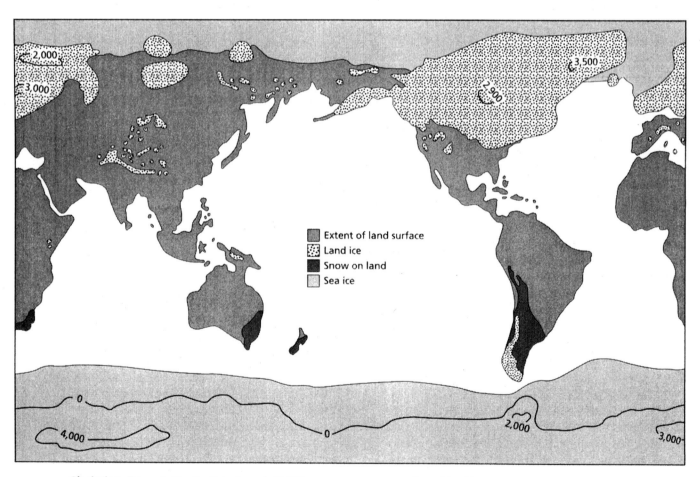

**Glaciation. Figure 4.** The Earth in August, 18,000 years ago, showing the land surface, including that exposed by a drop in sea level of 85 meters (shaded); oceans and seaways (unpatterned); and land ice, sea ice, and snow, with calculated contours of elevation (in meters). (Modified from CLIMAP, 1976.)

parts of what are now continental shelves. Antarctic ice probably had a similar configuration as at present, including extension of ice sheets into the sea as floating ice shelves. Other main bodies existed in Greenland, Iceland, eastern Siberia, the Himalayas, and the European Alps. Mass balance on these glaciers probably fluctuated during much of the Quaternary owing to catastrophic surges of ice into the sea, but the final decline of the ice sheets was preceded by a sizable increase in Northern Hemisphere Milankovitch radiation. Near the peak of radiation, there was a marked increase in atmospheric carbon dioxide (from 180–200 to ca. 280 parts per million by volume) and this was soon followed by increased production of North Atlantic Deep Water. The latter event resulted in marked increases in deep convection in the Atlantic. A further response to the melting, which led to the Holocene Epoch at about 10,000 years ago, was a gradual rise in sea level of about 120 meters.

A great deal is known about the character and genesis of glaciations of the recent past, although many questions remain unanswered. Knowledge about the more ancient episodes is less comprehensive. Because circumstances for each glaciation are different (for example, latitude of formation), it is difficult to apply conclusions from one to the other. At this stage, the fairest conclusion is that each glaciation may have had, and probably did have, a unique style of genesis.

[*See also* Climate Change; Cryosphere; Earth History; Earth Motions; Erosion; Hydrologic Cycle; Ice Sheets; Land Surface Processes; Natural Hazards; Sea Level; Younger Dryas; *and the biography of Agassiz.*]

## BIBLIOGRAPHY

Alley, R. B. "Resolved: The Arctic Controls Global Climate Change." *Arctic Oceanography: Marginal Ice Zones and Continental Shelves.* Special issue. *Coastal and Marine Studies* 49 (1995), 263–283. One side of a debate on the importance of glaciers in forcing climate change.

Barnola, J. M., D. Raynaud, Y. S. Korotkevitch, and C. Lorius. "Vostok Ice Core Provides 160,000-year Record of Atmospheric CO₂." *Nature* 329 (1987), 408–414. A summary of the variation of an important greenhouse gas during recent history.

Barron, E. J., W. H. Peterson, D. Pollard, and S. Thompson. "Past Climate and the Role of Ocean Heat Transport: Model Simulations for the Cretaceous." *Paleoceanography* 8 (1993), 785–798. Illustrates how the ocean can assist in overcoming potential low temperatures at high latitudes.

Bennett, M. R., and N. F. Glasser. *Ice Sheets and Landforms.* New York: Wiley, 1996.

Berger, A., and M. F. Loutre. "Insolation Values for the Climate of the Last 10 Million Years." *Quaternary Science Review* 10 (1991), 297–317. An excellent summary of the calculated insolation figures in recent times.

Berner, R. A. "Geocarb III: A revised model of atmospheric CO₂ over Phanerozoic Time." *American Journal of Science* 294 (1994), 56–91. Presents a comprehensive geochemical model of atmospheric carbon dioxide through geologic time.

Bond, G., W. Broecker, S. Johnsen, J. McManus, L. Labeyrie, J. Jouzel, and G. Bonani. "Correlation between Climate Records from North Atlantic Sediments and Greenland Ice." *Nature* 365 (1993), 143–147.

Boyle, E. A. "Quaternary Deepwater Paleoceanography." *Science* 249 (1990), 863–870. Relates ocean chemistry to glacial episodes.

Broecker, W. S., and G. H. Denton. "The Role of Ocean–Atmosphere Reorganizations in Glacial Cycles." *Geochimica Cosmochimica Acta* 53 (1989), 2465–2501. An important discussion of changes in the global system and how they interact.

Caputo, M. V., and J. C. Crowell. "Migration of Glacial Centers across Gondwana during the Paleozoic Era." *Geological Society of America Bulletin* 96 (1985), 1020–1036.

CLIMAP. "The Last Glacial Ocean." *Quaternary Research* 21 (1984), 123–224.

COHMAP. "Climatic Changes of the Last 18,000 Years: Observations and Model Simulations." *Science* 241 (1988), 1043–1052. Documents climate changes through data and modeling.

Eyles, N. "Earth's Glacial Record and Its Tectonic Setting." *Earth Science Reviews* 35 (1993), 1–248. Provides an excellent review of glacial history and offers a tectonic model to explain the distribution of glaciers.

Fairbanks, R. G. "A 17,000-Year Glacio-eustatic Sea Level Record: Influence of Glacial Melting Rates on the Younger Dryas Event and Deep-Ocean Circulation." *Nature* 342 (1989), 637–642.

Frakes, L. A., and J. E. Francis. "A Guide to Phanerozoic Cold Polar Climates from High-Latitude Ice-Rafting in the Cretaceous." *Nature* 333 (1988), 547–549. Summarizes the latitudinal distribution of seasonal and permanent ice during the Phanerozoic.

Hambrey, M. J., and W. B. Harland. *Earth's Pre-Pleistocene Glacial Record.* Cambridge: Cambridge University Press, 1981. A compendium of the Earth's glacial deposits.

Hughes, T. J. "Abrupt Climatic Change Related to Unstable Ice-Sheet Dynamics: Toward a New Paradigm." *Palaeogeography, Palaeoclimatology, Palaeoecology (Global and Planetary Change)* 97 (1992), 203–234. Models how ice sheets contribute to global change.

Imbrie, J., A. Berger, E. A. Boyle, S. C. Clemens, A. Duffy, W. R. Howard, G. Kukla, J. Kutzbach, D. G. Martinson, A. McIntyre, A. C. Mix, B. Molfino, J. J. Morley, L. C. Peterson, N. G. Pisias, W. L. Prell, M. E. Raymo, N. J. Shackleton, and J. R. Toggweiler. "On the Structure and Origin of Major Glaciation Cycles 1. Linear Responses to Milankovitch Forcing." *Paleoceanography* 7 (1992), 701–738. A thorough analysis of the myriad of data sets and their application to climate studies.

———. "On the Structure and Origin of Major Glaciation Cycles 2. The 100,000-year Cycle." *Paleoceanography* 8 (1993), 699–735. A broad-based study of possible controls on the 100,000-year cycle.

Johnsen, S. J., H. B. Clausen, W. Dansgaard, K. Fuhrer, N. Gundestrup, C. U. Hammer, P. Iversen, J. Jouzel, B. Stauffer, and J. P. Steffensen. "Irregular Glacial Interstadials Recorded in a New Greenland Ice Core." *Nature* (1992), 311–313. Documents rapid climate changes.

Kerr, A. "Topography, Climate and Ice Masses: A Review." *Terra Nova* 5 (1993), 332–342.

Lorius, C., J. Jouzel, D. Raynaud, J. Hansen, and H. Le Treut. "The Ice-Core Record: Climate Sensitivity and Future Greenhouse Warming." *Nature* 347 (1990), 139–145.

Manabe, S., and A. J. Broccoli. "The Influence of Continental Ice Sheets on the Climate of an Ice Age." *Journal of Geophysical Research* 90 (1985), 2167–2190. Tests the sensitivity of climate to variability in polar ice volume.

Miller, K. G., R. G. Fairbanks, and G. S. Mountain. "Tertiary Oxygen Isotope Synthesis, Sea Level History, and Continental Margin Erosion." *Paleoceanography* 2 (1987), 1–19. A landmark study setting the marine isotope limits for growth of ice sheets.

Nesje, A., and S. O. Dahl. *Glaciers and Environmental Change.* London: Edward Arnold, 2000.

Raymo, M. E., and W. F. Ruddiman. "Tectonic Forcing of Late Cenozoic Climates." *Nature* 359 (1992), 117–122.

Saltzman, B., and M. Y. Verbitsky. "Multiple Instabilities and Modes of Glacial Rhythmicity in the Plio-Pleistocene." *Climate Dynamics* 8 (1993), 147–161. Examines the possible sources of climate cycles.

Shackleton, N. J. "Oxygen Isotopes, Ice Volume, and Sea Level." *Quaternary Science Review* 6 (1987), 183–190. An important summary of relationships among these three variables.

—LAWRENCE A. FRAKES

## GLACKEN, CLARENCE J.

Clarence J. Glacken (1909–1989) was an American human geographer and historian of environmental thought. The place of humans in nature is central to our understanding of global change, and, in that investigation, the work of Glacken must stand very high. His 1956 essay, "Changing Ideas of the Habitable World," gave a hint of what was to follow in *Traces on the Rhodian Shore.* The latter must be, according to one obituarist, David Hooson, "one of the most scholarly books written by a geographer, or by a historian of ideas, in this century." The comprehensiveness and scope of its 763 pages are dazzling, and the care and range of the references are extraordinary.

The theme of *Traces* is that, from classic Greek times to the end of the eighteenth century, humans yearned to find purpose and order in their life and surroundings. Consequently, they constantly asked three basic questions that, in one form or another, still underpin much of the environmental debate: First, was the Earth designed by some supreme being, and were people merely one part of God's creation and equal to all other life, or were they something superior? Second, did nature influence or mold human action, or was the opposite true—that humans molded nature? Third, if humans were superior to nature and could alter it, then what was the role of humans on Earth? Were they to dominate nature or did they have a divine role in acting as stewards on Earth—tending and caring for the wonderful creation around them?

Ill health prevented Glacken from completing the companion volume for the nineteenth and twentieth centuries. However, three papers published between 1970 and 1973 give a hint of what it might have been like. They show how Glacken's deep understanding of humankind's predicament was moving from the academic (if it ever had been) to the practical in the age of burgeoning environmental concern. His rationale was disarmingly simple and straightforward: he linked the history of ideas to contemporary environmental concerns by giving them historical and intellectual underpinnings. "If the history of thought teaches us anything about culture and environment," he wrote in his essay "Man against Nature: an Outmoded Concept" (1970), "it is the importance of the conceptions which people have of both—whether these conceptions are religious, philosophical, scientific or utilitarian."

The nonappearance of his final volume is one of the great ironies of the history of global change and human society. As chair of Geography at Berkeley in the "swinging sixties," his work on early thought about the place of humans in nature was pilloried and derided as irrelevant, and his lectures disrupted by activist students who were too shortsighted to see that it was at the very core of many of their concerns. He suffered an acute crisis of confidence and a nervous breakdown. It is thought that he destroyed the nearly completed manuscript; it has certainly never been found. We are all the poorer for that.

[*See also* Global Change, *article on* History of Global Change.]

### BIBLIOGRAPHY

WORKS BY CLARENCE GLACKEN

Glacken, C. "Changing Ideas of the Habitable World." In *Man's Role in Changing the Face of the Earth*, edited by W. L. Thomas, Jr., pp. 70–92. Chicago: University of Chicago Press, 1956. A succinct foretaste of his later work.

———. *Traces on the Rhodian Shore: Nature and Culture in Western Thought from Ancient Times to the End of the Eighteenth Century.* Berkeley and Los Angeles: University of California Press, 1967. An erudite and scholarly investigation into western cultural history.

———. "Man against Nature: An Outmoded Concept." In *The Environmental Crisis*, edited by H. W. Helrich, Jr., pp. 127–142. New Haven: Yale University Press, 1970.

———. "Man and Nature in Recent Western Thought." In *This Little Planet*, edited by M. Hamilton, pp. 163–201. New York: Scribner, 1970.

———. "Environment and Culture." In *Dictionary of the History of Ideas: Studies of Selected Pivotal Ideas*, edited by P. P. Wiener, vol. 2, pp. 127–134. New York: Scribner's, 1973.

WORKS ON CLARENCE GLACKEN

Hooson, D. "Clarence Glacken, 1909–1989." *Annals of the Association of American Geographers* 81 (1991), 152–158. A useful account of Glacken's work and life.

—MICHAEL WILLIAMS

## GLOBAL CHANGE

[*To discuss the historical and human dimensions of global change, this entry comprises two articles. The first article focuses on the historical understanding of global impacts on the Earth, citing the seminal works of numerous writers. The second pre-*

*sents human responses to environmental change as reflected in the information revolution, the rise of popular culture, social movements, and population mobility.*]

## History of Global Change

This article deals with the history of ideas regarding human-induced changes in the biophysical environment that are global in their magnitude or distribution or significance.

The phrase "global environmental change" has come into wide use only within the past few decades. Interest in much of what it denotes, however, predates the term itself. Attention to such change has never been stronger or more widespread than it is today, yet it existed in more scattered and fragmentary form before talk about global change as such became common. But to what extent it existed, and how it differed from current awareness, are important questions that largely await satisfactory answers. Adequate histories of ideas regarding global environmental change still remain to be written.

Such ideas take two forms: (1) beliefs about what changes have taken place or might take place and how, and (2) evaluations of the changes (e.g., as being for the better or for the worse). A full intellectual history of global change would include many ideas of the first sort that would now be judged untenable, cases where human-induced change was incorrectly thought to have occurred. It would also include many past evaluations of most real or imagined human-induced changes as almost necessarily improvements upon nature and the more substantial the more so, part of the remodeling of the Earth from its bleak, hostile, and forbidding state into one much better suited to its human occupants. This point of view, though scarcely typical of contemporary discourse, is nevertheless part of the wider history to which it belongs. A history guided too closely by present-day concerns is apt to overlook it.

A history so guided is also apt to distort the significance of much that it does include. Many possibilities of misinterpretation arise when earlier ideas are examined mainly because of their relation to contemporary ones. One of the most common is that of giving undue attention to ideas similar to today's just because they are similar to today's and ascribing them more importance in the thought of the past than they actually had. In the field of global change, it is dangerously easy to class all past thinkers who expressed concern over any perceived human-induced change in the environment as having been "ahead of their time" and necessarily more insightful than contemporaries who did not share their concerns. At an extreme, it can mean lauding marginal figures whose views were absurd by the standards of

the best science of their time but whose errors led them to express worries in some way similar to those of the present.

Documenting what has been thought and said about global change is not made easier by the unevenness and unrepresentativeness of the primary source material. Though much of past and much of current awareness consists of popular as well as scientific beliefs, the character, extent, and distribution of those beliefs as they exist even today are little known. Scientific and explicit understandings of global change are far more thoroughly recorded than lay and tacit ones are. As a result, little is known or can be said with confidence about what views of global change prevailed in the past and among what groups. As they may have differed greatly and may differ today along such dimensions as location, ethnicity, age, class, race, and gender, one must be quite wary of generalizing about the beliefs of a population or a period without recognizing how narrowly a generalization may apply even within the domain for which it is made.

As the geographer Clarence Glacken has shown (1967), observations and discussions of human influence on the environment stretch back to classical antiquity and beyond. It is a difficult and somewhat arbitrary matter to fix the points at which awareness of environmental change could be said to have become awareness of global change. Much was said in classical texts about the decay or the improvement of the Earth generally that could be considered global environmental theorizing. Yet it was based on scant empirical knowledge of the Earth's surface. It reached global conclusions mainly by abstract reasoning rather than by thorough analysis and synthesis of observed phenomena. So too, many premodern Christian theologians viewed the Fall of Man and the Deluge as cases in which human misdeeds led to the physical transformation for the worse of the entire Earth, shaping its current conditions from its geomorphology to its climate. These conclusions about global change again derived more from reasoning or revelation than from detailed knowledge established by observation or report. To be sure, many such thinkers did appeal to the evidence of the current configuration of the Earth's surface, which they read to support their case, and they linked, as many environmentalists do today, human misdeeds with environmental degradation in a call for reform. Inasmuch as the mechanism they invoked to link misdeeds with degradation was an entirely supernatural one and the reforms for which they called were not in behavior toward the environment as such, it would be an error to emphasize too strongly their kinship with the mainstream of contemporary environmental thought. It would also be an error to ignore the extent to which similar views persist and even thrive today, though in areas, such as fundamentalist religion, generally little

**ONE THOUSAND YEARS AGO**

### The Known World

In the West, Ptolemy's concept of an Earth-centered universe prevailed; but his rational view of the world had been abandoned for a geography of the imagination, strongly influenced by religious concepts. Maps were schematic drawings, often circular, with distorted views of Europe, Africa, and Asia surrounded by the Ocean Sea. It would be another 150 years before the Arabic world produced the Idrisi map with its more realistic portrayal of the known regions. By the year 1000 there had been some contact with lands to the west of Europe; but there was no widespread concept of new continents there. It would be nearly 300 years before Marco Polo returned from his travels to recount the wonders of China.

### Events of the Time (After Asimov, 1991)
*950 to 1000*
- End of West Frankish kingdom and founding of Capetian line of French Kings
- Moorish Spain begins to decline
- Kingdoms of Denmark, Sweden, and Norway established
- Vikings discover and settle Greenland
- Leif Ericsson lands in North America
- Kingdoms of Hungary, Bohemia, and Poland established
- Russia and Serbia converted to Byzantine version of Christianity
- Fatimid Egypt at its peak
- China under the Sung dynasty
- Japan under the Shogunate

*1000 to 1050*
- Denmark, under Canute, conquers England
- Macbeth in Scotland
- Brian Boru drives Vikings out of Ireland
- Rise of Normandy
- Rise of Venice
- Christian drive to retake Spain from the Moors increases in intensity
- Kievan Russia at its peak
- Basil II brings Byzantine Empire to a peak, destroying Bulgarian kingdom
- Rise of Seljuk Turks
- Tale of the Genji written in Japan
- Mayans enter final decline

### Estimated World Population
- Between 250 and 500 million people, concentrated in Europe, China, India, and the highlands of Mexico and the Andes

### Cities over 250,000 Population (After Chandler, 1987)

- Cordova, Spain          450,000
- Kaifeng, China          400,000
- Constantinople          300,000
- Teotihuacán (northern valley of Mexico) with population around 250,000 collapsed around 800 CE

### Science and Technology
- Crossbow invented in France
- Arab physicist, Alhazen (965–1039), recognized basic principles of light and the use of lenses
- Horse collar first used. Along with moldboard plow and horseshoes invented earlier, this helped to increase food production

### Forests
- Pliny, the Elder, in his Natural History nearly 1,000 years earlier, had warned that indiscriminate cutting of forests would cause soils to wash away (see Thomas, 1956)

In the year 1000 CE:

- Significant deforestation was occurring in Europe and China
- North African forests had been extensively cut
- Substantial reforestation of large portions of tropical forests of Southern Mexico and Central America was underway because of demise of the classic lowland Mayan civilization

### Natural Change
- North America then was roughly 10 meters closer to Europe
- Hudson Bay shore elevation then was 10-15 meters lower
- Niagara Falls then was roughly 1000 meters farther downstream
- World average temperature then was slightly higher than now, as the Medieval Maximum was coming to a close

#### BIBLIOGRAPHY

Asimov, I. *Asimov's Chronology of the World.* New York: HarperCollins, 1991.
Chandler, T. *Four Thousand Years of Urban Growth. An Historical Census.* Lewiston, N.Y.: St. David's University Press, 1987.
Thomas, W. L., Jr., ed. *Man's Role in Changing the Face of the Earth.* Chicago: University of Chicago Press, 1956.

—DAVID J. CUFF

known to either intellectual historians or environmental activists.

Through the end of the nineteenth century, most discussions of human impact through means other than supernatural ones emphasized land-cover changes that were usually viewed as improvements: the clearing of forests, the draining of wetlands, and the cultivation of the soil. But the work of authors collating and comparing observations of these processes and their effects in different regions represents another point where concern over environmental change could be said to have attained a global scale. It is also awareness for which the term *environmental* in its modern sense becomes more appropriate than before, for it drew attention not only to the direct and deliberate effects of human action but to their possible secondary, inadvertent, and harmful consequences. The eighteenth century, as the historian Richard Grove has demonstrated (1995), saw the emergence of a school of thought incorporating scientists, travelers, and government officials who linked deforestation in Europe's far-flung tropical colonies to disastrous climatic change in the form of lessened rainfall. The thesis gained further adherents and credibility during the first half of the nineteenth century and helped to shape the forestry policies of many colonial governments. It marks an important point at which a coordinated and politically influential concern emerged over widespread and evidently harmful effects of human alteration of the land cover. Many of the connections that the desiccation theorists invoked, though, it must be noted, would be judged scientifically dubious by modern standards, and many of their claims about human impact overblown. They have some resemblances to current thought but no seamless continuity with it.

Such theorizing about deforestation and desiccation developed in tandem with Western growth in knowledge about other parts of the world, a growth hastened by the formal and informal efforts of governments and learned societies in promoting systematic exploration of the Earth and its human occupancy. The reports and descriptions that resulted included much observation and comment on the human alteration of the environment, among many other themes. In the latter part of the eighteenth century, the French savant Count Buffon (1707–1788) produced the period's most ambitious synthesis of global geography and natural history. His *Histoire Naturelle* (44 vols., 1749–1804) devoted much attention to the ways in which humankind had been a powerful force altering the Earth. Buffon judged most of these alterations, from forest clearance to species transplantation to soil drainage, to have been changes for the better. His Earth as affected by human action was largely one improved by it. He expressed an optimistic Enlightenment confidence in the steady improvement of the environment as part of the progressive perfection of the human race.

For Buffon and even for later compilers and interpreters of global geographic knowledge, human impact had not been the central theme, but only one, though an important one, among many. The first full-length work devoted entirely to a survey and inventory of the impacts of human activity on the Earth came in 1864. *Man and Nature, or Physical Geography as Modified by Human Action* was the work of the American diplomat, legislator, linguist, and historian George Perkins Marsh (1801–1882). The book's subtitle described the contents better than the title. Marsh more aptly called his later revisions of the work (published in 1874, in 1882, and posthumously in 1885) *The Earth as Modified by Human Action,* for his focus was not on all relations between man and nature or between society and the environment. Even with deliberate and successful changes, Marsh declared himself not concerned. Hence he offered only a partial assessment of human-induced change, and one further limited by the scantiness of knowledge about many regions of the world. He dealt chiefly with land-cover change, especially deforestation but also wetland drainage, changes in surface and underground water, the creation and spread of deserts, and the transfer of plant and animal species.

Its limitations notwithstanding, the book was a landmark in the history of global change. It was based factually on the best collection of evidence yet amassed, drawn from a wide reading in many languages and enriched by personal observations in North America and the lands surrounding the Mediterranean. Throughout, Marsh emphasized what he called the "secondary and collateral" effects of human actions, their unforeseen and often distant, surprising, and dismaying consequences. In places, he emphasized them beyond what modern scientific opinion would consider justified, ascribing too much of a role to land-cover modification in climate change, the spread of deserts, and the flooding of rivers.

Marsh, while challenging some of the assumptions current in his time, accepted others. He tended to view nature in the absence of human occupants as harmonious and unchanging or at most changing only slowly and gradually. He thus tended to suppose that any sudden or profound change occurring within the human period was necessarily the work of human action. Like many of his contemporaries, he took an anthropocentric view that placed humankind outside of and above nature and focused on the significance of the environment for human life and livelihood. He questioned,

however, the widespread optimism, the equation of environmental alteration with progress that may be said to have prevailed in his milieu, by documenting instead the damage that had been done and the mass of unintended environmental consequences that could flow from human actions. He emphasized the immense power of human agency in the environment and the havoc that it could produce if it were not restrained and informed by a better understanding of its possible effects on the complex web of natural processes.

Marsh's influence on later thought is not easy to define. He clearly and forcefully stated many of the themes that have become central to late twentieth-century concern about human impact on the environment. How far did he help to shape that concern, and how far has he simply been rediscovered because of his congeniality to a later climate of opinion? Certainly he was widely read in his day. Among the most important of several writers who closely followed Marsh in the topics with which they dealt and in the spirit in which they approached them was the Russian meteorologist and geographer Aleksandr Ivanovich Voeikov (1842–1916). In 1901 he published an extensive and impressive evaluation of human alteration of land and land cover and its further repercussions for such realms as climate, water balance, and sediment flows. Voeikov was one of a number of European scholars around the turn of the century who echoed Marsh in their concern with what the German geographer Ernst Friedrich termed *Raubwirtschaft,* or destructive occupation and exploitation of the land. Similar concerns took shape in many countries in a forest and soil conservation movement that received added impetus and much popular support as a result of the American Dust Bowl and similar disasters in other places in the 1930s. A 1955 conference held at Princeton University on the theme of "Man's Role in Changing the Face of the Earth" was defined by its organizers—the geographer Carl Sauer, the social theorist Lewis Mumford, and the biologist Marston Bates—as a "Marsh Festival" dedicated to the memory of the author of *Man and Nature.* Published in 1956, the papers and proceedings of the conference updated Marsh's analysis in light of nearly a century of further scholarship and of further human-induced environmental change.

They also dealt with some newer themes. One of them was a strand of global concern that had developed separately from the Marsh tradition. It had to do with the adequacy of natural resources for human needs and the possibility that waste and overuse would exhaust ones vital to human civilization, such as coal, timber, and minerals. The late nineteenth century had also seen the earliest serious attention to the ways in which coal consumption could affect human life on a global scale not only by depleting the supply but by releasing waste products into the environment. Work published in 1896 by the Swedish chemist Svante Arrhenius (1859–1927) suggested that the temperature of the Earth might be increased by several degrees K if releases from coal combustion were to double atmospheric concentrations of carbon dioxide. Arrhenius's work represents another landmark in the understanding of global change, introducing as it did the first credible forecast of substantial human impact on a fluid global system potentially affecting the entire planet.

It is also a landmark that is liable to be misinterpreted when viewed too much in relation to the concerns of the present. As his biographer Elisabeth Crawford points out (1997), Arrhenius was not the first to suggest that human activities might raise the atmospheric level of carbon dioxide and thereby affect the global climate. Moreover, he was centrally concerned with explaining the mechanism by which continental glaciation had occurred in the geological past. He was no more than incidentally concerned with anything to do with human impact in the future. Finally, Arrhenius, in common with contemporaries of his who took up the suggestion, looked on a carbon dioxide–induced warming as an unmixed good that would make agriculture more productive, life more comfortable, and the return of an Ice Age less likely. It might be added that throughout the first half of the twentieth century, Arrhenius's suggestions did not win general acceptance. Many meteorologists believed both that the oceans would promptly absorb any additional carbon dioxide released by fuel combustion and that any added to the atmosphere would mean only a minor further increase in the amount of heat trapped. What would come to be known as the greenhouse effect was not among the topics seriously discussed at Princeton in 1955.

More general in character than Arrhenius's, but almost as optimistic in tenor, were the speculations of the Russian chemist Vladimir Ivanovich Vernadsky (1863–1945), the most significant early theorist of biogeochemical change as a major element in global environmental transformation. Vernadsky regarded the central fact of modern Earth history as the emergence of humankind as a force comparable to many shaping the geological record in its transformation of the planet's material and energy flows. He described its central (and still continuing) process as the reshaping of the biosphere—the envelope of soil, air, and water at the Earth's surface capable of supporting life—into a noosphere dominated by the power of human consciousness. Language and cultural barriers meant that Vernadsky's ideas were slow to diffuse into other countries save in the somewhat altered form in which they were adapted by the French philosopher Teilhard de Chardin. There was little note of warning or pessimism in Vernadsky's work, which nonetheless was of fundamental importance in defining the human impact on the Earth's bio-

geochemical flows as comparable in importance to the land-cover transformations assessed by Marsh and his successors.

Detailed attention to the human impact on the biogeochemical cycles noted by Vernadsky flowered in the post–World War II decades under the stimulus of an important new development, a much-improved capacity to monitor and detect change in the environment. Technological advances in the gathering and analysis of data made it possible to record trends of change in a far wider range of environmental realms than before and with increasing precision. Year-by-year monitoring of the carbon dioxide content of the troposphere begun at the Mauna Loa observatory in Hawaii in the 1950s clearly demonstrated a steady increase as fossil fuel combustion and land-cover change released carbon, at the same time that scientists began to consider anew the idea that it and other trace gases could indeed significantly warm the global climate.

The data thus accumulated testified less to the increasing rational human control over the biosphere that Vernadsky had foreseen than to an increasingly varied collection of impacts produced as the chance and inadvertent product of novel human activities. The dark side of the picture was brought to wide popular and scientific attention in the 1950s through the discovery of radioactive fallout from atmospheric testing of nuclear weapons. It was further dramatized to great effect by Rachel Carson's (1907–1964) *Silent Spring*, published in 1962. By linking the widespread use of the synthetic pesticide dichlorodiphenyltrichloroethane (DDT) to declines in songbird populations, Carson vividly pointed up the way in which human interventions in the biosphere could produce consequences vastly more varied, wide-reaching, and undesirable than those that they were intended to have. Her polemic gave urgent statement and wide currency to a sense of the environment as Marsh had portrayed it, as a complex, interconnected system in which surprises and unforeseen effects were always possible and even likely, but one now threatened by a far wider array of stresses and impacts than in Marsh's time. It represents a landmark in the emergence of full-blown global environmental concern. The fears that it helped to arouse were kept awake by many further and well-publicized examples of increases in concentrations of metals and chemical compounds in air, water, soil, and biota often quite distant from their points of release. The most dramatic and far-reaching was the conclusion by scientists that the release of chlorofluorocarbons, synthetic organic chemicals used in refrigeration and other processes, was significantly depleting the ozone of the upper atmosphere that reduces the surface incidence of high-energy solar radiation.

One other undeniably global-scale human impact attracted even more scientific and public attention. From a change understood largely in terms of resource conservation as a net benefit to the globe, as it was in the early twentieth century, a climatic warming through greenhouse gas accumulation came more and more to be seen as a potential catastrophe. The new and more pessimistic assessment was the result of a greater awareness of climate change's manifold likely second- and third-order consequences. Those now seen as likely range from possible sea level rise from melting of the Antarctic and Greenland ice sheets to biodiversity loss through ecosystem disruption to impacts on human health through rising heat stress and changes in the distribution of infectious diseases.

One distinctive feature of modern global change thought is thus a much wider range of concerns than in the past. Another inventory of human impact, *The Earth as Transformed by Human Action* (1990), presented a catalog as much expanded over that of the 1955 Princeton conference as its own had been over Marsh's *Man and Nature*. Its core consisted of chapters on eighteen major realms and flows of the global environment, documenting the often multiple changes that had taken place in each, many unrecognized in 1956. It followed with a dozen chapters exploring the further combinations and interactions of changes occurring in particular regions and producing diverse syndromes of impact.

Another striking recent trend is the growth of avid discussion and debate in public forums of all sorts. Public opinion has always been an element of global change thought, but with the rapid popularization of the topic it has become a more important one than in the past. Research has shown that scientific and public beliefs about global change often differ profoundly. Just as the term "twentieth-century physics" is generally used to refer to concepts understood by only a tiny fraction of the human population of the twentieth century, at the same time that studies show that the model of dynamics held by most people is not even a Newtonian one, let alone a "post-Newtonian" one, but the "impetus model" of Aristotelian physics discredited centuries ago, so too the contemporary public understanding of almost every aspect of global climate change, from its physical mechanisms to its likely impacts, has been found to diverge widely from the models generally accepted among climate scientists and impact researchers. These differences are sometimes described as being between scientific understanding on the one side and a lay misunderstanding on the other, sometimes explained as the expression of different but equally legitimate concerns. However they are best characterized, and whatever the reasons for them, they raise the question of why expressed public concern over global change is so substantial if, as is apparently the case, the reason does not lie in the public assimilation of scientific reasons for thinking it a matter for concern.

But concern, certainly, is not universal, as witness the rancorous debates that surround almost every environmental issue of importance. One final striking characteristic of present thought is a lack of accord and consensus, both globally and within any country, on almost every issue. This absence of consensus is most of all apparent in differing evaluations of human impact and disagreements over the need to take action to regulate it. Influential perspectives in today's environmental debate span the range from radical Green and doomsday perspectives to a radical reaction questioning the reality of almost any environmental threat, and contain many shades and nuances in between.

[*See also* Climate Change; Environmental Movements; Introduction; *and the biographies of Carson, Glacken, Marsh, and Sauer.*]

## BIBLIOGRAPHY

Carson, R. *Silent Spring.* Boston: Houghton Mifflin, 1962.
Crawford, E. "Arrhenius's 1896 Model of the Greenhouse Effect in Context." *Ambio* 26 (1997), 6–11. An excellent discussion by the author of a longer biography of the scientist.
Glacken, C. *Traces on the Rhodian Shore: Nature and Culture in Western Thought from Ancient Times to the End of the Eighteenth Century.* Berkeley: University of California Press, 1967. The classic history of early Western environmental thought.
Grove, R. H. *Green Imperialism: Colonial Expansion, Tropical Island Edens, and the Origins of Environmentalism, 1600–1860.* Cambridge: Cambridge University Press, 1995.
Lowenthal, D. *George Perkins Marsh: Versatile Vermonter.* New York: Columbia University Press, 1958. The standard biography.
———. "Awareness of Human Impacts: Changing Attitudes and Emphases." In *The Earth as Transformed by Human Action,* edited by B. L. Turner II et al., pp. 121–135. New York: Cambridge University Press, 1990. Possibly the best short synthesis to date.
Marsh, G. P. *Man and Nature, or Physical Geography as Modified by Human Action,* edited by D. Lowenthal. Cambridge: Belknap Press, 1965. An annotated edition of the 1864 original.
Raumolin, J. "L'homme et la destruction des ressources naturelles: La *Raubwirtschaft* au tournant du siècle." *Annales: Economies, Sociétés, Civilizations* 39 (1984), 798–819. A good survey of an important episode in environmental thought.
Thomas, W. L., Jr., ed. *Man's Role in Changing the Face of the Earth.* Chicago: University of Chicago Press, 1956.
Thompson, M., and S. Rayner. "Cultural Discourses." In *Human Choice and Climate Change,* vol. 1, *The Societal Framework,* edited by S. Rayner and E. L. Malone, pp. 265–343. Columbus, Ohio: Battelle, 1998. A thoughtful inventory and analysis of differences between contemporary "expert" and "lay" perceptions of global climate change.
Turner, B. L., II, et al., eds. *The Earth as Transformed by Human Action.* New York: Cambridge University Press, 1990.
Vernadsky, V. I. *The Biosphere,* translated by D. B. Langmuir and edited by M. A. S. McMenamin. New York: Copernicus, 1998. The first full scholarly translation into English, with notes and introductory materials.

—WILLIAM B. MEYER

## Human Dimensions of Global Change

Human beings have played a crucial role in all the many phases of recent global change, either as agents engaged in modifying or redirecting natural terrestrial processes, while also carrying out certain economic, social, and cultural activities at the planetary scale, or as subjects experiencing the effects of such changes, be they good or ill. Obviously enough, complex interactions among people and places have been going on for many millennia, sometimes resulting in important local or regional transformations. But it is only with the dawning of the modern age that we can begin detecting some truly global dimensions in the behavior of our species. Consequently, this discussion is limited to developments taking place since the late fifteenth century.

**The Historical Perspective.** The following sections provide a brief historical perspective of the roles humans have played in global change.

*The Columbian encounter.* The pivotal event was the abrupt Columbian Encounter: the effective mutual discovery by the inhabitants of the two hemispheres of the existence of the transoceanic other, resulting in an immediate establishment of tight interconnections between the peoples, biotic, and other resources of the Americas on the one hand, Europe, Africa, and Asia on the other. Australia, New Zealand, and the Pacific Islands were to enter this extended community some two to three hundred years later.

We have sound botanical, linguistic, and archaeological evidence indicating sporadic human contacts across both the Atlantic and Pacific Oceans long before 1492, but the impacts of such interchanges were minor or local. The only actual colonizations we can be sure of, those of the Norse in Greenland, Newfoundland, and perhaps other localities in northeastern North America, were tiny in scale and failed to endure. But it was in that same general source of wanderers—northwestern and western Europe—a region that had long lagged behind the civilizations of Asia and the Mediterranean—that our great ongoing adventure in globalization was to take off. It was also there, as documented by Fernand Braudel (1981–1984), that, for whatever reasons, we find the earliest decisive stirrings of the capitalist system, arguably the most essential component of the entire modernization process. The fact of relative geographic and navigational proximity to the New World certainly did not hinder the ascent of western European power. In any event, it has been argued persuasively that the vigorous plundering of so much American wealth in the form of labor power, precious metals and gems, foodstuffs, furs, and industrial raw materials gave such European nations as Spain, Portugal, England, France, and the Netherlands, with their nascent capitalist economies, a

decisive boost over such competitors as China, India, and the Ottoman Empire.

As the first chapter in the chronicle of human-induced global change, the Columbian Exchange had many immediate and long-term effects, but perhaps none more sudden or dramatic than those in the biotic realm. On balance, the interhemispheric swapping of domesticated plants and animals was quite beneficial to the affected parties. The introduction of maize, the white potato, manioc (cassava), and peanuts into Eastern Hemisphere farms generated major demographic and economic change. Similarly, the acquisition of the tomato, cacao, sunflowers, pineapple, and various New World squashes and beans also had a significant, mostly positive impact upon the diet and welfare of their Old World cultivators and their communities. But perhaps the ideal exemplar of biological globalization—if of rather dubious benefit to human well-being—is the case of tobacco, which spread with amazing rapidity from its American homeland to even the remotest inhabited tracts of Eurasia and Africa during the sixteenth and seventeenth centuries.

Perhaps equally impressive was the westward flow of domesticated plants and animals into the Americas. The arrival, then veritable population explosion, of horses, cattle, sheep, goats, swine, and poultry meant not only the creation of totally novel rural economies but also some drastic impacts upon indigenous biota, forests, grasslands, and hydrology, even though the changes occurred at different times and rates in various regions. These effects were further compounded by the importation of virtually the entire repertory of cultivated Eurasian and African plants. The major grains—wheat, barley, oats, rye, rice, and sorghum—may have been the most consequential, but there were also many vegetables, fruits, nuts, and such plantation items as sugar cane, coffee, bananas, and various spices. But there were less welcome transatlantic passengers in the form of weeds and insects that proved to be quite successful in their aggression against native competition. For whatever reasons, American species were unable to retaliate in the Eastern Hemisphere.

Venturing beyond the biological category, we dare not ignore the transfer of Old World technologies that had major consequences for American ecosystems. Thus we have the addition of metallurgical knowledge to societies in which metal objects had been rare previously and not especially utilitarian. Among the more conspicuous results relating to the habitat was the universal availability of guns and iron-tipped plows.

The early intercontinental sharing of artificially controlled species inaugurated an era of genuine globalization of economic activity that has yet to run its full course. An ideal example of the process is the massive cultivation of sugar cane and the refining thereof in nearly every locality throughout the world where climate, terrain, soil, labor supply, and other economic considerations render the practice profitable. In parallel fashion, there has been a worldwide dispersal of wine grapes into virtually every feasible ecological niche. Both cultures were being pursued vigorously in the Americas by the early 1500s, although the sugar plantation was to have much the greater economic, social, and geopolitical impact. In subsequent periods, these two agricultural items, along with other globalizing plantation crops, were to locate in many of the more receptive portions of Africa, Asia, Australia, and the Pacific Islands as well as the Americas.

On balance, the Columbian Exchange of domesticated life forms and various technologies would seem to have benefited the affected populations. Quite a different verdict must be offered for the exchange of disease organisms between two previously isolated worlds. Indeed, for all too many indigenous American peoples the consequences were sudden and catastrophic. For reasons related to prehistoric patterns of migration of small bands of human beings along high-latitude routes from Eurasia into the Americas, the many millions of the pre-Columbian residents of the latter lacked immunity to the various infectious and parasitic ailments that had been acquired over many generations by Eurasians and Africans. [See Migrations.] The outcome was a series of devastating epidemics among susceptible populations, whose numbers were reduced by 50–90 percent in just a few years, a holocaust chronicled in detail by William Denevan (1992). Indeed, exotic pathologies seem to have been the principal factor in the total extinction of a good many communities. The phenomenon was to be repeated later among various isolated groups in the Pacific and Siberia, and is still being played out today in the remoter reaches of Amazonia.

Syphilis is the only likely candidate for retribution, a lethal disease apparently originating in the Americas and subsequently diffusing throughout all of humanity, but that claim is still subject to scholarly dispute. Horrific as the intercontinental exchange of certain microorganisms may have been, the resulting carnage was continental or regional in scope; these were not global pandemics. However, with the gradual universalization of immunities and low-level infections, the stage was set for the genuine worldwide pandemics that have become an inescapable feature of the later phases of modernization.

One of the more unfortunate results of universal intensification of contact among peoples and places is the occurrence of truly global pandemics. Thus far the greatest of these has been the influenza pandemic of 1918–1919 that accounted for more than twenty million deaths throughout the world, far exceeding the toll of later out-breaks of flu viruses or the earlier cholera episodes. The current acquired immune deficiency syndrome (AIDS) pandemic will almost certainly not be the last such event.

A full definition and treatment of modernization is not feasible at this time, but we can note a few key characteristics vital to its origin and eventual triumph. The most crucial development was the appearance of a new mind-set among a critical, interactive mass of human beings. It involved the notions of rapid, perpetual change and material progress, individualism, scientific rationality, a devout belief in technology, and a mechanistic vision of the world; a loosening of the claims of traditional religion, new ways of regarding space and time, and the ascendance of impersonal relationships. These new sensibilities are obviously compatible with the rise of capitalism and, later, state socialism. Such symptoms of the new regime as industrialization, urbanization, and advanced communication systems should be regarded not as causes of modernization but rather as concomitants or consequences of an unprecedented mode of dealing with the realities and challenges of a globalizing, humanized planet. [See Industrialization.] We must also associate the modern age with that extraordinarily important fifteenth century invention, the printing press; and with the massive, widespread exploitation, beginning some generations later, of fossil fuels. The emergent new order would have been unthinkable in the mental realm without an abundance of cheap printed matter, to be supplemented eventually by universal literacy and (often compulsory) mass education; and, in the material realm, by cheap and abundant energy.

***Europeanization of the world.*** From its opening episodes until quite recently, the modern age could be quite accurately characterized as the "Europeanization" of the world. From the mid-1400s onward, western European mariners and explorers availed themselves of steadily improving vessels, charts, and navigational aids to penetrate all the seas of the world and to scout their shores. They were followed soon after by explorers of interiors, soldiers, merchants, administrators, clerks, missionaries, teachers, curious scientists, and, where conditions looked inviting, European settlers. Eventually, by the early twentieth century, the entire inhabited world, with only a few exceptions such as China and Japan, was occupied or governed directly or indirectly by Europeans. A series of imperial regimes of truly global reach came into being—the Spanish, Portuguese, Dutch, French, and, most far-flung of all, the British. All were managed with a degree of sophistication and systematic exploitation and racism unknown in the premodern era.

The globalization of the spiritual realm was especially noteworthy. Proselytizing activity kept pace with advancing flag and weaponry, sometimes even preceding it. The claims of a universal (Western) Christian church were being realized, even though dominion was divided between Roman Catholics and any number of Protestant competitors. Although degrees of success in conversion varied greatly from place to place, only new localities were spared the attentions of European and North American missionaries. These evangelists often introduced Western languages, schooling, and medicine, and much else in the way of cultural baggage as well as the gospel. At a later date, the worldwide program of Christianity was to be emulated by Buddhists and Moslems, whose previous range had been regional.

From its outset, European hegemony brought about redistributions of population on a scale and over distances never experienced previously. Initially, the transfers overseas of Europeans in their various imperial capacities were much smaller than the forced movements of slaves and indentured laborers. The largest such stream of migrants was that of enslaved Africans shipped to plantations, mines, and other menial workplaces in the West Indies and portions of South and North America. This process began in the early 1500s and lasted well into the nineteenth century, involving an estimated fifteen million captives. Much smaller in magnitude, but also producing major alterations in the human geography of the receiving areas, was the recruitment, often under false pretenses, of laborers from India, Malaya, China, the Philippines, the Pacific Islands, and elsewhere for farm labor in various European colonies lacking adequate local pools of workers.

Beginning as a trickle in the 1600s but increasing phenomenally in volume thereafter, interrupted only by major wars and business depressions, an emigration of Europeans to attractive settlement areas climaxed in the early twentieth century. Reaching a total of perhaps sixty million persons, this was certainly the largest migration in human history. The United States absorbed a majority of these uprooted individuals, but Canada, Argentina, Uruguay, Brazil, South Africa, Australia, New Zealand, Siberia, and Central Asia also received substantial influxes of newcomers. In the process, the extermination or displacement of indigenous peoples was a common occurrence. Many of the pioneers pushed into tracts that had been thinly occupied by their predecessors, and all too frequently the ecological well-being of the colonized territories was imperiled by the new modes of land use.

The initial wave of European settlement into distant rural and urban places was generally followed by substantial natural population growth, so that, after several generations of the great European exodus and other movements instigated by Europeans, a radical transformation of the world's population has materialized. Paralleling the virtual globalization of European peoples, but on a more modest numerical scale, was a set of diasporas of certain ethnic groups who served the economy as middlemen in commerce or various crafts or services in places where neither Europeans nor the subordinated population could fill vital niches. Thanks to modern modes of transportation and communication, these scattered communities of Chinese, East Indians, Jews, Lebanese, Armenians, and Gypsies, to take the more prominent examples, were able to sustain their cultural integrity and globalized interconnections.

Among the unintended consequences of the often massive shifts of peoples incident upon the modernization process was a decided increase in miscegenation, that is, mating between members of different "races" and the bearing of hybrid progeny. This is a process that had gone on for millennia at a relatively low level, for even in premodern times it was a rare population that was totally isolated from other visibly different human beings. However, within the past few centuries, the thrusting together of previously widely separated groups has strongly accelerated the phenomenon, so that some completely novel mixed-blood populations have been formed: for example, the Mexican mestizos and South Africa's Cape Coloured as well as other unnamed combinations. Perhaps even more important ultimately is a recent, steady, well-documented increase in the frequency of interracial marriages in a wide range of countries. Unless the current trend should somehow reverse itself—and it is difficult to envision such a turn of events—one is free to speculate how, at some quite distant point in the future, the Earth will be populated by a human species with few, if any, group differences in visible physical attributes from place to place or even among classes—a genuine consummation of the globalization process.

***Demographic and settlement transformation.*** But much more immediately consequential for the welfare of *Homo sapiens* and the ecological health of the planet has been the enormous growth in human numbers during the modern period. It seems quite likely that, after a major spurt following the adoption of agriculture and animal husbandry in the Neolithic, there was a relatively slow, gradual buildup of total global population thereafter up through the Middle Ages, despite the occasional temporary setback. [*See* Agriculture and Agricultural Land.] But a new demographic era arrived once the modernization process was well under way. The result has been at least a twelve-fold increase in the world's population, from an estimated half-billion in the late 1600s to more than six billion at the close of the twentieth century. This genuine population explosion began with death control—the first phase of the Demographic Transition—initially among European populations, then spreading later to other communities, with improvements in food supply, hygiene, housing, and, quite late in the process, effective medical treatment or prevention of disease and injuries. The globalization of death control was responsible for the climactic decade of the 1960s, when world population was growing at 2 percent per annum. A universal fall in death rates was followed eventually by a parallel decline in birth rates, once again initiated among more developed countries and extending recently to the majority of developing countries. Currently a number of developed nations are reporting an excess of deaths over births, while others are approaching that situation. For the world as a whole, if present hopes and expectations of demographers are realized, global human numbers may stabilize by the end of the twenty-first century, but certainly at a level above ten billion.

Among the most significant aspects of modernization has been a worldwide extension of urbanization. In pre-modern times, cities were to be found only in those portions of the world containing complex civilizations. But the modern era has witnessed the creation of cities large and small over vast areas that had never before known such agglomerations, most notably North America, the West Indies, most of Central and South America, Africa south of the Sahara, Siberia, and Australia. Moreover, many new cities sprouted in such historically advanced areas as China, India, Russia, and Europe, even as many older places enjoyed substantial growth. Many older metropolises in such advanced societies as those in China, India, Russia, and Europe experienced substantial growth even as totally new cities appeared and grew rapidly. Almost everywhere, internal and international migrants flocked to the bright lights and economic opportunities of these centers. Although urban residence is not uniformly defined by various national agencies, it seems likely that, as of 2000 CE, more than half of the world's people will have become city dwellers. Inevitably associated with this widespread urbanization are commercial and manufacturing enterprises large and small, and a broad array of service industries and complex social and cultural activities.

The modernization of the world also entailed a novel reordering of its political–geographic arrangements. Once again, following a project originating in Western Europe, all of the Earth's land surface (except Antarctica) was parceled out into nominally sovereign states, now numbering more than 150, each with its carefully defined boundaries. The most avidly sought-after political situation, albeit one seldom fully achieved, is that of the perfected nation-state, an entity within which a political apparatus and a single national, or ethnic, group coexist in close symbiosis, to the exclusion of other groups or loyalties. After the decolonization of the European empires, which began haltingly after World War I, then accelerated rapidly after World War II, the former colonies have tried to emulate the European model, but frequently with questionable results.

**A Globalized Human Society Today.** The existence of a single globalized human world is the paramount social fact of the late twentieth century. The concept of globalization ties together two related phenomena: (1) a worldwide sharing of various political, economic, social, and cultural patterns and modes of behavior as suggested above; and (2) a close-knit interdependency among all the geographic and other components of humankind, a set of indispensable working interrelationships. Contrary to the general misconception, globalization cannot be equated with homogenization. Although there are many obvious examples of standardized items replicated all across the world, we are not undergoing simple convergence toward a universal cultural sameness, however inexorable the progress toward biological unity. In fact, the forces of globalization, while obliterating some place-to-place differences, are also breeding totally new varieties of places and people. Furthermore, the operation of the current world-system has yielded highly uneven levels of development, and the gaps between centers of power and dominance in the developed world and a huge less-developed periphery, show no signs of narrowing. Similarly, the developmental disparities within a number of countries—for example, Italy, Argentina, Brazil, or China—refuse to disappear during our current advanced state of capitalism.

The mechanisms immediately responsible for the present state of affairs are advanced modes of transportation and communication. The advent of railroads, automobiles, trucks, and aircraft has vastly reduced the friction of distance and costs for passengers and commodities, while the instantaneous transmission of information and money, thanks to the telegraph, telephone, recordings, radio, television, video cassettes, and rapidly evolving electronic and computer networks, has transformed the nature of work and leisure. Most conspicuously, these innovations have made possible the creation and growth of immensely powerful, spatially volatile multinational firms. [*See* Multinational Enterprises.] This information, travel, and business revolution has been greatly facilitated by the adoption in recent years of English as the world language. Completely routing such early competitors as Latin, French, or German, English in all its various dialects has become the first or second language of a majority of countries and the unrivaled medium for commerce, finance, diplomacy, science, and scholarship.

A uniquely modern development has been the emergence of popular culture, something that was manifested by the late nineteenth century as spectator sports and mass-circulation periodicals began to enthrall the populace. Subsequently, with the compression of space and time, there has been a worldwide sharing of movies, comics, television programs, gadgets, popular music and dance, dress, cuisines, other forms of titillation, and much else.

The recent boom in human numbers has been accompanied by a marked rise in territorial mobility. International migrations of the conventional type do persist, but the outward streaming of Europeans has dropped precipitously and has been largely replaced by flows of Asians, Latin Americans, West Indians, Middle Easterners, and Africans gravitating toward developed-world and other prosperous destinations. In addition, there is now a brisk circulation of temporary labor migrants among countries and a new series of transnational communities taking shape. The latter phenomenon, one with interesting political, economic, and social implications, involves both the working-class and elite strata of society, persons oscillating among two or more countries without definitive attachment to any. The expansion of tourism, at both the domestic and international levels, into a leading industry has been tremendous and shows no signs of abating. Few places fall outside the tourist circuits, and the social and environmental impacts, while undoubtedly substantial, are still poorly understood. A related latter-day development has been the growth of vacation trips and retirement migration. Less welcome is one of the twentieth century's worst chronic crises: the proliferation of refugees. At a conservative estimate, at least fifteen million refugees and asylum-seekers have been displaced from their homes by warfare, civil and ethnic strife, and natural and industrial disasters.

The massive growth of human numbers and the complex redistribution thereof certainly have implications for environmental change at the global scale, but they are difficult to spell out. At the metropolitan level, the impact of human presence and activity upon all aspects of the habitat is obvious enough, especially in those burgeoning cities and conurbations with ten million or more inhabitants. But it would be highly simplistic to equate environmental change at the planetary scale with body counts or population densities measured in regional or national terms. Patterns of activity, consumption, and organization are much more relevant. Thus, through its purchases and trips of all sorts, a single affluent household in Frankfurt or Los Angeles may inflict more ecological harm scattered throughout the world than a score of families in rural Bangladesh or the Philippines.

For some time it has been clear that even the most successful of nation-states lack the means to cope with economic, environmental, and other problems that transcend its borders. Whatever the strengths or weaknesses of individual states, the globalization of human affairs has necessitated intergovernmental treaties and other arrangements for postal, navigational, aeronautical, meteorological, medical, criminal, intellectual property, humanitarian, and other concerns. Along with the creation of a growing number of official international organizations since the late nineteenth century, a vast array of nongovernmental organizations (NGOs) and all manner of special-interest associations have come into being. The range of interests is seemingly endless, including sport, technology, science, hobbies, philanthropy, social issues, and much else.

As the ultimate recognition of the impotence of the sovereign state, the world has witnessed the founding of the League of Nations in the aftermath of World War I and its successor, the United Nations, a quarter-century later. Although neither institution has fully lived up to the hopes of its creators, their very existence testifies to the stubborn reality of an interactive, globalized human community.

In summary, then, at the end of the twentieth century the human species inhabits a planet radically transformed from the premodern past. It is truly a global community, however troubled by internal tensions and conflicts. Our contemporary world is one in which there is universal sharing of modes of production and consumption, of values, amusements, anxieties, and ways of behaving—even though the flow of influences may be much stronger from developed to less-developed lands than the reverse movement. It is a place in which distant events can prompt instantaneous perturbations everywhere throughout the system. It is, finally, also a world increasingly unified in its worries over the viability of an ecosystem so palpably modified by human activities that have both local and planetary consequences.

[*See also* Economic Levels; Environmental Movements; Global Economy; Human Populations; International Cooperation; International Human Dimensions Programme on Global Environmental Change; *and* Urban Areas.]

## BIBLIOGRAPHY

Appadurai, A. *Modernity at Large: Cultural Dimensions of Globalization.* Minneapolis: University of Minnesota Press, 1996. The nine chapters in this volume are varied in focus and uneven in value, but *in toto* this is a most significant exploration of transnationalism, diasporas, and the postnational idea.

Barber, B. R. *Jihad vs. McWorld.* New York: Times Books, 1995. A highly readable, popular survey of the apparent contradiction between global cultural convergence and localized, militant new social identities in the contemporary world.

Barnet, R. J., and J. Cavanagh. *Global Dreams: Imperial Corporations and the New World Order.* New York: Simon and Schuster, 1994. An excellent account of the rising power of multinational firms and their impact upon culture and demography as well as the economy.

Blaut, J. M. *The Colonizer's Model of the World: Geographical Diffusionism and Eurocentric History.* New York: Guilford Press, 1993. A persuasive polemic that not only attacks the conventional Eurocentric view of history and geography but also presents a plausible, if iconoclastic, hypothesis for the rise of Europe and the origin of the modern world-system.

Braudel, F. *Civilization and Capitalism, 15th to 18th Century*, pp. 181–184. 3 vols. New York: Harper and Row, 1981–1984. A rich, magisterial treatment of the many dimensions of the European genesis of capitalism.

Butzer, K. W., ed. "The Americas before and after 1492: Current Geographical Research." *Annals of the Association of American Geographers* 82 (1992), 343–568. A dozen articles that provide an excellent conspectus of what is known and what still remains to be discovered concerning this enormous hemispheric transformation.

Crosby, A. W., Jr. *The Columbian Exchange: Biological and Cultural Consequences of 1492*. Westport, Conn.: Greenwood, 1972. A pioneering study, one with special emphasis on the effects of the Columbian Encounter on the peoples and ecologies of the New World.

Crosby, A. W. *Ecological Imperialism: The Biological Expansion of Europe, 900–1900*. Cambridge: Cambridge University Press, 1986. A superb treatment of the changes wrought in the non-European world by the introduction of European organisms, with special attention to disease.

Denevan, W. M. *The Native Population of the Americas in 1492*. 2d ed. Madison: University of Wisconsin Press, 1992. A study of precontact human numbers and subsequent depopulation that is as nearly definitive as is feasible given difficult data problems.

Fishman, J. A., et al. *The Spread of English: The Sociology of English as an Additional Language*. Rowley, Mass.: Newbury House, 1977. The most extensive coverage of the topic to date.

Goudie, A. *The Human Impact on the Natural Environment*. 4th ed. Cambridge, Mass.: MIT Press, 1994. Details the kinds of biophysical impacts that humankind has made on Earth.

Grübler, A. *Technology and Global Change*. Cambridge: Cambridge University Press, 1998. Assesses the role of technological change in global environmental change, focusing on the history of that change in the twentieth century.

Hopkins, T. K., and I. Wallerstein, eds. *The Age of Transition: Trajectory of the World-System, 1945–2025*. London: Zed Books, 1997. Nine essays, looking before and after, that examine issues of politics, production, labor, and welfare at the world scale.

Kotkin, J. *Tribes: How Race, Religion, and Identity Determine Success in the New Global Economy*. New York: Random House, 1993. A breezy, semipopular account of how traditional "tribes" (Jewish, Chinese, Japanese, English, Armenian) and nascent ones (East Indian, Mormon, Palestinian) operate effectively at a global or quasi-global level. The author claims that tribalism will be a major ingredient in the new global order.

McCrum, R., et al. *The Story of English*, rev. ed. New York: Penguin, 1993. The final chapters of this excellent volume deal with the worldwide diffusion of the language.

McEvedy, C., and R. Jones. *Atlas of World Population History*. Penguin, 1978. Although far from definitive, this atlas—the only one of its kind—remains useful.

Meinig, D. W. *The Shaping of America: A Geographical Perspective on 500 Years of History*, vol. 1, *Atlantic America, 1492–1800*. New Haven: Yale University Press, 1986. This volume, the first of a planned tetralogy, deals masterfully with the genesis and development of the Atlantic World, the core of the emergent modern world-system.

Meyer, W. B. *Human Impact on the Earth*. Cambridge: Cambridge University Press, 1994. Traces the human impact on the biosphere, including causes and regional variations, all placed in historical context.

Meyer, W. B., and B. L. Turner II, eds. *Changes in Land Use and Land Cover*. Cambridge: Cambridge University Press, 1994. A set of essays documenting the role of human-induced land changes in global environmental change.

Schiller, H. I. *Mass Communications and the American Empire*. 2d ed. Boulder, Colo.: Westview, 1992. An acerbic, but well-informed, overview of the ways in which modern media have saturated the world and the effects and implications thereof, with special attention to the role of the United States.

Socolow, R., et al., eds. *Industrial Ecology and Global Change*. Cambridge: Cambridge University Press, 1994. A set of thirty-six essays offering a variety of useful perspectives on the ecological implications of worldwide industrialization.

Thomas, W. L., Jr., ed. *Man's Role in Changing the Face of the Earth*. Chicago: University of Chicago Press, 1956. A monumental, highly influential work consisting of fifty-two essays and complementary discussion. Still indispensable for people seeking understanding of the effect of human activities on major categories of the natural environment.

Turner, B. L., II, and K. W. Butzer. "The Columbian Encounter and Land-Use Change." *Environment* 43.8 (1992), 16–20. Connects the European conquest of the Western Hemisphere with current global environmental change.

Turner, B. L., II, et al. *The Earth as Transformed by Human Action: Global and Regional Changes in the Biosphere over the Past 300 Years*. Cambridge: Cambridge University Press, 1990. Essentially a sequel to the preceding item, this massive, richly informative volume devotes pp. 19–141 to "Changes in Population and Society."

Wallerstein, I. *The Modern World-System: Capitalist Agriculture and the Origins of the European World-Economy in the Sixteenth Century*. New York: Academic Press, 1976. A relatively brief summary of research by a highly influential author concerning a crucial period in the development of the modern world-system.

Wills, C. *Children of Prometheus: The Accelerating Pace of Human Evolution*. Reading, Mass.: Perseus Books, 1998.

—WILBUR ZELINSKY

# GLOBAL ECONOMY

Commercial interactions on a global scale have been an important feature of the growth of the capitalist system for half a millennium. Patterns of trade as much as three or four centuries ago involved shipping both luxury goods and bulk commodities, such as basic foodstuffs, across thousands of miles. The last century saw a rapid increase in international efforts to encourage and manage the global economy, and in the number of transnational and international organizations that participate in it. The last few decades have seen a rapid growth in globalization, with international trade, finance, investment, and production consistently growing much faster than total economic output.

International commerce has always had an impact on the environment. Long-distance trade and investment allowed resources to be extracted at far greater rates than local populations could have done, and agricultural production for international markets displaced native ecosystems. The impact of contemporary globalization is different in both scale and kind. The scale of the contemporary global economy magnifies the traditional environmental effects of economic production, while technologies create environmental hazards on a global scale, such as global warming and widespread extinctions, that were unknown to previous generations. Yet the process of globalization also helps to create the surplus wealth that is often necessary for efforts at environmental protection, and makes countries more willing to participate in international efforts to protect the environment.

This entry addresses both the process of economic globalization and the effect of the process on the natural environment. It does so in three sections. The first looks at the historical development of the global economy through the course of the twentieth century, with a particular focus on the institutional structure within which it has operated. The second looks at the contemporary process of globalization, and examines its particular patterns and effects. The third and final section points out some of the ways in which this process has affected and effected environmental change and management internationally.

**History and Institutions.** The global economy at the beginning of the twentieth century was already quite integrated. Levels of both international trade and international investment were higher than they had ever been, and, as a proportion of global economic output, were comparable to the situation in the late twentieth century. Most of the world's major currencies were part of the classical gold standard, which meant that exchange rates were stable and inflation a rarity. This era also saw a rapid expansion of the reach of the global economy. Foreign investment in infrastructure by European colonial powers, primarily in railways, opened up the interiors of much of the Americas, Asia, and Africa. This allowed regions that had previously been cut off by their geography from large-scale trade with the rest of the world to produce for global markets. In short, processes of globalization at the beginning of the twentieth century bore many similarities to equivalent processes at the end. This is true both of the benefits of globalization and the costs.

In important ways, though, economic globalization then was different from the more contemporary model. Both world economies were based to important degrees on the leadership of one country, Great Britain then and the United States now. At the beginning of the twentieth century, however, international economic activity was largely the product of patterns of interaction among governments and among national firms. In other words, agreements on international economic issues, from trade to the management of environmental issues, tended to be bilateral and specific, while commerce usu-

ally involved firms that operated primarily within one country, selling finished goods or services to their equivalents abroad. By the end of the century, international economic activity had become much more transnational in nature. Agreements on broad ranges of international economic issues tend now to be embedded in multilateral institutions, which in turn often take on a life of their own and have important independent effects on the rules governing international commerce. And international trade and investment is now more likely to happen within multinational corporations than between national firms. The story of the global economy in the twentieth century is largely one of how and why these changes occurred.

***The classical gold standard system.*** The primary features of the global economy at the beginning of the twentieth century, as suggested above, were the gold standard, a relatively open international trading system, and British international economic leadership. The gold standard meant that most major currencies were freely convertible both to gold and to each other. This allowed money to cross borders easily, lowering the costs of international trade and removing one of the biggest barriers to international investment and finance. Trade barriers such as tariffs were on average quite low by historical standards, but varied considerably across different countries. For example, Great Britain and much of its empire practiced free trade. Except for a few luxury items, there were no tariffs at all, no protection for domestically made goods. The United States, by contrast, levied tariffs that, while low by historical standards, were considerably higher than they are today.

This difference in tariff levels was one example of the central leadership role that Great Britain played in the global economy. Another example was the Bank of England's role in managing the gold standard. While there is some historiographical dispute on the issue, it is generally accepted that the Bank essentially acted as a central bank for the entire gold standard system, and thus to most of the world. It played this role for two reasons. The first is that a majority of the total foreign investment in the world at the time was denominated in British currency, and this gave the Bank of England, which managed the currency, a structurally central role. The second reason is that there was strong confidence in both the ability and willingness of Great Britain to stay on the gold standard, a confidence that existed only to much lesser degrees for other currencies. Confidence is a crucial element of monetary stability, because people value money as a store of wealth, and if they do not have confidence that the money will retain its value over time, they are less likely to hold it. If people are not willing to hold it, it will in fact begin to lose its value. Thus loss of popular confidence in a currency becomes a self-fulfilling prophecy. Because confidence in the British currency was higher than in other currencies, it tended to be the one most used for international commerce.

This structure of the global economy lasted until the outbreak of World War I in 1914. As a result of the war, patterns of international trade were disrupted because the combatants would not trade with each other and each side tried to interfere with the shipping of the other. The war also caused most of the world's major currencies, except the American dollar, to cease trading openly. After the war ended, many countries, including Great Britain, tried to recreate the global economic structure that had existed before the war, but with only limited success. Tariff levels were lowered from their wartime highs, but in most cases not to where they had been before the war. Most countries also tried to get back on to the gold standard, although for a variety of reasons this often took several years. Great Britain, for example, did not get back on to the standard until 1925. By then, though, it looked like the prewar system had been more or less recreated. This appearance would only last a few years. The system began to come apart in 1929, and by 1932 lay in ruins.

**The Great Depression and World War II.** The cause of this unraveling of the international economic system was the Great Depression, which began with the crash of the New York Stock Exchange in October 1929. The Depression had profound effects on both international finance and international trade. Because Great Britain was not as central to international finance as it had been, and because the Bank of England had never managed to restore to Sterling the confidence that it had commanded before the war, the gold exchange standard of the late 1920s had always been somewhat brittle. So much so, in fact, that it could not withstand the shock of the deepening depression that radiated from the United States to the rest of the world in 1930. As a result, the system collapsed. Most currencies were devalued over the next couple of years, and the global economy began to disintegrate into a group of currency blocs. At the same time, many governments imposed much higher tariffs on imported goods, the most infamous being the U.S. Smoot–Hawley tariff of 1930, which substantially increased American tariff levels across the board. The combination of higher tariffs and a dysfunctional international monetary system led to a downward spiral in international commerce; total world trade fell by two-thirds in less than four years from 1929 to 1933. The balkanization of the global economy into exclusive currency blocs is also seen by some historians (though not all) as a major cause of World War II; since the blocs centered on Germany and Japan could not trade for the primary resources that they needed, both countries ended up getting them through military expansion.

Some efforts were made in the mid-1930s to revive the global economy by negotiating new trade and mon-

etary agreements, particularly among the United States, Great Britain, and France. These efforts had borne little fruit, though, by the time of the outbreak of war in 1939. World War II had the same effects as World War I on international commerce, which was put on hold for the duration (except such commerce as was directly related to the war effort). One interesting effect of this process has been called the import substitution effect. In some of the larger developing countries, particularly in South America, the loss of traditional sources of manufactures resulting from the war led to the indigenous development of industrialization. Most of these import substitution industries, though, withered after the war, when imports from developed countries once again became available. During the 1960s and 1970s, many developing countries increased restrictions on imports of manufactured goods in an attempt to recreate this effect and thus promote domestic industrialization.

***The postwar reconstruction.*** The victorious powers got together as World War II ended to discuss the reconstruction of a global economic system that would be less fragile than the system that had failed so completely in 1929–1930. They came into these discussions perceiving three lessons from the Great Depression. The first was that some kind of formal organization was needed to stabilize the international monetary and financial structure. The second was that a formal set of rules needed to be created to stabilize international trade, to create a system in which tariff increases in some countries, and retaliations against them in other countries, would not threaten to create another downward spiral in international trade. The third was that capitalism should not be allowed to operate unfettered, that any new international economic structures that were created needed to allow states to become more actively involved in economic planning, as suggested by the new Keynesian economics.

These priorities resulted in the creation of a set of institutions that formalized a system of multilateral rule-making in the international political economy. The first institutions to be created, known as the Bretton Woods institutions after the town in New Hampshire where their creation was negotiated in 1944, were the International Monetary Fund (IMF) and the World Bank. The role of the IMF was to help countries experiencing temporary balance-of-payments problems and foreign reserve crises, so that the domino effect of the failure of one currency bringing others down, as happened in 1930, would not be repeated. The IMF was thus created to act in important ways as an international central bank. The World Bank was created to help countries in more long-term ways, to lend money to projects designed to reconstruct countries damaged in war, and to promote economic and industrial development. Both of these institutions are still central features of global economic management. They still fulfill roles similar to those for which they were created, although both now lend exclusively to developing countries.

After the war, negotiations began for the creation of an International Trade Organisation (ITO) that would be to trade what the IMF and the World Bank were to international money and finance. These negotiations were less successful, however, and the ITO never came into being. Salvaged from these negotiations was the General Agreement on Tariffs and Trade (GATT), a general set of rules for the conduct of international trade. The core of the GATT is the principle of nondiscrimination, which means that any country that is a member of the GATT has to be accorded equal treatment by all other members; countries are not allowed to treat any individual trading partner any worse than they treat their other trading partners. The agreement also set maximum tariff levels for many goods. The GATT continued to be the basic set of rules governing international trade; while the rules have remained quite consistent, average tariff levels within the agreement have been renegotiated several times, and have fallen dramatically. The goal of a more formal international trade organization was finally realized in 1994 with the creation of the World Trade Organization (WTO), which incorporates the GATT, a new General Agreement on Trade in Services (GATS), other new trade agreements, and a formal institutional structure and dispute-settlement mechanism. [See World Trade Organization.]

The final priority listed above, the greater role for the state in the economy and in the management of social welfare, was reflected by these institutions. The monetary system created at Bretton Woods was one of fixed exchange rates, as the classical gold standard had been, but in the newer system much less capital mobility was allowed. A major effect of this decrease in capital mobility was a concurrent decrease in pressure on governments by international financial markets, giving governments greater flexibility in their monetary and fiscal policies. As a corollary, however, the system prevented foreign investment on the scale that had been the norm before World War I. At the same time, the rules of the GATT were written to allow for a greater governmental role in economic and industrial management than had ever been the case before, and in particular allowed developing countries large loopholes through which their industries could be protected.

It is important to note here that this reconstruction of the international economy after World War II was not global: many countries, particularly those with Communist governments, did not participate. With the communist victory in China in 1949, almost half of the world's people lived in countries that were not part of the Western postwar economy. In fact, many Eastern Bloc countries participated in the Communist Economic

Community (COMECON), which provided an institutional structure for their international economic interactions that reflected socialist rather than market principles. With the collapse of communism forty years later, however, this structure disappeared, and the institutions discussed above have since become effectively global in scope. Many formerly communist countries have not yet been admitted to membership in the WTO, but most have applied.

To help along the process of recovery from World War II, and also to help convince western European voters of the value of the market rather than the socialist economic model, the United States engaged in a number of efforts to fund reconstruction. The most famous of these efforts was the Marshall Plan, under which the U.S. government gave U.S.$17 billion in grants, equivalent to 2.5 percent of American gross national product (GNP), to western European governments in 1947–1951. Similar aid was also given to Japan. Partially as a result of this aid, the two and a half decades after World War II were a period of rapid growth and low unemployment for most of the industrialized world. The international economic system designed in the 1940s seemed to be working.

The postwar period was also one of rapid expansion in the number of countries participating in the system, largely as a result of decolonization. Beginning with India in 1946 and continuing through much of Africa in the early 1960s, most of the colonial possessions of the European powers became independent, a process affecting roughly a third of the world's population. Both the economic conditions and the economic strategies of developing countries in this period varied greatly. Many states, though, did become heavily involved in various kinds of industrial policy, including import substitution industrialization (discussed above), often with considerable apparent success. Much of the developing world saw solid economic growth and industrialization through this period.

***Crises of the 1970s and 1980s.*** After two decades of relative stability and consistent growth, the first half of the 1970s saw two major changes in the structure of the global economy. The first was the collapse of the international monetary system created at Bretton Woods. The system had been based, formally as well as practically, on a strong American dollar. The dollar had, however, been growing gradually weaker since the early 1960s. Because of the mounting pressure of an overvalued dollar (and an undervalued West German mark and Japanese yen), the United States pulled out of the system in 1971. Once the United States had withdrawn there was little hope of maintaining an international system of fixed exchange rates, and the system was formally abandoned two years later. What replaced it was a system in which most major currencies traded freely on an international market. This system is a major innovation in that it is the first time that the values of major international currencies have not been set in terms of precious metals. The flexibility of the new system has allowed capital to flow much more freely across national borders; total capital flows have increased more than a hundredfold since 1971.

The second major change was the oil shock in 1973. In a brief period in that year the price of petroleum, the basic fuel of modern industry, tripled as a result of an embargo by the oil producers' cartel, the Organization of the Petroleum Exporting Countries (OPEC). The price rise had the effect of dampening economic growth and generating recessions in most industrialized economies. It also created huge new pools of wealth in the major petroleum-exporting countries, particularly those bordering the Persian Gulf. These new pools of wealth, denominated in dollars and thus known popularly as petrodollars, were invested abroad and gave a crucial boost to the new international capital markets developing in response to the end of the Bretton Woods monetary system. The combination of these two phenomena led to a major and rapid change in the global economy, from one that was fairly closed to international capital flows to one that was very open.

Partially as a result of this new capital openness, many developing countries, both oil exporters and nonexporters, had access to far larger loans from the developed countries than ever before. Many of them borrowed large amounts, which often generated impressive growth rates for much of the decade. Thus while the 1970s was a period of economic stagnation for many developed countries, much of the developing world was experiencing rapid industrialization. When interest rates went up in the early 1980s, however, many of the primary debtor countries could no longer pay the interest on their loans. This resulted in a major international debt crisis, as many developing countries and many of the world's biggest banks faced bankruptcy. The banks, for the most part, survived, but the fallout from the crisis meant that for many of the biggest developing economies the 1980s were a lost decade.

*The new liberalism.* One effect of the changes in the global economy in the last quarter of the twentieth century has been the end of the model of state involvement in the economy that developed after World War II. The Great Depression had left a generation of economic planners believing in an activist state, one that tempered market forces with an active welfare state and a certain amount of direct involvement by the government in industry. This kind of state seemed to work well through the 1950s and 1960s. However, the economic shocks of the 1970s, which led to both high inflation and high unemployment in the West, and the debt crisis of the 1980s, which led to a decade of little or no growth in much of

the developing world, undermined faith in the model. At the same time, the rapid increase in capital flows in the 1970s and 1980s made the model less viable, by undermining state autonomy in economic decision-making. A belief in pure market principles, and thus in a smaller state and less state interference in the marketplace, had been growing among some economists since the 1960s, but, with the election of Margaret Thatcher in Britain in 1979 and of Ronald Reagan in the United States in 1980, free marketeers came to control two of the leading liberal democracies. Since then, a free-market orthodoxy has come to replace the more statist Keynesian orthodoxy of the 1940s and 1950s. This trend has been reinforced by the demise of the Soviet economic model. By the end of the century, few governments or international organizations were willing to claim adherence to anything but free-market principles.

**Contemporary Globalization.** Globalization means an increase in economic integration that is different both in scale and kind from what has gone before. At its most straightforward, globalization is often taken to refer to increases in trade and capital flows across borders. Both of these are measures of scale, and while they suggest a level of integration higher than it was half a century ago, they can be read as suggesting a level not qualitatively different from what was the case a century ago. Yet there are a number of aspects to contemporary globalization that suggest basic structural changes in the international political economy.

*National incomes.* One of the most notable aspects of the global economy since World War II has been its continual expansion. Even through the crises of the 1970s and 1980s, average income per capita has continued to increase. At the same time, international trade has increased even faster. Parts of the world have seen their growth slow for significant periods of time, but, taken as a whole, global economic activity and output has increased steadily and shows every sign of continuing to do so. The rise in average per capita income has been so great that, by the middle of the 1990s, average global per capita GNP (which is to say, the average for all countries) is roughly the same as per capita GNP in the richest country in the world a century earlier.

Another notable feature of the contemporary global economy is that it is becoming increasingly postindustrial. It is a conventional wisdom that, as countries develop economically, agriculture as a proportion of the total economy decreases while industry's share increases. This is still the case in many lower-income and middle-income economies; between 1980 and 1995, agriculture's share of low-income economies decreased from 34 to 25 percent, while industry's share increased from 32 to 38 percent. In wealthier countries, however, a different trend can be seen. In these countries, the importance of industry is declining, and many of the wealthiest economies are becoming dominated by services. In the United States, for example, roughly three-quarters of the economy is generated by services. Because of this trend in the wealthier countries, the importance of services to the global economy as a whole is increasing rapidly, from 53 percent in 1980 to 63 percent in 1995. This trend is expected to continue.

The distinction between low-income and high-income countries highlights a disturbing feature of the contemporary global economy that is not apparent from figures on aggregate international economic growth, namely, the increasing disparity between the richest countries and the poorest. The per capita income in the richest countries in the world is four to five hundred times greater than the per capita income in the poorest. Most of the world's industrialized countries show slow but relatively consistent economic growth. Elsewhere, growth rates, both across and within regions, vary dramatically. A number of countries in sub-Saharan Africa are poorer, on a per capita basis, than they were two or three decades ago. At the same time, other countries, primarily in the Asia–Pacific region but to a lesser extent in South America as well, have transformed themselves from primarily agrarian to primarily industrial countries over the same period. This has had the effect of making generalizations like "third world countries" or "developing countries" to a significant degree meaningless; a newly industrializing country such as South Korea has far more in common economically with the United States than it does with countries that remain agrarian and poor (see Figure 1).

***Globalization and economic integration.*** Several other factors also contribute to contemporary processes of globalization. Some of these are purely economic, the most notable of which is the globalization of production. Whereas until the relatively recent past goods tended to be made in one country and then traded internationally as finished goods, it is increasingly the case that the various parts of a good will be made and assembled wherever it is most cost-efficient. Because of this globalization of production, it is increasingly meaningless to say of a particular product that it is made somewhere specific. For example, an automobile may be designed in the United States and assembled in Japan from parts manufactured throughout Southeast Asia. A related phenomenon is the increasing degree to which international commerce happens within companies rather than between them. In other words, trade between two countries now often consists of one part of a multinational corporation (MNC) "selling" goods or services to another part of the same corporation rather than one national firm selling a finished good or service to another.

This globalization of production allows MNCs to perform each specific economic function wherever it can be done most cost-effectively, but it also makes it much more difficult for governments to regulate production processes. At the same time, private international capital flows can now swamp the efforts of governments and central banks to manage their currency exchange rates. This means that government policy that is not interpreted as market-friendly by the international capital markets can result in rapid and sharp depreciations of national currencies, which is a major added cost of adopting such policies. Thus the globalization of both production and capital has the effect of limiting the effectiveness of states in managing their own economies, and limiting their ability to adopt policies that are not seen to be market-friendly. Global economic forces such as MNCs and capital markets are increasing in importance relative to states as sources of economic governance.

Along with this globalization of production and decision-making comes what might be called a globalization of consumption. Along with increased flows of trade and capital come increasing flows of technology, innovation, and culture. On the one hand this kind of globalization makes new technologies available worldwide, both improving living standards and increasing the pace of innovation as new ideas are spread much more quickly than at any time in the past. [*See* Technology.] On the other hand, it can also be seen as a homogenization of culture, as the cultural specificity of individual places is replaced by the same companies, goods, and services that can be found worldwide. This latter phenomenon is often referred to as cultural colonialism or cultural imperialism, because most of the culture being internationalized comes from a few developed countries, particularly the United States.

***Problems of globalization.*** The cultural critique, that globalization is creating a homogenized world culture based on the lowest common denominator of Western multinational exports, is one of several forms of criticism of the process of globalization. Other critiques relate to the effects of globalization on local governance. It was suggested above that increases in the globalization of finance and production have the effect of decreasing state decision-making autonomy. The broader phenomenon of the geographical separation of the production of goods and services from their consumption diminishes local empowerment and governance by increasing the extent to which local economic production is held captive to the effects of markets abroad. This phenomenon is exacerbated by modern advances in telecommunications and transportation. It has recently become possible, for the first time, for decisions on local production on a day-to-day basis to be made from abroad, further reducing the need for local input into decision-making. Continual decreases in the costs of transportation have the effect of tying local production to conditions elsewhere to an ever greater degree. At

**Global Economy. FIGURE 1.** Gross National Product Per Capita, 1997.

A country's gross national product is divided by its population. (From The World Bank, 1999, p. 189. With permission of The World Bank.)

the same time, much of the local decision-making power that is being lost is being gained by MNCs, whose priorities may often be very different from those of local communities. Because these communities often have far fewer resources than MNCs with which to lobby (or bribe) governments, they have little ability to compete with MNCs for governmental attention.

Another adverse effect of the separation of production from consumption engendered by the process of globalization is the protection of consumers from the effects of production. The most straightforward example

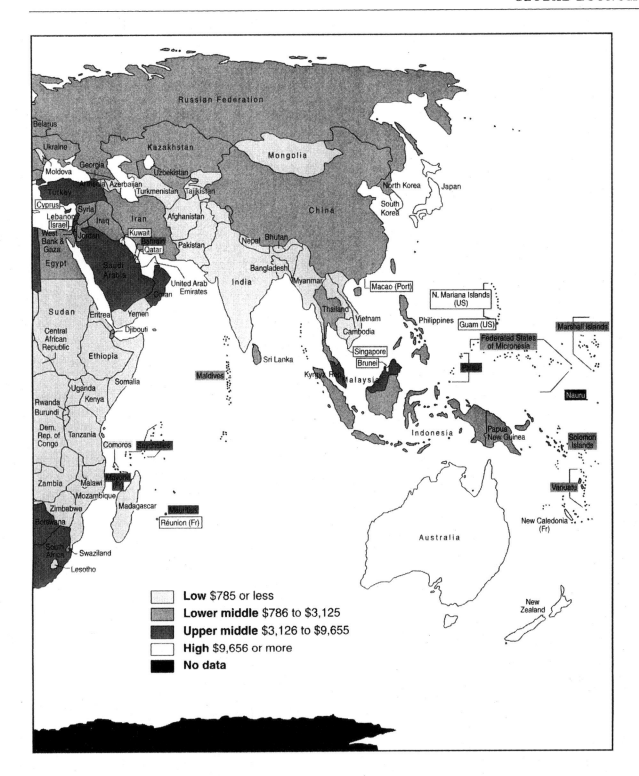

Low $785 or less
Lower middle $786 to $3,125
Upper middle $3,126 to $9,655
High $9,656 or more
No data

of this phenomenon is the demand from the industrialized world for resources from the developing world. The depletion of local resource bases has often provided a check to economic excesses, but, in today's world, overconsumption can deplete resources on a global rather than just a local scale. If a resource is depleted locally, it is lost only to that local community. If it is depleted globally, it is lost to humankind forever. A second ex-

ample is the international trade in pollution. If richer countries can export pollution to poorer countries, then a natural check on local pollution levels is lost, and problems of pollution threaten to become global rather than just local in scale. [*See* Basel Convention; *and* Hazardous Waste.]

Proponents of globalization argue that it allows more, and faster, economic growth than would otherwise be the case. Some critics make the same point, but suggest that such untrammeled growth is problematic. These critics suggest that from a social perspective, globalization can be seen as promoting a culture of consumption for consumption's sake, at the expense of the whole range of other potential social and cultural values. From an ecological perspective, unlimited consumption must necessarily eventually deplete available natural resources and the ability of the environment to absorb pollution. Opponents of globalization thus argue that it has the effect of speeding up the rate at which we approach natural limits to economic growth.

**The Global Economy and the Global Environment.** While the "limits to growth" approach is the most overarching indictment of the effects of economic globalization on the natural environment, there are a number of other ways in which the two are related. Resource depletion and the trade in pollution and other environmental "bads" are important phenomena that are dealt with in other entries in this encyclopedia. Also dealt with in separate entries are a range of issues relating to the effects of both international trade and MNCs on the environment. [*See* Multinational Enterprises.]

Other issues linking economic globalization to environmental management include questions of environmental regulation and competitiveness, debt, and issues of the global commons. All countries impose some degree of regulation on commerce and industry for environmental reasons, but the extent of such regulation varies widely, and in some developing countries is clearly inadequate. Some experts argue that the process of globalization has the effect of forcing better-regulated countries to deregulate to a lowest common denominator of environmental management. To the extent that environmental regulations impose costs on businesses, increased regulation makes production more expensive, and as such less competitive internationally, while decreasing regulation makes production cheaper, and thus more competitive. In a global economy in which MNCs can easily move production facilities to wherever costs are lowest, there is an incentive for governments to enforce weaker environmental regulation than they would otherwise prefer to, either to attract more businesses into the country, or to discourage businesses from moving to less-regulated countries. Thus they argue that globalization can have the effect of discouraging environmental regulation in favor of maintaining a country's international competitiveness.

Other experts argue, however, that while worrying about competitiveness makes sense for companies, it does not make sense for countries. By this logic, environmental regulations do create costs, as they would in a nonglobalized economy, but they do not result in a loss of national competitiveness. They suggest that there is little evidence that MNCs are being drawn to countries

with laxer environmental regulation. Furthermore, increased environmental regulation can create new environmental management industries in a country that would not have developed them otherwise. Also, globalization can give more environmentally conscious governments tools to encourage increased environmental regulation abroad, tools that they otherwise might not have.

Many governments, particularly of developing countries, are limited in their freedom to protect their environments by their levels of foreign debt. High levels mean that countries must ensure strong growth in exports to service their debt. This can result in governments promoting higher, and less sustainable, levels of resource extraction and a greater focus on cash crops than would otherwise be the case, so as to generate these needed exports. Such export industries are often damaging to the environment (e.g., unsustainable logging).

The global commons includes those areas of the planet that are not within national jurisdictions, those environmental goods and bads that cross national boundaries as a matter of course, and such elements of the ecosystem held to be part of the common heritage of humanity. Examples include the high seas and the upper atmosphere, migrating species, and biodiversity. The economic expansion associated with globalization has given the human economy a far greater ability than it has ever had in the past to exploit the global commons. It is in the nature of commons resources that they are difficult to manage, because even though the community of users has an incentive to manage the commons in a sustainable way, each individual user has an incentive to overuse it. This was not much of a problem on a planetary scale until well into this century, when new technologies gave humanity the ability to deplete fisheries on an oceanic scale, put enough chemicals into the air to deplete the ozone layer and have a measurable effect on a global scale, and cause species to become extinct at a geologically unprecedented rate.

Commons problems are difficult to deal with, but not impossible. Successful international action has been taken on ozone-depleting substances, on a number of fisheries issues, and on the trade in endangered species, all of which are discussed elsewhere in this encyclopedia. The climate change issue shows signs of meaningful international action, although whether this will lead to successful abatement of the problem remains to be seen. Yet continuing global economic expansion will inevitably exacerbate pressures on the global commons, and it is not at all clear that multilateral efforts to deal with them will be able to keep abreast of these continually mounting pressures.

The international community has increasingly in the recent past come to recognize that issues of economy and environment are intimately and necessarily related. The context within which such recognition has taken place is sustainable development. Most countries have signed the Rio Declaration, formally committing themselves to the principle of sustainable development, and many of the international organizations that govern the global economy, such as the World Trade Organization, officially recognize the centrality of the principle to the contemporary global economy. Skeptics might suggest that this recognition is meaningless in the absence of specific limitations on behavior, and that in any case it is by no means clear exactly what sustainable development means or how it might be put into operation. Optimists might suggest that the active adoption of the term indicates a broader trend toward the incorporation of environmental sensibilities into the management of global economic affairs. The truth probably lies somewhere in between.

[See also Economic Levels; European Community; Industrial Ecology; Industrialization; Nuclear Hazards; and Trade and Environment.]

## BIBLIOGRAPHY

Block, F. The Origins of International Economic Disorder: A Study of United States International Monetary Policy From World War II to the Present. Berkeley: University of California Press, 1977. A critical discussion of the unraveling of the Bretton Woods system.

Brecher, J., and T. Costello. Global Village or Global Pillage. Boston: South End Press, 1995. A radical contemporary critique of globalization.

Caves, R., J. Frankel, and R. Jones. World Trade and Payments: An Introduction. 7th ed. New York: Harper Collins, 1996. A classic introductory international economic text.

Foreman-Peck, J. A History of the World Economy: International Economic Relations since 1850. Totowa, N.J.: Barnes & Noble Books, 1983. A representative general economic history.

Gardner, R. Sterling–Dollar Diplomacy in Current Perspective: The Origins and Prospects of our International Economic Order. New York: Columbia University Press, 1980. The classic history of the creation of the Bretton Woods system.

Gilpin, R. The Political Economy of International Relations. Princeton: Princeton University Press, 1987. A good graduate-level survey.

Helleiner, E. States and the Reemergence of Global Finance: From Bretton Woods to the 1990s. Ithaca, N.Y.: Cornell University Press, 1994. A discussion of the globalization of finance.

Keohane, R., and J. Nye. Power and Interdependence: World Politics in Transition. Boston: Litttle, Brown, 1977. One of the earliest statements of the globalization hypothesis.

Kindleberger, C. The World in Depression, 1929–39. Berkeley: University of California Press, 1973. Useful both for the history of the Depression and for the discussion of international economic leadership.

Krasner, S., ed. International Regimes. Ithaca, N.Y.: Cornell University Press, 1983. A collection of several different approaches to studying international economic institutions and governance.

Krugman, P. The Age of Diminished Expectations: United States Economic Policy in the 1990s. 3d ed. Cambridge, Mass.: MIT Press, 1997. Collection of essays for nonspecialists.

Landes, D. The Unbound Prometheus: Technological Change and Industrial Development in Western Europe from 1750 to the Present. London: Cambridge University Press, 1969.

Samuelson, P., and W. Nordhaus. Economics. 15th ed. New York: McGraw-Hill, 1995. A basic introduction to the subject.

Spero, J., and J. Hart. The Politics of International Economic Relations. 5th ed. New York: St. Martin's Press, 1997. A good undergraduate survey.

World Bank. The World Development Report, various years. New York and Oxford: Oxford University Press. A source both for various analyses of contemporary global economic issues and for comparative statistics.

—J. SAMUEL BARKIN

# GLOBAL ENVIRONMENT FACILITY

The Global Environment Facility (GEF) was established in November 1990 as a pilot program to finance the incremental costs of developing-country action on four global environmental problems: climate change, biodiversity loss, pollution of international waters, and depletion of the ozone layer. In June 1992 it was designated as the interim financial mechanism for the Framework Convention on Climate Change (FCCC) and the Convention on Biological Diversity (CBD). After a highly politicized and contentious negotiating process over the GEF's mandate, governance, and administration, the restructured GEF was formally established in March 1994. Since then, the GEF's key stakeholders have concentrated on developing the Facility's operational strategy and operational programs to support implementation of the FCCC and CBD, and to a lesser extent to combat pollution of international waters and ozone depletion.

The GEF's governance structure and its allocation of roles and responsibilities among key stakeholders are the result of hard political bargaining between developed- and developing-country governments, the GEF's three implementing agencies, and international environmental nongovernmental organizations (NGOs). Membership in the GEF is open and voluntary for all countries; as October 1999, it had 166 members, including all major developed, developing, and transition countries. The Facility is governed by a thirty-two-member council (its members represent fourteen developed and eighteen developing- and transition-country constituencies); the council is accountable to the GEF Assembly, which represents all GEF member countries and meets once every three years. The Council is responsible for setting GEF policies (with guidance from the Conferences of the Parties of the FCCC and CBD and the Parties to the Montreal Protocol on Substances that Deplete the Ozone Layer), overseeing the GEF's operations, and approving work programs developed by the GEF's three implementing agencies. The GEF council's process for consultation with nongovernmental organizations (NGOs) is unique among international financial organizations: the council holds a formal NGO consultation before each council meeting, and allows NGO observers at its meetings. A significant number of NGOs are also involved in designing and implementing GEF projects.

The GEF's Secretariat and its three implementing agencies—the World Bank, the United Nations Development Programme, and the United Nations Environment Programme—are jointly responsible for managing the GEF's operations. At the international level, they are guided by the Council, assisted by a Scientific and Technical Advisory Panel (STAP), and receive advice (solicited and unsolicited) from a wide range of developed- and developing-country NGOs. At the national level in developing countries, the GEF's implementing agencies seek to coordinate their activities with each other, with other international agencies, and with national governments, NGOs, and the private sector.

The GEF's budget is determined primarily by its developed-country members, who make contributions in a separate negotiated burden-sharing process. During the GEF's 1990–1994 pilot phase it had a total budget of U.S.$1.1 billion; after the GEF's restructuring, developed countries committed U.S.$2 billion for the period 1994–1997; they committed U.S.$2.7 billion for the period 1998–2001. Climate change and biodiversity projects have each received roughly 40 percent of GEF funds; international-waters projects have received roughly 10 percent; ozone projects about 6 percent, and regional or global projects that address multiple focal areas the remaining 4 percent.

The GEF seeks to fulfill its mandate primarily by providing grants to cover the agreed incremental costs of projects that deliver net global benefits in each of its four focal areas. The agreed incremental cost is the difference between the cost of a project that would deliver some net national benefit and the cost of an alternative project that would produce an additional, global net benefit (in the form of biodiversity conservation, greenhouse gas reduction, protection of international waters, or reduction of ozone-depleting substances). The GEF is supposed to pay countries on what might be called a fee for service basis for the incremental cost of projects that deliver a net global benefit. It is also supposed to give priority to projects that are cost-effective (namely, those that deliver the greatest net global benefit at the lowest incremental cost). To date, the incremental-cost criterion has been difficult to use as a practical basis for allocating GEF funds, and the GEF's Council, Secretariat, and implementing agencies have gradually shifted in the direction of more flexible and common-sense interpretations of the incremental-cost criterion.

As a complement and alternative to the incremental-cost criterion, the GEF has developed operational programs in its two top-priority focal areas, climate change and biodiversity conservation, and in the international-waters focal area.

In consultation with the parties to the FCCC, the GEF has given priority to climate change projects that reduce barriers to the adoption of commercially viable energy conservation measures, energy-efficient technologies on the supply and demand sides, and renewable energy technologies; and to projects that promote the commercialization of new energy technologies that are low in greenhouse gas emissions. The GEF also funds enabling activities under the FCCC to build the capacity of states to analyze sources of greenhouse gas emissions, develop action plans to reduce the growth of emissions, and report to the Conferences of the Parties. GEF climate change funding has been heavily concentrated in the top ten greenhouse-gas-emitting developing countries. The major questions about the GEF's climate change projects are the extent to which they can catalyze change in national policies that discourage the adoption of energy-efficient and renewable technologies, and the extent to which "demonstration effects" of GEF projects actually reduce market, policy, and technology risk for energy producers and consumers.

In consultation with the Parties to the CBD, the GEF has sought to fund biodiversity projects across four types of ecosystem: arid and semiarid; coastal, marine, and freshwater (including wetlands); forest; and mountain. It funds projects oriented toward protection and sustainable use for each of these ecosystem types. As in the climate change area, the GEF also funds enabling activities under the CBD. The GEF has allocated most of its biodiversity funds to ecosystem protection, and has concentrated its funding in countries with high levels of species diversity and endemism. The major questions about the GEF's biodiversity projects are the extent to which they can influence national policies that encourage settlement, commercial use, and degradation of biodiversity-rich land and marine ecosystems, and the financial sustainability of biodiversity protection projects after GEF funding ends.

In the international-waters focal area, the GEF is not required to seek the guidance of the parties to any international convention. The GEF Secretariat has taken the lead in developing the GEF's operational strategies in this area. The strategies promote regional cooperation in analyzing land- and sea-based pollution sources and developing action plans to reduce pollution. The major questions about the GEF's work in this area are whether participating countries will act on the findings of their joint analyses and whether sufficient GEF funding will be available to support regional action, given the relatively low share of GEF funds that are dedicated to the international-waters focal area and the potentially high costs of projects.

In the ozone-depletion focal area, the GEF's actions are limited to supporting ozone-depleting substance phaseout projects and enabling activities in a few countries (primarily in central and eastern Europe) that are not developing countries and are also too wealthy to qualify for funding from the Multilateral Fund of the Montreal Protocol.

Independent evaluations of the GEF's structure and operations were carried out in 1994 and 1997. These evaluations, as well as analyses by other informed observers, suggest that the GEF has established itself as the leading source of public funding for developing-country projects that seek to reduce greenhouse gas emissions, protect biodiversity, and reduce pollution of international waters. The GEF's stakeholders have succeeded in evolving a governance structure and operational strategy that are politically viable in the hotly contested arena of global environmental politics, and have also demonstrated some ability to modify both structure and strategy in response to changing circumstances and new information. Nevertheless, the GEF's financial and human resources are and will remain very small relative to the problems they seek to address. Given these limits, the GEF must seek to leverage its resources in two ways: by raising the priority that its own implementing agencies give to global environmental issues, and by supporting analytic and institutional change to influence policy and market incentives for energy use and ecosystem conservation in developing countries.

[See also Convention on Biological Diversity; Framework Convention on Climate Change; Nongovernmental Organizations; United Nations Environment Programme; and World Bank, The.]

### INTERNET RESOURCE

Global Environment Facility. http://www.gefweb.org/. This is the best source of information about the GEF. It is regularly updated with information about the GEF's projects and policies, documentation from meetings of the GEF Council and Assembly, and GEF Working Papers.

### BIBLIOGRAPHY

Fairman, D. "The Global Environment Facility: Haunted by the Shadow of the Future." In *Institutions for Environmental Aid: Pitfalls and Promise*, edited by R. Keohane and M. Levy. Cambridge, Mass.: MIT Press, 1996, pp. 55–87. Analyzes the interests and interactions of the GEF's key stakeholders during the GEF's establishment, pilot phase, and restructuring negotiations.

Global Environment Facility. *Instrument for the Establishment of the Restructured Global Environment Facility*. Washington, D.C.: Global Environment Facility, 1994. As the GEF's "consti-tution," the Instrument defines the roles and responsibilities of the GEF Assembly, Council, Secretariat, Implementing Agencies, and Scientific and Technical Advisory Body, and the GEF's relationship to the FCCC and CBD.

———. *Operational Strategy of the Global Environment Facility*. Washington, D.C.: Global Environment Facility, 1996. Outlines guiding principles for the GEF's work, and establishes operational programs for climate change, biodiversity, and international waters.

Porter, G., R. Clémençon, W. Ofosu-Amaah, and M. Philips. *Study of GEF's Overall Performance*. Washington, D.C.: Global Environment Facility, 1997. A thorough and balanced examination of the implementation of the GEF's operational strategy and programs from 1994 to 1997; makes recommendations for strategic and administrative change.

Sjoberg, H. *From Idea to Reality: The Creation of the Global Environment Facility*. Working Paper Number 10. Washington, D.C.: Global Environment Facility, 1994. A carefully researched history of the GEF's establishment.

United Nations Environment Programme. *United Nations Development Programme and the World Bank, 1993: Report of the Independent Evaluation of the GEF Pilot Phase*. Washington, D.C.: Global Environment Facility, 1993. A comprehensive examination of the GEF's pilot phase operations. Makes extensive recommendations for strategic and administrative changes.

—DAVID M. FAIRMAN

## GLOBALIZATION.

**GLOBALIZATION.** *See* Global Economy; Information Technology; *and* Regional Assessment.

## GLOBAL MONITORING

Global monitoring refers to the collection, processing, interpretation, dissemination, and archiving of data about the state of the Earth, including both its natural environment and its human population. In recent years, the volume of data collected about the Earth has grown enormously as a result of increased use of automatic sensors in the oceans and atmosphere, increased use of satellite remote sensing, and increasing complexity in the reporting of statistical data by census and other agencies. Data are almost without exception available in digital form. Modern satellite sensors are capable of producing data and transmitting them to Earth at rates of at least one terabyte ($10^{12}$ bytes) per day.

Collection of data at these rates is essential if careful track is to be kept of changes in the Earth's environment. For example, the Antarctic ozone hole was first identified from satellite data. But effective monitoring requires frequent sampling and high spatial resolution, as well as high specificity to particular phenomena, and our ability to sense and create data now threatens to outstrip our ability to absorb, analyze, and interpret. The global change research community often uses the metaphor of "drinking from a fire hose" to describe the inability to cope with the flood of monitoring data.

Recently, therefore, substantial effort has gone into developing adequate data and information delivery systems that can deal simultaneously with the growing appetite for data by global change scientists and the growing acquisitions by sensors, avoiding the bottleneck shown in Figure 1.

**Historical Trends.** Comprehensive global monitoring is very recent; the first efforts to provide an accurate picture of the Earth's surface date back no further than the fifteenth century. The nineteenth century saw a massive expansion of global monitoring, as the colonial powers attempted to characterize the environments of their newly acquired territories (and, not incidentally, to use mapping and monitoring as a means to assert their dominion). International collaborations led eventually to the integration of national monitoring networks, through the World Meteorological Organization and comparable agencies in other fields. Thus the historical pattern has been to organize global monitoring by coordinating national efforts. This allows national data to be disseminated nationally, and international data to be obtained through cooperative arrangements. Moreover, each discipline has tended to develop its own, independent system of monitoring and dissemination.

Several trends have affected this set of arrangements in the latter half of the twentieth century. First, remote sensing has made it possible for data to be collected from above, without reference to national jurisdiction. For some early sensors, limited communication technology required the location of ground receiving stations in several countries, and this served as a way of preserving some sense of national ownership over data. But today, communications have improved significantly, and there are no longer any technical limitations on collection of data about one country by another country's satellites. Instead, agencies such as the United States National Aeronautics and Space Administration (NASA) operate in a highly internationalized environment, and data dissemination has become an international function also.

Second, almost all data are now handled in digital form. This has allowed for much greater integration of data dissemination and archiving: the World Data Center network, for example, handles data across a wide range of scientific disciplines, allowing scientists some degree of "one-stop shopping" for data. There is now widespread adherence to common data format and data description standards, allowing data to be transferred easily between systems and to be readily imported to analysis and modeling software, using widely recognized formats.

Third, there is increasing acceptance that an understanding of the Earth system can only come from research efforts that span disciplines, and from models that integrate both human and physical aspects of the Earth environment. The data to support these models must come from many disciplines, each of which has tended in the past to adopt its own methods of representation, format standards, and terminologies. Thus there is increasing pressure for integrated approaches that transcend the boundaries of disciplines.

Fourth, in recent years there has been very rapid progress in technologies for the dissemination of data. We have moved in the course of not much more than a decade from stand-alone computing systems, each serving many users, and exchange of data on magnetic tape, to an environment of high-bandwidth connectivity in which it is possible to transmit megabytes of information in seconds, at no cost, between computers located on opposite sides of the globe. This global network, the Internet, now provides its users with highly sophisticated search tools capable of finding information anywhere on its connected servers, of which there are currently of the order of ten million. The Internet and World Wide Web truly revolutionized the world of data dissemination in the 1990s.

Finally, arrangements for data dissemination have been affected in recent years by the political changes that have occurred in many countries. Traditionally, the dissemination of data relevant to global monitoring has been funded by national governments, either through national programs or through international collabora-

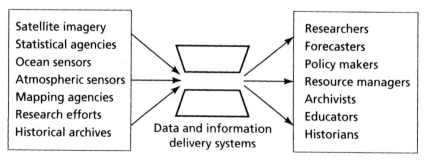

**Global Monitoring. Figure 1.** The Data and Information Bottleneck.

tion. Many national governments have been forced to curtail these efforts sharply, and to move them to a cost-recovery basis by charging users for the costs of data collection. The principle that data for scientific purposes should be freely available has been challenged, particularly in areas where data serve both scientific and other purposes, such as climatic data and base mapping. Budget cuts have also forced the thinning of global monitoring networks and the curtailment of many data-collection programs.

**Global Data Dissemination Today.** The current patchwork of arrangements for data dissemination reflects the legacy of past arrangements and the impact of rapid technological change. Data are available from large servers offering collections defined by regional interest (e.g., the International Centre for Antarctic Information and Research [ICAIR] server for scientific information on the Antarctic); by the data's sources (e.g., NASA's EOS Data and Information System server for information from the Earth Observing System satellites); or by disciplinary theme (e.g., the World Data Centers). The World Wide Web's search engines (e.g., Altavista, Yahoo) can be used in a hit-or-miss process to hunt for information by keyword. More ambitious arrangements allow searches over distributed collections of servers if each server agrees to adopt a standard method of data description. For example, the U.S. Federal Geographic Data Committee's FGDC National Geospatial Data Clearinghouse allows users to search over any of the several hundred servers in its international network, using FGDC's Content Standard for Digital Geospatial Metadata, which prescribes standard formats for information on geographic coverage and other key parameters.

To make effective use of this set of current arrangements, it is clear that a user needs a high level of experience and skill. The term *collection-level metadata* describes knowledge about the nature of a collection's contents: for example, knowledge about the information likely to be found in the National Geospatial Data Clearinghouse. Where, for example, should one look for information on the climate of Costa Rica? This problem was handled in the traditional research library by ensuring that every library possessed a copy of every important book. In the digital world of specialized data this is clearly impossible, but the arrangements that will evolve to replace it are not yet clear.

**Prospects.** The prospects for a single, monolithic server for all global monitoring data are clearly nil; such a server would be impossibly large, would conflict with innumerable special interests, and would be impossible to fund. Scientists and agencies will insist on maintaining some level of control over the data they have collected or created. But it is technologically possible to provide local control and, at the same time, to allow the user to see the distributed resource as a virtual whole, provided that appropriate protocols and standards can be adopted for description of data set contents, as the National Geospatial Data Clearinghouse experience indicates. Thus the problem of building an international system for storage and dissemination of global monitoring data is not so much technical as behavioral and institutional. On the behavioral side, there is always a cost associated with adopting any standard or protocol, and the perceived benefits to a server in joining a distributed network must therefore outweigh the perceived costs. On the institutional side, it will be necessary to identify an appropriate sponsoring organization, with access to sufficient funds to develop and sustain the network.

It is possible, however, that technical developments will make these problems much less significant in the future. The current generation of Web search engines works on the basis of keywords, relying on intelligent search of the textual contents of a Web site to identify the right words; no standards need to be adopted by servers. This hit-or-miss approach is clearly better suited to research in literature or history than global change science, where data sets are often characterized numerically (e.g., by latitude and longitude coverage limits). If a new generation of more intelligent search engines could be developed to recognize and catalog scientific data sets, then it would not be necessary to adopt the kinds of data description standards outlined earlier. At present, however, this remains an interesting problem for research.

[*See also* Geographic Information Systems; Information Management; International Cooperation; Public Policy; *and* Remote Sensing.]

### BIBLIOGRAPHY

Committee on Issues in the Transborder Flow of Scientific Data, United States National Committee for CODATA. *Bits of Power: Issues in Global Access to Scientific Data.* Washington, D.C.: National Academy Press, 1997. Includes commentary on the political and economic trends affecting data production and dissemination.

Federal Geographic Data Committee. *Content Standard for Digital Geospatial Metadata.* Washington, D.C.: Department of the Interior, 1998. http://www.fgdc.gov/. This Web site includes the text of the standard and much related information.

Mounsey, H., ed. *Building Databases for Global Science.* London: Taylor & Francis, 1988. A review of the issues before the advent of the WWW.

Onsrud, H. J., and G. Rushton, eds. *Sharing Geographic Information.* New Brunswick, N.J.: Center for Urban Policy Research, Rutgers University, 1995. A collection of essays on problems of sharing data, including scientific data.

Rhind, D. W., ed. *Framework for the World.* Cambridge: GeoInformation International, 1997. A recent collection on the issues impeding comprehensive global mapping.

—MICHAEL F. GOODCHILD

**GLOBAL POSITIONING SYSTEMS.** *See* Geographic Information Systems.

# GLOBAL WARMING

[*To explore the major environmental issue of global warming, this entry comprises two articles. The first article presents an overview of the increases in the concentration of greenhouse gases and their potential to bring about global warming; the second focuses on the index that qualitatively compares the warming effects of a given greenhouse gas relative to a reference.*]

## An Overview

The concept of global warming combines the observed rise in global mean temperatures since the beginning of the twentieth century with the measurable enhancement of the greenhouse effect over the same period. Unlike earlier periods of rising temperature in the climate record, current global warming appears to be the first to be created by human activities, its basic cause is seen as the rising anthropogenic contribution to levels of greenhouse gases in the atmosphere.

**The Greenhouse Effect.** The greenhouse effect is created by the ability of the atmosphere to be selective in its response to radiation. It is transparent to high-energy, shortwave radiation, such as that from the Sun, but partially opaque to the lower-energy, longwave radiation emanating from the Earth's surface. For example 10–30 percent of ultraviolet radiation with wavelengths between 0.28 and 0.32 micrometers and a major proportion of the radiation in the visible range of the spectrum, between 0.3 and 0.7 micrometers, is transmitted through to the surface without losing its high-energy content. Following absorption, the return radiation from the infrared end of the spectrum—with wavelengths between 1 and 30 micrometers—is captured and the temperature of the atmosphere rises. The capture of the outgoing terrestrial radiation is effected largely by water vapor and carbon dioxide, along with methane and traces of about twenty other gases, which together are called the *greenhouse gases*. The whole process was labeled the *greenhouse effect* because the gases, by trapping the heat, appeared to work in much the same way as the glass in a greenhouse.

Since the greenhouse effect depends upon carbon dioxide and the other gases in the atmosphere, it follows that any change in these gases, including their relative concentration, will impact on the intensity of the effect. Any increase in greenhouse gas levels, for example, should cause the retention of additional terrestrial radiation, which in turn should lead to a rise in global temperature. Current evidence suggests that the relationship is real. According to generally accepted estimates, the Earth's surface temperature has increased by between 0.3° and 0.6°C since 1900, at a rate broadly consistent with that expected from the measured rise in levels of greenhouse gases in the atmosphere. These gases differ in their ability to cause radiative forcing and contribute to global warming, however, and the concept of global warming potential (GWP) was incorporated in the first Scientific Assessment of the Intergovernmental Panel on Climate Change (IPCC) as a means of measuring the relative radiative effects of the different greenhouse gases. [*See* Intergovernmental Panel on Climate Change.] The GWP of a gas is a measure of the cumulative radiative forcing caused by unit volume of the gas over a given period of time, starting from the present. The values are expressed in the form of a comparison to carbon dioxide, the most commonly used reference gas. With the GWP for carbon dioxide over a one-hundred-year time horizon set at 1, for example, the GWP for methane over the same period is 21. Comparable values for nitrous oxide and CFC-11 are 290 and 3,500, respectively. The cumulative effects of the greenhouse gases and the relative contributions of individual gases to the overall warming can be estimated from such values by multiplying the specific GWP of each gas by the volume of gas emitted. The GWP concept has certain shortcomings, and values typically have an uncertainty factor of ±35 percent relative to their carbon dioxide reference index. However, they provide a simple, direct comparison of the effectiveness of the various greenhouse gases and can be used to estimate the potential impacts if individual gases are targeted for emission reductions.

Greenhouse gas levels in the past rose and fell as a result of natural processes, but, since the middle of the nineteenth century, levels have been rising as human activities have led to the release of higher volumes of carbon dioxide, methane, and other greenhouse gases into the atmosphere. By the last decade of the nineteenth century, the changing concentrations of atmospheric carbon dioxide and their potential to cause climate change were already being investigated by a number of researchers. Svante Arrhenius, a Swedish chemist, is usually credited with being the first to quantify the relationship and provide the first predictions of the temperature rise that might be expected with an increasing concentration of carbon dioxide. He published his findings in 1896, at a time when the environmental implications of the Industrial Revolution were just beginning to be appreciated, and although neither his prediction of the likely rate of carbon dioxide accumulation in the atmosphere nor that of the consequent temperature rise was particularly accurate, they were not bettered until the early 1960s. Little attention was paid to the issue during the first half of the twentieth century. Occasional papers appeared, but interest only began to increase in the early 1970s as part of a growing appreciation of the potentially dire consequences of human interference in the environment. [*See the biography of Arrhenius.*]

Atmospheric concentrations of carbon dioxide have varied considerably in the past. Analysis of air bubbles trapped in polar ice indicates that the lowest levels of atmospheric carbon dioxide occurred during the Quaternary glaciations. At that time, the atmosphere contained as little as 180–200 parts per million by volume (ppmv) of carbon dioxide, although there is some evidence that levels fluctuated by as much as 60 ppmv in periods as short as 100 years. Levels rose to 275 ppmv during the warm interglacial phases, and that level is also considered representative of the preindustrial era of the nineteenth century.

The first modern measurements at Mauna Loa in Hawaii in 1957 indicated that concentrations had risen to 310 ppmv, and they continued to rise by over 1 ppmv per year, reaching 335 ppmv in 1980. Increases of 2–4 ppmv per year brought the level to 345 ppmv by the middle of that decade, to 353 ppmv by 1990, and to 358 ppmv by 1994. The current volume of carbon dioxide is more than 25 percent higher than the preindustrial volume, and without precedent in the past 100,000 years of Earth history. If the present rate of increase continues until the year 2050, it is estimated that the atmospheric carbon dioxide concentration will then be 450 ppmv, and by the end of the twenty-first century it will be close to 500 ppmv, almost twice the 1800 level.

Carbon dioxide is the most abundant greenhouse gas, but it is not the most powerful nor is it the most rapidly increasing. That distinction belongs to methane, which is at least twenty times more powerful (molecule for molecule) than carbon dioxide and has an atmospheric concentration that has at times grown at twice the rate of carbon dioxide. As with carbon dioxide, the preindustrial concentration of methane in the atmosphere was relatively steady. Its level of about 0.8 ppmv was probably close to the average for the past 2,000 years. By the end of the nineteenth century, the concentration had increased to 0.9 ppmv, and in 1978, when atmospheric concentrations of methane were first measured directly, the average volume was 1.51 ppmv. The continued release of the gas from agricultural activities, coal mining, natural gas leaks, and possibly from sanitary landfill sites has pushed the concentration up at an average rate of 0.8 percent per year to a value of 1.7–1.75 ppmv. This is double the preindustrial level, and at the present growth rate a further doubling could take place in the next century. The IPCC Supplementary Report of 1992 noted some evidence that the growth in atmospheric methane concentration had already stabilized, but after 1993 growth began again at a rate of about 0.5

percent per year. The lifespan of methane in the atmosphere, at about ten years, is short. As a result, it responds relatively rapidly to changes in output or changes in the concentrations of substances such as the hydroxyl radicals that control it. It is difficult, therefore, to estimate the long-term level of methane in the atmosphere. [*See* Methane.]

After carbon dioxide and methane, the third highest concentration among the greenhouse gases is that of nitrous oxide. With an atmospheric concentration of only 310 parts per billion by volume (ppbv), it is about a thousand times less common than carbon dioxide, and with a growth rate of only 0.2–0.3 percent per annum, it is increasing less rapidly than carbon dioxide or methane. Nitrous oxide is released naturally into the atmosphere through the denitrification of soils, but the growth from preindustrial concentrations of 280–290 ppbv to a level of about 312 ppbv in 1994 was thought to have been brought on by the increased use of fossil fuels and the denitrification of nitrogen-based agricultural fertilizers. The first IPCC Scientific Assessment (1990) concluded, however, that past estimates of the contribution of fossil fuel combustion to the increase are too large—by perhaps as much as ten times—and nitrous oxide production rates during agricultural activity are difficult to quantify. As a result, the global nitrous oxide budget remains poorly understood and its future concentration is therefore difficult to predict. [*See* Nitrous Oxide.]

Although current carbon dioxide, methane, and nitrous oxide levels are being augmented by human activities, these gases are also supplied from natural sources. In contrast, the halocarbons, the fourth important group of greenhouse gases, are almost entirely anthropogenic in origin. Only methyl bromide and methyl chloride have natural sources. The others, such as chlorofluorocarbons (CFCs) and halons, have been released in increasing quantities from refrigeration units, insulating foams, aerosol spray cans, fire extinguishers, and industrial plants since the 1950s. [*See* Chlorofluorocarbons.] They are best known for their ability to damage the stratospheric ozone layer, but are also among the most potent greenhouse gases. For example, CFC-11 is about twelve thousand times more effective, molecule for molecule, than carbon dioxide as a radiative forcing agent. Concentrations of CFCs in the atmosphere are much lower than those of the other greenhouse gases, ranging from CFC-115 at 5 parts per trillion by volume (pptv) to CFC-12 at 484 pptv. Up to the early 1990s these concentrations were growing at rates between 4 and 10 percent per annum. Other halocarbons, such as Halon 1211 and Halon 1301, used mainly in fire extinguishers, were growing at rates as high as 15 percent per year from concentrations of 2 pptv. Recent international agreements, beginning with the Montreal Protocol in 1987, were aimed at reducing further damage to the ozone layer by reducing the production and use of CFCs, but they will also have some impact on the greenhouse effect. Already the rate of increase in the atmospheric concentration of halocarbons is stabilizing and the ultimate outcome will be a reduced contribution to the greenhouse effect and therefore global warming. However, CFCs have a long residence time in the atmosphere—up to four hundred years in the case of CFC-13 and CFC-115—and even as emission rates fall, they will continue to contribute to global warming for some time to come. Other halogenated compounds such as perfluorocarbons, sulfur hexafluoride, and hydrofluorocarbons are currently present in the atmosphere in very small quantities, but they may become significant in the future if concentrations continue to increase.

In its 1990 Scientific Assessment, the IPCC provided projections of future levels of carbon dioxide, methane, nitrous oxide, and CFCs. The projections were based on four different scenarios ranging from "business as usual," in which emissions continued at 1990 levels, through three other scenarios in which increasing controls were introduced to reduce the growth of emissions. Projections from these scenarios to the year 2100 indicated potential concentrations of carbon dioxide ranging from 400 to 825 ppmv. Projections for methane ranged from 1,500 to 4,000 ppbv, for nitrous oxide from 355 to 418 ppbv, and for CFC-12 from 350 to 1,400 pptv. These values were recalculated in 1992 and 1994 using revised scenarios, but the net differences in the results were generally small.

**Global Temperature Trends.** Increases in the concentrations of greenhouse gases contribute to positive radiative forcing in the Earth–atmosphere system and have the potential to bring about climate change in the form of warming. Analysis of the instrumental record of the last one hundred years appears to support such a relationship, with the succession of exceptionally warm years in the 1980s and 1990s widely regarded as an indication that the warming of the Earth, which had be-

gun some time toward the beginning of the century, is continuing and even increasing (Figure 1). Estimates obtained from the instrumental climate record indicate that mean global temperatures have risen between 0.3° and 0.6°C since 1900, but the change has not been steady. The main increase took place between 1910 and 1940, and again after 1975. Between 1940 and 1975, despite rising greenhouse levels, mean global temperatures declined, particularly in the Northern Hemisphere. In addition, analysis of the records suggests that relatively rapid warming prior to 1940 was probably of natural origin. Global surface temperatures have increased by about 0.2°–0.3°C since the 1950s. Warming was particularly evident in the 1980s, which was overall the warmest decade on record up to that time. Initially, the warming continued into the 1990s, with 1991 the second warmest year on record, but relatively cool conditions, close to the 1951–1980 average, returned in 1992 and 1993, following the eruption of Mount Pinatubo. Warmer temperatures returned in 1994 to make that year among the warmest 5 percent of all years since 1860.

Assessments of the reliability of such results are central to understanding global warming. Although they may be presented with what appears to be a high degree of certainty, their accuracy can be reduced by a number of factors. At the global level, gaps exist in the observational coverage in both time and place. [See Climate Change, *article on* Climate Change Detection.] Some areas such as the high Arctic and parts of Africa have records that are short and sparsely distributed; some have records that started in the mid-nineteenth century, but have been disrupted by war or by changing economic requirements. Over the oceans, the existence of long-term records is the exception. Even where meteorological stations have been reporting regularly for many years, the quality of the record may be compromised by changes in observing schedules or by the way in which the thermometers are exposed. Station locations have also changed with time. In the 1930s and 1940s, for example, many stations were moved from ur-

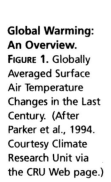

**Global Warming: An Overview.**

FIGURE 1. Globally Averaged Surface Air Temperature Changes in the Last Century. (After Parker et al., 1994. Courtesy Climate Research Unit via the CRU Web page.)

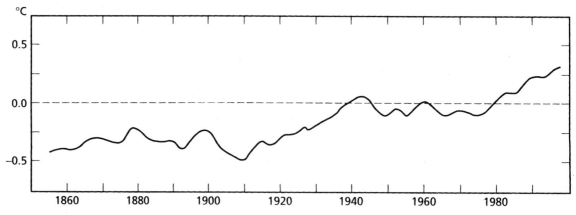

Average global temperature in 1998 was higher than at any time since the mid-1980s, exceeding 1997, the previous highest year, by a substantial margin. The year 1999 was not as warm as 1998, partly because of the cool La Niña phase that followed the 1997–98 El Niño. [See El Niño–Southern Oscillation.] Nevertheless, global mean temperature up to the end of July, 1999 is roughly 0.7°C above that at the end of the last century; and it is likely that 1999 will prove among the warmest ten years since global records began nearly 140 years ago. The decade of the 1990s clearly is the warmest on record, averaging 0.6°C above that of 1900 (see figure, right).

There is some discrepancy between temperatures at the surface and those in the upper troposphere (i.e., at altitude of 3–5 kilometers), especially in certain decades (see figure below, left). This lack of correspondence is leading some researchers to reconsider traditional models of greenhouse effect.

Records of Arctic sea ice during the last three decades (see figure below, right) show progressive shrinkage in the area covered, possibly as a reflection of global warming. [See Sea Ice.] Records for sea ice in Antarctica are relatively short, and do not show a significant trend.
—David J. Cuff

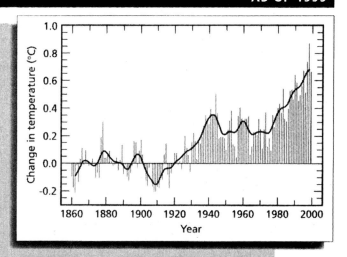

Global Mean Surface Air Temperature from 1860 to 1999, Relative to That in 1990.

The 1999 value includes observations through July. (Based on data from British Meteorological Office, 1999.)

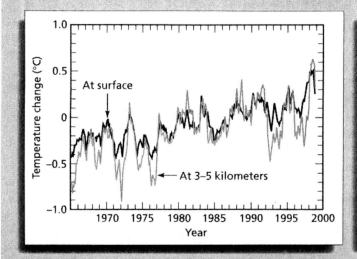

Change in Monthly World Average Air Temperature from 1965 to 1999, at the Surface and at the Height of 3–5 Kilometers. (Based on data from British Meteorological Office, 1999.)

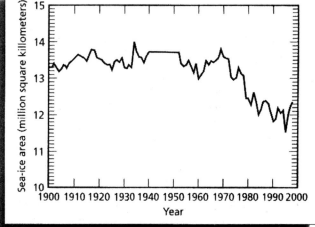

Observed Change in Area of Arctic Sea Ice from 1900 to 1998. (Based on data from British Meteorological Office, 1999.)

ban centers to rural or semirural airport locations to meet the needs of the expanding aviation industry. This may well have introduced an apparent but artificial cooling into the record, as a result of the removal of the station from the effects of the urban heat island. While some records have been adjusted to allow for such changes, there is no consensus on the significance of their impact on the global temperature record. In contrast, the impact of increasing urbanization around meteorological observing stations is generally considered

**GLOBAL HOT SPOTS**

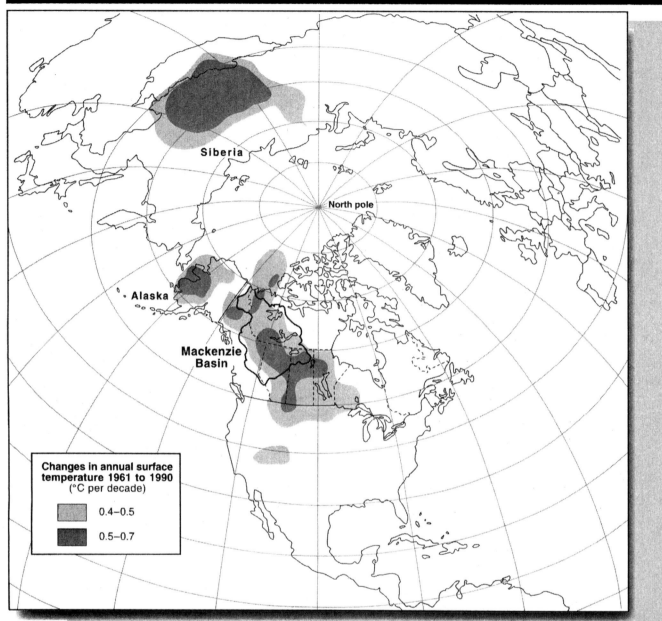

Global Hot Spots. The map illustrates changes in annual surface temperatures from 1961 to 1990 (°C per decade). (After Grescoe, 1998. With permission of Environment Canada.)

Global warming through the last century is usually expressed as an average for the world. But the warming is not evenly distributed. Three regions in the Northern Hemisphere show warming rates roughly triple the world average: one area is near Lake Baikal in Siberia; another is near Nome in western Alaska; and the third is an extensive area that includes parts of Canada's Prairie Provinces, North West (Canada) Territories, and Yukon Territory (see figure). This Canadian area includes the Mackenzie River watershed, which was the subject of a six-year study of the impacts of warming (Environment Canada, 1997). The Mackenzie Basin Impact Study (MBIS), sponsored by the federal and North West Territories (Canada) governments, B.C. Hydro, the University of Victoria, and Esso Resources Ltd., was released in August 1997. The study documents how rising temperatures have already melted permafrost, contributed to landslides, and increased the area burned by forest fires. The study then projects the future effects of global warming on the region's wildlife, vegetation, and humans, using a climate model provided by the Canadian Centre for

Climate Modeling and Analysis in Victoria, B.C. Expected effects include the following (Grescoe, 1998):

- Wheat farmers in the Peace River area of northwestern Alberta may benefit from longer and warmer growing seasons, but will have to deal with drier summers.
- The effect on forests will be mixed: a species like Douglas fir will be able to expand northward, but so will beetles such as the white pine weevil; and hotter summers will favor more forest fires, which could burn areas more than double the size of those now affected each year.
- Warmer summers and increased evaporation are expected to reduce wetlands in the Mackenzie delta. This will be detrimental to many shorebirds, but will favor those that prefer drier conditions.
- The warming is expected to have mixed effects on the region's caribou herds: warmer winters will bring heavier snowfall that will make grazing difficult; but the early spring season should increase the survival rates for calves. Hotter summers will increase the length of the season for mosquitoes and bot flies that harass the herds.
- Human occupants of the region will have to adapt to changes in wildlife and water levels, and to the progressive melting of permafrost that now supports many of the buildings in predominantly native communities.

—David J. Cuff

to be much more significant. With its impact on temperature through the creation of urban heat islands, urbanization is probably the element that has the greatest potential to introduce error into the climate record, but the magnitude of that error is difficult to evaluate. The effects of urbanization vary from region to region, being greater in the United States and parts of China than in Australia, for example, but all fall within the range 0.05°–0.1°C, with the latter considered most likely. When included in the longer-term record, the warming associated with urbanization may well be offset by the artificial cooling introduced when city center stations were relocated to rural sites.

The bulk of the research into global temperature change has involved the study of surface temperatures. In addition, temperature records are available for the troposphere and stratosphere, although they are commonly shorter, beginning only in the 1940s with data from instrument packages carried aloft by weather balloons (radiosonde ascents) and augmented from the 1970s on through satellite observations. Radiosonde data tend to be geographically sparse and of limited reliability because of changes in techniques and equipment over the period of the record. In contrast, the data from Microwave Sounding Units in weather satellites are much more precise and provide a more uniform global coverage of upper atmospheric temperatures.

The initial analysis of radiosonde and satellite observations indicated a slight cooling of the troposphere

between 1979 and 1993, at a time when surface temperatures were rising. Further investigation indicated that transient events such as the occurrence of an El Niño or a volcanic eruption accounted for much of the apparent difference between the tropospheric and surface results. After appropriate adjustments for these events, it was evident that a slight warming had occurred in the troposphere, continuing a trend of between +0.08° and +0.11°C per decade evident in data from 1958 on, and generally matching the rising trend in surface temperature over the same period. In contrast, temperatures in the lower stratosphere have been in decline at a rate of as much as 0.36°C per decade since the mid-1960s. By the mid-1990s, global stratospheric temperatures were at their lowest since radiosonde and satellite observations began.

In analyzing the instrumental record, even after adjustments are made for potential errors, the temperature changes that have taken place since the beginning of the century are well within the range of normal natural variations in global temperatures, but Hansen and Lebedeff (1988) calculated that the warming between the 1960s and the 1980s was more rapid than that between the 1880s and the 1940s, suggesting that the alleged greenhouse warming might be beginning to emerge from the general background "noise." Wigley and Barnett (1990) in their contribution to the first IPCC Scientific Assessment, noted that there was not at that time any evidence of an enhanced greenhouse effect in the observational record, but cautioned that it might be in part a function of the uncertainties and inadequacies in then-current investigative techniques.

In the five years between the first and second IPCC Scientific Assessments, sufficient progress was made toward a clearer understanding of these uncertainties—through the incorporation of anthropogenic effects in model experiments, a better definition of the natural variability of the climate system, and more powerful statistical techniques—to allow an anthropogenic contribution to be recognized in the recent global warming. That human signal is only gradually emerging from the natural background noise and it has not been possible to quantify it, but that does not detract from the significance of the IPCC conclusion that "the balance of the evidence suggests a discernible human influence on global climate."

**Earth's Climate History.** The study of climate change in the period prior to the development of the instrumental observational record in the mid-nineteenth century requires a reliance on a wide variety of proxy data. The indicators that provide the data take many forms. They may, for example, be biological, stratigraphical, archaeological, agricultural, glaciological, or historical in nature, but all reflect to a greater or lesser degree the climatic conditions that prevailed at the time they developed. They vary in quality: some, such as tree rings and fossil pollen, allow past conditions to be quantified with some precision, whereas others, such as some of the historical data, may provide only qualitative results. The calibration of the data is also variable. Tree rings and historical documents can provide specific dates for meteorological events, but other data may provide only relative dating or at best establish a range within which an event might have occurred. Gaps in the proxy record also create problems. Few long-term climate reconstructions depend entirely upon one source of proxy data. They incorporate information from a variety of sources, not only to fill gaps, but also to provide cross-references that improve the reliability of the results. Using the available proxy data, scientists have reconstructed the climatic history of the last ten thousand years with some reliability, and in places there is sufficient evidence to extend the record back further into glacial times. [*See* Climate Reconstruction.]

Such proxy data provide evidence that over the last several thousand years, during the ice ages and since they ended some ten thousand years ago, fluctuations in climate have produced alternating periods of increasing cold and increasing warmth (Figure 2). Ice cores from Greenland and the Antarctic and deep oceanic sediment cores, along with paleobotanical evidence, from the northern continents indicate a period of major warming during the immediate postglacial period between 10,000 and 5,000 BP. This event is known

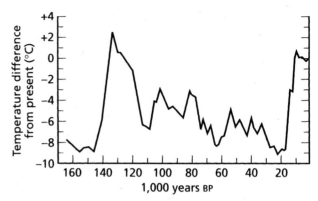

**Global Warming: An Overview. Figure 2.** Global Climate Change over the Past 160,000 Years, Based on Data from the Vostok Ice Core in Antarctica. (Adapted from data in Genthon et al., 1987.)

as the *Climatic Optimum*, when temperatures were perhaps 1°–3°C higher than at present. Changing oxygen isotope (oxygen-18/oxygen-16) ratios in ice and ocean sediments indicate rapid warming at about 10,000 BP or perhaps earlier. In contrast, the paleobotanical evidence, based on the changing distribution of plant communities obtained through pollen analysis, suggests that the warming peaked much later, between 7,000 and 5,000 BP. As with most reconstructions of early climate change, limitations in the distribution and precision of the proxy data ensure that the magnitude, extent, and timing of events such as the Climatic Optimum remain imprecise.

A much later period of warming identified through proxy data was the Medieval Climatic Optimum, which occurred between about 750 and 1200 CE and was marked by ameliorating climatic conditions in Europe and North America and probably elsewhere. Although it appears to have been best developed around the North Atlantic, that perception may be a reflection of the greater availability of proxy data from that area. Physical data from glacier fluctuations, tree-ring analysis, and pollen analysis have all provided evidence of warming, but the Medieval Climatic Optimum was the first in which proxy data from documentary sources also made a significant contribution to the evidence. Any assessment of the geographical extent and magnitude of the warming during the Medieval Climatic Optimum retains many uncertainties that tend to frustrate comparisons with the late-twentieth-century warming. [*See* Medieval Climatic Optimum.]

The climatic amelioration of the Medieval Climatic Optimum was followed by a gradual cooling that peaked in the Little Ice Age. This was a period of global cooling commonly accepted as having occurred in the three hundred years from the mid-sixteenth to the mid-nineteenth century, but with variations in time, place, and character. There is evidence, for example, that the cooling started as early as the mid-fourteenth century in some areas. The Little Ice Age included prolonged periods of particularly cool and wet conditions—as in the 1690s—but it was also characterized by considerable variability, with some periods, such as the 1730s, that were warm and dry. The final phases of the Little Ice Age overlapped slightly with the beginning of the observational record, but the bulk of the evidence has been supplied by proxy data from glaciological, paleobotanical, and documentary sources. Most of that information is from the Northern Hemisphere, but there are sufficient data from other areas to confirm that the Little Ice Age was a global event, although not all of the elements were completely synchronous. [*See* Little Ice Age in Europe.]

All of these events—the Climatic Optimum, the Medieval Climatic Optimum, and the Little Ice Age—occurred before the human impact on the environment was globally significant. In contrast, therefore, to the current warming linked to human activities, they were caused by natural processes in the Earth–atmosphere system. Given their past contribution to change, it seems unlikely that these processes will now remain quiescent while anthropogenic inputs contribute to change in the system, yet they have received only limited attention in planning for the consequences of global warming. [*See* Climate Models.]

**Climate Models.** The current global warming scenario, in which a doubling of atmospheric carbon dioxide would cause a temperature increase of between 1.0° and 3.5°C, has evolved from the results of numerous experiments with theoretical climate models. In these models, physical processes in the Earth–atmosphere system are represented by a series of fundamental equations, which, when solved repeatedly for a series of small but incremental changes, provide a forecast of the future state of the atmosphere. Climate models take various forms and involve various levels of complexity, depending upon the application for which they are designed, but general circulation models (GCMs) are most commonly used in the study of global warming.

GCMs provide full spatial analysis of the Earth–atmosphere system through the use of powerful computer programs capable of processing as many as 200,000 equations at tens of thousands of points in a three-dimensional grid covering the Earth's surface, and reaching through two to fifteen levels as high as 30 kilometers into the atmosphere. Despite the apparent intensity of such numbers, they provide a horizontal resolution of only about 250 kilometers, which is sufficient to simulate major atmospheric processes but too coarse to allow the incorporation of processes that operate at a local or regional scale. The latter requires

the use of parameterization, a technique that involves the establishment of statistical relationships between grid-scale variables of the basic model and the smaller-scale processes. Calculations at the grid scale can then provide estimates for the local or regional conditions. Cloudiness, net surface radiation, various elements of the hydrologic cycle, and land surface processes all require parameterization for inclusion in the GCMs. Regional or mesoscale conditions can also be simulated by using a model with a finer resolution nested or embedded in the global model. Such a nested model, driven by the global model that surrounds it, is programmed to provide much finer detail on regional climatic conditions than is possible with the standard model alone.

The first GCMs to be developed were component models capable of simulating conditions in one element of the Earth–atmosphere system. Atmosphere models, for example, were developed from weather forecasting models in the 1970s to simulate large-scale distributions of surface air temperature, precipitation, and mean sea level pressure, to be followed by oceanic models providing simulations of temperature, salinity, and sea ice distributions. Although climate modelers have continued to refine these component models, it has long been apparent that the constituents of the real environment do not operate in isolation, but as interrelated elements of the Earth–atmosphere system. In response to this, individual component models have been coupled in an attempt to emulate the integrated nature of that system (Figure 3). The most common approach is the coupling of atmospheric and oceanic models, with more sophisticated systems also including land surface schemes and the representation of sea ice. In theory, such models should provide a more accurate representation of the Earth's climate, but this is not always the case. The coupling of the models leads to the coupling of any errors in the individual models. Variations in heat, momentum, and freshwater fluxes between the atmosphere, ocean, and land surface, for example, can lead to a problem known as *model drift*, and although it can be treated it remains a constraint for coupled models.

Another major problem for coupled models is the difference in time scales over which the different components develop and respond to change. This is particularly true of coupled atmosphere–ocean models. The atmosphere generally responds within days, weeks, or months, while parts of the oceans—the ocean deeps, for example—may take centuries or even millennia to respond. Running a completely interactive coupled atmosphere–ocean model is therefore time-consuming and costly. As a result, the oceanic element in most coupled models is much less comprehensive than the atmospheric component. The ocean may be modeled as a slab, which represents only the uppermost layer of wa-

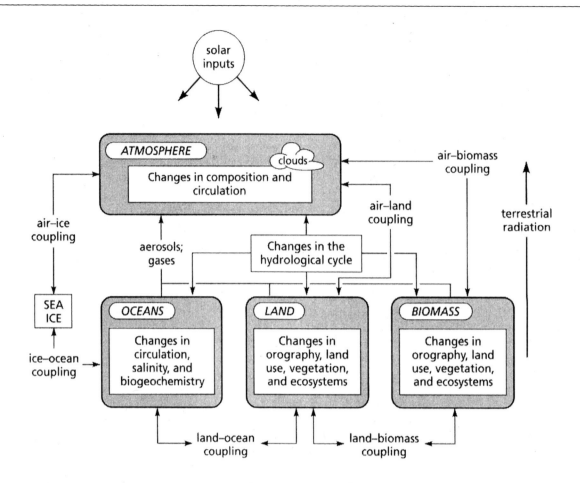

**Global Warming: An Overview. FIGURE 3.** The Elements and Processes that Must be Considered in Developing Models to Emulate the Integrated Nature of the Earth–Atmosphere System.

ter where the temperature is relatively uniform with depth. Oceanic heat storage is calculated only for the chosen depth of the layer, and other elements such as oceanic heat transport and exchanges with the deeper parts of the ocean are neglected or calculated indirectly. Mixed-layer and deep-ocean models have also been developed, incorporating data from fifteen to twenty layers to the ocean floor. However, the paucity of available data for the deeper oceans creates a problem with accuracy, and the relatively slow response rate of the oceans compared with the atmosphere creates major demands on computer time and costs. In the search for greater realism, the potential for sea ice models, carbon cycle models, and chemical models to contribute to climate simulation has also been explored. Sea ice simulation is a common component of many ocean models, for example. Carbon cycle models, particularly important in studies of global warming, have already been coupled to ocean models, and chemical models, involving

tropospheric ozone and methane, for example, are being developed to investigate the influence of other trace gases on the general circulation of the atmosphere.

Atmospheric aerosols have also received considerable attention in the development of climate models. They have long been recognized as being able to cause cooling in the Earth–atmosphere system, brought about by the backscattering of incoming solar radiation by the aerosol particles or by the increase in cloud cover when the aerosols act as condensation nuclei. Aerosols have a variety of forms and sources, but current attention has focused on sulfate aerosols, which are particularly effective in scattering incoming solar radiation. They are derived from emissions of sulfur dioxide from natural sources such as volcanoes, or increasingly from anthropogenic sources. Because they cause a negative radiative forcing, aerosols have the ability to counteract to some extent the positive forcing associated with the increases in greenhouse gas levels. They must therefore be incorporated in global warming models where possible, and although modeling of the radiative effects of aerosols remains complex and uncertain, improvements continue to be made.

In the development of GCMs, most of the effort was invested initially in producing equilibrium models. In these, instantaneous change is introduced into a model that represents existing climate conditions, and the model is then allowed to run until a new equilibrium is reached. The new model climate can then be compared with the original to establish the overall impact of the change. Most of the estimates of the global warming expected from a doubling of atmospheric carbon dioxide are of this type. They make no attempt to estimate changing climate conditions during the transient phase of the model run, although these conditions may well have important environmental impacts long before equilibrium is reached. The development of transient or time-dependent models has lagged behind that of equilibrium models. They require increased computer power and improved ocean observation data if they are to live up to their potential to simulate the progressive changes that take place in the real environment. Their potential is recognized, however, and ten transient experiments were included in the second IPCC Scientific Assessment, compared with only one in the first Assessment in 1990 (Figure 4).

GCMs are becoming increasingly complex and comprehensive, but a high level of sophistication is no guar-

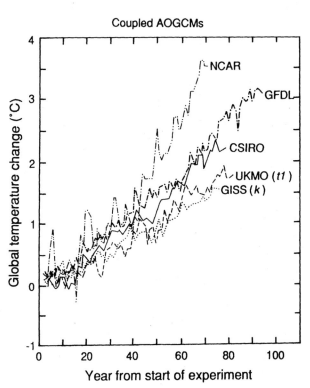

**Global Warming: An Overview.** FIGURE 4. Comparison of Selected Transient Experiments, Run for the IPCC 1995 Assessment of Climate Change, Using Coupled Atmosphere–Ocean GCMs.

Although the different experiments provide similar results in the short term, as the models continue to run they tend to diverge. (Based on data in Intergovernmental Panel on Climate Change, 1996, p. 300. With permission of Cambridge University Press.)

antee of perfection. Even state-of-the-art models include some degree of simplification, and certain variables are difficult to deal with whatever the level of model employed. Part of the difficulty in simulating the environment in models lies in a continuing incomplete understanding of the processes involved in the real environment, compounded by limited observational data on elements that are central to the models. Most of the uncertainties associated with model simulations arise from difficulties in these areas. The radiative properties of clouds, the links between the oceans and the atmosphere, interactions between the atmosphere and vegetation, and the role of the hydrologic cycle are all integral to the functioning of the Earth–atmosphere system, yet all are sources of uncertainty in climate models. GCMs also have difficulty in dealing with feedback mechanisms that act to augment or diminish change once it has been initiated (Figure 5). Deficiencies in the treatment of feedbacks involving clouds, surface radiation budgets, and the carbon cycle in the terrestrial biosphere, for example, were recognized in the 1995 IPCC Assessment as limiting the accuracy of global warming simulations.

Since individual models treat uncertainties in different ways, models using the same data often generate different results. When compared with the results obtained using observed data, the range of model projections is apparent (Figure 6). Physical and statistical components differ from model to model, as do the quality of the parameterization and the number and complexity of the feedbacks included. In addition, the spatial resolution of most models is such that although they may provide broadly similar results at the global or even the hemispheric scale, the quality of prediction at the regional scale remains low and quite variable from model to model (Figure 7). The Atmospheric Model Intercomparison Project was established to examine such discrepancies among models. It allowed the direct comparison of simulated and observed climatic conditions, by requiring participating modeling groups to simulate conditions over the period 1979–1988, using the same values for such elements as monthly averaged distributions of sea surface temperatures and sea ice, atmospheric carbon dioxide concentration, and the solar radiation flux at the outer edge of the atmosphere (solar constant). Following the model runs, the results were compared with the actual climate indicated by the observational record for the same ten-year period. The results confirmed that current models provide simulations of large-scale or global climatic conditions that resemble the observed conditions, but that substantial intermodel differences remain, particularly at the regional level.

The U.S. Global Change Research Program (USGCRP) has promoted the development of predictive models for

**Impact of temperature increase**

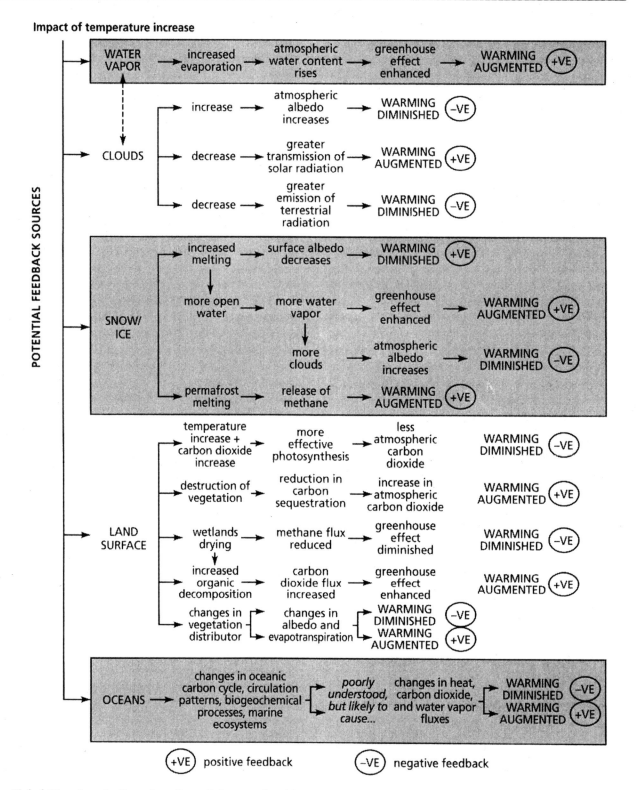

**Global Warming: An Overview. FIGURE 5.** Sources of Positive and Negative Feedback Mechanisms Initiated by a Temperature Increase.

Such mechanisms cannot easily be incorporated into current climate models, but they are essential for the accurate prediction of climate.

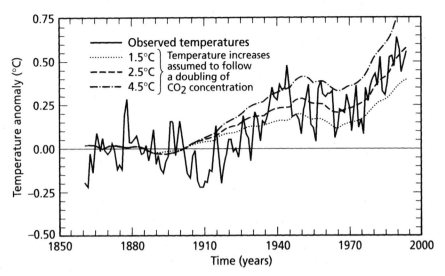

**Global Warming: An Overview. FIGURE 6.** Comparison of Observed Global Surface Temperature Change over the Past Century with Estimates Provided by Climate Models.

The curves provided by the models are smoother than that from the direct observations because of the simplification required by the models, but there is general agreement in the direction and range of the change. (After Wigley et al., 1997. Copyright 1997, National Academy of Sciences, U.S.A.)

research into climate change. Its Forum on Global Change Modeling attempted to assess the quality of the predictions produced by existing climate models and produced a qualitative classification that grouped the predictions into categories listed as virtually certain, very probable, probable, and uncertain. Of the predictions directly involving temperature, only one was considered virtually certain—the indication of a large stratospheric cooling occurring as a result of the increase in carbon dioxide concentration and ozone depletion. The start of the cooling has already been observed in the upper stratosphere. The Forum considered it very probable that global mean surface temperatures would increase by between 0.5° and 2°C above 1990 values by the middle of the twenty-first century, and that the consequence of a doubling of carbon dioxide would be a warming of 1.5°–4.5°C, with 2.5°C being the most probable estimate. This compares closely with a range of 1°–3.5°C and a best estimate of 2°C calculated for IPCC's second Scientific Assessment.

All of these increases represent global mean values. The USGCRP Forum recognized that regional warming was also to be expected, and cited wintertime warming over the arctic land areas as very probable. Regional warming over the North Atlantic and Antarctic ocean areas, although expected, was considered less certain, being designated only probable, and likely to be slower than the global average rate.

In almost all cases in which uncertainty remains high, the problems arise from the inability of models to simulate short-term temporal or geographic changes accurately. Changes in interannual variability depend upon such factors as midlatitude circulation patterns, El Niño–Southern Oscillation events, small-scale convective activity, and land–ocean temperature differences. Models are as yet unable to deal with the short-term complexity of these issues and generally cannot be expected to provide detailed estimates of changes on a time scale less than several decades. Similarly, although it is evident that regional-scale changes will differ from the global average, current models do not operate at a spatial scale that will allow projections of regional responses to change. These results tend to match the IPCC conclusion that models simulate large-scale features of the climate system reasonably well, but that discrepancies become much more common when regional-scale features are introduced (Figure 8).

Much of the effort in the study of global warming has been expended in predicting future change through the use of a variety of GCMs. In the 1980s and 1990s, models have become increasingly sophisticated and confidence in their predictions has increased. Despite this, all models represent a compromise between the complexities of the Earth–atmosphere system and the constraints imposed by such factors as data availability, computer size and speed, and the cost of model development and operation, which limit the accuracy of the simulations. Regional patterns of warming are poorly represented by global GCMs because of the limited spatial resolution allowed by such compromise. Inadequacies in the representation of complex climatic processes such as feedbacks involving clouds, oceans, sea ice and vegetation also limit the quality of simulations and require continued research. As the quality of the models

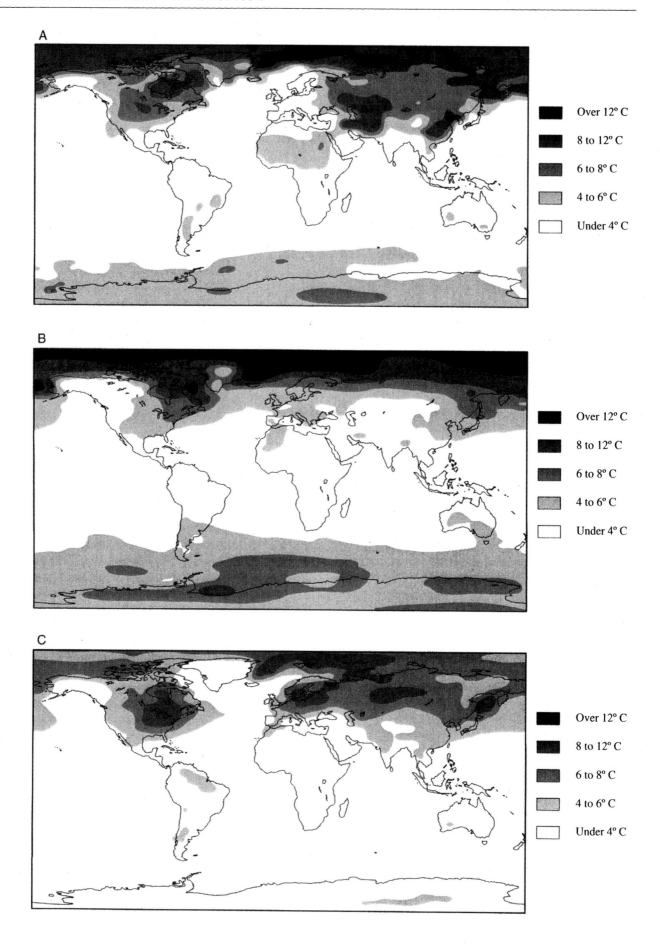

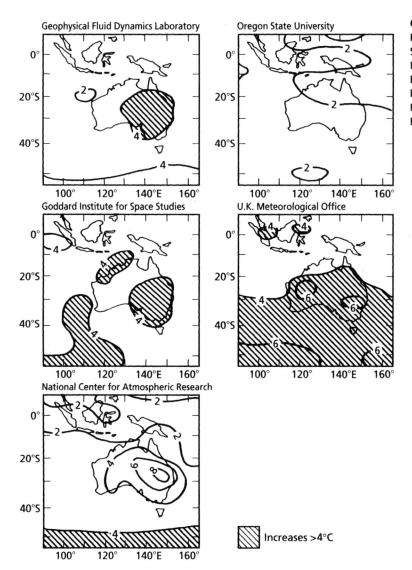

**Global Warming: An Overview. FIGURE 8.** Regional Projections of Climate Change Showing Different Estimates of Temperature Increases over Australia and Indonesia during December, January, and February following a Doubling of Carbon Dioxide. (After Henderson-Sellers, 1991. With permission of Edward Arnold Publishers.)

improves, it is important to continue the collection and analysis of preindustrial and paleoclimatic records, particularly those that provide information on system variables such as solar output, energy fluxes, hydrologic cycles, and land surface ecosystem changes. This will allow model testing against past conditions at both the global and the regional levels, providing an indication of the natural variability that must be taken into account when predicting future anthropogenically instigated changes. [*See* Modeling of Natural Systems.]

**Global Warming: An Overview. FIGURE 7.** Changes in December, January, and February surface air temperature due to a doubling of carbon dioxide as projected by three high-resolution models: (A) Canadian Climate Centre Model, (B) Geophysical Fluid Dynamics Laboratory Model, and (C) United Kingdom Meteorological Office Model. (After Intergovernmental Panel on Climate Change, 1990, p. 165. With permission of Cambridge University Press.)

Even with these uncertainties resolved, various non-climatic factors still have the potential to disrupt projections of change. Greenhouse gas emission rates, for example, often depend upon socioeconomic factors that may significantly alter the rates and mix of gases released over the period covered by a typical model run. In theory, the requirements of the Framework Convention on Climate Change (FCCC) will allow better representation of future emissions, but international agreements are not always easily achieved and can be difficult to enforce. [*See* Framework Convention on Climate Change.] Similarly, most simulations take into account the direct and indirect effects of anthropogenic aerosols. In terms of total volume, however, the aerosols produced by human activity cannot compare with the quantities produced by natural phenomena, such as volcanic activity. Major eruptions such as that of Mount Pinatubo have had a measurable impact on global warming, yet they happen with little or no warning and cannot therefore be fully accommodated in existing simulations.

Widely recognized as a major environmental issue, global warming enjoys a high profile both in the media and in the scientific community. It is central to the FCCC, for example, and receives ongoing attention from the Conference of the Parties, set up to deal with the provisions of the Convention. The Earth's greenhouse effect has been intensifying since the latter part of the nineteenth century, apparently as a result of human activities that have increased greenhouse gas concentrations. However, a 1989 survey of environmental scientists involved in the study of climate change revealed that more than 60 percent of those questioned were not completely confident that the current warming was beyond the range of normal variations in global temperatures. By 1995, the scientists of the IPCC had made sufficient progress in their attempts to distinguish between natural and anthropogenic influences on climate that they were able to conclude that there was a discernible human influence in recent climate change.

Research into global warming is continuing at a high level, and many of the uncertainties that reduce the quality of current predictions are being addressed. Even as better models are developed, however, and confidence in their results improves, there will be surprises associated mainly with the natural variability of the Earth–atmosphere system. Any decisions or proposals based on model projections of global warming will have to take that uncertainty into consideration.

[*See also* Aerosols; Atmosphere Dynamics; Atmospheric Chemistry; Carbon Dioxide; Catastrophist–Cornucopian Debate; Gaia Hypothesis; Greenhouse Effect; Sea Level; *and* Water Vapor.]

## BIBLIOGRAPHY

GLOBAL WARMING

Ball, T. "Historical Evidence and Climatic Implications of a Shift in the Boreal Forest–Tundra Transition in Central Canada." *Climatic Change* 8 (1986), 121–134. An example of the use of paleobotanical and historical proxy data for the estimation of climate change.

Barnola, J. M., D. Raynaud, Y. S. Korotkevitch, and C. Lorius. "Vostok Ice Core: A 160,000 Year Record of Atmospheric CO$_2$." *Nature* 329 (1987), 408–414. Results from the deepest and probably the best known of the Antarctic ice cores.

Blake, D. R., and F. S. Rowland. "Continuing Worldwide Increase in Tropospheric Methane." *Science* 293 (1988), 1129–1131.

Bolin, B. "On the Exchange of Carbon Dioxide between the Environment and the Sea." *Tellus* 12 (1960), 247–281.

Bradley, R. S., and P. D. Jones. *Climate Since A.D. 1500.* London: Routledge, 1992. A combination of records from paleobotanical, physical, historical, and observational sources.

British Meterological Office. *Climate Change and Its Impacts.* U.K. Department of the Environment, Transport and the Regions, October 1999.

Callendar, G. S. "The Artificial Production of Carbon Dioxide and its Influence on Temperature." *Quarterly Journal of the Royal Meteorological Society* 64 (1938), 223–237. Paper by the first of the modern researchers to link atmospheric carbon dioxide and climate change.

Charlson, R. J., S. E. Schwartz, J. M. Hales, R. D. Cess, J. A. Coakley, J. E. Hansen, and D. J. Hofmann. "Climate Forcing by Anthropogenic Aerosols." *Science* 255 (1992), 423–430.

Crane, A., and P. Liss. "Carbon Dioxide, Climate and the Sea." *New Scientist* 108 (1985), 50–54.

Cook, E. R. "Temperature Histories from Tree Rings and Corals." *Climate Dynamics* 11 (1995), 211–222. An examination of the ways in which temperature records can be derived from proxy data.

Gates, W. L. "AMIP: The Atmospheric Model Intercomparison Project." *Bulletin of the American Meteorological Society* 73 (1992), 1962–1970. An assessment of the ability of different models to simulate existing observed climates.

Genthon, C., J. M. Barnola, D. Reynaud, C. Lovius, J. Jouzel, N. I. Barkov, Y. S. Korokevich, and V. M. Kotlyakov. "Vostok Ice Core: Climate Resonse to CO$_2$ and Orbital Forcing Changes over the Last Climatic Cycle." *Nature* 329 (1987), 414–418.

Giorgi, F., C. S. Brodeur, and G. T. Bates. "Regional Climate Change Scenarios over the United States Produced with a Nested Regional Climate Model." *Journal of Climate* 7 (1994), 375–399. An example of regional-scale climate projections using a finer-resolution model nested or embedded in a global model.

Grove, J. M. *The Little Ice Age.* London: Routledge, 1988. The definitive work on the Little Ice Age, synthesizing evidence from every region of the world.

Grove, J. M., and R. Switsur. "Glacial Geological Evidence for the Medieval Warm Period." *Climatic Change* 26 (1994), 143–169.

Hansen, J., and S. Lebedeff. "Global Surface Air Temperatures: Update through 1987." *Geophysical Research Letters* 15 (1988), 323–326.

Hansen, J., H. Wilson, M. Sato, R. Ruedy, K. Shah, and E. Hansen. "Satellite and Surface Data at Odds?" *Climatic Change* 30 (1995), 103–117. An examination of the nature and causes of differences in surface and lower tropospheric temperature data.

Henderson-Sellers, A. "Global Climate Change: The Difficulties of Assessing Impacts." *Australian Geographical Studies* 29 (1991), 202–225.

Hengeveld, H. G. *Understanding Climate Change.* SOE Report 91–2. Ottawa: Environment Canada, 1991. A well-written account of the basics of current climate change for the nonexpert.

Hughes, M., and H. F. Diaz. "Was there a Medieval Warm Period, and if so, Where and When?" *Climatic Change* 26 (1994), 109–142.

Hulme, M. "Global Warming." *Progress in Physical Geography* 21.3 (1997), 446–453. One of a continuing series of reports on the development of studies in global warming.

Intergovernmental Panel on Climate Change. *Climate Change: The IPCC Scientific Assessment.* Cambridge: Cambridge University Press, 1990. The original attempt by the IPCC to assess available scientific information on climate change.

———. *Climate Change 1992: The Supplementary Report to the IPCC Scientific Assessment.* Cambridge: Cambridge University Press, 1992. An update of the 1990 report that generally confirmed the original findings.

———. *Climate Change 1994: Radiative Forcing of Climate Change and an Evaluation of the IPCC IS92 Emission Scenarios.* Cambridge: Cambridge University Press, 1995.

———. *Climate Change 1995: The Science of Climate Change.*

Cambridge: Cambridge University Press, 1996. A second comprehensive report made necessary by the rapid accumulation of data in the field of climate change.

Jones, M. D. H., and A. Henderson-Sellers. "History of the Greenhouse Effect." *Progress in Physical Geography* 14 (1990), 1–18.

Jones, P. D., P. M. Kelly, G. B. Goodess, and T. R. Karl. "The Effect of Urban Warming on the Northern Hemisphere Temperature Average." *Journal of Climate* 2 (1989), 285–290.

Kacholia, K., and R. A. Reck. "Comparison of Global Climate Change Simulations for a $2\times CO_2$-Induced Warming." *Climatic Change* 35 (1997), 53–69. An intercomparison of 108 temperature-change projections published between 1980 and 1985.

Lamb, H. H. *Climate, History and the Modern World.* 2d ed. London: Routledge, 1996. An authoritative examination of past climate records, how they were reconstructed, the causes of climatic variation, and its impact on human affairs.

Lorius, C., J. Jouzel, C. Ritz, I. Merlivat, N. I. Barkov, Y. S. Korotkevich, and V. M. Kotlyakov. "A 150,000 Year Climatic Record from Antarctic Ice." *Nature* 316 (1985), 591–596.

Parker, D. E., P. D. Jones, A. Bevan, and C. K. Folland. "Interdecadal Changes of Surface Temperature since the Late 19th Century." *Journal of Geophysical Research* 99 (1994), 14373–14399.

Slade, D. H. "A Survey of Informed Opinion Regarding the Nature and Reality of Global Greenhouse Warming." *Climatic Change* 16 (1990), 1–4. A survey indicating that, as late as the beginning of the decade, as many as 60 percent of environmental scientists involved in the study of climate change were not confident that the current global warming was beyond the normal range of global temperature variability.

Spenser, R. W., and J. R. Christy. "Precision Lower Stratospheric Temperature Monitoring with the MSU: Validation and Results 1979–91." *Journal of Climate* 6 (1993), 1194–1204. One of a series of papers providing upper atmospheric temperature data derived from satellite observations.

U.S. Global Change Research Program. *Forum on Global Change Modeling.* Washington, D.C.: USGCRP, 1995. An attempt to assess the quality of the predictions produced by existing climate models.

———. *Our Changing Planet.* Washington, D.C.: USGCRP, 1998.

Wigley, T. M. L., P. D. Jones, and S. C. B. Raper. "The Observed Global Warming Record. What Does It Tell Us?" *Proceedings of the National Academy of Sciences* 94 (1997), 8314–8320.

GLOBAL HOT SPOTS

Environment Canada. *Mackenzie Basin Impact Study.* Ottawa, 1997. Further reading on the Mackenzie Basin hot spot.

Grescoe, T. "Temperature Rising." *Canadian Geographic* (November–December 1998), 36–44.

Intergovernmental Panel on Climate Change (IPCC). *Aviation and the Global Atmosphere.* Cambridge: Cambridge University Press, 1999.

—DAVID D. KEMP

## Global Warming Potential

A Global Warming Potential (GWP) is an index that has been created in an attempt to compare quantitatively the warming effect of a given greenhouse gas (GHG) relative to a reference GHG. The most commonly used reference gas is carbon dioxide. Different GHGs differ in their effects on global warming for two main reasons: (1) different gases have a different inherent ability to alter the Earth's radiation balance (by absorbing infrared radiation in the case of GHGs), and (2) different GHGs, once emitted into atmosphere, will be removed from the atmosphere at different rates. Consequently, the relative heat-trapping ability (or radiative forcing) of two different gases that are emitted into the atmosphere today will change over time. The challenge motivating the development of GWPs is to come up with a single number or index that combines these two factors.

The most commonly used method of computing GWPs is to consider a pulse emission at time $t = 0$ of equal masses of the gas under consideration and of the reference gas. These gases have heat-trapping abilities per unit mass of $f_i(t)$ and $f_r(t)$, respectively, and the amounts remaining in the atmosphere change over time by $C_i(t)$ and $C_r(t)$, respectively. The heat trapping by a given amount of a given gas depends on the preexisting concentration of the gas, which can vary with time, so both $f_i$ and $f_r$ are written as functions of time. The relative heat trapping for the two gases at time $t$ is given by

$$(f_i(t) \times C_i(t)) / (f_r(t) \times C_r(t)).$$

Rather than comparing this ratio at some single time in the future, the GWP is based on the summation or integral of this ratio from the moment the pulses are emitted ($t = 0$) up to some arbitrary time horizon $T$. That is,

$$GWP_i(T) = \frac{\int_0^T f_i(t)C_i(t)dt}{\int_0^T f_r(t)C_r(t)dt} \qquad (1)$$

where the GPW is written as $GWP_i(T)$ because it depends on the gas under consideration and on the time horizon $T$.

The main purpose of GWPs is to compare the effect of a given emission of a given GHG with that of carbon dioxide, which accounts for over half of the enhancement in the natural greenhouse effect. Carbon dioxide is therefore the natural reference gas, but this raises a major difficulty for GWPs. The removal of carbon dioxide initially proceeds quickly, then more slowly, and, for all practical purposes, some of the emitted carbon dioxide is never removed from the atmosphere. Ideally, one would compute GWPs by integrating over all time (that is, by letting $T$ go to infinity), but this is not possible with carbon dioxide as the reference gas because some of the emitted carbon dioxide is never removed from the atmosphere. [*See* Impulse Response.] The GWP must then, of necessity, be computed over some finite time horizon. For gases that are removed more quickly than carbon dioxide, the GWP will be smaller the longer the time horizon, while for gases that are removed more slowly than the initial rate of removal of carbon dioxide, the GWP will be larger the longer the time horizon. Whatever time horizon is chosen for the GWP, the computed GWP will not accurately reflect the relative warming effects of different gases for other time horizons.

The Kyoto Protocol places restrictions on a basket of six gases or groups of gases: carbon dioxide ($CO_2$), methane ($CH_4$), nitrous oxide ($N_2O$), sulfur hexafluoride ($SF_6$), the HFCs (hydrofluorocarbons), and the PFCs (perfluorocarbons). These gases all have a sufficiently long lifetime that they can spread throughout the atmosphere (they are said to be "well mixed"). Table 1 shows the GWP as computed by the IPCC for time horizons of twenty, one hundred, and five hundred years. Also shown are the two ingredients that go into the computation of the GWPs—the heat-trapping ability on a mass basis and the atmospheric lifetime. Some gases, such as sulphur hexafluoride, are extraordinary powerful GHGs compared with carbon dioxide and have a very long atmospheric lifetime, resulting in a very large GWP. The Kyoto Protocol will use GWPs as computed by the IPCC for a time horizon of one hundred years. [See Kyoto Protocol.]

There are a number of significant uncertainties associated with GWPs: (1) the direct radiative forcing of some gases (the HFCs in particular) is still uncertain by up to ±30 percent, (2) some gases (the halocarbons and methane in particular) have important indirect radiative forcings that are very difficult to compute accurately, (3) the atmospheric lifetimes of some of the gases that are being compared with carbon dioxide are still uncertain and (in the case of methane) are expected to change over time, and (4) changes in climate could cause important changes in the net rate of removal of carbon dioxide from the atmosphere, and this would alter the computed GWPs for all gases. Other uncertainties, such as the assumed future concentrations of the GHGs (which alter the $f_i(t)$ and $f_r(t)$), are less important. The use of the GWP (as defined above) contains a number of implicit assumptions as well. Most importantly, it is assumed that the climatic effect of a given GHG is directly proportional to its radiative forcing or, in other words, that the climate sensitivity is the same for all gases. This assumption is valid to within ±20 percent when comparing the GHGs that are covered by the Kyoto Protocol. The concept of a GWP is completely inapplicable to gases such as carbon monoxide and nitrogen oxides (which affect climate through their role in the production of tropospheric ozone) and to aerosols, since the climate sensitivity to ozone and aerosol radiative forcing appears to be quite different from that for well-mixed GHGs.

A much less problematic application of GWPs is in comparing the impact on future climatic change of alternative replacements for CFC-11 and CFC-12. The CFCs are powerful GHGs but are being phased out because of their impact on stratospheric ozone. HFC replacements have no effect on the ozone layer, but most are powerful GHGs. In this case, CFC-11 can be used as a reference gas and the time horizon $T$ can be taken to infinity. The difficulties involving carbon dioxide as a reference gas can thus be avoided.

To summarize, the concept of GWP is an attempt to force the complexities of nature into a single number. This is not scientifically justifiable. However, the political process has demanded a single number to intercompare different GHGs. This is because current international agreements have ignored scientific reality by focusing on a "basket" of GHGs rather than by framing gas-by-gas restrictions. The latter would eliminate altogether the most problematic applications of GWPs.

[See also Carbon Dioxide; Chlorofluorocarbons; Global Warming; Greenhouse Effect; Methane; Nitrous Oxide; and Ozone.]

**Global Warming: Global Warming Potential. TABLE 1.** Relative Heat-Trapping Ability on a Mass Basis, Atmospheric Lifetimes, and GWPs for 20, 100, and 500-Year Time Horizons for Selected Gases Covered by the Kyoto Protocol

| GAS | HEATING TRAPPING COMPARED TO CARBON DIOXIDE PER UNIT MASS, (°F) | ATMOSPHERIC LIFESPAN (YEARS) | Global warming potential time horizon | | |
|---|---|---|---|---|---|
| | | | 20 YEARS | 100 YEARS | 500 YEARS |
| $CH_4$ | 58* | 12.3 ± 3 | 56 | 21 | 6.5 |
| $N_2O$ | 206* | 120 | 280 | 310 | 170 |
| $SF_6$ | 10,900† | 3,200 | 16,300 | 23,900 | 34,900 |
| HFC-23 | 6,310† | 264 | 9,100 | 11,700 | 9,800 |
| HFC-32 | 5,240‡ | 5.6 | 2,100 | 650 | 200 |
| HFC-134 | 4,290† | 10.6 | 2,900 | 1,000 | 310 |
| $CF_4$ | 2,900‡ | 50,000 | 4,400 | 6,500 | 10,000 |
| $C_2F_6$ | 4,130‡ | 10,000 | 6,200 | 9,200 | 14,000 |

Sources for (°F):

*Shine et al. (1995, Table 4.2).

†Shine et al. (1995, Table 4.2) for the CFC-11:$CO_2$ ratio (3,970); and Shine et al. (1995, Table 4.3) for the gas:CFC-11 ratio.

‡Shine et al. (1995, Table 4.2) to get the CFC-11:$CO_2$ (3,970); and Schimel et al. (1996, Table 2.7) to get the gas:CFC-11 ratio.

## BIBLIOGRAPHY

Harvey, L. D. D. "A Guide to Global Warming Potentials." *Energy Policy* 21 (1993), 24–34.

Schimel, D. "Radiative Forcing of Climate Change." In *Climate Change 1995: The Science of Climate Change*, edited by J. T. Houghton, L. G. M. Filho, B. A. Callander, N. Harris, A. Kattenberg, and K. Maskell, pp. 65–131. Cambridge: Cambridge University Press, 1996.

Shine, K. P., Y. Fouquart, V. Ramaswamy, S. Solomon, and J. Srinivasan. "Radiative Forcing." In *Climate Change 1994: Radiative Forcing of Climate Change and an Evaluation of the IPCC IS92 Emission Scenarios*, edited by J. T. Houghton, L. G. Meira Filho, J. Bruce, H. Lee, B. A. Callander, E. Haites, N. H. Harris, and K. Maskell, pp. 163–203. Cambridge: Cambridge University Press, 1995.

Smith, S. J., and T. M. L. Wigley. "Global Warming Potentials. 1. Climatic Implications of Emissions Reductions." *Climatic Change* 44 (2000a), 445–457.

———. "Global Warming Potentials. 2. Accuracy." *Climatic Change* 44 (2000b), 459–469.

—L. D. DANNY HARVEY

# GRASSLANDS

Temperate grasslands occur in middle-latitude regions where seasonal climate favors the dominance of perennial grasses. The most extensive area is the prairies of North America, which originally covered more than 350 million hectares of the Central Lowlands (Figure 1). In Eurasia the steppes cover some 250 million hectares of rolling plains that extend from Hungary to Manchuria. Temperate grasslands are represented in the Southern Hemisphere by the pampas of Argentina, Uruguay, and southeastern Brazil (70 million hectares); smaller areas are found in the drier parts of New Zealand (0.6 million hectares) with occasional patches in southeastern Australia. The veld of the high plains of southern Africa is also included in this biome. Vegetation cover in the temperate grasslands is relatively homogeneous, but important floristic and structural differences occur in response to regional and local conditions. Tall, perennial, sod-forming grasses are the dominant species in moister grasslands, but in drier sites they are replaced by shorter bunch grasses and annuals. Similarly, warm-season grasses are less abundant than the cool-season grasses at higher latitudes. Sedges and broad-leaved forbs are common associates, and in some areas woody plants are conspicuous. Much of the grasslands has been broken for cereals and other crops, while areas that remain in pasture have typically been improved by the introduction of more palatable grasses and legumes. Although native species are still present along roadsides and in areas that have not been plowed, they are subject to very different environmental forces compared to presettlement conditions.

**North America Formation.** The main grassland region in North America extends through the Central Lowlands from southern Canada into northern Mexico. Elevations generally range from two hundred to twelve hundred meters. The climate is one of recurring drought, and differences in the amount and reliability of precipitation are reflected in the diversity of the plant cover. Precipitation comes mostly from cyclonic storms and is usually greatest in the early summer. Later in the summer, thunderstorms are common, often accompanied by hail and, less frequently, by tornadoes. Most of the winter precipitation falls as snow. Precipitation decreases northward and westward, with values ranging from approximately 350 to 1,200 millimeters. More than 80 percent of the precipitation is lost through evaporation, and by the end of the growing season soil moisture levels are often below 25 percent of field capacity. Mean temperatures in January range from 5 to $-25°C$, rising to 18 to 28°C in July, and results in frost-free periods of about 250 days in the south and about 100 days in the north. Floristic diversity increases toward the south, where the growing season is longer, and to the east, where precipitation is higher. Structural changes are also apparent, with tall-grass, mixed-grass, and short-grass prairies distinguished according to the growth habits of the dominant grasses. Representative species of the eastern tall-grass prairies include big bluestem (*Andropogon gerardii*), Indian grass (*Sorghastrum nutans*), and switch grass (*Panicum virgatum*). The tall grasses exceed two meters in height by late summer, when they come into flower; they are deep-rooting, sod-forming species that typically spread by rhizomes and form a densely matted layer ten to fifteen centimeters below the soil surface (Kucera, 1992). Mixed-grass prairie occurs in central Canada and the northern Great Plains. Western wheat grass (*Agropyron smithii*) is a common sod-forming species in this region, while bunch grasses, such as needle-and-thread grass (*Stipa comata*), which propogate by tillering, are widespread. The mixed prairie canopy may reach one meter by the end of the growing season (Coupland, 1992). Drought-tolerant short grasses, including blue grama (*Bouteloua gracilis*) and buffalo grass (*Buchloe dactyloides*), become progressively more abundant in the western grasslands. Most species are shallow-rooting bunch grasses that reach twenty to fifty centimeters in height by the time growth ceases in late summer (Lauenroth and Milchunas, 1992).

**The Eurasian Formation.** The grasslands of Eurasia comprise four zonal types that reflect latitudinal gradients in aridity, temperature, and length of the frost-free period. Further differentiation on the basis of longitude incorporates the increasingly severe climate of the continental interior (Lavrenko and Karamysheva, 1993). Precipitation ranges from three hundred to six hundred millimeters in the western steppes, where approximately 25 percent comes as snow, which lies on the

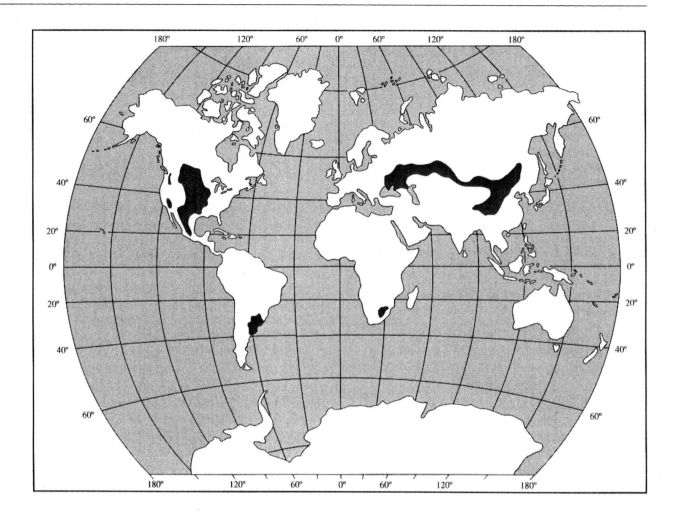

**Grasslands. FIGURE 1.** Worldwide Grasslands Distribution. (After Archibold, 1995, p. 204, Figure 7.1. With permission of Kluwer Academic Publishers.)

ground for 100–140 days. East of the Urals, annual precipitation decreases from about four hundred millimeters in the northern regions to two hundred millimeters in the south. Mean summer temperatures range from 18 to 24°C with daytime maxima often near 40°C; in winter, temperatures in the continental interior may drop to −50°C. The semihumid forest-steppe (also termed meadow-steppe) forms the northernmost zone; it is dominated by taller bunch grasses such as *Stipa joannis* and *S. pennata*, sedges (*Carex* spp.), and forbs, which flower mainly in the early summer. Groves of oaks (*Quercus robur*) occur in the western areas, while to the east birch (*Betula verrucosa*) and aspen (*Populus tremula*) are present. Further south is the semiarid true steppe, a bunch-grass formation comprising densely rooted, xerophillous grasses, including *Stipa lessingia* and *Stipa pulcherrima*, interspersed with occasional forbs and shrubs. Dwarf shrubs, such as worm-

wood (*Artemisia maritima*), increase in abundance in the drier areas and together with grasses (mainly *Stipa capillata* and *Festuca sulcata*) provide a sparse cover that stretches discontinuously from the Black Sea to the margins of the Gobi Desert. In China the main region of grassland occurs in Inner Mongolia and extends eastward into Heilongjiang, Jilin, and Liaoning Provinces (Zhu, 1993). The humid northern areas support a mixture of bunch grasses (*Stipa baicalensis, Filifolium sibiricum*) and rhizomatous species (*Leymus chinensis, Arundinella hirta*), growing to a height of sixty centimeters with occasional patches of halophytic meadow composed of *Polygonum sibiricum, Pucinellia tenuflora*, and *Suaeda corniculata*. In drier areas the cover changes to dry steppe dominated by species of *Stipa* and other caespitose grasses, together with forbs, legumes, and small shrubs. Where precipitation is below three hundred millimeters, desert steppe predominates. Here the drought-resistant grasses (*Stipa brevifolia, S. glareosa, S. gobica*) and shrubs (*Artemisia xerophytica, Caragana microphylla*) rarely exceed thirty centimeters and 30 percent cover. In the mountainous areas of the continental interior steppe, vegetation occurs at elevations from nine hundred to two thousand meters in response to increased orographic precipitation.

**The Grasslands of the Southern Hemisphere.** The pampas of South America is found in the undulating lowland around the Rio de la Plata, with regional subdivisions recognized on the basis of physiography, drainage, and soils (Soriano, 1992). Winter temperatures average 7°C with occasional frosts and in summer rise to about 22°C. Precipitation falls mostly in the spring and autumn and decreases from fifteen hundred millimeters in the northwest to less than four hundred millimeters in the southeast, where summer droughts can limit productivity. Tall and medium bunch grasses, including species of *Stipa*, *Poa*, *Piptochaetium*, and *Aristida*, dominate the plant cover. Numerous herbs and sedges are interspersed with the grasses, but shrubs are poorly represented. In New Zealand the temperate tussock grasslands occur in the rain shadow of the Southern Alps from sea level to about fifteen hundred meters. Precipitation generally exceeds eight hundred millimeters, but water demand is high, and droughts can limit productivity in low-elevation sites, where summer temperatures occasionally reach 40°C. Winter temperatures rarely fall below −15°C even at high elevations. The dominant grasses are xerophytic bunch grasses, but distinct types are recognized according to physiognomy and species composition (Mark, 1992). The short-tussock grasslands occur in lowland areas from sea level to one thousand meters. The dominant species are *Festuca nova-zealandica*, *Poa cita*, and *Elymus rectisetus*, which grow about 0.6 meter tall in association with many herbaceous and woody species. The upland tall-tussock grasslands are dominated by alpine species of *Chionochloa*, which grow from 0.1 to 1.5 meters tall. *C. rigida* is the most widespread species near tree line, where it is accompanied by shrubs and tall forbs. At higher elevations it is replaced by *C. crassiuscula* and *C. oreophila*. In the wetter parts of Australia, small patches of tall-grass communities are distributed throughout the temperate woodlands of New South Wales and Victoria, where they are mainly associated with hilltops or frost hollows. These communities typically consist of a low-density cover of *Themeda australis*, *Poa labillardieri*, and species of *Dichelachne*. Areas with annual precipitation of about five hundred millimeters support temperate short-grass communities dominated by species of *Danthonia* and *Stipa* (Moore, 1992). In southern Africa, the high veld occurs on gently undulating plains and high plateaus at elevations near fifteen hundred meters. Annual precipitation is generally above 750 millimeters, with a summer maximum, and mean temperatures range from about 5°C in winter to 20°C in summer. The plant cover is dominated by tussock grasses growing to a height of eighty centimeters. Common species include *Themeda triandra*, *Hetero-*

*pogon contortus*, *Cymbopogon plurinoides*, and *Digitaria eriantha*, together with cool-season species of *Bromus* and *Koeleria* and forbs such as *Helichrysum* and *Senecio* species (Tainton and Walker, 1992).

**Evolution of the Grasslands.** The flora of the North American grasslands has few endemic species, which suggests that it has developed comparatively recently and attained its present distribution in postglacial times (Axelrod, 1985). Abundant fossil grasses and herbs have been found in Miocene (5–24 million years BP) deposits together with the remains of hardwood species, indicating that the region supported a woodland cover 10–12 million years BP. The drier climate of the Pliocene (5–6 million years BP) favored expansion of the grasslands, but during the Pleistocene spruce forest was predominant. The forests persisted until 10–12,000 BP on the northern Great Plains, with grasslands reestablishing about 9,500 years BP. The Eurasian grasslands supported wooded steppe prior to the Pleistocene, with patches of steppe grassland further south. Drier conditions during periods of glacial advance favored expansion of the short-grass and wooded steppe formations, and by the end of the Pleistocene, this type of cover extended from northern France to the plains of Manchuria (Frenzel, 1968). In New Zealand, grasslands persisted in the South Island during glacial maxima and were replaced by forests as the climate improved. Subsequent development of grasslands is attributed mainly to fires set by the moa hunters. Paleobotanic evidence is not available for the pampas, but vertebrate fossils suggest that the climate during the Pleistocene was similar to present. In southern Africa the climate of the high plateaus is thought to have been wet and cold during periods of ice advance with temperatures 8°–10°C lower than present. This favored the spread of woodland, alternating with semidesert vegetation during interglacials. Here the temperate grasslands appear to have developed about 4,600 years BP. [*See* Climate Change.]

**Human Impact on Grasslands.** Very little of the temperate grassland biome remains in its natural state. In North America substantial changes in the grasslands occurred as European settlement expanded westward across the continent in the mid-1800s. The most direct effects included plowing and seeding, the alteration of grazing pressures as the bison herds were extirpated, and the introduction of a new fire regime. Much of the region is now under crop production, and areas that remain in pasture have generally been altered by various range management techniques to improve quality and resilience. Considerable changes in the composition of grasslands have occurred through widespread use of introduced species such as smooth brome (*Bromus inermis*) and crested wheatgrass (*Agropyron cristatum*), which are native to Eurasia. These species quickly replaced native species. Consequently, the present grazing lands in North America bear only a superficial resemblance to the native prairie. Similarly, the Eurasian steppes have undergone significant transformation through human activities. Much of the European steppe has been converted to cropland and pasture. Less disturbance has occurred in the Mongolian steppes, where burning and mowing are uncommon and grazing is moderate, while in China natural steppe vegetation is now restricted to refuges that represent 2–3 percent of the former area. In South America the natural population of grazing animals was small, but horses and cattle, which arrived with the Spaniards in the sixteenth century, rapidly increased in numbers after a few escaped. Sheep production further intensified the use of the grasslands. Introduced pasture species (e.g., *Lolium multiflorum*, *Trifolium repens*, and *Medicago* spp.) and infestation by many aggressive weeds such as thistles (e.g., *Cirsium vulgare*, *Carduus acanthoides*) have greatly altered the plant cover, while competition from animals has severely reduced the native populations of guanaco, deer, and rheas. Little indigenous grassland remains in New Zealand. Intensive grazing by sheep and other livestock and use of fire to encourage more palatable regrowth have affected the structure and composition of the grasslands. In relatively inaccessible areas pasture improvement has been achieved through aerial applications of fertilizer and seed of exotic grasses and legumes. The introduction of game animals such as red deer (*Cervus elaphus*) and chamois (*Rupicapra rupicapra*) has also affected the grassland ecosystem, especially through intense competition with the native avian grazers such as the takahe (*Notornis mantelli*) and kakapo (*Strigops habroptilus*). Australian grasslands support a large population of native herbivores, notably kangaroos and wallabies. Evidence suggests that these have increased in numbers and their ranges have changed as a result of modification of the grasslands by domestic livestock. Although efforts to control their numbers increase in times of drought, the native animals do not compete vigorously with livestock. Here the main scourge of the temperate grasslands is the rabbit (*Oryctolagus cuniculus*). In southern Africa human impact in the grasslands has been through sowing grasses (*Lolium* spp., *Festuca* spp., and *Dactylis* spp.) and legumes (*Trifolium pratense*, *T. repens*) and through frequent burning, especially in the wetter "sour" grasslands, to remove old growth. As in North America, the native herbivore population has been reduced considerably, and present grazing activity by nonmigratory livestock with different food preferences has introduced additional selective pressures on these ecosystems. [*See* Animal Husbandry.]

**Effects of Grazing on Rangeland Condition.** In natural ecosystems grazing animals respond to changes in forage supply through migration and population cycles. Animal densities also are regulated by climatic perturbations such as droughts and severe winter storms.

Rangeland soils and vegetation are relatively stable under moderate grazing, and forage quality improves if livestock are removed periodically. Overstocking was widespread during the early ranching period in North America, but following the devastating impact of the drought years of the 1930s there have been considerable advances in range management and conservation (Kothman, 1995). The plants available to grazing animals range from those that are highly palatable to those that are strictly avoided. The palatable species are most severely injured, and continual grazing can reduce the quality of the forage when these are replaced by less desirable plants. Species that decline in abundance as a result of grazing are termed decreasers (Dyksterhuis, 1949); increasers become more abundant, but eventually these too may be replaced by unpalatable invaders in severely overgrazed areas. The quality of the grassland is therefore related to its floristic composition rather than the amount of growth that occurs in a given season. In undulating areas the effect of grazing is not uniform. Moister lowland areas are generally more resilient than hillside and drier upland sites. Forage production is normally reduced by grazing, although an increase in aboveground production of 25–30 percent has been noted in grazed tall-grass prairie; this is due to less shade from dead shoots in the canopy.

All rangeland eventually deteriorates under severe and prolonged grazing. Soil moisture can be reduced in heavily grazed pastures as the soil becomes compacted by the animals. Infiltration is further reduced by the loss of actively growing roots and litter. Runoff rates can increase under poor grazing management and thus contribute to accelerated soil erosion. Despite considerable losses of water through interception and transpiration, soil moisture reserves are usually higher in ungrazed pastures. Similarly, reserves of soil nitrogen and organic matter decrease under grazing because less material is added as surface litter and from roots. Nitrogen added to the soil in animal wastes increases microbial breakdown of organic matter, as do the higher soil temperatures following reduction of the plant cover (Risser et al., 1981); this further reduces the amount of soil litter and increases the possibility of erosion and desiccation. Erosion by wind and water is uncommon in natural grassland, but can be a serious problem in grazed ecosystems and is especially common where land has been broken for crops. Similarly, leaching of soil nutrients is slight in natural grasslands compared to agricultural land. Productivity is often maintained by high inputs of nitrogenous fertilizer rather than use of legumes, and this has increased the likelihood of groundwater contamination in some areas. At the same time, substantial losses of gaseous ammonia and nitrous oxide have been reported from managed grasslands and has raised concerns about air quality and global warming (Jarvis and Pain, 1997).

**Response of Grasslands to Global Warming.** Climate change scenarios predict a rise of 2°–4°C in winter and 2°–3°C in summer over central North America, accompanied by a 0–15 percent increase in winter precipitation and drier summers. This is expected to cause significant extension of grasslands northward into the boreal forest, although drier conditions could also favor the spread of desert vegetation into the short-grass prairies. Similar changes may occur in Eurasia. The link between atmospheric pollution and potential global warming is well known and is attributed to the burning of fossil fuels, forest clearance, and agricultural production, which have increased the atmospheric concentrations of carbon dioxide, methane, and nitrous oxide. Carbon dioxide is expected to account for more than 50 percent of the predicted global warming, but uncertainties in the global carbon budget limit the accuracy of this prediction. Carbon dioxide enrichment of the atmosphere may increase net primary production by 8–40 percent (Gifford, 1994), although the effect varies between different species. $C_4$ (warm-season) grasses appear to be less responsive than $C_3$ (cool-season) species, although increased growth has been reported in well-watered *Andropogon gerardii* in tall-grass prairie (Nie et al., 1993). Thus the predicted spread of $C_4$ species under warmer conditions may be offset by the $C_3$ growth response. Increased nitrogen fixation has been noted in legumes grown under enhanced carbon dioxide levels. However, the accompanying changes in temperature and precipitation may limit gains in production arising from higher carbon dioxide concentrations. The long-term effects of global warming on grasslands are poorly understood. Some believe that large quantities of carbon dioxide will be released to the atmosphere through faster decomposition rates. It is more probable, however, that the carbon-to-nitrogen ratio in plants will widen, thereby reducing decomposition rates and increasing the inactive pool of carbon in the soil. [*See* Global Warming.]

**Grassland Reclamation and Conservation.** Few ecosystems have been altered by human activities as dramatically as the grasslands. The Great Plains ecosystem of North America is considered the most endangered of any on the continent. What grassland remains in this region is fragmented and subject to invasion by aliens and exotics. This has had a profound effect on the native plant and animal populations. Humans have caused the decline or disappearance of several species of grassland mammals through overhunting, deliberate extermination, and habitat modification. The extirpation of keystone species such as the bison and prairie dog (*Cynomys* spp.) has altered the landscape for many other animals. For example, the black-footed ferret (*Mustela nigripes*), a species that is essentially restricted to prairie dog burrows, was reduced to only twelve animals by 1985. The population of black-footed ferret was increased through captive breeding, and appropriately managed prairie dog colonies have been selected for its reintroduction. A similar predicament faced the steppe marmot (*Marmota bobac*) in Eurasia, although its numbers also have recovered as a result of conservation and reintroduction programs. Other North American animals that have suffered from human activities include the large grazing mammals, such as the wapiti (elk), and large predators, including the black bear, gray wolf, and mountain lion. Conversely, the populations of dietary and habitat generalists, such as raccoons, opossums, and coyotes, have increased during the period of settlement. Many prairie bird populations continue to decline. Most of these species are migratory, however, and some of these changes may be attributable to disturbances in their winter habitat outside the grasslands. Invertebrates are equally threatened by habitat loss and competition from indiscriminate use of chemicals. Conservation of remnant grasslands has increased in the past few decades, and many tracts are now preserved as ecological refugia. The creation of the Grasslands National Park in southern Saskatchewan and Alberta in 1999 attests to government efforts to preserve a national legacy. Many ranchers and farmers have become interested in prairie restoration, and several now harvest seed of native species for this purpose. Similarly, native species are used to revegetate pipeline right-of-ways and road corridors. Conservation efforts will continue to require active cooperation of governments, private agencies, and individuals and must necessarily focus on different spatial scales to ensure suitable conditions for all of the threatened grassland species.

[*See also* Biomes; Fire; *and* Savannas.]

## BIBLIOGRAPHY

Axelrod, D. L. "Rise of the Grassland Biome, Central North America." *Botanical Review* 51 (1985), 280–334.

Coupland, R. T. "Mixed Prairie." In *Ecosystems of the World*, vol. 8A, *Natural Grasslands: Introduction and Western Hemisphere*, edited by R. T. Coupland, pp. 151-182. Amsterdam: Elsevier, 1992.

Dyksterhuis, E. J. "Condition and Management of Range Land Based on Quantitiative Ecology." *Journal of Range Management* 2 (1949), 104–115.

Frenzel, B. "The Pleistocene Vegetation of Northern Eurasia." *Science* 161 (1968), 637–649.

Gifford, R. M. "The Global Carbon Cycle: A Viewpoint on the Missing Sink." *Australian Journal of Plant Physiology* 21 (1994), 1–15.

Jarvis, S. C., and B. F. Pain. *Gaseous Nitrogen Emissions from Grasslands*. Wallingford, Conn.: CAB International, 1997.

Kothman, M. M. "Rangeland Ecosystems in the Great Plains: Status and Management." In *Conservation of Great Plains Ecosystems: Current Science, Future Options*, edited by S. R. Johnson and A. Bouzaher, pp. 199–209. Dordrecht: Kluwer, 1995.

Kucera, C. L. "Tall-Grass Prairie." In *Ecosystems of the World*, vol. 8A, *Natural Grasslands: Introduction and Western Hemisphere*, edited by R. T. Coupland, pp. 227–268. Amsterdam: Elsevier, 1992.

Lauenroth, W. K., and D. G. Milchunas. "Short-Grass Steppe." In *Ecosystems of the World*, vol. 8A, *Natural Grasslands: Introduction and Western Hemisphere*, edited by R. T. Coupland, pp. 183–226. Amsterdam: Elsevier, 1992.

Lavrenko, E. M., and Z. V. Karamysheva. "Steppes of the Former Soviet Union and Mongolia." In *Ecosystems of the World*, vol. 8B, *Natural Grasslands: Eastern Hemisphere and Résumé*, edited by R. T. Coupland, pp. 3–59. Amsterdam: Elsevier, 1993.

Mark, A. F. "Indigenous Grasslands of New Zealand." In *Ecosystems of the World*, vol. 8B, *Natural Grasslands: Eastern Hemisphere and Résumé*, edited by R. T. Coupland, pp. 361–410. Amsterdam: Elsevier, 1993.

Moore, R. M. "Grasslands of Australia." In *Ecosystems of the World*, vol. 8B, *Natural Grasslands: Eastern Hemisphere and Résumé*, edited by R. T. Coupland, pp. 315–360. Amsterdam: Elsevier, 1993.

Nie, D., H. He, M. B. Kirkham, and E. T. Kanemasu. "Photosynthesis and Water Relations of a $C_4$ and a $C_3$ Grass under Doubled Carbon Dioxide." In *Proceedings XVII International Grassland Congress*, pp. 1139–1141, 1993.

Risser, P. G., E. C. Birney, and H. D. Blocker. *The True Prairie Ecosystem*. Stroudsburg, Pa.: Hutchinson Ross, 1981.

Soriano, A. "Rio de la Plata Grasslands." In *Ecosystems of the World*, vol. 8A, *Natural Grasslands: Introduction and Western Hemisphere*, edited by R. T. Coupland, pp. 367–407. Amsterdam: Elsevier, 1992.

Tainton, N. M., and B. H. Walker. "Grasslands of Southern Africa." In *Ecosystems of the World*, vol. 8B, *Natural Grasslands: Eastern Hemisphere and Résumé*, edited by R. T. Coupland, pp. 265–290. Amsterdam: Elsevier, 1993.

Zhu, T.-C. "Grasslands of China." In *Ecosystems of the World*, vol. 8B, *Natural Grasslands: Eastern Hemisphere and Résumé*, edited by R. T. Coupland, pp. 60–82. Amsterdam: Elsevier, 1993.

—O. W. ARCHIBOLD

**GRAZING LAND.** *See* Agriculture and Agricultural Land; Animal Husbandry; *and* Desertification.

**GREEN CHEMISTRY.** *See* Chemical Industry.

**GREEN DESIGN.** *See* Industrial Ecology.

## GREENHOUSE EFFECT

[*For an introduction to* Greenhouse Effect, *see articles on* Carbon Dioxide *and* Global Warming.]

### A Scientific Analysis

The term *greenhouse effect* refers broadly to the partial trapping by the atmosphere of radiation from the Earth's surface, leading to a surface temperature that is larger than would be the case without the atmosphere. While the atmosphere is relatively transparent to shortwave radiation (sunlight), it is nearly opaque to infrared radiation, owing to the presence of certain trace gases and of clouds. Much of the infrared radiation passing upward from the Earth's surface is absorbed and reradiated, both upward and downward. Because the surface therefore receives not just solar radiation but also infrared radiation from the atmosphere and clouds, it is much warmer than it would be in the absence of the atmosphere. (Actual greenhouses work mostly by preventing the upward *convection*—not radiation—of heat received from the sun, so that the term *greenhouse effect* is something of a misnomer.)

The most important greenhouse gas in the atmosphere is *water vapor*. Along with clouds, composed of water drops or ice crystals, water vapor plays a key role in trapping outgoing terrestrial radiation. But because water vapor responds rapidly to changing conditions, water is treated as a feedback in the climate system, not an external forcing. It is generally believed that water vapor and clouds are the most important feedbacks in the climate system, at least on time scales of thousands of years or less. [*See* Clouds; *and* Water Vapor.]

Next to water in all its phases, the important greenhouse substances in the atmosphere include *carbon dioxide, methane, nitrous oxide, ozone,* and various *chlorofluorocarbons* (CFCs). The CFCs are entirely of anthropogenic origin. Carbon dioxide is thought to respond to changes in sources and sinks over a time scale of from forty to several hundred years, while methane has an inherent time scale of about eight years. The ozone concentration peaks in the middle stratosphere, where it filters out most of the harmful, very shortwave (ultraviolet) radiation from the sun. Although its lifetime is short, ozone's concentration is affected by the presence of other trace gases, notably chlorine, which is in turn related to the chlorofluorocarbons, which have very long lifetimes. Analysis of gas bubbles trapped in polar ice shows that carbon dioxide and methane had somewhat lower concentrations during the ice ages but had quite stable concentrations from the end of the last major glaciation until the beginning of the Industrial Revolution. The carbon dioxide concentration of the atmosphere has been increasing since the early nineteenth century, owing to our consumption of fossil fuels and to deforestation, and is expected to reach twice its natural, postglacial value sometime in the twenty-first century. [*See* Deforestation.]

The concentration of methane increased even more rapidly over the past two centuries (but now seems to be stabilizing), for reasons that are less clear but possibly related to the influence of human activities, which are also affecting the concentrations of ozone, nitrous oxide, and the chlorofluorocarbons. Because of the important role these gases play in the greenhouse effect, it is feared that their increasing concentrations may lead to noticeable global warming. [*See* Carbon Dioxide;

Chlorofluorocarbons; Methane; Nitrous Oxide; *and* Ozone.]

The recognition that certain trace gases may play an important role in determining the temperature of the Earth's surface and atmosphere dates back two centuries, to the French scientist Jean Baptiste-Joseph Fourier (1768–1830), who described the surface warming in terms of a "hothouse." There followed many others who contributed to the understanding of the greenhouse effect, among them John Tyndall in England (1820–1893), who measured the absorption of infrared radiation by carbon dioxide and water vapor, and, most importantly, Svante Arrhenius (1859–1927), the Swedish Nobel laureate in chemistry who performed the first quantitative estimates of the surface temperature change that would result from changes in atmospheric carbon dioxide concentration (see Arrhenius, 1896). [*See the biography of Arrhenius.*]

**Basic Physics of the Greenhouse Effect.** Averaged over a year and over the whole area of the Earth, about 340 watts per square meter of solar radiant energy enters the top of the atmosphere, mostly in the form of visible light. Suppose there were no atmosphere, and that none of the incoming radiation were reflected back to space. Then, since in the long run the Earth must be in thermal equilibrium, it has to radiate as much energy back to space as it receives from the sun. According to the Stefan-Boltzmann equation, the amount of radiation emitted by a perfect radiator is given by $\sigma T^4$, where $\sigma$ is the Stefan-Boltzmann constant (about $5.67 \times 10^{-8}$ $Wm^{-2}\ K^{-4}$) and $T$ is the temperature of the emitter. Thus, in equilibrium, the surface temperature of the Earth would be given by $\sigma T^4 = 340\ Wm^{-2}$. Solving for $T$ gives a temperature of about 278 kelvins, or 5°C. This is considerably colder than the observed average temperature of the Earth's surface, which is more like 288 kelvins, or 15°C.

A photo of the Earth taken from space dramatically illustrates how clouds reflect a great deal of the incoming solar radiation. In fact, on average the clouds—and, less importantly, the Earth's surface itself—reflect about 30 percent of the incoming solar radiation, reducing the amount absorbed to about 240 $Wm^{-2}$. Solving the Stefan-Boltzmann equation again gives a surface temperature of about 255 kelvins, or $-18°C$, even colder than before. It is the greenhouse effect of trace gases and clouds that accounts for the much larger observed average temperature.

The basic idea of the greenhouse effect is illustrated by Figure 1. Suppose that the atmosphere and clouds can be represented by a single layer of gas and clouds at some temperature $T_a$ and that this layer of gas and clouds can be treated as a perfect emitter. The layer therefore emits radiant energy both upward and downward at the rate $\sigma T_a^4$, while the surface emits upward

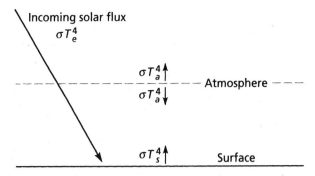

**Greenhouse Effect. FIGURE 1.** A Simple Model of the Greenhouse Effect.

Incoming solar radiation is absorbed at the surface, which, in the absence of an atmosphere, would have an equilibrium temperature $T_e$. Infrared radiation emitted by the surface is absorbed by the atmosphere, whose temperature is $T_a$. The atmosphere in turn radiates infrared radiation both upward to space and downward to the surface. $T_a = T_s = 2^{1/4}\ T_e$.

at the rate $\sigma T_s^4$, where $T_s$ is the surface temperature. At the top of the atmosphere, the total outgoing radiative flux, given by $\sigma T_a{}^4$, must balance the net incoming solar flux of 240 $Wm^{-2}$, giving $T_a = 255$ kelvins. But the surface receives energy from *both* the sun *and* the atmosphere, and the thermal equilibrium of the surface requires that

$$\sigma T_s^4 = 240\ Wm^{-2} + \sigma T_a^4 = 480\ Wm^{-2}.$$

Solving for $T_s$ gives a surface temperature now of 303 kelvins, or 30°C. Although this result is now much warmer than the observed temperature, this simple model illustrates the basic concept of the greenhouse effect. One aspect of the whole system that is of special interest is the contradictory role of clouds in climate. On the one hand, they reflect much solar radiation back to space, tending to cool the climate; on the other hand, they are important contributors to the greenhouse effect, which warms the climate.

The model of the greenhouse effect illustrated by Figure 1 is obviously a gross oversimplification. To perform an accurate estimate of the greenhouse effect, we have to divide the atmosphere into many layers, each with its own temperature and each radiating only a fraction of what the Stefan-Boltzmann law gives (this fraction is called the *emissivity*). This fraction depends crucially on the amount of greenhouse gases and clouds in the layer, and it determines not only how much radiation the layer can emit but also what fraction of the radiation passing through it from other layers can be absorbed. We also have to take into account that the atmosphere is not completely transparent to sunlight; clouds and water vapor absorb some of the incoming solar radiation. When all these things are accounted for in sophisticated radiative transfer models, the equilibrium surface temperature works out to be about 350 kelvins or about 80°C. But the equilibrium temperature decreases quite rapidly with altitude, reaching a minimum of around 210 kelvins (about −60°C) about eight miles above the surface.

The greenhouse effect is so large that, in the absence of other processes, it would make the surface of the Earth very hot indeed. But the rapid decrease of temperature with altitude in the radiative equilibrium state cannot be sustained because it is unstable to convective overturning, as first recognized by R. Emden in 1913 (see Goody and Yung, 1989 for details). Cool air aloft sinks and warm air from lower down rises until a new equilibrium is achieved, characterized by a vertical rate of decrease of temperature that is nearly neutrally stable to convective overturning. In the Earth's atmosphere, water usually condenses in the rising convective plumes, and much of the condensed water falls to the surface as precipitation. When the water condenses, the latent heat of vaporization is released to the air, making it warmer than it would be otherwise. The result of this moist convection is the establishment of a vertical rate of decrease of temperature that is nearly neutrally stable to the moist convection. This vertical rate of decrease of temperature is referred to as the *moist adiabatic lapse rate*, and it varies with both altitude and temperature. Figure 2 shows a pair of moist adiabats for the Earth's atmosphere.

Because of the greenhouse effect, acting in concert with cumulus convection, the actual state of much of the Earth's atmosphere is very nearly one of *radiative-convective equilibrium*. (Such a state does not exist, however, over middle- and high-latitude continents in winter.) In such a state, convection carries much of the heat flux from the surface to the lower atmosphere, but as one goes higher in the atmosphere, the radiative fluxes become progressively more important. At 10–12 kilometers above the surface, the convective fluxes cease, and above that altitude, the atmosphere is nearly in pure radiative equilibrium. In the tropics, almost all of the heat flux from the surface to the lowest layers of the atmosphere is carried by convection.

The most important consequence of the convective heat transfer is that the surface is much cooler than it would be in pure radiative equilibrium. The surface is still receiving about 240 $Wm^{-2}$ from the sun and an even larger amount from back-radiation from the atmosphere, but it is losing most of that energy by convection, not directly by radiation. In connection with the reduced emission from the surface, the surface temperature is lower. Conversely, the atmosphere is receiving energy from the convection and thus must be warmer than it would be in pure radiative equilibrium, so it is cooling radiatively. For example, the tropical tro-

posphere cools at an average rate of about 2°C per day by radiation.

The cooling effect of convection on the Earth's surface is illustrated in Figure 2. First, note that there is a particular altitude at which the temperature happens to be equal to the temperature the surface would have in the absence of an atmosphere (i.e., 255 kelvins). Most of the greenhouse gases are below this altitude. In the present climate, this altitude is around 6 km above the surface. In pure radiative equilibrium, the temperature would increase very rapidly downward from this level. But because of convection, the temperature actually increases downward much more slowly (i.e., along a moist adiabat), giving a much lower surface temperature. The actual operation of the greenhouse effect is more complicated and subtle than is usually recognized. However, excellent treatments in the literature have existed for many years (e.g., Goody and Yung, 1989).

There is one particularly interesting complication in this picture: It is the convection itself that largely determines the distribution of the two most important greenhouse substances, water vapor and clouds. Thus, real radiative-convective equilibrium is a highly interactive process.

**Response of the Greenhouse Effect to Increasing Concentrations of Trace Gases.** The basic physics underlying the sensitivity of global temperature to greenhouse gas concentration can also be seen in Figure 2. The key point is that the temperature of the atmosphere and surface is most sensitive to changes in the concentrations of greenhouse gases near the level at which $T = 255$ kelvins. Changing the concentration of greenhouse gases or clouds at and above this level will change the altitude where $T = 255$ kelvins. (Changing the amount of greenhouse gases lower in the atmosphere will have less effect.) For example, increasing the concentration of carbon dioxide or methane will move upward the altitude at which $T = 255$ kelvins. Following a moist adiabat down to the surface shows that the surface temperature will increase, but because of the divergence of moist adiabats with altitude, the surface temperature increase will be only about half the temperature increase in the upper troposphere (less in the tropics and more in polar regions).

If changes in greenhouse gas concentration were the only factor involved, it would be straightforward to calculate the resulting temperature change in the atmosphere and at the surface. For example, doubling the carbon dioxide concentration would increase the global average surface temperature by about 1°C. The complexity of the climate problem arises from the great number of feedbacks in the climate system. For example, raising the temperature of the atmosphere may increase the amount of water vapor—a greenhouse gas—thereby giving an even greater increase in temperature.

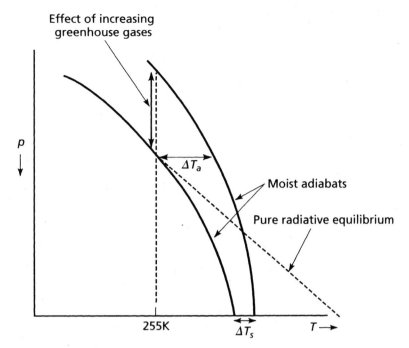

**Greenhouse Effect. FIGURE 2.** The Effect of Convection and of Increasing Concentration of Greenhouse Gases.

Pressure decreases upward and temperature increases to the right. The two thin, solid lines are moist adiabats; convection keeps the temperature profile close to curves of this shape. In the absence of convection, the temperature would be close to a state of radiative equilibrium, shown by the thick dashed line. Increasing greenhouse gases, particularly in the upper troposphere, near and above the level at which the temperature equals 225K, moves the altitude at which $T = 255K$ upward. This causes an increase in upper tropospheric temperature of $\mathring{a}T_a$ and a corresponding increase of $\mathring{a}T_s$ in the surface temperature.

**Greenhouse Feedbacks.** As is apparent from Figure 2 and from the discussion in the preceding section, the greenhouse effect is particularly sensitive to changes in greenhouse gases and clouds near and above the altitude at which the actual temperature is equal to the effective emission temperature (255 kelvins). It is well established from more detailed calculations (e.g., Lindzen, 1997) that tropospheric and surface temperature are somewhat more sensitive to changes in greenhouse gases and clouds in the upper troposphere and lower stratosphere. Thus one of the great challenges in predicting and understanding climate change is to understand the response of upper-tropospheric and lower-stratospheric variable greenhouse constituents—namely, water vapor and clouds—to changes in forcing. It is the uncertainty in this response that has driven much of the controversy surrounding the issue of global warming.

Two major physical processes control the amount of water vapor (and, indirectly, clouds) in the Earth's atmosphere: the circulation of the atmosphere and microphysical processes within the cloud. The latter determine how much cloud water (very small droplets or ice crystals effectively suspended in the air) is converted to precipitation and how much of the precipitation reevaporates before reaching the surface. We know that if all cloud water were converted to precipitation, all of which fell to the ground without reevaporation, the atmosphere would be far drier than observed; conversely, if precipitation did not occur at all, the atmosphere would become saturated with water vapor and filled with cloud. Reality lies somewhere in between. The great sensitivity of actual water vapor concentration to the details of cloud microphysics was first demonstrated by Emanuel (1991) and Renno et al. (1994).

The circulation of the atmosphere also exerts a strong influence on the distribution of water vapor. Over much of the Earth, particularly in the subtropics, air near the effective emission altitude is subsiding, having detrained from convective clouds thousands of kilometers away (Sun and Lindzen, 1993). Only within 1–2 kilometers of the Earth's surface in these regions is air directly influenced by the local sea surface. Some of this subsiding air is observed to be extremely dry, implying that little addition of water occurred from the time the air originally left the tops of tall cumulonimbus clouds (Spencer and Braswell, 1997). In other places, the descending air is more moist, implying that some mixing with other airstreams has occurred, or that the air has been moistened by evaporating precipitation or outflow from shallower clouds.

Satellite images demonstrate that a disproportionate amount of radiation leaving the Earth originates in these dry, subtropical "windows." Thus the net greenhouse effect of the planet as well as its sensitivity to changes in forcing can be sensitive to the relative size of these regions of subsidence and to the degree of mixing of moist air into such regions.

Unfortunately, there is little quantitative understanding of many of the processes that have been described here, and there is considerable reason to be skeptical of the ability of current global climate models to handle these processes correctly. Not a single existing climate model contains even the most rudimentary representation of the fundamental cloud microphysical processes operating within cumulus clouds (Emanuel and Zivkovic-Rothman, 1999). For example, it is well known that the efficiency of conversion of cloud water to precipitation increases strongly with temperature (e.g., see Sun and Lindzen, 1993), yet this is not contained in the convective representations used in climate models. Moreover, as demonstrated by Tompkins and Emanuel (2000), the vertical resolution of climate models is inadequate, resulting in incorrect prediction of water vapor and reduced sensitivity of water vapor content to cloud microphysics (Emanuel and Zivkovic-Rothman, 1999). These problems also result in an artificially strong coupling between water vapor fluctuations at the surface and aloft in climate simulations (Sun and Held, 1996). Problems of this nature will have to be tackled before we can have confidence in model simulations of global climate change.

**Summary.** Greenhouse gases work by absorbing some of the infrared radiation that would otherwise pass directly from the Earth's surface to space and reradiating part of this energy back down toward the surface, which thus receives radiation not only from the sun but from the atmosphere and clouds as well. The last two require warming in the atmosphere to operate. The most important greenhouse constituents of the atmosphere are water vapor and clouds, but water vapor cycles so quickly through the atmosphere that its concentration is usually regarded as a feedback, rather than a forcing, in the climate system. Next to water vapor and clouds, carbon dioxide, methane, nitrous oxides, and ozone are the most important greenhouse constituents. Rising levels of these constituents raise concern that the Earth's climate may warm appreciably. Just how much warming occurs depends crucially on the response of water vapor and clouds to changing climate, but the physics controlling these important feedbacks is still inadequately understood or modeled. Thus while the basic physics of the greenhouse effect is well understood, quantitative estimates of its sensitivity to climate change are hampered by poor understanding of certain key physical processes and by inadequate measurement of clouds and water vapor in the upper troposphere.

[*See also* Aerosols; Albedo; Atmosphere Dynamics; Atmosphere Structure and Evolution; Atmospheric Chemistry; Climate Change; Global Warming; Hydrologic Cycle; *and* Sun.]

## BIBLIOGRAPHY

Arrhenius, S. "On the Influence of Carbonic Acid in the Air upon the Temperature of the Ground." *Philosophical Magazine* 41 (1896), 237–276.

Emanuel, K. A. "A Scheme for Representing Cumulus Convection in Large-Scale Models." *Journal of Atmospheric Sciences* 48 (1991), 2313–2335.

Emanuel, K. A., and M. Zivkovic-Rothman. "Development and Evaluation of a Convection Scheme for Use in Climate Models." *Journal of Atmospheric Sciences* 56 (1999), 1766–1782.

Goody, R. M., and Y. L. Yung. *Atmospheric Radiation: Theoretical Basis.* New York: Oxford University Press, 1989.

Lindzen, R. S. "Can Increasing Atmospheric $CO_2$ Affect Global Climate?" *Proceedings of the National Academy of Sciences* 94 (1997), 8335–8342.

Renno, N. O., K. A. Emanuel, and P. H. Stone. "Radiative-Convective Model with an Explicit Hydrological Cycle. Part I. Formulation and Sensitivity to Model Parameters." *Journal of Geophysical Research* 99 (1994), 14429–14441.

Rogers, R. R., and M. K. Yau. *A Short Course in Cloud Physics.* 3d ed. New York: Pergamon Press, 1989.

Spencer, R. W., and W. D. Braswell. "How Dry is the Tropical Free Troposphere? Implications for Global Warming Theory." *Bulletin of the American Meteorological Society* 78 (1997), 1097–1106.

Sun, D.-Z., and I. Held. "A Comparison of Modeled and Observed Relationships between Variations of Water Vapor and Temperature on the Interannual Time Scale." *Journal of Climate* 9 (1996), 665–675.

Sun, D.-Z., and R. S. Lindzen. "Distribution of Tropical Tropospheric Water Vapor." *Journal of Atmospheric Sciences* 50 (1993), 1643–1660.

Tompkins, A. M., and K. A. Emanuel. "Simulated Equilibrium Tropical Temperature and Water Vapor Profiles and their Sensitivity to Vertical Resolution." *Quarterly Journal of the Royal Meteorological Society* (2000), in press.

—RICHARD S. LINDZEN AND KERRY A. EMANUEL

**GREENHOUSE GASES.** *See* Atmospheric Chemistry; Carbon Dioxide; Greenhouse Effect; *and* Nitrous Oxide.

# GREENING OF INDUSTRY

This is a research field concerned with the environmental performance of industry and the contribution of industry to sustainable development. The field is not defined by a particular theoretical approach or policy orientation. Most research seeks to identify ways to reduce the materials, waste, and energy intensity of production, and to promote the development of products and production processes that enhance the environmental sustainability of economic activity. At the macro level this involves an assessment of such public and private initiatives as pollution control, pollution prevention, clean production, eco-efficiency, and environmental technology development. At the level of individual organizations, researchers examine the mix of internal and external forces that promote improvements in the environmental performance of industry. Research in this field seeks to determine whether the values, strategies, and practices of business are compatible with the goals of sustainable development.

Why highlight industry as a core focus for research concerned with sustainable development and global change? Industry is a large and in some cases a dominant source of pollution, greenhouse gas emissions, and toxic waste. Of equal importance is the critical role of industry in shaping the environmental impacts of consumption through the characteristics of the products and services that are produced. Industry has both a direct and an indirect impact on the environment, as, for example, in emissions released in the production and the subsequent use of an automobile. Researchers are currently trying to determine what scale of environmental benefits can be achieved through improvements in the energy and materials efficiency of industry (eco-efficiency), combined with the development of greener products and production processes. Life cycle analysis is used to compare the environmental impacts associated with the production, use, and disposal of different products and services. Of particular concern is the extent to which improvements in the environmental performance of industry can be achieved at no additional cost over existing technologies and business practices, to provide so-called win–win solutions.

Much of the direct environmental impact of industry derives from the energy and materials used, transformed, stored, and released in the production of goods and services. Research into the greening of industry shares with industrial ecology and with industrial metabolism this interest in materials and energy flows within firms, industries, and economies. [See Chemical Industry.] But in contrast to industrial ecology and industrial metabolism, which have drawn heavily on engineering and materials science, the greening of industry builds principally upon ideas and theories from the social sciences. The typical research design varies from detailed case studies of individual firms to cross-sectional comparisons of industrial practice in different countries. An initial focus upon firms that have demonstrated leadership in greening, a series of highly visible success stories, has now given way to more systematic analysis, often organized by sector or by size of firm.

Three areas of research are especially prominent. One major focus is the way in which firms respond to market, public, and regulatory pressures to improve their environmental performance. How do firms recognize the need for change in their business practices, and then effect such change? Drawing upon organization and management theory, researchers have explored such issues as internal and external drivers of change in firms, processes of learning and adaptation, issues of leadership and of employee participation in the greening process, the relationship of firms to their customers, suppliers, research organizations, and other aspects of their external environment, and the forms of regulation that best support change in a firm's environmental performance. One critical research question is the likely success of various market-based and "soft" forms of regulation, ranging from performance disclosure to eco-labeling and voluntary environmental codes of conduct as a driver of industrial greening. A second core issue is that of technology change. The development of green products and production processes, be it an energy-efficient motor or biodegradable packaging, will be crucial to any attempt to meet the goals of sustainable development. Of particular concern are the opportunities for public policy intervention to enhance the rate of technology change, either through direct investment in research and development or through induced changes in demand for greener products and services. Here again, researchers have typically drawn upon an existing body of theory concerned with innovation and the management of research and development and then sought to use this theory to understand obstacles to the accelerated development, deployment, and use of green technology. In addition, there has been considerable interest in technology transfer and in the role of multinational corporations and multilateral funding agencies in this process of greening on an international scale.

The third research focus is that of business strategy and the environment. There is wide diversity in the strategies that industrial firms currently pursue with respect to the environment, ranging from a reactive stance focused upon compliance with existing regulatory standards to a proactive positioning of greening as a source of competitive advantage in the marketplace. The primary focus remains that of environmental management and the extension of existing management systems to include waste, materials, and energy optimization as a goal. For example, can the types of management innovation that have helped manufacturers improve product quality also be harnessed to improve environmental performance? At the same time, there is growing interest in the prospects for firms moving beyond environmental management to what some have labeled *management for sustainability*. This approach reflects a view that improvements in the environmental performance of industrial activity will be insufficient to meet societal goals of sustainable development, and that business shares responsibility with other parts of society to address such challenges as growth, consumption, and equity in development. Management for sustainability is a nascent field of study without a clearly dominant research paradigm.

Global change is of crucial importance to the greening of industry. The environmental performance of firms around the world is increasingly driven by a set of international forces, from international business standards such as ISO 14000 to the product requirements of customers in distant markets. And yet it is far from clear whether globalization of markets, investment, trade, information flows, and other dimensions of economic activity will facilitate, or act to undermine, the greening of industry on an international scale.

[See Energy Policy; Environmental Economics; Industrial Ecology; Industrialization; Industrial Metabolism; Market Mechanisms; Recycling; Sustainable Development; Technology; and Waste Management.]

### BIBLIOGRAPHY

Angel, D., and J. Huber, eds. "Greening of Industry Network Special Issue." *Business Strategy and the Environment* 5 (1996), 127–215. Special issue of the journal focusing on greening of industry.

Cairncross, F. *Green, Inc.* Washington, D.C.: Island Press, 1995. A nontechnical discussion of the need for greening.

Fischer, K., and J. Schott, eds. *Environmental Strategies for Industry*. Washington, D.C.: Island Press, 1993. One of the best introductions to the field.

Groenewegen, P., K. Fischer, E. Jenkins, and J. Schott, eds. *The Greening of Industry Resource Guide and Bibliography*. Washington, D.C.: Island Press, 1995. An annotated international bibliography of important studies.

Heaton, G. R., R. Repetto, and R. Sobin. *Back to the Future: U.S. Government Policy toward Environmentally Critical Technologies*. Washington, D.C.: World Resources Institute, 1992. An excellent introduction to the development of green technology.

Socolow, R., C. Andrews, F. Berkhout, and V. Thomas. *Industrial Ecology and Global Change*. Cambridge: Cambridge University Press, 1994. Covers the perspective of many disciplines.

Welford, R. *Environmental Strategy and Sustainable Development*. London: Routledge, 1995. A provocative introduction to management for sustainability.

—DAVID P. ANGEL

## GREEN PARTIES. *See* Environmental Movements.

## GREENPEACE

Greenpeace is a transnational activist group committed to protecting the global environment. Conceived in Vancouver, Canada, in 1971, Greenpeace was originally concerned with the ecological effects of nuclear testing. Its first campaign involved twelve people sailing a small boat into the United States atomic test zone off Amchitka in Alaska as a form of protest and public education. At the heart of that campaign was a strategy that still informs much of Greenpeace activity: the belief that bearing witness—providing an unwavering presence at the site of environmental abuse, whatever the risk—can draw public attention to issues and mobilize change. This strategy has been Greenpeace's signature among activist groups.

Since its inception, Greenpeace has expanded its agenda to concentrate on all forms of ecological abuse. It presently organizes public campaigns to stop climate change, protect the oceans, eliminate toxic chemicals, prevent the release of genetically modified organisms into nature, pursue nuclear disarmament, and end nuclear contamination. Greenpeace undertakes these campaigns worldwide by coordinating activities among its regional and national offices in forty countries, with a central office in Amsterdam. The entire network of offices, known as Greenpeace International, is financed almost entirely through contributions from 2.5 million supporters in 158 countries, and by sales of merchandise. In terms of membership and budget, it is one of the largest transnational environmental activist groups.

Greenpeace works for global environmental protection by pressuring states and international organizations to establish and enforce environmental regimes, and by influencing public opinion in general to support environmentally friendly practices. Greenpeace possesses observer status at the United Nations and is an accredited nongovernmental organization in most international environmental negotiations. This allows it to lobby governmental officials and international civil servants in the service of environmental protection. In addition, individuals from Greenpeace's national offices pressure private actors (for example, firms) and government officials to enhance domestic environmental policy. In terms of galvanizing public opinion, Greenpeace works to change codes of good conduct, social mores, and cultural understandings of human impact on the natural world. In this capacity, it undertakes education campaigns throughout the world.

Like other transnational organizations, Greenpeace's effectiveness stems partially from its nonterritorial status. Unlike states, Greenpeace can pursue environmental protection free from the task of preserving and enhancing the welfare of a given, geographically situated population. In other words, Greenpeace can effect a global perspective on issues because it is not charged with representing the interests of only one, territorially situated, segment of world society. This does not mean that Greenpeace somehow represents world opinion. It simply suggests that it is not territorially bound in its perspective.

Evaluation of Greenpeace's effectiveness is difficult

because of the complexities of measuring shifts in public opinion and tracing governmental policy changes directly back to efforts of particular activists, organizations, and campaigns. Nonetheless, there is some scholarly consensus that Greenpeace has had a significant influence on a number of issues. These include the establishment of the Basel Convention, which bans the exportation of toxic waste material from Organization for Economic Co-operation and Development (OECD) to non-OECD countries, the strengthening of the International Whaling Commission's moratorium on commercial whaling, and the enhancement of the London Dumping Convention to ban all waste disposal at sea.

[*See also* Basel Convention; Environmental Movements; London Convention of 1972; Nongovernmental Organizations; *and* Whaling Convention.]

### BIBLIOGRAPHY

Stairs, K., and P. Taylor. "Non-Governmental Organizations and the Legal Protection of the Oceans: A Case Study." In *The International Politics of the Environment: Actors, Interests and Institutions*, edited by A. Hurrell and B. Kingsbury. New York: Oxford University Press, 1992. A detailed study of Greenpeace's efforts to strengthen international commitments to protect against waste disposal at sea, and a general review of NGO political efforts.

Wapner, P. *Environmental Activism and World Civic Politics*. Albany: State University of New York, 1996. A detailed examination of Greenpeace, Friends of the Earth, and the World Wildlife Fund, and their influence on world environmental politics.

—PAUL KEVIN WAPNER

## GREEN POLITICS. *See* Environmental Movements.

## GREEN REVOLUTIONS. *See* Agriculture and Agricultural Land; Biotechnology; *and* Food.

## GROUND WATER

When rain falls, some of it percolates into the soil. While a proportion will be taken up by plant roots, some deeper infiltration will occur (under the influence of gravity) in the more permeable soils, whose pore spaces are both larger and better connected. This moisture will eventually accumulate above an impermeable bed, saturating the pore space of the ground. The water that remains underground is known as *ground water*. A layer of soil or rock that forms an underground reservoir and is both sufficiently porous to store substantial quantities of water and sufficiently permeable to allow this water to flow is called an *aquifer*. [*See* Hydrologic Cycle.]

The ground above the aquifer, through which the excess rainfall passed vertically, is termed the *vadose* (or *unsaturated*) *zone*, because its pore space contains both air and water. Its retained water is held below atmospheric pressure by the surface tension of soil particles. Over a period of time, this equilibrium is disturbed by evaporation (normally via plant roots) and/or by infiltration. The latter is the result of renewed rainfall which, after replenishing any soil moisture deficit resulting from evaporation, will cause water to drain downwards.

The level to which the ground is fully saturated is known as the *water table*. The water table is not horizontal, but slopes toward places where water leaves the aquifer system (such as some river valleys) and away from areas where water enters the aquifer system. In most climates, ground water discharges from aquifers continuously, but rainfall replenishes them for only a few days, weeks, or months of the year (Clarke et al., 1996). Thus the water table is not at a constant level, but rises and falls as the amount of water stored in the aquifer increases or decreases.

Where an aquifer dips beneath another layer that is much less permeable, its ground water becomes confined by the overlying layer. In such situations, if boreholes are drilled into the aquifer, water is encountered under pressure and rises until the column of water is enough to balance the pressure in the aquifer at the corresponding location. The imaginary pressure surface joining the water levels of boreholes in a confined aquifer is known as the *piezometric surface*. In the unconfined part of the aquifer system, this surface would be represented by the water table. In some areas, the ground water in the aquifer will be under sufficient pressure to overflow from the borehole, and such a borehole is known as an *artesian well*.

If ground water is pumped from an unconfined aquifer it will be derived from both pore drainage and interception of natural flow in the aquifer system. When water is pumped from a confined aquifer, however, none of its pore space is drained. In this case, some or all of the supply is said to come from elastic storage, because the natural pressure of ground water slightly increases the pore space volume, and water itself is to some degree compressed elastically, rather like the air in a car tire (Price, 1985).

**Types of Aquifer.** The pore space of aquifers is connected to form a system of tiny tubes in which ground water circulates, generally very slowly. All aquifers have two fundamental capacities—one for ground water storage and the other for ground water flow—but not all aquifers are of the same type.

Unconsolidated granular sediments, such as sands, contain pore space between the grains and thus their saturated water content can exceed 30 percent of their total volume. This, however, reduces progressively if cementation of the pores has occurred. In highly consolidated rocks, ground water is found only in fractures, whose porosity rarely exceeds 1 percent of the volume of the rock mass. However, over geologic time these fractures tend to become enlarged through dissolution by flowing ground water to form fissures and caverns, especially in limestone.

Aquifers vary in size and extent, ranging from small local ones to systems extending over huge areas and containing enormous volumes of water. The major aquifers in England, for example, are made of consolidated limestones or sandstones. They are extensive, covering areas of several thousand square kilometers, and in many places are more than 100 meters (300 feet) deep (Downing, 1998). The major limestone aquifer that underlies large areas of southern and eastern England, known as "the Chalk," is estimated to contain 2 billion cubic meters of water, very many times more than all of the major surface-water reservoirs of England and Wales (Price, 1985). The great aquifers of the world, such as those that underlie the Ganges and the northeastern China plains, are, of course, very much larger in terms of geographical extension and water storage.

**Natural Groundwater Quality.** Most ground water originates from water that has permeated first through the soil and then through the rock below it. It is normally of excellent quality because its slow passage through the subsurface acts as a natural self-purification system (Freeze and Cherry, 1979). Thus, under natural conditions, it is free from pathogenic microorganisms and undesirable organic matter. At the same time, it tends to take into solution small quantities of minerals such as calcium bicarbonate and magnesium sulfate, which are generally considered beneficial to human health in either continuous or occasional consumption. In consequence of these characteristics, some groundwater sources have become fashionable in the form of bottled mineral water.

The passage of water through different soils and rocks controls its pH (acidity) and its content of dissolved oxygen and other gases (Downing, 1998). In certain hydrogeological situations, the natural evolution of groundwater chemistry can result in excessive concentrations of certain mineral ions (such as the soluble forms of iron, manganese, arsenic, and fluoride). These can represent a nuisance in terms of water use or taste, or a significant threat to human health, and this condition is normally referred to as *natural contamination* of ground water.

**Groundwater Flow Regimes.** All fresh water found underground must have had a source of recharge. This is normally rainfall, but can also sometimes be seepage from rivers and lakes, or from man-made water storage and distribution systems, such as those used in irrigated agriculture or in urban water supply and wastewater disposal. [*See* Irrigation.] Average rates of diffuse ground water recharge depend on the climatic regime, the soil characteristics, and the degree of human intervention. They can range from a few millimeters per year to more than one meter per year under extreme conditions, but are normally between 50 and 500 millimeters per year.

Groundwater systems are dynamic, and water is continuously in slow motion from areas of recharge (in the unconfined parts of aquifers) to areas of discharge. The total volume of water in storage in aquifers is usually large relative to the rate of flow through them, and thus aquifer residence times can be very long. Tens, hundreds, or even thousands of years may elapse during the passage of water through this subterranean part of the hydrological cycle, since flow rates do not normally exceed 10 meters per day and are often as low as one meter per year, or less. The residence times of ground water in aquifers vary widely with their geologic origin. The overall hydrogeological environment, including the size of the aquifer system and the prevailing climatic regime, also exerts a major influence in this respect (Table 1).

**Social and Environmental Importance.** It is estimated that about 22 percent of the planet's total freshwater resources are stored underground in the form of ground water. Excluding water locked in the polar icecaps, ground water constitutes more than 95 percent of all available fresh water. The remainder is stored in lakes, rivers, and swamps. [*See* Water.]

Groundwater sources (boreholes, wells, and springs) have numerous advantages over surface-water sources for water supply (Figure 1). First, they are not nearly as susceptible to drought, flood, or earthquake impacts. Second, they are normally cheap to develop because of their natural storage and excellent quality (requiring little or no treatment before use) and because they can be developed stage by stage according to demand. Third, they can often be tapped near to where the water is needed, thus reducing the need for costly distribution systems.

Throughout human history, ground water has been used extensively on all continents. While some early populations settled by sources of surface water (such as rivers and lakes), many others based settlements around areas of ground water discharge (namely, springs) or of shallow water table (where wells could easily be dug or galleries excavated). During the twentieth century, drilling and pumping technology has developed rapidly, so that it has become easy to sink deep boreholes and to extract large volumes of water from them.

No reliable statistics exist on the percentage of the

**Ground Water. TABLE 1.** Classification of Hydrogeological Environments with (A) Their Susceptibility to Exploitation Side Effects and (B) Their Vulnerability to Groundwater Pollution

*(A)*

| HYDROGEOLOGICAL ENVIRONMENT AND AQUIFER TYPE | | SALINE INTRUSION OR UPWELLING | LAND SUBSIDENCE | INDUCED POLLUTION |
|---|---|---|---|---|
| | | *Type of Side Effect* | | |
| Major alluvial formations | Coastal | ** | (In some cases)** | ** |
| | Inland | (In few areas)* | (In few cases)* | ** |
| Intermontane valley-fill | With lacustrine deposits | (In some areas)** | (In most cases)** | * |
| | Without lacustrine deposits | (In few areas)* | In few cases)* | * |
| Consolidated sedimentary aquifers | | (In some areas)* | 0 | (In few cases)* |
| Recent coastal limestones | | ** | 0 | * |
| Glacial deposits | | (In few cases)* | (In few cases)* | * |
| Weathered basement | | 0 | 0 | * |
| Loess-covered plateau | | 0 | (In few cases)* | * |

*Occurrences known; **Major effects; 0 = no significant effects.

*(B)*

| HYDROGEOLOGICAL ENVIRONMENT AND AQUIFER TYPE | | TYPICAL TRAVEL TIMES TO WATER TABLE | ATTENUATION POTENTIAL OF ENVIRONMENT | POLLUTION VULNERABILITY OF AQUIFER |
|---|---|---|---|---|
| Major alluvial formations | Unconfined | Weeks/years | Moderate | Moderate |
| | Semi-confined | Years/decades | High | Low |
| Intermontane valley-fill* | Unconfined* | Months/years* | Moderate* | Moderate* |
| | Semi-confined | Years/decades | Moderate | Low |
| Consolidated sedimentary aquifers* | Porous sandstones* | Weeks/years* | Moderate* | Moderate* |
| | Karstic limestones | Days/weeks | Low | Extreme |
| Recent coastal limestones | Unconfined | Days/weeks | Low | Extreme |
| Glacial deposits* | Unconfined | Weeks/years* | Moderate* | Moderate* |
| Weathered basement* | Unconfined | Days/weeks | Low | High |
| | Semi confined* | Weeks/years* | Moderate* | Moderate* |
| Loess-covered plateau | Unconfined | Weeks/months | Moderate | Moderate |

*Environments and aquifers characterized by highest degree of intrinsic variability, for which generalization is more difficult.

world's water supply that is derived from ground water. It is estimated that about one-third of Asia's population (some 1,000–1,200 million people), some 150 million people in Latin America, and more than 100 million people each in western Europe and North America depend on ground water. Some individual countries, such as Denmark and the Netherlands, use ground water almost exclusively for public water supply, and many of the world's largest and most rapidly growing cities, including Mexico City, Lima, Dhaka, Beijing, Bangkok, and Lusaka, also are heavily dependent upon ground water (Foster et al., 1998).

Ground water is also now widely developed to provide rural communities with more reliable supplies through low-yielding boreholes, and it is often the pre-ferred source of industrial water supply, where high quality and/or reliable yield is critical. Where aquifers of sufficient size exist, ground water has also become widely exploited for irrigated agriculture. In some cases, aquifers are the sole source of irrigation water, but more often they supplement surface-water irrigation systems and increase their drought reliability, or act as drought insurance in the case of rain-fed agriculture.

To speak in terms of direct human use alone, however, is to omit one of the most important functions of ground water, namely that of sustaining some important elements of the surface-water environment (Clarke et al., 1996). The natural discharge from major aquifer systems provides most of the natural dry-weather flow (baseflow) in the upper reaches of many lowland rivers

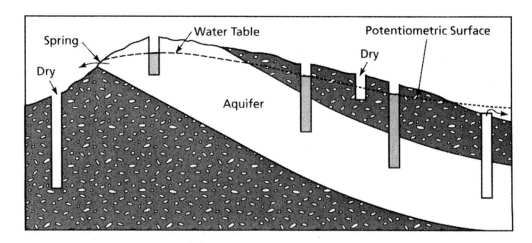

**Ground Water. Figure 1.** Typical Sedimentary Aquifer System Showing the Occurrence of Ground Water in Boreholes and as Springs.

(Downing, 1998). Moreover, the existence of a shallow water table or upward ground water flows to the surface is a critical element in sustaining many freshwater wetland habitats and in maintaining the salinity balance in some coastal brackish-water lagoons. Perhaps the classic wetland habitat sustained by ground water is that of the oases found in many deserts, including the Sahara.

**Groundwater Resource Degradation.** In numerous areas of the world, groundwater resources are under threat from human activities (Clarke et al., 1996). These threats take two main forms: (1) excessive or inadequately controlled pumping, and (2) pollution from activities at the land surface. Both processes can lead to damage to aquifers themselves, to the quality of their ground water, and to the surrounding environment. The damage will in some cases be irreversible, and in many other cases only reversible at very high cost and/or after very long periods. Sustainability issues arise since future generations may be deprived of the resource or of the environmental feature concerned.

*Aquifer overexploitation.* Ground water is being abstracted from many aquifers at rates greater than the average rate of recharge. The most immediate effect of this is a continuous lowering of the water table and a reduction of the volume of ground water in permanent storage. In the more extreme cases, this situation is sometimes referred to as *groundwater mining.* [*See* Ogallala Aquifer, Depletion and Restoration of the.]

In reality, however, the term *aquifer overexploitation* is not as easy to define as it might at first appear (Foster et al., 2000). First, it is necessary to specify the area under consideration; it is possible to have local pockets of severely depressed water table within an aquifer in which average recharge exceeds average abstraction overall. Second, it is necessary to specify the time period under consideration; in the case of more arid areas where major groundwater recharge occurs once a decade or even less frequently, the ability to support an abstraction rate in the long term will be more a function of the aquifer storage than of recent average recharge. Third, it is necessary to define how much of the average aquifer recharge and throughflow should be allowed to discharge naturally, since serious impacts in the surface water environment can occur at levels of abstraction well below those that would be termed overexploitation in simplistic terms.

In practice, when talking about aquifer overexploitation, we are mainly concerned with the internal consequences on the aquifer system and the external consequences on the surface environment (Foster et al., 2000). While all groundwater development may have some side effects, the main concern will be with serious consequences such as the following.

1. Large reductions in well yields and/or major increases in their pumping costs
2. Impacts on the surface-water environment, such as reductions in springflow and streamflow, or in lake and wetland water levels
3. Inflow or upwelling of saline water of coastal or other origin
4. Seepage of poor-quality ground water from overlying shallow aquifers
5. Land subsidence resulting from drainage and settlement of interbedded clays and silts, affecting urban buildings, infrastructure, and drainage

Most of the these consequences (except, to some extent, 1 and 2) are likely to be irreversible in practice. Many, if they proceed unchecked, will also give rise to increased social inequity because of reduced availability and/or increased cost of water supply. [*See* Desertification.]

Because of its elevation at 7,415 feet (2,260 meters) on a high plateau, Mexico City has no easy access to surface water. Its population of over 15 million persons relies mostly on one aquifer that supplies over 70 percent of the city's water. The aquifer has been overdrawn since the early 1900s, and in recent years the water table has been falling at roughly one meter per year.

Because of this overdraft, the aquifer is being compacted, and the land is subsiding: over the past one hundred years the central area of the Mexico City Metropolitan Area has fallen by an average of 7.5 meters, with serious damage to building foundations and the sewer system. Demand for water will continue to grow rapidly as the region develops: the per capita use of water in the urban area now is only half that of New York City.

To ameliorate the problem, new pricing and improved water metering systems are being undertaken, along with other conservation measures.

SOURCE: *World Resources 1996–97*, pp. 64–65. New York: Oxford University Press, 1996.

—DAVID J. CUFF

The extent to which a given aquifer will suffer adverse side effects from overpumping will depend on its hydrogeological characteristics and setting. Some consequences can be avoided or mitigated by exercising control over well construction and pumping rates. The control of groundwater abstraction is, however, not a straightforward task. Ground water (like fish stock) is a replenishable resource whose ownership is difficult to define. The preferred approach is to separate the legal right of land ownership from that of groundwater ownership, with licenses to construct wells and to abstract ground water granted from some form of regulatory agency. The involvement of representatives of water users in the activities of such agencies is a key element in the successful implementation of aquifer management policies (Foster et al., 2000).

**Groundwater pollution.** Groundwater pollution is becoming increasingly widespread. It occurs when the rate of discharge or leaching of a given water contaminant (normally at or near the land surface) exceeds the natural attenuation capacity of the soil vadose (unsaturated) zone profile and the dilution capacity of the aquifer system itself (Foster et al., 1998).

The list of human activities that can potentially generate a significant subsurface contaminant load and cause groundwater pollution is long, but most can be grouped under (1) urbanization and industrialization processes and (2) the intensification of agriculture (Table 2). A distinction is also usefully made between readily identifiable point sources and more diffuse sources. [*See* Pollution.]

A major source of groundwater pollution in and around urban areas is the inadequate design, operation, and maintenance of sanitation systems. In numerous instances, especially in rapidly developing cities, *in situ* sanitation systems (such as septic tanks and latrines) are deployed with inappropriate design and/or excessive density for the ground conditions, leading to widespread groundwater contamination by nitrate, and in some instances also by pathogenic microorganisms (Foster et al., 1998). Inadequately designed and/or badly sited waste disposal landfills (or tips) may also allow the infiltration of highly saline, ammonia-rich, and organically contaminated leachate. [*See* Sanitation.] Industry also pollutes ground water by discharging liquid wastes directly or indirectly to the ground, together with the leakage or spillage of hydrocarbon fuels and industrial solvents; the latter, known as *dense nonaqueous phase liquids*, are especially insidious groundwater pollutants. Mine drainage and spoil are another common source of groundwater pollution.

The intensification of agricultural cultivation, sustained through large (and often excessive) applications of inorganic fertilizers and pesticides often gives rise to diffuse groundwater pollution; the effects are most marked on permeable soils and/or where low-efficiency irrigation is practiced (Foster et al., 2000). This is especially the case where intensive cultivation of arable or horticultural crops over extensive areas is practiced, since there is no possibility of dilution from neighboring land that harvests groundwater of good quality.

Groundwater pollution is insidious and expensive: insidious because it takes many years to appear in water abstracted from deep boreholes, by which time it will

**Ground Water. TABLE 2.** Summary of Activities and Installations Potentially Generating a Subsurface Contaminant Load**

| ACTIVITY/INSTALLATIONS | DISTRIBUTION CATEGORY | *Character of Pollution Load* | | |
| --- | --- | --- | --- | --- |
| | | MAIN TYPES OF POLLUTION | HYDRAULIC SURCHARGE | SOIL ZONE BYPASS |
| **Urban Wastewater and Other Services** | | | | |
| UNSEWERED SANITATION | u/p/r | P-D | n f o | + | * |
| Leaking Sewers[a] | u | P-L | o f n | + | * |
| SEWAGE OXIDATION LAGOONS[†] | u/p | P | o f n | ++ | * |
| Sewage Sludge/Effluent Land Discharge[†] | u/p/r | P-D | n s o f | + | |
| SEWAGE TO INFILTRATING RIVER[†] | u/p/r | P-L | n o f | ++ | * |
| Leaching Refuse Landfill/ Tips[†] | u/p/r | P | o s h | | * |
| Fuel Storage Tanks | u/p/r | P-D | o | | * |
| Highway Drainage Soakaways | u/p/r | P-D | s o | + | * |
| **Industrial** | | | | |
| Leaking Tanks/Pipelines[‡] | u | P-D | o h | | * |
| Accidental Spillages | u | P-D | o h | + | * |
| PROCESS/EFFLUENT LAGOONS | u | P | o h s | ++ | * |
| EFFLUENT LAND DISCHARGE | u | P-D | o h s | + | |
| EFFLUENT TO LOSING RIVER | u | P-L | o h s | ++ | * |
| Leaching Residue Tips | u/p/r | P | o h s | | |
| Soakaway Drainage | u/p/r | P | o h | ++ | * |
| Aerial Fallout | u/p/r | D | s o | | |
| **Agricultural/Horticultural[§]** | | | | |
| a) SOIL CULTIVATION | | | | |
| • with AGROCHEMICALS | p/r | D | n o | | |
| and with IRRIGATION | p/r | D | n o s | + | |
| • with sludge/slurry | p/r | D | n o s | | |
| • with WASTEWATER IRRIGATION | p/r | D | n o s f | + | |
| b) Livestock Rearing/Crop Processing | | | | |
| • effluent lagoons | p/r | P | f o n | ++ | * |
| • effluent land discharge | p/r | P-D | n s o f | | |
| • effluent to losing river | p/r | P-L | o n f | ++ | * |
| **Mineral Extraction** | | | | |
| Hydraulic Disturbance | r/p | P-D | s h | | * |
| Drainage Water Discharge | r/p | P-D | h s | ++ | * |
| PROCESS WATER/SLUDGE LAGOON | r/p | P | h s | + | * |
| LEACHING RESIDUE TIPS | r/p | P | s h | | * |

**Block capitals indicate more common and serious sources of groundwater pollution.

[†]Can include industrial components.

[‡]Can also occur in nonindustrial areas.

[§]Intensification presents main pollution risks.

**f** = fecal pathogens; **h** = heavy metals; **n** = nutrient component; **o** = microorganic compounds and/or organic load; **P/L/D** = point/line/diffuse; **s** = salinity; and **u/p/r** = urban/periurban/rural.

+, ++ = the degree of significance of the associated hydraulic surcharge.

* = bypass of the soil zone through preferential flow is likely.

be too late to take preventive action; expensive because the costs of providing alternative water supplies or remediating polluted aquifers are extremely high (indeed, restoration to drinking-water standards may be technically impossible).

For these reasons, there is an urgent need to assess the vulnerability of aquifers to pollution and to estimate the capture areas of groundwater sources. The application of appropriate controls over land use and effluent discharge in the land surface zones thus defined is the best way of achieving a balance between groundwater resource and source protection. The highest priority will need to be attached to controlling subsurface contaminant loading in areas of high vulnerability (lowest natural attenuation capacity), especially where these coincide with the capture zones of key sources of drinking water.

[*See also* Desalination; Energy; Environmental Law; Erosion; Salinization; Water Quality; *and* Water Resources Management.]

## GROUND WATER RISE UNDER LONDON, ENGLAND

In some parts of the world, ground water is the main source of water for industrial, municipal, and agricultural use. Some rocks, including sandstones and limestones, have characteristics that enable them to hold and transmit large quantities of water, which can be reached by installing pumps and boreholes.

Considerable reductions in groundwater levels have been caused by abstraction. The rapid increase in the number of wells tapping ground water in the London area from 1850 until after World War II caused substantial changes in groundwater conditions. The piezometric surface in the confined chalk aquifer fell by more than 60 meters over hundreds of square kilometers.

In some industrial areas, recent reductions in industrial activity have led to less ground water being taken out of the ground. As a consequence, groundwater levels in such areas have begun to rise, a trend exacerbated by considerable leakage from ancient, deteriorating pipe and sewer systems. This is already happening in British cities including London, Liverpool, and Birmingham. In London, because of a 46 percent reduction in groundwater abstraction, the water table in the chalk and tertiary beds has risen by as much as 20 meters. Such a rise has numerous implications, both good and bad:

- increase in spring and river flows
- re-emergence of flow from dry springs
- surface water flooding
- pollution of surface waters and spread of underground pollution
- flooding of basements
- increased leakage into tunnels
- reduction in stability of slopes and retaining walls
- reduction in bearing capacity of foundations and piles
- increased hydrostatic uplift and swelling pressures on foundations and structures
- swelling of clays as they absorb water
- chemical attack on building foundations.

—ANDREW S. GOUDIE

### BIBLIOGRAPHY

Clarke, R., A. R. Lawrence, and S. S. D. Foster. *Groundwater: A Threatened Resource.* UNEP Environment Library 15, Nairobi: United Nations Environment Programme, 1996.

Downing, R. A. *Groundwater: Our Hidden Asset.* U.K. Groundwater Forum Publication. Keyworth-Notts: British Geological Survey, 1998.

Foster, S. S. D., P. J. Chilton, M. Moench, F. J. Cardy, and M. Schiffler. *Groundwater in Rural Development: Facing the Challenges of Supply and Resource Sustainability.* World Bank Technical Paper 463, Washington, D.C.: World Bank, 2000.

Foster S. S. D., A. R. Lawrence, and B. L. Morris. *Groundwater in Urban Development: Assessing Management Needs and Formulating Policy Strategies.* World Bank Technical Paper 390, Washington, D.C.: World Bank, 1998.

Freeze, R. A., and J. A. Cherry. *Groundwater.* Upper Saddle River, N.J.: Prentice-Hall, 1979.

Price M. *Introducing Ground Water.* London: Chapman and Hall, 1985.

—STEPHEN FOSTER

## GROWTH, LIMITS TO

All organisms are subject to some constraint on their numbers; no species can grow infinitely abundant. For organisms such as bacteria, this phenomenon can be observed in a microcosm. The bacteria on a culture plate will grow exponentially until growth is inhibited by exhaustion of resources, toxic-waste buildup, or other factors. In ecology, the idea that the population size and growth rate of a species are negatively associated is termed *density-dependent growth.* Many mechanisms can produce density-dependent growth, including resource and space depletion, disease outbreaks, and aggressive social interactions.

While density-dependent growth is well documented in organisms from bacteria to trees, a more controversial issue is whether human beings are subject to the same ecological phenomenon. So far in history, human beings have managed to escape a limit to global population growth. As resources have become scarce, technological innovation or, in some cases, resource substitution has allowed more people to be added to the planet with no apparent detriment to the global human growth rate. The question remains, however: if the human population continues to increase, will density-dependent mechanisms limit population growth?

Limits to an organism's population growth might be posed by many factors. Here, the discussion is restricted to biophysical resources that might limit human population numbers. In addition, the discussion is focused at a global scale; at a local level, limits will be set by different factors that will vary from place to place and time to time.

For human beings, resources are goods such as productive land, fresh water, and energy, and the ecosystem services that supply these goods. Examples of ecosystem services are purification of air and water, pollination, natural pest control, generation and renewal of soil fertility, stabilization of climate, and decomposition and recycling of wastes.

Although it is easy to identify general classes of key resources, it is difficult to determine which specific resources are truly essential. First, there is huge variation within and between societies in patterns of resource use. Second, technological innovation may greatly reduce the total amount of a resource needed and may even offer a complete substitute. Because of these inherent complexities, research on limits to human population growth has focused on a few resources that appear totally essential (for example, organic energy, fresh water, an ozone layer) or that are being depleted so rapidly that substitutes are not being developed fast enough to offset the economic and other consequences of exhaustion (such as fisheries and forests).

Many lines of evidence suggest that humanity may be approaching limits to essential resources and, therefore, limits to growth. In one of the first studies to attempt to quantify limits to growth, Vitousek et al. (1986) estimated the total annual net primary productivity (NPP) on Earth and the fraction thereof that people have appropriated. NPP is the amount of energy fixed biologically by primary producers (mainly plants) after subtracting the energy required for their own metabolic processes. In other words, it is the total amount of food resources on Earth for all organisms other than plants. The authors considered three classes of use of NPP by people: direct use (for example, food, timber, and fuel); indirect use (such as uneaten portions of crops, clearing of lands by fire); and forgone (conversion of land to less productive areas, such as from natural habitat to agriculture, a conversion that results in an increase in NPP edible by humans but typically a decrease in total NPP). They concluded that our species is using about 38 percent of the world's terrestrial annual NPP, leaving only 62 percent for all other terrestrial species. Moreover, the human population is likely to double in the near future, but capturing a second 40 percent will be more difficult than capturing the first 40 percent, simply because the most accessible NPP has already been co-opted.

A recent analysis (Pauly and Christensen, 1995) refined the estimate of human appropriation of aquatic NPP. The authors combined statistics on global fisheries catches with data on the primary production required to support these catches and estimated that humanity is co-opting 8 percent of the planet's aquatic NPP. Although this figure may seem low, there are many reasons why it would be difficult to increase this percentage much further. For instance, most primary productivity (phytoplankton and detritus) is unavailable to higher levels of the food chain because it too dispersed across the open ocean. Also, in coastal shelf areas, intensive blooms account for a substantial part of the primary production, but most of this available energy settles to the bottom as detritus before zooplankton can use it.

Furthermore, it is already evident that 8 percent is not sustainable given current practices. Major fish stocks are declining everywhere, at a cost of billions of dollars in economic losses and the loss of critical animal protein from the diets of some human populations. Also, people are gradually fishing down the food chain; in other words, the top predators have been so overharvested that fisherman are now catching more of the prey of these fish. In recent years, fisheries catches have declined an average of 0.1 trophic (food chain) level per decade without, so far, substantially increasing the amount harvested, suggesting that humans have already greatly altered the planet's aquatic ecosystems and their ability to sustain fishing pressure (Pauly et al., 1998). [See Fishing.]

Fresh water is another resource that is clearly essential to human survival. Less than 1 percent of the water on Earth is fresh and directly available to society; the remainder exists in the form of salt water or frozen in icecaps. Nonreplenishable ground water can be tapped, but this depletes reserves that require thousands of years to replenish. Only water that flows through the solar-powered hydrologic cycle is renewable. Postel et al. (1996) estimated that human beings are utilizing 26 percent of the renewable freshwater supply. Excluding amounts that are geographically and temporally inaccessible, human beings are using 54 percent of available fresh water. As with other resources, people have used the most accessible fresh water first. Postel and her colleagues projected that new dam construction could increase available runoff by 10 percent by 2025; however, human population is projected to grow by more than 30 percent in that same period (Population Reference Bureau, 1997). [*See* Dams.]

Biodiversity is also a resource that is crucial to human well-being, although it is unclear what fraction of biodiversity is essential to human survival. The term *biodiversity* encompasses many of the tangible resources mentioned above (for example, fisheries), as well as indirect resources, including genetic diversity for improving crops and medicines. Moreover, recent studies suggest that biodiversity plays a role in ecosystem functioning and stability and, therefore, in the delivery of ecosystem services (Tilman et al., 1996). Every hour, an estimated one to three species and eighteen hundred populations are driven to extinction by tropical forest destruction alone (Hughes et al., 1997). As the scale of human activities grows and natural habitats are destroyed, biodiversity loss may intensify. [*See* Biodiversity.]

In addition to nearing limits of tangible resources, society is approaching limits on the ability of ecosystems to supply services. For example, humans have disrupted the protective layer of ozone in the stratosphere by the widespread release of chlorofluorocarbons (CFCs). The release of carbon dioxide and other greenhouse gases (GHGs) into the atmosphere seems to be warming the Earth's surface. Thus it appears that we have passed the atmosphere's limit of being able to accommodate gases emitted by human activities without noticeably altering the regulation of the Earth's climate. Global warming threatens to cause more frequent and extreme storms and droughts and could impact all aspects of human civilization, from food production to human susceptibility to disease.

Natural nutrient cycles that we rely on to recycle wastes and purify air are also being disrupted. Humanity has doubled the amount of nitrogen fixation from the unreactive atmospheric pool to the biologically available forms on land (Vitousek et al., 1997). These extra sources of nitrogen are mainly nonorganic fertilizers, fossil fuel emissions, and nitrogen-fixing crops such as soybeans, peas, and alfalfa. The consequences of this disturbance include an increase of nitrous oxide ($N_2O$) (an effective GHG as well as a facilitator of stratospheric ozone destruction) in the atmosphere, changes in species diversity and composition in terrestrial ecosystems, eutrophication in lakes, estuaries, and coastal seas, and nitrate contamination in drinking water. Thus, "limits to growth" now involves a new kind or level of resources—the very functioning of the biosphere. [*See* Biogeochemical Cycles.]

The research on biophysical resources suggests that, at current rates of resource consumption and concomitant environmental change, global human population growth may be inhibited in the future. This may be attributable to specific resource losses, or to the cumulative stress that resource use places on the functioning of the biosphere. Locally, malnutrition, disease, and war have historically limited, and currently limit, population growth in some areas. Of course, technological innovation may increase the efficiency of essential resource use so that global population growth could continue for a time with no further impact. For instance, a shift from coal and oil to solar power would reduce GHG emissions per capita. Also, the level of consumption (or resource use) per capita will influence the number of people the Earth can support before encountering limits to growth. Certainly, the limits would be different if everyone adopted a lifestyle resembling that of the average person in Costa Rica rather than that of the average North American. Finally, many social factors (such as allocation and security of property rights and patterns of trade) affect patterns of resource use, so that changes in institutions could potentially forestall the exceeding of some resource limits.

[*See also* Brundtland Commission; Carrying Capacity; Catastrophist–Cornucopian Debate; Future Studies; Human Populations; IPAT; Population Dynamics; Scenarios; Sustainable Development; *and the biography of Schumacher.*]

## BIBLIOGRAPHY

Brown, L. R., ed. *The State of the World.* New York: Norton, 1984–1998. An annual series with up-to-date information on the progress of sustainable development.

Cohen, J. *How Many People Can the Earth Support?* New York: Norton, 1995. A well documented discussion of human growth limits.

Daily, G. C., ed. *Nature's Services: Societal Dependence on Natural Ecosystems.* Washington, D.C.: Island Press, 1997. Surveys the present state of knowledge of the value of ecosystem services.

Ehrlich, P. R., and A. H. Ehrlich. *The Population Explosion.* New York: Simon and Schuster, 1990. A readable account of the relationship between population and all global environmental problems.

Ehrlich, P. R., A. H. Ehrlich, and G. C. Daily. *The Stork and the Plow.* New York: Grosset/Putnam, 1995. A well-documented introduction to population and food issues.

Ehrlich, P. R., and J. Roughgarden. *The Science of Ecology.* New York: Macmillan, 1997. Provides a basic discussion of density-dependent growth.

Holdren, J. "Population and the Energy Problem." *Population and Environment* 12 (1991), 21–255. A detailed article on the relationship between population, energy, and a sustainable society.

Hughes, J. B., G. C. Daily, and P. R. Ehrlich. "Population Diversity: Its Extent and Extinction." *Science* 278 (1997), 689–692.

Meadows, D. H., et al. *Limits to Growth.* New York: New American Library, 1972. The classic book on human population limits.

Meadows, D. H., D. L. Meadows, and J. Randers. *Beyond the Limits.* White River Junction, Vt.: Chelsea Green Publishing, 1992. An update to the classic.

Myers, N. *The Primary Source.* New York: Norton, 1992. A readable documentation of the importance and loss of tropical biodiversity.

Pauly, D., and V. Christensen. "Primary Production Required to Sustain Global Fisheries." *Nature* 374 (1995), 255–257.

Pauly, D., V. Christensen, J. Dalsgaard, R. Froese, and F. Torres. "Fishing Down Marine Food Webs." *Science* 279 (1998), 860–863.

Postel, S. L., G. C. Daily, and P. R. Ehrlich. "Human Appropriation of Renewable Fresh Water." *Science* 271 (1996), 785–788.

Schneider, S. *Laboratory Earth: The Planetary Gamble We Can't Afford to Lose.* New York: Basic Books, 1997. An excellent general discussion of global environmental issues.

Tilman, D., D. Wedin, and J. Knops. "Productivity and Sustainability Influenced by Biodiversity in Grassland Ecosystems." *Nature* 379 (1996), 718–720.

Vitousek, P. M., P. R. Ehrlich, A. H. Ehrlich, and P. A. Matson. "Human Appropriation of the Products of Photosynthesis." *Bioscience* 36 (June 1986), 368–373.

Vitousek, P. M., J. D. Aber, R. W. Howarth, G. E. Likens, P. A. Matson, D. W. Schindler, W. H. Schlesinger, and D. G. Tilman. "Human Alteration of the Global Nitrogen Cycle: Sources and Consequences." *Ecological Applications* 7 (1997), 737–750.

—JENNIFER B. HUGHES

# H

**HABITAT LOSS.** *See* Deforestation; Extinction of Species; *and* Human Impacts, *article on* Human Impacts on Earth.

## HAZARDOUS WASTE

Hazardous waste can loosely be described as solid, semiliquid, or liquid waste that contains chemical substances that are toxic to humans, plants, or animals, are flammable, corrosive, or explosive, or have high chemical reactivity and as such require special handling, treatment, and disposal. Examples of hazardous wastes include waste oils, paint residues, solvents, resins, metal finishing sludges, and hospital wastes; these contain typically one or more potentially toxic chemicals such as strong acids and alkalis, metals (e.g., cadmium, mercury, chromium, lead), solvents (e.g., trichloroethylene), polychlorinated biphenyls (PCBs), polynuclear aromatic hydrocarbons, pesticides, and pathogenic organisms.

**Sources and Nature of Hazardous Waste.** Hazardous wastes are for the most part associated with industrial activity and are generated in the course of the production and consumption of goods and services. The processing of wastes arising from these activities and the end-of-pipe treatment of factory emissions and effluents generate further quantities of hazardous wastes.

Households are also responsible for generating hazardous waste, or more specifically, for contaminating municipal waste streams with potentially hazardous chemicals discarded in batteries, household cleaners, do-it-yourself and garden chemicals, (e.g., pesticides, wood treatments, paints), and car-care products. Small businesses, including photographic laboratories, paint and printing workshops, educational establishments, research laboratories, hospitals and the agricultural sector also produce significant quantities of similar wastes.

It is difficult to put a precise figure on just how much hazardous waste is generated globally, or even by individual countries. This is partly because of the low priority that historically has been given to the collection of waste statistics by national statistical offices and local authorities responsible for the management of solid wastes, and partly because, until relatively recently, there has been no universal agreement on what constitutes a hazardous waste. These difficulties notwithstanding, attempts have been made to quantify the hazardous waste burden. According to the Organisation for Economic Co-operation and Development (OECD),

some 338 million metric tons per annum of hazardous or special wastes are produced worldwide, most of which is generated by the industrialized countries (see Table 1). Other estimates of the global hazardous waste burden are somewhat higher, at 500 million metric tons per annum or more (United Nations Environment Programme [UNEP], 1993).

Although there is uncertainty about the precise amounts of hazardous waste generated, it is generally accepted that the volume has risen substantially everywhere in tandem with the growth in the world population, increased levels of industrialization and urbanization, and rising standards of living. Furthermore, advances in technology have led to the development of new materials and chemicals, which have increased both the diversity and the complexity of the waste streams that have to be dealt with. [*See* Industrialization.]

**Environmental and Human Health Impacts.** Hazardous wastes represent a potentially serious threat to both human health and the environment. Of primary concern is the risk posed by the improper disposal of hazardous waste, a practice that can introduce—and without doubt has introduced—toxic chemicals into the environment, contaminating surface waters, groundwater resources, land, and the air. The problem is not confined to the local scale but is global in extent; trace quantities of long-lived or persistent chemicals such as PCBs, the pesticide dichlorodiphenyltrichloroethane (DDT), and other chlorinated hydrocarbons have now been detected in all environmental media, especially in biota (wildlife) even in remote parts of the world once considered pristine and free from contamination. The environmental and human health implications of this type of contamination are only just, after decades of research, becoming apparent. More disconcerting still is the fact that the sheer number of chemicals in production, currently estimated to be over the 100,000 mark, means that for the vast majority of chemicals, no detailed toxicity or hazard assessment profiles are available. As a result, the environmental effects of many chemicals present in hazardous waste streams cannot be predicted with any degree of reliability. [*See* Chemical Industry.]

*Past mistakes: contaminated sites.* Until the middle to late 1970s most wastes, including industrial and hazardous wastes, were disposed of by simply dumping them in landfills, pits, ponds, or lagoons, with no regulation or control and with scant regard to any potential deleterious environmental effects. However, a

**Hazardous Waste. TABLE 1.** Industrial and Hazardous Waste Generation in Member States of the Organisation for Economic Co-operation and Development (OECD), Eastern Europe, and the Rest of the World, Late 1980s

| REGION | INDUSTRIAL WASTES (MILLION METRIC TONS PER YEAR) | HAZARDOUS AND SPECIAL WASTES (MILLION METRIC TONS PER YEAR) |
|---|---|---|
| World | 2,100 | 338 |
| OECD | 1,430 | 303 |
| North America | 821 | 278* |
| Europe | 272 | 24 |
| Pacific | 333 | <1 |
| Eastern Europe | 520 | 19 |
| Rest of World | 180 | 16 |

*The value for the United States (275 million metric tons per year) used to derive the regional total for North America includes liquid wastes that are classified as hazardous.

series of highly publicized incidents involving hazardous wastes, one of the most celebrated of which occurred in 1978 at Love Canal in New York State, where abandoned chemical wastes caused serious contamination of surface waters, have since demonstrated that such practices are totally inadequate and have provided the driving force for implementing increasingly stringent regulations to control waste disposal. [See Waste Management.]

The scale of the damage caused by past indiscriminate dumping is difficult to assess, as most countries have yet to carry out national-scale inventories of contaminated sites. What data are available do, however, indicate widespread problems. In the United States, for example, over thirty thousand potentially hazardous sites have been identified, twelve hundred of which are described as requiring immediate attention (Russel et al., 1992). The cleanup of polluted sites has proved to be a costly exercise; the bill for remedial action in the United States could be as high as U.S.$100 billion, and somewhere between 1.0 billion and 1.5 billion ECUs in OECD Europe.

***Controlled landfill.*** In most industrialized countries the uncontrolled dumping of hazardous wastes has been replaced by controlled disposal in "secure landfills," that is, in pits or trenches fitted with waterproof bottom liners (usually high-density polyethylene), a leachate collection system, and a rain-shedding cover. These measures are designed to prevent leachate containing trace metals and organic compounds from contaminating local surface and ground waters. In some countries the practice of codisposal is particularly popular; this entails leaving landfilled municipal waste to mature for some months and then placing a small quantity of hazardous waste within it, the theory being that biodegradation will reduce the toxicity of the hazardous waste.

Disposal in secure landfill has dominated hazardous

waste management programs over the last ten to fifteen years and is still the most popular option in most countries today. In Europe, for example, around 75 percent of hazardous wastes are currently disposed of in this way. Where properly managed, controlled landfill offers a relatively low-cost yet effective means of handling hazardous waste. In more recent years, however, landfill has come under increasing pressure. Reports of poorly designed and badly managed landfill operations, giving rise to local pollution problems and nuisance (e.g., odor, litter, dust, and birds), have been all too common in many countries. This has led to a loss of faith and greater opposition by a more environmentally sensitive public to the siting of new facilities. Furthermore, the methane generated by the decomposition of waste in landfills is increasingly perceived to be a problem; according to recent estimates, landfills globally are responsible for around 8 percent of the current total anthropogenic emission of methane to the atmosphere (UNEP, 1993).

Under recent, stricter regulation, many countries in the industrialized world are now subjecting their hazardous wastes to some form of pretreatment, a physical, chemical, or biological process that reduces the toxicity and volume of waste prior to landfilling. Some countries have gone even further and have introduced legislation that prohibits the landfilling of certain types of hazardous waste (OECD, 1991; Yakowitz, 1993). Taken together, these factors imply that, in some countries at least, landfill faces a difficult and uncertain future.

***Ocean disposal.*** As an alternative to burial in landfill, some countries have relied on dumping at sea as a disposal option for selected waste streams, including certain industrial wastes, sludges, and packaged low-level radioactive wastes. However, concerns regarding the potential impacts on the marine environment have lead to a gradual reduction and near ban on such practices in most world regions. In the Northern Hemisphere, for example, ocean disposal has been under the control of the International Convention for the Dumping of Wastes and Other Matter (more commonly known as the 1972 London Dumping Convention) and the Oslo Convention. Under the terms of these agreements, the incineration at sea of noxious liquid wastes was banned at the end of 1994 and the ocean dumping of all industrial wastes at the end of 1995. [See Ocean Disposal.]

*Incineration.* Incineration (i.e., high-temperature combustion under controlled conditions) is generally the main alternative to landfilling for waste disposal in most countries, at least in the developed world. For municipal solid wastes, incineration offers the advantage of substantially reducing the volume of waste requiring final disposal, and in many cases its cost effectiveness can be improved by making use of the heat produced during the incineration process in energy recovery schemes. In the case of hazardous wastes the main objective is to destroy any potentially toxic chemical substances and infectious materials and relative to other treatment technologies, well-designed incineration systems are capable of the highest overall degree of destruction for the broadest range of hazardous waste streams. Incineration is particularly suited to wastes contaminated with organic chemicals such as PCBs, chlorinated pesticides, and halogenated solvents.

On the downside, incineration is the more costly option; nor is it without its own environmental impacts. Where incinerator design and operation are below standard, it can result in the release of particulates and other air pollutants (trace metals, hydrogen chloride, sulfur dioxide, and trace organics) into the atmosphere. Moreover, the residues, the fly and bottom ash, require final disposal and can sometimes contain trace quantities of heavy metals and organic compounds.

Of the possible emissions from incinerators, public attention has focused on the polychlorinated dioxins and furans. These can be formed as a result of incomplete combustion during the incineration of wastes containing PCBs and other chlorinated hydrocarbons. Emissions of metals such as mercury, lead, and cadmium have also caused concern. In countries where data are available, evidence suggests that incinerators are currently responsible for a significant proportion of the national emissions of dioxins and selected trace metals, and thus may represent a growing source of human exposure to these pollutants.

The magnitude of the environmental impact of these emissions and associated human health risk, being difficult to quantify, continues to be a matter of some controversy. Evidence from the United States, however, would appear to indicate that the problem lies more with municipal incinerators than with the dedicated hazardous waste incinerators (Demsey and Oppelt, 1993).

At present, incineration takes care of a relatively small proportion of hazardous waste, accounting for around 10 percent in OECD Europe. Despite the problems outlined above, it is widely predicted that incineration will grow in importance in coming years as the pressures on landfill intensify and the cost differential narrows.

*Other disposal options.* Hazardous wastes that are neither incinerated nor landfilled may be disposed of by some form of physicochemical treatment process, recycled in some way, or exported to other countries. Stabilization (or solidification), the most widely practiced option in the first category, involves the addition of a binding agent (such as cement) to liquids or slurries containing hazardous chemicals in order to convert them to solid, relatively inert masses that are suitable for landfilling. Inorganic waste sludges containing heavy metals can be dealt with in this way.

The recycling of hazardous byproducts of industrial processes currently accounts for somewhere between 5 and 10 percent of hazardous waste in most industrialized nations and is clearly desirable from an environmental point of view. Less desirable has been the practice of exporting hazardous waste to other countries for disposal. Such movements could be construed as a viable management option in cases where there is a need to find a specialized facility where particular wastes can be disposed of safely. However, it is impossible to justify the much-publicized movement of hazardous waste from industrialized countries to the less developed regions, since the latter do not have the facilities for treating hazardous wastes. In most cases imported hazardous wastes are simply dumped, sometimes in open urban dumps, where they can pose a significant health risk to scavengers who obtain their livelihood from sorting refuse.

**Policy Instruments for Hazardous Waste Management.** In order to ensure proper and safe management of hazardous wastes and thus adequate environmental protection, all countries need to have enforceable legislative and regulatory frameworks for the control of waste generation, treatment, storage, and final disposal, ideally set within a national strategy for waste management and environmental protection. Today most countries do indeed have comprehensive legislative frameworks in place.

In the United States, legislation governing hazardous waste at the federal level is divided into two parts. The Resource Conservation and Recovery Act of 1976 (RCRA) and its subsequent amendments comprise "cradle-to-grave" provisions for the storage, transport, treatment, and final disposal of new hazardous waste, whereas the cleanup of contaminated sites created by past inept disposal practices is covered by the Comprehensive Environmental Response, Compensation and Liability Act (CERLA) of 1980 and its 1986 amendment. This act established a national fund, known as Superfund, to assist with the costs of cleanup. It is largely financed by taxes on industrial and chemical production. While many individual U.S. states have their own regulations, they are usually similar to the federal ones and can be no less stringent.

Similarly, in the European region an extensive body of legislation, in the form of directives, now exists under the auspices of the European Commission; these cover all aspects of hazardous waste generation and disposal and include provisions for reporting and monitoring transfrontier shipments. As in North America, European laws governing hazardous wastes have been progressively strengthened since their initial adoption in the mid-1970s.

In the wake of public outrage over the illegal trade in hazardous waste, legislation has also been introduced to control movements of hazardous waste across national boundaries. Based on principles laid down by OECD member countries in the mid-1970s, the Basel Convention on the Control of Transboundary Movements of Hazardous Wastes and Their Disposal, which came into force in 1992, represents a truly international effort to reduce the amount of hazardous wastes that are exported. Under the terms of the convention, export may take place only if the necessary facilities for disposal do not exist in the country where the waste is generated. Furthermore, prior written approval of the importing country—the principle of Prior Informed Consent—is necessary before export is initiated. [See Basel Convention; and Prior Informed Consent for Trade in Hazardous Chemicals and Pesticides.]

Although the details may differ, the philosophies that guide much current waste legislation in different parts of the world are fairly similar. The idea that the burden of responsibility for limiting the adverse impacts of hazardous wastes lies with the producer is a common theme, embodied in the European region in the "polluter pays" principle. Under current U.S. law, the federal government can require the responsible parties who generated the wastes, or owned or operated a waste site subsequently classified as hazardous as a result of past indiscriminate dumping, to clean up the site. Only if no willing or able responsible party can be found is cleanup financed by the Superfund program. However, the concept of joint and several liability that underpins the Superfund program has proved to be problematic; the administration and compliance costs have been unnecessarily high and the revenue collected relatively small (Probst et al., 1995).

Controls on hazardous wastes tend to operate through systems of licensing of operators and treatment, storage, and disposal facilities. These licenses or permits are used to impose operating conditions or standards, which determine, for example, the types and

quantities of waste that can be accepted by landfill sites, or the level of emissions from incinerators. The underlying principle for setting these standards is the same in both Europe and North America, that is, those achievable by the "best demonstrated available technology" or "best available technology not entailing excessive cost."

Tighter regulatory control of hazardous wastes has led not only to more precise definitions of hazardous wastes but also to much improved monitoring and reporting of the types and quantities of wastes generated. Hitherto, much of this type of information has been lacking, making assessments of the scale of hazardous waste impacts difficult. Now, following the lead of the U.S. Environmental Protection Agency and its Toxic Release Inventory, many more countries are planning to set up detailed Pollutant Release and Transfer Registers and Materials Accounting Data systems in order to track the flow of potentially toxic chemicals more accurately and so better assess the associated environmental risks.

**Concluding Comments.** Over the past two decades the imposition of progressively tighter regulatory controls aimed at mitigating the adverse environmental and human health impacts that can arise as a result of the generation and disposal of hazardous wastes has, not surprisingly, substantially increased the cost of waste disposal. Although prices vary markedly between countries, the processing of hazardous waste within the OECD now costs an average of fifty to sixty U.S. dollars per metric ton (OECD, 1991).

Rising costs, coupled with pressure from the public and environmental lobby groups, have led to significant changes in the perception and general attitude toward the management of wastes. Instead of a preoccupation with the "safe management" of wastes and investment in pollution control equipment (i.e., end-of-pipe technologies), waste managers are placing a much greater emphasis on strategies that reduce the amount and toxicity of hazardous waste generated in the first place and on recycling and reclamation.

The prevention of hazardous waste at the source, the idea of "waste minimization" as it has come to be called, is now regarded as the preferable approach to hazardous and indeed all waste management, and is the primary objective of many national hazardous waste management plans. By the same token, managers are increasingly looking at ways in which potentially hazardous chemicals can be recovered and reused in the production cycle. To date, recycling rates for hazardous chemicals have been quite low, tending to focus on solvents and metals. Nowadays, however, waste streams are increasingly being perceived as having a residual value as a secondary raw material. The change is not just in perception; evidence suggests that waste exchange schemes, in which one industry's waste is used as another's raw material, are on the increase.

Although significant progress has been made over the past two decades with regard to the control of hazardous waste, a number of key challenges and problems remain. First, despite the universal acceptance of the desirability of the waste minimization and cleaner production concepts, their adoption in practice has been slow. It is unlikely that the barriers are entirely technological—it is estimated by some that up to 50 percent of all hazardous waste could be eliminated with existing technology—but rather a reluctance by industry to make the necessary large-scale investment required to change or adapt their existing processes. Policy makers have thus come to the conclusion that regulatory measures alone are not going to be sufficient to move the hazardous waste management industry toward a more sustainable position. In recognition of this, many countries are now beginning to experiment with a more flexible and varied approach; that is, one that combines direct regulatory control with the use of economic incentives—green taxes—in order to try and improve the efficiency and effectiveness of their waste management programs. What is clear, however, is that until waste minimization is more widely practiced, the volume of hazardous waste is likely to continue to grow, placing an ever increasing burden on national economies.

Public perception of the risks associated with hazardous wastes represents a further ongoing challenge for waste managers. Public outrage about past mistakes has created an atmosphere of distrust and lack of confidence in the companies that operate hazardous waste treatment and disposal facilities, and in the government agencies that regulate them. Consequently, in public opinion surveys, the general public frequently ranks problems relating to hazardous waste high on lists of environmental problems, higher than is perhaps warranted on the current scientific evidence of the actual risks involved (Kunrenther and Patrick, 1991). Clearly, future policies will need to incorporate strategies for improving the communication between scientific experts and the public.

[*See also* Environmental Economics; Ethics; Industrial Ecology; Industrial Metabolism; Risk; *and* Sustainable Development.]

## BIBLIOGRAPHY

British Medical Association. *Hazardous Waste and Human Health.* Oxford: Oxford University Press, 1991.

Dempsey, C. R., and E. T. Oppelt. "Incineration of Hazardous Waste: A Critical Review Update." *Air and Waste* 43 (January 1993), 25–73.

Freeman, H. M. *Standard Handbook of Hazardous Waste Treatment and Disposal.* 2d ed. New York: McGraw-Hill, 1998. Detailed information on existing and emerging treatment and disposal technologies for hazardous chemical wastes, along with a description of relevant legislation from the U.S. perspective.

Kunreuther, H., and R. Patrick. "Managing the Risks of Hazardous Waste." *Environment* 33.3 (April 1991), 13–15, 31–35.

Organisation for Economic Co-operation and Development (OECD). *The State of the Environment.* Paris: Organisation for Economic Co-operation and Development, 1991.

Probst, K. N., et al. *Footing the Bill for Superfund Cleanups: Who Pays and How?* Washington, D.C.: Brookings Institution and Resources for the Future, 1995.

Royal Society of Chemistry. *Simple Guide on Management and Control of Wastes.* London: The Royal Society of Chemistry, 1996. Comprehensive overview of waste management practices, policies, and legislation in the member countries of the European Community.

Russell, M., F. W. Colazier, and B. E. Tonn. "The U.S. Hazardous Waste Legacy." *Environment* 34.6 (1992), 12–15, 34–39.

United Nations Environment Programme (UNEP). *United Nations Environment Programme Environmental Data Report, 1993–94.* 3d ed. Oxford: Blackwell, 1993.

Yakowitz, H. "Waste Management: What Now? What Next? An Overview of Policies and Practices in the OECD Area." *Resources, Conservation, and Recycling* 8 (1993), 131–178.

—ANN D. WILLCOCKS

**HAZARDS.** *See* Drought; Earthquakes; Insurance; Natural Hazards; Risk; *and* Sea Level.

## HEATHLANDS

In everyday language the word *heathland* may signify any of quite a wide range of open, largely treeless, uncultivated landscapes, usually on acidic soils of low fertility. Defined more precisely, it refers to a type of vegetation or biome in which the prominent plants are subshrubs, mostly evergreen, with small tough leaves (sclerophylls). Trees and tall shrubs, if present, are sparse, and grasses are either absent or play a role subordinate to that of the subshrubs. The dominant species belong to a rather small number of botanical families—examples are Ericaceae, Epacridaceae, Myrtaceae, Proteaceae, and Vacciniaceae—not necessarily closely related but sharing certain structural features described as "ericoid" or "heathlike." Thus they are usually much-branched, woody subshrubs, seldom more than about 1–1.5 meters tall and sometimes less than 10 centimeters, often forming a rather dense low canopy composed of the numerous small or very small, often narrow leaves.

Heathland soils are almost universally low in plant nutrients (notably phosphorus and calcium) and usually acidic. They include freely drained sands, gravels and podzols, on which dry heaths develop, and moist or seasonally saturated peats, gleys, and so on, which support humid or wet heaths. Where such soils occur under oceanic or seasonally moist climatic conditions, ranging from subarctic through cool temperate to Mediterranean-type climates, heathlands may develop. Hence they occur chiefly on the margins of continents (for example, western and eastern North America, the Cape Province of South Africa, southwest and southern Australia, and in much of the subarctic tundra). They are absent where continental climates prevail, but are found on mountains in many parts of the world (including tropical regions) under suitable combinations of soil and climatic conditions. However, arid lands, wet tropical and warm temperate lowlands, permanently waterlogged habitats, and very fertile soils are all unsuitable for heathlands.

Heathland vegetation is made up of a number of characteristic plant communities that include, in addition to the subshrubs, a variety of (mostly perennial) herbaceous species, bryophytes, and lichens. In each of the regions in which they are present, the floristic composition of these communities is distinctive and often of great ecological interest and visual attractiveness, especially when the heath plants are in bloom. Furthermore, some heathland areas, notably the *fynbos* of South Africa and *kwongan* of southwestern Australia, have become centers of spectacular evolutionary radiation, producing flora of amazing diversity and beauty. Heathlands also support important invertebrate and vertebrate fauna.

**Natural and Seminatural Heathlands.** The absence of a dominant tree or tall shrub stratum may be a result of purely natural causes (i.e., without human influence), as in the case of tundra heaths, heaths above the treeline on mountains, and heaths on exposed clifftops or on certain dry and extremely nutrient-deficient soils. Here tree growth may be inhibited by climatic factors, including very strong winds and low temperatures, or by edaphic conditions. Naturally occurring fire, when frequent, may also prevent forest growth and permit the development of heath.

Heathland also develops naturally as a stage in a plant succession prior to the entry of tree species, for example, on some acid sand dunes and some peat bogs subject to surface drying. Such heathland plant communities usually give place in time to scrub or woodland.

In addition to these types of heathland, however, there are also (or have been in the past) very extensive lowland or middle-altitude heathlands in locations entirely suitable for forest or woodland vegetation. Such heathlands may make a major contribution to the landscape, as in the heath region of west Europe (from western Norway, through southwestern Sweden, Denmark, the North German plain, the Low Countries, and northern and western France, to northern Spain, and including the British Isles). These heathlands have replaced former forest and owe their origin and maintenance largely to human activities, including clearance of trees and subsequent management by cutting or burning the vegetation, turf stripping, or the use of grazing animals.

However, the plant communities consist of naturally occurring, not cultivated, species and so the vegetation may be described as *seminatural.*

**Fire.** A factor very generally associated with heathland is periodic fire, which may be caused naturally, for example by lightning or sparks from falling rocks, or through human agency. Some heathland areas have long been subject to natural fires, but many have in addition a history of occasional or frequent burning related to human use. In the South African *fynbos*, the pine plains of New Jersey and Long Island (North America) and the Australian heaths, although fire has always been an integral component of the ecosystem, its frequency has been much increased by mankind from early times. Fire has also contributed in some regions to forest clearance, creating areas open to colonization by heathland vegetation.

Where frequent, fire has led to selection of a flora composed of species that regenerate vigorously after burning, either vegetatively from protected buds or subsurface perennating structures, or by seed from a soil seed-bank or contained in woody fruits which disperse their contents only when heated by a fire. [*See* Fire.]

**Origins of Heathlands.** Purely natural heathlands may have a long history dating, sometimes from the Pleistocene period or earlier. Few of them, however, have escaped modification by human influence from early times: for example, the South African *fynbos* was subject to "firestick farming" at least 10,000 years ago and in places has been used for grazing by domestic stock for some 2,000 years.

At many different times in the past, heathlands have evidently replaced former forest. In such instances, studies of quaternary forest history almost invariably point to direct links between human activity and reductions in forest cover, with its replacement by grassland in the more fertile areas and heathland on poorer soils. This is especially true of the western European heathlands, and because they were formerly extensive and their origins have been intensively researched, the following summary concentrates on this area.

At first, inroads into the forest, which date from late Neolithic times (more than four thousand years ago), were small in scale and only temporary, representing a form of shifting cultivation. After cropping for a few seasons, the cleared patches were abandoned, allowing the heath plants especially heather (*Calluna vulgaris*), already present as part of the forest flora on poor, acidic soils, to colonize from a buried seed bank or dispersal from nearby. If no further use was made of these patches, however, trees quickly reinvaded, and the area reverted to woodland.

Later, especially during the Bronze Age (about thirty-five hundred years ago) and Iron Age (about twenty-five hundred years ago), heathland acquired value as grazing for the increasing numbers of domestic herbivores (sheep, cattle, goats, horses), and return to forest was delayed or prevented. Heather proved to be a useful grazing plant where the more palatable and nutritious grasses were scarce, particularly in winter when other forage was unavailable. The herbivores themselves tended to stimulate the production of edible green shoots by the heather plants and to prevent the establishment of tree seedlings. Where grazing intensity was insufficient for these purposes, recourse was made from time to time to the use of fire as a means of management.

Hence the origin of most heathlands in western Europe was anthropogenic. Although only a few have an unbroken history extending over three thousand years or more, the process of forest clearance continued at intervals because of the demand for timber for buildings, ships, and fuel (including charcoal for iron smelting), and because of increasing needs for arable and grazing lands. Heather proved to have other uses, including thatch, animal bedding, resilient foundations for tracks and roads, and as a source of dyes and honey. The areas of open heathland grew, and they were maintained and managed until their extent reached a peak in most countries of the heathland region in the late nineteenth century. In Britain, extensive midaltitude and upland heathlands became important for hill sheep farming, particularly in the late eighteenth and early nineteenth centuries, and later for sporting purposes because they provided a habitat for large populations of the red grouse (*Lagopus scoticus scoticus*, a game bird) and the red deer (*Cervus elaphus*). Hill sheep and grouse rely on the young green shoots of heather for a large proportion of their diet, and they are also grazed by deer.

Rotational burning was adopted as the regular method of heathland management. This was carried out in small strips and patches, such that each patch was burned at intervals of, usually, between ten and fifteen years. The branching pattern of the heather bush is such that, so long as it is not allowed to get too old, it responds well to burning or cutting and will resprout from buds at the stem base, which are protected from damage by surface litter or vegetation.

**The Decline of Heathlands.** In many regions the area of heathland declined over the past one hundred to three hundred years owing to transformation for agriculture and other land uses. In places its disappearance has been rapid, so that there is now little left outside nature reserves or other protected areas. For example, it has been estimated that the extent of *fynbos* in South Africa has declined by some 60 percent and that of heath and scrub heath in the southwest botanical province of Australia by between 40 and 60 percent. Even greater losses have been incurred in the lowland heathland regions of western Europe: figures as high as 95 percent of the area covered by heath around 1850 have been

quoted for Denmark and the Netherlands, and 90 percent in the province of Halland in southern Sweden. Similar trends are reported in Belgium and parts of northern Germany and France. In southern England, six of the main heathland areas have decreased in extent in the same period by 72 percent overall, the figures ranging from 50 to 90 percent in different localities. In the uplands and north of Britain the decline has been rather less, but in Scotland between the 1940s and 1980s there has been a contraction in the area of "heather moorland" of at least 23 percent. In a few localities there has been limited gain, however, owing to change toward heathland on bog surfaces that have been drained, or where former forest has been cleared.

**The Cause of Loss of Heathland Vegetation.** In northern Europe many factors are contributing to this decline in the extent of heathland, most of them associated with the disappearance of traditional forms of management and the demands on land for alternative uses. Increased production on better land has made the use of heathland for grazing domestic herbivores relatively unprofitable, and there is now little interest in harvesting cut vegetation or turves for fuel or other purposes. Where management has ceased, heathland may quickly be invaded by shrubs and trees, or by other aggressive competitors such as bracken (*Pteridium aquilinum*). In Europe and elsewhere, however, most of the losses of heathland have been due to conversion to arable land, grassland, or plantation forestry, accompanied by soil improvement, including the use of fertilizers. For example, nearly 50 percent of the area of *fynbos* in South Africa has been transformed by afforestation or agriculture, mainly in the period following colonial settlement from the late seventeenth century onward, while much of the loss and fragmentation of southwestern Australian heaths has also been a consequence of land use change, mainly during the past one hundred years. In Europe, conversion of heath to plantation forest has been the major change in countries such as Sweden and Scotland, while "reclamation" for agriculture was predominant in Denmark, the Netherlands, Germany, and southern England.

In Australia, heathlands have suffered in many places from the effects of the introduced pathogen *Phytophthora cinnamomi*, while in northwestern Europe damage to heather has been caused by phytophagous insects such as the heather beetle (*Lochmaea suturalis*) and (very recently in northern Scotland) the winter moth (*Operophtera brumata*). Heathlands have also disappeared under urban expansion, industrial development, roads, airports, and other artifacts. Some have given place to invasive plants, especially trees, as for example in the *fynbos*, where some 36 percent of the total area of this vegetation type has changed in this way.

In many places, heathlands have survived only where

protected or where they retain some economic or social value, for example, as open countryside for hunting (e.g., in Britain, shooting red grouse or red deer), as grazing land for ponies (northern France, southern England), or as a source of cut flowers for export (South Africa). Pastoral use of heathlands continues in some areas, including to some extent the *fynbos* and, more especially, Scotland, where sheep and red deer are the chief grazers. In parts of these regions, grazing pressure has increased and become excessive in recent years, in contrast to its cessation elsewhere, and this too has damaged the vegetation, usually causing change to rough, relatively unproductive grassland.

In many heathlands there is a delicate balance between heath and grassland species. The heath vegetation belongs to infertile soils, and any significant increase in nutrient status leads to change toward grassland or other vegetation types. Traditional forms of management in Europe, such as grazing, burning, and turf stripping have always had the effect of periodically depleting the ecosystem of a proportion of the nutrient fund, thus maintaining conditions suitable for the heath. Abandonment of this management may allow some accumulation of nutrients in soil and vegetation, improving conditions for competitor species. Change of this kind has been greatly accelerated in recent times in certain areas such as the Netherlands, northern Germany and southeastern England by the input of nutrients, especially nitrogen and phosphorus, from atmospheric pollution. In the Netherlands, for example, from the 1980s inputs of nitrogen have exceeded 40 kilograms per hectare per year, and in places more than 60 kilograms. This development has been clearly linked to the decline of heath species such as heather (sometimes associated with increased attacks of heather beetle) and their replacement by dominant grasses such as wavy hair grass (*Deschampsia flexuosa*) or purple moor grass (*Molinia caerulea*). Throughout much of the heathland region of western Europe inputs of nitrogen now exceed 15 kilograms per hectare per year, which is said to be beyond the ecological tolerance of most heath species. These heathland ecosystems must therefore be regarded as endangered.

Climatic warming and increased carbon dioxide concentrations in the atmosphere are also possible causes of change, though their likely impact on heathland is unknown. In Europe, heather occurs over a wide temperature range and might therefore be affected by climatic warming only at the margins of its geographical area, though changes in precipitation and atmospheric humidity regimes might prove more important. Other species of heathland communities have narrower temperature ranges and in some cases might be significantly affected by global warming. There is experimental evidence of a positive growth response in heather to increased carbon dioxide levels, but a maximum is quickly reached in typical nutrient-poor habitats by limited nutrient supply. However, flowering is strongly enhanced and advanced in time under increased carbon dioxide treatments. The ecological significance of these responses, however, is unknown. In Australia, as elsewhere, it is a general view that land use changes are, at least at present and in the foreseeable future, of greater significance in regard to loss of heathland than climatic change.

**Is There a Future for Heathlands?** The total global extent of heathlands has never been large in comparison with other biomes, and its rate of decline in the past two hundred years is such as to make it an endangered type, except in a very few areas. The scientific and aesthetic value of heathlands, however, is great, especially where they are seminatural, managed ecosystems and where they have high biodiversity. Conservation of heathlands must therefore have high priority. With the disappearance of traditional uses of heathlands in Europe, this poses problems because conservation here depends on continued management, which, although not necessarily using traditional methods, must perpetuate their ecological effects (such as preventing nutrient accumulation). In view of the drastic reductions of area and fragmentation of heathland in many countries, attention is also turning now to the restoration of heath vegetation in places from which it has disappeared, and to the creation of corridors to link surviving fragments. Continuing and increasing interest in the conservation and restoration of heathland suggests that there is indeed a future for heathlands.

## BIBLIOGRAPHY

Cowling, R. M., ed. *The Ecology of Fynbos.* Capetown: Oxford University Press, 1992. An authoritative, multiauthor account of the Fynbos Biome Project.

Forman, R. T. T., ed. *Pine Barrens: Ecosystem and Landscape.* New York: Academic Press, 1979. A thorough coverage of all aspects of the ecology of these areas, including the heaths of the "pine plains."

Gimingham, C. H. *Ecology of Heathlands.* London: Chapman and Hall, 1972. Covers the origins, history, structure, ecology and uses of heathlands, with emphasis on the western European heathland region.

Gimingham, C. H., and J. T. de Smidt. "Heaths as Natural and Seminatural Vegetation." In *Man's Impact on Vegetation*, edited by W. Holzner, M. J. A. Werger, and I. Ikusima, pp. 185–199. The Hague: W. Junk, 1983. Reviews the origins, uses, and management of European heaths, with special reference to grazing, burning, and "plaggen" (turf stripping).

Hobbs, R. J., ed. *Biodiversity of Mediterranean Ecosystems in Australia.* Chipping Norton, New South Wales, Australia: Surrey Beatty, 1992. Includes much material on Australian heathlands.

Hobbs, R. J., and C. H. Gimingham. "Vegetation, Fire and Herbivore Interactions in Heathland." In *Advances in Ecological Research 16*, edited by A. Macfadyen and E. D. Ford, pp. 87–173.

London: Academic Press, 1987. A review of research and its bearing on management and conservation in northwestern Europe, especially Scotland and northern England.

Pate, J. S., and J. S. Beard. *Kwongan: Plant Life of the Sand Plain.* Nedlands, Western Australia: University of Western Australia Press, 1984. A comprehensive review of southwestern Australian heathlands.

Smidt, J. T. de. "Phytosociological Relations in the North West European Heath." *Acta Botanica Neerlandica* 15 (1967), 630–647. Analyzes the variation in floristic composition of European heathland vegetation.

Specht, R. L., ed. *Heathlands and Related Shrublands,* vols. A and B. Ecosystems of the World. Amsterdam: Elsevier, 1979. Indispensable source books of information on the world's heathlands, divided into (A) descriptive studies, and (B) analytical studies (biological, physiological, and ecological, including conservation).

Thompson, D. B. A., A. J. Hester, and M. B. Usher, eds. *Heaths and Moorland: Cultural Landscapes.* Edinburgh: H.M.S.O., 1995. Twenty-four contributions to a conference, providing an up-to-date source of information on dynamic processes and changes in heathlands in the U.K.

Webb, N. R. *Heathlands.* The New Naturalist Series. London: Collins, 1986. An excellent, readable account of the history, natural history, and ecology of heathlands, especially the lowland heaths of England.

Woodin, S. J., B. Graham, A. Killick, U. Skiba, and M. Cresser. "Nutrient Limitation of the Long Term Response of Heather [*Calluna vulgaris* (L.) Hull] to $CO_2$ Enrichment." *New Phytologist* 122 (1992), 635–642. An experimental study relevant to the possible effects of $CO_2$ enrichment and global warming on *Calluna* heathland.

—CHARLES H. GIMINGHAM AND
RICHARD J. HOBBS

**HEAVY METALS.** *See* Metals; Pollution; *and* Sanitation.

**HERBICIDES.** *See* Agriculture and Agricultural Land; *and* Water Quality.

## HOLOCENE IN THE SAHARA

*[This case study describes the tremendous climatic and environmental changes, relative to the present, undergone in northern Africa during the last 2°C increase in global temperature.]*

The tropical deserts in the Old World are vast, uninhabitable, and too dry to allow cultivation. Their present-day existence is due to a drastically negative water budget, resulting from scarce precipitation that is unable to match a high evaporation rate, which in turn is related, at tropical latitudes, to intense insolation and high wind strength and velocity. Consequently, these deserts are "lost land" for their overpopulated margins

as well as for global socioeconomic development. Considering the demographic expansion, it has become a scientific priority to understand whether, in past cooler or warmer periods, those deserts have varied in their surficial extent and degree of aridity, and, if so, why, and to what extent such mechanisms are liable to operate again in the future. [*See* Deserts.]

Over the last two decades, multidisciplinary studies of geologic records (paleolacustrine and eolian sediments, vegetal and animal remains, prehistoric settlements) have allowed scientists to attest to alternate expansions (during the glacial phases) or reductions (during the interglacials) of the arid and hyperarid areas, over the last climatic cycle, as summed up by Petit-Maire (1994). Some questions remain to be answered by more precise studies:

• What were the intensity and velocity of the recorded changes toward a more or less arid climate?
• Were the recorded major regional changes affected by human activities?
• Did mid- or short-term nonorbital events, due to unknown external or internal forcings, interrupt the major Milankovich trends?
• Can the observed data allow us better to understand present-day global warming and to anticipate the future evolution of tropical deserts?

The Sahara in northern Africa covers a surface area comparable to Australia or China: 9.5 million square kilometers (Figure 1). The study of its climatic evolution throughout the last climatic cycle is thus a priority. Between 1973 and 1993, over three thousand radiocarbon dates were performed from lake, animal, vegetal, and human remains dating from the Holocene period (ten thousand years ago to the present). As a result, a paleoenvironmental map may be drawn summarizing the climatic evolution of the present hyperarid core of the desert—that is, the area located between hundred-millimeter isohyets to the north and south that neither the monsoon nor the Atlantic cyclones now reach. The evolutionary pattern is strikingly similar from the Nile to the Atlantic, although the dates for the beginning and end of the humid phase may vary by about one thousand years according to latitude, altitude, and oceanic proximity. The eastern Sahara has always remained, as at present, somewhat more arid than the western Sahara.

Five evolutionary stages may be discerned:

1. By about 10,000 radiocarbon years BP, the precipitation and evaporation rate drastically changes. At 9,500 years BP, seasonal lakes or swamps appear in the topographical depressions, fed by runoff and rising ground water. The landscapes change from desert to flowing wadis and strings of shallow surface water bodies.

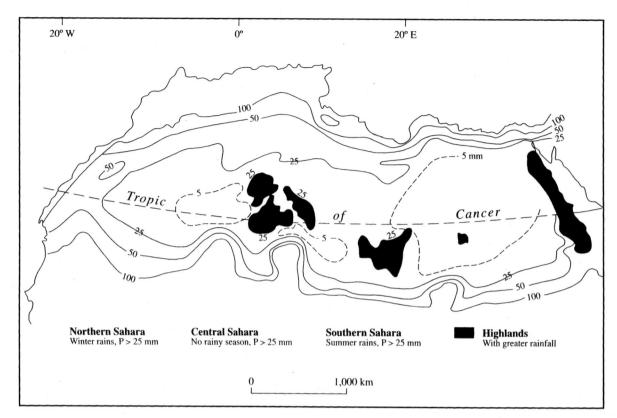

**Holocene in the Sahara. FIGURE 1.** Present-Day Location and Annual Precipitation (in millimeters) of the Sahara.

2. Between about 9,000 and 7,000 years BP, humidity is optimal. A grassland cover exists between the lakelets and swamps. As described in detail by paleozoologists in Petit-Maire and Riser (1983), abundant large mammals have migrated from the south up to the Tropic of Cancer: elephants, rhinoceros, large antelopes, and warthogs. The aquatic or semiaquatic sites are rich in freshwater foraminifera, ostracods, and mollusks, as well as in fish and crocodiles when channels existed between the observed sites and flowing rivers. Hippopotamus bones were found up to 22° north, five hundred kilometers to the north of the present-day location of this species. The presence of this large animal also testifies to plentiful vegetal food. Giraffes, inhabitants of wooded savannas, live up to 20° north. Small Neolithic groups have also moved into the new inhabitable territories. They live around the water bodies on large aquatic and terrestrial game and on vegetable resources, as shown by the abundance of grinders and millstones associated with brown soils. Pollen evidence confirms the existence up to 23° north of Sahelian vegetal species, now only living south of 18° north.

3. After about 6,800 years BP, a period of higher climatic variability is characterized by frequent, sometimes drastic drying up of the shallow lakes and salinization of others, as precisely shown in continuous records from northern Mali by Fabre and Petit-Maire (1988) and Petit-Maire et al. (1991). This episode lasted several centuries, at ages varying between 6,800 and 5,800 years BP, according to local geographical conditions. Vegetation progressively includes more Saharan species, and large game is scarcer.

4. This drier episode is followed by a short second humid optimum. The lakes are still permanent but become shallower and smaller. Progressive disappearance of Sahelian plants and animals from the southern Sahara matches the migration of Neolithic groups, following the regressive range of the rains.

5. Toward 4,500 years BP, an abrupt decline of rainfall is recorded from east to west. Surface fresh water, grasses, and the associated biotopes disappear, the aridification progressing latitudinally to the south (monsoons) and north (Atlantic depressions) relative to the Tropic of Cancer. At 3,000 years BP, the present-day pattern is established. In the western Sahara, prehistoric sites are found nearly exclusively along the oceanic borders, or south of 17° north.

Figure 2 sums up the preceding observations and allows us to answer some questions:

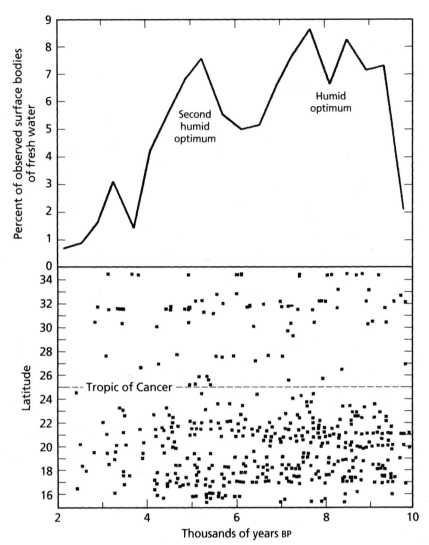

**Holocene in the Sahara. Figure 2.**
Distribution of 560 Holocene Freshwater Bodies throughout the Sahara, According to Radiocarbon Age (top) and Latitude (bottom).

- The distribution of radiocarbon ages from Holocene freshwater lacustrine and paludal deposits in the Saharan areas that are presently hyperarid and without any surface water indicates two optimal periods (early Holocene and about 5,000 years ago), a slightly drier episode between 6,700 and 5,500 years ago, and a very severe arid period around 4,000 years ago. This scheme is confirmed by the study of continuous lacustrine deposits near Taoudenni, in northern Mali.
- The latitudinal location of the deposits shows occurrence of monsoons up to the Tropic of Cancer and of Mediterranean rains down to 29° north. Rainfall over the intermediate area was scarcer and may be attributed to occasional tropical or Atlantic depressions.
- The intensity of the changes after the end of the Pleistocene is drastically high. In the early Holocene, average rainfall was at least fifty times its present values, from east to west.

- The velocity of the later changes is striking. In five hundred years, the landscape changes from swampy areas and flowing rivers to desert.

The small Neolithic groups then living in the Sahara cannot be held responsible for such extreme changes in atmospheric circulation. Moreover, no evidence for agriculture (and intentional bush fires) or cattle breeding have ever been found in western Sahara, even if well looked for in northern Mali, where the climatic changes are latitudinally well studied and dated.

The curve shows three "events" not attributable to orbital forcings. The short, drier event at about 8,000 years BP could be taken as a sampling bias if it did not coincide with an increase of salinity in ocean cores. A decrease in the methane atmospheric ratio, shown in 1991 to be related to tropical vegetation changes, is also registered in the Greenland Ice-Core Project (GISP) ice cores from Greenland, although not at the same moment: 8,200 *calendar* years BP.

The increase of aridity around 6,500 to 6,000 years ago also corresponds to several observations in other areas in the world and in some ocean cores. The Holocene climatic optimum is not 6,000 years ago, which fits the sea level rise after final deglaciation but not atmospheric circulation.

The event at 4,000 years ago is marked throughout the Old World by the decline of great civilizations, often attributed by historians to barbarian invasions: in fact, the Anatolia, Mesopotamia, Indus, and Tarim basin cultures were probably also devastated by the drastic droughts culminating between 4,000 and 3,800 years ago.

These observations on the evolution of the largest tropical desert in the world allow a careful guess: despite the slow neoglacial trend related to Milankovich forcings ($-0.01°$C per century), the recent enhancement of the natural greenhouse effect, due to anthropic pollution of the atmosphere, will increase the globe's surficial temperature quite quickly to an extent at least similar to the natural Holocene one. Even if not due to the same astronomical causes, the warming should have similar effects in the tropical world, namely, an enhancement of the monsoons. [See Earth Motions.]

Will the human-enhanced greenhouse effect green the tropical deserts? According to the data that we have considered here, the answer is yes, beginning along their margins. The demographic pressure in such areas is strong; it should not be allowed to destroy the first effects of increased rainfall. One must protect from anthropozoic pressure the regeneration of ground water and vegetation. Patience, in such a case, is the only key to a better future.

[See also Greenhouse Effect; and Younger Dryas.]

#### BIBLIOGRAPHY

Fabre, J., and N. Petit-Maire. "Holocene Climatic Evolution at 22–23°N from Two Palaeolakes in the Taoudenni Area (northern Mali)." *Palaeogeography, Palaeoclimatology, Palaeoecology* 65 (1988), 133–148. A geologic description and interpretation of several continuous sections through the deposits of two paleolakes with different geochemistry, in the most arid part of the present-day Sahara.

Neumann, K. "Vegetations Geschichte der ÖstSahara im Holozän." *Africa praehistorica* 2 (1989), 13–182.

Pachur, J. J., S. Kröpelin, P. Hoelzmann, H. Goshin, and H. Altmann. "Late Quaternary Fluvio-Lacustrine Environments of Western Nubia." *Berliner geowissenschaftlichen Abhandlungen* 120 (1990), 203–260.

Petit-Maire, N. "Paleoclimates in the Sahara of Mali: A Multidisciplinary Study." *Episodes* 9 (1986), 7–16. The first discoveries in the most unhospitable zone of the Sahara.

———. "Will Greenhouse Green the Sahara?" *Episodes* 13 (1990), 103–107. One of the most important papers linking the past with the future of the tropical deserts.

———. "Natural Variability of the Asian, Indian, and African Monsoons over the Last 130 ka." In *Global Precipitations and Climate Change*, edited by M. Desbois and F. Désalmand, pp. 3–26.

Heidelberg: Springer, 1994. A clear, short summary of paleoclimatic data from the Old World desert belt.

Petit-Maire, N., ed. *Paléoenvironnements du Sahara: Lacs holocènes à Taoudenni (Mali)*. Marseille: Centre National de la Recherche Scientifique, 1991. French and English detailed multidisciplinary study of continuous Holocene lacustrine bed.

Petit-Maire, N., and J. Riser, eds. *Sahara ou Sahel? Quaternaire récent du Bassin de Taoudenni (Mali)*. Marseille: Lamy, 1983. A detailed monograph especially rich in faunal descriptions of remains found between 1980 and 1983 over 500,000 square kilometers in northern Mali.

Petit-Maire, N., and N. Page. "Remote Sensing and Past Climatic Changes in Tropical Deserts: Example of the Sahara." *Episodes* 15 (1992), 113–117. A short review of the glacial and interglacial evolution of the Sahara over the last twenty thousand years.

Petit-Maire, N., M. Fontugne, and C. Rouland. "Atmospheric Methane Ratio and Environmental Changes in the Sahara and Sahel during the Last 130 kyrs." *Palaeogeography, Palaeoclimatology, Palaeoecology* 86 (1991), 197–204.

Petit-Maire, N., N. Page, and J. Marchand. *The Sahara in the Holocene*. Paris: UNESCO/CGMW, 1993. A striking, colorful visual summary of paleoenvironmental radiocarbon-dated observations in the Sahara during the Holocene.

Schulz, E. "The Taoudenni-Agorgott Pollen Record and the Holocene Vegetation of the Central Sahara." In *Paléoenvironnements du Sahara: Lacs holocènes à Taoudenni (Mali)*, edited by N. Petit-Maire, pp. 143–162. Marseille: Centre National de la Recherche Scientifique, 1991.

—NICOLE PETIT-MAIRE

## HUMAN ECOLOGY

Human ecology is defined as the relations between human beings and their biophysical environment, or between nature and society, though the term has also been and continues to be used in many narrower senses as well.

The word *ecology* is derived from *Oekologie*, introduced in the 1860s by the German biologist Ernst Haeckel. He coined the term to denote the relations of the organism with its environment, a topic he held no less deserving of scientific study than the organism's internal anatomy and physiology. In the late nineteenth and early twentieth centuries, *ecology* became the accepted English-language name of a subfield of biology examining the ways in which climate, soil, other environmental conditions, and interactions among organisms influence the distribution of plant and animal species and species assemblages. Biological ecology developed a number of highly influential terms and concepts. Among them, those of community and ecosystem emphasized the close relations and interdependencies of different components of the landscape. Those of succession and climax described an orderly evolution of species assemblages toward an end state determined largely by local conditions of climate and soil. [See Ecosystems.]

The approach and the concepts developed by ecologists proved attractive to some social scientists as well. A subfield of sociology that developed in the 1920s under the name of *human ecology* transferred them into the social realm to address the relations between human individuals and their social—economic, institutional, and technological—surroundings. Its name notwithstanding, sociological human ecology paid and has continued to pay little attention to human relations with the biophysical environment. But it is not the only school of thought to take the name that it did or a similar one, and the term *human ecology* and similar coinages have also been attached to the study of particular aspects of human interaction with the biophysical world. A subfield of medicine addressing the environmental relations of disease was dubbed human ecology; so was the study of the physiological and psychological influence of weather on human beings. Anthropologists and geographers interested in cultural ecology—a term that dates back to the 1940s—have focused on human use of resources in rural and especially premodern settings. More recently, the term political ecology has been used to describe similar work with a stronger emphasis on the role of class and power in shaping and constraining resource use. Still more recent coinages include those of feminist cultural ecology and feminist political ecology, denoting a particular interest in gender relations. Concern among economists for such matters as the material and energy dimensions of transactions has given rise to the field of ecological economics. "Industrial ecology" has recently been proposed as an organizing concept for the study of the entire set of relations between industry and the environment, from the consumption of resources to the emission of byproducts and wastes. A radical school of environmental ethics associated with the Norwegian philosopher Arne Naess has taken the name of deep ecology. [*See* Industrial Ecology.]

As used here, "human ecology" takes in many of these more partial senses. It was defined in its full breadth as early as 1922 by the American geographer Harlan H. Barrows as denoting "the mutual relations between man and his natural environment." It is used in this wider meaning by, among others, the U.S.-based journal *Human Ecology*, the International Society for Human Ecology, and a standard library classification for books dealing with human-environment relations.

Taken in this sense, *human ecology* is one of several possible umbrella terms under which to group the study of all relations between humankind and the biophysical world. Other terms that can be and often are used for that purpose are *human-environment relations, nature-society relations, environmentalism, environmental studies,* and *environmental science. Human*

*ecology* is the crispest and least clumsy of the available terms, and it carries few of the connotations of a political agenda that *environmentalism* may suggest. Not only is the literal meaning of the term apt, but because the concepts of integration, complexity, systems, and feedback central to biological ecology are basic to the contemporary understanding of the nature of human-environment relations and especially of global environmental change, its connotations are equally appropriate.

The name also has some possible disadvantages. It is sometimes argued that a human ecology, so-called, must tend to emphasize the elements that human beings share with plants and animals and thereby to reduce social science to some form of biological determinism; that it must focus on such variables as population numbers and means of subsistence while neglecting what is distinctive in human and social activities, including relations with the environment: reflexivity, values and ethics, institutions, and culture. It has also often been suggested that classical plant and animal ecology places such emphasis on the notions of equilibrium and climax states that any social science drawing upon it must have conservative or reactionary political implications. It has been suggested too that any importation of natural science into social analyses must have the effect of justifying whatever social relations happen to exist as "natural" and therefore beyond criticism. But many counterexamples suggest that such need not be the case, nor is there any necessary reason why the full range of distinctively human characteristics cannot be brought into a field defined as human ecology. The danger of reductionism indeed exists and must be carefully guarded against, though a social science that ignores the biophysical dimension of human life offers no less distorted a picture of reality than one that overemphasizes it. The objection that drawing upon ecology implies a bias toward stability carries less weight now than it once might have, for biological ecology has undergone such change in this regard that many now speak of a "new ecology" that has abandoned the steady-state and climax dogmas of its past.

If human ecology is defined as encompassing all human relations with the biophysical environment, it covers a much wider field than global environmental change or the human dimensions thereof. Global changes are one possible consequence of human use of the environment, and they are one possible source of change in human-environment relations. They are far from the only factor in those relations, nor, in any time and place, are they necessarily the most important one. Human ecology is the whole of which the human dimensions of global change form a part. Under any sensible organization of inquiry, therefore, an encyclopedia of human ecology would incorporate an entry or entries on global environmental change, not the other way around. Yet human ecology does not exist, under that or any other name, as an organized academic field incorporating all work on the topic. Global change and the human dimensions of global change, on the other hand, are increasingly approaching that status, with their own journals, academic organizations, professional meetings, and budget lines with funding agencies. How this state of affairs arose, and how the narrower focus has flourished at the expense of the broader—how the study of global environmental change has taken precedence over the wider study of environment—has much to do with the history of attitudes in twentieth-century social thought toward the biophysical dimensions of human existence.

The first important grand theory of human ecology in modern social science was that of environmental determinism, in which the biophysical environment also figured in a partial role, in this case chiefly as something affecting rather than affected by humankind. In the early twentieth century, determinism dominated the academic field of geography in many countries and influenced work in every other discipline as well. Its best-known proponents included the American geographers Ellen Churchill Semple and Ellsworth Huntington. Both emphasized the role of the biophysical environment in shaping individual behavior and collective social attributes alike. Semple, for instance, proposed that the inhabitants of mountains, mountain passes, islands, seacoasts, wetlands, and other physiographic zones had certain fixed psychological and social characteristics impressed upon them by their surroundings. Huntington sought the sources of such characteristics mainly in climate. He tried to identify the ideal and the substandard climates and seasons for civilization, mental and manual labor, and social progress, and he attempted to explain the rise and decline of ancient civilizations as the consequence of shifts in storm tracks that had stimulated or sapped human energies. In the strictest sense, not even Semple or Huntington was an environmental determinist. Both paid frequent lip service and sometimes more to ways in which social and cultural differences affected the significance of environmental features. Huntington in some of his writings emphasized race and diet as factors ranking with climate as forces in human life. But both did regard environment as a powerful and sometimes a dominant external and independent influence, and to call their work deterministic is not unjust or misleading.

By the 1920s, criticisms of environmental determinism in the social sciences were already widespread and effective. Within a few decades, it had almost entirely ceased to be a significant presence even within those disciplines, such as geography and anthropology, most closely concerned with human relations to the environ-

ment. The most constructive reaction against determinism took the form of the counterthesis that came to be known as possibilism. Expounded most influentially by the French historian Lucien Febvre, it held that environments never impose any characteristics or any particular way of life on those who dwell in them. Instead, they offer a broad range of possibilities. Any particular society, depending on its own characteristics and activities, will recognize and exploit only some of those possibilities as resources and encounter and suffer from only some of them as hazards or obstacles. Thus the environment is never an entirely independent variable in possibilist human ecology; its significance can never be defined except with relation to the goals, technological capacity, internal organization, and external relations typical of the society that occupies it. Over time, the possibilities offered by the environment can be altered by changes in the environment itself, whether natural or human-induced (and, if the latter, whether deliberate or inadvertent). They are likewise altered, even if the environment remains stable, by any changes occurring in human society, from technological innovation to cultural and political change to shifts in settlement patterns to the expansion of trade to the development or decay of property rights, exchange institutions, or social safety nets, that affect the range of possibilities and problems that the environment offers.

The possibilist approach to human ecology emphasized detailed attention to the full array of characteristics of both peoples and environments as necessary for understanding their interactions. It drew empirical support from research in anthropology, geography, and history contradicting the claims of environmental determinists and showing that similar environments had been and continued to be inhabited by very different societies and cultures and used by them in quite varied ways. A French-based approach to history associated with Febvre and his colleague Marc Bloch came to be known, after the name of the journal on which it centered, as the *Annales* school. It devoted much attention to climate, terrain, soils, biota, and other aspects of the human environment and to the changing conditions under which societies exploited them as resources or experienced them as hazards. Possibly the most influential single work it produced was Fernand Braudel's *La Méditerranée et le monde méditerranéen a l'époque de Philippe II* (orig. 1949). Braudel dealt at the outset with the human ecology of the Mediterranean region—the biophysical environment and the technology and institutions that governed its use—as intimately intertwined with its economic and political history.

Equally within the possibilist vein, the German-trained, American-based economist Erich Zimmermann enunciated what he called the functional theory of natural resources. Resources, Zimmermann emphasized, are defined as such not through their inherent physical qualities alone but through the roles that those qualities can play in particular human activities. The usefulness of a resource, Zimmermann argued, is therefore dependent on the social conditions, including technology and institutions and culture, that allow the activity in which it is useful to be profitably carried on. A natural substance can thus cease to be a resource, though itself remaining physically unchanged, when changes in activities put an end to its usefulness, or it may become a resource in a new way, as the chief use of petroleum shifted from lighting in the nineteenth century to transportation in the twentieth. Environmental change in the form of physical depletion or exhaustion owing to overuse is only one possible end to a resource among many. Mines are not abandoned only when the ore runs out or becomes too difficult to extract. Product substitution may render the ore valueless. Ruinous competition may arise from other and richer mines opened elsewhere. Declines in ore prices or in protective tariffs may make extraction uneconomical. Changes in labor relations may raise wages, or changes in markets raise energy prices beyond what extraction can profitably cover, and so on.

The American geographer Gilbert F. White pioneered an analogous social science approach to environmental hazards. Rather than seeing certain natural events or phenomena as inherently hazardous to human activity, White described losses from hazards as the result of the interaction between natural events and human activities that effectively invite or avoid loss from them. The former approach implied that only altering or controlling a feature of the environment could make it less hazardous. What White called the "range of choice" perspective, on the other hand, called attention to the many ways in which changes in human use and occupation of the environment could make natural agents more or less threatening and harmful without those agents themselves undergoing any change. Subsequent hazards research has defined the vulnerability of different groups, individuals, and activities exposed to the same natural agents as the differences in ability to cope with its effects that make losses greater for some than for others. [See Natural Hazards.]

Mid-twentieth-century research inquiring into the role of culture and technology in the adaptation of peoples to their habitats came to be known as cultural ecology through the writings of the American anthropologist Julian Steward. One of its key concerns has been the role of indigenous environmental knowledge in resource management and its relation to the pressures of the economic and political context. Similar interests, including a local scale of study and an emphasis on field research in rural settings, characterized the Berkeley school of Carl O. Sauer. Apart from natural hazards research, it represented the chief nature-society subfield in the discipline of geography. All of these approaches were broadly representative of a possibilist human ecology.

It was not possibilism, however, but rather a near-entire disregard for the biophysical environment that became the rule in the Western social sciences by mid-century. Questions of human ecology in any form disappeared from teaching and research save in a few subfields and the work of scattered individuals. This state of affairs came about for more than one reason. A strong contributing factor was a reaction against the excesses and the distasteful political affiliations of much of environmental determinism that sometimes went to the extreme of disregarding the biophysical world altogether. Many social scientists may also have supposed, consciously or not, that technology had become so powerful and culture and society so intricate as to deprive the environment of most of its significance for modern human life, however much it might have mattered in the past. The American sociologists William Catton and Riley Dunlap have described this belief as the "human exemptionalist paradigm" and found it reflected in most mainstream postwar social science. But it was the organization of academic research and instruction into separate realms devoted to natural and social phenomena respectively that perhaps did the most to discourage work on topics necessarily involving both. That separation allowed social scientists to establish their independence from the natural science disciplines and allowed and encouraged them to insist that social phenomena in their causes and consequences be discussed only in terms of other social phenomena. Equally, it diverted Earth and biological scientists from any close or sophisticated consideration of the possible human role in the patterns and processes they studied.

When human ecological concerns began to reappear widely in academic research and teaching on both sides of the divide, it was largely because of events occurring outside of academia. The rise of popular environmental consciousness and concern affected social and natural scientists no less than it did other members of the public. What has resulted from their reawakened interest is a body of research on the whole displaying several characteristics shaped by the circumstances under which it developed. For one thing, its origin in a sense of crisis and a wish to contribute to the solution has colored it in such a way as to leave it open to the suspicion of being less than dispassionate and even-handed. It is often accused of a predisposition to judge human impact to be harmful and its regulation to be necessary, allowing those who do not agree to dismiss it out of hand. For another, as the product not of a discipline of its own but of researchers trained in the established natural and social science disciplines, it lacks a unifying body of theory and methods. Most of the concepts used in it are simply the concepts central to various other, nonenvironmental fields transplanted and applied to environmental issues as if fully adequate to deal with them. As a result, environmental research is kept intellectually fragmented into subschools of thought originating in the various social and natural science disciplines. There exist few concepts and theories truly indigenous to human ecology. Last and most strikingly, arising as it did out of concern over human impact, it has tended to deal with the effects of environmental changes, and particularly global ones, to the exclusion of other aspects of human-environment relations. Yet because of the previous neglect of human ecology, it has been in the position of performing applied science before the basic science has been done; of asking, that is, what role environmental change might play in human life while knowing little about what role environment plays in human life. [*See* Human Impacts, *article on* Human Impacts on Earth.]

The fields of global change and the human dimensions of global change do not constitute a full-fledged human ecology because of their partial focus. Environmental change, global or otherwise, can make any feature of the environment more or less valuable as a resource and more or less threatening as a hazard, but so too can social change that leaves that feature physically unaffected. The increasingly large-scale and rapid social changes that contribute to an unprecedented rate of alteration in the global environment also make for an unprecedented rate of change in human relations with aspects of the environment in which the role of environmental change is modest. The most worrisome of the projected consequences of human-induced stratospheric ozone depletion is an increase in the incidence of skin cancer in the human population. But incidence has already increased dramatically during the twentieth century for other reasons: chiefly migration, especially of fair-skinned peoples of northern European descent, to lower latitudes and changes in dress and behavior increasing exposure to sunshine. Dry and sunny climates during the late twentieth century have become much more valuable a resource than ever before, not because of any changes in climate but thanks to factors increasing the potential for their exploitation by tourism and "sunbelt" migration, and as a result they have become more deadly a hazard. The role of stratospheric ozone depletion in the increase in skin cancer is a matter of the human dimensions of global change. The role of all factors together, including but not limited to ozone depletion, contributing to the increase is one of human ecology. It is the former—the human dimensions of global change—that have been the most closely studied, despite their small relative importance to date, because anything having to do with environmental change is guaranteed a degree of attention that other processes of human ecology are not. Likewise, agricultural land abandonment can sometimes be traced to environmental change, such as climatic change or soil degradation or pest infestation. But it often occurs for other reasons entirely: changing tastes, increases in labor and energy costs, and the abandonment of tariffs, import quotas, and other restrictions on trade that lessened competition from elsewhere, for instance. A complete understanding of the process will not be obtained from analyzing the role of stress from environmental change alone.

But as its relevance to the human dimensions of global change becomes clearer, such a wider perspective may well be developed. The urgency associated with many global change issues may finally overcome the barriers to environmental research more generally and make possible the emergence of a human ecology equally attentive to natural and social processes. Much of the research and many of the ideas developed in the global change field may form some of its building blocks. Studies prompted by concern about the effects of global climatic change have greatly, if incidentally, enlarged our understanding of society-climate relations more generally. The recently popularized concept of "natural services" has been elaborated by ecological economists to capture the point that ecosystems, climate, and other biophysical realms provide services that are vital or valuable to human society but are often taken for granted rather than being ascribed their real value and protected accordingly. The concept expands the notion of direct environmental value beyond the classical notion of "natural resources" to include many services that are useful without being consciously exploited and whose importance may be neglected until they disappear or decay through mismanagement. The parallel concept of natural disservices, broadening the classical one of natural hazards, could be developed to catalog and value the sources of human loss in nature-society relations. In either case, a key insight of earlier resource and hazards work must not be lost: that the services and disservices rendered by any aspect of the environment are not inherent in its physical qualities. Rather, they emerge from the interaction of those qualities with human arrangements and can be altered by a change on either side, or both.

[*See also* Environmental Movements.]

## BIBLIOGRAPHY

Barrows, H. H. "Geography as Human Ecology." *Annals of the Association of American Geographers* 13 (1923), 1–14.

Burton, I., R. W. Kates, and G. F. White. *The Environment as Hazard*. 2d ed. New York: Guilford Press, 1993. A standard work on the nature of natural hazards.

Daily, G., ed. *Nature's Services*. New York: Island Press, 1997.

Ellen, R. *Environment, Subsistence, and System: The Ecology of Small-Scale Social Formations*. Cambridge: Cambridge University Press, 1982. A solid assessment of anthropological human ecology.

Febvre, L. *La Terre et l'evolution humaine.* Paris: La Renaissance du Livre, 1922. (Translated into English as *A Geographical Introduction to History.* Boston, 1925.) The classic statement of possibilism.

*Human Ecology.* Journal, 1966–present. Features as wide a range as any journal of scholarly research on environment and society.

Huntington, E. *Mainsprings of Civilization.* New York: Wiley, 1945. A classic treatise of environmental determinism.

Worster, D. *Nature's Economy: A History of Ecological Ideas.* Cambridge: Cambridge University Press, 1985. An outstanding if controversial analysis of the emergence and evolution of ecological thought.

Zimmermann, E. W. *World Industries and Resources.* New York: Harper, 1951.

—WILLIAM B. MEYER

# HUMAN HEALTH

There are multiple determinants of health and well-being. Biological and psychological factors, ecological and global systems all play their roles, while economics and access to health care determine the social vulnerabilities to disease. These ingredients can interact, and our chief means of controlling infections—antibiotics and insecticides—have become drivers of new, resistant microbes and disease carriers. Meanwhile, the growing numbers of people with depressed immune systems and those with malnutrition help to select and disseminate emerging organisms.

Environmental influences interacting with internal biological factors can be profound. While land use patterns affect the distribution of disease carriers, climate affects their range, and weather affects the timing and intensity of outbreaks. The subject of this article is the interplay of social, local environmental, and global dynamics that influences our health. The discussion focuses primarily on the environment, for—given the scale and pace of change—environmental change may play an ever-increasing role in determining disease patterns in the future.

Ecosystems show enormous resilience, resistance, and flexibility, and can rebound to contain explosions of nuisance organisms. Human flexibility and new manufacturing and restorative technologies hold great promise for responding to environmental challenges. Greater surveillance, treatment, vaccines, and support for basic public health infrastructure are the first responses needed to address the resurgence of infectious disease in the latter part of this century. But, ultimately, our social system must resolve how we interact with other natural systems to maintain one of nature's essential "services"—the containment and control of pests, parasites, and pathogens.

The expected redistribution of infectious disease is but one of the biological consequences of global environmental change. The impacts of global environmental change on agriculture pests and crop yields, on livestock health and fisheries, and on human illness can be enormous; and the costs of epidemics can cascade through economies and ripple through societies. The resurgence of infectious disease thus poses a threat to food security, to biological security, and to economic development. [*See* Disease.]

**Background and History of Epidemics.** A 1996 report of the World Health Organization (WHO) notes that, since 1976, thirty diseases that are new to medicine have emerged. The reappearance of old ones—once thought under control—is of equal concern: drug-resistant tuberculosis, exacerbated by human immunodeficiency virus/acquired immune deficiency syndrome (HIV/AIDS), now causes three million deaths annually, while childhood diphtheria, whooping cough, and measles, also transmitted person to person, are also on the rise, particularly in those places where social systems have deteriorated. Malaria, dengue (or "breakbone fever," a severe, sometimes-deadly tropical disease transmitted by mosquitoes and characterized by headache and severe joint pain), yellow fever, cholera, and a number of rodent-borne viruses are also appearing with increased frequency. These latter diseases, which rely on animals or water as vehicles (or vectors) for transmission, are a reflection of both environmental and social change. In 1995, U.S. mortality figures from infectious diseases rose 58 percent above the levels of fifteen years before; 22 percent was attributable to causes other than HIV/AIDS.

Pandemics (affecting multiple continents) emerge out of social and environmental conditions, and they can propel changes in both. At times, these changes have been disruptive; at other times they have generated significant social reform.

***The rise and fall of infectious disease.*** From a long-term historical perspective, pandemics have often been associated with major social transitions and overtaxed infrastructures. Their impacts have been profound and lasting.

A pandemic of debated cause, but remembered as the Plague of Justinian, struck Europe in 541 CE as the Roman Empire was in decline, and raged for two centuries, claiming over forty million lives. Urban centers were abandoned, and the plague encouraged population resettlement into rural, feudal communities.

After a six-hundred-year hiatus, plague again appeared in 1346, when growing urban populations had once more outstripped the capabilities of cities to sustain sanitation and basic public health. Several other factors played compounding roles: human populations had migrated from east to west; the Medieval Climatic Optimum of the twelfth and thirteenth centuries may have contributed to the proliferation of rats and fleas that carried the bubonic plague; and cats had been killed in the belief that they were witches. In the ensuing five years of the Black Plague, about twenty-five million lives were taken—about one of every three people who lived in Europe at the time.

A third outbreak of widespread epidemic disease, almost three hundred years later, had a more positive outcome. In the early Industrial Revolution, urbanization led to a substantial decline in mortality from infectious disease. Then, in the 1830s, under the growing weight of industrialization and the growth of population (sevenfold in London from 1790 to 1850), the conditions in European cities that were described in the novels of Charles Dickens became breeding grounds for three major infectious diseases: cholera, smallpox, and tuberculosis. Growth and development had outgrown infrastructure, and infectious diseases rebounded.

This resurgence of infectious disease precipitated protests throughout the European continent, and ultimately led to constructive responses. In England, the sanitary and environmental reform movements began; and the field of epidemiology ushered in modern public health principles and eventually led to a national health program. The epidemics were abated over the ensuing decades, three-quarters of a century before the advent of the antimicrobials that are prescribed to counter the same diseases today.

***Recent history.*** By the 1960s, widespread improvements in hygiene, sanitation, and mosquito control led most public health authorities to believe that humans would soon conquer infectious diseases. In the 1970s, public health schools turned their attention instead to chronic ailments, such as heart disease, stroke, diabetes, and cancer. But the epidemiological transition to diseases of modernity never materialized in most developing nations. And, in the 1980s, the global picture shifted dramatically.

According to the World Health Organization's 1996 report, drug-resistant strains of bacteria and other microbes are having a deadly impact on the fight against tuberculosis, malaria, cholera, and pneumonia—which collectively killed more than ten million people in 1995. Antibiotic resistance has resulted from increasing pressure from overuse, selection of mutated microbes in vulnerable hosts, and the geographic movement of humans, insects, rodents, and microbes. Ironically, the very means of control over infectious disease—antibiotics and insecticides—are themselves rapidly driving the evolution of new and unaffected strains. And two-thirds of antibiotic use is for animal husbandry, agriculture, and aquaculture.

Diphtheria rose exponentially in the former Soviet Union as the public health system deteriorated following political and economic changes. The incidence rose from four thousand cases in 1992 to eight thousand in 1993 and forty-eight thousand in 1994, and the disease has killed more than four thousand residents of the former Soviet Union since 1990. It has spread through fifteen nations of Eastern Europe, but recent immunization campaigns have begun to control this infection.

Dengue fever, for which no vaccine is yet available, had essentially disappeared from the Americas by the 1970s, but has of late seen a resurgence in South America, and in 1995 infected over 240,000 people. According to a recent editorial in the British medical journal *Lancet*, "*Aedes aegypti* is now well established in all areas of the Americas except Canada and Chile." In the United States, the Asian tiger mosquito, *Aedes albopictus*, has been introduced in the past two decades, apparently imported from Asia in used tires. Fortunately, *A. albopictus* does not seem to be the most efficient carrier of dengue fever. In Asia, the resurgence of the disease in New Delhi in 1996 resulted in more than eight thousand cases and several hundred deaths. Settlements that surround India's large cities, where discarded nonbiodegradable containers can serve as excellent mosquito breeding sites, provide especially vulnerable settings, while mild cool seasons and weather extremes can precipitate large outbreaks in nonimmune populations. Meanwhile, previous exposure and a change in the circulating dengue viral type may lead to dengue hemorrhagic fever, which carries a 5–10 percent mortality.

In 1995 the largest epidemic of yellow fever since 1950—carried by the same mosquito that transmits dengue—hit the Americas. Peru and the Amazon basin were heavily impacted. There is a growing potential for urban yellow fever, and, while there is a vaccine, the supply is inadequate for present needs.

In 1996 the largest epidemic of meningitis ever recorded struck West Africa, associated with pervasive drought on that continent. (Dry mucus membranes may facilitate invasion of colonizing organisms.) Over 100,000 people contracted the disease, and more than ten thousand died.

Conditions conducive to the spread of epidemic infectious diseases now exist worldwide, according to the WHO report. Infectious diseases that are prevalent in the United States, such as Hantavirus Pulmonary Syndrome, Lyme Disease, and toxic *Escherichia coli*, were not imported from other nations. *E. coli*, for example, spreads in cattle raised in closed quarters. When put through a huge meat grinder, one infected cow can contaminate an entire batch. Domestic environmental and social conditions conducive to the spread of *E. coli* prevail in the United States today. The transmission of tuberculosis, to give another example, is facilitated in homeless shelters and in prisons. And, while poorer populations are at greater risk, outbreaks of infectious disease are not restricted to disadvantaged regions, for today's population movements facilitate "microbial traffic" between nations and economic groups.

**Malaria.** Malaria is an ancient, mosquito-borne disease that has played a significant role in the history of Africa. Through the ages, the presence of sickle cells and other types of red blood cells warded off the parasite's damage to cells and limited malaria's impact on native Africans. But the disease also served to ward off foreign colonizers, who lacked these evolved defenses, and until the latter part of the nineteenth century malaria helped deter the deeper penetration of the continent. In the first steps of the race for African territory, Europeans selectively colonized highland regions to escape from swampy areas, and this contributed to the separation of the races. For further protection, they drank gin in water flavored with quinine (one of nature's many useful remedies), discovered in Peru in the fifteenth century in the bark of the cinchona tree.

Control measures in Africa and the Americas (where malaria was also found) eventually included environmental improvements and the application of insecticides, and by the 1950s there were dramatic drops in the worldwide incidence of the disease. It was not conquered, however, but only held at bay. By the late 1970s, dwindling investments in public health programs, growing insecticide resistance, and prevalent environmental change, such as forest clearing, contributed to a widespread resurgence. By the late 1980s, large epidemics were once again the rule, often associated with warm, wet periods.

In the 1990s, the worldwide incidence of malaria has quadrupled. This increase in malaria has been influenced by changes in both land development and regional climate. In Brazil, satellite images depict a fishbone pattern where roads have opened the tropical forest to localized development. In these areas, malaria has seen a resurgence. Temperature changes have encouraged a redistribution of the disease: malaria is now found in higher elevations in Africa's central highlands, and could threaten cities such as Nairobi, Kenya (at about 1,500 meters, or the altitude of Denver, Colorado), as freezing levels have shifted higher in the mountains. In the summer of 1997, for example, hundreds of adult malaria deaths occurred in the Kenyan highlands in previously unexposed populations. An acute strain—falciparum malaria—is more temperature sensitive and is increasing disproportionately in highland regions of Papua New Guinea, for example. Anopheles mosquitoes are present in the United States, and there was transmission of malaria in the United States earlier this century. And locally transmitted malaria has reappeared in the 1990s in Texas, Georgia, Florida, Michigan, New Jersey, and New York—and again in California as in the 1980s—primarily during hot, wet spells. A persistence of conducive climatic conditions, combined with inadequate (or ineffective) control methods, could allow further outbreaks to occur in the United States.

Up to 500 million people—twice the population of the United States—contract malaria every year, and between one and a half and three million, mostly children, die. Africa is most affected. Mosquito resistance to insecticides and parasite resistance to many drugs are widespread, and there are no operational vaccines, nor any foreseen in the near future. Ecological changes and climate change and increased variability appear to be playing increasing roles in the spread of the disease.

**Global Change.** Several features of global change tend to reduce predators disproportionately and, in the process, release prey from their normal biological controls. Among the most widespread are: (1) fragmentation and loss of habitat, (2) dominance of monocultures in agriculture and aquaculture, (3) excessive use of toxic chemicals, (4) increased ultraviolet radiation, and (5) climate change and weather instability. The breaking up of large tracts of forest or other natural wilderness into smaller and more diverse patches, compounded by edge effects, fragments and reduces the available habitat for large predators, while favoring many pests. Habitat loss and climate change may act synergistically: the loss of habitat frustrates the migration a species requires to survive changed climatic conditions. Extensive deforestation and a climate anomaly (such as the delayed monsoon due to the 1997 El Niño) may also act synergistically and produce surprises: witness the massive blaze and haze covering Southeast Asia in the fall of 1997, causing acute and chronic respiratory damage and losses in trade and tourism. [*See* Fire, *article on* Indonesian Fires.]

The dedication of land to monoculture, that is, the cultivation of single crops with reduced genetic and species diversity, brings greater vulnerability to infection. Biological diversity buffers against infection and invasions of exotic species, because (1) multiple hosts retard spread by separating them, (2) genetic diversity prevents spread through entire populations, and (3) birds and some insects (for example, ladybugs and lacewings) provide generalized defenses to control invaders, by eating them. Simplified systems lack "insurance" species, and are thus vulnerable to climatic extremes and to outbreaks of pests. [*See* Biodiversity.]

Overuse of pesticides harms birds and beneficial insects, as noted in 1962 by Rachel Carson. The title of her book, *Silent Spring*, made reference to the absence of the chorus of birds in springtime, and the resulting resurgence of plant-eating insects with evolved resistance to the pesticides. The worldwide response to her message transformed agricultural policies and ushered in integrated pest management. But still today, the heavy application of pesticides carries risks to human health and ecosystems. Overuse of pesticides in Texas and Alabama to control the boll weevil has alarmed farmers; friendly insects such as spiders and ladybugs have died off and other plant pests have rebounded. Stratospheric ozone depletion and climate change have had global impacts. [*See the biography of Carson.*]

**Ecosystem Health.** One of nature's "services" is to keep opportunistic species under control. For this service to be maintained, ecosystem health and integrity must be nurtured and sustained. This requires genetic and species biodiversity, providing alternative hosts for disease organisms, and the relative stability among functional groups (and insurance species) that preserve coevolved cooperative interactions and predation of species that can become pests and carriers of pathogens. As examples, a landscape of stands of trees interspersed with agricultural fields supports birds to control bugs, clean ponds with healthy fish populations control mosquito larvae, and wetlands sop up excess nutrients, chemicals, and organisms. Restoration of wetlands and riverbed trees in Maryland, for example, can reduce the sediments, chemicals, and organisms flowing into the Chesapeake, thus reducing the risk of the duoflagellate *Pfeisteria piscida*, a type of phytoplankton. Maintenance of these environmental systems thus provides generalized defense systems against the proliferation of opportunist pests.

Population explosions of nuisance organisms, be they animals or plants or microbes, are often a sign of failing ecosystem health—of systems out of equilibrium, where the natural ratio of organisms performing essential functions has been altered. The damage done, moreover, can be cumulative, for multiply stressed systems are less able to resist and rebound when impacted by new stresses. Rodents, insects, and algae are key biological indicators of ecosystem health. Their populations and species compositions change rapidly in response to environmental change, and particularly to an increase in their food supply or a drop in the number of their natural predators. Indeed, these factors can act synergistically, resulting in "biological surprises" such as sudden explosions in the populations of pests or pathogens.

The present rate of species extinction around the world is a potential threat to human health when one considers the role that predators play in containing infectious disease. From the largest to the smallest scales, an essential element in natural systems for countering stress is a diversity of defenses and responses. Thus, groups of animals that seem redundant may serve as "insurance" species in a natural ecosystem, providing a backup layer of resilience and resistance in the face of diseases, a shortage of food or water, or environmental change.

In 1996 the World Conservation Union found that one quarter of all species of mammals—and similar propor-

tions of reptiles, amphibians, and fish—are threatened. The current rate of extinction (estimated to be one hundred to one thousand times prehuman levels) falls most heavily on large predators and "specialists," and would appear to favor the spread of opportunistic, pioneering species. Throughout the long history of the Earth, times of mass extinctions—the upheavals or punctuations that have interrupted long periods of gradual evolutionary change—have been followed by the emergence of new species, and, at a more general level of taxonomic classification, of new phyla with different body forms.

**Climate and Health.** Persistent spells of extremely hot weather can threaten the health of people who live in temperate latitudes. Farm animals are also adversely affected when air temperature remains uncommonly high throughout the night. During the summer of 1995, excess deaths in Chicago and other large cities around the world were directly associated with heat waves (social isolation was also a factor). In many instances, according to meteorologists, the key factor may be the lack of relief at night. Models of climate change, based on global greenhouse warming, project more prolonged and intense heat waves and a disproportionate rise in minimum temperatures, in either daily or seasonally averaged measurements. Since the 1940s, the timing of the seasons has been altered, and, in the latter half of this century, minimum temperatures over land areas have risen at a rate of 1.86°C per one hundred years, while maximum temperatures have risen at 0.88°C per one hundred years. Warm winters and recurrent thawing can damage forests and allow harmful insects to survive the winter. [*See* Global Warming.]

Social factors such as the accelerating growth of "megacities" and ecological changes such as deforestation create conditions that aid the spread of infectious diseases. But climate restricts the range in which vector-borne diseases (VBDs) can spread, and weather affects the timing of their outbreaks. Insect biting rates and the growth of insect-borne microorganisms that are transmissible to humans are temperature dependent, and both increase when the air warms. Warming also increases the number of insects, provided there is adequate moisture. Excessive warmth, on the other hand, can decrease the survival of microorganisms or their hosts. Between limits of too hot and too cold is an optimum range of temperature in which warmer air will further the chances of transmission of a given organism or disease.

Most insects are highly sensitive to temperature change. Ants even run faster in warmer weather. Insects are R-strategists, which reproduce rapidly, have large broods, good dispersal mechanisms, and easily colonize disturbed environments. Warming speeds up their rate of reproduction, their biting rates, and the development time for parasites and viruses within them. (Excessive

## IMPACTS OF GLOBAL CHANGE IN MONTANE REGIONS

Both insects and insect-borne diseases (including malaria and dengue fever) are today being reported at higher altitudes in Africa, Asia, and Latin America. Highland malaria is becoming a problem for rural areas in Papua New Guinea and for urban centers in Central Africa. In 1995, dengue fever blanketed Latin America, and the diseases or its mosquito vector, *Aedes aegypti*, are appearing at increasing altitudes. In addition, the migration of plants to higher altitudes has been documented on thirty peaks in the European Alps, and has also been observed in Alaska, the Sierra Nevada range in the United States, and in New Zealand. These botanical trends, indicative of decadal warming, accompany other widespread physical changes: for example, montane glaciers are in retreat in Argentina, Peru, Alaska, Iceland, Norway, the Swiss Alps, Kenya, the Himalayas, Indonesia, and New Zealand.

Since 1970, the lowest level at which freezing occurs climbed about 160 meters higher in the mountains around the world (from 30° north to 30° south latitude), based on radiosonde data analyzed at U.S. National Oceanic and Atmospheric Administration (NOAA's) Environmental Research Laboratory. The shifts to higher levels on mountainsides correspond to a warming at these altitudes of about 1°C, which is nearly twice the average global warming that has been documented over the Earth as a whole. Notably, atmospheric models that incorporate observed trends in stratospheric ozone, sulfate aerosols, and greenhouse gases predict that, at least in the Southern Hemisphere, the warming trend at high mountain elevations should indeed exceed that at the Earth's surface. In fact, mountain regions—where shifts in isotherms are especially apparent—can serve as sentinel areas for monitoring more widespread climate change.

—PAUL R. EPSTEIN

heat shortens their longevity. Warming increases transmission only within the viable range.)

Findings from paleoclimatic (fossil) studies demonstrate that changes in temperature (and especially in minimum temperatures) were closely correlated with geographic shifts of beetles near the end of the last ice age, about 10,000 BP. Indeed, fossil records indicate that, when changes in climate occur, insects shift their range far more rapidly than do grasses, shrubs, and forest, and move to more favorable latitudes and altitudes hundreds of years before larger animals do. "Beetles," said one climatologist, "are better climatic indicators than bears."

Computer models of global greenhouse warming project increased temperatures, which will in turn favor the spread of VBDs to higher altitudes and more temperate latitudes. While 42 percent of the globe presently offers conditions that can sustain the transmission of malaria, the fraction could rise to 60 percent with a global increase of a few degrees Celsius.

Mosquitoes are hot-weather insects that have rigid thresholds for survival, and some are far more restricted than the common garden varieties that survive temperate winters. Anopheline mosquitoes and falciparum malaria transmission are sustained only where the winter temperature never falls below 16°C (61°F), while the

organism that transmits dengue fever, *Aedes aegypti*, is limited by the 10°C (50°F) winter isotherm. Shifts in the geographic limits of equal temperature (isotherms) that accompany global warming may extend the areas that are capable of sustaining the transmission of these and other diseases. The transmission season may also be extended in regions that now lie on the margins of the temperature and moisture conditions that allow vector spread. Similar considerations apply to cold-blooded agricultural pests, called *stenotherms*, that require specific temperatures for their survival. Agricultural entomologists estimate that 42 percent of yields and stored food are currently lost to weeds, pests, and pathogens, amounting to U.S.$247 billion annually.

These projected changes may already be under way, for there are now reports from several continents of new outbreaks of VBDs in mountainous regions—findings that are consistent with the recorded temperature increase, the general retreat of alpine glaciers, and the reported upward migration of temperature-sensitive plants. [See the vignette on Impacts of Global Change in Montane Regions.]

This consistency among varied physical and biological indicators agrees with the most recent (1996) consensus findings of the Intergovernmental Panel on Climate Change (IPCC) that climate appears to be changing, that some of the anticipated impacts are now observable, and that the changes are probably due to human activities. There may be some positive impacts (fewer winter deaths, or a drop in schistosomiasis in areas where excessive heat kills off snails), but, overall, the current evaluation is that the impacts on human health are likely to be overwhelmingly negative.

**Climate Variability and Human Health.** Another significant climate trend that has been linked to systematic changes in temperature and precipitation is an increase in the variability, or extremes, of climate. Such a change could affect not only the intensity of individual events, such as storms and floods, but the timing and spatial patterns of weather as well. Since the mid-1800s the average surface temperature of the globe has risen by about 0.4°–0.6°C, and periods of warming can in general be associated with increased variability. The IPCC projects that more intense heatwaves and altered drought and rainfall patterns may accompany the warming trend. [See Extreme Events.]

Data from the National Climatic Data Center—the main U.S. repository for meteorological data—indicate that, since the 1970s, extreme weather events have indeed increased in the continental United States. On average, periods of drought are systematically longer and bursts of precipitation (greater than 50 millimeters of rain over twenty-four hours) are more frequent. A warmed atmosphere accelerates evaporation and holds more water. One consequence is that we are now receiving a greater percentage of precipitation in the form of sudden, intense bursts that are more typical, for example, in the tropics. Longer droughts and more heavy bursts of rain (accompanied by flash floods) were more common in the 1980s than in the 1970s, and more so in the 1990s than in the 1980s.

Extreme events—floods, storms, droughts, and uncontained fires—can be devastating for agriculture, for human settlements, and for health. Both heatwaves and winter storms bring an increase in cardiac deaths. Floods spread bacteria, viruses, and chemical contaminants, foster the growth of fungi, and contribute to the breeding of insects. Prolonged droughts, interrupted by heavy rains, favor population explosions of both insects and rodents. Extreme weather events (most often associated with the recurring climatic conditions known as El Niño/La Niña events) have been accompanied by harmful algal blooms in Asia and North America, and in Latin America and Asia by malaria outbreaks and various waterborne diseases, such as typhoid, hepatitis A, bacillary dysentery, and cholera. [See El Niño–Southern Oscillation.]

In August 1995 the eastern, tropical region of the Pacific Ocean surface turned cold, initiating a La Niña event that would last until late 1996. Along the Caribbean coast of Colombia a summer 1995 heatwave was followed by the heaviest August rainfall in fifty years, ending a long drought that accompanied the preceding, prolonged El Niño of 1990–1995. The heat and flooding precipitated a cluster of diseases involving mosquitoes (Venezuelan equine encephalitis and dengue fever), rodents (leptospirosis), and toxic algae (killing 350 metric tons of fish).

Prolonged anomalous conditions of the sort that applied in 1990–1995 can also have biological consequences. In New Orleans, for example, a period of five years without frost was associated with an explosion of mosquitoes, termites, and cockroaches. Termites persisted inside trees into the cold winter of 1995–1996, and now threaten to destroy stands of New Orleans' fabled "mighty oaks."

**Rodent-Borne Diseases.** Rodents are today a growing problem in the United States, Latin America, Africa, Europe, Asia, and Australia. These preeminent opportunists are believed to be the fastest reproducing mammal; they eat everything that humans do (and much else besides), thrive on contaminated water and food, and are extremely capable swimmers. Meadow voles, for example, can have up to seventeen broods a year, each of half a dozen offspring (predatory marsh hawks keep their numbers from exploding). Rodents consume 20 percent of the world's grain, including almost a seventh in the United States, and up to three-quarters of what is grown and stored in some African nations. Rodents can also carry diseases.

A controlled experiment with Canadian snowshoe hares depicts how multiple factors can act synergistically to greatly increase the number of rodents. Exclusion of predators by confining the rabbits to cages led to a doubling of population compared with a similar number of animals in the wild. Augmentation of food tripled the hare density. Together, the interventions in the controlled experiment resulted in more than a tenfold population explosion.

*Rats, mice, and the hantavirus.* The story of the hantavirus illustrates a similar synergy in the case of microbial agents. A prolonged drought in the southwestern United States in the early 1990s reduced the populations of animals such as owls, coyotes, and snakes that prey on rodents.

When the drought yielded to intense rains in 1993, the grasshoppers and pinyon nuts on which rodents feed became more abundant. The result, when combined with the drop in predators, was a tenfold increase in rats and mice by June 1993. One outcome was the emergence of a new disease—called Hantavirus Pulmonary Syndrome—stemming from a virus transmitted through rodent droppings that was perhaps already present, but dormant, and was now transmitted through rodent saliva, urine, and droppings.

Again, we note the resilience of natural systems: after the initial delay, the animals that prey on rodents returned by September 1993 to keep them in check once more. Upsurges of pests and pathogens may be short-lived, but these events can serve as early warnings of distress in natural ecosystems.

Rodent-borne hantaviruses have resurged in several European nations, particularly in the former Yugoslavia. In late 1996, hantavirus infection emerged in western Argentina; at least ten deaths resulted, frightening off tourists and threatening the economic livelihood of the region. The recent evidence for person-to-person spread is of great concern.

***Other rodent-borne diseases.*** Another rodent-borne disease—leptospirosis—is increasingly reported in U.S. urban centers where the disposal of sewage and other measures to protect public health have declined. In 1995, there were major outbreaks of leptospirosis in Central America and Colombia, as heavy rains following drought (La Niña after the five-year El Niño period) drove rodents from their burrows. Leptospirosis is treatable with antibiotics, but fatalities occurred before the diagnosis was established.

A combination of stresses have contributed to the sudden appearances of several viral hemorrhagic fevers in rural Latin America in the past few decades: *Junin* in Argentina (1953), *machupo* in Bolivia (1962 and 1996), *guaranito* in Venezuela, and *sabia* in Brazil. In Bolivia, systematic clearing of trees apparently shifted populations of a variety of disease-carrying mice, known as *Calomys*, from forest to field settings where they became dominant. Heavy applications of dichlorodiphenyltrichloroethane (DDT)—meant to eradicate malaria—helped reduce their natural predators. When cats were reintroduced to the area in 1962, the epidemic of Bolivian hemorrhagic fever was abated, although not until it had killed 10–20 percent of the inhabitants of the small villages in which the disease was present.

In southern Africa, rodent populations exploded as a consequence of the heavy rains of 1993 and 1994 that followed six years of prolonged drought. When the rains came, rodents found themselves in a world in which avian and land predators were virtually absent. Moreover, because so many draft animals had also succumbed, there was little tillage of the land, and the underground burrows in which the rodents lived went largely undisturbed. After an initial successful harvest in 1993, the maize crop in Zimbabwe was decimated by rodents. Soon afterwards, human plague broke out in Zimbabwe and on the borders of neighboring Malawi and Mozambique, carried—as in the Black Plague of fourteenth-century Europe—by fleas and rats. Subsequently, a rodent-borne virus took the lives of eighty-one elephants in South Africa's Kruger Park. Also, following the prolonged El Niño of the early 1990s and five years of drought, rodents emerged in Australia in 1995 as serious crop pests.

Current land use practices and the overuse of chemicals to control pests may increase the chances for such nasty synergies. Climate variability is a key element in upsurges of pests—and, were climate to become more unstable, it could exert greater influence on the patterns of infectious diseases in the future. A disturbance in one factor can be destabilizing; but multiple perturbations can affect the resistance and the resilience of a entire system.

**Interventions.** Interventions and approaches to solutions may be divided into three categories.

1. Improved surveillance and response capability for the public health sector, along with vaccine development, better treatments, and support for basic public health measures.
2. Integration of health surveillance into environmental monitoring, employing remote sensing and climate forecasting to develop early-warning systems for conditions conducive to outbreaks.
3. Evaluation of environmental and energy policies in light of the growing threats. Maintenance of ecosystems integrity (e.g., forest habitat and wetlands) can provide defense against outbreaks of opportunists, and buffer against climatic vagaries, whether or not the climate regime changes.

Today the activities of one species—humans—are affecting the global environment and reducing the diversity of all others. Environments that experience stresses of the sort described above become more susceptible to the emergence, invasion, and spread of opportunistic species. When subject to multiple stresses, natural environments can exhibit symptoms that signal reduced resilience, resistance, and regenerative capabilities. Ecosystems are not fragile, and their flexibility and survival strategies can be strengthened by systematic stress (the alternating battering of seasonal changes in temperate ecosystems, for example). But their tolerance for abuse has its limits.

Water, food, and health are three of our basic needs. These issues are interrelated, and global change threatens to change all three. Maintenance of health entails providing clean water, safe food, clean air, and clean energy for cooking, transportation, and development. Thus these basic public health infrastructure issues (water and rural electrification) cannot be separated from confronting regional and global environmental and energy issues.

There are increasing indications of climate change as our influence upon natural systems deepens. We are changing atmospheric chemistry, which is altering the world's heat budget, and these changes—along with changes in coastal and ocean chemistry—have begun to affect biological systems. Underlying these chemical, physical, and biological changes, we are using the Earth's resources and generating wastes at rates beyond those at which biogeochemical systems can adequately recycle them. Moreover, our patterns of consumption come at costs that are real, often very high, and not acknowledged by current systems of economic accounting. Practices affecting forestry, fisheries, petrochemicals, and fossil fuels all need to be examined in light of their costs across the full range of their ultimate impacts, including their effects on biodiversity, climate, and the global resurgence of infectious diseases.

[*See also* Air Quality; Drought; Hazardous Waste; Natural Hazards; Ozone; Risk; Water Quality; *and* World Health Organization.]

### BIBLIOGRAPHY

Epstein, P. R. "Is Global Warming Harmful to Health?" *Scientific American* (August 2000), 50–57.
Epstein, P. R., H. F. Diaz, S. Elias, G. Grabherr, N. E. Graham, W. J. M. Martens, E. Mosley-Thomson, and J. Susskind. "Biological and Physical Signs of Climate Change: Focus on Mosquito-Borne Disease." *Bulletin of the American Meteorological Society* 78 (1998), 409–417.
Grifo, F., and J. Rosenthal. *Biodiversity and Human Health*. Washington, D.C.: Island Press, 1997.
Harvell, C. D., K. Kim, J. M. Burkholder, R. R. Colwell, P. R. Epstein, J. Grimes, E. E. Hofmann, E. Lipp, A. D. M. E. Osterhaus, R. Overstreet, J. W. Porter, G. W. Smith, and G. Vasta. "Diseases in the Ocean: Emerging Pathogens, Climate Links, and Anthropogenic Factors." *Science* 285 (1999), 1505–1510.
Karlan, A. *Man and Microbes: Disease and Plagues in History and Modern Times*. New York: Simon and Schuster, 1995.
Levins, R., T. Auerbuch, V. Brinkmann, I. Eckardt, P. Epstein, T. Ford, N. Makhoul, C. A. dePossas, C. Puccia, A. Spielman, and M. E. Wilson. "New and Resurgent Diseases: The Failure of Attempted Eradication." *The Ecologist* 25 (1995), 21–26.
McMichael, A. J., A. Haines, R. Slooff, and S. Kovats, eds. *Climate Change and Human Health*. Geneva: World Health Organization, 1996.
Rapport, D., R. Costanza, P. R. Epstein, C. Gaudet, and R. Levins, eds. *Ecosystem Health*. Malden, Mass.: Blackwell Science, 1998.
Trenberth, K. E. "The Extreme Weather Events of 1997 and 1998." *Consequences* 5 (1999), 3–15.

—PAUL R. EPSTEIN

# HUMAN IMPACTS

[*This entry consists of two articles*, Human Impacts on Biota *and* Human Impacts on Earth. *The first article explains the ways that human beings have and do impact the biota, and discusses why this is important globally. The second article reviews the major kinds of changes that humankind has made on the biosphere as well as the underlying attitudes toward nature that supported these changes.*]

## Human Impacts on Biota

Human impacts on biota became significant during the late Pleistocene and contributed to the loss of many animal species, including the mastodon and woolly mammoth. In North America this period of mass extinction coincides with the spread of humans from Asia and led to the disappearance of about 75 percent of the large mammalian species about eleven thousand years ago. Overhunting is usually cited as the cause, but natural factors are also implicated, as this period is contemporaneous with rapid changes in climate and vegetation (Pielou, 1991). The human element is less in doubt on previously isolated landmasses. The arrival of humans in Australia about fifty thousand years ago is contemporaneous with the extinction of many species of large mammals, giant snakes and reptiles, and several species of large flightless birds. Colonization of the Hawaiian Islands in the fourth and fifth centuries CE coincided with the extinction of 50 percent of the endemic bird species. In New Zealand the entire Moa family was exterminated by the end of the eighteenth century, with a similar loss of ratites on Madagascar. Island populations generally are more vulnerable than continental populations and 75 percent of recorded animal extinctions since 1600 CE are associated with these fragile ecosystems. The restricted geographical range of island endemics makes them innately vulnerable, and most have evolved in the absence of terrestrial predators. Elsewhere, habitat destruction and fragmentation is the most serious threat to biodiversity, but significant numbers of species have also been lost through overexploitation for subsistence and commercial trade, accidental or deliberate introduction of competitive exotic species, eradication of species, considered to be pests, incidental take in commercial harvests, and diseases introduced with livestock and crops (Groombridge, 1992). The variety, intensity, and extent of human impacts are accelerating as a result of population growth and the rate of resource consumption. Consequently, direct and readily observed changes are now augmented by several indirect and unintentional effects at the planetary scale.

The first humans are distinguished from other animals mainly by the fashioning of primitive tools and the use of fire, but the key step in development came with the domestication of animals and plants. Archaeological evidence suggests a date of 8500 BP for the domestication of cattle, 7000 BP for goats and sheep, and 5000 BP for horses. Einkorn, emmer, barley, peas, and lentils appeared about 9000 BP (Goudie, 1999). The development of reliable food supplies made it possible to adopt a more sedentary way of life. Domestication was followed by new agricultural technologies. The first irrigation projects began in Egypt about 5000 BP, about the same time as the plow was invented. Plant species were redistributed as new crops were introduced; of particular significance was the spread of vigorous weedy species that began to compete with indigenous floras. The rapid spread of agriculture was accompanied by an increasing demand for land and initiated the process of forest clearance. Wood was needed for sleds and rafts, and later, carts and ships. A considerable amount of wood was also consumed in metal smelters, and this continued until the use of coal heralded the Industrial Revolution. Traditional industries place heavy demands on the world's nonrenewable resources and impact ecosystems in various ways. Water and air are especially sensitive to industrial pollution; in the past few decades the deleterious effects of such pollution have become important public issues both locally and internationally.

**Climate Change's Impact on Terrestrial Ecosystems.** Atmospheric pollution is implicated in climate change mainly through perturbations in the carbon cycle. The concentration of atmospheric carbon dioxide is estimated to have increased from 280 to nearly 370 parts per million since the beginning of the Industrial Revolution, with half of this increase recorded since the mid-1960s. The principal source is the consumption of fossil fuels. Because of its relative abundance in the atmosphere, carbon dioxide accounts for about 60 percent of potential global warming, compared to 15 percent for methane, 12 percent for chlorofluorocarbons, 8 percent for ozone, and 5 percent for nitrous oxide. At current rates of emission, atmospheric concentrations of carbon dioxide could double by 2050, leading to a predicted rise in global temperatures of 2.0°C. The effects of global climate change are expected to be most pronounced in higher northern latitudes, where winter temperatures may rise by more than 8°C and would be accompanied by changes in precipitation regimes. In central North America winter precipitation may increase by 15 percent, with summer precipitation decreasing by 10 percent. A similar response is expected in southern Europe (Intergovernmental Panel on Climate Change, 1990). [*See* Global Warming].

The impact of climate change is speculative. Computer simulations predict that 40–55 percent of the Earth's land area would experience a change in natural vegetation in response to human-induced climate change. The world's forests are especially sensitive because they are not intensively managed and so must adapt through slow migration or physiological change. The change in forest distribution is expected to be especially significant at high latitudes where potential warming trends are greatest (Lenihan and Nelson, 1995; Smith et al., 1995). The boreal forest could expand slowly into the tundra and polar regions, the rate being controlled mainly by the development of suitable soils. Northward migration of grassland and shrubland could ultimately reduce the total area of forest by as much as 35 percent. The economic impact would be considerable, as most of the reduction would occur in areas of high productivity. Another consequence of global warming is a potential increase in the extent and severity of fires in boreal forests. The present area of tropical forest is predicted to remain relatively stable but would be supplemented by newly established forests that theoretically could double the current area.

However, the growing demand for agricultural land would counteract any potential increase. The nature of the forest cover would also change. Rapid transition from savanna to dry forest could occur in areas where there is an increase in precipitation. Similarly, a general decrease in desert and semiarid regions is predicted because of reduced evapotranspiration. Elsewhere the present dry forest cover would be replaced slowly by tropical moist forests, while some alpine communities could be displaced by vegetation presently growing at lower elevations.

**Human Impact on Aquatic Ecosystems.** Continued climatic warming is expected to cause heavier precipitation and increase the probability of severe floods and droughts. Changes in water temperature, water level, and flow regimes will have several direct and indirect effects on the biota of streams, lakes, and wetlands. As well as general changes in species distributions, climate warming may lead to reduced biodiversity in streams as temperatures in the headwaters and downstream reaches become less variable. Similar effects are predicted for temperate lakes, where the thermal environment is determined by the timing of freeze and thaw, the degree of stratification, and the depth of the thermocline. Some species will be at risk of extinction because of changes in their thermal niches or through competition from species that were previously excluded (Graves and Reavey, 1996). Wetlands are also sensitive to factors that affect water flow, and this sensitivity has important implications for resident species as well as the downstream ecosystems, which are influenced by rates of groundwater recharge, nutrient retention, and flood control.

Saltwater marshes, mangrove ecosystems, and coastal wetlands are particularly vulnerable to climate change and rising sea levels. In some locations, sea level is expected to rise faster than sediment deposition rates, thereby restricting coastal ecosystems to estuaries with suitable substrates. Coastal wetlands are vulnerable to inundation and erosion by rising sea levels and to salt water intrusion. Coral reefs are especially sensitive to an increase in seawater temperature. A sustained increase in water temperature of 3°–4°C above long-term average seasonal maxima over a six-month period can cause significant coral mortality; short-term increases of only 1°–2°C can cause "bleaching," which leads to reef destruction (Smith and Buddemeier, 1992). Changes in water temperatures in the deep oceans would alter the geographical distribution of species with narrow thermal tolerances and modify the dynamics of marine ecosystems. [*See* Ocean Life; *and* Reefs.]

**Toxic Metals and Acid Rain.** In addition to greenhouse gases, many other chemicals are emitted to the atmosphere. Such emissions may originate from point sources such as metal smelters or from nonpoint sources ranging from broad tracts of eroding land to numerous, smaller sources such as vehicles. Heavy metals such as zinc, lead, and copper emitted from smelters can be detected in precipitation many kilometers from their source (Boutron et al., 1995), causing soil contamination and subsequent accumulation in vegetation. Mosses are good bioindicators of airborne contamination by heavy metals. Because they also absorb sulfur dioxide to their detriment, their density and frequency can be used to assess long-term air quality around industrial sites. Lichens also are sensitive to sulfur dioxide, and distinctive species zonations occur around industrial areas that correspond to the mean gas concentrations.

Sulfur dioxide dissolves in water to form acid rain. This has had a significant effect on regional forest covers in many parts of the world. In North America, the vulnerability of sugar maple (*Acer saccharum*) to acid rain has caused concern because it is used in the production of maple syrup. Conifers are especially sensitive to pollutants, and forest damage is particularly severe around smelters, although general forest decline because of poor ambient air quality is also reported (Innes, 1992). In Scandinavia, acid precipitation is commonly below pH 5.0; under these conditions aluminum and other toxic elements can be leached from soil and bedrock. This can affect various organisms and lead to changes in ecosystem diversity as sensitive species begin to decline. [*See* Acid Rain and Acid Deposition.]

The problem of acidification of fresh waters was first identified in Scandinavia, where drainage basins have little natural buffering capacity. Similar conditions are encountered in glaciated regions of Canada and the United States. Acidification of lakes and rivers is typically accompanied by a decline in the algal community and an increase in acid-tolerant macrophytes such as *Sphagnum* moss and rushes. In some acidified lakes, filamentous algae may become abundant. Microbial activity falls sharply below pH 5.0, the decomposer bacteria are replaced by fungi, and litter decay may be five to twenty times slower. Corresponding changes occur in the phytoplankton and invertebrate populations. A threshold pH of 5.0 is also characteristic for many species of freshwater fish. Young migratory fish are especially vulnerable to the surge in acidity that may occur during snowmelt or during periods of heavy rain. During these episodes metals are flushed from bryophytes, and concentrations of aluminum and manganese can exceed the lethal thresholds of sensitive species such as salmon. Nonmigratory species decline in acidified lakes because of high egg and fry mortalities, and ultimately the fish disappear from the water bodies, even though conditions may not be overly toxic to the adults (Rosseland, 1986).

**Eutrophication and Contamination of Lakes and Rivers.** Most lakes and rivers now are affected by excessive inputs of nutrient and organic matter and by contamination from potentially toxic metals and organics. Phytoplankton production increases rapidly as more phosphorus and nitrogen become available, and the undesirable growths of algae associated with polluted nutrient-rich lakes are now synonymous with eutrophication. The discharge of organic wastes is the most common disturbance to aquatic ecosystems. The main source of organic enrichment is domestic sewage, but other materials are added from various industries associated with food processing, tanning, textiles, paper making, and petrochemicals (Hellawell, 1986). The principal effect of organic discharge is the depletion of dissolved oxygen by the large population of heterotrophic decomposers. Toxic substances, which rarely occur in natural water bodies, also are added through industrial processes. Inert solids from coal washeries and similar operations increase turbidity of the water and smother benthic organisms. The discharge of heated water used in industrial cooling also affects the distribution and activities of aquatic organisms. Polluted rivers show a characteristic change in physical and biological properties downstream from the point of discharge, which reflects the degree of dilution and dispersion that occurs in flowing water. The response of macroinvertebrates to differing degrees of pollution is so distinctive that it is commonly used as a standard measure of water quality. Fish are equally sensitive to organic pollution mainly because of the reduction in dissolved oxygen. In addition, high concentrations of ammonia in organic wastes can be toxic to fish, and suspended solids affect feeding efficiency and smother spawning gravels. Rivers are self-cleaning systems, but until recently this rarely occurred because of the varied and heavy usage they received. Nevertheless, conservation measures and enforced legislation in many parts of the world has greatly improved water quality for humans and wildlife. [*See* Pollution.]

**Dams, Reservoirs, and Channel Modification.** A more permanent effect on the ecology of rivers occurs when the hydrologic regime and sediment transport is altered by the construction of dams and reservoirs and by channel modification. Important changes occur in the upstream ecosystem when migratory fish are unable to pass a dam; for this reason fish ladders sometimes are provided. The effect downstream depends on the design of the dam, which determines the quantity and quality of the discharged water. If water is released from the bottom of the structure, as is the case for most hydroelectric installations, it is cold and low in dissolved oxygen but will usually carry bacteria, plankton, and nutrients out of the reservoir. If the water is discharged over a spillway, the effects are similar to a natural lake; nutrients are trapped and heat is exported. Rapid fluctuations in discharge, which is typical of dams used for power generation, and unusually low flow, such as occurs with irrigation, are disruptive for biota. [*See* Dams.] Dredging and channel straightening also tend to have adverse effects on aquatic communities in that they create unnaturally uniform conditions. Several species have extended their ranges by way of canals and other waterways; the introduction of the sea lamprey (*Petromyzon marinus*) and alewife (*Alosa pseudoharengus*) into the Great Lakes via the Erie and Welland Canals in 1921 had a devastating effect on lake trout and other indigenous species. More recently, the zebra mussel (*Dreissena polymorpha*) has already become a major problem around Lake Superior. In tropical regions the creation of lakes and waterways is notorious for spreading waterborne diseases such as schistosomiasis. Another effect is the explosive growth of water weeds; water fern (*Salvinia molesta*), water lettuce (*Pistia stratiotes*) and water hyacinth (*Eichhornia crassipes*) are common problem species.

Most of the sediment formerly carried by a dammed river is deposited in the reservoir, and turbidity decreases downstream. The silt that is carried in turbid flood waters is an important source of nutrients for riparian communities. The decline in cottonwoods along dammed rivers in North America is attributed to alterations of the flood regime. Changes can be more extreme when water is diverted for irrigation; this is the case with the Aral Sea. Rivers that previously flowed into this large inland water body were used to support an intensive cotton monoculture. Since 1960 the water level in the Aral Sea has dropped by more than fifteen meters, and the area of the lake has decreased from about 68,000 to about 33,000 square kilometers (Micklin and Williams, 1996). Salt, sand, and dust from the exposed seabed now blows across the region, and contamination by pesticides and fertilizers has affected the remaining fish stocks. [*See* Aral Sea, Desiccation of the.]

**Desertification.** The desert surrounding the Aral Sea has expanded considerably in the past twenty-five to thirty years. Equally tragic is the damage that has occurred in the sub-Saharan region of Africa, where the area of desert increased by approximately 1.3 million square kilometers between 1980 and 1984. The traditional pastoralists in this semiarid area reduced the stressful impact on their animals by moving the herds with the rains. Their hardy domestic animals are well suited to the environment and can travel for several days between waterholes. However, a change toward a more sedentary way of life has led to larger herds and problems of overgrazing. As grazing pressure increased, the perennial grasses were replaced by less palatable annual species, and with declining productivity the soil became susceptible to erosion by wind and water. Intensification of agriculture, increased use of marginal, drought-prone areas, and poorly managed irrigation schemes have exacerbated the problem in other parts

of the world. Desertification now affects the southwest United States and parts of Australia, southern Africa, southern Europe, and China. Globally, desertification threatens about 3.6 billion hectares, or nearly twenty-five percent of the total land area of the Earth, and impacts about one-sixth of the world's population (Tolba and El-Kholy, 1992). [*See* Desertification.]

**The Impact of Agriculture.** The amount of land used in agriculture is greater than for any other activity. Approximately 11 percent of the ice-free land area of the world is currently under cultivation, with a further 24 percent under permanent pasture. More than 50 percent of harvested land is devoted to wheat, rice, and corn. Much of the remaining area is used for barley, oats, rye, millet, and sorghum. Weeds are equally ubiquitous; the most aggressive are usually introduced aliens that thrive in areas where the natural vegetation cover is broken and the soil is exposed. Many of these troublesome species were carried from Europe mixed with grains or on the fleeces of animals, while species such as the South American barbed-wire plant (*Xanthium spinosum*) and poisonous thorn apple (*Datura stramonium*) have been returned to Europe through wool exports. [*See* Agriculture and Agricultural Land].

The cultivated lands of the world are concentrated in the temperate regions, where typically they have replaced forest and grassland ecosystems. In Europe and Asia, forest clearance has been going on for centuries, while in North America intensive agriculture began to spread across the Great Plains in the early nineteenth century. Deliberate eradication of native species, such as the bison in North America, by hunting, trapping, and poisoning has had considerable ecological impact. Many nontarget species have also become locally extinct, allowing other species to extend their ranges. Some parts of the tropics still have substantial natural areas that are suitable for conversion, but these often coincide with regions of high biodiversity.

Significant changes in agriculture occurred in the 1960s and 1970s, when traditional farming practices were replaced by modern techniques that depended on heavy inputs of energy and agrochemicals and improved crop varieties grown in monocultures on large fields. A principal feature of this Green Revolution was the development of new genetic strains of wheat and rice that provided higher yields and faster growth. This was achieved with increased use of fertilizers, pesticides, and herbicides and greater use of irrigation water. Despite obvious benefits in terms of food supply, there have been several adverse effects. Large-scale plantings consisting of a single genetic strain are more vulnerable to short-term perturbations in weather and disease than the native varieties they replaced. In some areas native varieties are in danger of extinction. In India, for example, thirty thousand varieties of rice were cultivated in 1980, but by the end of the century it is estimated that twelve varieties may account for 75 percent of production (Tolba and El-Kholy, 1992).

Intensive agriculture has proven to be nonsustainable in some areas, partly for economic reasons and partly because of adverse environmental effects such as soil erosion and loss of soil fertility. Chemicals are needed to combat pests and diseases, but these chemicals also reduce their natural enemies. These beneficial species are eliminated directly by contact with the pesticide, or indirectly through biomagnification when the contaminant is accumulated in other species in the food chain. The creation of large fields has removed hedgerows and other sites that afforded protection to birds and other predators. Consequently, there has been renewed interest in ecological control in which the use of chemicals is greatly reduced in favor of various biological and cultural alternatives. The concept of integrated pest management is based on broad ecological principles and incorporates a range of control methods (El-Hinnawi and Hashmi, 1987). Control methods include simple tillage to destroy pests in the soil through exposure or injury, crop rotation and mixed cropping to reduce the buildup of pest populations that infest the root zone, and the use of trap crops to divert pests to an area where they can be controlled more efficiently. Male sterilization and chemicals that alter insect behavior, such as feeding deterrents, also have been developed. The most successful examples of biological control have involved the reduction of introduced pests using natural enemies imported from the pests' native areas.

**The Impact of Forestry.** It is estimated that by 1970 forests and woodlands had been reduced by about 15 percent of their preagricultural area. Much of this occurred in temperate forests where clearing began about seven thousand years ago. By the early 1900s over 600,000 square kilometers of forests were cleared in the United States, with a further 400,000 square kilometers in Canada, New Zealand, Australia, and South Africa (Tolba and El-Kholy, 1992). The total area of forest in temperate zones has changed relatively little in recent decades because losses in places such as Australia and the United States have been offset by afforestation programs elsewhere. In tropical areas, however, the current rate of loss is estimated at approximately 1 percent yearly and is mainly to meet the growing demand for agricultural land. Resettlement schemes, such as those in Brazil, have caused extensive forest clearance, soil degradation, and conversion to grazing lands. Forest clearance for agriculture or simply to gain revenue from the timber is viewed by many as a wasteful use of an irreplaceable resource. On purely economic grounds, the short-term benefits derived from timber sales may be insignificant compared to the potential income that might accrue from the unscreened genetic resources of the for-

est. At a global scale, widespread deforestation in the tropics could lead to significant changes in the global circulation of the atmosphere. Changes in the energy balance of the Earth have also been postulated as less carbon dioxide is removed from the atmosphere by the dwindling forest cover (Apps and Price, 1996). In temperate regions, much of the deciduous forest has been cleared for agriculture. What remains is mostly secondary growth dispersed as small patches in the landscape. Forest clearance has resulted in the extirpation of many species of animals, although others have adapted to the new conditions and so have become much more widely distributed (Burgess and Sharpe, 1981). [*See* Deforestation; Forestation; *and* Forests.]

Concerns about habitat change have been raised in boreal forests where logging has affected much of the plant cover. The traditional harvest method is clear cutting, in which all of the economically useful trees are removed and the area is then replanted with commercial species or left to regenerate naturally. Accelerated erosion is a common problem in logged areas. Erosion is especially important on steeper terrain and increases exponentially with the size of the clear cuts. In most cases erosion is triggered by improper logging practices such as poor road construction and culvert design, the use of stream beds to extract logs, and from dragging logs downslope and thereby creating channels for surface runoff. To alleviate these problems, the customary practice is to leave uncut buffer strips fifteen to twenty meters wide besides streams and lakes. These not only function as wildlife corridors but also help to reduce erosion and maintain other natural qualities of the streams. Multiple-use forestry, in which factors such as wildlife and recreation are considered as well as timber harvesting, has led to a change in forestry practices (Hunter, 1990). The preference for a greater diversity of stand conditions is being incorporated into smaller, irregularly shaped clear cuts that are more hospitable to a range of wildlife. Important microhabitats, such as overmature live trees and dead snags, were often destroyed during logging; only recently have the long-term noncommercial benefits of old-growth forests come to be appreciated.

**Conclusion.** The past few decades have seen considerable advances in understanding environmental processes and the role of humans in the biosphere. The development of global monitoring systems and powerful modeling programs has provided data and techniques to predict long-term impacts with some accuracy. At the same time, the growth of environmental organizations has increased the pressure on industry and government to formulate and enforce environmental protection policies. The current emphasis on sustainable development is an attempt to ensure the long-term security of the environment. The growth in public awareness of the problems of human impacts on biomes can be assessed by the increase in membership in environmental organizations and the growing influence that they have in national environmental policies. Although environmental issues are addressed at the national and international level, governments are under pressure to maintain economic growth and viability. In most cases, this is achieved at the expense of the environment. Consequently, there is continuing stress on the environment and the resources on which all living things depend. Many species are in urgent need of protection. Numerous captive breeding and reintroduction programs and translocation projects have been initiated to help maintain viable populations. Many of these programs have met with limited success because the pressures on these populations have not been relaxed. Loss of habitat is a major problem facing conservation efforts, but even in less disturbed environments vigorous aliens and exotics also pose significant threats to endemic species. Human society has begun to see the global benefits of natural ecosystems, and world opinion is changing in favor of sustainable development and the establishment of protected areas and habitat reserves.

[*See also* Animal Husbandry; Biological Diversity; Convention on Long-Range Transboundary Air Pollution; Electrical Power Generation; Fishing; Hazardous Waste; Mining; Nuclear Hazards; Ocean Disposal; Pest Management; Petroleum Hydrocarbons in the Ocean; *and* Transportation.]

### BIBLIOGRAPHY

Apps, M. J., and D. T. Price, eds. *Forest Ecosystems, Forest Management, and the Global Carbon Cycle.* Berlin: Springer, 1996.

Boutron, C. F., J. P. Candelone, and S. M. Hong. "Greenland Snow and Ice Cores: Unique Archives of Large-Scale Pollution of the Troposphere of the Northern Hemisphere by Lead and Other Heavy Metals." *Science of the Total Environment* 161 (1995), 233–241.

Burgess, R. L., and D. M. Sharpe, eds. *Forest Island Dynamics in Man-dominated Landscapes.* New York: Springer, 1981.

El-Hinnawi, E., and M. H. Hashmi. *The State of the Environment.* London: Butterworth, 1987.

Goudie, A. S. *The Human Impact on the Environment.* 5th ed. Oxford: Blackwell, 1999.

Graves, J., and D. Reavey. *Global Environmental Change: Plants, Animals, and Communities.* Harlow, U.K.: Longman, 1996.

Groombridge, B., ed. *Global Biodiversity: Status of the Earth's Living Resources.* London: Chapman and Hall, 1992.

Hellawell, J. M. *Biological Indicators of Freshwater Pollution and Environmental Management.* London: Elsevier, 1986.

Hunter, M. L. *Wildlife, Forests, and Forestry: Principles of Managing Forests for Biological Diversity.* Englewood Cliffs, N.J.: Prentice-Hall, 1990.

Innes, J. L. "Forest Decline." *Progress in Physical Geography* 16 (1992), 1–64.

Intergovernmental Panel on Climate Change (IPCC). *Scientific Assessment of Climate Change.* Geneva: Intergovernmental Panel on Climate Change, WMO/UNEP, 1991.

Lenihan, J. M., and R. P. Nelson. "Canadian Vegetation Sensitivity to Projected Climatic Change at Three Organizational Levels." *Climate Change* 30 (1995), 27–56.

Micklin, P. P., and W. D. Williams. *The Aral Sea Basin.* NATO Advanced Science Institute Series. Berlin: Springer, 1996.

Pielou. E. C. *After the Ice Age: The Return of Life to Glaciated North America.* Chicago: University of Chicago Press, 1991.

Rosseland, B. O. "Ecological Effects of Acidification on Tertiary Consumers: Fish Population Responses." *Water, Air and Soil Pollution* 30 (1986), 451–460.

Smith, S. V., and R. W. Buddemeier. "Global Change and Coral Reef Ecosystems." *Annual Review of Ecology and Systematics* 23 (1992), 89–118.

Smith, T. M., P. N. Halpin, H. H. Shugart, and C. M. Secrett. *Global Forests.* In *As Climate Changes: International Impacts and Implications*, edited by K. M. Strzepack and J. B. Smith, pp. 146–179. Cambridge: Cambridge University Press, 1995.

Tolba, M. K., and O. A. El-Kholy, eds. *The World Environment 1972–1992: Two Decades of Challenge.* London: Chapman and Hall, 1992.

—O. W. ARCHIBOLD

## Human Impacts on Earth

This article deals with changes produced in the terrestrial environment as a result of human activity. Though the use of the word *impacts* for such changes may imply that they are, on the whole, more likely to be harmful than beneficial, it is used here in an entirely neutral sense to designate all anthropogenic changes in the biophysical environment, with no implication as to whether any change represents improvement or damage.

Whether for the better or the worse on the whole, human impacts have assuredly been extensive, profound, and far-reaching. Few of the major systems of the biosphere have not been substantially altered by them. The current scale and variety of changes and of the activities directly driving them suggest that still larger and more varied impacts are yet to come, but a 1990 inventory and assessment of impacts to that date concluded that we live already on an Earth fundamentally transformed by human action (Turner et al., 1990).

It has not necessarily been transformed, however, by deliberate human action or according to any particular conscious purpose. Human impacts are of two sorts. Some are the intended results of human actions, some are the incidental and inadvertent results of actions taken for other purposes altogether. Some losses in biodiversity, for example, are the result of deliberate efforts to exterminate unwanted species, but by far the larger share are the unforeseen consequences of other practices. Some forms of outdoor weather modification have been practiced with modest and local success, but most human impacts on weather and climate, documented or suspected, are entirely inadvertent. It is in the nature of human-environment relations that the unintended impacts are numerous and frequent. The biosphere and the overlapping ecosystems of which it is composed are such complexly woven systems of interacting and interdependent elements—of climate, soil, and biota and of water, material, and energy flows—that it is nearly impossible for any of the parts to be altered without at the same time many others being affected. The ways in which they are affected are often—at least at first—unexpected and even counterintuitive. These secondary impacts are, on the whole, less likely than the deliberate and intentional changes to be of benefit. If human-induced environmental change overall tends to be viewed as harmful today more often than it was previously—as "impact" rather than as improvement—one of the reasons is that much more is now known than in the past about the wider results of human actions, the repercussions that may spread far and wide from their point of origin.

A second difference between the present and the past offers a second reason for greater concern. As human activities expand in scale and variety, so do the environmental impacts that they may produce on a planet that has not itself grown larger. As a result, it has become more conceivable and even more plausible than ever before that the integrity of the biosphere as a whole and the necessary conditions for human existence on Earth may be disrupted by the sum total of human impacts. Throughout most of human history, excessive demands on resources might place local and regional environments under stress, but the Earth overall remained so vast in comparison as never to seem fundamentally threatened by it. The Earth's impacts on humans were, and justifiably, more of a cause for worry than those of humans on the Earth.

The change in the scale and type of human actions has led to the emergence of impacts that can reasonably be described as global environmental change. *Global change* and *human impact* are overlapping but not synonymous terms. Many significant human impacts on the environment, past and present, do not constitute global change. Those that do are of two different sorts. Some changes reach a global magnitude and character, if they do, by affecting a realm of the environment that operates as a fluid global system. Examples are the release of carbon dioxide and other stable trace gases into the atmosphere and possible change in sea level resulting from climatic change produced by increased concentrations of such gases. These are the sort that have been designated as globally systemic changes. What have been called globally cumulative changes, on the other hand, are those that reach a global magnitude either by occurring so widely across the Earth's surface as to attain a worldwide character or by significantly affecting the total stock of some resource, however spatially concentrated it may be.

Some human impacts can also be described as cumulative in a second sense, temporal rather than spatial. They are impacts that are essentially irreversible and that therefore steadily add up, as opposed to ones that can be reversed by natural or human-aided processes of recovery. Species extinction and nonrenewable resource extraction are cumulative in this sense. Other changes, such as the pollution or depletion of ground water, may be reversible, but on so long a time scale that for most human purposes they are not. Cumulative impacts of this sort are a particular cause for concern in that they represent a permanent alteration and perhaps impoverishment of the terrestrial environment.

Human impacts on individual aspects of the environment are dealt with in detail in other entries in this encyclopedia, as are questions about their significance and consequences for society and their root causes. The focus here is on general themes common to many of them and to their overall character and magnitude. The present entry briefly discusses some of the chief difficulties and sources of uncertainty in documenting human impacts; reviews the chief realms of the environment affected by human impact; and describes the overall character, magnitude, spatial pattern, and chronology of human impact.

**Documenting Human Impact.** It is often no simple matter to assess the degree of human impact in any sphere of the environment. Doing so requires knowing both the previous and the current state of affairs. It also requires knowing what contribution, if any, natural processes made to any change that has occurred. These requirements are much more easily satisfied in forms of change and some aspects of the environment than in others.

The original, preimpact state of the environment can be reconstructed from documentary sources, on the indirect evidence of what are called "proxy" indicators, or on the even more indirect evidence of what theories and assumptions suggest must have been the case. Documentary sources offering information on past environmental conditions range from government tax and land survey records to maps and detailed scientific observations. But such sources do not exist in most settings for most of the human past. Even where they do, they are often far from accurate, reliable, or comprehensive in their coverage. Much of the information that they offer must be treated with caution, and much extrapolation may be needed to fill the gaps left by available written sources.

To supplement them, scientists have exploited sources of data that offer useful indirect evidence regarding the past state of the phenomenon in question. Proxy indicators as varied as oxygen isotopes present in the sediments of the deep sea floor and pollen in those of the land surface have been used with much success in reconstructing climates of the geological and recent past, and their use has been steadily refined. It has proven possible to reconstruct long-term changes in many marine fish populations by measuring the abundance of fish scales deposited and preserved in ocean sediments and changes in land use from the patterns of soil erosion and deposition in streambeds. Human activities also may offer proxies for environmental change. Harvest times and traces of the abandonment of human settlements have been used to document climatic conditions during the past millennium. Such data must always be used with great caution, though, for such changes may have occurred for purely social reasons having nothing to do with environmental shifts. [*See* Climate Reconstruction; *and* Erosion.]

Uncertainty about the past remains a great obstacle to impact assessment. In some areas, human-induced environmental changes have turned out to have a much longer history than was once assumed. As a result, a pristine, preimpact state has proven much harder to define. In other realms, the knowledge of the preimpact baseline from which change can be measured is seriously incomplete because knowledge is particularly costly to obtain. The extent of known deposits of many mineral and fossil fuel resources in the Earth's crust depends heavily on the economic incentives that exist to search for such deposits. Shortages that arise as a result of the depletion of known sources tend to encourage the discovery of more. Over the course of human history, known stocks of such nonrenewable resources have steadily grown as a result of exploration at the same time that they have been reduced by extraction and use. The proportion of the total resource judged usable for human wants also depends on ever-changing conditions of technology and economics. In such cases, it is very difficult to define with any precision the size of an initial resource that has subsequently been depleted and how much of it remains, even though it may be a reasonably simple task to calculate roughly what amount has been mined and used to date. It cannot be assumed that discovery has come to an end and that what is currently known is the total of what exists. The gaps can be filled by estimates made using theoretical assumptions, but the results tend to be controversial.

Calculations based on theoretical models of species size and distribution are often used in assessing human impact on plants and animals. The knowledge of many species and populations is far sketchier than that of ore and energy deposits because little or no direct use is made of them and little or no effort has ever been invested in taking stock of their numbers. It is generally agreed that more species exist that have not been identified and named than that have been, but how many more remains a matter of immense uncertainty. Biogeographical theory has necessarily been invoked to

provide estimates, but with results that are inevitably less compelling and more controversial than ones established by observation would be. The resulting uncertainty affects both the original total baseline number of species and the number that can be presumed to have been extinguished by human action.

Present-day numbers of species and rates of extinction by no means represent the only areas where the knowledge of current conditions, as of past ones, is far from ideal. In the area of land use and land cover change, documentary sources are much more abundant than for the past, statistics especially, but many are of questionable accuracy, and standard meanings and measures that would permit comparison of figures from different places or periods are rare. Nonetheless, much progress has been made. Remote sensing technologies have greatly expanded data archives and the possibilities for reliably assessing land cover. Observing stations have been multiplied for the regular collection of data on atmospheric composition, climate, streamflow, and many other variables of interest. Techniques of chemical analysis have become so sophisticated as to permit the detection of trace substances in tiny quantities in the environment.

Establishing how much change has occurred between some earlier and some later state of the environment still is not enough to specify how much of it was due to human impact. In many realms, natural change occurring during the same period must also be considered. The last ice age, ending roughly ten thousand years ago, was one of tremendous natural upheavals in climate, biotic patterns, hydrology, and landforms across the globe at a time when *Homo sapiens* was already a significant presence in many regions. Natural changes, though less dramatic and profound than continental glaciation, have not ceased to occur in the subsequent period of accelerated human impact.

The problems of separating anthropogenic from natural change are nowhere greater than in the area of climate and other realms closely linked to it. The reconstructed record of global climate throughout geological times shows numerous fluctuations owing to forces that have certainly continued to operate even as human activities such as the release of greenhouse gases have reached a scale capable of affecting the climate system. Natural climatic variations are a possible explanation, in whole or in part, for a number of environmental changes that are often attributed in whole or in part to human impact. Vegetation decline in arid and semiarid zones, one of the phenomena sometimes collectively called desertification, can be the work of either human actions or of climatic changes or variations. [*See* Desertification.] Abrupt declines in marine fish populations may be caused by overharvesting, but evidence exists that such declines also occurred, probably triggered by climate, long before the advent of any significant human impacts. What especially complicates the task of separating human and natural processes is that there may be important synergies at work between them, making causality difficult to attribute. The two forces, moreover, may work at cross purposes, as may be the case with some forms of climatic change, thereby masking each other's effects.

In certain environmental realms, the question of natural change is not a major one. Human impacts may be far more rapid or powerful than similar processes known to occur in nature, or the possibility of nonanthropogenic variations can be dismissed because the changes are entirely novel ones that do not occur in nature. Changes that fall into the former category include certain kinds of land-cover alteration, such as forest clearance and wetland drainage, which do occur naturally but where human impact is generally not difficult to identify, and such modern processes as the large-scale mining and release of heavy metals, such as lead, mercury, and cadmium, which are mobilized by human action in quantities far exceeding their flows in nature. The latter category includes the presence in the environment of nonnatural substances such as synthetic organic chemicals, plastics in the oceans, and artificial radionuclides.

**Human Impacts.** The Earth's land cover—the soil, vegetation, and structures on the land surface—displays the most obvious marks of human impact and many that are not so obvious. Cultivated land now accounts for roughly 15 million square kilometers of the Earth's surface. It has expanded at the expense of other land-cover types. The global area of forest has diminished by some 15–20 percent since preagricultural times. Grassland has experienced a smaller net decline. It has grown through the conversion of forest to pasture and shrunk as it has been converted to cultivation and other uses. The global area of coastal and freshwater wetlands has declined, mostly as a result of drainage for cultivation and construction. Human settlements have come to occupy several percent of the world's land area and a much higher proportion in densely populated regions. There has also been significant modification within the major classes of land cover. Forests have been extensively thinned and altered in composition. Vegetation decline has been substantial in many grasslands and drylands as a result of human activity, especially livestock grazing. Humans have enjoyed a steadily increasing flow of food and other outputs from the lands of the globe, which in recent decades has been the result less of the cultivation of new lands than of the more intensive use of lands already farmed. The expansion of cultivated area has thus slowed even in a period of rapidly rising overall output. In some regions of the world there has been a net reversion of area from crop to forest cover. Among the in-

advertent consequences of land use, soil degradation in various forms, particularly salinization and nutrient depletion, has been significant over a large part of the Earth's cultivated area. [*See* Salinization.] Soil erosion by wind and running water has in many places been accelerated by human action, though to a degree difficult to quantify globally or regionally. Erosion has further impacts of siltation and sedimentation downwind and downstream, where the eroded material is deposited.

The stocks and flows of surface fresh water have been more profoundly altered than the land surface. Both total and per capita human withdrawals from the hydrologic cycle have risen rapidly during the past several centuries and especially during the twentieth. They now claim a substantial fraction of its mean annual volume, by some estimates as much as half of the readily available flow. Globally, irrigation of crops has been and remains the chief activity for which water is withdrawn, but other uses are dominant in many basins and regions, especially those in humid climates. Transfers of water between basins have increased, but they remain minor in the global and regional aggregate. The impounding and regulation of rivers and the construction of reservoirs across the globe have made overall flows of water significantly more even than in the past. In some places, however, the trend is partially offset by an increased spread between peak and low flows due to land-cover changes such as deforestation, drainage, and urbanization. As water use has increased, so has the volume of polluted waste water returned to the hydrologic cycle. The variety of pollutants that it contains has expanded as well, with consequences for water composition, biota, and human health. In turn, wastewater treatment and the provision of safe supplies have also expanded, though much more in the developed than in the developing world. [*See* Ground Water.]

The list of changes in the global biota begins with the great increase in the size and distribution of the human population itself since glacial times and its particularly rapid growth during the past several centuries. The expansion of human numbers and activities has in turn greatly affected the numbers and distribution of other creatures. Species extinctions are the most dramatic form of impact, but sizes of populations and genetic diversity within them have also been greatly affected and for the most part reduced. Losses of biodiversity to date have been heaviest in island and freshwater ecosystems and in tropical forests rich in endemic forms of life. Habitat destruction as a result of land-cover change has been the most powerful human cause of decline. Hunting, pollution, and competition from introduced species have been secondary forces overall, though each has been dominant in some cases and regions. Estimates of terrestrial plant and animal extinctions are controversial, but it is generally agreed that the rate of loss as a result of human action far exceeds the natural rate. Extinctions of marine animals have been few, but populations of large marine mammals, especially whales, were drastically reduced by past hunting, and many marine fish stocks have collapsed as an apparent result of overharvesting possibly combined with natural, climate-related variations in numbers. Species transfer, whether deliberately or accidentally effected by humankind, is both a source of extinctions and declines and a major change in its own right. The environmental history of bacterial and viral pathogens represents a distinctive case of human impact on biota. Some social changes have magnified their effects. Once-isolated outbreaks of illness have acquired a worldwide reach as the rate and scale of trade, travel, and migration have increased, leading to what is sometimes called the disease unification of the globe. Other trends, such as improvements in sanitation, nutrition, and medical care, have lessened—though quite unevenly across the world—the burden of infectious illness. Yet some interventions have not taken place without further unforeseen consequences: the rapid evolution by selection of strains resistant to common antibiotics, for example, a phenomenon also seen in the evolutionary response of agricultural pests to chemical biocides. [*See* Disease; *and* Human Health.]

Human impacts have not been limited to the Earth's surface. They have included substantial extraction of ground water from subsurface aquifers; of the fossil fuels of coal, oil, and natural gas; and of metal ores and other mineral resources. In each case, the result has been a significant depletion of the resource stock, though generally a difficult one to quantify. Global groundwater resources and the net changes that they have undergone are not well documented, but many local and regional cases of serious depletion by withdrawal far exceeding rates of recharge have occurred. The chief known deposits of high-quality ores of more than one metal have been exhausted and production shifted to much lower-quality ores.

Materials extracted from the surface or the subsurface are released during production, use, or disposal into the environmental compartments of air, water, and soil. As a result, many of the chemical flows of the biosphere have been significantly altered and some entirely new ones have been created. Releases from fossil fuel combustion, added to those from land-cover changes such as deforestation, have increased the annual flow of carbon through the biosphere by roughly 10 percent. Human releases of sulfur to water and atmosphere, mostly from fossil fuels and metal ore processing, have approximately doubled the natural global flow, with effects including the acidification of precipitation, soil, and surface water. Nitrogen flows have been increased chiefly through the manufacture of synthetic fertilizer

and fossil fuel combustion, raising agricultural yields while contributing to acidification akin to that resulting from sulfur, to the eutrophication of lakes and coastal waters, and to certain forms of air pollution. The flow of phosphorus, chiefly mined as another agricultural fertilizer, has also been greatly increased and has also contributed to eutrophication. Besides altering these great flows of the biosphere, human impact has been particularly noteworthy on those constituents of the biosphere that are naturally present in quite small concentrations. Human action annually mobilizes many metals—copper, lead, mercury, cadmium, and numerous others—in quantities much larger than the natural releases from weathering, volcanoes, and other sources. The current average natural and anthropogenic doses of ionizing radiations, at least in the United States, have been estimated as roughly equal. Finally, chemistry has introduced into the biosphere many synthetic substances that did not previously exist. Important compounds created by industrial organic chemistry include pesticides, solvents, and chlorofluorocarbons (CFCs), used as coolants and propellants.

Many of these emissions have contributed to changes in the chemistry of the oceans. Such changes have been most pronounced off populous coastal areas and in enclosed seas, such as the Mediterranean and the Baltic, where human activity is highly concentrated and diluting capacity is small. Nitrogen and phosphorus enrichment through sewage discharge and fertilizer runoff have degraded near-shore waters in many parts of the world. Oil is added to the oceans naturally through seeps, but tanker discharges have multiplied the amount many times over, especially along main shipping routes. Much wider but less pronounced increases in many trace substances from atmospheric deposition have been registered in surface waters across the globe. The accumulation in the ocean of decay-resistant plastic debris of various sorts, including litter, discarded fish nets, and production residues, is a recent but rapidly expanding phenomenon.

Emissions have also changed the chemistry of the atmosphere. Highly stable CFCs have accumulated in the stratosphere, where their decay products catalyze the breakdown of ozone molecules. The result has been a decline in high-altitude ozone concentrations of some 5–10 percent on global average, an uneven decline that is especially marked over the high latitudes and in winter. Because stratospheric ozone blocks ultraviolet solar rays from the Earth's surface, its depletion increases surface exposure to such high-energy radiation. In the lower atmosphere, there have been many localized impacts. Various reactive trace gases have increased or decreased as human activities have developed. Particulate pollution has thickened with industrialization and in some places, especially in the developed world, has been reduced by management. Nitrogen oxides released by fossil fuel combustion have widely promoted the formation of ozone smog, especially in sunny low-latitude metropolitan regions where conditions are most favorable. Globally, the atmospheric concentration of carbon dioxide has been increased by about 30 percent over the preindustrial level. That of methane, released chiefly from land use, has approximately doubled during the past several centuries. [See Air Quality.]

Carbon dioxide, methane, CFCs, and several other trace components of the atmosphere are "greenhouse gases." They have the property of absorbing and retaining heat radiated from the Earth. The likely net outcome of an increase in their concentrations in the atmosphere is an increase in global average temperatures, a change that might be further amplified or dampened by its own effects on greenhouse gas concentrations. Climatic changes thus inadvertently produced by human action would not be globally uniform but might vary greatly across regions. They would bring about many further impacts in climate and in other realms of the environment: in precipitation, storm patterns, sea level, biota, and land cover. But precisely what climatic changes and what further consequences could be expected from existing and projected levels of greenhouse gas concentrations remain matters of intense debate. They are largely forecast from models of doubtful reliability or from the use of analogies with other situations that may not be similar enough to be useful. A clear signal of human-induced change has been difficult to detect so far against a background of considerable natural fluctuation. Many local-scale changes in climate are far better documented. Though the atmosphere, like the oceans, is a globally fluid system, no more than in the oceans are all climatic changes of the globally systemic sort. The most substantial local-level change is the elevated temperature typical of cities and their surroundings known as the urban heat island, which may in turn affect precipitation, storminess, and other processes.

**Patterns and Trends of Human Impacts.** Human impacts on the environment are so varied that generalizations about them must be made with caution, but several can be offered.

Both impacts in general and those that are significant at a global scale have expanded greatly in variety over time as human activities have done the same. Such impacts once involved chiefly alterations in the Earth's surface features but have come to include many changes in the biospheric flows of energy and material as well. Preagricultural impacts were modest and were confined chiefly to the land surface and the biota. They were greatly intensified and expanded by the development of agriculture, but as late as the beginning of the nineteenth century, the chief forms of human impact were forest clearing, land cultivation, and the extinction, depletion,

domestication, and transfer of plant and animal species. Biotic impacts aside, they were largely reversible processes. Natural recovery could take place if allowed to. A few of these activities affected global material flows. Preindustrial carbon dioxide release from forests and soils and methane releases from rice and livestock were already substantial fractions of current levels. Impacts of regional and global significance on realms affected by extractive and industrial processes are largely a phenomenon of the twentieth century. [See Chemical Industry.]

Among them are most global-scale alterations of the major biospheric cycles. There has been an expansion in the spatial scale of impact as many changes once substantial in particular regions have become so cumulatively in the global aggregate. Yet most impacts, save for the purest examples of globally systemic change—atmospheric mixing of relatively stable and long-lived gases—remain highly differentiated across the Earth's surface. In many respects, there are vast differences between advanced industrial societies and less industrialized ones and between the urban and rural areas in each—differences in land cover, in the availability and quality of fresh water, in air quality, and in many other respects. Global trends and patterns of change are only poorly mirrored in most subglobal areas of the world.

There has been a striking acceleration at the global scale in many, probably most, long-term trends of change as human activity has expanded in scale and scope. Annual freshwater withdrawals, for example, have more than doubled since the end of the World War II. More than half of all anthropogenic sulfur releases, nitrogen fixation, and phosphorus extraction has occurred since 1960 or later. Large-scale synthetic organic chemical manufacturing is chiefly a development of the twentieth century.

On the other hand, a few human impacts have decelerated regionally and even globally as compared with their earlier trajectories, whether as the incidental result of a shift in activities taken for reasons of economic or technological change or because of purposeful environmental management. Forest regrowth on once cultivated or grazed land has been extensive in much of the developed world, but it has probably had more to do with the economics of agricultural land abandonment than with deliberate measures for environmental restoration. So too, the clearance of tropical forests has slowed in periods of economic depression, as have global releases of carbon and sulfur from fossil fuel combustion. But some examples of lessened impact can be credited directly to managerial intervention. They include the decline of radioactive fallout following a international treaty banning atmospheric nuclear testing in the early 1960s, the recovery of many marine mammal populations following restrictions on harvesting, the declining use of lead additives in gasoline, and the restriction of persistent pesticides such as dichloro-diphenyltrichloroethane (DDT). International arrangements have been negotiated to curb chlorofluorocarbon emissions and protect the stratospheric ozone layer, and early agreements have been reached to slow carbon emissions in order to avoid dangerous interference with the global climate system.

Several of these problems addressed by management, however—particularly the impacts of CFCs on stratospheric ozone—are among the most striking examples to date of another trend, the more frequent emergence in recent times of surprise impacts where some human action has had consequences of a sort not foreseen when it was undertaken. Human activities in earlier centuries had their unexpected environmental consequences, such as inadvertent transfers of plant and animal species, or planned transfers that got out of hand, but to a much larger degree today the unintended effects may exceed the intended ones. As novel human activities multiply and technological innovations are produced ever faster and diffuse much more rapidly than before, many new direct and indirect impacts on the biosphere are created. Substances are mobilized and released in large quantities long before their full range of possible consequences, some of which may be subtle, surprising, or slow but cumulative, can be assessed. Other modern examples have included the worldwide spread of lead as a result of its use in gasoline and that of DDT residues following its development and adoption as a pesticide. Prevailing trends suggest that there will be many more surprise environmental impacts of this sort to come during the foreseeable future. The same trends in innovation and diffusion, however, may also mean that remedies for both novel and long-standing environmental problems and substitutes for activities causing damage can be fashioned and deployed much more rapidly than they could in the past.

Finally, the increasing variety of human activities and of impacts that they produce can greatly complicate the task of dealing with any of them. A measure that alleviates one problem may worsen or create another. When DDT itself was introduced, it was most of all as a means of countering an environmental hazard in some ways worsened by human actions, the prevalence of mosquito-borne malaria in many regions of the world. As its harmful effects on other biota became known, it was widely replaced by more reactive pesticides that did not persist in the environment to such a damaging degree upon release, but that posed all the more of a danger to the health of the farmworkers applying them. The replacement of one form of energy generation by another generally replaces one set of environmental impacts with a second. Nuclear power, for example, does not release greenhouse gases, but creates a different set of risks that must be weighed against those of fossil fuels.

There is every reason to suppose that the trade-offs of this sort involved in managing environmental change will only become more perplexing and challenging in the future.

[See also Anthropogeomorphology; Climate Change; Fire; Global Change, article on History of Global Change; Greenhouse Effect; and Introduction.]

## BIBLIOGRAPHY

Consequences: The Nature & Implications of Environmental Change, 1995–. A quarterly journal presenting articles written and reviewed by specialists that sum up in accessible fashion the current state of knowledge in key areas.

Goudie, A. S. The Human Impact on the Natural Environment. 5th ed. Cambridge, Mass.: Blackwell, 1999. A standard textbook on the subject.

———, ed. The Human Impact Reader: Readings and Case Studies. Oxford: Blackwell, 1997. A useful collection of research papers that illustrate the range of human impacts and the methods used in their study.

Goudie, A. S., and H. Viles. The Earth Transformed: An Introduction to the Human Impact on the Environment. Oxford: Blackwell, 1997. A comprehensive recent survey.

Simmons, I. G. Environmental History: A Concise Introduction. Oxford: Blackwell, 1993. A readable and comprehensive synthesis of the broad sweep of human–nature interactions.

Turner, B. L., II, et al., eds. The Earth as Transformed by Human Action. New York: Cambridge University Press, 1990. The most detailed and comprehensive inventory of human impact.

Vitousek, P. M., H. A. Mooney, J. Lubchenco, and J. M. Melillo. "Human Domination of Earth's Ecosystems." Science 277 (1997), 494–499. A useful short recent review.

—WILLIAM B. MEYER

# HUMAN POPULATIONS

Human populations are among both the causes and consequences of global change. The size of the population, the ways in which it changes (the balance of births and deaths, there being no migration at the global scale), and its internal distribution and redistribution will all affect the type and nature of global change; and the economy, the environment, the state, and society will in turn affect population size, distribution, and rate of change. The broad course of global trends in population over the approximately three thousand millennia of human history, and the relationships between fertility and mortality, are the starting point for an examination of the causes and implications of a fundamental transition in human population that began only within the last three hundred years and is likely to continue into the next millennium: from an equilibrium of high mortality and high fertility, with crude death and birth rates of thirty-five to forty per thousand, to an equilibrium of low mortality and fertility and rates of ten to fifteen per thousand. This has been a demographic revolution of dramatic proportions.

**Sources of Information on Human Populations.**
Evidence for the presence of human populations, if not their size, at any given time comes from archaeology. Human remains—typically bones—are the most familiar archaeological source, while deoxyribonucleic acid (DNA) evidence is used increasingly to study relationships between human groups, and in particular for identifying long-run patterns of human migration. There is evidence of the volume and type of human activity from its imprint on the land, for example in ancient irrigation and terracing systems. Further evidence is provided by artifacts such as ornaments and military and domestic tools, and by urban and manufacturing remains, especially from metal and stone sources. Archaeological evidence is of necessity partial and geographically patchy, much dependent on where archaeologists have been active (many developing countries remain archaeologically under-investigated), and also on the fact that some materials are less durable than others, especially in hot and wet climates where bacterial decay is rapid and evidence from wood and other organic materials does not survive.

Estimates of population numbers rely heavily on documentary sources. Where early centralized civilizations, mostly associated with the Neolithic revolutions of some six to eight thousand years ago, developed a written culture, as in ancient Egypt, there were early estimates of human populations. Rather later, in the Roman Empire for example, population counts in history were primarily associated with taxation, as in the case of the census in Judaea in the late first century BCE, during the reign of Herod the Great. In China, Japan, and India, and in Europe and the Middle East, the primary sources for reconstructing the size and rates of change of populations in history have been national administrative records, such as the Domesday Book for England (1086 CE).

The growth of the economic and political cohesion of European and North American states from the eighteenth century was associated with the first formal national compilations of population data, and the state is now everywhere the prime agency for population enumeration. This takes two forms. First, the establishment of national registration systems for births, deaths and marriages, often as a direct derivative of ecclesiastical records of baptisms and burials. In most developed countries, registration has been a continuous source for monitoring population change, and in a few cases (namely the Netherlands, Sweden) includes registration of migrants. Registration data are used not only to examine contemporary population trends, but also to analyze historical registers; knowledge of the patterns and causes of the sharp decline in mortality in Europe in the nineteenth century has been largely derived from age and cause-of-death entries in the parish or town registers.

Second, the establishment of regular censuses, usually at ten-year intervals. The essence of any census is that it is cross-sectional, universal, simultaneous, and normative (that is, it asks questions that are relevant to all, such as age, sex, occupation, and place of residence). Census data are valuable in their own right as providing a "snapshot" of the size, distribution, and essential characteristics of a population, but are also used comparatively: successive censuses allow identification of patterns and rates of change between census dates.

Population censuses have been the principal source of population data worldwide over the last fifty years. They have proved relatively simple to collect, even in colonial societies with largely illiterate peoples and no tradition of population counts. The United Nations (UN) Population Division has been instrumental in generalizing the census experience, and all countries have now had at least one reasonably reliable national population census within the last twenty years. The results of these censuses and other survey sources permit international comparison, and the UN has become the principal source of estimates of past global trends and projections of future trends. These are available in the annual UN Demographic Yearbook and associated publications.

**Premodern Populations.** Global population change is the resultant of the balance between births and deaths. Although there is constant redistribution of the global population through migration, both of individuals and of groups, over long and short distances, most populations throughout most of human history have experienced a balance of births and deaths, at least in the medium term (perhaps twenty-five to fifty years). In the short term, whether seasonally, annually, or longer, there may be some growth or decline, but in the very long term, over a period of centuries or millennia, growth of the global population has generally been very slow. It is likely that for most of the period before the last two millennia, the population at the end of any millennium was no more than 5 percent greater than at the beginning of that millennium. Over this long period there has been slow expansion from the earliest foci of human existence in the East African Rift Valley in Tanzania, Kenya, and Ethiopia, some 1.5–3 million years ago, to all habitable areas of the Earth's surface.

During most of this period, life expectancy at birth for most populations was probably less than thirty years, which may be compared with a figure of more than forty years in the poorest countries today. Life for most people in most groups was nasty, brutish, and short. In particular, in the absence of systematic medical knowledge and with the low and erratic levels of nutrition that were and still are characteristic of hunter–gatherer modes of life, survival rates for infants and children were very low. As food supply became more secure with settled agriculture after the Neolithic revolution, beginning some ten thousand years ago, particularly in the riparian civilizations of the Middle East (based on the Nile, the

Tigris–Euphrates, and the Indus) between about 5,000 and 3,000 BP and rather later in the rice civilizations of eastern and southern Asia, it is likely that long-term mortality levels improved. Even with settled agriculture, rates remained liable to fluctuate from year to year because of the random occurrence of epidemics or famines and the associated phenomenon of famine mortality. In many cases there will have been an excess of deaths over births, often by a considerable margin, followed by population decline. Some declines may have been spectacular: the most notorious of these was the Black Death, an outbreak of bubonic plague that killed an estimated one-third of the population of Europe in the middle of the fourteenth century. The nearest twentieth-century equivalent was the global influenza pandemic in 1918–1919, which killed an estimated thirty million people in a twelve-month period (more than three times as many as were killed in World War I, 1914–1918).

The high and fluctuating mortality of the premodern period was accompanied by equally high but much less volatile levels of natural fertility. In theory, women could have at least twenty pregnancies during their biologically productive years—roughly between the ages of fifteen and fifty—but the highest recorded mean fertilities for whole groups are between twelve and fourteen children per woman. Some 3–5 percent of women in most populations are naturally infertile, a further and much more variable proportion become infertile through infection and disease, especially sexually transmitted disease, and there are periods of biological infertility for all women for several months after childbirth. These infertile periods are extended by social constraints on fertility, notably those associated with marriage and sexual abstinence before, during, or after marriage. These biological and social proximate determinants of fertility reduced the number of children to between four and eight children per woman for most preindustrial societies. Even where there is a strong tradition of early marriage, as in contemporary India, Bangladesh, and Pakistan, there is not necessarily a correlation with high fertility because of an equally strong tradition of long birth intervals, which are good for the health of mothers as well as for their newborn infants. The demographic benefits of traditional practices associated with long birth intervals, especially breastfeeding, are integral to contemporary maternal and child-health programs in developing countries.

Unlike mortality, fertility is not governed principally by factors beyond individual control. Even in early historical periods, social norms—notably marriage—are likely to have given relative stability to fertility rates, for these norms evolve only slowly. E. A. Wrigley and R. S. Schofield's classic analysis, *The Population History of England, 1541–1871* (Wrigley and Schofield, 1981) showed, for example, that marital fertility was relatively stable in England between the sixteenth and nineteenth

centuries, rising gradually, with some fluctuations, over a period of substantial annual fluctuations in mortality, especially before 1700—with upturns in fertility during some periods of relative economic prosperity.

Although it is not normally possible to reconstruct directly the demographic balance in premodern populations, the apparent outcome of very slow growth for most populations suggests that high but relatively stable fertility, and high but annually and seasonally fluctuating mortality, have been the norm for the greater part of human history. This "natural" regime seems to have been underpinned by social institutions and economic and political systems, and also by generally fragile food-production systems, resulting in inadequate nutrition, variable from place to place and from season to season, as manifested in the universal evidence for famine mortality (although the likelihood of famine over large areas was itself also a function of environmental conditions, notably variable rainfall). The long-term trend was for the slow overall growth to be incorporated *in situ* or else to fuel migrations to colonize new land.

**The Demographic Transition and Modern Populations.** This long-established high-level global demographic equilibrium has altered very fundamentally and rapidly over the last third of the present millennium. There are now much lower levels of fertility and mortality in developed countries, where there are greater short-term fluctuations in fertility than in mortality. The net global effect of the period of the changing balance between fertility and mortality has been a much higher and, for the most part, continually rising rate of overall population growth, to the extent that the world population has grown from an estimated 300 million, two thousand years ago, to approximately 1 billion by 1800, to 1.65 billion by 1900, and to nearly 6 billion at the century's close (Figure 1).

There has clearly been a fundamental shift from a high-level equilibrium of births and deaths to a lower level of both that seems to mark a new low-level equilibrium in some countries. The transition from high to low demographic equilibrium was first evident in Europe. Over a period of two hundred years, from 1700 to 1900, there was very rapid population growth throughout western Europe as mortality fell but fertility remained at its pre-eighteenth-century level. The sharp and consistent falls in mortality seem to have been triggered in the first instance by improved nutrition. Medieval and Renaissance Europe experienced economic prosperity, with urban growth and relative political stability. The agricultural revolution brought increased local availability of food, and greater inter-regional and international trade in food was able to reduce greatly the incidence of famine mortality. After 1500, new foods brought from the New World, notably the potato, began to enter the European diet. By the nineteenth century, medical science had also advanced, with direct effects

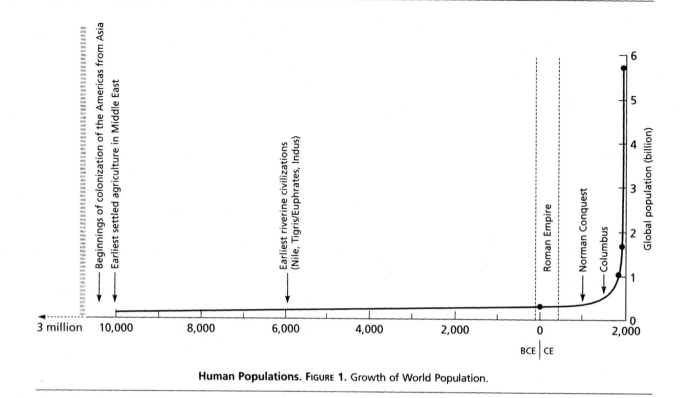

**Human Populations. Figure 1.** Growth of World Population.

on infant and childhood mortality, especially through vaccinations. Private and public incomes were also rising at this time of rapid urbanization and industrialization, creating a greater purchasing power in the mass of the population, allowing greater food "entitlements," to use Amartya Sen's terminology (Sen, 1981). An "urban penalty" in mortality had emerged throughout Europe, whereby rural areas, with less crowded and less insanitary conditions for the mass of the population, experienced lower levels of mortality than urban and industrial populations. The quality of the environment in overcrowded and poorly managed towns had deteriorated seriously, but, during the second half of the nineteenth century in particular, substantial public resources were directed to improvements in urban housing and sanitation. By the early years of the twentieth century, rural and urban mortality levels were low and continued to decline in developed countries.

For most of this period, fertility remained high in Europe, with a growing gap between fertility rates and mortality rates that resulted in high rates of natural increase. Some of that increase fed the large-scale migrations to the New World, but most of it was absorbed in the growing urban populations in Europe, with substantial rural–urban internal migrations that continued well into the twentieth century. Only by the last quarter of the nineteenth century and into the twentieth century did fertility begin to fall in Europe and North America, a fall that was fairly continuous until it approached levels of

mortality, and therefore produced very low—or even negative—rates of natural population growth. This occurred from the 1930s in a few countries (for example, Britain and France), but in most of them by the end of the twentieth century.

Why did fertility fall in these countries at the time it did? It happened in the absence of two features currently associated with contemporary developing-world fertility decreases. First, there was no sense of national population policy, and no direct intervention by governments to facilitate fertility decline. Second, there was no general use of artificial means of contraception: neither barrier methods, including condoms (in widespread use only from the 1920s), nor pharmacological methods, principally the contraceptive pill (which became available only in the 1960s). It happened spontaneously: as a result of decisions of individual couples to limit their family size, either through sexual abstinence or by such means as the rhythm method or *coitus interruptus.*

Thus, while mortality in industrial populations fell as a result of structural causes, such as health improvements or economic and environmental conditions beyond individual control but affecting everyone (though not necessarily equally), fertility fell as a result of individual choice, a human response to changing economic and social conditions. Life in towns meant overcrowded housing and children who were much more of an economic burden (attending school and, after the enforcement of legislation limiting the employment of children, not direct contributors to the household economy) than they were in rural areas, creating effective incentives for limiting family size. Couples seem to have responded to these new social and economic conditions by reducing mean family size to levels that would allow low or even negative population growth: a lagged response to earlier mortality decline.

This historical experience of western as well as some eastern populations (including Japan) of moving from high-level equilibrium to low-level equilibrium has been generalized into the demographic transition model: an empirical generalization based on historical experience, used to describe the general course of population change (Figure 2). All populations began in the high-equilibrium stage, with fluctuating mortality and rather more stable fertility, and remained thus until recently. The underlying presumption of the demographic transition model is that all countries will progress through these stages to the low-level equilibrium now evident in the developed countries, but will do so at different rates and perhaps even for different reasons. Thus there is a predictive implication: all populations will progress through the transition, and populations with high rates of growth with currently excess fertility are merely lagging in the process of global change.

Twentieth-century development has largely sought to replicate elsewhere the structural conditions that were associated with falling mortality in nineteenth-century Europe: industrialization and associated urbanization and rising personal incomes; the application of medical science and healthcare systems; environmental improvements in clean water supply, sanitation and better housing; and social improvements such as modern education. Global development has taken place through the diffusion of these structural features out of western Europe and North America, first into eastern Europe and the Mediterranean countries, and subsequently into the less developed areas of Africa, Asia, and Latin America. This was implicit in the colonial objectives of European powers in bringing "civilization" to the colonized populations, and it brought evident—though not necessarily immediate—success as mortality fell almost universally in the twentieth century.

Global modernization had been presumed by the development strategies of the major bilateral and multilateral development agencies, notably The World Bank, which, in its 1984 *World Development Report* (World Bank, 1984), famously argued an explicitly diffusionist explanation of the global demographic transition. There have been very effective medical interventions, as in the eradication of smallpox (the last recorded case was in 1974) and in immunization programs urged by such agencies as United Nations Children's Fund (UNICEF) and Save the Children. There is much less malnutrition globally: the proportion of the world's population living in countries with mean dietary energy supplies below the World Health Organization (WHO) recommended level fell from 83 percent in 1961–1963 to 44 percent in 1990. Food is now much better distributed, both internally in most countries as well as globally, and, with bet-

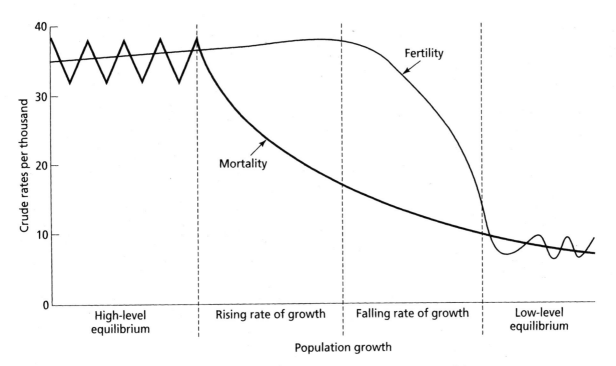

**Human Populations. FIGURE 2.** The Demographic Transition Model.

ter food security and much greater warning of potential problems, the occurrence of famine mortality is much reduced. Although there is still a long way to go in the poorest countries, mortality rates, especially in children under five, have fallen in all countries of the world in the twentieth century. In many developing countries they are beginning to approach the levels and stability of developed countries.

As we have seen, fertility levels are the result of individual choice and cultural factors rather than the structural conditions associated with development. There are, however, important similarities in the factors affecting fertility in nineteenth-century populations of countries now defined as developed and in contemporary developing-country populations, notably the increasing proportion of the population that is urban and the increasing number of women who have had a modern education and have access to modern health care and contraception. However, there are also important differences in the cultural and economic roles of children, such that many developing-country populations may be more resistant to reduced family size. A higher proportion of such populations remains predominantly rural, winning a marginal living from the land. Children are more likely to be net contributors to the household economy from an early age, whether on the farm in weeding, harvesting, bird scaring or cattle herding, or in the domestic sphere in fetching water or caring for younger siblings. In such conditions, large families bring more hands to work rather than more mouths to feed, and weaken economic incentives for lower fertility. Furthermore, in societies that are tied to the land, large families cement inheritance systems and social obligations.

In sub-Saharan Africa, with its strong cultural legacy of traditional landholding and land management systems, the evidence for rural fertility decline has been most elusive, but even here most recent data do indicate falling fertility in all countries (Table 1). The de-

**Human Populations. TABLE 1.** Total Fertility Rates, 1970, 1980, and 1999

| (A) BY MAJOR REGION | 1970 | 1980 | 1999* |
|---|---|---|---|
| Sub-Saharan Africa | 6.6 | 6.7 | 5.8 |
| East Asia and Pacific | 5.7 | 3.1 | 2.05[†] |
| South Asia | 5.8 | 5.3 | 3.6[‡] |
| Middle East and North Africa | 6.7 | 6.1 | 3.6 |
| Latin America and Caribbean | 5.2 | 4.1 | 3.2 |
| Low- and middle-income Countries | 5.4 | 4.1 | 3.2 |
| High-income countries | 2.3 | 1.9 | 1.5 |

| (B) 15 LARGEST COUNTRIES (65 PERCENT OF WORLD'S POPULATION) | 1999 POPULATION* | 1970 | 1980 | 1999 |
|---|---|---|---|---|
| China | 1254 m | 5.8 | 2.5 | 1.8 |
| India | 987 m | 5.5 | 5.0 | 3.4 |
| United States | 272 m | 2.2 | 1.8 | 2.0 |
| Indonesia | 212 m | 5.3 | 4.2 | 2.8 |
| Brazil | 168 m | 4.9 | 3.9 | 2.3 |
| Russia | 146 m | 2.0 | 1.9 | 1.2 |
| Pakistan | 146 m | 7.0 | 7.0 | 5.6 |
| Japan | 127 m | 2.0 | 1.8 | 1.4 |
| Bangladesh | 126 m | 7.0 | 6.1 | 3.3 |
| Nigeria | 114 m | 6.4 | 6.9 | 6.2 |
| Mexico | 100 m | 6.5 | 4.5 | 3.0 |
| Germany | 82 m | 1.9 | 1.6 | 1.3 |
| Vietnam | 79 m | 5.9 | 5.0 | 2.7 |
| Philippines | 75 m | 5.7 | 4.8 | 3.7 |
| Iran | 66 m | 6.7 | n.a. | 3.0 |

*Population Reference Bureau, 1999.

[†]Average of East Asia and Oceania, 1999.

[‡]North Africa only, 1999.

SOURCE: World Development Reports, various.

cline is small in most cases, although much larger in some of the richer countries (South Africa, Botswana, Zimbabwe, Kenya) than in poorer countries with recurring political instability, but overall the annual rate of population growth in this region is still over 2.5 percent (Figure 3). This would suggest that even sub-Saharan Africa is now exhibiting a lagged fertility response, proceeding through a transition to levels of fertility that approximate to its levels of mortality, themselves considerably higher than elsewhere.

Stronger support for a universal demographic transition would certainly be offered by the experience of Latin America and Asia. Latin America is a much more urbanized and industrialized continent that Asia or Africa, and much more exposed to Western economic and cultural attributes and values. There were substantial falls in mortality in the region between 1930 and 1970, and there have also been substantial reductions in fertility, especially since the 1970s. National annual rates of growth, rising to over 2 percent in the 1970s, are now mainly between 1 and 2 percent, and fertility continues to fall more rapidly than mortality, such that the achievement of a low-level equilibrium seems likely. The Asian experience has been even more dramatic. Mortality declines have proceeded slowly, especially in India and China, but they have been general, such that by the 1970s the Asian population was growing at a rate of over

2 percent each year, with little evidence of fertility decline. Since then, however, there have been sharp fertility declines, characteristically in this region associated with direct population policies. The case of China, with its "one child" policy backed by strongly coercive measures and severe economic penalties for exceeding individual or group targets, is the most extreme example of government intervention. Most interventions in the region have taken the form of family-planning programs, with success in relatively poor countries—Bangladesh, Indonesia, and Thailand—as well as in the Asian "tigers" of rapid economic growth—Singapore, Hong Kong, South Korea, Taiwan, and Malaysia.

The rationale for direct intervention in family planning, principally to increase the prevalence of contraception, has been underpinned by the empirical evidence of an apparent universal relationship between total fertility rate (TFR—an index of the number of children each woman can expect to bear) and contraceptive prevalence rate (CPR): as contraceptive prevalence rises, so fertility falls. The most effective means of reducing fertility available to governments, therefore, is to raise contraceptive usage through effective family-planning programs. Governments of developing countries have been provided with massive international assistance to expand contraceptive availability and knowledge, usually in conjunction with maternal and child-health pro-

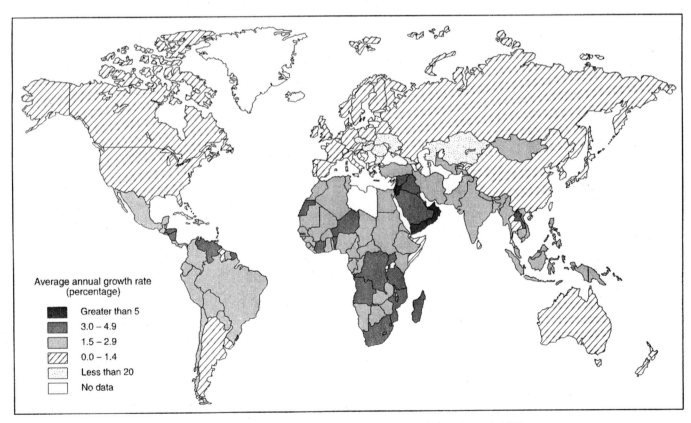

**Human Populations. Figure 3.** Annual Rate of Population Growth (Percentage), 1995.

grams that target birth spacing as much as the use of contraceptives to prevent further conception.

There seems little doubt that there has been a global reproductive revolution in recent decades. Contraceptive use is everywhere rising rapidly, even in sub-Saharan Africa, and, although this is not the sole factor in fertility decline, it is most effective when applied in economically and politically stable societies. Where women and couples have assured contraception and a high likelihood of survival of existing children, together with a higher income, they usually choose to reduce their family size.

**Age and Sex Structure.** A major effect of these recent trends in fertility and mortality has been the differentiation of age and, to a lesser extent, sex structures of populations. On the global scale, differentials in sex (as a biological characteristic) have not been of major significance, for in all populations the numbers of males and females have been roughly the same. Females have a longer life expectancy (roughly 5–10 percent longer than males in most societies), but about 5 percent more boys than girls have typically been born. There is now, however, considerable potential for the sex ratio at birth to be altered significantly by technology that enables couples to ascertain the sex of unborn children in the womb, and to exercise what is usually a preference for male children either by differential abortion, increasingly common in India, or else by differential infanticide at birth, a practice said to have been fairly widespread in China at the height of its one-child campaigns in the 1980s, resulting in a reported sex ratio at birth of 118 males per 100 females in 1991.

As a result of these trends in fertility and mortality, intercontinental variability in age structures has become large (Figure 4). In Europe and North America, after at least three generations of low fertility and mortality, the proportion of the population under fifteen years of age is much lower (North America, 21 percent; Europe, 18 percent) than the global average of 32 percent, but the proportion aged over sixty-five is 18 percent in these regions, more than double the global average of 6 percent. In these populations, the proportion in the economically active age groups, commonly defined as fifteen to sixty-four (but note that this age range loses meaning as an increasing number of people below sixty-five are retired, and many people over sixty-five remain in employment), is higher than the global average of 62 percent. This is in clear contrast to Asia and Africa, where continuing high fertility, especially in Africa, keeps the proportion of under-fifteens very high (over 40 percent), and the poor quality of healthcare will keep the proportion over sixty-five low.

While it is certainly the case that, on these continents, age is a far less rigid determinant of economic activity, the fact that the proportion aged fifteen to sixty-four

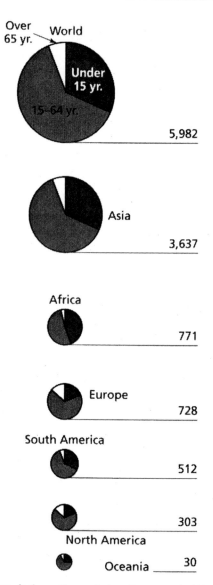

**Human Populations. Figure 4.** Age Structures by World Region, 1995.

remains low, and dependency ratios (the ratio of the dependent population, the young and the old, to the nondependent population) are adverse, has clear implications for the productive potential to generate economic activities to support development expenditures—including the large educational and health expenditures consumed disproportionately by the young and the old. However, as fertility falls, the proportion of under-fifteens will fall, and they will progress into the labor force. In China, the absolute number of fifteen-year-olds entering the labor force is already less than it is in India, with its higher fertility but smaller overall population. These children will also be entering the reproductive age groups, creating a population momentum: although fertility continues to fall, the reproductive population will keep on growing, albeit at a declining rate. By contrast, improved health care and medical knowledge are allowing adults to live longer. Even in the poorest countries, the proportion of old people is rising.

**Population Distribution and Density.** As a result of the joint effects of differential population growth and migration, the global distribution of the human population has been in constant change. Humans have spread to occupy all continents, but with very different continental as well as local population densities. Particularly in the period of very rapid growth in the last three hundred years, the distribution has altered substantially. A. M. Carr-Saunders, in his book *World Population* (Carr-Saunders, 1936), estimated that, in 1650, just before major growth of the European population, just over half the world's population was in Asia, and over 20 percent in each of Europe and Africa. By 1900, Africa's share had more than halved, but Europe's share had risen to nearly 25 percent. Growth in the present century has raised the proportion in the Southern Hemisphere to over 80 percent, with further rises to over 90 percent within one hundred years anticipated by current population projections (Table 2). The current global distribution of population by size of national population and national population density remains highly uneven.

Within each country, the most important global feature of recent redistribution has been a dramatic change in the proportion of rural to urban populations. The rapid increases in population in nineteenth-century Europe were associated with increases in the proportion of the population living in urban areas, and urbanization has been a major feature in all countries within the present century. It is estimated that, in 1995, 45 percent of the global population lived in urban areas (Figure 5). This proportion was still as low as 26 percent in Asia

and 31 percent in Africa, but these are where recent urban growth has been most rapid. By the middle of the twenty-first century, it is likely that, for the first time in human history, the global urban population will exceed the rural population.

**World Population Conferences.** Global concern for the growth, structure, and distribution of population, first voiced in the League of Nations before World War II, has been a matter of concern for the UN throughout its existence. The critical technical role of the UN Population Division in collecting and analyzing population data (discussed above) has in recent years tended to be overshadowed by its broader role in considering global population issues through a series of World Population Conferences, the most recent of which took place in 1974, 1984, and 1994.

At Bucharest in 1974, attitudes to "the population problem" were ambivalent. Although this conference took place just after the first UN Conference on the Human Environment (held in Stockholm in 1972) and the publication of the strongly neo-Malthusian *The Limits to Growth* (1972), it was also a time when many developing countries were still experiencing economic optimism after independence, believing that development would come quickly, especially when associated with a global redistribution of wealth. It was generally, although not universally, recognized that population growth rates (then typically over 2 percent per annum) were too high in developing countries, but that "development is the best contraceptive": direct intervention in population control is generally unnecessary because fertility falls spontaneously with development, as it did in the developed countries. Furthermore, the Green Revolution was by then having an effect on crop yields in Asia, increasing the food supply and reducing the

**Human Populations. TABLE 2.** Shares of World Population by Continent, 1650, 1900, 1995, and 2150

|  | 1650* | 1900* | 1995[†] | 2150[‡] |
|---|---|---|---|---|
| Europe | 21.5 | 24.2 | 12.8 | 4 |
| North America | 0.2 | 2.4 | 5.1 | 3 |
| Oceania | 0.4 | 0.4 | 0.5 | 1 |
| South America | 2.6 | 3.9 | 8.4 | 8 |
| Africa | 21.5 | 7.4 | 12.6 | 19 |
| Asia | 53.8 | 58.3 | 60.8 | 65 |
| Global North | 22 | 27 | 19 | 8 |
| Global South | 78 | 73 | 81 | 92 |
| Total (billions) | 0.465 | 1.6 | 5.7 | 11.5 |

*Willcox, estimated in Carr-Saunders (1936, p. 42). Includes Central America in South America.

[†]United Nations Demographic Yearbook (1995). Includes Central America in North America.

[‡]United Nations Population Projections (1998).

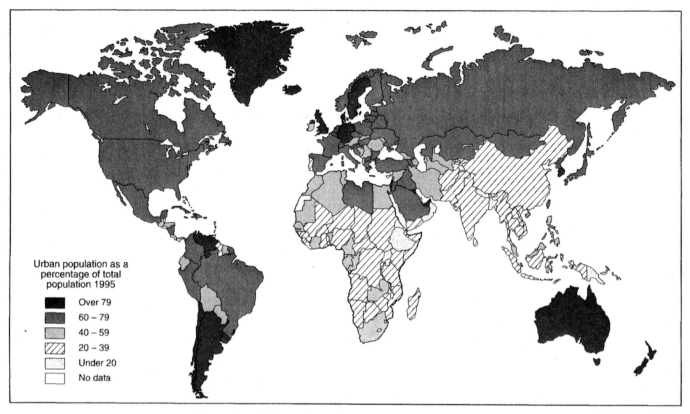

**Human Populations. FIGURE 5.** Urban Proportion of National Populations.

threat of local and regional famines. There was, therefore, an implicit acceptance of the demographic transition model.

By 1984, however, the global economy had deteriorated sharply after the oil shocks of the 1970s, and environmental concerns were becoming more evident. There were droughts and famines in Africa, where annual rates of population growth were already over 3 percent in some countries and generally continuing to rise. There were recurring food crises in China. Green Revolution technologies had been important in increasing global food supplies, but there were clearly limits to the technological improvement and agricultural intensification. National population growth was now seen as a major problem for many governments, even those previously hostile to direct intervention, as in Africa. The Mexico Conference in 1984 saw the acceptance by the majority of governments of the need for direct intervention to control population growth, and especially through family-planning programs. As indicated above, the close statistical correlation between national TFR and CPR had been used to justify intervention, but now governments, supported by international agencies, would be able to allocate additional resources to fertility reduction. Fertility began to fall in almost all devel-

oping countries in the 1980s, and in some with great speed, even to below replacement level (TFR of 2.1), as in China (Table 2).

Coming after the Rio Conference on Environment and Development (1992), at which many population issues were raised directly, the Cairo Conference (1994) was styled, unlike its predecessors, a conference on *Population and Development*. Nevertheless, the main issue remained direct fertility control, and particularly the extent to which the contraceptive revolution that had allowed much greater choice over reproductive outcomes could be generalized, and in what form. There was criticism of the fairly crude application of family planning in the 1980s, often equated with fairly coercive "pill-pushing." Support was directed toward a more sensitive approach in which family planning could be part of maternal and child-health programs aimed at meeting the expressed needs of the majority of women for smaller families.

These world conferences have been particularly important in two respects: they have encouraged a global view of population questions, in which national governments have been able to place their own policies in the broader context of international population trends; and they have broadened the recognition that population seldom constitutes a problem in itself, but that the problem arises when population size, growth, and/or distribution interact with particular economic, political, and environmental circumstances.

**Population Projections.** It is clear that the universal downturn in fertility, resulting in a fall in the estimated annual increment to the global population from eighty-seven million in the late 1980s to eighty-one million by the late 1990s, has encouraged some limited optimism. The unprecedented growth rate of the last hundred years has now peaked, and a consistent fall through the early years of the next millennium is confidently expected. Current UN projections of global population change suggest that the population will continue to grow for the next 150 years. High, medium, and low projections are estimated, with the medium variant identifying a population stabilizing at approximately eleven billion people, roughly double its present level, by 2150. Then, some four hundred years after the first major signs of mortality decline in Europe, a new low-level global equilibrium will be reached. The transition to the low equilibrium will be complete, but the world population will have risen more than tenfold.

Like all population projections, the UN projections are inherently open to a range of outcomes, even though only two variables are modeled, namely, birth trends and mortality trends. Of the two, mortality trends are the easier to predict. As we have seen, the joint impact of medical science, of better living conditions, and of better nutrition has been to bring mortality levels to less than ten per thousand in developed countries, and to reduce the large short-term fluctuations. Mortality levels are low and stable in the richest countries, and development policies elsewhere seek to reduce mortality to these levels. The World Health Organization has set targets for mortality, especially in its *Health Care for All to 2000*, that assume this downward trend. However, even in developed countries, there may be a limit to improvement in life expectancy, although what that limit might be remains a matter of some dispute. In developing countries, mortality levels still have a long way to fall if they are to reach developed-country levels, and in poor tropical countries, with high levels of bacterial activity and many conditions still beyond medical control, there is no guarantee of the triumph of medical science—even where, unfortunately in a minority of countries, increased resources are likely to be available for health provision. Furthermore, new diseases are appearing, and further global epidemics may occur, now much more easily spread in a world of easy personal transport. The global human immunodeficiency virus/acquired immune deficiency syndrome (HIV/AIDS) pandemic is the most obvious case. It has already been responsible for an estimated 18.5 million deaths from 1980 to 1995, with an estimated one million AIDS deaths per year by 2000—thus growing in absolute as well as relative importance, especially in developing countries.

The unpredictability of fertility projections is acknowledged by the UN in the presentation of high, medium, and low projections. Since fertility is inherently more volatile than mortality in contracepting populations, the experience of the last two decades of negative or near-zero population growth and fertility below replacement levels in developed countries does not necessarily provide assurance of global trends over the next two centuries. For developing countries, the pace and extent of any future fertility decline must remain uncertain, although the projections assume that fertility will fall everywhere to replacement level by the end of the twenty-first century. The recent experience of sharp fertility declines in eastern Asia would certainly support an optimistic scenario, although even here there may be some possibility for reversals if economic performance continues to slump after the recent decades of unprecedented economic growth that was able to finance intensive family-planning programs as well as rising personal incomes and increasing urbanization.

However uncertain these projections might be, they do seem to suggest continuous growth in human populations, perhaps a doubling overall in the next 100–150 years, with a very sharp increase in the proportion in developing countries. Is such a population, with such a distribution, sustainable? In an absolute sense it probably is: 11 billion people could probably survive at the low levels of consumption of those currently in India or Ethiopia. However, fertility and mortality declines have been associated with rising levels of wealth and education, and, by implication, rising levels of consumption and production. With Western energy-consumption levels and lifestyles in Africa and Asia, can the global environment support 11 billion people? Furthermore, can this be sustained where most of that population increment remains in the developing countries, or might there be a necessity for substantial intercontinental migration? Clearly the problems associated with this projected population growth are larger than those of population size alone. [*See* Catastrophist–Cornucopian Debate.]

Human populations have been successful in reproducing themselves while achieving rising living standards, but does that success inherently contain the seeds of their own destruction in the longer term? A comparison with change in other plant and animal populations might suggest that it does. However, the biological analogy is inappropriate to a fertility revolution based on the general use of contraception that offers human populations, for the first time, an immediate choice about levels of fertility not available to other species. The future of population may be less about Malthusian checks—war, famine, disease, although these will surely remain—than about human choices in family size and family location and about broader societal choices concerning production and consumption.

[*See also* Agriculture and Agricultural Land; Carrying Capacity; Climate Change; Economic Levels; Growth, Limits to; Human Ecology; Industrialization; IPAT; Migrations; Population Dynamics; Resources; Urban Areas; *and the biography of Malthus*.]

### BIBLIOGRAPHY

Bongaarts, J. "A Framework for Analyzing the Proximate Determinants of Fertility." *Population and Development Review* 4 (1978), 105–132. This journal, published by the Population Council, is one of the main sources of discussion of population change, and contains many of the most authoritative discussions of global population change, both general, such as this reference, and country-specific, such as Martine (1996), see below.

———. "Global Trends in AIDS Mortality." *Population and Development Review* 22 (1996), 21–45.

Carr-Saunders, A. M. *World Population: Past Growth and Present Trends*. Oxford: Clarendon Press, 1936.

Cleland, J. "Population Growth in the 21st Century: Cause for Crisis or Celebration?" *European Journal of Tropical Medicine and International Health* 18 (1996), 15–26.

Coleman, D., and R. Schofield. *The State of Population Theory: Forward from Malthus*. New York and Oxford: Oxford University Press, 1986. A collection of essays by leading authors on population trends in a range of historical and contemporary contexts.

Douglas, R. M., G. Jones, and R. M. D'Souza, eds. "The Shaping of Fertility and Mortality Declines: The Contemporary Demographic Transition." *Health Transition Review*. Supplement to vol. 6, 1996, Canberra, Australia.

Gould, W. T. S., and M. S. Brown. "A Fertility Transition in Sub-Saharan Africa?" *International Journal of Population Geography* 2 (1996), 1–22. Discusses in full detail the controversies surrounding the relevance of the demographic transition model to explanations of population trends in Africa. This relatively new journal also contains many contributions on the various themes explored in this entry: e.g., Thomas (1995), see below.

Grigg, D. *The World Food Problem*. 2d ed. New York and Oxford: Oxford University Press, 1993. Grigg's studies, also available in the journal literature (e.g., Grigg, 1997, see below), consistently relate population growth to food production and consumption.

———. "The World's Hunger, A Review: 1930–1990." *Geography* 82 (1997), 197–206.

Hardin, G. *The Ostrich Factor: Our Population Myopia*. New York: Oxford University Press, 1998.

Jones, H. R. *Population Geography*. 2d ed. London, 1991. A very readable and informative introduction to a range of population issues.

Martine, G. "Brazil's Fertility Decline, 1965–95." *Population and Development Review* 22 (1996), 47–95.

Meadows, D. H., et al. *The Limits to Growth: A Report on the Club of Rome Project on the Predicament of Mankind*. New York: Universe Books, 1972.

Robey, B., S. O. Rutstein, L. Morris, and R. Blackburn. "The Reproductive Revolution: New Survey Findings." *Population Reports* Series M 11 (1992).

Ross, J. R., ed. *International Encyclopedia of Population*. 2 vols.

New York, 1982. A definitive exploration of all aspects of population change.

Sen, Amartya. *Poverty and Famines*. New York and Oxford: Oxford University Press, 1981.

Simon, J. L. *The Ultimate Resource 2*. Princeton: Princeton University Press, 1998.

Thomas, N. "The Ethics of Population Control in China." *International Journal of Population Geography* 1 (1995), 3–18.

United Nations. *Demographic Yearbook*. An annual compilation of national statistics, with a standard format, but with special features each year (e.g., a companion on "Ageing" in 1994). This is the most widely used of a very large range of summary and technical publications of the UN Population Division.

———. *World Population Projections to 2050*. New York: UN Population Division, 1998.

United Nations Children's Fund (UNICEF). *The State of the World's Children*. New York, annual.

United Nations Fund for Population Activities. Readings in Population Research Methodology. 8 vols. New York, 1993. Technical manuals and readings for all aspects of population analysis.

Whitmore, T. M., B. L. Turner, D. L. Johnson, R. W. Kates, and T. R. Gottschang. "Long-Term Population Change." In *The Earth as Transformed by Human Action*, edited by B. L. Turner, W. C. Clark, R. W. Kates, J. F. Richards, J. T. Matthews, and W. B. Meyer, pp. 26–39. Cambridge: Cambridge University Press, 1990.

World Bank. *World Development Report*. New York, annual. This publication has a large data appendix of population and other data, and the main text is devoted to a particular theme each year. The 1984 edition was concerned with "Population Change and Development."

World Health Organization. *Health Care for All to 2000*. Geneva: World Health Organization, 1990.

Wrigley, E. A., and R. S. Schofield. *The Population History of England, 1541–1871*. London: Edward Arnold, 1981.

—W. T. S. GOULD

## HUNTING AND POACHING

[*This article surveys the effects of hunting and poaching on biodiversity.*]

Since its origin, humankind has fed itself by hunting. Humans shifted to agriculture and animal domestication a scant seven thousand to eight thousand years ago. As hunters, early humans successfully competed with other mammalian carnivores by being fleet-footed and coordinating their actions. The development of weapons such as stone-bladed spears and throwing sticks enabled humans to kill prey as large as mammoths and mastodons (Ward, 1997).

Humans in historic times have sometimes engaged in the wanton killing of prey animals, exceeding the hunters' needs. The slaughter of 60 million American bison between 1865 and 1884 is a classic example (Matthiessen, 1959). Capricious killing by other mammals is rare, but if it is characteristic of *Homo sapiens*, such behavior could account in part for the massive Pleistocene extinctions in North America ten thousand to twelve thousand years ago when about twenty to forty bird species and forty large mammal species disappeared.

The Pleistocene overkill theory blames human hunters from the Old World for the extermination of all species of horses and North American camels, mammoths, mastodons, giant ground sloths, and wooly rhinos; these animals lacked time to develop an effective defense to a newly arrived, intelligent predator. With their prey gone, carnivores and scavengers such as saber-toothed tigers, dire wolves, and giant condors also died out (Van Valkenburgh, 1994). If such excessive hunting occurred, it would have been the greatest human-caused mammal extinction ever. The evidence is sketchy, but man's arrival in the New World coincides with the disappearance of these animals. Climatic and other factors may have contributed to these New World extinctions, but human intervention was likely an important element (Flannery, 1999).

The effect of these extinctions on the biological diversity of North and South America was almost immeasurable, because along with each mammal and bird species that disappeared went its species-specific parasites and ecological dependents. The loss of seed-disseminating animals affected plant and tree distribution, which then influenced the survival of the invertebrates dependent on them for their life cycle. Scientists can only surmise the ecological changes that occurred with the extirpation of so many birds and mammals.

Today humans hunt for both sustenance and sport, with the former having the greater impact on the hunting area's biodiversity. Sustenance hunting generally concentrates on species that are easily harvestable. For example, many West African forest species such as duikers and primates that evolved unthreatened by hunters with dogs and guns are now heavily harvested for bush meat.

Sustenance hunting can be wasteful and may have caused the extinction of species such as the great auk, Atlantic gray whales, dodos, and Steller's sea cow. These animals were generally slow-moving and easily killed. If they were already dying out when they first came in contact with humans, then a sudden reduction in their fragile populations could have accelerated their extermination. The long human history of sustenance hunting has contributed to the elimination of prey species and their parasites and symbionts, thereby resulting in reduced biodiversity.

Sport hunting thrived under Assyrian kings and Egyptian pharaohs, but so few engaged in the sport that the killing had little effect on biodiversity. By the seventeenth century, rulers set aside large natural areas as exclusive hunting reserves and maximized quarry populations. Local farmers, however, now precluded from sustenance hunting, began poaching, and their illegal harvests eventually affected biodiversity. Today, European royal hunting reserves are credited with maintaining such large ungulates (hoofed mammals) as the wisent (European bison). Similar efforts in India have helped the Asian rhino survive, and in China, Pére David's deer.

Hunting organizations are actively involved in conservation. Ducks Unlimited promotes breeding of waterfowl in the United States and Canada by saving and restoring wetlands. Anglers, too, have increased the use of catch-and-release fishings to maintain quantities of heavily caught game fish. Finally, hunting and fishing license fees are used to restore and protect wild populations and habitats.

Despite hunters' conservation efforts to save prey species, modern poachers thwart these attempts by hunting and selling endangered species. A large-scale, profitable, commercial trade exists in poached rhino horn, tiger parts, ivory, and even caviar from Caspian Sea sturgeon. To curtail this illegal harvest, producer and consumer nations of protected animals and their products have signed the Convention on International Trade in Endangered Species of Wild Fauna and Flora (CITES), which has significantly reduced the international market for these commodities. [See Convention on International Trade in Endangered Species.]

Illegal trade has been curbed by the efforts of the Trade Records Analysis of Flora and Fauna in Commerce, a program of the World Wildlife Fund that monitors trade in wild plants and animals and their products through an international network, and by the U.S. Customs Service and Fish and Wildlife Service. Negative publicity about poaching is also an effective deterrent to illegal trade in animal parts and stimulates calls for legislative action. Poaching has decimated populations of tigers, elephants, rhinos, and other mammals to the extent that some of them may be so scattered already that the critical minimum population necessary for long-term survival no longer exists.

Policy conflicts arise when a country with a well-managed elephant herd wants to sell ivory from culled animals to help finance its management, while its neighbor suffers from a declining, heavily poached population. Advocates believe that ivory sales will encourage management and legitimize its international trade, while opponents believe that such sales will encourage poaching because accurate identification of an ivory's source is not yet practical. In time a compromise solution may be found, as it was in the trade of crocodile hides. Poaching declined when entrepreneurs started farming crocodiles to produce better quality skins and even developing hybrid strains for rapid growth and easy hide identification (Webb et al., 1987; Thorbjarnarson, 1992).

Another conflict arises when federal authorities forbid hunting endangered species whose products play a

## BODY PARTS AND TRADITIONAL MEDICINE IN ASIA

Significant advances in wildlife conservation have been made recently throughout much of Asia: there is increasing awareness of the value of forests and wildlife; more protected areas have been set aside than ever before; and governments have enacted laws and signed international conventions that discourage trade in endangered species. Yet a strong market for body parts used in medicines continues to motivate the hunting and trapping of many large animals and threatens many with extinction.

| ANIMAL | BODY PART | MEDICINAL USE |
|---|---|---|
| **Native to Asia** | | |
| Musk deer | Musk gland | Malaria, convulsions, and general tonic |
| Malayan sun bear | Bile | Liver disease, blood disorders, and digestive ailments |
| Malayan tapir | Skin | Boils, infections |
| Pig-tailed macaque | Flesh | Malaria, general tonic |
| Saiga antelope | Horn | Colds, fever, liver |
| Pangolin (anteater) | Scales | Infections, limb stiffness |
| Tiger | Bone | Rheumatism, arthritis, muscles |
| | Many other parts | Variety of complaints |
| **Asia and Africa** | | |
| Leopard | Bone | Substitute for tiger bone |
| Rhinoceros | Horn | Fever, headache, delerium |
| **Asia and North America** | | |
| Black bear | Gall bladder | Fevers, convulsions, skin lesions |
| | Paw | General tonic |
| **Pacific Coasts** | | |
| Seal | Penis, testes | Impotence, kidney |

**BIBLIOGRAPHY**

Gaski, A. L., and K. Johnson. *Prescription for Extinction: Endangered Species and Patented Oriental Medicine in Trade.* Washington, D.C.: TRAFFIC USA (World Wildlife Fund), 1994.
McCracken, C., D. Rose, and K. Johnson. *Status, Management, and Commercialization of the American Black Bear.* Washington, D.C.: TRAFFIC USA (World Wildlife Fund), 1995.
Rabinowitz, A. "Killed for a Cure." *Natural History* (April 1998), 22–24.
Robbins, C. *An Overview of World Trade in Cervid (Deer) Antler with an Emphasis on United States.* Washington, D.C.: TRAFFIC USA (World Wildlife Fund), 1997.

—DAVID J. CUFF

religious or cultural role. Bald eagle feathers, for example, have traditionally adorned Native American objects and clothing. The U.S. government decided that Native Americans could use feathers from confiscated, illegally killed eagles or those found dead from natural causes. A more complicated conflict is the Inuits' insistence on hunting rare bowhead whales off Alaska's north coast. Annual quotas are now set and when reached, whaling must cease. As a result, quotas are seldom reached and the bowhead population is increasing (Marine Mammal Commission Report, 1999).

As nations become politically and economically stable, they begin to husband their natural resources. Legislation is enacted to protect endangered species, and reserves are established. Governments regulate hunting and forbid poaching, but enforcement of game laws is often erratic. As the Earth's wildlife competes with humans for a declining habitat, we have an increasing responsibility to preserve the Earth's biodiversity.

[*See also* Animal Husbandry; Biological Diversity; Biological Productivity; Carrying Capacity; Deforestation; Extinction of Species; Fishing; Food; *and* Trade and Environment.]

**BIBLIOGRAPHY**

Flannery, T. F. "Debating Extinction." *Science* 283 (1999), 182–183.
Marine Mammal Commission. *Annual Report to Congress 1999*, pp. 23–26. Bethesda: Marine Mammal Commission, 2000.
Matthiessen, P. *Wildlife in America.* New York: Viking Press, 1959.
Thorbjarnarson, J. *Crocodiles—An Action for Their Conservation*, edited by H. Messel, F. W. King, and J. P. Ross, pp. 51–52. IUCN/SSC. Crocodile Specialist Group. Gland, Switzerland: IUCN, 1992.

Van Valkenburgh, B. "Tough Times in the Tar Pits." *Natural History* 103.4 (1994), 84–85.

Ward, P. D. *The Call of the Distant Mammoths.* New York: Springer-Verlag, 1997.

Webb, G. J. W., S. C. Manolis, and P. J. Whitehead, eds. *Wildlife Management: Crocodiles and Alligators*, pp. 369–371. Chipping Norton, NSW, Australia: Surrey Beatty & Sons Pty. Ltd., 1987.

—DAVID CHALLINOR

**HYDROCARBONS.** *See* Fossil Fuels; Petroleum Hydrocarbons in the Ocean.

**HYDROELECTRICITY.** *See* Dams; *and* Renewable Energy Sources.

# HYDROGEN

Hydrogen is a high-quality energy carrier that can be used with high conversion efficiency and essentially zero pollutant emissions. Hydrogen can be made from a variety of widely available primary sources such as natural gas, biomass, wastes, coal, solar energy, wind, or nuclear power. Hydrogen transportation, heating, and power generation systems have been demonstrated technically, and, in principle, hydrogen could replace current fuels in all their present end uses. If hydrogen were made from renewable or decarbonized fossil sources, it would be possible to produce and use energy on a large scale with no emissions of pollutants or greenhouse gases. Although hydrogen energy systems offer environmental benefits, however, there are significant technical, cost, and infrastructure challenges that must be met before they could be widely used.

**Sources of Hydrogen.** Hydrogen is the most abundant element in the universe, and is found in hydrocarbons (e.g., fossil fuels or biomass) and in water. Since free hydrogen does not occur naturally on Earth in large quantities, it must be produced from hydrogen-containing substances.

A number of well-known methods for making hydrogen are in commercial use in the chemical industries today. Most commonly, hydrogen is made thermochemically by processing hydrocarbons (such as natural gas, coal, biomass, or wastes) in high-temperature chemical reactors to make a synthetic gas containing hydrogen, carbon monoxide, carbon dioxide, water, and methane. This gas can be processed further, and pure hydrogen extracted. A process diagram for making hydrogen thermochemically from natural gas is shown in Figure 1.

In areas where inexpensive electricity is available, water electrolysis is sometimes used to make hydrogen. In this process, electricity is passed through a conduct-

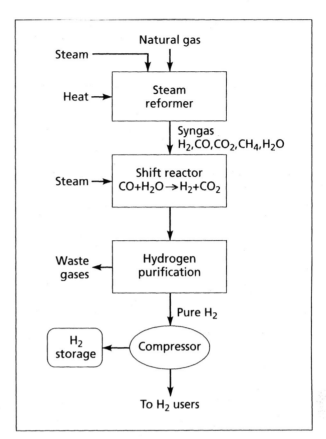

**Hydrogen. FIGURE 1.** Thermochemical Hydrogen Production from Natural Gas.

ing aqueous electrolyte, breaking water down into its constituent elements, hydrogen and oxygen (Figure 2). The hydrogen can be compressed and stored for later use. Any source of electricity can be used, including intermittent (time-varying) sources such as solar or wind energy. Hydrogen provides an energy storage medium for intermittent renewables and a means of transmitting energy from remote locations where wind power, for example, is abundant.

Basic research is being conducted on experimental methods of hydrogen production, including direct conversion of sunlight to hydrogen in electrochemical cells and hydrogen production by biological systems such as algae or bacteria.

Unlike fossil energy resources such as oil, natural gas, and coal, which are unevenly distributed geographically, there are one or more local resources for hydrogen production virtually everywhere in the world. The lowest-cost and most widely used method for hydrogen production today is thermochemical processing of natural gas. Biomass, wastes, or coal could also be used, and could be significant sources of energy. Wind and solar power are potentially huge resources for electrolytic hydrogen production that could meet future fuel

**Hydrogen. Figure 2.** Hydrogen Production Via Water Electrolysis.

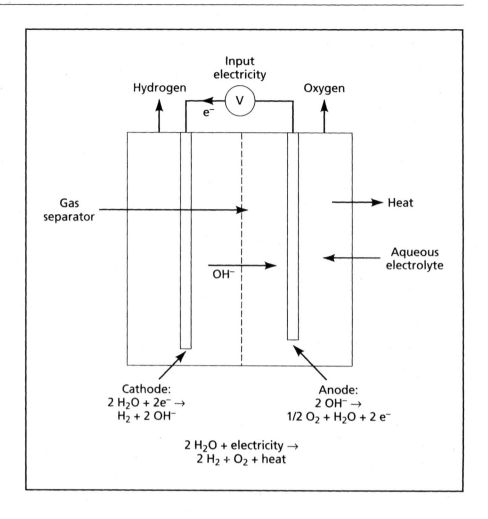

Cathode:
$2 H_2O + 2e^- \rightarrow$
$H_2 + 2 OH^-$

Anode:
$2 OH^- \rightarrow$
$1/2 O_2 + H_2O + 2 e^-$

$2 H_2O + electricity \rightarrow$
$2 H_2 + O_2 + heat$

demands, although other methods are likely to be less costly in most locations.

**Hydrogen Storage and Distribution.** The technologies for storing hydrogen and delivering it to users have been developed in the chemical industry, where large quantities of hydrogen are handled routinely. Hydrogen can be liquefied at low temperature ($-253°C$) and delivered by cryogenic tank truck or compressed to high pressure and delivered by truck or gas pipeline. If a large demand developed in the future, hydrogen could be distributed via a gas pipeline network, as natural gas is today.

Storage of hydrogen on vehicles poses special problems. Unlike gasoline or alcohol fuels, which are easily handled liquids at ambient conditions, hydrogen is a lightweight gas, and has the lowest volumetric energy density of any fuel at ambient temperature and pressure. Hydrogen can be stored on a vehicle as a compressed gas (in high-pressure gas cylinders), as a cryogenic liquid at $-253°C$ (in a special insulated vessel), or in a hydrogen compound (such as a metal hydride) where the hydrogen is easily removed by applying heat.

Onboard storage systems for hydrogen are heavier, bulkier, and more costly than those for liquid fuels or compressed natural gas, although they are lighter, more compact, and less costly than electric batteries. With high-efficiency hydrogen cars, it appears that onboard storage weights, costs, and volumes would be acceptable, yielding a range of 500 kilometers (300 miles) or more. Innovative storage methods such as hydrogen adsorption in carbon nanostructures are being researched. If a "breakthrough" hydrogen storage technology were successfully developed, it might speed the introduction of hydrogen as a fuel.

**Hydrogen for Transportation, Heat, and Power.** Hydrogen can be converted to useful heat and power in engines or fuel cells. Hydrogen engines resemble those that use more familiar fuels such as natural gas, gasoline, or diesel, but have substantially lower emissions, and can have somewhat higher efficiency. The hydrogen fuel cell offers even greater benefits.

A fuel cell is an electrochemical device that converts the chemical energy in a fuel (hydrogen) and an oxidant (oxygen in air or pure oxygen) directly to electricity, water, and heat. The only emission is water vapor. The electrical conversion efficiency can be quite high (up to 60 percent of the energy in the hydrogen is converted to electricity).

The fuel cell works as follows (Figure 3). Hydrogen is introduced at one electrode (the anode) and oxygen at the other electrode (the cathode). The two reactants (hydrogen and oxygen) are physically separated by an electrolyte, which can conduct hydrogen ions (protons), but not electrons. The electrodes are loaded with a catalyst, usually platinum, which allows the hydrogen to dissociate into a proton and an electron. To reach the oxygen, the hydrogen "splits up," the proton travels across the electrolyte, and the electron goes through an external circuit, doing work. The proton, electron, and oxygen combine at the cathode to produce water.

Although the principle of fuel cells has been known since 1838, the first practical applications were in the space program, where fuel cells powered the Gemini and Apollo spacecraft. In the 1960s and 1970s, fuel cells were used in space and military applications such as submarines. More recently, fuel cells have been developed for low-polluting cogeneration of heat and power in buildings. In the past few years, there has been a growing worldwide effort to commercialize fuel cells for zero-emission vehicles. As of January 2000, most of the major automobile companies (Daimler-Chrysler, Ford, General Motors, Honda, Mazda, Nissan, and Toyota) had declared their intention to commercialize fuel cell automobiles by 2003–2005.

**Environmental and Safety Considerations.** Several environmental and safety considerations must be taken into account in the design of hydrogen energy systems.

*Emissions from a hydrogen energy system.* Hydrogen can be used with zero or near-zero emissions at the point of use. When hydrogen is burned in air, the main combustion product is water, with traces of nitrogen oxides, which can be controlled to very low levels. No particulates, carbon monoxide, unburned hydrocarbons, or sulfur oxides are emitted. With hydrogen fuel cells, water vapor is the only emission. Moreover, the total fuel cycle emissions (i.e., all the emissions involved in producing, transmitting, and using an alternative fuel) of pollutants and greenhouse gases can be much reduced compared with conventional energy systems.

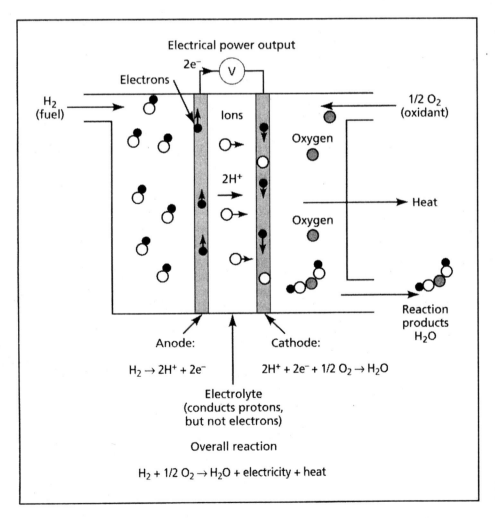

**Hydrogen. Figure 3.** A Hydrogen-Powered Fuel Cell.

*Hydrogen safety.* When hydrogen is proposed as a future fuel, safety concerns often arise. The proposed use of hydrogen in vehicles has raised the question of whether the excellent hydrogen safety record in the chemical industries could be replicated in consumer energy systems.

A 1997 study by Ford Motor Company addressed this question, finding that the safety of a hydrogen fuel cell vehicle would be potentially better than that of a gasoline or propane vehicle. A 1994 hydrogen vehicle safety study by researchers at Sandia National Laboratories stated, "There is abundant evidence that hydrogen can be handled safely, if its unique properties—sometimes better, sometimes worse and sometimes just different from other fuels—are respected."

**Developing a Hydrogen Energy System.** The technical building blocks for a future hydrogen energy system already exist. The technologies for producing, storing, and distributing hydrogen are well known and widely used in the chemical industries today. Hydrogen end-use technologies—fuel cells, hydrogen vehicles, power and heating systems—are undergoing rapid development. Still, the costs and logistics of changing the present energy system mean that building a large-scale hydrogen energy system would probably take many decades.

Because hydrogen can be made from many different sources, a future hydrogen energy system could evolve in many ways. In industrialized countries, hydrogen might get started by "piggybacking" on the existing energy infrastructure. Initially, hydrogen could be made locally, avoiding the need to build an extensive hydrogen pipeline distribution system. For example, in the United States, where low-cost natural gas is widely distributed, hydrogen will probably be made initially from natural gas, in small reformers located near the hydrogen demand (e.g., at refueling stations). As demand increased, centralized production with local pipeline distribution would become more economically attractive. Eventually, hydrogen might be produced centrally and distributed in local gas pipelines, as natural gas is today. A variety of sources of hydrogen might be brought in at this time. In developing countries, where relatively little energy infrastructure currently exists, hydrogen from local resources such as biomass may be more important from the beginning.

Hydrogen offers many potential benefits in terms of energy efficiency, greatly reduced emissions of pollutants and greenhouse gases, and diversity of energy supply. The costs of hydrogen energy systems are significantly higher than the alternatives today. However, if hydrogen technologies such as fuel cells are successfully commercialized on a large scale, the cost of energy services with zero or near-zero emission might become comparable to today's energy costs.

[*See also* Electrical Power Generation; Energy; Fossil Fuels; *and* Renewable Energy Sources.]

## BIBLIOGRAPHY

Appleby, A. J., and F. R. Foulkes. *Fuel Cell Handbook.* New York: Van Nostrand Reinhold, 1989.

Ford Motor Company. "Direct Hydrogen Fueled Proton Exchange Membrane Fuel Cell System for Transportation Applications: Hydrogen Vehicle Safety Report." Contract No. DE-AC02-94CE50389, May 1997.

Ogden, J. M., and J. Nitsch. "Solar Hydrogen." In *Renewable Energy Sources of Electricity and Fuels,* edited by T. Johannsson et al. Washington, D.C.: Island Press, 1993.

Ogden, J. M., and R. H. Williams. *Solar Hydrogen.* Washington, D.C.: World Resources Institute, 1989.

Ogden, J., et al. "Hydrogen as a Fuel for Fuel Cell Vehicles." Proceedings of the 8th National Hydrogen Association Meeting, Arlington, Va., March 11–13, 1997.

Ringland, J. T., et al. "Safety Issues for Hydrogen-Powered Vehicles." Sandia National Laboratories, March 1994.

Winter, C.-J., and J. Nitsch. *Hydrogen as an Energy Carrier.* New York: Springer, 1988.

"The Future of Fuel Cells" Special Edition. *Scientific American* (July, 1999).

—JOAN M. OGDEN

# HYDROLOGIC CYCLE

Water is both commonplace and unique. It is found everywhere at and near the surface of the Earth, but is the only naturally occurring inorganic liquid and is the only chemical compound that occurs in normal conditions as a gas (water vapor), a liquid (water), and a solid (ice). The hydrologic cycle describes the global circulation of water as it moves unendingly through these three states. Water is indestructible, so the total quantity of water in the hydrologic cycle cannot be diminished as it changes from water vapor, to liquid or solid, and back again. Nor can it be increased, except through the occasional release of minute quantities of fossil ground water or possible additions from comets. Instead the processes of streamflow, groundwater flow, and evaporation ensure the never-ending transfer of water between land, ocean, and atmosphere, followed by its return as precipitation to the Earth's surface. Some 97 percent of global water occurs as a saline liquid in the seas and oceans, which therefore play a dominant role in the global water and energy budgets. The vast majority (72 percent) of the remaining global fresh water occurs in the form of ice sheets and glaciers, including, in the case of Antarctica, enormous reservoirs of liquid water, which are now known to occur near the base of the ice at depths of more than 3,000 meters (Hawkes, 1996). However, the main focus of the hydrologic cycle is normally on the relatively small amount of fresh water occurring as lakes, rivers, soil water, and shallow ground water, and in the vegetation cover and the atmosphere.

The essential nature of the hydrologic cycle is depicted in Figure 1. Water vapor in the atmosphere condenses and may give rise to precipitation. In the terrestrial portion of the cycle, not all of this precipitation will reach the ground surface because some will be intercepted by the vegetation cover or by the surfaces of buildings and other structures, and will from there be evaporated back into the atmosphere. The precipitation that reaches the ground surface may be stored in the form of pools, puddles, and surface water, which are usually evaporated into the atmosphere quite quickly or be stored as snow and ice before melting or sublimation occurs, possibly after many years or even centuries. It may also flow over the surface into streams and lakes, from where it will move either by evaporation into the atmosphere, or by seepage toward the ground water, or by further surface flow into the oceans. Precipitation may also infiltrate through the ground surface to join existing soil water. This may be removed either by evaporation from the soil and vegetation cover, by throughflow toward stream channels, or by downward percolation to the underlying ground water, where it may be held for periods ranging from weeks to millennia. The groundwater component will eventually be removed ei-

ther by upward capillary movement to the soil surface or to the root zone of the vegetation cover, whence it will be returned by evaporation to the atmosphere or by groundwater seepage and flow into surface streams and into the oceans. [See Ground Water.]

The uninterrupted, sequential movement of water implied in this simplified description of the hydrologic cycle is rarely achieved, however. For example, falling or newly fallen precipitation may be returned to the atmosphere by evaporation without becoming involved in streamflow, soil water, or groundwater movement. Or precipitation may fall upon the oceans and be evaporated without touching the land surface at all, or it may fall upon the land and percolate to the main groundwater body within which it moves slowly toward a discharge point such as a spring, which it may not reach for centuries or even millennia. In hot deserts, rainfall is spasmodic and so too are other processes in the cycle, such as evaporation and streamflow, which can take place only for a short period during and after rainfall. Accordingly, a short burst of hydrologic activity for a week or so may be followed by a long period of virtual inactivity, apart from a slow redistribution of ground water at some depth below the surface. In cold climates the time delay between snowfall and the active involvement of the precipitated moisture, after melting,

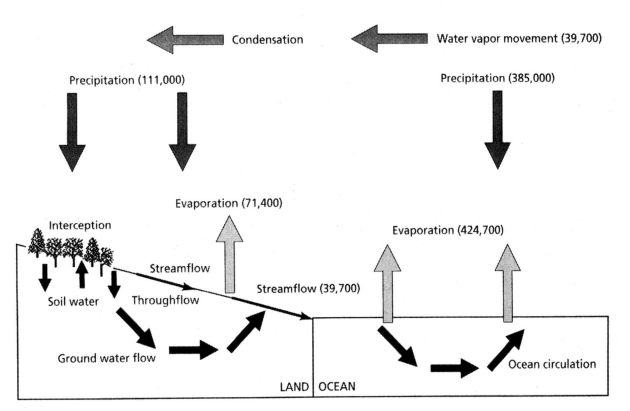

**Hydrologic Cycle. Figure 1.** The Hydrologic Cycle Showing Annual Estimates of Some Water Fluxes (in cubic kilometers).

See also Table 3.

in the subsequent phases of the hydrologic cycle may range from months (seasonal snowpacks) through centuries (valley glaciers) to millennia (Antarctic ice sheet). Also, increasingly, the cycle is interrupted and modified by human activities.

The essential continuity of water circulation through the hydrologic cycle has been recognized since ancient times, as exemplified in the well-known verse from Ecclesiastes 1:7,

> All the rivers run into the sea, yet the sea is not full; unto the place from which the rivers come thither they return again . . .

and in some of the writings of Aristotle (384–322 BCE).

However, a detailed understanding of the mechanisms and relative magnitudes of the hydrologic processes involved, especially the relationships between evaporation, precipitation, and streamflow, developed only slowly. Not until the late seventeenth century were explanations of the hydrologic cycle based upon reliable scientific evidence such as that of Perrault and Mariotte, whose work on the Seine drainage basin in northern France demonstrated that, contrary to earlier assumptions, rainfall was more than adequate to account for river flow; and Halley, the English astronomer, who showed that total streamflow into the Mediterranean Sea could be more than accounted for by oceanic evaporation.

**Components of the Hydrologic Cycle.** The fundamental components of the hydrologic cycle, evaporation and precipitation, are inextricably linked. The replenishment of atmospheric water vapor by evaporation can take place from any part of the Earth's surface, although in practice there are significant spatial and temporal variations. At its simplest, 71 percent of the Earth's surface is ocean and approximately one-third of the land surface is desert, so that evaporation from the sea surface dominates the supply of water vapor to the atmosphere, contributing nearly 86 percent of the water vapor used in the global precipitation process, compared with just over 14 percent contributed from the land areas. However, for both sea and land surfaces, the rate of evaporation depends on three factors: energy supply, diffusion processes, and water supply. Energy, either directly from the Sun or indirectly from the atmosphere itself, is necessary to convert liquid water to water vapor; and molecular and turbulent diffusion processes control the rate at which water vapor is diffused away from the Earth's surface. In addition, the evaporation rate from land surfaces also depends on the supply of water to be evaporated, which varies from time to time, but especially seasonally. In many areas, for example, the ground surface is wetter in the spring than in the fall.

The combination of these factors means that some areas of the Earth's surface act as the principal sources of water vapor, and others as the primary sinks. Redistribution, or advection, from sources to sinks occurs via the atmospheric, hydrologic, and oceanic circulations. For example, globally the oceans lose more water by evaporation than they gain by precipitation. Hence the land areas, over which precipitation exceeds evaporation, act as sinks for water vapor and the return flow is made up by runoff from the land areas into the oceans. Oceans cover a larger proportion of the total surface in the Southern Hemisphere (80.9 percent), than in the Northern Hemisphere (60.6 percent), so that, although precipitation is very similar in both hemispheres, evaporation exceeds precipitation in the Southern Hemisphere (see Table 1). This results in a net transfer of water vapor into the Northern Hemisphere, a significant part of which probably occurs during the summer monsoon season in the Indian Ocean (Sellers, 1965).

The greater prevalence of ocean in the Southern Hemisphere is also reflected in the latitudinal variations of evaporation. As Table 1 and Figure 2 show, evaporation is highest in the subtropics and decreases slightly toward the equator and markedly toward the poles. However, the evaporation peak is more pronounced in the oceanic Southern Hemisphere than in the Northern Hemisphere, where extensive subtropical deserts reduce the evaporation total. Precipitation exceeds evaporation polewards of 40° in both hemispheres and in the equatorial zone between 10° north and 10° south latitude. These zones are therefore sinks for water vapor that are replenished, via the atmospheric circulation, from the subtropical source areas in both hemispheres. The maximum precipitation surpluses occur in the latitude zones 0°–10° north (677 millimeters per year) and 50°–60° south (510 millimeters per year). Precipitation surpluses are redistributed between latitude zones by a combination of river flow and ocean currents. However, the precipitation excess of the Northern Hemisphere must flow into the Southern Hemisphere as ocean currents since the Nile and Amazon are the only major rivers to cross the equator, and both have their headwaters in the Southern Hemisphere (Sellers, 1965).

Other than by evaporation, precipitation falling on to the land surface is redistributed through the terrestrial segment of the hydrologic cycle in different ways and at different speeds. Redistribution is most rapid as river flow, which may reach average rates of more than 200 kilometers per day in flood conditions. Shallow ground water occurring within the upper few hundred meters of the Earth's surface, which is an important contributor to the baseflow of streams and rivers, may flow at rates of up to 2 meters per day (Todd, 1980). Deep groundwater movement is much slower, and Todd suggested 2 meters per year as a typical flow rate, although this will vary greatly depending on the permeability of strata. Usually

**Hydrologic Cycle. TABLE 1.** Mean Annual Precipitation (P) and Evaporation (E) by 10° Zones of Latitude

| LATITUDE ZONE | OCEAN AREA (% OF ZONE) | PRECIPITATION (MILLIMETERS) | EVAPORATION (MILLIMETERS) | P–E (MILLIMETERS) |
|---|---|---|---|---|
| 80–90°N | 93.4 | 116 | 41 | 76 |
| 70–80 | 71.3 | 179 | 141 | 39 |
| 60–70 | 29.4 | 402 | 323 | 79 |
| 50–60 | 42.8 | 765 | 455 | 310 |
| 40–50 | 47.5 | 879 | 621 | 258 |
| 30–40 | 57.2 | 845 | 971 | −126 |
| 20–30 | 62.4 | 766 | 1,208 | −442 |
| 10–20 | 73.6 | 1,115 | 1,346 | −231 |
| 0–10 | 77.2 | 1,874 | 1,197 | 677 |
| 0–90°N | 60.6 | 978 | 915 | 63 |
| 0–10°S | 76.4 | 1,400 | 1,264 | 137 |
| 10–20 | 78.0 | 1,097 | 1,493 | −396 |
| 20–30 | 76.9 | 831 | 1,372 | −542 |
| 30–40 | 88.8 | 903 | 1,217 | −314 |
| 40–50 | 97.0 | 1,188 | 867 | 321 |
| 50–60 | 99.2 | 1,014 | 504 | 510 |
| 60–70 | 89.6 | 405 | 169 | 236 |
| 70–80 | 24.6 | 79 | 44 | 36 |
| 80–90 | 0.0 | 29 | 0 | 29 |
| 0–90°S | 80.9 | 969 | 1,031 | −62 |
| Global total | 70.8 | 973 | 973 | 0 |

SOURCE: Based on data adapted from Sellers (1965), and other sources.

the slowest movement of all is associated with valley glaciers and ice sheets, which therefore act as long-term storages for the precipitation falling on to them.

**The Hydrologic Cycle Quantified.** The hydrologic cycle may be considered as a series of water reservoirs or storages that are recharged and discharged largely by the relatively rapid flows or fluxes of precipitation, evaporation, and streamflow.

The main storages within the hydrologic cycle are illustrated and quantified in Figure 3 and Table 2. These estimates must be treated with caution since, in most cases, monitoring and exact quantification at the macro scale are difficult. For example, the volumes of the ocean basins and of the major ice sheets depend upon seabed and sub-ice topography, which it has only recently become possible to map with reasonable accuracy. Reserves of deep ground water are notoriously difficult to assess but (like those of fossil fuel) tend to be subject to periodic and usually upward revision. Shallow groundwater storage is more accessible and therefore easier to estimate with accuracy, although the proportion of usable, nonsaline water is still far from certain. Atmospheric water vapor content is normally

monitored either by radiosonde balloons released daily from a mere fifteen hundred locations around the world, or from infrared (IR) spectrometers in the main weather satellite systems of, for example, U.S. National Oceanic and Atmospheric Administration (NOAA) and Endangered Species Act (ESA). Unfortunately, IR spectrometry is more problematic in the air layers closest to the Earth's surface, where water vapor values are highest, because of the presence of clouds (Boucher, 1997). It is not surprising, therefore, that estimates of the main water storages vary widely, depending upon sources of information used and assumptions made. Speidel and Agnew (1988) quoted ten, mainly American and Russian, sources (see Bibliography) whose estimates for water storage in the ocean basins varied from 1,320 million to 1,370 million cubic kilometers; estimates of atmospheric water vapor varied from 10,500 to 14,000 cubic kilometers; estimates of water storage in icecaps and glaciers varied in the same sources from 16.5 million to 29.2 million cubic kilometers; and estimates of ground water varied from 7 million to 330 million cubic kilometers. Another valuable source of data on the global hydrologic cycle is that presented by Shiklomanov (1993).

**Hydrologic Cycle. Figure 2.** Mean Annual Precipitation and Evaporation by 10° Zones of Latitude. (Based on data in Table 1.)

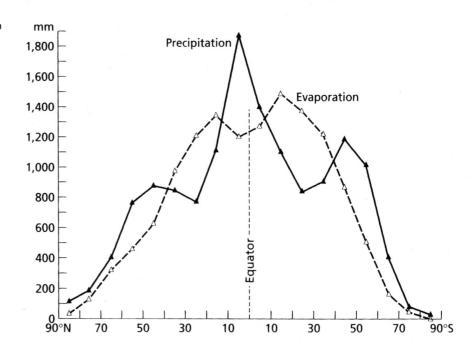

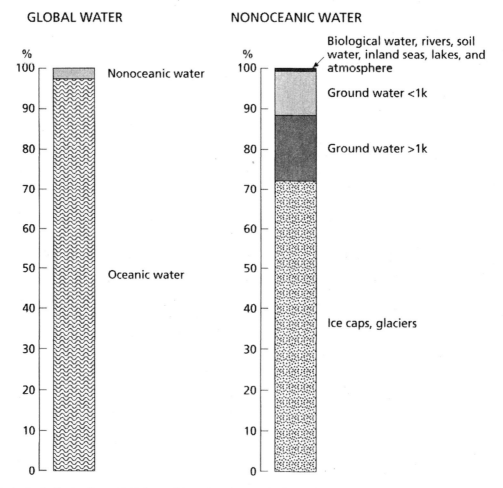

**Hydrologic Cycle. Figure 3.** Estimated Storages of Water in the Hydrologic Cycle. (Based on data in Table 2.)

**Hydrologic Cycle. TABLE 2.** Estimated Storages of Water in the Hydrologic Cycle

| STORAGE | PHASE | WATER VOLUME (CUBIC KILOMETERS × 1,000) | NONOCEANIC WATER (%) | GLOBAL WATER (%) | TYPICAL RESIDENCE TIME (YEARS) |
|---|---|---|---|---|---|
| Atmosphere | Gas | 13 | 0.035 | 0.001 | 0.026 |
| Land | Solid and liquid | 37,275 | 99.965 | 2.697 | 0.019->4000 |
| Oceans | Liquid | 1,345,000 | | 97.302 | 200->3,000 |
| Total Global Water | | 1,382,288 | | 100 | |
| Land | | | | | |
|   Biological water | Liquid | 1 | 0.003 | 0.000 | 0.019 |
|   Rivers | Liquid | 2 | 0.005 | 0.000 | 0.045 |
|   Soil water | Liquid | 65 | 0.174 | 0.005 | 0.038–1 |
|   Inland seas, saline | Liquid | 100 | 0.268 | 0.007 | 200 |
|   Freshwater lakes | Liquid | 107 | 0.287 | 0.008 | 0.040–20 |
|   Ground water, <1k | Liquid | 4,000 | 10.727 | 0.289 | 0.038->500 |
|   Ground water, >1k | Liquid | 6,000 | 16.091 | 0.434 | >1,000 |
|   Ice caps, glaciers | Solid | 27,000 | 72.409 | 1.953 | >10->4,000 |
| Total Land Water | | 37,275 | 99.965 | | |
| Total Nonoceanic Water | | 37,288 | 100.000 | | |

SOURCE: Based on data from various sources; see text for further details.

Despite the uncertainties implicit in the wide variation of values used to construct Table 2 and Figure 3, the estimates of storage show clearly that the hydrologic cycle is primarily concerned with less than 3 percent of the global water resource. Of this relatively small amount of nonoceanic water, more than 72 percent occurs in solid form in high-latitude and high-altitude areas, and a further 16 percent occurs as deep ground water, more than 1 kilometer below the Earth's surface. This deeper ground water normally forms part of a slow circulation, may be relatively inaccessible, and is sometimes highly mineralized. By far the largest accessible reservoir of nonoceanic water is that stored as shallow ground water within 1 kilometer of the ground surface. Although some of this water also has a high mineral content, its large volume exceeds by more than twenty times the combined total of all other accessible fresh water in the atmosphere, rivers, freshwater lakes, soil, and plants. It is therefore an extremely valuable water resource.

However, the resource value of the various storages of nonoceanic water is not simply a question of total volume. Just as important is the rate of turnover, or the residence time, of water in each of the reservoirs. Rough estimates of typical residence times are shown in the right-hand column of Table 2. The wide range of values shown for some of the storages reflects the fact that these have been calculated in different ways. Some of the estimates are likely to be more valid than others. For example, the value for precipitation may be quite accurate since it is known that the mean annual global rainfall of 496,100 cubic kilometers is approximately thirty-eight times greater than the volume of water vapor stored in the atmosphere at any one time; that is, atmospheric water vapor must be turned over thirty-eight times each year, giving an average residence time of 9.6 days or 0.026 years. The value for rivers is probably quite accurate as well, being largely a function of mean water velocity and channel length. The residence time of soil water, however, depends upon a complex interaction between soil matrix characteristics, shallow groundwater depth, and the intensity and variability of precipitation and evaporation; it may vary, therefore, between a few days and one year. Estimates for large lakes and inland seas and, of course, the oceans themselves will vary with the assumptions made about the internal circulation of the water bodies and the proportion of the water body that links actively with the other components of the hydrologic cycle. In the case of the oceans, for example, the annual evaporation of 424,700 cubic kilometers from a reservoir of 1,345 million cubic kilometers represents an average residence time of more than three thousand years. However, if it is assumed that most of the evaporation/precipitation interchange between ocean and atmosphere occurs from the upper 100 meters or so below the ocean surface, then the residence time for that layer would be only a few hundred years. Clearly, therefore, the residence time of ocean water varies enormously, with deep bottom water probably remaining within the ocean mass for many thousands of years.

By dividing the volume of the individual nonoceanic

storages by their residence times, it can readily be shown that the more dynamic storages make a resource contribution out of all proportion to their size. For example, the 13,000 cubic kilometers of atmospheric water makes an annual contribution of 500,000 cubic kilometers, and rivers contribute 44,000 cubic kilometers. If it is assumed that soil water is turned over four times a year, then its annual contribution is 260,000 cubic kilometers. The combined volume of the atmospheric, fluvial, and soil water annual contributions to the hydrologic cycle (804,000 cubic kilometers) is therefore not far short of the contribution of shallow ground water (one million cubic kilometers), if it is assumed that the latter has an average residence time of four years.

Estimated values of the annual fluxes are summarized in Table 3 and are illustrated schematically in Figure 4. As with the storage values discussed previously, difficulties of measurement mean that these estimates must be treated with some caution. For example, water vapor advection, especially in the lower layers of the atmosphere, is difficult to monitor with accuracy because concentrations change on a short-term (e.g., hourly) basis, particularly in the latitude zones dominated by rapidly moving cyclonic systems. Again, although the accurate measurement of precipitation is considerably easier than that of evaporation, there are many possible sources of error in the measurement of precipitation at a point and in the calculation of representative areal depths of precipitation from a network of point measurements; these have been well documented and extensively discussed (Ward and Robinson, 2000). Rather few countries have rain gauge networks as dense as those in the United Kingdom (60 square kilometers per gauge) or the United States (600 square kilometers per gauge). Elsewhere, networks involving several thousand square kilometers per gauge are quite common, and over the oceans the gauging network is even sparser. In addition, snowfall is much more difficult to measure accurately than rainfall. The growing use of weather radar and satellite imagery has greatly improved our ability to monitor the amounts and distribution of precipitation. For example, FRONTIERS (Forecasting Rain Optimised using New Techniques of Interactively Enhanced Radar and Satellite) is one of two systems in use in the United Kingdom that combine the use of digital radar data from the U.K. network with Meteosat weather satellite data (Austin and Moore, 1996). The use of radar and satellite measurements has also been developed via SHARP in Canada, PROFS in

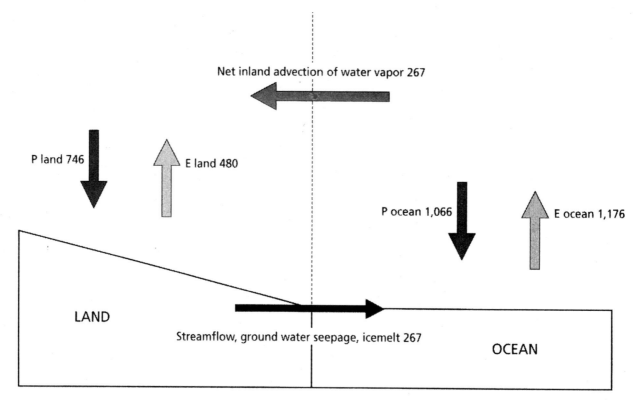

**Hydrologic Cycle. FIGURE 4.** Estimated Annual Fluxes in the Hydrologic Cycle.

Values are expressed in millimeters precipitation equivalent. See also Table 3, in which these fluxes are expressed both in terms of precipitation equivalent and as volumes (i.e., precipitation equivalent depth multiplied by total land or ocean area). (Based on data in Speidel and Agnew, 1988, and other sources.)

**Hydrologic Cycle. TABLE 3.** Estimated Fluxes of Water in the Hydrologic Cycle

| FLUX | OCEANS (CUBIC KILOMETERS) | LAND (CUBIC KILOMETERS) | GLOBAL TOTAL (CUBIC KILOMETERS) | OCEANS* (MM) | LAND* (MM) | GLOBAL TOTAL* (MM) |
|---|---|---|---|---|---|---|
| Precipitation | 385,000 | 111,100 | 496,100 | 1,066 | 746 | 973 |
| Evaporation | −424,700 | −71,400 | −496,100 | −1,176 | −480 | −973 |
| Runoff, including ground water seepage and icemelt | 39,700 | −39,700 | 0 | 110 | −267 | 0 |
| Net inland advection of water vapor | −39,700 | 39,000 | 0 | −110 | 267 | 0 |
| Total | −39,700 | 39,700 | 0 | −110 | 266 | 0 |

*Figures on these three columns are calculated assuming a total land area of 148,904,000 square kilometers.

SOURCE: Based on data from Speidel and Agnew (1988), and other sources; see text for further details.

the United States, and PROMIS 600 in Sweden. Even so, it is easy to see why estimates of global precipitation have varied widely. Evaporation is notoriously difficult to measure at a point, although regional values can be calculated with greater accuracy using a combination of empirical and physical approaches. The Thornthwaite–Holzman equation is widely used throughout the world for this purpose, as also is the Penman–Monteith equation, which forms the basis of MORECS (Meteorological Office Rainfall and Evaporation Calculation System) in the United Kingdom.

The ten sources used by Speidel and Agnew (1988) varied in their precipitation and evaporation estimates from 446,000 to 577,000 cubic kilometers, and their estimates of runoff to the oceans varied from 33,500 to 47,000 cubic kilometers. Such uncertainty about the magnitude of major fluxes of water and energy makes it difficult to identify and interpret global change within the hydrologic, atmospheric, and oceanic systems or to predict the likely impact of that change upon human beings and their activities. In turn, these difficulties provided a major impetus to the development of the Global Energy and Water Cycle Experiment (GEWEX; World Meteorological Organisation, 1993). GEWEX was initiated in the late 1980s to rectify the absence of reliable global data on precipitation, evaporation, and water vapor advection. The program continues beyond the end of the twentieth century and includes new Earth-observing satellites to complement existing hydrologic networks (Boucher, 1997). As the name of the experiment makes explicit, there is a need to understand better the role of water fluxes in both the hydrologic cycle and the global energy cycle. The aim of GEWEX is, therefore, not only to improve estimates of the global water fluxes shown in Figure 4 but also to investigate the important role of water in the warming and cooling of the Earth's atmosphere and surface, including the

oceans, given the release of 0.0251 joules per kilogram of latent heat on condensation and 0.0034 joules per kilogram on freezing, and the absorption of similar amounts of latent heat on evaporation and melting respectively.

The net inland advection of water vapor, illustrated in Table 3 and Figure 4, involves a complex interchange of water vapor between ocean and land areas. Although a significant quantity of water vapor generated by evaporation from the land is eventually precipitated over the oceans, an even greater quantity of water vapor generated by evaporation from the oceans is precipitated over the land areas. However, the compensating advective flux, involving the movement of liquid water from the land areas into the oceans, is, at the same time, both simpler and more complex. It is simpler because it is unidirectional, that is, apart from the drift of sea spray and the occasional flooding of limited coastal areas by storm surges or tsunamis, flow occurs only from the land to the oceans and not in the reverse direction. It is more complex because the quantities of the separate components involved are more difficult to ascertain. There is, for example, no simple separation between ground water and streamflow. For much of the year, streamflow in humid environments is maintained largely by the flow of shallow subsurface water, or throughflow, and ground water from various depths and from all parts of the river basin. It is in this sense, primarily, that rivers are regarded as synonymous with their drainage basins and it is by this means that the largest proportion of groundwater flow is removed from the land areas into the oceans. In near-coastal areas, however, and particularly where river systems are sparse and widely separated, a significant direct movement of ground water into the oceans will take place. Direct groundwater flow into the oceans is, therefore, an important feature of the hydrologic system in many arid and semiarid areas

**Hydrologic Cycle. TABLE 4.** Runoff Contributed Annually by the World's Biggest Rivers

| RIVER | BASIN AREA (SQUARE KILOMETERS × 1,000) | RUNOFF (CUBIC KILOMETERS) | GLOBAL RUNOFF |
|---|---|---|---|
| Amazon | 7,050 | 3,767.8 | 9.5 |
| Congo | 3,691 | 1,255.9 | 3.2 |
| Yangtze | 1,959 | 690.8 | 1.7 |
| Mississippi | 3,221 | 556.2 | 1.4 |
| Yenisei | 2,598 | 550.8 | 1.4 |
| Mekong | 811 | 538.3 | 1.4 |
| Orinoco | 907 | 538.2 | 1.4 |
| Parana | 3,103 | 493.3 | 1.2 |
| Lena | 2,424 | 475.5 | 1.2 |
| Brahmaputra | 935 | 475.5 | 1.2 |
| Total 10 rivers | 26,699 | 9,342.3 | 23.5 |
| Irrawaddy | 431 | 443.3 | 1.1 |
| Ganges | 489 | 439.6 | 1.1 |
| Mackenzie | 1,766 | 403.7 | 1.0 |
| Ob | 3,706 | 395.5 | 1.0 |
| Amur | 1,843 | 349.9 | 0.9 |
| St. Lawrence | 1,010 | 322.9 | 0.8 |
| Indus | 963 | 269.1 | 0.7 |
| Zambezi | 1,330 | 269.1 | 0.7 |
| Volga | 1,380 | 256.6 | 0.6 |
| Niger | 1,502 | 224.3 | 0.6 |
| Total 20 rivers | 41,119 | 12,716.3 | 32.0 |

SOURCE: Based on data in Speidel and Agnew (1988).

and in areas where seasonal low temperatures restrict streamflow to a comparatively short melt season. It is, however, difficult to estimate with precision, and the inevitable uncertainty about its magnitude largely accounts for the wide discrepancies in global groundwater flow values that were noted earlier.

Although streamflow is much easier to monitor, its uneven distribution, both spatially and seasonally, complicates the assessment of its role in the hydrologic cycle. Table 4 illustrates, for example, the dominant role played by a small number of major rivers. Nearly 10 percent of all global runoff flows out of the Amazon basin into the Atlantic Ocean. Furthermore, the ten largest rivers account for nearly a quarter of the global flow from land to oceans, and the twenty largest rivers together contribute nearly one-third. The input of land water into the ocean basins is, therefore, concentrated at a relatively small number of locations.

Viewed from the broader continental scale, it is clear from Table 5, and indeed from the list of rivers in Table 4, that some continents are bigger sources of water than others. Thus Asia and South America together contribute 59 percent of the global flow from land to oceans. It is equally clear from Table 5 that, despite its overwhelming size (about 181 million square kilometers), the Pacific Ocean receives only 30.6 percent of the water leaving the land areas each year, whereas the Atlantic Ocean (about 82 million square kilometers) receives 48.7 percent.

Because of these inequalities and the marked seasonal variations of river flow, both within climatic regions, especially with monsoon and meltwater regimes, and between the Northern and Southern Hemispheres, the advective flow of liquid water is achieved only partly by runoff from land to oceans. The remainder of this advective flow takes place via ocean currents. These redistribute not only the excess runoff received by some ocean basins compared with others, as shown in Table 5, but also the imbalances between evaporation and precipitation that are indicated in Table 1. For example, in

**Hydrologic Cycle. TABLE 5.** Runoff Sources (Continents) and Main Destinations (Oceans) Annual Estimates

| CONTINENT | RUNOFF (CUBIC KILOMETERS) | GLOBAL TOTAL (%) | Destination (%) | | | |
|---|---|---|---|---|---|---|
| | | | ATLANTIC | PACIFIC | INDIAN | ARCTIC |
| Asia | 12,467 | 31.4 | 1.7 | 51.9 | 28.3 | 18.1 |
| South America | 11,039 | 27.8 | 87.5 | 12.5 | | |
| North America | 5,840 | 14.7 | 61.8 | 32.8 | | 5.4 |
| Africa | 3,409 | 8.6 | 82.7 | | 17.3 | |
| Europe | 2,564 | 6.5 | 98.6 | | | 1.4 |
| Australia | 2,394 | 6.0 | | 75.7 | 24.3 | |
| Antarctica | 1,987 | 5.0 | 26.4 | 28.2 | 45.4 | |
| Total | 39,700 | | 19,351 | 12,137 | 5,601 | 2,611 |

SOURCE: Based on data in Baumgartner and Reichel (1975).

the Atlantic Ocean there is a deficit in the precipitation–evaporation balance of 36,500 cubic kilometers, which is not made up by the, albeit large, inflow of land water (19,351 km$^3$) and which must therefore be compensated by flow from the arctic seas, and even more so from the Pacific, where there is a surplus in the precipitation–evaporation balance (Speidel and Agnew, 1988).

**Human Interactions with the Hydrologic Cycle.** The hydrologic cycle impacts upon human existence and behavior both directly through its influence on the disposition of water resources and indirectly through its inextricable meshing with the atmospheric circulation and with climate. In turn, humans have a powerful influence on the hydrologic cycle, modifying systems and changing flows and storages of water.

Human dependence upon the components of the hydrologic cycle is absolute. Drinking water supplies are literally vital and are normally obtained by collecting or storing fresh water (rainfall, streamflow, ground water) or, to a minor degree, by desalting brackish, saline, or highly mineralized water. Rainfall and an adequate store of soil water are essential for crop growth and in water-deficient areas are often supplemented by irrigation and the diversion of river flows. Streams, rivers, and lakes are used for navigation, power generation, and the dilution of effluents, as a result of which discharges, water levels, and even the channels themselves may be substantially modified. [See Dams.]

The intermeshing of the hydrologic cycle with the global atmospheric circulation, though less obvious, is equally important. As has been emphasized earlier, evaporation at the Earth's surface is a major source of cooling. The subsequent release of the latent heat of evaporation, usually after significant vertical and horizontal displacement of the water vapor, is a principal cause of the redistribution of heat energy in the global atmosphere as well as a primary factor in the formation of rainfall-producing cyclones and convective systems. Particularly important in this respect are the roles of sea surface temperatures and El Niño–Southern Oscillation (ENSO) phenomena. [See El Niño–Southern Oscillation.]

Human influences upon the hydrologic cycle are widespread and may operate either directly on the components of the cycle or indirectly through climatic processes. At the present time the three most important of these relate to: (1) direct modifications of channel flow and storage through dam construction, and large-scale indirect modifications by means of surface changes such as afforestation, deforestation, and urbanization (which affect surface runoff, including the incidence or magnitude of flooding, and water quality); (2) the widespread development of irrigation and land drainage (which affects salination and groundwater levels); and (3) the large-scale abstraction of ground water and surface water for domestic and industrial uses (which affects river flow, groundwater levels and, in some areas, groundwater quality).

Other modifications that may become increasingly important in the future include artificial recharge of ground water; interbasin transfers of surface and ground water, including channel flow reversal; the artificial stimulation of precipitation; and the use of transpiration suppressants. Some of these human interactions with the hydrologic cycle operate at a local scale, but others, such as large-scale deforestation, river reversal, large-scale irrigation (and salination), and climate change, are regional or even global in their potential impact.

[See also Climate Change; Clouds; Deforestation; Glaciation; Global Warming; Ice Sheets; Snow Cover; Soils; Water; and Water Vapor.]

## BIBLIOGRAPHY

Austin, R. M., and R. J. Moore. "Evaluation of Radar Rainfall Forecasts in Real-Time Flood Forecasting Models." *Quaderni di Idronomia Montana* 16 (1996), 19–28. Special issue, *Integrating Radar Estimates of Rainfall in Real Time Flood Forecasting*, edited by M. Borga and R. Casale.

Baumgartner, A., and E. Reichel. The *World Water Balance*. Amsterdam: Elsevier, 1975. Translated by R. Lee. Quoted by Speidel and Agnew (1988).

Boucher, K. "Hydrological Monitoring and Measurement Methods." In *Contemporary Hydrology*, edited by R. L. Wilby, pp. 107–149. Chichester, U.K.: Wiley, 1997.

Garrels, R. M., and F. T. Mackenzie. *Evolution of Sedimentary Rocks*. New York: Norton, 1971. Quoted by Speidel and Agnew (1988).

Hawkes, N. "Long Lost Lake." *The Times* (London), May 20, 1996.

Korzoun, U. I., ed. *Atlas of World Water Balance*. Paris: UNESCO, 1978a. Quoted by Speidel and Agnew (1988).

———. *World Water Balance and Water Resources of the Earth*. Paris: UNESCO, 1978b. Quoted by Speidel and Agnew (1988).

Lvovich, M. I. "The Global Water Balance." *EOS* 54 (1973), 28–42. Quoted by Speidel and Agnew (1988).

———. "World Water Resources Present and Future." *Ambio* 6 (1977), 13–21. Quoted by Speidel and Agnew (1988).

———. *World Water Resources and Their Future*. Washington, D.C.: American Geophysical Union, 1979. English translation edited by R. L. Nace. Quoted by Speidel and Agnew (1988).

Menard, H. W., and S. M. Smith. "Hypsometry of Ocean Basin Provinces." *Journal of Geophysical Research* 71 (1966), 4305–4325. Quoted by Speidel and Agnew (1988).

Nace, R. L. "Water Resources: A Global Problem with Local Roots." *Environmental Science and Technology* 1 (1967), 550–560. Quoted by Speidel and Agnew (1988).

———. "World Water Inventory and Control." In *Water, Earth and Man*, edited by R. J. Chorley, pp. 31–42. London: Methuen, 1969. Quoted by Speidel and Agnew (1988).

Sellers, W. D. *Physical Climatology*. Chicago: University of Chicago Press, 1965. Despite its publication date contains a useful compilation of data, especially chapters 2, 7, and 8.

Shiklomanov, I. A. "World Fresh Water Resources." In *Water in Crisis: A Guide to the World's Freshwater Resources*, edited by P. H. Gleick, pp. 13–24. New York and Oxford: Oxford University Press, 1993. A valuable and up-to-date source of global hydrologic data.

Speidel, D. H., and A. F. Agnew. "The World Water Budget." In *Perspectives on Water*, edited by D. H. Speidel, L. C. Ruedisili, and A. F. Agnew, pp. 27–36. New York and Oxford: Oxford University Press, 1988. Another concise and valuable source of data.

Todd, D. K. *Groundwater Hydrology*. 2d ed. New York: Wiley, 1980.

Ward, R. C., and M. Robinson. *Principles of Hydrology*. 4th ed. Maidenhead, U.K.: McGraw-Hill, 2000. A comprehensive and readable discussion of the components and processes of the hydrologic cycle.

World Meteorological Organization. "The Global Energy and Water Cycle Experiment (GEWEX)." *WMO Bulletin* 42 (1993), 20–27.

—ROY C. WARD

**HYDROTHERMAL VENTS.** *See* Ocean Life.

**ICE AGES.** *See* Climate Change; Glaciation; *and* Little Ice Age in Europe.

**ICE CAPS.** *See* Cryosphere; Glaciation; *and* Ice Sheets.

## ICE SHEETS

The Greenland and Antarctic ice sheets cover 10 percent of the Earth's land area and contain 77 percent of the world's fresh water and 99 percent of all the glacier ice. Their average thicknesses are both approximately 2,100 meters, but the Antarctic Ice Sheet (14 million square kilometers in area) contains about ten times the ice volume of the Greenland Ice Sheet (1.7 million square kilometers in area; Figure 1). Unlike the small mountain glaciers, which flow through channels bounded on their sides by land, ice sheets rest on land that is relatively flat in comparison with the ice thickness. Icecaps are similar in structure to ice sheets, but much smaller. Although the Earth's crust is depressed by hundreds of meters by the tremendous weight of the ice, the bases of the ice sheets are on average close to sea level. However, the West Antarctic Ice Sheet, which is the portion (about 12 percent) of the antarctic ice that lies mostly in the Western Hemisphere, is grounded as much as 2,500 meters below sea level. If all of the ice were to melt, sea level would rise by nearly 80 meters.

During the last few million years, ice sheets have waxed and waned with major climate changes occurring about every 100,000 years in response to changes in the Earth's position relative to the Sun. Before the end of the last ice age approximately fifteen thousand years ago, huge ice sheets also covered parts of Eurasia and much of North America, extending as far south as Pennsylvania. As the climate warmed, sea level rose by about 125 meters at an average rate of 2.5 centimeters per year for roughly five thousand years. As the Northern Hemisphere ice sheets melted, several episodes of massive outbreaks of ice from the Hudson Bay region spread many icebergs over the North Atlantic Ocean. During this time, the Antarctic Ice Sheet decreased by only about 10 percent. The last interglacial warm period approximately 120,000 years ago was even warmer than today's climate; sea level may have been 6 meters higher, and the West Antarctic Ice Sheet may have disintegrated or much of Greenland may have melted.

Although the current rate of sea level rise is as much as 2 millimeters per year, it is not known whether the present ice sheets are growing or shrinking. Each year, about 8 millimeters of water from the entire surface of the Earth's oceans accumulates as snow on Greenland and Antarctica. The average ice accumulation is about 26 centimeters per year on Greenland and 16 centimeters per year on Antarctica (measured in centimeters of water equivalent, or about five times greater in terms of snowfall). However, we do not know to better than ±25 percent whether the amount of water returned to the oceans in icebergs and meltwater runoff equals the snow accumulation. This large uncertainty exists because there is little direct information on ice sheet volume change.

**Mass Balance and the Formation of Ice Sheets.** The term *mass balance* refers to the difference between the mass input to a glacier or ice sheet and the mass loss, and it is this parameter that governs the development and decay of these bodies of ice. Glacier formation occurs when, for a sustained period of time, the snow deposition in an area consistently exceeds the amount of snow that is lost (that is, the mass balance is positive). In the case of ice sheets, this occurs over prolonged cold periods on the order of ten thousand years. Snow falls on a large area of land, such as Greenland and Antarctica, at various times of the year (depending on local climate conditions; Figure 2). During the summer months, some, but not all, of this snow is removed by melting and evaporation. When snow remains beyond one melt season, it becomes denser as the grains bond to each other and increase in size; such snow is referred to as *firn*. As the cycle is repeated year after year, the previously fallen snow becomes buried deeper and deeper relative to the surface. Under the increased pressure of the overlying snow, the grains become larger and more aggregated, and the density continues to increase. Eventually, the process reaches a point where the pressure and sintering are so great that the air between particles is closed off, and the ice is formed. As this process continues, older ice is pushed deeper and outward by virtue of the increasing pressure of the overlying ice and firn. As a result, the ice sheets are generally thicker, with higher surface elevations, near their center, and thinner, with lower surface elevations, near their edges.

While the only mass gain to the ice sheets as a whole comes from the accumulation of snow on the surface, the mechanisms of mass loss are somewhat more complicated. As ice flows along the elevation gradient

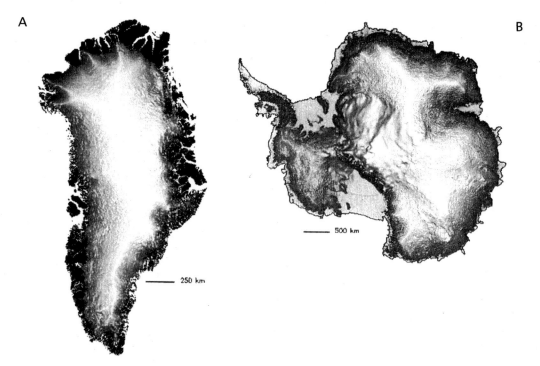

**Ice Sheets. Figure 1. Topography of Greenland and Antarctica.**

(A) Greenland is characterized by distinct ridges that define hydrologic and climatologic boundaries, and with undulating surfaces in between. (B) Antarctica is characterized by these ice ridges and undulations as well, but also contain ice shelves (the large light gray areas without texture in the figure). Surface elevations for these maps are measured by radar altimeters on the European ERS-1 and 2 satellites (except near the center of the Antarctic, where only aircraft measurements have been made). Ice sheet thicknesses are measured by airborne radars that penetrate as much as 4,000 meters of ice to map the bedrock. In the future, ice thickness changes as small as 1 centimeter will be measured by a laser altimeter on NASA's ICESat (Ice Cloud and Land Elevation Satellite).

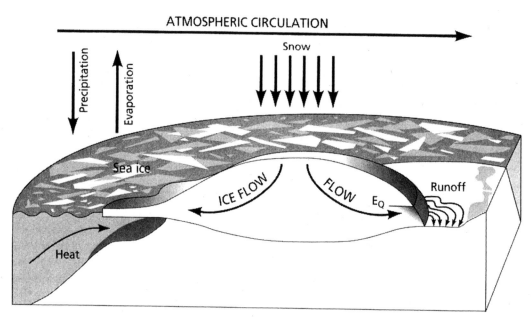

**Ice Sheets. Figure 2. Schematic Diagram of an Ice Sheet.**

Snow accumulates in the higher central regions, sinks, and is transformed to ice over time. The figure shows the flow processes (at and below the surface as well as in ice streams), iceberg calving, flow in ice streams, ice shelves, the ablation and accumulation areas of an ice sheet, and the equilibrium line.

toward the edge of the ice sheet, it is transported to the lower elevations, and in most cases lower latitudes, where temperatures are warmer. In the case of Greenland, the temperatures are sufficiently warm over about half of the ice sheet that surface melting occurs. In Antarctica, typically less than 5 percent experiences melting. The difference is a result of Antarctica's more polar location. When the snow melts, the water that forms may either evaporate, be discharged as surface or subsurface runoff, or percolate into the firn, where it refreezes at greater depths. The first two processes clearly result in mass loss by the system, but the last is simply a redistribution, representing no net decrease in mass. There is also some additional loss due to sublimation (the direct transition from dry snow to water vapor). This occurs because the air is very dry and there is sustained solar radiation during the summer, which provides the energy for the conversion.

Finally, as the remaining ice approaches the outer portions of the ice sheet, the most dramatic and most significant form of loss occurs—calving. The topography and characteristics of the underlying bedrock are such that the ice tends to flow in major streams, or drainage basins, where it is carried out to the surrounding seas. Because of the mechanical strength of the ice, as well as its buoyancy, it can extend well beyond the water–land boundary as a solid unit. In the case of ice discharged through singular narrow channels or ice streams, as is generally the case in Greenland, the floating ice is referred to as a glacier tongue. In instances where the discharge from several ice streams merges into a single frozen table of ice, or where ice is discharged over wide areas, the floating ice is referred to as an ice shelf, as is generally the case in Antarctica. Eventually, the stresses on the floating ice that result from its weight and buoyancy, and the erosion from warm water in contact with its base, cause it to break into icebergs that are carried out to sea. The size of these icebergs can range from a few meters to tens of kilometers across. Iceberg calving is believed to account for roughly 40 percent of the mass loss in Greenland and 80–90 percent in Antarctica. In addition, the underside of the floating ice experiences melt as a result of its contact with the warmer water. This basal melting also contributes substantially to mass loss and affects ocean circulation by adding cold fresh water to the saline seas.

The relative magnitudes of the accumulation and mass loss components determine the sizes and shapes of the ice sheets, as well as their impact on sea level and their role in the changing climate.

**Ice Sheet History.** Of particular importance in the understanding of ice sheets and climate, past and present, is the record contained beneath its surface. Changes in the atmospheric conditions affect the physical and chemical structure of the firn. These changes can be ob-

served in ice cores and related to temperature, accumulation, and atmospheric circulation, among other phenomena. Physically, the changes appear as horizontal bands in the core of varying clarity and texture that are related to an annual signal. Temperature, radiation, and wind characteristics vary considerably from summer to winter, and they affect the structure of the surface snow differently in different seasons. During the summer, the temperature gradients in the snow can be very high, which can cause snow grains to become very large and faceted, a condition known as hoar formation. These low-density, faceted crystals metamorphose differently than the rest of the snow pack, so that, when they transform into ice, the resulting layer appears more cloudy than the rest. In this way, they provide annual markers, and the ice between these layers represents the year-to-year snow accumulation. Additional information is contained in the chemical makeup of the snow. Various chemical compounds and isotopes vary with temperature at the time of ice formation (oxygen-18 and deuterium), source of moisture (salt compounds), and volcanic activity (sulfur dioxide). These amounts vary seasonally and interannually, and, through their combined analyses, a past climate history can be reconstructed. [*See* Climate Reconstruction.]

To date, approximately ten deep ice cores 2–3 kilometers in length have been extracted from Greenland and Antarctica. These span a period of tens and even hundreds of thousands of year. The oldest record is from Vostok in Antarctica, which dates back 220,000 years. These cores provide an invaluable record of ice sheet histories. [*See* Aerosols *and the vignette on* GRIP and GISP2 *in the article on* Climate Reconstruction.]

**Ice Sheets and Climate Change.** The Greenland and Antarctic ice sheets are of great importance to sea level and global climate, both through their impact on and their response to climate changes. However, the precise role of the ice sheets in the climate system and their interaction with global change are complex. In the short term, changes in atmospheric temperature, winds, clouds, precipitation, and radiation balance affect the surface snow accumulation, temperature, and meltwater runoff from the ice sheets, which in turn have long-term effects on the dynamics of the ice sheet and the rates of ice flow and iceberg discharge into the ocean. Changes in ocean temperatures and circulation beneath the antarctic ice shelves also have long-term effects on the rates of ice discharge from the grounded ice sheets.

Recent estimates suggest that, in Greenland, the mass loss associated with increased melting and calving will exceed the increased input caused by enhanced accumulation in the warmer environment. It is expected that the Greenland Ice Sheet, which already experiences melt over roughly half of its surface, will shrink, contributing to an increase in sea level. Antarctica, on the other hand, which is situated much closer to the pole than Greenland, has very few areas that experience temperatures that are even close to the melting point. As a result, increased global temperatures are not expected to change significantly the amount of melt on the ice sheet, and the increased accumulation is presumed to be the dominant phenomenon. Consequently, the Antarctic Ice Sheet is expected to increase in size and help lower sea level. The extent to which these estimates are true and the relative magnitudes of each are currently the subject of much scientific investigation. With the wide application of satellite technology, improved ice models, and a number of comprehensive field campaigns, these issues are being addressed, and scientists are converging on the answers.

[See also Cryosphere; Glaciation; Global Warming; Hydrologic Cycle; and Sea Level.]

### BIBLIOGRAPHY

Bindschadler, R. "Monitoring Ice Sheet Behavior from Space." *Reviews of Geophysics* 36.1 (1998), 79–104.

Colbeck, S. C., ed. *Dynamics of Snow and Ice Masses.* New York: Academic Press, 1980.

Hambrey, M., and J. Alean. *Glaciers.* Cambridge: Cambridge University Press, 1994.

Paterson, W. S. B. *The Physics of Glaciers.* 3d ed. Oxford: Pergamon, 1994.

Van der Veen, K. C. "Land Ice and Climate." In *Climate System Modeling*, edited by K. Trenberth, pp. 437–450. Cambridge: Cambridge University Press, 1992.

Warrick, R. A., et al. "Changes in Sea Level." In *Climate Change 1995: The Science of Climate Change*, edited by J. T. Houghton et al., pp. 361–405. Cambridge: Cambridge University Press, 1996.

—H. JAY ZWALLY AND WALEED ABDALATI

**IMMIGRATION.** *See* Migrations.

**IMPACT.** *See* IPAT.

**IMPACT ASSESSMENT.** *See* Environmental Impact Assessment.

## IMPACTS BY EXTRATERRESTRIAL BODIES

Our solar system is a dynamic environment that has undergone change throughout geologic time. [See Earth History.] One of the agents for rapid changes is the impact of interplanetary bodies, such as asteroids and comets. On the Earth, a variety of possible effects have been ascribed to impacts. To date, approximately 150 impact craters have been identified on Earth, and several new structures are found each year. The basic types of impact structure are (1) simple structures, up to 4 kilometers in diameter, with uplifted and overturned rim rocks, surrounding a bowl-shaped depression, partially filled by breccia, and (2) larger complex impact structures and basins, with a distinct central uplift in the form of a peak and/or ring, an annular trough, and a slumped rim. The interiors of these structures are partially filled with broken and mixed rocks known as breccia and rocks melted by the impact.

**Crater Identification.** A reliable set of criteria are required to determine that a particular craterlike feature on Earth resulted from the impact of an extraterrestrial body. Meteorite fragments are found only at the smallest craters and are quickly destroyed by weathering. For impact events that form craters larger than about 1 kilometer across, the pressures and temperatures upon impact are sufficient to melt and even vaporize the impacting body and some of the target rocks. For example, the peak pressure produced by the impact of a stony (chondritic) body into granite, a common terrestrial rock, at 25 kilometers per second (an average impact velocity for an asteroid impact on the Earth) is 900 gigapascals, or 9 million times atmospheric pressure. In such cases, the recognition of a unique suite of rock and mineral deformations, termed shock metamorphism, is indicative of an impact origin. Examples include conical fractures known as shatter cones, microscopic deformation features (particularly the development of planar deformation features in silicate minerals such as quartz), various glasses and high-pressure minerals, and rocks melted by the intense heat of impact. In some cases, a distinct geochemical anomaly in the form of relative enrichment in platinum group elements can be detected in lithologies such as impact melt rocks. This represents the admixture of a small amount of meteorite material from primitive meteoritic bodies such as chondrites.

**Impact and Earth Evolution.** As the impact flux has varied throughout time, the potential for impact to act as an agent for change has varied. Simulations of planetary accretion envisage the collision and growth of 100 million to 1 trillion ($10^{12}$) bodies with diameters of the order of 10 kilometers. These asteroidal-sized bodies grow rapidly by impact, in less than 100,000 years, to about 20–30 embryonic planets with masses of the order of 10 percent of the mass of the Earth. This early stage of planetary accretion is believed to be followed by a stage involving truly giant impacts between the embryos. During this stage, a few planet-sized bodies are formed in 1 million to 100 million years.

In the case of the Earth, the last embryonic planetary collision may have resulted in the formation of the Moon. Numerical simulations of an impact of a Mars-sized body with the Earth indicate the formation of an

Earth-orbiting accretionary disk from which a Moon-like body can form. The key is that the material that forms the accretionary disc is originally impact-produced vapor—either jetted terrestrial mantle or projectile material. While the collision hypothesis for the origin of the Moon is consistent with the constraints on the nature of the Moon from lunar missions, it is still only a model. The consequences of such a giant impact for the proto-Earth would have been severe. They would have included massive remelting of the Earth and the loss of the original atmosphere.

Following planetary formation, the subsequent high rate of bombardment by the remaining tail of accretionary debris is recorded on the Moon and other terrestrial planets that have preserved portions of their earliest crust. In the case of the Moon, a minimum of six thousand craters with diameters greater than 20 kilometers formed during this early period. In addition, there were roughly forty-five impacts producing impact basins ranging in diameter from Schrödinger, at 320 kilometers, to the South Pole Aitken basin at 2,500 kilometers and the putative Procellarum basin at 3,500 kilometers.

The Earth has received a higher cratering flux than the Moon throughout geologic time because of its greater gravitational cross section. Most of the record of this flux, however, is not preserved on the Earth because of the high level of geologic activity and its relatively young surface. From the lunar cratering rate, it is estimated that as many as two hundred impact basins greater than 1,000 kilometers in diameter may have been formed on the Earth from 4.5 to about 3.8 billion years ago. There has been speculation as to the potential effects of these basin-sized impacts, but the cumulative effect of such a bombardment on the surface and upper crustal rocks of the Earth is unknown. The Earth's earliest crust was either not voluminous or was destroyed by impact and tectonic forces. By analogy with the lunar case, it is likely that few terrestrial surface rocks survived intact through this period of heavy bombardment.

Basin-sized impacts must also have affected the atmosphere, the hydrosphere, and the development of the biosphere. The plume of high-velocity ejecta and vapor in large impacts can blow off the atmospheric mass above the point of impact. The primordial atmosphere on the early Earth could have been reduced in density by a factor of five through such a process. The impact on Earth of a body in the 500-kilometer size range, similar to the present-day asteroids Pallas and Vesta, would release sufficient energy to vaporize the present world's oceans. Such an event would have sterilized the surface. The Earth would have been enveloped by a hot rock vapor atmosphere that would radiate heat downward on to the surface, with a resultant temperature of over 2,000°C. It is estimated that it would take three thousand years for the water-saturated atmosphere to rain out and re-form the oceans. Impacts by bodies in the 200-kilometer size range would be sufficient to evaporate the zone where photosynthesis takes place in the oceans. Life could survive in a deep marine setting as early as 4.2–4.0 billion years ago. Smaller impacts, however, would continue to make the surface inhospitable until 4.0–3.8 billion years ago.

As the debris from accretion were removed from the solar system by ejection or collision, the impact flux began to stabilize. While major basin-forming impacts were no longer occurring, there were still occasional impacts resulting in craters hundreds of kilometers in diameter. For example, the Precambrian terrestrial record contains remnants of the Sudbury (Canada) and Vredefort (South Africa) structures, which have estimated original crater diameters of 200–300 kilometers and ages of about 2 billion years.

A number of large craters have been preserved on Earth from more recent Phanerozoic time (the past 550 million years). Examples are the Manicouagan (Canada), 100 kilometers diameter, age $212 \pm 2$ million years, and the Popigai (Russia), 100 kilometers diameter, age $35 \pm 5$ million years, which are temporally close to transitions in the biostratigraphic record (Figure 1). These and other known large impact events, as well as the many similar-sized impact events not discovered or preserved in the record, must have had effects at least on a continental, if not a global, scale. Impact events resulting in craters in the 100-kilometer size range release energies of $10^{21}$–$10^{22}$ joules, or the equivalent of the explosion of a million megatons of trinitrotoluene (TNT), of the same order of magnitude as the annual output of internal energy of the Earth, that is, all the energy that comes out of the Earth from earthquakes, volcanic eruptions, and heat flow.

**The Cretaceous–Tertiary Extinction.** An extensive literature exists concerning the evidence for a major impact event at the Cretaceous–Tertiary (K/T) boundary and its association with a mass extinction in the terrestrial biosphere 65 million years ago. [See Extinction of Species.] The physical evidence consists of the following: shock-produced microscopic planar deformation features in the minerals quartz, feldspar, zircon, and chromite; the occurrence of stishovite, a high-pressure form of quartz; high-temperature magnesioferrite spinels believed to be vapor condensates; and various spherules, generally altered but including glass spherules. The chemical evidence consists primarily of a geochemical anomaly in the K/T boundary deposits, indicative of an admixture of meteoritic material.

The original mechanism for destruction of life suggested for the K/T impact was global darkening and cessation of photosynthesis due to ejecta in the atmosphere. Early atmospheric evolution models of a K/T-

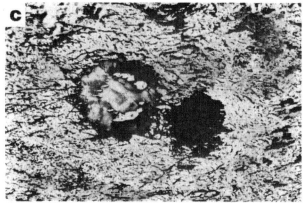

**Impacts By Extraterrestrial Bodies. FIGURE 1.** Some Examples of Terrestrial Complex Impact Craters.

(a) Oblique aerial photograph of Steinheim, Germany (3.8 kilometers diameter, 15.1 ± 1.0 million years old), with a central peak and flat floor. (b) Oblique aerial photograph of Gosses Bluff, Australia (22 kilometers diameter, 142.5 ± 0.5 million years old), which has been heavily modified by erosion. The prominent ring of hills is 5 kilometers in diameter and is an erosional remnant of a central peak within a much larger structure. (c) Shuttle photograph of the twin craters in Clearwater Lake, Canada (22 kilometers and 32 kilometers diameter, respectively, 290 ± 20 million years old). A central peak is submerged in the East crater, and a small central peak occurs within the ring of islands at the West crater. (d) Landsat photograph of Manicouagan, Canada (200 kilometers diameter, 214 ± 1 million years old). The original rim lay outside the annular lake, which is 65 kilometers in diameter.

sized impact event indicated no photosynthesis for about three months, and below-freezing temperatures on land for about six months. They also indicated similar effects for dust loadings one hundred times lower. Such loadings are achieved in the formation of impact structures in the 50- to 100-kilometer size range, that is, corresponding to known impact structures that are not known to be associated with mass extinctions. More recent, three-dimensional atmospheric general circulation models suggest that the heat capacity of the oceans can offset land surface cooling. They indicate more rapid cooling than in earlier models, but global freezing of the land surface is not achieved; rather, below-zero temperatures occur only in some areas.

Models of impact-induced global dust loading of the atmosphere and its potential effects also stimulated studies of nuclear winter scenarios. [*See* Nuclear Winter.] In the latter, global darkening and cooling is much more efficient because of the soot and particulate matter from the combustion of materials from the high thermal pulse associated with nuclear explosions. Soot, however, has also been identified in K/T boundary deposits, and its origin has been ascribed to global wildfires. The impact-related soot, therefore, may enhance or even overwhelm the effects produced by global dust clouds. (On the other hand, the heat input per unit area of wildfires is probably not enough to loft the soot into the stratosphere, while precipitation would wash it out of the troposphere in roughly two weeks.)

In the study of the K/T event, increasing emphasis has been placed on understanding the effects of vaporized and melted ejecta on the atmosphere. This hot expanding cloud provides a mechanism for igniting wildfires. Models of the thermal radiation produced by the ballistic reentry of ejecta condensed from the vapor plume of a K/T-sized impact indicate a massive thermal radiation pulse, up to 50–150 times the present solar input, on the Earth's surface for as long as several hours. Calculations indicate that the thermal radiation produced by the K/T event is at the lower end of that required to ignite living plant materials. Other calculations indicate that the prompt and direct ignition of wildfires would have been continental rather than global in scale. If global wildfires are required to produce mass extinction effects, in addition to the effect of atmospheric dust loadings, then this may explain why impacts that result in 50- to 100-kilometer craters do not result in mass extinctions. The accompanying thermal radiation is insufficient to result in the global ignition of plant materials.

Despite the growing, self-consistent evidence, a sector of the geologic community has argued against the impact interpretation for the K/T boundary, preferring rather to ascribe it to more traditional geologic processes, such as volcanism. [See Volcanoes.] One of the early arguments of the opponents was that there was no known K/T crater. Without this smoking gun, they argued, the physical and chemical evidence is open to various interpretations. It is now generally accepted that an 180-kilometer feature known as Chicxulub, on the Yucatán Peninsula, Mexico, is the K/T crater (Figure 2). It is a buried structure, and evidence for impact consists of shock-produced planar deformation features in quartz and feldspar from deposits interior and exterior

to the structure, as well as impact melt rocks. Variations in the concentration and size of shocked quartz grains and the thickness of K/T boundary deposits point toward a source in the Americas. In addition, the geochemistry of K/T glasses from Haiti matches the mixture of lithologies found at the Chicxulub site. Isotopic ages for the melt rocks at Chicxulub of $64.98 \pm 0.05$ million years are indistinguishable from the K/T glasses at $65.07 \pm 1.00$ million years.

A considerable thickness of anhydrite ($CaSO_4$) occurs in the target rocks at Chicxulub. Impact heating would drive off sulfur dioxide, producing about 1 trillion ($10^{12}$) metric tons of sulfur aerosols in the atmosphere. It is not clear, however, what percentage of these aerosols would rapidly recombine to solids in the immediate area of the impact. From the study of major volcanic eruptions, the effects of sulfate aerosols on atmospheric cooling are known to be greater than those of dust particles, and, ultimately, the aerosols will rain out as sulfuric acid, which will create further severe damage to the biosphere. It may be, therefore, that the devastating effects of the K/T impact compared with other known large impacts is due to the unusual composition of the target rocks at Chicxulub. In any case, the temporal association of an extremely large impact crater with a worldwide ejecta layer and a global mass extinction event is well established. The cause-and-effect relationship is less well established, however. Modeling of the atmospheric and biospheric effects of large impact events is in its infancy.

The terrestrial cratering rate indicates that the fre-

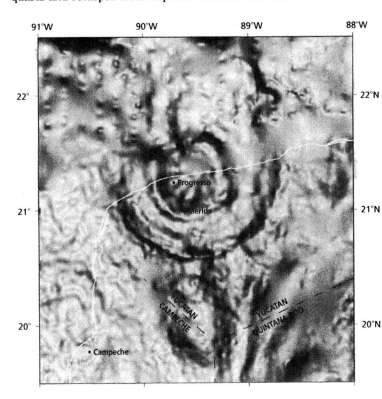

**Impacts By Extraterrestrial Bodies. Figure 2.**
Black and White Rendition of Shaded Relief of Circular Gravity Gradient Features in the Area of Chicxulub: the K/T Crater.

The technique displays subtle features in the gravity field due to changes in density of the rocks brought about by crater formation. The crater is buried by 1 kilometer of sediments and is not expressed at the surface. Fuzzy areas correspond to areas with limited data.

quency of K/T-sized events on Earth is of the order of one every $100 \pm 50$ million years (Figure 3). Smaller, but still significant, impact events occur on shorter time scales. Although such events will not produce mass extinctions, they will affect the global climate and biosphere to varying degrees. Model calculations suggest that dust loadings from the formation of impact craters as small as 20 kilometers in diameter could produce light reductions and temperature disruptions. Such impacts occur on Earth with a frequency of two or three every million years. The most recent known structure in this size range is Zhamanshin in Kazakhstan, with a diameter of 15 kilometers and an age of about 700,000 years. Impacts of this scale are not likely to have a serious effect upon the biosphere, because the model climatic disruptions are of limited severity and duration. The most fragile component of the present environment, however, is human civilization, which is highly dependent on an organized and technologically complex infrastructure for its survival. While we seldom think of human civilization in terms of millions of years, there is little doubt that if civilization lasts long enough, it will suffer severely or may even be destroyed by an impact event.

While we take some comfort in the fact that civilization-threatening impact events occur only every half million years or so, it must be remembered that impact is a random process in time. "Impact" events do occur on human time scales. [See Extreme Events.] For example, the Tunguska event in 1908 was due to the atmospheric explosion of a relatively small body (several hundred meters in diameter) at an altitude of 10 kilometers or less. The energy released, based on the energy of the observed seismic disturbances, has been estimated to be $10^{17}$–$10^{18}$ joules, which is the equivalent of 10–100 megatons of TNT. Although the air blast resulted in the devastation of about 2,000 square kilometers of Siberian forest, there was no recorded loss of human life because of the very sparse population. Events such as Tunguska occur on a time scale of hundreds of years. The next time a Tunguska-sized event occurs, it may be over an urban center.

The study of impact events on Earth is a relatively recent scientific endeavor. It was the preparations for and, ultimately, the results of the lunar and planetary exploration programs that provided the impetus and rationale for their study. It is apparent, however, that such impacts can no longer be considered a process of interest only to the planetary community. This is a process that has affected terrestrial evolution in fundamental ways. For example, without the K/T impact and the extinction of the dominant dinosaurs, the present-day biosphere may not have included mammals, and, in particular, humans, as an important component. Similarly, without a Mars-sized impact forming the Moon with its accompanying tidal forces, one can only speculate as to how the littoral zone, one of the most important areas in the terrestrial ecosystem, would have evolved and been populated.

Impacts are a natural consequence of the character

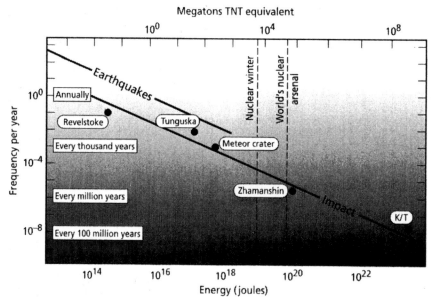

**Impacts By Extraterrestrial Bodies. FIGURE 3.** Frequency of Impact Plotted against Energy Released, in Joules and Megatons of TNT Equivalent.

A number of specific impact and atmospheric explosion events are indicated. For comparison, the frequency and energy of earthquakes are shown. Earthquakes are more common than impacts but are limited in the amount of energy they release. Also shown (dotted vertical lines) are the energies required to produce a global "nuclear winter" and in the world's nuclear arsenal.

## THE RISK OF IMPACTS WITH PLANET EARTH

The larger the asteroid or comet, the less likely we are to be struck by it. Smaller objects, of 3-meter diameter, strike the Earth on average once a month. At the other end of the scale are objects 10 kilometers in diameter—the size of the asteroid thought to have made the Chicxulub Crater near Yucatan roughly 65 million years ago. Such a destructive invader is expected only once in 100 million years.

| AVERAGE FREQUENCY | DIAMETER (METERS) | IMPACT ENERGY (A-BOMBS) |
|---|---|---|
| Month | 3 | 0.05 |
| Year | 6 | 0.5 |
| Decade | 15 | 10 |
| Century | 30 | 100 |
| Millennium | 100 | 2,500 |
| 10,000 years | 200 | 50,000 |
| 1,000,000 years | 2,000 | 50,000,000 |
| 100,000,000 years | 10,000 | 5,000,000,000 |

SOURCE: de Grasse Tyson, 1997.

**BIBLIOGRAPHY**

de Grasse Tyson, N. "Asteroid Collisions Made Us What We Are Today. So Why Worry About the One That Will Do Us in Tomorrow?" *Natural History* (September 1997), 82–88.

—DAVID J. CUFF

of the solar system. They have happened on Earth throughout geologic time and will happen again. Although the occurrence of a large impact event has a low probability on the time scale of human lifetimes, its consequences will be globally disastrous.

[*See also* Atmospheric Chemistry; Climate Change; Evolution of Life; Fire; *and* Planets.]

### BIBLIOGRAPHY

Alvarez, W. *T. Rex and the Crater of Doom*. Princeton, N.J.: Princeton University Press, 1997. An easily read anecdotal account of the development of the impact hypothesis for the K/T boundary and mass extinctions and the discovery of the Chicxulub crater. Written by the son of a Nobel laureate, L. Alvarez, who was the primary author of the initial scientific article presenting chemical evidence for the involvement of impact at the K/T boundary.

Chapman, C. R., and D. Morrison. *Cosmic Catastrophes*. New York: Plenum, 1989. A review of cosmic catastrophes, mainly impact, for the general reader, now slightly dated.

Gehrels, T., ed. *Hazards Due to Comets and Asteroids*. Tucson: University of Arizona Press, 1994. A compendium of scientific papers by 120 authors dealing with a range of subjects related to impact hazards.

Hatmann, W. K., et al., eds. *Origin of the Moon*. Lunar and Planetary Institute, Houston: 1986. A compendium of scientific papers presented at the Conference on the Origin of the Moon, held in Kona, Hawaii, 1984. Includes a general review of hypotheses concerning the origin of the Moon, and presents one of the first series of papers dealing with its impact origin.

Remo, J. L., ed. *Near-Earth Objects: The United Nations International Conference*. New York: New York Academy of Sciences, 1997. A compendium of scientific papers dealing with near-Earth objects. Can be considered a follow-on and update to some of the topics covered in Gehrels (1994).

Ryder, G., et al., eds. *The Cretaceous–Tertiary Event and Other Catastrophes in Earth History*. Geological Society of America Special Paper 307. Washington, D.C., 1996. The latest compendium of scientific papers dealing with impacts, the K/T boundary, and other boundaries. Makes an interesting comparison with Silver and Schultz (1982).

Sharpton, V. L., and P. D. Ward, eds. *Global Catastrophes in Earth History: An Interdisciplinary Conference on Impacts, Volcanism and Mass Mortality*. Geological Society of America Special Paper 247. Washington, D.C., 1990. A compendium of scientific papers published ten years after the original suggestion that the K/T boundary was due to an impact event. It is a follow-on from Silver and Schultz (1982) and indicates the growing strength and consensus regarding the origin of the K/T boundary.

Silver, L. T., and P. H. Schultz, eds. *Geological Implications of Impacts of Large Asteroids and Comets with the Earth*. Geological Society of America Special Paper 190. Washington, D.C., 1982. A compendium of scientific papers dealing with the effects of large-scale impact. Published soon after the original suggestion that the K/T boundary was due to an impact event, this book is a historical snapshot of the scientific debate at the time.

—RICHARD A. F. GRIEVE

**IMPLEMENTATION.** *See* Environmental Law; International Cooperation; *and* Public Policy.

## IMPULSE RESPONSE

Atmospheric gases are cycled through the atmosphere through a variety of emission sources and removal processes. In a balanced state, the total rates of emission and of removal will be equal, so that there is an unchanging concentration of the gas in the atmosphere. If a sudden, additional pulse of gas is emitted into the atmosphere (as a result of human activities, for example), then the natural removal processes will gradually remove this extra gas, tending to restore the atmospheric concentration to the value that existed before the pulse was added. The term *impulse response* refers to the variation in the amount of gas remaining in the atmosphere after a sudden injection of a given amount of gas into the atmosphere. The impulse response is computed via computer simulation models that incorporate the various removal processes.

For gases such as methane ($CH_4$) and nitrous oxide ($N_2O$), which are removed largely through chemical reactions within the atmosphere itself, the rate of removal varies directly with the concentration of the remaining gas. The concentration thus decreases exponentially toward zero according to

$$C(t) = C_0 e^{-t/\tau} \qquad (1)$$

where $C(t)$ is the concentration of the remaining pulse at a time $t$ after the pulse injection, $C_0$ is the concentration of the initial pulse, and $t$ is the time required for the concentration to decrease to $1/e$ of the initial concentration. The time $t$ is also referred to as the life span of the gas in the atmosphere, since it is equal to the average length of time that molecules of the pulse remain in the atmosphere before being removed. Since the concentration of the remaining pulse decreases exponentially to zero, the impulse response can be characterized by the single time constant, $t$. The larger $t$ is, the more slowly the pulse is removed.

In the case of carbon dioxide ($CO_2$), the removal processes involve absorption by the terrestrial biosphere and by the oceans. Impulse responses can be computed for either process acting alone or for both processes acting together. When absorption by the oceans only is considered, a pulse injection is not removed with a single time constant and the amount of the pulse remaining in the atmosphere does not decrease all the way to zero. Rather, we can think of the pulse as being divided into a number of fractions, each of which is absorbed with its own time constant. For example, the first 20 percent of the pulse is removed with a time constant of less than

ten years. Successive increments take progressively longer to be removed, and the final 10–15 percent is not removed at all, which is equivalent to giving it an infinitely long time constant. When absorption by the terrestrial biosphere alone is considered, the pulse concentration initially falls rapidly as extra carbon dioxide is taken up through an increase in the rate of photosynthesis, but then the excess concentration begins to *increase* as some of the extra biomass is transferred to soil carbon and returned to the atmosphere through increased respiration. Figure 1A shows the impulse response when accounting for absorption by the oceans alone, the terrestrial biosphere alone, and when the two act together. Different models of the carbon cycle will give somewhat different impulse responses, so the impulse response serves as a convenient way to compare different models. Also shown, for purposes of contrast, is the impulse response for nitrous oxide ($N_2O$), which has a single time constant $t$ of 120 years.

A single time constant for carbon dioxide is frequently quoted, usually a value of around one hundred years. However, since the life span for some of the excess carbon dioxide molecules in the atmosphere is infinity, the average life span is also infinity. It is more accurate to speak of the average life span of that portion of the injected carbon dioxide that is eventually removed. As can be seen from the comparison of the $CO_2$ and $N_2O$ impulse responses in Figure 1A, this is still a rather crude representation of the behavior of carbon dioxide, since its rate of removal is initially much more rapid than that of $N_2O$, but later is much slower.

Carbon exists as three different isotopes: carbon-12 (about 99 percent), carbon-13 (about 1 percent), and carbon-14 (about one molecule in $10^{12}$ in the atmosphere). Figure 1B shows the separate impulse responses for carbon-12 and carbon-14 if the pulse input contains some carbon-14. The decay in the perturbation of the atmospheric carbon-14:carbon-12 ratio is much faster than the decrease in the total amount of carbon in the atmosphere and, unlike the total carbon pulse, the carbon-14 pulse decays essentially to zero within a few hundred years. The isotope ratio decreases much faster than the pulse of total carbon because, as carbon-14 diffuses into the ocean, some of it is replaced by carbon-12 that diffuses out (this oceanic carbon-12 can be thought of as being dislodged by the excess carbon-14). As a result, the atmospheric carbon-14 content has recovered quickly from the injection of carbon-14 due to the atmospheric testing of nuclear bombs (from 1953 to 1986), but this behavior provides no guidance concerning how the carbon cycle responds to anthropogenic injections of carbon dioxide from deforestation or the burning of fossil fuels.

The impulse response is at the core of the calcula-

A

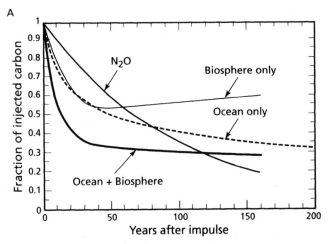

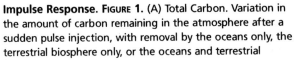

B

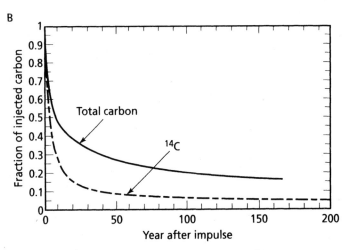

**Impulse Response. FIGURE 1.** (A) Total Carbon. Variation in the amount of carbon remaining in the atmosphere after a sudden pulse injection, with removal by the oceans only, the terrestrial biosphere only, or the oceans and terrestrial biosphere. The impulse response for a sudden injection of $N_2O$ is also given, for comparison. (B) Atmosphere. Variation in the amount of total carbon and of carbon-14 in the atmosphere following a sudden injection of carbon-14.

tion of the global warming potential, which takes into account the change in the relative concentrations over time in the pulses of two different gases. The impulse response can also be used to estimate the buildup of atmospheric carbon dioxide from an arbitrary $CO_2$-emission scenario. In this case, a continuous emission can be broken down into a series of closely spaced, small emission pulses, each of which then decays according to the given impulse response. The amount of carbon dioxide remaining in the atmosphere at some time in the future is given by the sum of the carbon dioxide remaining from each of the small pulses emitted from the start of the scenario up to the time in question. This procedure assumes that the carbon cycle is linear, that is, that each pulse decays in exactly the same way. The true carbon cycle is not linear, in that successive emission pulses will decay slightly more slowly than the preceding pulses, because of saturation effects. However, this can be accounted for by changing the time constants in the impulse response function based on the cumulative emission. The impulse response thus remains as a powerful tool for characterizing the behavior of different carbon cycle models and for quickly estimating the carbon dioxide buildup associated with different emission scenarios.

[*See also* Global Warming, *article on* Global Warming Potential; *and* Modeling of Natural Systems.]

### BIBLIOGRAPHY

Harvey, L. D. D. "Managing Atmospheric $CO_2$." *Climatic Change* 15 (1989), 343–381.

O'Neill, B. C., S. R. Gaffin, F. N. Tubiello, and M. Oppenheimer. "Reservoir Timescales for Anthropogenic $CO_2$ in the Atmosphere." *Tellus* 46B (1994), 378–389.

Tubiello, F. N., and M. Oppenheimer. "Impulse-Response Functions and Anthropogenic $CO_2$." *Geophysical Research Letters* 22 (1995), 413–416.

—L. D. DANNY HARVEY

**INCINERATION.** *See* Waste Management.

## INDUSTRIAL ECOLOGY

Industrial ecology, or IE, has been defined as the study of materials and energy flows and transformations, not just within an economic system but also, and more important, across its boundaries with the natural world with which the industrial and economic activities of humankind must coexist. According to this definition, IE is a very broad, encompassing concept. On the other hand, IE is evidently a metaphor, intended to evoke the idea that an industrial system of diverse specialized firms could (in principle) imitate a natural ecosystem in the specific sense of utilizing and recycling all material resources, just as a natural ecosystem does.

This metaphor is more appealing to the imagination than realistic. A natural ecosystem consists of primary producers (photosynthetic green plants), first-level grazers, second- and third-level predators, plus assorted parasites and decay organisms. The industrial system in its present form has some primary producers (agriculture and forestry). It has a rough resemblance to the food chain, in the sense that firms higher on the chain "consume" the output of firms lower on the chain. Reprocessors and recyclers are only roughly analogous to parasites and decay organisms. But the human industrial system differs fundamentally from an ecosystem in that it depends essentially on exploiting natural capital—minerals and fossil fuels—put in place eons ago, whereas an ecosystem utilizes only current or recent primary production.

Because of this confusion the term industrial ecology is not uncontroversial. Nor is it universal. In Europe the term *industrial metabolism* (IM) is more common, although there are also significant differences in nuance between the two terms. [*See* Industrial Metabolism.] Some environmentalists object to the ecosystem metaphor on the grounds that it is being promoted by industry as a way to shift emphasis away from the need to treat and dispose of wastes harmlessly, or (more important) to reduce waste emissions at the source by innovating and adopting cleaner production technology. This objection (insofar as it is based on imputed motivations) seems overblown. But the IM metaphor conveys better than IE the notions of ingestion (of raw materials), digestion, energy conversion, and excretion of wastes. The latter aspect, and the environmental implications, are not central to IE. By the same token, there are some who prefer the ecosystem metaphor precisely because it does convey the importance, or at least the theoretical possibility, of closing the materials cycle.

Both IE and IM are essentially multidisciplinary. They are holistic and systems-oriented, rather than reductionist. They focus on the long term rather than the immediate. They are applicable to several scales, ranging from industrial parks or complexes to regional and global scales. While IE and IM are largely descriptive, they also include a variety of analytic and normative elements. Analytic tools include life-cycle analysis, mass flow analysis, and substance flow analysis as well as thermodynamic (exergy) and thermoeconomic analysis tools from engineering. Both use input-output economic models. Insofar as they deal with resources and residuals, both IE and IM overlap with resource and environmental economics, respectively. At the interface with ecology they overlap with ecological economics. [*See* Environmental Economics.]

Normative aspects arise from a concern with "sustainable development" in the sense of long-term compatibility with ecological systems. [*See* Sustainable Development.] The authors of the first textbook in the field of IE defined it as "the means by which humanity can deliberately and rationally approach and maintain a desirable carrying capacity, given continued economic cultural and technological evolution. . . It is a systems view which seeks to optimize the total materials cycle from virgin material, to finished material, to components, to product, to obsolete product, and to ultimate disposal. Factors to be optimized include resources, energy, and capital" (Graedel and Allenby, 1995).

**IE and Economics.** Cross-boundary impacts of human industrial activity on the biosphere have important economic implications. This is because they affect (usually adversely) vital environmental services to humans. The services in question range from physical—climatic stabilization, nutrient (carbon, nitrogen, and sulfur) cycling, food supply, and waste assimilation—to recreational and aesthetic. Needless to say, the environment is our source of raw materials for production, both renewable and nonrenewable. IE concerns itself with identifying and analyzing situations where human activity perturbs natural systems to an extent that overwhelms or threatens to overwhelm natural stabilizing tendencies. It also concerns itself with developing and implementing engineering and economic approaches to ameliorating these problems.

Having said this, a caveat is needed. To the extent that the optimization of the materials cycle, as advocated by the "systems view" noted above, is more than the ideal outcome of some modernized "invisible hand," the underlying mechanism—the "fingers of the hand"—must be articulated. To put it more bluntly, this optimization is unlikely to occur in the absence of a suitable profit motive. A number of firms have implied the existence of such a motive in choosing memorable acronymic slogans, such as 3M's PPP for "Pollution Prevention Pays," Dow's WRAP for "Waste Reduction Always Pays," and Chevron's SMART for "Save Money and Reduce Toxics."

The problem is that, while there are certainly many opportunities for saving money and reducing pollution at the same time—called "double dividends" in the current jargon—to assert that double dividends are always possible flies in the face of logic and common sense.

Waste reduction can be profitable if—and only if—it can be accomplished by reducing costly inputs without much additional capital outlay. However, end-of-pipe waste treatment is rarely, if ever, profitable. It can be profitable only if the treated waste can find a productive use. One of the few examples of such a use is found in the case of the famous Kalundborg industrial complex in Denmark. [*See the vignette on* Examples of Industrial Ecosystems.] Other such examples surely do exist; indeed, the history of technology is replete with examples of former wastes that have become valuable feedstocks or products. Examples range from gasoline (once a waste from petroleum refineries) to coal tar, coke oven gas, natural gas (originally a waste from petroleum drilling), and blast furnace slag.

Nevertheless, many such treatment wastes still remain essentially valueless. Four of the most persistent examples are phospho-gypsum waste (from phosphate rock processing), "red mud" from bauxite processing, wet ferrous sulfate from steel pickling, and "black liquor"

## EXAMPLES OF INDUSTRIAL ECOSYSTEMS

The industrial system in Kalundborg, Denmark, has been evolving since the 1970s. Waste heat from a coal-burning power plant provides energy for a pharmaceutical company and warms more than fifty commercial fish ponds. The plant also sells fly ash from coal combustion to an enterprise that makes cement and road aggregate. An associated oil refinery uses steam from the power plant, contributes fuels to that power plant, and sends sulfur to an acid plant. Both power plant and refinery pipe their excess heat to district heating systems and send other wastes to a gypsum wallboard plant. Sludge from the pharmaceutical plant is used as fertilizer on nearby farms.

Since 1993, more than twenty U.S. cities have initiated plans to develop similar industrial parks. In October 1997 the U.S. National Science Foundation and Lucent Technologies foundation jointly awarded eighteen grants, totaling U.S.$1.2 million, to support industrial ecology research across the country. The resulting projects have taken various forms. One of them, the Port of Cape Charles Sustainable Technologies Industrial Park in Cape Charles, Virginia, is self-contained like Kalundborg. A second demonstration project, in Fairfield, Maryland, not only has established an industrial park but also coordinates waste exchanges among 160 companies, most of which are beyond the park boundaries. Taking the idea further, a third demonstration project, in Brownsville, Texas, has no central site: instead it has targeted about thirty companies within a fifteen-mile radius that could profit from exchanges of byproducts.

The Industrial Ecosystem at Kalundborg, Denmark. (After Kneese, 1998, p. 11. With permission.)

**BIBLIOGRAPHY**

Dwortzan, M. "The Greening of Industrial Parks." *Technology Review* (January–February 1998), 18–19.

—DAVID J. CUFF

from kraft paper mills. The latter is normally burned for its fuel value and for chemical recovery, but the economics of doing this would be unfavorable if not for the need to satisfy environmental pollution regulations at the same time. This also applies to flue gas desulfurization (FGD) waste, which is recycled in Kalundborg, Denmark, but not in the United States, where disposal in ponds or impoundments is still cheaper than recycling.

Apart from this issue, it is also important to realize that if a firm makes a profit by waste reduction through input reduction (e.g., by internal recycling), this will simultaneously reduce the sales of an upstream supplier. One firm's economic gain, in this case, is another's loss. The game among competing firms may well be zero sum, not positive sum. A possibility for real economic gains at the societal level does remain, however. The more material products are reused, remanufactured, and recycled, the less these materials need be extracted from the environment. Because processing steps higher on the value-added chain generate less pollution than pri-

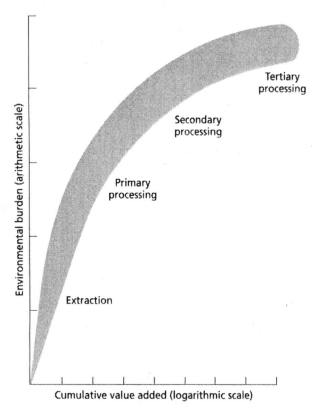

**Industrial Ecology. FIGURE 1.** Pollution and Wastes versus Value Added.

mary processing, this means a reduction in environmental pollution and resulting damage and social costs. The essential relationship is indicated in Figure 1.

A subtler but no less important economic implication is the following. The real wealth of society consists of a stock of natural capital plus a stock of tangible produced capital in the form of durable goods and structures. These forms of capital either generate services directly to consumers or perform an indirect but vital role in the economic production process. The latter, in turn, generates "final" goods and services for consumers. Durable consumer goods such as housing, automobiles, television sets, and books are in the first category, along with air, water, soil, forests, fisheries, landscapes, and biodiversity. Producer durables (e.g., machines, factory buildings, infrastructure) are in the second category. It follows that societal wealth can be increased by any strategy that (1) extends the life of existing durable goods or (2) reduces their rate of depreciation. [*See* Manufacturing.]

**Materials Processes in the Value-Added Chain.** As suggested already, the industrial ecosystem is not at present a closed cycle. It can be described schematically as a sequence of processing stages between extraction and ultimate disposal, with a number of actual or hypothetical intermediate loops that would permit the sys-

tem to be more nearly closed, in principle, with respect to mass flows (Figure 2). The IE system cannot, of course, be closed with respect to exergy flows. (Energy is conserved, but exergy, which is roughly equivalent to the "useful" component of energy, is used up.)

The extent to which mass-flow closure would be desirable from an environmental perspective depends on the nature of the materials being lost or discarded. It also depends very much on the ultimate source of energy; if hydrocarbon fuels are being consumed, the resulting pollution imposes a definite limit on the extent to which energy-intensive recycling would be justifiable. On the other hand, if solar energy can be used for recycling, the environmental cost-benefit calculation could be very different.

Some kinds of mass flows, such as soil displaced by erosion, dredging, road building, mining, or construction, are quantitatively large but qualitatively unimportant except in a few special cases. (In fact, "clean fill" often has economic value.)

Physical separation wastes from mineral concentration processes and agricultural and forestry activities constitute a second category that is slightly less benign but quantitatively smaller (Figure 2). These are intermediate in quantity (compared to the first category) and variable in harmfulness. Some, such as crop and animal wastes and wood ash, are actually beneficial, if properly recycled to the land. Others, such as flotation wastes from nonferrous metal and phosphate rock mining, are contaminated by chemicals, some of which (like cyanides) are extremely toxic. These must be impounded. [*See* Waste Management.]

The next stage of raw materials processing consists of chemical separation. Examples include metallic ore reduction (smelting) and alumina and phosphate rock processing. Smelting produces a low-grade solid waste, slag, which can be used for road paving. Nonferrous metals are usually found in nature as sulfides, so nonferrous ore processing also generates significant quantities of sulfur dioxide. In principle this can be economically recovered for sulfuric acid production; in the United States roughly 3.5 million metric tons of recovered sulfuric acid was obtained as a byproduct of 2.3 million metric tons of primary copper, lead, and zinc in 1993. In many parts of the world, however, sulfur dioxide is still simply emitted into the air without treatment because of the lack of a local market for the byproduct acid.

Other examples of wastes from the reduction-refining stage include FGD, sludge (from coal-burning power plants), phospho-gypsum (from phosphoric acid plants), and "red mud" from alumina producers. Generally, these are comparatively inert, chemically, and harmful—if at all—only in very large quantities, although extensive impoundments near processing facilities re-

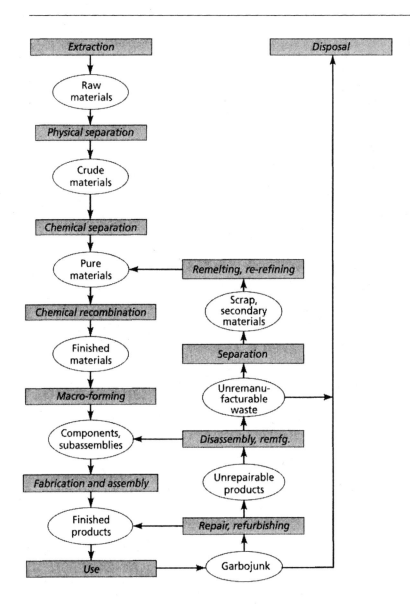

**Industrial Ecology.** FIGURE 2. Processing Stages between Extraction and Ultimate Disposal.

semble moonscapes where nothing will grow. At the extreme are toxic processing wastes, such as arsenic- or cadmium-containing sludges and particulate emissions from copper, lead, and zinc smelters; fluoride emissions from aluminum smelters; mercury-containing sludges from chlor-alkali plants; and contaminated heavy metal-containing ash from incinerators.

Wastes from shaping, forming, cutting, and assembly stages of production are much less dangerous in general. Minor localized pollution problems arise in connection with surface treatments: the use of acids, detergents, or chlorinated solvents for surface cleaning (as in the semiconductor industry), spray painting, electroplating, and so on. Most of these problems have been or can be solved by process modification, if not by adequate collection and treatment.

Another general category consists of wastes arising from the dissipative one-time use of intermediate or final materials that are not fully recoverable or recyclable,

even in principle. Examples include fuels, lubricants, additives, catalysts, explosives, surfactants (wetting agents), antifreeze, desiccants, detergents, flocculants, solvents (in some applications), acid and alkali neutralizing agents, foam blowing agents, fertilizers, pesticides, and many packaging materials. These can occur at any stage of the value-added chain. Most of them are, effectively, chemically changed by the process of use. Fuels and explosives are chemically converted to combustion products; neutralizing agents react with acids or alkalis; fertilizers are dissolved or volatilized, taken up by plants or microorganism, or immobilized by reaction with some environmental component. Biocides react with living organisms, by definition; most organic biocides are either degraded by photolysis or accumulated in the food chain. (The latter type, such as dichlorodiphenyltrichloroethane (DDT) and polychlorinated biphenyls (PCBs), are supposedly being phased out.) Toxic metals can best be immobilized by ionic bonding with clay and similar substances. Unfortunately, the ability of clay to bond with toxic metals is a function of environmental acidity (pH). In an acid environment, toxic elements that were previously immobilized may be mobilized. This phenomenon has been called "the toxic time bomb."

In some cases, toxic or carcinogen compounds are actually produced by chemical reactions with the environment at the point of use. The most obvious examples are nitrogen oxides, which are generated in "lean" combustion processes where excess oxygen reacts with nitrogen from the air and "burns" it. Photochemical smog is another example of such an atmospheric process. There is growing evidence that dioxins are also produced in fires of any sort where trace quantities of chlorine-containing compounds are available, especially in the presence of copper (which seems to catalyze the reaction).

Those that are merely dispersed but not changed, such as lubricants, solvents, surfactants, and blowing agents, can be recovered in principle, although doing so may require a major change in the production process (and more capital equipment). Analysis of tradeoff questions like these is very much at the core of IE. Ingenious schemes have been implemented in some cases to recycle materials that were formerly discarded. For instance, Dow-Europe has created a small joint venture to collect, recover, and redistill chlorinated solvents for the electronics industry. Dow has also initiated a venture to collect, redistill, and redistribute automotive antifreeze and brake fluids that would otherwise be burned or dumped into sewers.

More generally, IE concerns itself with mass flow and life-cycle analysis of industrial subsystems. It is helpful, but not sufficient, for purposes of identifying linkages and tradeoffs between waste flows and for prioritizing opportunities for intervention. The long-run goal of closing the materials cycle by finding and implementing practical uses for most wastes will be accomplished only very slowly. Indeed, there are a number of large-volume industrial and consumer wastes—carbon dioxide for one—for which there is no conceivable practical possibility of collection and use in significant quantities.

Nevertheless, there are many future opportunities for increased recycling, especially of metals. It is easy to verify that the ferrous metals sector is much bigger than the nonferrous sector in terms of the quantity of finished product. On the other hand, the comparison is reversed in terms of mining activity, waste overburden moved, tailings generated, and water and air pollution. Indeed, the process wastes from nonferrous metallurgy are in principle much more likely to be harmful. This would seem to be a powerful a priori argument for sharply increasing recycling in the nonferrous sector. Yet, ironically, it is in the ferrous metals sector that the process

is most advanced (except in the case of lead). [See Recycling.]

The future use of coal is also disturbing. Even now, coal mining disturbs huge land areas and generates enormous quantities of overburden. It also causes methane emissions, subsidence (in the case of underground mines), water pollution by acid mine drainage, and fire and explosion hazards for the miners. Coal combustion is a major source of carbon dioxide, sulfur dioxide, nitrogen oxides, and particulates (smoke). These are all health hazards. Coal washing, to reduce sulfur content, generates huge piles of sulfurous waste refuse near the mines. Sulfur not removed by washing must eventually be removed from flue gases of electric power-generating stations. This process, as currently implemented, also consumes large quantities of limestone and generates larger quantities of FGD waste and additional carbon dioxide.

During the last century and a half, coal consumption has been dramatically reduced as compared to what it would have been otherwise, thanks to the availability of better fuels, petroleum and natural gas. These are extraordinary natural resources that have been available in large quantities, at low cost, in nearly pure (or at least usable) form. Even the sulfur contained in natural gas is easily recovered and constitutes a valuable byproduct. Apart from emissions of small amounts of volatile organic compounds associated with petroleum refining and moderate quantities of saline water waste from drilling operations, the only pollutants associated with the extraction and processing of these resources are combustion products and some minor downstream pollutants from the petrochemical sector.

As this is written, however, there is increasingly convincing evidence that we are approaching the end of the age of cheap oil. Global production has been monotonically increasing for many decades, but the rate of discovery of new petroleum resources has been falling for a long time, and if the trend continues it will soon fall below the rate of consumption. When this happens, the point of peak production will have arrived. That point was predicted for the United States in 1957 by M. K. Hubbert, and his prediction of the peak date (1969–1979) was borne out in practice. U.S. production did peak in those years. Currently a number of petroleum geologists are making a similar prediction for global peak production. There are some data uncertainties, such as the immediate economic outlook for Asia, but the most probable range appears to be 2010–2015. [See Fossil Fuels.]

The peak year of global oil production implies much more than straightforward substitution of other forms of energy. Obviously the end will not occur suddenly. It would be more accurate to think in terms of the end of stable reserve capacity/production ratios and price stability followed by an accelerating decline in reserves and rise in prices (World Resources Institute, 1996). What will end, rather more suddenly, is the long period of declining global energy prices. Thus the "end of the age of oil" really means the beginning of rising fossil energy prices. With regard to pollution, admittedly, the immediate substitute for petroleum is likely to be natural gas, which is even cleaner than petroleum. However, this substitution would also hasten the end of the Age of Gas, which, on present estimates, could come as soon as two or three decades later.

Gas consumption can be increased for a while, but not without limit; much of current gas production is associated with oil wells. Nuclear power could not be brought onstream rapidly (a plant takes ten years to design and build) even were the public to accept it. If the short-term economic substitute for gas and oil is coal, as would seem likely based on current conditions and institutions, the period after 2015 also portends a vast increase in extraction and processing wastes.

**Economic Implications of the Industrial Ecology Perspective.** Many of the economic implications of IE have been touched upon, at least glancingly, in the previous sections. But perhaps it is worthwhile to reiterate some of them more explicitly. The major implication of both the IE and IM perspectives is that the economic system is embedded in a larger biogeochemical system by material and energy (exergy) flows. Thus, the economic system is constrained by physical laws, such as the laws of thermodynamics. The first law, conservation of mass and energy, is the basis of the mass balance principle, which has been cited repeatedly. The second law, often called the entropy law, states that all processes in isolated systems are irreversible.

The economic implications of the first law are fairly obvious. For instance, it is a consequence of the law of conservation of mass that the total quantity of materials extracted from the environment and processed through the economic system will ultimately return to the environment as some sort of waste residuals or "garbojunk." Of course, the mass balance principle is valid at every level of aggregation, from individual chemical processes to national accounts.

The relevance of the second law of thermodynamics to resource depletion and resource economics is equally obvious. From a thermodynamic perspective it is quite evident that the biogeochemical environment is a dissipative system, in the terminology of Ilya Prigogine, far from thermodynamic equilibrium. It is maintained in this state by a flux of exergy, that is, available energy, from the sun, some of which drives the hydrological cycle and some of which drives the carbon and nitrogen cycles. However, the deeper implications for resource scarcity are less clear.

The resource scarcity argument was thrashed out to some extent—and at a very abstract level—by neoclassical economists in the 1970s. They explored variations of abstract growth models in which resources (meaning exergy) were given some explicit role. The general conclusion was that resource scarcity would not limit economic growth in the long run, given continued capital investment and technological progress. In most of these models, it was assumed that human capital and natural capital, that is, resources, are inherently substitutable and interchangeable without limit. This assumption implies that resource inputs can be reduced to arbitrarily small levels by correspondingly large capital and labor inputs. It is evident that the real economic system today is a materials-processing system in which both intermediate commodities and final products are material in nature.

But, a perfectly acceptable neoclassical answer to this critique would seem to be that in the distant future the economic system need not produce significant amounts of material goods at all. In principle, it could produce final services from very long-lived capital goods, with very high information content, using nonscarce renewable sources of energy, such as sunlight. At the end of its useful life, a capital good in this hypothetical economy would be repaired, upgraded, and remanufactured, but rarely discarded entirely.

In short, it can be argued that there is no physical limit in principle to the economic output that can be obtained from a given resource input. Another way of saying the same thing is that there is no limit in principle to the degree of dematerialization that can be achieved in the very long run. This does not mean that no virgin materials need be processed at all. Nor does it imply that recovery, remanufacturing, and recycling can be 100 percent efficient. No such claim need be made. It is sufficient merely to claim that nobody can define a finite absolute minimum material input requirement to produce a unit of economic welfare.

It is clear that this formulation of the resource scarcity/resource recovery controversy is quite straightforward within the IE/IM framework. The same can be said for a number of the emerging tools of environmental analysis and assessment, including the notion of materials-process-product chains, chain management, life-cycle analysis, and substance flow analysis. Any of them could have developed without the framework, and some did, but these tools fit very naturally into it.

Actually, the introduction of exergy as a general quality measure for heterogeneous material flows, mentioned in the last section, has interesting potential economic implications. As noted in the previous section, by means of this measure it is possible to construct meaningful dimensionless measures of materials-processing efficiency and waste effluent loss per unit input, at the sectoral level. The first of these measures, over time, can be interpreted as a direct measure of technological progress for the sectors that extract, refine, and process materials into finished forms (e.g. steel, paper, concrete, plastic, rubber) and even material products.

This, in turn, suggests several new measures of productivity that could be constructed without great difficulty once the exergy calculations have been made and that might provide important new insights. One of them might be the ratio of value added per unit of exergy loss in processing (input less output), over time. Another might be the ratio of value to exergy content of the output, over time. A third might be the ratio of gross domestic product to exergy input to the whole economy, over time. One might even speculate that an improved theory of technical progress, and ultimately of economic growth, could eventually be built around these concepts.

All this having been said, one is still left with an unresolved problem with regard to the problem of measuring ecosustainability. Sustainability at the global biogeochemical system level has some aspects with which economics, the science of resource allocation among human activities *at the margin*, cannot begin to cope. At this level, economics must take a subordinate role to natural science.

[*See also* Greening of Industry.]

### BIBLIOGRAPHY

Ayres, R. U., and A. V. Kneese. "Production, Consumption, and Externalities." *American Economic Review* 59.3 (1969), 282–297.

Graedel, T., and B. R. Allenby. *Industrial Ecology.* Englewood Cliffs, N.J.: Prentice-Hall, 1995.

Hawken, P., A. Lovins, and L. H. Lovins. *Natural Capitalism: Creating the Next Industrial Revolution.* New York: Little, Brown, 1999.

Kneese, A. V. "Industrial Ecology and Getting the Prices Right." *Resources* 130 (1998), 10–13.

World Resources Institute (WRI). *World Resources, 1996–97.* New York: Oxford University Press, in collaboration with UNEP, UNDP, and the World Bank, 1996.

—R. U. AYRES

# INDUSTRIALIZATION

Industrialization is the second of the great transformations of the human economy. The first was the beginning of agriculture some two thousand years before, a change that led to the first permanent settlements. When industrialization began in Europe in the middle to late 1700s, it initiated an era of unprecedented use of energy and materials and a drastic change in the character and magnitude of human impact on resources and the environment in general.

Crucial to the beginnings of industry was the more effective use of nonanimate sources of energy. Wind and water power had been used for centuries, powering some mills and pumping operations at locations dictated by the energy flow; but when invention and refinement of the steam engine exploited the chemical energy in coal and wood, that allowed entrepreneurs to power mills and manufacturing operations at any locations convenient to the transport of the fuels.

As one industry spawned another related industry, the demand for raw materials and for coal increased rapidly: thus the mining of coal, the mining and processing of metals, the quarrying of rock, and the transportation of both raw materials and manufactured products all expanded and were accompanied by the disruption of land surface, pollution of air and water, and generation of solid waste that have been environmental hallmarks of the industrial age. In the twentieth century, petroleum fueled the vehicles that have transformed the landscapes of industrial nations, altered the dynamics of urban regions, and introduced new forms of air pollution. Through large-scale combustion of fuels, the industrial era coincides with serious urban air pollution, regional acid rain and ozone problems, and increased concentrations of carbon dioxide in the atmosphere worldwide. It is an era that has seen the proliferation of synthetic materials—plastics, herbicides, pesticides, fertilizers, solvents, and propellants—many of which have been accompanied by unforeseen environmental complications. [*See* Energy and Human Activity; *and* Fossil Fuels.]

At the same time, the beginnings of industry coincided with a scientific revolution that brought sweeping advances in medicine, agriculture, sanitation, transportation, and communication that collectively have improved the lives of millions of people. The medical and sanitation practices that first aided and bolstered European and colonial populations were exported after World War II to Asian, Latin American, and African nations, sharply reducing death rates and fostering rapid population growth. Currently, improved communication and education are altering the roles of women and reducing birth rates in some of those same nations.

Arguably, we now are entering a postindustrial age in which services, information management, and creativity will become the dominant mechanisms through which humans gain wealth and well-being. This *sustainability transition* will be spurred by a combination of changing consumption behavior, more environmentally and socially centered management objectives, and adjustments in market pricing and regulation to stimulate more frugal and environmentally clean technology.

This article summarizes four important trends in industrialization vis-à-vis environmental management:

- The sustainability revolution in the form of a more ecoefficient industry
- The shift to consumption as the main cause of environmental damage

- The new alliances between industry and the environmental movement on a global basis
- The dilemma of industrialization for developing countries

By way of providing an overview, Figure 1 summarizes the relationship between industrialization and the planetary life-support system.

**The Sustainability Revolution.** We are witnessing the early stages of this important transformation in industrial change. The process that we are observing is a reduction in industrial metabolism, or the total amount of materials, energy, and waste generated per unit of production. As the world's population continues to increase and consumption thereby continues to rise, the amount of environmental damage and social description associated with industrialization continues to grow, despite the trend toward a reduction in the extraction and waste of resources per unit of wealth created.

So we observe two powerful trends. One is the huge global effort to make the industrial process more environmentally and socially accountable. The other is the

continued increase of industrial activity throughout the world, with all its demands for water, wood, fertile soil, coastal margins, toxic chemical transformation, and widespread disruption of global ecosystems. To date the second is outstripping the first the world over. As the developing countries open up their economies to markets and freer trade, so this imbalance will continue to widen, despite heroic efforts to improve technological efficiency and upgrade management systems.

**Consumption and Environmental Damage.** The balance of this environmental onslaught is shifting from the production phase to the consumption and waste generation phases. Though the extractive industries (in metals, in timber, and in water) still take their toll, developments in international agreements and the opening up of ecolabeling and source inventories (i.e., the audit of what is being extracted from where) are creating the scope for more environmental accountability in the extraction sector. Already there is a forest products stewardship scheme designed to ensure that wood is taken only from forests managed to replace the timber at the rate at which is removed. Soon there is to be a

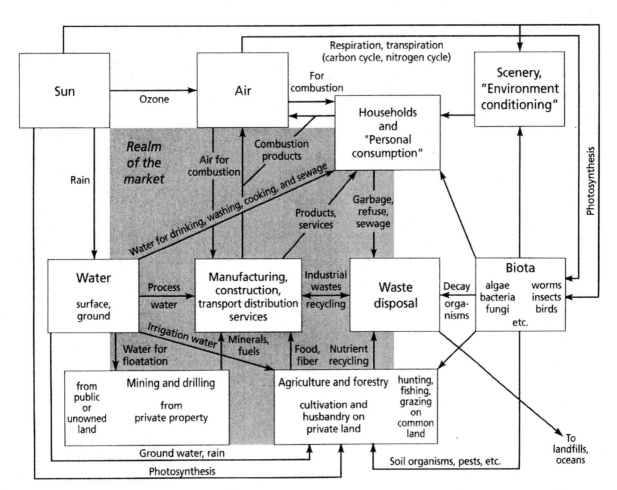

**Industrialization. FIGURE 1.** The Relationship between Ecology and Economics in the World of the Market.

(After Ayres and Simonis, 1996, p. 4. Copyright 1996 by the United Nations University. All rights reserved.)

marine stewardship scheme with the same purpose for the fish resource in particular, but also for corals and other marine life in a decade or so. These schemes are immensely important, for they are validated by global environmental organizations, carried on the Internet for ready reference, and provided with a logo that is placed on all products traded under the scheme.

However, the extraction stewardship approach remains confined to a narrow range of products and a few participating companies. So far the consumption element remains largely unregulated from the point of view of sustainable use. Yet goods are often associated with toxic materials that are released in use (e.g., chlorofluorocarbons) or in disposal (e.g., polyvinyl chloride products such as plastic bags and window frames). Similarly, many cosmetics contain traces of heavy metals or pharmaceutical products, or antibiotic substances that can accommodate in a host of pathways throughout global ecosystems. [See Chemical Industry.]

The contamination of the consumption end of the industrial cycle is proving a barrier to waste management. For example, sewage sludge nowadays contains so many heavy metals from household products and food wastes (via the body) that it cannot be placed on agricultural land as a fertilizer. Similarly, in closed water cycles—for example, on cruise liners—waste water cannot be cleansed of antibiotic and trace toxic chemicals associated with health regimes and cosmetics so as to be reusable on board. Throughout the world, local residents, because of the threat of toxic leakage, increasingly resist landfill sites, while incinerators are also opposed, despite advances in removal technology, because local residents fear the contamination of potentially toxic emissions in their neighborhoods. [See Waste Management.]

In sum, the industrialization process is a major transformer of the planet. Because it is intimately associated with global extraction, transportation, and waste disposal, and because it is highly dependent on energy and notably electricity, industrialization is a major cause of global climate change, acid rain, urban air pollution, and the removal of biological diversity. Despite the shift to the more eco-friendly knowledge economy, these transformational aspects of industrialization will be evident for generations to come.

**Industrial Metabolism.** The metabolism of industry is the integrated collection of physical processes that convert raw material and energy, plus labor, into finished products and wastes. This process is illustrated in Figure 1. The notion of metabolism applies also to the unit of industrial organization, namely, the firm. The economy is essentially the collectivity of firms together with regulatory institutions and workers who are also consumers (along with their families), using a common currency and a political structure.

In the process of manufacture, firms use energy and move materials around. Roughly 40 percent of all global energy consumption is associated with industrial activity, and about half the emissions of greenhouse gases is linked to industrialization. Table 1 shows the scale of the flux of key nutrients linked to industrialization. Table 2 provides an estimate of global emissions of trace metals, many of which are toxic and accumulative in biological pathways (such as the food chain and in the pattern of marine concentrations for dispersed entry). [See Metals.]

In an ideally self-sustaining world, all industrial products would be either reused or recycled. In practice, the laws of thermodynamics preclude this because of the energy required to reconvert used materials at a higher

**Industrialization. TABLE 1.** Anthropogenic Nutrient Fluxes (in teragram per year)

| | Carbon | | Nitrogen | | Sulfur | | Phosphorus | |
|---|---|---|---|---|---|---|---|---|
| | TG/YR | % | TG/YR | % | TG/YR | % | TG/YR | % |
| To atmosphere, total | 7,900 | 4 | 55.0 | 12.5 | 93 | 65.5 | | 12.5 |
| Fossil fuel combustion and smelting | 6,400 | | 45.0 | | 92 | | 1.5 | |
| Land clearing, deforestation | 1,500 | | 2.6 | | 1 | | | |
| Fertilizer volatilization* | | | 7.5 | | | | 1.5 | |
| To soil, total | | | 112.5 | 21 | 73.3 | 23.4 | 15 | 7.4 |
| Fertilization | | | 67.5 | | 4.0 | | 15 | |
| Waste disposal† | | | 5.0 | | 21.0 | | | |
| Anthropogenic acid deposition | | | 30.0 | | 48.3 | | | |
| Anthropogenic $NH_3$, $NH_4$ deposition | | | 10.0 | | | | | |
| To rivers and oceans, total | | | 72.5 | 25 | 52.5 | 21 | 5 | 10.3 |
| Anthropogenic acid deposition | | | 55.0 | | 22.5 | | | |
| Waste disposal | | | 17.5 | | 30.0 | | 5 | |

*Assuming 10 percent loss of synthetic ammonia-based fertilizers applied to land surface 75 teragrams per year (tg/yr).

†Total production less fertilizer use, allocated to landfill. The remainder is assumed to be disposed of via waterways.

SOURCE: Ayres and Simonis (1996, p. 10).

**Industrialization.** Table 2. Worldwide Atmospheric Emissions of Trace Metals (1,000 metric tons per year)

| ELEMENT | ENERGY PRODUCTION | SMELTING, REFINING, AND MANUFACTURING | MANUFACTURING PROCESSES | COMMERCIAL USES, WASTE INCINERATION, AND TRANSPORTATION | TOTAL ANTHROPOGENIC CONTRIBUTIONS | TOTAL CONTRIBUTION BY NATURAL ACTIVITIES |
|---|---|---|---|---|---|---|
| Antimony | 1.3 | 1.5 | — | 0.7 | 3.5 | 2.6 |
| Arsenic | 2.2 | 12.4 | 2.0 | 2.3 | 19.0 | 12.0 |
| Cadmium | 0.8 | 5.4 | 0.6 | 0.8 | 7.6 | 1.4 |
| Chromium | 12.7 | — | 17.0 | 0.8 | 31.0 | 43.0 |
| Copper | 8.0 | 23.6 | 2.0 | 1.6 | 35.0 | 6.1 |
| Lead | 12.7 | 49.1 | 15.7 | 254.9 | 332.0 | 28.0 |
| Manganese | 12.1 | 3.2 | 14.7 | 8.3 | 38.0 | 12.0 |
| Mercury | 2.3 | 0.1 | — | 1.2 | 3.6 | 317.0 |
| Nickel | 42.0 | 4.8 | 4.5 | 0.4 | 52.0 | 2.5 |
| Selenium | 3.9 | 2.3 | — | 0.1 | 6.3 | 3.0 |
| Thallium | 1.1 | — | 4.0 | — | 5.1 | 29.0 |
| Tin | 3.3 | 1.1 | — | 0.8 | 5.1 | 10.0 |
| Vanadium | 84.0 | 0.1 | 0.7 | 1.2 | 86.0 | 28.0 |
| Zinc | 16.8 | 72.5 | 33.4 | 9.2 | 132.0 | 45.0 |

SOURCE: Ayres and Simonis (1996, p. 11).

state of dissipation and disorder (i.e., high entropy) to a lower state of disorder (i.e., reconstitution of the product). Because energy, including transportation, is so environmentally unfriendly, the process of creating reusable products actually defeats the exercise. For example, virtually all sulfur mined is ultimately dissipated in use, in the form of fertilizers and pigments or in waste acids or gypsum (the residue of stripping sulfur dioxide out of coal-fired generating plant emissions). Table 3 summarizes key industrial products where the original materials move from a natural state of low entropy to a final, consumption-oriented state of high entropy.

The point here is that, up to a point, no amount of regulatory pressure or price incentives will guarantee sustainable reabsorption of certain industrial products. Even if a heroic effort is made, the full environmental burden analysis of the effort involved will result in uncompensated environmental and social costs. This is why it is all but impossible to visualize an industrial economy that is truly self-sustaining. In practice, therefore, various ways forward are offered.

***Product stewardship.*** Industrial products may be redesigned so that they explicitly avoid environmental damage. This action involves proactive investment in less environmentally burdensome sources and marketing strategies that involve the consumer in a sense of sustainability commitment. Examples include the removal of ozone-depleting chemicals from refrigerants and foam-filled goods (furniture and insulated materials) and the substitution by recyclable materials. For example, the McDonald's organization has phased out all

products involving foam-filled materials. On a global scale, this is a major commitment.

***Materials intensity analyses.*** The use of energy and materials may be assessed in terms of their environmental burden and their contribution to wealth creation. Analysts talk of the materials "rucksack," namely, the load of damage and disruption carried per unit of usable material. This kind of analysis shows more clearly the full costs associated with production and consumption, as well as providing a basis for reassembling the production-consumption relationship, for example, by designing recyclability into the production line. Already the top car manufacturers have created vehicles that are fully recyclable in their materials. The trick is to ensure that the costs of materials wastage are incorporated into the primary production of goods and services so that these "closed loop" production cycles are sufficiently profitable for technological investment.

***Social audits.*** The impact of production on the societies for which goods are created and removed is rarely given full weight. The reasons for this are fairly obvious. To begin with, businesses rarely feel a responsibility for the social context of their actions, though they may be regulated, or feel morally responsible, for the ecological impacts. Then, businesses like to believe that social matters of benefit payments, education, and health care are policy issues for communities or governments rather than the private sector. Third, the cultural impact of damage or disruption to a family or community is difficult to place into figures or equivalent compensatory measures. The loss of a way of life is not

**Industrialization. TABLE 3.** Examples of Dissipative Use (Global)

| SUBSTANCE | METRIC TONS ($\times 10^6$) | DISSIPATIVE USES |
| --- | --- | --- |
| Heavy metals | | |
| Copper sulfate ($CuSO_4$, $5H_2O$) | 0.10 | Fungicide, algaecide, wood preservative, catalyst |
| Sodium bichromate | 0.26 | Chromic acid (for planting), tanning, algaecide |
| Lead oxides | 0.24 | Pigment (glass) |
| Lithopone (ZuS) | 0.46 | Pigment |
| Zinc oxides | 0.42 | Pigment (tires) |
| Titanium oxides ($TiO_2$) | 1.90 | Pigment |
| TEL | ? | Gasoline additive |
| Arsenic | ? | Wood preservative, herbicide |
| Mercury | ? | Fungicide, catalyst |
| Other chemicals | | |
| Chlorine | 25.9 | Acid, bleach, water treatment, poly(vinyl)chloride (PVC) solvents, pesticides, refrigerants |
| Sulfur | 61.5 | Acid ($H_2SO_4$), bleach, chemicals, fertilizer, rubber |
| Ammonia | 93.6 | Fertilizers, detergents, chemicals |
| Phosphoric acid | 24.0 | Fertilizers, nitric acid, chemicals (nylon, acrylics) |
| NaOH | 35.8 | Bleach, soap, chemicals |
| $Na_2CO_3$ | 29.9 | Chemicals (glass) |

easily quantified or substituted. When firms try to provide any compensation—for example, as Shell has sought out to do for the Ogoni people of southern Nigeria—there is usually plenty of anger and mistrust because of the history of social abuse and dislocation.

One example is the attempt by retail firms to take into account the well-being of local communities supplying food or products such as flowers for the supermarket showrooms. In the past, the disruption caused by the change of food-producing practices—from, say, traditional organic to high-tech agrichemical—has often resulted in the displacement of the poor farmers and the substitution of capital-rich entrepreneurs. They in turn sometimes employ the former farmers as cheap labor for agricultural production that is exposed to toxic materials. The original farmers thus lose out on their livelihood and their health. Social audits are supposed to deal with all this but rarely do so for the reasons given. This is why the whole issue is controversial and extremely difficult to put into practice, even for a well-intentioned firm. The most promising solution is some sort of governmental–transgovernmental–private sector partnership of cooperative ventures in community training and support initiatives to ensure fairness of treatment and dignity and reallocation of jobs and land. Until consumers really press for this kind of approach, it will be long delayed. And until the market analysts try to promote the ethical basis of social audits in their evaluation of a firm's potential, again the likelihood of this more radical but just approach to business and sustainable development will remain an ideal.

***Trade deals and sustainability audits.*** The opening up of the free-trading era following the completion of the negotiations on international trading barriers and opportunities has placed the spotlight on the environmental and social repercussions of open, competitive trade markets. [*See* Trade and Environment.] The World Trade Organization has failed to provide adequate safeguards for environmental and social protection. There is nominally a filter of scrutiny on the basis of the soundness of production processes. But trading nations are suspicious that any country that uses an environmental or social safeguard as a basis for controlling trade is doing so for covert protectionist reasons. Despite many attempts to ensure greater safeguards for "fair trade"—namely, the guarantee of social and environmental justice in the extraction and manufacturing processes—this effort has hardly got off the ground. Nevertheless, both enlightened governments and nongovernmental organizations are trying hard to bring these safeguards into the trading area. So far, the forces of competitive advantage are winning. Again, the most likely forces of change are the consumers and the market analysts, though neither are as yet enlightened on the sustainability theme.

**Industrialization and the Developing World.** Before World War II, the pace of industrialization was extremely uneven. The reasons for this can be put down to:

- Inflexibility of political systems to the needs to mobilize labor and employment practices

- Inability to accumulate capital to ensure sufficient leverage for investment, notably in new technology and communications
- Lack of a disciplined industrial proletariat
- Inability to provide or create a market at home and abroad

Consequently, the global corporation emerged in the privileged and competitively advantageous existing industrial countries to colonize the developing world. This process took place systematically in the period 1950–1980, mostly through capital-accumulating corporations with the power to control prices of energy, goods, and labor. The result was a pattern of industrial hegemony that created huge amounts of both environmental damage and an industrial laboring class. The consequences are still with us today. Over 200 million children are forced out of school to work at low wages to make ends meet for their impoverished families. Many of these same families are manipulated and abused by their own social neighbors in the murky politics of community-level adjustment to poverty, disease, and carefully manipulated economic opportunity. Governments are forced to subsidize the price of energy and raw materials, even below nominal market rates, simply to keep a firm in place. The "true" cost of resource extraction, that is, taking into account the environmental and social disruption of present and future generations, is never remotely calculated in these highly subsidized arrangements. [See the vignette on Perverse Subsidies in the article on Environmental Economics.]

This is why this essay has concentrated on the changing role of company accountability by means of ethical accounting, ecological economics, and social audits. The major international corporations are at least sensitive to these matters nowadays. But the boardroom remains largely unconvinced, while the more junior managers, who are aware of the issues, fret in anguish over the practices that they feel forced to pursue. The way forward will come through a combination of factors, namely:

- The rise of internal regulations to ensure equality of treatment regarding environmental and social matters by firms operating worldwide
- The growth of national regulations in both the consumer and environmental arenas that begin to standardize company costs
- The coalescence of consumer and environmental nongovernmental organizations, first with enlightened industry and subsequently with governments (forced or shamed to act) to ensure product stewardship
- The entry of an environmentally concerned junior managerial class seeking high standards and prepared to join firms only with a good track record

- The interest of market analysts in ecoefficiency not only as a good management tool but also as a precursor to ethical investments more generally
- The rise of consumer power, buoyed by a better education, more precise media coverage, and a sense of concern for the future, and also assisted by product stewardship ecolabels
- A general trend toward ecoefficiency, created by a demand for more efficient materials production and the application of the knowledge economy

Industrialization is on the verge of a profound revolution. In principle it is possible to create and sustain wealth. In practice it will be a long time before all the pressures listed above have any demonstrable effect. And in any case, the result will no doubt be uneven and inequitable across the globe.

[See also Economic Levels; Greening of Industry; Human Health; Human Populations; Industrial Metabolism; Land Use, article on Land Use Planning; Resources; Sustainable Development; Technology; and Water Resources Management.]

## BIBLIOGRAPHY

Ayres, R. U. Turning Point: The End of Growth Paradigm. London: Earthscan, 1998.

Ayres, R. U., and U. E. Simonis, eds. Industrial Metabolism: Restructuring for Sustainable Development. New York: United Nations University Press, 1996.

Carley, M., and P. Spapens. Sharing the Earth: Sustainable Living and Global Equity in the Twenty-first Century. London: Earthscan, 1998.

Fussler, C., and P. James. Driving Eco Innovation. London: Pitman, 1996.

Kemp, T. Industrialization and the Nonwestern World. London: Longman, 1997.

Sachs, N., R. Loske, and M. Linz. Greening the North: A Post-Industrial Blueprint for Ecology and Equity. London: Zed Books, 1998.

Socolow, R., R. C. Andrews, F. Berkhout, and V. Thomas, eds. Industrial Ecology and Global Change. Cambridge: Cambridge University Press, 1996.

Von Weiszacker, E., A. Lovins, and H. Lovins. Factor 4. London: Earthscan, 1997.

—TIMOTHY O'RIORDAN

# INDUSTRIAL METABOLISM

Industrial metabolism (IM) is a metaphor for calling attention to the resemblance between the materials processing functions of an industry—defined broadly—and those of a biological organism or population of organisms. There are obvious industrial analogues of many metabolic functions involving the processing of materials, including ingestion of food (intake of raw materials), digestion (processing of materials), and excretion of wastes, as well as energy conversion, maintenance, and repair. This metaphor is applicable to individual factories, firms, industries, or even regions or political units, such as cities or nations.

There is also a closely related industrial analogue for the reproductive processes of the life cycle of an individual organism: conception (invention, design), birth (production), maturity (use), and death (disposal). This analogy has given rise to the terminology of life cycle analysis (LCA), of a product from "cradle to grave." However, it is also evident that the LCA metaphor is applicable to specific materials, products, or services, whereas the IM metaphor is more applicable to organizations, regions, or the Earth as a whole. The two metaphors are closely related and complementary but distinct.

A third related metaphor is industrial ecology (IE), which evokes the similarities between a natural ecosystem and an industrial complex, consisting of specialized firms that utilize each other's products and byproducts. LCA is an essential tool of both IM and IE. Both perspectives are holistic. This is an important distinction in viewpoint compared with the reductionist perspective of most scientific disciplines and government agencies.

Industrial metabolism is the systematic study of the interactions between human economic (that is, industrial) activity and the natural systems within which our economy is embedded and from which it obtains many important services. In particular, IM is the study of mass flows and transformations of materials, both within and across the boundaries of firms, industries, and regional units. It focuses especially on the linkages between energy and material flows within the economic system and the environment. These relationships can be explored at the plant, firm, sectoral, regional, or global level.

**Energy and Material Flows.** Major energy and material flows result from raw material extraction, refining, transformation, and consumption. One of the important insights of IM is that all materials extracted from the environment, of which fossil fuels are only the most obvious example, eventually become wastes and emissions. Of course, not all emissions are equally harmful. But it is not clear that waste treatment or abatement activities can compensate for the increased quantities of wastes and emissions that result from quantitative growth. A second key insight is that dissipative use of process intermediates, such as fuels, lubricants, solvents, neutralizing agents, catalysts, flotation agents, flocculants, wetting agents, detergents, fertilizers, and biocides, constitutes by far the most important source of waste pollution associated with economic activity. Consumption-related wastes are increasingly important compared with wastes resulting from conversion losses or inefficiencies, or the elimination and disposal of contaminants. A further insight is that environmentally

harmful waste flows associated with materials extraction processes may be extremely large in relation to the quantities that end up in commerce, as will be seen subsequently.

Fossil fuels constitute by far the largest flow of materials in commerce, with coal, petroleum, and gas roughly equal in mass terms (Table 1). However, coal involves by far the greatest mass flow. Coal mining is one of the most material-intensive human activities, with current global annual production of around 5 billion metric tons (nearly 1 billion metric tons in China alone). Coal mining contributes approximately 6 metric tons of overburden per metric ton of coal, based on U.S. data. Coal mining (and, in some countries, coal washing) also generates acid mine drainage wastes that pollute many nearby rivers. Coal combustion is the largest single global source of the greenhouse gas carbon dioxide, as well as sulfur dioxide, nitrogen oxides, and particulates. Coal mining is also a significant source of the second most important greenhouse gas, methane. Furthermore, coal ash—more than 10 percent of the mass of coal—constitutes a significant solid waste from electric power plants, as well as a major source of airborne particulates. [See Fossil Fuels.]

Important atmospheric emissions from fossil fuel extraction and consumption include carbon dioxide, carbon monoxide, sulfur dioxide, nitrogen oxides, methane (from coal mining, gas drilling, and gas distribution), unburned hydrocarbons, tropospheric ozone and lead (mainly from motor vehicles), small particulates, and fly ash (the latter mostly from coal burning). As much as 80 percent of all greenhouse gas emissions are associated with fossil fuel usage (the remainder are mostly related to agriculture). Environmental acidification, from sulfuric and nitric acids produced in the atmosphere by emissions of sulfur and nitrogen oxides, is also almost entirely associated with the use of fossil fuels. [See Acid Rain and Acid Deposition.]

Other resource extraction activities involve even greater material mobilization per unit of product, albeit less in aggregate terms. Gold and platinum mining are by far the most notable examples, each involving the displacement and processing of the order of 1 million metric tons of overburden and crude ore per ton of metal recovered. Gold mining is probably the second most material-intensive mining activity in the world, after coal. Next in order of concentration requirements are silver and uranium, although copper mining is third on the list in terms of total material mobilized globally. Up to 300 metric tons of ore and overburden must be moved and/or mined to yield 1 ton of copper. Iron and aluminum ore mining, by contrast, involve displacing only about 4 metric tons (or less) of crude ore per ton of metal. So, in both cases, less material is displaced, even though more metal is produced.

**Industrial Metabolism. TABLE 1.** Fossil Fuel Production and Consumption 1991/1992*

| | Hard Coal | | Brown Coal | | PETROLEUM | GAS |
|---|---|---|---|---|---|---|
| REGION | PRODUCTION | CONSUMPTION | PRODUCTION | CONSUMPTION | PRODUCTION | PRODUCTION |
| World | 3,790 | 1,138 | 905 | 66 | 3,215 | 1,753 |
| Non-OECD Total | 2,344 | 765 | 278 | 38 | 2,331 | 957 |
| Africa (except Middle East) | 226 | 26 | | | 339 | 81 |
| Latin America | 43 | 8 | | | 314 | 73 |
| Asia—Other | 393 | 125 | 54 | 17 | 166 | 146 |
| Asia—China | 1,373 | 548 | | | 161 | 19 |
| Asia—Former USSR) | 305 | 55 | 104 | 13 | 358 | 496 |
| Middle East (Asia and Africa) | 1 | 1 | | | 989 | 129 |
| Non-OECD Europe | 2 | 1 | 120 | 8 | 9 | 13 |
| OECD Total | 1,446 | 137 | 627 | 28 | 885 | 796 |
| OECD Europe | 274 | 71 | 444 | 22 | 304 | 211 |
| OECD Pacific | 219 | 33 | 58 | | 23 | 30 |
| OECD North America | 954 | 32 | 124 | 6 | 558 | 556 |
| United States | 910 | 30 | 78 | 5 | 318 | 404 |
| EU | 118 | 27 | 251 | 2 | 146 | 167 |

*Units are in million metric tons.

SOURCE: International Energy Agency (1999a and b).

Needless to say, there are enormous waste flows from some of the mining and concentration processes, not to mention significant wastes associated with smelting and refining. Copper concentration involves grinding and flotation of copper minerals from the gangue, or dross (what remains after the valuable minerals have been removed), most of which end up in open ponds. Phosphate rock—for fertilizer—is another large-scale mining activity, involving very large quantities of waste phosphogypsum that is contaminated with fluorides and radioactive metals (notably, uranium and thorium). This waste is normally impounded behind dikes or dumped at sea. Aluminum ore (bauxite) processing generates a corrosive alkaline waste called *red mud*, which is also left in impoundments. [*See* Mining.]

**Regional Aspects.** It is informative to compare materials and energy productivities at the sectoral level between countries at different stages of economic development. These flows arise from industrial concentrations that tend to be regionally specialized. Thus heavy industry, based on coal and steel, developed historically in South Wales, the Ruhr–Rhine region of Germany, the Saar–Meuse–Moselle region of Germany, Belgium, Luxembourg, France, and the Monongahela–Ohio valley of western Pennsylvania. These industries have gradually migrated to other coal-rich regions such as the Great Lakes region of the United States, Bohemia–Silesia, the Dnieper–Donets basin of the Ukraine and Russia and, more recently, to Australia, Japan, Korea, and China. Petroleum and natural gas played a major role in the industrial development of the southern and western United States (Texas, Louisiana, and California), Azerbaijan and the other countries bordering the Caspian Sea, and the entrepôt cities of Hamburg, Rotterdam, Antwerp, Le Havre, and Marseilles. These industries are now expanding mainly in hydrocarbon-rich places such as Saudi Arabia, the Persian Gulf emirates, the western Urals of Russia, and Indonesia.

Pulp and paper production is generally collocated with forest resources, especially in North America and Scandinavia. Forest products industries are now expanding in Indonesia, Malaysia, and Brazil. Economic geography suggests that humid tropical regions will increasingly dominate the world's supply of forest products, including paper and lumber. The dry desert regions, on the other hand, could become major sources of exportable electric energy, either directly or as products of electrolysis (especially hydrogen), as photovoltaic electric generating technology becomes competitive with conventional fossil fuels in another decade or two.

In the future, material-intensive processing industries will tend to move from consuming countries, such as Europe and the other advanced industrial countries, into the resource-rich developing regions. These trends are the natural consequence of value-added economic development strategies, which dictate a shift from raw material exports to intermediate material exports. Based on considerations of economic geography alone, the most likely pattern of future global industrialization would see much of the basic petrochemical industry moving to the Persian Gulf and North Africa, where most of the world's gas and oil reserves are located. Ammonia, urea, nitrate fertilizers, and methanol-based chemicals will also be produced near sources of cheap gas, rather than near the consumers. Similar considerations suggest that primary aluminum operations will be concentrated in Canada, Australia, Brazil, Russia, and Norway, to exploit cheap hydroelectric power. Copper and other nonferrous metal smelting and refining will increasingly be concentrated in southern Africa, western South America, and parts of Russia. Phosphate fertilizers should increasingly be produced in North Africa from locally mined phosphate rock, rather than being produced in Europe from imported concentrate. Russia and the United States will continue to be major suppliers for a long time, however. Economic development also implies increasing local and regional demand for capital goods and building materials, especially for infrastructure. The latter implies an increasing emphasis, especially in China, on "heavy" industry.

By the same logic, the integrated steel mills still operating in the Midwestern United States and in the Ruhr area of Germany are likely to be closed down gradually in coming decades. By 2020 or so, ferrous metallurgy in North America and western Europe will likely be almost exclusively based on recycling scrap through "mini-mills." The large iron and steel complexes in Japan and Korea will probably also be phased out a decade or two later. Western Europe and the United States would also increasingly depend on secondary recovery of aluminum. Similarly, the heavy concentration of petrochemical operations in Texas–Louisiana and western Europe are likely to be phased out gradually as the oil and gas resources of these regions are exhausted in coming decades.

The shift of industry from region to region implies a corresponding shift of waste generation and disposal activities. Countries in which mining activities are located not only suffer the consequences of land despoliation at and near the mine, but also the concentration and smelting or refining wastes. All of these can, in turn, pollute both air and local streams and ground water. Air pollution consists mainly of sulfur dioxide (from nonferrous metal operations), fluoride emissions from aluminum smelters, and mercury vapor emissions from small-scale gold recovery operations based on the old amalgam process. Particulates from smelters carry heavy metals such as arsenic, cadmium, lead, mercury, and zinc. Water pollution consists mainly of soluble leachates from waste heaps and impoundments. These include cyanide (from gold and silver heap leaching) as well as sulfuric acid from copper ore heap leaching, not to mention toxic heavy metals. All of these can be damaging to local agriculture, as well as to human health.

**Global Aspects.** The subtle effects of uncontrolled material flows on natural systems and cycles may be equally or even more devastating than localized pollution in the long run. Exponential economic growth, following traditional materials- (and energy-) intensive patterns, is utilizing exponentially increasing amounts of raw materials and fossil fuels.

From the biochemical perspective, the balance of nature is a remarkable evolutionary accident. The Earth is not in thermodynamic equilibrium. On the contrary, it is a system very far from equilibrium, maintained in a relatively stable state by a continuous flow of energy from the Sun. Solar energy drives the hydrologic cycle that waters the land and warms the polar regions; it drives the trade winds, the storms, the jet stream, and the ocean currents. The rains and surface runoff erode and weather the rocks and carry soluble salts into the ocean. Life began on Earth billions of years ago, as a result of some accidental confluence of natural phenomena that we still cannot fully comprehend. But it is quite clear that the existing patterns of life on Earth today are only possible because of climatic and chemical conditions created and maintained by life itself.

To mention only the most obvious example, our oxygen atmosphere, which is essential to all animal life, would not exist without prior eons of photosynthetic activity by plants, and could not continue to exist if that activity were to cease. If life on Earth were destroyed by some cosmic event, the oxygen in the atmosphere would gradually react with nitrogen (to form nitric acid) or with carbon (to form carbon dioxide). The ocean would soon be too acidic to sustain life, and the greenhouse effect of the carbon dioxide and water vapor buildup in the atmosphere would heat the surface of the Earth to a temperature well above the boiling point of water. In short, if life on Earth were destroyed, it could not re-create itself.

Apart from climatic stabilization, living systems perform other essential services. Among them is the cycling of nutrients. A nutrient is an element essential to life that would not be chemically available to living organisms (in significant quantities) if it were not concentrated and recycled by living systems. Carbon, oxygen, nitrogen, and perhaps sulfur are the major examples. The carbon–oxygen cycle is maintained by a balance between plant photosynthesis and animal respiration. Plants capture carbon from carbon dioxide to form carbohydrates and release oxygen in the process. Animals consume oxygen to burn carbohydrates for metabolic purposes. The two processes are roughly balanced in undisturbed nature. [See Biogeochemical Cycles.]

But human industrial activity is tilting the balance by burning large amounts of formerly buried fossil carbon and converting it to carbon dioxide. This gas gradually builds up in the atmosphere, where it absorbs and rera-

**ECOLOGICAL FOOTPRINTS**

An ecological footprint is an accounting tool for ecological resources in which various categories of human consumption are translated into areas of productive land required to provide resources and assimilate waste products. It is thus a measure of how sustainable the lifestyles of different population groups are. The great differences in the ecological footprints of different nations is brought out in the table, which shows that the fooprint of the average American is more than twelve times that of the average Indian.

**Comparing People's Average Consumption in the United States, Canada, India, and the World**

| CONSUMPTION PER PERSON IN 1991 | CANADA | UNITED STATES | INDIA | WORLD |
|---|---|---|---|---|
| $CO_2$ emission (metric tons per year) | 15.2 | 19.5 | 0.81 | 4.2 |
| Purchasing power ($U.S.) | 19,320 | 22,130 | 1,150 | 3,800 |
| Vehicles per 100 persons | 46 | 57 | 0.2 | 10 |
| Paper consumption (kilograms per year) | 247 | 317 | 2 | 44 |
| Fossil energy use (gigajoules per year) | 250 (234) | 287 | 5 | 56 |
| Freshwater withdrawal (in cubic meters per year) | 1,688 | 1,868 | 612 | 644 |
| **Ecological Footprint (hectares per person)** | 4.3 | 5.1 | 0.4 | 1.8 |

BIBLIOGRAPHY

Wackernagel, M., and W. Rees. *Our Ecological Footprint. Reducing Human Impact on the Earth.* Gabriola Island, British Columbia: New Society Publishers, 1995.

—ANDREW S. GOUDIE

diates thermal radiation from the Earth, thus triggering the greenhouse climate-warming effect. Up to a point this buildup may be self-limiting through its stimulation of faster plant growth—known as the carbon dioxide fertilization effect—which increases the rate of carbon dioxide removal. But climate warming could become self-sustaining, rather than self-limiting, if it results in increased evapotranspiration of water vapor and/or increased emissions of methane from anaerobic decay processes (for example, in arctic tundra). Both water vapor and methane are also greenhouse gases.

The nitrogen cycle is more complex but equally important. This is because nitrogen is a component of all amino acids, which are the building blocks of proteins. Nitrogen from the atmosphere is fixed by certain bacteria and fungi, which make it available to plants via their roots. But organic nitrogen does not accumulate indefinitely, thanks to other denitrifying bacteria that return nitrogen to the atmosphere. Again, the nitrification and denitrification processes are roughly balanced in undisturbed nature. But human industrial activity has already altered the nitrogen cycle significantly by fixing large amounts of nitrogen (as synthetic ammonia) for plant fertilizers and also by generating nitrogen oxides in combustion processes. As in the carbon case, this buildup may be self-limiting up to a point, since atmospheric nitrogen oxides eventually wash out in rainfall as nitric acid or nitrates. These are partly deposited on land, where they may have a fertilizing effect (for example, on forests) resulting in an increase in the amount of organic nitrogen that is in circulation, and probably an increase in the local rate of denitrification. However, other less desirable consequences are also possible and may already be occurring. [*See* Carbon Cycle; *and* Nitrogen Cycle.]

**Trends and Indicators.** A question of some importance is whether the world economy is evolving toward, or away from, environmental compatibility. If overall mobilization of materials is the indicator, the answer must obviously be negative. The more important question is whether aggregate mass displacement is equivalent to, or proportional to, environmental damage or "ecotoxicity." There are some influential arguments in favor of this proposition, mainly based on the rather negative point that a more complex indicator would inevitably lose transparency while not gaining significantly in credibility. Indeed, it must be acknowledged that various composite indicators have been proposed, but they vary widely from each other.

It is also true that the damage attributable to some mass flows, per unit mass, is negligible compared with others. For instance, the damage due to damming and thus diverting a river is likely to be extremely small per ton of water displaced compared with the damage done by releasing a ton of cadmium or arsenic into the environment. Clearly, some sort of damage-weighting scheme is needed. But, equally clearly, no such weighting scheme has yet been widely accepted.

Nevertheless, it is also fairly clear that environmental problems are getting worse, by almost any test. These problems are due partly to human-induced land use changes and partly to uncontrolled chemical reactions or physical changes triggered by materials released into the environment, either in normal usage by industrial and economic activity (including agriculture) or as process wastes. The formation of photochemical smog in the atmosphere is one example. Stratospheric ozone depletion is another. Climate warming is a third. Acidification of rain and soils is a fourth. An indirect environmental impact of acidification is the buildup of toxic elements, such as heavy metals, in soils and ground water. In undisturbed nature, the heavy metals are tightly bound to clay particles. But many metals are mobilized by acidification of the environment.

The only way to reduce the severity of these problems is to reduce the materials intensity of the economic system. The way to reduce waste flows and emissions is to reduce extraction. This, in turn, implies a sharp future increase in the productivity of materials. In ordinary language, this means much more reuse, renovation, and remanufacturing of durable goods. It also implies reducing the dissipative (namely, wasteful) use of process intermediates and consumables, especially fossil fuels, but also solvents, lubricants, cleaning agents, fertilizers, pesticides, and so forth. These profound changes imply a major economic transformation in the direction of more and more sophisticated services and away from the current emphasis on consumption of goods.

[*See also* Air Quality; Energy; Environmental Economics; Industrialization; Mass Consumption; Pollution; Recycling; *and* Technology.]

### BIBLIOGRAPHY

International Energy Agency. *Energy Balances of OECD Countries 1960–1997*. Paris: International Energy Agency, Organisation for Economic Co-operation and Development, 1999a.
———. *Energy Statistics of OECD Countries 1971–1997*. Paris: International Energy Agency, Organisation for Economic Co-operation and Development, 1999b.

—R. U. AYRES

**INDUSTRIAL REVOLUTION.** *See* Human Impacts, *article on* Human Impacts on Earth; Human Populations; *and* Industrialization.

**INFECTIOUS DISEASE.** *See* Disease; *and* World Health Organization.

## INFORMATION MANAGEMENT

Earth, space, and environmental scientists have become increasingly interested in issues related to climate change, decreases in plant and animal biodiversity, significant changes in land use, and human dimensions of global change. To address these issues, scientists are increasingly working at very broad spatial and temporal scales. Many attempts to scale research to the region, continent, and globe require unprecedented collaboration among scientists, data sharing across international borders, and ready access to high-quality, well-documented data that have been preserved in data archives.

**Sources of Data and Information.** The primary source for information about data relevant to global change is the Committee on Earth Observation Satellites (CEOS) International Directory Network (IDN), which represents an international effort to assist researchers in locating information on available data sets. The CEOS IDN is supported by CEOS as a service to the Earth science and space science communities. The CEOS IDN provides free online access via the World Wide Web (WWW; see Internet Resources at the end of this article) to information on worldwide scientific data, including Earth sciences (e.g., geoscience, hydrospheric science, biospheric science, satellite remote sensing, atmospheric sciences), space physics, solar physics, planetary science, and astronomy/astrophysics. The CEOS IDN describes data held by universities, government agencies, and other organizations.

There are four coordinating nodes of the CEOS IDN for the Asian, American, European, and African continents. Three of these maintain a copy of the Global Change Master Directory (GCMD; see Table 1), which is a comprehensive directory of descriptions of data sets relevant to global change research. The GCMD is maintained by the American node of CEOS IDN, which is operated by the U.S. National Aeronautics and Space Administration's (NASA) Global Change Data Center at the NASA/Goddard Space Flight Center. The European coordinating node is operated by the European Space Agency at the Earthnet Programme Office in Frascati,

**Information Management.** TABLE 1. Online Sources for Data Relevant to Global Change

| DATA CENTER | DATA CENTER UNIFORM RESOURCE LOCATOR (URL) | DATA CENTER FOCUS |
|---|---|---|
| Alaska SAR Facility—DAAC | http://www.asf.alaska.edu | Synthetic aperture radar data |
| The Center for International Earth Science Information Network—Socioeconomic Data and Applications Center | http://www.ciesin.org | Integrated social and natural science data |
| EROS Data Center—DAAC | http://edcwww.cr.usgs.gov/landdaac/landdaac.html | Land processes and characteristics data |
| Global Change Master Directory | http://gcmd.gsfc.nasa.gov | Comprehensive directory of descriptions of data sets relevant to global change research |
| Goddard Space Flight Center—DAAC | http://daac.gsfc.nasa.gov | Upper atmosphere, atmospheric dynamics, and the global biosphere data |
| Jet Propulsion Lab—DAAC | http://podaac.www.jpl.nasa.gov | Data about the physical state of the oceans |
| Langley Research Center—DAAC | http://eosweb.larc.nasa.gov | Radiation budget, clouds, aerosols, and tropospheric chemistry |
| Marshall Space Flight Center—DAAC | http://ghrc.msfc.nasa.gov | Global hydrology and climate data |
| National Oceanic and Atmospheric Administration—Satellite Active Archive | http://www.saa.noaa.gov/ | Real-time and historical satellite data (NOAA's Polar-Orbiting Operational Environmental Satellites) |
| National Snow and Ice Data Center—DAAC | http://eosims.colorado.edu | Snow and ice data |
| Oak Ridge National Lab—DAAC | http://www-eosdis.ornl.gov | Biogeochemical dynamics (i.e., biological, geologic, and chemical components of the Earth's environment) data |
| World Data Centers | http://www.wdc.rl.ac.uk/wdcmain | Collection of five international data centers that archive a variety of geologic, atmospheric, and ocean processes data |
| World Health Organization | http://www.who.int/whosis/ | Data related to infectious diseases, consolidated on a country-by-country basis |

Italy. The Asian coordinating node is operated by the National Space Development Agency in Saitama, Japan. The African coordinating node is operated by the United Nations Environment Programme.

Each of the four coordinating nodes has cooperating affiliated directories or cooperating nodes that provide a path for researchers within a country or region to participate in the CEOS IDN. Cooperating nodes may support directories specializing in a specific subject or may maintain the complete GCMD database. Automated transfer of new or revised entries takes place among the coordinating nodes on a monthly schedule. This information can optionally be received by the cooperating nodes. These procedures assure that data set descriptions and supplementary information (collectively referred to as metadata) obtained from various parts of the world are exchanged with other areas, thereby expanding the base of information available to researchers worldwide. The GCMD employs a user-friendly search interface that allows individuals to search easily for particular types of data. Resulting metadata records provide information on the nature of the data (e.g., param-

eters measured, geographic location) and where the data are stored. It is also possible to register data easily in the GCMD via a Web-based registration form.

The actual data and metadata related to aspects of global change are stored by universities, government agencies, and other organizations in locations that are variously referred to as data repositories, digital libraries, data clearinghouses, data centers, or data archives. Various online sources of data are presented in Table 1. Examples of particularly comprehensive data sources include the eight distributed active archive centers (DAACs) that are funded as part of the NASA Earth Observing System (EOS) program.

**Types of Data.** Many different types of data are of potential interest to the global change research community. For instance, the GCMD database includes descriptions of data sets covering climate change, the biosphere (e.g., land use and forest cover), hydrosphere and oceans (e.g., water quality, sea surface temperature), geology (e.g., soils), geography, and human dimensions of global change (e.g., disease outbreaks, resource inventories; Figure 1).

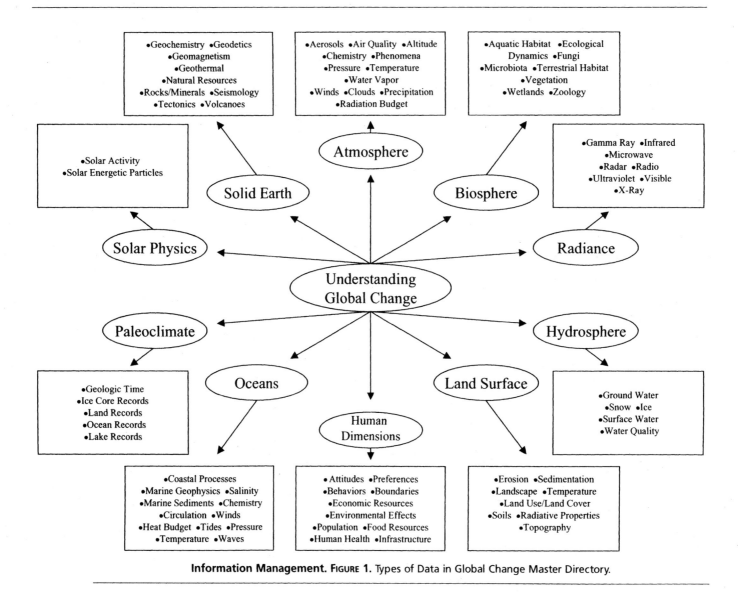

**Information Management.** Figure 1. Types of Data in Global Change Master Directory.

Available global change data cover many spatial, thematic, and temporal scales of resolution. Two relevant examples related to soils and land cover data are included in Figure 2. Both examples indicate some of the challenges and potential associated with broad-scale data. For instance, in many cases the data that are available at the global scale will be limited in terms of the number of parameters that are included. The Global Soils Data Set and the Global Vegetation Data Set contain twenty-six major soil units and thirty-two vegetation types, respectively. Both data sets were compiled at 1° latitude by 1° longitude and are currently stored at the Goddard Institute for Space Studies. At the midrange scale, U.S. Geological Survey Land Use and Land Cover Data (LULC) for the conterminous United States and Hawaii are classified into nine general categories (e.g., urban, agricultural, rangeland, forest), each of which is subdivided into detailed subclasses. The State Soil Ge-

ographic (STATSGO) Database (1:250,000) is based on generalization of the Soil Survey Geographic (SSURGO) Database, which is the most detailed (1:12,000 to 1:63,360) level of mapping by the Natural Resources Conservation Service. The State of Georgia land cover database is based on a system of U.S. land-surface observation satellites (Landsat) Thematic Mapper (TM) data (30 meters × 30 meters) that were obtained in 1988–1990. These data are grouped into fifteen major classes and have a much higher spatial resolution than data in the nationwide LULC database. In addition to scale, it should be emphasized that the amount of field validation and the timeliness (the degree to which the data represent recent conditions) will vary significantly among the various data sets.

Spatial resolution, the size of an area that can be discriminated, will often be limited in data sets that include many parameters and cover a broad spatial extent. Data

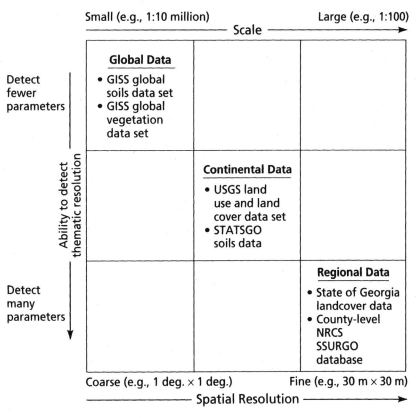

**Information Management. FIGURE 2.** Spatial Data Sets.

relevant to human dimensions of global change often fall in this category. For example, data related to infectious diseases are consolidated on a country-by-country basis by the World Health Organization (see Table 1) and included in the "Global Health for All Database." In addition to mortality and morbidity rates by disease category, the database includes a broad range of parameters related to economic, demographic, and social trends, food and water supply, and human and financial resources for health. Compilation of these countrywide data and subsequent analyses often represent a considerable investment in time and resources.

In contrast to the previous example, other types of data may undergo little or no postprocessing or analysis and may be available in raw form. For example, data collected by automated weather stations, ocean buoys, and other continuously recording sensors may be accessible in "real time" via direct Internet connections or following frequent uploading to a Web site. Very large volumes of data can be collected at buoys and weather stations. These data may be extremely useful for documenting conditions at a limited number of sites, but may require considerable processing and analysis effort when expanding to a regional, continental, or global scale. Clustering of the stations or buoys near the shore or near cities because of ease of maintenance and other

factors must also often be figured into subsequent analyses and interpretation.

***Spatial data.*** Although all data are spatial in the sense that they are collected at one or more points in space, scientists often consider spatial data to be those that explicitly contain a spatial component. Thus, maps, satellite images, aerial photos, and related data that can be geographically related to coordinates on the Earth's surface represent good examples of spatial data. However, data that are collected at a network of sites (e.g., climate stations throughout a country) also constitute a viable spatial data set according to this definition.

Data collected by sensors associated with a variety of satellite platforms constitute the largest proportion of data that are available for global change studies. Examples of some of the satellites that may be relevant to global change studies include:

1. Landsat Multispectral Scanner (MSS), a U.S. satellite launched in 1972 with four spectral bands that sense the visible and near infrared every sixteen days. Spatial characteristics include an 80-meter instantaneous field of view (IFOV) and a 185-kilometer scene width.
2. U.S. National Oceanic and Atmospheric Administration (NOAA) Advanced Very High Resolution Radiometer (AVHRR), a U.S. satellite launched in 1978

with five spectral bands that sense the visible, near infrared, and thermal infrared every twelve hours. Spatial characteristics include a 1–4-kilometer IFOV and a 2,400-kilometer scene width.

3. Systeme Pour l'Observation de la Terre (SPOT) High-Resolution Visible (HRV) Multispectral (XS), a French satellite launched in 1986 with three spectral bands that sense the visible and near infrared every twenty-six days. Spatial characteristics include a 20-meter IFOV and a 117-kilometer scene width.

4. Japanese Earth Resource Satellite-1 (JERS-1) Synthetic Aperture Radar (SAR), a Japanese satellite launched in 1992 with one spectral band that senses radar every forty-four days. Spatial characteristics include an 18-meter IFOV and a 75-kilometer scene width.

Spectral data collected from satellite platforms must generally undergo one or more postprocessing steps so that the data are accurately georeferenced (related to Earth coordinates) and corrected for atmospheric and other types of distortion. Generally, additional analyses are necessary to relate the "corrected" spectral data to parameters of interest such as land cover, biomass, or vegetation vigor. For example, multispectral data from Landsat TM, SPOT, and AVHRR are often transformed into a Normalized Difference Vegetation Index or related index that is closely related to vegetation biomass or vigor. In addition, spatial data sets covering a large areal extent must frequently be condensed to facilitate subsequent processing and visualization. New satellite data acquisition programs will soon add more than a terabyte of data each day to national and commercial data repositories, providing a valuable resource for global change research.

Historically, it has been difficult to relate data collected at point locations to coordinates, such as latitude and longitude, on the Earth's surface. This process, known as *georeferencing*, has required expensive surveying equipment, expertise, and ready access to benchmarks (points with known geographic coordinates). Consequently, it has frequently been difficult to relate data collected from aerial and satellite platforms to spectral values, vegetation type or condition, or other environmental factors that are recorded on the ground surface, an activity known as field or ground validation. Another satellite program, the Navstar Global Positioning System (GPS), represents a relatively new solution to this problem, greatly facilitating ground validation efforts. Navstar consists of twenty-four satellites that are continuously orbiting the Earth in very precise orbits. GPS receivers triangulate their position on the ground using radio signals (timing and satellite position information) that are sent from each of the satellites. Three satellites are required to calculate a position in two-dimensional space, whereas at least four satellites are required when accurate elevation (i.e., three-dimensional space) is desired. [*See* Remote Sensing.]

Spatial data can also be generated from interpretation of aerial photographs, on-the-ground surveys using GPS or conventional surveying equipment, spatial interpolation of data collected at multiple discrete points, and other sources. For example, many graphical and data analysis programs offer algorithms that can contour data collected at multiple points. Regardless of the source of spatial data, the types of questions being addressed in global change studies typically require management and analysis of large volumes of different types of data that represent many different parameters of interest. Analysis of these large, complex spatial databases frequently necessitates the use of specialized technology.

*Geographic information systems.* One type of specialized tool that can be used to increase our understanding of global change is the Geographic Information System (GIS). This is a technology that is designed to capture, store, manipulate, analyze, and visualize the diverse set of georeferenced data that are required to support accurate modeling of the Earth's environmental processes. A GIS consists of computer hardware (e.g., computer, digitizer, plotter), computer software, data, and personnel. During the past two decades, GIS technology has moved from large mainframe computers to small desktop machines with increased power and decreased costs. Furthermore, GIS can now interface with many related technologies including computer-assisted design (CAD), remote sensing, database management systems (DBMSs), video, and terrain modeling systems.

Several characteristics distinguish GIS as a powerful tool for global change research and other applications, including the ability to easily update or modify existing geospatial databases, the rapid synthesis of information contained in multiple data layers, and a range of analytic algorithms for supporting complex spatial analyses. Analytic operations routinely performed in GIS-based applications include renaming and reclassification of mapped parameters, geometric operations (e.g., rotation, translation, scaling, rectification, registration), Boolean operations ("and," "or," "not," etc.), spatial overlay of one data layer onto another, spatial coincidence, proximity, measurements (e.g., length, area, perimeter, fractal dimension), and statistical operations. GIS is being linked increasingly with simulation models to examine broad-scale questions related to primary productivity, sea level rise, nonpoint source pollution, and global climate change. [*See* Geographic Information Systems.]

**Data and Information Management Procedures.** High-quality, well-documented, securely preserved, and accessible data are essential for addressing long-term and broad-scale environmental issues associated with global change. Access to high-quality data requires a strong commitment to the implementation of effective information management procedures. The absence of such procedures impairs our ability to use data over long periods of time. For instance, the loss of information content associated with data through the degradation of the raw data or the metadata is unavoidable and has been referred to as "information entropy" (Michener et al., 1997).

Many processes can lead to information entropy. Processes that operate continuously, such as the gradual deterioration of storage media, as well as discrete events, such as retirement or death of the scientist(s) and technician(s) involved in data collection and loss of storage media through catastrophic events, lead to information entropy. Adherence to recommended data management practices, especially the development of comprehensive metadata and the submission of both data and metadata to data archives, greatly slows the progression of information entropy.

Data and information management represents a process that often starts with project design and extends beyond the data analysis and publication phases. For example, information management includes design of paper and digital data forms for data entry, quality assurance and quality control (QA/QC), various types of data processing that can occur within or outside a database management system (e.g., subsetting, merging), metadata development, and submission of data and metadata to a data center (Brunt, 2000). Procedures for data entry and QA/QC vary considerably among the various scientific disciplines. For example, the data relevant to global change may be digitized (e.g., GIS data layers), stored as a large matrix (e.g., remotely sensed data), or manually typed into word processors, DBMSs (e.g., dBASE, ORACLE), or spreadsheets (e.g., Lotus 1-2-3, Excel). The type of data entry is related to data volume as well as to the availability of specific software programs. Except for GIS data, which are often stored in a vendor-specified format (e.g., ARC/INFO Export file), data that are of potential interest to scientists from throughout the world are often stored in ASCII format to facilitate access. ASCII format is very flexible (e.g., data can be fixed column, vertical bar delimited, tab delimited, or comma separated) and can be understood by most software programs. Other storage formats such as the wide variety of spreadsheet formats can become outdated as companies upgrade their products or, in some cases, go out of business. Consequently, it is advisable to store data in internationally accepted or industry standard formats whenever possible.

Quality assurance varies markedly depending on accepted practices within a discipline, data type and volume, and how the data were acquired. Ideally, every effort should first be made to prevent errors from occurring. Error prevention can be facilitated through training efforts, routine equipment maintenance, well-designed data entry forms, and other approaches. "Data filter" programs are often used to identify keypunch errors and illegal data (e.g., sex = "Z" instead of "F" or "M"). It should be noted, however, that it is important to define "illegal" data thoughtfully. For example, many businesses and governmental agencies now wish that "2000" (and higher numbers) had not been declared illegal year values. Additional quality assurance procedures range from simple graphical approaches to more sophisticated statistical programs that identify outliers in regression data (Edwards, 2000).

Many of the data sets available through commercial and governmental data centers are stored in a generic format (e.g., ASCII). Although it is often not clear how the data arrived in their final format, it is frequently the case that some type of DBMS or GIS was employed. There is a dizzying array of DBMS and GIS software packages available on the market, and there are no established standards that are accepted across all scientific disciplines. Since the DBMS packages on the market today may not be here tomorrow, it is especially important to use software that provides easy export capabilities (i.e., the ability to translate from proprietary or software-specific formats to generic formats like ASCII). Furthermore, increased reliance on the WWW for data acquisition highlights the need for user-friendly interfaces between the DBMS and the Web.

*Metadata.* The higher-level data necessary to understand and effectively use primary data, including documentation of the data set contents, context, quality, structure, and accessibility, are known as metadata (Michener et al., 1997). In essence, metadata describe the *who, what, when, where,* and *how* of every aspect of the data. More specifically, metadata address the five basic questions that normally arise when a scientist attempts to acquire and use a specific data set:

1. What potentially relevant data exist?
2. Why were those data sets collected and are they suitable for my particular use?
3. How can these data sets be obtained?
4. How are the data organized and structured?
5. What additional information is available that would facilitate data use and interpretation?

At least three major benefits may be derived as a consequence of investing adequate time and energy into metadata development. First, information entropy is delayed and, correspondingly, data set longevity is increased. Development and maintenance of comprehensive metadata counteracts the natural tendency for data to degrade in information content through time. Second, data reuse by the originator and data sharing with others are facilitated. Scientists often find that a data set they previously collected for a specific purpose can be "mined" later to answer new questions. In such cases, adequate documentation of sampling and analytical procedures, data quality, and data set structure will help ensure that the data can be correctly interpreted or reinterpreted. Third, well-documented data may be used to expand the scale of ecological inquiry. Many of the new types of questions being addressed in global change research require far more data than could feasibly be collected, managed, and analyzed by a single investigator or project (group of investigators). Consequently, scientists must increasingly rely upon data collected by other scientists from numerous disciplines, often for different purposes. Metadata provide critical information for expanding the scales at which ecologists work. For example, field validation data from multiple sites are frequently used to calibrate (or, in some cases, are merged with) remotely sensed data, thereby expanding the spatial domain from the site to broader scales. Cross-site comparative studies depend heavily upon the availability of sufficient metadata. For cross-site comparisons, it is especially important that both methods and instrumentation calibration and intercalibration (measurements of similar parameters by different methods or instruments) be well documented to confirm data integrity, proper use of experimental methods, and data acquisition.

Metadata are receiving increased attention from the scientific community. The geographic sciences community, for example, has devoted considerable resources to developing, adopting, and implementing spatial metadata standards. Most metadata standardization efforts have focused, thus far, on data with a strong geospatial component. In 1994, the U.S. Federal Geographic Data Committee completed the Content Standards for Digital Geospatial Metadata. These standards were developed as part of both the ongoing evolution of the National Biological Information Infrastructure (NBII) in the United States and the standardization of geographical data collected by the federal government. The Content Standards contain more than two hundred metadata fields that are categorized into seven classes of metadata descriptors: identification, data quality, spatial data organization, spatial reference, entity and attribute, distribution, and metadata. Efforts are under way to add extensions to the Content Standards, creating metadata supersets relevant to vegetation classification, cultural, demographic, and other types of data. International Standards Organization (ISO) geospatial metadata standards will be forthcoming in the near future.

Metadata standards for nongeospatial ecological data do not exist currently in any accepted format beyond individual studies, projects, or organizations. Environ-

mental studies often require large amounts of a diverse array of data related to the chemical and physical attributes of the environment, as well as to the individual organisms, populations, communities, and ecosystems composing the biotic portion of the environment. It is therefore unlikely that a single metadata standard, no matter how comprehensive, could encompass all types of environmental data that are relevant to global change studies.

Consequently, a generic set of nongeospatial metadata descriptors for environmental data was recently introduced for the ecological sciences (Michener et al., 1997). The list of metadata descriptors was proposed as a template that could serve as the basis for more refined subdiscipline- or project-specific metadata guidelines. Five classes of metadata descriptors were delineated:

1. *data set descriptors*: basic attributes of the data set (e.g., data set title, associated scientists, abstract, and keywords);
2. *research origin descriptors*: all relevant metadata that describe the research leading to the genesis of a particular data set (e.g., hypotheses, site characteristics, experimental design, and research methods);
3. *data set status and accessibility descriptors*: the status of the data set and associated metadata, as well as information related to data set accessibility;
4. *data structural descriptors*: all attributes related to the physical structure of the data file; and
5. *supplemental descriptors*: all other related information that may be necessary for facilitating secondary usage, publishing the data set, or supporting an audit of the data set.

Numerous other efforts are under way in a variety of scientific disciplines to define metadata content. For instance, metadata content guidelines for climate monitoring, emphasizing those descriptors necessary for determining data fitness for use, have been proposed in support of the Global Climate Observing System (Miller et al., 1996). Also, special types of metadata associated with large statistical databases, including logical or arithmetic expressions (e.g., weighting, aggregation, error), data quality indicators, and statistical summary data are discussed by McCarthy (1982). Ziskin and Chan (1997) further emphasize the importance of including data statistical summaries in metadata to support content-based data selection from large scientific databases, such as those associated with the NASA Earth Observing System.

A major impediment to metadata implementation has been the lack of software tools that can facilitate metadata entry. Many organizations have independently developed word processing or DBMS forms that can be filled in to meet in-house metadata requirements. In other cases, metadata development is an *ad hoc* process performed by the data originators. One exception has been the development of a personal computer (PC)–based program for geospatial metadata. As part of the NBII in the United States, the NBII MetaMaker was developed to support geospatial metadata generation in a format that conforms to U.S. Federal Geographic Data Committee (FGDC, 1994) guidelines (see Internet Resources at the end of this entry).

***Data archives.*** A data archive consists of a collection of digital data sets and metadata that are organized and stored so that users can easily locate, acquire, and utilize data that meet a particular objective (Olson and McCord, 2000). Furthermore, data in an archive are generally stored in multiple locations so that they are secure against natural and anthropogenic disasters. With the primary exception of the GCMD, which acts as a pointer to data that reside elsewhere, the online data centers listed in Table 1 qualify as data archives. Many of the data archives focus on a restricted set of themes. One good example of a data archive is the National Climatic Data Center (NCDC), which was established in 1951 by the U.S. National Oceanographic and Atmospheric Administration (NOAA). The NCDC has one of the largest environmental data archives in relation to length (data date back to the nineteenth century), volume (55 gigabytes are added daily), and users (more than 170,000 requests annually).

A major objective for every data archive, as well as institutional data centers, is the secure storage of the data and metadata. Many different approaches to data storage can be taken, depending on the size of the data holdings, the anticipated rate of access by users, and other factors. For example, a relatively small volume of data can be stored on a single computer's hard disk and made available via the WWW. In this case, some form of manual or automated backup scheme will be required. Moderate data holdings (e.g., 10–100 gigabytes) may be stored online in a series of disks (e.g., a redundant array of independent devices, RAID) that can be configured to support various levels of redundancy. Extremely large data holdings, such as those maintained by the NCDC in the previous example, may be stored in a large mass storage system consisting of multiple RAID units and automated tape libraries. In this latter case, data are either online or near-line, and data backup may be performed offsite (at a mirror location).

One of the primary benefits of focusing on a particular theme is the ability of the data archive to add value to data more easily. For instance, the Carbon Dioxide Information Analysis Center (CDIAC), which is funded by the U.S. Department of Energy, emphasizes the value-added component of data sets that results from the participation of scientists and users in metadata preparation, rigorous QA/QC processing, peer review of data and metadata, beta testing of data sets prior to general

release, and incorporation of user feedback into data packages. In addition to extensive metadata, CDIAC data packages contain examples of data applications and copies of important associated literature. By focusing on a few types of data, a data archive can establish and maintain high standards (e.g., QA/QC) and develop the requisite pool of experts for data and metadata peer review.

The process of data submission will vary from one data archive to another. Different data archives may have different data structure, QA/QC, and metadata content standards that must be adhered to. Frequently, data and metadata are reviewed by data archive staff for internal consistency and completeness. In some cases, the quality assurance procedures that are documented in the metadata will be reviewed. After going through the review process, data and metadata may then be incorporated into the archive database and made publicly accessible.

Following submission to an archive, the availability of a data set may be "publicized" in a data directory such as the GCMD. Listing a data set in the GCMD is rela-

**Information Management. Table 2.** Summary of the Global Change Master Directory (GCMD) Online Data Set

| CONTENTS | DESCRIPTION |
| --- | --- |
| Title | Data set title |
| Summary | Short description of the data set<br>Resolution<br>General attribute information<br>Data source (e.g., digitized from paper<br>    maps, derived from satellite data, etc.) |
| Coverage | Temporal coverage<br>    start and stop date<br>Geographic coverage<br>    southwest and northeast extent<br>Location keywords |
| Attributes | Parameters<br>    category<br>    topic<br>    term<br>    variable<br>    detailed variable<br>Discipline/subdiscipline<br>General keywords<br>Entry ID/originating center |
| Distribution | Data center contact<br>Storage media |
| Personnel | Technical contact<br>Directory Interchange Format (DIF)<br>    author |
| Reference | Data set citation<br>International Directory Network (IDN)<br>    node<br>Revision date<br>Review date |

tively straightforward and primarily requires filling out a form, known as a Directory Interchange Format (DIF), that describes the data. A DIF consists of several fields that describe the data and essentially represent a subset of the metadata for that particular data set (Table 2). Several fields are required, including the entry ID, title, parameters, originating center, data center, and summary. Other fields that provide location keywords and describe the spatial and temporal coverage of the data are deemed critical for data set selection (i.e., searching) and user understanding of the data. The GCMD provides a number of guidelines to facilitate DIF writing, such as recommended keywords and definitions of what constitutes a good title and summary. Furthermore, there are several free, downloadable tools that can be used to author DIFs for UNIX (using the Emacs editor), PCs (both DOS and Microsoft Windows), and the WWW. Once entered, the GCMD can perform very effective searches to match users with data that meet their particular objectives.

[*See also* Global Monitoring; *and* Information Technology.]

### INTERNET RESOURCES

Committee on Earth Observation Satellites International Directory Network (CEOS IDN). http://gcmd.gsfc.nasa.gov/ceosidn/info.html/.

National Biological Information Infrastructure (NBII). http://www.nbii.gov/. Visit this site for the most recent version of the NBII MetaMaker program.

[*See also* Table 1.]

### BIBLIOGRAPHY

American Society for Testing and Materials. *Standard Practice for Dealing with Outlying Observations.* Philadelphia: ASTM, 1994. An authoritative reference for data quality assurance.

Brunt, J. W. "Data Management Principles, Implementation, and Administration." In *Ecological Data: Design, Management and Processing,* edited by W. K. Michener and J. W. Brunt. Oxford: Blackwell Science, 2000.

Chambers, J. M., et al. *Graphical Methods for Data Analysis.* Boston: Duxbury Press, 1983. One of many excellent references related to graphical analysis.

Deming, W. E. *Out of the Crisis.* Cambridge: MIT Center for Advanced Engineering Study, 1986. A landmark reference for quality assurance and quality control.

Easterling, D. R., et al. "On the Development and Use of Homogenized Climate Databases." *Journal of Climate* 9 (1996), 1429–1434. A must-read for those interested in using or developing large comprehensive databases.

Edwards, D. "Data Quality Assurance." In *Ecological Data: Design, Management and Processing,* edited by W. K. Michener and J. W. Brunt. Oxford: Blackwell Science, 2000.

Federal Geographic Data Committee. "Content Standards for Digital Spatial Metadata (June 8 Draft)." Federal Geographic Data Committee. Washington, D.C., 1994. The reference for geospatial metadata content standards, pending adoption of an international standard.

Flournoy, N., and L. B. Hearne. "Quality Control for a Shared Multidisciplinary Database." In *Data Quality Control: Theory and*

*Pragmatics,* edited by G. E. Liepins and V. R. R. Uppuluri. New York: Marcel Dekker, 1990. One of only a few comprehensive references for quality control in large databases.

Goodchild, Michael F., et al. *GIS and Environmental Modeling: Progress and Research Issues.* Fort Collins, Colo.: GIS World Books, 1996. An excellent source of information about linking GIS to a variety of statistical and simulation models.

Heery, R. "Review of Metadata Formats." *Program: Automated Library and Information Systems* 30.4 (1996), 345–373.

Hogan, R. *A Practical Guide to Data Base Design.* Englewood Cliffs, N.J.: Prentice-Hall, 1990. A very readable account of database design.

Ingersoll, R. C., et al. "A Model Information Management System for Ecological Research." *Bioscience* 47 (1997) 310–316. A good example of a site-based information management system.

Inmon, W. H. *Building the Data Warehouse.* New York: Wiley, 1992. An authoritative reference to this exploding subject.

Jensen, John R. *Introductory Digital Image Processing: A Remote Sensing Perspective.* Englewood Cliffs, N.J.: Prentice-Hall, 1996. The authoritative reference for processing and analyzing digital images.

Johnston, Carol A. *Geographic Information Systems in Ecology.* Oxford: Blackwell Science, 1998. An up-to-date reference dealing with all aspects of GIS.

Justice, C. O., et al. "Recent Data and Information System Initiatives for Remotely Sensed Measurements of the Land Surface." *Remote Sensing and the Environment* 51 (1995), 235–244.

McCarthy, J. L. "Metadata Management for Large Statistical Databases." In *Proceedings of the International Conference on Very Large Databases,* New York: IEEE, 1982, pp. 234–243.

Michener, W. K., et al. *Environmental Information Management and Analysis: Ecosystem to Global Scales.* London: Taylor and Francis, 1994. A comprehensive reference for most aspects of data and information management.

———. "Non-Geospatial Metadata for the Ecological Sciences." *Ecological Applications* 7 (1997), 330–342.

Michener, W. K., and J. W. Brunt. *Ecological Data: Design, Management and Processing.* Oxford: Blackwell Science, 2000. A comprehensive reference to managing and processing environmental data.

Miller, C., et al. Documenting Climatological Data Sets for GCOS: A Conceptual Model. In *Proceedings of the First IEEE Metadata Conference.* Silver Spring, Md.: IEEE Computer Society, 1996.

National Research Council. "Bits of Power: Issues in Global Access to Scientific Data." Washington, D.C.: National Academy Press, 1997. A very readable account of the technological and sociological challenges to data sharing.

———. *Finding the Forest in the Trees.* Washington, D.C.: National Academy Press, 1995. A valuable source of information on how different organizations have managed data and information.

Olson, R. J., and R. A. McCord. "Archiving Ecological Data and Information." In *Ecological Data: Design, Management and Processing,* edited by W. K. Michener and J. W. Brunt. Oxford: Blackwell Science, 2000.

Strebel, D. E., et al. "The FIFE Data Publication Experiment." *Journal of Atmospheric Sciences* 55 (1998), 1277–1282.

———. "Scientific Information Systems: A Conceptual Framework." In *Environmental Information Management and Analysis: Ecosystem to Global Scales,* edited by W. K. Michener et al., pp. 59–85. London: Taylor and Francis, 1994. An excellent account of how data "mature" within an organization.

Webster, F. *Solving the Global Change Puzzle: A United States Strategy for Managing Data and Information.* Report by the Committee on Geophysical Data Commission on Geosciences, Environment and Resources, National Research Council. Washington, D.C.: National Academy Press, 1991.

Ziskin, D. C., and P. Chan. "Innovations in Response to Floods of Data." In *1997 International Geoscience and Remote Sensing Symposium.* pp. 1255–1256. New York: IEEE, 1997.

—WILLIAM K. MICHENER

# INFORMATION TECHNOLOGY

Information technology (IT) is a driver of global change, both through the changes in economies and through more direct environmental effects. It is also a tool for studying, managing, and responding to global change. Understanding information technology from both of these perspectives requires an understanding of some of the history of information technology and its consequences for society and the environment, as well as its existing applications in the study of global change through the collection, storage, analysis, and dissemination of environmental information.

**Origins of Information Technology.** Since its origins in the early 1940s, electronic digital information technology—the sensing, communication, storage, analysis, and display of information by digital electronic means—has influenced the global course of events and reshaped everyday life in much of the world. Military applications have driven much of the development of this information technology. IT has played a significant role in conflicts from World War II, in which the English created the special-purpose Colossus to decrypt German military messages prior to D-day, through the cold war, to the decisive victory of the United States–led coalition in the Persian Gulf war. The U.S. government has supported a substantial fraction of all U.S. research conducted in computing, especially in the early years of computing. In 1950, U.S. government funding for research and development exceeded all industrial research and development spending on computing by a factor of three.

However, the subsequent ubiquitous application of information technology in countless domains has tremendously exceeded the expectations of early analysts. Early pessimism about the prospects for electronic stored-program computers held sway for a while even at International Business Machines (IBM) Corporation, which would dominate the computer industry for the next three decades. In the late 1940s and early 1950s, the head of IBM, Thomas J. Watson Sr., was reluctant for his firm to venture into electronic computing, believing, as did his marketing organization, that there was an insufficiently broad market for computers. Moreover, he was reluctant to introduce products that might compete with existing IBM product lines. Nonetheless, his son, Thomas J. Watson Jr., was able to secure IBM's entry into the electronic computer market with the IBM 701, originally known as the Defense Calculator, in part by documenting that a single design for a general-purpose computer could meet many of the needs of the U.S. defense establishment for the Korean War. IBM's lack of foresight in the early years of computing would be repeated in its failure (along with other mainframe computer manufacturers) to recognize early on the importance of the personal computer, as well as in Microsoft Corporation's failure to recognize the importance of the Internet promptly.

Although the 1950s saw the expansion of commercial uses of computers, national security issues continued to be key drivers of IT development. Over time, the U.S. Department of Defense (DOD) has been the largest funder of computing and communications research. In particular, the Information Processing Techniques Office of the Defense Advanced Research Projects Agency (DARPA) has played a dominant role. DARPA was established in 1958 after the launch of Sputnik with the mission of maintaining U.S. leadership in military technology, and it has provided more support for computer science research than all other U.S. agencies combined. DARPA's funding for computer science and electrical engineering research, along with that of other U.S. agencies such as the National Science Foundation (NSF), totaled $1 billion to $1.7 billion annually between 1976 and 1995.

A number of important industrial products have resulted from technologies developed using this support. In the 1950s, the U.S. Air Force's Semi-Automatic Ground Environment (SAGE) command-and-control early-warning system pioneered developments in real-time computing and core memory. The cold war and concerns about communications in the event of an enemy attack ultimately gave rise to a robust communication network called ARPANET, the forerunner of today's Internet. ARPANET, named for the Advanced Research Projects Agency, was first demonstrated publicly in 1972, the same year that Intel Corporation released the milestone 8008 microprocessor. Federally sponsored research helped make these and other technologies such as time-sharing computers, relational database management systems, and the graphical user interface (GUI) available for commercial use.

Numerous software advances and dramatic increases in available computational and telecommunications power, coupled with decreases in the physical requirements and cost of computers and telecommunications, have allowed information technology to transform manufacturing, government and law enforcement, banking, shipping, travel, entertainment, the practice of medicine, and scientific research throughout much of the world. By some accounts, the information technology industry is now over 10 percent of the U.S. gross national product and is responsible for over 5 percent of the total employment in the United States. The advent of the World Wide Web is likely to increase information technology's transformative power and sheds light on what the continuing merger of computer and communications technologies can achieve.

Early prospective assessments of information technology could not have richly captured even its intentional consequences, let alone the unintentional. However, there has been an observable trend to the growth in computing power that has fueled the spread of this technology. In 1965, Gordon Moore, who would later found Intel Corporation with Robert Noyce, predicted that the capacity of integrated circuits would double yearly. In fact, the trend of exponential growth in computing power has continued, and, on average, computing power has doubled every year and a half. This trend became known as Moore's law, and information technology optimists such as Microsoft's Bill Gates not only project that it will hold for the next twenty years but also see similar trends in computer storage and communications.

**Societal Consequences of Information Technology.** For many, automated teller machines, computer-based entertainment in the form of games and movie special effects, word processors, spreadsheets, electronic mail, and the World Wide Web provide the greatest evidence of the changes wrought by information technologies. These are but elements of a broader pattern of societal transformation.

For a number of years, futurists and others have discussed the transformation of economies from industrial toward information and knowledge-based services, driven in part by information technology. The provision of services comprises 75 percent of the U.S. gross national product and employs 70 percent of its labor force. By making factories more productive, information technologies have reduced industrial labor requirements. Moreover, information technology creates the possibility of new classes of valuable services, such as credit cards, worldwide reservation systems, and overnight package delivery. Not only has IT made new industries possible, but it has also allowed the operation of old industries (from airline reservations to mail order stores) at scales, and with response times, that would simply not have been possible with traditional technologies.

A snapshot of the scope of technology at the United Parcel Service (UPS) provides an example of the magnitude of information technology's impact in the service economy. UPS is an American firm with estimated 1996 revenues of $22.5 billion. Its 339,000 employees deliver 12 million parcels and documents daily to over two hundred countries and territories using a fleet of 147,000

ground vehicles and hundreds of aircraft. UPS lauds information technology as its means for maintaining efficiency and price competitiveness and the provision of new customer services. UPS has a global electronic data communications network that consists of 500,000 miles of communications lines and a communications satellite. The network links more than thirteen hundred distribution sites in forty-six countries. Customers can also obtain instant tracking information about their barcoded shipments through the Web, electronic mail, or special tracking software. UPS also makes use of computerized flight planning, scheduling, and load handling. Between 1986 and 1991, UPS spent $1.5 billion on technology improvements, and it plans to spend billions more.

In addition to contributing to the rise of the service economy, information technology has contributed to the globalization of commerce. In 1992, about 25 percent of the U.S. gross national product depended on international trade, including both imports and exports of goods and services of all kinds. It is difficult to imagine the management of global enterprises and intensive international trade without the communications power enabled by information technology. In the case of UPS, we can see the contribution of IT to globalization both directly, by allowing efficient global operations, and indirectly, by allowing UPS to provide rapid, inexpensive global shipping. Air freight, in combination with facsimile transmission and electronic mail, have contributed to a shift in textile manufacturing for the North American and European markets to Asia. In addition to manufacturing, knowledge work such as data entry or programming for these markets has been moving to South Asia as well.

Another set of organizational trends that some forecasters see coming from information technology are a decrease in the importance of large organizations relative to the importance of small organizations and individual entrepreneurial knowledge workers. Recent statistics indicated that 53 percent of the labor force works for firms of fewer than one hundred employees. Large firms are transforming as well. For instance, managers at a large financial services organization in the United States tell employees that they should not expect to maintain traditional long-term employment but instead should seek chances to develop a useful set of skills to take with them to their next position. Akin to this emphasis on smaller organization units is a trend of disintermediation, or elimination of intermediate business, functions, and people that do not create new value.

To sum up many of these ideas, technology optimists such as Bill Gates apply the label "friction-free capitalism" to many of the ways that IT may transform economies globally. In doing so, they refer to decreased transaction costs thought to be brought about by IT. They also hearken to the assumption of perfect information that underlies claims of the economic efficiency of markets and the fact that decreases in the cost of information and improvements in its availability should lead to corresponding increases in economic efficiency.

These supposed gains in efficiency are not certain and do not come without costs. Even though computer power and memory capacity have improved even more rapidly than optimists predicted, tasks to which we would have liked to apply computers, such as the translation of natural languages, have proven to yield slowly to our efforts. Information technology has held surprising costs as well. For instance, the introduction of computers into the bookkeeping practices of businesses in the 1950s decreased the need for accountants as anticipated, but business needed to hire more programmers than they had projected. Moreover, as computers moved to workers' desktops, workers have lost time to the "PC futz factor," the tendency of computer users to spend time playing games, installing the latest versions of applications and operating systems, and working with printouts to get colors, fonts, and alignments just so. A study of twenty cases in five large corporations concluded that the use of computers may actually pull professionals and middle managers away from their most important work and into work that support staff would have handled previously.

Perhaps these disappointments can be understood through parallels between the early results of microcomputer use and the early use of electric motors in industries previously dependent on water and steam power. Although the advantages of electric motors were known in the 1880s, the transformation of industrial processes by this new technology required forty years or more, in part because of existing capital investment but also because of the time and effort required to shift entire manufacturing structures. Perhaps some of the discouraging findings about the impacts of information technology are simply indicators that society is still transforming in order to take advantage of what information technology has to offer. However, these disappointments may also be a result of the difficulty of measuring productivity gains from office automation, especially since our notions of productivity may be somewhat crude compared to the complexity of production in real organizations.

**Environmental Consequences of Information Technology.** Information technology seems upon first consideration to be a very clean technology. Because it comprises 10 percent of the gross national product of the United States, however, we would be surprised if computers did not have some consequences for the global environment. For instance, office equipment uses about 7 percent of commercial-sector electricity, amounting to 58 terawatt hours (TWh) in 1990. This amount is expected to grow, perhaps to as much as 78

TWh in 2010. This, combined with the electricity used by computers in homes and the increased air-conditioning load created by this equipment, results in a nonnegligible contribution to the quantity of electricity demanded. Much of this electricity is generated by the combustion of fossil fuels, and thus in one way information technology directly contributes to increasing greenhouse gas concentrations.

In years past, the manufacture of computers contributed to stratospheric ozone depletion because some steps depended on chlorofluorocarbons and 1,1,1 trichloroethane, although the use of these chemicals has declined considerably since the 1980s. Manufacturers had used these chemicals as cleaning solvents for printed circuit boards and other components of computers. Following the life cycle of the computer beyond manufacturing and operating leads to another environmental concern—disposal. Computers are complex, heterogeneous artifacts, and some estimates suggest that by the year 2005, 150 million personal computers will be in landfills, requiring perhaps 300 million cubic feet of landfill. The presence of lead, nickel, cadmium, and mercury in computers has given rise to concerns about the dispersal of these toxic materials.

Of course, information technology has had a number of environmental benefits. In addition to the efficiency gains already noted in businesses such as parcel services, computer monitoring and control have allowed cleaner and more efficient industrial processes in countless industries.

Some observers have noted or anticipate more surprising environmental consequences of information technology. IT would seem to allow "moving bits instead of people," through telecommuting and telecollaboration. In the past, though, advances in telecommunications have been a complement rather than a substitute for transportation. Although societal transformation may one day allow a net reduction in transportation needs through the use of information technology, this is by no means certain.

Increasing the consumption of travel is one of many possible examples of unanticipated consequences of information technology, and this is only one way that IT may remove constraints to the consumption of resources. In this case, IT contributes to the spread of knowledge of consumption opportunities and increases affluence. IT may also decrease other costs of resource consumption. For instance, the growth of overnight shipping services is in part due to declines in the cost of these services, and IT has contributed to these cost declines.

Another well-known instance of perverse environmental consequences of information technology is its effect on paper use. The long heralded "paperless office" has been anything but, with obvious consequences for land use patterns, although future technological and societal transformations could reverse this effect as well.

**Addressing Global Change with Information Technology.** In addition to being a driver of global change both directly and indirectly, information technology is one of the most important tools for understanding global change. Efforts in this area date back at least to 1946, when John von Neumann, a titan of early computing and decision sciences, recognized parallels between his work in the simulation of nuclear weapons and possible numerical approaches to weather prediction. Aiming to pursue weather modeling in order to develop the ability to control the weather for national security, von Neumann received grants from the U.S. Weather Bureau, the navy, and the air force to form a Meteorology Group at Princeton's Institute for Advanced Study (IAS). This group ran its first computerized forecast on the ENIAC computer in 1950. The forecast covered North America and used finite-difference methods and a two-dimensional grid with 270 points roughly seven hundred kilometers apart and a time step of three hours. In the next few years, the Royal Swedish Air Force Weather Service in Stockholm began regular numerical weather forecasting based on a model developed by the Institute of Meteorology at the University of Stockholm. It was not until much later, however, that these computer-based methods began to compare favorably with results from human forecasters. Less concerned with immediate weather forecasts, theoretical meteorologists were able to achieve some successes in the ten-year period starting in 1955. In particular, Norman Phillips created a two-layer, hemispheric, quasi-geostrophic computer model in 1955. This model is considered to be the first general circulation model, or GCM. GCMs play a key role in modern climate change research and policy analysis, especially as increases in computer power (see Table 1) have obviated the need for a number of simplifying assumptions and have allowed considerable decreases in the cell sizes of atmospheric simulations. [*See* Climate Models; *and* Modeling of Natural Systems.]

More recent advances in high-performance computers such as the development of parallel vector processor (PVP), massively parallel processor (MPP), and distributed shared memory (DSM) architectures have allowed the combination of atmospheric models with ocean models. Investigators are also able to employ available processing power in order to provide a treatment of biological responses and feedback mechanisms in climate change, and Geographic Information Systems (GIS) provide new opportunities to study ecological and human dimensions of global change. [*See* Geographic Information Systems.]

Not only have advances in information technology allowed new kinds of analyses, but they have also allowed the storage and dissemination of global change information in ways that would not have been possible in previous decades. Beginning with the establishment of world data centers and worldwide communications networks for the International Geophysical Year of 1957, information technology has transformed geophysics research. Stolarski et al. (1986) provided an example of this transformation by confirming ground-based measurements of the springtime decrease in stratospheric ozone over Antarctica by using archived Nimbus 7 satellite measurements. This work illustrated not only the importance of attention to data management practices but also how investigators not immediately involved in an experiment or data collection effort may make use of the resulting data to explore new hypotheses.

More recently, the U.S. Global Change Research Program (USGCRP) has placed a great deal of emphasis on

**Information Technology. TABLE 1.** Increases in Computer Power during the First Three Decades of General Circulation Modeling—Computers Used at the Geophysical Fluid Dynamics Laboratory of Princeton University

| COMPUTER | TIME PERIOD | RELATIVE POWER |
|---|---|---|
| IBM 701 | 1956–1957 | 1 |
| IBM 704 | 1958–1960 | 3 |
| IBM 7090 | 1961–1962 | 20 |
| IBM 7030 | 1963–1965 | 40 |
| CDC 6600 | 1965–1967 | 200 |
| UNIVAC 1108 | 1967–1973 | 80 |
| IBM 360/91 | 1969–1973 | 400 |
| IBM 360/195 | 1974–1975 | 800 |
| Texas Instruments X4ASC | 1974–1982 | 3,000 |

SOURCE: "A History of Atmospheric General Circulation Models," http://www.stanford.edu/group/STS/sloan.project/intro.html and United States Department of Commerce (1981).

information technology in general and in the dissemination of information in particular. Early in the history of the USGCRP, the United States Executive Office of the President, through the Office of Science and Technology Policy, put forth a set of policy statements on data management for global change research. These statements had the purpose of facilitating "full and open access to quality data for global change research."

Among the policy statements were a "commitment to the establishment, maintenance, validation, description, accessibility, and distribution of high-quality, long-term data sets," a requirement for the preservation of all data needed for long-term global change research, storage of detailed information about the data holdings (metadata), and a requirement that the data be provided at the lowest possible cost to researchers.

In response to these policy statements, the USGCRP established the Global Change Data and Information System (GCDIS) and one of its largest elements, NASA's Earth Observation System Data and Information Systems (EOSDIS). EOSDIS received $248.0 million in fiscal year 1997, making it second among NASA's Mission to Planet Earth (MTPE) programs only to Earth Observation System Flight Development, which received $427.4 million. This amount is not entirely devoted to information systems per se—the flight operations segment of EOSDIS will perform spacecraft and instrument planning and scheduling as well as command and control, for instance. However, when considered with information technology elements of other components of USGCRP-sponsored research, from the high-performance computing facilities at the National Center for Atmospheric Research to the data collection and archiving systems of the National Climate Data Center, it is clear that a significant portion of the USGCRP budget ($1.81 billion in fiscal year 1997) is in fact spent on information technology.

EOSDIS itself is a distributed data archival and distribution system. Its multiple data centers are connected via an Information Management System (IMS), allowing searching and ordering of data from multiple centers at once. The centers have data on or from sea ice and polar processes, synthetic aperture radar, human interactions in the environment, land processes, global hydrology, the upper atmosphere, the global biosphere, atmospheric dynamics, geophysics, physical oceanography, the radiation budget, tropospheric chemistry, clouds, aerosols, comprehensive satellite data, snow and ice, the cryosphere and climate, and biogeochemical dynamics. [See Information Management.]

The storage and dissemination of the volume and variety of data associated with EOSDIS would not be possible without the simultaneous advances in computation, storage, and communication technologies of recent decades. Among the many information technol-

ogy efforts outside the United States are applied efforts such as Japan's Global Observation Information Network and research programs such as the Telematics Applications for the Environment Programme of the European Commission.

The formulation of policies to address global change is also facilitated by advances in information technology. Because uncertainty pervades global change decisions, policies must be developed that are robust across uncertainty, and computationally demanding techniques such as Monte Carlo simulation and exploratory modeling allow policy analysts to systematically consider thousands of global change scenarios.

The applications of information technology described above tend to be large-scale research-oriented efforts. Information technology also influences understanding, managing, and responding to global change on smaller scales. To the extent that information technology makes societies wealthier, it can in that way increase the ability of individuals and local governments to adapt to global change and to cope with extreme events. Moreover, information technology is generally thought to increase the access to information of individuals who are not geographically or sociologically near centers of power and is thought to help provide a voice for the voiceless. This may lead to greater interest and understanding in global change on the part of individuals, a larger and richer role for individuals in global change decision processes, and a greater ability for individuals to mitigate pernicious global changes and to adapt to those changes that we do not mitigate.

Individuals with access to the World Wide Web are currently able to examine immediately the texts and status of international environmental agreements, all bills before the Congress of the United States, tutorials for laypeople on issues such as global climate change, and advice on how to minimize one's contribution to undesirable global change. Electronic mail, distribution lists, and the creation of Web pages by individuals and small groups also can aid grassroots education, organization, and involvement in global change issues. A number of observers attribute to information technology the power to enable individuals to influence globally significant events. These observers point to the spread of democratic values in the former Soviet Union, global information dissemination and discussions about Chinese demonstrations in 1989, and the 1996 uprisings in Serbia.

The consequences for global change of broad access to information technology are of course difficult to predict, as are other consequences of information technology. Any such predictions should be tempered with a recollection of the contrast between the current pervasiveness of information technology throughout much of the world with IBM's early assessments of the market for computers, and as with so many technological assessments, usually neither the real benefits nor the real perils are the ones that we expect.

[*See also* Global Monitoring.]

### BIBLIOGRAPHY

Aspray, W. *John von Neumann and the Origins of Modern Computing*. Cambridge: MIT Press, 1990. Contains a discussion of von Neumann's early work in weather prediction by computer.

Berners-Lee, T. "WWW: Past, Present, and Future" *Computer* 29, no. 10 (October 1996), 69–77. A description of the World Wide Web by its original developer.

Committee on Environment and Natural Resources Research. *The U.S. Global Change Data and Information System Implementation Plan*. USGCRP-95-02. 1995. Although more current information is available via the World Wide Web, this report is a starting point for understanding GCDIS, its components (such as EOSDIS), and its organizational context.

Computer Science and Telecommunications Board, National Research Council. *Funding a Revolution: Government Support for Computing Research*. Washington, D.C.: National Academy Press, 1999. A comprehensive and authoritative report on the role of the U.S. government in funding research in computer science and related fields.

Department of Engineering and Public Policy, Carnegie Mellon University. *Design Issues in Waste Avoidance*. Pittsburgh: Department of Engineering and Public Policy, Carnegie Mellon University, 1991. A project report with projections of the scope and nature of the problem of computer disposal.

Flamm, K. *Targeting the Computer: Government Support and International Competition*. Washington, D.C.: Brookings Institution, 1987. A dated but important study of the role of the U.S. federal government in the development of computer-related technologies.

Gates, W. H., III. *The Road Ahead*. New York: Viking, 1995. An optimistic view of the future of information technology from the head of Microsoft Corporation.

Kittel, T. G. F., N. A. Rosenbloom, T. H. Painter, D. S. Schimel, and VEMAP Modeling Participants. "The VEMAP Integrated Database for Modeling United States Ecosystem/Vegetation Sensitivity to Climate Change." *Journal of Biogeography* 22 (1995), 857–862.

Koomey, J. G., M. Cramer, M. Piette, and J. H. Eto. *Efficiency Improvements in U.S. Office Equipment: Expected Policy Impacts and Uncertainties*. Ernest Orlando Lawrence Berkeley National Laboratory LBL-37383, December 1995. Includes reports on the magnitude of electricity usage by office equipment.

Laudon, K., C. Traver, and J. Laudon. *Information Technology and Society*. Belmont, Calif.: Wadsworth, 1994. A comprehensive undergraduate textbook containing an introduction to information technology and its societal aspects.

Mokhtarian, P. L. "Now That Travel Can Be Virtual, Will Congestion Virtually Disappear?" *Scientific American* 277.4 (October 1997), 93. A suggestion that decreases in commuting and travel that some anticipate from information technology may be overstated.

Schaller, R. R. "Moore's Law: Past, Present and Future." *IEEE Spectrum* 34.6 (June 1997), 53–59. An interesting history of Moore's law and the microelectronics advances that are behind much of the information technology advance of the last few decades.

Semtner, A. J. "Ocean and Climate Modeling on Advanced Parallel Computers: Progress and Prospects." Paper delivered at SC98: High Performance Networking and Computing Conference, Orlando, Florida, 7–13 November 1998. Available at http://www.oc.nps.navy.mil/~braccio/ACM/acm.html.

Shurkin, J. N. *Engines of the Mind: The Evolution of the Computer from Mainframes to Microprocessors*. New York: Norton, 1996. A history of the computer, focusing especially on early developments.

Stolarski, R. S., A. J. Krueger, M. R. Schoeberl, R. D. McPeters, P. A. Newman, and J. C. Alperr. "Nimbus-7 Satellite Measurements of the Springtime Antarctic Ozone Decrease." *Nature* 322 (1986), 808–811.

Subcommittee on Global Change Research, Committee on Environment and Natural Resources of the National Science and Technology Council. *Our Changing Planet: The FY 1998 U.S. Global Change Research Program*. Washington, D.C.: National Academy Press, 1997.

Sullivan, W. "The International Geophysical Year." In *International Conciliation*. New York: Carnegie Endowment for International Peace, 1959.

Tapscott, D. *The Digital Economy: Promise and Peril in the Age of Networked Intelligence*. New York: McGraw-Hill, 1996. Observations and projections about the consequences of information technology for the world's economies.

Tenner, E. *Why Things Bite Back: Technology and the Revenge of Unintended Consequences*. New York: Knopf, 1996. Includes discussion of the parallels drawn by P. A. David between the adoption of microcomputers and the previous adoption of electric motors. Tenner also summarizes a set of case studies by P. G. Sassone on adverse effects productivity caused by computer use.

Toffler, A. *Powershift: Knowledge, Wealth, and Violence at the Edge of the 21st Century*. New York: Bantam Books, 1990. The well-known futurist includes a discussion of the shift toward the importance of knowledge power.

United States Department of Commerce. "Geophysical Fluid Dynamics Laboratory: Activities—FY80, Plans—FY81." Princeton, N.J.: U.S. Department of Commerce, 1981.

U.S. Global Change Research Program. *Policy Statements on Data Management for Global Change Research*. DOE/EP-001P, 1991. Contains the guiding principles for data systems for the USGCRP.

Yourdon, E. *Decline and Fall of the American Programmer*. Englewood Cliffs, N.J.: Yourdon Press, 1993. In addition to containing a critical examination of the state of software development practice, this book devotes an appendix to the rising software industry in India.

—CHARLES D. LINVILLE

**INSECTICIDES.** *See* Pest Management; *and* Water Quality.

# INSURANCE

Discussions of the enhanced greenhouse threat, often known as *global warming*, began substantively in the financial sector only in 1990. Concern is now proliferating rapidly in the financial institutions, fueled by a crop of billion-dollar losses from suspiciously extreme windstorms, wildfires, and floods. The first book specifically on this subject, published by a publishing affiliate of the giant German insurer Gerling, did not appear until 1996 (Leggett, 1996). In it, senior financial-sector managers and executives from a broad range of institutions express their concerns. Among the insurers, the chief manager of operations at General Accident concludes that "there is no doubt to me that weather patterns are changing. There is also no doubt from historical study that small climate changes have very big effects on society. . . . Looking forward, I am sure that climate change will speed up. I am also sure there will be major consequences for insurers. . . . And I am absolutely certain there will be long term international effects." Among the bankers, the head of environmental affairs at the Union Bank of Switzerland describes global warming as the most serious of all environmental problems: " . . . we must recognize that the financial markets will be affected by climate change . . . site contamination and lender liability are only the tip of the banker's 'environmental iceberg.' Climate change is from my perspective the mass underneath the water line and we can't ignore this part . . . [it] will probably surface soon."

The threat pertains at two levels: the baseline and the worst case. The baseline threat involves the impacts of midrange estimates of global average temperature rise and sea level rise by the world's climate scientists, as codified by the authoritative Intergovernmental Panel on Climate Change (IPCC) in its various reports to governments. What this means is that, for insurers, the past is no longer a guide to the future. This makes actuarial work, on which premium rates are set, difficult. For bankers, the concerns that arise are somewhat different and relate to the security of investments. A September 1995 British Bankers' Association paper on climate change (Blackman, 1995) spells this out: "as climate change takes place it will have significant adverse effects upon the activities of banks' customers; their operations will fail or become not viable . . . within the lifetime of loans granted today climate change is forecast to have a dramatic impact on industrial operations within 20 to 40 years."

The worst-case threat has been summarized best on the public record by the president of the U.S. Insurance Institute for Property Loss Reduction. His view of the threat to the insurance industry in 1993 makes especially grim reading (Lecomte, 1993): "Despite the fact that the industry is financially healthy, and has some $160 billion in surplus, in two events you could take 70 billion, or maybe 80 billion of that surplus away, and you'd cripple the industry. It wouldn't be able to take on new risks. It wouldn't have the capacity to underwrite the business of the future. We'd have massive, massive availability problems." This is a global insurance crash, in other words. The kinds of events referred to include potential megacatastrophes such as a category-five hurricane on, say, Miami and New Orleans, a typhoon hitting Tokyo,

a drought-related wildfire burning out of control in an urban center such as Los Angeles or Sydney, or a failure of London's Thames barrier under a storm surge. The critical point is that such events become ever more likely the longer heat-trapping greenhouse gases are emitted into the atmosphere in the quantities they are today. Nor does the risk of a potential wipeout of global surplus (the amount kept by the industry to hedge against property catastrophe losses in any one year, globally) extend just to unlucky hits on cities. As one analyst for the reinsurance group Swiss Re put it in 1993, the industry could suffer "a machine-gun fire" of smaller catastrophes in a warming world, which would have the same effect (Leggett, 1993).

This scope for a global wipeout of property catastrophe reserves is backed up by a growing number of on-the-record, worst-case prognoses within the industry. As the president of the Reinsurance Association of America has bluntly quoted to *Time* magazine (1994), "the insurance business is first in line to be affected by climate change . . . it could bankrupt the industry." As an unnamed official from the Marine and Fire Association of Japan has put it (Shimbun, 1993), "if more disasters like [Typhoon Mireille] follow, it could affect the industry's very existence." More recently, Munich Re wrote in a report on the 1996 catastrophe record (Munch Re, 1997) that "according to current estimates, the possible extent of losses caused by extreme natural catastrophes in the one of the world's metropolises or industrial centers would be so great as to cause the collapse of entire countries' economic systems and could even bring about the collapse of the world's financial markets."

In the Second Assessment Report of the Intergovernmental Panel on Climate Change (1995), a majority of the world's best climate scientists concluded that the first faint signal of an enhanced greenhouse effect is already appearing in recent unusually hot years and other climatic patterns. (The First IPCC Assessment, in 1990, forced governments to begin negotiations for the Convention on Climate Change, signed at the Earth Summit in Rio in 1992. At that time, the climate science community concluded that it would be ten years before a signal became apparent.) The IPCC judged, however, that there is not yet a detectable signal in the pattern of extreme events. A chapter in the IPCC report written by insurers and bankers on the impacts of climate change on their industry concludes, among other things, that "there is a need for increased recognition by the financial sector that climate change is an issue which could affect its future at the national and international level. This could require institutional change."

Deep cuts in greenhouse gas emissions provide the only route to meaningful risk abatement and loss reduction for financial institutions. Changes to business practice can buy an insurer interim risk management. The insurance industry would have no future if, as the global greenhouse crisis deepened, it simply retreated serially from areas of perceived threat. Neither would it be allowed to; this is a heavily regulated industry. After the full extent of the losses from Hurricane Andrew became clear, thirty insurance companies tried to pull out of Florida altogether. Had they been permitted to do so, economic growth in the state would probably have come to a halt. But the state legislated and forced the companies to stay, permitting them to retreat only at the rate of 5 percent of capacity per year. In Japan, the situation is even more stringent. Insurers are bound to offer coverage if it is requested and are required to seek the agreement of the Ministry of Finance for every rate increase. The message is clear: if the financial services industry wants to cut the risk from global warming, its interests lie in cutting global warming itself.

Recently, it has become clear that the first serious business interest is arising, both in the financial and energy sectors, in the opportunities involved in greenhouse risk abatement and loss reduction. Significant reductions in global greenhouse gas emissions will inescapably have to involve the progressive replacement of fossil fuel energy markets, and energy profligacy, with burgeoning new markets in renewable energy and energy-efficiency technologies. The first major players are aware of this, and also the implications for investments in the years ahead. For example, the Chartered Insurance Institute Society of Fellows (the U.K. insurance industry's premier professional body) concluded on climate change that "the industry has a limited breathing space in which to gather its wits, and plan in a truly long-term timeframe." It recommended many options, one of which was as follows: "all investment managers should modify their investment policies to take account of the potential direct and indirect effects of global warming" (Chartered Insurance Institute, 1994). Swiss Re is among the financial giants that seem to agree, having written in a report on global warming that "there is no shortage of practical suggestions, especially with regard to a drastic reduction of greenhouse gases" (Swiss Re, 1994).

Some effort has been made by financial institutions toward risk abatement at source. In November 1995, insurance companies from Europe and Japan signed the United Nations Environment Programme (UNEP) Declaration of Environmental Commitment by the Insurance Industry, featuring climate change high on their list of concerns and reasons to pledge action. Over sixty companies have now signed up. However, this initiative has been criticized by environmentalists as too little, too late.

[*See also* Extreme Events; Global Warming; Natural Hazards; Nuclear Hazards; *and* Risk.]

## BIBLIOGRAPHY

Blackman, P. "Environmental Issues: Liability, Climate Change." British Bankers Association paper for the Annual Meeting of Officers, 20–23 September 1995, Oiso, Japan.

"Burned by Warming." *Time*, 14 March 1994. An article based on the U.S. College of Insurance conference referred to in the present article. The *Time* correspondent concluded his article by observing that "these risks and the crucial role played by the $1.41 trillion insurance industry in the world economy could change the dynamic of the debate about global warming." Chartered Insurance Institute. "The Impact of Changing Weather Patterns on Property Insurance." Chartered Insurance Institute Special Report, May 1994.

"Insurance Industry Concerned about the Earth." *Asahi Shimbun*, 3 February 1993.

Intergovernmental Panel on Climate Change. IPCC Second Assessment Report: Climate Change, 1995.

Lecomte, Eugene. Speech given at the U.S. College of Insurance, November 1993.

Leggett, J. K., ed. "Climate Change and the Financial Sector: The Growing Threat, the Solar Solution." Zurich: Gerling Akademie Verlag, 1996.

Leggett, J. K. "Climate Change and the Insurance Industry: Solidarity among the Risk Community?" Greenpeace International Special Publication, February 1993.

Munich Re. "Topics: Annual Review of Natural Catastrophes 1996." Munich Re Special Publication, 1997.

Swiss Re. "Element of Risk." Swiss Re Special Publication, 1994.

—JEREMY LEGGETT

# INTEGRATED ASSESSMENT

An *assessment* may be defined as a collection of knowledge on related topics prepared by a group of interested parties. A parliamentary debate, this encyclopedia, and the scientific assessments by the Intergovernmental Panel on Climate Change (IPCC) are all examples of assessments.

For an assessment to be *integrated*, the interactions between its various elements need to be specified explicitly. This necessitates the coordination of assumptions underlying each element of the assessment. This emphasis on coherent narratives is a hallmark of integrated assessments. Rarely is it possible to impose such constraints on essays prepared by a group of scientists or on discussions among stakeholders about climate change or any other problem. This limitation can make it more difficult to arrive at concrete conclusions from such deliberations. Nevertheless, such informal integrated assessments offer greater latitude to the imagination of participants. When we are faced with considering issues far from our past experience, this extra degree of freedom can lead to the exploration of alternatives beyond the diagnostic or prognostic capability of formal integrated assessment models.

*Integrated assessment models* are used to explore the quantitative as well as qualitative interactions between various elements of a problem domain. These models can be developed as prognostic or diagnostic tools. The choice of how to develop and use the model is largely determined by the state of knowledge about the relevant processes.

Integrated assessments can be designed to follow a *vertical integration* path or a *horizontal integration* path. In vertical integrations, the goal is to explore and close the chain of causation linking the various elements of change impacting the issue in question. For example, a comprehensive vertical integration of climate change would include explicit interactions between demographic change and economic development; development and resource mobilization; resource mobilization and biosphere/climate change; and the impact of this biosphere/climate change back on demographics, economic development, and resource mobilization. Such studies often limit themselves to consideration of the direct interactions between different processes. Horizontal integrations involve understanding the key processes interacting across each of the elements of a vertical study. For example, a horizontal integration of water resources would involve the mutual interactions between water quality, on the one hand, and, on the other hand, demographics, economics, energy and food production, land use and land cover, surface and groundwater flows, local and regional climate, transportation, pollution, diseases, and sanitation.

Integrated assessments can be developed to shed light on many issues. For issues with confined temporal and spatial extent, the interactions between various elements of the system are relatively easy to characterize and prognostic analysis is feasible. Short-lived transboundary pollution, in general, and acid rain, in particular, are examples of such well-defined issues. Prognostic integrated assessment models such as RAINS, developed by Alcamo et al. (1990), have been used successfully to develop regional policies for control of nitrogen oxides and sulfur oxides emissions on a regional basis. [*See* Regional Assessment.]

For a number of compelling reasons, prognostic approaches to century-scale policy issues, such as climate change, are unlikely to be successful or useful. In such extensive problems, there are significant uncertainties about the dynamics of each element of the problem, and gaping holes riddle our understanding of the first- and higher-order interactions between these elements. These challenges are a reflection of how much of our scientific knowledge is about marginal changes in relatively isolated systems. For example, in both the social and natural sciences we have yet to develop theories linking the dynamics of system elements observed at different scales. We need to understand these cross-scale phenomena in order to understand nonmarginal regime changes. We need to characterize these processes to specify how path dependencies affect the interactions and dynamics defining the Earth system. Embracing this simple truth in the design of integrated assessments of problems such as climate change can deliver powerful diagnostic tools. The emphasis of analysis in such efforts falls on characterization of uncertainties, speculation on dynamics, and exploration of possible interactions. Such diagnostic tools are not capable of forecasting the future of the climate state, or identifying an optimal policy from an economic or other objective. They are, however, capable of informing disciplinary scientists about the interstices of current knowledge. They can be used to identify the type of observations that can shed light on uncertain aspects of system dynamics and interactions. They can also help stakeholders and policy makers visualize the opportunities and constraints they face in meeting their objectives. This paradigm is rarely the guiding philosophy behind integrated assessments of climate change.

**Objectives of Climate Integrated Assessments.** The consequences of rapid and substantial global climate change could be disastrous. The impact of stringent emission control programs could be enormous, and the efficacy of such action is uncertain. This issue is especially relevant if abrupt climate change could be triggered at a threshold close to the present concentration of greenhouse gases.

There are four, possibly overlapping, options available for dealing with the climate problem:

1. If we accept that significant climate change is caused by anthropogenic emissions of greenhouse gases, we may undertake actions to reduce these emissions.
2. If we accept that climate is going to change (from a combination of natural and anthropogenic causes), we may undertake actions to adapt to such change.
3. If we believe that processes leading to climate change can be "managed," we may undertake actions designed to control the climate system (geoengineering).
4. If we believe that we simply do not know enough to know what to do, we can undertake research to improve our understanding of how humans perturb the environment, the climate system, its interaction with the biosphere, and feasible strategies for implementing the options noted above.

If the anticipated impacts of climate change are judged to be unacceptable, the cost of abating greenhouse gas emissions must be borne long before we know the actual magnitude and severity of climate change and its consequences. This stems from the long lag time between the release of greenhouse gases and equilibrium response of the biosphere and climate systems. A further cause for concern lies in uncertainty about the reversibility of the biosphere-climate system. Through anthropogenic activity we may force the Earth system into a new quasi equilibrium of biosphere-climate interactions. If we judge this new system state to be undesirable after arriving there, there is no guarantee that we can recover the current quasi-equilibrium state.

The climate issue's characteristic of prompt costs and delayed benefits has resulted in the focus of early policy research on cost-effectiveness analyses of various greenhouse gas abatement strategies. These models help decision makers identify not climate change policy objectives but merely the cost of meeting various abatement targets and the efficacy of different strategies. Concurrently, scientific research has been focused on explorations of the Earth's environment if the atmospheric concentrations of greenhouse gases continue to increase. Despite a significant effort to understand the impacts of climate change, the results of this effort has been less well received than the economic analyses of the costs of climate policy. This stems from a significant asymmetry between our concerns and knowledge. For example:

- Climate change will impact us at a local level. However, we are unable to project future local climate change with sufficient confidence to assess whether we are going to be better or worse off.
- Impacts of climate change are believed to be most severe in agrarian societies. However, we are particularly ignorant of their state and vulnerability in the distant future.
- Impacts of climate change on unmanaged ecosystems are believed to dominate all others in our sense of well-being and happiness. However, we know that aesthetics are labile and that humans are capable of rapid psychological adaptation to favorable and unfavorable outcomes.
- Impact assessments are often stated in terms of social costs, which reflect whether society as a whole will gain or lose from climate change. However, governance is about management of the incidence of costs and benefits on individuals and society as a whole.

Such asymmetries notwithstanding, there is an immediate need for policy decisions on how to prevent or adapt to climate change and how to allocate scarce funds for climate research. We need to move beyond isolated studies of the various parts of the problem. Analysis frameworks are needed that incorporate our knowledge about precursors to, processes of, and consequences from climate change. This framework also needs to represent the reliability with which the various pieces of the climate puzzle are understood and to be able to propagate uncertainties through the analysis reflecting them in the conclusions.

**Pioneering Studies.** The first serious attempt at integrated assessment of climate change was the Integrated Model to Assess the Greenhouse Effect (IMAGE) model developed in the Netherlands by Rotmans (1990). IMAGE has a rich description of the physical world and biogeochemical processes. In IMAGE the Edmonds and Reilly model (1985), long established as a benchmark for predicting future greenhouse gas emissions, was used to drive the inputs to the biogeochemical cycle.

Most early economic models focused on the cost-effectiveness of meeting a particular emissions target. An important model within this group is Global 2100, developed by Manne and Richels (1992). Because such models lack benefit estimates, however, they are often not counted among the integrated assessments of climate change. Economic models incorporating the costs and benefits of climate change fall into two categories: optimization models and simulation models. The DICE model, by Nordhaus (1994), and CETA, by Peck and Teisberg (Peck et al., 1992), are early exemplars of the optimization paradigm. IMAGE-1, by Rotmans (1990); PAGE, by Hope et al. (1993); and ICAM-1 by Dowlatabadi and Morgan (1993) are exemplars of integrated assessment models designed to explore the outcome of prescribed policy initiatives through simulations.

Since the early 1990s, the number of integrated assessment models of climate change has mushroomed to over fifty. Recent notable achievements in this arena are the AIM model, focused on the Asian and Pacific Rim countries, by Morita et al. (1995), and IMAGE-2, by Alcamo et al. (1994).

Continuing with his pioneering practice, Rotmans has spearheaded the development of the first integrated assessment of global change—the TARGETS model (Rotmans and de Vries, 1997). At the time of writing this overview, many new efforts are being launched. These offer the promise of progress toward identifying critical scientific gaps and promoting international cooperation among disciplinary experts.

Efforts to develop integrated science models for climate change are also noteworthy. These are exemplified by the research undertaken by the University of Illinois and the Massachusetts Institute of Technology, where atmospheric chemistry, climate dynamics, and land cover are being coupled in dynamic interactive frameworks.

**Good Practice.** In an ideal world where computers are infinitely fast and cheap, an integrated assessment would incorporate the most detailed available representations of each element of the climate problem. It would incorporate a calculable nonequilibrium model of the world economy, three-dimensional models of atmospheric chemistry and dispersal, coupled ocean–atmosphere global circulation models, general coupled ecological systems models, and models of social preferences and dynamics. Each of these represents the holy grail of its particular discipline. To date, none has been attained with sufficient satisfaction. All are too large for conventional uncertainty analysis using today's computation engines. It will probably never be appropriate to incorporate such models directly in an integrated assessment framework. As the quality of these large models improves, however, integrated assessments should try to capture their most salient features, in reduced-form or metamodels. [*See* Climate Models.]

While the arguments for integrated assessment are intellectually compelling, current understanding of the natural and social science of the climate problem is so incomplete that today it is not possible to build traditional analytical models incorporating all the elements, processes, and feedbacks that are likely to be important. Faced with similar problems in the past, the policy research community has typically modeled what was understood and waved their hands at the rest. The result has often been that the policy discussion has focused on what we know, rather than on what is important. To avoid this difficulty in the climate problem it will be necessary to evolve a new class of policy models that allow an integration of subjective expert judgment about poorly understood portions of the problem with formal analytical treatments of the well-understood portions of the problem. Preliminary work on such hybrid models that close the loop on the climate problem is under way at several institutions.

The design of an integrated assessment should start with the identification of outcomes that matter to key policy makers—not with the science. Different outcome measures will be important to different actors. With each actor's decision rule and an integrated assessment model that incorporates the relevant science, we can choose among alternative policy objectives. For example, an environmentalist in the developed world, using the precautionary principle (i.e., minimize the worst possible outcome), might identify stringent reduction in greenhouse gas emissions as the appropriate policy objective. An industrialist, making decisions on the basis of expected values, might identify a per capita emission limit as the appropriate policy objective (Lave et al., 1993).

When a policy objective has been identified as desirable, sensitivity analysis of the integrated model can identify the most effective policy strategy for meeting that objective. For example, environmentalists might identify population control as the strategy with the greatest potential for achieving their objective. The industrialist might find improvement in energy efficiency as the strategy with the greatest potential. Once an integrated framework is able to simulate these preferences among key actors, it should become possible to systematically explore opportunities for trade-off and cooperation among key actors.

At times, uncertainty in an integrated assessment model will make policy choice ambiguous. In this case, uncertainty analysis of the model can identify key areas where better information is needed. Research prioritization cannot be based on value of information alone. Expert judgment of returns to policy-motivated research investment are also needed. Work to combine these two is at the frontier of current policy-modeling practice.

**Summary.** The challenge of performing integrated assessment is enormous. First, there is the challenge of casting the problem from the perspective of a variety of key decision makers. Second, there is the challenge of information management, development of suitable reduced-form representations of key elements of the climate problem, and the successful incorporation and parametric analysis of expert judgments. Third, there is the challenge of developing operational methods for estimating value of information and setting research priorities. Finally, there is an institutional/political challenge of using integrated assessments to evaluate various policy objectives, identify better policy strategies for meeting these objectives, clarify opportunities for cooperation and trade-offs between various key actors, and set priorities in policy-motivated research.

[*See also* Climate Change; Climate Impacts; Global Warming; Public Policy; *and* Surprise.]

## BIBLIOGRAPHY

Alcamo, J., ed. *IMAGE 2.0: Integrated Modeling of Global Climate Change.* Dordrecht: Kluwer, 1994.

Alcamo, J., R. Shaw, and L. Hordijk. *The RAINS Model of Acidification: Science and Strategies in Europe.* Dordrecht: Kluwer, 1990.

Dowlatabadi, H., and M. G. Morgan. "A Model Framework for Integrated Studies of the Climate Problem." *Energy Policy* 21, no. 3 (1993), 209–221.

Edmonds, J. A., and J. M. Reilly. *Global Energy: Assessing the Future.* New York: Oxford University Press, 1985.

Hope, C. W., J. Anderson, and P. Wenman. "Policy Analysis of the Greenhouse Effect: An Application of the PAGE Model." *Energy Policy* 21, no. 3 (1993), 327–338.

Lave, L. B., and H. Dowlatabadi. "Climate Change Policy: The Effects of Personal Beliefs and Scientific Uncertainty." *Environmental Science and Technology* 27, no. 10 (1993), 1962–1972.

Manne, A. S., and R. G. Richels. *Buying Greenhouse Insurance: The Economic Costs of $CO_2$ Emission Limits.* Cambridge, Mass.: MIT Press, 1992.

Morita, T., M. Kainuma, H. Harasawa, K. Kai, L. Dong-Kun, and Y. Matsuoka. *The Asian Pacific Integrated Model: What Can It Predict? A Collection of AIM Simulation Results.* Tsukuba, Japan: National Institute for Environmental Studies, 1995.

Nordhaus, W. D. *Managing the Global Commons: The Economics of Climate Change.* Cambridge: MIT Press, 1994.

Peck, S. C., and T. J. Teisberg. "CETA: A Model for Carbon Emissions Trajectory Assessment." *Energy Journal* 13, no. 1 (1992), 55–77.

Rotmans, J. *IMAGE: An Integrated Model to Assess the Greenhouse Effect.* Dordrecht: Kluwer, 1990.

Rotmans, J., and B. de Vries, eds. *Perspectives on Global Change: The TARGETS Approach.* Cambridge: Cambridge University Press, 1997.

—HADI DOWLATABADI

**INTEREST GROUPS.** *See* Nongovernmental Organizations.

**INTERGLACIAL PERIOD.** *See* Climate Change; Glaciation; *and* Global Warming.

## INTERGOVERNMENTAL PANEL ON CLIMATE CHANGE

The Intergovernmental Panel on Climate Change (IPCC) is a specialized agency of the United Nations that is responsible for providing scientific and technical information on climate change, its potential implications, and response options. The IPCC was established in 1988, under the joint sponsorship of the World Meteorological Organization and the United Nations Environment Programme. [*See* World Meteorological Organization; *and* United Nations Environment Programme.] The mission of the IPCC is to review the scientific and technical literature in disciplines as diverse as physics, ecology, agronomy, and economics and provide assessments of major scientific and technical issues related to climate change on the basis of this literature. The distinguishing characteristic of the IPCC is the elaborate process used to prepare and review the panel's reports, a process closer to that used in peer-reviewed research than in most governmental institutions. The IPCC's First Assessment Report, issued in 1990, served as a primary basis for negotiations leading to establishing the United Nations Framework Convention on Climate Change (UNFCCC) in 1992. [*See* Framework Convention on Climate Change.] The Second Assessment Report (SAR), issued in 1995, as well as a Special Report on regional impacts of climate change and a series of Technical Papers, served as primary sources of information during negotiations leading to the Kyoto Protocol in 1997. IPCC Guidelines for National Greenhouse Gas Inventories provide the basis for National Communications required under the UNFCCC. Additional Special Reports on Aviation and the Global Atmosphere, Emissions Scenarios, and Technology Transfer were planned for 1999, while a comprehensive Third Assessment Report (TAR) is planned to be completed in 2001.

**Principles.** Because of the complexity and broad scope of the climate change issue as well as the potential for controversy, the IPCC process was developed around a set of principles meant to ensure accuracy and wide acceptance of its assessments of the state of knowledge. These principles include:

- Openness and transparency: Make draft reports freely and widely available to experts, policymakers, and the public for review and comment
- Comprehensiveness: Provide information on all relevant aspects of the climate change issue
- Inclusiveness and neutrality: Assess the full range of scientifically valid views on any given issue, establishing consensus when possible, highlighting scientific disagreements when contradictory evidence cannot be reconciled, and characterizing uncertainties and gaps in knowledge. Maintain a balance of views and avoid taking an advocacy position.

**The IPCC Process.** The IPCC works at two overlapping levels: as a formal intergovernmental body, and as a scientific/technical assessment body. IPCC reports are prepared and reviewed in a rather lengthy process of interaction between these two levels. The process requires more than two years for each of the comprehensive assessments, which are conducted about every five years. It is designed to ensure that all relevant information is assessed and that all scientifically grounded views are included in the final reports. The IPCC process includes five main components:

- Commissioning of reports
  Official government representatives approve the terms of reference and focus of IPCC reports and other IPCC activities (e.g., workshops, expert meetings) to ensure their relevance to decision making needs

- Author nomination and selection process
  Governments and organizations are invited to nominate scientists, researchers, and other experts to be lead authors or contributing authors for IPCC reports. Substantively and geographically balanced author teams are appointed by the IPCC Bureau, after careful review of the professional credentials of the nominees

- Drafting of technical materials
  The chapters of the reports—technical in nature—are prepared by the lead authors, based on information openly available in the scientific and technical literature. Lead authors do not "vote" on the content of these chapters; rather, findings are reached through careful collective evaluation of peer-reviewed research articles published in major academic journals and collections

- Review
  Reports undergo a three-stage review process: first by experts only, then by government representatives and experts, and finally by governments

- Acceptance and approval

The final stage is "acceptance" by governments of the technical material, and verbatim "approval" of the Summaries for Policymakers (SPMs). The SPMs are subject to this approval process because, by their nature, they cannot include all of the conclusions of the technical chapters, and the process of selecting information for inclusion is political. However, the lead authors participate in drafting and reviewing the summaries, which is intended to ensure that the information is consistent with the underlying technical chapters and represents a fair balance of available information.

**Controversies and Evolution of the IPCC Process.** During its first ten years, the IPCC evolved into one of the primary sources of scientific and technical information on climate change. Several of its conclusions resulted in controversy, and have spurred the evolution of the organization's procedures. The conclusion of the SAR that "the balance of evidence suggests a discernible human influence on global climate" produced a great deal of controversy in public and media circles, although it was not particularly controversial within the informed scientific community. Another controversy in the SAR concerned the use of economic methods to estimate potential climate change damages; these methods were based on the assumption that the (statistical) value of human life was higher for developed than for developing countries. Finally, the finding that there were many "no-regrets" options for reducing greenhouse gas emissions also generated significant debate.

In response to these controversies, the IPCC process was refined to enhance transparency and credibility of the panel's reports. Review editors were added to ensure that all review comments were treated fairly and that the political biases of the authors or members of the Bureau did not inject bias into the reports. In addition, in recognition of the diverse interests potentially affected by climate change and by potential adaptation and mitigation responses, the panel took steps to increase participation of researchers from developing countries and to encourage individuals from nongovernmental organizations to participate as contributors and reviewers. In this latter category, special efforts were made to entrain the business community in evaluation of the feasibility of options for adaptation and emissions mitigation.

[*See also* Climate Change; *and* Global Warming.]

### BIBLIOGRAPHY

Boehmer-Christiansen, S. "Global Climate Protection Policy: The Limits of Scientific Advice." *Global Environmental Change* 4.3 (1994), 140–159.

Intergovernmental Panel on Climate Change (IPCC). *IPCC Procedures for Preparation, Review, Acceptance, Approval, and Publication of Its Reports.* Approved by the panel at its ninth session (Geneva, 29–30 June 1993).

Moss, R. H. "The IPCC: Policy Relevant (Not Driven) Scientific Assessment." *Global Environmental Change* 5.3 (1995), 171–174.

Pace Energy Project. "The Great Global Warming Debate." Released June 16, 1997. New York: Pace University School of Law's Center for Environmental Legal Studies, Global Warming Central. http://www.law.pace.edu/env/energy/debateintro.html#Believe. (Introduction by Ed Smeloff, Director, The Pace Energy Project.)

—RICHARD H. MOSS

# INTERGOVERNMENTAL PANEL ON FORESTS

The Intergovernmental Panel on Forests (IPF) was established by the United Nations Commission on Sustainable Development (CSD) in 1995 to formulate options for action to support the management, conservation, and sustainable development of all types of forests. At the 1992 United Nations Conference on Environment and Development (UNCED), forestry issues were hotly debated, and agreement was possible on only a loosely worded Statement of Forest Principles. The IPF is a direct descendant of that debate.

After UNCED, governments established a number of forums that began to build bridges between the developed and developing countries and to provide the basis for international cooperation on forests. These forums included the Intergovernmental Working Group on Forests (Malaysian/Canadian Initiative), the Workshop on Reporting Guidelines (U.K./India Initiative), two intergovernmental processes to develop criteria and indicators for sustainable forest management (Helsinki and Montreal Processes), and the Centre for International Forestry Research/Indonesia Policy Dialogue on Science, Forests and Sustainability. Yet, by 1995, it was clear that the political debate on forests would need more attention and should focus on priority issues; IPF was created for that purpose.

The IPF met four times between September 1995 and February 1997 and submitted final conclusions and policy recommendations to the CSD in April 1997. The IPF's agenda was defined to consider the following: local, national, regional, and global dimensions of priority issues; the complexity and cross-sectoral aspects of forest issues; and fostering of partnerships between local communities, forest dwellers, the private sector, governments, experts, and institutions at all levels. The Panel, which was noted for its transparent and inclusive process, received support from the UN system, nongovernmental organizations, and eleven country-sponsored initiatives that produced over two hundred reports.

Among its conclusions, the Panel noted the critical need to understand the underlying causes of deforestation and forest degradation. The importance of traditional forest-related knowledge was stressed. There was also emphasis on national forest inventories and the development of criteria and indicators to help formulate effective national forest programs. The Panel recognized the importance of the multiple benefits provided by forests, such as wood and nonwood forest products and a range of services, including those related to biological diversity and global climate regulation. The Panel also urged countries to consider the value of voluntary certification and labeling schemes, provided they are not used as a form of disguised protectionism.

Nevertheless, everything was not resolved. Governments remained divided over such issues as national control of natural resources versus international oversight or regulation of global environmental concerns. Developing countries continued to call for new and additional financial resources and transfer of technology. Finally, there was divergence of opinion on action regarding a future international forest treaty. In the end, the Panel left open the question of what the international community's next steps would be and forwarded a range of options to the CSD, which was also unable to reach agreement. Finally, the Nineteenth Special Session of the General Assembly in June 1997 agreed to establish an Intergovernmental Forum on Forests, which would consider the need for or build the necessary consensus for a legally binding instrument on all types of forest. This Forum began its work in October 1997 and, after four sessions, forwarded a new set of recommendations to the CSD in February 2000.

[*See also* Carbon Cycle; Deforestation; Forestation; *and* Forests.]

### BIBLIOGRAPHY

Carpenter, C., et al. "Summary of the Fifth Session of the Commission on Sustainable Development: 8–25 April 1997." *Earth Negotiations Bulletin* 5.82 (28 April 1997). http://www.mbnet. mb.ca/linkages/vol05/0582000e.html. Contains information on the treatment of the report of the IPF.

———. "Summary of the Nineteenth United Nations General Assembly Special Session to Review Implementation of Agenda 21: 23–27 June 1997." *Earth Negotiations Bulletin* 5.88 (30 June 1997). http://www.mbnet.mb.ca/linkages/vol05/0588000e.html. Contains information on the treatment of the report of the IPF.

Davenport, D., et al. "Summary of the Fourth Session of the Intergovernmental Panel on Forests: 11–21 February 1997." *Earth Negotiations Bulletin* 13.34 (24 February 1997). http://www. iisd.ca/linkages/vol13/1334000e.html. Summary of the work of the IPF at its fourth session.

United Nations. "Forests in the Global Political Debate: The Intergovernmental Panel on Forests: Its Mandate and How It Works." New York, 1996. gopher://gopher.un.org:70/00/esc/cn17/ipf/ipf-fly.txt. A summary of the history and purpose of the IPF.

———. "Report of the Ad Hoc Intergovernmental Panel on Forests on its Fourth Session." (E/CN.17/1997/12). 20 March 1997. gopher://gopher.un.org:70/00/esc/cn17/ipf/session4/97—12.EN. Final report on the work of the IPF.

—Pamela S. Chasek

## INTERNATIONAL ATOMIC ENERGY AGENCY

The International Atomic Energy Agency (IAEA) was first proposed by U.S. President Dwight D. Eisenhower in his "Atoms for Peace" speech to the United Nations on 8 December 1953. It came into existence as an "autonomous agency under the aegis of the United Nations" when its statute entered into force on 29 July 1957; in December 1999, it had 130 members. Eisenhower's motivation was to balance the extensive and growing military applications of nuclear energy with a peaceful applications program that would expand the peaceful uses of nuclear energy and, at the same time, present the United States in the intensifying cold war as the leader in its technological applications. It was assumed that the uranium needed to power nuclear reactors was in short supply (in the event quite wrong) so that providing an international mechanism to make uranium available for peaceful uses under controlled conditions would be an effective way to prevent diversion for military purposes. It was also assumed, incorrectly as well, that nuclear power would be practical and competitive as an energy source within a relatively short time.

Equally important, the proposal reflected a desire to reopen consideration of a nuclear arms control agreement after the failure some years earlier of the Baruch Plan for control of atomic arms. The concept was that, in time, both the United States and the Soviet Union would turn over uranium to the IAEA for peaceful applications in other countries; that uranium would thus not be used for armaments and, perhaps, would eventually come from decommissioned weapons.

Although the hope of an agreement to reduce the military use of nuclear energy was illusory, the U.S. government did believe that the Soviet Union would at least share the goal of controlling the spread of nuclear material and technology. But the Soviets were resistant for a number of years. In 1963 they changed their position and gradually became staunch supporters of the IAEA, in particular, its safeguards program, which carries out inspections in member states to detect and deter clandestine diversions of nuclear fuel. The Agency also has a substantial technical assistance program to aid in peaceful applications in developing countries.

The IAEA might have had a major role in mitigating global climate change if nuclear power, which yields practically no emissions of greenhouse gases, had become the inexpensive technology expected. Although widely used in several Western countries, future prospects for nuclear power in those countries are bleak, for economic and safety reasons. In Asia its prospects are better, at least in the short term, but are limited by high capital costs. [*See* Nuclear Industry.]

The Agency is also concerned about the physical

safety of nuclear power and has been instrumental in monitoring and studying the effects of the one large nuclear accident so far, that at Chernobyl, Ukraine, in 1986. Clearly, any future nuclear accidents would have a significant global effect, either through direct physical effects or through the psychological impact on countries dependent on nuclear energy for generation of electricity. [*See* Nuclear Hazards.]

Perhaps the IAEA's greatest contribution to global change will not be in nuclear power directly but in demonstrating that international organizations can successfully monitor and supervise stringent international conventions. Its safeguards record is good and is getting better as its experience, especially in North Korea and Iraq, has become more extensive. That may be a very large contribution indeed.

[*See also* Electrical Power Generation; *and* Nuclear Accident and Notification Conventions.]

### BIBLIOGRAPHY

Aspen Institute for Humanistic Studies. *Proliferation, Politics, and the IAEA: The Issue of Nuclear Safeguards.* Parts I and II. Berlin, 1987.
Greenwood, T. *Nuclear Power and Weapons Proliferation.* London: International Institute for Strategic Studies, 1976.
International Atomic Energy Agency. Annual Yearbooks. Vienna.
Office of Technology Assessment. *Nuclear Safeguards and the International Atomic Energy Agency.* Washington, D.C., 1995.
Scheinman, L. *The Nonproliferation Role of the International Atomic Energy Agency: A Critical Assessment.* Washington, D.C.: Resources for the Future, 1985.
———. *The International Atomic Energy Agency and World Nuclear Order.* Washington, D.C.: Resources for the Future, 1987.
Smart, I., ed. *World Nuclear Energy: Toward a Bargain of Confidence.* Baltimore: Johns Hopkins University Press, 1982.

—EUGENE B. SKOLNIKOFF

## INTERNATIONAL COOPERATION

International cooperation refers to processes of policy coordination by which states and other entities (such as multinational corporations or nongovernmental organizations) adjust their behavior to the actual or anticipated preferences of other states, or of entities based in other countries. Cooperation should be sharply differentiated from harmony. In harmonious situations, actors support one another merely by pursuing their own interests: no discord occurs, and there is no need for cooperation. Cooperation arises not from harmony but from discord: in the absence of policy coordination, conflict would arise. For example, cooperation was necessary to reduce the production of ozone-depleting chemicals (in the Montreal Protocol on Substances that Deplete the Ozone Layer of 1987, and subsequent agreements), since many states would not independently have reduced their production and use of such substances

without reciprocal commitments by others. [*See* Montreal Protocol.]

Some international cooperation is self-enforcing. That is, when agreements are reached, everyone has an incentive to comply with the terms. In such coordination situations, bargaining may be difficult because actors have different preferences about what agreement should be reached: what language air traffic controllers and pilots should speak, what standard should be used for high-definition television, whose tanks should be purchased by an alliance, which state should pay more, which less. But once agreement is reached, the common preference for a uniform standard overrides these preferences, and everyone can be expected to conform to the established norm.

Unfortunately, much international cooperation, including all of the major environmental issues that have been the subject of negotiation during the last quarter of the twentieth century, does not lead to self-enforcing agreements. Even after negotiations have yielded rules, states (and nongovernmental actors as well) may seek to reinterpret them or even to evade them—trying to shift the burdens of regulation onto others, while retaining the benefits of collective action. Collaboration of this type requires ongoing monitoring and, potentially, enforcement of agreements.

International cooperation typically passes through several conceptually distinct but often temporally overlapping phases, around which this article will be organized. First, interdependence generates discord, which produces a demand for some sort of international regulation. Through such a contentious process, some issues are placed on the international agenda. Second, the nature of the problem is defined, or framed, through a political process involving advocacy and discourse. Third, negotiations take place to reach an international decision on solutions, typically involving the construction or adaptation of international institutions. Negotiations take place in the shadow of concern about ratification by major governments. Finally, cooperation has to be implemented: nominal policies have to be put in place, and each country's actions need to be monitored so that others are assured that compliance is mutual. As just noted, in collaboration situations these issues of implementation are not trivial; indeed, they are often quite difficult.

As this sketch suggests, international cooperation is a complex, multilevel process. States are key actors, since it is they who have to agree to policies and to devise measures to enforce them within their jurisdictions. But other actors play important roles, ranging from issue-advocacy networks and nongovernmental organizations, which generate demands for policy change, to professional networks of scientists who evaluate environmental risks, to multinational corporations and transgovernmental coalitions of bureaucrats from a variety of countries, who may implement, or resist, policy change. As discussed at the end of this article, attempts at international cooperation can therefore vary widely in effectiveness.

**Interdependence, Sovereignty, and Regulation.** The discord that generates efforts at international cooperation arises from a combination of transnational interdependence and dilemmas of collective action created by sovereignty. On environmental issues, increasing transnational interdependence is evident, as illustrated by issues ranging from pollution of international rivers to trade in endangered species to climate change. But states remain in control of policy making and implementation: this is what sovereignty means. Sovereign governments respond principally to domestic pressures, through their political systems. Hence there is a disjunction between the scope of problems (increasingly transnational and even global) and the scope of political systems.

The widening scope of problems means that states find it difficult or impossible to achieve their policy objectives unless other states pursue compatible policies. On environmental issues, the essential problem is that each actor (state or firm) has incentives to capture the benefits of industrial action for itself while pushing many of the negative environmental effects onto others. In economic language, this problem is referred to as one of "externalities," since the costs of these actions are not borne by the state or firm in question. For example, a number of firms could rationally decide to engage in industrial activities that produce large quantities of sulfur dioxide, and the countries in which they operate could tolerate or even encourage these polluting activities—if much of the pollution were blown to neighbors' territories. Even if these activities generated large amounts of acid rain whose costs for the region as a whole were much greater than the benefits from the industrial activities, the offending firms and their governments would have little reason to cease their pollution, on the basis of their own interests.

Negative externalities are common. On an international basis, furthermore, the effects of actions that affect the environment are rarely symmetrical. Countries that are downriver or downwind will be environmentally damaged by their neighbors' pollution more than vice versa. Discord will lead to diplomatic protests; and, if these protests are not heeded, they may get louder. But protests do not necessarily lead to issues appearing on the international agenda. Powerful states, such as the United States, Japan, and the larger members of the European Union, can usually get their concerns inscribed quickly, since they can promise rewards from cooperation and threaten punishment for recalcitrance. Weak states will have more difficulty getting their concerns

heard, and the problems will be even more severe for groups that do not command the support of their own governments. Their discordant voices are likely to need the assistance of issue-advocacy networks, with access to the mass media and to public opinion in the advanced industrialized countries—as illustrated by campaigns to save the Brazilian and Malaysian rainforests.

Some important issues may never get onto the agenda; others will be slow to do so. There is no guarantee that the issues on which international cooperation takes place will be objectively the most important issues from the standpoint of world welfare or environmental quality. On the contrary, since power and voice are so important, we can expect that the international agenda will disproportionately reflect the interests of the powerful. It is not surprising, therefore, that stratospheric ozone depletion and climate change—issues that concern people in advanced industrialized countries—are higher on the international agenda than desertification, access to healthy water, and forest destruction in the poorest parts of the world.

**Framing the Problem.** Framing refers to the way in which a particular issue is viewed; that is, how it is placed in a social context by the actors concerned with it. How problems are framed is crucial in determining the paths that attempts at cooperation will later follow. Whether the framing is benign or malign can profoundly affect the degree of concern people express about it: is surgery to remove the clitoris "female circumcision" (which may seem benign) or "female genital mutilation" (malign)? The environmental issue of climate change was often framed early on as "global warming," which might suggest a benign process to people in colder climes, whereas "climate change" conveys feelings of increased uncertainty and threat.

How issues are framed depends in large measure on the constituency that dominates discussion of the issue. Acid rain was not a problem of dying lakes for the Scandinavian countries, but of forests for Germany. The nuclear safety issue after the 1986 explosion at Chernobyl was framed by many environmental groups and some Western governments as a problem of quickly ending reliance on nuclear energy, but it was framed by the nuclear industry and other governments (East and West) as one of devising and implementing technical measures to increase the safety of nuclear power plants. Rhine River pollution could have been framed broadly as a multiple-source, multiple-chemical issue, but by the 1960s a definition of the problem had been institutionalized as one of chloride pollution from a small number of sources. Early in a process of cooperation, one of the most promising ways to achieve one's desired results is to influence the framing of a problem, in the public as well as among policy makers. At this stage, nongovernmental actors such as Greenpeace may have a large voice, quite disproportionate to their material resources, if they can articulate themes that resonate with the public or with policy makers. [*See* Greenpeace.]

Framing is done by institutionalized groups of scientists as well as by advocacy groups. Indeed, a recent development in international cooperation—especially evident in the environmental area—is the institutionalization of scientific groups, whose function is to assess changes occurring in physical processes, determine their impact, and discuss social and political factors that may affect their severity for human life. Consider, as an example, the Intergovernmental Panel on Climate Change (IPCC). In 1988 two United Nations (UN) agencies, the World Meteorological Organization (WMO) and the United Nations Environment Programme (UNEP), created the IPCC to assess whether human actions were having significant effects on the global climate, and what those impacts might be. When the Rio Conference of 1992 established the Framework Convention on Climate Change (FCCC), the work of the IPCC was linked to the FCCC. Although the IPCC is an intergovernmental body, most of its work is carried out through three scientific working groups, run by scientists using extensive peer review, with twenty to sixty reviewers per chapter. In other words, the IPCC institutionalizes scientific assessment in such a way as to produce comprehensive, systematic statements of what is scientifically known about climate; and it does so repeatedly. When Working Group I concluded in 1995 that the evidence, on balance, suggested that human activities were altering the global climate, this finding became the basis for a major intergovernmental conference at Kyoto in December 1997. [*See* Kyoto Protocol.]

How a problem is framed reflects a combination of public concern, political calculation (by governments and others), and scientific knowledge. Truth, interests, and power are all involved. How problems are defined affects what is negotiated, how urgently action is taken, and the types of solution proposed.

**Negotiating Agreements.** Many people probably equate international cooperation with the negotiation of international agreements, such as the Montreal Protocol or the Framework Convention on Climate Change. These negotiations are indeed an indispensable part of most international cooperation: tacit or spontaneous cooperation is possible, but, on complex issues ranging from trade to banking to environmental issues, negotiated agreements are by far the most important.

In international negotiations, reciprocity is always a central principle: providing benefits in return for benefits and imposing costs in retaliation for recalcitrance. In an elegant analysis, Robert Axelrod (1984) has shown how reciprocity can generate cooperation among egoists. In world politics, reciprocity is both an informal convention of behavior and formally institutionalized in

organizations such as the World Trade Organization (WTO).

Reciprocity, however, does not imply symmetry. As indicated above, discord arises asymmetrically: those that are hurt by externalities generated by others lead the fight to put issues on an intergovernmental agenda. As a result, negotiations are typically characterized by a dynamic between leaders and laggards. Leaders put issues on the table and press for action; laggards raise objections, propose less extensive or less binding rules, and generally drag their feet. Different countries play different roles in different situations: the United States, for instance, was a leader on ozone depletion but is a laggard on climate change.

Beyond the public pronouncements, bargaining proceeds through offers of benefits on other issues, implicit or explicit threats of recalcitrance if sufficient accommodation from others is not forthcoming, and—in the case of the United States—hints that the Senate will not approve the agreement unless it meets certain U.S. terms. But in public, emphasis is typically on principles and world social welfare. As a diplomat once told the author, "At the UN, the meal of self-interest is served up on a platter of universal principle." Interests and power play a key role, but democratic governments are constrained by the way in which the problem has been framed and how their actions will appear to their publics. The administration of U.S. President George Bush, for instance, was very reluctant to impose strong environmental regulations, particularly at the behest of other countries; but it nevertheless reluctantly agreed to the process leading to the Rio Conference, since it did not wish to be labeled as blatantly antienvironmental.

Some negotiations are merely *ad hoc*, called by a government or set of governments—for example, the Canadian Government convened a conference on banning land mines in 1997, under the influence of a nongovernmental issue-advocacy network. Most negotiations, however, take place within an institutionalized context of rules and organizations. For example, the United Nations Conference on Environment and Development of 1992 (the Rio Conference) was formally authorized by the United Nations General Assembly, after previous activity by the UNEP.

Most negotiations, however, are sponsored by formal international organizations and take place within a firmly established institutional context, against the background of previous agreements and with the expectation of future negotiations on similar issues. These institutions are conventionally referred to as *international regimes*—sets of rules, procedures, norms, practices, and organizations that affect behavior, and expectations of others' behavior, in an issue area. The international environmental regime includes rules and procedures such as those contained in the Montreal Protocol, in the European Convention on Long-Range Transboundary Air Pollution (LRTAP), and in the Convention on International Trade in Endangered Species of Wild Fauna and Flora (CITES). It also involves organizations, such as UNEP and the WMO. Actual political processes within these formal institutions also entail informal practices, based on procedural principles such as reciprocity and substantive ones such as the norm of sustainability, that help to shape which negotiating tactics are acceptable and which are not.

International institutions are important partly because they reduce the difficulties and costs of negotiating agreements: their secretariats and facilities and established procedures make it easier to get together and to organize complex global or regional conferences. Beyond this facilitating role, the rules and procedures of international institutions constrain states' bargaining strategies, limiting legitimate tactics and helping to coordinate expectations about likely pathways, or focal points, for agreement. By grouping a number of issues together, they make linkages or trade-offs among issues more feasible, thus creating negotiating space for mutual benefits—political gains from exchange. Having multiple issues involved also provides leverage for inducing compliance when a state finds that a given rule is costly to it: if it benefits from other aspects of an agreement, it may be reluctant to renege too blatantly where it is a loser, for fear that the agreement as a whole will unwind. Finally, virtually all international institutions play a role in monitoring agreements—either through international organizations established partly for that purpose (as in the LRTAP conventions on acid rain in Europe) or by establishing rules facilitating decentralized monitoring by states.

One way to summarize the role of institutions is to say that well-designed formal institutions help to make states' commitments more credible. Governments are unlikely to take costly actions that they would not take independently, unless other states make credible commitments to take actions that are helpful to others. By providing information about other states' behavior, and creating expectations that agreements will, on the whole, be implemented, international institutions increase the credibility of agreements. In view of the functions they perform, it is not surprising that international institutions are almost always associated with sustained patterns of international cooperation.

Nevertheless, by the domestic standards of well-functioning societies, these institutions are weak. Decisions usually require unanimous consent (with the notable exception of the European Union, where qualified majority voting applies on issues covered under Pillar One of the Maastricht Treaty). Voting is unusual in these meetings; instead, negotiations continue until a consensus appears. Consensus does not imply unanimity on every point, but it does require that all states with the ability to block the implementation of an agreement be sufficiently satisfied to accept it. Furthermore, since centralized enforcement is usually weak or nonexistent, negotiators have to recognize that nominal agreement may not translate into real action. Coalitions that oppose strong regulation—such as the Climate Change Coalition sponsored by industrial firms in industries such as petroleum—may be able to muster considerable support through public-relations campaigns and lobbying within key countries, as well as at conferences themselves. Although it may be relatively easy for issue-advocacy networks to get issues on the international agenda—especially if the opposition is slow to mobilize—it is much more difficult to engineer the passage of costly regulatory agreements.

As a result of these weaknesses, international negotiations are frustrating to advocates of strong measures. Observers have commented on the "law of the least ambitious program," by which agreements are diluted to meet the demands of laggard states. Some negotiations, such as those for a Law of the Sea Convention that lasted throughout the 1970s, failed to reach formal agreements accepted by all major states (the United States has not ratified the Law of the Sea Convention).

Nevertheless, despite these frustrations, many cooperative agreements have been reached during the last half century. Indeed, the period since 1944 has been unprecedented in the number and scope of international agreements reached, and in the extent and significance of the international regimes that have been established. Well over half the environmental treaties adopted since 1921 have been concluded since the United Nations Conference on Human Environment (the Stockholm Conference) in 1972. Max Weber defined *politics* as "the boring of hard boards." Certainly, this characterization would apply to international negotiations. But in a long-run perspective, what is more striking than the difficulty of reaching agreements is the number and importance of the agreements that have been reached and the regimes that have been established. International cooperation is difficult, but its existence is an unchallengeable fact.

**Implementation and Monitoring of Agreements.** The implementation record of major international agreements is mixed. On the positive side, most international agreements are adhered to by most countries most of the time. States are disinclined to enter into agreements that they cannot implement, since inconsistency between nominal assent and action can be embarrassing and can adversely affect a state's reputation, making other states reluctant to make further agreements with the offender. Occasionally, noncompliance can even generate retaliation. Furthermore, in pluralistic democracies, officials assigned to carry out agreements usually believe that they should be implemented. Fairness and good faith are deeply embedded principles—and, among friendly countries, essential to managing common problems. Norms of compliance thus undoubtedly exist and affect the behavior of governments.

One implication of this tendency of governments toward compliance, however, is a reluctance to accept agreements that are too demanding. The "law of the least ambitious program" is the other side of the compliance norm. For a democratic government, the worst result of a negotiation is to find itself committed to a demanding program that has high domestic political costs—that is, jeopardizes its chances at reelection. In such a case, no option is very appealing. Compliance jeopardizes reelection; noncompliance jeopardizes reputation. In this event, we often find governments seeking to reinterpret their commitments to fit their political exigencies, so that implementation becomes a process of argument and loophole seeking as much as one of rule following. From the government's standpoint, however, it is better to anticipate these problems and negotiate an undemanding standard. If noncompliance were more acceptable, there might be more highly demanding international agreements.

As noted above, a major function of international institutions is to monitor adherence to international agreements. Success in monitoring depends heavily on the types of rules that are enacted. For example, attempts to limit deliberate dumping of oil sludge by tankers at sea were frustrated between 1954 and 1978 by the impossibility of monitoring a rule against discharges. The oceans are large and the nights are black: tanker captains ignored the rules. But new rules, enacted in 1978, required specific equipment on tankers, whose existence could be readily monitored in ports. Furthermore, the new rules, unlike the old ones, were largely self-enforcing, since insurance companies demanded the equipment and, once the equipment existed, tanker owners and captains had every incentive to use it.

As this example suggests, it is often nonstate actors—such as tanker captains and shipping firms—that violate agreements, while states either are incapable of taking effective action or their officials look the other way. A notable example is CITES, which seeks to regulate trade in endangered species. The borders of many states are porous, and customs officials may either be poorly trained or corrupt. Russia since the breakup of the Soviet Union is a notable example. Other countries that could regulate the trade may profit from it, and so deliberately allow illegal traffic to occur.

When states fail to implement provisions of agreements, others sometimes resort to the threat and (more rarely) the use of sanctions, particularly economic sanctions. But sanctions are a blunt instrument, often falling on the innocent rather than the guilty; and only in ex-

treme cases (such as Iraq's attack on Kuwait and its manufacture of chemical and biological weapons) do a sufficient number of countries actively cooperate to make sanctions effective. More important are combinations of encouragement, technical assistance ("capacity building"), systematic monitoring, and public pressure—designed, on the one hand, to make implementation easier and less costly, and, on the other, to discourage blatant noncompliance by penalizing it, in more or less formal or informal ways.

For example, the Montreal Protocol regime includes a Non-Compliance Procedure (NCP) with a standing committee that makes general recommendations, and an *ad hoc* component that investigates complaints. These arrangements have established general reporting requirements and have brought to light instances of noncompliance. States that seek to comply with the rules but have difficulties doing so have been assisted. But when deliberate noncompliance has occurred, the reconciliation processes of the NCP have not been effective by themselves: they have had to mobilize pressure on the offenders from powerful governments.

The overall pattern of implementation could be summarized as follows. States need to be persuaded to accept and implement rules, since, outside the European Union, there is very little centralized enforcement. Since noncompliance is embarrassing, states tend to be conservative in the rules they will accept, leading to the law of the least ambitious program. In the vast majority of cases, states do implement their agreements, certainly in the environmental area. However, there are many instances of known noncompliance, often involving the failure of states to prevent illegal action by nonstate actors under their jurisdiction. Some of these problems are intractable, particularly when effective enforcement (as with respect to illegal drugs or endangered species) tends to drive up the price of the product and therefore increase incentives to evade the rules. The most effective way to deal with them seems to be through systems of implementation review, which combine monitoring, capacity building, appeal to norms, publicity, and, only as a last resort, threats of sanctions.

**The Effectiveness of International Cooperation.** Ascertaining the effectiveness of international cooperation is difficult, because it is hard to know what policies states would have followed, and what practice nonstate actors would have engaged in, in the absence of the agreements that we observe. Often states only agree to actions that they would in any case have taken. However, there is some rather clear evidence about the effects of efforts at cooperation.

Apart from recalcitrant states that are willing to be international pariahs, it is probably the leading states—those that press for international action—whose regulatory actions are least likely to be affected by cooper-

| DATE | AGREEMENT (FULL NAME) | REFER TO ARTICLE |
|---|---|---|
| 1946 | International Convention for the Regulation of Whaling (ICRW) | Whaling Convention |
| 1959 | Antarctic Treaty | Antarctica, *article on* Antarctic Treaty System |
| 1972 | Stockholm Declaration on the Human Environment | Stockholm Declaration |
| 1972 | Convention for Conservation of Antarctic Seals | Antarctica, *article on* Antarctic Treaty System |
| 1972 | London Dumping Convention | London Convention of 1972 |
| 1973 | Convention on International Trade in Endangered Species (CITES) | Convention on International Trade in Endangered Species |
| 1973 | International Convention for Prevention of Pollution from Ships (MARPOL 73/78) | Marine Pollution Convention |
| 1975 | Convention on Wetlands of International Importance | Ramsar Convention |
| 1979 | Convention on Long-Range Transboundary Air Pollution | Convention on Long-Range Transboundary Air Pollution |
| 1980 | Convention on Conservation of Antarctic Marine Living Resources | Antarctica, *article on* Antarctic Treaty System |
| 1982 | United Nations Convention on the Law of the Sea | Law of the Sea |
| 1985 | Vienna Convention for Protection of the Ozone Layer | Montreal Protocol |
| 1986 | Convention on Assistance in Case of a Nuclear Accident or Radioactive Emergency | Nuclear Accident and Notification Conventions |
| 1986 | Convention on Early Notification of a Nuclear Accident | Nuclear Accident and Notification Conventions |
| 1987 | Montreal Protocol on Substances that Deplete the Ozone Layer | Montreal Protocol |
| 1988 | Convention on Regulation of Antarctic Mineral Resource Activities | Antarctica, *article on* Antarctic Treaty System |
| 1989 | Prior Informed Consent in Hazardous Chemicals and Pesticides | Prior Informed Consent for Trade in Hazardous Chemicals and Pesticides |
| 1989 | Convention for Prevention of Fishing with Long Driftnets in the South Pacific | Driftnet Convention |
| 1989 | Basel Convention on the Control of Transboundary Movements of Hazardous Wastes, and Their Disposal | Basel Convention |
| 1991 | Antarctic Environment Protocol | Antarctica, *article on* Antarctic Treaty System |
| 1992 | Rio Declaration on Environment and Development | Rio Declaration |
| 1992 | United Nations Framework Convention on Climate Change | Framework Convention on Climate Change |
| 1992 | Convention on Biological Diversity (CBD) | Convention on Biological Diversity |
| 1992 | Agenda 21 | Agenda 21 |
| 1994 | United Nations Convention to Combat Desertification in Countries Experiencing Serious Drought and/or Desertification | Desertification, *article on* Desertification Convention |
| 1995 | United Nations Agreement on Straddling Fish Stocks and Highly Migratory Fish Stocks | Straddling and Migratory Stocks Agreement |
| 1997 | Kyoto Protocol | Kyoto Protocol |
| 1998 | Convention for Prior Informed Consent for Trade in Hazardous Chemicals and Pesticides (binding convention versus voluntary of 1989) | Prior Informed Consent for Trade in Hazardous Chemicals and Pesticides. (*Note:* Convention not in force as of 1999.) |

—DAVID J. CUFF

ation. Often these states have set domestic standards that are higher than those negotiated internationally. They see their task as getting others to move toward their standards, rather than as using international negotiations to toughen their own policies.

However, the process of publicity and negotiation can even affect the policies of leaders, particularly if they have democratic governments. In the first place, one major effect of international assessment procedures is to generate scientific knowledge. If such knowledge increases the certainty that environmental harm is occurring, it may lead governments to act more vigorously. In democratic states, publicity can also have a significant effect. On the issue of European acid rain dealt with by LRTAP, for instance, knowledge of damage was enhanced by LRTAP, leading some governments to take a stronger stance. Furthermore, international negotiations raised the salience of issues to publics, providing incentives for politicians to seek to take stronger action so as to gain credit with "green" electorates.

International agreements are probably more likely to raise standards in laggard states—those that would otherwise do nothing or follow even weaker policies than those enacted in treaties. For example, the United Kingdom's policy on acid rain was affected by issue linkages. As an upwind state, it had little direct incentive to take strong action; but its early recalcitrance led to the label of "dirty man of Europe," which hindered its relationship with its European partners on other issues. The regime on oil pollution at sea induced tanker firms that would have otherwise continued to dump sludge to reform their behavior, since they had to have tank-cleaning equipment if they were to enter major ports. Russia and other successor states of the Soviet Union have not complied well with the provisions of the Montreal Protocol, but arguably the situation would have been much worse in the absence of the Protocol and its systems of implementation review.

Another problem with effectiveness is that international regimes in different areas may conflict with one another. For instance, by any standard the rules of the WTO are remarkably strong, and the WTO contains provisions for third-party settlement of disputes. There is a genuine trade regime, which has demonstrably reduced protectionism and curbed intergovernmental trade disputes. Yet this same trade regime makes it more difficult for states to protect the environment through the use of trade restrictions, hence arguably reducing political pressures for multilateral environmental regimes as alternatives to such unilateral measures. Effectiveness in one issue area may create difficulties for effective actions on other issues.

In summary, there is clear evidence that international environmental regimes are sometimes effective, at least to some extent. But their effectiveness should not be exaggerated. Deadlock can prevent any effective agree-

ment, as is still the case with respect to such issues as protection of tropical forests and climate change. Agreements are typically not very ambitious, so not all states that comply actually experience policy change. Rules in adjacent issue areas may compete with one another. And some noncompliance occurs. Implementation, like negotiation, involves "the boring of hard boards."

For international cooperation, to paraphrase Charles Dickens, it is the best of times, and the worst of times. It is the best of times because the last few decades have seen unprecedented levels of cooperation, on issues ranging from European security to trade and finance to environmental protection. Some of these efforts have led to highly institutionalized regimes, to ongoing programs of scientific research, and to the creation of international organizations and issue-advocacy networks. Not only does it become more difficult for governments to "backslide," but a dynamic of negotiations for more extensive and stronger rules is created. This dynamic is clearest in the trade area, but one sees it on some environmental issues, such as the ozone regime, in which the Montreal Protocol was successively strengthened and timetables for phasing out production of ozone-depleting substances shortened.

The dynamic character of international cooperation implies feedback between the stages of action around which this article has been organized. Cooperation is not a linear process. Negotiations may reframe issues, leading to different definitions of the problem. Failed negotiations, or inadequate action, may lead to increased discord and pressure for action. New scientific knowledge generated as a result of international cooperation may lead issues to be reframed, or may generate pressure for more specific or demanding agreements—as in the process that led from the vague and general Vienna Convention for Protection of the Ozone Layer (1985) to the Montreal Protocol (1987) and the amendments negotiated in London (1990) and Copenhagen (1992). As discussed above, failure to implement earlier conventions on the dumping of oil wastes at sea led to more effective action in 1978.

If it is the best of times for international cooperation, it is also the worst of times, because our awareness of the need to regulate action at the global level has risen faster than our capacity to do so. Partly because of international scientific cooperation, we are more aware of environmental dangers—of which global climate change is only the most dramatic—than we were only one or two decades ago. International cooperation proceeds at a snail's pace compared with our upgrading of the magnitude of ecological risks. In a world of intensified ecological, material, and security interdependence, advocates and practitioners of international cooperation need to run ever faster just to stand still.

[*See also* Commons; Environmental Law; Environmental Movements; Nongovernmental Organizations;

United Nations; *and* United Nations Environment Programme.]

## BIBLIOGRAPHY

Axelrod, R. *The Evolution of Cooperation.* New York: Basic Books, 1984. A brilliant demonstration, through computer simulation and with illustrations from international relations, of how reciprocity can generate cooperation.

Chayes, A., and A. H. Chayes. *The New Sovereignty: Compliance with International Regulatory Agreements.* Cambridge: Harvard University Press, 1995. An argument for a "managerial" rather than an "enforcement" model of compliance.

Downs, G. W., D. M. Rocke, and P. N. Barsoom. "Is the Good News about Compliance Good News about Cooperation?" *International Organization* 50.3 (Summer 1996), 379–406. A challenge to the "managerial model," pointing out incentives for governments to agree only to shallow cooperation.

Fearon, J. D. "Bargaining, Enforcement and International Cooperation." *International Organization* 52.2 (Spring 1998), 269–306.

Haas, E. B. *When Knowledge Is Power: Three Models of Change in International Organizations.* Berkeley: University of California Press, 1990. A highly original theoretical analysis of how knowledge, power, and policy are linked in international organizations.

Haas, P. M. *Saving the Mediterranean: The Politics of International Environmental Cooperation.* New York: Columbia University Press, 1990. Haas develops the concept of "epistemic community" in exploring the impact of scientific networks on pollution control in the Mediterranean.

———, ed. "Knowledge, Power and International Policy Coordination." *International Organization* 46.1 (Winter 1992). An important compilation of work on issues relating science to international cooperation.

Haas, P. M., R. O. Keohane, and M. Levy, eds. *Institutions for the Earth: Sources of Effective International Environmental Protection.* Cambridge: MIT Press, 1993. The articles by Marc Levy and Ronald Mitchell are particularly outstanding exemplars of work that demonstrates the effects of international environmental regimes.

Jasanoff, S. *The Fifth Branch: Science Advisors as Policymakers.* Cambridge: Harvard University Press, 1990. A major study of the interaction between science and policy, important for understanding international cooperation on issues involving scientific knowledge.

Keck, M. E., and K. Sikkink. *Activists beyond Borders: Advocacy Networks in International Politics.* Ithaca, N.Y.: Cornell University Press, 1998. An impressive study of how nongovernmental activists affect international policy on human rights and environmental issues.

Keohane, R. O. *After Hegemony: Cooperation and Discord in the World Political Economy.* Princeton, N.J.: Princeton University Press, 1984. This work defines cooperation and shows how international institutions can promote it.

Keohane, R. O., and M. Levy, eds. *Institutions for Environmental Aid.* Cambridge: MIT Press, 1996. This joint work uses a consistent analytical framework to compare problem definition, negotiations, and implementation on seven environmental issues involving financial transfers between countries.

Keohane, R. O., and E. Ostrom, eds. *Local Commons and Global Interdependence: Heterogeneity and Cooperation in Two Domains.* London: Sage, 1995. An attempt to show the parallels between international cooperation and cooperation on commons issues in small communities.

Litfin, K. T. *Ozone Discourses: Science and Politics in Global Environmental Cooperation.* New York: Columbia University Press, 1994. An exploration of the relationship between science and policy with respect to international cooperation, through an analysis of the international ozone regime.

Mitchell, R. B. *Intentional Oil Pollution at Sea: Environmental Policy and Treaty Compliance.* Cambridge: MIT Press, 1994. A rigorous analysis of compliance, using the oil pollution regime as the domain of analysis.

Ostrom, E. *Governing the Commons: The Evolution of Institutions for Collective Action.* Cambridge: Cambridge University Press, 1990. A major theoretical and empirical statement about how small communities can govern the commons without encountering tragedy.

Victor, D. G., K. Raustiala, and E. B. Skolnikoff, eds. *The Implementation and Effectiveness of International Environmental Commitments.* Cambridge: MIT Press, 1998. Part I of this work develops the valuable concept of systems for implementation review.

Young, O. R. *International Cooperation: Building Regimes for Natural Resources and the Environment.* Ithaca, N.Y.: Cornell University Press, 1989. A pioneering work on international environmental cooperation.

———. *International Governance: Protecting the Environment in a Stateless Society.* Ithaca, N.Y.: Cornell University Press, 1994. A continuation of Young's major research program on environmental cooperation.

—ROBERT O. KEOHANE

# INTERNATIONAL COUNCIL FOR SCIENCE (ICSU)

This organization was known as the International Council of Scientific Unions during most of its history. The name was changed to International Council for Science in 1998 but the acronym ICSU was retained. The roots of the International Council for Science go back a hundred years. The innate desire of scientists to converse with one another, to work together, and to take science beyond the limitations of a solo effort prompted Eduard Suess, president of the Austrian Academy of Sciences, in 1897 to invite the Royal Society of London to attend a meeting of the Academies of Göttingen, Leipzig, Vienna, and Munich. The purpose of the meeting was to explore the creation of an association of academies.

The International Association of Academies (IAA) was established in 1899 by twelve members from Austria, England, France, Germany, Italy, Russia, and the United States. Its main purpose was "to initiate and otherwise to promote scientific undertakings . . . and to facilitate scientific intercourse between different countries." The governance included a Council and a General Assembly consisting of a "section on Natural Science and a Literary and Philosophical Section."

IAA did not survive the hostilities of 1914–1917, but

its motivation did. It arose, phoenixlike, in 1919, as an interim International Research Council (IRC), largely through the efforts of G. Hale in the United States, C. Picard in France, and A. Schuster in England. In the aftermath of World War I, membership from nations was restricted to Allied Powers, but a "sunset clause" limited IRC's lifetime to ten years. An important step was taken in 1922 when membership was extended to include international, disciplinary, scientific Unions (where *Union* was an alternative term for *Society* or *Association*).

In 1928, the transition began from a caucus of academies of the Allied Powers to a truly international system. On 11 July 1931, the International Council of Scientific Unions came into being with forty National and eight Union Members. By 1999, ICSU had ninety-eight National Members, twenty-six Union Members, twenty-eight Scientific Associates (who support the objectives and principles of the ICSU but do not have voting rights), and more than two dozen standing, special, and scientific committees. Hundreds of international congresses, symposia, and scientific meetings are organized each year. The annual budget of the total system is over U.S.$15 million, with many times that amount contributed on a volunteer basis by thousands of scientists.

The context for ICSU's central role in global change studies was provided in its sponsorship of the 1957–1958 International Geophysical Year (IGY). Eighty thousand scientists and technicians from sixty-seven countries gathered data from eight thousand stations around the world, adding to the storehouse of knowledge concerning the habitat of humanity. The pattern of international cooperation established by the IGY stimulated a half century of wide-ranging global scientific programs. The Antarctic Treaty and the Space Treaty were part of the geopolitical heritage of the IGY.

ICSU first became associated with global change research through an initiative in meteorological research that became the Global Atmospheric Research Program (GARP). At the urging of individuals from the ICSU family, President John F. Kennedy called on the United Nations in 1961 to address the issues of weather prediction and climate change. ICSU was invited to participate and quickly took the lead in developing the research plans for an interagency GARP that evolved into the World Climate Research Program (WCRP).

The illumination of the role of biogeochemical cycles in global change by ICSU's Scientific Committee on Problems of the Environment (SCOPE) led ICSU to establish, in 1986, the International Geosphere–Biosphere Programme (IGBP) as the centerpiece of its global change studies. [*See* International Geosphere–Biosphere Programme; *and* Scientific Committee on Problems of the Environment.]

Since 1966, ICSU has had a committee (COSTED) pursuing the application of science and technology for economic development in developing countries. In preparation for the 1992 United Nations Conference on Environment and Development, ICSU developed a program in science and technology to support the concept of sustainable development—meeting the needs of the present generation without compromising the capability of future generations to meet their needs. A major program in human-capacity building was launched in 1996 as one step toward implementation of that program. [*See* Sustainable Development.]

The emerging recognition that knowledge, broadly construed and properly framed, will become the organizing principle for society in the twenty-first century is leading ICSU to examine integration of the social, health, and engineering sciences and the humanities into its constituency and activities in the natural sciences.

[*See also* World Meteorological Organization.]

### INTERNET RESOURCE

Web site: http://www.icsu.org.

### BIBLIOGRAPHY

Boyer E. *Scholarship Reconsidered*. Princeton, N.J.: Princeton University Press, 1990. The seminal work establishing the discovery, integration, communication, and application of knowledge as the basic framework for scholarship.

Dooge, J., et al. *An Agenda for Science for Environment and Development in the 21st Century*. Cambridge: Cambridge University Press, 1992. Contribution of ICSU to the 1992 Earth Summit in Rio de Janeiro.

Greenway, F. *Science International: A History of the International Council of Scientific Unions*. Cambridge: Cambridge University Press, 1996.

Malone, T. *A New Agenda for Science and Technology in the Twenty-First Century*. Seoul: The Korea Science and Engineering Foundation, 1997. The role of science and technology in the pursuit of a new vision for society in the knowledge age.

Malone, T., and J. Roederer, eds. *Global Change*. Cambridge: Cambridge University Press, 1985. Proceedings of the symposium that initiated IGBP.

—THOMAS F. MALONE

## INTERNATIONAL GEOSPHERE–BIOSPHERE PROGRAMME

A cooperative research program that became the International Geosphere–Biosphere Programme (IGBP) was proposed by the Executive Board of the International Council for Science (ICSU) at its meeting in Stockholm, Sweden, in February 1983. It was envisioned as the centerpiece of a number of initiatives under way in the world scientific community to develop and to integrate the knowledge required for sound decision making in an era in which human activity on planet Earth had already begun to reach a scale that introduced perturbations in the physical, chemical, and biological processes that had been driving global environmental change.

Global environmental change, in turn, was bringing into jeopardy the life-supporting capabilities of the environment. Assessments by ICSU's Scientific Committee on Problems of the Environment had underscored the need to interrelate these processes and determine the impact of human activity on them.

IGBP was launched in 1986 as a result of an international symposium in Ottawa, Canada, in 1984. Its mission is to describe and understand the interactive physical, chemical, and biological processes that regulate the total Earth system, the unique environment that system provides for life, the changes that are occurring in this system, and the manner in which these changes are influenced by human action.

IGBP has inaugurated eight core projects in the atmospheric sciences, terrestrial ecology, oceanography, and hydrology, and three cross-cutting activities on data, modeling, and regional networks linking analysis, research, and training. Each initiative is guided by a scientific steering committee. By 1997, forty-two major reports had been published through the collective efforts of over a thousand scientists in over seventy-six countries.

IGBP concluded in 1990 that research focused on the natural and social sciences must be carried out in regions characterized by diverse climates and biogeography, particularly in developing countries. This step was necessary to achieve the understanding of global change processes required to develop and utilize the practical predictive capability essential for sound policy responses.

An array of regional networks under the rubric of a Global Change SysTem for Analysis, Research and Training (START) was established for these purposes. START is now under the joint sponsorship of IGBP, the interinstitutional World Climate Research Programme (WCRP), and the International Human Dimensions Programme on Global Environmental Change (IHDP), the latter jointly sponsored by ICSU and the International Social Science Council.

After the first ten years of IGBP operations, major new scientific results have emerged, front-ranking networks formed, and liaison strengthened with social scientists. The emphasis in all IGBP program activities is shifting toward synthesis, publication, and communication of research results and their implications for public and private policy.

Secretariats for IGBP and START, respectively, are maintained at the Royal Swedish Academy of Sciences in Stockholm and at the American Geophysical Union in Washington, D.C. Both secretariats publish quarterly newsletters.

[See also Biosphere; International Council for Science (ICSU); and Scientific Committee on Problems of the Environment.]

## INTERNET RESOURCE

Information on the International Geosphere–Biosphere Program can be accessed by visiting http:www.icsu.org.

## BIBLIOGRAPHY

Fuchs, R. "START: The Road from Bellagio." *Global Environmental Change* 5.3 (1995), 397–404. The background for START.

International Geosphere–Biosphere Programme. *A Study of Global Change: The Initial Core Projects.* IGBP Report No. 12. Stockholm, 1990. Available from the IGBP secretariat.

Malone T. F., and J. G. Roederer, eds. *Global Change.* Proceedings of the ICSU Symposium, Ottawa, Canada, 25 September 1984. Cambridge: Cambridge University Press, 1985. The background for IGBP.

Turner, B. L., W. C. Clark, R. W. Kates, J. F. Richards, J. T. Mathews, and W. B. Meyer, eds. *The Earth as Transformed by Human Action.* Cambridge: Cambridge University Press, 1990. The overall context.

—THOMAS F. MALONE

# INTERNATIONAL HUMAN DIMENSIONS PROGRAMME ON GLOBAL ENVIRONMENTAL CHANGE

The International Human Dimensions Programme on Global Environmental Change (IHDP) is an international, interdisciplinary, social science program to promote and coordinate research aimed at describing, analyzing, and understanding the human dimensions of global environmental change (GEC); that is, the way people and societies (1) contribute to GEC, (2) are influenced by GEC, and (3) mitigate and adapt to GEC. The growing recognition of the complex economic, social, political, and cultural implications of possible impacts and responses to environmental change have pushed human dimensions issues higher on the agenda of both the research and the policy communities.

In 1996, the International Council of Scientific Unions (ICSU, now known as the International Council for Science) and the International Social Science Council (ISSC) combined their scientific forces to cosponsor the IHDP. While strengthening the IHDP itself, this has also facilitated collaboration with the International Geosphere–Biosphere Programme (IGBP) and the World Climate Research Programme (WCRP). The IHDP, supported by a strengthened secretariat in Bonn, Germany, is guided by a Scientific Committee representing different geographic and scientific backgrounds, but with the common objective of strengthening links and synergies between individual researchers and research institutions working on human dimensions issues. National Human Dimensions Committees are an essential component of the IHDP's research strategy, emphasizing a bottom-up approach, and the IHDP is presently working with over twenty national committees and programs at various levels of development and activity. These activities are facilitated by the establishment of a quarterly newsletter, *IHDP Update*, and a Web site.

As defined within the IHDP, the human dimensions of GEC encompass the full range of social and natural sciences disciplines necessary to analyze and understand humans' role as both the possible cause and the target of global environmental change. IHDP Science Projects are a key mechanism used to generate IHDP research activities in priority areas, to promote international collaboration of researchers, and to link policy makers and researchers. Projects evolve from proposals by the Scientific Committee or via proposals to the IHDP made directly by one or more National Human Dimensions Committees or programs. There are presently four IHDP Science Projects: Land Use and Land Cover Change, Global Environmental Change and Human Security, Institutional Dimensions of Global Change, and Industrial Transformation.

Land Use and Land Cover Change (LUCC), which is jointly sponsored with IGBP, is carrying out a wide range of research projects within three research foci: first, land use dynamics—comparative case study analysis; second, land cover dynamics—empirical observations and diagnostic models; and third, regional and global integrated models.

Global Environmental Change and Human Security (GECHS) focuses on five themes: (1) conceptual and theoretical issues; (2) environmental change, resource use, and human security; (3) population, environment, and human security; (4) modeling of regions of environmental stress and human vulnerability; and (5) institutions and policy development in environmental security.

Institutional Dimensions of Global Change (IDGC) will develop a response to the following research questions: What roles do institutions play in causing and confronting GEC? How effective are institutional innovations intended to confront GEC? What are the prospects for (re)designing institutions to confront environmental challenges?

The Industrial Transformation (IT) research agenda focuses on the relationship between changes in the industrial systems and changes in environment and is expected to carry out research in three areas: first, macrosystems and incentive structures; second, industrial ecology; and third, consumers and consumer choice.

[*See also* Global Change, *article on* Human Dimensions of Global Change.]

### BIBLIOGRAPHY

Lambin, E. F., et al. "Land-Use and Land-Cover Change: Implementation Strategy." International Human Dimensions Programme Report No. 10/International Geosphere–Biosphere Programme Report No. 48. Stockholm: IHDP/IGBP, 1999.

Lonergan, S., et al. "Global Environmental Change and Human Security: Science Plan." International Human Dimensions Programme Report No. 11. Bonn: IHDP, 1999.

Turner II, B. L., D. Skole, S. Sanderson, G. Fischer, L. Fresco, and R. Leemans. "Land-Use and Land-Cover Change: Science/Research Plan." International Human Dimensions Programme Report No. 7/International Geosphere–Biosphere Programme Report No. 35. Stockholm: IHDP/IGBP, 1995.

Vellinga, P., and N. Herb, eds. "Industrial Transformation: Science Plan." International Human Dimensions Programme Report No. 12. Bonn: IHDP, 1999.

Young, O., et al. "Institutional Dimensions of Global Environmental Change: Science Plan." International Human Dimensions Programme Report No. 9. Bonn: IHDP, 1999.

—LARRY R. KOHLER

## INTERNATIONAL INSTITUTE FOR APPLIED SYSTEMS ANALYSIS

The International Institute for Applied Systems Analysis (IIASA) was founded in 1972 by nongovernmental national member organizations (NMOs) from twelve countries: the United States, the Soviet Union, Bulgaria, Canada, Czechoslovakia, France, the Federal Republic of Germany (FRG), the German Democratic Republic (GDR), Italy, Japan, Poland, and the United Kingdom. Although formally nongovernmental, all NMOs had the clear support and financing of their governments. The idea had first been proposed by U.S. President Lyndon B. Johnson in 1966 as one of several bridge-building initiatives between the United States and the Soviet bloc. The substantive focus would be systems analysis and modern management techniques, fields in which the United States led the world and where many believed great gains could be achieved through improving and disseminating American methods.

Through the end of the cold war, IIASA maintained a broad research portfolio bringing scholars together for joint studies on demography, energy and natural resources, management, and applied mathematics. In 1991, members negotiated a new Strategic Plan focused on environmental, technological, and economic issues of global change and emphasizing long-term connections between regional policies and global considerations.

**Activities Related to Global Change.** IIASA published the first truly global study of energy supply and demand, including initial estimates of greenhouse gas emissions and impacts (Häfele, 1981), and conducts long-term studies analyzing alternative energy resources, technologies, and environmental policies (Nakićenović et al., 1998). The Institute is a major contributor to the work of the Intergovernmental Panel on Climate Change on energy scenarios. IIASA's work on transboundary air pollution provided the principal scientific support for negotiations on implementing the 1979 Convention on Long-Range Transboundary Air Pollution (LRTAP). Reduction targets for sulfur dioxide in the 1994 Second Sulfur Protocol were based on IIASA calculations. IIASA's models have been extended to include tropospheric ozone and to support negotiations on the second $NO_x$ Protocol. In agricultural research, IIASA is at the forefront of estimating impacts of global warming, carbon dioxide fertilization, and sulfur emissions on agricultural production, trade, and consumption (see, for example, Parry et al., 1988a, b). In population research, the Institute published the first fully probabilistic world population projections (Lutz, 1996) and conducts studies on population aging, international interactions among national social security policies, impacts of changing household and family structures on greenhouse gas emissions, and linkages among demographic, environmental, and economic development models. [*See* Convention on Long-Range Transboundary Air Pollution; *and* Intergovernmental Panel on Climate Change.]

IIASA's agenda also covers the effectiveness of international environmental agreements, technological change, boreal forests, land use and land cover change, water resources, economic transitions in the formerly centrally planned economies, and analytic methods valuable to the Institute's policy-related research. IIASA coordinates several research networks connecting scholars around the globe and each summer brings approximately fifty doctoral students to the Institute for its three-month Young Scientists Summer Program.

**Structure and Governance.** IIASA is located in Laxenburg, Austria, outside Vienna, with fifteen NMOs from Austria, Bulgaria, Finland, Germany, Hungary, Japan, Kazakhstan, the Netherlands, Norway, Poland, Russia, the Slovak Republic, Sweden, Ukraine, and the United States. Contributions are organized according to a four-tier schedule, with Russia and the United States providing the most. The chief executive officer of the Institute is the director. The Institute's Governing Council, with one representative from each NMO, appoints the director, establishes the Institute's overall policy and direction, provides scientific guidance, and approves the annual budget and research program. Of a total staff of 190 people, there are approximately eighty-five research scholars recruited primarily from NMO countries. The sole working language is English.

[*See also* Climate Models; Energy; Modeling of Natural Systems; *and* Population Dynamics.]

### BIBLIOGRAPHY

Fischer, G., K. Frohberg, M. A. Keyzer, and K. S. Parikh. *Linked National Models: A Tool for International Food Policy Analysis.* Dordrecht: Kluwer, 1988.
Fischer, G., K. Frohberg, M. L. Parry, and C. Rosenzweig. "Climate Change and World Food Supply, Demand and Trade: Who Ben-

efits, Who Loses?" *Global Environmental Change* 4.1 (1994), 7–23.
Fischer, G., and C. Rosenzweig. *The Impacts of Climate Change, $CO_2$, and $SO_2$ on Agricultural Supply and Trade: An Integrated Assessment.* Laxenburg, Austria: International Institute for Applied Systems Analysis, 1996.
Häfele, W. *Energy in a Finite World: A Global Systems Analysis.* Cambridge, Mass.: Ballinger, 1981.
Lutz, W. *The Future Population of the World: What Can We Assume Today?* Rev. ed. London: Earthscan, 1996.
Nakićenović, N., A. Grübler, and A. McDonalds, eds. *Global Energy Perspectives.* Cambridge: Cambridge University Press, 1998.
Parry, M. L., T. R. Carter, and N. T. Konijn, eds. *The Impact of Climatic Variations on Agriculture,* vol. 1, *Assessments in Cool Temperate and Cold Regions.* Dordrecht: Kluwer, 1988a.
———. *The Impact of Climatic Variations on Agriculture,* vol. 2, *Assessments in Semi-Arid Regions.* Dordrecht: Kluwer, 1988b.

—ALAN MCDONALD

**INTERNATIONAL LAW.** *See* Environmental Law; European Community; *and* Montreal Protocol.

**INTERNATIONAL UNION FOR THE CONSERVATION OF NATURE AND NATURAL RESOURCES.** *See* World Conservation Union, The.

**INTERNET.** *See* Information Technology.

**INTRODUCED SPECIES.** *See* Exotic Species.

## IPAT

IPAT is an acronym for a mathematical identity, $I = P \times A \times T$, that describes the environmental impact ($I$) of human activities as the product of three factors: population size ($P$), affluence ($A$), and technology ($T$). An identity is an equation that is true by definition; in this case, affluence is often represented by production (or consumption) of goods per person, and technology by environmental impact per unit of production (or consumption), so that

$$\text{Impact} = \text{Population} \times \frac{\text{Production}}{\text{Population}} \times \frac{\text{Impact}}{\text{Production}}.$$

The IPAT identity was developed in the early 1970s during the course of a debate between Barry Commoner, who argued that environmental impacts in the United States were due primarily to post–World War II changes in production technology, and Paul Ehrlich and John Holdren, who argued that all three factors were important and emphasized in particular the role of population growth. Ehrlich and Holdren originally proposed the equation in 1970 in prose form as a framework to rebut assertions by Commoner and others that population growth was unimportant to rising pollution levels in the United States. Ehrlich and Holdren pointed out that population has no effect on pollution only when the technology factor becomes zero; that is, when technology is so benign that economic activity produces no pollutants whatsoever. Commoner then applied Ehrlich and Holdren's formula to several decades of data on various pollutants in the United States in an attempt to quantify the proportion of growth in pollution levels due to each of the three driving forces, concluding that postwar changes in production technology far outweighed the influence of either population growth or rising affluence. At the same time, Ehrlich and Holdren used an equation of the form $I = P \times F$, where $F$ was a combination of $A$ and $T$ measuring impact per person, to determine that population was indeed an important driver of environmental degradation.

The publication of Commoner's book, *The Closing Circle* (Commoner, 1971), which elaborated on his previous results, set off a heated exchange between the two sides. In a critique, Ehrlich and Holdren condensed Commoner's equation into the now familiar IPAT form, using it as a tool to argue that Commoner's results were based on misleading mathematical analysis and selective use of data. Commoner's response, and a subsequent rejoinder from Ehrlich and Holdren, also relied heavily on IPAT to defend vigorously their respective points of view. The debate was never settled, and both sides have since held to their respective points of view, often using IPAT analyses to buttress their arguments.

**Identity versus Explanation.** Since the early 1970s, IPAT has been employed by many researchers in the analysis of a wide range of environmental issues in all regions of the world, including deforestation and other kinds of land use/cover change, food and agriculture, climate change, energy use, air and water quality, and biodiversity loss. As an identity, it has been most useful as an orienting perspective on environmental degradation processes, simplifying their conceptualization by dividing the forces driving environmental impacts into a small number of broad categories. For example, land under cultivation, a proxy for a number of impacts on the environment, can be expressed as the product of population, food production per person, and area of land required per unit of food produced. This formulation identifies three classes of primary influences on food-related land use: population size, diet, and agricultural productivity. Similarly, emissions of the greenhouse gas carbon dioxide resulting from energy use can be expressed as the product of population, per capita economic production, and the amount of carbon

dioxide emitted per unit of economic production. This breakdown organizes factors influencing emissions into three simple categories. The last, known as carbon intensity, can be further subdivided into two terms reflecting the energy intensity of the economy and the mix of energy sources in overall supply.

The IPAT identity also illustrates an important consequence of the multiplicative relationship between driving forces: that each variable amplifies changes in any other. As a result, a given change in technology may have only a small effect on the environment in a society with a small, low-income population, while the same change would have a much greater effect in a populous, affluent society. Likewise, a given increment in population would have a much greater impact in affluent societies than in low-income countries, assuming levels of technology were similar.

On the other hand, IPAT has been strongly criticized for a number of perceived flaws. Although as an identity it is always true by definition, when it is used as an explanatory model it implicitly assumes that there are only three relevant variables and that they are related in a simple linear fashion. Critics assert that its lack of social science content, particularly the influence of policies and institutions on environmental outcomes, render it misleading at best. Some researchers have suggested that the $P$, $A$, and $T$ variables be thought of as proximate (direct) causes of environmental impact, which are themselves influenced by a wide range of indirect, but more fundamental, ultimate causes including income distribution, land management practices, urban–rural settlement patterns, prices, political empowerment, trade relations, attitudes and preferences, wars, and so on. These factors may be critical to environmental outcomes and differ widely across settings, making the equation ill suited to analysis at the micro level, and casting doubt on the results of larger-scale studies. Nonetheless, the fact that data for the $P$, $A$, and $T$ variables are generally available across a range of settings has invited widespread use of IPAT for cross-site studies and studies at large spatial and temporal scales.

**Calculating Blame.** The IPAT model was developed for the purpose of quantifying arguments seeking to apportion blame for environmental impacts among contributing factors, and it has been widely used for this purpose ever since. The goal in such exercises is to rank the importance of the $P$, $A$, and $T$ variables, usually to prioritize policy recommendations for reducing impacts. However, such exercises suffer from a long list of mathematical ambiguities inherent to the decomposition of index numbers (such as the $I$ in IPAT) that make results difficult, if not impossible, to compare. These ambiguities have also allowed attacks on methods of quantitative analysis to fuel larger debates without bringing them closer to resolution.

In the initial Ehrlich–Holdren–Commoner exchanges, for example, the decomposition method used was a comparison of the ratios of final to initial values of each of the variables over a given period. Suppose impact increased by 500 percent, while population increased only by 50 percent. One interpretation might be that the $A$ and $T$ variables combined are ten times as important as population. Yet because the factors multiply each other, the combined increase in $A$ and $T$ in this example is actually only 300 percent, implying that they are six (300/50), not ten, times more important than $P$. Further, if the changes in the variables are expressed as ratios rather than percentages, then $I$ increased sixfold while $P$ increased 1.5-fold and $AT$ fourfold. Now $AT$ may be said to be less than three (4/1.5) times as important as $P$. Clearly, the method allows a wide range of interpretations of results.

An additional problem with the original method is that, whether expressed in percentages or ratios of absolute values, the changes in the driving forces do not add up to the change in impact, confounding attempts to divide up blame for $I$ among $P$, $A$, and $T$. In the example above, one might conclude on the basis of percentage increases that population was responsible for 10 percent (50/500) of the increase in impact while $AT$ accounted for 60 percent (300/500). Yet this leaves 30 percent of the change in impact unaccounted for; it is due not to any single variable but to the multiplicative effect of each on the others. To circumvent this problem, first Commoner, and later Holdren and many subsequent researchers, converted the multiplicative IPAT relation into an additive one based on average annual growth rates over a given period. To a good approximation, if a prime ($'$) indicates annual percentage change, then

$$I' = P' + A' + T'.$$

Assuming the example above applies to a span of thirty years, the $I$, $P$, and $AT$ variables would have grown at 6.2, 1.4, and 4.7 percent per year. The contribution of each factor is then expressed as the ratio of its own growth rate to the growth rate of $I$. In this case $P$ would be assigned 23 percent (1.4/6.2) of the blame for the increase in $I$, and $AT$ would be responsible for the remaining 77 percent (4.7/6.2).

Neither approach is completely satisfactory, and several others are possible. There is, in fact, no single correct method, and each will yield different results. Decomposition of IPAT belongs to a larger class of problems related to index numbers whose results are influenced by several factors.

*Choice of variables.* Aside from the question of whether the IPAT framework itself captures the relevant variables, it is not always clear what quantities to use for $P$, $A$, and $T$. Although population size usually serves as the demographic variable, other units of measure

such as numbers of households or urban population may also be relevant. Because these quantities have generally grown faster than population size itself, the choice of demographic variable can have an important effect on results. Defining affluence appropriately was a point of contention in the early debates and continues to be problematic in some applications. While per capita gross national product is often taken as a convenient proxy, at times only particular sectors of the economy are relevant. The choice of variable for $T$ is generally determined by the choice of $I$, since $T$ is often not measured but solved for as $I/(PA)$; as a result, it can be thought of not only as a technology variable, but as a variable that includes all influences, including technology, not captured by $P$ and $A$. Choosing $I$ is itself not straightforward. Environmental impact is often difficult to measure, so proxies such as release of pollutants, rates of resource use, or land devoted to agriculture are commonly substituted. Finally, the number of variables included will also determine the outcome of decomposition exercises. The $T$ variable, for example, can often be broken down into further subunits or combined with $A$ to form a single per capita impact variable.

*Heterogeneity.* The level of aggregation at which an analysis is performed can strongly influence the results. If population growth and per capita environmental impact are negatively correlated, as is the general case with greenhouse gas emissions, then calculations at the global level will generally assign a larger proportion of the blame to population growth than more disaggregated analyses. The global view misses the fact that population growth is generally fastest where per capita impact is lowest. On the other hand, if per capita impact is positively correlated with population growth, as for example in the case of land degradation due to fuelwood harvesting, then a large-scale analysis will underestimate the role of population. Even analyses at the country level mask inequalities in the distribution of per capita impact across national populations. Disaggregation by income level is probably more appropriate, but more difficult in practice because of constraints of data availability.

*The offset problem.* Decomposition exercises become particularly difficult to interpret when not all variables move in the same direction. For example, in many regions of the world the productivity of agricultural land, or the carbon intensity of the economy, has improved over time, so that the $T$ variable has been declining while $P$ and $A$ have risen. Suppose $I$ increases at 1 percent per year while $P$, $A$, and $T$ change at $+2$ percent, $+1$ percent, and $-2$ percent per year, respectively. The decomposition method based on annual rates of change described above assigns blame of 200 percent, 100 percent, and $-200$ percent to the three driving forces—a confusing result. A number of fixes are possible, but each method of handling the offset problem has its own set of difficulties and leads to different conclusions.

***Interaction between variables.*** Even the earliest users of the IPAT equation recognized that it made the simplifying assumption that $P$, $A$, and $T$ behave independently. Arguments have been made for the existence of bidirectional relationships among all of the variables. Ehrlich and Holdren, for example, argued that increases in $P$ were likely to increase $T$ as population growth induced resource scarcity, which would necessitate ever more environmentally destructive means of extraction. On the other hand, other schools of thought view population growth as an impetus to environmentally beneficial technological change either through scarcity-driven incentives to innovate or simply through the additional capacity for innovation provided by a larger population. Relationships between $P$ and $A$ have received even more attention, although empirical data are inconclusive on whether a significant effect of population growth on growth in per capita income exists. Finally, a strong case has been made for relationships between $A$ and $T$ in which greater affluence drives an "environmental transition": first an increase, and later a reduction, in the impact of technology on the environment.

***Implausible scenario comparison.*** All decomposition exercises suffer from the shortcoming that they do not take account of how much change in a particular variable is plausible. Instead, they estimate the contribution of one variable to total impact by implicitly comparing a particular set of data to a hypothetical scenario in which that variable remains constant at its initial level. While they may shed some light on the narrow question of the source of absolute changes in environmental impact, results of decompositions do not necessarily translate directly into priorities for intervention. From a policy point of view, it is much more relevant to ask how much a realistic scenario for one variable, relative to a baseline path, would change impact over a given period of time. Plausible alternative scenarios account for, among other things, momentum built into population age structures that make immediate population stabilization impossible, as well as momentum in technological systems and patterns of consumption.

**Recent Developments.** Because of its simplicity, applicability to a wide range of impacts, and convenience as a conceptual framework, IPAT continues to be widely used. While early applications focused on resource depletion and emissions of local pollutants, the growth of concern over global issues—in particular, climate change, has led to a number of IPAT-based analyses of the forces driving greenhouse gas emissions. This recent work has attempted to extend previous approaches by introducing new variables, recasting IPAT in nonlinear form, and placing it in a larger theoretical context, all of which are likely to improve its usefulness. Growing recognition of the factors that it leaves out and the shortcomings of decomposition exercises should discourage its misapplication. Its use as a starting point for more complex models in sensitivity analyses should also improve its applicability to questions of environmental policy.

[*See also* Carrying Capacity; Catastrophist–Cornucopian Debate; Economic Levels; Environmental Economics; Global Change, *article on* Human Dimensions of Global Change; Growth, Limits to; Human Impacts, *article on* Human Impacts on Earth; Human Populations; Mass Consumption; Population Policy; Sustainable Development; *and* Technology.]

## BIBLIOGRAPHY

Ang, B. W. "Decomposition Methodology in Industrial Energy Demand Analysis." *Energy* 20.11 (1995), 1081–1095. Provides an overarching framework for decomposition methodologies, demonstrating that results are method dependent.

Commoner, B. *The Closing Circle.* New York: Knopf, 1971. This work set off a heated exchange with Ehrlich and Holdren.

———. "Rapid Population Growth and Environmental Stress." Summary Report of the United Nations Expert Group Meeting on Consequences of Rapid Population Growth in Developing Countries. New York: United Nations, 1988. Extends Commoner's analyses of the role of population growth in environmental impact in the United States to the developing countries, concluding again that the contribution from population is minor relative to the influence of technology.

Commoner, B., et al. "The Causes of Pollution." *Environment* 13.3 (1971), 2–19. Along with Ehrlich and Holdren's 1971 *Science* paper, this paper contained the first quantitative analysis explicitly based on an IPAT-type equation, concluding that changes in production technology were the dominant causes of environmental degradation in the United States.

Dietz, T., and E. A. Rosa. "Effects of Population and Affluence on Carbon Dioxide Emissions." *Proceedings of the National Academy of Sciences* 94 (1997), 175–179. A good example of recent extensions of the IPAT equation presenting a nonlinear form of IPAT estimated from panel data on national carbon dioxide emissions, population size, and per capita income.

Ehrlich, P. R., and A. H. Ehrlich. *The Population Explosion.* New York: Simon and Schuster, 1990. An updated and expanded version of the original argument presented by Ehrlich and Holdren that slowing of population growth is essential to alleviating environmental impact. Uses the IPAT identity as a touchstone throughout.

Ehrlich, P. R., and J. Holdren. "Impact of Population Growth." *Science* 171 (1971), 1212–1217. Along with Commoner's 1971 *Environment* paper, this paper contained the first quantitative analysis explicitly based on an IPAT-type equation. It concluded that population was a significant contributor to environmental impact.

———. "The People Problem." *Saturday Review* (4 July 1970), 42–43. A short piece that first introduced a prose version of the IPAT equation.

Harrison, P. *The Third Revolution: Environment, Population and a Sustainable World.* London: I. B. Taurus, 1992. A readable introduction to the multifaceted debate over the influence of population growth on environmental degradation, with examples from several locations around the world. Employs the IPAT equation to quantify population's role.

Mackellar, F. L., et al. "Population and Climate Change." In *Human Choice and Climate Change*, vol. 1, *The Societal Framework*, edited by S. Rayner and E. L. Malone. Columbus, Ohio: Battelle Press, 1998. Thoroughly documents the shortcomings of IPAT decomposition exercises, focusing on those that have been applied to greenhouse gas emissions.

Marden, P. G., and D. Hodgson, eds. *Population, Environment, and the Quality of Life*. New York: Wiley, 1975. A collection that includes the key papers in the Ehrlich–Holdren–Commoner debate, including the exchange over Commoner's *The Closing Circle* that appeared in the 1971 *Bulletin of the Atomic Scientists*, with some interesting background on events surrounding their publication.

United Nations Population Fund. *Population, Resources and the Environment: The Critical Challenges*. London: Banson, 1991. An examination of the role of population growth in a wide range of environmental issues, making frequent reference to IPAT analyses. A good example of how IPAT exercises subject to many of the shortcomings described in the text are nonetheless used to draw strong policy conclusions.

Wexler, L. "Decomposing Models of Demographic Impact on the Environment." Working Paper WP-96-85, International Institute for Applied Systems Analysis. Laxenburg, Austria, 1996. Available through IIASA (see http://www.iiasa.ac.at). A technical review of various methods of IPAT decomposition that have been proposed in the literature, using historical and projected future greenhouse gas emissions as an example.

—BRIAN C. O'NEILL

# IRRIGATION

Irrigation is any human-induced change in the natural flow of water for the purpose of growing plants. This includes water conservation measures such as bunding (impounding water in a field using small earthen dykes), mulching, and furrowing that increase the quantity of water available and the reliability of the supply, whether from rainfall or runoff. Irrigation water is usually delivered to farm fields through canal networks or pipes. Common sources of water include rivers and streams (either through direct diversion or after storage in a natural or artificial lake) and underground aquifers (lifted from open wells or from tube wells). Drainage is the removal of excess water from the land to reduce yield losses due to flooding, waterlogging, and salinity.

Irrigation water made it possible for the earliest forms of civilization to develop in the arid lands of the Middle East and Asia along the major rivers—the Nile, Euphrates, Tigris, Indus, and Yellow—by producing the food surpluses needed for these civilizations to survive and flourish. More recently, the colonial powers developed irrigation not only to increase food production and stave off famine, but also to produce crops for export—cotton, sugar cane, and rice. The rapid growth in food production during the Green Revolution, from the mid-

1960s to the present, was accomplished in large part by expanding irrigated area. During this period, a major portion of the increase in global food production came from increased yields on an expanded area of irrigated land.

Rice is grown on approximately one-third of the world's irrigated cropland and 50 percent of Asia's irrigated cropland. Most rice is grown in areas dominated by a monsoon climate. In the main rice-growing season (the wet season), heavy rains often result in an excess of water. Rice is extremely tolerant of flooding—hence its widespread cultivation in such climatic zones. In these areas, drainage facilities are needed to remove excess water and prevent yield losses due to waterlogging and salinity. In the dry season and in drier climates, most irrigated agricultural production comprises crops other than rice, ranging from grain crops such as wheat and maize to cash crops such as cotton, vegetables, and fruit.

**Worldwide Expansion of Irrigation.** In the half century after World War II, the irrigated area of the world tripled from approximately 90 million hectares to 270 million hectares, an annual compound growth rate of over 2.5 percent. Most of the irrigated land is in the developing countries; over 50 percent is in Asia (Table 1). There is a wide variation among regions in the fraction of cropland irrigated. Almost all of the cropland in North Africa and the Middle East is irrigated, over 20 percent in Asia, but 12 percent or less in the rest of the world.

The growth of irrigated areas in the developing world during this period was strongly influenced by two periods of severe drought: (1) the drought in the Indian subcontinent in the mid-1960s and (2) the more general drought and shortfall in grain production accompanied by the energy crisis and fertilizer shortages in the mid-1970s. The Green Revolution—the development and spread of higher-yielding, shorter-duration, fertilizer-responsive varieties—was closely associated with the expansion of irrigation.

The expansion of irrigation has been based mostly on the construction of new storage dams and reservoirs or, where ground water is readily available, through the use of tube wells. These reservoirs help to assure water supplies and allow carry-over of surplus flows in the rainy season for use in the dry season when water is scarce. Of the more than forty thousand large dams (defined by the International Commission on Large Dams as measuring 15 meters from foundation to crest), all but five thousand have been built since 1950. [*See* Dams.]

The World Bank has been the largest foreign financier of irrigation projects, having lent over U.S.$58 billion for more than six hundred dams in ninety-three countries. However, in most countries, domestic resources dominated the financing of irrigation development. In India, for example, which alone has more than three thousand

**Irrigation. TABLE 1.** Net Irrigated Area as a Percentage of Primary Crop Area and World Irrigated Area

| REGION | PRIMARY CROPS— HARVESTED AREA (1,000 HECTARES)* | IRRIGATED AREA (1,000 HECTARES) | IRRIGATED AREA (AS PERCENTAGE OF PRIMARY CROPS AREA) | IRRIGATED AREA (AS PERCENTAGE OF WORLD IRRIGATED AREA) |
|---|---|---|---|---|
| Asia | 762,244 | 166,860 | 22 | 63 |
| India | 209,605 | 57,000 | 27 | 22 |
| China | 188,212 | 49,880 | 27 | 19 |
| Rest of Asia† | 364,426 | 59,980 | 16 | 23 |
| North Africa and Middle East‡ | 25,208 | 22,405 | 89 | 9 |
| Sub-Saharan Africa | 211,667 | 6,096 | 3 | 2 |
| Europe | 266,595 | 25,077 | 9 | 10 |
| North America | 192,875 | 22,110 | 11 | 8 |
| Latin America and Caribbean | 150,685 | 17,749 | 12 | 7 |
| Oceania | 63,149 | 2,605 | 4 | 1 |
| World | 1,672,422 | 262,902 | 16 | 100 |

*Figures exclude pasture range and hay areas. All data except for North America are for 1996. Data for North America Irrigation Area are for 1995; world irrigation area adjusted accordingly.

†Includes Pakistan, Bangladesh, Sri Lanka, all Southeast Asian countries, all East Asian countries except the People's Republic of China, and all West Asian countries except those noted for North Africa and the Middle East.

‡Includes Morocco, Algeria, Tunisia, Libya, Egypt, Jordan, Israel, Saudi Arabia, Syria, Iraq, Iran, and Afghanistan.

SOURCE: The Food and Agriculture Web site for statistical data (FAOSTATS): http://www.apps.fao.org.

large dams, foreign funding of irrigation accounted for only about 10 percent of total expenditure in any year. Since the 1960s, net irrigated area served by wells has grown more rapidly than that served by reservoirs and canals, and wells now account for over half of the total area in India and China.

The growth of irrigation has slowed in the last decade. At least three factors have influenced the recent decline in the rate of new irrigation developments. First, and perhaps foremost, has been the steady decline in cereal grain prices, which in real terms are about 50 percent of 1960 levels—in large measure because of the very success of the irrigation strategy of previous decades. Second, costs of new dam and canal construction have been rising steadily—several countries have reported cost increases of 50–100 percent over the past two decades. This has been due in large measure to the fact that the more easily irrigated areas have already been brought into production, and suitable new dam sites are not readily available. Finally, there has been growing pressure from environmental groups and local communities against the construction of new large dams because of concerns about their ecological and social impacts.

For most of modern history, the world's irrigated area has grown faster than the population. Since 1980, however, the irrigated area per person has declined, and per capita cereal grain production has stagnated (Figure 1). The debate regarding the world's capacity to feed a growing population, brought to the fore in the writings

of Malthus two centuries ago, continues unabated. But the growing scarcity and competition for water adds a new element to this debate over food security. Most of the gains in crop production over the past three decades have come from higher yields on expanding irrigated land area. While irrigated area grew at about 2 percent per annum during much of this period, the period of major construction of new systems has come to an end. Future growth in irrigated cropland is projected to be less than 1 percent per annum, and, with losses in irrigated land from salinity, urbanization, and other factors, the net irrigated area in the world may already be declining.

**A New Paradigm.** To meet the challenge posed by the growing scarcity and competition for water, a new paradigm of water resource management is gradually emerging from the work of researchers in the field. This paradigm has two components. First, there is increased recognition of the need to recycle and reuse water in river basins. When water is withdrawn for a particular use, only a fraction of it is used up, or lost to the river basin, through evaporation or contamination. The balance flows to other surface and subsurface areas, where it may be recaptured and reused. Because of this recycling, a "water multiplier effect" occurs, whereby the sum of all the withdrawals in a river basin may exceed the initial amount of water entering the basin by several times. Recognition of the fact of water recycling has profound implications for water resource planning and policies.

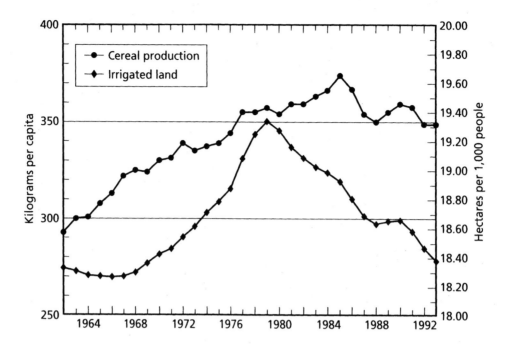

**Irrigation. Figure 1.** World Cereal Production per Capita and Irrigated Land per Thousand People. (From The Food and Agriculture Web site for statistical data [FAOSTATS]: http://www.apps.fao.org.)

A second component of the paradigm concerns recent advances in information technologies. The rapid development of techniques such as remote sensing and geographic information systems will enable water managers to determine water needs and availability, both in the surface and subsurface areas, at the scale of the river basin. Through observation of actual evapotranspiration via remote sensing, it will be possible to determine how well irrigation water is being managed, thus increasing yields and reducing water needs. Further development of this paradigm and of information technologies combined with a better understanding of the social, institutional, and economic aspects of water management can result in dramatic increases in water productivity and equity of water use.

**Environmental Concerns.** The benefits of irrigation included lower food prices, higher employment, and more rapid agricultural and economic development. But irrigation development has also led to social and environmental problems such as salinization of soils, pollution of aquifers by agrochemicals, loss of wildlife habitats, and the forced resettlement of those previously living in areas submerged by reservoirs. The result has been a growing conflict between those who see the potential benefits of continued irrigation development and those who view further development as a threat to the environment and even to human health.

There are valid arguments to support the views of both the promoters and the detractors of further irrigation development, and both the positive and negative effects must be considered in the development of any project. Careful analysis of the benefits and costs of alternatives is needed to assist policy makers in making informed judgments. The long-term, diverse, and complex nature of the impacts of irrigation development make it especially hard to balance these views within a simple cost-benefit framework. For example, the stability of employment for agricultural laborers is hard to weigh against real or potential loss of wetlands or species.

Environmental problems arise largely as a consequence of efforts to intensify agricultural production in irrigated areas. Most forms of environmental degradation represent a cost to society that is borne neither by the suppliers nor by the users of irrigation water. This reinforces the rationale for a new focus on water basin planning and management in which these external or offsite costs are taken into account.

**Competition for Water.** Of the world's total water resources, over 97 percent is in the oceans and seas and is too salty for productive use. Two-thirds of the remainder is locked in icecaps, glaciers, permafrost, swamps, and deep aquifers. The remainder, less than 100,000 cubic kilometers, is found in the rivers and lakes that constitute the bulk of usable supply. [See Hydrologic Cycle.]

Global demand for water has grown at 2.4 percent per annum since 1970. Agriculture currently accounts for approximately 70 percent of global water withdrawals. However, on a smaller geographic scale, this figure ranges from as high as 90 percent in the lowest-income developing countries to less than 50 percent in some developed countries. But the demand for water for industrial and municipal use is expected to double in the next twenty-five years, leading to growing competition with agricultural water needs. Agriculture's share may decline as a result.

Many countries are entering a period of severe water shortage. None of the projections of the global food situation, such as those done for the World Bank, the Food and Agriculture Organization, and the International Food Policy Research Institute, has explicitly incorporated water as a constraint. There will be an increasing number of water-deficit countries and regions, including not only the Middle East and North Africa, but also some of the major breadbaskets of the world such as the Indian Punjab and the north central plain of China. A recent study by Seckler et al. (1998) estimates that about 50 percent of the increase in demand for water by the year 2025 might be met by increasing the effectiveness of irrigation. This study suggests, however, that the remainder must be met by the expansion of irrigated area, which will almost certainly necessitate the construction of some new dams.

**Increasing Productivity.** The productivity of water can be increased by any one or a combination of the following: (1) increasing output per unit of consumed water, (2) reducing water losses to sinks (including the ocean), (3) reducing the pollution of water, and (4) reallocating water from lower- to higher-valued crops or uses.

There are a wide range of farming practices and technologies available to increase irrigation water productivity, including recharge of aquifers, more intensive control and management of water in canal systems, use of efficient sprinklers and drip irrigation, and recycling. The suitability of any given technology or practice will vary according to the particular physical, economic, and institutional environment.

Eventually, however, as demand increases, the fresh water available in the river basin is fully utilized: that is to say, all remaining flows are of sufficiently poor quality, the cost of recovery is too high, or they are committed to meet environmental needs. When this happens, since there is no more surplus water to be developed, water-related activities in the river basin become wholly competitive. More water consumed upstream makes less water available downstream, and more pollution upstream reduces quality for all downstream users. At this stage, the basin is "closed"—a phenomenon that is increasingly common, either at the low-flow season or throughout the year, for many of the most populated river basins in the world.

**Demand.** As the demand for water increases and its value rises, there is increasing pressure to treat water as an economic good, subject to market forces, with the price to be determined by some form of market. While there are situations where this is desirable, irrigation water is not a good that can be easily traded or bought and sold. So-called market failure calls for government intervention. Furthermore, it must be recognized that minimum levels of water, like food, are a necessity of life to which even the poorest should have access. In the design and management of systems, close attention should be paid to cost-effectiveness and economic efficiency. As suggested above, greater efficiency in planning and management is best achieved at the river basin level, where overall effects on productivity and the environment can be taken into account.

**Privatization.** Private systems include individually owned wells and various forms of canal irrigation managed by user groups or by private entities that own systems and provide services on behalf of users. Despite the fact that much of the recent expansion of irrigation has involved the use of public funds for the construction and operation of large irrigation systems, a major portion of the irrigated area in the world is privatized.

It is well known that some public systems have not performed well and that their productivity of water use could be increased. Privatization of these publicly managed systems is now often viewed as the best strategy for improving performance. The aim of privatization is to create markets for the management and allocation of water. Elements of a privatized system might include one or more of the following: (1) charges to users for services, (2) pricing that varies with volume of use, (3) water markets, (4) tradable water rights, and (5) management of irrigation systems by local user groups. The small farm size found in most developing countries and the absence of established water rights presents a major constraint to volumetric pricing and development of water markets in most developing countries. Hence, the current trend is to turn over the management of public irrigation systems to local entities. While this typically results in reduced government expenditures in the short run, there is no widespread evidence that significant productivity increases are achieved, and there is growing concern that the private sector pays insufficient attention to support services and long-term expenditures for major rehabilitation and repairs.

**The Future of Irrigation.** Until very recently, increased food production needs have been met by applying modern technology to an expanding area of irrigated cropland, but opportunities for further expansion are increasingly limited. The growing scarcity and competition among users for finite renewable water resources means that less water per capita may be available for agriculture in the future than in the past.

Increased food production and maintenance of food security will thus depend increasingly on the more productive use of existing water resources. As the new paradigm indicates, water resource management must include a river basin perspective, taking into account social objectives in allocating water among competing uses. Efficient irrigation technologies (for example, drip irrigation), biological technologies, and agronomic practices that save water and/or improve its quality will be adopted more widely in the future. Information technologies such as remote sensing, geographic information systems, and better models can assist in management and intersector water resource allocation at the basin level. Much more work needs to be done to assess the relative costs of these various alternatives and their potential for increasing the productivity of irrigated agriculture.

[*See also* Agriculture and Agricultural Land; Animal Husbandry; Erosion; Famine; Food; Land Use, *article on* Land Use and Land Cover; Methane; Nitrous Oxide; Pest Management; Salinization; Soils; Water; *and* Water Quality.]

### INTERNET RESOURCE

The Food and Agriculture Organization Web site for statistical data (FAOSTATS) can be found at http://apps.fao.org/.

### BIBLIOGRAPHY

Falkenmark, M., et al. "Macro-Scale Water Scarcity Requires Micro-Scale Approaches: Aspects of Vulnerability in Semi-Arid Development." *Natural Resource Forum* 13.4 (1997), 258–267. A pioneering work in the definition of measurement of water scarcity.

Food and Agriculture Organization. *Crop Water Requirements.* FAO Irrigation and Drainage Paper 24. Rome, 1984. Still considered a standard text for determining crop water requirements and their application in planning, design, and operation of irrigation projects.

Gleick, P. H., ed. *Water in Crisis: A Guide to the World's Fresh Water Resources.* New York and Oxford: Oxford University Press, 1993. This important reference contains, in Part I, a series of essays on freshwater issues and, in Part II, a comprehensive set of data covering irrigation, agricultural production, environment, human health, and other related issues.

Molden, D. "Accounting for Water Use and Productivity." Systemwide Initiative for Water Management, Paper No. 1, International Irrigation Management Institute. Colombo, Sri Lanka, 1997. Of interest principally to researchers, this paper presents a conceptual framework for water accounting and provides generic terminologies and procedures to describe the status of water resource use and consequences of actions related to water resources. The framework applies to water resource use at three levels: a use level such as an irrigated field or household, a service level such as an irrigation or water supply system, and a water basin level that may include several uses.

Pingali P. L., et al. *Asian Rice Bowls: The Returning Crisis.* Wallingford, U.K.: CAB International, 1997. There are two excellent chapters on the environmental problems associated with intensification of agriculture through irrigation. Unlike many works dealing with the environment, the authors present data and research findings to support their arguments.

Postel, S. *Last Oasis: Facing Water Scarcity.* New York: Norton, 1992. A very readable account of the growing scarcity and competition for water, its implications for the production of food, and related issues.

———. *Pillar of Sand: Can the Irrigation Miracle Last?* Washington, D.C.: Worldwatch Institute, 1999.

Rosegrant, M. *Water Resources in the Twenty-First Century: Challenges and Implications for Action.* Food, Agriculture and the Environment Discussion Paper 20. Washington, D.C.: International Food Policy Research Institute, 1997. A comprehensive look at the growing competition for water for irrigation and other purposes, and its implications for policy.

Seckler, D. *The New Era of Water Resources Management: From "Dry" to "Wet" Water Savings.* Research Report No. 1, International Irrigation Management Institute. Colombo, Sri Lanka, 1996. This report contains the rationale for a new paradigm emphasizing a water basin perspective for water resource and irrigation management.

Seckler, D., et al. *World Water Demand and Supply, 1990 to 2025: Scenarios and Issues.* Research Report No. 19, International Irrigation Management Institute. Colombo, Sri Lanka, 1998. This report establishes the methodological framework and the basic set of data for analyzing the present and future water supply situations of major countries of the world. The work to date highlights the national and regional disparities in water resources and provides a basis from which to begin to assess the future supply and demand for this vital natural resource.

—RANDOLPH BARKER AND DAVID SECKLER